COLD SPRING HARBOR SYMPOSIA
ON QUANTITATIVE BIOLOGY

VOLUME LXXI

www.cshl-symposium.org

Institutions that have purchased the hardcover edition of this book are entitled to online access to the Symposium Web site. Please contact your institution's library to gain access to the Web site. The site contains the full text articles from the 2006 Symposium and the Symposia held in 1998–2005 as well as archive photographs and selected papers from the 62-year history of the annual Symposium.

If a token number is given above, you will need to activate the token to gain access. If you have previously registered and activated your online account for any prior volume without the use of a token, then you do not have to register again.

If you do not have an account number or a token number, or are experiencing access problems, please contact Kathy Cirone, CSHL Press Subscription Manager, at 1-800-843-4388, extension 4044 (Continental U.S. and Canada), 516-422-4100 (All other locations), cironek@cshl.edu, or subscriptionfeedback@symposium.org.

COLD SPRING HARBOR SYMPOSIA
ON QUANTITATIVE BIOLOGY

VOLUME LXXI

Regulatory RNAs

www.cshl-symposium.org

Meeting Organized by Bruce Stillman and David Stewart
COLD SPRING HARBOR LABORATORY PRESS
2006

COLD SPRING HARBOR SYMPOSIA ON QUANTITATIVE BIOLOGY VOLUME LXXI

©2006 by Cold Spring Harbor Laboratory Press
International Standard Book Number 978-0-87969-817-1 (cloth)
International Standard Book Number 978-0-87969-818-8 (paper)
International Standard Serial Number 0091-7451
Library of Congress Catalog Card Number 34-8174

COLD SPRING HARBOR SYMPOSIA ON QUANTITATIVE BIOLOGY
Founded in 1933 by
REGINALD G. HARRIS
Director of the Biological Laboratory 1924 to 1936
Previous Symposia Volumes

I (1933) Surface Phenomena
II (1934) Aspects of Growth
III (1935) Photochemical Reactions
IV (1936) Excitation Phenomena
V (1937) Internal Secretions
VI (1938) Protein Chemistry
VII (1939) Biological Oxidations
VIII (1940) Permeability and the Nature of Cell Membranes
IX (1941) Genes and Chromosomes: Structure and Organization
X (1942) The Relation of Hormones to Development
XI (1946) Heredity and Variation in Microorganisms
XII (1947) Nucleic Acids and Nucleoproteins
XIII (1948) Biological Applications of Tracer Elements
XIV (1949) Amino Acids and Proteins
XV (1950) Origin and Evolution of Man
XVI (1951) Genes and Mutations
XVII (1952) The Neuron
XVIII (1953) Viruses
XIX (1954) The Mammalian Fetus: Physiological Aspects of Development
XX (1955) Population Genetics: The Nature and Causes of Genetic Variability in Population
XXI (1956) Genetic Mechanisms: Structure and Function
XXII (1957) Population Studies: Animal Ecology and Demography
XXIII (1958) Exchange of Genetic Material: Mechanism and Consequences
XXIV (1959) Genetics and Twentieth Century Darwinism
XXV (1960) Biological Clocks
XXVI (1961) Cellular Regulatory Mechanisms
XXVII (1962) Basic Mechanisms in Animal Virus Biology
XXVIII (1963) Synthesis and Structure of Macromolecules
XXIX (1964) Human Genetics
XXX (1965) Sensory Receptors
XXXI (1966) The Genetic Code
XXXII (1967) Antibodies
XXXIII (1968) Replication of DNA in Microorganisms
XXXIV (1969) The Mechanism of Protein Synthesis

XXXV (1970) Transcription of Genetic Material
XXXVI (1971) Structure and Function of Proteins at the Three-dimensional Level
XXXVII (1972) The Mechanism of Muscle Contraction
XXXVIII (1973) Chromosome Structure and Function
XXXIX (1974) Tumor Viruses
XL (1975) The Synapse
XLI (1976) Origins of Lymphocyte Diversity
XLII (1977) Chromatin
XLIII (1978) DNA: Replication and Recombination
XLIV (1979) Viral Oncogenes
XLV (1980) Movable Genetic Elements
XLVI (1981) Organization of the Cytoplasm
XLVII (1982) Structures of DNA
XLVIII (1983) Molecular Neurobiology
XLIX (1984) Recombination at the DNA Level
L (1985) Molecular Biology of Development
LI (1986) Molecular Biology of *Homo sapiens*
LII (1987) Evolution of Catalytic Function
LIII (1988) Molecular Biology of Signal Transduction
LIV (1989) Immunological Recognition
LV (1990) The Brain
LVI (1991) The Cell Cycle
LVII (1992) The Cell Surface
LVIII (1993) DNA and Chromosomes
LIX (1994) The Molecular Genetics of Cancer
LX (1995) Protein Kinesis: The Dynamics of Protein Trafficking and Stability
LXI (1996) Function & Dysfunction in the Nervous System
LXII (1997) Pattern Formation during Development
LXIII (1998) Mechanisms of Transcription
LXIV (1999) Signaling and Gene Expression in the Immune System
LXV (2000) Biological Responses to DNA Damage
LXVI (2001) The Ribosome
LXVII (2002) The Cardiovascular System
LXVIII (2003) The Genome of *Homo sapiens*
LXIX (2004) Epigenetics
LXX (2005) Molecular Approaches to Controlling Cancer

Front Cover (*Paperback*): Petunia images courtesy of R. Jorgensen, University of Arizona. Selected images with permission of the American Society of Plant Biologists. Design by Cat Eberstark, Cold Spring Harbor Laboratory.

All Cold Spring Harbor Laboratory Press publications may be ordered directly from Cold Spring Harbor Laboratory Press, 500 Sunnyside Boulevard, Woodbury, NY 11797-2924. Phone: 1-800-843-4388 in Continental U.S. and Canada. All other locations: (516) 422-4100. FAX: (516) 422-4097. E-mail: cshpress@cshl.edu. For a complete catalog of all Cold Spring Harbor Laboratory Press publications, visit our World Wide Web Site http://www.cshlpress.com/

Web Site Access: Institutions that have purchased the hardcover edition of this book are entitled to online access to the companion Web site at www.cshl-symposium.org. For assistance with activation, please contact Kathy Cirone, CSHL Press Subscription Manager, at cironek@cshl.edu.

Symposium Participants

ABAD, XABIER, Dept. of Gene Therapy and Hepatology, Centro de Investigación Médica Aplicada, Universidad de Navarra, Pamplona, Spain

ACH, ROBERT, Agilent Laboratories, Agilent Technologies, Palo Alto, California

ADAMS, CHRISTOPHER, Dept. of Gene Expression Profiling, Invitrogen Life Technologies, Carlsbad, California

AFONINA, IRINA, Dept. of Research and Development, Nanogen, Bothell, Washington

AHN, JEONGHYUN, Dept. of Microbiology, University of Ulsan College of Medicine, Seoul, South Korea

AIBA, HIROJI, Div. of Biological Science, Graduate School of Science, Nagoya University, Nagoya, Japan

AIGNER, STEFAN, Lab. of Genetics, The Salk Institute, La Jolla, California

AIZAWA, YASUNORI, Center for Biological Resources and Informatics, Tokyo Institute of Technology, Yokohama, Japan

AKSOY, CANAN, Cold Spring Harbor Laboratory, Cold Spring Harbor, New York

ALVAREZ-SAAVEDRA, EZEQUIEL, Dept. of Biology, Massachusetts Institute of Technology, Cambridge, Massachusetts

AMAN, PIERRE, Dept. of Pathology, Sahlgrenska Academy, Göteborg University, Gothenburg, Sweden

AMARAL, PAULO, Dept. of Biochemistry, Institute of Chemistry, University of São Paulo, São Paulo, Brazil

AMARIGLIO, NINETTE, Institute of Hematology, Sheba Medical Center, Tel Hashomer, Israel

AMERES, STEFAN, Max F. Perutz Laboratories, Medical University of Vienna, Vienna, Austria

ANDERSON, EMILY, Dept. of Biology Research and Development, Dharmacon, Lafayette, Colorado

ANDERSSON, MATTIAS, Dept. of Pathology, Lundberg Laboratory for Cancer Research, Göteborg University, Gothenburg, Sweden

ANDINO, RAUL, Dept. of Microbiology and Immunology, University of California, San Francisco

ANDRUSS, BERNARD, Dept. of Discovery, Asuragen, Austin, Texas

ANSEL, K. MARK, CBR Institute for Biomedical Research, Harvard Medical School, Boston, Massachusetts

ARAVIN, ALEXEI, Cold Spring Harbor Laboratory, Cold Spring Harbor, New York

BAK, MADS, Dept. of Medical Biochemistry and Genetics, Panum Institute, University of Copenhagen, Copenhagen, Denmark

BALZER, ERICA, Dept. of Molecular Biology, School of Medicine, University of Medicine and Dentistry of New Jersey

BANFI, SANDRO, Telethon Institute of Genetics and Medicine, Fondazione Telethon, Naples, Italy

BARBASH, ZOHAR, Dept. of Hemato-Oncology, Sheba Medical Center, Tel Aviv University, Ramat Gan, Israel

BARFORD, DAVID, Section of Structural Biology, Chester Beatty Laboratories, Institute of Cancer Research, London, United Kingdom

BARTA, ANDREA, Max F. Perutz Laboratories, Medical University of Vienna, Vienna, Austria

BARTEL, DAVID, Dept. of Biology, Howard Hughes Medical Institute, Whitehead Institute, Massachusetts Institute of Technology, Cambridge, Massachusetts

BASS, BRENDA, Dept. of Biochemistry, Howard Hughes Medical Institute, School of Medicine, University of Utah, Salt Lake City

BATEY, ROBERT, Dept. of Chemistry and Biochemistry, University of Colorado, Boulder

BAULCOMBE, DAVID, Sainsbury Laboratory, John Innes Centre, Norwich, United Kingdom

BÄURLE, ISABEL, Dept. of Cell and Developmental Biology, John Innes Centre, Norwich, United Kingdom

BEAZLEY, KIM, Dept. of Agricultural Biotechnology, Monsanto, Chesterfield, Missouri

BEEMON, KAREN, Dept. of Biology, Johns Hopkins University, Baltimore, Maryland

BEREZIKOV, EUGENE, Dept. of Functional Genomics, Hubrecht Laboratory, Utrecht, The Netherlands

BERNARDS, RENÉ, Dept. of Molecular Carcinogenesis, Netherlands Cancer Institute, Amsterdam, The Netherlands

BERNHOFF, EVA, Dept. of Microbiology and Infectious Control, University Hospital of North Norway, Tromsø, Norway

BERNSTEIN, EMILY, Lab. of Chromatin Biology, Rockefeller University, New York, New York

BHATTACHARYYA, SUVENDRA, Dept. of Epigenetics, Friedrich Miescher Institute for Biomedical Research, Basel, Switzerland

BIELKE, WOLFGANG, Dept. of Cell Biology, QIAGEN GmbH, Hilden, Germany

BIEMAR, FREDERIC, Dept. of Molecular and Cell Biology, University of California, Berkeley

BLACKBURN, ELIZABETH, Dept. of Biochemistry and Biophysics, University of California, San Francisco

BOBEK, JAN, Institute of Microbiology, Czech Academy of Sciences, Prague, Czech Republic

BOCCARA, MARTINE, Unité de Recherche en Génétique Végétale, Atelier de BioInformatique, Université Paris, Evry, France

BOREL, CHRISTELLE, Dept. of Genetic Medicine and Development, University of Geneva Medical School, Geneva, Switzerland

BOUHOUCH, KHALED, Laboratoire de Génétique Moléculaire, Unité Mixte de Recherche, Centre National de la

Recherche Scientifique, Ecole Normale Supérieure, Paris, France

BOWDEN, MICHAELA, Dept. of Genome and Proteome Sciences, Novartis Institute for Biomedical Research, Cambridge, Massachusetts

BOZZONI, IRENE, Dept. of Genetics and Molecular Biology, University of Rome, La Sapienza, Rome, Italy

BRIZUELA, LEONARDO, Dept. of Life Sciences and Chemical Analysis, Agilent Technologies, Cambridge, Massachusetts

BROCK, GRAHAM, Cancer Center, Ordway Research Institute, Albany, New York

BÜHLER, MARC, Dept. of Cell Biology, Harvard Medical School, Boston, Massachusetts

BUKER, SHANE, Dept. of Cell Biology, Harvard Medical School, Boston, Massachusetts

BUSHELL, MARTIN, School of Pharmacy, University of Nottingham, Nottingham, United Kingdom

BUSKIRK, ALLEN, Dept. of Chemistry and Biochemistry, Brigham Young University, Provo, Utah

BUTLER, MAURICE, Dept. of Genetics, Harvard Medical School, Boston, Massachusetts

CAIKOVSKI, MARIAN, Department of Sciences, University of Geneva, Geneva, Switzerland

CALCIANO, MARGARET, Dept. of Genetics and Developmental Biology, Health Center, School of Medicine, University of Connecticut, Farmington

CAMIOLO, MATTHEW, Cold Spring Harbor Laboratory, Cold Spring Harbor, New York

CARMELL, MICHELLE, Cold Spring Harbor Laboratory, Cold Spring Harbor, New York

CARMICHAEL, GORDON, Dept. of Genetics and Developmental Biology, Health Center, School of Medicine, University of Connecticut, Farmington

CARTHEW, RICHARD, Dept. of Biochemistry, Molecular and Cell Biology, Northwestern University, Evanston, Illinois

CASTOLDI, MIRCO, Dept. of Pediatric Oncology, Hematology and Immunology, Heidelberg University, Heidelberg, Germany

CAYOTA, ALFONSO, Dept. of Medicine, Facultad de Medicina, Universidad de la República, Montevideo, Uruguay

CECH, THOMAS, Howard Hughes Medical Institute, Chevy Chase, Maryland

CERNILOGAR, FILIPPO, Dept. of Epigenetics and Genome Reprogramming, Dulbecco Telethon Institute, Institute of Genetics and Biophysics, Consiglio Nazionale delle Ricerche, Naples, Italy

CHALKER, DOUGLAS, Dept. of Biology, Washington University, St. Louis, Missouri

CHARTRAND, PASCAL, Dept. of Biochemistry, Université de Montréal, Montréal, Québec, Canada

CHAVES, DANIEL, Program in Molecular Medicine, Medical School, University of Massachusetts Worcester

CHELOUFI, SIHEM, Cold Spring Harbor Laboratory, Cold Spring Harbor, New York

CHEN, CHUN-HONG, Dept. of Biology, California Institute of Technology, Pasadena, California

CHEN, JACK, Cold Spring Harbor Laboratory, Cold Spring Harbor, New York

CHEN, JICHAO, Dept. of Molecular Biology and Genetics, Johns Hopkins University, Baltimore, Maryland

CHEN, KEVIN, Dept. of Biology, Center for Comparative Functional Genomics, New York University, New York

CHEN, XIANMING, Dept. of Biochemical and Molecular Biology, College of Medicine, Mayo Clinic, Rochester, Minnesota

CHOOKAJORN, THANAT, Dept. of Organismic and Evolutionary Biology, Harvard University, Cambridge, Massachusetts

CHOU, CHEN-KUNG, Dept. of Life Science, Chang Gung University, Kwei-Shan, Tao-Yuan, Taiwan, Republic of China

CHOW, YEN-HUNG, Vaccine Research and Development Center, Taiwan National Health Research Institutes, Chu-Nan, Miao-Li, Taiwan, Republic of China

CIAUDO, CONSTANCE, Dept. of Development Biology, Centre National de la Recherche Scientifique, Pasteur Institute, Paris, France

CLARK, ALEJANDRA, The Wellcome Trust, Cancer Research UK Gurdon Institute, University of Cambridge, Cambridge, United Kingdom

CONNELL, LAUREEN, *Genes & Development*, Cold Spring Harbor Laboratory Press, Woodbury, New York

COOKE, ROBERT, Stow, Massachusetts

CORTES-CROS, MARTA, Dept. of Platform and Chemical Biology, Novartis Institutes for Biomedical Research, Basel, Switzerland

COSTA, ANA, Warwick Horticultural Research International, University of Warwick, Warwick, United Kingdom

COUGHLAN, SEAN, Agilent Technologies, Wilmington, Delaware

CULLEN, BRYAN, Dept. of Genetics, Howard Hughes Medical Institute, Duke University, Durham, North Carolina

DAHLBERG, JAMES, Dept. of Biomolecular Chemistry, University of Wisconsin, Madison

DARNELL, ROBERT, Dept. of Molecular Neuro-Oncology, Howard Hughes Medical Institute, Rockefeller University, New York, New York

DAUGHERTY, MATTHEW, Dept. of Biochemistry and Biophysics, University of California - San Francisco

DAWSON, ELLIOTT, BioVentures, Murfreesboro, Tennessee

DEGUIRE, VINCENT, Dept. of Biochemistry, Université de Montréal, Montréal, Québec, Canada

DE JONG, DEBORAH, Dept. of Molecular and Cellular Biology, Harvard University, Cambridge, Massachusetts

DENLI, AHMET, Lab. of Genetics, The Salk Institute, La Jolla, California

DIAZ-MARTINEZ, LAURA, Dept. of Genetics, Cell Biology and Development, University of Minnesota, Minneapolis

DIBENEDETTO, ANGELA, Dept. of Biology, Villanova University, Villanova, Pennsylvania

DIK, WILLEM, Dept. of Immunology, Erasmus Medical Center, Rotterdam, The Netherlands

DING, YE, Dept. of Developmental Genetics and Bioinformatics, Wadsworth Center, Albany, New York

DITTMAR, KIMBERLY, Dept. of Biochemistry and Molecular Biology, University of Chicago, Chicago, Illinois

DORNAN, DAVID, Dept. of Molecular Oncology, Genentech, South San Francisco, California

DORNER, SILKE, Dept. of Molecular Biology and Genetics, School of Medicine, Johns Hopkins University, Baltimore, Maryland

DOUDNA, JENNIFER, Dept. of Molecular and Cell Biology, Howard Hughes Medical Institute, University of California, Berkeley

DREYFUSS, GIDEON, Dept. of Biochemistry and Biophysics, Howard Hughes Medical Institute, School of Medicine, University of Pennsylvania, Philadelphia

DUBOWITZ, VICTOR, Dept. of Paediatrics, Imperial College London, London, United Kingdom

DUMITRU, AMALIA, Dept. of Biology, FORTH, Institute of Molecular Biology and Biotechnology, Heraklion, Crete, Greece

DUS, MONICA, Cold Spring Harbor Laboratory, Cold Spring Harbor, New York

EBERT, MARGARET, Dept. of Biology, Massachusetts Institute of Technology, Cambridge, Massachusetts

EDDY, SEAN, Dept. of Genetics, Howard Hughes Medical Institute, School of Medicine, Washington University, St. Louis, Missouri

EGANA, ANA, Div. of RNA Biology, New England Biolabs, Ipswich, Massachusetts

EGGLESTON, ANGELA, *Nature*, Nature Publishing Group, Cambridge, Massachusetts

EHM, SEBASTIAN, Lab. of RNAi, Institute of Molecular Biotechnology, Vienna, Austria

ELLIOT, MARIE, Dept. of Biology, McMaster University, Hamilton, Ontario, Canada

ELMÉN, JOACIM, Dept. of Drug Disovery and Manufacturing, Santaris Pharma, Hørsholm, Denmark

ESAU, CHRISTINE, Dept. of Antisense Drug Discovery, Isis Pharmaceuticals, Carlsbad, California

ESUMI, NORIKO, Dept. of Ophthalmology, School of Medicine, Johns Hopkins University, Baltimore, Maryland

FABIAN, MARC, Dept. of Biology, York University, Toronto, Ontario, Canada

FAGEGALTIER, DELPHINE, Dept. of Developmental Biology, Centre National de la Recherche Scientifique, Pasteur Institute, Paris, France

FAGHIHI, MOHAMMAD ALI, Dept. of Neurobiology, The Scripps Research Institute, Jupiter, Florida

FARABAUGH, PHILIP, Dept. of Biological Sciences, University of Maryland, Baltimore

FEIG, ANDREW, Wayne State University, Detroit, Michigan

FELICE, KRISTIN, Dept. of Molecular, Microbial, and Structural Biology, Health Center, School of Medicine, University of Connecticut, Farmington

FERBEYRE, GERARDO, Dept. of Biochemistry, Université de Montréal, Montréal, Québec, Canada

FILIPOWICZ, WITOLD, Dept. of Epigenetics, Friedrich Miescher Institute for Biomedical Research, Basel, Switzerland

FISCHER, SYLVIA, Dept. of Molecular Biology, Massachusetts General Hospital, Harvard Medical School, Boston, Massachusetts

FLINTOFT, LOUISA, *Nature Reviews Genetics*, Nature Publishing Group, London, United Kingdom

FONTANA, LAURA, Dept. of Hematology, Oncology and Molecular Medicine, Istituto Superiore di Sanità, Rome, Italy

FORTIN, KRISTINE, Div. of Neuropathology, University of Pennsylvania, Philadelphia

FOX, RICHARD, Dept. of Pathology, University of Washington, Seattle

FREEMAN, KATIE, Dept. of Pathway Genomics, GlaxoSmithKline, Collegeville, Pennsylvania

FRENDEWEY, DAVID, Velocigene, Regeneron Pharmaceuticals, Tarrytown, New York

FRENSTER, JOHN, Activator RNA Research, *Physicians' Educational Series*, Atherton, California

FRIEDMAN, LILACH, Dept. of Human Molecular Genetics and Biochemistry, Sackler School of Medicine, Tel Aviv University, Tel Aviv, Israel

FRITZ, BRIAN, Div. of Human Biology, Fred Hutchinson Cancer Research Center, Seattle, Washington

GABEL, HARRISON, Dept. of Molecular Biology, Massachusetts General Hospital, Boston, Massachusetts

GANEM, DON, Dept. of Microbiology, Howard Hughes Medical Institute, University of California, San Francisco

GANN, ALEXANDER, Cold Spring Harbor Laboratory Press, Woodbury, New York

GARY, SYDNEY, Banbury Center, Cold Spring Harbor Laboratory, Cold Spring Harbor, New York

GAUR, RAJESH, Dept. of Molecular Biology, Beckman Research Institute of the City of Hope, Duarte, California

GEORGES, MICHEL, Dept. of Animal Genomics, University of Liège, Liège, Belgium

GHOSHAL, KALPANA, Dept. of Molecular and Cellular Biochemistry, Ohio State University, Columbus, Ohio

GIEGERICH, ROBERT, Faculty of Technology, Bielefeld University, Bielefeld, Germany

GILADI, HILLA, Dept. of Gene Therapy, Hadassah University Hospital, Jerusalem, Israel

GILBERT, SUNNY, Dept. of Chemistry and Biochemistry, University of Colorado, Boulder

GINGERAS, THOMAS, Transcription Mapping and Regulation Group, Affymetrix, Santa Clara, California

GIRALDEZ, ANTONIO, Dept. of Molecular and Cellular Biology, Harvard University, Cambridge, Massachusetts

GIRARD, ANGELIQUE, Cold Spring Harbor Laboratory, Cold Spring Harbor, New York

GLOVER-CUTTER, KIRA, Dept. of Biochemistry and Molecular Genetics, Health Sciences Center, University of Colorado, Aurora

GOFF, LOYAL, Dept. of Cell and Developmental Biology, Rutgers University, Piscataway, New Jersey

GOTTESMAN, SUSAN, Lab. of Molecular Biology, Center for Cancer Research, National Cancer Institute, National Institutes of Health, Bethesda, Maryland

GOTTWEIN, EVA, Dept. of Molecular Genetics and Microbiology, Duke University, Durham, North Carolina

GRABOWSKI, PAULA, Dept. of Biological Sciences, University of Pittsburgh, Pittsburgh, Pennsylvania

GRACHEVA, ELENA, Dept. of Biology, Washington University, St. Louis, Missouri

GRAVELEY, BRENTON, Dept. of Genetics and Developmental Biology, Health Center, School of Medicine, University of Connecticut, Farmington

GREEN, RACHEL, Dept. of Molecular Biology and Genetics, Howard Hughes Medical Institute, School of Medicine, Johns Hopkins University, Baltimore, Maryland

GREIDER, CAROL, Dept. of Molecular Biology and Genetics, School of Medicine, Johns Hopkins University, Baltimore, Maryland

GREWAL, SHIV, Lab. of Molecular Cell Biology, National Cancer Institute, National Institutes of Health, Bethesda, Maryland

GREY, FINN, Dept. of Molecular Microbiology and Immunology, Oregon Health and Sciences University, Portland, Oregon

GRIMAUD, CHARLOTTE, Institute of Human Genetics, Centre National de la Recherche Scientifique, Montpellier, France

GRODZICKER, TERRI, Cold Spring Harbor Laboratory Press, Woodbury, New York

GROISMAN, EDUARDO, Dept. of Molecular Microbiology, Howard Hughes Medical Institute, School of Medicine, Washington University, St. Louis, Missouri

GROISMAN, REGINA, Oncogenese, Differenciation et Transduction, Unité Propre de Recherche, Centre National de la Recherche Scientifique, Villejuif, France

GUIL, SONIA, Dept. of Chromosomes and Gene Expression, Medical Research Council Human Genetics Unit, Western General Hospital, Edinburgh, Scotland, United Kingdom

GUNDERSON, SAMUEL, Dept. of Molecular Biology and Biochemistry, Rutgers University, Piscataway, New Jersey

GUTIERREZ-NAVA, Maria de la Luz, Dept. of Genetic Discovery, DuPont, Wilmington, Delaware

HAISER, HENRY, Dept. of Biology, McMaster University, Hamilton, Ontario, Canada

HALE, CARYN, Dept. of Biochemistry and Molecular Biology, University of Georgia, Athens

HAMMELL, CHRISTOPHER, Dept. of Biochemistry, Dartmouth Medical School, Hanover, New Hampshire

HAN, JINJU, School of Biological Sciences, Seoul National University, Seoul, South Korea

HANNON, GREGORY, Cold Spring Harbor Laboratory, Cold Spring Harbor, New York

HARVEY, JAGGER, Sainsbury Laboratory, John Innes Centre, Norwich, United Kingdom

HATZIGEORGIOU, ARTEMIS, Dept. of Genetics, University of Pennsylvania, Philadelphia

HAVECKER, ERICKA, Sainsbury Laboratory, John Innes Centre, Norwich, United Kingdom

HE, LIN, Cold Spring Harbor Laboratory, Cold Spring Harbor, New York

HE, XINGYUE, Cold Spring Harbor Laboratory, Cold Spring Harbor, New York

HEARD, DAVID, Dept. of Genomic Sciences Bioinformatics, Sanofi Aventis Pharma, Vitry-sur-Seine, France

HEARD, EDITH, Mammalian Developmental Epigenetics Group, Unité Mixte de Recherche, Centre National de la Recherche Scientifique, Curie Institute, Paris, France

HEIDRICH, NADJA, AG Bacteriagenetik, Friedrich-Schiller University, Jena, Germany

HEIMSTAEDT, SUSANNE, Sainsbury Laboratory, John Innes Centre, Norwich, United Kingdom

HENKIN, TINA, Dept. of Microbiology, Ohio State University, Columbus, Ohio

HINAS, ANDREA, Dept. of Molecular Biology, Swedish University of Agricultural Sciences, Uppsala, Sweden

HIRSCH, HANS, Dept. of Clinical and Biological Sciences, University of Basel, Basel, Switzerland

HOBERT, OLIVER, Dept. of Biochemistry, Columbia University College of Physicians & Surgeons, New York, New York

HOECHSMANN, MATTHIAS, Faculty of Technology, Bielefeld University, Bielefeld, Germany

HONG, SUN WOO, Dept. of Chemistry, School of Molecular Science, Pohang University of Science and Technology, Pohang, South Korea

HOTTA, IKUKO, Cold Spring Harbor Laboratory, Cold Spring Harbor, New York

HU, ZONGLIN, Lab. of Bacterial Diseases, Bacterial Toxins and Therapeutics Section, National Institute of Allergy and Infectious Diseases, National Institutes of Health, Bethesda, Maryland

HUANG, JULIE, Dept of Molecular Biology, Massachusetts General Hospital, Harvard Medical School, Boston, Massachusetts

HUNTER, CRAIG, Dept. of Molecular and Cellular Biology, Harvard University, Cambridge, Massachusetts

HUR, INHA, School of Biological Sciences, Seoul National University, Seoul, South Korea

HÜTTENHOFER, ALEXANDER, Div. of Genomics and RNomics, Innsbruck Biocenter, Innsbruck Medical University, Innsbruck, Austria

HYUN, SEOGANG, School of Biological Sciences, Seoul National University, Seoul, South Korea

IBARRA, INGRID, Cold Spring Harbor Laboratory, Cold Spring Harbor, New York

INGLIS, JOHN, Cold Spring Harbor Laboratory Press, Woodbury, New York

IRVINE, DANIELLE, Cold Spring Harbor Laboratory, Cold Spring Harbor, New York

IZAURRALDE, ELISA, Dept. of Biochemistry, Max-Planck-Institute for Developmental Biology, Tübingen, Germany

JACOBSEN, STEVE, Dept. of Molecular, Cell, and Developmental Biology, Howard Hughes Medical Institute, University of California, Los Angeles

JAN, CALVIN, Dept. of Biology, Whitehead Institute, Massachusetts Institute of Technology, Cambridge, Massachusetts

JAYASENA, SUMEDHA, Dept. of Cancer Biology, Amgen, Thousand Oaks, California

JAZDZEWSKI, KRYSTIAN, Comprehensive Cancer Center, Ohio State University, Columbus, Ohio

JEFFRIES, CLARK, School of Pharmacy, University of North Carolina, Chapel Hill

JENSEN, KEVIN, Dept. of Molecular Biology and Biochemistry, Health Center, School of Medicine, University of Connecticut, Farmington

JI, XINJUN, Department of Genetics, School of Medicine, University of Pennsylvania, Philadelphia

JOHNSTON, BRIAN, Dept. of Pediatrics, School of Medicine, Stanford University, Stanford, California

JORGENSEN, RICHARD, Dept. of Plant Sciences, University of Arizona, Tucson, Arizona

JOSHUA-TOR, LEEMOR, Cold Spring Harbor Laboratory, Cold Spring Harbor, New York

JOVANOVIC, MARKO, Institute of Molecular Biology, University of Zürich, Zürich, Switzerland

JOZSI, PETER, Dept. of Gene Regulation, Invitrogen Life Technologies, Carlsbad, California

JUNG, JUN EUN, Dept. of Microbiology, University of Ulsan College of Medicine, Seoul, South Korea

KALYNA, MARIA, Institute of Medical Biochemistry, Medical University of Vienna, Vienna, Austria

KARGINOV, FEDOR, Cold Spring Harbor Laboratory, Cold Spring Harbor, New York

KATAYAMA, SHINTARO, Lab. for Genome Exploration Research, Genomic Sciences Centre, RIKEN Yokohama Institute, Yokohama, Japan

KAUPPINEN, SAKARI, Dept. of Medical Biochemistry and

Genetics, University of Copenhagen, Copenhagen, Denmark

KAWANO, MITSUOKI, National Institute of Child Health and Human Development, National Institutes of Health, Bethesda, Maryland

KENNEDY, SCOTT, Dept. of Pharmacology, University of Wisconsin, Madison

KIERMER, VERONIQUE, *Nature Methods*, Nature Publishing Group, New York, New York

KIM, DANIEL, Dept. of Molecular Biology, Beckman Research Institute of the City of Hope, Duarte, California

KIM, V. NARRY, School of Biological Sciences, Seoul National University, Seoul, South Korea

KIM, YOUNG-KOOK, School of Biological Sciences, Seoul National University, Seoul, South Korea

KISHORE, SHIVENDRA, Institute for Biochemistry, University of Erlangen, Erlangen, Germany

KISS, TAMÁS, Laboratoire de Biologie Moléculaire Eucaryote, Centre National de la Recherche Scientifique, Toulouse, France

KLATTENHOFF, CARLA, Program in Molecular Medicine, Medical School, University of Massachusetts, Worcester

KLAUSEN, MIKKEL, Dept. of Clinical Biochemistry, Rigshospitalet, University of Copenhagen, Copenhagen, Denmark

KO, JAE-HYEONG, Dept. of Molecular, Cellular, and Developmental Biology, Yale University, New Haven, Connecticut

KONFORTI, BOYANA, *Nature Structural & Molecular Biology*, Nature Publishing Group, New York, New York

KONG, YI WEN, School of Pharmacy, University of Nottingham, Nottingham, United Kingdom

KONSTANTINOVA, PAVLINA, Dept. of Human Retrovirology, Academic Medical Center, University of Amsterdam, Amsterdam, The Netherlands

KOZU, TOMOKO, Research Institute for Clinical Oncology, Saitama Cancer Center, Saitama, Japan

KUMAR, MADHU, Dept. of Biology, Center for Cancer Research, Massachusetts Institute of Technology, Cambridge, Massachusetts

KURODA, MITZI, Harvard-Partners Center for Genetics and Genomics, Harvard University, Boston, Massachusetts

KUZNETSOV, VLADIMIR, Institute for Mathematical Sciences, Genome Institute of Singapore, Singapore, Republic of Singapore

LAI, ERIC, Dept. of Developmental Biology, Sloan-Kettering Institute, New York, New York

LAMBERTZ, IRINA, Dept. of Molecular Biomedical Research, Flanders Interuniversity Institute for Biotechnology, Ghent University, Ghent, Belgium

LANDTHALER, MARKUS, Dept. of RNA Molecular Biology, Rockefeller University, New York, New York

LAPIDOS, KAREN, Dept. of Biochemistry, Molecular and Cellular Biology, Northwestern University, Evanston, Illinois

LAU, NELSON, Dept. of Molecular Biology, Massachusetts General Hospital, Boston, Massachusetts

LECHMAN, ERIC, Dept. of Cell and Molecular Biology, University Health Network, Toronto, Ontario, Canada

LEE, DONG-KI, Dept. of Chemistry, School of Molecular Science, Pohang University of Science and Technology, Pohang, South Korea

LEE, HEUIRAN, Dept. of Microbiology, University of Ulsan College of Medicine, Seoul, South Korea

LEE, JEANNIE, Dept. of Molecular Biology, Massachusetts General Hospital, Harvard Medical School, Boston, Massachusetts

LEPÈRE, GERSENDE, Laboratoire de Génétique Moléculaire, Unité Mixte de Recherche, Centre National de la Recherche Scientifique, Ecole Normale Supérieure, Paris, France

LEUNG, ANTHONY, Center for Cancer Research, Massachusetts Institute of Technology, Cambridge, Massachusetts

LI, CHENGJIAN, Dept. of Biochemistry and Molecular Pharmacology, Medical School, University of Massachusetts, Worcester

LI, LONG-CHENG, Dept. of Urology, University of California, San Francisco

LI, PAN, Dept. of Chemistry, University of California, Berkeley

LIANG, CHEN, Dept. of Medicine, Lady Davis Institute, Jewish General Hospital, Montréal, Québec, Canada

LIPOVICH, LEONARD, Dept. of Information and Mathematical Sciences, Genome Institute of Singapore, Singapore, Republic of Singapore

LIU, JIDONG, Cold Spring Harbor Laboratory, Cold Spring Harbor, New York

LORENZ, CHRISTINA, Dept. of Biochemistry, University of Vienna, Vienna, Austria

LOU, HUA, Dept. of Genetics, Case Western Reserve University, Cleveland, Ohio

LUND, ELSEBET, Dept. of Biomolecular Chemistry, University of Wisconsin, Madison

LUSSIER, JACQUES, Dept. of Biomedicine, Animal Reproduction Research Center, Université de Montréal, Saint-Hyacinthe, Québec, Canada

MA, JIN-BIAO, Structural Biology Program, Memorial Sloan-Kettering Cancer Center, New York, New York

MACARI, MARISA, Cold Spring Harbor Laboratory, Cold Spring Harbor, New York

MACRAE, IAN, Dept. of Molecular and Cell Biology, University of California, Berkeley

MAHER, CHRISTOPHER, Dept. of Biomedical Engineering, State University of New York, Stony Brook

MAITY, TUHIN, Dept. of Chemistry, University of North Carolina, Chapel Hill

MAKEYEV, EUGENE, Dept of Molecular and Cellular Biology, Harvard University, Cambridge, Massachusetts

MANIATAKI, ELISAVET, Dept. of Pathology, University of Pennsylvania, Philadelphia

MARAIA, RICHARD, Lab. of Molecular Growth Regulation, National Institutes of Health, Bethesda, Maryland

MARTIENSSEN, ROBERT, Cold Spring Harbor Laboratory, Cold Spring Harbor, New York

MATSKEVICH, ALEXEY, Institute of Medical Virology, University of Zürich, Zürich, Switzerland

MATTICK, JOHN, Institute for Molecular Bioscience, University of Queensland, St. Lucia, Brisbane, Australia

MATTIE, MICHAEL, Dept. of Quantitative Expression, Glaxo SmithKline, Research Triangle Park, North Carolina

MATUKUMALLI, LAKSHMI, Dept. of Bioinformatics and Computational Biology, George Mason University, Manassas, Virginia

MATZKE, MARJORI, Gregor Mendel Institute of Molecular Plant Biology, Austrian Academy of Sciences, Vienna, Austria

MAZUREK, ANTHONY, Cold Spring Harbor Laboratory, Cold Spring Harbor, New York

MCMULLAN, LAURA, Dept. of Virology and Infectious Disease, Rockefeller University, New York, New York

MCSWIGGEN, JAMES, Dept. of Biochemistry and Bioinformatics, Nastech Pharmaceutical, Bothell, Washington

MEGRAW, MOLLY, Dept. of Genetics, Genomics and Computational Biology Graduate Group, University of Pennsylvania, Philadelphia

MEHLE, ANDREW, Dept. of Molecular and Cell Biology, Howard Hughes Medical Institute, University of California, Berkeley

MEISTER, GUNTER, Dept. of RNA Biology, Max-Planck-Institute for Biochemistry, Munich, Germany

MEYER, ERIC, Laboratoire de Génétique Moléculaire, Centre National de la Recherche Scientifique, Ecole Normale Supérieure, Paris, France

MICA, ERICA, Dept. of Biomolecular Sciences and Biotechnology, University of Milan, Milan, Italy

MICHALIK, STEVE, Dept. of Research and Development, Sigma Aldrich, St. Louis, Missouri

MILLS, NICHOLAS, Dept. of Chemistry and Chemical Biology, University of California, San Francisco

MIRIAMI, ELANA, Dept. of Genetics, The Hebrew University of Jerusalem, Jerusalem, Israel

MISCHO, HANNAH, Sir William Dunn School of Pathology, University of Oxford, Oxford, United Kingdom

MOAZED, DANESH, Dept. of Cell Biology, Harvard Medical School, Boston, Massachusetts

MOELLING, KARIN, Institute of Medical Virology, University of Zürich, Zürich, Switzerland

MOORE, TROY, Open Biosystems, Hudson Alpha Institute for Biotechnology, Huntsville, Alabama

MORILLON, ANTONIN, Centre de Génétique Moléculaire, Centre National de la Recherche Scientifique, Gif-sur-Yvette, France

MORITA, TEPPEI, Div. of Biological Science, Graduate School of Science, Nagoya University, Nagoya, Japan

MORRIS, KEVIN, Dept. of Molecular and Experimental Medicine, The Scripps Research Institute, La Jolla, California

MOSCHENROSS, DARCY, Center for Vascular Biology, School of Medicine, University of Connecticut, Farmington

MUENCHOW, SONJA, Adolf-Butenandt-Institute, Ludwig Maximilians University, Munich, Germany

MULLINAX, BECKY, Dept. of Research and Development, Stratagene, La Jolla, California

MURCHISON, ELIZABETH, Cold Spring Harbor Laboratory, Cold Spring Harbor, New York

NEWMAN, MARTIN, Dept. of Cell and Developmental Biology, University of North Carolina, Chapel Hill

NG KWANG LOON, STANLEY, Dept. of Systems Biology, Bioinformatics Institute, Singapore, Republic of Singapore

NILSEN, TIMOTHY, Center for RNA Molecular Biology, Case Western Reserve University, Cleveland, Ohio

OBERNOSTERER, GREGOR, Lab. of RNAi, Institute of Molecular Biotechnology, Vienna, Austria

OCHSENREITER, TORSTEN, Global Infectious Disease Program, Marine Biological Laboratory, Woods Hole, Massachusetts

OHLSON, JOHAN, Dept. of Molecular Biology and Functional Genomics, Stockholm University, Stockholm, Sweden

OHMAN, MARIE, Dept. of Molecular Biology and Functional Genomics, Stockholm University, Stockholm, Sweden

OONO, KIYOHARU, Research Center for Environmental Genomics, Kobe University, Kobe, Japan

ORLANDO, VALERIO, Dept. of Epigenetics and Genome Reprogramming, Dulbecco Telethon Institute, Institute of Genetics and Biophysics, Consiglio Nazionale delle Ricerche, Naples, Italy

PACKER, ALAN, *Nature Genetics*, Nature Publishing Group, New York, New York

PADALON, GILLY, Dept. of Molecular Genetics and Biotechnology, The Hebrew University-Hadassah Medical School, Jerusalem, Israel

PAN, TAO, Dept. of Biochemistry and Molecular Biology, University of Chicago, Chicago, Illinois

PARK, SEUNG-WON, Dept. of Cell Biology and Molecular Medicine, University of Medicine and Dentistry of New Jersey, Newark

PARK, SUNG-YEON, School of Biological Sciences, Seoul National University, Seoul, South Korea

PARKER, JAMES, Section of Structural Biology, The Institute of Cancer Research, London, United Kingdom

PAROO, ZAIN, Dept. of Biochemistry, Southwestern Medical Center, University of Texas, Dallas

PARROTT, ANDREW, Dept. of Biochemistry, University of Medicine and Dentistry of New Jersey, Newark

PARRY, DEVIN, Dept. of Molecular Biology, Massachusetts General Hospital, Harvard Medical School, Boston, Massachusetts

PATEL, DINSHAW, Dept. of Structural Biology, Memorial Sloan-Kettering Cancer Center, New York, New York

PATEL, REKHA, Dept. of Biological Sciences, University of South Carolina, Columbia

PATTON, JAMES, Dept. of Molecular Biology, Vanderbilt University, Nashville, Tennessee

PAWLICKI, JAN, Dept. of Pharmacology, Yale University, New Haven, Connecticut

PAZ, NURIT, Dept. of Genetics, Sheba Medical Center, Tel-Aviv University, Ramat-Gan, Israel

PEDERSEN, JAKOB, School of Engineering, Center for Biomolecular Science and Engineering, Santa Cruz, California

PENNELL, ROGER, Dept. of Trait Development, Ceres, Thousand Oaks, California

PERKINS, JOHN, Dept. of Biotechnology Research and Development, DSM Nutritional Products, Basel, Switzerland

PERRIMON, NORBERT, Dept. of Genetics, Harvard Medical School, Boston, Massachusetts

PERSSON, HELENA, Dept. of Oncology, Lund University, Lund, Sweden

PETERS, LASSE, Dept. of RNA Biology, Max-Planck-Institute for Biochemistry, Munich, Germany

PEZIC, DUBRAVKA, Lab. of RNAi, Institute of Molecular Biotechnology, Vienna, Austria

PIKAARD, CRAIG, Dept. of Biology, Washington University, St. Louis, Missouri

PLASTERK, RONALD, Dept. of Functional Genomics,

Hubrecht Laboratory, Utrecht, The Netherlands

POETHIG, SCOTT, Dept. of Biology, University of Pennsylvania, Philadelphia

POLLOCK, MILA, Cold Spring Harbor Laboratory Library, Cold Spring Harbor, New York

PRASANTH, KANNANGANATTU, Cold Spring Harbor Laboratory, Cold Spring Harbor, New York

PRASANTH, SUPRIYA, Cold Spring Harbor Laboratory, Cold Spring Harbor, New York

PRESNAIL, JAMES, Dept. of Agriculture and Nutrition, DuPont Pioneer Hi-Bred International, Wilmington, Delaware

PROUDFOOT, NICHOLAS, Sir William Dunn School of Pathology, University of Oxford, Oxford, United Kingdom

QI, YIJUN, Cold Spring Harbor Laboratory, Cold Spring Harbor, New York

RAJEWSKY, NIKOLAUS, Dept. of Biology, Center for Comparative Functional Genomics, New York University, New York

RAPICAVOLI, NICOLE, Dept. of Neuroscience, School of Medicine, Johns Hopkins University, Baltimore, Maryland

RASMUSSEN, SOREN, Dept. of Research and Development, Exiqon, Vedbaek, Denmark

REBATCHOUK, DMITRI, Dept. of Bioinformatics, Sanofi-Aventis, Bridgewater, New Jersey

RECHAVI, GIDEON, Cancer Research Center, Sheba Medical Center, Tel Hashomer, Israel

REDDI, PRABU, Dept. of Pathology, University of Virginia, Charlottesville

REHMSMEIER, MARC, Center for Biotechnology, Bielefeld University, Bielefeld, Germany

REIJNS, MARTIN, Wellcome Trust Centre for Cell Biology, University of Edinburgh, Edinburgh, Scotland, United Kingdom

REINKE, CATHERINE, Dept. of Biology, Carleton College, Northfield, Minnesota

REIS, EDUARDO, Dept. of Biochemistry, University of São Paulo, São Paulo, Brazil

RICHARD, PATRICIA, Laboratoire de Biologie Moléculaire Eucaryote, Centre National de la Recherche Scientifique, Toulouse, France

RIDDIHOUGH, GUY, *Science*, Science Magazine, Washington, D.C.

RIGOUTSOS, ISIDORE, Dept. of Bioinformatics and Pattern Discovery, IBM T.J. Watson Research Center, Yorktown Heights, New York

RINALDO, CHRISTINE, Dept. of Microbiology and Infectious Control, University Hospital of North Norway, Tromsø, Norway

RIVAS, FABIOLA, Cold Spring Harbor Laboratory, Cold Spring Harbor, New York

RODRIGUEZ, ANTONY, Lab. of Mouse Genomics, The Wellcome Trust Sanger Institute, Cambridge, United Kingdom

ROGLER, CHARLES, Dept. of Medicine, Albert Einstein College of Medicine, Bronx, New York

ROMFO, CHARLES, Dept. of Discovery, PTC Therapeutics, South Plainfield, New Jersey

ROUNBEHLER, ROBERT, Dept. of Biochemistry, St. Jude Children's Research Hospital, Memphis, Tennessee

ROVIRA, CARLOS, Dept. of Oncology, Lund University, Lund, Sweden

RUVKUN, GARY, Dept. of Molecular Biology, Massachusetts General Hospital, Harvard Medical School, Boston, Massachusetts

SACCHI, NICOLETTA, Dept. of Cancer Genetics, Roswell Park Cancer Institute, Buffalo, New York

SACHS, ALAN, Dept. of Molecular Profiling, Rosetta Inpharmatics, Merck Research Laboratories, Seattle, Washington

SALEH, MARIA CARLA, Dept. of Microbiology and Immunology, University of California, San Francisco

SALOTTI, JACQUELINE, Dept. of Biochemistry, Institute of Chemistry, University of São Paulo, São Paulo, Brazil

SALZMAN, DAVID, Dept. of Molecular, Microbial, and Structural Biology, Center for Vascular Biology, Health Center, School of Medicine, University of Connecticut, Farmington

SANBONMATSU, KEVIN, Dept. of Theoretical Biology and Biophysics, Los Alamos National Laboratory, Los Alamos, New Mexico

SANDER, CHRIS, Computational Biology Center, Memorial Sloan-Kettering Cancer Center, New York, New York

SANDVIK, KJERSTI, Dept. of Microbiology and Infectious Control, University Hospital of North Norway, Tromsø, Norway

SARNOW, PETER, Dept. of Microbiology and Immunology, School of Medicine, Stanford University, Stanford, California

SAUGSTAD, JULIE, R.S. Dow Neurobiology Laboratories, Legacy Research, Portland, Oregon

SCHIER, ALEXANDER, Dept. of Molecular and Cellular Biology, Harvard University, Cambridge, Massachusetts

SCHOENBERG, DANIEL, Dept. of Molecular and Cellular Biochemistry, Ohio State University, Columbus, Ohio

SCHUMACHER, HEIKO, Dept. of Plant Sciences, Institute of Molecular Biology and Biotechnology, Heraklion, Crete, Greece

SEBAT, JONATHAN, Cold Spring Harbor Laboratory, Cold Spring Harbor, New York

SEILA, AMY, Center for Cancer Research, Massachusetts Institute of Technology, Cambridge, Massachusetts

SERGANOV, ALEXANDER, Dept. of Structural Biology, Memorial Sloan-Kettering Cancer Center, New York, New York

SETHUPATHY, PRAVEEN, Dept. of Genetics, Center for Bioinformatics, University of Pennsylvania, Philadelphia

SETO, ANITA, Dept. of Molecular Biology, Massachusetts General Hospital, Harvard Medical School, Boston, Massachusetts

SHARMA, CYNTHIA, Dept. of RNA Biology, Max-Planck-Institute for Infection Biology, Berlin, Germany

SHARP, PHILLIP, Center for Cancer Research, Massachusetts Institute of Technology, Cambridge, Massachusetts

SHAW, ROBERT, The Wellcome Trust, Cancer Research UK Gurdon Institute, University of Cambridge, Cambridge, United Kingdom

SHEA, CATHY, Dept. of Research, New England Biolabs, Ipswich, Massachusetts

SHEFER, KINNERET, Dept. of Genetics, The Hebrew University of Jerusalem, Jerusalem, Israel

SHIBATA, KAZUHIRO, Dept. for Coordination Program of Science and Technology Projects, Japan Science and Technology Agency, Tokyo, Japan

SHIELDS, ROBERT, *Trends in Genetics*, Current Trends, Elsevier Science, London, United Kingdom

SHIN, JINWOOK, National Creative Research Initiative Center, Seoul National University, Seoul, South Korea

SHINOHARA, FUMIKAZU, BioFrontier Laboratories, Kyowa Hakko Kogyo Co., Tokyo, Japan

SIGOVA, ALLA, Dept. of Biochemistry and Molecular Pharmacology, Medical School, University of Massachusetts, Worcester

SIJEN, TITIA, Dept. of Functional Genomics, Hubrecht Laboratory, Utrecht, The Netherlands

SIM, SOYEONG, Dept. of Cell Biology, Yale University School of Medicine, New Haven, Connecticut

SIOLAS, DESPINA, Cold Spring Harbor Laboratory, Cold Spring Harbor, New York

SIU, FAI, Dept. of Molecular and Cell Biology, University of California, Berkeley

SLACK, FRANK, Dept. of Molecular, Cellular and Developmental Biology, Yale University, New Haven, Connecticut

SMIT, MARIA, *Genome Research*, Cold Spring Harbor Laboratory Press, Woodbury, New York

SMITS, GUILLAUME, Dept. of Developmental Genetics and Genomic Imprinting, The Babraham Institute, Cambridge, United Kingdom

SNYDER, MICHAEL, Dept. of Molecular, Cellular and Developmental Biology, Yale University, New Haven, Connecticut

SONENBERG, NAHUM, Dept. of Biochemistry, McGill University, Montréal, Québec, Canada

SOUTSCHEK, JÜRGEN, Alnylam Pharmaceuticals, Kulmbach, Germany

SPECTOR, DAVID, Cold Spring Harbor Laboratory, Cold Spring Harbor, New York

STAMM, STEFAN, Institute for Biochemistry, University of Erlangen, Erlangen, Germany

STEIN, PAULA, Dept. of Biology, University of Pennsylvania, Philadephia

STEINER, FLORIAN, Dept. of Functional Genomics, Hubrecht Laboratory, Utrecht, The Netherlands

STEITZ, JOAN, Dept. of Molecular Biophysics and Biochemistry, Yale University School of Medicine, New Haven, Connecticut

STERN, BODO, *Cell*, Cell Press, Cambridge, Massachusetts

STEWART, DAVID, Meetings and Courses Programs, Cold Spring Harbor Laboratory, Cold Spring Harbor, New York

STILLMAN, BRUCE, President and CEO, Cold Spring Harbor Laboratory, Cold Spring Harbor, New York

STORZ, GISELA, Cell Biology and Metabolism Branch, National Institute of Child Health and Human Development, National Institutes of Health, Bethesda, Maryland

STRECK, RANDAL, Dept. of Investigative Developmental Toxicology, Pfizer, Groton, Connecticut

STRICKER, STEFAN, Max F. Perutz Laboratories, Institute for Microbiology and Genetics, Center for Molecular Medicine, University of Vienna, Vienna, Austria

SUESS, BEATRIX, Department of Microbiology, University of Erlangen, Erlangen, Germany

SUHASINI, AVVARU, Centre for Cellular and Molecular Biology, Hyderabad, India

SUNWOO, HONGJAE, Cold Spring Harbor Laboratory, Cold Spring Harbor, New York

SUSSMAN, HILLARY, *Genome Research*, Cold Spring Harbor Laboratory Press, Woodbury, New York

SYLVESTRE, YANNICK, Dept. of Biochemistry, Université de Montréal, Montréal, Québec, Canada

TAKEDA, HARUKO, Dept. of Genetics, University of Liège, Liège, Belgium

TERNS, MICHAEL, Dept. of Biochemistry and Molecular Biology, University of Georgia, Athens

THEUNISSEN, JAN-WILLEM, *Nature Biotechnology*, Nature Publishing Group, New York, New York

THEURKAUF, WILLIAM, Program in Molecular Medicine, Medical Center, University of Massachusetts, Worcester

THOMSON, J. MICHAEL, Dept. of Cell and Developmental Biology, University of North Carolina, Chapel Hill

TIAN, YUAN, Dept. of Structural Biology, Memorial Sloan-Kettering Cancer Center, New York, New York

TIMMERMANS, MARJA, Cold Spring Harbor Laboratory, Cold Spring Harbor, New York

TOLIA, NIRAJ, Cold Spring Harbor Laboratory, Cold Spring Harbor, New York

TOPS, BASTIAAN, Dept. of Functional Genomics, Hubrecht Laboratory, Utrecht, The Netherlands

TOREN, GINAT, Dept. of Human Molecular Genetics and Biochemistry, Sackler School of Medicine, Tel Aviv University, Tel Aviv, Israel

TROTTA, CHRISTOPHER, Dept. of Discovery, PTC Therapeutics, South Plainfield, New Jersey

TSAI, HSIN-YUE, Dept. of Molecular, Cellular and Developmental Biology, Ohio State University, Columbus, Ohio

TSAI, NIEN-PEI, Dept. of Pharmacology, University of Minnesota, Minneapolis

TSIRIGOS, ARISTOTELIS, Dept. of Bioinformatics, IBM Research, Astoria, New York

TSUKADA, KOJI, Dept. of Biotechnology, Osaka University, Osaka, Japan

TU, KIMBERLY, Dept. of Molecular Biology, Princeton University, Princeton, New Jersey

TU, ZHIJIAN, Dept. of Biochemistry, Virginia Polytechnic Institute and State University, Blacksburg, Virginia

TZERTZINIS, GEORGE, Dept. of Research, New England Biolabs, Ipswich, Massachusetts

URBAN, JOHANNES, Dept. of RNA Biology, Max-Planck-Institute for Infection Biology, Berlin, Germany

USHIDA, CHISATO, Dept. of Biochemistry and Biotechnology, Hirosaki University, Hirosaki, Japan

VAGIN, VASILY, Dept. of Biochemistry and Molecular Pharmacology, Medical School, University of Massachusetts, Worcester

VAN DER MEIJDEN, CAROLINE, Cold Spring Harbor Laboratory, Cold Spring Harbor, New York

VAN RIJ, RONALD, Dept. of Microbiology and Immunology, University of California, San Francisco

VASUDEVAN, SHOBHA, Dept. of Molecular Biophysics and Biochemistry, Yale University School of Medicine, New Haven, Connecticut

VAUGHN, MATTHEW, Cold Spring Harbor Laboratory, Cold Spring Harbor, New York

VIKESAA, JONAS, Dept. of Clinical Biochemistry, Rigshospitalet, University of Copenhagen, Copenhagen, Denmark

VOGEL, JOERG, Dept. of RNA Biology, Max-Planck-Institute for Infection Biology, Berlin, Germany

VOLFOVSKY, NATALIA, Advanced Biomedical Computing Center, SAIC-Frederick, National Cancer Institute, Frederick, Maryland

WACHTMANN, TIM, Cardiovascular Multimedia Information Network, Pfizer, Groton, Connecticut

WADE, NICHOLAS, *The New York Times*, Science Department, New York, New York

WAHLSTEDT, HELENE, Dept. of Molecular Biology and Functional Genomics, Stockholm University, Stockholm, Sweden

WANG, HUI, Agilent Laboratories, Agilent Technologies, Palo Alto, California

WANG, I.-FAN, Neuroscience Research Institute, University of California, Santa Barbara,

WANG, JINHUA, Hartwell Center for Bioinformatics and Biotechnology, St. Jude Children's Research Hospital, Memphis, Tennessee

WARE, DOREEN, Cold Spring Harbor Laboratory, Cold Spring Harbor, New York

WEHNER, KAREN, Dept. of Microbiology and Immunology, School of Medicine, Stanford University, Stanford, California

WESTERHOUT, ELLEN, Dept. of Human Retrovirology, Academic Medical Center, University of Amsterdam, Amsterdam, The Netherlands

WHITMARSH, BARBARA, Center for Scientific Review, National Institutes of Health, Bethesda, Maryland

WINKLER, WADE, Dept. of Biochemistry, Southwestern Medical Center, University of Texas, Dallas

WISHART, WILLIAM, Dept. of Genome and Proteome Sciences, Novartis Institutes for Biomedical Research, Basel, Switzerland

WITKOWSKI, JAN, Banbury Center, Cold Spring Harbor Laboratory, Cold Spring Harbor, New York

WITTRUP, ANDERS, Dept. of Oncology, Lund University, Lund, Sweden

WOLIN, SANDRA, Dept. of Cell Biology, Howard Hughes Medical Institute, Yale University School of Medicine, New Haven, Connecticut

WOLLMANN, HEIKE, Dept. of Molecular Biology, Max-Planck-Institute for Developmental Biology, Tübingen, Germany

WOMBLE, KRISTIE, Dept. of Innovative Projects, BioVentures, Murfreesboro, Tennessee

WU, XIAOYUN, Dept. of Molecular Biology, Massachusetts General Hospital, Boston, Massachusetts

XU, NING, Dept. of Medical Biochemistry and Microbiology, Uppsala University, Uppsala, Sweden

XUAN, ZHENYU, Cold Spring Harbor Laboratory, Cold Spring Harbor, New York

YAO, GANG, Dept. of Biochemistry and Molecular Biology, University of New Hampshire, Durham

YAO, JIE, Dept. of Applied and Engineering Physics, Cornell University, Ithaca, New York

YEO, GENE, Crick-Jacobs Center for Computational and Theoretical Biology, The Salk Institute, La Jolla, California

YILDIRIM, EDA, Lab. of Signal Transduction, National Institute of Environmental Health Sciences, National Institutes of Health, Research Triangle Park, North Carolina

YOU, SHIHYUN, Center for the Study of Hepatitis C, Rockefeller University, New York, New York

YOUNGMAN, ELAINE, Dept. of Molecular Biology and Genetics, School of Medicine, Johns Hopkins University, Baltimore, Maryland

YU, YANG, Dept. of Biochemistry, RNA Center, Case Western Reserve University, Cleveland, Ohio

YU, ZHENBAO, Biotechnology Research Institute, National Research Council Canada, Montréal, Québec, Canada

YUAN, CHIH-CHI, Cold Spring Harbor Laboratory, Cold Spring Harbor, New York

YUAN, YU-REN, Structural Biology Program, Memorial Sloan-Kettering Cancer Center, New York, New York

ZAMORE, PHILLIP, Dept. of Biochemistry and Molecular Pharmacology, Medical School, University of Massachusetts, Worcester

ZATECHKA, STEVEN, Dept. of Biochemistry, Howard Hughes Medical Institute, St. Jude Children's Research Hospital, Memphis, Tennessee

ZHANG, CHI, Dept. of Biochemistry and Molecular Biology, Massachusetts General Hospital, Harvard Medical School, Boston, Massachusetts

ZHANG, ELISA, Dept. of Biochemistry, Molecular and Cellular Biology, Harvard University, Cambridge, Massachusetts

ZHANG, WEIGUO, Div. of Life Sciences, Lawrence Berkeley National Laboratory, Berkeley, California

ZHANG, YINHUA, Dept. of Biochemistry and Parasitology, New England Biolabs, Ipswich, Massachusetts

ZHANG, YUANJI, Dept. of Biochemistry and Bioinformatics, Monsanto, St. Louis, Missouri

ZHAO, MENG, Institute for Cell and Molecular Biology, University of Texas, Austin

ZHENG, ZHI-MING, HIV and AIDS Malignancy Branch, National Cancer Institute, National Institutes of Health, Bethesda, Maryland

ZHOU, QIANG, Dept. of Biochemistry, Molecular and Cell Biology, University of California, Berkeley

ZHOU, RUI, Dept. of Biochemistry and Genetics, Harvard Medical School, Watertown, Massachusetts

ZIEVE, GARY, Dept. of Biochemistry and Pathology, State University of New York, Stony Brook

First row: P. Aman, Z.-M. Zheng; R. Maraia, Q. Zhou
Second row: S. Wolin, A. Seto; P. Zamore; J. Doudna, L. Joshua-Tor
Third row: M. Jovanovic, M. Landthaler; P. Farabaugh, D. Schoenberg; P. Sarnow, A. Mehle
Fourth row: N. Rajewsky, D. Stewart; M. Matzke, A. Barta

First row: N. Sonenberg, P. Sharp, B. Stillman, T. Cech; S. Eddy, C. Hunter
Second row: S. Stamm, P. Grabowski; T. Nilsen, B. Bass, A. Feig
Third row: S. Grewal, D. Moazed; J. Elmén, M. Bak
Fourth row: P. Sethupathy, A. Hatzigeorgiou, J. Brennecke; R. Bernards, R. Plasterk

First row: J. Perkins, V. Dubowitz; S. Jayasena, J. McSwiggen
Second row: K. Beemon presenting her poster; G. Carmichael, A. Krainer
Third row: C. van der Meijden, J. Inglis, S. Locke; M.C. Saleh, R. Andino
Fourth row: S. Poethig, E. Lund; B. Whitmarsh, C. Pikaard

First row: A. Denli, Y. Qi; M. Rehmsmeier, B. Seuss; E. Makeyev, V. Vagin
Second row: E. Izaurralde, T. Nilsen; F. Slack, S. Kennedy, C. Hammell
Third row: Dining on the terrace overlooking Cold Spring Harbor

First row: N. Sacchi, J. Watson; J. Mattick, E. Blackburn; S. Gilbert
Second row: E. Heard, R. Martienssen; M. Snyder, J. Perkins, D. Baulcombe
Third row: E. Berezikov, C. Hunter, G. Ruvkun, C. Esau; M. Terns, J. Dahlberg, J. Steitz
Fourth row: J. Hicks, N. Wade, H. Sussman; G. Hannon; S. Gundersen, N. Proudfoot
Fifth row: K. Moelling, B. Stillman; V. Narry Kim; Lunch al fresco

First row: M. Timmermans, L. Zhang, M. Guo; J. Salotti
Second row: E. Dawson, J. Witkowski; A. Hüttenhofer, T. Kiss
Third row: I. Bozzoni, W. Filipowicz; G. Zieve, G. Dreyfuss; R. Shields, M. Pollock
Fourth row: C. Trotta, J. Vikesaa; C. Borel, E. Yildirim; D. Schoenberg, W. Winkler
Ffth row: Picnic at Blackford

Foreword

The role of DNA as the primary information carrier and RNA as the interpretative system for that information has been the key tenet of modern biology— the central dogma of molecular biology. This was modified by the discovery of reverse transcription and presented at this Symposium in 1970. The large majority of enzymatic reactions are carried out by proteins, although the discovery of catalytic RNA and its emerging centrality in certain highly conserved cellular processes such as protein synthesis and eukaryotic mRNA splicing have increasingly suggested that RNA plays a more diverse and complex role in the cell than initially proposed. The discovery of RNA interference and the ubiquity of microRNAs in diverse systems have clearly established that this polymer regulates gene transcription and translation in hitherto unexpected ways, as well as providing cellular defense mechanisms against viruses. How many of these regulatory processes are relics of an ancient RNA-based world, as has been suggested for the riboswitches, remains unclear, but their growing importance in both prokaryotic and eukaryotic biology is fuelling research and new ideas in laboratories around the world. Transcriptional and translational control by RNA, epigenetic phenomena such as RNA-directed DNA methylation and RNAi-directed heterochromatin formation, meiotic silencing, and cellular defense mechanisms against viruses are just some of the current areas under intense investigation through a variety of genetic, genomic, biochemical, cell biological, and structural approaches.

It was therefore deemed timely to focus the annual Symposium broadly on the theme of regulation of cellular processes by RNA-based mechanisms. It is increasingly clear that many of these processes may be harnessed to perturb biological function in a variety of biological systems, work that has reached its most widespread application with the use of large-scale RNAi screens. The Symposium therefore explored how the application of these technologies, combined with a growing understanding of the molecular basis of many of these processes, is yielding new insights into the treatment of many human diseases.

In organizing this Symposium with considerable help from Terri Grodzicker, we relied on the assistance of our colleagues Greg Hannon and Adrian Krainer for suggestions for speakers, while Susan Gottesman provided useful guidance in the prokaryotic field. We also thank the first evening speakers, Susan Gottesman, David Baulcombe, Gary Ruvkun, and Greg Hannon for providing an overview of the areas to be covered. This year's Reginald Harris Lecture was delivered by Elizabeth Blackburn on telomere ribonucleoprotein biology. We particularly thank Joan Steitz, with help from Greg Hannon, for delivering a thoughtful and realistic summary of the current state of the field, and Ronald Plasterk, who ably managed to convey the excitement and surprises of the emerging world of small RNAs in his Dorcas Cummings Lecture to the local community and the attending scientists.

This Symposium was attended by 469 scientists from over 25 countries, and the program included 70 oral presentations and 209 poster presentations. Essential funds to run this meeting were obtained from the National Cancer Institute, a branch of the National Institutes of Health. In addition, financial help from the corporate benefactors, sponsors, affiliates, and contributors of our meetings program is essential for these Symposia to remain a success, and we are most grateful for their continued support.

We thank Val Pakaluk and Mary Smith in the Meetings and Courses Program office for their efficient help in organizing the Symposium. Joan Ebert and Patricia Barker in the Cold Spring Harbor Laboratory Press, headed by John Inglis, ensured that this volume would be produced. We thank them for their dedication to producing high-quality publications.

Bruce Stillman
David Stewart
January 2007

Sponsors

This meeting was funded in part by the **National Cancer Institute,** a branch of the **National Institutes of Health.**

Contributions from the following companies provide core support for the Cold Spring Harbor meetings program.

Corporate Patron

Pfizer, Inc.

Corporate Benefactors

Amgen, Inc.
GlaxoSmithKline

Novartis Institutes for BioMedical Research

Corporate Sponsors

Abbott Laboratories
Applied Biosystems
BioVentures, Inc.
Bristol-Myers Squibb Company
Diagnostic Products Corporation
Forest Laboratories, Inc.
GE Healthcare Bio-Sciences
Genentech, Inc.
Hoffmann-La Roche, Inc.

Johnson & Johnson Pharmaceutical Research & Development, LLC
Kyowa Hakko Kogyo Co., Ltd.
Merck Research Laboratories
New England BioLabs, Inc.
OSI Pharmaceuticals, Inc.
Pall Corporation
Sanofi-Aventis
Schering-Plough Research Institute

Plant Corporate Associates

ArborGen
Monsanto Company
Pioneer Hi-Bred International, Inc.

Corporate Affiliates

Abcam, Ltd.
Agilent Technologies

Corporate Contributors

Aviva Systems Biology
Bethyl Laboratories
Cell Signaling Technology
Epicentre Biotechnologies

Illumina
inGenious Targeting Laboratory, Inc.
IRx Therapeutics, Inc.

Foundations

Albert B. Sabin Vaccine Institute, Inc.
Hudson Alpha Institute for Biotechnology

Contents

Control of Gene Expression by Noncoding RNAs

Heterochromatin

Quality Control, Messenger RNA Turnover, and Translational Control

Small RNA Regulators and the Bacterial Response to Stress

S. Gottesman,* C.A. McCullen,* M. Guillier,* C.K. Vanderpool,*† N. Majdalani,*
J. Benhammou,* K.M. Thompson,* P.C. FitzGerald,‡ N.A. Sowa,* and D.J. FitzGerald*

*Laboratory of Molecular Biology, Center for Cancer Research, National Cancer Institute,
Bethesda, Maryland 20892; ‡Genome Analysis Unit, Center for Cancer Research,
National Cancer Institute, Bethesda, Maryland 20892

Recent studies have uncovered dozens of regulatory small RNAs in bacteria. A large number of these small RNAs act by pairing to their target mRNAs. The outcome of pairing can be either stimulation or inhibition of translation. Pairing in vivo frequently depends on the RNA-binding protein Hfq. Synthesis of these small RNAs is tightly regulated at the level of transcription; many of the well-studied stress response regulons have now been found to include a regulatory RNA. Expression of the small RNA can help the cell cope with environmental stress by redirecting cellular metabolism, exemplified by RyhB, a small RNA expressed upon iron starvation. Although small RNAs found in *Escherichia coli* can usually be identified by sequence comparison to closely related enterobacteria, other approaches are necessary to find the equivalent RNAs in other bacterial species. Nonetheless, it is becoming increasingly clear that many if not all bacteria encode significant numbers of these important regulators. Tracing their evolution through bacterial genomes remains a challenge.

The bacterial genomes of organisms such as *E. coli* contain the information to allow the bacteria to thrive under a variety of conditions, both inside mammalian hosts and in the external environment. This requires systems to sense, respond to, and recover from a variety of stressful treatments and changes in nutrient availability. Our understanding of these systems has expanded rapidly, enhanced by the sequencing of multiple bacterial genomes. Transcriptional regulators and changes in the basic transcription machinery via use of alternative σ factors provide appropriate regulated expression of a variety of repair and recovery genes.

In recent years, it has become increasingly clear that in addition to these transcriptional regulatory programs, stress responses also involve small regulatory RNAs that have important roles in the posttranscriptional regulation of many genes. These RNAs are frequently regulated at the level of synthesis, as part of larger regulons, and may have roles in the immediate response to stress and/or the recovery from stress. As with eukaryotic microRNAs (miRNAs) and small interfering RNAs (siRNAs), these bacterial regulatory RNAs, called sRNAs here, frequently act by pairing with specific target mRNAs to change their translation and/or stability. Other sRNAs act to influence the activity of proteins; the function of many others is still unknown (for review, see Gottesman 2004; Storz et al. 2005; Storz and Gottesman 2006) and not discussed further here. Global searches for sRNAs in *E. coli*, expanding studies on these RNAs and their function, and extension of findings to other bacterial species have provided us with an understanding of how these sRNAs contribute to bacterial physiology and stress responses.

PAIRING RNAs: MODE OF ACTION OF A "TYPICAL" BACTERIAL SMALL RNA

The most studied and thus far most numerous family of regulatory sRNAs act by base-pairing with specific mRNAs and, in *E. coli*, are almost always associated with the RNA chaperone Hfq (see below). The steps in the function of sRNAs of this family are outlined in Figure 1 and discussed in some detail below.

Control of Synthesis as a Primary Regulatory Step

For any regulator, an important issue is how regulation occurs at the right time and place. For regulatory sRNAs, the primary response to environmental cues occurs at the level of synthesis. When the regulatory sRNA is made at high levels, biological effects rapidly follow. Promoters for bacterial regulatory RNAs do not differ qualitatively from promoters for protein-coding genes. However, the strength of the sRNA promoter when fully expressed and the extent of regulation may both be particularly stringent. sRNAs belong to well-characterized regulons such as those responding to oxidative stress (Altuvia et al. 1997), osmotic stress (Guillier and Gottesman 2006), and iron starvation (Massé and Gottesman 2002). Thus, the complex network of bacterial transcriptional regulators and their mechanisms for sensing cues are put to use to synthesize these sRNAs.

Some of the regulatory proteins that have been demonstrated to regulate sRNAs are summarized in Table 1. It seems likely that other global regulators will also prove to include noncoding RNAs in their regulons. Only SgrR, newly discovered via its role in regulating the sRNA SgrS (Vanderpool and Gottesman 2004), has not been found to control a large regulon in addition to the sRNA (C. Vanderpool and S. Gottesman, in prep.). The genes encoding SgrR, as well as the LysR family regulators

---

†Present address: Department of Microbiology, University of Illinois, Urbana-Champaign, Ilinois 61801.

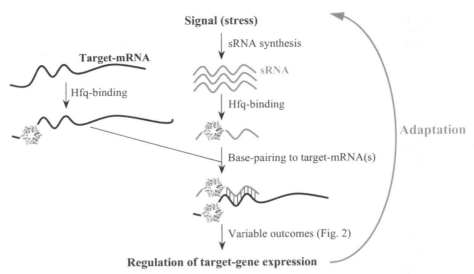

Figure 1. Cycle of activity of Hfq-dependent regulatory RNAs.

GcvA and OxyR, are next to but transcribed divergently from the sRNA that they regulate. The intergenic regions encoding the promoters and regulatory sites for the transcriptional regulator and sRNA thus overlap, allowing coordinate regulation of the regulatory protein and the sRNA, and this suggests the possibility of coevolution during any horizontal transfer events. In all the other cases listed in Table 1, the regulatory proteins and regulated noncoding RNAs are encoded far from each other.

A given regulatory protein can regulate more than one sRNA. σ^E controls expression of the apparently unrelated sRNAs MicA and RybB (Table 1). There are a number of examples of closely related RNAs, presumably evolved through duplication events, whose synthesis is controlled

Table 1. Regulatory Circuits with Hfq-dependent Small RNAs

Regulatory protein family	Regulatory protein	Inducing signal	Small RNAs[a]	Targets[b]	References
LysR	OxyR	oxidative stress	OxyS	*fhlA*; *yobF*, *wrbA*, *ybaY*	Altuvia et al. (1997); Argaman and Altuvia (2000); Tjaden et al. (2006)
	GcvA		GcvB	*dpp*, *opp*	Urbanowski et al. (2000); McArthur et al. (2006)
Two-component	OmpR	osmotic shock	OmrA, OmrB	*ompT*, *cirA*, other cell surface genes	Guillier and Gottesman (2006)
	RcsB	cell surface stress	RprA	*rpoS*	Majdalani et al. (2001)
	LuxO	quorum sensing	Qrr1-4 (Vc)	*hapR*	Lenz et al. (2004)
Sigma factor	Sigma E	periplasmic stress	MicA	*ompA*	Udekwu et al. (2005); Johansen et al. (2006)
			RybB	*sigE*	Johansen et al. (2006); K. Thompson and S. Gottesman (in prep.); Vogel and Papenfort (2006)
Fur repressor	Fur	iron limitation	RyhB (Ec) (Vc) PrrF (Pa)	*sodB*, *sdh*, iron-binding proteins	Massé and Gottesman (2002); Wilderman et al. (2004); Davis et al. (2005); Mey et al. (2005)
Mar family	Mar, SoxS, Rob	oxidative stress, antibiotic stress	MicF[c]	*ompF*	Delihas and Forst (2001)
Sugar binding, novel SgrR		glucose-phosphate accumulation	SgrS	*ptsG*	Vanderpool and Gottesman (2004)
CRP	CRP	glucose limitation	Spot 42	*galK*	Møller et al. (2002a)

[a]Small RNAs were identified in *E. coli* (Ec) unless otherwise indicated. (Vc) *Vibrio cholerae*; (Pa) *Pseudomonas aeruginosa*.
[b]Targets are negatively regulated except for *rpoS*, which is positively regulated.
[c]MicF has also been reported to be regulated by OmpR under some conditions (Ramani et al. 1994).

by the same regulatory protein, including the four Qrr RNAs in *Vibrio cholerae*, the duplicate PrrF RNAs in *Pseudomonas*, and OmrA and B in *E. coli*. Duplicated RNAs may have evolved somewhat different targets, or their synthesis may respond with different thresholds to inducing signals.

The primary transcript of the sRNA is generally active for regulation, without any requirement for processing, in contrast to the essential processing steps for microRNAs in eukaryotic cells (for review, see Cullen 2004; Gottesman 2005). Although processed transcripts are sometimes seen (Argaman et al. 2001; Repoila and Gottesman 2001; Sledjeski et al. 2001; Vogel et al. 2003; Opdyke et al. 2004), there is little information that addresses whether these are active or, if active, whether processing is essential for activity. In immunoprecipitation experiments with the RNA chaperone Hfq, the primary transcript is found to bind this protein (see below) (Wassarman et al. 2001; Zhang et al. 2003).

Finding the regulators of a small RNA. Given that the major regulatory step for most sRNAs is at the level of transcription, identifying the transcriptional regulators that control sRNA expression is a key step to understanding the physiological importance of the sRNA. Analysis of expression of the RNA by northern blot under various growth conditions is frequently the first step; computational approaches or other methods may provide clues to conditions to test (see, e.g., Massé and Gottesman 2002; Johansen et al. 2006). For more precise information, or in the absence of clues, the same sorts of approaches that can be used to define the promoters of any gene can be used to find regulators of these RNAs—deletion analysis of promoters using transcriptional fusions, and genetic screens and selections using the fusions to define *trans*-acting regulators (Majdalani et al. 2002; Guillier and Gottesman 2006).

Hfq Binding and the Pairing of the Regulatory Small RNAs with mRNAs

In vivo roles of Hfq. The RNA-binding protein Hfq was first identified biochemically by its role in the in vitro replication of the RNA phage Qβ (Blumenthal and Carmichael 1979). Later studies of *hfq* mutants demonstrated that cells which are devoid of Hfq grow slowly and have very low levels of RpoS (Tsui et al. 1994; Muffler et al. 1996). The finding that Hfq was also involved in the action of some sRNAs provided a possible explanation for these phenotypes (Zhang et al. 1998; Sledjeski et al. 2001).

In immunoprecipitation experiments using an anti-Hfq antibody, the sRNAs that bind and use Hfq are significantly enriched in the immunoprecipitate and can be detected even when not specifically induced (Wassarman et al. 2001; Zhang et al. 2003). This tight binding to Hfq defines the family of Hfq-binding sRNAs. In many cases, the Hfq-binding sRNAs are significantly less stable in the absence of Hfq and consequently accumulate to lower levels (Sledjeski et al. 2001; Møller et al. 2002b; Massé et al. 2003; Zhang et al. 2003; Antal et al. 2005). Thus, it is generally assumed that Hfq rapidly binds and protects

sRNAs of this class and that it is the Hfq-bound form which is active in vivo for pairing with target mRNAs (Fig. 1), but this has not been directly demonstrated.

The biological effects of these Hfq-binding sRNAs are absent in *hfq* mutants (see, e.g., Zhang et al. 1998; Sledjeski et al. 2001; Massé and Gottesman 2002; Møller et al. 2002a). For instance, translation of the stationary-phase σ factor, RpoS, is dramatically reduced in *E. coli hfq* mutants; mutations that increase translation disrupt an inhibitory hairpin in the *rpoS* leader mRNA (Brown and Elliott 1997). At least two sRNAs, and probably others, stimulate translation of *rpoS* by interacting with and opening the inhibitory hairpin (for review, see Repoila et al. 2003). In an *hfq* mutant, these sRNAs can no longer stimulate translation and RpoS is not made. Another example is provided by the phenotype of *fur* mutants of *E. coli*. Fur represses RyhB, and RyhB down-regulates the genes encoding succinate dehydrogenase (see below). As a result, *fur* mutants cannot grow on succinate. Mutations in either *ryhB* or *hfq* can restore growth (Massé and Gottesman 2002). *hfq* mutants of *Vibrio* and *Brucella* are avirulent (Robertson and Roop 1999; Ding et al. 2004), and *hfq* mutants of *Vibrio* and *Pseudomonas* have defects in the quorum sensing pathway (Lenz et al. 2004; Sonnleitner et al. 2006), although it has not been demonstrated in all of these cases that the phenotypes are due to loss of function of sRNAs and not some other role of Hfq.

In vitro activities of Hfq. Rapid turnover of the regulatory sRNAs in vivo in the absence of Hfq might be a sufficient explanation for loss of function of these sRNAs, but a variety of in vitro experiments suggest that Hfq has a more direct role as an RNA chaperone. Specifically, interactions of an sRNA and target mRNA are promoted by the presence of Hfq in vitro (Møller et al. 2002b; Zhang et al. 2002; Lease and Woodson 2004; Kawamoto et al. 2006). In these and other experiments, Hfq binds the target mRNA as well as the regulatory RNA (Geissmann and Touati 2004). What proportion of the population of a given mRNA or sRNA is bound at any time has not been examined in vivo.

How does binding of Hfq to both mRNA and regulatory RNA promote pairing? Hfq subunits assemble into a hexameric ring (Fig. 1); a model oligonucleotide has been shown to bind to the inner surface of the ring, and evidence for secondary RNA-binding sites have also been described (Schumacher et al. 2002; Sauter et al. 2003; Mikulecky et al. 2004; Valentin-Hansen et al. 2004). Models depending on pairing of Hfq rings or binding of both RNAs to a single ring have been suggested (Storz et al. 2004) and may be important in stabilizing RNA:RNA interactions once they have been initiated. However, most bacterial species, including *E. coli*, encode only one *hfq* gene; thus, Hfq must be able to promote interactions between many sRNAs that pair with different mRNAs. This implies that Hfq does not itself provide specificity of pairing. Instead, pairing may proceed more rapidly when the RNA secondary or tertiary structure is remodeled by Hfq binding (Moll et al. 2003a; Geissmann and Touati 2004; Lease and Woodson 2004; Antal et al. 2005), and, once pairing is initiated, it may be extended and/or stabilized by interactions with Hfq. Thus, although complementarity between the sRNA and the target mRNA may be essential for pairing (Kawamoto et al. 2006), it is apparently not sufficient in most cases. High-affinity Hfq-binding sites on both the sRNA and the target mRNA provide additional necessary elements. Computational approaches that do not take this into account may predict that pairing between sRNAs and target mRNAs would not occur in vivo.

Outcomes of Pairing

Pairing of the sRNA to its target mRNA can result in positive or negative effects on translation. In the best-studied case of stimulation of translation, two different sRNAs, DsrA and RprA, bind to counteract the formation of an inhibitory hairpin in the mRNA of *rpoS* that blocks translation (Fig. 2A) (Majdalani et al. 1998, 2002). In many cases, negative effects are associated with the sRNA binding close to the ribosome-binding site. Binding may both inhibit translation (under conditions where mRNA degradation is blocked) and lead to rapid mRNA degradation (Morita et al. 2006) (Fig. 2B). It is not yet clear whether mRNA degradation is secondary to blocking ribosome access.

Translation Inhibition and mRNA Degradation

For most of the cases of sRNA-dependent mRNA degradation, the target message disappears rapidly (within 2–3 minutes) and completely after induction of the regulatory RNA (Massé and Gottesman 2002; Massé et al. 2005; Rasmussen et al. 2005; Udekwu et al. 2005; Guillier and Gottesman 2006); in some cases, much smaller degradation products can be detected (Morita et al. 2003; Vanderpool and Gottesman 2004). Degradation is generally due to the activity of RNase E (Massé et al. 2003; Afonyushkin et al. 2005; Morita et al. 2005). RNase E is an essential endonuclease that cleaves at single-stranded AU-rich RNA (for review, see Kennell 2002).

Presumably, endonucleolytic cleavage by RNase E serves as a primary step, followed by degradation by a variety of exonucleases. Absence of translation can uncover RNase E cleavage sites internal to a coding region that are otherwise inefficiently cleaved (Joyce and Dreyfus 1998; Baker and Mackie 2003; for review, see Deana and Belasco 2005). Thus, an sRNA might promote rapid degradation of a message by blocking ribosome access, thereby allowing access of RNase E to recognition sites internal to the mRNA normally masked by translating ribosomes. If this is the primary cause of degradation of mRNAs targeted by sRNAs, the absence of an internal RNase E site on a given mRNA would allow regulation of translation without degradation of that mRNA. Regulation of translation without significant mRNA degradation has been described for Spot 42 regulation of translation in the *gal* operon (Fig. 2B) (Møller et al. 2002a), although the presence or absence of RNase E sites has not been addressed. Alternatively, the act of pairing may by itself make the mRNA accessible to RNase E, by changing the structure of the RNA. There is evidence that RNase E activity is affected by secondary structure (for review, see Kennell 2002); possibly, the paired RNAs can mimic secondary structure elements in stimulating recognition or cleavage of the mRNA. Finally, recent work suggests that Hfq and RNase E are physically associated in the cell (Morita et al. 2005). If so, the Hfq binding, which is also found to be at AU-rich single-stranded regions (for review, see Valentin-Hansen et al. 2004), may in itself recruit RNase E, bypassing the need for a high-affinity site for RNase E binding to promote efficient degradation. However, Hfq binds to both mRNA and regulatory RNA in the absence of pairing, whereas degradation is only triggered upon pairing and does not occur in the case of positive regulation. Thus, a number of aspects of the mechanism remain to be explained.

Fate of small RNA. A second issue must be considered in thinking about how sRNAs stimulate RNase-E-depen-

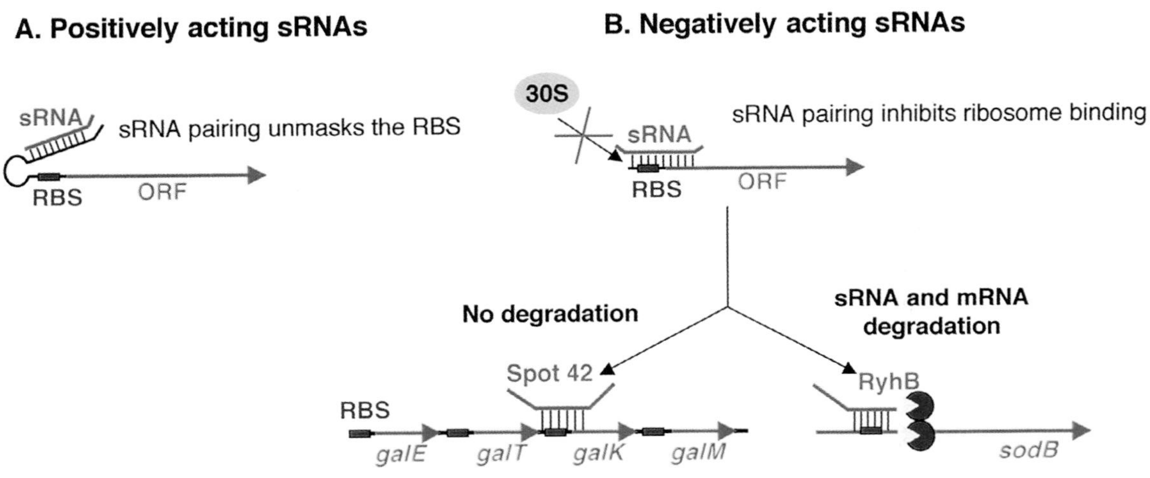

A. Positively acting sRNAs   B. Negatively acting sRNAs

Figure 2. Outcomes of pairing for Hfq-dependent regulatory RNAs. (*A*) Positively acting RNAs are exemplified by DsrA and RprA action in stimulating *rpoS* translation (for review, see Repoila et al. 2003). (*B*) Negatively acting sRNAs generally pair with target mRNAs near the ribosome-binding site, but it is not yet clear what distinguishes cases where there is degradation of the mRNA (RyhB and *sodB* in the figure) and those where there is no degradation (Spot 42 and *galK* in the figure).

dent cleavage. How does this pairing stimulate degradation of the regulatory RNA itself? The evidence that there is coupled degradation of the mRNA and the regulatory RNA is not yet complete, but it is nevertheless fairly compelling. A variety of sRNAs are relatively stable when turnover is measured in the presence of a general inhibitor of transcription such as rifampicin (Sledjeski et al. 2001; Vogel et al. 2003); however, the same regulatory RNAs are significantly more unstable when synthesis is shut down in a specific fashion (i.e., by turning down the promoter for the sRNA) (Massé et al. 2003). This has been interpreted to mean that the sRNA is degraded upon pairing to its target, which will be rapidly depleted after rifampicin treatment. The recruitment of RNase E by Hfq binding would provide part of the explanation for such coupled degradation, assuming, as noted above, that the act of pairing in itself helps to stimulate the degradation. RNase E endonuclease activity has been shown to be stimulated by a 5′ monophosphate end in vitro (Jiang and Belasco 2004). Such a 5′ monophosphate would be the expected primary product of cleavage by RNase E (Kennell 2002); thus, it is possible that a cut within the mRNA may directly stimulate RNase E to cleave the paired sRNA. However, in the absence of Hfq, and therefore presumably in the absence of pairing, the sRNAs are also rapidly degraded and RNase E seems to be critical for this degradation (Massé et al. 2003), suggesting that pairing and initial cleavage of the mRNA is not the sole mechanism for RNase E to obtain access to these sRNAs. It is intriguing that Hfq and RNase E both share a preference for single-stranded AU-rich RNA. Thus, the absence of Hfq binding may uncover RNase E cleavage sites. This has been directly demonstrated in vitro for the *sodB* message and has been suggested in other cases (Moll et al. 2003b; Zhang et al. 2003; Afonyushkin et al. 2005).

Further work will be necessary to identify the basis for mRNA degradation and how translational inhibition can lead to mRNA degradation in some cases but not others. However, based on work done thus far, rapid degradation of a target mRNA is a useful and easily detectable outcome of pairing by many sRNAs, allowing the use of approaches, such as microarrays, reverse transcriptase–polymerase chain reaction (RT-PCR), as well as northern blots, to define targets for these sRNAs (discussed further below).

Defining Targets

Hfq-binding sRNAs thus far all act by complementary base-pairing to target mRNAs. Finding these targets continues to pose a challenge. Our ability to predict targets based on the expected complementarity is improving, although it still fails to find many targets (Tjaden et al. 2002). The level of false-positive predictions (good pairing predicted, but no regulation found) is also high; possibly, these potential targets do not contain Hfq-binding sites, a characteristic that is not yet integrated into search programs. It is also unclear how many targets a given sRNA will have. At least one sRNA, RyhB, has been found to regulate dozens of target mRNAs (Massé et al. 2005), whereas others have been described as regulating only one or a small number of targets (for examples of known targets, see Table 1).

One experimental approach that has been widely used is examination of the effects of sRNA expression by microarray; the success of this method depends on both detectable expression levels for the mRNA and the degradation of target mRNAs upon pairing (Davis et al. 2005; Massé et al. 2005; Mey et al. 2005; Guillier and Gottesman 2006; Tjaden et al. 2006). This approach is most likely to uncover direct rather than indirect effects if changes in mRNA levels are examined a short time (5 minutes or less) after expression of the regulatory RNA (Massé et al. 2005). Other approaches require direct interactions to capture target mRNAs, but they may be most appropriate for well-expressed messages. Affinity purification of mRNAs binding to an sRNA immobilized on a column has been successful in a number of cases (Antal et al. 2005; Douchin et al. 2006).

Regulatory Steps Beyond Synthesis

As we learn more about how the Hfq-binding sRNAs act, other possible steps for regulation of their activity, in addition to regulation of synthesis, can be identified, although most have not been fully explored. Competition for Hfq has been suggested as a mechanism for negative regulation of RpoS translation by OxyS (Zhang et al. 1998). The intrinsic stability of the sRNA will of course contribute to its accumulation, and this can be affected by Hfq binding as well. Competition between target mRNAs for a given regulatory RNA may be an important point of control if the sRNA is limiting and used stoichiometrically, which is probably the case if degradation of the sRNA is coupled with pairing. Under the strongest induction conditions, sRNA levels may not be a limiting factor. However, sRNA levels may well be limiting when there is only a basal level of induction or during recovery from an inducing stress, when synthesis of the sRNA has slowed or stopped.

FINDING SMALL RNAs AND THEIR TARGETS: EXPANDING OPTIONS

The existence of some species of sRNAs in bacteria has been known since the early days of identifying novel stable RNAs, but in many cases, their function took many years to uncover. In recent years, a combination of the availability of sequence information for many bacterial genomes and the development of microarray-based approaches for studying gene expression has allowed both computational and experimentally based discovery of many additional sRNAs and the growing understanding of their specific expression and function. Approaches for finding sRNAs have been reviewed recently (Gottesman 2004; Huttenhofer and Vogel 2006). The result of these genome-wide approaches has defined more than 80 noncoding RNAs in *E. coli*, with at least 20 of them binding to and stabilized in vivo by Hfq (Zhang et al. 2003). Of these RNAs, studies on 14 have now been published (see Table 1).

The methods for searching for regulatory RNAs in *E. coli* have been extrapolated to other organisms. Start and stop sites are useful landmarks for some of these searches. Flanking an sRNA, one can expect to find promoter elements and/or Rho-independent terminators with a characteristic stem-loop followed by a run of U's as a stop signal. Combining conservation and the presence of a promoter and Rho-independent terminator has been developed into a program, sRNAPredict2, recently used to predict sRNAs in a variety of pathogens; a number of these predictions have been confirmed (Livny et al. 2006).

More specific predictions can be made by assuming that a stress and the physiologic response mediated by an sRNA are conserved from one species to another. Under such conditions, one can search for the regulatory sites, in combination with a terminator, for instance, within a given intergenic region. Quorum sensing sRNAs were identified in *Vibrio* by this approach (Lenz et al. 2004), as were iron-regulated sRNAs, described below.

Regulating Iron Homeostasis with Small RNAs

All organisms, from bacteria to mammals, need to carefully regulate their intracellular iron (Fe) pools. Iron is an essential cofactor for many enzymes; at the same time, iron can be quite toxic, causing damage to proteins and nucleic acids (Imlay 2003). Therefore, organisms generally need to minimize the accumulation of free iron. Iron acquisition may be regulated so that intracellular iron is kept to the minimum needed to satisfy requirements, and iron storage systems that sequester excess iron exist in all organisms. It has been known for decades that iron storage is regulated in mammalian cells by an intriguing post-transcriptional mechanism, in which the aconitase enzyme, an iron-binding enzyme of the TCA cycle, also acts as a regulator, binding an RNA element at the 3′ end of some mRNAs to both positively and negatively regulate translation (Rouault 2002).

Bacteria inside mammalian hosts need to obtain iron from their host, therefore requiring that they develop methods for iron acquisition that can effectively compete for bound iron with the host. The Fur repressor is used by many bacteria, both gram-positive and gram-negative, to keep iron acquisition systems under control until they are needed (Hantke 2001). Fur directly binds Fe^{2+}, and acts as a repressor only in the iron-bound form. RyhB, a small noncoding RNA made in *E. coli*, acts to regulate internal iron use by targeting for degradation the mRNAs of as many as 16 operons encoding iron-binding proteins (Massé and Gottesman 2002; Massé et al. 2005). RyhB is directly repressed by the Fur repressor and rapidly induced when iron is removed from Fur. It was proposed that RyhB helps to prioritize intracellular iron use under iron-limiting conditions by sparing iron for essential functions (Massé et al. 2005; Jacques et al. 2006), although genes involved in biofilm formation, acid resistance, and others have also been found to be regulated by RyhB in other organisms (Davis et al. 2005; Mey et al. 2005; Oglesby et al. 2005).

The biology underlying the use of Fur and RyhB to regulate intracellular iron utilization is likely to extend to many other bacteria as well. Is a sRNA also used in these bacteria? RyhB homologs can easily be found in other enterobacteria, including *Salmonella, Shigella, Klebsiella, Photorhabdus*, and *Yersinia* (Fig. 3, top) (Massé and Gottesman 2002). *Yersinia, Photorhabdus*, and *Salmonella* have two copies of an RNA similar to RyhB; one is in the context of the *E. coli* gene (location A), whereas the second copy is in a different chromosomal context (location B for *Yersinia* and *Photorhabdus* and location C for *Salmonella*) (Fig. 3, bottom). Evolutionary trees of these RyhB molecules suggest a possible early duplication event, to give the A and B locations; subsequently, the B copy may have been lost in *E. coli* and *Salmonella*. The RyhB at the C locus in *Salmonella* is the most divergent of this set and might have been acquired from elsewhere by horizontal transfer. In all cases, the RyhB homolog is preceded by a predicted Fur-binding site (not shown). *Vibrio cholerae* and other *Vibrio* species have a single *ryhB* gene with some distinct characteristics. Although still regulated by Fur, the *ryhB* gene is in yet another genome context and is significantly longer than the *E. coli* RNA (Davis et al. 2005; Mey et al. 2005).

In organisms more evolutionarily distant from *E. coli*, RyhB cannot be found by sequence homology. For instance, a search of *Pseudomonas aeruginosa* sequences failed to reveal a sequence related to RyhB. However, pseudomonads contain a Fur repressor and have a similar requirement to regulate iron homeostasis. A computational search for a Fur-regulated sRNA (searching intergenic regions for RNA secondary structure within an appropriate distance from a Fur-binding site) revealed a pair of closely related RNAs that were shown to have a role in *Pseudomonas* similar to that of *E. coli* RyhB and have been called PrrF1 and PrrF2 (*Pseudomonas* regulatory RNA involving Fe) (Wilderman et al. 2004). Comparison to other pseudomonads demonstrated conservation of PrrF-like RNAs. In fact, all sequenced pseudomonads have two PrrF sRNAs, although only *P. aeruginosa* has two in tandem; the others have one at the same position as the *P. aeruginosa* sRNAs, whereas the other is elsewhere (Fig. 4). An examination of the evolution of PrrF in the pseudomonads would suggest that as for *E. coli* and its close relatives, PrrF duplicated at some point in the past. The second copy (Context B in Fig. 4) of PrrF may have been lost from *P. aeruginosa;* a later duplication of the remaining *prrF* gene yielded tandem genes that are more similar to each other than they are to any of the other pseudomonad PrrFs. No similarity in sequence or genome context is obvious between the PrrF sRNAs and RyhB sRNAs in *E. coli* or its close relatives. At least one target, *sodB*, is shared by the PrrF and RyhB sRNAs (Wilderman et al. 2004).

As with RyhB in *Yersinia* and *Salmonella*, it would thus seem that multiple independent duplication events have occurred during evolution, suggesting a requirement for more than one *ryhB* gene. Exactly why this would be is not yet clear. Mutation of one or the other *P. aeruginosa prrF* genes gives a partial phenotype in down-regulation of target mRNAs, suggesting that two may be necessary to achieve enough PrrF RNA for full control under severe iron depletion conditions (Wilderman et al. 2004). Alternatively, subtle differences in expression patterns and differences in sequences, and therefore pairing, might be driving the duplication of these loci.

It seems highly likely that the *E. coli* and pseudomonad RyhB-like RNAs are evolved from a common, unidentified ancestor. The lack of sequence homology in these RNAs between *Pseudomonas* and *E. coli,* coupled with the conservation of promoter sequences, suggests that

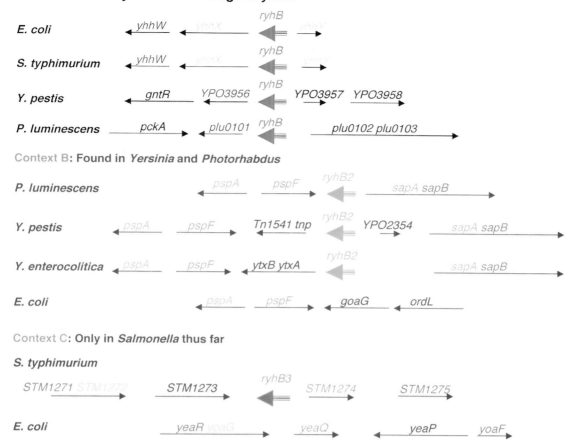

Figure 3. RyhB homologs in *Enterobacteriaceae*. Homologous genes are similarly color-coded. The three contexts shown are far from each other on the bacterial chromosome. We have arbitrarily named those RNAs in the same context as *E. coli ryhB* and second genes elsewhere *ryhB*2 or *ryhB*3 (depending on the context). These are not meant to be permanent names. Information on genes is adapted from information provided on the Comprehensive Microbial Resource of TIGR (The Institute for Genome Research [http://cmr.tigr.org/tigr-scripts/CMR/CmrHomePage.cgi; see Peterson et al. 2001).

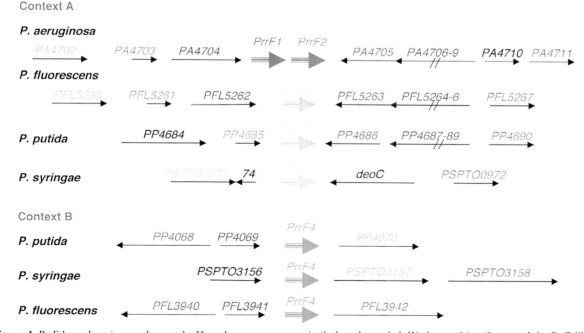

Figure 4. PrrF homologs in pseudomonads. Homologous genes are similarly color-coded. We have arbitrarily named the PrrF-like RNAs in Context A PrrF3 and those in Context B PrrF4.

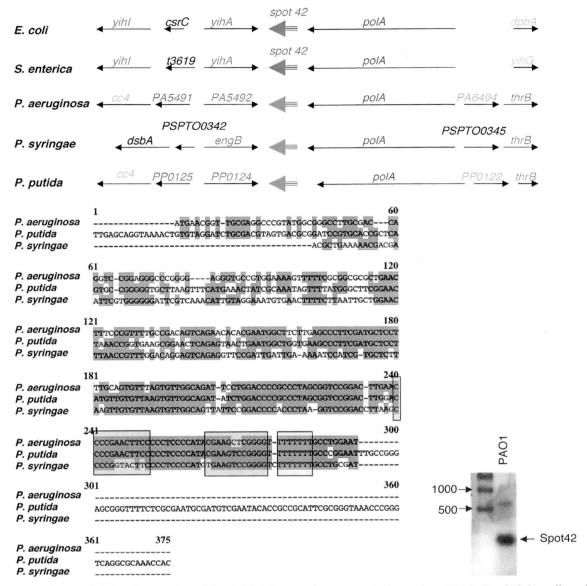

Figure 5. A possible *Pseudomonas* Spot 42 RNA. (*Top*) Conserved genome context around an sRNA in *E. coli, Salmonella,* and pseudomonads. Homologous genes are similarly color-coded. (*polA*) DNA polymerase I; (*engB*) GTP-binding protein. (*Bottom left*) Multiple sequence alignment of *Pseudomonas* intergenic region between *polA* and *engB*. The sequences overlapping the putative sRNAs are shown, with conserved regions in gray and the putative terminator stem-loop boxed in color. Sequences are from Genbank files AE004091 (*P. aeruginosa* PA01), AE015451 (*P. putida* KT2440), and AE016853 (*P. syringae* pv tomato str. DC3000). Other sequenced pseudomonads are also conserved in this intergenic region. (*Bottom right*) Detection of a Spot 42 sRNA in *P. aeruginosa.* RNA extracted from *P. aeruginosa* strain PA01 was probed for an RNA of the sequence predicted in Figure 5 (*left*). The RNA isolation and northern blot procedure were carried out as described previously (Wilderman et al. 2004).

RNA genes can change relatively rapidly, possibly as their targets change or new targets are acquired. As more bacterial genomes are sequenced, it should become possible to fill in some of the missing steps and better understand how these RNAs are evolving.

Genome Context as a Landmark for Small RNA Discovery

Given the relatively rapid evolution of the RyhB-like noncoding RNAs, tracing these RNAs through evolution

can be easier if a protein-coding gene can be used as a genome marker for the RNA. For instance, the *E. coli* 6S RNA, a regulator of RNA polymerase, is processed from a longer message that includes an open reading frame (*ygfA*) in many organisms, and by identifying the ORF, the linked but not always homologous 6S RNA was identified (and found to have a conserved structure and function) (Barrick et al. 2005; Trotochaud and Wassarman 2005).

The Hfq-binding sPRNAs have generally been found to be free-standing transcripts, not processed from a longer RNA. However, as noted above, for a few sRNAs (SgrS,

OxyS, and GcvB), the transcriptional regulator is encoded next to and divergently from the RNA (Table 1). Thus, this protein can serve as a marker for the small RNA (McArthur et al. 2006).

A Unique Genome Neighborhood for Spot 42

A more general investigation of the genome context of the sRNAs has suggested that at least in a few cases, genes for sRNAs flanked by genes for highly conserved proteins can be found in a similar neighborhood, even when the sRNAs themselves have diverged. In particular, this appears to be the case for the genome context of the Hfq-binding Spot 42 RNA.

In *E. coli*, Spot 42 has been found to be responsible for polarity in the *gal* operon; it negatively regulates the translation of the third gene in the operon (Fig. 2B) (Møller et al. 2002a). Sequence comparisons show that Spot 42 is also present in *Salmonella, Erwinia, Yersinia,* and *Vibrio* (see RFAM; Griffiths-Jones et al. 2005).

What is notable is the genome location of this RNA (Fig. 5, top). It is found between *polA*, encoding DNA polymerase I, a protein found widely in bacteria, and *yihA* (*engB*), a conserved GTPase of unknown function. These two genes are also near each other in *P. aeruginosa*, and the sequences in the intergenic region are conserved in other pseudomonads and are reminiscent of those expected for a noncoding RNA (Fig. 5, bottom left). In fact, an sRNA is expressed from this locus (Fig. 5, bottom right).

As with the RyhB-like RNAs, no obvious sequence similarity can be found between Spot 42 from enterobacteria and the *Pseudomonas* small RNA. In addition, the *gal* operon, the major demonstrated target for Spot 42 in *E. coli* (Møller et al. 2002a), is not found in *P. aeruginosa*, suggesting either that this is an RNA with divergent function in the same site or that the major (conserved) function of Spot 42 has not yet been described. Further analysis of its function in *Pseudomonas* as well as further examination of the evolution of this and other sRNAs will clearly be fruitful.

PROSPECTS FOR THE FUTURE

Our understanding of the role of small regulatory RNAs in bacterial physiology has grown significantly over the last decade. Discoveries and approaches developed initially in *E. coli* are now being applied to a variety of bacteria, including pathogens. It seems highly likely that this once ignored level of regulation will provide new insights into bacterial growth and adaptation to stress and new targets for future antibiotics.

ACKNOWLEDGMENTS

The work described in this manuscript was supported by the Intramural Research Program of the National Institutes of Health, National Cancer Institute, Center for Cancer Research.

REFERENCES

Afonyushkin T., Vecerek B., Moll I., Blasi U., and Kaberdin V.R. 2005. Both RNase E and RNase III control the stability of *sodB* mRNA upon translational inhibition by the small regulatory RNA RyhB. *Nucleic Acids Res.* **33:** 1678.

Altuvia S., Weinstein-Fischer D., Zhang A., Postow L., and Storz G. 1997. A small stable RNA induced by oxidative stress: Role as a pleiotropic regulator and antimutator. *Cell* **90:** 43.

Antal M., Bordeau V., Douchin V., and Felden B. 2005. A small bacterial RNA regulates a putative ABC transporter. *J. Biol. Chem.* **280:** 7901.

Argaman L. and Altuvia S. 2000. *fhlA* repression by OxyS RNA: Kissing complex formation at two sites results in a stable antisense-target RNA complex. *J. Mol. Biol.* **300:** 1101.

Argaman L., Hershberg R., Vogel J., Bejerano G., Wagner E.G.H., Margalit H., and Altuvia S. 2001. Novel small RNA-encoding genes in the intergenic region of *Escherichia coli. Curr. Biol.* **11:** 941.

Baker K.E. and Mackie G.A. 2003. Ectopic RNase E sites promote bypass of 5′-end-dependent mRNA decay in *Escherichia coli. Mol. Microbiol.* **47:** 75.

Barrick J.E., Sudarsan N., Weinberg Z., Ruzzo W.L., and Breaker R.R. 2005. 6S RNA is a widespread regulator of eubacterial RNA polymerase that resembles an open promoter. *RNA* **11:** 774.

Blumenthal T. and Carmichael G.G. 1979. RNA replication: Function and structure of Qβ-replicase. *Annu. Rev. Biochem.* **48:** 525.

Brown L. and Elliott T. 1997. Mutations that increase expression of the *rpoS* gene and decrease its dependence on *hfq* function in *Salmonella typhimurium. J. Bacteriol.* **179:** 656.

Cullen B.R. 2004. Transcription and processing of human microRNA precursors. *Mol. Cell* **16:** 861.

Davis B.M., Quinones M., Pratt J., Ding Y., and Waldor M.K. 2005. Characterization of the small untranslated RNA RyhB and its regulon in *Vibrio cholerae. J. Bacteriol.* **187:** 4005.

Deana A. and Belasco J.G. 2005. Lost in translation: The influence of ribosomes on bacterial mRNA decay. *Genes Dev.* **19:** 2526.

Delihas N. and Forst S. 2001. *MicF:* An antisense RNA gene involved in response of *Escherichia coli* to global stress factors. *J. Mol. Biol.* **313:** 1.

Ding Y., Davis B.M., and Waldor M.K. 2004. Hfq is essential for *Vibrio cholerae* virulence and downregulates σ^E expression. *Mol. Microbiol.* **53:** 345.

Douchin V., Bohn C., and Bouloc P. 2006. Down-regulation of porins by a small RNA bypasses the essentiality of the RIP protease RseP in *E. coli. J. Biol. Chem.* **281:** 12253.

Geissmann T.A. and Touati D. 2004. Hfq, a new chaperoning role: Binding to messenger RNA determines access for small RNA regulator. *EMBO J.* **23:** 396.

Gottesman S. 2004. The small RNA regulators of *Escherichia coli:* Roles and mechanisms. *Annu. Rev. Microbiol.* **58:** 273.
———. 2005. Micros for microbes: Non-coding regulatory RNAs in bacteria. *Trends Genet.* **21:** 399.

Griffiths-Jones S., Moxon S., Marshall M., Khanna A., Eddy S.R., and Bateman A. 2005. Rfam: Annotating non-coding RNAs in complete genomes. *Nucleic Acids Res.* **33:** D121.

Guillier M. and Gottesman S. 2006. Remodelling of the *Escherichia coli* outer membrane by two small regulatory RNAs. *Mol. Microbiol.* **59:** 231.

Hantke K. 2001. Iron and metal regulation in bacteria. *Curr. Opin. Microbiol.* **4:** 172.

Huttenhofer A. and Vogel J. 2006. Experimental approaches to identify non-coding RNAs. *Nucleic Acids Res.* **2:** 635.

Imlay J.A. 2003. Pathways of oxidative damage. *Annu. Rev. Microbiol.* **57:** 395.

Jacques J.-F., Jang S., Prévost K., Desnoyers G., Desmarais M., Imlay J., and Massé E. 2006. RyhB small RNA modulates the free intracellular iron pool and is essential for normal growth during iron limitation in *Escherichia coli. Mol. Microbiol.* (in press).

Jiang X. and Belasco J.G. 2004. Catalytic activation of multimeric RNase E and RNase G by 5′-monophosphorylated RNA. *Proc. Natl. Acad. Sci.* **101:** 9211.

Johansen J., Rasmussen A.A., Overgaard M., and Valentin-Hansen P. 2006. Conserved small non-coding RNAs that belong to the σ^E regulon: Role in down-regulation of outer membrane proteins. *J. Mol. Biol.* (in press).

Joyce S.A. and Dreyfus M. 1998. In the absence of translation, RNase E can bypass 5′ mRNA stabilizers in *Escherichia coli*. *J. Mol. Biol.* **282:** 241.

Kawamoto H., Koide Y., Morita T., and Aiba H. 2006. Base-pairing requirement for RNA silencing by a bacterial small RNA and acceleration of duplex formation by Hfq. *Mol. Microbiol.* **61:** 1013.

Kennell D. 2002. Processing endoribonucleases and mRNA degradation in bacteria. *J. Bacteriol.* **184:** 4645.

Lease R.A. and Woodson S.A. 2004. Cycling of the Sm-like protein Hfq on the DsrA small regulatory RNA. *J. Mol. Biol.* **344:** 1211.

Lenz D.H., Mok K.C., Lilley B.N., Kulkarni R.V., Wingreen N.S., and Bassler B.L. 2004. The small RNA chaperone Hfq and multiple small RNAs control quorum sensing in *Vibrio harveyi* and *Vibrio cholerae*. *Cell* **118:** 69.

Livny J., Brencic A., Lory S., and Waldor M.K. 2006. Identification of 17 *Pseudomonas aeruginosa* sRNAs and prediction of sRNA-encoding genes in 10 diverse pathogens using the bioinformatic tool sRNAPredict2. *Nucleic Acids Res.* **34:** 3484.

Majdalani N., Hernandez D., and Gottesman S. 2002. Regulation and mode of action of the second small RNA activator of RpoS translation, RprA. *Mol. Microbiol.* **46:** 813.

Majdalani N., Chen S., Murrow J., St. John K., and Gottesman S. 2001. Regulation of RpoS by a novel small RNA: The characterization of RprA. *Mol. Microbiol.* **39:** 1382.

Majdalani N., Cunning C., Sledjeski D., Elliott T., and Gottesman S. 1998. DsrA RNA regulates translation of RpoS message by an anti-antisense mechanism, independent of its action as an antisilencer of transcription. *Proc. Natl. Acad. Sci.* **95:** 12462.

Massé E. and Gottesman S. 2002. A small RNA regulates the expression of genes involved in iron metabolism in *Escherichia coli*. *Proc. Natl. Acad. Sci.* **99:** 4620.

Massé E., Escorcia F.E., and Gottesman S. 2003. Coupled degradation of a small regulatory RNA and its mRNA targets in *Escherichia coli*. *Genes Dev.* **17:** 2374.

Massé E., Vanderpool C.K., and Gottesman S. 2005. Effect of RyhB small RNA on global iron use in *Escherichia coli*. *J. Bacteriol.* **187:** 6962.

McArthur S.D., Pulvermacher S.C., and Stauffer G.V. 2006. The *Yersinia pestis gcvB* gene encodes two small regulatory RNA molecules. *BMC Microbiol.* **6:** 52.

Mey A.R., Craig S.A., and Payne S.M. 2005. Characterization of *Vibrio cholerae* RyhB: The RyhB regulon and role of *ryhB* in biofilm formation. *Infect. Immun.* **73:** 5706.

Mikulecky P.J., Meenakshi K., Brescia C.C., Takach J.C., Sledjeski D., and Feig A.L. 2004. *Escherichia coli* Hfq has distinct interaction surfaces for DsrA, *rpoS* and poly(A)RNAs. *Nat. Struct. Mol. Biol.* **11:** 1206.

Moll I., Leitsch D., Steinhauser T., and Blasi U. 2003a. RNA chaperone activity of the Sm-like Hfq protein. *EMBO Rep.* **4:** 284.

Moll I., Afonyushkin T., Vytvytska O., Kaberdin V.R., and Blasi U. 2003b. Coincident Hfq binding and RNase E cleavage sites on mRNA and small regulatory RNAs. *RNA* **9:** 1308.

Møller T., Franch T., Udesen C., Gerdes K., and Valentin-Hansen P. 2002a. Spot 42 RNA mediates discoordinate expression of the *E. coli* galactose operon. *Genes Dev.* **16:** 1696.

Møller T., Franch T., Hojrup P., Keene D.R., Bachinger H.P., Brennan R., and Valentin-Hansen P. 2002b. Hfq: A bacterial Sm-like protein that mediates RNA-RNA interaction. *Mol. Cell* **9:** 23.

Morita T., Maki K., and Aiba H. 2005. RNase E-based ribonucleoprotein complexes: Mechanical basis of mRNA stabilization mediated by bacterial noncoding RNAs. *Genes Dev.* **19:** 2176.

Morita T., Mochizuki Y., and Aiba H. 2006. Translational repression is sufficient for gene silencing by bacterial small noncoding RNAs in the absence of mRNA destruction. *Proc. Natl. Acad. Sci.* **103:** 4858.

Morita T., El-Kazzaz W., Tanaka Y., Inada T., and Aiba H. 2003. Accumulation of glucose 6-phosphate or fructose 6-phosphate is responsible for destabilization of glucose transporter mRNA in *Escherichia coli*. *J. Biol. Chem.* **278:** 15608.

Muffler A., Fischer D., and Hengge-Aronis R. 1996. The RNA-binding protein HF-I, known as a host factor for phage Qβ RNA replication, is essential for *rpoS* translation in *Escherichia coli*. *Genes Dev.* **10:** 1143.

Oglesby A.G., Murphy E.R., Iyer V.R., and Payne S.M. 2005. Fur regulates acid resistance in *Shigella flexneri* via RyhB and *ydeP*. *Mol. Microbiol.* **58:** 1354.

Opdyke J.A., Kang J.-G., and Storz G. 2004. GadY, a small-RNA regulator of acid response genes in *Escherichia coli*. *J. Bacteriol.* **186:** 6698.

Peterson J.D., Umayam L.A., Dickinson T.M., Hickey E.K., and White O. 2001. The comprehensive microbial resource. *Nucleic Acids Res.* **29:** 123.

Ramani N., Hedeshian M., and Freundlich M. 1994. *micF* antisense RNA has a major role in osmoregulation of OmpF in *Escherichia coli*. *J. Bacteriol.* **176:** 5005.

Rasmussen A.A., Eriksen M., Gilany K., Udesen C., Franch T., Petersen C., and Valentin-Hansen P. 2005. Regulation of *ompA* mRNA stability: The role of a small regulatory RNA in growth phase-dependent control. *Mol. Microbiol.* **58:** 1421.

Repoila F. and Gottesman S. 2001. Signal transduction cascade for regulation of RpoS: Temperature regulation of DsrA. *J. Bacteriol.* **183:** 4012.

Repoila F., Majdalani N., and Gottesman S. 2003. Small noncoding RNAs, co-ordinators of adaptation processes in *Escherichia coli*: The RpoS paradigm. *Mol. Microbiol.* **48:** 855.

Robertson G.T. and Roop R.M.J. 1999. The *Brucella abortus* host factor I (HF-1) protein contributes to stress resistance during stationary phase and is a major determinant of virulence in mice. *Mol. Microbiol.* **34:** 690.

Rouault T.A. 2002. Post-transcriptional regulation of human iron metabolism by iron regulatory proteins. *Blood Cells Mol. Dis.* **29:** 309.

Sauter C., Basquin J., and Suck D. 2003. Sm-like proteins in eubacteria: The crystal structure of the Hfq protein from *Escherichia coli*. *Nucleic Acids Res.* **31:** 4091.

Schumacher M.A., Pearson R.F., Møller T., Valentin-Hansen P., and Brennan R.G. 2002. Structures of the pleiotropic translational regulator Hfq and an Hfq-RNA complex: A bacterial Sm-like protein. *EMBO J.* **21:** 3546.

Sledjeski D.D., Whitman C., and Zhang A. 2001. Hfq is necessary for regulation by the untranslated RNA DsrA. *J. Bacteriol.* **183:** 1997.

Sonnleitner E., Schuster M., Sorger-Domenigg T., Greenberg E.P., and Blasi U. 2006. Hfq-dependent alterations of the transcriptome profile and effects on quorum sensing in *Pseudomonas aeruginosa*. *Mol. Microbiol.* **59:** 1542.

Storz G. and Gottesman S. 2006. Versatile roles of small RNA regulators in bacteria. In *The RNA world*, 3rd edition (ed. R.F. Gesteland et al.), p. 567. Cold Spring Harbor Laboratory Press, Cold Spring Harbor, New York.

Storz G., Altuvia S., and Wassarman K.M. 2005. An abundance of RNA regulators. *Annu. Rev. Biochem.* **74:** 199.

Storz G., Opdyke J.A., and Zhang A. 2004. Controlling mRNA stability and translation with small, noncoding RNAs. *Curr. Opin. Microbiol.* **7:** 140.

Tjaden B., Saxena R.M., Stolyar S., Haynor D.R., Kolker E., and Rosenow C. 2002. Transcriptome analysis of *Escherichia coli* using high density oligonucleotide probe arrays. *Nucleic Acids Res.* **30:** 3732.

Tjaden B., Goodwin S.S., Opdyke J.A., Guillier M., Fu D.X., Gottesman S., and Storz G. 2006. Target prediction for small, noncoding RNAs in bacteria. *Nucleic Acids Res.* **34:** 2791.

Trotochaud A.E. and Wassarman K.M. 2005. A highly conserved 6S RNA structure is required for regulation of transcription. *Nat. Struct. Mol. Biol.* **12:** 774.

Tsui H.-C.T., Leung H.-C.E., and Winkler M.E. 1994. Characterization of broadly pleiotropic phenotypes caused by an *hfq* insertion mutation in *Escherichia coli* K-12. *Mol. Microbiol.* **13**: 35.

Udekwu K.I., Darfeuille F., Vogel J., Reimegard J., Holmqvist E., and Wagner E.G. 2005. Hfq-dependent regulation of OmpA synthesis is mediated by an antisense RNA. *Genes Dev.* **19**: 2355.

Urbanowski M.L., Stauffer L.T., and Stauffer G.V. 2000. The *gcvB* gene encodes a small untranslated RNA involved in expression of the dipeptide and oligopeptide transport systems in *Escherichia coli*. *Mol. Microbiol.* **37**: 856.

Valentin-Hansen P., Eriksen M., and Udesen C. 2004. The bacterial Sm-like protein Hfq: A key player in RNA transactions. *Mol. Microbiol.* **51**: 1525.

Vanderpool C.K. and Gottesman S. 2004. Involvement of a novel transcriptional activator and small RNA in post-transcriptional regulation of the glucose phosphoenolpyruvate phosphotransferase system. *Mol. Microbiol.* **54**: 1076.

Vogel J. and Papenfort K. 2006. Small non-coding RNAs and the bacterial outer membrane. *Curr. Opin Microbiol.* **9**: 1.

Vogel J., Bartels V., Tang H.H., Churakov G., Slagter-Jager J.G., Huttenhofer A., and Wagner E.G.H. 2003. RNomics in *Escherichia coli* detects new sRNA species and indicates parallel transcriptional output in bacteria. *Nucleic Acids Res.* **31**: 6435.

Wassarman K.M., Repoila F., Rosenow C., Storz G., and Gottesman S. 2001. Identification of novel small RNAs using comparative genomics and microarrays. *Genes Dev.* **15**: 1637.

Wilderman P.J., Sowa N.A., FitzGerald D.J., FitzGerald P.C., Gottesman S., Ochsner U.A., and Vasil M.L. 2004. Identification of tandem duplicate regulatory small RNAs in *Pseudomonas aeruginosa* involved in iron homeostasis. *Proc. Natl. Acad. Sci.* **101**: 9792.

Zhang A., Wassarman K.M., Ortega J., Steven A.C., and Storz G. 2002. The Sm-like Hfq protein increases OxyS RNA interaction with target mRNAs. *Mol. Cell* **9**: 11.

Zhang A., Altuvia S., Tiwari A., Argaman L., Hengge-Aronis R., and Storz G. 1998. The *oxyS* regulatory RNA represses *rpoS* translation by binding Hfq (HF-1) protein. *EMBO J.* **17**: 6061.

Zhang A., Wassarman K.M., Rosenow C., Tjaden B.C., Storz G., and Gottesman S. 2003. Global analysis of small RNA and mRNA targets of Hfq. *Mol. Microbiol.* **50**: 1111.

Short Silencing RNA: The Dark Matter of Genetics?

D.C. Baulcombe

The Sainsbury Laboratory, Norwich, United Kingdom, NR4 7UH

Plants and animals have single-stranded silencing RNAs (sRNAs) of 21–25 nucleotides in length that are derived from a double-stranded (ds)RNA precursor by Dicer (DCL) processing. These RNAs are the guide RNA for nucleases of the AGO class that cleave targeted RNA in a nucleotide sequence-specific manner. The cleaved RNAs are then degraded further or they are the template for an RNA-dependent RNA polymerase (RDR) that generates a dsRNA. In this paper, I discuss the possibility that this RDR-generated dsRNA initiates a cascade in which there are multiple rounds of secondary sRNA production. I propose that these secondary sRNAs feature in mechanisms that can either buffer mRNA populations against change or, in certain circumstances, mediate extensive changes in mRNA populations. The RNA cascades may also have RNA-mediated epigenetic characteristics in addition to the DNA and chromatin transcriptional silencing potential that has been previously linked with RNA silencing.

In 1993, the Ruvkun and Ambros laboratories described two different regulatory short RNAs affecting the timing of developmental transitions in *Caenorhabditis elegans* (Lee et al. 1993; Wightman et al. 1993). They referred to these RNAs as short temporal (st) RNAs, and they discovered that they mediate gene silencing by base-pairing to a complementary nucleotide sequence in the 3′UTR of target mRNAs. For many years, they were the only known example of regulatory RNA acting in this manner. However, in 2001 and 2002, when short RNA was sequenced from animals and plants, it became apparent that the stRNAs are typical of a large class of short regulatory RNAs that are now referred to as micro (mi)RNAs (Lagos-Quintana et al. 2001; Lau et al. 2001; Lee and Ambros 2001; Reinhart et al. 2002). Ruvkun described this new family of RNAs using a dark matter metaphor because, like the hypothetical dispersed mass in the physical universe, they were recently discovered, abundant, and functionally significant (Ruvkun 2001). This dark matter metaphor remains appropriate because, as discussed below, there are many previously unidentified families of these RNAs, in addition to miRNAs, that mediate gene silencing. I refer generically to these RNAs as silencing RNAs (sRNAs). Correspondingly, I refer to silencing mechanisms based on these RNAs as examples of "RNA silencing."

The sRNAs are processed from a longer dsRNA precursor by an RNase III homolog known as dicer or dicer-like (DCL) (Bernstein et al. 2001; Schauer et al. 2002). DCL cleaves these precursor RNAs into a 20- to 24-nucleotide dsRNA with 5′ phosphate and 3′ hydroxyl groups that is recruited by a second nuclease known as slicer (Fig. 1) (Liu et al. 2004). Slicer, a member of the Argonaute (AGO) family of proteins, cleaves one of the two strands, and the cleavage products are released (Matranga et al. 2005). The remaining single-stranded sRNA is then available for a targeting mechanism in which slicer is guided by Watson-Crick base-pairing to a complementary target RNA or DNA.

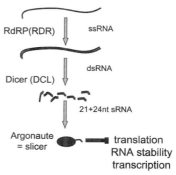

Figure 1. The core mechanism of RNA silencing in plants. Single-stranded RNA (ssRNA) is converted into a double-stranded (ds) form by an RDR protein. The dsRNA is cleaved by DCL and the single-stranded sRNA is retained in an AGO ribonucleoprotein complex where it has the role as a guide RNA.

The consequence of the RNA targeting mechanism is either RNA cleavage by AGO, translational arrest, or both (Schwab et al. 2005). The DNA-targeted mechanism leads to heterochromatin formation and, in some instances, DNA methylation at the target locus (Lippman and Martienssen 2004). The DNA-targeted mechanism is less well understood than RNA-targeted silencing, although it does require an AGO protein (Volpe et al. 2002; Zilberman et al. 2003; Li et al. 2006). In principle, the DNA-targeted mechanism could involve base-pairing of an sRNA with one of the DNA strands at the target locus. However, in cells in which nuclear DNA is not replicating, this mechanism would only succeed if there was unwinding of the DNA double helix. A more plausible mechanism involves indirect targeting of DNA: The direct target of the sRNA interaction could be a nascent RNA that is still attached to the transcribed chromatin, rather than DNA. In fission yeast, it is likely that sRNA is targeted at a nascent RNA (Bayne and Allshire 2005), but in plant systems, the evidence is not conclusive.

There are two variations of the sRNA biogenesis mechanism. In the first of these mechanisms, the sRNA precursor is an RNA with partially complementary internal repeats (Fig. 2). The inverted repeat regions anneal to form an incompletely double-stranded RNA in which mismatched regions in the RNA precursor are likely to provide structural cues that guide Dicer to a single site in the precursor RNA. It is also possible that these mismatch sites exclude DCL from the flanking regions so that each precursor normally gives rise to only a single sRNA (Han et al. 2006). The sRNAs generated by this process are the miRNAs referred to above.

In the second variation, the DCL substrate is completely double-stranded, and several sRNAs are generated from each dsRNA precursor (Fig. 2). Presumably, the absence of unpaired bases means that there are no structural cues to guide DCL to any one region of its substrate RNA. Consequently, the cleavage of dsRNA is less targeted than with the miRNA precursors. The dsRNA may be produced either by annealing of complementary single-stranded RNAs, by intramolecular base-pairing of self-complementary inverted repeat regions, or by the action of an RNA-dependent RNA polymerase (RDR) on a single-stranded RNA template. The sRNAs produced by the cleavage of the completely double-stranded RNA are referred to as short interfering (si) RNAs (Carmell and Hannon 2004).

In this paper, I speculate about the possible biological roles and mechanisms of sRNA-mediated silencing. My discussion is based primarily on the sRNAs of plants, although some of the mechanisms proposed may also be relevant to animal and fungal systems. First, I summarize the genetic and molecular approaches to characterize the sRNAs in *Arabidopsis*. I then describe how these sRNAs may form interaction cascades and networks affecting genetic and epigenetic systems in growth and development.

GENETICS AND THE HIGH-THROUGHPUT SEQUENCING REVOLUTION

In *Arabidopsis*, the analysis of biogenesis and targeting of sRNAs has benefited from a classic mutation analysis combined with a more recent approach—the Hubble tele-

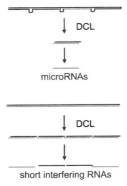

Figure 2. The difference between miRNA and siRNA. miRNAs (*upper diagram*) are generated from precursor dsRNAs in which sequence mismatches provide structural cues so that DCL releases only a single short RNA after cleavage. The precursors of siRNAs, in contrast (*lower diagram*), are completely dsRNA, and the absence of structural cues means that DCL releases siRNAs from many positions.

scope of the RNA universe—that uses novel high-throughput cDNA sequencing technologies (Lu et al. 2005; Henderson et al. 2006). The picture emerging from these analyses, as described below, is of multiple variations on the core RNA silencing mechanisms.

The mutation analysis involving both forward and reverse genetics reveals the existence of multigene families encoding the three key proteins—RDR, DCL, and AGO—in RNA silencing (Baulcombe 2004). Many, if not all, of these isoforms are involved in variations of RNA silencing. DCL1, for example, generates miRNAs (Baulcombe 2004), whereas DCL2, 3, and 4 feature in the siRNA biogenesis pathways (Dunoyer et al. 2005; Gasciolli et al. 2005). DCL3 has the most distinct function because it generates longer siRNAs—24 nucleotides rather than 21 or 22 nucleotides—that are associated with heterochromatin silencing and DNA modification (Xie et al. 2004). RDR2 and AGO4 are similarly distinct from their homologs in that they also are primarily involved in heterochromatin rather than mRNA targeting (Xie et al. 2004; Zilberman et al. 2004). Mutants at four of the six RDR loci and eight of the ten AGO loci have yet to be characterized in terms of their RNA silencing phenotypes.

High-throughput sequencing technologies are still being refined, but at present, the most useful method involves a mechanical cloning procedure and an automated pyro-sequencing of DNA (Margulies et al. 2005). Other suitable techniques use single-molecule DNA sequencing. The power of these methods derives from the absence of a requirement for the labor-intensive manipulations associated with cloning of genomic and complementary DNA in plasmid and phage vectors. It is therefore relatively straightforward to generate sequence data from hundreds of thousands of sRNAs. The current versions of these technologies have a limitation for analysis of genomic DNA in that the sequence data are only short: 20–100 nucleotides. However, this limitation is not relevant to the analysis of short sRNAs.

These sequence data reveal that the *Arabidopsis* sRNAs are predominantly in the size range of 20–26 nucleotides with prominent size groups of 21 and 24 nucleotides (Lu et al. 2005; Henderson et al. 2006). Alignment of the sRNA sequences shows that they are not randomly distributed copies of genomic DNA, and I refer to the regions of the genome with a match to many individual sRNA sequences as sRNA loci. Some of these loci correspond to single-copy genome sequences, and they are the genes from which the sRNAs are produced. Other sRNAs correspond to repeated DNA sequences from which the sRNA genes cannot be identified with certainty. Any of these repeated sRNA loci could be either a source gene of sRNAs or a target of sRNA-mediated silencing. It could also be that they are completely inactive DNA with the same sequence as transcribed regions.

At present, the classification of these loci is an ongoing task in several laboratories using data from various tissues in wild type and from RNA silencing mutants at RDR, DCL, AGO, and other genes (Lu et al. 2005; Henderson et al. 2006). The absence or hyper-accumulation of sRNA in these mutant plants may be informative about the function of sRNA encoded at the locus. My laboratory and others have also used the association of sRNAs with an AGO protein as a parameter in the classification process (Baumberger and Baulcombe 2005; Qi et al. 2005, 2006). Epitope-tagged versions of the AGO proteins are immunoprecipitated, and the associated sRNAs are characterized using the high-throughput sequencing technology. The information produced by the high-throughput sequencing approaches is a major advance because it is informative about sRNAs and sRNA loci that would not be accessible through simple genetics and phenotype analysis or by using smaller-scale sequencing technology.

The extensive sRNA profile from the high-throughput sequence analysis includes all of the recognized classes of sRNA, including miRNAs and various types of siRNA. However, a new point to emerge is that these sRNA classes are represented by more loci than had been recognized previously (Lu et al. 2005; Henderson et al. 2006). At present, we do not have precise information about the numbers of each type of sRNA because the analysis of the high-throughput sequence data is at an early stage: The numbers change as new data are produced or when the parameters of the computational analysis are modified. However, for the purposes of this paper, it is sufficient to know that there are many thousands of sRNA loci in the *Arabidopsis* and presumably other genomes. Nor is it necessary, for the discussion, to have complete knowledge of all types of sRNA. My interpretation requires only the knowledge that there are many different types of sRNA and that the targets include both mRNAs and heterochromatic loci.

CASCADES AND NETWORKS OF RNA SILENCING

sRNA interactions may lead to AGO-mediated silencing of mRNA or DNA as described above. However, these interactions can also lead to secondary sRNA production through a process involving RDR proteins (Vaistij et al. 2002; Xie et al. 2005; Yoshikawa et al. 2005). In this section, I first discuss secondary siRNAs. I then assess the implications of these two outcomes of sRNA interactions—silencing or secondary sRNA production—for the potential to form cascades and networks of RNA silencing.

Secondary siRNA production in plants and *C. elegans* involves RDR proteins (Sijen et al. 2001; Vaistij et al. 2002). A single-stranded RNA is targeted by an sRNA–slicer complex and is then converted into a dsRNA form by an RDR (Fig. 3). In plants, one of the best-understood secondary siRNA systems leads to the production of trans-acting (ta)siRNAs (Xie et al. 2005; Yoshikawa et al. 2005). The primary sRNAs are miRNAs that direct the cleavage of a single-stranded noncoding RNA. The cleavage process involves the same AGO-mediated mechanism that leads to mRNA degradation, but the cleaved RNA becomes a substrate for RDR-mediated dsRNA production, DCL cleavage, and production of secondary sRNAs—the tasiRNAs.

Other mechanisms of secondary siRNA production have also been proposed. It could be, for example, that the

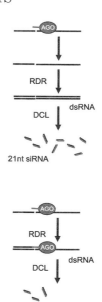

Figure 3. Secondary siRNA production. An AGO slicer with primary sRNA cleaves a target RNA. The cleaved RNA is then an RDR substrate so that dsRNA and sRNA are produced from the whole length of the targeted RNA (*upper diagram*). sRNA-primed synthesis of dsRNA would generate sRNAs only on the 5′ side of the primary sRNA target (*lower panel*) and is not consistent with observations from plant systems.

primary siRNA anneals to its target RNA and is used as a primer by the RDR rather than as a guide for a slicer nuclease (Fig. 3) (Sijen et al. 2001; Makeyev and Bamford 2002). This priming mechanism has been invoked as an explanation for secondary sRNA production in *C. elegans*. However, it is not likely that this priming method is involved in plants because the secondary sRNAs are produced on both the 3′ and 5′ sides of the primary sRNA target (Voinnet et al. 1998; Vaistij et al. 2002).

Another possible mechanism of secondary siRNA production involves the heterochromatin-targeted siRNAs and a recently discovered RNA polymerase: pol IV. pol IV is a structural homolog of the well-known DNA-dependent RNA polymerases I–III, and genetic analysis indicates that there are two isoforms—pol IVa and pol IVb (Herr et al. 2005; Kanno et al. 2005; Onodera et al. 2005). These isoforms share a common second-largest subunit, but they have largest subunits with different carboxy-terminal domains. From the siRNA phenotype of RNA silencing mutants, it seems that pol IV provides a template RNA that is converted into dsRNA by an RDR and then into siRNA by a DCL (Herr et al. 2005; Pontier et al. 2005).

These mutant phenotypes also indicate that sRNA production by pol IVb is dependent on pol IVa but not vice versa (Pontier et al. 2005). A plausible mechanism that explains this interdependence is similar to the amplification of tasiRNAs: A primary siRNA produced in a pol IVa pathway would cleave a pol IVb transcript so that it becomes a template for dsRNA production by an RDR. siRNA would then be produced in the normal way by a DCL. A second possible mechanism for amplification of pol-IV-dependent siRNAs involves the RNA-directed heterochromatinization process and requires that pol IVb preferentially transcribe heterochromatin. According to this model, the pol-IVa-dependent sRNAs would direct heterochromatinization of a locus so that it then has the potential to be transcribed by pol IVb. In this scenario, the pol-IVb-generated precursor of secondary siRNAs would be dependent on the primary siRNAs.

The above descriptions involve mechanisms in which primary sRNAs initiate a single round of secondary sRNA production. However, there is no reason, in principle, that there should be just a single round of secondary sRNA production: There could be many layers in a cascade of secondary sRNA production. As a framework for analysis of RNA silencing cascades, I describe a hypothetical situation in which there are component RNA species referred to as initiators, nodes, and end points (Fig. 4). In the simplest situation, the initiator would be an RNA, such as a miRNA precursor with an inverted repeat region. Primary 21- to 24-nucleotide sRNA would be produced by DCL-mediated cleavage of the self-complementary region in this RNA. End point RNAs, exemplified by most miRNA targets in plants, would be cleaved and degraded by interaction with the sRNA (Fig. 4). These end point RNAs would not be substrates for RDR, and they would not direct secondary sRNA production.

The node RNAs would be recruited into the silencing network if they are targeted and cleaved by AGO slicer enzymes associated with siRNAs derived from the initiator (Fig. 4). These node RNAs would be complementary to the primary sRNAs according to miRNA targeting rules (Schwab et al. 2005, 2006), and after sRNA-directed cleavage, they could be substrates for RDR and further rounds of sRNA production. As each primary sRNA would have generated multiple secondary sRNAs, this involvement of node RNAs results in amplification of

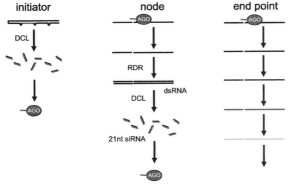

Figure 4. The components of RNA silencing cascades. The three parts of the figure illustrate the processes associated with initiator, node, and end point RNAs. The siRNAs produced by DCL from the initiator RNA are the primary sRNAs. The primary siRNA then directs AGO-mediated cleavage of a node RNA, and the RDR-mediated synthesis of dsRNA and secondary sRNA then follows. The nucleoprotein complex at the bottom of the node RNA column is the AGO slicer protein associated with a secondary sRNA. The AGO slicer at the top of the end point RNA column could be associated either with primary or secondary sRNAs.

silencing pathways. The production of secondary siRNA is an amplification process in silencing pathways.

Systems with initiator, node, and end point RNAs would expand with time because additional nodes would be recruited into the cascade. In addition, they could form a closed network if downstream secondary sRNAs are targeted at initiator or node RNAs involved at upper levels of the cascade. An RNA silencing network would influence the cell in which it was initiated because the initiator, node, and end point RNA may be protein-coding mRNAs. In addition, through the DNA targeting mechanism, the RNA silencing systems would directly affect heterochromatinization of certain regions and have an indirect effect on gene expression and chromosome behavior. The cascade could also operate between cells because sRNAs or sRNA precursors have the potential to spread through plasmodesmata (Voinnet et al. 1998; Palauqui and Balzergue 1999).

Analysis of high-throughput sRNA sequence data indicates that there may be many thousands of node RNAs including hundreds of mRNAs (Lu et al. 2005; Henderson et al. 2006). The hallmark of these node RNAs is that they would match multiple secondary sRNAs and be RDR dependent (Lu et al. 2006). However, although there may be many of these RNAs, the analysis of sRNA profiles indicates that there are fewer mRNA nodes than would be expected on the basis of a simple statistical analysis in which it is assumed that secondary sRNA production requires targeting by a single sRNA with a sequence match in 15/21 positions or more. This value corresponds approximately to the match between bona fide miRNAs and their targets (Schwab et al. 2005, 2006). To explain lower than expected frequency of mRNA nodes, I propose that a single sRNA targeting event is not sufficient to initiate secondary sRNA production: There must be additional factors. This requirement for additional factors would explain the absence of transitivity and secondary sRNA production on endogenous mRNAs targeted by transgenic sRNAs or miRNAs (Vaistij et al. 2002; Parizotto et al. 2004).

In some situations, an additional factor that predisposes an RNA to behave as a node could be structural "aberrancy." "Aberrant" RNA was invoked originally as an explanation for transgene RNA silencing: It was proposed that transgene RNAs would be more aberrant than endogenous RNAs, although it was difficult originally to define aberrancy. Recently, based on analysis of *Arabidopsis* mutants, there is evidence that aberrancy could be perturbation of 5′ cap or 3′ end mRNA features that would influence the ribonucleoprotein (RNP) status of an mRNA (Gazzani et al. 2004; Herr et al. 2006). An interpretation of these findings is that the normal messenger ribonucleoprotein (RNP) structure would inhibit RDR. With the disrupted RNP of an aberrant RNA, the enzyme would have fuller access to its template.

By extrapolation from these findings, I propose that an RNA with aberrancy at either the 5′ or 3′ end would behave as a node if it is targeted by a single sRNA. It could also be, as indicated in a very recent report, that a "normal" RNA with two sRNA target sites could also become a source of secondary sRNAs and, consequently, a node (Axtell et al. 2006). RNAs with aberrancy at both ends might also generate sRNAs in the absence of targeting by a primary sRNA: The presence of the second aberrancy would substitute for sRNA targeting. In this scenario, the doubly aberrant RNA would be an initiator RNA rather than a node because the sRNA production would be independent of an sRNA. Correspondingly, the sRNAs from this doubly aberrant RNA would be primary rather than secondary. One of the major current challenges in sRNA studies is to find out how much of the RDR-dependent sRNA in *Arabidopsis* is primary or secondary. This information will be informative about the extent of sRNA networks and cascades.

BIOLOGICAL IMPLICATIONS OF sRNA CASCADES

The existence of RNA silencing cascades and networks would have profound implications for cellular regulation for several reasons. These systems would have inherent feedback characteristics, and they would provide a mechanism through which a change in the expression level of a node RNA would influence secondary sRNAs targeted at other nodes or end points in the cascade or network. In addition, the RNA silencing networks could have epigenetic characteristics. In this section, I consider the biological implications that are related and consequent to these characteristics of RNA silencing cascades and networks.

The simplest manifestation of negative feedback in RNA silencing cascades and networks would operate if the transcription of a node RNA was reduced as a consequence of an external stimulus or due to mutation of a promoter or transcription factor. The amount of RNA available for secondary sRNA production would also be reduced as a result of this change and, correspondingly, there would be less RNA silencing targeted by this sRNA. Since the targets of this sRNA would include the original node RNA, the reduced silencing activity would compensate, at least partially, for the reduced transcription. The converse would apply if the transcription of the node RNA was increased: The increased RNA silencing would offset the effects of the increased transcription (Baulcombe 2004). In addition, there could be more complex negative feedback systems, as discussed previously, whenever sRNAs target mRNAs for proteins that are required for silencing or for transcription factors that are required for production of sRNAs (Achard et al. 2004; Vaucheret et al. 2004, 2006).

These examples of negative feedback illustrate how RNA silencing loci and cascades would provide a buffer against the influence of the environment or mutation on the expression of node RNAs. In some respects, this system could provide an RNA-level equivalent of the heat shock protein (HSP)-mediated buffering mechanism that protects against the effect of misfolded proteins caused by mutation or environmental stress (Queitsch et al. 2002).

The RNA silencing could also mediate large-scale changes to the regulatory state of a cell if perturbations result in the introduction of a novel source of sRNAs. The new sRNAs would feed into a new cascade and, consequently, there could be silencing of multiple node and end point RNAs. Situations in which cascades could be activated by novel sources of sRNA could involve inducible genes or transposons. Transposons are abundant sources

of sRNA (Lu et al. 2005; Henderson et al. 2006) and they may be the major components of RNA silencing cascades either as initiators or nodes. Phenotypes of transposon-induced silencing could be manifested, for example, in hybrids between related plants with different transposon populations. Such transposon-induced silencing could explain, in part, the unexpected phenotypes that frequently arise in hybrid plants (Tanksley and McCouch 1997).

Viruses might also initiate RNA silencing cascades or networks. If a virus-induced cascade targets mRNAs, there could be viral symptoms that are caused by silencing. There are several examples of viral or viroid RNAs that have properties predicted of silencing RNAs in that they induce symptoms but do not encode proteins (Devic et al. 1989; Jaegle et al. 1990; Wang et al. 2004). However, there is no reason, in principle, that coding sequence viral RNAs could not also induce symptoms by silencing.

If viral RNAs have inverted repeats, they could serve as initiators of silencing cascades. However, even without inverted repeat regions, they could participate in cascades as nodes if they have endogenous sRNA target sites or a combination of sRNA target sites and aberrancy, as discussed above. In such a situation, a cascade that includes viral sRNAs would have been initiated by an endogenous sRNA. Since the viral sRNAs would have an antiviral role, this type of process illustrates how the endogenous sRNAs would, in effect, be the initiators of an immune response. A precedent for the antiviral role of endogenous sRNAs exists in animal cells producing miRNA targeted at a retrovirus, although there does not appear to be secondary sRNA produced in this system (Lecellier et al. 2005).

The potential for RNA silencing cascades to have an epigenetic characteristic involves the well-characterized mechanisms involving RNA-directed DNA methylation and heterochromatin formation (Lippman and Martienssen 2004). There is also the potential for a second epigenetic mechanism at the RNA level. To illustrate this second mechanism, I consider a silencing system in which the initiator RNA is expressed transiently and in which the node RNAs would be expressed continuously (Fig. 5). The epigenetic silencing would be initiated when these primary sRNAs interact with and direct cleavage of node RNAs. Secondary sRNAs would be produced in these initiator rounds of silencing, and they would have the potential to target other node RNAs. However, they would also have the potential to target further "maintenance" rounds of sRNA production on the original node RNAs (Fig. 5). The property of plant silencing systems, that secondary sRNAs are produced on both the 5′ and 3′ sides of the sRNA target, means that these maintenance rounds of silencing and the downstream cascade would persist, even in the absence of the primary sRNA. If secondary sRNAs were produced only on the 5′ side of an sRNA target, the sRNAs would be progressively more restricted in each maintenance round and the silencing would not persist (Bergstrom et al. 2003) unless there are direct sequence repeats in the node RNA, as discussed by Martienssen (2003).

An experimental example of an RNA-mediated epigenetic mechanism involves viruses inoculated to plants carrying a viral transgene. Some of these lines are not ini-

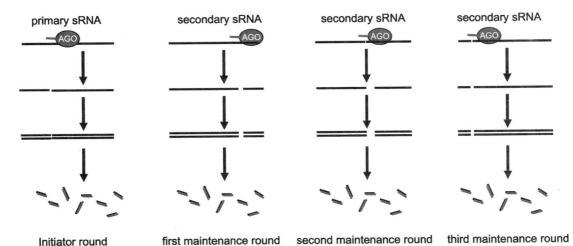

primary sRNA	secondary sRNA	secondary sRNA	secondary sRNA
Initiator round	first maintenance round	second maintenance round	third maintenance round

Figure 5. RNA-mediated epigenetic silencing. An RNA-mediated epigenetic mechanism is shown as a series of rounds involving primary and secondary sRNAs. In the initiator round, the primary sRNA triggers the production of secondary sRNAs on a node RNA. These secondary sRNAs then target other RNAs in the cascade in processes not shown in the diagram. However, they can also target the node RNA again and supplant the role of the primary sRNA in the subsequent rounds of secondary sRNA production. These subsequent rounds are referred to as maintenance rounds because they perpetuate the silencing mechanism even if the primary sRNA is no longer being produced. In effect, these maintenance rounds of silencing provide a molecular memory of the primary sRNA.

tially resistant to the virus, despite the presence of complete sequence identity between the virus and the transgene. The plants manifested viral symptoms on the inoculated leaves and the first leaves to become systemically infected. However, on the upper leaves of the plant, the virus was absent and the transgene was silent (Lindbo et al. 1993). It is likely in this situation that the viral RNA was the initiator producing sRNAs that targeted a transgene node RNA. This node RNA remained silent and, presumably, was a source of secondary siRNAs even after elimination of the initiator viral RNA. In a second example, the initiator of GFP RNA silencing was transiently expressed in one leaf of a plant. Once a signal of silencing had spread out of this leaf, the silencing persisted for the lifetime of the plant, even after the leaf had been removed (Voinnet and Baulcombe 1997; Voinnet et al. 1998). This effect was dependent on an RDR (Schwach et al. 2005). These examples illustrate the principle that there can be an RNA-mediated epigenetic mechanism although they do not involve cascades involving multiple nodes.

CONCLUSION: FUTURE ANALYSES OF CASCADES

In this paper, I speculated on the existence of sRNA cascades. I have justified this speculation by describing data to support the existence of the component molecules and processes in these cascades, including primary and secondary sRNA, transitivity, and end point RNAs. To support my speculation, I describe the limited evidence that primary sRNAs initiate secondary sRNA production. There is also evidence for enormous numbers of sRNAs in plants, and their dependency on RDR activity is consistent with, but not proof that, the idea that many of them are secondary. These sRNA cascades would have the buffering, integrative, and epigenetic characteristics described above. However, their extent is unknown. Do they encompass all

or most of the RNA in a cell either as initiator, node, or end point molecules—or just a minority of the cellular RNAs? Are there a few cascades or many, and do the cascades interact with each other? Fortunately, we are now in a position to address these questions using the full genome sequence of *Arabidopsis* and the power of the high-throughput sequencing technologies. For the first time, using these resources, we have the capability to determine the sequence profile of sRNA populations associated with AGO proteins, in RNA silencing mutants, in *Arabidopsis* genotypes, and after biotic and abiotic stimuli. We also have genetic tools—mutants in many RNA silencing genes and the potential to isolate insertion mutants in sRNA loci so that the small RNA profiles can be dissected in detail.

ACKNOWLEDGMENTS

I am grateful to the Gatsby Charitable Foundation and the Biotechnology and Biological Sciences Research Council for supporting research in my laboratory. Most of all, I acknowledge past and present members of the lab for discussions that have prompted the speculation in this paper.

REFERENCES

Achard P., Herr A.J., Baulcombe D., and Harberd N.P. 2004. Modulation of floral development by a gibberellin-regulated microRNA. *Development* **131:** 3357.
Axtell M.J., Jan C., Rajagopalan R., and Bartel D.P. 2006. A two-hit trigger for siRNA biogenesis in plants. *Cell* **127:** 565.
Baulcombe D. 2004. RNA silencing in plants. *Nature* **431:** 356.
Baumberger N. and Baulcombe D.C. 2005. *Arabidopsis* ARGONAUTE1 is an RNA Slicer that selectively recruits micro RNAs and short interfering RNAs. *Proc. Natl. Acad. Sci.* **102:** 11928.
Bayne E.H. and Allshire R.C. 2005. RNA-directed transcriptional gene silencing in mammals. *Trends Genet.* **21:** 370.

Bergstrom C.T., McKittrick E., and Antia R. 2003. Mathematical models of RNA silencing: Unidirectional amplification limits accidental self-directed reactions. *Proc. Natl. Acad. Sci.* **100:** 11511.

Bernstein E., Caudy A.A., Hammond S.M., and Hannon G.J. 2001. Role for a bidentate ribonuclease in the initiation step of RNA interference. *Nature* **409:** 363.

Carmell M.A. and Hannon G.J. 2004. RNase III enzymes and the initiation of gene silencing. *Nat. Struct. Mol. Biol.* **11:** 214.

Devic M., Jaegle M., and Baulcombe D.C. 1989. Symptom production on tobacco and tomato is determined by two distinct domains of the satellite RNA of cucumber mosaic virus (strain Y). *J. Gen. Virol.* **70:** 2765.

Dunoyer P., Himber C., and Voinnet O. 2005. DICER-LIKE 4 is required for RNA interference and produces the 21-nucleotide small interfering RNA component of the plant cell-to-cell silencing signal. *Nat. Genet.* **37:** 1356.

Gasciolli V., Mallory A.C., Bartel D.P., and Vaucheret H. 2005. Partially redundant functions of *Arabidopsis* DICER-like enzymes and a role for DCL4 in producing *trans*-acting siRNAs. *Curr. Biol.* **15:** 1494.

Gazzani S., Lawrenson T., Woodward C., Headon D., and Sablowski R. 2004. A link between mRNA turnover and RNA interference in *Arabidopsis. Science* **306:** 1046.

Han J., Lee Y., Yeom K.-H., Narm J.-W., Heo I., Rhee J.-K., Sohn S.Y., Cho Y., Zhang B.-T., and Kim V.N. 2006. Molecular basis for the recognition of primary micro-RNAs by the Drosha-DGCR8 complex. *Cell* **125:** 887.

Henderson I.R., Zhang X., Lu C., Johnson L., Meyers B.C., Green P.J., and Jacobsen S.E. 2006. Dissecting *Arabidopsis thaliana* DICER function in small RNA processing, gene silencing and DNA methylation patterning. *Nat. Genet.* **38:** 721.

Herr A.J., Jensen M.B., Dalmay T., and Baulcombe D. 2005. RNA polymerase IV directs silencing of endogenous DNA. *Science* **308:** 118.

Herr A.J., Molnar A., Jones A., and Baulcombe D.C. 2006. Defective RNA processing enhances RNA silencing and accelerates flowering in *Arabidopsis. Proc. Natl. Acad. Sci.* **103:** 14994.

Jaegle M., Devic M., Longstaff M., and Baulcombe D.C. 1990. Cucumber mosaic virus satellite RNA (Y strain): Analysis of sequences which affect yellow mosaic symptoms on tobacco. *J. Gen. Virol.* **71:** 1905.

Kanno T., Huettel B., Mette M.F., Aufsatz W., Jaligot E., Daxinger L., Kreil D.P., Matzke M., and Matzke A.J. 2005. Atypical RNA polymerase subunits required for RNA-directed DNA methylation. *Nat. Genet.* **37:** 761.

Lagos-Quintana M., Rauhut R., Lendeckel W., and Tuschl T. 2001. Identification of novel genes coding for small expressed RNAs. *Science* **294:** 853.

Lau N.C., Lim L.P., Weinstein E.G., and Bartel D.P. 2001. An abundant class of tiny RNAs with probable regulatory roles in *Caenorhabditis elegans. Science* **294:** 858.

Lecellier C.-H., Dunoyer P., Arar K., Lehmann-Che J., Eyquem S., Himber C., Saib A., and Voinnet O. 2005. A cellular MicroRNA mediates antiviral defense in human cells. *Science* **308:** 557.

Lee R. and Ambros V. 2001. An extensive class of small RNAs in *Caenorhabditis elegans. Science* **294:** 862.

Lee R., Feinbaum R.L., and Ambros V. 1993. The *C. elegans* heterochronic gene lin-4 encodes small RNAs with antisense complementarity to lin-14. *Cell* **75:** 843.

Li C.F., Pontes O., El-Shami M., Henderson I.R., Bernatavichute Y.V., Chan S.W., Lagrange T., Pikaard C., and Jacobsen S.E. 2006. An ARGONAUTE4-containing nuclear processing center colocalized with Cajal bodies in *Arabidopsis thaliana. Cell* **126:** 93.

Lindbo J.A., Silva-Rosales L., Proebsting W.M., and Dougherty W.G. 1993. Induction of a highly specific antiviral state in transgenic plants: Implications for regulation of gene expression and virus resistance. *Plant Cell* **5:** 1749.

Lippman Z. and Martienssen R. 2004. The role of RNA interference in heterochromatic silencing. *Nature* **431:** 364.

Liu J., Carmell M.A., Rivas F.V., Marsden C.G., Thomson M.,

Song J.J., Hammond S.M., Joshua-Tor L., and Hannon G.J. 2004. Argonaute2 is the catalytic engine of mammalian RNAi. *Science* **305:** 1437.

Lu C., Tej S.S., Luo S., Haudenschild C.D., Meyers B.C., and Green P.J. 2005. Elucidation of the small RNA component of the transcriptome. *Science* **309:** 1567.

Lu C., Kulkarni K., Souret F.F., MuthuValliappan R., Tej S.S., Poethig R.S., Henderson I.R., Jacobsen S.E., Wang W., Green P.J., and Meyers B.C. 2006. MicroRNAs and other small RNAs enriched in the *Arabidopsis* RNA-dependent RNA polymerase-2 mutant. *Genome Res.* **16:** 1276.

Makeyev E.V. and Bamford D.H. 2002. Cellular RNA-dependent RNA polymerase involved in posttranscriptional gene silencing has two distinct activity modes. *Mol. Cell* **10:** 1417.

Margulies M., Egholm M., Altman W.E., Attiya S., Bader J.S., Bemben L.A., Berka J., Braverman M.S., Chen Y., Chen Z., et al. 2005. Genome sequencing in microfabricated high-density picolitre reactors. *Nature* **437:** 376.

Martienssen R. 2003. Maintenance of heterochromatin by RNA interference of tandem repeats. *Nat. Genet.* **35:** 1.

Matranga C., Tomari Y., Shin C., Bartel D.P., and Zamore P.D. 2005. Passenger-strand cleavage facilitates assembly of siRNA into Ago2-containing RNAi enzyme complexes. *Cell* **123:** 1.

Onodera Y., Haag J.R., Ream T., Nunes P.C., Pontes O., and Pikaard C.S. 2005. Plant nuclear RNA polymerase IV mediates siRNA and DNA methylation-dependent heterochromatin formation. *Cell* **120:** 613.

Palauqui J.-C. and Balzergue S. 1999. Activation of systematic acquired silencing by localised introduction of DNA. *Curr. Biol.* **9:** 59.

Parizotto E.A., Dunoyer P., Rahm N., Himber C., and Voinnet O. 2004. In vivo investigation of the transcription, processing, endonucleolytic activity, and functional relevance of the spatial distribution of a plant miRNA. *Genes Dev.* **18:** 1.

Pontier D., Yahubyan G., Vega D., Bulski A., Saez-Vasquez J., Hakimi M.-A., Lerbs-Mache S., Colot V., and Lagrange T. 2005. Reinforcement of silencing at transposons and highly repeated sequences requires the concerted action of two distinct RNA polymerases IV in *Arabidopsis. Genes Dev.* **19:** 2030.

Qi Y., Denli A.M., and Hannon, G.J. 2005. Biochemical specialization within *Arabidopsis* RNA silencing pathways. *Mol. Cell* **19:** 421.

Qi Y., He X.H., Wang X., Kohany O., Jurka J., and Hannon G.J. 2006. Distinct catalytic and non-catalytic roles of ARGONAUTE4 in RNA-directed DNA methylation. *Nature* **443:** 1008.

Queitsch C., Sangster T.A., and Lindquist S. 2002. Hsp90 as a capacitor of phenotypic variation. *Nature* **417:** 618.

Reinhart B.J., Weinstein E.G., Rhoades M., Bartel B., and Bartel D.P. 2002. MicroRNAs in plants. *Genes Dev.* **16:** 1616.

Ruvkun G. 2001. Glimpses of a tiny RNA world. *Science* **294:** 797.

Schauer S.E., Jacobsen S.E., Meinke D.W., and Ray A. 2002. DICER-LIKE1: Blind men and elephants in *Arabidopsis* development. *Trends Plant Sci.* **7:** 487.

Schwab R., Ossowski S., Riester M., Warthmann N., and Weigel D. 2006. Highly specific gene silencing by artificial microRNAs in *Arabidopsis. Plant Cell* **18:** 1121.

Schwab R., Palatnik J.F., Riester M., Schommer C., Schmid M., and Weigel D. 2005. Specific effects of MicroRNAs on the plant transcriptome. *Dev. Cell* **8:** 517.

Schwach F., Vaistij F.E., Jones L., and Baulcombe D.C. 2005. An RNA-dependent RNA-polymerase prevents meristem invasion by potato virus X and is required for the activity but not the production of a systemic silencing signal. *Plant Physiol.* **138:** 1842.

Sijen T., Fleenor J., Simmer F., Thijssen K.L., Parrish S., Timmons L., Plasterk R.H.A., and Fire A. 2001. On the role of RNA amplification in dsRNA-triggered gene silencing. *Cell* **107:** 465.

Tanksley S.D. and McCouch S.R. 1997. Seed banks and molecular maps: Unlocking genetic potential from the wild. *Science* **277:** 1063.

Vaistij F.E., Jones L., and Baulcombe D.C. 2002. Spreading of RNA targeting and DNA methylation in RNA silencing requires transcription of the target gene and a putative RNA-dependent RNA polymerase. *Plant Cell* **14:** 857.

Vaucheret H., Mallory A.C., and Bartel D.P. 2006. AGO1 Homeostasis entails coexpression of *MIR168* and *AGO1* and preferential stabilization of miR168 by AGO1. *Mol. Cell* **22:** 129.

Vaucheret H., Vazquez F., Crete P., and Bartel D.P. 2004. The action of ARGONAUTE1 in the miRNA pathway and its regulation by the miRNA pathway are crucial for plant development. *Genes Dev.* **18:** 1187.

Voinnet O. and Baulcombe D.C. 1997. Systemic signalling in gene silencing. *Nature* **389:** 553.

Voinnet O., Vain P., Angell S., and Baulcombe D.C. 1998. Systemic spread of sequence-specific transgene RNA degradation is initiated by localised introduction of ectopic promoterless DNA. *Cell* **95:** 177.

Volpe T., Kidner C., Hall I.M., Teng G., Grewal S.I.S., and Martienssen R. 2002. Regulation of heterochromatic silencing and histone H3 lysine-9 methylation by RNAi. *Science* **297:** 1833.

Wang M.-B., Bian X.-Y., Wu L.-M., Liu L.-X., Smith N.A., Isenegger D., Wu R.-M., Masuta C., Vance V.B., Watson J.M., et al. 2004. On the role of RNA silencing in the pathogenicity and evolution of viroids and viral satellites. *Proc. Natl. Acad. Sci.* **101:** 3275.

Wightman B., Ha I., and Ruvkun G. 1993. Posttranscriptional regulation of the heterochronic gene lin-14 by lin-4 mediates temporal pattern-formation in *C. elegans*. *Cell* **75:** 855.

Xie Z., Allen E., Wilken A., and Carrington J.C. 2005. DICER-LIKE 4 functions in *trans*-acting small interfering RNA biogenesis and vegetative phase change in *Arabidopsis thaliana*. *Proc. Natl. Acad. Sci.* **102:** 12984.

Xie Z., Johansen L.K., Gustafson A.M., Kasschau K.D., Lellis A.D., Zilberman D., Jacobsen S.E., and Carrington J.C. 2004. Genetic and functional diversification of small RNA pathways in plants. *PLoS Biol.* **2:** 642.

Yoshikawa M., Peragine A., Park M.Y., and Poethig R.S. 2005. A pathway for the biogenesis of *trans*-acting siRNAs in *Arabidopsis*. *Genes Dev.* **19:** 2164.

Zilberman D., Cao X., and Jacobsen S.E. 2003. *ARGONAUTE4* control of locus specific siRNA accumulation and DNA and histone methylation. *Science* **299:** 716.

Zilberman D., Cao X., Johansen L.K., Xie Z., Carrington J.C., and Jacobsen S.E. 2004. Role of *Arabidopsis ARGONAUTE4* in RNA-directed DNA methylation triggered by inverted repeats. *Curr. Biol.* **14:** 1214.

Misexpression of the *Caenorhabditis elegans* miRNA *let-7* Is Sufficient to Drive Developmental Programs

G.D. Hayes and G. Ruvkun

Department of Molecular Biology, Massachusetts General Hospital and Department of Genetics,
Harvard Medical School, Boston, Massachusetts 02114

The *Caenorhabditis elegans* microRNAs (miRNAs) *lin-4* and *let-7* promote transitions between stage-specific events in development by down-regulating the translation of their target genes. Expression of *let-7* is required at the fourth larval stage for the proper transition from larval to differentiated, adult fates in the hypodermis; however, it was not known whether expression of *let-7* is sufficient to specify these adult fates. To test this, we created fusion genes between *lin-4* and *let-7* that direct the expression of *let-7* two stages early, at the L2 stage. We find that animals bearing the fusion genes show precocious adult development at the L4 stage, indicating that temporal misexpression of *let-7* is sufficient to direct the larval-to-adult transition. Additionally, an RNA interference (RNAi)-based screen for enhancers of the precocious phenotype identified the *period* ortholog *lin-42*, among other genes, which are candidate modulators of the effects of *let-7* expression. *let-7* is conserved throughout bilaterian phylogeny, and orthologs of its targets have roles in vertebrate development, suggesting the importance of understanding how *let-7* promotes terminal differentiation in *C. elegans* and other organisms.

The first two miRNAs discovered, *lin-4* and *let-7*, regulate the timing of certain developmental events that occur in particular stages of *C. elegans* development. *lin-4* is required for the transition from L1 to L2 and from L2 to L3 developmental programs in the hypodermal seam cells, among other tissues. Animals lacking *lin-4* reiterate in each larval stage and, at the nominal adult stage, seam cell division patterns that are normally confined to the L1 stage (Chalfie et al. 1981; Ambros and Horvitz 1984). *lin-4* promotes the normal progression through developmental programs by down-regulating translation of its target genes, *lin-14* and *lin-28* (Lee et al. 1993; Wightman et al. 1993; Moss et al. 1997). Inactivation of the *lin-4* targets causes the opposite heterochronic phenotype as loss of *lin-4*: Developmental programs appropriate to later stages occur precociously.

Similarly, *let-7* acts to trigger a transition from larval to adult patterns of development. In *let-7* mutants, the hypodermal seam cells continue to divide in adults, rather than fusing and secreting the adult-specific cuticular structure known as alae (Reinhart et al. 2000). *let-7* acts by inhibiting the translation of its targets *lin-41* and *hbl-1* (Slack et al. 2000; Abrahante et al. 2003; Lin et al. 2003), and down-regulation of these targets in turn allows synthesis of the transcription factor encoded by *lin-29* (Rougvie and Ambros 1995), the most downstream gene discovered in the heterochronic pathway, the set of miRNAs and their protein-coding targets that regulate developmental timing (Ambros and Horvitz 1984). At the larval-to-adult transition, LIN-29 directs expression of genes (such as collagens) involved in synthesis of the adult cuticle and inhibits transcription of those required for synthesis of larval cuticle (Liu et al. 1995; Rougvie and Ambros 1995). Recently, *let-7* was shown to target additional genes at the larval-to-adult transition, including the gene encoding the nuclear hormone receptor DAF-12 as well as several other transcription factors (Grosshans et al. 2005).

Although a number of other genes that regulate developmental timing in *C. elegans* have been identified, it remains unclear how transitions between developmental programs following the L1 stage are promoted. Work on the *let-7* paralogs *mir-48, mir-84,* and *mir-241* has suggested that they have a role in the transition to L3 developmental programs (Abbott et al. 2005; Hayes 2005; Li et al. 2005). It was not clear whether any outputs of the heterochronic pathway that are independent of *let-7* expression are required to promote the larval-to-adult transition.

Understanding how *let-7* promotes the transition to terminal differentiation in *C. elegans* is important because of the roles that *let-7* has in many animals. *let-7* is perfectly conserved throughout bilaterian phylogeny, and roles for its target, *lin-41*, has also been identified in vertebrate development (Pasquinelli et al. 2000; Kloosterman et al. 2004; Lancman et al. 2005; Schulman et al. 2005). Analogously to its role in *C. elegans*, up-regulation of *let-7* is associated with transition from the larval to pupal stage in *Drosophila* (Sempere et al. 2002, 2003; Bashirullah et al. 2003). Reduced expression of *let-7* has also been observed in a variety of tumors, suggesting it may have a broader role in promoting terminal differentiation (Takamizawa et al. 2004; Johnson et al. 2005; Karube et al. 2005).

Here, we show that precocious expression of *let-7* under the control of the *lin-4* promoter is sufficient to drive premature execution of adult developmental programs. At the fourth larval stage, animals carrying a *lin-4::let-7* fusion gene show precocious adult differentiation of the hypodermis that is dependent on *lin-29* activity. We also report results of an RNAi-based screen to identify enhancers of the precocious phenotype that results from early *let-7* expression. Inactivation of certain genes involved in developmental timing and RNA metabolism causes an enhanced precocious phenotype in animals bearing the *lin-4::let-7* fusion gene.

SECTION THEMES

miRNA Fusion Genes

To address whether expression of the miRNA *let-7* is sufficient to direct the larval-to-adult transition in the lateral hypodermal seam cells of *C. elegans*, we expressed *let-7* in the first larval stage, two stages earlier than it is normally expressed. We designed two constructs, both placing the *let-7* precursor and downstream sequence immediately 3′ to the putative *lin-4* promoter and precursor (Fig. 1A). We chose to use the *lin-4* promoter because mutations in *lin-4* and *let-7* affect the same tissues, suggesting that they are expressed in the same places. In addition, although it has now been shown that miRNA-coding genes are transcribed by RNA polymerase II (Cai et al. 2004; Lee et al. 2004), it was not originally clear how

miRNA expression is controlled. We included the *lin-4* precursor because of the possibility that elements regulating *lin-4* transcription could be embedded in the precursor sequence. Construct A (*ExA*) included 573 nucleotides upstream of the *lin-4* precursor, which was shown to be sufficient to rescue a *lin-4* mutant (Lee et al. 1993). Construct B (*ExB*) was nearly identical, except that it included 1668 nucleotides upstream of the *lin-4* precursor, and the sequence 3′ of the *lin-4* precursor was extended by 8 nucleotides before the fusion point to *let-7*.

The *lin-4* Promoter Can Drive Precocious Expression of *let-7*

Northern blot analysis demonstrated that both constructs *Ex[lin-4A::let-7]* (data not shown) and *Ex[lin-*

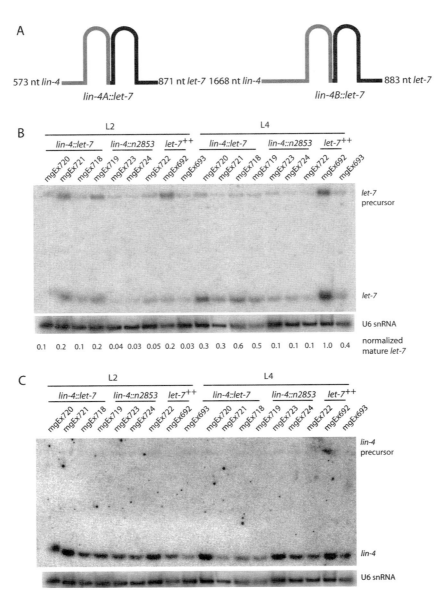

Figure 1. Fusions to *lin-4* sequence drive precocious expression of *let-7*. (*A*) Schematics of the *lin-4A::let-7* and *lin-4B::let-7* fusion constructs. (*B*) Blot of RNA isolated from *lin-4B::let-7* and control lines at the L2 and L4 stages probed for *let-7*. Information on the transgenic arrays is presented in Table 1. (*C*) The same blot reprobed for *lin-4*.

4B::let-7] (Fig. 1B) express the mature *let-7* miRNA by the beginning of the L2 stage. This was also the case with *Ex[lin-4::let-7(n2853)]*, in which the *n2853* point mutation was included. Surprisingly, when *let-7* was overexpressed under its own promoter from a 2459-bp genomic fragment beginning 1576 nucleotides 5′ of the *let-7* precursor, it was also detected by early L2. The method used here normally reveals *let-7* expression at a low level in the L3 stage in wild-type *C. elegans* (Reinhart et al. 2000), although a more sensitive method has detected low-level expression at the L2 stage (Abbott et al. 2005). It is possible that a negative regulatory element was not included in the *let-7* overexpression construct or that increased expression from the multicopy array reveals *let-7* earlier than is the case when it is expressed from the endogenous *let-7* locus. The temporal response element described by Johnson et al. (2003) as necessary for proper regulation of *let-7* was included. Quantitation of the signal from the 22-nucleotide band in each lane and normalization of loading relative to U6 small nuclear RNA (snRNA) signal revealed that *let-7* was expressed at similar levels in the L2 stage in all lines, although the highest level of expression was achieved under the *lin-4* promoter (Fig. 1B). The apparently slightly lower expression level of the *n2853* control construct could be due to imperfect hybridization between the probe and the mutated target sequence. In all

cases, the abundance of the mature form of the *let-7* miRNA increased at the L4 stage. Interestingly, in each line, an accumulation of the precursor form of *let-7* relative to the abundance of the mature product was observed in the L2, but this accumulation was resolved by the L4 stage (Fig. 1B and data not shown). This suggests the possibility that processing of *let-7* by the Dicer-ALG-1/2 complex (Grishok et al. 2001; Hutvagner et al. 2001), as well as the transcription of *let-7*, could be temporally regulated. No similar L2-stage accumulation of *lin-4* precursor was observed, demonstrating that miRNA processing in general was not disrupted (Fig. 1C).

Both of the *lin-4::let-7* fusion constructs drove production of functional *let-7*, as demonstrated by their ability to rescue a *let-7* null mutant (Table 1). However, all of the lines that were tested showed at most weak rescue of a *lin-4* null (data not shown). In addition, when RNA derived from animals carrying the *lin-4B::let-7* transgenes *mgEx719* and *mgEx720* in a *lin-4* null background was probed, no clear *lin-4* expression was detected (data not shown). These observations suggest that the *lin-4* precursor produced from these fusion genes may be degraded rather than processed to mature *lin-4*. Many miRNAs are found in genomic clusters (such as *mir-35–mir-41* in *C. elegans* and *let-7* and the *lin-4* homolog, *miR-125*, in *Drosophila*) and are processed from a single primary transcript (Lau et al.

Table 1. Phenotypes Generated by *let-7* Precocious- and Overexpression Transgenes

Strain/Array	Construct	Injection concentration (ng/µl)	Precocious alae at L4 stage (% [n])	Rescue of *let-7* (% [n])
mgEx704	*lin-4A::let-7*	40	0 (5)	n.d.
mgEx705	*lin-4A::let-7*	40	n.d.	20 (10)
mgEx706	*lin-4A::let-7*	20	n.d.	77 (22)
mgEx707	*lin-4A::let-7*	20	n.d.	80 (10)
mgEx708	*lin-4A::let-7*	20	n.d.	100 (26)
mgEx709	*lin-4A::let-7*	20	6 (17)	71 (14)
mgEx710	*lin-4A::let-7*	40	0 (7)	<5 (>20)
mgEx711	*lin-4B::let-7*	15	80 (5)	n.d.
mgEx712	*lin-4B::let-7*	15	33 (12)	n.d.
mgEx713	*lin-4B::let-7*	15	40 (10)	40 (10)
mgEx714	*lin-4B::let-7*	10	0 (3)	59 (12)
mgEx715	*lin-4B::let-7*	10	29 (7) [b]	80 (5)
mgEx716	*lin-4B::let-7*	10	69 (26)	n.d.
mgEx717[a]	*lin-4B::let-7*	5	0 (7)	100 (9)
mgEx718[a]	*lin-4B::let-7*	5	35 (17)	n.d.
mgEx719[a]	*lin-4B::let-7*	5	43 (7)	n.d.
mgEx720[a]	*lin-4B::let-7*	1	38 (8)	100 (6)
mgEx721[a]	*lin-4B::let-7*	1	45 (31)	n.d.=
mgEx722[a]	*lin-4B::n2853*	5	0 (12)	n.d.=
mgEx723[a]	*lin-4B::n2853*	1	0 (20)	n.d.=
mgEx724[a]	*lin-4B::n2853*	1	4 (24)	0 (9)
mgEx692[a]	*let-7++*	1	0 (14)	100 (8)
mgEx693[a]	*let-7++*	1	0 (16)	n.d.=
lin-41(ma104)	n.a.	n.a.	50 (12)	n.d.
mgEx719[a]	*lin-4B::let-7*	5	78 (9)	n.d.
mgEx719; lin-29 (RNAi)[a]	*lin-4B::let-7*	5	0 (9)	n.d.
mgEx720; vector (RNAi)	*lin-4B::let-7*	1	26 (19)[b]	n.d.
mgEx720; lin-29(RNAi)	*lin-4B::let-7*	1	0 (14)[b]	n.d.

[a] All observations reported on this line of the table were done in a *wIs54[scm::gfp]* V; *let-7(mn112) unc-3(e151)* X background.
[b] Scored in a *let-7(mn112) unc-3(e151)* background.
= This transgene rescued *let-7(mn112)* (only at 15°C for *lin-4::n2853* lines), but the penetrance of rescue was not determined. n.d. indicates not determined; n.a. indicates not applicable.

2001; Bashirullah et al. 2003). However, the mature *lin-4* and *let-7* sequences are positioned more closely (7–15-bp separation) in our fusion genes than the miRNAs in these naturally occurring clusters (>30-bp separation).

Early Expression of *let-7* Generates Precocious Development

It was expected that early expression of *let-7* would cause precocious down-regulation of its target, LIN-41, causing a phenotype similar to that seen in animals bearing loss-of-function mutations in *lin-41*. Animals lacking LIN-41 are dumpy (Dpy) and partly sterile. In wild-type *C. elegans*, the lateral hypodermal seam cells fuse and secrete a cuticular structure known as alae at the L4-to-adult molt (Sulston and Horvitz 1977). In *lin-41* loss-of-function mutants, both seam cell fusion and production of alae occur one stage early, at the L3-to-L4 molt (Slack et al. 2000). Animals expressing both *Ex[lin-4A::let-7]* and *Ex[lin-4B::let-7]* were Dpy and had a partially penetrant egg-laying defect (data not shown). However, animals expressing *Ex[lin-4A::let-7]* only very rarely produced alae precociously, whereas those expressing *Ex[lin-4B:: let-7]* produced alae with a penetrance similar to that of *lin-41(ma104)*; 29–80% of animals bearing *Ex[lin-4B:: let-7]* and 50% of *lin-41* mutants produced alae covering at least one of the lateral hypodermal seam cells at the L3-to-L4 molt (Table 1). As shown in Figure 2, both *lin-41(ma104)* animals and those expressing *Ex[lin-4B:: let-7]* often produced "patchy" alae during the L4 stage, with some seam cells precociously expressing the adult fate, while others continued to express the larval fate. The "patchy alae" phenotype has also been observed for other heterochronic mutants (Ambros 1989).

Consistent with the results for alae formation, animals bearing *lin-4B::let-7* transgenes showed precocious expression of a reporter for *col-19*, which encodes a collagen that is a component of the adult cuticle. *col-19*, which is positively regulated by LIN-29, is expressed only in the adult stage in wild-type *C. elegans* (Abrahante et al. 1998); 75% (*n* = 12) of animals carrying *mgEx719[lin-4B::let-7]* expressed the *veIs13[col-19::gfp]* reporter in the L4 stage. Animals bearing *lin-4B::let-7* transgenes only rarely expressed *col-19::gfp* as early as the L3 stage (Fig. 3).

Precocious Alae Result from Early Expression of *let-7*

To test whether the precocious phenotype generated by the *Ex[lin-4B::let-7]* transgenes was due to expression of *let-7*, we introduced the *n2853* point mutation into the *lin-4B::let-7* construct. Introduction of the *n2853* mutation was expected to block *let-7* function, because *let-7(n2853)* mutant worms have a temperature-sensitive lethal phenotype similar to that of *let-7(mn112)* null mutants. In fact, animals bearing the *Ex[lin-4B:: let-7(n2853)]* constructs only very rarely (0–4%) produced alae precociously (Table 1). Those animals bearing *mgEx[lin-4B::let-7(n2853)]* that did produce alae precociously produced only faint alae, not distinct patches as observed with animals bearing the *Ex[lin-4B::let-7]*

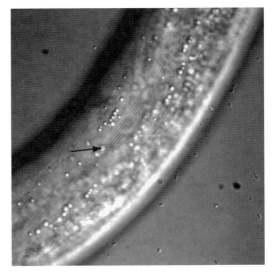

Figure 2. A typical L4 larva bearing the *lin-4B::let-7* transgene showing precocious alae formation. This animal expressed a mix of larval and adult fates in the hypodermis, indicated by the break in the alae (*arrow*).

transgenes (data not shown). The fact that the introduction of the *n2853* point mutation abrogates the precocious phenotype induced by expression of *Ex[lin-4B::let-7]* demonstrates that the precocious phenotype results from early expression of *let-7*, not overexpression of *lin-4*.

The precocious phenotype generated by the *lin-4::let-7* constructs was not a product of simple overexpression of *let-7*. Animals bearing the *Ex[let-7++]* transgenes never formed alae precociously (Table 1), even though they produced *let-7* at the L2 stage at a level comparable to that of animals bearing the *lin-4::let-7* constructs (Fig. 1B).

The Heterochronic Pathway Mediates the Effects of Precocious *let-7* Expression through *lin-29*

To confirm that the precocious phenotype produced by expression of *Ex[lin-4B::let-7]* is mediated by the established heterochronic pathway, we inactivated the terminal gene in this pathway, *lin-29*, by feeding RNAi. Animals bearing either of two *lin-4B::let-7* transgenes produced alae at the L3-to-L4 molt when fed vector RNAi, but they never produced alae when fed *lin-29(RNAi)* (Table 1), indicating that the precocious phenotype produced by expression of the *lin-4::let-7* fusion gene is mediated by *lin-29*.

RNAi Screen for Enhancers of the Precocious *lin-4::let-7* Phenotype

Although the *lin-4B::let-7* construct drives precocious production of *let-7* two stages early, by the beginning of the L2 stage, its expression results in production of alae and expression of *col-19::gfp* only one stage early, at the L3-to-L4 molt. It seemed possible that some factor might inhibit the early action of *let-7* or that some other factor necessary for *let-7* function might not normally be present as early as the L2 stage.

To identify such factors, we conducted a feeding RNAi screen using a library of 438 clones selected because they were expected to have a role in RNA metabolism, RNAi, genome stability and transposon silencing, miRNA function, or control of developmental timing. Animals bearing *mgEx719[lin-4B::let-7]* in a *col-19::gfp*; *let-7(mn112) unc-3(e151)* background were fed each clone, and they and their progeny were scored for expression of *col-19:: gfp* earlier than the L4 stage; 128 clones emerged that either appeared to cause expression of *col-19::gfp* earlier than the L4 stage or that caused other phenotypes (such as sickness or sterility) specifically in the animals bearing the *lin-4::let-7* transgene (data not shown). None of the seven vector clones included and scored blindly was noted to cause precocious *col-19::gfp* expression nor any other phenotypes in larvae. 25 of the 128 clones appeared to enhance precocious expression of *col-19::gfp* in animals bearing the *lin-4::let-7* transgene upon a single retest (Table 2; data not shown). Four of the 25 positive clones that caused the most potent or penetrant precocious expression of *col-19::gfp* were examined by Nomarski microscopy to determine the age of the green animals. The clones corresponding to *lin-42* (*period*), Y57G11A.5 (which encodes a predicted RNA-binding protein that may be involved in splicing), and Y77E11A.7 and T26A5.5, which are required for genome stability, were all confirmed to cause *col-19::gfp* expression by the L3 stage in animals bearing *mgEx719[lin-4B::let-7]* (Fig. 3A; Table 2).

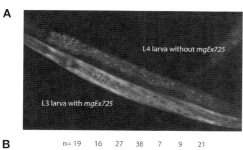

A

B

n= 19   16   27   38   7   9   21

Figure 3. Enhancement of the *mgEx725[lin-4B::let-7]* precocious *col-19::gfp* expression phenotype by RNAi. (*A*) *lin-42(RNAi)* enhanced the precocious *col-19::gfp* expression phenotype of animals bearing the *lin-4B::let-7* transgene. On the bottom is an L3 animal bearing *mgEx725[lin-4B::let-7]* and expressing *col-19::gfp* precociously. Above it is an L4 animal without the *mgEx725* transgene and showing only background fluorescence. (*B*) Effect of feeding RNAi against the indicated clone on the age distribution of animals expressing *col-19::gfp* in the F$_1$ generation.

Table 2. Possible Enhancers of the Precocious *col-19::gfp* Expression in Animals Bearing *mgEx719[lin-4B::let-7]*

Developmental timing	Nucleases
F47F6.2 (*lin-42*)	R17.2
F11A1.3 (*daf-12*)	Y39G8C.1 (*xrn-1*)
Argonaute orthologs	F31E3.4
F55A12.1	C50C3.7
C35D6.3	ZK1098.3
Nucleic acid binding	**Helicases**
C08H9.2	Y54E2A.6
C56G2.1	**Genome stability /**
C44E4.4	**transgene silencing**
R05H10.2	R06C7.7 (*lin-61*)
M4.2 (*puf-4*)	T26A5.5
Y57G11A.5	Y77E11A.7
Y116A8C.32	AC8.1 (*pme-6*)
C30G7.1 (*hil-1*)	**Unknown function**
mRNA nuclear export	C28A5.1
F10G8.3 (*npp-17*)	Y54E2A.7

The 25 apparent enhancers were tested a final time with a strain bearing a red fluorescent protein (*rfp*)-marked *lin-4B::let-7* transgene that eliminated background green fluorescence from the original *tub-1::gfp*-marked *lin-4B:: let-7* transgene. Animals bearing the *lin-4B::let-7* transgene showed *col-19::gfp* expression by the L3 stage when grown in synchrony and fed RNAi against the following genes beginning at the L1 stage: *lin-42*, 7 of 22; C08H9.2, 3 of 23; C56G2.1, 3 of 27; *hil-1*, 4 of 29; R17.2, 2 of 22; Y54E2A.6, 2 of 18; Y54E2A.7, 3 of 18; *lin-61*, 6 of 21. No animals lacking the *lin-4B::let-7* transgene showed expression of *col-19::gfp* before the L4 stage. One of 20 *lin-4B::let-7* transgene-bearing animals fed vector RNAi dimly expressed *col-19::gfp* at the L3 stage. For clones that induced relatively penetrant precocious expression of *col-19::gfp* in the P$_0$ generation, the F$_1$ progeny were washed from the well, and the age of *col-19::gfp*-expressing animals that bore the *lin-4::let-7* transgene was determined (Fig. 3B). Inactivation of *lin-42* (*period*) and R17.2 (which encodes a nuclease) in particular increased the proportion of L3 animals that expressed *col-19::gfp* as compared to vector. Targeting of Y54E2A.6 (which encodes a helicase) and C08H9.2 (which encodes a KH-domain protein) had less effect at the L3 stage, but increased the proportion of animals expressing *col-19::gfp* before the adult stage.

A final category of clone that emerged from the screen was those that allowed *let-7* null animals to live in the absence of the *lin-4::let-7* transgene. Among the six suppressors was *lin-41*, a known suppressor of *let-7* and one of its established targets (Slack et al. 2000). Three of the six clones (those targeting F55F8.3, K07C5.4, and F17C11.9) corresponded to genes required for assembly of the ribosome or translation. The significance of this finding is unclear, but it may be that the lethality of *let-7* derives from derepression of the translation of *lin-41* or other *let-7* targets and that this lethality is prevented by defects in translation. Two additional suppressors were *crn-1*, which encodes an endonuclease (Parrish et al. 2003), and *puf-3*, which encodes an ortholog of Pumilio, an RNA-binding protein implicated in posttranscriptional regulation (Wickens et al. 2002).

CONCLUSIONS

The data presented here demonstrate that early expression of the *let-7* miRNA is sufficient to drive adult differentiation of the hypodermis and that this precocious phenotype is mediated by the heterochronic gene *lin-29*. It is perhaps surprising that animals bearing the *Ex[let-7⁺⁺]* transgenes failed to produce alae precociously, given that animals expressing both *Ex[lin-4B::let-7]* and *Ex[let-7⁺⁺]* produced similar levels of mature *let-7* miRNA by the L2 stage. One explanation is that the *lin-4* and *let-7* promoters may drive *let-7* expression in different tissues and that expression of *let-7* in the L2 stage in tissues where the *lin-4* promoter is active is required to yield the precocious phenotype. It could be that some factor prevents the early activity of *let-7* in the tissues where it is expressed under its own promoter but that this factor is not present in additional tissues where *Ex[lin-4B::let-7]* drives *let-7* expression. Candidates for such a factor include molecules such as LIN-42 and others identified in the enhancer screen.

The *period* homolog *lin-42* has been previously shown to cause precocious development of the hypodermis when it is inactivated (Jeon et al. 1999). Loss of other genes that cause precocious heterochronic phenotypes, such as *lin-28*, has been shown to up-regulate levels of *let-7* (Johnson et al. 2003). It seemed possible that the precocious phenotype of *lin-42* mutants could also depend on such up-regulation of *let-7*. However, the fact that inactivation of *lin-42* strongly enhances the precocious phenotype of animals bearing *lin-4B::let-7* transgenes suggests that up-regulation of *let-7* is not primarily responsible for the precocious phenotype of *lin-42* mutants.

It was recently shown that the abundance of mature *let-7* and many other miRNAs in vertebrates is controlled at the level of processing of the primary transcript by Drosha (Thomson et al. 2006). No similar regulation has been shown in *C. elegans*; indeed, transcription of a *let-7* reporter (Johnson et al. 2003) and processing of the primary transcript (Bracht et al. 2004) are closely correlated with the appearance of mature *let-7*. However, regulation of processing, at the level of either the primary transcript or the hairpin precursor, is one reason the precocious phenotypes generated by *let-7* expression under the *lin-4* promoter could be weaker than anticipated. Indeed, the accumulation of the hairpin precursor form of *let-7* at the L2 stage (Fig. 1B) suggested that such regulation might occur.

Additional candidates for factors that could act together with *let-7* to direct the larval-to-adult transition were the *let-7* paralogs *mir-48, mir-84,* and *mir-241* (Lau et al. 2001; Lim et al. 2003). Indeed, *let-7* and *mir-84* act together to promote cessation of molting in adults, and overexpression of *mir-84* causes precocious execution of the L3-stage pattern of seam cell divisions in the L2 stage (Hayes 2005). However, when *lin-4::mir-48* and *lin-4::mir-84* constructs analogous to the *lin-4B::let-7* construct were injected, they caused no obvious precocious phenotypes (data not shown). Animals injected with combinations of these constructs and the *lin-4B::let-7* construct showed precocious development, but not to a greater extent than animals bearing *Ex[lin-4B::let-7]* alone (data

not shown). Similarly, no effect on the L2-stage seam cell divisions was noted for the *lin-4B::let-7* transgenes, even in combination with the enhancer clones identified here (data not shown). These results suggest that despite their sequence similarity, *let-7* and its paralogs have at least partly distinct targets in promoting developmental progression. This finding is in agreement with the observation that the effects of *mir-84* overexpression are similar to but distinct from those of *mir-48* overexpression (Hayes 2005; Li et al. 2005) and also with the observation by Leaman et al. (2005) that inactivation of individual *miR-310* family members caused similar, but not identical, phenotypes in *Drosophila* larval development.

The results of the RNAi screen for enhancers of the precocious phenotype caused by expression of *lin-4B::let-7* point to possible roles in the regulation of *let-7* (at least as expressed from the fusion transgene) for molecules implicated in various aspects of RNA metabolism. One caveat is that the verification of these observations remains incomplete. Retesting with the *rfp*-marked *lin-4::let-7* transgene failed to detect precocious *col-19::gfp* expression in the P₀ generation of animals for many of the clones. Greater penetrance of precocious phenotypes is expected upon screening of the F₁ progeny of animals fed each clone. Determining whether these clones enhance precocious *col-19::gfp* expression in heterochronic mutant backgrounds (such as *lin-14, lin-28, lin-42,* and *hbl-1*) will point to roles the corresponding genes might have in the heterochronic pathway. It would also help clarify the relationship of the enhancers to the regulation of endogenous *let-7*.

ACKNOWLEDGMENTS

John Kim and Sylvia Fischer generously shared the sublibrary of RNAi clones they assembled from the library created in Julie Ahringer's laboratory. *col-19::gfp* was a kind gift from Ann Rougvie. The assistance of Jinling Xu, Snjezana Joksimovic, and Li Xue with microinjections is gratefully acknowledged. Sylvia Lee, Ho Yi Mak, Amy Pasquinelli, Brenda Reinhart, and other members of the Ruvkun lab all contributed advice and helpful discussion throughout the project. This work was supported by National Institutes of Health grant GM44619 to G.R.

REFERENCES

Abbott A.L., Alvarez-Saavedra E., Miska E.A., Lau N.C., Bartel D.P., Horvitz H.R., and Ambros V. 2005. The *let-7* microRNA family members *mir-48, mir-84,* and *mir-241* function together to regulate developmental timing in *Caenorhabditis elegans. Dev. Cell* **9**: 403.

Abrahante J.E., Miller E.A., and Rougvie A.E. 1998. Identification of heterochronic mutants in *Caenorhabditis elegans.* Temporal misexpression of a collagen::green fluorescent protein fusion gene. *Genetics* **149**: 1335.

Abrahante J.E., Daul A.L., Li M., Volk M.L., Tennessen J.M., Miller E.A., and Rougvie A.E. 2003. The *Caenorhabditis elegans* hunchback-like gene *lin-57/hbl-1* controls developmental time and is regulated by microRNAs. *Dev. Cell* **4**: 625.

Ambros V. 1989. A hierarchy of regulatory genes controls a larva-to-adult developmental switch in *C. elegans. Cell* **57**: 49.

Ambros V. and Horvitz H.R. 1984. Heterochronic mutants of the nematode *Caenorhabditis elegans*. *Science* **226**: 409.

Bashirullah A., Pasquinelli A.E., Kiger A.A., Perrimon N., Ruvkun G., and Thummel C.S. 2003. Coordinate regulation of small temporal RNAs at the onset of *Drosophila* metamorphosis. *Dev. Biol* **259**: 1.

Bracht J., Hunter S., Eachus R., Weeks P., and Pasquinelli A.E. 2004. Trans-splicing and polyadenylation of *let-7* microRNA primary transcripts. *RNA* **10**: 1586.

Cai X., Hagedorn C.H., and Cullen B.R. 2004. Human microRNAs are processed from capped, polyadenylated transcripts that can also function as mRNAs. *RNA* **10**: 1957.

Chalfie M., Horvitz H.R., and Sulston J.E. 1981. Mutations that lead to reiterations in the cell lineages of *C. elegans*. *Cell* **24**: 59.

Grishok A., Pasquinelli A.E., Conte D., Li N., Parrish S., Ha I., Baillie D.L., Fire A., Ruvkun G., and Mello C.C. 2001. Genes and mechanisms related to RNA interference regulate expression of the small temporal RNAs that control *C. elegans* developmental timing. *Cell* **106**: 23.

Grosshans H., Johnson T., Reinert K.L., Gerstein M., and Slack F.J. 2005. The temporal patterning microRNA *let-7* regulates several transcription factors at the larval to adult transition in *C. elegans*. *Dev. Cell* **8**: 321.

Hayes G.D. 2005. "Control of developmental timing by microRNAs in *C. elegans*." Ph.D. thesis, Harvard University, Cambridge, Massachusetts.

Hutvagner G., McLachlan J., Pasquinelli A.E., Balint E., Tuschl T., and Zamore P.D. 2001. A cellular function for the RNA-interference enzyme Dicer in the maturation of the *let-7* small temporal RNA. *Science* **293**: 834.

Jeon M., Gardner H.F., Miller E.A., Deshler J., and Rougvie A.E. 1999. Similarity of the *C. elegans* developmental timing protein LIN-42 to circadian rhythm proteins. *Science* **286**: 1141.

Johnson S.M., Lin S.Y., and Slack F.J. 2003. The time of appearance of the *C. elegans let-7* microRNA is transcriptionally controlled utilizing a temporal regulatory element in its promoter. *Dev. Biol.* **259**: 364.

Johnson S.M., Grosshans H., Shingara J., Byrom M., Jarvis R., Cheng A., Labourier E., Reinert K., Brown D., and Slack F.J. 2005. RAS is regulated by the *let-7* microRNA family. *Cell* **120**: 635.

Karube Y., Tanaka H., Osada H., Tomida S., Tatematsu Y., Yanagisawa K., Yatabe Y., Takamizawa J., Miyoshi S., Mitsudomi T., and Takahashi T. 2005. Reduced expression of Dicer associated with poor prognosis in lung cancer patients. *Cancer Sci.* **96**: 111.

Kloosterman W.P., Wienholds E., Ketting R.F., and Plasterk R.H. 2004. Substrate requirements for *let-7* function in the developing zebrafish embryo. *Nucleic Acids Res.* **32**: 6284.

Lancman J.J., Caruccio N.C., Harfe B.D., Pasquinelli A.E., Schageman J.J., Pertsemlidis A., and Fallon J.F. 2005. Analysis of the regulation of *lin-41* during chick and mouse limb development. *Dev. Dyn.* **234**: 948.

Lau N.C., Lim L.P., Weinstein E.G., and Bartel D.P. 2001. An abundant class of tiny RNAs with probable regulatory roles in *Caenorhabditis elegans*. *Science* **294**: 858.

Leaman D., Chen P.Y., Fak J., Yalcin A., Pearce M., Unnerstall U., Marks D.S., Sander C., Tuschl T., and Gaul U. 2005. Antisense-mediated depletion reveals essential and specific functions of microRNAs in *Drosophila* development. *Cell* **121**: 1097.

Lee R.C., Feinbaum R.L., and Ambros V. 1993. The *C. elegans* heterochronic gene *lin-4* encodes small RNAs with antisense complementarity to *lin-14*. *Cell* **75**: 843.

Lee Y., Kim M., Han J., Yeom K.H., Lee S., Baek S.H., and Kim V.N. 2004. MicroRNA genes are transcribed by RNA polymerase II. *EMBO J.* **23**: 4051.

Li M., Jones-Rhoades M.W., Lau N.C., Bartel D.P., and Rougvie A.E. 2005. Regulatory mutations of mir-48, a *C. elegans let-7* family microRNA, cause developmental timing defects. *Dev. Cell* **9**: 415.

Lim L.P., Lau N.C., Weinstein E.G., Abdelhakim A., Yekta S., Rhoades M.W., Burge C.B., and Bartel D.P. 2003. The microRNAs of *Caenorhabditis elegans*. *Genes Dev.* **17**: 991.

Lin S.Y., Johnson S.M., Abraham M., Vella M.C., Pasquinelli A., Gamberi C., Gottlieb E., and Slack F.J. 2003. The *C. elegans* hunchback homolog, hbl-1, controls temporal patterning and is a probable microRNA target. *Dev. Cell* **4**: 639.

Liu Z., Kirch S., and Ambros V. 1995. The *Caenorhabditis elegans* heterochronic gene pathway controls stage-specific transcription of collagen genes. *Development* **121**: 2471.

Moss E.G., Lee R.C., and Ambros V. 1997. The cold shock domain protein LIN-28 controls developmental timing in *C. elegans* and is regulated by the *lin-4* RNA. *Cell* **88**: 637.

Parrish J.Z., Yang C., Shen B., and Xue D. 2003. CRN-1, a *Caenorhabditis elegans* FEN-1 homologue, cooperates with CPS-6/EndoG to promote apoptotic DNA degradation. *EMBO J.* **22**: 3451.

Pasquinelli A.E., Reinhart B.J., Slack F., Martindale M.Q., Kuroda M.I., Maller B., Hayward D.C., Ball E.E., Degnan B., Muller P., et al. 2000. Conservation of the sequence and temporal expression of *let-7* heterochronic regulatory RNA. *Nature* **408**: 86.

Reinhart B.J., Slack F.J., Basson M., Pasquinelli A.E., Bettinger J.C., Rougvie A.E., Horvitz H.R., and Ruvkun G. 2000. The 21-nucleotide *let-7* RNA regulates developmental timing in *Caenorhabditis elegans*. *Nature* **403**: 901.

Rougvie A.E. and Ambros V. 1995. The heterochronic gene *lin-29* encodes a zinc finger protein that controls a terminal differentiation event in *Caenorhabditis elegans*. *Development* **121**: 2491.

Schulman B.R., Esquela-Kerscher A., and Slack F.J. 2005. Reciprocal expression of lin-41 and the microRNAs *let-7* and *mir-125* during mouse embryogenesis. *Dev. Dyn.* **234**: 1046.

Sempere L.F., Dubrovsky E.B., Dubrovskaya V.A., Berger E.M., and Ambros V. 2002. The expression of the *let-7* small regulatory RNA is controlled by ecdysone during metamorphosis in *Drosophila melanogaster*. *Dev. Biol.* **259**: 9.

Sempere L.F., Sokol N.S., Dubrovsky E.B., Berger E.M., and Ambros V. 2003. Temporal regulation of microRNA expression in *Drosophila melanogaster* mediated by hormonal signals and broad-Complex gene activity. *Dev. Biol.* **259**: 9.

Slack F.J., Basson M., Liu Z., Ambros V., Horvitz H.R., and Ruvkun G. 2000. The *lin-41* RBCC gene acts in the *C. elegans* heterochronic pathway between the *let-7* regulatory RNA and the LIN-29 transcription factor. *Mol. Cell* **5**: 659.

Sulston J.E. and Horvitz H.R. 1977. Post-embryonic cell lineages of the nematode, *Caenorhabditis elegans*. *Dev. Biol.* **56**: 110.

Takamizawa J., Konishi H., Yanagisawa K., Tomida S., Osada H., Endoh H., Harano T., Yatabe Y., Nagino M., Nimura Y., et al. 2004. Reduced expression of the *let-7* microRNAs in human lung cancers in association with shortened postoperative survival. *Cancer Res.* **64**: 3753.

Thomson J.M., Newman M., Parker J.S., Morin-Kensicki E.M., Wright T., and Hammond S.M. 2006. Extensive post-transcriptional regulation of microRNAs and its implications for cancer. *Genes Dev.* **20**: 2202.

Wickens M., Bernstein D.S., Kimble J., and Parker R. 2002. A PUF family portrait: 3′UTR regulation as a way of life. *Trends Genet.* **18**: 150.

Wightman B., Ha I., and Ruvkun G. 1993. Posttranscriptional regulation of the heterochronic gene *lin-14* by *lin-4* mediates temporal pattern formation in *C. elegans*. *Cell* **75**: 855.

Function and Localization of MicroRNAs in Mammalian Cells

A.K.L. Leung and P.A. Sharp

Center for Cancer Research, Massachusetts Institute of Technology, Cambridge, Massachusetts 02139

microRNAs (miRNAs) represent a large set of master regulators of gene expression. They constitute 1–4% of human genes and probably regulate 30% of protein-encoding genes. These small regulatory RNAs act at a posttranscriptional level—mediating translational repression and/or mRNA degradation—through their association with Argonaute protein and target mRNAs. In this paper, we discuss various mechanisms by which miRNAs regulate posttranscriptionally, including their subcellular localization. Recent results indicate that the majority of miRNA-targeted and thus translationally repressed mRNA is probably distributed in the diffuse cytoplasm, even though a small fraction is concentrated in subcellular compartments, such as processing bodies or stress granules; notably, the stress granule localization of Argonaute depends on the presence of miRNAs. Here we discuss the structural requirement of these subcellular compartments in light of their potential miRNA functions.

The miRNAs are a class of noncoding RNAs found in animals and plants, whose mature products are approximately 21–22 nucleotides in length (Bartel 2004; Kim 2005). These RNAs are predicted to regulate about 30% of mammalian mRNAs through interactions with their 3′UTRs (untranslated regions) (Lewis et al. 2005; Rajewsky 2006). The binding of miRNAs can result in translational repression and/or degradation of mRNAs (Filipowicz et al. 2005; Valencia-Sanchez et al. 2006). We discuss here the various functions of miRNAs and their possible cytoplasmic locale in mammalian cells.

MiRNA-MEDIATED TRANSLATIONAL REPRESSION

The role of miRNA was first demonstrated as the translational repressor by its founding member *lin-4* in the nematode *Caenorhabditis elegans* (Lee et al. 1993; Wightman et al. 1993). Genetic analysis showed that the gene product of *lin-4* acted as a negative regulator of LIN-14 expression after larval stage L1, and the 3′UTR of the *lin-14* mRNA was required for this regulation (Arasu et al. 1991). After the discovery of the gene product of *lin-4* as a 22-nucleotide RNA and the potential binding sites at the 3′UTR of *lin-14* mRNA, the miRNA:mRNA base-pairing was realized to be crucial in regulating *lin-14* expression (Lee et al. 1993; Wightman et al. 1993). The 3′UTR of *lin-14* mRNAs contains seven *lin-4* miRNA-binding sites, which are conserved in *Caenorhabditis briggsae*. *lin-14* mRNAs with mutations in all of these binding sites can no longer be down-regulated by *lin-4* (Ha et al. 1996). It was initially found that the steady-state level of *lin-14* transcript was not reduced more than twofold by the *lin-4* (although a recent report detected a fivefold reduction; Bagga et al. 2005; and see below), whereas its protein level was reduced by approximately tenfold from larval stage L1 to L2 (Olsen and Ambros 1999). Furthermore, neither the transcription rate nor the status of polyadenylation for *lin-14* mRNA changed from L1 to L2 (Olsen and Ambros 1999). This evidence points to the role of *lin-4* as a translational repressor of *lin-14* mRNA.

The mechanisms by which *lin-4* suppresses the expression of *lin-14* mRNA were investigated. Biochemical analysis suggested that both *lin-4* miRNA and *lin-14* mRNA cosedimented with polyribosomes (Olsen and Ambros 1999). A similar polyribosomal association of miRNA is also observed in mammalian cells (Kim et al. 2004; Nelson et al. 2004; Petersen et al. 2006). However, another substantial fraction of miRNAs do not cosediment with the polyribosomes, but with a fraction slower than monosomes (Nelson et al. 2004). This more slowly sedimenting fraction is typically devoid of target mRNA, probably representing miRNA-containing ribonucleoparticles (miRNPs) in excess to target mRNAs or those miRNAs that dissociated from their target mRNAs during isolation. Evidence showing that the polyribosome-associated miRNA target mRNAs are engaged in some form of active translation has been reported, for example, human K-*ras* and *lin-28* mRNAs that are targeted by *let-7* miRNA (Nelson et al. 2004; Maroney et al., this volume) or a model luciferase mRNA targeted by a partially complementary small interfering RNA (siRNA) (Petersen et al. 2006). In both cases, short pulses of the peptidyl transferase inhibitor puromycin released the repressed mRNAs from the polyribosomes. Conversely, miRNA-mediated translational repression was proposed to act at the stage of initiation. When targeted by *let-7* miRNA or directly tethered with Ago2, a model luciferase mRNA in the polysome profile was significantly shifted toward the monosome fraction from the polyribosomes (Pillai et al. 2005). Thus, miRNA-mediated translational repression can probably occur both at a step postinitiation or at initiation (Fig. 1).

Repression Postinitiation

On the basis of the following observations, our laboratory proposed that miRNA might act to promote abortive termination, resulting in "ribosome drop-off" during elongation of translation (Petersen et al. 2006): First, the readthrough rate of a termination codon artificially introduced between two open reading frames is less efficient in the presence of a miRNA that targets their shared

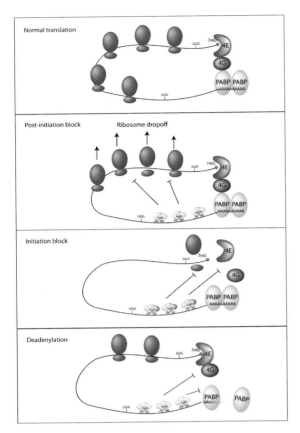

Figure 1. Different modes of miRNA-mediated translational repression.

3′UTR, suggesting that more termination events occur in the presence of miRNA. Second, miRNA-targeted mRNAs dissociate from ribosomes more readily than nontargeted control mRNA when translation initiation is inhibited, suggesting an enhanced rate of termination or release of ribosomes. Third, miRNA can still repress the translation of mRNAs that bypass any requirement for a typical initiation, including cap binding, scanning, and binding of multiple initiation factors; this involved the use of an internal ribosome entry site (IRES) from a cricket paralysis virus intergenic region (CrPV-IGR). These data, together with the puromycin sensitivity of the polyribosome profile of the target mRNA, suggest that miRNA acts at a step postinitiation. In particular, we propose that it might modulate the processivity of translation, that is, an enhancement of the abortive termination rate. This abortive translational mechanism ostensibly wastes energy in a nonproductive manner; however, when relieved, these types of inhibitions with continual translation could result in rapid modulation of the translation rate, and hence increase in the protein yield within a short period of time.

Postinitiation repression is not restricted to miRNA-type translational repression. For proper embryonic development, 4% of *Nanos* mRNAs are exclusively localized to the posterior pole of the *Drosophila* embryo, whereas the remaining 96% of unlocalized mRNAs must be translationally repressed elsewhere (Gavis and

Lehmann 1994); the repression is mediated by an element at the 3′UTR. These unlocalized *Nanos* mRNAs are associated with polyribosomes in a puromycin-sensitive manner (Clark et al. 2000). Interestingly, the *Nanos* mRNA-associated polyribosomes are capable of completing the elongation phase of the translation cycle in in vitro extracts made from a specific embryo strain where the mRNAs are all unlocalized and translationally repressed. This suggests a regulatory mechanism that does not involve a stable arrest of the translation cycle, similar to what we observed in miRNA-mediated translational repression.

Repression at Initiation

Most translational initiation begins with the recognition of the mRNA cap, where initiation factors are assembled together with Met-tRNA and a 40S ribosome to form a preinitiation complex (Pestova et al. 2001; Sonenberg and Dever 2003). This preinitiation complex then scans across the message to the initiation codon, resulting in the joining of other initiation factors and finally the 60S ribosome. A luciferase reporter with three artificial bulged *let-7*-binding sites, when transfected into HeLa cells in which *let-7* miRNA is abundant, results in approximately 30% reduction at the mRNA level and tenfold reduction in the protein level (Pillai et al. 2005). Polysome profiling of the reporter mRNA showed that a major fraction shifts toward the top of the gradient as compared to a control where the anti-*let-7* 2′O-Me oligonucleotide was cotransfected into the cells. This suggests that the *let-7* miRNA is responsible for the shift in the polysome profile of the target mRNA. This shift in polysome profile with miRNA-mediated repression is similarly observed when Ago2 is tethered to the 3′UTR of the luciferase reporter (Pillai et al. 2005). In both cases, the shift in the profile results in an increase of 25% of reporter mRNA to the top of the gradient; however, there are still about 50% of the signals notably remaining in the polyribosome fraction in the experiment. The moderate shift in profile and the remaining fraction in the polyribosomes suggest a decrease in efficiency, rather than a total block, at translation initiation.

Other experiments using a target mRNA with defined initiation requirements have been used to address which steps of translation are repressed by miRNAs (Humphreys et al. 2005; Pillai et al. 2005). Pillai et al. (2005) reported that miRNA repression did not occur when translation was driven by tethering initiation factors, such as the cap-binding protein eIF4E or its binding partner eIF4G, upstream of an internal open reading frame, suggesting that miRNA might act at initiation, most likely upstream of the step where eIF4E recruits eIF4G. However, in these cases, it is not known whether artificially tethering multiple copies of initiation factor will outcompete the repression machinery or if the rate of artificial initiation is so strong that it masks an otherwise subtle effect such as a decrease in processivity. Another way to modulate the initiation dependence on cap is to use different viral IRES sequences, which differ by their initiation factor requirement (Humphreys et al. 2005; Pillai et al. 2005). When transfecting in-vitro-synthesized RNA with CrPV or hepatitis C virus (HCV) IRES, miRNAs no longer repress their translation. However, miRNA-mediated translational repression still occurs when DNA constructs encoding messages with the CrPV or HCV IRES were transfected into cells (Petersen et al. 2006), potentially suggesting that the "nuclear history" of an mRNA might be important for miRNA-mediated repression and/or that the protein associated with the in-vivo-synthesized mRNA may be different from those transfected after in vitro synthesis. In either case, if miRNA-mediated repression acts at the initiation step, it is yet to be reconciled with the fact that the bulk of target-associated miRNA cosediments with polyribosomes (Kim et al. 2004; Nelson et al. 2004).

Both the cap and poly(A) tail seem to be required, but are not sufficient by themselves, for the full repression mediated by miRNAs (Humphreys et al. 2005; Pillai et al. 2005). Factors binding at the cap and poly(A) tail are known to synergistically interact with each other to enhance translation. Such interactions are commonly modulated by elements at the 3′UTR that result in the lowering of the initiation efficiency (Gebauer and Hentze 2004). If miRNAs can alter the polyadenylation status of the message (see below), the disruption of the interaction between the cap and poly(A) tail might provide yet another mechanism for miRNA-mediated translational repression (Fig. 1, bottom panel).

miRNA-INDUCED DEGRADATION OF mRNA

Numerous examples demonstrate that miRNAs repress translation with little or no change of steady-state mRNA levels (Brennecke et al. 2003; Saxena et al. 2003; Zeng et al. 2003; Chen et al. 2004; Doench and Sharp 2004; Poy et al. 2004; Cimmino et al. 2005; Pillai et al. 2005). Yet, a growing number of examples show that miRNAs also cause a reduction in mRNA levels (Yekta et al. 2004; Bagga et al. 2005; Davis et al. 2005; Jing et al. 2005; Wu and Belasco 2005; Behm-Ansmant et al. 2006; Giraldez et al. 2006; Gupta et al. 2006; Rehwinkel et al. 2006; Wu et al. 2006). In most cases, the reduction in RNA level is approximately 1.5- to 2-fold as measured by microarrays after transfection with miRNAs in the form of an siRNA duplex (Lim et al. 2005). Consistent with these transfection results, recent bioinformatic comparisons of miRNA expression and predicted target mRNA levels show an inverse relationship (Farh et al. 2005). Currently, there are three reported types of miRNA-induced degradations (Fig. 2): The first requires extensive complementarity throughout the miRNA, the second requires complementarity particularly at the seed region (i.e., 2nd–8th nucleotides) of miRNA, and the third is based on the complementarity at a central AU-rich region of miRNA (Fig. 2).

Endonucleolytic Cleavage

Unlike the situation in plants, mammalian miRNAs rarely have extensive complementarity with their targets. So far, there are two reported cases: The first case involves *miR-196*, where the whole 22-nucleotide sequence has conserved complementarity to a region in the 3′UTR of the *HoxB8* mRNA; the complementarity

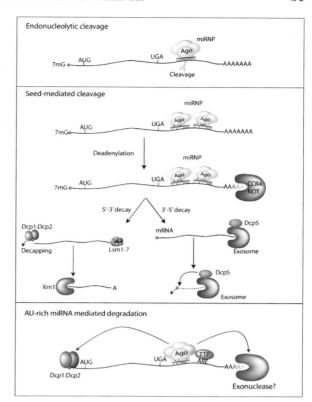

Figure 2. Different modes of miRNA-mediated mRNA degradation.

includes a non-Watson-Crick base-pair G:U wobble at the fifth position (Yekta et al. 2004; Hornstein et al. 2005). The other case involves several miRNAs encoded in the *Dlk1-Gtl2* imprinted domain region. These miRNAs are processed from a transcript that is antisense to a paternally expressed *Rtl1/Peg11* mRNA; therefore, the antisense transcript, by definition, possesses multiple perfectly complementary binding sites for these miRNAs. The presence of these miRNAs results in the cleavage of their targets between the corresponding 10th and 11th positions of the miRNA, reminiscent of an RNA interference (RNAi)-like cleavage event supposedly by Argonaute 2 (Ago2). Currently, Ago2 is the only known Argonaute family member in mammalian cells that possesses endonucleolytic (or "slicer") capability (Liu et al. 2004). However, when complexed with other Argonaute family members, these miRNAs are not thought to mediate direct cleavage, but their associations with particular targets should mediate translational repression and perhaps degradation of the transcripts.

"Seed"-mediated Degradation

miRNA and siRNA silencing pathways have been shown to have some common activities. For example, endogenous miRNAs can direct cleavage of exogenously expressed targets that contain sites of extensive complementarity, and siRNAs can function as miRNAs in mediating translational repression (Hutvagner and Zamore 2002; Doench et al. 2003; Saxena et al. 2003; Zeng et al.

2003). Another aspect of siRNA behaving as an miRNA pathway was revealed when it was observed that there is 1.5- to 2-fold degradation of multiple mRNAs (the so-called "off" target) that does not have a perfect complementary binding site with the siRNA (Jackson et al. 2003). The major determinant for this type of degradation turns out to be the propensity of the 5′ end of the siRNA pairing with these "off-targets" (Lin et al. 2005; Birmingham et al. 2006; Jackson et al. 2006a). This is strikingly similar to the seed region requirement for miRNA-mediated mRNA degradation (Lim et al. 2005). This mode of degradation does not involve RNAi-like endonucleolytic cleavage activity because the sequences of mRNA degradation products are not ended at the corresponding 11th position of miRNA (Bagga et al. 2005; Wu et al. 2006).

The underlying mechanism for this type of degradation is still not clear, but we can draw parallels from the study of the "off-target" effect: (1) This effect is observed at a low level of transfected siRNA and is probably the result of a direct interaction between siRNA and target mRNA, as it is observed shortly after transfection (Jackson et al. 2003, 2006a). The degree of off-target degradation can be partially inhibited by adding a 2′O-methyl ribosyl substitution at position 2, suggesting that this substituted position might be crucial in some steps leading to mRNA degradation (Fedorov et al. 2006; Jackson et al. 2006b). (2) This effect is not observed in cells depleted of either nonslicer Ago1 or slicer Ago2, suggesting that endonucleolytic capability of Argonaute protein might not be a prerequisite (Fedorov et al. 2006). (3) The degree of off-target degradation is greater with multiple seed sites and is also dependent on sequence context surrounding the complementary site (Lin et al. 2005). (4) This effect is observed more often if the complementary site is in the 3′UTR, rather than the coding sequence or 5′UTR (Birmingham et al. 2006; Jackson et al. 2006a).

Other lines of evidence concerning miRNA-induced degradation come from genetic studies in *Drosophila* and zebra fish (Behm-Ansmant et al. 2006; Giraldez et al. 2006; Rehwinkel et al. 2006). When depleted of the primary component, such as dAgo1 or GW182, of the miRNA pathway in *Drosophila* S2 cells, the RNA level of predicted miRNA targets increases as measured by microarrays (Behm-Ansmant et al. 2006); therefore, these targets are most likely degraded by miRNA in addition to being translationally repressed. When these predicted miRNA targets are fused to the 3′UTR of luciferase, these targets are rapidly degraded with cotransfection of the corresponding miRNA due to an increase of the deadenylation rate. Subsequent studies have shown that this type of mRNA degradation requires both CCR4:NOT deadenylase and DCP1:DCP2 decapping complexes (Behm-Ansmant et al. 2006). Tethering of GW182 to a luciferase mRNA alone can bypass the requirement of dAgo1 in promoting decay, suggesting that the binding of GW182 to the message through the dAgo1/miRNA complex marks the transcript as a target for degradation (Behm-Ansmant et al. 2006). In many systems, deadenylation is often the first step toward degradation of the transcript, which is followed by two pathways (Fig. 2)—one involving decapping followed by 5′→3′ degradation by an exoribonuclease such as Xrn1, and the other beginning with 3′→5′ decay by the exosome, followed with cap removal by the scavenger DcpS (Parker and Song 2004; Fillman and Lykke-Andersen 2005). The idea of accelerated deadenylation by miRNA is reinforced in the study of zebra fish (Giraldez et al. 2006). *miR-430*, a highly expressed miRNA in early zebra fish embryogenesis, increases the deadenylation rate of its targets at the onset of zygotic gene expression. Given that most of the *miR-430* targets are maternal mRNAs, the resultant approximately fivefold reduction of these targets underlies the clearance of maternal mRNAs. Interestingly, miRNA-induced deadenylation still occurs in the absence of translation, suggesting that deadenylation is not necessarily a secondary effect of nonproductive translation (Giraldez et al. 2006). It should, however, be stressed that in both cases, deadenylation/degradation was only studied for those miRNA targets that displayed measurable changes in mRNA level, and the phenomenon might not hold true for all miRNA targets.

Wu et al. (2006) also described similar miRNA-induced deadenylation in mammalian cells in a study of *miR125b* and its target *lin-28*. These authors reported that the down-regulation of *lin-28* expression is due to both translational repression and degradation by accelerated deadenylation. The degradation requires that the miRNA-binding site be located prior to the polyadenylation signal on the message (Wu and Belasco 2005). As observed in zebra fish, the miRNA-induced deadenylation of the target occurs even in the presence of a large stem-loop in the 5′UTR that abolished 99% of protein synthesis. Conversely, translational repression still occurs in mRNA containing a histone stem-loop that replaces the poly(A) tail, suggesting that translational repression and deadenylation induced by this process are not mutually dependent.

"AU-rich miRNA"-mediated Decay

A genome-wide RNAi screen revealed that AU-rich element (ARE)-mediated decay requires Dicer in *Drosophila* S2 cells, and this requirement was also demonstrated in human HeLa cells (Jing et al. 2005). This ARE-mediated decay involves *miR-16*, which has previously been shown to mediate translational repression (Cimmino et al. 2005). However, this miRNA contains an UAAAUAUU sequence between the 10th and the 17th position, which is complementary to a typical ARE. The mRNA decay requires both the ARE-binding protein tristetraprolin (TTP) and *miR-16*. However, TTP does not directly bind *miR-16*, but instead interacts through protein–protein interaction with human Ago2/Ago4 that binds the miRNA. It is noteworthy that the binding site in the target mRNA is complementary to the central region of the miRNA, rather than the 5′ region, as observed in the seed-mediated cleavage. The significance of this differential binding mode remains to be determined.

WHERE ARE THESE FUNCTIONS LOCALIZED?

Where are these various functions of miRNAs localized in the cells? Are they related to subcellular compartments in the cytoplasm? Interestingly, miRNAs, miRNA-targeted mRNAs, and proteins associated with their activities have been found to localize to specific subcellular structures. We first examine the evidence of the localization of miRNAs, miRNA targets, and their associated proteins and discuss the significance of colocalization of these factors in specific structures.

Localization of miRNAs

In animals, the primary transcript of miRNA (pri-miRNA) is transcribed by RNA polymerase II and sequentially processed to an approximately 70-nucleotide pre-miRNA intermediate in the nucleus by class III ribonuclease Drosha (Kim 2005). The pre-miRNA intermediate can be folded in silico into a stem-loop hairpin with a short (~2 nucleotides) 3′ overhang. The resulting "minihelix" structure of the pre-miRNA intermediate is recognized by Exportin 5 and exported to the cytoplasm in a RanGTP-dependent manner. The double-stranded pre-miRNA intermediate is then processed by another class III ribonuclease enzyme Dicer in the cytoplasm, where it generates the mature single-stranded miRNA (miR) and the opposite strand (miR*). The opposite strand presumably undergoes rapid degradation, supported by its usual absence in small RNA cloning or northern blot analysis. Subcellular fractionation studies have confirmed that the active single-stranded form of miRNA is primarily localized in the cytoplasm (Fig. 3).

Several methods have been used to identify the specific location of mature miRNAs in the cytoplasm (Jakymiw et al. 2005; Pillai et al. 2005; Bhattacharyya et al. 2006). Pillai et al. (2005) microinjected an in-vitro-transcribed, fluorescently labeled artificial precursor of *let-7a* into nuclei and found that 21.8% of the cytoplasmic fluorescent signals localized at (13.7%) or near (8.1%) process-

ing bodies (PBs), a known site for mRNA decay. However, in this approach, both the active miR strand and the miR* strand of miRNA were equally labeled and therefore cannot easily be distinguished from each other. The signal observed at PBs could represent the degradation site for miR*, rather than the active strand. Furthermore, it is not possible to determine the fraction of the labeled intracellular RNA that is processed to a mature miRNA. Jakaymiw et al. (2005) labeled the antisense strand of an siRNA duplex and found that the labeled strand accumulated in PBs. In this case, it is similarly difficult to be certain that the labeled antisense strand is the one loaded into the RNA-induced silencing complex (RISC) as contrasted to the one targeted for decay. Whether PB is the site for decay of the miR* strand or the passenger strand remains unclear. In situ hybridization using locked nucleic acid (LNA) probes against a highly abundant miRNA, *miR122*, followed by signal amplification, also revealed the accumulation of the signals at PBs in hepatocytes (Bhattacharyya et al. 2006). Thus, several lines of evidence suggest that miRNAs might associate with PBs; the extent of this association is not clear. Furthermore, since PBs are free of ribosomal subunits (Anderson and Kedersha 2006), it is difficult to reconcile this subcellular location with several observations showing that a substantial fraction of miRNA associates with polyribosomes (Kim et al. 2004; Nelson et al. 2004).

To study miRNA localization, we aimed to avoid the ambiguous detection of inactive miR/miR* duplexes or the miR* strand and to quantitate the signal in the linear range. To this end, we have modified a well-characterized siRNA duplex that has been shown to function as a miRNA. The antisense strand is labeled with tetramethylrhodamine (TAMRA) at its 3′ end, and the 5′ phosphate group of the sense strand is substituted with black hole quencher 2 (BHQ2). In this way, the fluorescence is quenched in the siRNA duplex form, thereby enhancing the signal detection from the active single strand. Since the 5′ phosphate group is a prerequisite for RISC loading (Tomari et al. 2004; Rivas et al. 2005), the substitution of the phosphate group with the quencher presumably favors the selection of the labeled antisense strand for loading into the RISC complex. Using this modified siRNA, we did not observe any signal at PBs that was higher than the cytoplasmic background; instead, the signal was diffusively distributed throughout the cytoplasm (Leung et al. 2006).

Localization of Target mRNA

To localize repressed targets, Liu et al. (2005a) used green fluorescent protein (GFP)-tagged MS2-binding proteins to track luciferase mRNA that has 24 copies of MS2-binding sites behind either two tandemly arranged *let-7*-binding sites or six CXCR4-binding sites. The target was localized to PBs in a miRNA-dependent manner. Furthermore, Pillai et al. (2005) quantitated the in situ hybridization signal of *let-7*-targeted luciferase mRNA and, similar to *let-7*, these authors found that 20.7% localized at (13.0%) or near (7.7%) PBs. We also found similar localization of the mRNA targets at or near PBs. Diffuse signal was also measured throughout the cytoplasm using the in situ hybridization method (Leung et al. 2006).

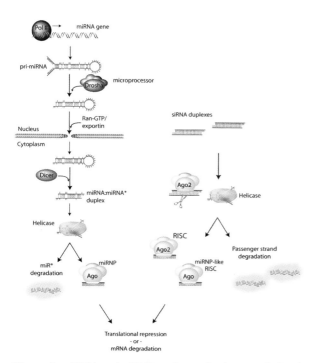

Figure 3. miRNA and siRNA pathways lead to translational repression or mRNA degradation.

Localization of Protein Components

Argonaute proteins bind to the miRNA forming the core component of RISC or miRNP for translational repression, endonucleolytic cleavage, and other types of mRNA degradation (Carmell et al. 2002). In human cells, there are four different Argonaute family members (Ago1–4), all of which accumulate at PBs (Jakymiw et al. 2005; Liu et al. 2005a,b; Pillai et al. 2005; Sen and Blau 2005; Chu and Rana 2006; Leung et al. 2006). Together with the colocalization of miRNA and its target, these data suggest a potential role of the PB as a site for miRNA function. This potentially new role of the PB, apart from being the site for mRNA decay, was reinforced by the fact that depletion of another PB component GW182, which physically associates with Argonaute through a protein–protein interaction, impaired PB formation and miRNA-mediated translational repression activity (Liu et al. 2005a).

To assess the extent of localization of Argonaute protein to PBs, we established a HeLa cell line that stably expresses GFP-tagged human Argonaute 2 (Leung et al. 2006). Consistent with the published data, we found that the enhanced GFP (EGFP)-Ago2 signal intensity is on average tenfold higher at PBs than the neighboring cytoplasm. By quantitation, however, we found that approximately only 1% of the total cytoplasmic signal of Argonaute is localized to PBs, whereas the majority is diffuse throughout the cytoplasm. Moreover, using photoactivation and photobleaching techniques (Lippincott-Schwartz et al. 2003), we found that the off-rate of EGFP-Ago2 at PBs is very slow. Similar observations of a slow rate of exchange was observed at PBs for both Argonaute-associated protein GW182 and the cap-binding protein eIF4E (Andrei et al. 2005; Kedersha et al. 2005). This is in contrast to the fast exchange rate observed for other PB components such as decapping enzymes Dcp1a/2 and decapping activator component Lsm6, suggesting that PBs comprise heterogeneous mRNPs with differential rates of exchange (Andrei et al. 2005; Kedersha et al. 2005; Leung et al. 2006). These quantitative analyses suggest that only small amounts of Argonaute proteins localize to PBs, and they rarely exchange with the cytoplasm once localized there. So, apart from PBs, could miRNA function elsewhere?

CYTOPLASMIC STRUCTURES FOR miRNA FUNCTIONS?

At least two cytoplasmic structures have been reported to associate with repressed mRNAs—processing bodies (PBs) and stress granules (SGs) (Anderson and Kedersha 2006).

Processing Bodies

PBs are well characterized in yeast, but much less studied in mammalian cells. Yeast and mammalian PBs share several characteristics: (1) They contain components that are involved in mRNA decapping and $5' \rightarrow 3'$ exonucleolytic decay, but lack ribosomal subunits (Sheth and Parker 2003; Cougot et al. 2004; Andrei et al. 2005; Anderson and Kedersha 2006), (2) their integrities depend on the presence of nontranslating mRNAs (Cougot et al. 2004; Teixeira et al. 2005), and (3) their structures dissociate in the presence of drugs that stabilize the association of mRNA with ribosomes, such as emetine and cycloheximide (Sheth and Parker 2003; Cougot et al. 2004; Kedersha et al. 2005; Teixeira et al. 2005). On the other hand, yeast and mammalian PBs differ in terms of composition, and in their response to growth and stress conditions. For example, only mammalian PBs contain the deadenylase CCR4 (Sheth and Parker 2003). Mammalian PBs increase in size and number when cells are proliferating or during the S/G_2 phase of the cell cycle, whereas yeast PBs increase when starved or grown at high density (Yang et al. 2004; Teixeira et al. 2005). With stress, yeast PBs, but not mammalian PBs, increase in size.

Stress Granules

In mammalian cells, under stress conditions, newly assembled structures known as stress granules (SGs) are formed where translationally stalled messages accumulate (Kedersha et al. 2005; Teixeira et al. 2005). Translationally repressed mRNAs localized in yeast PBs can leave and reenter the polyribosome pool (Brengues et al. 2005), but a similar phenomenon has not been demonstrated for mammalian PBs. Instead, SGs are thought to function as triage sites where mRNAs are sorted for future degradation, storage, or translation (Kedersha and Anderson 2002). SGs are known to be sites where translationally stalled poly(A)$^+$ mRNAs accumulate when cells experience stress, such as heat shock, osmotic stress, or oxidative stress, or when translation initiation is specifically inhibited. PBs and SGs are known to be dynamically associated with, and often juxtaposing, each other. Although they share components, such as the cap-binding protein eIF4E and the translational repressor rck/p54, not all PB components are found in SGs, or vice versa (Anderson and Kedersha 2006). The close interaction observed between SGs and PBs in mammalian cells suggests that transcripts may be exported from SGs to PBs for degradation (Kedersha et al. 2005). Whether miRNAs, target mRNAs, and Argonaute proteins localize to SGs has not been determined. We report here the first evidence for the localization of miRNA and its target to SGs and that Argonaute protein localization to SGs is miRNA-dependent (Leung et al. 2006).

miRNA, mRNA Target, and Argonaute also Localize to Stress Granules

We found that upon limiting translation initiation with an eIF4A-specific inhibitor, Hippuristanol, EGFP-Ago2 remained localized to PBs at the same level, but approximately 2–3% of the total fluorescence signal now localized to SGs. Quantitatively, there was about threefold more EGFP localized to SGs than to PBs after Hippuristanol treatment. Given that PB localization did not change with treatment, additional Argonaute proteins were redistributed from the diffuse cytoplasm to these distinct structures. In contrast to the relative immobility of

Argonaute proteins demonstrated at PBs, Argonaute proteins at SGs were more dynamic as shown by photobleaching experiments. The fluorescence at a single photobleached spot at an SG can be fully recovered to its initial intensity level within 6 minutes, whereas the photobleached spot at a PB never recovered during this period. Furthermore, these preformed, Argonaute-positive SGs dissociate in the presence of Emetine, a drug that stabilizes the association of mRNA and polyribosomes, suggesting a dynamic relationship between SGs and polyribosomes. These treatments do not dissociate EGFP-Ago2 from PBs.

Upon SG induction, we detected a 1.5-fold enrichment of miRNAs at SGs as compared with the neighboring cytoplasmic signal, but no enrichment was observed at PBs, using our quencher-TAMRA-based detection method. Interestingly, miRNA targets also accumulated in SGs. These observations prompted us to postulate that Argonaute protein localization to SG is dependent on miRNA. To test this hypothesis, we transfected EGFP-Ago2 into three clonal mouse embryonic stem (ES) cell lines that contained a mutant nonfunctional form of Dicer and hence lacked mature miRNAs. We found that EGFP-Ago2, although still observed in PBs, no longer associated with SGs. However, in the presence of exogenously supplied *let-7a* miRNA transfected in the form of siRNA, EGFP-Ago2 association with SGs was restored. This strongly suggests that the Argonaute localization to SGs is miRNA-dependent.

Do We Need Any Structure for miRNA Function?

Given that microscopically visible SGs are not observed in cells when translation initiation is not limiting, these macrostructures themselves are most likely not required for miRNA functions. However, upon stress, the accumulation of miRNA and its target at SGs and the miRNA-dependent localization of Argonaute proteins to SGs might reflect a potential relationship between miRNAs and SGs. SGs could possibly represent the aggregation of submicroscopic miRNPs that were formerly involved in translational repression of the target mRNA. Interestingly, multiple SG components FMRP, TTP, and HuR are known to interact with the Argonaute protein (Caudy et al. 2002; Jing et al. 2005; Bhattacharyya et al. 2006). Therefore, another possibility is that these newly assembled SGs preferentially recruit particular miRNA targets through protein–protein interaction for specific functions during stress.

Several lines of evidence indicate that microscopically visible PBs are also not required for translational repression, endonucleolytic activity, and ARE-mediated mRNA decay. For example, depletion of the PB component Lsm1 impairs PB formation but does not affect miRNA-mediated translational repression activity or endonucleolytic activity (Chu and Rana 2006). Therefore, the miRNA and siRNA pathways appear to be intact in these unstressed cells, where no microscopically visible PBs and SGs are observed. Similarly, no negative effects on ARE-mediated decay or deadenylation were observed in the absence of visible PBs when p54/rck or Dcp2 was overexpressed (Fenger-Gron et al. 2005; Lykke-Andersen and Wagner

2005), suggesting that these processes do not require the presence of visible PBs. Furthermore, in vitro extract that presumably lacks visible structures can recapitulate miRNA-mediated endonucleolytic activity and translational repression (Wang et al. 2006).

Neither PBs nor SGs are confined by membranes, and their components continually exchange with the cytoplasm (Gilks et al. 2004; Yang et al. 2004; Andrei et al. 2005; Kedersha et al. 2005). Given that these proteins are not confined to specific structures and the structures are not required for the translational repression, endonucleolytic activity, or deadenylation/degradation of target mRNA, miRNAs most likely act predominately in the diffuse cytoplasm. This is consistent with the current quantitation data showing that the majority of miRNAs, miRNA targets, and Argonaute proteins are not localized to PBs, but are diffusively distributed in the cytoplasm (Pillai et al. 2005; Leung et al. 2006). Notably, Argonaute family members (dAgo1 and Alg1/2) were also found to be diffusely distributed in the cytoplasm of *Drosophila* S2 cells and *C. elegans* epidermal/neuronal cells (Ding et al. 2005; Behm-Ansmant et al. 2006). Only when GW182 homologs were overexpressed did these Argonaute members localize to punctate PB-like structures in the cytoplasm (Ding et al. 2005; Behm-Ansmant et al. 2006). Moreover, GW182 is not always concentrated in mammalian PBs throughout the cell cycle (Yang et al. 2004), and this is also the case for endogenous Argonaute protein (A. Leung, unpubl.). Therefore, the existence of these microscopically visible structures is most likely not required for miRNA function.

So What Are These Structures For?

The appearance of a specific structure, such as PB, where multiple proteins colocalize, most likely represents complex formation from interactions between different components for a specific pathway or function (Misteli 2001), even though equivalent interactions can probably occur elsewhere in the cytoplasm. As discussed earlier, components in PBs or SGs are continually exchanging with the cytoplasm. In turn, these differential dynamic exchange rates of the protein components should determine their concentration at such structures. Increased local concentration of factors that participate in related steps of a particular process at a physical structure can increase the overall functional efficiency, but the structure itself may not be a prerequisite for the function. Furthermore, visible structures like PBs most likely represent sites for multiple pathways where shared components localize. However, these colocalizations do not necessarily imply functional relationship/dependency. For example, components of the NMD pathway (UPF1), RNAi (dAgo2), and miRNA function (dAgo1) are colocalized in a PB-like structure in *Drosophila* S2 cells, and knockdown of individual components of one pathway does not affect the function of others (Rehwinkel et al. 2005). These results suggest that a visible structure could be composed of heterogeneous mRNPs. Therefore, structure itself does not imply exclusive functional location, nor does colocalization imply the same functional pathway.

PB association with miRNA components most likely reflects its role linked to degradation/deadenylation. Previous immunoprecipitation data showed that the associations between Argonaute proteins and other PB components, such as Dcp1a (Liu et al. 2005a), rck/p54 (Chu and Rana 2006), and GW182 (Liu et al. 2005b), were due to protein–protein interactions, rather than through a common RNA scaffold. Consistent with this, we observed that the PB association with Argonaute protein does not require miRNAs. Since PBs do not contain either ribosomal subunits or initiation factors, mRNAs interacting with miRNAs at PBs cannot be undergoing repression either at postinitiation or at arrest of initiation processes postbinding of the 40S subunit. Therefore, the ribosome-free PBs are most likely not the sites where active miRNA-mediated translational repression occurs. However, it is possible that the observed stable association of Argonaute protein and PBs, through protein–protein interactions, reinforces a repressive state by keeping the bound message in a translation-incompetent environment. This may be the case for other structures that could simply sequestrate miRNAs from active function in polarized cell types. For example, neurons, which have long cytoplasmic extensions, are known to require specialized machineries to transport mRNAs to dendrites. It remains unclear whether localization of miRNA at dendrites, such as those reported by Schratt et al. (2006), requires specialized structures or is a consequence of miRNA binding to mRNA that is targeted to dendrites.

In light of these results, miRNA-mediated translational repression is likely associated with other complexes in the diffuse cytoplasm. By limiting translation initiation, we identify a pool of Argonaute protein in the cytoplasm that continually exchanges with ribosomes. In this new steady state, the ribosome-associated Argonaute pool dynamically relocates to newly assembled SGs, which contain translationally stalled mRNAs, in a miRNA-dependent manner. We propose here that the pool localized to SGs most likely represents a subset of Argonaute proteins formerly involved in miRNA-mediated translational repression.

FINAL WORDS

In summary, the majority of Argonaute protein is diffusively distributed in the cytoplasm. Argonaute proteins localized at PBs and SGs are not static but are continually exchanging with the cytoplasm at different rates. Translational repression, endonucleolytic cleavage, and deadenylation remain functional in the absence of microscopically visible PBs and SGs. These results suggest that miRNA-mediated processes can occur in submicroscopic complexes, but not necessarily in distinct subcellular structures, in the cytoplasm.

ACKNOWLEDGMENTS

We thank Mary Wernett for preparing the illustrations and Alla Grishok, Joel Neilson, Amy White, and Lourdes Aleman in the Sharp Lab for comments on the manuscript, and Tun-Han Leung for proofreading. A.K.L.L. is a recipient of a Human Frontier Science Program long-term fellowship. This work was supported by Program Project grant PO1 CA42063 from the National Cancer Institute, by National Science Foundation grant 0218506 to P.A.S., and partially by Cancer Center Support (core) grant P30-CA14051 from the National Cancer Institute.

REFERENCES

Anderson P. and Kedersha N. 2006. RNA granules. *J. Cell Biol.* **172:** 803.

Andrei M.A., Ingelfinger D., Heintzmann R., Achsel T., Rivera-Pomar R., and Luhrmann R. 2005. A role for eIF4E and eIF4E-transporter in targeting mRNPs to mammalian processing bodies. *RNA* **11:** 717.

Arasu P., Wightman B., and Ruvkun G. 1991. Temporal regulation of lin-14 by the antagonistic action of two other heterochronic genes, lin-4 and lin-28. *Genes Dev.* **5:** 1825.

Bagga S., Bracht J., Hunter S., Massirer K., Holtz J., Eachus R., and Pasquinelli A.E. 2005. Regulation by *let-7* and *lin-4* miRNAs results in target mRNA degradation. *Cell* **122:** 553.

Bartel D.P. 2004. MicroRNAs: Genomics, biogenesis, mechanism, and function. *Cell* **116:** 281.

Behm-Ansmant I., Rehwinkel J., Doerks T., Stark A., Bork P., and Izaurralde E. 2006. mRNA degradation by miRNAs and GW182 requires both CCR4:NOT deadenylase and DCP1:DCP2 decapping complexes. *Genes Dev.* **20:** 1885.

Bhattacharyya S.N., Habermacher R., Martine U., Closs E.I., and Filipowicz W. 2006. Relief of microRNA-mediated translational repression in human cells subjected to stress. *Cell* **125:** 1111.

Birmingham A., Anderson E.M., Reynolds A., Ilsley-Tyree D., Leake D., Fedorov Y., Baskerville S., Maksimova E., Robinson K., Karpilow J., et al. 2006. 3′ UTR seed matches, but not overall identity, are associated with RNAi off-targets. *Nat. Methods* **3:** 199.

Brengues M., Teixeira D., and Parker R. 2005. Movement of eukaryotic mRNAs between polysomes and cytoplasmic processing bodies. *Science* **310:** 486.

Brennecke J., Hipfner D.R., Stark A., Russell R.B., and Cohen S.M. 2003. bantam encodes a developmentally regulated microRNA that controls cell proliferation and regulates the proapoptotic gene hid in *Drosophila*. *Cell* **113:** 25.

Carmell M.A., Xuan Z., Zhang M.Q., and Hannon G.J. 2002. The Argonaute family: Tentacles that reach into RNAi, developmental control, stem cell maintenance, and tumorigenesis. *Genes Dev.* **16:** 2733.

Caudy A.A., Myers M., Hannon G.J., and Hammond S.M. 2002. Fragile X-related protein and VIG associate with the RNA interference machinery. *Genes Dev.* **16:** 2491.

Chen C.Z., Li L., Lodish H.F., and Bartel D.P. 2004. MicroRNAs modulate hematopoietic lineage differentiation. *Science* **303:** 83.

Chu C.Y. and Rana T.M. 2006. Translation repression in human cells by microRNA-induced gene silencing requires RCK/p54. *PLoS Biol.* **4:** e210.

Cimmino A., Calin G.A., Fabbri M., Iorio M.V., Ferracin M., Shimizu M., Wojcik S.E., Aqeilan R.I., Zupo S., Dono M., et al. 2005. miR-15 and miR-16 induce apoptosis by targeting BCL2. *Proc. Natl. Acad. Sci.* **102:** 13944.

Clark I.E., Wyckoff D., and Gavis E.R. 2000. Synthesis of the posterior determinant Nanos is spatially restricted by a novel cotranslational regulatory mechanism. *Curr. Biol.* **10:** 1311.

Cougot N., Babajko S., and Seraphin B. 2004. Cytoplasmic foci are sites of mRNA decay in human cells. *J. Cell Biol.* **165:** 31.

Davis E., Caiment F., Tordoir X., Cavaille J., Ferguson-Smith A., Cockett N., Georges M., and Charlier C. 2005. RNAi-mediated allelic trans-interaction at the imprinted Rtl1/Peg11 locus. *Curr. Biol.* **15:** 743.

Ding L., Spencer A., Morita K., and Han M. 2005. The developmental timing regulator AIN-1 interacts with miRISCs and may target the argonaute protein ALG-1 to cytoplasmic P bodies in *C. elegans*. *Mol. Cell* **19:** 437.

Doench J.G. and Sharp P.A. 2004. Specificity of microRNA target selection in translational repression. *Genes Dev.* **18:** 504.

Doench J.G., Petersen C.P., and Sharp P.A. 2003. siRNAs can function as miRNAs. *Genes Dev.* **17:** 438.

Farh K.K., Grimson A., Jan C., Lewis B.P., Johnston W.K., Lim L.P., Burge C.B., and Bartel D.P. 2005. The widespread impact of mammalian microRNAs on mRNA repression and evolution. *Science* **310:** 1817.

Fedorov Y., Anderson E.M., Birmingham A., Reynolds A., Karpilow J., Robinson K., Leake D., Marshall W.S., and Khvorova A. 2006. Off-target effects by siRNA can induce toxic phenotype. *RNA* **12:** 1188.

Fenger-Gron M., Fillman C., Norrild B., and Lykke-Andersen J. 2005. Multiple processing body factors and the ARE binding protein TTP activate mRNA decapping. *Mol. Cell* **20:** 905.

Filipowicz W., Jaskiewicz L., Kolb F.A., and Pillai R.S. 2005. Post-transcriptional gene silencing by siRNAs and miRNAs. *Curr. Opin. Struct. Biol.* **15:** 331.

Fillman C. and Lykke-Andersen J. 2005. RNA decapping inside and outside of processing bodies. *Curr. Opin. Cell Biol.* **17:** 326.

Gavis E.R. and Lehmann R. 1994. Translational regulation of nanos by RNA localization. *Nature* **369:** 315.

Gebauer F. and Hentze M.W. 2004. Molecular mechanisms of translational control. *Nat. Rev. Mol. Cell Biol.* **5:** 827.

Gilks N., Kedersha N., Ayodele M., Shen L., Stoecklin G., Dember L.M., and Anderson P. 2004. Stress granule assembly is mediated by prion-like aggregation of TIA-1. *Mol. Biol. Cell* **15:** 5383.

Giraldez A.J., Mishima Y., Rihel J., Grocock R.J., Van Dongen S., Inoue K., Enright A.J., and Schier A.F. 2006. Zebrafish MiR-430 promotes deadenylation and clearance of maternal mRNAs. *Science* **312:** 75.

Gupta A., Gartner J.J., Sethupathy P., Hatzigeorgiou A.G., and Fraser N.W. 2006. Anti-apoptotic function of a microRNA encoded by the HSV-1 latency-associated transcript. *Nature* **442:** 82.

Ha I., Wightman B., and Ruvkun G. 1996. A bulged *lin-4/lin-14* RNA duplex is sufficient for *Caenorhabditis elegans lin-14* temporal gradient formation. *Genes Dev.* **10:** 3041.

Hornstein E., Mansfield J.H., Yekta S., Hu J.K., Harfe B.D., McManus M.T., Baskerville S., Bartel D.P., and Tabin C.J. 2005. The microRNA miR-196 acts upstream of Hoxb8 and Shh in limb development. *Nature* **438:** 671.

Humphreys D.T., Westman B.J., Martin D.I., and Preiss T. 2005. MicroRNAs control translation initiation by inhibiting eukaryotic initiation factor 4E/cap and poly(A) tail function. *Proc. Natl. Acad. Sci.* **102:** 16961.

Hutvagner G. and Zamore P.D. 2002. A microRNA in a multiple-turnover RNAi enzyme complex. *Science* **297:** 2056.

Jackson A.L., Burchard J., Schelter J., Chau B.N., Cleary M., Lim J., and Linsley P.S. 2006a. Widespread siRNA "off-target" transcript silencing mediated by seed region sequence complementarity. *RNA* **12:** 1179.

Jackson A.L., Bartz S.R., Schelter J., Kobayashi S.V., Burchard J., Mao M., Li B., Cavet G., and Linsley P.S. 2003. Expression profiling reveals off-target gene regulation by RNAi. *Nat. Biotechnol.* **21:** 635.

Jackson A.L., Burchard J., Leake D., Reynolds A., Schelter J., Guo J., Johnson J.M., Lim L., Karpilow J., Nichols K., et al. 2006b. Position-specific chemical modification of siRNAs reduces "off-target" transcript silencing. *RNA* **12:** 1197.

Jakymiw A., Lian S., Eystathioy T., Li S., Satoh M., Hamel J.C., Fritzler M.J., and Chan E.K. 2005. Disruption of GW bodies impairs mammalian RNA interference. *Nat. Cell Biol.* **7:** 1267.

Jing Q., Huang S., Guth S., Zarubin T., Motoyama A., Chen J., Di Padova F., Lin S.C., Gram H., and Han J. 2005. Involvement of microRNA in AU-rich element-mediated mRNA instability. *Cell* **120:** 623.

Kedersha N. and Anderson P. 2002. Stress granules: Sites of mRNA triage that regulate mRNA stability and translatability. *Biochem. Soc. Trans.* **30:** 963.

Kedersha N., Stoecklin G., Ayodele M., Yacono P., Lykke-Andersen J., Fitzler M.J., Scheuner D., Kaufman R.J., Golan D.E., and Anderson P. 2005. Stress granules and processing bodies are dynamically linked sites of mRNP remodeling. *J. Cell Biol.* **169:** 871.

Kim J., Krichevsky A., Grad Y., Hayes G.D., Kosik K.S., Church G.M., and Ruvkun G. 2004. Identification of many microRNAs that copurify with polyribosomes in mammalian neurons. *Proc. Natl. Acad. Sci.* **101:** 360.

Kim V.N. 2005. MicroRNA biogenesis: Coordinated cropping and dicing. *Nat. Rev. Mol. Cell Biol.* **6:** 376.

Lee R.C., Feinbaum R.L., and Ambros V. 1993. The *C. elegans* heterochronic gene lin-4 encodes small RNAs with antisense complementarity to lin-14. *Cell* **75:** 843.

Leung A.K.L., Calabrese J.M., and Sharp P.A. 2006. Quantitative analysis of Argonaute protein reveals microRNA-dependent localization to stress granules. *Proc. Natl. Acad. Sci.* **103:** 18125.

Lewis B.P., Burge C.B., and Bartel D.P. 2005. Conserved seed pairing, often flanked by adenosines, indicates that thousands of human genes are microRNA targets. *Cell* **120:** 15.

Lim L.P., Lau N.C., Garrett-Engele P., Grimson A., Schelter J.M., Castle J., Bartel D.P., Linsley P.S., and Johnson J.M. 2005. Microarray analysis shows that some microRNAs downregulate large numbers of target mRNAs. *Nature* **433:** 769.

Lin X., Ruan X., Anderson M.G., McDowell J.A., Kroeger P.E., Fesik S.W., and Shen Y. 2005. siRNA-mediated off-target gene silencing triggered by a 7 nt complementation. *Nucleic Acids Res.* **33:** 4527.

Lippincott-Schwartz J., Altan-Bonnet N., and Patterson G.H. 2003. Photobleaching and photoactivation: Following protein dynamics in living cells. *Nat. Cell Biol.* (suppl.) S7.

Liu J., Valencia-Sanchez M.A., Hannon G.J., and Parker R. 2005a. MicroRNA-dependent localization of targeted mRNAs to mammalian P-bodies. *Nat. Cell Biol.* **7:** 719.

Liu J., Rivas F.V., Wohlschlegel J., Yates J.R., III, Parker R., and Hannon G.J. 2005b. A role for the P-body component GW182 in microRNA function. *Nat. Cell Biol.* **7:** 1261.

Liu J., Carmell M.A., Rivas F.V., Marsden C.G., Thomson J.M., Song J.J., Hammond S.M., Joshua-Tor L., and Hannon G.J. 2004. Argonaute2 is the catalytic engine of mammalian RNAi. *Science* **305:** 1437.

Lykke-Andersen J. and Wagner E. 2005. Recruitment and activation of mRNA decay enzymes by two ARE-mediated decay activation domains in the proteins TTP and BRF-1. *Genes Dev.* **19:** 351.

Misteli T. 2001. The concept of self-organization in cellular architecture. *J. Cell Biol.* **155:** 181.

Nelson P.T., Hatzigeorgiou A.G., and Mourelatos Z. 2004. miRNP:mRNA association in polyribosomes in a human neuronal cell line. *RNA* **10:** 387.

Olsen P.H. and Ambros V. 1999. The *lin-4* regulatory RNA controls developmental timing in *Caenorhabditis elegans* by blocking LIN-14 protein synthesis after the initiation of translation. *Dev. Biol.* **216:** 671.

Parker R. and Song H. 2004. The enzymes and control of eukaryotic mRNA turnover. *Nat. Struct. Mol. Biol.* **11:** 121.

Pestova T.V., Kolupaeva V.G., Lomakin I.B., Pilipenko E.V., Shatsky I.N., Agol V.I., and Hellen C.U. 2001. Molecular mechanisms of translation initiation in eukaryotes. *Proc. Natl. Acad. Sci.* **98:** 7029.

Petersen C.P., Bordeleau M.E., Pelletier J., and Sharp P.A. 2006. Short RNAs repress translation after initiation in mammalian cells. *Mol. Cell* **21:** 533.

Pillai R.S., Bhattacharyya S.N., Artus C.G., Zoller T., Cougot N., Basyuk E., Bertrand E., and Filipowicz W. 2005. Inhibition of translational initiation by Let-7 microRNA in human cells. *Science* **309:** 1573.

Poy M.N., Eliasson L., Krutzfeldt J., Kuwajima S., Ma X., Macdonald P.E., Pfeffer S., Tuschl T., Rajewsky N., Rorsman P., and Stoffel M. 2004. A pancreatic islet-specific microRNA regulates insulin secretion. *Nature* **432:** 226.

Rajewsky N. 2006. microRNA target predictions in animals. *Nat. Genet.* (suppl. 1) **38**: S8.

Rehwinkel J., Behm-Ansmant I., Gatfield D., and Izaurralde E. 2005. A crucial role for GW182 and the DCP1:DCP2 decapping complex in miRNA-mediated gene silencing. *RNA* **11**: 1640.

Rehwinkel J., Natalin P., Stark A., Brennecke J., Cohen S.M., and Izaurralde E. 2006. Genome-wide analysis of mRNAs regulated by Drosha and Argonaute proteins in *Drosophila melanogaster. Mol. Cell. Biol.* **26**: 2965.

Rivas F.V., Tolia N.H., Song J.J., Aragon J.P., Liu J., Hannon G.J., and Joshua-Tor L. 2005. Purified Argonaute2 and an siRNA form recombinant human RISC. *Nat. Struct. Mol. Biol.* **12**: 340.

Saxena S., Jonsson Z.O., and Dutta A. 2003. Small RNAs with imperfect match to endogenous mRNA repress translation. Implications for off-target activity of small inhibitory RNA in mammalian cells. *J. Biol. Chem.* **278**: 44312.

Schratt G.M., Tuebing F., Nigh E.A., Kane C.G., Sabatini M.E., Kiebler M., and Greenberg M.E. 2006. A brain-specific microRNA regulates dendritic spine development. *Nature* **439**: 283.

Sen G.L. and Blau H.M. 2005. Argonaute 2/RISC resides in sites of mammalian mRNA decay known as cytoplasmic bodies. *Nat. Cell Biol.* **7**: 633.

Sheth U. and Parker R. 2003. Decapping and decay of messenger RNA occur in cytoplasmic processing bodies. *Science* **300**: 805.

Sonenberg N. and Dever T.E. 2003. Eukaryotic translation initiation factors and regulators. *Curr. Opin. Struct. Biol.* **13**: 56.

Teixeira D., Sheth U., Valencia-Sanchez M.A., Brengues M., and Parker R. 2005. Processing bodies require RNA for assembly and contain nontranslating mRNAs. *RNA* **11**: 371.

Tomari Y., Matranga C., Haley B., Martinez N., and Zamore P.D. 2004. A protein sensor for siRNA asymmetry. *Science* **306**: 1377.

Valencia-Sanchez M.A., Liu J., Hannon G.J., and Parker R. 2006. Control of translation and mRNA degradation by miRNAs and siRNAs. *Genes Dev.* **20**: 515.

Wang B., Love T.M., Call M.E., Doench J.G., and Novina C.D. 2006. Recapitulation of short RNA-directed translational gene silencing in vitro. *Mol. Cell* **22**: 553.

Wightman B., Ha I., and Ruvkun G. 1993. Posttranscriptional regulation of the heterochronic gene *lin-14* by *lin-4* mediates temporal pattern formation in *C. elegans. Cell* **75**: 855.

Wu L. and Belasco J.G. 2005. Micro-RNA regulation of the mammalian *lin-28* gene during neuronal differentiation of embryonal carcinoma cells. *Mol. Cell. Biol.* **25**: 9198.

Wu L., Fan J., and Belasco J.G. 2006. MicroRNAs direct rapid deadenylation of mRNA. *Proc. Natl. Acad. Sci.* **103**: 4034.

Yang Z., Jakymiw A., Wood M.R., Eystathioy T., Rubin R.L., Fritzler M.J., and Chan E.K. 2004. GW182 is critical for the stability of GW bodies expressed during the cell cycle and cell proliferation. *J. Cell Sci.* **117**: 5567.

Yekta S., Shih I.H., and Bartel D.P. 2004. MicroRNA-directed cleavage of HOXB8 mRNA. *Science* **304**: 594.

Zeng Y., Yi R., and Cullen B.R. 2003. MicroRNAs and small interfering RNAs can inhibit mRNA expression by similar mechanisms. *Proc. Natl. Acad. Sci.* **100**: 9779.

The RNAi Pathway Initiated by Dicer-2 in *Drosophila*

K. Kim,* Y.S. Lee,† D. Harris,* K. Nakahara,‡ and R.W. Carthew*

*Department of Biochemistry, Molecular Biology and Cell Biology, Northwestern University, Evanston,
Illinois 60208; †Division of Biotechnology, College of Life Sciences and Biotechnology, Korea University,
Seoul, South Korea 136-713; ‡Graduate School of Agriculture, Hokkaido University, Sapporo 060-8589, Japan

Injection or expression of double-stranded RNA (dsRNA) in *Drosophila* serves as a trigger that causes cells to specifically cleave homologous mRNA transcripts. Our approach is to identify essential components of the RNA interference (RNAi) mechanism by isolating and characterizing mutations that cause the RNAi response to be abnormal. These studies have thus far led to the identification of seven genetic loci that encode proteins acting at various steps in the RNAi process. We have molecularly identified several of these proteins. Two are members of the Dicer family. Dicer-1 and Dicer-2 are required for short interfering RNA (siRNA)-directed mRNA cleavage by facilitating distinct steps in the assembly of the RNA-induced silencing complex (RISC). AGO2 is a RISC component that both carries out transcript cleavage and facilitates RISC maturation. Other factors appear to function as regulators of RISC assembly rather than as core factors for RNAi.

SECTION THEMES

Small regulatory RNAs have important roles in a variety of physiological and developmental processes (Zamore 2001). How the generation of such RNAs imparts changes in cell physiology and gene expression is still not well understood. The intracellular machinery used to interpret and implement the effects conveyed by the siRNAs are being actively investigated in a wide variety of experimental systems, using both biochemical and genetic approaches. Biochemical studies with *Drosophila* and mammalian extracts have led to the identification of proteins that bind to or modify siRNAs (Caudy et al. 2002; Martinez et al. 2002; Caudy and Hannon 2004). Although some of these interactions suggest potential mechanisms of gene regulation, the identities and functions of numerous factors are still unclear.

Genetic screens in *Caenorhabditis elegans*, *Arabidopsis thaliana*, and *Neurospora crassa* have yielded numerous mutant strains with defects in RNAi, and in some cases, the individual mutations were shown to impair specific phases of the RNAi response (Cogoni and Macino 1999; Tabara et al. 1999; Fagard et al. 2000). Biochemical experiments with tissue culture cell extracts have identified several components of the RNAi machinery (Harborth et al. 2001). *Drosophila melanogaster* embryo and cell extracts have also been employed to dissect the in vitro RNAi pathway (Haley et al. 2003; Lee et al. 2004). Despite these advances, no system has been developed that has fully harnessed the combined power of large-scale genetics and in vitro mechanistic analysis. Only *Drosophila* currently offers the combination of sophisticated genetic tools and a well-developed in vitro RNAi system. Accordingly, in vitro analyses using extracts from RNAi-defective *Drosophila* strains present a unique opportunity to more precisely define the pathway that leads to gene regulation in response to siRNAs.

Gene expression is silenced by siRNAs in *Drosophila* through cytoplasmic mRNA transcript degradation, typical of the classically defined RNAi response (Kennerdell and Carthew 1998, 2000). In some cases, as in the maturing oocyte, RNAi activity can be correlated with the translational state of the target transcript (Kennerdell et al. 2002). It suggests that the RNAi machinery is limited to detecting and/or degrading only actively translated mRNAs or that both RNAi and translation independently rely upon some common mechanism of transcript licensing. Although a targeted transcript apparently disappears in toto, the first step in its destruction is a single cleavage of the transcript phosphodiester backbone at a site complementary to one strand of an effector siRNA. This event is followed by degradation of the 5′ and 3′ cleavage products by the exosome and other degradation machineries (Orban and Izaurralde 2005). In this paper, we focus on recent progress in understanding the means by which transcripts are targeted for cleavage and, in particular, the use of genetic analysis to reveal aspects of the intracellular pathway that transforms an siRNA into a gene regulator.

A Way to Identify RNAi Factors through Genetics

The simplest way to screen for RNAi mutants is to isolate recessive mutants that are viable and fertile but are defective for RNAi. This approach has been used for many genetic screens to isolate RNAi mutants in various model systems. However, much of the RNAi mechanism could have critical links to other processes that are essential for viability or fertility. If so, then the majority of mutants that perturb RNAi would be recessive-lethal or sterile, and would be missed in a screen for viable mutants.

To get around this problem, we created a situation in which flies are heterozygous for a mutation, but their compound eyes are composed of cells that are homozygous mutant. We then looked for mutations that perturb RNAi activity in the compound eye (Fig. 1). We first carried out construction of a *Drosophila* strain in which eye-

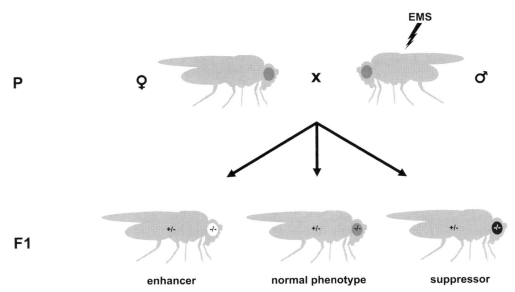

Figure 1. Summary of the genetic screen for mutants with perturbed RNAi activity. RNAi-mediated silencing of the *white* gene is triggered by the presence of a transgene *GMR-wIR*, which expresses *white* dsRNA. The result is an adult fly with partially pigmented eyes (shown here as gray eyes). Males of the parental (P) generation are mutagenized with ethanemethane sulfonate (EMS), and then mated to nonmutagenized females. Their offspring, the F₁ generation, are heterozygous mutant and carry one copy of *GMR-wIR*. However, eye cells carry one chromosome arm that has recombined between sister chromosomes such that the arm is homozygous mutant. If the homozygous-mutant eye cells have normal RNAi, the eye color appears silenced (*gray*). If the mutant cells have impaired RNAi, the eye color appears dark, and if the mutant cells have enhanced RNAi, the eye color appears white.

specific RNAi is easily monitored (Lee and Carthew 2003; Lee et al. 2004). We chose the *white* gene as an RNAi target based on several criteria. The *white* gene encodes an ABC transporter required cell-autonomously to pigment the adult compound eye. A *white*⁺ eye is dark red in color, whereas a null *white* mutant is completely white. This phenotype is easily scored, affording easy and rapid phenotype screening. Moreover, a semiquantitative relationship exists between expression level and phenotype such that activity can be ordered by a phenotypic series: red → brown → orange → yellow → white. To generate RNAi against *white*, a single exon of *white* was placed in a tail-to-tail orientation downstream from the eye-specific GMR promoter, and this snapback construct was transformed into *white*⁺ flies to create transgenic lines that had a *white* loss-of-function phenotype (Lee and Carthew 2003). *GMR-wIR* lines bearing one copy of the transgene had pale orange eyes.

To create flies with homozygous mutant eyes, we used the FLP/FRT system to induce site-specific mitotic recombination (Golic and Lindquist 1989). FLP recombinase was expressed under the control of the *eyeless* promoter (*eyFLP*), which is exclusively active throughout the proliferative phase of eye development (Newsome et al. 2000). *EyFLP* induces recombination between FRT elements on sister chromosomes during cell division, so almost all eye tissue becomes homozygous by the end of the proliferative phase. The nonmutagenized FRT chromosome was marked with the cell-death transgene *GMR-hid*, which kills retinal cells during metamorphosis (postproliferative). Any retinal cell possessing even one copy of *GMR-hid* will die, thus leaving only retinal cells homozygous for the mutage-

nized chromosome. Since the remaining cells compensate in number for the missing ones by further division, the size and morphology of eyes engineered in this fashion are basically normal.

Animals with one copy of *GMR-wIR* were examined for mutant eyes that had variant *white* gene activity. Seven genes were identified in the screen. The products of four of the seven loci were identified by other means as being core RNAi factors.

Dicer-2: An RNase III Enzyme with RISC Assembly Functions

One of the four loci identified in our screen corresponds to the *Dicer-2* gene. The *Dicer-2*-encoded protein (Dicer-2) is synthesized as a 200-kD protein. The amino-terminal region contains a DExH helicase domain, whereas the carboxy-terminal region contains two tandem RNase III catalytic domains and a single dsRNA-binding domain. The Dicer-2 protein has been shown to have RNase-III-type activity in vitro, and this activity appears to produce siRNAs from dsRNA substrates (Bernstein et al. 2001). Moreover, this activity appears to be essential for Dicer-2 function in vivo. Three lines of evidence suggest that Dicer-2 has enzymatic preference to process siRNAs rather than microRNAs (miRNAs). First, purified Dicer-2 protein exhibits strong siRNA-producing activity and weak miRNA-producing activity (Provost et al. 2002; Myers et al. 2003). Second, *Dicer-2* mutants show specific deficits in siRNA production in vivo. Third, *Dicer-2* mutants are completely viable and fertile, whereas loss of miRNAs results in complete loss of viability (Lee et al. 2004).

What domains in Dicer-2 are responsible for siRNA production? Point mutation of key catalytic residues within each RNase III domain abolishes siRNA production (Lee et al. 2004). Thus, these domains are necessary. In addition, the DExH helicase domain is important for siRNA production; point mutation of residues within the ATP-binding pocket greatly reduces activity. Why is a helicase activity essential? It might be needed to displace the enzyme from the dsRNA substrate upon cleavage, thus enabling multiple rounds of enzymatic activity. Interestingly, the other Dicer in *Drosophila*, Dicer-1, lacks a DExH helicase domain and has poor siRNA productivity in vivo.

Dicers act as catalysts for processing small regulatory RNAs including siRNA and miRNA. Do Dicers have any other function in the RNAi reaction? We found that Dicer-2 is required in vivo for transcript destruction in response to artificially supplied siRNA (Lee et al. 2004). This implies that Dicer-2 functions in a second step in addition to its known function as a small RNA processor. We took advantage of the robust biochemistry of RNAi afforded by *Drosophila* to characterize the RNAi defects of the *Dicer-2* mutant. We found that extracts derived from *Dicer-2* mutants exhibit severe defects in mRNA target cleavage activity triggered by an added siRNA. This suggests that Dicer-2 protein functions in the effector phase of RNAi, again consistent with in vivo results. To explore the biochemical basis for the target mRNA cleavage defect in the mutant extract, we examined the assembly of RISC complexes with labeled siRNA.

When siRNA is incubated with extract derived from *Drosophila* embryos, several discrete complexes are seen by native gel electrophoresis (Pham et al. 2004). Dicer-2 protein directly contacts siRNA in these complexes, as determined by UV cross-linking experiments. Moreover, the complexes fail to form when siRNA is incubated with extract derived from *Dicer-2* mutant embryos (Pham et al. 2004). Thus, Dicer-2 associates with siRNA in vitro to form stable complexes. What are the relationships of the different complexes to each other? Pulse-chase experiments show that one of the complexes serves as a precursor to all of the other complexes. Further analysis has revealed that the precursor complex is a heterotrimer composed of siRNA, Dicer-2, and the R2D2 protein (Pham and Sontheimer 2005). For this reason, we call this complex the R2D2–Dicer-2 initiator (RDI). RDI is thought to initiate the assembly of the other complexes. One of the other complexes has many features expected of the RISC effector, although it is far larger (~80S) than any RISC yet described. The complex contains siRNA in a single-stranded state, it copurifies with known RISC factors (TSN, FMR, VIG, and AGO2), and it binds and cleaves targeted mRNAs (Pham et al. 2004). For these reasons, we refer to this complex as holo-RISC. Interestingly, holo-RISC also contains Dicer-2 protein, although unlike RDI, it does not have Dicer-2 in intimate association with siRNA. The RDI and holo-RISC appear to be the beginning and end of an ordered biochemical pathway for RISC assembly (Fig. 2). Dicer-2 does not simply transfer siRNAs to

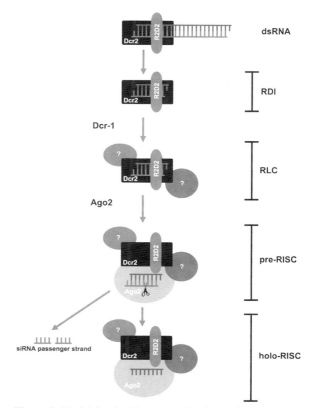

Figure 2. Model for the Dicer-2-mediated RNAi pathway (see text for details).

a distinct RISC complex, but rather assembles into RISC along with the siRNAs, indicating that its role extends beyond the initiation phase of RNAi.

R2D2: A dsRNA-binding Protein Partner of Dicer-2

The Dicer-2 protein is not required for viability or fertility; its loss appears to affect only RNAi activity. We were therefore encouraged to seek mutations in other genes that would lead to the loss of RNAi and, like Dicer-2, would not lead to lethal or sterile phenotypes. One such gene that emerged from our mutagenesis screen encodes the R2D2 protein. R2D2 contains two tandem dsRNA-binding domains (dsRBDs). Loss of *R2D2* function results in strong derepression of *white* gene silencing by *GMR-wIR*. The R2D2 protein heterodimerizes with Dicer-2, but Dicer-2 does not require R2D2 to cleave dsRNA into siRNA (Liu et al. 2003; Liu et al. 2006). What is R2D2 doing in association with Dicer-2? When *R2D2* is mutated, siRNA is not efficiently loaded into holo-RISC, and the assembly of the RDI complex is not observed in the mutant. This observation correlates with the detection of R2D2 as a subunit of RDI. Therefore, R2D2 and Dicer-2 seem to connect the initiator and effector phases of RNAi. The *C. elegans* RDE-4 protein has a structure similar to that of R2D2, and RDE-4 associates with the nematode Dicer in vitro (Tabara et al. 2002). Similar orthologs have been detected in complex with Dicers of vertebrate species. Thus, the role of R2D2 is probably conserved in other organisms.

AGO2 Has an Essential Role in RISC Assembly and Function

Our screen identified a viable mutation that partially derepressed *white* gene activity. On the basis of a variety of genetic data, we have identified the mutation as an allele of *Ago2*. Several previous studies have indicated a role for *Ago2* in RNAi. AGO2 protein is a physical subunit of RISC, and loss of *Drosophila Ago2* gene activity results in a failure to form holo-RISC (Okamura et al. 2004). Furthermore, structural studies of an archaebacterial AGO protein found that it resembles RNase H, an RNase that cleaves the RNA strand of RNA/DNA hybrid duplexes (Lingel et al. 2003; Song et al. 2004). These studies suggested that AGO proteins might directly cleave the target RNA strand of a target:siRNA hybrid duplex in RISC. Indeed, purified human AGO2 protein possesses target RNA cleavage activity, and mutation of key catalytic residues results in loss of this activity (Liu et al. 2004).

There is a sequence motif, GxDV, within the PIWI domain of eukaryotic AGO proteins that is highly conserved (Parker et al. 2004). The aspartate residue is critical to coordinate the nucleophile that catalyzes target RNA cleavage. The mutation $Ago2^{V966M}$ found in our screen changes the GxDV motif into GxDM (Kim et al. 2006). This valine residue is highly conserved among all AGO proteins. Modeling of a GxDM variant of the AGO structure with PyMol reveals a steric change in the integrity of the structure at the catalytic site. We suspect that the V-to-M mutation in AGO2 alters nucleophile coordination required for normal catalysis. Consistent with this notion, western blot analysis reveals that there is a normal level of AGO2 protein present in the mutant even though target RNA cleavage activity is reduced 16-fold in the mutant.

To test if $Ago2^{V966M}$ is nevertheless able to form holo-RISC, we analyzed RISC assembly by native gel electrophoresis of siRNA–protein complexes. Typically, native gels exhibit the RDI and holo-RISC complexes, as well as two intermediate complexes (Kim et al. 2006). One of these is the RISC-loading complex (RLC) and the other is the pre-RISC complex. The assembly pathway in vitro follows RDI to RLC to pre-RISC to holo-RISC (Fig. 2). Formation of each complex appears to have discrete requirements for different protein factors. For example, Dicer-2 is required to form RDI, and hence RLC, pre-RISC, and holo-RISC (Pham et al. 2004). In contrast, AGO2 protein is required to form pre-RISC and holo-RISC but not RDI and RLC (Kim et al. 2006). When we tested the $Ago2^{V966M}$ mutant, both pre-RISC and holo-RISC are present, but there is 2-fold less holo-RISC and an apparent buildup of pre-RISC. The fact that $Ago2^{V966M}$ has a 2-fold decrease in the level of holo-RISC and a 16-fold decrease in holo-RISC activity suggests that the mutant protein has impaired catalytic activity, as predicted. However, why should the mutant show a partial block in the pre-RISC to holo-RISC assembly step?

The answer to this question requires us to understand the differences between pre-RISC and holo-RISC. When we visualize the conformation of siRNA associated with various complexes formed in the assembly reaction, RDI,

RLC, and pre-RISC complexes contain duplex siRNA (Kim et al. 2006). In contrast, holo-RISC contains single-stranded siRNA, the other strand presumably lost from the complex. Thus, key steps in converting pre-RISC to holo-RISC are siRNA strand unwinding and preferential strand retention (Fig. 2). However, $Ago2^{V966M}$ complexes predominantly contain duplex siRNA. Thus, AGO2 catalytic activity appears to stimulate conversion of siRNA duplex into single-stranded siRNA within higher-order RISC complexes (Kim et al. 2006), possibly by strand unwinding or facilitating strand loss. We found that the mutant shows defective siRNA-unwinding comparable to its defective holo-RISC-forming activity. Recent in vitro studies have shown that unwinding of the siRNA duplex is accompanied by cleavage of the ejected siRNA strand during holo-RISC formation (Matranga et al. 2005; Rand et al. 2005). In the $Ago2^{V966M}$ mutant, the rate of siRNA strand cleavage is strongly reduced. These data support a model where siRNAs are initially loaded as duplexes into AGO2-containing pre-RISC. The catalytic activity of AGO2 is then required to cleave one strand of the duplex, stimulating it to dissociate from the uncleaved strand and ultimately dissociate from RISC. The end result is a RISC retaining one siRNA strand, fully competent to down-regulate transcript targets.

If AGO2-dependent catalysis is obligatory for complex conversion, then we might have expected a more profound assembly block in the $Ago2^{V966M}$ mutant. One explanation is that conversion of pre-RISC to holo-RISC occurs by an independent mechanism when AGO2 is unable to cleave siRNA. Alternatively, conversion is promoted by several factors working in parallel, each factor augmenting conversion. For example, it is conceivable that AGO2-dependent cleavage of one siRNA strand acts to facilitate an active unwindase or to prevent stable reassociation of the strands after an unwindase has acted. The identity of such an unwindase remains unknown. The RNA helicase Armitage is required for RISC assembly (Tomari et al. 2004), but it seems to act in the RLC to pre-RISC step of the pathway before unwinding has occurred (data not shown).

RISC Assembly and Dicer-1

Formation of the RLC complex from RDI is poorly understood. It is a rapid ATP-independent step, forming a relatively unstable complex. RLC contains duplex siRNA in association with Dicer-2 and R2D2. The identities of other RLC subunits are yet unknown. Interestingly, another *Drosophila* Dicer protein has been implicated in RLC formation. Dicer-1 is a 220-kD protein that lacks the amino-terminal DExH helicase domain typically found in Dicer proteins. The Dicer-1 protein has RNase-III-type activity in vitro, but this activity appears to be specific for miRNA synthesis and not siRNA synthesis (Lee et al. 2004). Consistent with this observation, *Dicer-1* mutants show specific deficits in miRNA production in vivo, resulting in complete loss of viability.

Despite this preference for miRNA processing, Dicer-1 is also active in the siRNA pathway. *Dicer-1* mutants show derepressed *white* gene silencing in the presence of the

GMR-wIR trigger, and these mutants are also defective in silencing targets that are triggered by exogenous siRNA. Thus, Dicer-1 is required downstream from siRNA production. Our biochemical analysis indicates that Dicer-1 acts as an activator of RLC formation. Extracts from Dicer-1 mutants fail to form stable RLC complexes in vitro. Moreover, UV cross-linking experiments implicate Dicer-1 protein as a factor in close association with siRNA in RLC complexes (Pham et al. 2004). Further biochemical studies will be required to determine whether Dicer-1 is actually an obligatory subunit of RLC. The available data do, however, permit the conclusion that the level of Dicer-1 can be a limiting step in RLC formation. A similar complex to RLC has been identified in *C. elegans*. It consists of Dcr-1, Rde-4, the AGO protein Rde-1, and one or two RNA helicases, Drh-1 and Drh-2 (Tabara et al. 2002).

Regulation of RISC Assembly: Links between RLC and RISC

The RLC consists of Dicer-2/R2D2 closely associated with siRNA duplex and yet unknown factors. The siRNA must undergo a radical transformation in that it changes to a single-stranded form that is closely associated with AGO2 rather than Dicer-2. Conversion of RLC to pre-RISC involves this latter step. Several lines of evidence suggest that this step might be rate-limiting and subject to regulation both in vitro and in vivo. First, in vitro, the rate of pre-RISC formation is very slow relative to RDI and RLC formation. Second, formation is stimulated by ATP, suggesting that the process is more complicated than mere intermolecular association/dissociation. Third, genetic analysis has identified putative regulators of RISC assembly that appear to affect the RLC to pre-RISC step.

Whereas *Dicer-1*, *Dicer-2*, *R2D2*, and *Ago2* mutations show impaired silencing of a *white* target gene in vivo, mutations in two other genes show enhanced silencing of *white* by *GMR-wIR*. Our data implicate these genes as encoding repressors of the RNAi response (Y.S. Lee et al., unpubl.). Indeed, biochemically both repressors appear to inhibit production of holo-RISC by attenuating the rate of pre-RISC formation. These data have several implications. They argue that RNAi is under some endogenous negative control for reasons that are as yet unclear. They argue that the RLC to pre-RISC step is an effective step in which regulation can be exerted, and they argue that RNAi can be engineered to be stronger using facile methods to manipulate fly strains.

CONCLUDING REMARKS

The experimental advantages of being able to examine mutant phenotypes in a tissue that is dispensable for viability and fertility ensure that new insights will continue to emerge from studies of RNAi in *Drosophila*. In particular, the utility of genetic screens for enhancers and suppressors of RNAi to identify components of the pathway has been established and has encouraged us to start similar endeavors for other pathways. We believe that the success of our previous screens depended on our ability to create a situation where the RNAi-responsive tissue

(the eye) contained completely mutant cells while the rest of the animal contained heterozygous cells. This ability to identify mutations in heterozygous individuals permits not only the screening of large numbers of individuals, but also the identification of genes with other roles during development whose total loss of function would be lethal. Finally, the advantage of performing biochemical assays with extracts from normal and mutant strains in parallel has permitted an unprecedented ability to pinpoint the molecular activities of those products identified in our screens.

ACKNOWLEDGMENTS

We thank the National Institutes of Health, FRAXA Foundation, and the Damon Runyon Foundation for their support.

REFERENCES

Bernstein E., Caudy A.A., Hammond S.M., and Hannon G.J. 2001. Role for a bidentate ribonuclease in the initiation step of RNA interference. *Nature* **409:** 363.

Caudy A.A. and Hannon G.J. 2004. Induction and biochemical purification of RNA-induced silencing complex from *Drosophila* S2 cells. *Methods Mol. Biol.* **265:** 59.

Caudy A.A., Myers M., Hannon G.J., and Hammond S.M. 2002. Fragile X-related protein and VIG associate with the RNA interference machinery. *Genes Dev.* **16:** 2491.

Cogoni C. and Macino G. 1999. Gene silencing in *Neurospora crassa* requires a protein homologous to RNA-dependent RNA polymerase. *Nature* **399:** 166.

Fagard M., Boutet S., Morel J.B., Bellini C., and Vaucheret H. 2000. AGO1, QDE-2, and RDE-1 are related proteins required for post-transcriptional gene silencing in plants, quelling in fungi, and RNA interference in animals. *Proc. Natl. Acad. Sci.* **97:** 11650.

Golic K.G. and Lindquist S. 1989. The FLP recombinase of yeast catalyzes site-specific recombination in the *Drosophila* genome. *Cell* **59:** 499.

Haley B., Tang G., and Zamore P.D. 2003. In vitro analysis of RNA interference in *Drosophila melanogaster*. *Methods* **30:** 330.

Harborth J., Elbashir S.M., Bechert K., Tuschl T., and Weber K. 2001. Identification of essential genes in cultured mammalian cells using small interfering RNAs. *J. Cell Sci.* **114:** 4557.

Kennerdell J.R. and Carthew R.W. 1998. Use of dsRNA-mediated genetic interference to demonstrate that *frizzled* and *frizzled 2* act in the *wingless* pathway. *Cell* **95:** 1017.

———. 2000. Heritable gene silencing in *Drosophila* using double-stranded RNA. *Nat. Biotechnol.* **18:** 896.

Kennerdell J.R., Yamaguchi S., and Carthew R.W. 2002. RNAi is activated during *Drosophila* oocyte maturation in a manner dependent on *aubergine* and *spindle-E*. *Genes Dev.* **16:** 1884.

Kim K., Lee Y.S., and Carthew R.W. 2006. Conversion of pre-RISC to holo-RISC by Ago2 during assembly of RNAi complexes. *RNA* (in press).

Lee Y.S. and Carthew R.W. 2003. Making a better RNAi vector for *Drosophila:* Use of intron spacers. *Methods* **30:** 322.

Lee Y.S., Nakahara K., Pham J.W., Kim K., He Z., Sontheimer E.J., and Carthew R.W. 2004. Distinct roles for *Drosophila* Dicer-1 and Dicer-2 in the siRNA/miRNA silencing pathways. *Cell* **117:** 83.

Lingel A., Simon B., Izaurralde E., and Sattler M. 2003. Structure and nucleic-acid binding of the *Drosophila* Argonaute 2 PAZ domain. *Nature* **426:** 465.

Liu J., Carmell M.A., Rivas F.V., Marsden C.G., Thomson J.M., Song J.J., Hammond S.M., Joshua-Tor L., and Hannon G.J. 2004. Argonaute2 is the catalytic engine of mammalian RNAi. *Science* **305:** 1437.

Liu Q., Rand T.A., Kalidas S., Du F., Kim H.E., Smith D.P., and Wang X. 2003. R2D2, a bridge between the initiation and effector steps of the *Drosophila* RNAi pathway. *Science* **301:** 1921.

Liu X., Jiang F., Kalidas S., Smith D., and Liu Q. 2006. Dicer-2 and R2D2 coordinately bind siRNA to promote assembly of the siRISC complexes. *RNA* **12:** 1514.

Martinez J., Patkaniowska A., Urlaub H., Luhrmann R., and Tuschl T. 2002. Single-stranded antisense siRNAs guide target RNA cleavage in RNAi. *Cell* **110:** 563.

Matranga C., Tomari Y., Shin C., Bartel D.P., and Zamore P.D. 2005. Passenger-strand cleavage facilitates assembly of siRNA into Ago2-containing RNAi enzyme complexes. *Cell* **123:** 607.

Myers J.W., Jones J.T., Meyer T., and Ferrell J.E. 2003. Recombinant Dicer efficiently converts large dsRNAs into siRNAs suitable for gene silencing. *Nat. Biotechnol.* **21:** 324.

Newsome T.P., Asling B., and Dickson B.J. 2000. Analysis of *Drosophila* photoreceptor axon guidance in eye-specific mosaics. *Development* **127:** 851.

Okamura K., Ishizuka A., Siomi H., and Siomi M.C. 2004. Distinct roles for Argonaute proteins in small RNA-directed RNA cleavage pathways. *Genes Dev.* **18:** 1655.

Orban T.I. and Izaurralde E. 2005. Decay of mRNAs targeted by RISC requires XRN1, the Ski complex, and the exosome. *RNA* **11:** 459.

Parker J.S., Roe S.M., and Barford D. 2004. Crystal structure of a PIWI protein suggests mechanisms for siRNA recognition and slicer activity. *EMBO J.* **23:** 4727.

Pham J.W. and Sontheimer E.J. 2005. Molecular requirements for RNA-induced silencing complex assembly in the *Drosophila* RNA interference pathway. *J. Biol. Chem.* **280:** 39278.

Pham J.W., Pellino J.L., Lee Y.S., Carthew R.W., and Sontheimer E.J. 2004. A Dicer-2-dependent 80s complex cleaves targeted mRNAs during RNAi in *Drosophila*. *Cell* **117:** 83.

Provost P., Dishart D., Doucet J., Frendewey D., Samuelsson B., and Radmark O. 2002. Ribonuclease activity and RNA binding of recombinant human Dicer. *EMBO J.* **21:** 5864.

Rand T.A., Petersen S., Du F., and Wang X. 2005. Argonaute2 cleaves the anti-guide strand of siRNA during RISC activation. *Cell* **123:** 621.

Song J.J., Smith S.K., Hannon G.J., and Joshua-Tor L. 2004. Crystal structure of Argonaute and its implications for RISC slicer activity. *Science* **305:** 1434.

Tabara H., Yigit E., Siomi H., and Mello C.C. 2002. The dsRNA binding protein RDE-4 interacts with RDE-1, DCR-1, and a DExH-box helicase to direct RNAi in *C. elegans*. *Cell* **109:** 861.

Tabara H., Sarkissian M., Kelly W.G., Fleenor J., Grishok A., Timmons L., Fire A., and Mello C.C. 1999. The rde-1 gene, RNA interference, and transposon silencing in *C. elegans*. *Cell* **99:** 123.

Tomari Y., Du T., Haley B., Schwarz D.S., Bennett R., Cook H.A., Koppetsch B.S., Theurkauf W.E., and Zamore P.D. 2004. RISC assembly defects in the *Drosophila* RNAi mutant armitage. *Cell* **116:** 831.

Zamore P.D. 2001. RNA interference: Listening to the sound of silence. *Nat. Struct. Biol.* **8:** 746.

Molecular Mechanism of Target RNA Transcript Recognition by Argonaute-Guide Complexes

J.S. PARKER, S.M. ROE, AND D. BARFORD

Section of Structural Biology, Institute of Cancer Research, Chester Beatty Laboratories,
London, SW3 6JB, United Kingdom

Argonaute proteins participate in conferring all known functions of RNA-mediated gene silencing phenomena. However, prior to structural investigations of this evolutionarily conserved family of proteins, there was little information concerning their mechanisms of action. Here, we describe our crystallographic analysis of the PIWI domain of an archaeal Argonaute homolog, *Af*Piwi. Our structural analysis revealed that the Argonaute PIWI fold incorporates both an RNase-H-like catalytic domain and an anchor site for the obligatory 5′ phosphate of the RNA guide strand. RNA-*Af*Piwi binding assays combined with crystallographic studies demonstrated that *Af*Piwi interacts with RNA via a conserved region centered on the carboxyl terminus of the protein, utilizing a novel metal-binding site. A model of the PIWI domain of Argonaute in complex with a small interfering RNA (siRNA)-like duplex is consistent with much of the existing biochemical and genetic data, explaining the specificity of the RNA-directed RNA endonuclease reaction and the importance of the 5′ region of microRNAs (miRNAs) (the "seed") to nucleate target RNA recognition and provide high-affinity guide–target interactions.

Argonautes are the central component of RNA-induced silencing complex (RISC)-like ribonucleoprotein (RNP) complexes that mediate all known effector functions associated with RNA interference (RNAi) (Hammond et al. 2001). Target RNA transcripts are selected by a guide strand–Argonaute complex. Depending on the degree of complementarity between guide and target strands, the subclass of Argonaute protein, and the origin of the guide RNA, the RNA-programmed RISC-like complex mediates either cleavage of the target, translation repression, or heterochromatin remodeling. Moreover, in addition to conferring these RNAi-based effector functions, Argonautes participate in selecting the single-stranded RNA guide from the double-stranded siRNA/miRNA duplex. Findings that a highly purified form of *Drosophila* RISC was dmAgo2 (Rand et al. 2004) and that recombinant hAgo2 was capable of selecting and cleaving target RNA transcripts (Rivas et al. 2005) provided the crucial insights that the critical biological functions of RISC are performed by Argonaute.

Our interests have focused on understanding the molecular mechanisms of Argonaute proteins. From a combination of structural, biochemical, and biophysical approaches, we addressed the following questions: (1) How does the guide RNA interact with Argonaute? (2) How does the Argonaute–guide RNA complex recognize its complementary RNA transcript selectively? (3) What structural features of the Argonaute–guide–target RNA complex determine whether the target strand is cleaved?

Argonaute proteins, originally discovered in plants (Bohmert et al. 1998), are indigenous to the three domains of life (Carmell et al. 2002). They are characterized by an amino-terminal PAZ domain (of ~150 residues) and a carboxy-terminal PIWI domain of about 400 residues linked by a region of variable length and sequence. PAZ domains also feature in Dicer, the endonuclease that generates siRNA and miRNAs from longer double-stranded RNA precursors. In both Argonaute and Dicer, the PAZ domain interacts with the 3′ end of an RNA molecule. However, intriguingly, the PIWI domain is unique to the Argonautes. An Argonaute homolog from the archaea *Archaeoglobus fulgidus* is the only known member of this family to consist solely of a PIWI domain. The protein of 427 amino acids comprises the conserved motifs that are characteristic of the PIWI domain and shares approximately 20% sequence identity with the PIWI domains of eukaryotic Argonautes. The protein therefore serves as a useful model system to understand the function of the PIWI domain within Argonautes.

PIWI IS COMPOSED OF RNASE H AND LAC REPRESSOR-LIKE DOMAINS

We determined the crystal structure of *Af*Piwi to 2.0 Å resolution, revealing important insights into the functions of the PIWI component of Argonaute (Parker et al. 2004). *Af*Piwi is organized into two major domains that we defined as A and B (Fig. 1). The combination of these two domains creates the PIWI fold that includes all conserved sequence motifs that were initially used to define the PIWI domain. The intimate association of the A and B domains also suggests that the functional and structural unit of the PIWI fold is generated from these two domains.

Domain A is formed from a simple α-β-α sandwich fold with weak structural resemblance to a subdomain of the Lac repressor. However, differences in the chain topology and structural context of this domain relative to its counterpart in the Lac repressor suggest that the two proteins share no functional similarity. The structure of domain B, on the other hand, revealing an architectural similarity to the RNase H family of endonucleases, provided significant insights into its function. Domain B

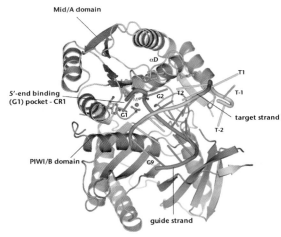

Figure 1. Ribbons representation of the crystal structure of *Af*Piwi in complex with siRNA. (*Green*) Guide strand; (*yellow*) target; (*salmon*) domain A (mid); (*blue*) domain B (PIWI); (*cyan*) residues of the G1-binding pocket. The 5′ nucleotide (G1) of the guide strand is located within the G1-binding pocket (CR1). (M) Metal ion.

(defined as the PIWI domain by Joshua-Tor and colleagues from their crystal structure of the *Pyrococcus furiosus* Argonaute [*Pf*Ago]) (Song et al. 2004) shares structural characteristics of the RNase H family members: archaeal RNase HII, *Escherichia coli* and human immunodeficiency virus (HIV) RNase HI, ASV- and HIV-encoded integrases, and *E. coli* RuvC, namely a central 5-stranded β-sheet surrounded on both sides by a pair of α-helices (Fig. 2). In RNase H enzymes, two aspartic acid residues situated on adjacent β-strands contribute to divalent metal coordination at the catalytic site. Although neither of these aspartic acids is conserved in *Af*Piwi, their presence in eukaryotic Argonautes, *Pf*Ago, an archaeal Argonaute, and the eubacterial Argonaute *Aquifex aeolicus*, suggested a potential catalytic function and implicated Argonaute as the Slicer activity of RISC (Song et al. 2004). Substantial evidence now supports the notion that

Argonautes are indeed responsible for the Slicer activity of RISC. Mutation of either catalytic aspartic acid in hAgo2 abolishes Slicer activity in vivo (Liu et al. 2004) and in vitro (Rivas et al. 2005), and similarly to RNase H, Slicer is an Mg^{2+}-dependent enzyme that utilizes metal-activated water molecules to cleave its phosphodiester bond substrate, yielding 3′OH groups and 5′ phosphates (Martinez and Tuschl 2004; Schwarz et al. 2004; Nowotny et al. 2005).

CONSERVED REGION 1 (CR1) DEFINES A PUTATIVE RNA-BINDING SITE

To gain insight into functional regions of the PIWI domain, we mapped regions of sequence conservation, shared between all three kingdoms of life, onto the molecular surface of *Af*Piwi. This analysis revealed three conserved regions: CR1, CR2, and CR3. All three conserved regions map to the same face of the PIWI domain surface. CR3 is unique to the eukaryotic Argonautes, as residues at this site are not conserved in eubacteria or archaea. In eukaryotes, this conserved region constitutes the "PIWI box," a region of the PIWI domain of Argonautes implicated in interactions with Dicer (Tahbaz et al. 2004). CR2, within the RNase-H-like domain B, corresponds to the RNase-H-like catalytic site, and its presence in eubacteria and archaea, in addition to eukaryotes, suggests that Argonaute-associated endonuclease activity is evolutionarily conserved. Strikingly, the region of highest structural conservation shared among all Argonaute sequences is CR1, a region at the interface of the A and B domains (Fig. 1). This interface features a deep cleft and is linked to CR2 by a channel located within domain B (see Fig. 4). The carboxy-terminal carboxylate group of the polypeptide chain, exposed on the molecular surface, defines the center of the interface that in *Af*Piwi is coordinated to a divalent cation (cadmium in our crystal structure). The aliphatic side chain of Leu-427 and preceding four amino acids are buried at the domain interface, ensuring that the carboxy-terminal carboxylate is rigidly located at the molecular surface. Metal coordination by a carboxy-terminal carboxyl group is unusual and suggests an interesting structural feature. Compared with the side chains of aspartic acid and glutamic acid residues, a carboxy-terminal carboxyl group is allowed reduced rotational freedom, and because Leu-427 is clamped at the interface of the A and B domains, it is likely that the metal ion site in *Af*Piwi is quite rigid. In our structure, the divalent cation also contacts the amide side chain of Gln-159, a residue conserved in the PIWI sub-family. The divalent cation-binding site is surrounded by four residues, contributed by domain A, almost universally invariant in Argonaute sequences (Fig. 3). These residues, two lysines, glutamine, and tyrosine, are characteristic of RNA-binding proteins and suggested the exciting possibility that CR1 may represent an RNA-binding site. Geometric arguments, based on the distance between CR1 and CR2, and the property of RISC to cleave the target strand at a defined site, namely, the phosphodiester bond bridging the two nucleotides that base-pair with the 10th and 11th nucleotides of the guide strand (Elbashir et al. 2001a,b;

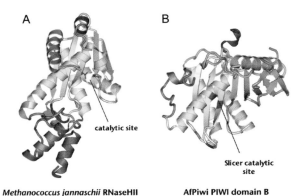

Figure 2. The PIWI domain (domain B) features an RNase-H-like fold (*cyan*). (*A*) structure of RNase HII from *Methanococcus jannaschii* (Lai et al. 2000). (*B*) *Af*Piwi domain B. (*Red*) Positions of residues of the catalytic triad.

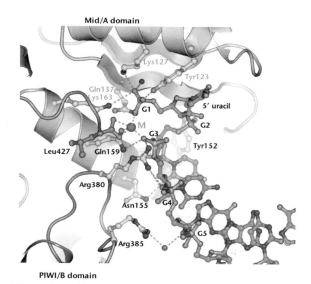

Figure 3. G1-binding pocket anchors the 5′ end of guide strand (target strand omitted for clarity). Details of protein–RNA guide interactions are at the G1-binding pocket. (M) Metal ion.

discussed in Parker et al. 2004), provided a strong indication that CR1 may anchor the obligatory 5′ phosphate of the guide strand (Nykänen et al. 2001; Chiu and Rana 2002; Schwarz et al. 2002), placing the scissile phosphate of the target strand within CR2. Although other studies had demonstrated that the PAZ domain of Argonaute binds RNA (Song et al. 2003; Yan et al. 2003; Lingel et al. 2004; Ma et al. 2004), it was not previously known whether PIWI acts as an RNA-binding domain. To test the possibility that *Af*Piwi interacts with RNA, in a manner that is dependent on metal binding to the carboxy-terminal carboxylate group, we performed protein–RNA-binding experiments, monitoring RNA–*Af*Piwi interactions by means of electrophoretic mobility-shift assays and UV-induced protein–RNA cross-linking. These studies, employing an siRNA-like molecule of 16 nucleotides with 5′ phosphate groups and 3′ dinucleotide overhangs, demonstrated that the high-affinity *Af*Piwi–RNA interactions required divalent cations and were severely diminished by addition of a glycine residue to Leu-427, a modification expected to disrupt the geometry of the divalent cation binding site of CR1 (Parker et al. 2004).

To understand the molecular mechanism of PIWI–RNA interactions, we cocrystallized *Af*Piwi with a 16-nucleotide siRNA-like duplex (Parker et al. 2005). The resultant structure represents a model of either a guide-passenger duplex or a guide-target duplex associated with the PIWI domain and provided the critical insight that CR1 functions as an anchor site for the 5′ phosphate of the guide strand. The 5′ end of the RNA duplex (defined as the 5′ end of the guide strand), which participates in extensive interactions with the protein, is highly ordered, but the duplex becomes progressively less well ordered toward its 3′ end. In our crystal structure, we observed well-resolved electron density for the nine 5′ nucleotides, assigned G1–G9 for the guide strand and

T1–T9 for the complementary nucleotides of the target strand (Fig. 1). The remainder of the RNA duplex is not clearly resolved in the electron density and is assumed to be mobile.

*Af*Piwi binds to the 5′ end of the siRNA-like duplex, engaging the RNA in an L-shaped channel (Fig. 1). Nucleotides 2–9 on each strand base-pair, adopting an A-form helix that extends from the domain interface along the channel of domain B to CR2. A striking feature of the structure is the profound distortion of the 5′ end of the RNA duplex phosphodiester backbone, resulting in the unwinding of the 5′ complementary nucleotides G1 (uridine) and T1 (adenosine). G1, which carries the obligatory 5′ phosphate of the guide, is displaced into a pocket surrounding the carboxyl terminus of the PIWI domain, CR1, now defined as the G1-binding pocket. T1 passes through the short "exit" portion of the channel, allowing the two 3′ overhanging nucleotides, T(-1) and T(-2), to fold back along the protein surface. The unpaired G1 and T1 nucleotides are physically divided by the αD helix of domain A, which contributes two aromatic residues (Phe-151 and Tyr-152) capping the first 5′ base pair of the RNA helix (U2–A2), therefore substituting for the unpaired nucleotides at the extreme 5′ end of the duplex (Fig. 1).

Distortion of the 5′ end of the guide strand is facilitated by RNA–phosphate interactions with conserved residues of the G1-binding pocket, particularly to the divalent cation (manganese in the RNA–*Af*Piwi complex) coordinated to the carboxylate of Leu-427 (Fig. 3). Via interactions that require distortion of the 5′ end of the guide, both the phosphate groups of G1 and G3 form direct contacts to the metal ion. In addition to the metal-mediated RNA contacts within the G1-binding pocket, the side chains of four invariant residues in Argonaute proteins (Tyr-123, Lys-127, Gln-137, and Lys-163) also contact the 5′ phosphate group of G1. Tyr-123 further stabilizes the flipped conformation of G1 by stacking with the unpaired uracil base. The 5′ phosphate group forms further contacts to the backbone amide of Phe-138, fixed in position by the adjacent Gln-137.

Interactions between *Af*Piwi and the guide are also mediated by the phosphate groups of G2–G5. A second key anchor site is the phosphate group of G3. In addition to its coordination by the metal, the phosphate interacts with Gln-159 and Arg-380, residues conserved in the PIWI and Ago subfamilies, respectively. The phosphate of G2 contacts Tyr-152, conserved as tyrosine or threonine in the Ago subfamily, and the main-chain amide of Arg-140. The phosphate of G4 also contacts Arg-380, and the phosphate of G5 interacts with Arg-385, via a water molecule. The only protein–base contact occurs between Asn-155, conserved in the Ago subfamily, and the 2′ oxygen of the G2 uracil. Hydrogen bond acceptors occupy equivalent positions in cytosine and purines. Therefore, as anticipated, none of the interactions provide RNA sequence specificity. Interactions with G1–G5 account for all conserved residues of CRI, strongly indicating a conserved mode of RNA 5′-end binding within the Argonaute family. In contrast to the extensive interactions between PIWI and the guide, few contacts are formed to the target strand, consistent with its requirement to dissociate after mRNA cleavage.

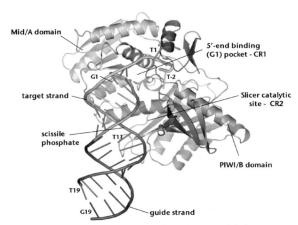

Figure 4. Model of a 19-nucleotide siRNA–*Af*Piwi structure. Guide strand nucleotides G1–G9 (and complementary target nucleotides and 3′ dinucleotide overhang) determined from the crystal structure are in light green and yellow, respectively. The modeled regions of both strands are in darker green and yellow. (*Red*) Positions of catalytic triad residues, and the scissile bond of the target strand.

MODEL FOR TARGET STRAND RECOGNITION AND CLEAVAGE

Nucleotides G10–G16 of the guide and T10–T14 of the target are less clearly defined in the electron density maps. To visualize the trajectory of a full-length (19-nucleotide) siRNA guide bound to an mRNA target, we modeled the remaining nucleotides, assuming an A-form helix contiguous with the visible duplex (Fig. 4). The weaker electron density for nucleotides G10 and G11 is compatible with this conformation. This model of a PIWI domain–siRNA complex provided significant insights into mechanisms of target strand recognition and mechanisms of target strand cleavage.

Distortion and displacement of G1 into the conserved binding pocket result in the unwinding of the cognate G1–T1 base pair, whereas the bases of nucleotides G2 to G8 are ordered and exposed to solvent. Bases 3′ of G9 are turned in toward the molecular surface of the protein and are therefore less accessible for base-pairing (Fig. 4). The nucleotides with exposed bases, accessible for base-pairing with a target transcript, as revealed by our crystal structure, correlate with the definition of a "seed" sequence identified in animal miRNAs. Bioinformatics analysis of the complementarity between miRNAs and their target recognition sequences within the 3′UTRs (untranslated regions) of RNA transcripts revealed that nucleotides 2–7 of the miRNA were complementary to the target sequence. The 5′ nucleotide and nucleotides 3′ of the seed region shared little to no sequence homology with their target. This implies that the seed sequence is sufficient for target recognition, suggesting that regions of the miRNA outside of the seed region would not base-pair with their target sequence (Lewis et al. 2003, 2005; Stark et al. 2003). The notion that a seed sequence dominates miRNA–target RNA interactions was strongly supported by empirical data of Doench and Sharp (2004).

Introducing mismatches between an miRNA and its target within the seed region profoundly reduced miRNA-mediated repression of translation, whereas mismatches 3′ to the seed region had little to no effect on translation repression. It is likely that an analogous "seed" region may operate for siRNAs because in RISC-mediated mRNA cleavage, bases at the 5′ end of the siRNA guide dominate its affinity for target RNA (Haley and Zamore 2004).

The model of the extended siRNA in complex with *Af*Piwi provides insights into the catalytic mechanism of Argonaute. Toward the middle of the guide RNA, the trajectory of the RNA directs the target strand into the RNase H catalytic site, placing the scissile phosphate, linking T10 and T11 (which base-pair with G10 and G11, respectively) in close proximity to the putative catalytic site. Modeling the position of the RNase H catalytic aspartic acid residues into the *Af*Piwi–siRNA complex provides a plausible explanation for the specificity of target RNA cleavage. With the 5′ phosphate of the guide strand anchored at the G1-binding pocket, an RNA duplex with complementary guide and target strands forming an A-form RNA duplex will adopt the correct geometry to position the target strand within the catalytic site, resulting in cleavage between the 10th and 11th nucleotides. Since an A-form RNA duplex is the preferred substrate of RISC (Chiu and Rana 2003; Haley and Zamore 2004), this could explain why some animal miRNAs containing mismatches between guide and target strand within the central region do not direct target strand cleavage, whereas plant miRNAs, which are normally homologous to their targets over their entire length, direct target strand cleavage. Argonaute proteins lacking either of the catalytic aspartic acids (e.g., human Ago4 and Hiwi2) would be predicted to lack the capacity to cleave RNA. In addition to the aspartic acid diad, a third catalytic residue (histidine in human Ago2), located on the αH helix, most likely contributes to metal binding at the catalytic site (Nowotny et al. 2005; Rivas et al. 2005).

MODEL FOR GUIDE STRAND SELECTION

The *Af*Piwi–RNA complex potentially also represents the initial phase of RISC association with an siRNA duplex; here, the target strand would represent the passenger strand. Argonautes are required for efficient unwinding of the siRNA duplex (Okamura et al. 2004). The relative stabilities of the ends of an siRNA duplex determine strand fate: The strand whose 5′ end is located at the less stable end becomes the guide, the other (passenger) strand is discarded (Khvorova et al. 2003; Schwarz et al. 2003). Specific recognition of the siRNA-like 5′ end, and partial unwinding of the duplex revealed from the *Af*Piwi–RNA complex, suggests that binding to PIWI may be the first step in siRNA duplex unwinding in RISC; PIWI could therefore participate in strand selection.

The guide–passenger duplex bound to *Af*Piwi is the exact structural analogy of the guide–target complex, predicting that the passenger strand would enter the RNase-H-like catalytic site, reminiscent with that of a target strand and hence subject to cleavage. Consistent with this notion, studies from Zamore (Matranga et al. 2005) and Wang (Rand et al. 2005) have recently revealed that passenger strand cleavage is associated with RISC assembly, guide strand binding, and dissociation of the passenger strand. Similar to the situation for target strand release following cleavage, cleavage of the passenger strand near its center would significantly reduce its affinity for the guide (since nucleic acid duplex interactions are highly cooperative) promoting its dissociation. However, there is also evidence that an ATP-dependent RNA helicase also participates in passenger strand unwinding.

CONCLUSIONS

A role for the PIWI domain in mediating interactions with the 5′ end of the guide RNA complements the function of the PAZ domain in recognizing RNA 3′ overhangs. Thus, Argonaute functions as a scaffold protein, engaging guide RNA to optimize recognition of target RNA transcripts, particularly those that are homologous to the seed region of the guide strand. On binding the target strand, we envisage, based on the low affinity for blunt-ended strands (Ma et al. 2004), that the PAZ domain dissociates from the 3′ end of the guide. The 5′ anchor site of the G1-binding pocket therefore provides the fundamental interaction between Argonaute and RNA, orienting and ordering the 5′ region of the guide strand for interactions with its RNA target. Our findings that both the 5′ phosphate receptor site and the Slicer catalytic site are located within the PIWI domain is consistent with the result of Miyoshi et al. (2005) that the PIWI domain of DmAgo2 alone incorporates the capacity for RNA-directed RNA endonuclease activity.

The study of prokaryotic Argonautes, homologs of their eukaryotic counterparts, has provided considerable insights into molecular mechanisms of RNA silencing phenomena. Both the G1-binding pocket and RNase-H-like catalytic site are conserved throughout evolution. Data from our laboratory showed that disrupting the invariant carboxy-terminal carboxylate position of *Af*Piwi significantly diminished *Af*Piwi–RNA interactions (Parker et al. 2004), whereas Tsuchl, Patel, and colleagues found that mutating residues in hAgo2 whose counterparts form the G1-binding pocket of *Af*Piwi (Parker et al. 2005) abolished hAgo2-mediated-RNA cleavage (Ma et al. 2005). However, it should be noted that prokaryotes are not known to utilize RNA silencing mechanisms, and the roles of prokaryotic Argonautes are not currently understood. Future studies will be directed at obtaining crystal structures of eukaryotic Slicer Argonautes in complex with RNA duplex substrates and in providing a thermodynamic explanation for the finding that the 5′ seed region of miRNAs/siRNAs dominates interactions with the target mRNA transcript.

ACKNOWLEDGMENTS

The work in the author's laboratory was funded by Cancer Research UK and the Wellcome Trust.

REFERENCES

Bohmert K., Camus I., Bellini C., Bouchez D., Caboche M., and Benning C. 1998. AGO1 defines a novel locus of *Arabidopsis* controlling leaf development. *EMBO J.* **17:** 170.

Carmell M.A., Xuan Z., Zhang M.Q., and Hannon G.J. 2002. The Argonaute family: Tentacles that reach into RNAi, developmental control, stem cell maintenance, and tumorigenesis. *Genes Dev.* **16:** 2733.

Chiu Y.L. and Rana T.M. 2002. RNAi in human cells: Basic structural and functional features of small interfering RNA. *Mol. Cell* **10:** 549.

———. 2003. siRNA function in RNAi: A chemical modification analysis. *RNA* **9:** 1034.

Doench J.G. and Sharp P.A. 2004. Specificity of microRNA target selection in translational repression. *Genes Dev.* **18:** 504.

Elbashir S.M., Lendeckel W., and Tuschl T. 2001a. RNA interference is mediated by 21- and 22-nucleotide RNAs. *Genes Dev.* **15:** 188.

Elbashir S.M., Martinez J., Patkaniowska A., Lendeckel W., and Tuschl T. 2001b. Functional anatomy of siRNAs for mediating efficient RNAi in *Drosophila melanogaster* embryo lysate. *EMBO J.* **20:** 6877.

Haley B. and Zamore P.D. 2004. Kinetic analysis of the RNAi enzyme complex. *Nat. Struct. Mol. Biol.* **11:** 599.

Hammond S.M., Boettcher S., Caudy A.A., Kobayashi R., and Hannon G.J. 2001. Argonaute2, a link between genetic and biochemical analyses of RNAi. *Science* **293:** 1146.

Khvorova A., Reynolds A., and Jayasena S.D. 2003. Functional siRNAs and miRNAs exhibit strand bias. *Cell* **115:** 209.

Lai L., Yokota H., Hung L.W., Kim R., and Kim S.H. 2000. Crystal structure of archaeal RNase HII: A homologue of human major RNase H. *Struct. Fold. Des.* **8:** 897.

Lewis B.P., Burge C.B., and Bartel D.P. 2005. Conserved seed pairing, often flanked by adenosines, indicates that thousands of human genes are microRNA targets. *Cell* **120:** 15.

Lewis B.P., Shih I.H., Jones-Rhoades M.W., Bartel D.P., and Burge C.B. 2003. Prediction of mammalian microRNA targets. *Cell* **115:** 787.

Lingel A., Simon B., Izaurralde E., and Sattler M. 2004. Nucleic acid 3′-end recognition by the Argonaute2 PAZ domain. *Nat. Struct. Mol. Biol.* **11:** 576.

Liu J., Carmell M.A., Rivas F.V., Marsden C.G., Thomson J.M., Song J.J., Hammond S.M., Joshua-Tor L., and Hannon G.J. 2004. Argonaute2 is the catalytic engine of mammalian RNAi. *Science* **305:** 1437.

Ma J.B., Ye K., and Patel D.J. 2004. Structural basis for overhang-specific small interfering RNA recognition by the PAZ domain. *Nature* **429:** 318.

Ma J.B., Yuan Y.R., Meister G., Pei Y., Tuschl T., and Patel D.J. 2005. Structural basis for 5′-end-specific recognition of guide RNA by the *A. fulgidus* Piwi protein. *Nature* **434:** 666.

Martinez J. and Tuschl T. 2004. RISC is a 5′ phosphomonoester-producing RNA endonuclease. *Genes Dev.* **18:** 975.

Matranga C., Tomari Y., Shin C, Bartel D.P., and Zamore P.D. 2005. Passenger-strand cleavage facilitates assembly of siRNA into Ago2-containing RNAi enzyme complexes. *Cell* **123:** 607.

Miyoshi K., Tsukumo H., Nagami T., Siomi H., and Siomi M.C. 2005. Slicer function of *Drosophila* Argonautes and its involvement in RISC formation. *Genes Dev.* **19:** 2837.

Nowotny M., Gaidamakov S.A., Crouch R.J., and Yang W. 2005. Crystal structures of RNase H bound to an RNA/DNA hybrid: Substrate specificity and metal-dependent catalysis. *Cell* **121:** 1005.

Nykänen A., Haley B., and Zamore P.D. 2001. ATP requirements and small interfering RNA structure in the RNA interference pathway. *Cell* **107:** 309.

Okamura K., Ishizuka A., Siomi H., and Siomi M.C. 2004. Distinct roles for Argonaute proteins in small RNA-directed RNA cleavage pathways. *Genes Dev.* **18:** 1655.

Parker J.S., Roe S.M., and Barford D. 2004. Crystal structure of a PIWI protein suggests mechanisms for siRNA recognition and slicer activity. *EMBO J.* **23:** 4727.

———. 2005. Structural insights into mRNA recognition from a PIWI domain-siRNA guide complex. *Nature* **434:** 663.

Rand T.A., Ginalski K., Grishin N.V., and Wang X. 2004. Biochemical identification of Argonaute 2 as the sole protein required for RNA-induced silencing complex activity. *Proc. Natl. Acad. Sci.* **101:** 14385.

Rand T.A., Petersen S., Du F., and Wang X. 2005. Argonaute2 cleaves the anti-guide strand of siRNA during RISC activation. *Cell* **123:** 621.

Rivas F.V., Tolia N.H., Song J.J., Aragon J.P., Liu J., Hannon G.J., and Joshua-Tor L. 2005. Purified Argonaute2 and an siRNA form recombinant human RISC. *Nat. Struct. Mol. Biol.* **12:** 340.

Schwarz D.S., Tomari Y., and Zamore P.D. 2004. The RNA-induced silencing complex is a Mg^{2+}-dependent endonuclease. *Curr. Biol.* **14:** 787.

Schwarz D.S., Hutvágner G., Haley B., and Zamore P.D. 2002. Evidence that siRNAs function as guides, not primers, in the *Drosophila* and human RNAi pathways. *Mol. Cell* **10:** 537.

Schwarz D.S., Hutvágner G., Du T., Xu Z., Aronin N., and Zamore P.D. 2003. Asymmetry in the assembly of the RNAi enzyme complex. *Cell* **115:** 199.

Song J.J., Smith S.K., Hannon G.J., and Joshua-Tor L. 2004. Crystal structure of Argonaute and its implications for RISC slicer activity. *Science* **305:** 1434.

Song J.J., Liu J., Tolia N.H., Schneiderman J., Smith S.K., Martienssen R.A., Hannon G.J., and Joshua-Tor L. 2003. The crystal structure of the Argonaute2 PAZ domain reveals an RNA binding motif in RNAi effector complexes. *Nat. Struct. Biol.* **10:** 1026.

Stark A., Brennecke J., Russell R.B., and Cohen S.M. 2003. Identification of *Drosophila* MicroRNA Targets. *PLoS Biol.* **1:** 397.

Tahbaz N., Kolb F.A., Zhang H., Jaronczyk K., Filipowicz W., and Hobman T.C. 2004. Characterization of the interactions between mammalian PAZ PIWI domain proteins and Dicer. *EMBO Rep.* **5:** 189.

Yan K.S., Yan S., Farooq A., Han A., Zeng L., and Zhou M.M. 2003. Structure and conserved RNA binding of the PAZ domain. *Nature* **426:** 468.

Drosha in Primary MicroRNA Processing

Y. Lee, J. Han, K.-H. Yeom, H. Jin, and V.N. Kim

School of Biological Sciences, Seoul National University, Seoul, 151-742, Korea

MicroRNA (miRNA)-mediated gene silencing is one of the major regulatory pathways in eukaryotes. Much effort has been made to identify the factors involved in the pathway, and our understanding of RNA silencing has significantly advanced in recent years. Our group has been working on some of the issues regarding miRNA biogenesis and, in this paper, we summarize what we and other workers in the field have learned thus far. The focus remains on the role of Drosha and DGCR8 in the early events of miRNA biogenesis in animals.

miRNAs are single-stranded RNAs 19–25 nucleotides in length that are generated from endogenous hairpin-shaped transcripts (Ambros et al. 2003b; Bartel 2004; Cullen 2004). miRNAs act as guide molecules in post-transcriptional gene silencing by base-pairing with the target mRNAs, which leads to translational repression and/or mRNA degradation. Through specific mRNA targeting, miRNAs form an extensive layer of gene regulatory networks. More than one-third of human genes are predicted to be targeted directly by miRNAs (Bartel and Chen 2004; Lewis et al. 2005). Recent studies have revealed the key roles of miRNAs in diverse regulatory pathways, including developmental timing control, organ development, cell differentiation, apoptosis, cell proliferation, and tumorigenesis. To unravel these networks, it is necessary to determine which target sets are controlled by a given miRNA in a certain cellular context. It is also important to elucidate how the expression of miRNA itself is regulated in those cells. Thus, understanding the miRNA biogenesis pathway is an important part of miRNA biology.

CURRENT MODEL FOR miRNA BIOGENESIS

In their seminal paper published in 1993 (Lee et al. 1993), Ambros and colleagues made the first report on a miRNA (lin-4 RNA) and provided important clues toward the biogenesis mechanism. They described the precursor of lin-4 RNA, which is approximately 70 nucleotides in length and is predicted to fold into a hairpin structure (Lee et al. 1993). It was later found that this precursor (referred to as "pre-miRNA") is processed by an RNase III type enzyme, Dicer, to generate approximately 22-nucleotide miRNA (Grishok et al. 2001). The subsequent discovery of abundant miRNAs in 2001 offered a plethora of information that paved the way for further studies on miRNA (Lagos-Quintana et al. 2001; Lau et al. 2001; Lee and Ambros 2001). One of the interesting findings was that approximately 50% of miRNAs are located in close proximity to other miRNAs (Lagos-Quintana et al. 2001; Lau et al. 2001; Mourelatos et al. 2002), which raised the possibility that these clustered miRNAs may be transcribed from a single polycistronic TU. Indeed, our detailed analysis of miRNA gene expression demonstrated that the clustered miRNAs are generated as polycistronic primary transcripts (pri-miRNAs) (Lee et al. 2002). Because pri-miRNAs are much longer than the approximately 70-nucleotide pre-miRNAs, it became clear that there must be an additional processing event prior to Dicer processing. We were able to establish an in vitro miRNA processing assay using human cell extract and proved that the additional processing indeed takes place. We also demonstrated that the catalytic activities for the first and the second processing are compartmentalized into the nucleus and the cytoplasm, respectively (Lee et al. 2002). Thus, pre-miRNA are made in the nucleus and should be transported out of the nucleus for the cytoplasmic processing to occur. These observations led to the current model for miRNA maturation (Fig. 1) (Lee et al. 2002).

Genesis of miRNA begins with transcription by RNA polymerase II (pol II) (Cai et al. 2004; Lee et al. 2004; Kim 2005). Like other pol II transcripts, pri-miRNAs are both capped and polyadenylated, although these structures do not appear to be important for miRNA processing. The main advantage of pol II-dependent transcription is that miRNA gene transcription can be modulated by a variety of pol II-associated transcription factors so that miRNA expression can be controlled elaborately at the transcription level. For instance, myogenic transcription factors such as MyoD, serum response factor, Mef2, and Myogenin were shown to regulate the transcription of muscle-specific miRNAs such as miR-1 and miR-206 (Zhao et al. 2005; Rao et al. 2006).

pri-miRNAs are usually more than several kilobases long and often exceed 10 kilobases. Thus, precise processing is pivotal to ensure the production of mature miRNA. The initial processing that occurs in the nucleus is known as "cropping" and is catalyzed by a member of the ribonuclease III family (RNase III) named Drosha (Lee et al. 2003). Drosha requires a cofactor known as DGCR8 for its activity (Gregory et al. 2004; Han et al. 2004; Landthaler et al. 2004) (known as Pasha in *Drosophila melanogaster* and *Caenorhabditis elegans*; Denli et al. 2004). Pre-miRNAs then exit the nucleus via the action of exportin-5 (Exp5) (Yi et al. 2003; Bohnsack et al. 2004; Lund et al. 2004) and are processed by Dicer, a cytoplasmic RNase III

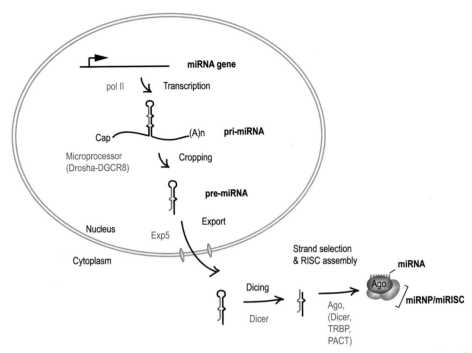

Figure 1. Model for miRNA biogenesis. miRNA genes are transcribed by a RNA polymerase II to generate the primary transcripts (pri-miRNAs). The initiation step (cropping) is mediated by the Drosha–DGCR8 complex (also known as Microprocessor complex). Drosha, as well as DGCR8, is located mainly in the nucleus. The product of the nuclear processing is ~70-nucleotide pre-miRNA, which possesses a short stem plus ~2-nucleotide 3′ overhang. This structure may serve as a signature motif that is recognized by the nuclear export factor, Exportin 5 (Exp5). pre-miRNA constitutes a transport complex together with Exp5 and its cofactor Ran (the GTP-bound form). Upon export, the cytoplasmic RNase III Dicer participates in the second processing step (dicing) to produce miRNA duplexes. The duplex is separated, and usually one strand is selected as the mature miRNAs, whereas the other strand is degraded. In *Drosophila*, R2D2 forms a heterodimeric complex with Dicer and binds to one end of the siRNA duplex. It thereby selects one strand of the duplex.

nuclease, yielding the aproximately 22-nucleotide miRNA duplex (Bernstein et al. 2001; Grishok et al. 2001; Hutvagner et al. 2001; Ketting et al. 2001; Knight and Bass 2001). One strand of the duplex is subsequently loaded onto the RNA induced silencing complex (RISC), which actively executes the RNA silencing procedure (Khvorova et al. 2003; Schwarz et al. 2003).

DROSHA AND OTHER RNASE III PROTEINS

Drosha was originally described as a gene that is located next to the RNase H gene in *Drosophila* (Filippov et al. 2000). An early study on the human Drosha homolog (initially referred to as human RNase III protein) showed that Drosha is a nuclear protein and that the knockdown of Drosha by antisense oligonucleotides results in the accumulation of rRNA precursors (Wu et al. 2000). The main substrates for Drosha, however, were later found to be pri-miRNAs (Lee et al. 2003). We have not been able to detect significant accumulation of rRNA precursors when Drosha was depleted by RNAi (J. Han and V.N. Kim, unpubl.). Thus far, no other RNA species apart from pri-miRNAs has been found to be processed in vivo by Drosha.

Drosha is conserved only in animals (Fig. 2) (Filippov et al. 2000; Wu et al. 2000; Fortin et al. 2002). Although plants express diverse miRNAs, they do not possess a

Drosha homolog, suggesting that the miRNA biogenesis pathway is distinct in plants. In *Arabidopsis* there are four Dicer homologs, and Dicer-like 1 is known to be responsible for miRNA processing in the nucleus (Kurihara and Watanabe 2004).

Drosha is a large nuclear protein of 130–160 kD (160 kD in human) with multiple domains (Fig. 2). Drosha contains two tandem RNase III domains (RIIIDs) and a double-stranded RNA-binding domain (dsRBD). The amino-terminal portion of Drosha contains a proline-rich region as well as a serine/arginine-rich region. To investigate the biochemical role of each domain, we carried out mutagenesis studies (Fig. 2). Deletions of the amino-terminal region that remove the P-rich region (ΔN220) and most of the RS-rich region (ΔN390) did not affect the enzymatic activity (Fig. 2). The RS-rich region appears to possess the nuclear localization signal because the mutant ΔN390 fails to localize to the nucleus (J. Han and V.N. Kim, unpubl.). The remainder of the mutants (ΔN490, ΔC432, and ΔC114) turned out to be inactive in our assay, indicating that the middle region, as well as the RIIIDs and the dsRBD, is required for pri-miRNA processing. Deletion of the middle region (ΔN490) and the RNase III domains (ΔC432) affected the binding activity toward DGCR8, indicating that DGCR8 interacts with Drosha through the middle region and the RIIIDs of the Drosha protein (Han et al. 2004).

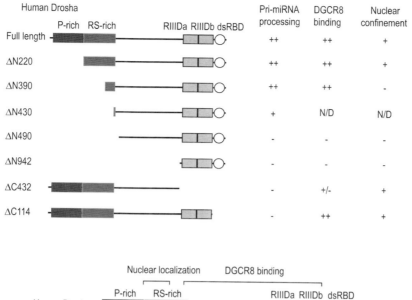

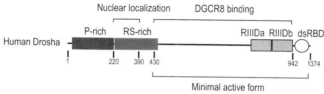

Figure 2. Domain organization of the roles of the domains of Drosha. The RIIID is the catalytic domain that executes the endonucleolytic reaction. The dsRBD is a well-conserved motif in many dsRNA-binding proteins of diverse functions. The "P-rich" indicates a proline-rich region, whose biochemical significance remains unknown. The "RS-rich" indicates a region that is abundant in arginine and serine. The Drosha mutants were assayed for pri-miRNA processing, DGCR8 binding, and nuclear confinement, N/D, not determined.

Both RIIID and dsRBD are common in all RNase III type enzymes, although domain organizations are different in other RNase III proteins. The differences in domain organization are often used to classify RNase III type proteins. Class I proteins include RNase III proteins found in bacteria and yeasts. Each of them contains one RIIID and one dsRBD. Drosha homologs constitute class II and possess two RIIIDs and a dsRBD. Class III includes Dicer homologs that are conserved in *Schizosaccharomyces pombe*, plants, and animals. Dicer homologs are approximately 200 kD and contain multiple domains apart from two RIIIDs and a dsRBD. Dicer has a long amino terminus containing a DExH RNA helicase/ATPase domain, as well as DUF283 and the PAZ domain. The PAZ domain is also found in a group of highly conserved proteins, referred to as Argonaute proteins. Structural and biochemical studies suggest that the PAZ domain binds to the 3′ protruding end of small RNA (Song et al. 2003; Yan et al. 2003; Lingel et al. 2004). Deletion mutagenesis analyses showed that the amino-terminal region containing the DExH RNA helicase/ATPase domain is dispensable for its catalytic activity, whereas the DUF283 domain whose function remains unknown appears to be critical for processing, as the mutant lacking this domain (ΔDUF) lost activity in vitro (Lee et al. 2006).

RNase III proteins show a high degree of conservation in their basic action mechanisms. Two RIIIDs interact with each other to make a dimer. In the interface between two RIIIDs, a single processing center is formed, which contains two closely placed catalytic sites. Each of the two catalytic sites cleaves each strand of dsRNA (Blaszczyk et al. 2001; Zhang et al. 2004). Both Drosha and Dicer form an intramolecular dimer of two RIIIDs, whereas class I enzymes containing a single RIIID make intermolecular dimers (Han et al. 2004; Zhang et al. 2004). The carboxy-terminal RIIID (RIIIDb), proximal to the dsRBD, cleaves the 5′ strand of the hairpin, whereas the other RIIID (RIIIDa) cleaves the 3′ strand. This "single processing center" model is supported by a recent structural study on a Dicer homolog (Macrae et al. 2006).

MICROPROCESSOR COMPLEX: DROSHA AND DGCR8/PASHA

Drosha forms a large complex that weighs about 500 kD in *Drosophila* (Denli et al. 2004) or about 650 kD in humans (Gregory et al. 2004; Han et al. 2004). In this complex, known as the "Microprocessor" (Denli et al. 2004; Gregory et al. 2004), Drosha interacts with its cofactor, DGCR8/Pasha (Denli et al. 2004; Gregory et al. 2004; Han et al. 2004; Landthaler et al. 2004). Neither recombinant DGCR8 nor Drosha alone is active in pri-miRNA processing, whereas combining these two proteins restores pri-miRNA processing activity, indicating that both proteins play essential roles in pri-miRNA processing (Gregory et al. 2004; Han et al. 2004). Both Drosha and DGCR8 are capable of oligomerization, although the functional significance of oligomerization remains unclear (Han et al. 2004). The size of the complex suggests that Microprocessor may contain multiple copies of the components. Another non-mutually exclusive possibility is that Microprocessor may have other uncharacterized components that may regulate the enzyme activity. In fact, immunoprecipitation of Drosha protein yielded a number of other proteins apart from DGCR8 (Gregory et al. 2004; Han et al. 2004).

The human DGCR8 gene (DiGeorge Syndrome Critical Region Gene 8) is located on chromosome 22q11 and is expressed ubiquitously from fetus to adult (Shiohama et al. 2003). Monoallelic deletion of this genomic region is associated with DiGeorge syndrome (Goldberg et al. 1993). It remains to be determined whether DGCR8 plays a role in DiGeorge syndrome.

DGCR8/Pasha is an approximately 120-kD protein that contains two dsRBDs at the carboxyl terminus and a putative WW domain (also termed Rsp5/wwp) in its middle region (Sudol 1996). To determine the functional domains of DGCR8, we generated deletion mutants of DGCR8 (Fig. 3) (Yeom et al. 2006). A relatively small region of the DGCR8 protein (residues 484–750) was found to be sufficient to facilitate pri-miRNA processing in vitro. This minimal region of DGCR8 contains both the dsRBDs and the carboxy-terminal Drosha-binding domain. The amino-terminal region was not critical for processing but was shown to be important for nuclear localization of DGCR8. Because the WW domain is known to interact with proline-rich motifs (Macias et al.

2002) and Drosha is proline-rich at its amino terminus, it was initially believed that the WW domain of DGCR8 may interact with the P-rich region of Drosha. However, it turned out that the domain responsible for Drosha binding is located at the carboxyl terminus of DGCR8 (Fig. 3).

DGCR8 is capable of direct interaction with pri-miRNA (Han et al. 2006; Yeom et al. 2006). On the contrary, we could not detect any significant pri-miRNA-binding activity for Drosha in any of our assays (gel mobility shift assays, UV-crosslinking experiments, immunoprecipitations, and GST pull-downs) (Han et al. 2006). Thus, DGCR8 may be the primary factor that recognizes the pri-miRNA structure, whereas Drosha interacts with the RNA substrates only transiently during catalysis. In addition, DGCR8 appears to stabilize the Drosha protein through its carboxy-terminal Drosha-binding domain (Yeom et al. 2006).

HOW DOES MICROPROCESSOR RECOGNIZE ITS SUBSTRATES?

Despite the similarities in their basic modes of action, RNase III proteins are different in their substrate specificities. Human Dicer tends to act on any dsRNA with a simple preference toward the terminus of the molecule and to produce approximately 22-nucleotide fragments progressively from the terminus (Bernstein et al. 2001;

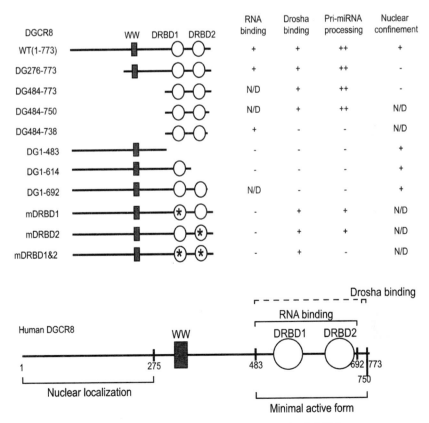

Figure 3. Domain mapping of human DGCR8. DGCR8 contains two dsRBDs. The DGCR8 mutants were assayed for pri-miRNA binding, Drosha binding, pri-miRNA processing, and nuclear confinement activity. Asterisks represent the sites of point mutagenesis. N/D, not determined.

Provost et al. 2002; Zhang et al. 2002). The PAZ domain of Dicer interacts with the 3′ overhang at the terminus and determines the processing site in a ruler-like fashion that measures approximately 22-nucleotide segments away from the terminus (Song et al. 2003; Yan et al. 2003; Lingel et al. 2004; Macrae et al. 2006).

Microprocessor also acts like a molecular ruler to determine the cleavage site (Han et al. 2006). A typical metazoan pri-miRNA consists of a stem of about 33 bp, with a terminal loop and flanking segments. Drosha cleaves both natural and artificial substrates at a site about 11 bp away from the ssRNA–dsRNA junction (SD junction) (Fig. 4). Manipulating the length of the outer stem affects cleavage-site selection, implicating the existence of a molecular device that measures the distance from the SD junction. A series of mutagenesis and biochemical analyses revealed that the flanking ssRNA segments are critical for processing (Han et al. 2006). Our data also show that the terminal loop is rather irrelevant to pri-miRNA processing (Han et al. 2006). We note, however, that pri-miRNA with a small loop is less active in processing. A large terminal loop may act as a flexible ssRNA and may contribute to tight binding to Microprocessor. It is also conceivable that a small terminal loop may impose structural constraints on the stem and affect processing.

DGCR8 interacts with pri-miRNAs both directly and specifically as determined by UV-cross-linking experiments (Han et al. 2006). The flanking ssRNA segments are vital for this binding to occur and the stem of over 33 bp is also required for efficient binding. Thus, DGCR8 may function as the molecular anchor that measures the distance from the dsRNA–ssRNA junction.

Pri-miRNA processing may consist of two sequential steps; substrate recognition and catalytic reaction (Fig. 4). First, DGCR8 may recognize the substrate by tight anchoring at the SD junction and interact with the approximately 33-bp stem. Drosha may not be in direct contact with RNA at this stage. After this "pre-cleavage" complex is made, Drosha may interact transiently with the stem,

and the processing center of the enzyme may be located at about 11 bp from the SD junction.

Because some large terminal loops can be seen as unstructured ssRNA segments, pri-miRNAs may be considered to be of "ssRNA–dsRNA (~3 helical turns)–ssRNA" structure. Because this structure is rather symmetrical, Microprocessor appears to recognize the terminal loop as ssRNA and binds to the stem-loop in an opposite orientation. In this case, cleavage can occur at an alternative site (at ~11 bp from the terminal loop). The processing at this alternative site would be "abortive" because the cleavage product does not contain miRNA sequences in full. The efficiency of abortive processing is usually much lower than that of productive processing in pri-miRNA processing.

CONCLUSIONS

A number of expression profiling studies have been conducted in recent years, and they indicate that most miRNAs are under the control of developmental and/or tissue-specific signaling. miRNA expression may be regulated at multiple steps of RNA biogenesis. Imminent questions concern which step is controlled and how this control is achieved. Some miRNAs seem to be controlled at the posttranscriptional level. For instance, *C. elegans* miR-38 is expressed only in the embryo, whereas the precursor (pre-miR-38) is ubiquitously detected, indicating that the maturation of pre-miR-38 may be temporally regulated (Ambros et al. 2003a). It is possible that the nuclear export of pre-miR-38 may be controlled by a certain developmental signal. Alternatively, Dicer processing may be repressed until a certain stage.

A more recent study indicated that posttranscriptional regulation of miRNA may be more extensive than previously anticipated (Thomson et al. 2006). For instance, let-7 RNAs are expressed only in differentiated cells although their primary transcripts are abundant even in undifferentiated cells. In addition, the analysis of gene

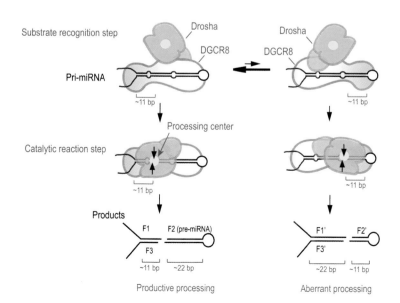

Figure 4. A ssRNA–dsRNA junction anchoring model for the processing of pri-miRNA. DGCR8 may play a major role in substrate recognition by directly anchoring at the ssRNA–dsRNA junction. DGCR8 also interacts with the stem of ~33 bp and the terminal loop for a full activity, although the terminal loop structure is not critical for DGCR8 binding and cleavage reaction. After the initial recognition step, Drosha may transiently interact with the substrate for catalysis. The processing center (*yellow circle*) of Drosha is placed at ~11 bp from the basal segments.

expression in human tumors showed that a number of miRNAs are down-regulated in cancer, although their pri-miRNAs are expressed at a relatively high level. These data suggest that the cropping step may be controlled dynamically during cell differentiation and tumorigenesis. It will be important to identify additional factors involved in the biogenesis pathway. For instance, the Drosha complex (Microprocessor) may contain additional components apart from DGCR8/Pasha, and the uncharacterized factors may regulate the activity and/or specificity of Drosha.

ACKNOWLEDGMENTS

We are grateful to members of our laboratory. This work was supported by a Molecular and Cellular BioDiscovery Research Program grant (2004-01749) from the Ministry of Science and Technology, and a Korean Research Foundation Grant (KRF-2005-005-J16001).

REFERENCES

Ambros V., Lee R.C., Lavanway A., Williams P.T., and Jewell D. 2003a. MicroRNAs and other tiny endogenous RNAs in C. elegans. Curr. Biol. 13: 807.
Ambros V., Bartel B., Bartel D.P., Burge C.B., Carrington J.C., Chen X., Dreyfuss G., Eddy S.R., Griffiths-Jones S., Marshall M., Matzke M., Ruvkun G., and Tuschl T. 2003b. A uniform system for microRNA annotation. RNA 9: 277.
Bartel D.P. 2004. MicroRNAs: Genomics, biogenesis, mechanism, and function. Cell 116: 281.
Bartel D.P. and Chen C.Z. 2004. Micromanagers of gene expression: The potentially widespread influence of metazoan microRNAs. Nat. Rev. Genet. 5: 396.
Bernstein E., Caudy A.A., Hammond S.M., and Hannon G.J. 2001. Role for a bidentate ribonuclease in the initiation step of RNA interference. Nature 409: 363.
Blaszczyk J., Tropea J.E., Bubunenko M., Routzahn K.M., Waugh D.S., Court D.L., and Ji X. 2001. Crystallographic and modeling studies of RNase III suggest a mechanism for double-stranded RNA cleavage. Structure 9: 1225.
Bohnsack M.T., Czaplinski K., and Gorlich D. 2004. Exportin 5 is a RanGTP-dependent dsRNA-binding protein that mediates nuclear export of pre-miRNAs. RNA 10: 185.
Cai X., Hagedorn C.H., and Cullen B.R. 2004. Human microRNAs are processed from capped, polyadenylated transcripts that can also function as mRNAs. RNA 10: 1957.
Cullen B.R. 2004. Transcription and processing of human microRNA precursors. Mol. Cell 16: 861.
Denli A.M., Tops B.B., Plasterk R.H., Ketting R.F., and Hannon G.J. 2004. Processing of primary microRNAs by the Microprocessor complex. Nature 432: 231.
Filippov V., Solovyev V., Filippova M., and Gill S.S. 2000. A novel type of RNase III family proteins in eukaryotes. Gene 245: 213.
Fortin K.R., Nicholson R.H., and Nicholson A.W. 2002. Mouse ribonuclease III. cDNA structure, expression analysis, and chromosomal location. BMC Genomics 3: 26.
Goldberg R., Motzkin B., Marion R., Scambler P.J., and Shprintzen R.J. 1993. Velo-cardio-facial syndrome: A review of 120 patients. Am. J. Med. Genet. 45: 313.
Gregory R.I., Yan K.P., Amuthan G., Chendrimada T., Doratotaj B., Cooch N., and Shiekhattar R. 2004. The Microprocessor complex mediates the genesis of microRNAs. Nature 432: 235.
Grishok A., Pasquinelli A.E., Conte D., Li N., Parrish S., Ha I., Baillie D.L., Fire A., Ruvkun G., and Mello C.C. 2001. Genes and mechanisms related to RNA interference regulate expression of the small temporal RNAs that control C. elegans developmental timing. Cell 106: 23.

Han J., Lee Y., Yeom K.H., Kim Y.K., Jin H., and Kim V.N. 2004. The Drosha-DGCR8 complex in primary microRNA processing. Genes Dev. 18: 3016.
Han J., Lee Y., Yeom K.H., Nam J.W., Heo I., Rhee J.K., Sohn S.Y., Cho Y., Zhang B.T., and Kim V.N. 2006. Molecular basis for the recognition of primary microRNAs by the Drosha-DGCR8 complex. Cell 125: 887.
Hutvagner G., McLachlan J., Pasquinelli A.E., Balint E., Tuschl T., and Zamore P.D. 2001. A cellular function for the RNA-interference enzyme Dicer in the maturation of the let-7 small temporal RNA. Science 293: 834.
Ketting R.F., Fischer S.E., Bernstein E., Sijen T., Hannon G.J., and Plasterk R.H. 2001. Dicer functions in RNA interference and in synthesis of small RNA involved in developmental timing in C. elegans. Genes Dev. 15: 2654.
Khvorova A., Reynolds A., and Jayasena S.D. 2003. Functional siRNAs and miRNAs exhibit strand bias. Cell 115: 209.
Kim V.N. 2005. MicroRNA biogenesis: Coordinated cropping and dicing. Nat. Rev. Mol. Cell Biol. 6: 376.
Knight S.W. and Bass B.L. 2001. A role for the RNase III enzyme DCR-1 in RNA interference and germ line development in Caenorhabditis elegans. Science 293: 2269.
Kurihara Y. and Watanabe Y. 2004. Arabidopsis micro-RNA biogenesis through Dicer-like 1 protein functions. Proc. Natl. Acad. Sci. 101: 12753.
Lagos-Quintana M., Rauhut R., Lendeckel W., and Tuschl T. 2001. Identification of novel genes coding for small expressed RNAs. Science 294: 853.
Landthaler M., Yalcin A., and Tuschl T. 2004. The human DiGeorge syndrome critical region gene 8 and its D. melanogaster homolog are required for miRNA biogenesis. Curr. Biol. 14: 2162.
Lau N.C., Lim L.P., Weinstein E.G., and Bartel D.P. 2001. An abundant class of tiny RNAs with probable regulatory roles in Caenorhabditis elegans. Science 294: 858.
Lee R.C. and Ambros V. 2001. An extensive class of small RNAs in Caenorhabditis elegans. Science 294: 862.
Lee R.C., Feinbaum R.L., and Ambros V. 1993. The C. elegans heterochronic gene lin-4 encodes small RNAs with antisense complementarity to lin-14. Cell 75: 843.
Lee Y., Jeon K., Lee J.T., Kim S., and Kim V.N. 2002. MicroRNA maturation: Stepwise processing and subcellular localization. EMBO J. 21: 4663.
Lee Y., Hur I., Park S.Y., Kim Y.K., Suh M.R., and Kim V.N. 2006. The role of PACT in the RNA silencing pathway. EMBO J. 25: 522.
Lee Y., Kim M., Han J., Yeom K.H., Lee S., Baek S.H., and Kim V.N. 2004. MicroRNA genes are transcribed by RNA polymerase II. EMBO J. 23: 4051.
Lee Y., Ahn C., Han J., Choi H., Kim J., Yim J., Lee J., Provost P., Radmark O., Kim S., and Kim V.N. 2003. The nuclear RNase III Drosha initiates microRNA processing. Nature 425: 415.
Lewis B.P., Burge C.B., and Bartel D.P. 2005. Conserved seed pairing, often flanked by adenosines, indicates that thousands of human genes are microRNA targets. Cell 120: 15.
Lingel A., Simon B., Izaurralde E., and Sattler M. 2004. Nucleic acid 3'-end recognition by the Argonaute2 PAZ domain. Nat. Struct. Mol. Biol. 11: 576.
Lund E., Guttinger S., Calado A., Dahlberg J.E., and Kutay U. 2004. Nuclear export of microRNA precursors. Science 303: 95.
Macias M.J., Wiesner S., and Sudol M. 2002. WW and SH3 domains, two different scaffolds to recognize proline-rich ligands. FEBS Lett. 513: 30.
Macrae I.J., Zhou K., Li F., Repic A., Brooks A.N., Cande W.Z., Adams P.D., and Doudna J.A. 2006. Structural basis for double-stranded RNA processing by Dicer. Science 311: 195.
Mourelatos Z., Dostie J., Paushkin S., Sharma A., Charroux B., Abel L., Rappsilber J., Mann M., and Dreyfuss G. 2002. miRNPs: A novel class of ribonucleoproteins containing numerous microRNAs. Genes Dev. 16: 720.

Provost P., Dishart D., Doucet J., Frendewey D., Samuelsson B., and Radmark O. 2002. Ribonuclease activity and RNA binding of recombinant human Dicer. *EMBO J.* **21:** 5864.

Rao P.K., Kumar R.M., Farkhondeh M., Baskerville S., and Lodish H.F. 2006. Myogenic factors that regulate expression of muscle-specific microRNAs. *Proc. Natl. Acad. Sci.* **103:** 8721.

Schwarz D.S., Hutvagner G., Du T., Xu Z., Aronin N., and Zamore P.D. 2003. Asymmetry in the assembly of the RNAi enzyme complex. *Cell* **115:** 199.

Shiohama A., Sasaki T., Noda S., Minoshima S., and Shimizu N. 2003. Molecular cloning and expression analysis of a novel gene DGCR8 located in the DiGeorge syndrome chromosomal region. *Biochem. Biophys. Res. Commun.* **304:** 184.

Song J.J., Liu J., Tolia N.H., Schneiderman J., Smith S.K., Martienssen R.A., Hannon G.J., and L. Joshua-Tor. 2003. The crystal structure of the Argonaute2 PAZ domain reveals an RNA binding motif in RNAi effector complexes. *Nat. Struct. Biol.* **10:** 1026.

Sudol M. 1996. Structure and function of the WW domain. *Prog. Biophys. Mol. Biol.* **65:** 113.

Thomson J.M., Newman M., Parker J.S., Morin-Kensicki E.M., Wright T., and Hammond S.M. 2006. Extensive post-transcriptional regulation of microRNAs and its implications for cancer. *Genes Dev.* **20:** 2202.

Wu H., Xu H., Miraglia L.J., and Crooke S.T. 2000. Human RNase III is a 160-kDa protein involved in preribosomal RNA processing. *J. Biol. Chem.* **275:** 36957.

Yan K.S., Yan S., Farooq A., Han A., Zeng L., and Zhou M.M. 2003. Structure and conserved RNA binding of the PAZ domain. *Nature* **426:** 468.

Yeom K.H., Lee Y., Han J., Suh M.R., and Kim V.N. 2006. Characterization of DGCR8/Pasha, the essential cofactor for Drosha in primary miRNA processing. *Nucleic Acids Res.* (in press).

Yi R., Qin Y., Macara I.G., and Cullen B.R. 2003. Exportin-5 mediates the nuclear export of pre-microRNAs and short hairpin RNAs. *Genes Dev.* **17:** 3011.

Zhang H., Kolb F.A., Brondani V., Billy E., and Filipowicz W. 2002. Human Dicer preferentially cleaves dsRNAs at their termini without a requirement for ATP. *EMBO J.* **21:** 5875.

Zhang H., Kolb F.A., Jaskiewicz L., Westhof E., and Filipowicz W. 2004. Single processing center models for human Dicer and bacterial RNase III. *Cell* **118:** 57.

Zhao Y., Samal E., and Srivastava D. 2005. Serum response factor regulates a muscle-specific microRNA that targets Hand2 during cardiogenesis. *Nature* **436:** 214.

Substrate Selectivity of Exportin 5 and Dicer in the Biogenesis of MicroRNAs

E. LUND AND J.E. DAHLBERG

Biomolecular Chemistry, University of Wisconsin-Madison, Madison, Wisconsin 53706

Each step in the biogenesis of microRNAs (miRNAs) depends on recognition of a correct substrate and efficient transport or processing of that RNA. Exportin5 (Exp5) and Dicer are proteins that mediate two key steps in this cascade, the nuclear export and cytoplasmic processing of microRNA precursor (pre-miRNAs). *Xenopus laevis* oocytes, eggs, and embryos constitute convenient experimental systems in which to study the substrate specificity of these proteins because specific RNAs or proteins can be injected directly into different subcellular compartments. We have used the *Xenopus* system and in vitro processing to define and compare the specificities of Exp5 and Dicer. Although both proteins act on many of the same substrates, we show that they recognize different structure elements of these RNAs. Our studies also revealed several unexpected activities. For example, Exp5 can mediate export of unspliced pre-mRNAs and excised lariat introns if these RNAs contain an aptamer sequence that itself is an Exp5 export substrate. Finally, we demonstrate that maturation of *Xenopus* oocytes into eggs leads to a large increase in Dicer activity, suggesting that miRNA biogenesis is subject to developmental control.

microRNAs (miRNAs) function in posttranscriptional control of gene expression through RISC (RNA-induced silencing complex)–mediated inhibition of translation, often accompanied by destabilization of the targeted mRNAs (for review, see Filipowicz et al. 2005; Valencia-Sanchez et al. 2006). miRNA biogenesis involves several coordinated processing and intracellular transport steps, some of which appear to be controlled during development and differentiation or in response to environmental stimuli (for review, see Bartel 2004; Du and Zamore 2005). These steps include (1) synthesis of primary transcripts containing one or more miRNAs, (2) processing of these transcripts by the RNase-III-like endonuclease Drosha (and associated proteins) to produce approximately 65- to 75-nucleotide-long incompletely base-paired hairpin RNAs (pre-miRNAs), (3) export of pre-miRNAs from the nucleus to the cytoplasm by the export factor exportin-5 (Exp5), (4) cleavage of the pre-miRNAs in the cytoplasm by a second double-strand-specific RNase-III-like processing enzyme, Dicer, into duplexes about 22 nucleotides long, and (5) selection of one strand of the product duplex as the miRNA and incorporation into a ribonucleoprotein complex, the RISC (for review, see Murchison and Hannon 2004; Kim 2005). Here we describe our studies on the activities, substrate specificities, and developmental control of two of the key proteins that participate in this process, Exp5 and Dicer (steps 3 and 4).

We have asked what features of pre-miRNA substrates contribute to their export by Exp5 and their processing by Dicer. These hairpin RNAs have imperfectly base-paired stem regions about 22 nucleotides long, flanked by free 5′ and 3′ ends and an unpaired loop region. To test the importance of these structural elements in export and processing, we prepared ^{32}P-labeled RNA substrates (by in vitro transcription; Grimm et al. 1997) in which these elements were either altered or deleted. We then monitored the fates of the variant RNAs in vivo, using microinjected *Xenopus* oocytes and embryos, or in vitro, using human recombinant Dicer.

EXPORTIN 5

We and other investigators previously demonstrated that Exp5 is responsible for export of pre-miRNAs from the nucleus to the cytoplasm of both *X. laevis* oocytes and mammalian cultured cells (Yi et al. 2003; Bohnsack et al. 2004; Lund et al. 2004), and that Exp5 binds directly to its RNA cargo in a RanGTP-dependent manner. Preferred export substrates resemble Drosha processing products (Lund et al. 2004; Zeng and Cullen 2004). Pre-miRNAs and other RNAs that are recognized by this export receptor (e.g., tRNA) all have a high degree of secondary structure, including short helices, indicating that double-stranded RNA structure is important to the interaction (Bohnsack et al. 2002; Calado et al. 2002; Gwizdek et al. 2003). Here, we asked what other features of the RNA contribute to its ability to serve as an export substrate for Exp5 in *Xenopus* oocytes.

Pre-miRNA-related Export Substrates

To determine whether Exp5 interacts with the free ends of the RNA cargos, we generated a pre-miRNA-like molecule that lacked free ends. This was done by ligating the ends of *Drosophila melanogaster* pre-let-7 RNA to each other, making a circular RNA (Fig. 1, left; top), which presumably maintained its secondary structure (Carrara et al. 1995). Upon injection into *Xenopus* oocyte nuclei, the gel-purified circular RNA was exported as efficiently as the linear pre-miRNA (Fig. 1, right; top panel). Likewise, a short imperfect miRNA–miRNA* duplex, generated by treatment of a human pre-miR-31 with Dicer, was exported as rapidly as was pre-miR-31

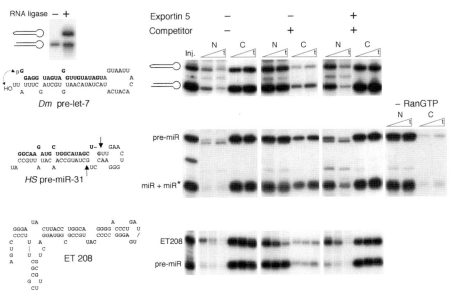

Figure 1. Substrate specificities of *Xenopus* Exp5. Structures on the left side show RNA substrates used in nuclear export experiments. The circular form of *Dm* pre-let-7 (*top*) was generated by treatment with RNA ligase (Carrara et al. 1995). Pre-miR-31 was digested with human recombinant Dicer, to produce the short miR-31/miR-31* duplex (indicated by *arrows*). ET-208 RNA is an aptamer selected for its ability to be exported in the presence of the M-protein of VSV (Grimm et al. 1997). Polyacrylamide gels display ^{32}P-labeled RNAs that had been injected into *Xenopus* oocyte nuclei in the absence or presence of 0.5 pmole of unlabeled pre-miR-31 (*top* and *middle panels*) or ET-208 RNA (*bottom panel*) competitor RNAs and recombinant Exp5, or into oocytes lacking RanGTP (**middle panel*, *right lanes*) (Lund et al. 2004). Inj. denotes the injected RNAs, and N (nucleus) and C (cytoplasm) show the intracellular distributions of the RNAs with time (*t*); the times were 25 and 40 min, 30 and 60 min, and 30, 60, and 120 min for the top, middle, and bottom panels, respectively.

(Fig. 1, right; middle panel). Export of both the circular RNA and the miRNA–miRNA* duplex was mediated by Exp5, as shown by the ability of an excess of unlabeled pre-miR-31 to inhibit it, and the ability of exogenous human recombinant Exp5 to reverse this inhibition (Fig. 1, right; top and middle panels).

We confirmed that the appearance of the miRNA–miRNA* duplex RNA in the cytoplasm was due to active transport by depleting oocytes of RanGTP, an essential cofactor in Exp5-mediated nuclear export. This was accomplished by nuclear injection of recombinant RanGAP (to activate the GTPase function of Ran) and RanT24N (to inhibit RanGEF, the GTP–GDP exchange factor) (Izaurralde et al. 1997). In the absence of RanGTP, more than 95% of the short RNA duplex, as well as pre-miR-31, remained in the nucleus for at least 60 minutes (Fig.1, far right of middle panel), demonstrating that, like pre-miRNAs, the short duplex RNA is exported in a RanGTP-dependent active process. Thus, neither the free ends nor the terminal loop region of a pre-miRNA is required for export by Exp5. This finding explains how Exp5 keeps duplex siRNAs out of the nucleus (Ohrt et al. 2006).

Other Structured Exp5 Substrates

Structured RNAs that are unrelated to pre-miRNAs can also be recognized and exported by Exp5. A novel example is ET-208 RNA (Fig. 1, left; bottom), an artificial, highly structured RNA aptamer that we selected from a library of sequences because of its ability to be exported from oocyte nuclei in the presence of the matrix protein of vesicular

stomatitis virus (M protein) (Grimm et al. 1997). Although M protein potently inhibits export of both snRNAs and mRNAs (Petersen et al. 2000), it does not affect export of pre-miR-31 (not shown), suggesting that its export receptor, Exp5, could also mediate export of ET-208 RNA. Indeed, when co-injected into the same oocyte nuclei, ET208 RNA and pre-miR-31 strongly competed with each other for export, and this competition was obviated by injection of exogenous Exp5 (Fig. 1, bottom panel and data not shown). Moreover, we found that Exp5 binds directly to ET-208 RNA in a RanGTP-dependent manner (not shown).

Similarly, other workers have shown that the highly structured small RNAs, adenovirus VA1 RNA and human Y1 RNA (hY1) (Gwizdek et al. 2003), are export substrates for Exp5, further substantiating the versatility of this RNA-binding export receptor. However, we have found that some Exp5-mediated export (e.g., hY1 RNA) is subject to inhibition by M protein (Grimm et al. 1997; Rutjes et al. 2001; data not shown). This differential sensitivity of Exp5-mediated export to M protein may reflect the fact that ET-208 RNA was selected solely for its ability to be exported in the presence of the inhibitor, whereas hY1 RNA is complexed with specific RNA–binding protein(s) that might act as nuclear retention factors.

Both ET-208 RNA and pre-miR-31 are exported even when they lack free 5′ or 3′ ends (Fig. 1) (Grimm et al. 1997), suggesting that Exp5 recognizes internal structure(s) of the RNAs. Therefore, we asked whether these RNAs could serve as *cis*-acting elements to promote nuclear export of other RNA sequences. For this, we inserted ET-208 RNA or pre-miR-31 sequences into the intron of an AdML pre-mRNA model substrate. After injection of the chimeric pre-mRNAs into oocyte nuclei, we monitored the intracellular distribution of the unspliced pre-mRNAs and excised intron lariats. As with other splicing/export systems, and as previously shown in oocytes (Pasquinelli et al. 1997), the AdML pre-mRNA and the excised intron lariat that lacked inserts were retained in the nucleus while spliced mRNA was exported (Fig. 2, left panel). However, both the chimeric pre-mRNA and the excised intron lariat were exported if they contained ET-208 RNA sequences (center panel); this export was dependent on Exp5, as it was inhibited by excess unlabeled competitor pre-miRNA and restored by exogenous Exp5 (data not shown). In contrast, the inserted pre-miR-31 sequence was unable to promote export of either the pre-mRNA or the excised intron lariat (Fig. 2, right panel). Thus, unlike ET-208 RNA, the structure of pre-miR-31 apparently is not recognized when embedded within a larger transcript, even though free ends are not needed for its export.

Thus, Exp5 is a versatile export receptor that recognizes RNA structure rather than the sequences or free ends of the substrate. The ability of Exp5 to recognize and export chimeric RNAs containing the ET-208 RNA sequence raises the possibility that this exportin-adapter pair could be used for nuclear export of other molecules.

DICER

Upon being exported to the cytoplasm by Exp5, pre-miRNAs are processed by Dicer to produce an imperfect RNA duplex (one strand of which will become the miRNA) with a 2-nucleotide single-stranded 3' extension on each of its approximately 21- to 22-nucleotide-long strands (Bernstein et al. 2001; Lee et al. 2002; Zhang et al. 2002, 2004). We investigated features of potential pre-miRNA substrates that might affect their ability to be processed by Dicer both in vitro, using human recombinant

Dicer, and in vivo, using *Xenopus* oocytes and early embryos. The substrates, which resembled *Drosophila* pre-let-7 or human pre-miR-31, differed from each other in terms of the character of their ends, as well as the lengths of their duplex regions and terminal loops. In all of these experiments, the indicated cleavage sites were determined by direct nucleotide sequence analysis of the Dicer digestion products (not shown).

Cleavage of Pre-miRNAs by Dicer In Vitro

Free ends of Dicer substrates. We first asked whether free ends of the pre-miRNAs are needed for cleavage, by incubating the linear and circularized forms of pre-let-7 RNA (see Fig. 1) with human recombinant Dicer and analyzing the cleavage products by polyacrylamide gel electrophoresis. As expected, the control RNA with free ends was efficiently digested to yield fragments about 22 and 23 nucleotides long, arising from the stem, and a shorter fragment, about 14 nucleotides long, generated by cleavages near the end of the stem, at either side of the loop. In contrast, the circularized RNA was completely resistant to cleavage by Dicer (Fig. 3, top left panel), showing that at least one end of the RNA stem-loop structure is required for cleavage by Dicer. An earlier report indicating that free ends were not needed for Dicer cleavage used a substrate whose ends were base-paired but not ligated together (Zhang et al. 2002).

To study the effects of ends on the cleavage, we made several variants of pre-let7 RNA and pre-miR-31 (Fig. 3, right). First, we compared the products generated by Dicer digestion of pre-miR-31 with those of similar RNAs but with extra unpaired nucleotides at either the 5' or 3' end (Fig. 3, RNAs 8, 9, and 10). The substrate with an extended 5' end yielded the wild-type array of products (Fig. 3, bottom), whereas the substrate with a similar extension at its 3' end yielded a complex mixture of products, and digestion was less efficient. The observed major sites of cleavage (denoted by horizontal bars on structures shown in Fig. 3, right) indicate that Dicer interacts with the free 3' end and cuts the duplex at a fixed distance (~22 nucleotides away) (see Zhang et al. 2004).

Effects of stem length and sequence on Dicer cleavage. If the sites of cleavage are determined solely with reference to the 3' extension at the base of the stem, pre-miRNA substrates with different length duplexes should be processed with similar efficiencies and yield products of predictable size. To test whether that was the case, we analyzed a series of synthetic substrates based on the sequence of *Drosophila* pre-let-7 (Fig. 3, right top) that were identical in and around the loop region but differed from each other by insertion of extra base pairs in various positions of the stem. All substrates with longer stems (except RNA 2) were cleaved less efficiently and, in most cases, yielded a heterogeneous array of products (Fig. 3, top; right panel), showing that the length of the stem also influences cleavage by Dicer. Moreover, digestion of substrates that had the same length but differed in the location of base-pair insertions resulted in different patterns of products (e.g., compare lanes 5 and 6), indicating that the

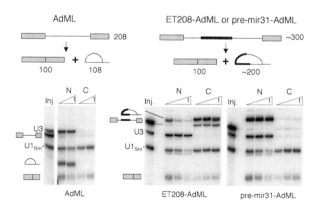

Figure 2. Exp5-mediated export of unspliced AdML pre-mRNA and its excised intron. Oocytes were injected in the nucleus with mixtures of control U3 and U1$_{Sm}$- RNAs and ^{32}P-labeled pre-AdML RNA (Pasquinelli et al. 1997) with either no insert (*left panels*) or with ET-208 RNA or pre-miR-31 sequences (*black bar*) within the intron (*middle* and *right panels*, respectively). After 105 or 180 min incubation (*left panel*) or 105, 180, and 360 min incubation (*middle* and *right panels*), the nucleo-cytoplasmic distributions of precursors and splicing products were determined. Abbreviations are as in Fig. 1.

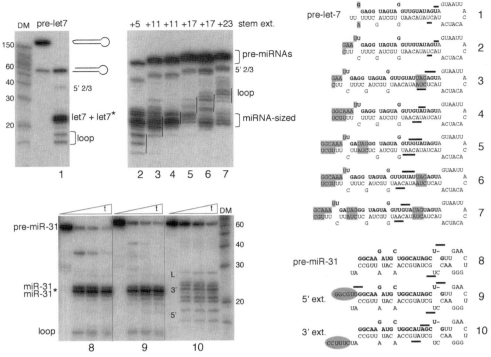

Figure 3. Role of the free ends of pre-miRNA and duplex lengths in cleavage by Dicer in vitro. Lanes 1–7 show products generated by incubation of ~0.5 pmole of ^{32}P-labeled circularized or linear pre-let-7 RNA (Fig. 1), or variants of pre-let-7 (*right side*) with 1 unit Dicer (Gene Therapy Systems) for 60 min at 37°C. Panels 8, 9, and 10 show the digestion products generated by incubation of pre-miR-31 and variant RNAs (*right side*) with Dicer for 0, 15, 30, and 60 min. Products were fractionated by electrophoresis in denaturing 20% polyacrylamide gels. Because the in vitro synthesized substrates were synthesized using either T7 or SP6 RNA polymerase, they all contain an extra non-encoded A at their 3′ ends (contributing or generating 3′ overhangs) which is not indicated in the figures.

sequence or structure within the stem region also could influence the site of cleavage.

Vertical lines in Figure 3 highlight the gel positions of fragments generated from the loop regions of the substrates. Apart from the short loop fragments, the shortest and most abundant cleavage products were miRNA-sized, in keeping with an initial cleavage at a fixed distance from the 3′ end. The products of all digests also included distinct, much larger fragments (5′ 2/3 molecules) that extended from the 5′ end of the stem through the loop to the cleavage site in the 3′ side of the stem. The existence of these "dead-end" partial digestion products indicates that Dicer cleaves first at the 3′ site, and that the duplex structure, which is needed for cleavage of the site in the 5′ side, has been disrupted in some of the substrate molecules.

Effects of loop size on Dicer cleavage. Although the site of cleavage clearly is determined in large part by measuring from the 3′ end of the stem, the effects of the length of the stem raise the possibility that interactions with the loop might also influence cleavage. Comparison of several variants of pre-let-7 showed that the size of the loop did not affect the site of cleavage (Fig. 4, top). However, loop size did affect the efficiency of cleavage, as shown by the reduced rate of digestion of the substrate with only 4 nucleotides (right panel). This reduction in efficiency may reflect distortion of the cleavage site near the end of the duplex, due to constraints imposed by the tight bend of the RNA backbone within the small loop.

As an RNase-III-like enzyme, Dicer cleaves RNAs in duplex regions, but it also fixes the site of cleavage at about 22 nucleotides from the 3′ end of the stem of the substrate (Fig. 3) (Zhang et al. 2004; Vermeulen et al. 2005). Therefore, duplexes shorter than 21 or 22 nucleotides would not be expected to be substrates for this enzyme. We tested this prediction by incubating Dicer with short hairpin RNAs (shRNAs) that had perfectly complementary stems 22, 21, or 19 bp long (RNAs 4, 5, and 6 at the bottom of Fig. 4). Surprisingly, the shRNA substrate with 22 bp (RNA 4) was cleaved much less efficiently than the corresponding pre-miRNA of similar length (RNA 3). This difference indicates that the unpaired nucleotides in the stems of pre-miRNAs may facilitate the positioning of the scissile bonds in appropriate RNase III active sites of the enzyme. As predicted, shortening the stem to 21 or 19 bp (RNAs 5 and 6) while maintaining the overall length of the shRNA (49 nucleotides) further reduced or completely abolished cleavage by Dicer.

shRNA substrates. Using similar substrates in their pioneering study on siRNA expression, Brummelkamp et al. (2002) reported that an shRNA with a stem of only 19 bp and a loop of 9 nucleotides (19/9) could be processed to make functional siRNAs, whereas substrates with shorter loops (19/7 or 19/5) could not. That led them to conclude that a 19-bp stem was sufficient for a substrate for Dicer, if the adjacent loop was 9 nucleotides long.

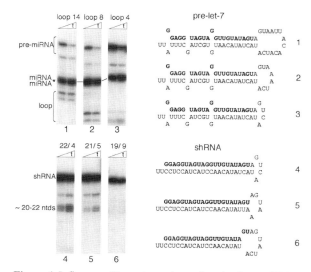

Figure 4. Influence of loop size and stem length of pre-miRNAs and shRNAs on cleavage by Dicer in vitro. ³²P-labeled variants of pre-let-7 and shRNAs (*right side*) were incubated with Dicer for 5 or 10 min (*top panels*) or 10 and 30 min (*bottom panels*) as in Fig. 3. Numbers above the bottom panels indicate the number of base pairs in the stem and the number of nucleotides in the loop of the shRNAs, respectively.

However, we note that two additional base pairs could be formed by nucleotides at the ends of the 9-nucleotide loop that they used, which would extend the stem to 21 pairs (and reduce the loop to 5). In the 19/9 substrate used here, extension of the stem was not possible and cleavage did not occur, showing that more than 19 bp are required for cleavage by Dicer. We conclude that the minimum stem length of an shRNA (and very likely a pre-miRNA) is 21 or 22 nucleotides.

Taken together, our data on the substrate requirements for cleavage of hairpin RNAs indicate that Dicer fixes the site of cleavage in a duplex region by measuring from the free 3′ end at the base of a stem, and that binding of the enzyme to the substrate is increased by interaction with the loop and perhaps the backbone of the stem. This interpretation is consistent with recent in vivo and in vitro studies of other workers (Zhang et al. 2004; Vermeulen et al. 2005; Chang et al. 2006), and it fits the model for substrate recognition based on the crystal structure of Dicer from *Giardia intestinalis* (MacRae et al. 2006). In that model, binding of the short 3′ overhang in the PAZ domain pocket at one end of the enzyme positions the duplex near the RNase III domains, allowing cleavage about 22 bp away. A mechanism for recognition of the loop by the minimal enzyme used in that study is unclear.

Non-pre-miRNA substrates. Finally, we asked whether other structured RNAs that are exported by Exp5 could also bind to, and be cleaved by, Dicer. As an indicator of binding, we measured the ability of an RNA to compete for cleavage of pre-let-7 RNA. hY1 RNA, which has a very short terminal stem (see Rutjes et al. 2001), had no effect on cleavage of pre-miR-31 (not shown). However, VA1 RNA of adenovirus 2, which has a very long, inter-

rupted stem (Fig. 5, left), was an effective inhibitor, as has recently been reported (Lu and Cullen 2004; Andersson et al. 2005). Likewise, ET-208 RNA competed for cleavage of pre-miRNA, albeit less efficiently (Fig. 5, top panels). However, neither VA1 RNA nor ET-208 RNA was a good substrate for cleavage; we detected only a very low level of VA1 cleavage, in agreement with other studies (Andersson et al. 2005; Sano et al. 2006), but we never observed any cleavage of ET-208 RNA (Fig. 5, lower panels). We propose that these RNAs bind via their 3′ overhangs to the PAZ domain and via their stem and loops to other structures of human Dicer including the dsRBD (Zhang et al. 2004; MacRae et al. 2006). Their ability to act as competitors without being cleaved indicates that they can do so without placing scissile bonds near the RNase III processing center of the enzyme. In vivo, none of these structured RNAs appeared to interfere with Dicer activity in *Xenopus* oocytes or eggs (not shown), perhaps because of interactions with other, more abundant or more avidly binding proteins. This is unlike the suppression of Dicer activity observed in VA1 RNA-producing mammalian cells (Lu and Cullen 2004; Andersson et al. 2005).

Processing by Dicer In Vivo

To study Dicer activity in vivo, we injected several synthetic pre-miRNAs and stem-loop RNAs into either the nucleus or cytoplasm of *Xenopus* oocytes or early embryos. As shown above (Fig. 1) and reported previously (Lund et al. 2004), pre-miRNAs injected into nuclei are rapidly exported into the cytoplasm where they appear to remain stable. However, these pre-miRNAs actually were processed in the cytoplasm, albeit slowly (Fig. 6, left

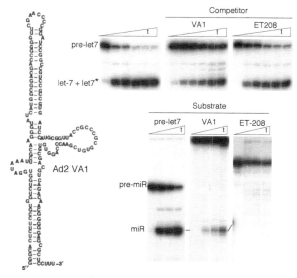

Figure 5. Interactions of structured RNAs with Dicer in vitro. The structure of VA1 RNA was adapted from Sano et al. (2006). The top panels show the products of Dicer cleavage of ³²P-labeled pre-let-7 upon incubation for 0, 5, 15, 30, 60, and 90 min in the absence or presence of 20 pmoles (~40-fold molar excess) of the indicated competitor RNAs. The bottom panels show the products of Dicer digestion of ³²P-labeled pre-let7, VA1, or ET-208 RNAs for 0, 5, 15, and 30 min, as in Fig. 3.

panel), to yield miRNA–miRNA* duplexes identical to the cleavage products generated in vitro. After 2 hours, only a small percentage of the injected pre-miRNA was cleaved, and even after overnight incubation, no more than 10–15% of the pre-miRNA substrates had been processed to miRNA. Thus, stage VI oocytes have detectable but low levels of Dicer activity.

To rule out that Dicer digestion products might have been produced in the nucleus and then rapidly exported to the cytoplasm by Exp5, we blocked nuclear export by depleting the oocytes of RanGTP (cf. Fig. 1). Under these conditions, the injected pre-miR-31 remained largely in the nucleus, but no processed miRNA was detected there (Fig. 6, second panel); thus, nuclei of stage VI oocytes are devoid of detectable amounts of Dicer activity, in agreement with the intracellular localization of Dicer in mammalian cells (Billy et al. 2001). Furthermore, the same extent of processing was observed regardless of whether the pre-miRNA substrate was injected into the nucleus or the cytoplasm (third panel).

Changes in Dicer Activity during Early Development

Incubation of stage VI oocytes with progesterone triggers their maturation into eggs. Upon breakdown of the germinal vesicle (nucleus), many hitherto dormant activities can be observed (Mendez and Richter 2001). When oocytes were matured, we observed a large increase in Dicer activity (Fig. 6, far right panel), with more pre-miR-31 RNA processed in matured oocytes within 30 minutes than during 4 hours in stage VI oocytes. Quantification of processing in several batches of oocytes showed a consistent five- to eightfold increase in Dicer activity upon maturation. Both miRNA and miRNA* products were detected, indicating that selection of one strand into a mature RISC might be inefficient. Comparable results were observed upon microinjection of several different pre-miRNAs (not shown).

It is unclear whether the large increase in Dicer activity that occurs upon breakdown of the germinal vesicle results from activation of preexisting enzyme or removal of an inhibitor, or from de novo synthesis of either Dicer or a cofactor(s) such as Xlrbpa (Eckmann and Jantsch 1997), the likely *Xenopus* homolog of TRBP (Haase et al. 2005; Maniataki and Mourelatos 2005) or PACT (Lee et al. 2006). A similar increase in Dicer activity was observed in a comparison of whole-cell extracts of stage VI oocytes, matured oocytes, and embryos (not shown), indicating that activation was not due simply to release of a sequestered cofactor from the nucleus.

We then asked whether the endogenous Dicer present in matured oocytes had the same substrate specificity as the human enzyme that we used in vitro. To make this comparison, matured oocytes were injected with several of the variant substrates that had been tested in vitro (Figs. 3 and 4). In general, processing in vivo matched the cleavage observed in vitro, but increased stringency in vivo was observed in several cases (Fig. 7). For example, the variant with only a 4-nucleotide-long loop that was cleaved well in vitro (RNA 3 in Fig. 4) was a poor substrate in vivo (middle panel, right lanes). Likewise, the RNA with a 14-nucleotide loop and 23-nucleotide stem extension (RNA 7 in Fig. 3) was not processed at all in vivo (Fig. 7, middle panel; left lanes). The shRNAs with 22 and 21 base-paired stems (RNAs 4 and 5 of Fig. 4) appeared to be better substrates in vivo than in vitro, but the one with only 19 bp (RNA 6) was not (Fig. 7, right panel). Comparable results were obtained in stage VI oocytes, but the extent of processing (and thus the signal) was considerably lower (data not shown). Changes in stringency of substrate recognition, while not great, could significantly influence the fidelity of processing events in vivo. The changes presumably result from interactions with one or more cofactors that facilitate Dicer activity.

The great increase in Dicer activity upon germinal vesicle breakdown (Fig. 6) presumably could support production of large quantities of miRNAs immediately after fertilization, during the early cleavage stage of *Xenopus* embryogenesis. However, we and other workers have observed that biogenesis of novel miRNAs is coincident with the mid-blastula transition (MBT), when zygotic transcription is activated (Newport and Kirshner 1982). At that time (~7 hours after fertilization), abundant accumulation of miR-427 is observed (Watanabe et al. 2005; E. Lund et al., in prep.).

We asked whether the early cleavage embryos, despite their very short cell cycle (~25–30 minutes), were still able

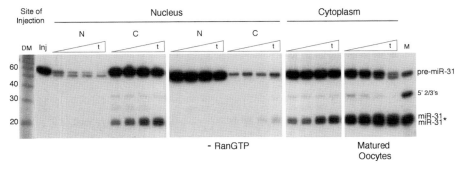

Figure 6. Processing of pre-miRNA in stage VI and matured *Xenopus* oocytes. ^{32}P-labeled pre-miR-31 was injected into the nuclei (*left two panels*) or cytoplasms (*third panel*) of stage VI oocytes without or with prior depletion of RanGTP (–RanGTP) (Lund et al. 2004). After 0.5, 1, 2, or 4 hours incubation at 23°C, the nucleo-cytoplasmic distributions of RNA products were determined as in Fig. 3. The far right panel shows the products resulting from injection into oocytes that had been matured by prior incubation (16 hours) with 10 μg/ml progesterone.

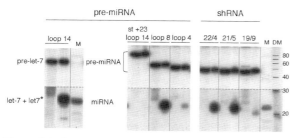

Figure 7. Influence of loop size and stem length on cleavage by Dicer in vivo. ^{32}P-labeled pre-miRNA and shRNA substrates (structures shown in Figs. 3 and 4) were injected into matured oocytes and analyzed 0 and 4 hours later for evidence of cleavage by endogenous *Xenopus* Dicer. Because several of the RNAs tested were poor substrates (despite the increased Dicer activity observed upon oocyte maturation), detection of the products was enhanced by ~10 times longer exposure of the bottom portions of the gels (below the dashed line).

to fully process pre-miRNAs and assemble them into RISCs. For this, we injected in vitro synthesized pre-miR-427 RNA into one- or two-cell embryos and observed its processing during normal development (Fig. 8). Mature miR-427 was visible already at the 32-cell stage (~1.5–2 hours after injection, or 3.5 hours post fertilization) and continued to accumulate through MBT and gastrulation (12 hours). The rate of disappearance of the injected pre-miR-427 appeared to accelerate in the approximately 2 hours prior to MBT, but that was not matched by a concomitant increase in the levels of mature miR-427, suggesting that processing by Dicer may have been saturated (although increased flux of material through mature miRNA cannot

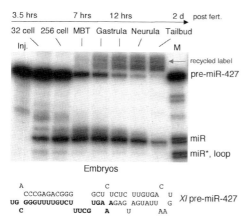

Figure 8. Processing of pre-miR-427 during early *Xenopus* embryogenesis. Between 1.5 and 2 hours after fertilization of *Xenopus* eggs (Newport and Kirshner 1982), ^{32}P-labeled pre-miR-427 (structure at the bottom) was injected into one- or two-cell embryos. Embryonic development was monitored as indicated, and at the indicated stages of embryonic development, RNAs were isolated from pools of four embryos and one embryo-equivalents of total RNAs were analyzed in a 20% denaturing polyacrylamide gel. After the mid-blastula transition (MBT), some of the labeled nucleotides from the injected RNA were recycled into newly synthesized tRNA molecules. The time line at the top indicates times after fertilization. The gel mobilities of Dicer digestion products (miR-427 and miR-427* and the loop) are indicated on the right.

be excluded). It is likely that the amount of accumulated miRNA represents a steady state that was achieved by simultaneous production and degradation of both the injected (^{32}P-labeled) and endogenous (unlabeled) miR-427. We note that only miR-427, but not its complementary strand, miR-427* (which migrates faster in the gel), accumulated, strongly indicating that the Dicer digestion products had been incorporated into mature RISCs. Importantly, this suggests that miRNA-mediated repression of translation may be operable during the earliest stages of embryonic development (E. Lund et al., in prep.).

CONCLUSIONS

Exp5 and Dicer are two proteins that are essential for the generation of microRNAs. The results discussed here demonstrate that, although both factors are able to accommodate a good deal of variability in their cargos or substrates, they also exhibit some strict substrate preferences or requirements. Undoubtedly, this level of stringency functions to reduce the level of spurious RNAs that could act like miRNAs, with detrimental consequences. Our data also demonstrate that the activity of at least one key enzyme, and probably several others (Thomson et al. 2006), involved in miRNA biogenesis is developmentally controlled. These findings highlight the importance of controlling the amounts of miRNAs that can be produced and function at various stem cell and embryonic stages, and probably in adult tissues, as well.

ACKNOWLEDGMENTS

This work was supported by National Institutes of Health grant GM 30220. We thank Susanne Imboden Blaser, Christian Bille Jendresen, Philipp Müller, Lei Tong, and David Trittle for help with RNA substrate preparation and in vitro Dicer assays.

REFERENCES

Andersson M.G., Haasnoot P.C., Xu N., Berenjian S., Berkhout B., and Akusjärvi G. 2005. Suppression of RNA interference by adenovirus virus-associated RNA. *J. Virol.* **79:** 9556.

Bartel D.P. 2004. MicroRNAs: Genomics, biogenesis, mechanism, and function. *Cell* **116:** 281.

Bernstein E., Caudy A.A., Hammond S.M., and Hannon G.J. 2001. Role for a bidentate ribonuclease in the initiation step of RNA interference. *Nature* **409:** 363.

Billy E., Brondani V., Zhang H., Muller U., and Filipowicz W. 2001. Specific interference with gene expression induced by long, double-stranded RNA in mouse embryonal teratocarcinoma cell lines. *Proc. Natl. Acad. Sci.* **98:** 14428.

Bohnsack M.T., Czaplinski K., and Görlich D. 2004. Exportin 5 is a RanGTP-dependent dsRNA-binding protein that mediates nuclear export of pre-miRNAs. *RNA* **10:** 185.

Bohnsack M.T., Regener K., Schwappach B., Saffrich R., Paraskeva E., Hartmann E., and Görlich D. 2002. Exp5 exports eEF1A via tRNA from nuclei and synergizes with other transport pathways to confine translation to the cytoplasm. *EMBO J.* **21:** 6205.

Brummelkamp T.R., Bernards R., and Agami R. 2002. A system for stable expression of short interfering RNAs in mammalian cells. *Science* **296:** 550.

Calado A., Treichel N., Muller E.C., Otto A., and Kutay U. 2002. Exportin-5-mediated nuclear export of eukaryotic elongation factor 1A and tRNA. *EMBO J.* **21:** 6216.

Carrara G., Calandra P., Fruscoloni P., and Tocchini-Valentini G.P. 1995. Two helices plus a linker: A small model substrate for eukaryotic RNase P. *Proc. Natl. Acad. Sci.* **92:** 2627.

Chang K., Elledge S.J., and Hannon G.J. 2006. Lessons from Nature: microRNA-based shRNA libraries. *Nat. Methods.* **3:** 707.

Du T. and Zamore P.D. 2005. microPrimer: The biogenesis and function of microRNA. *Development* **132:** 4645.

Eckmann C.R. and Jantsch M.F. 1997. Xlrbpa, a double-stranded RNA-binding protein associated with ribosomes and heterogeneous nuclear RNPs. *J. Cell Biol.* **138:** 239.

Filipowicz W., Jaskiewicz L., Kolb F.A., and Pillai R.S. 2005. Post-transcriptional gene silencing by siRNAs and miRNAs. *Curr. Opin. Struct. Biol.* **15:** 331.

Grimm C., Lund E., and Dahlberg J.E. 1997. Selection and nuclear immobilization of exportable RNAs. *Proc. Natl. Acad. Sci.* **94:** 10122.

Gwizdek C., Ossareh-Nazari B., Brownawell A.M., Doglio A., Bertrand E., Macara I.G., and Dargemont C. 2003. Exportin-5 mediates nuclear export of minihelix-containing RNAs. *J. Biol. Chem.* **278:** 5505.

Haase A.D., Jaskiewicz L., Zhang H., Laine S., Sack R., Gatignol A., and Filipowicz W. 2005. TRBP, a regulator of cellular PKR and HIV-1 virus expression, interacts with Dicer and functions in RNA silencing. *EMBO Rep.* **6:** 961.

Izaurralde E., Kutay U., von Kobbe C., Mattaj I.W., and Görlich D. 1997. The asymmetric distribution of the constituents of the Ran system is essential for transport into and out of the nucleus. *EMBO J.* **16:** 6535.

Kim V.N. 2005. MicroRNA biogenesis: Coordinated cropping and dicing. *Nat. Rev. Mol. Cell Biol.* **6:** 376.

Lee Y., Jeon K., Lee J.T., Kim S., and Kim V.N. 2002. MicroRNA maturation: Stepwise processing and subcellular localization. *EMBO J.* **21:** 4663.

Lee Y., Hur I., Park S.Y, Kim Y.K., Suh M.R., and Kim V.N. 2006. The role of PACT in the RNA silencing pathway. *EMBO J.* **25:** 522

Lu S. and Cullen B.R. 2004. Adenovirus VA1 noncoding RNA can inhibit small interfering RNA and MicroRNA biogenesis. *J. Virol.* **78:** 12868.

Lund E., Güttinger S., Calado A, Dahlberg J.E., and Kutay U. 2004. Nuclear export of microRNA precursors. *Science* **303:** 95.

MacRae I.J., Zhou K., Li F., Repic A., Brooks A.N., Cande W.Z., Adams P.D., and Doudna J.A. 2006. Structural basis for double-stranded RNA processing by Dicer. *Science* **311:** 195.

Maniataki E. and Mourelatos Z. 2005. A human, ATP-independent, RISC assembly machine fueled by pre-miRNA. *Genes Dev.* **19:** 2979.

Mendez R. and Richter J.D. 2001. Translational control by CPEB: A means to the end. *Nat. Rev. Mol. Cell Biol.* **2:** 521.

Murchison E.P. and Hannon G.J. 004. miRNAs on the move:

miRNA biogenesis and the RNAi machinery. *Curr. Opin. Cell. Biol.* **16:** 223.

Newport J. and Kirschner M. 1982. A major developmental transition in early *Xenopus embryos*. I. Characterization and timing of cellular changes at the midblastula stage. *Cell* **30:** 675.

Ohrt T., Merkle D., Birkenfeld K., Echeverri C.J., and Schwille P. 2006. In situ fluorescence analysis demonstrates active siRNA exclusion from the nucleus by Exportin 5. *Nucleic Acids Res.* **34:** 1369.

Pasquinelli A.E., Ernst R.K., Lund E., Grimm C., Zapp M.L, Rekosh D., Hammarskjöld M.L., and Dahlberg J.E. 1997. The constitutive transport element (CTE) of Mason-Pfizer monkey virus (MPMV) accesses a cellular mRNA export pathway. *EMBO J.* **16:** 7500.

Petersen J.M., Her L.S, Varvel V., Lund E., and Dahlberg J.E. 2000. The matrix protein of vesicular stomatitis virus inhibits nucleocytoplasmic transport when it is in the nucleus and associated with nuclear pore complexes. *Mol. Cell. Biol.* **20:** 8590.

Rutjes S.A., Lund E., van der Heijden A., Grimm C., van Venrooij W.J., and Pruijn G.J. 2001. Identification of a novel cis-acting RNA element involved in nuclear export of hY RNAs. *RNA* **7:** 741.

Sano M., Kato Y., and Taira K. 2006. Sequence-specific interference by small RNAs derived from adenovirus VAI RNA. *FEBS Lett.* **580:** 1553.

Thomson J.M., Newman M., Parker J.S., Morin-Kensicki E.M., Wright T., and Hammond S.M. 2006. Extensive post-transcriptional regulation of microRNAs and its implications for cancer. *Genes Dev.* **20:** 2202.

Valencia-Sanchez M.A., Liu J., Hannon G.J., and Parker R. 2006. Control of translation and mRNA degradation by miRNAs and siRNAs. *Genes Dev.* **20:** 515.

Vermeulen A., Behlen L., Reynolds A., Wolfson A., Marshall W.S., Karpilow J., and Khvorova A. 2005. The contributions of dsRNA structure to Dicer specificity and efficiency. *RNA* **11:** 674.

Watanabe T., Takeda A., Mise K., Okuno T., Suzuki T., Minami N., and Imai H. 2005. Stage-specific expression of microRNAs during *Xenopus* development. *FEBS Lett.* **579:** 318.

Yi R., Qin Y., Macara I.G., and Cullen B.R. 2003. Exportin-5 mediates the nuclear export of pre-microRNAs and short hairpin RNAs. *Genes Dev.* **17:** 3011.

Zeng Y. and Cullen B.R. 2004. Structural requirements for pre-microRNA binding and nuclear export by Exportin 5. *Nucleic Acids Res.* **32:** 4776.

Zhang H., Kolb F.A., Brondani V., Billy E., and Filipowicz W. 2002. Human Dicer preferentially cleaves dsRNAs at their termini without a requirement for ATP. *EMBO J.* **21:** 5875.

Zhang H., Kolb F.A., Jaskiewicz L., Westhof E., and Filipowicz W. 2004. Single processing center models for human Dicer and bacterial RNase III. *Cell* **118:** 57.

The Argonautes

L. Joshua-Tor

Cold Spring Harbor Laboratory, Cold Spring Harbor, New York 11724

RNA interference (RNAi) has been greatly exploited in recent years as an increasingly effective tool to study gene function by gene silencing. The introduction of exogenous double-stranded RNA (dsRNA) into a cell can trigger this gene silencing process. An RNase III family enzyme, Dicer, initiates silencing by releasing approximately 20 base duplexes, with 2-nucleotide 3′ overhangs called siRNAs. The RNAi pathway also mediates the function of endogenous, noncoding regulatory RNAs called miRNAs. Both miRNAs and siRNAs guide substrate selection by similar if not identical effector complexes called RISCs. These contain single-stranded versions of the small RNA and additional protein components. Of those, the signature element, at the heart of all RISCs, is a member of the Argonaute family of proteins. Our structural and biochemical studies on Argonaute identified this protein as Slicer, the enzyme in RISC that cleaves the mRNA as directed by the siRNA. The role of the Argonautes as Slicers and non-Slicers is discussed.

RNAi has made an enormous impact on biology in a very short period of time. It became an extraordinarily useful and simple tool for gene silencing, even as its mechanism was gradually unraveling. Deciphering the mechanism of RNAi-related pathways, including both transcriptional and posttranscriptional gene silencing, has benefited from an incredible marriage of genetics, biochemistry, molecular biology, and, of course, structural biology.

RNAi EFFECTOR COMPLEXES

The various pathways of RNAi can be broken down into the initiation steps in which the dsRNA triggers are processed into siRNAs or miRNAs, and the effector steps in which silencing is realized. This is, of course, an over-simplified view of these related pathways; however, in each case, the effector step involves an effector complex called the RNA-induced silencing complex (RISC) (Hammond et al. 2000). RISCs of varying compositions and sizes have been purified from both human and fly cells (Hammond et al. 2001; Nykanen et al. 2001; Caudy et al. 2002, 2003; Ishizuka et al. 2002; Martinez et al. 2002). However, the minimal defining features of RISC include the presence of an Argonaute family member and the guide strand of a short RNA. The short RNA (si- or miRNA) then acts to guide the RISC to its target through base complementarity. The best-characterized pathway, and the one that is predominantly used for gene knock-down technology, is a posttranscriptional silencing (PTGS) pathway called Slicing. Here the RISC is targeted to the mRNA and produces an endonucleolytic cut in the mRNA target, thus preventing gene expression from proceeding (for a recent review, see Zamore and Haley 2005). This cut is made 10 bases from where the 5′ end of the guide siRNA hybridizes to the target mRNA (Elbashir et al. 2001a, b). Other RNAi silencing pathways such as translational inhibition and transcriptional gene silencing (TGS) are also mediated through RISCs. In the fission yeast, *Schizosaccharomyces pombe*, there is a specialized RISC, called RITS, that is composed of Ago1, the sole

Argonaute protein in *S. pombe*; Tas3; and a chromo-domain protein, Chp1 (Verdel et al. 2004). This complex is involved in regulation of silencing of centromeric repeats and other heterochromatic loci. The crystal structure of Argonaute from *Pyrococcus furiosus* (PfAgo), described below, revealed an RNase H-like PIWI domain capable of RNA cleavage (Song et al. 2004). This established Argonaute as the catalytic core of the effector complex, RISC (Liu et al. 2004; Song et al. 2004).

ARGONAUTE—THE CATALYTIC CORE OF RISC

As mentioned above, the Argonaute proteins are at the heart of RISC. They have two characteristic domains: the PAZ domain, which is shared with another key RNAi enzyme, Dicer; and a unique domain called the PIWI domain. Both PAZ and PIWI domains appear to be unique to proteins involved in RNAi-type processes. The PAZ domain is approximately 140 amino acids in length and is named after three proteins—Piwi, Argonaute, and Zwille. Three PAZ domain structures were determined and published concurrently: the NMR structure of the PAZ domain of *Drosophila melanogaster* Ago1 by the Zhou group (Yan et al. 2003), the NMR structure of the PAZ domain of *D. melanogaster* Ago2 by the Sattler group (Lingel et al. 2003), and our crystal structure of the same PAZ domain *D. melanogaster* Ago2 (Song et al. 2003). The structures are all similar to each other and contain two subdomains with a prominent cleft between. The larger subdomain has a fold that is reminiscent of an OB-fold domain, although with slightly different topology, and the second subdomain, called by some "the appendage," is composed of a β-hairpin followed by an α-helix. OB fold stands for oligonucleotide-oligosaccharide binding, since a majority of OB-fold proteins bind single-stranded oligonucleotides or oligosaccharides. This suggested a role for the PAZ domain in nucleic acid binding. A series of crosslinking and mutagenesis experiments showed that the siRNAs bind to the PAZ intersubdomain cleft through the 2-nucleotide overhang at their 3′ end (Song et al. 2003). This mode of PAZ–siRNA recognition was confirmed shortly after by structures of the PAZ in complex with nucleic acids (Lingel et al. 2004; Ma et al. 2004) that also showed binding to be sequence independent. Highly conserved aromatic residues—two tyrosines and a histidine—contact the phosphate between the two bases in the overhang.

The structure of the full-length Argonaute from *P. furiosus* (PfAgo) was critical in defining the function of Argonaute (Song et al. 2004). It showed the Argonautes to be composed of four domains: the amino-terminal, PAZ, middle, and PIWI domains. Overall, the protein has a crescent-shaped base comprising the amino, middle, and PIWI domains. The region following the amino-terminal domain forms a "stalk" that holds the PAZ domain above the rest of the structure (Fig. 1). An interdomain connector cradles the four domains of the molecule. This architecture forms a large, positively charged groove between the PAZ domain and the crescent base, and a smaller one between the amino and PIWI domains.

The PAZ domain of PfAgo is very similar in structure to all the other PAZ domain structures that were determined previously, despite a less than 6% sequence identity. Importantly, the side chains of the aromatic residues in the PAZ intersubdomain cleft that were shown to bind the 3′ end of the siRNA fall in the same positions in space. Y212, Y216, H217, and Y190 of PfAgo are equivalent to residues Y309, Y314, H269, and Y277 of human Ago1 (hAgo1) (Ma et al. 2004; Song et al. 2004). However, in the case of the histidine of PfAgo, it comes from a very different location on the backbone than the equivalent one from hAgo1. Indeed, the sequences of the PfAgo and human Ago1 can only be aligned using the two structures. The middle

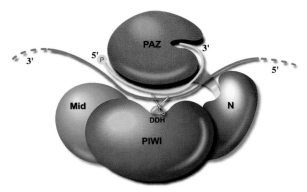

Figure 2. Schematic depiction of the model for siRNA-guided mRNA cleavage. The domains are colored as in Fig. 1. The siRNA, shown in yellow, binds with its 3′ end in the PAZ cleft. Its 5′ phosphate is important for enzyme stability and fidelity. The mRNA is depicted in brown, comes in between the amino-terminal and PAZ domains and out between the PAZ and middle domain. The active site in the PIWI domain, depicted as scissors, with its DDH motif, cleaves the mRNA opposite the middle of the siRNA guide.

domain is similar in structure to sugar-binding proteins such as the sugar-binding domain of Lac repressor.

The big surprise came upon examination of the PIWI domain. It is at the carboxyl terminus of Argonaute and located across the large groove from the PAZ domain. Not much was known about it from sequence alone, and it appeared to reside only in Argonaute proteins. The crystal structure revealed that it clearly belongs to the RNase H family of enzymes (Song et al. 2004). Apart from the RNase H fold at its core, it also has the two highly conserved aspartates always present on adjacent β-strands in these types of enzymes. The sequence of the PIWI domain of PfAgo aligns well with other members of the Argonaute family. Modeling of siRNA and target (mRNA) binding starting from the PAZ 3′ end binding site revealed that the large groove could accommodate the duplex positioning the predicted scissile phosphate at the active site near the two aspartates. Directed by the structure, mutation of each of these aspartates that are conserved in human Argonaute2 eliminates Slicer activity (Liu et al. 2004; Rivas et al. 2005). These findings identified Argonaute as being the long-sought catalytic enzyme of RISC, dubbed Slicer, that slices the mRNA substrate guided by the siRNA (Fig. 2). Moreover, the dependence of RISC Slicer activity on the presence of Mg^{2+} ions and the generation of similar products to RNase H family enzymes further underscores this conclusion.

The crystal structure of AfPIWI (Parker et al. 2004), which possesses both middle and PIWI domains but lacks the amino-terminal and PAZ domains, also demonstrated the similarity to Lac repressor and RNase H. Because this protein does not contain the catalytic aspartates, it does not bind a divalent metal ion in the active site. However, a Cd^{2+} ion is bound to the carboxy-terminal carboxylate of the protein at a conserved region of the protein. The carboxyl terminus of both AfPiwi and PfAgo lies at the interface between the mid and Piwi domains. Structures of a complex of this protein with dsRNA were later deter-

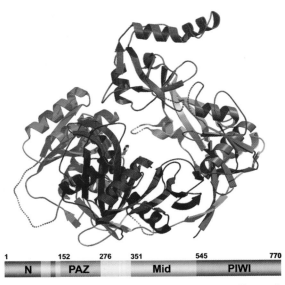

1	152	276	351	545	770
N	PAZ		Mid		PIWI

Figure 1. Crystal structure of *P. furiosus* Argonaute. Stereo ribbon representation of Argonaute with the amino-terminal domain shown in blue, the "stalk" in light blue, the PAZ domain in red, the middle domain in green, the PIWI domain in purple, and the interdomain connector in yellow. The active-site aspartates and histidine residues are drawn in stick representation. Disordered loops are drawn as dotted lines. Shown below is a schematic diagram of the domain borders.

mined that suggest a mode of recognition for the 5′ phosphate of the guide RNA (Ma et al. 2005; Parker et al. 2005). The 5′ nucleotide of the siRNA is unpaired, and the base and phosphate interact with residues of the middle domain. This is consistent with the observation that only nucleotides 2–8 of the guide RNA (the "seed" sequence) play a role in target recognition and specificity (Lewis et al. 2003; Stark et al. 2003; Doench and Sharp 2004). Coordination of the 5′ phosphate is also dependent on the Cd^{2+} metal ion. However, the metal ion is not observed in the apo and manganese-soaked crystals of PfAgo (Rivas et al. 2005) and may require the presence of the guide strand for coordination if the mode of siRNA binding is similar between these proteins.

More recently, the structure of a eubacterial Argonaute protein from *Aquifex aeolicus* (AaAgo) was determined and shows identical domain architecture as PfAgo (Yuan et al. 2005). However, the relative orientation of the PAZ domain is quite different. Interestingly, this protein was shown to have RNase-H-like activity, whereby it shows preference for a DNA/RNA hybrid as a substrate.

ACTIVE SITE OF SLICER

A combination of crystallographic and mutational analyses showed that Argonaute contains two aspartates and a histidine at its active site in a conserved "DDH" motif (Liu et al. 2004; Song et al. 2004; Rivas et al. 2005). A crystal structure of PfAgo soaked with manganese chloride showed that the Mn^{2+} ion bound these three residues (Rivas et al. 2005). This is equivalent to metal ion binding in other RNase H family enzyme structures lacking substrate. A second metal ion is present in structures of Tn5 transposase (Lovell et al. 2002) and RNase H (Nowotny et al. 2005) with bound substrate. It is reasonable to suggest that Argonaute uses a similar two-metal-ion mechanism for catalysis.

The two aspartates appear to be critical to convey slicer activity; however; it appears that the histidine might be replaced by another aspartate, a glutamate, or even a lysine, resulting in a more relaxed active site requirement of DD(H/D/E/K) (Table 1). Of the four immunopurified Argonautes tested (hAgo1-4), the only one with slicer activity is hAgo2 (Liu et al. 2004; Meister et al. 2004). hAgo1 has an arginine in place of the histidine, and hAgo4 has a glycine in place of one of the aspartates. Three of the four *D. melanogaster* Argonautes have been tested for activity—DmAgo1 (Miyoshi et al. 2005) and DmAgo2 (Rand et al. 2005) with DDH, and DmPiwi (Saito et al. 2006) with DDK, and all have slicer activity. *Arabidopsis* has 10 family members with either DDH or DDD as active-site residues, although only AtAgo1 and AtAgo4 have been shown to exhibit slicer activity (Baumberger and Baulcombe 2005; Qi et al. 2005, 2006). *Caenorhabditis elegans* has as many as 24 Argonaute proteins. Although none of these, to our knowledge, has been shown in vitro to have slicer activity, some, such as CeRDE-1, function via slicing (Tabara et al. 1999). The sole Argonaute in *S. pombe*, SpAgo1, has a complete DDH, and we have now shown that it is active for slicing in vitro (Irvine et al. 2006).

The presence of the catalytic residues is necessary for slicer activity, but it is not the whole story, hAgo3 being a case in point. hAgo3 has a complete DDH motif similar to hAgo2, but the protein immunopurifed from 293 or HeLa cells does not appear to have slicer activity, even though it is able to bind siRNA (Liu et al. 2004; Meister et al. 2004).

Previously, a phylogenic analysis of Argonaute proteins showed that they can be divided into two classes—the Argonaute-like Argonautes that are similar to AtAgo1 and the Piwi-like Argonautes that are similar to DmPiwi. We performed a new analysis including more Argonaute proteins from *Homo sapiens, Arabidopsis thaliana, C. elegans, D. melanogaster*, and *S. pombe* (Yigit et al. 2006). This suggests the existence of a third clade which we termed the Group III Argonautes (Table 1). Interestingly, this clade is worm-specific and predominantly contains non-Slicer-like Argonautes, meaning they do not have an intact catalytic motif. In contrast, the Ago-like and Piwi-like clades contain a high proportion of proteins from all species that are either Slicers, which have been shown to have slicer activity, or are Slicer-like, those which contain the three catalytic residues. Thus, we postulate that the worm has evolved a large set of non-Slicer-like Argonautes. Inclusion of 21 Argonautes from *Caenorhabditis briggsae* in a similar analysis showed that *C. briggsae* also possesses Group III Argonautes. Thus, the divergence of the Group III Argonautes arose prior to the *C. elegans–C. briggsae* split. Interestingly, CeRde-1, CePpw-1, and CePpw-2 also cluster in this clade. CeRde-1 was the first Argonaute family member to be genetically linked to silencing (Tabara et al. 1999), as CeRde-1 mutants have a penetrant loss of RNAi. CePpw-1 and CePpw-2 appear to be involved in RNAi in the germ line of *C. elegans* (Tijsterman et al. 2002; Vastenhouw et al. 2003).

TGS AND SLICING

Slicing of the substrate mRNA has long been established for PTGS. Recently, we have shown that SpAgo1 is an active Slicer that is required for TGS in *S. pombe* (Irvine et al. 2006). Mutation of active-site aspartate residues of SpAgo1 results in a defective enzyme in vitro. In vivo, this results in a derepression of silencing at heterochromatic repeats and a reduction in histone H3 lysine 9 dimethylation as well as an increase in histone H3 lysine 4 methylation. SpAgo1 levels at these repeats were observed at similar levels for the Slicer mutant strains, although RNA-dependent RNA polymerase (RdRP) levels were decreased, implicating slicing and the production of a free 3′OH as a requirement for recruitment of the RdRP complex (RDRC) to target RdRP activity, which in turn would initiate spreading.

MINIMAL RISC AND THE SLICING CYCLE

Because hAgo2 and hAgo3 were immunopurified from RNAi-competent cells (Liu et al. 2004; Meister et al. 2004), the differences in activity between them may have been due to a difference in purity between the two preparations, giving hAgo2 its activity. To eliminate this trivial

Table 1. Catalytic Residues of Argonautes from *H. sapiens* (Hs), *A. thaliana* (At), *D. melanogaster* (Dm), *S. pombe* (Sp), and *C. elegans* (Ce).

HsAgo1	VIFLGA	D	V	IFYR	D	GVP	IPAPAYYA	R	LVAF	Ago
HsAgo2	VIFLGA	D	V	IFYR	D	GVS	IPAPAYYA	H	LVAF	Ago
HsAgo3	VIFLGA	D	V	IFYR	D	GVS	IPAPAYYA	H	LVAF	Ago
HsAgo4	VIFLGA	D	V	IYYR	G	GVS	IPAPAYYA	R	LVAF	Ago
HsHIWI / PIWIL1	VMIVGI	D	C	IVYR	D	GVG	VPAPCQYA	H	KLAF	PIWI
HsHILI / PIWIL2	LMVIGM	D	V	VVYR	D	GVS	VPAPCKYA	H	KLAF	PIWI
HsPIWIL3	TMFVGI	D	C	IVYR	D	GVG	VPAPCHYA	H	KLAY	PIWI
HsHIWI2 / PIWIL4	LMVVGI	D	V	IVYR	A	GVG	VPAPCQYA	H	KLTF	PIWI
AtAgo1	TIIFGA	D	V	IFYR	D	GVS	IVPPAYYA	H	LAAF	Ago
AtAgo2	VMFIGA	D	V	VIFR	D	GVS	LVPPVYYA	D	MVAF	Ago
AtAgo3	VMFIGA	D	V	VIFR	D	GVS	LVPPVSYA	D	KAAS	Ago
AtAgo4	TIILGM	D	V	IIFR	D	GVS	VVAPICYA	H	LAAA	Ago
AtAgo5	TIIMGA	D	V	IFYR	D	GVS	IVPPAYYA	H	LAAF	Ago
AtAgo6	TLILGM	D	V	IIFR	D	GVS	SVAPVRYA	H	LAAA	Ago
AtAgo7	VIFMGA	D	V	IFFR	D	GVS	IVPPAYYA	H	LAAY	Ago
AtAgo8	TIIIGM	D	V	IFYR	D	GVS	VVAPVCYA	H	LAAA	Ago
AtAgo9	TIIVGM	D	V	IIFR	D	GVS	VVAPVCYA	H	LAAA	Ago
AtZll/Pnh	TIIFGA	D	V	IFYR	D	GVS	IVPPAYYA	H	LAAF	Ago
DmAgo1	VIFLGA	D	V	ILYR	D	GVS	IPAPAYYA	H	LVAF	Ago
DmAgo2	TMYIGA	D	V	IYYR	D	GVS	YPAPAYLA	H	LVAA	Ago
DmAUB	LMTVGF	D	V	LFFR	D	GVG	VPAVCHYA	H	KLAF	PIWI
DmPIWI	LMTIGF	D	I	VFYR	D	GVS	VPAVCQYA	K	KLAT	PIWI
SpAgo1	TLILGG	D	V	IYFR	D	GTS	LVPPVYYA	H	LVSN	Ago
CeZK757.3	TMVVGI	D	V	IVYR	D	GVS	IPTPVYYA	D	LVAT	Ago
CeT22B3.2	TMVVGI	D	V	IVYR	D	GVS	IPTPVYYA	D	LVAT	Ago
CeT23D8.7	VLFIGC	H	L	IIYR	A	GIA	IPSPVYYA	K	LVAQ	Ago
CeAlg-1	VIFFGC	D	I	VVYR	D	GVS	IPAPAYYA	H	LVAF	Ago
CeAlg-2	VIFLGC	D	I	VVYR	D	GVS	IPAPAYYA	H	LVAF	Ago
CeR09A1.1	TLVLGI	D	V	VVYR	D	GLS	LPAPVLYA	H	LAAK	PIWI
CePRG-1	TMIVGY	D	L	ILYR	D	GAG	VPAPCQYA	H	KLAF	PIWI
CePRG-2	TMIVGY	D	L	ILYR	D	GAG	VPAPCQYA	H	KLAF	PIWI
CeRde-1	TMYVGI	D	V	VVYR	D	GVS	LPVPVHYA	H	LSCE	III
CeF20D12.1	TFVIGM	D	V	IIFR	D	GVS	IPTPVYVA	H	ELAK	III
CeC04F12.1	TLIISY	D	V	VILR	D	GVS	LPESIYAA	D	EYAK	III
CeM03D4.6	LLLIGL	S	T	VIYL	C	GMS	LPAPLYLT	A	EMAE	III
CeK12B6.1	RLIIGF	E	T	LIYF	S	GVS	LPIPLHIA	G	TYSE	III
CeF56A6.1	RLIVGF	V	T	LLYF	N	GVS	LPVPLYIA	D	RYSQ	III
CePpw-1	RLIVGF	V	T	LLYF	N	GVS	LPVPLYIA	D	RYSQ	III
CeZK1248.7	QLIIGV	G	V	IIYR	S	GAS	IPTPLYVA	N	EYAK	III
CeF58G1.1	HLIIGV	G	I	IVYR	T	GTS	LPTPLYVA	N	EYAK	III
CeC06A1.4	HLIIGV	G	I	IIYR	T	ETS	LPTPLYVA	N	EYAK	III
CePpw-2	HLIIGV	G	I	TIYR	S	GSS	IPTPLYVA	N	EYAK	III
CeF55A12.1	QLIIGV	G	V	IIYR	S	GAS	IPTPLYVA	N	EYAK	III
CeR06C7.1	QLIIGV	G	V	IIYR	S	GAS	IPTPLYVA	N	EYAK	III
CeY49F6A.1	TQFIGF	E	M	VIYR	T	GAG	VPHILYAA	D	NLAK	III
CeR04A9.2	TQFIGF	E	M	VVYR	V	GSG	IPNVSYAA	Q	NLAK	III
CeT22H9.3	VQFIGF	D	I	VIYR	I	GAG	FPDVLYAA	E	NLAK	III
CeC16C10.3	VQFIGF	E	I	VIYR	V	GAG	VPDVLYAA	E	NLAK	III
CeC14B1.7	VQFIGF	E	I	VIYR	V	GAG	VPDVLYAA	E	NLAK	III

Critical aspartates are colored in red, and histidines are colored in dark blue. Substitutions for the DDH motif such as aspartate to glutamate or histidine to either aspartate or glutamate are colored in light red. Substitution of the histidine to a lysine is colored in light blue.

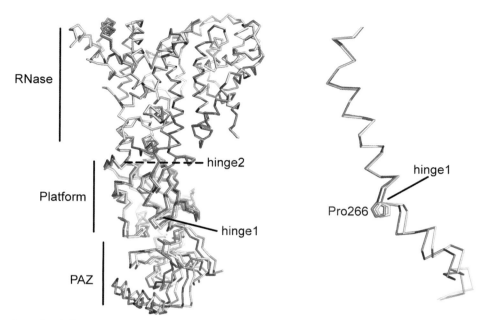

Figure 2. Conformational diversity in *Giardia* Dicer crystal structures. (*Left*) *Giardia* Dicer is composed of three structurally rigid regions termed RNase, Platform, and PAZ. These regions are connected by two flexible hinges in the protein. (*Right*) Close-up view of the connector helix. Hinge1 resides at a kink in the connector helix that is likely formed by Pro-266.

reveals that Dicer is composed of three structurally rigid regions connected by flexible hinges. The "RNase region" is made up of both RNase III domains and the bridging domain. The "platform region" is composed of the platform domain and the connector helix, and the third region is the PAZ domain (Fig. 2, left).

The most dramatic difference between the two structures is the shift in position of the PAZ domain, which moves about 5 Å. The hinge for this movement (hinge 1) can be traced to a single point in the connector helix where there is a distinct kink in the helix (Fig. 2, right). The kink is most likely induced by the presence of Pro-266, which sits in the apex of the kink. Comparison of Dicer sequences from a variety of organisms shows that proline is strictly conserved at this position in the primary sequence (MacRae et al. 2006). Thus, a kinked connector helix is likely a structural feature of all Dicer enzymes. The strict conservation of proline at this position argues that the connector helix kink and the hinge it forms between the platform and PAZ regions have an important role in Dicer function, most likely facilitating conformational flexibility.

The second hinge in Dicer resides between the platform and RNase regions. Unlike the connector helix kink, there is no distinct point in the polypeptide chain that forms the hinge between the two regions. Instead, the positional shift is diffused along the platform loop and the loop following the connector helix. The movement between the two structures in this region is a 5° rotation of the RNase region around the long axis of the molecule, which approximately coincides with the connector helix.

In addition to comparing static crystal structures of *Giardia* Dicer, we also explored possible conformational flexibility in the protein by normal mode analysis (Suhre and Sanejouand 2004). This analysis predicts similar structurally rigid regions of Dicer connected by hinges in

approximately the same locations as described above. The largest positional shifts in the analysis were between the RNase and platform regions, with the RNase predicted to both rotate and bend with respect to the platform (Fig. 3).

Structural analysis of *Giardia* Dicer reveals that the enzyme possesses a substantial degree of conformational flexibility. We propose that the three structural regions of Dicer flex and rotate relative to each other as the enzyme engages its dsRNA substrates. This is analogous to the "induced fit" classically seen in enzyme-active sites when bound to small-molecule substrates, except that in the case of Dicer, which has a substrate more than 80 Å long, the induced fit spans the length of the protein.

Flexibility is likely a critical feature of Dicer because it allows the enzyme to adjust its shape to accommodate structural diversity in its dsRNA substrates. This is particularly important in the processing of pre-microRNAs (pre-miRNAs), which are typically riddled with RNA helical imperfections (Pasquinelli et al. 2000; Lau et al. 2001). The ability of Dicer to cope with deviations from ideal RNA duplex structure allows pre-miRNAs to function as imperfect duplexes. Evolutionarily, this is an important point because it means that mutations on both sides of miRNA precursors will be tolerated, allowing nature to create new miRNAs from old as well as to adjust thermodynamic asymmetry across the pre-miRNA and thereby optimizing the guide:passenger strand ratio that is incorporated into miRNPs. Thus, conformational flexibility in Dicer's structure allows evolutionary flexibility in miRNA function.

ROLES FOR ACCESSORY DOMAINS IN DICER

The structure of *Giardia* Dicer represents the functional core structure that is likely to be common to all Dicer enzymes (Zhang et al. 2004). This relatively small

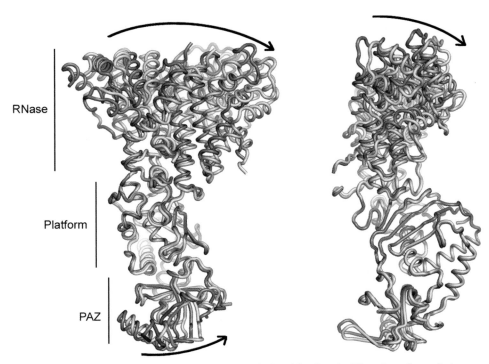

RNase

Platform

PAZ

Figure 3. Comparison of the two lowest-frequency normal modes calculated for *Giardia* Dicer. In each panel, the crystal structure is shown in blue, and the yellow structure represents the conformational change predicted by the analysis. Structures were aligned in their Platform regions, and arrows indicate direction of relative movements of RNase and PAZ regions.

and simple protein is fully capable of dicing dsRNA in vitro and is sufficient to support RNAi in vivo (MacRae et al. 2006). However, several of the domains commonly found in Dicer proteins of higher eukaryotes are not present in the *Giardia* enzyme. Most notably, *Giardia* Dicer lacks the amino-terminal DExD helicase present in human, fly, worm, plant, and fission yeast Dicers. Although the presence of the Dicer helicase is broadly conserved among higher eukaryotes, no function has yet been assigned to this domain. Possible activities of the Dicer helicase domain include unwinding of product siRNAs, translocating Dicer along dsRNA substrates, and facilitating the handoff of substrates or products with other components of the RNAi machinery.

We examined the role of the helicase in dicing by producing a version of human Dicer that lacks the entire helicase. The truncated protein was stable and well behaved, indicating that the core structure of Dicer was not perturbed by the removal of the helicase. Furthermore, the truncated Dicer displayed activity comparable to that of the wild-type protein in an in vitro dicing assay in both the presence and absence of ATP (Fig. 4). Thus, in the absence of other cellular factors, the helicase domain does not facilitate processing of dsRNA by human Dicer. This finding is consistent with previous results which showed that mutation of the ATP-binding Walker A motif did not impair the activity of purified recombinant human Dicer (Zhang et al. 2002).

To investigate the in vivo function of the Dicer helicase domain, we produced a mutant version of *Schizosaccharomyces pombe* Dicer in which Lys-38, the helicase Walker A motif lysine, is changed to alanine

(named *dcr1*-K38A). Mutation of the analogous lysine residue eliminates ATP binding and hydrolysis in the DExD helicase eIF4AIII (Shibuya et al. 2006). This mutation should therefore render the Dicer helicase catalytically inactive while leaving the helicase domain structurally intact. We also created a version of *S. pombe* Dicer lacking the entire helicase domain (named *dcr1*-hd for helicase deletion).

pREP2 expression plasmids carrying either the wild-type *S. pombe* Dicer or one of the two mutants were introduced into a Dicer deletion strain of *S. pombe*. The *S. pombe* strain used also contained the marker gene *ade6* at the heterochromatic *otr* region of the centromere. Transformants were isolated and plated in serial dilutions on Edinburgh minimal medium (EMM) without adenine. Because silencing of the marker gene *ade6* was disrupted, the *dcr1Δ* strain grew well on the plate. As expected, the strain transformed by the vector with wild-type Dicer displayed poor growth, indicating that ectopic expression of Dicer rescues the *dcr1Δ* centromeric silencing defect (Fig. 5, top). The two strains carrying the helicase mutants, *dcr1*-K38A and *dcr1*-hd, displayed a slower growth rate than the *dcr1* deletion strain, although not as slow as the wild type, indicating that even the Dicer lacking a helicase domain imparts partial functional complementation to the *dcr1Δ* mutant.

To further characterize the *S. pombe* Dicer mutants, sensitivity to the microtubule-destabilizing drug, thiabendazole (TBZ), was investigated. *dcr1Δ* cells display a strong sensitivity to TBZ, presumably due to increased chromosome missegregation. The vector with wild-type *dcr1+* fully rescued the TBZ sensitivity

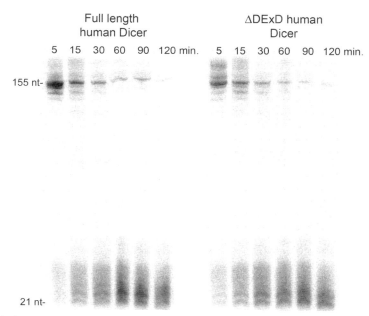

Figure 4. Truncation of the human Dicer helicase domain does not disrupt dicing activity. Full-length human Dicer and the truncated Dicer lacking its helicase (ΔDExD) were assayed in vitro.

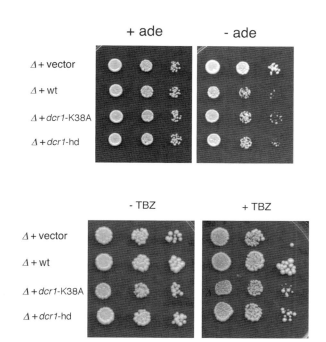

Figure 5. Functional complementation of *dcr1*Δ heterochromatin defect by *dcr1* helicase mutants, *dcr1*-K38A and *dcr1*-hd. (*Top*) *dcr1*Δ cells carrying a marker gene for adenine inserted in centromeric heterochromatin were transformed with a vector expressing *S. pombe* wild-type *dcr1*⁺, *dcr1*-K38A, *dcr1*-hd, or empty vector. Transformants were grown on selective media without supplementation of adenine to assay loss of silencing in the centromeric region. Each panel contains a tenfold serial dilution of each strain. The control plate (*left panel*) contains nonselective media. (*Bottom*) Sensitivity to the microtubule-destabilizing drug, TBZ, in *dcr1*Δ cells is partly rescued by expression of *dcr1* helicase mutants *dcr1*-K38A and *dcr1*-hd. Serial dilutions of indicated cultures were plated onto YES medium (*left panel*) or YES medium supplemented with 10 μg/ml TBZ.

phenotype of the *dcr1*Δ mutant. A partial functional rescue of *dcr1*Δ cells was also achieved by episomal expression of the *dcr1* helicase mutants, consistent with the results from the silencing assay (Fig. 5, bottom). Taken together, these results demonstrate that the helicase domain is not absolutely required for Dicer function in vivo. However, the helicase may still have a role in increasing efficient Dicer activity since overexpression of *dcr1* helicase mutants gave only partial rescue of the *dcr1*Δ mutant.

The finding that the Dicer helicase is not required for RNAi suggests that this domain does not have an essential role or that it is redundant with another factor. It is therefore unlikely that the helicase is solely responsible for a critical step in the process such as unwinding of the short interfering (siRNA) duplex. Consistent with this notion, we were unable to observe siRNA duplex unwinding activity using purified recombinant full-length human Dicer in vitro (data not shown).

We favor a model in which the Dicer helicase domain functions by facilitating the action of the enzyme in vivo. Two possible roles by which the helicase could increase Dicer activity are as a molecular motor for processivity or as an aid to the handoff of dsRNA substrates or products with other components of the RNAi machinery. Our finding that removal of the helicase domain from human Dicer does not impair dicing activity in a purified in vitro assay argues against a processivity function for the helicase. Consistent with this notion, we were unable to detect processivity during dsRNA processing with recombinant human Dicer (data not shown). We therefore suspect that the Dicer helicase functions in vivo to facilitate the handoff of dsRNA substrates or products with other components of the RNAi machinery. A likely possibility is that the dsRNA substrates of Dicer are not naked in the cell but

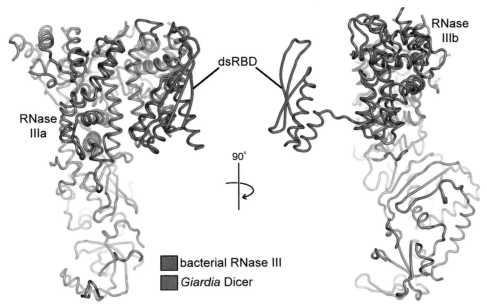

Figure 6. Modeling of a dsRBD onto the *Giardia* Dicer structure. The crystal structure of bacterial RNase III was truncated to contain only one dsRBD and then superimposed onto *Giardia* Dicer by aligning the RNase III domains of the two proteins.

are bound to a variety of cellular factors. The helicase domain of human Dicer may help remove or reposition these factors and allow Dicer to bind and process the RNA. Indeed, the DExD RNA helicases NPH-II and DED1 have been shown to catalyze protein displacement from RNA (Jankowsky et al. 2001; Fairman et al. 2004). This model is also consistent with the finding that in HeLa cell extracts, Dicer requires ATP to bind exogenous siRNAs (Pellino et al. 2005), whereas the purified recombinant human Dicer readily binds siRNAs in the absence of ATP (Zhang et al. 2002; Pellino et al. 2005). Furthermore, ATP has been shown to stimulate Dicer activity in crude mouse embryonal carcinoma cell lysates (Billy et al. 2001). On the other hand, *Drosophila* Dcr-2 requires ATP to process dsRNA, even when purified to near homogeneity (Liu et al. 2003). Likewise, immunopurified Dcr-1 from *Caenorhabditis elegans* shows ATP-dependent dicing activity (Ketting et al. 2001). It therefore appears that the Dicer helicase may serve multiple functions that are emphasized differentially in Dicer enzymes from different organisms.

In addition to lacking an amino-terminal helicase, *Giardia* Dicer does not contain the carboxy-terminal dsRBD common to most Dicer proteins. Although the function of this domain has never been directly tested, it is likely to be involved in binding dsRNA substrates in a fashion similar to that of the dsRBDs of bacterial RNase III. To help visualize how the dsRBD might fit into the structure of Dicer, we superimposed the crystal structure of bacterial RNase III (Gan et al. 2006) into *Giardia* Dicer by aligning their RNase III domains (Fig. 6). This modeling exercise reveals that the *Giardia* Dicer structure could easily accommodate a carboxy-terminal dsRBD and that in the larger Dicer proteins of higher eukaryotes, the dsRBD is likely to sit adjacent to

the RNase IIIb domain, outside of the nuclease core structure.

IMPLICATIONS FOR RISC LOADING

In addition to illuminating the structural basis of dsRNA processing, the structure of Dicer offers insight into another function of the enzyme: RISC loading. After hydrolysis of substrate RNAs, Dicer aids in loading its small dsRNA product into RISC. In *Drosophila*, Dicer-2 and its partner R2D2 are critical components of the protein complex responsible for loading small dsRNAs into RISC (Liu et al. 2003; Tomari et al. 2004b). Likewise, humans contain a similar RISC-loading complex, which is composed of Dicer and TRBP (TAR RNA-binding protein) bound to Argonaute (Gregory et al. 2005). The structure of Dicer, together with the structure and known biochemical properties of Argonaute (Ago) proteins, reveals the likely overall architecture of the RISC-loading complex.

The prevalent view of RISC loading is that thermodynamic asymmetry along the siRNA or miRNA duplex determines which RNA strand will be retained as the "guide" and which RNA will be discarded as the "passenger." Specifically, the RNA strand with its 5′ end at the thermodynamically less stable end of the siRNA duplex is preferentially loaded into RISC as the guide strand (Khvorova et al. 2003; Schwarz et al. 2003). Thermodynamic stability is thought to be sensed by R2D2, which binds to the more stable end of siRNAs as they are loaded into RISC (Tomari et al. 2004a). However, in addition to thermodynamic parameters, the orientation of the small RNA as it is produced by Dicer, or the processing polarity, can also have a role in guide strand selection (Rose et al. 2005).

The importance of processing polarity was uncovered by Rose et al. (2005), who were working to improve

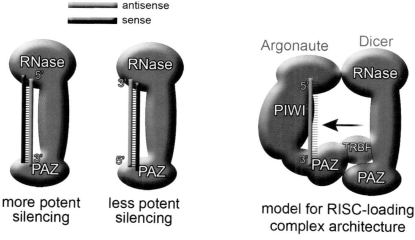

Figure 7. A model for the architecture of the RISC-loading complex. (*Left*) Silencing of a target gene is more effective when the 3′ end of the antisense strand in the siRNA duplex is bound to the Dicer PAZ domain. (*Right*) A model of the RISC-loading complex with Dicer and Argonaute PAZ domains adjacent to each other could explain the observed relationship between processing polarity and guide strand selection.

siRNA design for commercial research purposes. It had previously been shown that upon transfection into mammalian cells, 27-nucleotide dsRNAs are processed by Dicer into approximately 21-nucleotide products as they enter the RNAi pathway (Kim et al. 2005). These Dicer-substrate small RNAs are 100 times more potent inducers of gene silencing than the "prediced" 21-nucleotide siRNAs traditionally used in gene knockdown experiments, presumably because dsRNA cleavage by Dicer is coupled to RISC loading (Gregory et al. 2005; Maniataki and Mourelatos 2005). However, a disadvantage of 27-nucleotide dsRNAs is that they can be processed by Dicer from either end of the duplex, resulting in a variety of possible 21-nucleotide siRNA products and a higher chance for off-target effects. Rose et al. (2005) found that blunt-ended dsRNAs with DNA bases on the last two nucleotides of the 3′ ends are poorly processed by human Dicer. This finding allowed the design of 27-nucleotide dsRNA substrates that are blocked at one end, forcing Dicer cleavage to result in a single desired product.

The motivation for this work was to improve rational design of Dicer-substrate reagents, but the finding also allowed inspection of the relationship between dsRNA processing polarity and guide strand selection. Rose et al. (2005) prepared pairs of 27-nucleotide dsRNA Dicer substrates designed to produce the same final 21-nucleotide siRNA duplex, but with opposite processing polarity. In seven out of nine cases, silencing of a reporter gene was more efficient when the 5′ end of the strand antisense to the reporter mRNA was positioned at the blocked end of the 27-nucleotide Dicer substrate. Because the two substrates result in identical siRNA duplexes upon cleavage by Dicer, differences in silencing efficiency are independent of thermodynamic asymmetry and most likely due entirely to processing polarity. These results are consistent with the early studies of Elbashir et al. (2001) in *Drosophila* extracts, which also correlated dsRNA processing polarity with selective target strand cleavage.

These results have several interesting implications when considered together with the structure of Dicer. First, the observation that Dicer processing can be blocked by adding two DNA bases to a blunt-ended dsRNA suggests that the human Dicer PAZ domain cannot bind dsRNA modified in this way. Second, and more importantly, these results reveal that in the absence of thermodynamic asymmetry, the RNA strand that has its 3′ end bound to the PAZ domain of Dicer after cleavage is much more likely to be incorporated as the guide strand into RISC (Fig. 7, left). Along with the knowledge that in RISC, the 3′ end of the guide strand is bound to the PAZ domain of Argonaute (Song et al. 2003; Ma et al. 2004), this suggests that the siRNA duplex is passed from the PAZ domain of Dicer to the PAZ domain of Argonaute (Fig. 7, right). Thus, a likely model for the RISC-loading complex is one in which the PAZ domains of Dicer and Argonaute sit adjacent to one another and the RNase III domains of Dicer sit opposite the PIWI domain in Argonaute (Fig. 7). TRBP, which has been shown to interact with both Dicer and Argonaute (Chendrimada et al. 2005; Lee et al. 2006), may reside between the two where it could function in a manner similar to that of its *Drosophila* ortholog R2D2 in sensing siRNA thermodynamic asymmetry as the duplex is passed from Dicer to Argonaute.

CONCLUSIONS

The crystal structure of Dicer has revealed how this family of specialized ribonucleases measures and produces dsRNAs of specific lengths that function in various gene-silencing pathways. Furthermore, the structure has enabled detailed models for Dicer recognition of imperfectly base-paired substrates and for its interactions with other key proteins involved in RNAi. These models can now be tested experimentally, and provide a framework for investigating the activities of different Dicer enzymes in vivo.

ACKNOWLEDGMENTS

The authors thank the Life Sciences Research Foundation for a postdoctoral fellowship to I.J.M.; this work was supported in part by a grant from the National Institutes of Health to J.A.D.

REFERENCES

Bernstein E., Caudy A.A., Hammond S.M., and Hannon G.J. 2001. Role for a bidentate ribonuclease in the initiation step of RNA interference. *Nature* **409:** 363.

Billy E., Brondani V., Zhang H., Muller U., and Filipowicz W. 2001. Specific interference with gene expression induced by long, double-stranded RNA in mouse embryonal teratocarcinoma cell lines. *Proc. Natl. Acad. Sci.* **98:** 14428.

Chendrimada T.P., Gregory R.I., Kumaraswamy E., Norman J., Cooch N., Nishikura K., and Shiekhattar R. 2005. TRBP recruits the Dicer complex to Ago2 for microRNA processing and gene silencing. *Nature* **436:** 740.

Elbashir S.M., Lendeckel W., and Tuschl T. 2001. RNA interference is mediated by 21- and 22-nucleotide RNAs. *Genes Dev.* **15:** 188.

Fairman M.E., Maroney P.A., Wang W., Bowers H.A., Gollnick P., Nilsen T.W., and Jankowsky E. 2004. Protein displacement by DExH/D "RNA helicases" without duplex unwinding. *Science* **304:** 730.

Gan J., Tropea J.E., Austin B.P., Court D.L., Waugh D.S., and Ji X. 2006. Structural insight into the mechanism of double-stranded RNA processing by ribonuclease III. *Cell* **124:** 355.

Gregory R.I., Chendrimada T.P., Cooch N., and Shiekhattar R. 2005. Human RISC couples microRNA biogenesis and post-transcriptional gene silencing. *Cell* **123:** 631.

Hammond S.M., Bernstein E., Beach D., and Hannon G.J. 2000. An RNA-directed nuclease mediates post-transcriptional gene silencing in *Drosophila* cells. *Nature* **404:** 293.

Hammond S.M., Boettcher S., Caudy A.A., Kobayashi R., and Hannon G.J. 2001. Argonaute2, a link between genetic and biochemical analyses of RNAi. *Science* **293:** 1146.

Hannon G.J. 2002. RNA interference. *Nature* **418:** 244.

Hutvagner G., McLachlan J., Pasquinelli A.E., Balint E., Tuschl T., and Zamore P.D. 2001. A cellular function for the RNA-interference enzyme Dicer in the maturation of the let-7 small temporal RNA. *Science* **293:** 834.

Jankowsky E., Gross C.H., Shuman S., and Pyle A.M. 2001. Active disruption of an RNA-protein interaction by a DExH/D RNA helicase. *Science* **291:** 121.

Ketting R.F., Fischer S.E., Bernstein E., Sijen T., Hannon G.J., and Plasterk R.H. 2001. Dicer functions in RNA interference and in synthesis of small RNA involved in developmental timing in *C. elegans*. *Genes Dev.* **15:** 2654.

Khvorova A., Reynolds A., and Jayasena S.D. 2003. Functional siRNAs and miRNAs exhibit strand bias. *Cell* **115:** 209.

Kim D.H., Behlke M.A., Rose S.D., Chang M.S., Choi S., and Rossi J.J. 2005. Synthetic dsRNA Dicer substrates enhance RNAi potency and efficacy. *Nat. Biotechnol.* **23:** 222.

Lau N.C., Lim L.P., Weinstein E.G., and Bartel D.P. 2001. An abundant class of tiny RNAs with probable regulatory roles in *Caenorhabditis elegans*. *Science* **294:** 858.

Lee Y., Hur I., Park S.Y., Kim Y.K., Suh M.R., and Kim V.N. 2006. The role of PACT in the RNA silencing pathway. *EMBO J.* **25:** 522.

Liu Q., Rand T.A., Kalidas S., Du F., Kim H.E., Smith D.P., and Wang X. 2003. R2D2, a bridge between the initiation and effector steps of the *Drosophila* RNAi pathway. *Science* **301:** 1921.

Lucast L.J., Batey R.T., and Doudna J.A. 2001. Large-scale purification of a stable form of recombinant tobacco etch virus protease. *Biotechniques* **30:** 544.

Ma J.B., Ye K., and Patel D.J. 2004. Structural basis for overhang-specific small interfering RNA recognition by the PAZ domain. *Nature* **429:** 318.

MacRae I.J., Zhou K., Li F., Repic A., Brooks A.N., Cande W.Z., Adams P.D., and Doudna J.A. 2006. Structural basis for double-stranded RNA processing by Dicer. *Science* **311:** 195.

Maniataki E. and Mourelatos Z. 2005. A human, ATP-independent, RISC assembly machine fueled by pre-miRNA. *Genes Dev.* **19:** 2979.

Moreno S., Klar A., and Nurse P. 1991. Molecular genetic analysis of fission yeast *Schizosaccharomyces pombe*. *Methods Enzymol.* **194:** 795.

Myers J.W., Jones J.T., Meyer T., and Ferrell J.E., Jr. 2003. Recombinant Dicer efficiently converts large dsRNAs into siRNAs suitable for gene silencing. *Nat. Biotechnol.* **21:** 324.

Pasquinelli A.E., Reinhart B.J., Slack F., Martindale M.Q., Kuroda M.I., Maller B., Hayward D.C., Ball E.E., Degnan B., Muller P., et al. 2000. Conservation of the sequence and temporal expression of let-7 heterochronic regulatory RNA. *Nature* **408:** 86.

Pellino J.L., Jaskiewicz L., Filipowicz W., and Sontheimer E.J. 2005. ATP modulates siRNA interactions with an endogenous human Dicer complex. *RNA* **11:** 1719.

Pillai R.S., Bhattacharyya S.N., Artus C.G., Zoller T., Cougot N., Basyuk E., Bertrand E., and Filipowicz W. 2005. Inhibition of translational initiation by Let-7 MicroRNA in human cells. *Science* **309:** 1573.

Rose S.D., Kim D.H., Amarzguioui M., Heidel J.D., Collingwood M.A., Davis M.E., Rossi J.J., and Behlke M.A. 2005. Functional polarity is introduced by Dicer processing of short substrate RNAs. *Nucleic Acids Res.* **33:** 4140.

Schwarz D.S., Hutvagner G., Du T., Xu Z., Aronin N., and Zamore P.D. 2003. Asymmetry in the assembly of the RNAi enzyme complex. *Cell* **115:** 199.

Shibuya T., Tange T.O., Stroupe M.E., and Moore M.J. 2006. Mutational analysis of human eIF4AIII identifies regions necessary for exon junction complex formation and nonsense-mediated mRNA decay. *RNA* **12:** 360.

Song J.J., Smith S.K., Hannon G.J., and Joshua-Tor L. 2004. Crystal structure of Argonaute and its implications for RISC slicer activity. *Science* **305:** 1434.

Song J.J., Liu J., Tolia N.H., Schneiderman J., Smith S.K., Martienssen R.A., Hannon G.J., and Joshua-Tor L. 2003. The crystal structure of the Argonaute2 PAZ domain reveals an RNA binding motif in RNAi effector complexes. *Nat. Struct. Biol.* **10:** 1026.

Suhre K. and Sanejouand Y.H. 2004. ElNemo: A normal mode web server for protein movement analysis and the generation of templates for molecular replacement. *Nucleic Acids Res.* **32:** W610.

Tomari Y., Matranga C., Haley B., Martinez N., and Zamore P.D. 2004a. A protein sensor for siRNA asymmetry. *Science* **306:** 1377.

Tomari Y., Du T., Haley B., Schwarz D.S., Bennett R., Cook H.A., Koppetsch B.S., Theurkauf W.E., and Zamore P.D. 2004b. RISC assembly defects in the *Drosophila* RNAi mutant armitage. *Cell* **116:** 831.

Verdel A., Jia S., Gerber S., Sugiyama T., Gygi S., Grewal S.I., and Moazed D. 2004. RNAi-mediated targeting of heterochromatin by the RITS complex. *Science* **303:** 672.

Zhang H., Kolb F.A., Brondani V., Billy E., and Filipowicz W. 2002. Human Dicer preferentially cleaves dsRNAs at their termini without a requirement for ATP. *EMBO J.* **21:** 5875.

Zhang H., Kolb F.A., Jaskiewicz L., Westhof E., and Filipowicz W. 2004. Single processing center models for human Dicer and bacterial RNase III. *Cell* **118:** 57.

Structural Biology of RNA Silencing and Its Functional Implications

D.J. Patel,* J.-B. Ma,* Y.-R. Yuan,*‡ K. Ye,*¶ Y. Pei,† V. Kuryavyi,*
L. Malinina,*§ G. Meister, †** and T. Tuschl†
*Structural Biology Program, Memorial Sloan-Kettering Cancer Center, New York, New York 10021; †Howard
Hughes Medical Institute, Laboratory of RNA Biology, The Rockefeller University, New York, New York 10021

We outline structure–function contributions from our laboratories on protein–RNA recognition events that monitor siRNA length, 5′-phosphate and 2-nucleotide 3′ overhangs, as well as the architecture of Argonaute, its externally bound siRNA complex, and Argonaute-based models involving guide-strand-mediated mRNA binding, cleavage, and release.

Small RNAs mediate many gene regulatory events, including posttranscriptional gene silencing in plants, RNA interference (RNAi) in animals, the assembly and function of heterochromatin, as well as microRNA (miRNA) pathways at the genome level (for review, see Hannon 2002; Dykxhoorn et al. 2003; Bartel 2004; Baulcombe 2004; Meister and Tuschl 2004; Verdel et al. 2004; Filipowicz et al. 2005; Hammond 2005; Voinnet 2005; Zamore and Haley 2005; Kim and Nam 2006). These processes are collectively known as RNA silencing, which represents a conserved biological response to expressed double-stranded RNA molecules. Dicer-like enzymes (RNase III family nucleases) recognize and process these double-stranded RNAs (dsRNAs) into small interfering RNAs (siRNAs) (Bernstein et al. 2001), characterized by 21- to 25-nucleotide-long sequences (Hamilton and Baulcombe 1999; Hammond et al. 2000; Parrish et al. 2000; Zamore et al. 2000), composed of duplex (19–23 bp) and 2-nucleotide 3′-overhang segments, with 5′-phosphate and 3′-hydroxyl termini (Elbashir et al. 2001a, b). siRNAs mediate the silencing by guiding a nuclease complex, RNA-induced silencing complex (RISC), to recognize and cleave target mRNAs, with the targeting specificity dictated by base-pairing interaction between guide and target RNA strands (Hammond et al. 2000; Zamore et al. 2000; Elbashir et al. 2001a).

Structural biology approaches provide a unique perspective into the molecular events associated with specific recognition of siRNA by protein domains and the precise processing of mRNA targets mediated by Ago within the RISC (for review, see Collins and Cheng 2005; Filipowicz et al. 2005; Hall 2005; Lingel and Sattler 2005; Parker and Barford 2006; Song and Joshua-Tor 2006). We describe below the structure–function contributions from our laboratories on protein–RNA recogni-

tion events that monitor siRNA length, 5′-phosphate and 2-nucleotide 3′-overhangs, as well as the architecture of Argonaute (Ago) proteins, their externally bound siRNA complexes, and Ago-based models involving guide-strand-mediated mRNA binding, cleavage, and release. We conclude by putting our results into perspective by summarizing the structural results from other laboratories on dsRNA processing by Dicer and outlining future challenges associated with attempts to determine the structures of ternary protein–RNA complexes associated with three key steps in the biogenesis of miRNAs.

VIRAL SUPPRESSION THROUGH siRNA LENGTH MEASUREMENT

In plants, one of the major functions of RNA silencing is to act as a defense against invading viruses by sequence-specifically targeting their mRNAs for degradation in response to production of siRNAs from dsRNA replication intermediates (for review, see Baulcombe 2004; Wang and Metzlaff 2005). Many viruses, as a counter-defense against RNA silencing, encode proteins that can specifically inhibit the silencing machinery (for review, see Li and Ding 2001; Silhavy and Burgyan 2004; Voinnet 2005), including an inhibitor encoded by an animal virus (Li et al. 2002). This inhibition could occur through sequestration of the siRNA by the viral suppressor, and a molecular understanding of this process emerged following the successful crystal structure determination of the viral suppressor p19 (Silhavy et al. 2002; Lakatos et al. 2005) bound to siRNA residues from the Traci Tanaka-Hall (Vargason et al. 2003) and our (Ye et al. 2003) laboratories. Studies of viral suppressors are important for our understanding of RNA silencing mechanisms, given a recent result that a cellular miRNA mediates antiviral defense in human cells (Lecellier et al. 2005).

Viral Suppressor p19–siRNA Complex

Our efforts first focused on defining the siRNA structural element(s) involved in p19 recognition. Electrophoretic mobility-shift assays (EMSA) established that p19 bound

Present addresses: ‡Structural Biology Group, Temasek Life Sciences Laboratory, National University of Singapore, Singapore, 117604; ¶National Institute of Biological Sciences, 7 Science Park Road, Zhongguancun Life Science Park, Beijing, 102206, China; §CIC biGune, Technological Park of Bizkaia, 48160 Derio, Spain; **Max Planck Institute of Biochemistry, Am Klopferspitz 18, D-82152 Martinsried, Germany.

21-mer siRNAs, but not their corresponding DNA counterparts of the same length and sequence. Furthermore, although the duplex characteristics of siRNA were a prerequisite for p19 recognition, the 2-nucleotide 3′ overhangs could be deleted with little impact on complex formation. A systematic study of p19 recognition of siRNA duplexes ranging in length from 16 bp to 25 bp established that optimal p19 binding spanned duplexes in the 19- to 21-bp length range, with binding attenuated for either shorter or longer lengths. These results established that the viral suppressor p19 measures the length of siRNAs to a resolution of 20 ±1 bp (Fig. 1A) (Ye et al. 2003).

The crystal structure of p19 from *Tomato Bush Stunt virus* (*TBSV*) in complex with a 21-nucleotide (19-bp) siRNA at 1.85 Å resolution is shown in Figure 1B (Ye et al. 2003). p19 forms a symmetric dimer in the complex where the RNA-binding folds of each monomer align to form an antiparallel eight-stranded β-sheet. The siRNA duplex, which adopts an A-helix conformation, is cradled within the concave face of this β-sheet platform, burying 1300 Å^2 of solvent-accessible surface area.

Intermolecular protein–RNA contacts span the entire length of the electrostatically positive-charged surface of the p19 homodimer, thereby interacting with the central minor groove and two adjacent partial major grooves of the siRNA. The observed direct and water-mediated intermolecular contacts are restricted to the backbone phosphates and 2′-hydroxyl groups consistent with sequence-independent siRNA recognition by the p19 protein. The RNA minor-groove-binding face of p19 is rich in serine and threonine residues, thereby forming an unusual hydroxyl group network for hydrogen bonding with 2′-hydroxyl groups of the siRNA. DNA lacks 2′-hydroxyl groups, and this readily explains the ability of p19 to distinguish between siRNA and DNA counterpart duplexes.

The unique feature of p19 is highlighted by the presence of α-helical reading heads attached to the β-sheet-containing RNA-binding fold of each monomer. These α-helical reading heads project from opposite ends of the p19 homodimer, thereby positioning pairs of tryptophans that stack over the terminal base pairs; in essence, measuring and bracketing both ends of the siRNA duplex. The observed stacking of the 3′- and the 5′-terminal bases with W39 and universally conserved W42, respectively, essentially extends the terminal base pairs at either end by an additional step (Fig. 1C). This bracketing is further anchored by the stacking of the long hydrophobic side chain of conserved R43 over W39.

The peptide segments linking the α-helical reading heads to the β-sheet-containing RNA-binding fold are disordered in the crystal. Such structural plasticity could account for the ability of p19 to accommodate 19- to 21-mer siRNAs, its caliper-like action restricted to siRNAs spanning ± 1 bp in length.

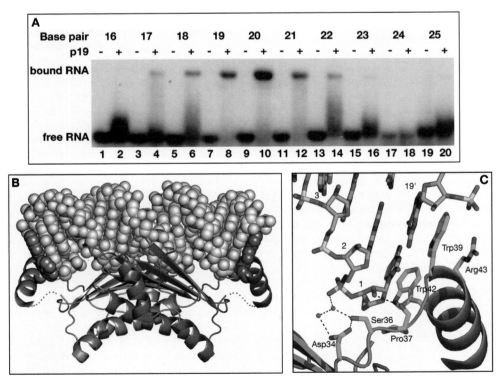

Figure 1. (*A*) Electrophoretic mobility-shift assays defining the length dependence of self-complementary siRNA duplexes for complex formation with viral suppressor p19. (*B*) The crystal structure of the *TBSV* p19–21-mer siRNA duplex (19-bp and 2-nucleotide 3′-overhang) (Ye et al. 2003). The RNA is shown in a space-filling representation with phosphorus atoms in red. The 2-nucleotide 3′-overhang bases are disordered and not shown. The protein is shown in a ribbon representation except for the pair of bracketing Trp residues at either end, which are shown in a space-filling representation. (*C*) Stick (RNA) and ribbon (protein) representation of stacking and hydrogen-bonding recognition of one RNA end in the complex. (Reprinted, with permission, from Ye et al. 2003 [Nature Publishing Group].)

Tanaka-Hall and colleagues focused their efforts on p19 from *Carnation Italian Ringspot virus* (*CIRV*) and have reported on the crystal structure of the complex of this p19 with a 21-mer siRNA (contains 19 bp) at 2.5 Å resolution (Vargason et al. 2003). There is excellent agreement between the structures of the *TBSV* (Ye et al. 2003) and *CIRV* (Vargason et al. 2003) p19–siRNA complexes, with both studies highlighting the principle that p19 acts as a molecular caliper for sequestering siRNAs based on their length.

The structural studies on the *CIRV* p19–siRNA complex were complemented by in vitro binding assays and in vivo suppression assays that monitored the impact of p19 mutations and siRNA variants on complex formation (Vargason et al. 2003). These studies established the importance of the bracketing residues W39 and W42, since substitution of both tryptophans simultaneously adversely affected complex formation. The binding constant was estimated to be 0.17 nM from EMSA-binding studies for *CIRV* p19 binding to a 21-mer siRNA containing a 5′-phosphate and a 2-nucleotide overhang at the 3′ end. Removal of the 5′-phosphate resulted in a 23-fold loss in binding, consistent with the 5′-phosphate forming a hydrogen bond with the indole NH of W42.

Viral Suppressor B2–dsRNA Complex

Recently, a crystal structure has also been reported for the complex of viral suppressor B2 from *insect Flock House virus* (*FHV*) and dsRNA at 2.6 Å resolution (Chao et al. 2005). *FHV* B2 suppressor has been shown to bind dsRNA independent of length with 1 nM affinity, but not dsDNA, single-stranded RNA (ssRNA), or DNA–RNA hybrids. The structure of the complex of FHV B2 with an 18-mer dsRNA has established that it does not use the unprecedented caliper-like principle highlighted above for the p19–siRNA complex (Vargason et al. 2003; Ye et al. 2003). Rather, B2 forms a dimer whose four-α-helix bundle topology binds to one face of the RNA duplex, thereby potentially inhibiting both cleavage of dsRNA by Dicer and incorporation of dsRNAs into RISC (Chao et al. 2005).

Viral Suppressor p21

One other viral suppressor, p21 (Chapman et al. 2004), has been characterized structurally in the free state, but not yet in its complex with siRNA. The crystal structure of the viral suppressor p21 from *Beet Yellow virus* (*BSV*) solved at 3.3 Å resolution in our laboratory exhibits an octameric ring architecture with a large central cavity of 90 Å in diameter (Ye and Patel 2005). Each monomer within the *BSV* p21 octamer is composed of amino- and carboxy-terminal α-helical domains, which associate with their neighboring counterparts through symmetric head-to-head and tail-to-tail interactions. In vitro binding studies establish that p21 binds both ssRNA and dsRNA, independent of length. There is a putative RNA-binding surface composed of conserved, positive charged amino acids, which line the inner surface of the octameric ring (Ye and Patel 2005).

siRNA 3′-OVERHANG RECOGNITION BY THE PAZ DOMAIN

The PAZ domain is an RNA-binding module found in Ago and some Dicer proteins. Despite initial suggestions regarding the PAZ domain's potential for mediating protein–protein interactions, it became clear as soon as its structure was determined and binding studies were undertaken that it is an RNA-binding module. Given this background, much effort has focused on elucidating the principles behind protein–RNA recognition involving the PAZ domain.

Architecture of the PAZ Domain

The structure of the PAZ domain has been determined in the free state and is composed of a central five-stranded β-barrel flanked by two α-helices, together with a flap-like segment composed of a β-hairpin and an α-helix (Lingel et al. 2003; Song et al. 2003; Yan et al. 2003). NMR chemical shift and mutation studies have mapped the nucleic-acid-binding region to a hydrophobic cleft containing conserved residues between its β-barrel core and αβ flap elements.

Structure of PAZ–siRNA–mimic-like Complex

We have solved the crystal structure of the PAZ domain from human Argonaute1 (*h*Ago1) bound to both ends of a 9-mer siRNA-like duplex. The RNA forms a 7-bp duplex with 2-nucleotide 3′-overhangs at either end, which are targeted by independent non-interacting PAZ modules (Ma et al. 2004). The two strands of the siRNA-like duplex contact specific PAZ domains in a highly asymmetric manner, such that the 3′-overhang-containing strand interacts with the PAZ domain along its entire 9-nucleotide length, whereas the complementary strand's interactions are restricted to its 5′-terminal residue (Fig. 2A). The bound RNA adopts a 5′ to 3′ stacked helical trajectory, except for a sharp clockwise 100° rotation in the phosphodiester backbone between the duplex and 2-nucleotide 3′-overhang segments (Fig. 2B), resulting in the insertion of the 2-nucleotide overhang into the preformed hydrophobic pocket between the central β-barrel and the αβ flap module.

The two overhang nucleotides at the 3′-end retain a stacked A-helical conformation, with the terminal residue buried deep within the pocket. Strikingly, the internucleotide phosphate linking the overhang bases is targeted through hydrogen bonding by the hydroxyls of three tyrosines, one of which is water-mediated, and a histidine, all conserved residues (Fig. 2C) (Ma et al. 2004). The sugar ring of the terminal overhang residue is anchored in place through van der Waals packing, augmented by hydrogen bonds of its 2′- and 3′-OH groups with the amide backbone, while portions of its base and sugar stack over a conserved Phe ring. Mutations of the aromatic and hydrophobic residues that line the binding pocket and contact the terminal overhang base, sugar, and internucleotide linkage, adversely affect binding affinity. In contrast, the penultimate overhang base and sugar make fewer contacts with PAZ residues.

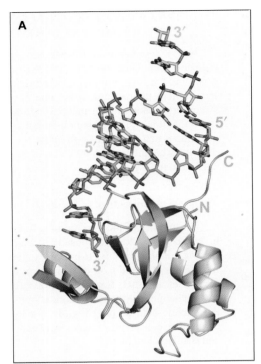
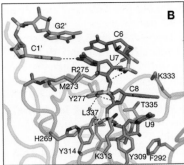
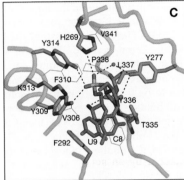

Figure 2. (*A*) The structure of one PAZ domain interacting with an siRNA-like end in the crystal structure of the human Ago1 PAZ-9-mer siRNA duplex (7-bp and 2-nucleotide 3′-overhang) (Ma et al. 2004). The complex is shown in stick (RNA) and ribbon (protein) representation. The RNA strand bound by its 3′-end is colored blue, whereas the strand bound by the 5′-end is colored yellow. (*B*) A stick representation of the interaction between the PAZ domain with the duplex terminus and 3′-overhang segments. Note the change in the RNA phosphodiester backbone between duplex and overhang segments. (*C*) Interactions between the 2-nucleotide 3′-overhang of the RNA and the walls of the conserved binding pocket on the PAZ domain. The residues directly contacting the RNA are shown in a stick representation, and other residues contributing to the hydrophobic pocket are shown as thin lines. Key hydrogen-bonding interactions from protein side chains to the internucleotide phosphate and the sugar hydroxyls are highlighted by dashed lines. (Reprinted, with permission, from Ma et al. 2004 [Nature Publishing Group].)

The stacked overhang bases are not involved in hydrogen bonding and face outward within an open pocket, suggesting that both pyrimidines and the larger purines can be equally well accommodated in a sequence-independent manner. The overhang-containing strand tracks the surface of the PAZ domain through electrostatic contacts between its phosphate backbone and five basic arginine and lysine side chains. In contrast, the base edges positioned within either groove of the RNA helix are not contacted by the PAZ domain, consistent with sequence-independent recognition.

The 2-nucleotide overhang bases contribute critically to binding, since removal of one or both overhang nucleotides results in 85-fold and >5000-fold reduction in binding affinity as measured by surface plasmon resonance. The binding pocket is incompatible with bulky 3′-end modifications, but accommodates a 2′-OCH$_3$ with an 18-fold reduction in affinity. The PAZ domain also binds ssRNA, but with 50-fold reduction in binding affinity (Ma et al. 2004).

The above structural studies complemented by binding assays on the *h*Ago1 PAZ domain bound to a siRNA-like duplex (Ma et al. 2004) establish that PAZ could serve as an siRNA-end-binding module for siRNA transfer in the RNA silencing pathway, and as an anchoring site for the 3′-end of guide RNA within silencing effector complexes.

The next challenge will be to solve the structure of Dicer PAZ bound to siRNA and identify similarities and differences with the known structure of the *h*Ago1 PAZ–siRNA complex.

Structure of PAZ-ssRNA Complex

The above studies on the crystal structure of *h*Ago1 PAZ–siRNA-like complex have been complemented by NMR structural studies of a uridine nucleotide bound to *Drosophila melanogaster* Ago1 (*Dm*Ago1) PAZ (Yan et al. 2003) and a 5-mer ssRNA bound to *Dm*Ago2 PAZ (Lingel et al. 2004). The terminal nucleotide binds in the hydrophobic binding pocket without sequence specificity. There are differences in detail between the stacked overhang bases in the crystal structure of the *h*Ago1 PAZ–siRNA-like complex (Ma et al. 2004) and the partially stacked overhang bases in the *Dm*Ago2–ssRNA complex (Lingel et al. 2004), perhaps suggestive of conformational plasticity within the open binding pocket.

siRNA 5′-PHOSPHATE RECOGNITION BY THE Piwi PROTEIN

The Piwi protein from the archaen *Archaeoglobus fulgidus* (*Af*) has played a key role in our understanding of siRNA 5′-phosphate end recognition at the molecular level. The Mid and PIWI domains of the *Af*Piwi protein also constitute the carboxy-terminal half of Ago proteins and hence structure–function studies of *Af*Piwi–siRNA complexes directly affect our functional understanding of guide-strand-mediated mRNA cleavage by Ago protein within RISC. Indeed, the 5′-phosphate end of siRNA is a prerequisite for both the stability and slicing fidelity during the processing steps within RISC (Rivas et al. 2005).

Architecture of a Bacterial Piwi Protein

The crystal structure of *Af*Piwi protein in the free state at 1.85 Å resolution established its two-domain architecture, with a substantial interface between its component Mid and PIWI domains (Parker et al. 2004). Strikingly, the PIWI domain adopted an RNase H scaffold, imparting the Piwi protein, as was first outlined for the Ago protein (Song et al. 2004), with potential mRNA cleavage or "slicer" activity. In addition, a basic pocket was identified within the Mid domain that was lined by pairs of conserved lysine and glutamine residues, supplemented by insertion of the carboxy-terminal carboxylate, whose charge was neutralized by an accompanying divalent cation (Parker et al. 2004).

Structure of a Bacterial Piwi–siRNA Complex

Biochemical studies in *Dm* lysates have demonstrated that the 5′-phosphate on the guide RNA is essential for its assembly into RISC (Nykanen et al. 2001). 5′-Phosphorylated single-stranded siRNAs exhibit more potent gene silencing activity than their non-phosphorylated counterparts in HeLa cells (Martinez et al. 2002). In addition, the position of the cleavage site within the mRNA that is paired to the guide RNA in RISC is defined by the 5′-end of the guide RNA strand (Elbashir et al. 2001a,b). A major advance in the field occurred with the reporting of the crystal structures of *Af*Piwi protein bound to siRNA simultaneously from the David Barford laboratory (Parker et al. 2005) and our laboratory (Ma et al. 2005). These studies provided the structural basis for guide-strand 5′-end recognition and insights into guide-strand-mediated mRNA target cleavage.

Our group has solved the 2.5 Å crystal structure of *Af*Piwi protein bound to a 5′-phosphate-containing 21-mer (19-bp) siRNA, capable of self-complementary duplex formation (Ma et al. 2005). We could trace the electron density for the Mid and PIWI domains of the Piwi protein, but we were restricted to a short 4-bp segment of the RNA duplex, positioned adjacent to the 5′-phosphate and its attached unpaired nucleotide (Fig. 3A). The identifiable RNA duplex segment is positioned within a basic channel spanning the Mid–PIWI interdomain interface of the protein, whereas the 5′-phosphate inserts into the conserved basic pocket composed mainly of residues from the Mid domain and the carboxy-terminal carboxylate of the PIWI domain. The 5′-phosphate is anchored within this pocket through a network of hydrogen bonds involving the side chains of Y123, K127, K163, and Q137, and a bound divalent cation (Fig. 3B).

The divalent cation is octahedrally coordinated to two nucleic acid phosphate oxygens, three protein-based oxygens, and a water molecule in the complex. The similarities between the free (Parker et al. 2004) and siRNA-bound (Ma et al. 2005; Parker et al. 2005) *Af*Piwi protein structures is consistent with formation of a preformed 5′-phosphate-binding pocket.

We propose that it is the guide RNA strand whose phosphate is anchored within the basic pocket of the Piwi protein. The base (A1 in our case) at the 5′-end of the guide RNA strand is unpaired and stacks over the aromatic ring of invariant Y123 in the complex (Fig. 3B). The two paired RNA strands bind the Piwi protein in a strongly asymmetric manner, such that the 5′-end of the guide strand (colored red in Fig. 3A) interacts extensively with the protein, whereas minimal contacts are made with the mRNA strand (colored green, Fig. 3A).

The wild-type *Af*Piwi protein binds the siRNA with an apparent K_d of 0.19 μM as monitored from filter-binding studies in our laboratory. Mutations of K127, Q137, and K163 that hydrogen-bond to the 5′-phosphate and Y123 that stacks on the unpaired base adjacent to the 5′-phosphate, either one at a time or in pairs to alanine, result in less than an order of magnitude loss in binding affinity (Ma et al. 2005). The 5′-phosphate stabilization involves multiple hydrogen-bonding interactions with conserved residues that line the binding pocket, and it appears that disruption of either single or pairs of interactions result in a modest decrease in binding affinity.

5′-End Specific Recognition of Guide Strand

A significant portion of the bound 21-mer is disordered in the crystal structure of the complex. We therefore constructed a model in which the trajectory of the 4-bp duplex was extended to form two turns of an A-form helix that spans one face of both the Mid and PIWI domains (Fig. 3C). In this model, the proposed mRNA cleavage site, between positions 10 and 11 as defined from the 5′-end of the guide strand (Elbashir et al. 2001a,b), is positioned opposite the putative catalytic residues associated with the RNAse H fold of the PIWI domain.

We next used sequence alignments to identify the corresponding amino acids that would line the 5′-end-binding pocket of *h*Ago2, the RISC component that mediates guide-strand-mediated cleavage of target mRNA (Liu et al. 2004; Meister et al. 2004). This allowed us to investigate whether mutation of key residues lining the 5′-end-binding pocket would affect *h*Ago2 mRNA cleavage activity. Although efficient cleavage of radiolabeled target mRNA was observed for wild-type *h*Ago2, single alanine mutants of the *h*Ago2-binding pocket residues K533, Q545, and K570 showed reduced cleavage activity, whereas double alanine mutants showed more severe defects in cleavage capacity (Ma et al. 2005). These results establish that the 5′-end-binding pocket identified in *Af*Piwi protein is conserved in *h*Ago2 and is critical for its small RNA-guided cleavage activity.

Given that there are no specific intermolecular contacts directed to the 2′-OH groups of the RNA in the crystal structure of the complex, it appears likely that the *Af*Piwi

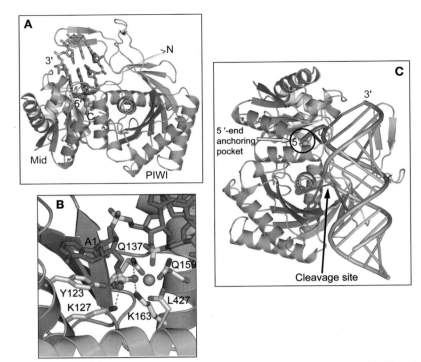

Figure 3. (*A*) The crystal structure of the *A. fulgidus* Piwi-21-mer siRNA duplex (19-bp and 2-nucleotide 3′-overhang) (Ma et al. 2005). The Piwi protein, shown in a ribbon representation, consists of a yellow-colored amino-terminal element (1–37), a magenta-colored Mid domain (38–167), and a cyan-colored PIWI domain (168–427). The segment of bound RNA that can be monitored in the structure is shown in a stick representation, with the protein interacting strand containing the 5′-phosphate colored red and its complementary partner strand colored green. (*B*) The 5′-phosphate-binding site in the complex. The 5′-phosphate is positioned in a basic pocket lined by K127, K163, Q137, and Q159 of the Mid domain and the carboxy-terminal carboxylate from the PIWI domain, and a bound divalent cation in orange. Bases A1 and G2 are splayed apart, with unpaired A1 stacked on Y123. (*C*) Model of the *A. fulgidus* Piwi protein bound to a 5′-phosphate-containing siRNA duplex complex. The 5′-end nucleotide and 4-bp duplex (colored in *red* for the guide strand and *green* for the target strand) observed in the crystal structure of the complex were extended by an A-form duplex (colored *tan* for the guide strand and *light green* for the target strand). The 5′-end anchoring pocket and putative catalytic site are circled and labeled by an arrow, respectively, and the phosphate at the cleavage site is marked by a yellow ball. (Reprinted, with permission, from Ma et al. 2005 [Nature Publishing Group].)

protein may also target other types of nucleic acids. Surprisingly, the results of double filter-binding studies establish that the binding is 20-fold tighter for ssDNA than for ssRNA, with binding affinity decreasing on proceeding from DNA–DNA and RNA–DNA to RNA–RNA duplexes (Ma et al. 2005). Related assays also established that 5′-phosphate-containing RNA binds the *Af*Piwi protein an order of magnitude more tightly than its non-phosphorylated counterpart.

The structure of the complex revealed that the stacked bases 2 to 5 at the 5′-end of the guide RNA are accessible for pairing with their complementary counterparts on the mRNA. Biochemical and bioinformatics analyses of miRNA genes have indicated a high degree of sequence conservation restricted to positions 2–8, indicative of a significant contribution of the 5′-end to activity and specificity (Doench and Sharp 2004; Mallory et al. 2004; Lewis et al. 2005). This concept receives support from the structure of our complex, in which the mRNA initially nucleates with the 5′-end of the protein-bound guide RNA strand (Stark et al. 2003; Haley and Zamore 2004), followed by the remaining segment zippering-up to form the guide RNA–mRNA duplex. Our observation that the first base of the guide RNA strand is not available for pairing with the mRNA in the structure of the complex is consistent with previously

reported non-sequence-specific recognition of the first nucleotide within the 5′-end-binding pocket (Stark et al. 2003; Doench and Sharp 2004; Lewis et al. 2005) and that disruption of this pair has no effect on cleavage activity, and under certain conditions can even facilitate target cleavage (Haley and Zamore 2004).

The related crystal structure of the *Af*Piwi protein bound to a 16-nucleotide siRNA was solved at 2.2 Å resolution (Parker et al. 2005), simultaneously with our contribution (Ma et al. 2005). This group could monitor 8 bp of duplex, together with the 5′-phosphate and its adjacent unstacked base positioned within the binding pocket. In addition, they could also follow the 2-nucleotide overhang at the 3′-end, which was shown to pass through a short exit channel. In essence, both groups have observed similar structures for the complexes and reached similar conclusions related to their functional impact. In summary, the structures of the *Af*Piwi protein–siRNA complexes (Ma et al. 2005; Parker et al. 2005) have provided the first molecular insights into 5′-end recognition of the guide RNA strand, aspects of the nucleation step associated with guide RNA–mRNA recognition, and the basis for distance-based identification of the mRNA cleavage site. Most importantly, perturbation of the 5′-end-binding pocket affects on mRNA cleavage efficiency (Ma et al. 2005).

ARGONAUTE—THE CATALYTIC ENGINE WITHIN RISC

RISC plays a key role in modulating eukaryotic gene expression through either chromatin remodeling, sequence-specific mRNA cleavage, or translational repression (for review, see Meister and Tuschl 2004; Sontheimer 2005; Tomari and Zamore 2005). RISC-associated Ago proteins (Cerutti et al. 2000; Hammond et al. 2001; Carmell et al. 2002), with their PAZ and PIWI domains, provide both architectural and catalytic functionalities associated with this scaffold serving as a platform for siRNA guide-strand selection and subsequent guide-strand-mediated cleavage of complementary mRNAs (Hutvanger and Zamore 2002; Martinez et al. 2002).

Architecture of Bacterial Argonautes

Our initial understanding of Ago architecture emerged following the report of the 2.25 Å crystal structure of *Pyrococcus furiosus* Ago (*Pf*Ago) in the free state (Song et al. 2004). In contrast to the *Af*Piwi protein that is composed of only two domains, the archaeal *Pf*Ago is composed of four domains labeled N, PAZ, Mid, and PIWI,

with the PIWI domain shown to adopt a RNase H fold composed of a conserved DD-containing catalytic motif with potential for target RNA cleavage activity. Subsequently, our group reported the 2.9 Å crystal structure of *Aquifex aeolicus* Ago (*Aa*Ago) and showed that the eubacterial *Aa*Ago adopts a bilobal architecture composed of PAZ-containing (PAZ and N domains and linkers L1 and L2) and PIWI-containing (Mid and PIWI domains) lobes (Fig. 4A) (Yuan et al. 2005).

The N, L1, and PAZ domains within the PAZ-containing lobe of *Aa*Ago are placed in a triangular arrangement (Fig. 4B), with each domain interacting with the other two domains. The linker L1 and L2 are in close proximity and, together with the N and PAZ domains, form a compact globular fold. The PAZ domain within the *Aa*Ago context contains the same nucleic-acid-binding pocket as reported previously for isolated PAZ domains (Lingel et al. 2003, 2004; Song et al. 2003; Yan et al. 2003) and its siRNA complex (Ma et al. 2004). The *Aa*Ago PAZ pocket is lined by a similarly spatially positioned cluster of hydrophobic and aromatic amino acids, implying a similar functional role in binding the 3′-end and in orienting the backbone of the guide RNA. One surface of the PAZ-containing lobe, the one directed

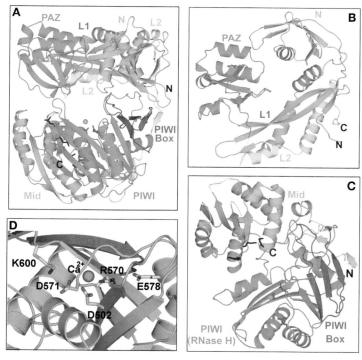

Figure 4. The crystal structure of *A. aeolicus* Argonaute (Yuan et al. 2005). The protein was crystallized in the presence of rU$_8$, but the RNA was disordered and could not be traced. The view emphasizes the bilobal topology of Ago, with the amino-terminal PAZ-containing lobe on top and the carboxy-terminal PIWI-containing lobe on the bottom. The various domains and linkers are numbered and color-coded. (*B*) Relative arrangements of the magenta-colored N (1–108), green-colored linker L1 (108–165), cyan-colored PAZ (166–262), and segment of yellow-colored linker L2 (263–311) within the PAZ-containing lobe. A Trp ring lining the PAZ-binding pocket is colored orange. (*C*) Relative arrangements of segment of the yellow-colored linker L2 (312–334), magenta-colored Mid domain (335–488), and cyan-colored PIWI domain (489–706) within the PIWI-containing lobe. Lys residues lining the 5′-phosphate-binding pocket on the Mid domain are colored dark blue. The DDE triad residues lining the catalytic binding pocket on the PIWI domain are colored red. The PIWI box (622–650) is colored red. (*D*) Relative positioning of invariant catalytic acidic D502, D571, and E578 residues and bound Ca cation on the surface of the RNase H fold of the PIWI domain. Invariant acidic R570 is also positioned in the catalytic pocket, whereas conserved basic K600 is directed toward the catalytic pocket. The Ca cation is also coordinated by D683, which is a His residue in *h*Ago2. (Reprinted, with permission, from Yuan et al. 2005 [© Elsevier].)

toward the PIWI-containing lobe, exhibits a pronounced basic electrostatic surface.

The Mid and PIWI domains within the PIWI-containing lobe of AaAgo interact with each other through an extensive interface (Fig. 4C). This lobe contains the 5′-phosphate recognition pocket identified earlier for the AfPiwi protein (Parker et al. 2004) and its siRNA complexes (Parker et al. 2004; Ma et al. 2005). As reported previously for AfPiwi, this pocket in AaAgo is also positioned near the interface between the Mid and PIWI domains and is lined by conserved aromatic and basic residues positioned on the surface of the Mid domain and the inserted carboxy-terminal carboxylate from the PIWI domain. The PIWI domain of AaAgo adopts an RNase H fold containing a catalytic DD-containing motif, with spatial characteristics similar to that which emerged from the crystal structures of PfAgo (Song et al. 2004) and AfPiwi (Parker et al. 2004). The RNase-H-like scaffold of the PIWI domain of AaAgo projects highly conserved (D502, D571, and E578) and nonconserved (D683) acidic residues, a basic residue (R570) and a bound Ca^{2+} cation (Fig. 4D). The spatial organization of the DD-containing motif projecting from the RNase H scaffold of the AaAgo PIWI-containing lobe has similarities with the corresponding DDE-containing RNase H motifs in RNase HII (Lai et al. 2000), as well as related motifs in integrases, reverse transcriptases, and transposases (Yang and Steitz 1995). The AaAgo PIWI-containing lobe also contains a PIWI box (Fig. 4A) (Cerutti et al. 2000), composed of three β-strands, and is positioned close to a pivot point linking the two lobes, and given its surface position, is available for recognition by Dicer (Tahbaz et al. 2004). One surface of the PIWI-containing lobe, the one directed toward the PAZ-containing lobe, exhibits a pronounced basic electrostatic surface, with its most basic segment lining the periphery of the composite 5′-phosphate-binding pocket.

Comparison of the PfAgo (Song et al. 2004) and AaAgo (Yuan et al. 2005) crystal structures highlights some differences in architecture and also perspectives as to relative alignments of individual domains. Thus, the AaAgo PAZ domain adopts a "closed" architecture with an α-helical pair subdomain positioned over the central β-barrel (Yuan et al. 2005), whereas the PfAgo PAZ domain adopts an "open" architecture, with the α-helical pair subdomain flipped out of the central β-barrel (Song et al. 2004). The AaAgo structure has a narrow channel between the N and PAZ domains and a wide channel between the PAZ and PIWI domains (Yuan et al. 2005), with the opposite trend observed in the PfAgo structure (Song et al. 2004). Finally, the AaAgo structure adopts a bilobal architecture defined by PAZ-containing (N and PAZ domains) and PIWI-containing (Mid and PIWI domains) lobes connected by a short hinge segment (Fig. 4A) (Yuan et al. 2005). In contrast, the PfAgo structure adopts an architecture where the PAZ domain is positioned over a crescent-shaped scaffold composed of the remaining three domains (Song et al. 2004). These contrasting architectures identify different linkers and hinge elements as potential contributors to inter-domain flexibility.

Bacterial Argonautes as Site-specific DNA-guided Endoribonucleases

The binding affinity and specificity of AaAgo for nucleic acids was tested in our laboratory using a double filter-binding assay (Wong and Lohman 1993). Unexpectedly, AaAgo bound most tightly to a 21-mer ssDNA (0.01 μM), with a tenfold reduction on binding to the corresponding 21-mer ssRNA. AaAgo also bound more tightly to a DNA–RNA hybrid duplex (0.64 μM) compared to a dsRNA duplex (>10 μM) (Yuan et al. 2005). Our laboratory also measured the same trend in binding affinities with PfAgo and *Thermus thermophilus* (Tth) Ago (J.-B. Ma et al., unpubl.).

The catalytic cleavage activity of AaAgo for radiolabeled target mRNA was tested in our laboratory as a function of 5′-phosphorylated guide DNA and RNA strands of varying length (18, 21, and 24 nucleotides), divalent ions (Ca, Mg, Mn) and temperature (35°C and 55°C). Cleavage by thermophilic AaAgo was most efficient for guide DNA strands (independent of length), in the presence of Mn as divalent cation, and at elevated temperatures (55°C) (Yuan et al. 2005). These unanticipated findings open new avenues for future investigation of the role of guide DNAs in mediating the functional activities of archaeal and eubacterial Ago proteins.

The number of divalent cations required for Ago-mediated slicing activity (Schwarz et al. 2004) needs further investigation. The catalytic triad residues are coordinated to a divalent ion in the AaAgo structure (Yuan et al. 2005), and Mn can be incorporated into the same site following soaking of this divalent cation into crystals of PfAgo (Rivas et al. 2005). One anticipates that the "slicer" activity of Ago is associated with a catalytic triad involving two coordinated divalent cations, similar to what has been recently observed for RNase H complexes with a DNA/RNA hybrid (Nowotny et al. 2005). Perhaps the second divalent cation accompanies the bound nucleic acid and will only be observable structurally in Ago complexes containing bound nucleic acid.

Although it was previously proposed that the guide strand was selectively loaded onto Ago within RISC (Schwarz et al. 2003), more recent research is consistent with Ago receiving the siRNA duplex (Matranga et al. 2005; Rand et al. 2005), and using either its catalytic potential to cleave the passenger strand or to expel the passenger strand with an RNA helicase action in case the Ago protein has no intrinsic cleavage activity.

Model of Guide Strand-mediated mRNA Binding, Cleavage, and Release

We have proposed a four-step catalytic cycle model of guide-strand-mediated mRNA binding, cleavage, and release within the context of the Ago scaffold within RISC (Fig. 5A) (Yuan et al. 2005), with similar perspectives into the mechanism of guide-strand-mediated mRNA cleavage also outlined in published reviews on RNA interference (Filipowicz 2005; Tomari and Zamore 2005). Conformer I, which corresponds to guide-strand-bound Ago, involves anchoring of both the 5′-phosphate and 2-nucleotide 3′-overhang ends of the guide strand in the basic pocket of the

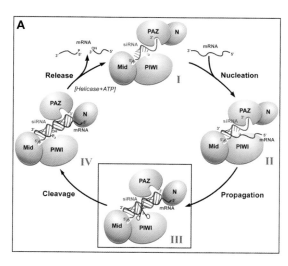

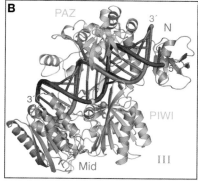

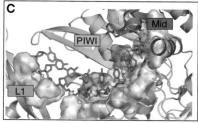

Figure 5. (*A*) Schematic of the catalytic cycle involving guide-RNA-dictated mRNA loading, cleavage, and product release within the context of the Ago scaffold (Yuan et al. 2005). The states represented by conformers I, II, III, and IV are described in the text. The mRNA nucleation step corresponds to the transition from conformer I to II, the mRNA propagation step corresponds to the transition from conformer II to III, the mRNA cleavage step corresponds to the transition from conformer III to IV, and the product release step corresponds to the transition from conformer IV back to conformer I. (*B*) A view of the model of the *Aa*Ago–DNA/RNA hybrid complex. The orientation of Ago is rotated along the vertical axis relative to the one shown in Fig. 4A. The color coding and labeling of domains and linkers, as well as key amino acids, are the same as listed in Fig. 4A. The hybrid duplex between the guide DNA strand (colored *red*) and the mRNA strand (colored *blue*) is shown in a tubular representation, with a thicker diameter for the sugar-phosphate backbone and thinner diameter for the bases. The cleavable phosphate positioned between residues 10 and 11 on the mRNA strand (as counted from the 5′-end of the guide strand) is shown by a yellow ball. (*C*) The phosphodiester backbone corresponding to positions 2–8 from the 5′-end of the guide strand are positioned within a trough-like segment of the Mid and PIWI domains in the model of the complex. The guide strand is shown in red, with phosphorus atoms as yellow balls. The trough is shown in a surface representation and exhibits surface complementarity with the sugar-phosphate backbone of the 5′-end region of the guide strand. (Reprinted, with permission, from Yuan et al. 2005 [© Elsevier].)

Mid domain and the aromatic-lined pocket of the PAZ domain, respectively. The bases are restricted to a stacked helical conformation toward the 5′-end due to constraints imposed by the binding channel, and their Watson-Crick edges are available for pairing with the target strand. Conformer II, which corresponds to the nucleation step, involves annealing of the mRNA with the accessible Watson-Crick base edges of the 5′-end of the guide strand, thereby maximizing pairing interactions spanning residues 2–8 of the guide strand. Conformer III, which corresponds to the propagation step, involves zippering-up of the mRNA to form a full-length bound guide strand–mRNA duplex. It is anticipated that steric constraints associated with this conformer will require release of the 2-nucleotide 3′-overhang from its binding pocket within the PAZ domain. Conformer IV, which corresponds to the cleavage step, involves cleavage of the mRNA phosphodiester bond between residues 10 and 11, as measured from the 5′-end of the guide strand, by the precisely positioned catalytic DD-containing residues of the RNase H motif of the PIWI domain. Release of mRNA fragments occurs on transition from conformer IV back to conformer I, perhaps facilitated by ATP-dependent RNA helicases. The catalytic cleavage models (Filipowicz 2005; Tomari and Zamore 2005; Yuan et al. 2005) incorporate concepts that have emerged from structural studies of

PAZ–siRNA (Ma et al. 2004) and Piwi–siRNA (Ma et al. 2005; Parker et al. 2005) complexes, as well as ideas that relate to preorganization of the 5′-end (Haley and Zamore 2004; Martinez and Tuschl 2004) and concepts that have emerged from miRNA target predictions (Lewis et al. 2003, 2005; Mallory et al. 2004), together with functional (Doench et al. 2003; Doench and Sharp 2004) and kinetic (Haley and Zamore 2004) studies on the contribution of 3′ bases of small RNAs to catalytic rate.

Sterochemically Robust Model of *Aa*Ago Bound to Guide DNA–mRNA

Our modeling efforts have focused on conformer III of the *Aa*Ago complex with a bound duplex composed of a guide DNA and mRNA strands (Yuan et al. 2005), given the findings of the double filter-binding and cleavage assays. The guide DNA–mRNA duplex was successively docked into a stereochemically compatible basic channel within the *Aa*Ago scaffold by anchoring the 5′-phosphate of the guide DNA strand within the 5′-end recognition pocket of the Mid domain, while positioning the scissible mRNA phosphate near the Ca-coordinated DD-containing motif of the RNase H fold of the PIWI domain, with maintenance of the DNA–RNA duplex in a standard helical form. Steric clashes could be relieved by interactive modeling and required repositioning and reorientation of L1 linker and the PAZ domain. A stereochemically robust model of the complex was obtained following molecular dynamics calculations. A view of the model of conformer III is shown in Figure 5B, with the guide DNA and mRNA strands colored red and blue, respectively, positioned between the basic electrostatic surfaces of the mutually facing lobes of the *Aa*Ago bilobal scaffold (Yuan et al. 2005). The phosphodiester backbone corresponding to positions 2–8 from the 5′-end of the guide strand are positioned in a stacked alignment within a trough-like segment of the Mid and PIWI domains (Fig. 5C), with their Watson-Crick edges available for recognition with complementary residues of the mRNA strand.

Structure of *Aa*Ago with Externally Bound siRNA

Considerable effort has gone into attempts at generating complexes of *Aa*Ago with single- and double-stranded nucleic acids containing RNA and/or DNA strands for structural characterization of one or more of the conformers in the proposed catalytic cycle outlined in Figure 5A. These efforts have not succeeded to date but, unexpectedly, have resulted in the crystallographic characterization of *Aa*Ago with an externally bound siRNA. Our group has solved the 3.0 Å crystal structures of 22-mer and 26-mer siRNAs bound to *Aa*Ago, where one 2-nucleotide 3′-overhang of the siRNA inserts into a cavity positioned on the outer surface of the PAZ-containing lobe of the bilobal *Aa*Ago architecture in both complexes (Fig. 6A) (Yuan et al. 2006). The first overhang nucleotide stacks over a tyrosine ring of Y119 (Fig. 6B), whereas the second overhang nucleotide, together with the intervening sugar-phosphate backbone, inserts into a preformed surface cavity (Figs. 6B, C).

Photochemical cross-linking experiments on complexes of 5-iodouridine-labeled siRNA in our laboratory provide support for this externally bound *Aa*Ago–siRNA complex (Yuan et al. 2006). Specifically, cross-linking was only observed when 5-iodouridine was incorporated at the

penultimate 3′-overhang position, but not at the 5′-end or in the center of the siRNA duplex, with the extent of cross-linking strongly attenuated for the Y119A mutant. The structure and biochemical results on the externally bound *Aa*Ago–siRNA complex together provide insights into a protein–RNA recognition event that could potentially be associated with the RISC-loading pathway.

Cleavage Specificity of Human Argonautes

There are four human Ago proteins labeled 1 to 4. Functional studies have established that only *h*Ago2 is capable of guide-strand-mediated cleavage of a complementary RNA target (Hutvagner and Zamore 2002; Martinez et al. 2002; Liu et al. 2004; Meister et al. 2004). The catalytic residues responsible for target cleavage within the PIWI domain of *h*Ago2 have been identified from mutation studies. Thus, Ala substitution of D597 and D669 resulted in retention of binding but loss of cleavage activity in both in vitro and in vivo experiments (Liu et al. 2004). The catalytic triad contributing to phosphodiester cleavage chemistry in *h*Ago2 has been identified as a DDH motif based on the results of mutation studies of H807 (Rivas et al. 2005). The catalytic residues are DDR for *h*Ago1, DDH for *h*Ago3, and DGR for *h*Ago4, none of which exhibit mRNA cleavage activity.

Domain-swapping experiments between *h*Ago2 and *h*Ago1 in our laboratory have also contributed to our understanding of the unique structural features of *h*Ago2 required for mRNA cleavage (Yuan et al. 2005). Only chimeral *h*Ago proteins containing the PIWI domain of *h*Ago2, either alone or paired with *h*Ago2 Mid domain, were capable of small RNA-guided cleavage of target RNA. These studies highlight the unique microenvironment restricted to the PIWI domain of *h*Ago2 and also reinforce earlier observations that the DD-containing motif within the PIWI domain of *h*Ago2 is the prominent determinant for its unique catalytic activity (Liu et al. 2004; Song et al. 2004; Rivas et al. 2005).

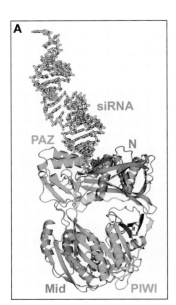
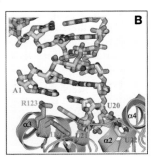
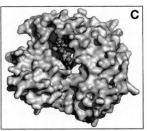

Figure 6. (*A*) The crystal structure of *A. aeolicus* Ago bound externally to a 22-mer siRNA (20-bp and 2-nucleotide 3′-overhang) (Yuan et al. 2006). The Ago protein is shown in a ribbon representation with the color coding and labeling of domains and linkers the same as listed in Fig. 4A. The externally bound siRNA, shown in a stick representation, is shown in beige with backbone phosphorus atoms in yellow, except for the 2-nucleotide overhang at the bound end, which is colored red. (*B*) A view of interactions between the 2-nucleotide overhang at one end of the externally bound 22-mer siRNA and the cavity positioned on the outer surface of the PAZ-containing lobe of *Aa*Ago. The overhang base U21 stacks on the aromatic ring of orange-colored Y119, whereas the overhang base U22 is inserted into a cavity whose walls involve segments of N domain, linker L1, and the PAZ domain. (*C*) Surface representation highlighting the cavity located on the outward-pointing face of the PAZ-containing lobe of the siRNA-bound *Aa*Ago. The bound 2-nucleotide 3′-overhang is shown in a space-filling representation. (Reprinted, with permission, from Yuan et al. 2006 [© Elsevier].)

An improved understanding of *h*Ago2 function will affect our understanding of mechanistic issues related to RNA silencing and in turn aid efforts focused on using this methodology for therapeutic intervention against virus-based diseases, and in the longer run, human genetic diseases.

New Developments within the Ago and Piwi Family

Recent functional studies have identified a new class of short regulatory RNAs that affected RNA silencing. These include Piwi-interacting RNAs (piRNAs) (Aravin et al. 2006; Girard et al. 2006; Grivna et al. 2006; Lau et al. 2006; Watanabe et al. 2006) and repeat-associated small interfering RNAs (rasiRNAs) (Aravin et al. 2003; Saito et al. 2006) that are associated with the Piwi sub-family. The piRNAs are in the 26- to 31-nucleotide range and the rasiRNAs are in the 24- to 29-nucleotide range, with both having a strong preference for U at the 5′-position. These newly identified small RNAs appear to have important implications for sperm developmental regulation and stem cell biology.

DOUBLE-STRANDED RNA PROCESSING BY DICER

We next briefly summarize recent biochemical (Zhang et al. 2004) and structural (Gan et al. 2006; MacRae et al. 2006) efforts that have considerably affected our mechanistic understanding of events associated with dsRNA processing by bacterial RNase III and Dicer (for review, see Carmell and Hannon 2004). Human Dicer is composed of an amino-terminal RNA helicase domain, followed by a domain of unknown function (DUF 283), a PAZ domain, two RNase III domains, and a carboxy-terminal RNA-binding domain (RBD).

The Witold Filipowicz group undertook a systematic mutation study of catalytic residues implicated in dsRNA cleavage within *Escherichia coli* RNase III and human Dicer. These studies have identified a single processing center generated by intramolecular dimerization of its two RNase III domains that is capable of simultaneously asymmetrically cutting partner strands of dsRNA (Zhang et al. 2004). This group also proposed that the PAZ domain in conjunction with the processing center measures the distance from the ends of the dsRNA to the cleavage site, thus generating appropriate-sized siRNAs.

The Xinhua Ji group has solved the crystal structure of a catalytically competent complex of *Aa*RNase III bound to its 2-nucleotide 3′-overhang-containing dsRNA product, which defines the protein–RNA interactions associated with substrate specificity and divalent cation-mediated scissible bond cleavage (Gan et al. 2006). This structure provides direct experimental support at the structural level for a single processing center and the role of induced fit in RNase III protein–RNA recognition and cleavage. The potential role of divalent ions in the catalytic cleavage mechanism has been clarified following comparison between the structures of *Aa*RNase III bound to dsRNA which contains one bound divalent cation at the cleavage site (Gan et al. 2006) and

Bacillus halodurans RNase H (*Bh*RNaseH) bound to a DNA/RNA hybrid which contains two bound divalent cations at the cleavage site (Nowotny et al. 2005; Yang et al. 2006).

The Jennifer Doudna laboratory has solved the crystal structure of a catalytically active *Giardia* Dicer composed of PAZ and two RNase III domains (MacRae et al. 2006). Strikingly, the linkers connecting the individual domains in Dicer adopt folded conformations, such that the overall molecule adopts a hatchet conformation, with the PAZ domain and linkers (a connector helix and a platform domain) forming the handle, while the two RNase III domains and a linker (the bridging domain) form the blade. This structure, which is of free Dicer, establishes that the PAZ module, which can bind to 2-nucleotide 3′-overhangs at the end of dsRNA, is separated by approximately 65 Å from the catalytic centers of the two ribonuclease III domains, by a flat surface, a length that spans 25 bp of dsRNA. Indeed, dsRNA can be modeled onto this flat surface, thereby suggesting that Dicer is a molecular ruler capable of cleaving dsRNA at a specified distance from its helical end (MacRae et al. 2006). A future challenge will be to structurally characterize either Dicer or its relevant domains with bound dsRNA, in efforts to further understand processing events associated with these RNase III enzymes that play a critical role in RNA silencing. Our laboratory has made progress toward this goal through ongoing crystallographic structural studies on human Dicer PAZ domain and flanking linkers bound to an siRNA-like duplex (J.-B. Ma et al., unpubl.).

MICRORNA BIOGENESIS

We conclude by outlining an emerging challenge for structural biologists in their efforts at structural characterization of the protein–RNA recognition events associated with the RNA silencing pathway. This challenge involves a molecular level elucidation of the steps associated with the processing of primary microRNAs (pri-miRNAs) into mature miRNA, which in turn are incorporated into RISC (for review, see Du and Zamore 2005; Kim 2005; Kim and Nam 2006). In humans, a protein–RNA complex involving the RNase III nuclease Drosha and DGCR8, a protein containing two dsRBDs, is responsible for converting pri-miRNAs into precursor microRNAs (pre-miRNAs) in the nucleus. The pre-miRNAs are then transported from the nucleus to the cytoplasm by a complex involving exportin-5 (Exp5) and RanGTP. Next, a protein–RNA complex involving the RNase III nuclease Dicer and TRBP, a protein containing three dsRBDs, is responsible for converting pre-miRNAs into mature miRNA duplexes in the cytoplasm. Thus, miRNA biogenesis involves three critical complexes, each involving two proteins and a RNA, whose structural elucidation could greatly advance our understanding of miRNA biogenesis at the molecular level. Because human Drosha and Dicer are large proteins that are both difficult to express and solubilize in *E. coli*, it will be necessary to identify minimalist domains and their interaction partners, which still retain their essential catalytic activity.

ACKNOWLEDGMENTS

The structural research was supported by National Institutes of Health (NIH) grant AI068776 to Dinshaw J. Patel and the functional research by NIH grant GM068476 to Thomas Tuschl. Yi Pei is supported by a Ruth Kirschstein NIH Postdoctoral Fellowship. The authors declare that they have no financial interests. The structural, mutation, and binding research on the *TBSV* p19–siRNA complex (Keqiong Ye with assistance from Lucy Malinina), *h*Ago1 PAZ–siRNA complex (Jin-Biao Ma with assistance from Keqiong Ye), *A. fulgidus* Piwi–siRNA complex (Jin-Biao Ma with assistance from Yu-Ren Yuan), *A. aeolicus* Argonaute (Yu-Ren Yuan and Jin-Biao Ma) and *A. aeolicus* Ago externally bound siRNA complex (Yu-Ren Yuan), and the modeling research on the *A. aeolicus* Ago-DNA/RNA hybrid complex (Vitaly Kuryavyi) was undertaken under the supervision of Dinshaw J. Patel. The functional research associated with target RNA cleavage assays (Yi Pei and Gunter Meister) was undertaken under the supervision of Thomas Tuschl. This contribution was written by Dinshaw J. Patel with input from Thomas Tuschl.

REFERENCES

Aravin A.A., Lagos-Quintana M., Yalcin A., Zavolan M., Marks D., Snyder B., Gaasterland T., Meyer J., and Tuschl T. 2003. The small RNA profile during *Drosophila melanogaster* development. *Dev. Cell* **5:** 337.

Aravin A., Gaidatzis D., Pfeffer S., Lagos-Quintana M., Ladgraf P., Iovino N., Morris P., Brownstein M.J., Kuramochi-Miyagawa S., Nakano T., et al. 2006. A novel class of small RNAs bind to MLL1 protein in mouse testis. *Nature* **442:** 203.

Bartel D.P. 2004. MicroRNAs: Genomics, biogenesis, mechanism and function. *Cell* **116:** 281.

Baulcombe D. 2004. RNA silencing in plants. *Nature* **431:** 356.

Bernstein E., Caudy A.A., Hammond S.M., and Hannon G.J. 2001. Role for a bidentate ribonuclease in the initiation step of RNA interference. *Nature* **409:** 363.

Carmell M.A. and Hannon G.J. 2004. RNase III enzymes and the initiation of gene silencing. *Nat. Struct. Mol. Biol.* **11:** 214.

Carmell M.A., Xuan Z., Zhang M.Q., and Hannon G.J. 2002. The Argonaute family: Tentacles that reach into RNAi, developmental control, stem cell maintenance, and tumorigenesis. *Genes Dev.* **16:** 2733.

Cerutti L., Mian N., and Bateman A. 2000. Domains in gene silencing and cell differentiation proteins: The novel PAZ domain and redefinition of the Piwi domain. *Trends Biochem. Sci.* **25:** 481.

Chao J.A., Lee J.H., Chapados B.R., Debler E.W., Schneemann A., and Williamson J.R. 2005. Dual modes of RNA-silencing suppression by Flock House virus protein B2. *Nat. Struct. Mol. Biol.* **12:** 952.

Chapman E.J., Prokhnevsky A.I., Gopinath K., Dolja V.V., and Carrington J.C. 2004. Viral RNA silencing suppressors inhibit the microRNA pathway at the intermediate step. *Genes Dev.* **18:** 1179.

Collins R.E. and Cheng X. 2005. Structural domains of RNAi. *FEBS Lett.* **579:** 5841.

Doench J.G. and Sharp P.A. 2004. Specificity of microRNA target selection in translational repression. *Genes Dev.* **18:** 504.

Doench J.G., Petersen C.P., and Sharp P.A. 2003. siRNAs can function as miRNAs. *Genes Dev* **17:** 438.

Du T. and Zamore P.D. 2005. microPrimer: The biogenesis and function of microRNA. *Development* **132:** 4645.

Dykxhoorn D.M., Novina C.D., and Sharp P.A. 2003. Killing the messenger: Short RNAs that silence gene expression. *Nat. Rev. Mol. Cell Biol.* **4:** 457.

Elbashir, S.M., Lendeckel W., and Tuschl T. 2001a. RNA interference is mediated by 21- and 22-nucleotide RNAs. *Genes Dev.* **15:** 188.

Elbashir S.M., Martinez J., Patkaniowska A., Lendeckel W., and Tuschl T. 2001b. Functional anatomy of siRNAs for mediating efficient RNAi in *Drosophila melanogaster* embryo lysate. *EMBO J.* **20:** 6877.

Filipowicz W. 2005. RNAi: The nuts and bolts of the RISC machine. *Cell* **122:** 17.

Filipowicz W., Jaskiewicz L., Kolb F.A., and Pillai R.S. 2005. Post-transcriptional gene silencing by siRNAs and miRNAs. *Curr. Opin. Struct. Biol.* **15:** 331.

Gan J., Tropea J.E., Austin B.P., Court D.L., Waugh D.S., and Ji X. 2006. Structural insight into the mechanism of double-stranded RNA processing by ribonuclease III. *Cell* **124:** 355.

Girard A., Sachidanandam R., Hannon G.J., and Carmell M.A. 2006. A germline-specific class of small RNAs binds mammalian Piwi proteins. *Nature* **442:** 199.

Grivna S.T., Beyret E., Wang Z., and Lin H. 2006. A novel class of small RNAs in mouse spermatogenic cells. *Genes Dev.* **20:** 1709.

Haley B. and Zamore P.D. 2004. Kinetic analysis of RNAi enzyme complex. *Nat. Struct. Mol. Biol.* **11:** 599.

Hall T.M. 2005. Structure and function of argonaute proteins. *Structure* **13:** 1403.

Hamilton A.J. and Baulcombe D.C. 1999. A species of small antisense RNA in posttranscriptional gene silencing in plants. *Science* **286:** 950.

Hammond S.M. 2005. Dicing and slicing: The core machinery of the RNA interference pathway. *FEBS Lett.* **579:** 5822.

Hammond S.M., Bernstein E., Beach D., and Hannon G.J. 2000. An RNA-directed nuclease mediates post-transcriptional gene silencing in *Drosophila* cells. *Nature* **404:** 293.

Hammond S.M., Boettcher S., Caudy A.A., Kobayashi R., and Hannon G.N. 2001. Argonaute2, a link between genetic and biochemical analysis of RNAi. *Science* **293:** 1146.

Hannon G.J. 2002. RNA interference. *Nature* **418:** 244.

Hutvagner G. and Zamore P.D. 2002. A microRNA in a multiple turnover RNAi enzyme complex. *Science* **297:** 2056.

Kim V.N. 2005. MicroRNA biogenesis: Coordinating cropping and dicing. *Nat. Rev. Mol. Cell Biol.* **6:** 376.

Kim V.N. and Nam J.-W. 2006. Genomics of microRNA. *Trends Genet.* **22:** 165.

Lai L., Yokota H., Hung L.W., Kim R., and Kim S.H. 2000. Crystal structure of archael RNase Hii: A homologue of human major RNase H. *Structure* **8:** 897.

Lakatos L., Csorba T., Pantaleo V., Chapman E.J., Carrington J.C., Liu Y.-P., Dolja V.V., Calvino L.F., Lopz-Moya J.J., and Burgyan J. 2005. Small RNA binding is a common strategy to suppress RNA silencing by several viral suppressors. *EMBO J.* **25:** 2768.

Lau N.C., Seto A.G., Kim J., Kuramochi-Miyagawa S., Nakano T., Bartel D.P., and Kingston R.E. 2006. Characterization of the piRNA complex from rat testes. *Science* **313:** 363.

Lecellier C.H,. Dunoyer P., Arar K., Lehmann-Che J., Eyquem S., Himber C., Saib A., and Voinnet O. 2005. A cellular microRNA mediates antiviral defense in human cells. *Science* **308:** 557.

Lewis B.P., Burge C.B., and Bartel D.P. 2005. Conserved seed pairing, often flanked by adenosines, indicates that thousands of human genes are microRNA targets. *Cell* **120:** 15.

Lewis B.P., Shih I.-H., Jones-Rhodes M.W., Bartel D.P., and Burge C.B. 2003. Prediction of mammalian microRNA targets. *Cell* **115:** 787.

Li H., Li W.X., and Ding S.W. 2002. Induction and suppression of RNA silencing by an animal virus. *Science* **296:** 1319.

Li W.X. and Ding S.W. 2001. Viral suppressors of RNA silencing. *Curr. Opin. Biotechnol.* **12:** 150.

Lingel A. and Sattler M. 2005. Novel modes of protein-RNA recognition in the RNAi pathway. *Curr. Opin. Struct. Biol.* **15:** 107.

Lingel A., Simon B., Izaurralde E., and Sattler M. 2003. Structure and nucleic acid-binding of the *Drosophila* Argonaute2 PAZ domain. *Nature* **426:** 465.

————. 2004. Nucleic acid 3′-end recognition by the Argonaute2 PAZ domain. *Nat. Struct. Mol. Biol.* **11:** 576.

Liu J., Carmell M.A., Rivas F.V., Marsden C.G., Thompson J.M., Song J.J., Hammond S.M., Joshua-Tor L., and Hannon G.J. 2004. Argonaute2 is the catalytic engine of mammalian RNAi. *Science* **305:** 1437.

Ma J.-B., Ye K., and Patel D.J. 2004. Structural basis for over-hang-specific small interfering RNA recognition by the PAZ domain. *Nature* **429:** 318.

Ma J.-B., Yuan Y.-R., Meister G., Pei Y., Tuschl T., and Patel D.J. 2005. Structural basis for 5′-end specific recognition of the guide RNA strand by *A. fulgidus* PIWI protein. *Nature* **434:** 666.

MacRae I.J., Zhou K., Li F., Repic A., Brooks A.N., Cande W.Z., Adams P.D., and Doudna J.A. 2006. Structural basis for double-stranded RNA processing by Dicer. *Science* **311:** 195.

Mallory A.C., Reinhardt B.J., Jones-Rhoades M.W., Tang G., Zamore P.D., Barton M.K., and Bartel D.P. 2004. MicroRNA control of PHABULOSA in leaf development: Importance of pairing to the microRNA 5′ region. *EMBO J.* **23:** 3356.

Martinez J. and Tuschl T. 2004. RISC is a 5′ phosphomo-noester-producing RNA endonuclease. *Genes Dev.* **18:** 975.

Martinez J., Patkaniowska A., Urlaub H., Luhrmann R., and Tuschl T. 2002. Single-stranded antisense siRNAs guide tar-get RNA cleavage in RNAi. *Cell* **110:** 563.

Matranga C., Tomari Y., Shin C., Bartel D.P., and Zamore P.D. 2005. Passenger-strand cleavage facilitates assembly of siRNA into Ago2-containing RNAi enzyme complexes. *Cell* **123:** 607.

Meister G. and Tuschl T. 2004. Mechanisms of gene silencing by double-stranded RNA. *Nature* **431:** 343.

Meister G., Landthaler M., Patkaniowska A., Dorsett Y., Teng G., and Tuschl T. 2004. Human Argonaute2 mediates RNA cleavage targeted by miRNA and siRNA. *Mol. Cell.* **15:**185.

Nowotny M., Gaidamakov S.A., Crouch R.J., and Yang Y. 2005. Crystal structures of RNase H bound to an RNA/DNA hybrid: Substrate specificity and metal-dependent catalysis. *Cell* **121:** 1005.

Nykanen A., Haley B., and Zamore P.D. 2001. ATP require-ments and small interfering RNA structure in the RNA inter-fering pathway. *Cell* **107:** 309.

Parker J.S. and Barford D. 2006. Argonaute: A scaffold for the function of short regulatory RNAs. *Trends Biochem. Sci.* **31:** 622.

Parker J.S., Roe S.M., and Barford D. 2004. Crystal structure of a PIWI protein suggests mechanisms for siRNA recognition and slicer activity. *EMBO J.* **23:** 4727.

————. 2005. Structural insights into mRNA recognition from a PIWI-domain-siRNA-guide complex. *Nature* **434:** 663.

Parrish S., Fleenor J., Xu S., Mello C., and Fire A. 2000. Functional anatomy of a dsRNA trigger: Differential require-ment for the two trigger strands in RNA interference. *Mol. Cell.* **6:** 1077.

Rand T.A., Petersen S., Du F., and Wang X. 2005. Argonaute2 cleaves the anti-guide strand of siRNA during RISC activa-tion. *Cell* **123:** 621.

Rivas F.V., Tolia N.H., Song J.J., Aragon J.P., Liu J., Hannon G.J., and Joshua-Tor L. 2005. Purified Argonaute2 and an siRNA form recombinant human RISC. *Nat. Struct. Mol. Biol.* **12:** 340.

Saito K., Nishida K.M., Mori T., Kawamura Y., Miyoshi K., Nagami T., Siomi H., and Siomi M.C. 2006. Specific associ-ation of Piwi with rasiRNA derived from retrotransposon and heterochromatic regions in the *Drosophila* genome. *Genes Dev.* **20:** 2214.

Schwarz D.S., Tomari Y., and Zamore P.D. 2004. The RNA-induced silencing complex is a Mg^{2+}-dependent endonucle-ase. *Curr. Biol.* **14:** 787.

Schwarz D.S., Hutvagner G., Du T., Xu Z., Aronin N., and Zamore P.D. 2003. Asymmetry in the assembly of the RNAi enzyme complex. *Cell* **115:** 199.

Silhavy D. and Burgyan J. 2004. Effects and side-effects of viral RNA silencing suppressors on short RNAs. *Trends Plant Sci* **9:** 76.

Silhavy D., Molnar A., Lucioli A., Sziiya G., Hornyik C., Tavazza M., and Burgyan J. 2002. A viral protein suppresses RNA silencing and binds silencing-generated, 21- to 25-nucleotide double-stranded RNAs. *EMBO J.* **21:** 3070.

Song J.J. and Joshua-Tor L. 2006. Argonaute and RNA-getting into the groove. *Curr. Opin. Struct. Biol.* **16:** 5.

Song J.J., Smith S.K., Hannon G.J., and Joshua-Tor L. 2004. Crystal structure of Argonaute and its implications for RISC slicer activity. *Science* **305:** 1434.

Song J.J., Liu J., Tolia N.H., Schneiderman J., Smith S.K., Martiensen R.A., Hannon G.J., and Joshua-Tor L. 2003. The crystal structure of the Argonaute2 PAZ domain reveals an RNA-binding motif in RNAi effector complexes. *Nat. Struct. Biol* **10:** 1026.

Sontheimer E.J. 2005. Assembly and function of RNA silencing complexes. *Nat. Rev. Mol. Cell Biol.* **6:** 127.

Stark A., Brennecke J., Russel R.B., and Cohen S.M. 2003. Identification of *Drosophila* microRNA targets. *PLoS Biol.* **1:** 397.

Tahbaz N., Kolb F.A., Zhang H., Jaronczyk K., Filipowicz W., and Hobman T.C. 2004. Characterization of the interactions between mammalian PAZ PIWI domain proteins and Dicer. *EMBO Rep.* **5:** 1.

Tomari Y. and Zamore P.D. 2005. Perspective: Machine for RNAi. *Genes Dev.* **19:** 517.

Vargason J.M., Szittya G., Burgyan J., and Tanaka-Hall T.M. 2003. Size selective recognition of siRNA by an RNA silenc-ing suppressor. *Cell* **115:** 799.

Verdel A., Jia S., Gerber S., Sugiyama T., Gygi S., Grewal S.I., and Moazed D. 2004. RNAi-mediated targeting of hete-rochromatin by the RITS complex. *Science* **303:** 672.

Voinnet O. 2005. Induction and suppression of RNA silencing: Insights from viral infections. *Nat. Rev. Genet.* **6:** 206.

Wang M.B. and Metzlaff M. 2005. RNA silencing and antiviral defense in plants. *Curr. Opin. Plant Biol.* **8:** 216.

Watanabe T., Takeda A., Tsukiyama T., Mise K., Okuno T., Sasaki H., Minami N., and Imai H. 2006. Identification and characterization of two novel classes of small RNAs in the mouse germline: Retrotransposon-derived siRNAs in oocytes and germline small RNAs in testes. *Genes Dev.* **20:** 1732.

Wong I. and Lohman T.M. 1993. A double-filter method for nitrocellulose-filter binding: Application to protein-nucleic acid interactions. *Proc. Natl. Acad. Sci.* **90:** 5428.

Yan K.S., Yan S., Farooq A., Han A., Zeng L., and Zhou M.M. 2003. Structure and conserved RNA binding of the PAZ domain. *Nature* **426:** 468.

Yang W. and Steitz T.A. 1995. Recombining the structures of HIV integrase, RuvC and RNase H. *Structure* **3:** 131.

Yang W., Lee J.Y., and Nowotny M. 2006. Making and break-ing nucleic acids: Two-Mg^{2+}-ion dependent catalysis and substrate specificity. *Mol. Cell* **22:** 5.

Ye K. and Patel D.J. 2005. RNA silencing suppressor p21 of beet yellow virus forms an RNA binding octameric ring structure. *Structure* **13:** 1375.

Ye K., Malinina L., and Patel D.J. 2003. Recognition of small interfering RNA by a viral suppressor of RNA silencing. *Nature* **426:** 874.

Yuan Y.R., Pei Y., Chen H.Y., Tuschl T., and Patel D.J. 2006. A potential protein-RNA recognition event along the RISC-loading pathway from the structure of *A. aeolicus* Argonaute with externally bound siRNA. *Structure* **14:** 1557.

Yuan Y.R., Pei Y., Ma J.-B., Kuryavyi Y., Zhadina M., Meister G., Chen H.-Y., Dauter Z., Tuschl T., and Patel D.J. 2005. Crystal structure of *A. aeolicus* Argonaute, a site-specific DNA-guided endoribonuclease, provides insights into RISC-mediated mRNA cleavage. *Mol. Cell* **19:** 405.

Zamore P.D. and Haley B. 2005. Ribo-gnome: The big world of small RNAs. *Science* **309:** 1519.

Zamore P.D., Tuschl T., Sharp P.A., and Bartel D.P. 2000. RNAi: Double-stranded RNA directs the ATP-dependent cleavage of mRNA at 21 to 23 nucleotide intervals. *Cell* **101:** 25.

Zhang H., Kolb F.A., Jakiewicz L., Westhof E., and Filipowicz W. 2004. Single processing center models for human Dicer and bacterial ribonuclease III. *Cell* **118:** 57.

Systemic RNAi in *Caenorhabditis elegans*

C.P. Hunter, W.M. Winston, C. Molodowitch, E.H. Feinberg, J. Shih,
M. Sutherlin, A.J. Wright, and M.C. Fitzgerald
Department of Molecular and Cellular Biology, Harvard University, Cambridge, Massachusetts 02138

RNA interference (RNAi) in *Caenorhabditis elegans* induced by ingestion or injection of double-stranded RNA (dsRNA) spreads throughout the organism and is even transmitted to the progeny. We have identified two proteins required for spreading of RNAi, SID-1 and SID-2, whose structure, subcellular localization, and expression pattern have been informative for how dsRNA can be transported into and between cells. SID-1 is a transmembrane protein that functions as a pore or channel that transports dsRNA into and out of cells. Proteins homologous to SID-1 are present in a wide range of invertebrate and vertebrate animals but are absent from plants. SID-2 is a small transmembrane protein that is expressed in the gut and localizes strongly to the luminal membrane where it appears to act as a receptor for uptake of dsRNA from the environment. Characterization of SID-2 activity in a variety of *Caenorhabditis* nematodes indicates that *C. elegans* SID-2 may have a novel activity.

The initial report that dsRNA triggers posttranscriptional gene silencing in *C. elegans* was accompanied by the observation that identical phenotypes were elicited regardless of the site of dsRNA injection (Fire et al. 1998). This result implies that dsRNA or a gene-specific silencing signal derived from the injected dsRNA is transported between cells and tissues (systemic RNAi). Similar observations of systemic gene silencing had been described in plants, where acquired viral resistance and transgene silencing information spread between old and new tissue and can even be transmitted across scions grafted onto the host plant (Baulcombe 2004). Building on the observation of systemic RNAi in *C. elegans*, two reports were published showing that RNAi could be triggered by exposure to environmental sources of dsRNA, either by soaking animals in concentrated solutions of dsRNA or by feeding animals *Escherichia coli* engineered to express gene-specific dsRNA (Tabara et al. 1998; Timmons and Fire 1998). How and why dsRNA or gene-specific signals are transported into animals and between cells was unknown.

THE SCREEN

To isolate mutations that specifically disrupt systemic RNAi, we designed a green fluorescent protein (GFP) transgene-based screen that monitored autonomous and systemic RNAi, separately (Winston et al. 2002). To monitor autonomous RNAi, we used a pharyngeal muscle promoter to express both GFP and a *gfp* hairpin construct (Fig. 1). To monitor systemic RNAi, we expressed a nuclear localized GFP in all the body wall muscle cells. The pharyngeal GFP hairpin efficiently reduced GFP expression in the pharynx in all animals and reproducibly reduced nuclear localized GFP in anterior but not posterior body wall muscle cells, consistent with a pharynx-derived gradient of silencing signal. Systemic silencing does not require coexpression of GFP in the pharynx,

indicating that the systemic silencing signal is produced independently of the target mRNA, although the hairpin construct may serve that role. Mutations in genes such as *rde-1* eliminated silencing of GFP in pharyngeal and body wall muscle cells, showing that the local and systemic RNAi observed in this strain require the same core RNAi machinery. Tabara et al. (1999) showed that *rde-1* function is not required to produce or transmit a systemic signal.

To identify systemic RNAi-defective (Sid) mutants, this triply transgenic strain was grown on bacteria that expressed a *gfp* hairpin RNA, which effectively silenced GFP in all body wall muscle cells. We then isolated 200 candidate Sid mutants that reexpressed GFP in all body wall muscle cells but not in the pharyngeal muscle. Thus, these mutants were either defective for uptake of dsRNA from the bacteria and/or spreading of RNAi silencing from the pharynx to the body wall muscle cells. These 200 alleles identified three large complementation groups (20–100 alleles each) and several smaller uncharacterized complementation groups. *sid-1(V)*, *sid-2(III)*, and *sid-3(X)* were mapped, and *sid-1* and *sid-2* cloned by DNA transformation rescue (Winston et al. 2002 and in prep.). *sid-1* and likely *sid-2* alleles were also isolated in independent screens (Timmons et al. 2003; Tijsterman et al. 2004).

SID-1

The predicted SID-1 protein encodes 11 potential transmembrane (TM) domains and a large extracellular amino-terminal domain (Fig. 2A). Experiments to determine the topology of individual TM domains confirmed predicted TM domains 1–4 and 6; however, the orientation of TM4 and TM6 indicates that TM5 cannot pass through the bilayer and requires that at least one additional predicted TM (7–11) also cannot pass through the bilayer (Feinberg and Hunter 2003). Thus, SID-1 contains 5, 7, or 9 TM domains.

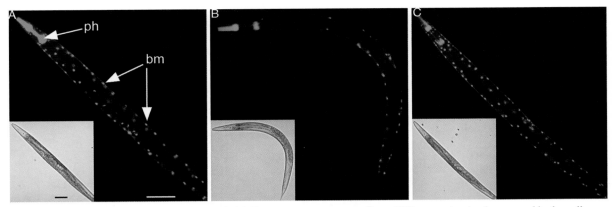

Figure 1. Direct visualization of systemic RNAi. (*A*) Two separate transgenes direct GFP expression in pharynx and body wall muscle. (*B*) The addition of a third transgene that expresses a *gfp* hairpin to produce dsRNA in the pharynx silences the GFP reporter in the pharynx (autonomous RNAi) and silences GFP in anterior body wall muscles (systemic RNAi). (*C*) A *sid-1* mutation in the background of the transgenes described for panel *B*. (ph) Pharynx muscle; (bm) body wall muscle. Insets are differential interference contrast images. (Reprinted in part, with permission, from Winston et al. 2002 [© AAAS].)

GFP driven by 700 bp of *sid-1* upstream sequences is detected in embryos and all larval stages. In adults, GFP is detected in all nonneuronal cells, perhaps explaining the refractoriness of neurons to systemic RNAi. To determine where SID-1 is localized within these cells, we examined the localization of a rescuing GFP-tagged SID-1 (SID-1C::GFP). SID-1C::GFP rescued the systemic RNAi defect, suggesting that expression and localization would be representative of SID-1. SID-1C::GFP localized throughout cells with strong enrichment at the periphery of cells, consistent with plasma membrane association (Fig. 2B, upper panel). SID-1C::GFP was detected at a much reduced level compared to the promoter fusion construct and was detected in a reduced subset of cell types; however, *sid-1* function is detected in cell types in which the fusion protein was not detected (e.g., body wall muscle). Curiously, SID-1C::GFP is most highly expressed in the cells and tissues that are directly exposed to the

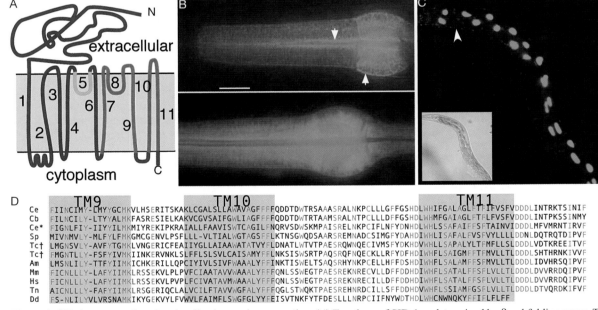

Figure 2. SID-1 structure, function, localization, and conservation. (*A*) Topology of SID-1 as determined by β-gal folding assay. The dark blue portions were demonstrated to be extracellular or intracellular constraining the orientation of the respective TM domains. The location and orientation of the light blue and red portions are undetermined. (*B*) Full-length functional SID-1::GFP (*top panel*) is enriched at the cell periphery (*arrows*) compared to GFP (*bottom panel*). Bar, 10 μm. (*C*) Injection of *gfp* dsRNA into a single intestinal cell (*arrowhead*) in a *sid-1* mutant animal that expresses nuclear-localized GFP in all cells demonstrates cell-autonomous RNAi. (*D*) Alignment of carboxy-terminal invertebrate and vertebrate SID-1 TM domains (*gray*). (*Red*) Amino acids that are identical among a majority of proteins; (*blue*) amino acids that are conserved among vertebrates. SID-1 homologs: (Ce) *C. elegans*; (Cb) *C. briggsae*; (Ce*) *C. elegans* ZK721.1; (Sp) *Strongylocentrotus purpuratus*; (Tc) *Tribolium castaneum*; (Am) *Apis mellifera*; (Mm) *Mus musculus*; (Hs) *Homo sapiens*; (Tn) *Tetraodon nigroviridis*; (Dd) *Dictyostelium discoideum*; (Tc†) the two *Tribolium* homologs are similar but distinct. (Reprinted in part, with permission, from Feinberg and Hunter 2003 [© AAAS].)

environment: pharynx, intestine, sphincter muscles, excretory cell, and phasmids, plus the gonad, which is indirectly exposed to the environment via the vulva. The significance of this observation remains unknown, but it portends a role in the uptake of environmental molecules.

SID-1 homologs are readily detected in genome sequence databases (Fig. 2D). The most conserved region is the carboxy-terminal region, which contains the TM domains, although all homologs are predicted to contain a large extracellular domain and a similar number of TM domains. At least one and usually two SID-1 homologs are present in all sequenced vertebrate genomes. In invertebrates, SID-1 homologs are present in some but not all sequenced genomes. For example, *Drosophila melanogaster*, *Anopheles gambiae* (mosquito), and *Ciona* (sea squirt) do not appear to have SID-1 homologs, whereas clear homologs are readily detected in *Apis mellifera* (honeybee), *Tribolium castaneum* (flour beetle), *Strongylocentrotus purpuratus* (sea urchin), and *Dictyostelium discoideum* (slime mold). Systemic RNAi has been reported in honeybee and flour beetle (Bucher et al. 2002; Amdam et al. 2003). The possible function of SID-1 homologs is discussed below.

To characterize the function of *C. elegans* SID-1, RNAi targeting a variety of endogenous and reporter genes was initiated by tissue-specific transgenes expressing RNA hairpins, by injecting dsRNA into specific cells, tissues, and body compartments, and by feeding and soaking animals in dsRNA. By these assays, RNAi is effective in *sid-1* mutants only within the cell(s) in which dsRNA is directly expressed or injected. For example, expression of *gfp* hairpin RNA in the pharynx or body wall muscles silences GFP expression in those tissues only. Injection of *gfp* dsRNA into the large injection-accessible intestinal cells leads to silencing of GFP only in the injected cell (Fig. 2C). Similarly, injection directly into the syncytial germ line of dsRNA that targets maternally expressed genes results in highly penetrant phenotypes. However, silencing is never observed in adjacent cells or tissues and is not observed in the *sid-1⁻* progeny of germ-line-injected mothers. These results are consistent with a requirement for *sid-1* in the import or export of a silencing signal. To test the role of *sid-1* in import of the silencing signal, we generated *sid-1* genetic mosaics (animals composed of *sid-1⁺* and *sid-1⁻* cells) and scored RNAi silencing of GFP in individual muscle cells, which was initiated by feeding the animals bacteria expressing a *gfp* hairpin. In all scored animals, GFP expression was silenced in *sid-1⁺* cells and was not affected in any *sid-1⁻* cells (Fig. 3). This result shows that *sid-1* is required cell-autonomously for the import or processing of a silencing signal. In analogous experiments, a requirement for *sid-1* for the export of a silencing signal has also been demonstrated (J. Shih and C.P. Hunter, unpubl.).

This phenotypic analysis is consistent with SID-1-dependent uptake and export of the systemic silencing signal. To investigate the nature of this systemic silencing signal and the molecular mechanism of SID-1 activity, we developed a heterologous cell-based assay (Feinberg and Hunter 2003). Published reports indicated that *Drosophila* was capable of robust cell-autonomous RNAi but appeared

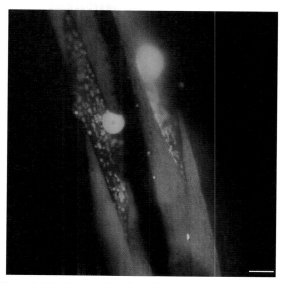

Figure 3. Mosaic analysis of *sid-1* function. *sid-1*(+) body wall muscle cells are marked by the coexpression of DsRED2 (*red*). The two non-red muscle cells are resistant to bacteria-mediated RNAi and retain expression of GFP (*green*), whereas *sid-1*(+) muscle cells (*red*) lose GFP expression. Bar, 10 μm. (Reprinted, with permission, from Winston et al. 2002 [© AAAS].)

to mount at best inefficient systemic RNAi, perhaps consistent with the lack of a detectable SID-1 homolog in the fly genome. When briefly serum-starved and soaked in a high concentration of dsRNA, phagocytic *Drosophila* S2 cells can efficiently initiate RNAi via a dsRNA uptake mechanism that requires endocytosis (Ulvila et al. 2006). It is unknown how the dsRNA crosses the membrane to initiate RNAi. In this S2 cell system, we developed both a quantitative silencing assay targeting luciferase and a radiolabeled dsRNA uptake assay using transient transfection of either wild-type SID-1 (wt) or a missense allele SID-1 (qt2) as a negative control. In the luciferase-silencing assay, expression of SID-1(wt) enabled cells to mount an RNAi response in the presence of approximately 10^5-fold less dsRNA than cells expressing the mutant SID-1 protein (Fig. 4A). This assay is quite sensitive, allowing us to detect silencing initiated at dsRNA concentrations below one dsRNA molecule per transfected cell. Using this assay, we determined that longer dsRNA was more effective at triggering *sid-1*-dependent silencing, either because it was taken up more efficiently or because it more efficiently interacted with the silencing machinery in the cell. The effective size cutoff is between 50 and 100 bp of dsRNA, with siRNAs effective only at the highest concentrations tested (Fig. 4B). This experiment suggested that systemic RNAi in the worm might be sensitive to the size of the trigger dsRNA. To address this possibility, we injected either a 50- or 100-bp dsRNA targeting the essential *mex-3* gene directly into the syncytial germ line or the intestine. Injection of either dsRNA into the germ line yields high embryonic lethality, whereas injection into the intestine and requisite transport to the germ line is only effective with the 100-bp dsRNA. To confirm that this result does not merely reflect varying efficiency of the dsRNA sequences in initiating RNAi, we generated a 100-bp

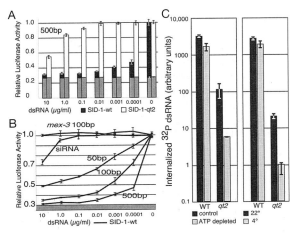

Figure 4. Characterization of SID-1 activity in *Drosophila* S2 cells. Wild-type (WT) or mutant (qt) SID-1 was transfected with a luciferase-expressing construct for activity assays. (*A*) Luciferase levels shown are relative to luciferase levels in cells exposed to no dsRNA. (*Gray bars*) Luciferase levels that remain at the end of the experiment in samples treated with the translation inhibitor cycloheximide (50 μg/ml) to control for protein stability. (*B*) Summary of SID-1-WT data showing dsRNA size-dependent silencing. (*C*) SID-1 mediates passive uptake of dsRNA in S2 cells. SID-1-mediated dsRNA uptake is resistant to ATP depletion (*left panel*) and to reduced temperature (*right panel*). (Reprinted in part, with permission, from Feinberg and Hunter 2003 [© AAAS.])

dsRNA containing the 50-bp *mex-3* dsRNA fused to 50 bp from luciferase. Intestinal injection of this dsRNA elicits efficient RNAi, confirming that dsRNA size, not merely sequence, determines systemic RNAi efficiency and explains the relatively poor systemic silencing capacity of shorter dsRNA (Parrish et al. 2000). Thus, the SID-1-dependent silencing assay in S2 cells revealed specific properties of systemic RNAi in *C. elegans*.

To gain further mechanistic insight into SID-1 function, we modified the S2 cell system to demonstrate SID-1-dependent intracellular accumulation of radiolabeled dsRNA and thereby showed that SID-1 is a dsRNA transporter (Fig. 4C) (Feinberg and Hunter 2003). In this assay, 100- and 500-bp dsRNAs accumulate in cells to at least fivefold higher levels than 50-bp dsRNA, consistent with the phenotypic effects observed in vitro and in vivo (M.C. Fitzgerald et al., unpubl.). Similar experiments have failed to detect the efficient transport of other

nucleic acids, although this may reflect the lack of retention within cells. The ability of SID-1 to function in a heterologous system indicates that it either functions alone or accommodates *Drosophila* variants of protein cofactors. Remarkably, depleting ATP or performing uptake assays at 4°C has little effect on the extent or rate of uptake, arguing for a passive uptake mechanism most consistent with a function as a dsRNA channel (Fig. 4D). This presents a still-unsolved problem of how SID-1 activity is regulated, as a channel large enough to admit dsRNA could allow essential cellular components to exit the cell.

The wide phylogenetic conservation of SID-1 homologs and the demonstration of systemic RNAi in other invertebrates (Bucher et al. 2002; Amdam et al. 2003; Newmark et al. 2003; Turner et al. 2006) and in mammals (Soutschek et al. 2004) suggest that the function of SID-1 as a dsRNA transporter may be conserved and may even be the selected activity for this protein family. In fact, Duxbury et al. (2005) obtained evidence that supports this idea: These authors report that overexpressing a human SID-1 homolog in human cells enabled efficient passive siRNA uptake and silencing. Although other substrates were not tested, the human and *C. elegans* SID-1 proteins have similar activities. We eagerly await further evidence of dsRNA transport functions for vertebrate and invertebrate SID-1 homologs.

SID-2

The second largest complementation group recovered in the original screen identified SID-2, a single-pass transmembrane protein strongly expressed in the gut (Winston et al. 2002 and in prep.). This restricted expression pattern corresponds to the phenotype observed in *sid-2* mutants, which are only defective for environmental RNAi, i.e., RNAi triggered by soaking in dsRNA or ingesting bacteria expressing dsRNA. Consequently, *sid-2* mutants are fully sensitive to systemic RNAi triggered by transgene expression or microinjection into any tissues. Topology-determining experiments show that the predicted TM domain separates an extracellular amino-terminal domain from an intracellular carboxy-terminal domain. Fusing GFP to the intracellular carboxyl terminus of SID-2 (SID-2-C::GFP) produces a fusion gene that fully rescues the *sid-2* environmental RNAi defect and strongly localizes GFP to intracellular and surface-localized membrane structures (Fig. 5).

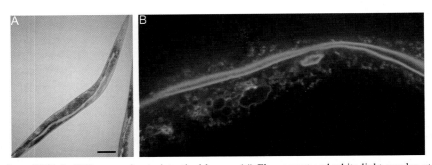

Figure 5. Localization of SID-2::GFP expression to intestinal lumen. (*A*) Fluorescent and white light overlay at low magnification and (*B*) confocal section at high magnification. (Confocal image courtesy of Daniel Schott.)

In contrast to the wide phylogenetic distribution of *sid-1*, the only readily detected *sid-2* homologous genes are from *Caenorhabditis briggsae* and *C. remanei*. The degree of amino acid sequence conservation among the three domains of SID-2 from all three species is similar: 23% amino-terminal (190 amino acids), 86% TM (21 amino acids), and 53% carboxy-terminal (100 amino acids), despite the relative recent divergence of *C. briggsae* and *C. remanei*. Among these three nematodes, only *C. elegans* efficiently initiates RNAi in response to environmental exposure to dsRNA. Given the relative lack of sequence conservation in the amino-terminal domain of SID-2, we suspected that *C. elegans* had gained this activity or *C. briggsae* and *C. remanei* had lost this activity. A Cb-SID-2::GFP fusion protein expressed in *C. briggsae* showed intestinal expression localized to the luminal membrane, indicating that changes in gene expression and protein localization do not account for the different sensitivities to environmental dsRNA. Furthermore, expression of the Cb-SID-2::GFP in *C. elegans* failed to rescue the *sid-2* mutant phenotype; conversely, expression of Ce-SID-2-C::GFP did confer environmental RNAi sensitivity on *C. briggsae*. To address the possibility that the amino acid changes in the highly divergent extracellular domain underlie the differential response to environmental dsRNA, we fused the Ce-SID-2 extracellular and TM domains to the Cb-SID-2 intracellular domain fused to GFP. This construct rescued the *C. elegans sid-2* mutant, thus localizing the relevant activity to the extracellular/TM domains. Further analysis of these domains is in progress.

To determine whether environmental RNAi sensitivity arose in *C. elegans* or had been lost in *C. briggsae* and *C. remanei*, we undertook a broad analysis of systemic RNAi among extant *Caenorhabditis* species (Fig. 6). We produced a set of species-specific RNA polymerase II subunit dsRNA that produced a similar early embryonic arrest phenotype among the progeny embryos (cell cycle arrest around gastrulation) when injected into the gonad of each species, showing that the dsRNA effectively initiated RNAi in each species. To test for systemic RNAi, we measured embryonic lethality among the progeny of intestine-injected mothers, and to test for environmental RNAi, we measured embryonic lethality among the progeny of mothers soaked in dsRNA overnight. All but one species was sensitive to intestine-injected dsRNA, showing that systemic RNAi is broadly conserved among *Caenorhabditis*. It will be interesting to learn whether *sid-1* has been conserved in this systemic RNAi-defective species. In contrast, only *C. elegans* and one distantly related and unnamed species were sensitive to environmental RNAi. This result indicates that sensitivity to environmental RNAi is rare and is not a general property of nematodes. Thus, this property of SID-2 likely arose within *C. elegans*. Ongoing investigations into the natural ecology of *Caenorhabditis* species may suggest a role, for example, in acquiring sequence-specific resistance to pathogens, for such an activity (Fitch 2005; Hong and Sommer 2006).

CONCLUSIONS AND PROSPECTUS

The initial genetic screen identified three large complementation groups with strong effects on systemic RNAi. Two of these, *sid-1* and *sid-2*, represent two easily mutated and recoverable mutants that disrupt the uptake and distribution of dsRNA in *C. elegans*. All other recovered alleles, which represent at least five additional genes, were recovered at 1–10% of the frequency of *sid-1*. Thus, new screens designed to avoid or eliminate the recovery of *sid-1* and *sid-2* alleles should identify new genes and pathways that function with *sid-1* or *sid-2*. The identification of genes that regulate the activity of SID-1 and SID-2 may provide an insight into the function of these pathways. For example, perhaps the SID proteins provide an antiviral defense similar to the role of systemic silencing in plants (Baulcombe 2004). Although data supporting an antiviral role for systemic RNAi have not been reported, it is clear that RNAi is antiviral in *C. elegans* (Lu et al. 2005; Schott et al. 2005; Wilkens et al. 2005). Alternatively, microRNAs or mRNAs may be generally or selectively transported between cells, tissues, or animals. Finally, it should not be assumed that dsRNA is the only substrate: A channel large enough to permit passage of long dsRNA through a membrane could accommodate many different large molecules.

The SID-1 protein family is ancient, with SID-1 homologs present in *Dictyostelium*, many invertebrates, and all sequenced vertebrates. However, it is unknown whether any of these homologs function in dsRNA transport. In *C. elegans*, ZK721.1 encodes a protein slightly more similar to the vertebrate SID-1 homologs than SID-1 itself. However, deletion of ZK721.1 does not have a detectable systemic RNAi defect in any tested *C. elegans* tissue. In contrast, overexpression of a human SID-1 homolog enables efficient RNAi in human cells. This hints

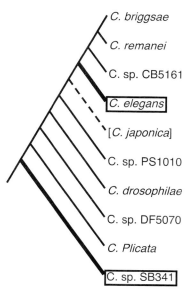

Figure 6. Phylogenetic relationship of *Caenorhabditis* species (Kiontke et al. 2004) and systemic and environmental RNAi proficiency. *C. japonica* (*dashed line*) was not tested. Only *C.* sp. CB5161 was deficient for systemic RNAi, whereas only *C. elegans* and *C.* sp. SB341 (*boxed*) were proficient at environmental RNAi.

that the evolutionarily selected function for some of the SID-1 homologs is indeed dsRNA transport. Investigation of the functions and substrate selectivity of these proteins will be important for the development of effective therapeutic applications of RNAi to human health.

ACKNOWLEDGMENTS

We thank Antony Jose and Daniel Schott for comments on the manuscript. This project was initiated with funds provided by the Beckman Young Investigator Foundation and continues with the support of the National Science Foundation (MCB-0417102) and the National Institutes of Health (GM06989).

REFERENCES

Amdam G.V., Simoes Z.L., Guidugli K.R., Norberg K., and Omholt S.W. 2003. Disruption of vitellogenin gene function in adult honeybees by intra-abdominal injection of double-stranded RNA. *BMC Biotechnol.* **3:** 1.

Baulcombe D. 2004. RNA silencing in plants. *Nature* **431:** 356.

Bucher G., Scholten J., and Klingler M. 2002. Parental RNAi in *Tribolium* (*Coleoptera*). *Curr. Biol.* **12:** R85.

Duxbury M.S., Ashley S.W., and Whang E.E. 2005. RNA interference: A mammalian SID-1 homologue enhances siRNA uptake and gene silencing efficacy in human cells. *Biochem. Biophys. Res. Commun.* **331:** 459.

Feinberg E.H. and Hunter C.P. 2003. Transport of dsRNA into cells by the transmembrane protein SID-1. *Science* **301:** 1545.

Fire A., Xu S., Montgomery M.K., Kostas S.A., Driver S.E., and Mello C.C. 1998. Potent and specific genetic interference by double-stranded RNA in *Caenorhabditis elegans. Nature* **391:** 806.

Fitch D.H. 2005. Evolution: An ecological context for *C. elegans. Curr. Biol.* **15:** R655.

Hong R.L. and Sommer R.J. 2006. *Pristionchus pacificus:* A well-rounded nematode. *Bioessays* **28:** 651.

Kiontke K., Gavin N.P., Raynes Y., Roehrig C., Piano F., and Fitch D.H. 2004. *Caenorhabditis* phylogeny predicts convergence of hermaphroditism and extensive intron loss. *Proc. Natl. Acad. Sci.* **101:** 9003.

Lu R., Maduro M., Li F., Li H.W., Broitman-Maduro G., Li W.X., and Ding S.W. 2005. Animal virus replication and RNAi-mediated antiviral silencing in *Caenorhabditis elegans. Nature* **436:** 1040.

Newmark P.A., Reddien P.W., Cebria F., and Sanchez Alvarado A. 2003. Ingestion of bacterially expressed double-stranded RNA inhibits gene expression in planarians. *Proc. Natl. Acad. Sci.* (suppl. 1) **100:** 11861.

Parrish S., Fleenor J., Xu S., Mello C., and Fire A. 2000. Functional anatomy of a dsRNA trigger: Differential requirement for the two trigger strands in RNA interference. *Mol. Cell* **6:** 1077.

Schott D.H., Cureton D.K., Whelan S.P., and Hunter C.P. 2005. An antiviral role for the RNA interference machinery in *Caenorhabditis elegans. Proc. Natl. Acad. Sci.* **102:** 18420.

Soutschek J., Akinc A., Bramlage B., Charisse K., Constien R., Donoghue M., Elbashir S., Geick A., Hadwiger P., Harborth J., et al. 2004. Therapeutic silencing of an endogenous gene by systemic administration of modified siRNAs. *Nature* **432:** 173.

Tabara H., Grishok A., and Mello C.C. 1998. RNAi in *C. elegans:* Soaking in the genome sequence. *Science* **282:** 430.

Tabara H., Sarkissian M., Kelly W.G., Fleenor J., Grishok A., Timmons L., Fire A., and Mello C.C. 1999. The *rde-1* gene, RNA interference, and transposon silencing in *C. elegans. Cell* **99:** 123.

Tijsterman M., May R.C., Simmer F., Okihara K.L., and Plasterk R.H. 2004. Genes required for systemic RNA interference in *Caenorhabditis elegans. Curr. Biol.* **14:** 111.

Timmons L. and Fire A. 1998. Specific interference by ingested dsRNA. *Nature* **395:** 854.

Timmons L., Tabara H., Mello C.C., and Fire A.Z. 2003. Inducible systemic RNA silencing in *Caenorhabditis elegans. Mol. Biol. Cell* **14:** 2972.

Turner C.T., Davy M.W., Macdiarmid R.M., Plummer K.M., Birch N.P., and Newcomb R.D. 2006. RNA interference in the light brown apple moth, *Epiphyas postvittana* (Walker) induced by double-stranded RNA feeding. *Insect Mol. Biol.* **15:** 383.

Ulvila J., Parikka M., Kleino A., Sormunen R., Ezekowitz R.A., Kocks C., and Ramet M. 2006. Double-stranded RNA is internalized by scavenger receptor-mediated endocytosis in *Drosophila* S2 cells. *J. Biol. Chem.* **281:** 14370.

Wilkins C., Dishongh R., Moore S.C., Whitt M.A., Chow M., and Machaca K. 2005. RNA interference is an antiviral defence mechanism in *Caenorhabditis elegans. Nature* **436:** 1044.

Winston W.M., Molodowitch C., and Hunter C.P. 2002. Systemic RNAi in *C. elegans* requires the putative transmembrane protein SID-1. *Science* **295:** 2456.

Zimmermann T.S., Lee A.C., Akinc A., Bramlage B., Bumcrot D., Fedoruk M.N., Harborth J., Heyes J.A., Jeffs L.B., John M., et al. 2006. RNAi-mediated gene silencing in non-human primates. *Nature* **441:** 111.

Transcriptional Landscape of the Human and Fly Genomes: Nonlinear and Multifunctional Modular Model of Transcriptomes

A.T. Willingham, S. Dike, J. Cheng, J.R. Manak, I. Bell, E. Cheung, J. Drenkow, E. Dumais, R. Duttagupta, M. Ganesh, S. Ghosh, G. Helt, D. Nix, A. Piccolboni, V. Sementchenko, H. Tammana, P. Kapranov, the ENCODE Genes and Transcripts Group,* and T.R. Gingeras

Affymetrix, Inc., Santa Clara, California 95051

Regions of the genome not coding for proteins or not involved in *cis*-acting regulatory activities are frequently viewed as lacking in functional value. However, a number of recent large-scale studies have revealed significant regulated transcription of unannotated portions of a variety of plant and animal genomes, allowing a new appreciation of the widespread transcription of large portions of the genome. High-resolution mapping of the sites of transcription of the human and fly genomes has provided an alternative picture of the extent and organization of transcription and has offered insights for biological functions of some of the newly identified unannotated transcripts. Considerable portions of the unannotated transcription observed are developmental or cell-type-specific parts of protein-coding transcripts, often serving as novel, alternative 5' transcriptional start sites. These distal 5' portions are often situated at significant distances from the annotated gene and alternatively join with or ignore portions of other intervening genes to comprise novel unannotated protein-coding transcripts. These data support an interlaced model of the genome in which many regions serve multifunctional purposes and are highly modular in their utilization. This model illustrates the underappreciated organizational complexity of the genome and one of the functional roles of transcription from unannotated portions of the genome.

WIDESPREAD RECOGNITION OF THE PHENOMENA OF EXTENSIVE AND COMPLEX PATTERNS OF TRANSCRIPTION THROUGHOUT THE GENOMES OF MANY SPECIES

Within the past 5 years, multiple large-scale, unbiased experimental approaches have identified surprisingly large amounts of RNA transcription far exceeding that estimated to be required for the production of messenger RNA for known proteins. This transcriptional "dark matter" has been observed in (1) large-scale full-length cDNA sequencing (Okazaki et al. 2002; Carninci et al. 2005), (2) mapping of 3' ends with serial analysis of gene expression (Chen et al. 2002; Saha et al. 2002), (3) mapping of 5' ends by cap analysis of gene expression (Carninci et al. 2005), and (4) analysis of expression by massively parallel signature sequencing (Jongeneel et al. 2005). Whereas these approaches have been key to analyzing widespread transcription, genomic tiling arrays have made substantial contributions by being both unbiased in their interrogation coverage (i.e., not limited to annotated regions) and sensitive with detection of low-copy-number transcripts (for review, see Johnson et al. 2005). In 2002, the first systematic and unbiased analysis of transcription across human chromosomes 21 and 22 was carried out. Strikingly, the sites of transcription across these chromosomes was determined to be at least an order of magnitude greater than that observed for annotated protein-coding genes (Kapranov et al. 2002). Implicit in this discovery was that a significant proportion of transcribed RNA was likely noncoding, suggesting an unanticipated degree of RNA complexity and possible novel function roles for such noncoding (nc)RNAs.

Careful review of earlier studies of the complexity and characteristics of transcribed RNA in eukaryotic cells finds evidence of widespread transcription heralded decades ago (for discussion, see Willingham and Gingeras 2006). These original studies were focused on large RNAs (i.e., >200–300 nucleotides) and arrived at the common conclusion that the complexity of transcripts made by organisms ranging from sea urchins to humans not only seemed to be inexplicably sizable and complex, but also contained non-polyadenylated (poly(A)⁻) RNAs that were more numerous than the standard polyadenylated (poly(A)⁺) RNAs associated with protein-coding functions.

Current efforts are focused on mapping transcription at very high nucleotide resolutions across the entire human genome and further dissecting this RNA into classes and cellular compartments (T.R. Gingeras et al., in prep.), searching for detectable patterns of stable RNA structures (Washietl et al. 2005; Torarinsson et al. 2006) and cross-comparing transcriptional complexity between evolutionarily distant genomes (Khaitovich et al. 2006; Manak et al. 2006). These studies have come to challenge the relatively straightforward protein-coding-centric view of genome organization in which genes are structured in discrete loci with a few transcript isoforms being made in the locus in a linear and mostly nonoverlapping fashion. Such loci are canonically seen to contain regulatory promoters immediately adjacent to annotated 5' ends of the encoded isoforms. As discussed in this paper, this "beads on a string" linear model is gradually being replaced by a more complicated interlaced architecture in which many discrete genomic loci (e.g., exons) serve a multitasking func-

*Group members listed in Acknowledgments.

tion in which sequences comprising such loci may also serve as promoters, and introns may serve as exons for overlapping transcripts on both strands (Gingeras 2006).

BRIEF BACKGROUND ABOUT TILING MICROARRAYS

In the work described here, we employed four different types of tiling arrays (for general review, see Mockler et al. 2005). The first contains a medium interrogation resolution of 20 nucleotides designed to interrogate the 44 regions of the human genome selected for the Encyclopedia of DNA Elements (ENCODE) project (ENCODE Project Consortium 2004). Transcription for 11 cell lines and 12 tissues was profiled using this ENCODE tiling array. Second, a higher-resolution tiling array with an interrogation resolution of 5 nucleotides covering 10 human chromosomes (~30% of the genome) was developed (Cheng et al. 2005). This 98-array set contains 380 Mb of probe-coverage and, because of its size, sample analysis was limited to 8 developmentally diverse cell lines. However, RNA samples were fractionated based on cellular compartments (cytosolic versus nuclear) and physical structure (poly(A)$^+$ vs. poly(A)$^-$). Very recently, a third tiling array set was developed that allowed creation of a 91-array set containing about 1300 Mb of probe-coverage, which translates to 100% of the nonredundant human genome tiled at 5 nucleotides resolution (T.R. Gingeras et al., in prep.). Finally, a single-array, low-resolution (35 nucleotides, 3.2 Mb of coverage) tiling array was created for the *Drosophila* genome and, taking advantage of the wealth of developmental biology for this model organism, was used to profile the first 24 hours of embryogenesis in 2-hour time increments (Manak et al. 2006).

Analysis of data gathered from tiling arrays indicates detected sites of transcription (transcribed fragments) which are termed "transfrags" (TFs). The labeled targets hybridized to the tiling arrays are double-stranded (ds) cDNAs made, in most instances, from processed (e.g., capped, polyadenylated, and/or spliced) RNAs isolated from cells. Thus, the detected TFs represent the sum of all transcripts mapping to the interrogated positions. Since the target is ds-cDNA, there is no strand information present in the tiling array data obtained from such labeled samples. Intensity thresholds are determined by calculating the intensity distribution of probes that originate from bacterial negative control regions. This allows for an estimated 5% false-positive rate to be used to determine where sites of transcription are located (see Kampa et al. 2004).

MOST OF THE HUMAN GENOME IS TRANSCRIBED AS NUCLEAR PRIMARY (UNPROCESSED) RNAs

This transcriptional complexity with overlapping sense and antisense transcripts significantly complicates interpretations of genome organization and annotation. For example, a 500-kb region selected by the ENCODE project (ENr 233) on chromosome 15 (Fig. 1) contains only 15 RefSeq annotated genes (top track), and yet, significant amounts of overlapping transcription as detected by tiling arrays (see Affymetrix and Yale tracks) far exceed these limited annotations. Messenger RNAs cloned by the Mammalian Gene Collection (MGC), H-Invitational Gene Database (HInv-DB), and submitted to GenBank also highlight the large degree of alternative splice isoforms. Paired-end ditag (PET) sequencing by the Genome Institute of Singapore has identified the starts and ends of hundreds of thousands of mRNAs, and similar large-scale 5′ cap analysis of gene expression (CAGE) by the RIKEN has predicted scores of gene start sites. Taken together with expression data from tiling arrays generated by Yale and Affymetrix, this 500-kb region appears to be entirely transcribed in a series of overlapping and intertwined transcripts. Furthermore, such transcriptional complexity is not limited to this example region, but rather appears to be the case for most of the ENCODE regions in the human genome (ENCODE Project Consortium, in prep.). In fact, by taking all the annotated mRNAs and empirically detected processed RNAs present in the ENCODE regions and then extrapolating the amount of primary nuclear transcription required to produce these RNAs, these analyses have inferred that >90% of the genomic sequence can be transcribed as nuclear primary sequences when looking at the totality of all examined biological samples (ENCODE Project Consortium, in prep.). This is illustrated by the last two tracks of Figure 1.

GENOMIC TILING ARRAYS DETECT PROCESSED, CYTOSOLIC TRANSCRIPTS COVERING FIVE TIMES MORE OF THE GENOME THAN SEQUENCES OF ANNOTATED PROTEIN-CODING GENES

When the polyadenylated cytosolic RNA present in each of eight analyzed cell lines was mapped to the non-repeat portion of 10 human chromosomes (Cheng et al. 2005), approximately 4–5% of interrogated nucleotides were observed to be transcribed (Table 1). Furthermore, when the nonredundant union of all detected poly(A)$^+$ RNA present in the eight cell lines was created, the detected transcribed non-repeat portions of the genome increased to 10.14%, suggesting that a significant portion of this transcription is cell-line-specific. Given that all of the nucleotides present in the exons of protein-coding genes amounts to 1–2% of the total (including repeated portions) human genome, the observed amount of transcription derived from poly(A)$^+$ cytosolic RNAs in eight cell lines is conservatively estimated to be more than five-fold greater.

The RNA products of this transcription are not uniformly distributed within a cell, nor are poly(A)$^+$ RNAs the only transcripts produced. To gain additional information on cellular compartment and structure of transcribed RNAs, one cell line (HepG2) was selected, and RNA was fractionated into nuclear and cytosolic portions. Further characterization was made by selecting poly(A)$^+$ and poly(A)$^-$ RNAs from each compartment (Cheng et al. 2005). Approximately 10.2% and 51.3% of detected sequences were found exclusively in the cytosolic and

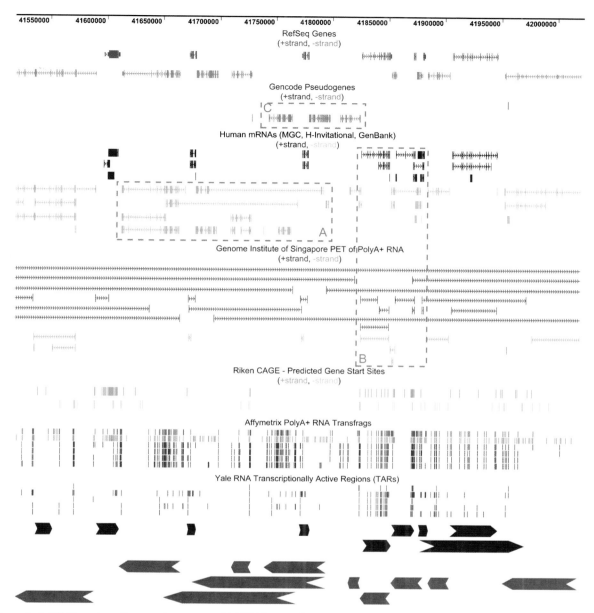

Figure 1. A 500-kb region of human chromosome 15 (ENCODE region ENr233) highlights the complexity of transcription observed across the genome. This region contains only 15 RefSeq annotated genes and yet contains (*A*) a variety of intertwined isoforms of annotated genes, (*B*) significant overlapping sense and antisense transcription, and (*C*) expressed pseudogenes (see Affymetrix transfrags). These examples are highlighted with red boxes. General areas of transcription are indicated by black (+ strand) and gray (– strand) arrow bars. Depicted annotations are based on the May 2004 human genome assembly and the UCSC Genome Browser (www.genome.ucsc.edu), where additional details on the represented data sets are available (Kent et al. 2002).

nuclear compartments (Table 2). This fivefold difference is striking, since it clearly indicates that most of the transcriptional products analyzed under these steady-state conditions never exit the nuclear compartment. These data also suggest that the sequences comprising the difference observed between the nuclear and cytosolic compartments are either synthesized and degraded, or that these products of widespread transcription have hitherto unknown nuclear functions. A total 19.4% and 43.7% of all detected transcription in both the nucleus and cytosol of HepG2 cells are exclusively poly(A)$^+$ or poly(A)$^-$, respectively (Table 2). The remainder of the detected transcription is bimorphic (detected as both poly(A)$^+$ and poly(A)$^-$). These data underscore the underappreciated prevalence (twofold greater overall) of poly(A)$^-$ transcription; for example, 84% of the cytosolic poly A$^-$ transcribed regions are located in unannotated genome regions. Furthermore, there is a renewed appreciation that many protein-coding transcripts exist within cells both with and without (or with shortened) polyadenylated 3′ termini. Although this bimorphic state has been previously described for some individual protein-coding transcripts (for list of genes, see Cheng et al. 2005), the extent of the number of genes exhibiting a bimorphic state was not understood.

Table 1. Poly(A)$^+$ RNA Detected in Eight Cell Lines

Sample	Coverage (bp)	% of all interrogated base pairs on the arrays
A-357	19, 330, 720	5.07
FHs 378Lu	18, 579, 012	4.87
Jurkat	18, 886, 873	4.96
NCCIT	21, 662, 254	5.68
PC-3	18, 242, 956	4.79
SK-N-AS	19, 088, 926	5.01
U-87 MG	14, 864, 583	3.90
HepG2	18, 899, 552	4.96
All cell lines (†)	38, 656, 627	10.14

Polyadenylated RNA map of the non-repeat portion 10 chromosomes representing ~30% of the human genome. For each cell line assayed, expressed poly(A)$^+$ RNA is listed as the number of nucleotides detected (coverage) and as a corresponding percentage of the total number of interrogated base pairs present on the microarrays. The nonredundant union of all detected poly(A)$^+$ RNA in the eight tested cell lines exceeds 10% of interrogated bases (†). Based on data in Cheng et al. (2005).

Table 2. Classes of RNAs Detected in HepG2 Cells

Sample	Coverage (bp)	% of all transcription detected in HepG2
Only in cytoplasmic fraction (‡)	6, 032, 310	10.25
Only in cytosolic poly(A)$^+$	1, 835, 709	3.12
Only in cytosolic poly(A)$^-$	3, 847, 281	6.53
Only in nuclear fraction (‡)	30, 207, 724	51.31
Only in nuclear (A)$^+$	5, 706, 194	9.69
Only in nuclear poly(A)$^-$	18, 237, 769	30.98
Only in poly(A)$^+$ RNA	11, 432, 433	19.42
Only in poly(A)$^-$ RNA	25, 747, 796	43.73
Both A$^+$ and A$^-$ RNA (†)	21, 693, 884	36.85

Percentages of HepG2 transcription detected in nuclear and cytosolic compartments. RNA from HepG2 cells was fractionated based on cellular compartment (nuclear vs. cytosolic) and polyadenylation (plus and minus). Total summing of transcription in each compartment (e.g., only in cytoplasmic fraction) includes the nonredundant contribution from both poly(A)$^+$ and poly(A)$^-$ fractions (‡). Sequences detected in both poly(A)$^+$ and (A)$^-$ are the nonredundant union (†). The number of nucleotides detected for poly(A)$^+$ RNA from nuclear (30,162,893 bp) and cytosolic (18,899,552 bp) fractions includes 15,936,128 bp present in both fractions. Based on data in Cheng et al. (2005).

However, is all this noncoding RNA biologically functional? Comparative genomic analysis of RNA structural elements has yielded intriguing clues and predicted thousands of functional ncRNAs. Conserved genomic sequences from several vertebrates were comparatively analyzed for base-pairing and thermodynamic stability contributions to structural conservation (Washietl et al. 2005). More than 30,000 RNA elements were identified in human, and nearly 1,000 were found conserved across vertebrates; furthermore, half of the structured RNA elements are located distant from known genes. In a separate study, Torarinsson et al. (2006) examined the approximately one-third of the nonrepetitive human genome that is not alignable with mouse, and a subset of these approximately 100,000 regions were analyzed for the presence of RNA structural elements. A significant number of "nonconserved" regions were found to have common RNA structure and, surprisingly, were twice as likely to overlap expressed transfrags as not to be expressed. Together, these studies begin to illustrate the value of profiling whole-genome transcription in multiple species and point the way to the genomic areas that will be the focus of additional studies.

UNANNOTATED ncRNA IS EXTENSIVELY UTILIZED DURING EARLY EMBRYOGENESIS OF *DROSOPHILA MELANOGASTER*

Within the first 24 hours of *Drosophila* embryogenesis, one oocyte and 15 nurse cells pattern and develop the complete musculature and nervous system for a larva which hatches about 24 hours postfertilization. This period of incredible transcriptional and developmental activity (e.g., early cell cycles are 7–8 minutes long) was interrogated with tiling arrays using samples gathered from 2-hour time points (Manak et al. 2006). In total over this 24-hour period, 27.6% of the 105.9-Mb nonrepetitive portion of the *Drosophila* genome is detected as transcribed RNAs with about 70% overlapping annotated gene structures (Fig. 2A). The 30% which is unannotated

is almost half that observed for the human transcriptome and, if analysis is further restricted to intergenic transcription, the 7% observed in *Drosophila* is approximately fourfold lower. These differences are likely attributable to the significantly more compact fly genome (106 Mb, nonrepetitive) having a more complete annotation than the approximately tenfold larger human genome (1255 Mb, nonrepetitive). Interestingly, in *Drosophila* the unannotated regions are often expressed at specific and discrete time points, similar to what was observed between differing human cell lines. A comparison of the overall expression levels across the developmental time points, measured as the number of transcribed nucleotides, finds minimum total expression (8.2%) at the 4- to 6-hour time point which roughly corresponds with the lull between the degradation of the maternally contributed RNA and the initiation of zygotically produced transcripts (Fig. 2B). Conversely, maximal expression (16.1%) is observed at the 10- to 12-hour time point.

Direct comparisons of human and fly intergenic genomic sequences are made complicated by their evolutionary distances; however, a study comparing human and chimpanzee intergenic transcription highlights the value of broadening the scope of comparative genomics to include noncoding regions. Chimpanzee and human transcription were compared in three tissues and one cell line across 1% of their genomes (ENCODE regions) using tiling arrays (Khaitovich et al. 2006). Intergenic transcripts show patterns of tissue-specific conservation of expression comparable to protein-coding exons of known genes, suggesting intergenic transcripts evolved under similar levels of positive selection and functional constraint. Additionally, about half of observed intergenic transcription is differentially expressed between humans and chimps.

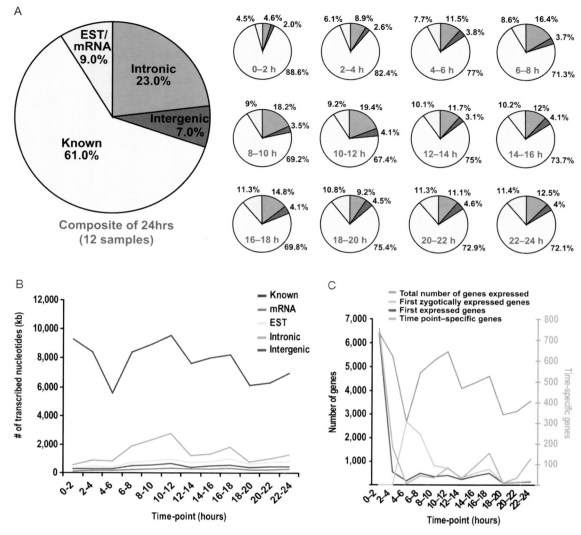

Figure 2. Analysis of transcription during *Drosophila* embryogenesis (0–24 hr). (*A*) Summary of annotated and unannotated (intronic and intergenic) transcription for each of 12 2-hr time points across the first 24 hrs of embryogenesis. (*B*) Comparison of the variation in overall expression levels (measured as number of transcribed kilobases of nucleotides) across time points for indicated classes of RNAs. (*C*) Relationship of expressed RefSeq genes: total number of expressed genes (*blue*), genes first expressed at a given time point (*dark green*), zygotically expressed genes (first 4 hrs disregarded due to maternal contributions of RNA) (*light green*), and genes expressed only in a specific time point (*orange*). (Based on data in Manak et al. 2006.)

Focusing on annotated fly protein-coding genes, a total of 9,808 genes are expressed during embryogenesis: 52.4% of a total of 18,716 known RefSeq genes (Fig. 2C). An average of approximately 4,400 are expressed in each time point, with an average duration of 10.8 hours. Despite the fact that average gene expression numbers in many hours, a significant number of genes (1,563; 16% of expressed genes) were found expressed at only one specific time point. It is worth noting that gene expression was computationally defined as the requirement that >70% of probes within an annotated gene be determined to be positive. Thus, it is likely that using this high level of stringency overlooks splice forms or other forms of partial gene expression and therefore underestimates overall transcription during embryogenesis.

ONE BIOLOGICAL FUNCTION OBSERVED FOR UNANNOTATED TRANSCRIPTION

A careful analysis of time-point-specific expression of genes and their correlation with adjacent unannotated transcription has proven a powerful method to understand the biological function of some unannotated transfrags. A convergence of a number of lines of evidence suggested that gene architectures are significantly more complicated than consensus annotations would suggest (see Fig. 1). Thus, we explored the possibility that this newly appreciated complex gene architecture could be related to some unannotated transcription and begin to address the important question concerning the biological function of the detected unannotated transcription. Specifically, one possibility was that some of the unannotated transfrags were previously unidentified parts of protein-coding genes. Interestingly, as suggested by the overlap of CAGE and Ditag data sets with transfrag data (Fig. 1), it is possible that some of the intergenic unannotated transfrags are novel 5′ transcriptional start sites (TSSs) which would allow the identification of novel regulatory regions, immediately 5′ to these transfrags. Therefore, two strategies were employed for the large-scale identification of alternative 5′ TSSs for protein-coding genes. For this study, both human and *Drosophila* genes were analyzed. Because of the limited number of protein-coding genes present within the ENCODE regions, a comprehensive empirical method relying on the use of the combination of 5′ RACE and tiling arrays was adopted, whereas in *Drosophila*, a whole-genome computational method was first developed to search for distal unannotated, developmental-stage-specific 5′ TSSs and upstream regulatory regions followed by the use of RT-PCR, cloning, and sequencing.

The hybridization of 5′ RACE reactions to tiling arrays forgoes time-consuming intermediate cloning and sequencing steps and permits the high-throughput assessment of gene structures. Products of the RACE reactions (RACEfrags) are resolved on a tiling array, thus providing all portions of a transcript that is connected to the site (index position) where the RACE primers are positioned (Kapranov et al. 2005). Thus, 5′ RACE reactions starting from a protein-coding exon proximal to the annotated 5′ end of a transcript would allow the identification of any novel exons that are upstream of the annotated end of the interrogated gene.

This strategy was applied to 399 genes contained within the 44 ENCODE regions that were subjected to 5′ RACE using RNA isolated individually from 12 human tissues. For the 359 loci with successful RACE products, almost half of the RACEfrags detected did not overlap with annotated exons. Nearly 80% of the genes were found to either have novel internal exons (between the index exon and the annotated 5′ end of the gene) or new 5′ extensions, with many of these new exons being tissue specific: 65.7% had 5′ extensions in at least one tissue, and 59.9% had new internal exons (ENCODE Project Consortium, in prep.). The distances to the tissue-specific unannotated 5′ extensions were surprisingly quite large, averaging about 108 kb for the new first intron, with 23% being ≥200 kb in size (Fig. 3A). These tissue-specific distal and often previously unannotated transcripts traverse several well-characterized protein-coding gene regions to reach the parent coding gene.

These findings are consistent with a large-scale analysis of TSSs conducted by Carninci et al., where sequencing of hundreds of thousands of CAGE tags culled from 145 mouse and 41 human libraries has permitted the quantitative analysis of the differential usage of promoters in numerous tissues (Carninci et al. 2006). Bidirectional promoters are found to be common, as are TSSs associated with internal exons, and the majority of protein-coding genes (58%) are shown to have alternative promoters. All told, TSSs are found to be abundant, tissue-specific, and bidirectional, further supporting emerg-

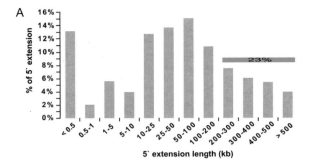

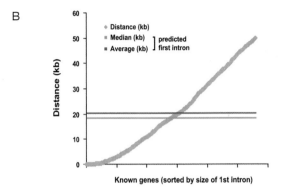

Figure 3. Many human and *Drosophila* genes have extensive, previously unannotated 5′ exons. (*A*) Summary of the size distribution of the longest 5′ RACE extensions per gene and tissue for 350 human genes within the ENCODE regions. The average size of these newly identified first introns is 108 kb; with 23% ≥200 kb in size. From data produced by the ENCODE Genes & Transcripts Analysis Group (ENCODE Project Consortium, in prep.). (*B*) Distance of most distal predicted 5′ site from the nearest co-regulated RefSeq gene (50-kb window) expressed in *Drosophila* embryonic development. Average size of predicted first introns was found to be 20.4 kb; ~12-fold larger than the average size of RefSeq annotated first introns. Median size of predicted new introns was 18.4 kb. (Based on data in Manak et al. 2006.)

ing models for a complex overlapping network of transcription across the genome.

What are the consequences of such extensive 5′ extensions? Sequencing of clones containing a subset of 5′ RACE extensions finds that approximately 45% likely add sequences to the protein open reading frame. In one striking example, a chimeric transcript isoform of DONSON, a gene of unknown function, was identified (Fig. 4A). The formation of such fusion transcripts has previously been reported (Kapranov et al. 2005). Tissue-specific RACEfrags were identified in brain, heart, kidney, and lung; but it was a small-intestine-specific product that was most intriguing, since it appeared to directly overlap exons from ATP5O about 330 kb upstream. A RT-PCR product was cloned and sequenced, confirming that three exons from ATP5O including the DNA-binding domain were being fused to at least two exons from DONSON. Independent paired end ditag (PET) data also confirm the presence of such an mRNA species. Such a combinatorial approach to building a protein takes full advantage of the modular potential of a genome's architecture and may prove to be a novel method for increasing complexity of the proteome.

For *Drosophila*, the availability of a developmental time course of whole-genome transcription provided the high-resolution temporal data that were used for predicting 5′ extensions based on the transcriptional co-regulation of unannotated regions with proximal (within 50 kb) protein-coding loci (Manak et al. 2006). More than 1000 genes were predicted to have 5′ extensions, and the distribution of lengths to the predicted 5′ sites for each are graphed in Figure 3B. As was found in the analysis of the human protein-coding genes, the average size of the predicted first introns was quite large: averaging 20.4 kb versus 1.7 kb for RefSeq annotated first introns. A subset of 180 co-regulated transfrags were then tested by RT-PCR and sequencing with a total verification rate of 78% for predicting new 5′ exons. Conservatively, 29% of the unannotated transcribed sequences function as misannotated constitutive exons or alternative exons of protein-coding genes. A total of 15.6% of the unannotated transcribed sequences appear to be TSSs used by 11.4% of the genes expressed during embryogenesis.

In one striking example, the RhoGAP gene was shown to be significantly larger than its annotated form, with novel 5′ exons adding more than 1300 amino acids to the protein and a TSS over 50 kb upstream of the annotated start site (Fig. 4B). This new annotation of RhoGAP now includes three P-element transposon insertions near the TSS and one lethal insertion within the gene, which adds a new mutant allele interrupting the transcript to the repertoire for future characterization of RhoGAP function. Given the abundance of intergenic P-element insertions mapped in the *Drosophila* genome, a continuation of this type of analysis would potentially allow the connection of "orphan" transposons with adjacent protein-coding loci. In summary, novel 5′ extensions are predicted to "convert" 16 Mb genomic sequence from intergenic to intronic and, when summed with all observed annotated and unannotated transcription, strongly suggest that >85% of the *Drosophila* genome is expressed as nuclear primary transcripts. This observation again underscores the observation made for the nuclear primary transcripts in the ENCODE regions.

Overall, the abundance of unannotated 5′ TSSs is surprising. As discussed, these novel 5′ extensions can result in novel protein open reading frames and even "chimeric" transcripts composed of protein-coding exons contributed by several different genes. Alternative TSSs can introduce new promoters with different regulatory timing and expression strengths, some of which appear to be well-characterized promoters used by upstream genes. It is possible that the long-range transcription afforded by these 5′ extensions marks the genomic interval as transcriptional capable for DNA and chromatin modifying machineries. This may allow transcription of independently regulated coding and noncoding RNAs from the intronic regions or even processing of active RNAs from the transcribed intron itself. Last, large-scale analysis of TSSs has found that 28% (human) to 36% (mouse) of TSSs appear to be associated with noncoding transcripts (Carninci et al. 2006), adding an additional layer of transcriptional complexity.

IMPLICATIONS AND CONCLUSIONS

The emerging consensus from whole-genome studies of human and *Drosophila* transcription is that almost all of the non-repeat portions of both genomes are transcribed as nuclear primary RNAs: *Drosophila* >85% and human ENCODE regions about 93%. Additionally, cytosolic processed transcribed regions of the nonrepetitive human genome are also pervasive. As consequence of such widespread transcription, many regions of the genome consist of overlapping networks of intertwined, independently regulated transcripts originating from both strands. In *Drosophila*, approximately 9% (29% are previously missed unannotated exons of the total of 30% of the total annotated detected transcribed sequences) of the unannotated transcribed sequences appear to function as parts of protein-coding transcripts. In human, estimates of the percent of unannotated transcription which are likely to be portions of protein-coding transcripts are yet undetermined. Furthermore, ncRNAs are increasingly emerging as novel functional regulators in cellular processes such as microRNAs regulating cancers (Esquela-Kerscher and Slack 2006), NFAT signaling (Willingham et al. 2005), heat-shock sensing (Shamovsky et al. 2006), and multiple other instances (for review, see Mattick 2004; Storz et al. 2005; Zamore and Haley 2005; Goodrich and Kugel 2006). Especially interesting is the recent recognition that the most rapidly evolving human gene yet identified is in fact a ncRNA specifically expressed in the neocortex during a critical period of neurodevelopment where it is potentially involved in specifying structural aspects of the human cortex (Pollard et al. 2006).

Beyond their role as unannotated parts of mRNAs and functional ncRNAs, noncoding transcripts are likely acting on the level of gene regulation and genome architecture. Small RNAs have been implicated in transcriptional silencing and chromatin formation (for review, see Andersen and Panning 2003; Matzke and Birchler 2005). Noncoding transcription appears to encompass a number of locus control regions (LCRs), and the contributions of ncRNA transcription are investigated for the LCR of the human growth hormone (hGH) (Ho et al. 2006). The hGH LCR is 14.5 kb from the gene, and levels of hGH expression directly correlate with the levels of LCR transcription. This LCR transcription is bidirectional, does not seem to be directly connected to hGH as a 5′ extension, and, if inhibited by the insertion of a transcriptional terminator, results in reduced hGH transcription. RNA may play a role in the interchromosomal interactions of a single enhancer element with select odorant receptors (OR). Within single neurons, the *trans*-acting H-enhancer element on chromosome 14 colocalizes with only one of 1300 possible OR promoters present in the mouse genome and likely serves as a selection method for expression of a single receptor gene in individual sensory neurons (Lomvardas et al. 2006). Interestingly, H-enhancer DNA is shown to colocalize with OR receptor RNA, raising the possibility that transcribed RNA might contribute to stabilizing H-enhancer association with a single locus.

Of particular interest is the prevalence of small RNAs such as microRNAs, snoRNAs, and other regulatory RNAs

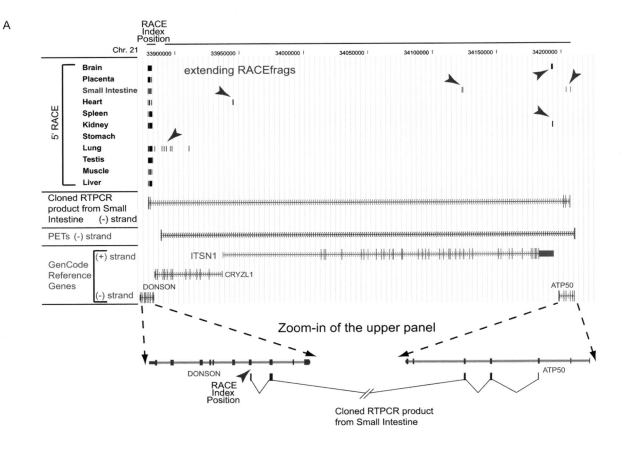

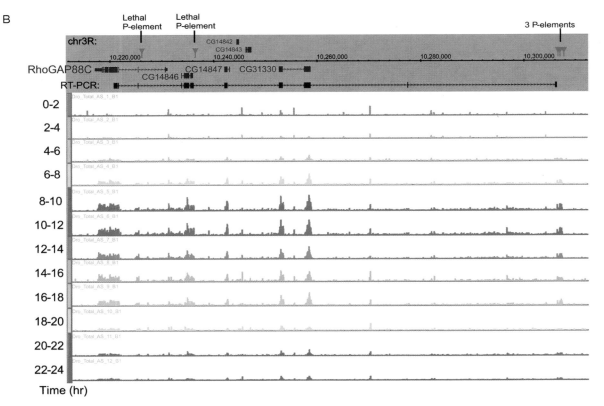

Figure 4. (*See facing page for legend.*)

(for review, see Mattick and Makunin 2005; Zamore and Haley 2005) as well as the surprising abundance of a newly discovered class of testes-specific approximately 30-bp RNAs which associate with the Piwi family and number in the tens of thousands (for review, see Kim 2006). Our group has begun to characterize the small RNA transcriptome on a whole-genome level using 5-bp resolution microarrays (T.R. Gingeras et al., in prep.). Known small RNA species such as miRNAs and snoRNAs are detected well, as are an astonishing number of novel small RNAs which could number over 400,000, averaging one small RNA every 3 kb. Ongoing investigations are exploring (1) small RNA biogenesis by perturbing proteins known to process small RNAs, (2) the relationship of these small RNAs to known gene annotations, and (3) the evolutionary conservation of these small RNAs. Clearly, as new technical approaches begin to answer old questions, they also elicit numerous new questions and insights into the organization and regulation of genomes.

ACKNOWLEDGMENTS

We are indebted to the ENCODE Genes & Transcripts Analysis Group for their invaluable computational, experimental contributions and discussions: Josep F. Abril, Tyler Alioto, Stylianos E. Antonarakis, Robert Baertsch, Peter Bickel, Ewan Birney, James B. Brown, Piero Carninci, Robert Castelo, Kuo Ping Chiu, Siew Woh Choo, Chiou Yu Choo, Jacqueline Chrast, Taane Clarke, France Denoeud, Emmanouil T. Dermitzakis, Mark C. Dickson, Olof Emanuelsson, Christoph Flamm, Paul Flicek, Sylvain Foissac, Adam Frankish, Claudia Fried, Mark Gerstein, James Gilbert, Roderic Guigó, Jörg Hackermüller, Jennifer Harrow, Yoshihide Hayashizaki, Charlotte N. Henrichsen, Jana Hertel, Heather Hirsch, Ivo L. Hofacker, Nancy Holroyd, Tim Hubbard, Chikatoshi Kai, Jun Kawai, Damian Keefe, Jan Korbel, Julien Lagarde, Zheng Lian, Jin Lian, Manja Lindemeyer, Todd M. Lowe, Caroline Manzano, Elliott H. Margulies, Nicholas Matthews, John S. Mattick, Kristin Missal, Zarmik Moqtaderi, Richard M. Myers, Ugrappa Nagalakshmi, Peter Newburger, Hong Sain Ooi, Sandeep Patel, Jakob S. Pedersen, Alexandre Reymond, Jane Rogers, Joel Rozowsky, Yijun Ruan, Albin Sandelin, Edward A. Sekinger, Atif Shahab, Michael Snyder, K.G. Srinivasan, Peter F. Stadler, Kevin Struhl, Wing-Kin Sung, David Swarbreck, Andrea Tanzer, Ruth Taylor, Daryl J. Thomas, Catherine Ucla, Stefan Washietl, Chia-Lin Wei, Matthew T. Weirauch, Sherman M. Weissman, Jiaqian Wu, Carine Wyss, Annie Yang, Xueqing Zhang,

Xiao-Dong Zhao, Deyou Zheng, and Zhou Zhu. This project has been funded in part with federal funds from the National Cancer Institute, National Institutes of Health, under Contract No. N01-CO-12400, and from the National Human Genome Research Institute, National Institutes of Health, under Grant No. U01 HG003147, and by Affymetrix, Inc.

REFERENCES

Andersen A.A. and Panning B. 2003. Epigenetic gene regulation by noncoding RNAs. *Curr. Opin. Cell Biol.* **15:** 281.

Carninci P., Kasukawa T., Katayama S., Gough J., Frith M.C., Maeda N., Oyama R., Ravasi T., Lenhard B., Wells C., et al. 2005. The transcriptional landscape of the mammalian genome. *Science* **309:** 1559.

Carninci P., Sandelin A., Lenhard B., Katayama S., Shimokawa K., Ponjavic J., Semple C.A., Taylor M.S., Engstrom P.G., Frith M.C., et al. 2006. Genome-wide analysis of mammalian promoter architecture and evolution. *Nat. Genet.* **38:** 626.

Chen J., Sun M., Lee S., Zhou G., Rowley J.D., and Wang S.M. 2002. Identifying novel transcripts and novel genes in the human genome by using novel SAGE tags. *Proc. Natl. Acad. Sci.* **99:** 12257.

Cheng J., Kapranov P., Drenkow J., Dike S., Brubaker S., Patel S., Long J., Stern D., Tammana H., Helt G., et al. 2005. Transcriptional maps of 10 human chromosomes at 5-nucleotide resolution. *Science* **308:** 1149.

ENCODE Project Consortium. 2004. The ENCODE (ENCyclopedia Of DNA Elements) Project. *Science* **306:** 636.

Esquela-Kerscher A. and Slack F.J. 2006. Oncomirs—microRNAs with a role in cancer. *Nat. Rev. Cancer* **6:** 259.

Gingeras T.R. 2006. The multitasking genome. *Nat. Genet.* **38:** 608.

Goodrich J.A. and Kugel J.F. 2006. Non-coding-RNA regulators of RNA polymerase II transcription. *Nat. Rev. Mol. Cell Biol.* **7:** 612.

Ho Y., Elefant F., Liebhaber S.A., and Cooke N.E. 2006. Locus control region transcription plays an active role in long-range gene activation. *Mol. Cell* **23:** 365.

Johnson J.M., Edwards S., Shoemaker D., and Schadt E.E. 2005. Dark matter in the genome: Evidence of widespread transcription detected by microarray tiling experiments. *Trends Genet.* **21:** 93.

Jongeneel C.V., Delorenzi M., Iseli C., Zhou D., Haudenschild C.D., Khrebtukova I., Kuznetsov D., Stevenson B.J., Strausberg R.L., Simpson A.J., and Vasicek T.J. 2005. An atlas of human gene expression from massively parallel signature sequencing (MPSS). *Genome Res.* **15:** 1007.

Kampa D., Cheng J., Kapranov P., Yamanaka M., Brubaker S., Cawley S., Drenkow J., Piccolboni A., Bekiranov S., Helt G., et al. 2004. Novel RNAs identified from an in-depth analysis of the transcriptome of human chromosomes 21 and 22. *Genome Res.* **14:** 331.

Kapranov P., Cawley S.E., Drenkow J., Bekiranov S., Strausberg R.L., Fodor S.P., and Gingeras T.R. 2002. Large-scale transcriptional activity in chromosomes 21 and 22. *Science* **296:** 916.

Figure 4. Examples of how unannotated transcriptional complexity can affect the landscape and composition of human and *Drosophila* genes. (*A*) 5′-RACE upstream of the DONSON gene identified tissue-specific 5′ exons (RACEfrags) spanning ~330 kb. A product specific to small intestine was shown by RT-PCR to be a novel fusion transcript incorporating three exons including the DNA-binding domain from the upstream ATP50 gene and at least two exons from DONSON. From data produced by the ENCODE Genes & Transcripts Analysis Group (ENCODE Project Consortium, in prep.). Gene annotations based on the May 2004 version of the human genome presented in the UCSC browser (Kent et al. 2002). (*B*) The annotated *Drosophila* RhoGAP gene is shown by transcriptional co-regulation and RT-PCR to have additional 5′ regions encompassing three separately in-silico annotated genes and a new unannotated 5′ exon. This extension includes four new P-element insertions and adds a new mutant allele interrupting the transcript for RhoGAP (this transposon was previously thought to disrupt an enhancer element). (Based on data in Manak et al. 2006.)

Kapranov P., Drenkow J., Cheng J., Long J., Helt G., Dike S., and Gingeras T.R. 2005. Examples of the complex architecture of the human transcriptome revealed by RACE and high-density tiling arrays. *Genome Res.* **15:** 987.

Kent W.J., Sugnet C.W., Furey T.S., Roskin K.M., Pringle T.H., Zahler A.M., and Haussler D. 2002. The human genome browser at UCSC. *Genome Res.* **12:** 996.

Khaitovich P., Kelso J., Franz H., Visagie J., Giger T., Joerchel S., Petzold E., Green R.E., Lachmann M., and Paabo S. 2006. Functionality of intergenic transcription: An evolutionary comparison. *PLoS Genet.* (in press).

Kim V.N. 2006. Small RNAs just got bigger: Piwi-interacting RNAs (piRNAs) in mammalian testes. *Genes Dev.* **20:** 1993.

Lomvardas S., Barnea G., Pisapia D.J., Mendelsohn M., Kirkland J., and Axel R. 2006. Interchromosomal interactions and olfactory receptor choice. *Cell* **126:** 403.

Manak J.R., Dike S., Sementchenko V., Kapranov P., Biemar F., Long J., Cheng J., Bell I., Ghosh S., Piccolboni A., and Gingeras T.R. 2006. Biological function of unannotated transcription during the early development of *Drosophila melanogaster*. *Nat. Genet.* **38:** 1151.

Mattick J.S. 2004. RNA regulation: A new genetics? *Nat. Rev. Genet.* **5:** 316.

Mattick J.S. and Makunin I.V. 2005. Small regulatory RNAs in mammals. *Hum. Mol. Genet.* **14:** R121.

Matzke M.A. and Birchler J.A. 2005. RNAi-mediated pathways in the nucleus. *Nat. Rev. Genet.* **6:** 24.

Mockler T.C., Chan S., Sundaresan A., Chen H., Jacobsen S.E., and Ecker J.R. 2005. Applications of DNA tiling arrays for whole-genome analysis. *Genomics* **85:** 1.

Okazaki Y., Furuno M., Kasukawa T., Adachi J., Bono H., Kondo S., Nikaido I., Osato N., Saito R., Suzuki H., et al. (FANTOM Consortium; RIKEN Genome Exploration Research Group Phase I & II Team). 2002. Analysis of the mouse transcriptome based on functional annotation of 60,770 full-length cDNAs. *Nature* **420:** 563.

Pollard K.S., Salama S.R., Lambert N., Lambot M.A., Coppens S., Pedersen J.S., Katzman S., King B., Onodera C., Siepel A., et al. 2006. An RNA gene expressed during cortical development evolved rapidly in humans. *Nature* **443:** 167.

Saha S., Sparks A.B., Rago C., Akmaev V., Wang C.J., Vogelstein B., Kinzler K.W., and Velculescu V.E. 2002. Using the transcriptome to annotate the genome. *Nat. Biotechnol.* **20:** 508.

Shamovsky I., Ivannikov M., Kandel E.S., Gershon D., and Nudler E. 2006. RNA-mediated response to heat shock in mammalian cells. *Nature* **440:** 556.

Storz G., Altuvia S., and Wassarman K.M. 2005. An abundance of RNA regulators. *Annu. Rev. Biochem.* **74:** 199.

Torarinsson E., Sawera M., Havgaard J.H., Fredholm M., and Gorodkin J. 2006. Thousands of corresponding human and mouse genomic regions unalignable in primary sequence contain common RNA structure. *Genome Res.* **16:** 885.

Washietl S., Hofacker I.L., Lukasser M., Huttenhofer A., and Stadler P.F. 2005. Mapping of conserved RNA secondary structures predicts thousands of functional noncoding RNAs in the human genome. *Nat. Biotechnol.* **23:** 1383.

Willingham A.T. and Gingeras T.R. 2006. TUF love for "junk" DNA. *Cell* **125:** 1215.

Willingham A.T., Orth A.P., Batalov S., Peters E.C., Wen B.G., Aza-Blanc P., Hogenesch J.B., and Schultz P.G. 2005. A strategy for probing the function of noncoding RNAs finds a repressor of NFAT. *Science* **309:** 1570.

Zamore P.D. and Haley B. 2005. Ribo-gnome: The big world of small RNAs. *Science* **309:** 1519.

Novel Transcribed Regions in the Human Genome

J. Rozowsky,* J. Wu,[†] Z. Lian,[‡] U. Nagalakshmi,[†] J.O. Korbel,* P. Kapranov,**
D. Zheng,* S. Dyke,** P. Newburger,[††] P. Miller,[§,¶] T.R. Gingeras,**
S. Weissman,[‡] M. Gerstein,*,[¶] and M. Snyder*,[†]

*Molecular Biophysics & Biochemistry Department, [†]Molecular, Cellular & Developmental Biology Department,
[‡]Department of Genetics, [§]Center for Medical Informatics, and [¶]Program in Computational Biology and
Bioinformatics, Yale University, New Haven, Connecticut 06520; **Affymetrix, Inc., Santa Clara, California 92024;
[††]University of Massachusetts Medical School, Children's Medical Center, North Worcester, Massachusetts 01605

We have used genomic tiling arrays to identify transcribed regions throughout the human genome. Analysis of the mapping results of RNA isolated from five cell/tissue types, NB4 cells, NB4 cells treated with retinoic acid (RA), NB4 cells treated with 12-O-tetradecanoylphorbol-13 acetate (TPA), neutrophils, and placenta, throughout the ENCODE region reveals a large number of novel transcribed regions. Interestingly, neutrophils exhibit a great deal of novel expression in several intronic regions. Comparison of the hybridization results of NB4 cells treated with different stimuli relative to untreated cells reveals that many new regions are expressed upon cell differentiation. One such region is the Hox locus, which contains a large number of novel regions expressed in a number of cell types. Analysis of the trinucleotide composition of the novel transcribed regions reveals that it is similar to that of known exons. These results suggest that many of the novel transcribed regions may have a functional role.

The human genome project has revealed the DNA sequence that governs all biological processes in humans (Lander et al. 2001; Venter et al. 2001). However, understanding how the sequence is interpreted to carry out cellular and developmental processes in humans is somewhat limited. In particular, we would like to identify the genes and the protein products they encode, the regulatory information that controls the level of expression of each RNA and protein, and how the different components function together to carry out complex molecular, cellular, developmental, and behavioral processes.

An important step in the process of characterizing the human genome is the identification of genes. This problem is particularly acute in mammals because the RNA coding segments of genes are split into relatively short exons of average length approximately 140 bp separated by often very large introns, which can sometimes be greater than 100,000 bp in length (Lander et al. 2001; Venter et al. 2001). In addition, the large number of alternatively spliced RNAs can make it difficult to determine which regions code for proteins. For these reasons, it can be extremely challenging to identify exons and genes, and computational approaches have generally been only partially successful (Burge and Karlin 1997; Guigo et al. 2006).

Other approaches for gene identification include (1) comparison of sequences among related species; in general, protein-coding genes are more conserved than noncoding regions (Cawley et al. 2003; Parra et al. 2003) and (2) mapping transcribed regions. For the latter case, sequences of cDNAs (ideally full length) and ESTs allow one to assign transcribed regions to segments of the genome. In general, ESTs have been less useful because of concerns that the sequences are often derived from unspliced RNAs and/or contaminating DNA in the starting RNA preparations. In contrast, full-length cDNAs

have been quite successful for gene and exon identification (Carninci et al. 2005). However, many genes are expressed at a low level, and the large number of alternative spliced RNAs precludes a comprehensive analysis of transcribed regions using only cDNA information.

MAPPING TRANSCRIBED REGIONS USING TILING ARRAYS REVEALS EXTENSIVE TRANSCRIPTION IN THE HUMAN GENOME

We have been mapping transcribed regions using genomic tiling arrays. Tiling arrays enable the detection of transcribed regions unbiased by genome annotation. Initially, arrays containing 21,000 800-bp PCR products from human chromosome 22 were used to cover the nonrepetitive portions of the chromosome (Rinn et al. 2003). Subsequently, we built a 36-base oligonucleotide array that tiled the nonrepetitive sequence of both strands of the entire genome at a resolution of one oligonucleotide every 46 bp (Bertone et al. 2004). By probing the chromosome-22 array and whole-genome arrays using poly(A)[+] RNA isolated from placenta and liver, respectively, we found that approximately 60% of the transcribed regions lay outside of annotated exons. Validation by PCR confirmed that the majority of these regions are expressed in poly(A)[+] RNA. Thus, there is at least twice as much of the genome expressed as processed poly(A)[+] transcripts as previously identified.

More recently, we have probed an Affymetrix oligonucleotide array covering the nonrepetitive DNA of the ENCODE regions using RNAs isolated from five different cell types or tissues: NB4 cells (a lymphoid cell line); NB4 cells treated with retinoic acid (RA), NB4 cells treated with TPA, neutrophils (isolated from ten different patients), and placenta. NB4 cells treated with RA are thought to differentiate toward neutrophils; NB4 cells treated with TPA are thought to differentiate toward monocytes. The placental sample was poly(A)$^+$ RNA, whereas the remaining samples used total RNA. The ENCODE regions comprise 44 genomic regions varying in size from 0.5 to 2 Mbp which collectively total 1% of the human genome (ENCODE Project Consortium 2004). The oligonucleotide probes are 25 nucleotides in length and overlap slightly such that they start on average every 20 bp. Hybridizing segments were scored using a sliding window approach described in Kampa et al. (2004) and Royce et al. (2005). Genomic sequences that are detected as transcribed are called either transcriptionally active regions, "TARs" (Rinn et al. 2003), or transcribed fragments, "transfrags" (Kapranov et al. 2002).

The results of these probings are revealed in Table 1. A large number (2,046) of novel transcribed regions were identified from all RNA samples. These new hybridizing regions lie in intergenic regions proximal (within 5 kb) or distal (greater than 5 kb) to annotated genes as well as in introns, proximal (within 5 kb) or distal (greater than 5 kb) to annotated exons or regions previously identified as ESTs which are not annotated as genes. In total, over twice as many transcribed regions were apparent as compared to previous annotation.

EXTENSIVE INTRONIC TRANSCRIPTON IN NEUTROPHILS

Although novel transcription was apparent in RNAs from each source, careful inspection of the results revealed that the intronic regions were often extensively expressed in neutrophils (see Table 1 and Fig. 1). Over 700 novel transcribed regions were expressed in neutrophil RNA; this pattern of expression was not generally observed for the other RNA samples. Most of the novel transcribed regions expressed in neutrophils were concentrated in a few ENCODE regions, in particular the HOXA locus.

COMPARISON OF EXPRESSED REGIONS REVEALS EXTENSIVE EXPRESSION CHANGES DURING CELL DIFFERENTIATION

Analysis of NB4 cells treated with different agents, RA and TPA, allowed us to compare transcriptional changes that occur upon cell differentiation. We found that NB4 cells treated with RA had transcriptional patterns similar to those of neutrophils, as expected. One such example in the TRIM22 region is shown in Figure 2. One region that exhibited particularly extensive transcription upon cell differentiation was the HOXA locus (Fig. 3). We observe that whereas a novel intergenic transcript is expressed in both the NB4 cells treated with RA and neutrophils, the HOXA1 gene is expressed in the RA-treated cells but not the neutrophils. Extensive intergenic transcription is apparent in this region in many cell types, such as NB4 cells treated with either RA or TPA. These results suggest that many novel transcribed regions are transcribed upon cell differentiation.

To gain further insight into the cell-type specificity of TARs, we determined the fraction of novel transcribed regions expressed in placental RNA that are expressed in other cell types and correspond to the same genomic coordinates. As shown in Figure 4, almost all of the novel transcribed regions expressed in placental RNA either entirely overlap with novel transcribed regions detected in other cell lines (ENCODE Project Consortium 2004) or do not overlap at all. These results demonstrate that novel transcribed regions are reproducible genomic regions which show varying expression profiles across different cell lines and tissues, similar to exons of known genes.

An initial attempt has been made to classify novel transcribed regions, using genomic location and expression profile across the mapped ENCODE RNAs, into those that are likely parts of alternative isoforms of known genes as well as those that correspond to novel transcribed loci (Rozowsky et al. 2006). Approximately 14% of the novel

Table 1. Distribution of Transcribed Regions (TARs) in the ENCODE Regions

	NB4 CTRL	NB4 RA	NB4 TPA	Neutrophil	Placenta
Count					
Gencode Exonic	727	552	728	880	2175
Intergenic Distal	52	29	69	61	147
Intergenic Proximal	55	35	70	81	99
Intronic Distal	26	15	36	211	80
Intronic Proximal	137	109	190	554	367
Other ESTs	60	46	79	299	168
Nucleotides					
Gencode Exonic	82,376	60,870	81,080	122,412	370,426
Intergenic Distal	3,564	2,009	4,544	4,353	10,668
Intergenic Proximal	3,499	2,157	4,468	7,185	6,638
Intronic Distal	1,991	1,283	2,598	35,631	5,171
Intronic Proximal	11,086	9,383	14,720	61,363	26,181
Other ESTs	3,958	4,247	7,050	38,209	16,723

Locations of transcribed sequences or TARs detected within the ENCODE regions for RNA from the following 5 cell lines/tissues: untreated NB4 cells, NB4 cells treated with RA, NB4 cells treated with TPA, neutrophils and placenta (poly(A)$^+$). Distributions are either the number of regions transcribed or the number of base pairs detected as being transcribed.

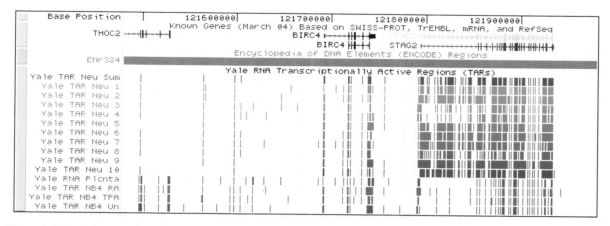

Figure 1. Plot of the genomic region surrounding the STAG2 locus on chromosome X. We observe that the entire STAG2 locus, including the intronic regions, is expressed in the RNA extracted from neutrophils.

TARs appear to connect with annotated loci, and approximately 21% of the remaining novel TARs are likely to be components of new annotated genes.

THE NOVEL TRANSCRIBED REGIONS HAVE A TRINUCLEOTIDE DISTRIBUTION SIMILAR TO EXONS

Our studies have revealed extensive transcription in the human genome. It is possible that this material represents random transcription throughout the human genome and that some fraction is maintained in stable poly(A)$^+$ RNA. To investigate this possibility, we analyzed the frequency of trinucleotide sequences of novel transcribed regions in intergenic and intronic regions and compared them with transcribed regions of exons, all GENCODE annotated exons (Harrow et al. 2006), and DNA from random selected nonrepetitive genomic intervals. As shown in Figure 5, transcribed regions and annotated exons show a nonrandom trinucleotide sequence composition. The composition of intergenic and intronic novel expressed regions is similar to that of transcribed annotated exons and all GENCODE exons, and very different from that of random DNA. Thus, novel transcribed regions have a sequence distribution similar to that of known genes, consistent with a functional role for these sequences.

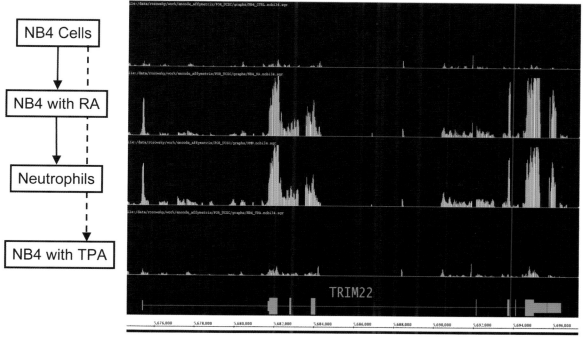

Figure 2. Plot of the expression profile of the TRIM22 locus on chromosome 11 for four of the RNAs mapped. This transcript is not expressed in the untreated NB4 cells; however, the transcript is present for both the NB4 cells treated with RA as well as the neutrophils, consistent with the hypothesis that RA differentiates NB4 cells toward neutrophils.

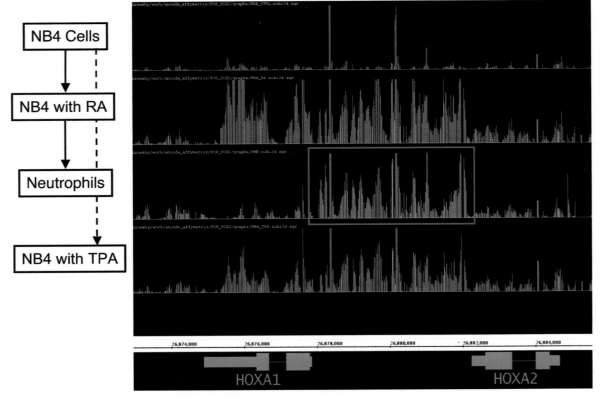

Figure 3. Plot of expression profiles of the genomic region from a portion of the HOXA locus on chromosome 7 for four of the RNAs mapped. The region displayed is not transcribed in the untreated NB4 cells; however, HOXA1 is expressed in the NB4 cells treated with RA or TPA. In addition, the intergenic region between HOXA1 and HOXA2 (shown with a *red box*) is transcribed in the NB4 cells treated with RA and TPA as well as the neutrophils. This transcript has been confirmed by RNA blot analysis.

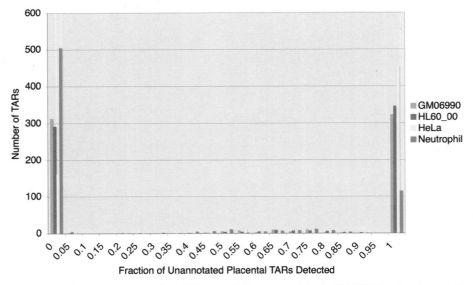

Figure 4. Plot of the number of novel TARs from four different cells lines mapped by the ENCODE project that overlap with the novel TARs detected in placental poly(A)$^+$ RNA. The horizontal axis shows the fraction of overlap between the novel TARs from each of the four cell lines vs the placental novel TARs. We observe that the novel TARs compared between the different cell/tissue types either overlap entirely or do not overlap at all. This is evidence that novel TARs are discrete transcribed regions similar to exons of known genes.

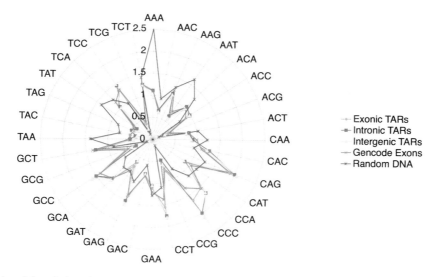

Figure 5. Radial plot of the relative trinucleotide frequency for the sequences of TARs that overlap exons of known genes, TARs that overlap introns of known genes, TARs that are in intergenic regions, exons of known genes, and DNA from randomly selected non-repetitive genomic regions. We observe that all the TARs exhibit a trinucleotide frequency distribution similar to that for exons of known genes and which is unlike the distribution for randomly selected nonrepetitive genomic regions.

DISCUSSION

We demonstrate that there are many more expressed sequences in the human genome than previous appreciated. Neutrophils in particular have extensive intronic transcription compared to other cell types. The reasons for this are not clear. This may be because not all of the transcribed RNAs from neutrophils are fully processed RNAs. A fraction of RNAs from some loci may remain in the cell as unspliced primary transcripts.

By analyzing RNA samples derived from cells treated with different stimuli, we examined the pattern of transcribed regions upon cell differentiation. We found many new transcribed regions are expressed in cell types upon differentiation, suggesting that they may contribute to new cell identities. Of particular interest is the extensive transcription throughout the HOXA region. It is likely that there are many more transcribed in this region than previously known. Consistent with this, RNA blot analysis reveals at least one new transcript encoded in the intergenic region in this locus (data not shown).

Although it is possible that much of the new transcription is due to random transcription throughout the genome, analysis of the trinucleotide composition of the novel expressed regions reveals that it is similar to annotated regions and not random DNA, similar to that reported previously (Bertone et al. 2004). This suggests that the new transcribed regions are either likely to be derived from similar genomics regions and/or that they have functional roles. Further characterization of the novel transcribed RNAs and the regions that encode them is likely to determine the function of these transcripts.

ACKNOWLEDGMENTS

This work was supported by grants from the National Institutes of Health.

REFERENCES

Bertone P., Stolc V., Royce T.E., Rozowsky J.S., Urban A.E., Zhu X., Rinn J.L., Tongprasit W., Samanta M., Weissman S., et al. 2004. Global identification of human transcribed sequences with genome tiling arrays. *Science* **306:** 2242.

Burge C. and Karlin S. 1997. Prediction of complete gene structures in human genomic DNA. *J. Mol. Biol.* **268:** 78.

Carninci P., Kasukawa T., Katayama S., Gough J., Frith M.C., Maeda N., Oyama R., Ravasi T., Lenhard B., Wells C., et al. 2005. The transcriptional landscape of the mammalian genome. *Science* **309:** 1559.

Cawley S., Pachter L., and Alexandersson M. 2003. SLAM web server for comparative gene finding and alignment. *Nucleic Acids Res.* **31:** 3507.

ENCODE Project Consortium. 2004. The ENCODE (ENCyclopedia Of DNA Elements) Project. *Science* **306:** 636.

Guigo R., Flicek P., Abril J.F., Reymond A., Lagarde J., Denoeud F., Antonarakis S., Ashburner M., Bajic V.B., Birney E., et al. 2006. EGASP: The human ENCODE Genome Annotation Assessment Project. *Genome Biol.* (suppl. 1) **7:** S2.1.

Harrow J., Denoeud F., Frankish A., Reymond A., Chen C.K., Chrast J., Lagarde J., Gilbert J.G., Storey R., Swarbreck D., et al. 2006. GENCODE: Producing a reference annotation for ENCODE. *Genome Biol.* (suppl. 1) **7:** S4.1.

Kampa D., Cheng J., Kapranov P., Yamanaka M., Brubaker S., Cawley S., Drenkow J., Piccolboni A., Bekiranov S., Helt G., et al. 2004. Novel RNAs identified from an in-depth analysis of the transcriptome of human chromosomes 21 and 22. *Genome Res.* **14:** 331.

Kapranov P., Cawley S.E., Drenkow J., Bekiranov S., Strausberg R.L., Fodor S.P., and Gingeras T.R. 2002. Large-scale transcriptional activity in chromosomes 21 and 22. Science 296: 916.

Lander E.S., Linton L.M., Birren B., Nusbaum C., Zody M.C., Baldwin J., Devon K., Dewar K., Doyle M., FitzHugh W., et al. (International Human Genome Sequencing Consortium). 2001. Initial sequencing and analysis of the human genome. *Nature* **409:** 860.

Parra G., Agarwal P., Abril J.F., Wiehe T., Fickett J.W., and Guigo R. 2003. Comparative gene prediction in human and mouse. *Genome Res.* **13:** 108.

Rinn J.L., Euskirchen G., Bertone P., Martone R., Luscombe N.M., Hartman S., Harrison P.M., Nelson F.K., Miller P., Gerstein M., et al. 2003. The transcriptional activity of human Chromosome 22. *Genes Dev.* **17:** 529.

Royce T.E., Rozowsky J.S., Bertone P., Samanta M., Stolc V., Weissman S., Snyder M., and Gerstein M. 2005. Issues in the analysis of oligonucleotide tiling microarrays for transcript mapping. *Trends Genet.* **21:** 466.

Rozowsky J., Newburger D., Sayward F., Wu J., Jordan G., Korbel J.O., Nagalakshmi U., Yang J., Zheng D., Guigo R., et al. 2006. The DART classification of unannotated transcription within the ENCODE regions: Associating transcription with known and novel loci. *Genome Res.* (in press).

Venter J.C., Adams M.D., Myers E.W., Li P.W., Mural R.J., Sutton G.G., Smith H.O., Yandell M., Evans C.A., Holt R.A., et al. 2001. The sequence of the human genome. *Science* **291:** 1304.

Computational Analysis of RNAs

S.R. Eddy

Howard Hughes Medical Institute and Department of Genetics, Washington University School of Medicine, Saint Louis, Missouri 63108

Genome sequence analysis of RNAs presents special challenges to computational biology, because conserved RNA secondary structure plays a large part in RNA analysis. Algorithms well suited for RNA secondary structure and sequence analysis have been borrowed from computational linguistics. These "stochastic context-free grammar" (SCFG) algorithms have enabled the development of new RNA genefinding and RNA homology search software. The aim of this paper is to provide an accessible introduction to the strengths and weaknesses of SCFG methods and to describe the state of the art in one particular kind of application: SCFG-based RNA similarity searching. The INFERNAL and RSEARCH programs are capable of identifying distant RNA homologs in a database search by looking for both sequence and secondary structure conservation.

A fundamental goal of genomics is to compile a comprehensive parts list for every organism: a catalog of all genes, regulatory elements, and other functional sequences in the genome (ENCODE Project Consortium 2004). But words such as "all" and "comprehensive" are terms of art in genomics. They mean as many as possible, for a reasonable cost and in reasonable time, of the kinds of functional sequences we know how to identify. For some kinds of sequence elements, we are only beginning to be able to take genome-wide approaches. Functional noncoding RNA elements are a striking example.

There are many known functional RNAs, ranging from catalysts in the ribosome and RNase P, to guide RNAs for RNA editing, to structural RNAs in the spliceosome, and more (Eddy 2001; Szymanski et al. 2003; Mattick and Makunin 2005). A series of discoveries in the past decade made it clear how incomplete our knowledge of functional RNA still is, including the discoveries of several large families of RNA genes, such as microRNAs (miRNAs) involved in posttranscriptional regulation of mRNAs (Bartel 2004), C/D small nucleolar RNAs (snoRNAs) directing site-specific 2′-O-methylation of target RNAs, and H/ACA snoRNAs directing site-specific pseudouridylation of target RNAs (Bachellerie et al. 2002; Brown et al. 2003). Even in *Escherichia coli*, many small new RNA genes have been discovered, at least some of which are posttranscriptional regulators (Storz et al. 2005), and numerous *cis*-regulatory elements called riboswitches have been identified in bacteria as well (Tucker and Breaker 2005; Winkler 2005).

For molecular biologists to discover whole new families of RNA elements in well-studied organisms is both embarrassing and exciting. These discoveries serve to remind us that unbiased discovery methods do not exist. Consider the recent explosion of papers on miRNAs (Bartel 2004). Up until 2001, tiny 21–25-nucleotide miRNA genes were not within the parameters of what we expected genes to look like, aside from two oddball *Caenorhabditis elegans lin-4* and *let-7* genes. miRNAs are often biochemically abundant, but they are only noticed if tiny RNAs are not run off the end of the gel. miRNAs are readily cloned and sequenced, but not when RNA samples are enriched for capped poly(A)$^+$ mRNA to eliminate the "uninteresting" background of poly(A)$^-$ rRNAs and tRNAs. miRNA genes show mutant genetic phenotypes, but if the mutation maps to an interval that contains no protein-coding genes, it takes intestinal fortitude to persevere and find the gene (as Victor Ambros's lab did with *lin-4*; Lee et al. 1993) as opposed to giving up. Specialized computational genefinding programs readily predict miRNA genes, but standard genefinders are looking for open reading frames and codon bias, which noncoding RNA genes do not have.

We probably do not know the full extent to which organisms use RNA regulatory motifs and noncoding RNA genes. We need better systematic genome-wide approaches for identifying functional RNA elements. One approach is to map complete transcriptomes, including both mRNA and noncoding RNA populations, by cDNA sequencing and tiled whole-genome microarrays (Okazaki et al. 2002; Carninci et al. 2005; Cheng et al. 2005). In mammalian genomes, these approaches have resulted in claims of thousands of putative noncoding RNA transcripts (Okazaki et al. 2002; Numata et al. 2003; Furuno et al. 2006; Ravasi et al. 2006). However, it remains unclear how many of these cDNA transcripts represent functional noncoding RNAs, as opposed to being artifacts of cryptic low-level promoters, pre-mRNA contamination, missplicing, unannotated alternative splicing, unrecognized small protein-coding genes, RNA degradation intermediates, and other sources of apparently noncoding RNA one should expect to find in a total cellular RNA population (Wang et al. 2004; Hüttenhofer et al. 2005; Babak et al. 2005; Lee et al. 2006). Additional analysis is required to distinguish functional noncoding RNAs from other transcripts. Additionally, although transcriptome mapping can identify novel independent transcripts, it does not help in identifying new *cis*-regulatory RNA elements contained in known mRNA transcripts.

Another approach is to systematically identify evolutionarily conserved elements by comparative genome sequence analysis. More than half of the conserved sequence in mouse/human genome comparisons appears to be in noncoding regions (Waterston et al. 2002). Large numbers of comparative genome sequences are enabling higher-resolution identification of short and/or weakly conserved elements (Eddy 2005; Stone et al. 2005). An advantage of comparative genome sequence analysis is that it can identify both conserved noncoding RNA genes and conserved *cis*-regulatory RNA elements. A disadvantage is that many other kinds of functional elements show sequence conservation, not just functional RNAs. Sequence conservation suggests that a genomic region is functional, but some kind of additional analysis is required to distinguish whether that function is at the level of DNA, RNA, or protein.

A focus of my lab has been on the development of computational analysis methods for identifying functional RNAs. The heart of our work is a general class of statistical models called "stochastic context-free grammars" (SCFGs), which we use to create computational methods that treat RNA as both primary sequence and base-paired secondary structure (Durbin et al. 1998). With different kinds of SCFGs, we have developed strong RNA similarity search methods (Eddy and Durbin 1994; Eddy 2002; Klein and Eddy 2003), reasonable noncoding RNA genefinding methods (Rivas and Eddy 2001; Rivas et al. 2001), and promising prototypes of RNA structure prediction methods (Dowell and Eddy 2004 and in prep.). Despite the off-putting jargon "stochastic context-free grammar," which we inherited from the field of computational linguistics where SCFGs were first developed (Lari and Young 1990), SCFGs are in fact a natural extension from familiar primary sequence analysis methods to RNA secondary structure methods.

RNA SEQUENCE ANALYSIS OUGHT TO MODEL RNA SECONDARY STRUCTURE

Computational tools are an essential part of comprehensive genome annotation. We rely on *similarity search* programs such as BLAST (Altschul et al. 1997) to identify informative protein homologies and to deduce gene structures by mapping cDNA and EST (expressed sequence tags) sequences onto genome sequence. *Genefinding* programs are used to identify novel genes, by looking for general statistical properties of that class of feature, such as the presence of an open reading frame and codon bias in protein-coding genes. *Motif identification* programs try to identify *cis*-regulatory elements, such as transcription-factor-binding sites, by identifying short conserved and/or overrepresented DNA sequences.

Most tools only look at linear primary sequence, scoring one residue (or aligned pair or column of homologous residues) at a time. In RNA analysis, linear sequence models are inadequate. Many (although not all) functional RNAs conserve a base-paired secondary structure. We want RNA computational analysis tools to be able to model both sequence and RNA secondary structure.

Why are we mostly satisfied with primary sequence analysis tools like BLAST for proteins, but not for RNA? Surely, *any* computational sequence analysis method would be more powerful if it took structural constraints into account. Both RNAs and proteins fold into three-dimensional structures composed of stereotyped secondary structure elements, and these structures constrain primary sequence evolution. That is, in general, if artificial sequences are produced with good primary sequence similarity to a known protein or RNA, few will fold properly (Socolich et al. 2005).

In making a practical computational tool, it is not sufficient to know that structure imposes constraints on sequence. These constraints also have to have predictable effects on sequence, and these effects have to be computable with time-efficient algorithms. Additionally, one uses the simplest tool that gets the job done. A simple linear sequence model is preferred over a more biologically realistic model if the simple model does just as well in a fraction of the time.

In the case of proteins, for the task of similarity searching, BLAST analysis has substantial power, routinely identifying significant homologies down to 20–30% amino acid sequence identity. Many proteins are conserved at this level across billions of years of divergence. Moreover, although protein structure clearly constrains primary sequence, we do not really understand *how* (i.e., we cannot yet predict very well which sequences will fold into active structures), nor do we know how to compute efficiently with what we do know (most existing protein folding or "threading" algorithms are very compute-intensive). Higher-order tools for protein analysis do exist (Godzik 2003), but they gain relatively little power at a high computational cost.

In the case of RNA, BLAST analysis is often unsatisfactory. Significant nucleic acid sequence alignments are only detected down to about 60–70% nucleotide sequence identity, largely as a consequence of the smaller nucleotide alphabet. Although some RNAs are highly conserved (notably ribosomal RNAs), many conserved RNAs will diverge below a 60–70% identity threshold in just tens or hundreds of millions of years. Thus, BLAST comparisons of RNAs are often unable to see reliably across important evolutionary divergences, such as across different animal phyla. The contrast between protein and RNA similarity searches is perhaps most striking when one looks at genome annotations of conserved homologs of the components of well-studied ribonucleoprotein (RNP) complexes. Often, the protein components of RNPs are annotated and the RNA components are not. If BLAST is used to search the fly, nematode, or yeast genome for homologs of human RNase P RNA, for example, no significant hits are seen, whereas conserved RNase P protein components are readily detectable. The presence of small nucleolar RNA homologs in Archaea was suspected based on BLAST detection of homologs of snoRNA-associated proteins, but detection of Archaeal snoRNAs required a combination of experiment and more sophisticated computational modeling (Omer et al. 2000).

Moreover, RNA structure is dominated by base pairs, and base pairs induce highly predictable patterns of long-distance pairwise residue complementarity in RNA primary sequence (Gutell et al. 2002). These patterns of structure-induced complementarity are so obvious in aligned RNA sequences that human analysts are often capable of accurately deducing the conserved secondary structure of an RNA sequence family solely from observed pairwise correlations in sequence alignments (*comparative sequence analysis*) (Pace et al. 1989). Robin Gutell and coworkers, for example, correctly predicted 97–98% of the conserved base pairs in ribosomal RNAs, essentially by eye (Gutell et al. 2002).

Thus, we need more powerful methods of RNA analysis than primary sequence analysis, and the constraints that conserved base-pairing imposes on RNA sequences are understood and easily predictable—by humans, at least. But are base-pairing constraints something we can use in efficient computer programs?

SCORING BOTH SEQUENCE AND RNA STRUCTURE

We want to be able to use a combination of sequence and structure information for a variety of RNA analysis problems, but for clarity, it will be useful to focus on a specific problem. Consider the problem of identifying homologs of a known RNA sequence family. Given a multiple sequence alignment of a family of homologous RNAs, and a consensus secondary structure for that family, we want to build a position-specific scoring model and use that model to search a sequence database and identify more homologs. (The model is called the *query*, and each database sequence is considered one at a time as a *target*.)

In a standard linear sequence profile, we assign 4 scores (for A, C, G, U) at each aligned column. If residue a occurs at some aligned position with probability p_a, compared to its average overall background frequency f_a, we calculate the score for that residue as:

$$S_a = \log_2 \frac{p_a}{f_a}$$

(The base two on the logarithm is an arbitrary and traditional choice, which makes scores in units of "bits.") For example, a completely conserved adenine ($p_A = 1.0$) gets a score of +2 (assuming uniform background expectation of 25% for each base). A position that allows either purine ($p_A = p_G = 0.5$) scores +1 for either purine. All standard sequence alignment methods (BLAST, Smith–Waterman, profile hidden Markov models) use essentially this same additive *log-odds scoring* method, which is well-grounded in statistics (Durbin et al. 1998).

Now, imagine that we have a perfectly conserved Watson–Crick base pair, but no primary sequence conservation in either column individually. That is, one column can be any of A, C, G, U with uniform probability, but when the residue there is an A, the residue in the other column is a U, and so on, with the two aligned columns maintaining a complementary Watson–Crick base pair. A primary sequence method assigns a score of 0 to any base in both columns, regardless of whether the two bases can base-pair or not. Substantial information is lost.

To capture that information, we must be able to score pairs of residues simultaneously. Log-odds scores are readily applied to pairs of positions. We obtain the score s_{ab} for residue pair a,b from the joint probability p_{ab} we expect to see that pair occur and the expected frequency that we would see a,b occur by chance independently, the product $f_a f_b$:

$$s_{ab} = \log_2 \frac{p_{ab}}{f_a f_b}$$

A perfectly conserved Watson–Crick pair could thus get a score of +2, when the individual positions are freely varying. That is, one base pair potentially conveys as much information as one completely conserved residue in a sequence profile.

Importantly, the pairwise score s_{ab} contains information about *both* primary sequence and base-pairing conservation. If both positions were completely conserved residues (say an A/U), s_{AU} would be +4, the same as if we scored the two columns independently by sequence as +2, +2. It is also important that the score s_{ab} is a general pairwise residue score, and it does not restrict the two residues to canonical Watson–Crick pairs. The pairwise residue score s_{ab} can deal with any pairwise correlation, including GU and noncanonical RNA pairs.

The amount of extra information in pairwise residue correlations induced by base-pairing is significant (Fig. 1). Generally speaking, in a typical structural RNA sequence family, about 50–60% of residues are involved in base pairs. Empirically, a scoring model that captures base-pairing typically has about 50% more information content as a sequence-only model (albeit with wide variation, depending on the RNA).

The point here is that formally grounded, statistical scoring of conserved RNA base pairs is straightforward and well understood. There is no reason for any proposed RNA alignment method to use arbitrary scores. The real difficulty is not in scoring residues, but in how to align a model to a target sequence when insertions and deletions are allowed. If we allow insertions and deletions, we do not know which target residues to score as which consensus base pairs and consensus singlet positions. We need the scoring system, and we also need an optimization algorithm that can look at all possible predicted structures and alignments of the target sequence, and find the best-scoring one(s). There are an astronomical number of possible solutions for typical alignment problems, unfortunately. Brute force enumeration of all possible solutions is not feasible. Either we need to simplify the problem, or we need a clever efficient algorithm.

If we restrict alignments to a limited number of possible gaps—by assuming that individual helices behave as ungapped blocks, for instance—then it becomes possible to exhaustively enumerate all possible alignments, as in Gautheret and Lambert's RNA profile search program ERPIN (Gautheret and Lambert 2001). But restricting where gaps are allowed is worrisome. It falls short of the general alignment methods we are accustomed to in primary sequence analysis.

This is where stochastic context-free grammars come in. SCFGs give us efficient and general alignment algorithms

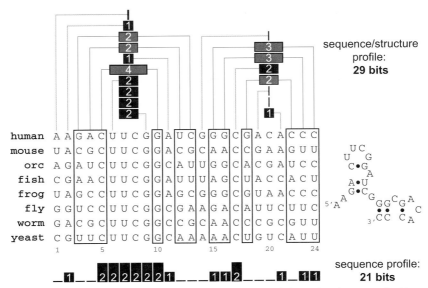

Figure 1. An example of how an RNA sequence/structure profile captures more information than a standard sequence profile. A multiple alignment is shown for eight homologous RNAs, all of which share the same consensus secondary structure (shown at right, for the "human" sequence). Below the alignment is a bar graph of the average expected score (information content) per position, in bits, for a sequence profile. In this contrived example, the only values are 2 bits (a 100% conserved residue), 1 bit (2 possible residues, 50% each), or 0 bits (no conservation, 25% each for all 4 residues). Above the alignment is a sequence/structure profile, a binary tree (instead of a linear array of columns) in which the six consensus base pairs are captured as six pairwise states (*red bars*) instead of as 12 single uncorrelated columns.

for base-paired RNA structure (Eddy and Durbin 1994; Sakakibara et al. 1994). To understand the generality of SCFGs, it is useful to make a brief digression into the current state of the art in linear sequence analysis.

PROBABILISTIC MODELS OF BIOLOGICAL SEQUENCES

Historically, sequence alignment algorithms were developed independently of statistical scoring methods. Different analysis applications typically used their own special algorithm(s) and parameterization methods. However, today, many primary sequence analysis methods, including similarity search, motif identification, and genefinding applications, are viewed by many computational biologists in a single formal framework called *hidden Markov models* (HMMs) (or more generally, *stochastic regular grammars*, and related models) (Durbin et al. 1998). Using HMM formalisms, every score is based on a probabilistic model—not just residue scores, but also any insertion/deletion penalties. HMMs allow straightforward creation of more complex yet consistent models that combine multiple information sources (protein-coding genefinders, for example [Burge and Karlin 1997], or position-specific profiles of conserved protein domains [Krogh et al. 1994]).

A key point is that any HMM, no matter how complicated, can be optimally aligned to a target sequence using a general algorithm called the Viterbi algorithm. For the special case of simple HMMs for sequence alignment, the Viterbi algorithm becomes identical to the well-known Smith/Waterman or Needleman/Wunsch sequence alignment algorithms (Needleman and Wunsch 1970; Smith and Waterman 1981).

Thus, adopting an explicit HMM framework has the advantage of splitting a sequence analysis problem into three pieces, two of which are standardized. The first piece is specifying the structure of the HMM. This is the interesting bit, where one decides what biological information to capture. The second piece is calculating the probability parameters (scores) of an HMM, which is standard probability theory. The third piece is how to align HMMs to target sequences, which is done with the standard Viterbi algorithm (and related HMM algorithms). By focusing specialization effort on the design of new models for different biological problems, rather than on the shared computational and statistical foundation, probabilistic models like HMMs give us powerful, biologically intuitive, and general toolkits for building a wide variety of sequence analysis methods of varying complexity and realism (Eddy 2004).

An HMM, however, is still "just" a primary sequence analysis method, scoring linear sequence one (or a few) residue at a time. HMMs cannot efficiently capture the long-distance pairwise correlations in RNA secondary structure.

STOCHASTIC CONTEXT-FREE GRAMMARS

In computational linguistics, HMMs are *stochastic regular grammars*, at the lowest level of a hierarchy of formal grammars originally defined by Noam Chomsky (1956) for the purpose of understanding the structure of natural languages. The next level in Chomsky's hierarchy are the so-called *context-free grammars,* or in probabilistic form as opposed to pattern-matching form, *stochastic context-free grammars* (SCFGs). One thing SCFGs can do that HMMs cannot is to efficiently model nested, long-

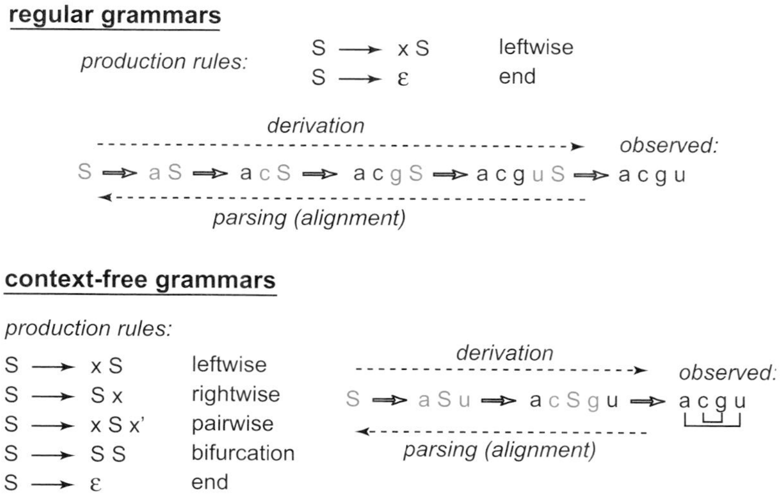

regular grammars

production rules:

$$S \longrightarrow x\, S \quad \text{leftwise}$$
$$S \longrightarrow \varepsilon \quad \text{end}$$

derivation

$S \Rightarrow a\,S \Rightarrow a\,c\,S \Rightarrow a\,c\,g\,S \Rightarrow a\,c\,g\,u\,S \Rightarrow a\,c\,g\,u$ *observed:*

parsing (alignment)

context-free grammars

production rules:

$$S \longrightarrow x\, S \quad \text{leftwise}$$
$$S \longrightarrow S\, x \quad \text{rightwise}$$
$$S \longrightarrow x\, S\, x' \quad \text{pairwise}$$
$$S \longrightarrow S\, S \quad \text{bifurcation}$$
$$S \longrightarrow \varepsilon \quad \text{end}$$

derivation

$S \Rightarrow a\,S\,u \Rightarrow a\,c\,S\,g\,u \Rightarrow a\,c\,g\,u$ *observed:*

parsing (alignment)

Figure 2. Comparison of regular grammars (linear sequence models) and context-free grammars, which can capture nested pairwise correlations. Productions of possible residues are shown as single rules generating a generic x (or ε, for end productions that generate a null symbol), rather than enumerating all possible 4 residues or 16 residue pairs. See text for more explanation.

distance pairwise correlations in strings of symbols—rare in natural languages, as it happens, but exactly what we need for RNA analysis. Shortly after HMMs were introduced into computational biology as general models of primary sequence analysis, SCFGs were brought into the field as general models of RNA sequence and structure (Eddy and Durbin 1994; Sakakibara et al. 1994).

Figure 2 briefly sketches the salient features of linear sequence algorithms (using regular grammars) and RNA sequence/structure algorithms (using context-free grammars). Both are *generative models*, consisting of a set of *production rules* that generate good sequences (those that belong to a homologous family, or align to a homologous query, or fit a gene model) with higher probability than other sequences. Production rules consist of *nonterminal symbols* (also called states) and *terminal symbols* (the observed A, C, G, U residues). Production rules describe the probabilistic expectation for what residues are favored in different places and for what states follow others. In essence, we use one state or production rule for each different way that we might want to score a residue. For example, a linear sequence profile of a multiple alignment might consist of a linear array of one state per consensus alignment column. An RNA structure profile might consist of one state per consensus base pair and one state per consensus single-stranded position. A genefinder might use two or more states to describe different residue compositions in exons versus introns.

Regular grammars are linear models. Production rules simply generate a symbol and move to a new state, from left to right. Still, regular grammar rules can capture a fair amount of complexity. For instance, to deal with insertions and deletions in a profile, we might move to the next consensus state, or move to an insertion state (and possibly stay there for a few self-transitions) to model a traditional gap-open/gap-extend penalty, or skip one or more of the next consensus states to model a deletion penalty.

Context-free grammars (CFGs) generate sequence outside-in, rather than left to right. For RNA, the crucial property of a CFG is that one production rule can generate a correlated pair of residues, then another correlated pair inside that. In an SCFG, a base-pair production is associated with a 4 × 4 probability table for all possible residue pairs, including Watson–Crick as well as noncanonical pairs. CFGs also allow one to fork off two or more substructures, so they can describe complex RNA secondary structures with multiple stem-loops.

In an actual application, the problem is not to generate simulated sequences, but to align a model to a given target sequence and assign it a score. The alignment problem is called *parsing* in linguistics. We aim to determine the optimal (highest probability) series of production rules that would have generated the target sequence. Just as all HMMs have a common parsing algorithm (the Viterbi algorithm), all SCFGs have a common parsing algorithm, the CYK algorithm.

The CYK algorithm identifies the best-scoring manner in which the model can generate the target sequence. The resulting so-called *parse tree* is the RNA secondary structure analog of a sequence alignment (Fig. 3),

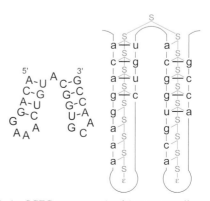

Figure 3. An SCFG parse tree (*right*) corresponding to a small example RNA secondary structure (*left*), for the set of five types of production rules in Fig. 2.

describing an assignment of residues in the target sequence to states of the query SCFG. A parse tree essentially specifies how to factorize an RNA alignment into a sum of additive scoring units, such as base pairs and singlet residues, or in more complicated models, base stacks and different lengths and types of loops.

Different SCFGs can be drawn for different problems (structure prediction, similarity search, or genefinding), depending on what statistical information one wants to capture in the model. Like HMMs for sequence analysis, SCFGs are a general toolkit for probabilistic modeling of RNA sequence and secondary structure.

LIMITATIONS OF SCFGs

The most important limitation of SCFGs is their computational complexity. Sequence analysis algorithms such as BLAST typically take time and memory proportional to L^2 for comparing two sequences of length L in residues, because the set of all possible alignments must try every residue in the query against every residue in the target. SCFG-based RNA analysis algorithms require time and memory proportional to at least L^3, because every possible pair of residues (L^2) must be tried against up to $L/2$ base-pairing states in the model (and in most RNA SCFGs, the time required more typically scales as L^4.) Thus, roughly speaking, for a typical RNA of length 100–1000 residues, compared to linear sequence analysis, SCFG-based algorithms take 100–1000-fold more memory and 10^4–10^6-fold more time. Although this is high, it should be noted that this kind of compute complexity is not foreign to biological sequence analysis. Well-known RNA secondary structure prediction programs such as the Zuker MFOLD program (Zuker 2000) have the same computational complexity (not a coincidence, because the MFOLD folding algorithm is essentially a special case of the SCFG CYK algorithm). Nonetheless, computational complexity is a very serious problem for SCFGs, especially in database-searching applications. Although the standard CYK algorithm is a useful general starting point for all SCFGs, much of the work in my lab is devoted toward finding more efficient algorithms.

A second limitation of SCFGs is that they can only model *nested* pairwise correlations. RNA pseudoknots, which involve nonnested interactions, cannot be described by SCFG formalisms. More powerful grammar classes (so-called "mildly context-sensitive grammars") can model RNA pseudoknots, but their computational complexity is currently prohibitive for many applications (Rivas and Eddy 1999). Likewise, base triples are usually prohibited, because they also usually involve nonnested pairing. These losses are unfortunate and would be especially serious if three-dimensional RNA structure prediction were the goal. But for many objectives, including homology search, motif identification, and genefinding, this information loss is an acceptable tradeoff. RNA pseudoknots typically account for something like 5–10% of the base pairs in most RNA structures. There is still a large gain in being able to capture most of the pairwise correlation information in an RNA structure in an efficient computational model.

SCFG-BASED RNA SIMILARITY SEARCH PROGRAMS

My lab has been exploring the use of SCFGs for a variety of tasks, including RNA structure prediction (Dowell and Eddy 2004 and in prep.) and noncoding RNA genefinding (Rivas and Eddy 2001). Rather than survey all this work, I continue to focus here on RNA similarity searching applications. So far, I have discussed the formal benefits of SCFGs from a mathematical and computational point of view. This leaves an important question—How well do they actually work?

Again, our problem is, given either a single RNA sequence and its secondary structure, or a RNA multiple alignment and a known consensus structure, we want to search a sequence database for similar sequences. A position-specific SCFG is constructed which has a set of 4 scores for each single-stranded position, 16 scores for each base pair, and appropriate extra states and state transitions that allow for insertions and deletions. Because there are many ways to deal with insertions and deletions, in terms of where to allow them and how to score them, there are many ways one can convert an RNA structure query into SCFG production rules. I adopted one particular general convention for building SCFGs for similarity searching, called "covariance models" (CMs) (Eddy and Durbin 1994; Eddy 2002). The conventions in CMs follow, as closely as possible, conventions in linear sequence analysis. Insertions and deletions are allowed anywhere, and are assigned gap-open and gap-extend penalties (affine gap penalties). Figure 4 shows an example of a CM alignment and parse tree for a tRNA profile aligned to yeast phenylalanine tRNA.

Essentially the same CM structure and alignment algorithms are the basis of three software packages from my lab. Two are for consensus profiles of multiple alignments—COVE (Eddy and Durbin 1994) (now obsolete) and COVE's replacement INFERNAL (Eddy 2002). The third, Robbie Klein's RSEARCH, is for searching with single RNA sequence/structure queries (Klein and Eddy 2003). INFERNAL is the RNA secondary structure analog of the HMMER profile HMM software for sequence analysis (Eddy 1998), and RSEARCH is the analog of Smith/Waterman single query sequence alignment (Smith and Waterman 1981).

One of the first practical applications of CMs for similarity search demonstrated both the power and the limitations of SCFG-based approaches. Todd Lowe in my lab developed a program for tRNA gene identification, TRNASCAN-SE, as a wrapper script around a CM built from a large alignment of known tRNAs (Lowe and Eddy 1997). At the time, the best tRNA gene identification programs had false-positive rates of about 0.2–0.3 per megabase and sensitivities of about 95–99% for known tRNAs (Fichant and Burks 1991; Pavesi et al. 1994). These programs were quite adequate for small genomes, such as *E. coli* or *Saccharomyces cerevisiae*, where they identified only a small number of false positives (≤10), but we realized that for large genomes like the 3000 Mb human genome, these false positive rates would become a problem. We expected only about 500 or so true tRNA

yeast tRNA-Phe: **CM parse tree:**

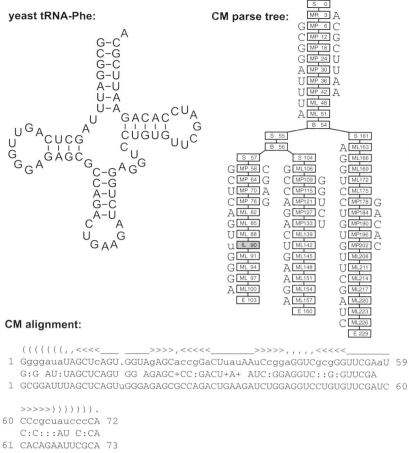

CM alignment:

```
     ((((((((,,<<<<____  ____>>>>,<<<<<_____>>>>>,,,,,<<<<<_____
  1  GgggauaUAGCUcAGU.GGUAgAGCaccgGaCUuauAAuCcggaGGUCgcgGGUUCGAaU 59
     G:G AU:UAGCUCAGU GG AGAGC+CC:GACU+A+ AUC::GGAGGUC::G:GUUCGA
  1  GCGGAUUUAGCUCAGUuGGGAGAGCGCCAGACUGAAGAUCUGGAGGUCCUGUGUUCGAUC 60

     >>>>>)))))))).
 60  CCcgcuaucccCA 72
     C:C:::AU C:CA
 61  CACAGAAUUCGCA 73
```

Figure 4. A real example of a CM parse tree (*top right*) for yeast phenylalanine tRNA (*top left*), for a CM constructed from a tRNA alignment, similar to the CM used by the TRNASCAN-SE program. Below, the two-dimensional parse tree is represented by the program as a four-line sequence alignment between the consensus of the query CM (*second line*) and the target yeast tRNA sequence (*fourth line*), in a format akin to BLAST output format. The output format is augmented in two ways to indicate RNA secondary structure. First, : symbols in the identity line (*third line*) indicate positive-scoring compensatory base pairs. (+ symbols indicate positive-scoring single residues as in BLAST). Second, an extra line of annotation (first line) annotates the base pairs in the consensus secondary structure with <> and ( ) pairs (and other symbols annotating different single-stranded residues).

genes in the human genome, but the programs were going to predict more than 1000 false positives. We needed to increase the specificity of tRNA gene identification by orders of magnitude if we were going to be able to annotate the tRNA gene family in large genomes.

Lowe showed that a CM built automatically by the COVE software from a large tRNA sequence alignment database (Steinberg et al. 1993) achieved 99.8% sensitivity with less than 0.002 false positives per megabase (Lowe and Eddy 1997). On the other hand, the search speed was far too slow for whole-genome analysis. We estimated needing 10 CPU (central processing unit)-years for a human genome, on 1997 CPUs. Since the existing programs were fast and sensitive, just not specific enough, Lowe solved the speed problem by using a combination of two existing programs as prefilters (Fichant and Burks 1991; Pavesi et al. 1994), and only passing their proposed tRNAs to COVE and the tRNA CM. The combination of programs resulted in TRNASCAN-SE, which showed 99.5% average sensitivity on tRNA genes, a false positive rate below the detection limit of our simulations

(<1 per 15 gigabases), and we could process the human genome in about 2 CPU-days (on 1997 CPUs).

Today, TRNASCAN-SE seems to still be the standard for whole-genome annotation of tRNA genes, although a newer heuristic program, ARAGORN, now has comparable performance (Laslett and Canback 2004). TRNASCAN-SE has, on occasion, even produced interesting new tRNA biology. For example, the CM in TRNASCAN-SE detected the noncanonical tRNA for the "22nd amino acid," pyrrolysine, in the *Methanosarcina barkeri* genome (Srinivasan et al. 2002). The principal failings of the program are largely unrelated to the similarity detection power of its CM. It has trouble distinguishing tRNA pseudogenes from true tRNA genes, and some genomes, such as the rat, have thousands of tRNA pseudogenes (Gibbs et al. 2004). It also has trouble correctly annotating some tRNA isoacceptor types in cases where a tRNA anticodon is posttranscriptionally modified.

TRNASCAN-SE was a nice demonstration of SCFGs. However, tRNAs are an unusually ideal case, in several ways:

1. tRNAs are small (~75 nucleotides), so the $O(L^3)$ memory requirement of CM alignment algorithms did not have serious impact. tRNA-sized RNAs can be aligned in about 1 MB RAM with the standard CYK algorithm. However, for larger RNAs, memory requirements could become prohibitive. We estimated that RNase P alignments (~400 nucleotides) would take 340 MB RAM, and LSU rRNA alignments (~2900 nucleotides) would take 150 GB RAM.

2. We had a highly reliable, deep, manually curated alignment of 1415 tRNAs to use to estimate the tRNA CM's parameters (Steinberg et al. 1993). Thus, we could use observed frequencies of single-stranded residues, base pairs, and indels as parameters, without any need for more sophisticated parameterization methods, such as the use of mixture Dirichlet priors in profile HMM parameter estimation (Sjölander et al. 1996), or the determination of general RNA substitution score matrices akin to the use of BLOSUM matrices in BLAST.

3. Structural variation in tRNAs is minimal. Almost all tRNAs adopt a four-stemmed cloverleaf consensus structure. Most structural RNAs—even ribosomal RNA—show much more substantial structural variation over evolutionary time, with stems or even whole substructural domains coming and going. Global alignment to a single consensus model works fine for tRNA, but fails for most other RNAs.

4. The $O(L^4)$ time requirement of the CM search algorithm is punitive, even for RNAs as small as tRNA. The only reason we could make a practical tRNA search program based on CM algorithms is that fast rule-based search programs were already available, which we could use as effective prescreens.

Several lines of research in my lab have focused on addressing each of these four problems in order to make RNA similarity searching practical.

IMPROVED MEMORY USAGE

The memory requirement of CMs was at first the most serious barrier. This issue was essentially solved in 2002 (Eddy 2002). Using an approach analogous to approaches already in common use in sequence alignment, I developed a "divide and conquer" (Hirschberg 1975) variant of the CYK algorithm that reduces the cubic $O(L^3)$ dynamic programming lattice for CMs to $O(L^2 \log L)$ while still guaranteeing a mathematically optimal alignment (Eddy 2002). The price is a relatively negligible extra factor in time (an extra 20% on average, as measured empirically). tRNA alignments now cost 0.1 megabytes, RNase P costs 4 MB, and LSU rRNA costs 270 MB, all well within the capabilities of standard current computers.

IMPROVED PARAMETERIZATION

Techniques for parameterizing probabilistic models are well understood. The same methods that are used for sequence alignment scores work for CMs, so once the memory problem was no longer limiting, it just required applying known methods to CMs.

For CMs built from single RNA query structures, Robbie Klein developed the RIBOSUM substitution matrices, 4 × 4 matrices for scoring single-residue alignments and 16 × 16 for scoring base-pair alignments (Klein and Eddy 2003). Klein implemented these in a program called RSEARCH. RSEARCH uses CM formalisms and the INFERNAL CM alignment engines, but parameterizes the models with RIBOSUM substitution scores and arbitrary gap penalities, much like standard sequence alignment algorithms.

For CMs built as profiles of a multiple RNA sequence alignment of known consensus structure, Eric Nawrocki estimated informative mixture Dirichlet priors from known RNA structural alignments and implemented these priors in INFERNAL (E.P. Nawrocki and S.R. Eddy, unpubl.). INFERNAL profiles built from only a few aligned sequences (five, for example) now seem to perform reasonably. (Quantifying "reasonably" would require a digression into how similarity search programs are benchmarked, which I will forego here.)

LOCAL STRUCTURAL ALIGNMENT

In most structural RNAs, not all of the secondary structure is conserved over evolutionary time. A computational model that requires a global RNA structural alignment is not ideal. We extended CMs to allow local structure alignment (Eddy 2002; Klein and Eddy 2003). The basic idea is in two parts. One is to allow a parse tree to start at any consensus position, instead of always starting at the root of the tree. The second is to allow the model to end prematurely from any consensus state and generate zero or more nonhomologous random residues before stopping. These rules allow large deletions and truncations of structural subdomains, where that deletion is consistent with the rest of the conserved secondary structure. For example, Figure 5 shows an INFERNAL alignment of a gamma-proteobacterial RNase P RNA profile to the *Bacillus subtilis* RNase P RNA, in which four substructural domains have to be deleted or inserted to see the homology.

The combination of the above three improvements has made it possible to routinely see remote RNA homologies that were previously below the radar of existing sequence-based approaches. RNase P RNAs are an interesting example. RNase P RNA is one of the best-studied catalytic RNAs and is thought to be nearly universally conserved in all domains of life. However, few metazoan RNase Ps had been identified until recently (Marquez et al. 2005; Piccinelli et al. 2005), even in sequenced model organisms like *Caenorhabditis elegans* and *Drosophila melanogaster*, because many RNase P RNA homologs are not detected by BLAST searches. A single RSEARCH search, using the human RNase P RNA structure as a query, cleanly identifies single RNase P RNA homologs in *C. elegans*, *D. melanogaster*, and several other eukaryotic genomes, with significant E-values. The predicted *C. elegans* RNase P is shown in Figure 6, along with the structure of the human RNase P query and the alignment output from RSEARCH.

```
       {{{{{{{{{{{{{{{{{{{,<<<<<<<<<<<<-<<<<<____>>>>>>>>->>>>>>>
     1 ggAGuggGgcaGgCaguCGCugcuucggccuuGuucaguuaacugaaaaggAccgaagga
       +: :::G::C:GG:A:UCGCU+C::::                U+         :::G+A
     4 CUUAACGUUCGGGUAAUCGCUGCAGAUC-----------UUG---------AAUCUGUA
          P1        P2        P3                                P3'
       >,,,,,,,,,,,,,[[[ [--------[[[[[        ((---(((((,,,,       )
    61 GAGGAAAGUCCGGGCUC.CACAGGGCAgGGUG        GGAAAGUGCCACAG       G
       GAGGAAAGUCC GCUC C A GG  :G G          :GAAAGUGCCACAG       G
    43 GAGGAAAGUCCAUGCUCgC--ACGGUGCUGAG*[102]*UGAAAGUGCCACAG*[37]*G
          P5          [P7]       [P10]    P11
       ))--))))]]]]]] ]]],,,        ,,,,,,,,,,,}}}}}}}--
   230 GUAAACCCCACCcG.GAGCAA        CuAGAUGAAUGacuGcCCA.............
       GUAAACC:C C: G GAG AA        UAGAU++AUGA:U:CC
   227 GUAAACCCCUCGAGcGAGAAA*[64]*GUAGAUAGAUGAUUGCC--gccugaguacgagg
       P11' P10' P7' P5'                       P2'
                                  ----------------}-}}}}}}}}}}}....
   345 ..........................CGACAGAACCCGGCUUAuagcCccaCUccucuu
                               ACA AAC  GGCUUA:AG::C::: :+ C
   343 ugaugagccguuugcaguacgaugga--ACAAAACAUGGCUUACAGAACGUUAGACCAC
```

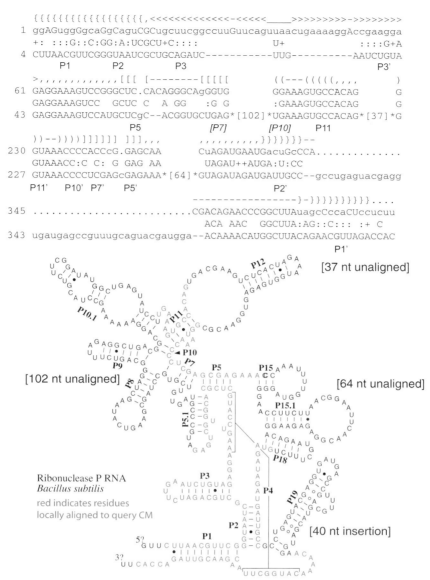

Figure 5. Example of output for a local structural alignment, when a CM built from five homologous gamma-proteobacterial RNase P RNAs (including *E. coli*) successfully detects a homologous alignment in the RNase P RNA of *B. subtilis* RNase P with a significant score of 49.4 bits. (*Top*) In the maximum likelihood alignment, three substructural domains of 102 (P8, P9, P10.1 helices), 37 (P12), and 64 (P15, P15.1, P18) nucleotides in the *B. subtilis* structure are treated as nonhomologous by the CM's local alignment rules, and an additional 40-nucleotide domain (P19) is handled as an insertion. The predicted P1–P11 stem regions are annotated beneath the output lines, with italics indicating mispredictions of the P7 and P10 stems as the result of the large structural variations in these regions. (*Bottom*) The accepted secondary structure of the *B. subtilis* RNase P is shown (Brown 1999), with red indicating which residues were aligned to the query model. These aligned residues roughly correspond to the conserved core of the RNase P three-dimensional structure.

SPEED IMPROVEMENTS

The remaining problem is the slow speed of CM searches. For example, whereas a BLAST search of the *C. elegans* genome with a mammalian RNase P RNA query takes CPU-seconds, an INFERNAL or RSEARCH search takes CPU-months. On the other hand, the BLAST search does not find anything significant, whereas SCFG searches do. Our first priority has been to get the right answer (however slowly), but now it is time to worry about speed. Up until now, we have been able to address the computational speed problem by brute force, by parallelizing our search

programs and running them on a cluster (~300 Linux processors in the St. Louis lab), but this is not a satisfactory long-term solution. We and other investigators are working on accelerated CM search algorithms. Zasha Weinberg and Larry Ruzzo at the University of Washington in Seattle have developed a clever "rigorous filter" approach, which Diana Kolbe in my lab has incorporated into the INFERNAL codebase (Weinberg and Ruzzo 2004, 2006). Eric Nawrocki and I have developed a complementary method, query-dependent banding, a banded dynamic programming algorithm specific to CMs. We think that the combination of these acceleration methods

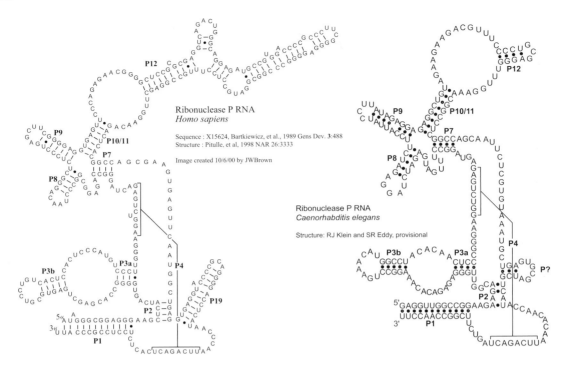

Figure 6. The RSEARCH program, with a human RNase P RNA query structure shown at top left (Brown 1999), detects one significant alignment in the *C. elegans* genome (*bottom*) with an E-value of 1.656e-5. We believe that this is the *C. elegans* RNase P RNA. The same sequence was detected by Steve Marquez and Norman Pace in 2005 (Marquez et al. 2005). Our tentative structure prediction, with a few manual corrections from the RSEARCH alignment, is shown in the upper right. The structure shown for the P2/P? region is uncertain (and different from that predicted by the RSEARCH alignment).

should soon give us about 10–100 times improvement in speed in INFERNAL's publicly distributed code. CM approaches will still be much more compute-intensive than BLAST, but they might start to become feasible on single desktop CPUs.

THE Rfam DATABASE

In collaboration with us, Sam Griffiths-Jones and coworkers at the Wellcome Trust Sanger Institute have developed a database called Rfam, which contains curated multiple alignments and CMs for known RNA sequence families (Griffiths-Jones et al. 2005). The current Rfam 7.0 release contains 503 families. Rfam is an RNA analog of the Pfam protein domain database, using INFERNAL software where Pfam uses the HMMER profile HMM software. Rfam makes it possible to automatically detect and annotate homologs of known RNA structures in genome sequences. At present, for speed reasons, Rfam processing relies on BLAST prefilters, however, so some of the added sensitivity that full CM searches could

provide is sacrificed. We hope to gain this back as CM search speed increases.

CONCLUSION

Developing better computational sequence analysis tools is like building better telescopes. With more and more powerful tools, we are trying to peer into the genome and discern the subtle signals left by functional elements that have diverged by billions of years of evolution. As our resolution power goes up, features come into sharper focus. True breakthroughs are relatively rare, because most features have been seen already, albeit at lower resolution and in less detail. Nonetheless, over time, steady incremental advances in technology can amount to surprising overall gains in power.

The advent of SCFGs for RNA sequence analysis was a significant *theoretical* advance, making it possible to harness RNA secondary structure constraints in almost arbitrarily complex, fully automated computational methods while still using formally well-grounded probabilistic modeling. However, converting the promise of SCFG formalisms to *practice* in RNA analysis has been a more usual case of incremental progress, requiring practical software implementations and a lot of work. We have almost reached the point of routine practical applications in RNA similarity search, with only computational time as our remaining barrier. In other areas, such as SCFG-based noncoding RNA genefinders, SCFG-based RNA structure prediction by comparative analysis, and SCFG-based RNA structural motif discovery in unaligned sequences, even more serious barriers still remain, but practical applications are developing in those areas as well.

ACKNOWLEDGMENTS

I am grateful to Elena Rivas, Ariane Machado-Lima, and Eric Nawrocki for comments on the manuscript. I thank the National Institutes of Health National Human Genome Research Institute, the Howard Hughes Medical Institute, and Alvin Goldfarb for their financial support of my group.

REFERENCES

Altschul S.F., Madden T.L., Schaffer A.A., Zhang J., Zhang Z., Miller W., and Lipman D.J. 1997. Gapped BLAST and PSI-BLAST: A new generation of protein database search programs. *Nucleic Acids Res.* **25:** 3389.

Babak T., Blencowe B.J., and Hughes T.R. 2005. A systematic search for new mammalian noncoding RNAs indicates little conserved intergenic transcription. *BMC Genomics* **6:** 104.

Bachellerie J.P., Cavaille J., and Hüttenhofer A. 2002. The expanding snoRNA world. *Biochimie* **84:** 775.

Bartel D.P. 2004. MicroRNAs: Genomics, biogenesis, mechanism, and function. *Cell* **116:** 281.

Brown J.W. 1999. The ribonuclease P database. *Nucleic Acids Res.* **27:** 314.

Brown J.W., Echeverria M., and Qu L.H. 2003. Plant snoRNAs: Functional evolution and new modes of gene expression. *Trends Plant Sci.* **8:** 42.

Burge C. and Karlin S. 1997. Prediction of complete gene structures in human genomic DNA. *J. Mol. Biol.* **268:** 78.

Carninci P., Kasukawa T., Katayama S., Gough J., Frith M.C., Maeda N., Oyama R., Ravasi T., Lenhard B., Wells C., et al. 2005. The transcriptional landscape of the mammalian genome. *Science* **309:** 1559.

Cheng J., Kapranov P., Drenkow J., Dike S., Brubaker S., Patel S., Long J., Stern D., Tammana H., Helt G., et al. 2005. Transcriptional maps of 10 human chromosomes at 5-nucleotide resolution. *Science* **308:** 1149.

Chomsky N. 1956. Three models for the description of language. *IRE Trans. Inf. Theory* **2:** 113.

Dowell R.D. and Eddy S.R. 2004. Evaluation of several light-weight stochastic context-free grammars for RNA secondary structure prediction. *BMC Bioinformatics* **5:** 71.

Durbin R., Eddy S.R., Krogh A., and Mitchison G.J. 1998. *Biological sequence analysis: Probabilistic models of proteins and nucleic acids.* Cambridge University Press, Cambridge, United Kingdom.

Eddy S.R. 1998. Profile hidden Markov models. *Bioinformatics* **14:** 755.

———. 2001. Non-coding RNA genes and the modern RNA world. *Nat. Rev. Genet.* **2:** 919.

———. 2002. A memory-efficient dynamic programming algorithm for optimal alignment of a sequence to an RNA secondary structure. *BMC Bioinformatics* **3:** 18.

———. 2004. What is a hidden Markov model? *Nat. Biotechnol.* **22:** 1315.

———. 2005. A model of the statistical power of comparative genome sequence analysis. *PLoS Biol.* **3:** e10.

Eddy S.R. and Durbin R. 1994. RNA sequence analysis using covariance models. *Nucleic Acids Res.* **22:** 2079.

ENCODE Project Consortium. 2004. The ENCODE (ENCyclopedia of DNA Elements) project. *Science* **306:** 636.

Fichant G.A. and Burks C. 1991. Identifying potential tRNA genes in genomic DNA sequences. *J. Mol. Biol.* **220:** 659.

Furuno M., Pang K.C., Ninomiya N., Fukuda S., Frith M.C., Bult C., Kai C., Kawai J., Carninci P., Hayashizaki Y., et al. 2006. Clusters of internally primed transcripts reveal novel long noncoding RNAs. *PLoS Genet.* **2:** e37.

Gautheret D. and Lambert A. 2001. Direct RNA motif definition and identification from multiple sequence alignments using secondary structure profiles. *J. Mol. Biol.* **313:** 1003.

Gibbs R.A., Weinstock G.M., Metzker M.L., Muzny D.M., Sodergren E.J., Scherer S., Scott G., Steffen D., Worley K.C., Burch P.E., et al. 2004. Genome sequence of the Brown Norway rat yields insights into mammalian evolution. *Nature* **428:** 493.

Godzik A. 2003. Fold recognition methods. *Methods Biochem. Anal.* **44:** 525.

Griffiths-Jones S., Moxon S., Marshall M., Khanna A., Eddy S.R., and Bateman A. 2005. Rfam: Annotating non-coding RNAs in complete genomes. *Nucleic Acids Res.* **33:** D121.

Gutell R.R., Lee J.C., and Cannone J.J. 2002. The accuracy of ribosomal RNA comparative structure models. *Curr. Opin. Struct. Biol.* **12:** 301.

Hirschberg D.S. 1975. A linear space algorithm for computing maximal common subsequences. *Commun. ACM* **18:** 341.

Hüttenhofer A., Schattner P., and Polacek N. 2005. Non-coding RNAs: Hope or hype? *Trends Genet.* **21:** 289.

Klein R.J. and Eddy S.R. 2003. RSEARCH: Finding homologs of single structured RNA sequences. *BMC Bioinformatics* **4:** 44.

Krogh A., Brown M., Mian I.S., Sjölander K., and Haussler D. 1994. Hidden Markov models in computational biology: Applications to protein modeling. *J. Mol. Biol.* **235:** 1501.

Lari K. and Young S.J. 1990. The estimation of stochastic context-free grammars using the inside-outside algorithm. *Comput. Speech Lang.* **4:** 35.

Laslett D. and Canback B. 2004. ARAGORN, a program to detect tRNA genes and tmRNA genes in nucleotide sequences. *Nucleic Acids Res.* **32:** 11.

Lee L.J., Hughes T.R., and Frey B.J. 2006. How many new genes are there? *Science* **311:** 1709.

Lee R.C., Feinbaum R.L., and Ambros V. 1993. The *C. elegans* heterochronic gene *lin-4* encodes small RNAs with antisense complementarity to *lin-14*. *Cell* **75:** 843.

Lowe T.M. and Eddy S.R. 1997. tRNAscan-SE: A program for improved detection of transfer RNA genes in genomic sequence. *Nucleic Acids Res.* **25:** 955.

Marquez S.M., Harris J.K., Kelley S.T., Brown J.W., Dawson S.C., Roberts E.C., and Pace N.R. 2005. Structural implications of novel diversity in eucaryal RNase p RNA. *RNA* **11:** 739.

Mattick J.S. and Makunin I.V. 2005. Small regulatory RNAs in mammals. *Hum. Mol. Genet.* **14:** R121.

Needleman S.B. and Wunsch C.D. 1970. A general method applicable to the search for similarities in the amino acid sequence of two proteins. *J. Mol. Biol.* **48:** 443.

Numata K., Kanai A., Saito R., Kondo S., Adachi J., Wilming L.G., Hume D.A., Hayashizaki Y., Tomita M., RIKEN GER Group, and GSL Members. 2003. Identification of putative noncoding RNAs among the RIKEN mouse full-length cDNA collection. *Genome Res.* **13:** 1301.

Okazaki Y., Furuno M., Kasukawa T., Adachi J., Bono H., Kondo S., Nikaido I., Osato N., Saito R., Suzuki H., et al. (FANTOM Consortium; RIKEN Genome Exploration Research Group Phase I & II Team). 2002. Analysis of the mouse transcriptome based on functional annotation of 60,770 full-length cDNAs. *Nature* **420:** 563.

Omer A.D., Lowe T.M., Russell A.G., Ebhardt H., Eddy S.R., and Dennis P.P. 2000. Homologs of small nucleolar RNAs in Archaea. *Science* **288:** 517.

Pace N.R., Smith D.K., Olsen G.J., and James B.D. 1989. Phylogenetic comparative analysis and the secondary structure of ribonuclease P RNA: A review. *Gene* **82:** 65.

Pavesi A., Conterlo F., Bolchi A., Dieci G., and Ottonello S. 1994. Identification of new eukaryotic tRNA genes in genomic DNA databases by a multistep weight matrix analysis of transcriptional control regions. *Nucleic Acids Res.* **22:** 1247.

Piccinelli P., Rosenblad M.A., and Samuelsson T. 2005. Identification and analysis of ribonuclease P and MRP RNA in a broad range of eukaryotes. *Nucleic Acids Res.* **33:** 4485.

Ravasi T., Suzuki H., Pang K.C., Katayama S., Furuno M., Okunishi R., Fukuda S., Ru K., Frith M.C., Gongora M.M., et al. 2006. Experimental validation of the regulated expression of large numbers of non-coding RNAs from the mouse genome. *Genome Res.* **16:** 11.

Rivas E. and Eddy S.R. 1999. A dynamic programming algorithm for RNA structure prediction including pseudoknots. *J. Mol. Biol.* **285:** 2053.

———. 2001. Noncoding RNA gene detection using comparative sequence analysis. *BMC Bioinformatics* **2:** 8.

Rivas E., Klein R.J., Jones T.A., and Eddy S.R. 2001. Computational identification of noncoding RNAs in *E. coli* by comparative genomics. *Curr. Biol.* **11:** 1369.

Sakakibara Y., Brown M., Hughey R., Mian I.S., Sjölander K., Underwood R.C., and Haussler D. 1994. Stochastic context-free grammars for tRNA modeling. *Nucleic Acids Res.* **22:** 5112.

Sjölander K., Karplus K., Brown M., Hughey R., Krogh A., Mian I.S., and Haussler D. 1996. Dirichlet mixtures: A method for improving detection of weak but significant protein sequence homology. *Comput. Appl. Biosci.* **12:** 327.

Smith T.F. and Waterman M.S. 1981. Identification of common molecular subsequences. *J. Mol. Biol.* **147:** 195.

Socolich M., Lockless S.W., Russ W.P., Lee H., Gardner K.H., and Ranganathan R. 2005. Evolutionary information for specifying a protein fold. *Nature* **437:** 512.

Srinivasan G., James C.M., and Krzycki J.A. 2002. Pyrrolysine encoded by UAG in Archaea: Charging of a UAG-decoding specialized tRNA. *Science* **296:** 1459.

Steinberg S., Misch A., and Sprinzl M. 1993. Compilation of tRNA sequences and sequences of tRNA genes. *Nucleic Acids Res.* **21:** 3011.

Stone E.A., Cooper G.M., and Sidow A. 2005. Trade-offs in detecting evolutionarily constrained sequence by comparative genomics. *Annu. Rev. Genomics Hum. Genet.* **6:** 143.

Storz G., Altuvia S., and Wassarman K.M. 2005. An abundance of RNA regulators. *Annu. Rev. Biochem.* **74:** 199.

Szymanski M., Barciszewska M.Z., Zywicki M., and Barciszewski J. 2003. Noncoding RNA transcripts. *J. Appl. Genet.* **44:** 1.

Tucker B.J. and Breaker R.R. 2005. Riboswitches as versatile gene control elements. *Curr. Opin. Struct. Biol.* **15:** 342.

Wang J., Zhang J., Zheng H., Li J., Liu D., Li H., Samudrala R., Yu J., and Wong G.K. 2004. Mouse transcriptome: Neutral evolution of "non-coding" complementary DNAs. *Nature* **431:** 757.

Waterston R.H., Lindblad-Toh K., Birney E., Rogers J., Abril J.F., Agarwal P., Agarwala R., Ainscough R., Alexandersson M., An P., et al. (Mouse Genome Sequencing Consortium). 2002. Initial sequencing and comparative analysis of the mouse genome. *Nature* **420:** 520.

Weinberg Z. and Ruzzo W.L. 2004. Exploiting conserved structure for faster annotation of non-coding RNAs without loss of accuracy. *Bioinformatics* (suppl. 1) **20:** I334.

———. 2006. Sequence-based heuristics for faster annotation of non-coding RNA families. *Bioinformatics* **22:** 35.

Winkler W.C. 2005. Riboswitches and the role of noncoding RNAs in bacterial metabolic control. *Curr. Opin. Chem. Biol.* **9:** 594.

Zuker M. 2000. Calculating nucleic acid secondary structure. *Curr. Opin. Struct. Biol.* **10:** 303.

and Filipowicz W. 2006. Relief of microRNA-mediated translational repression in human cells subjected to stress. *Cell* **125:** 1111.

Brengues M., Teixeira D., and Parker R. 2005. Movement of eukaryotic mRNAs between polysomes and cytoplasmic processing bodies. *Science* **310:** 486.

Brennecke J., Stark A., Russell R.B., and Cohen S.M. 2005. Principles of microRNA-target recognition. *PLoS Biol.* **3:** e85.

Brown V., Jin P., Ceman S., Darnell J.C., O'Donnell W.T., Tenenbaum S.A., Jin X., Feng Y., Wilkinson K.D., Keene J.D., et al. 2001. Microarray identification of FMRP-associated brain mRNAs and altered mRNA translational profiles in fragile X syndrome. *Cell* **107:** 477.

Chalfie M., Horvitz H.R., and Sulston J.E. 1981. Mutations that lead to reiterations in the cell lineages of *C. elegans*. *Cell* **24:** 59.

Chen X. 2004. A microRNA as a translational repressor of APETALA2 in *Arabidopsis* flower development. *Science* **303:** 2022.

Cimmino A., Calin G.A., Fabbri M., Iorio M.V., Ferracin M., Shimizu M., Wojcik S.E., Aqeilan R.I., Zupo S., Dono M., et al. 2005. miR-15 and miR-16 induce apoptosis by targeting BCL2. *Proc. Natl. Acad. Sci.* **102:** 13944.

Ding L., Spencer A., Morita K., and Han M. 2005. The developmental timing regulator AIN-1 interacts with miRISCs and may target the argonaute protein ALG-1 to cytoplasmic P bodies in *C. elegans*. *Mol. Cell* **19:** 437.

Dinkova T.D., Keiper B.D., Korneeva N.L., Aamodt E.J., and Rhoads R.E. 2005. Translation of a small subset of *Caenorhabditis elegans* mRNAs is dependent on a specific eukaryotic translation initiation factor 4E isoform. *Mol. Cell. Biol.* **25:** 100.

Doench J.G. and Sharp P.A. 2004. Specificity of microRNA target selection in translational repression. *Genes Dev.* **18:** 504.

Dreyfuss G., Kim V.N., and Kataoka N. 2002. Messenger-RNA-binding proteins and the messages they carry. *Nat. Rev. Mol. Cell Biol.* **3:** 195.

Farh K.K., Grimson A., Jan C., Lewis B.P., Johnston W.K., Lim L.P., Burge C.B., and Bartel D.P. 2005. The widespread impact of mammalian microRNAs on mRNA repression and evolution. *Science* **310:** 1817.

Galban S., Fan J., Martindale J.L., Cheadle C., Hoffman B., Woods M.P., Temeles G., Brieger J., Decker J., and Gorospe M. 2003. von Hippel-Lindau protein-mediated repression of tumor necrosis factor alpha translation revealed through use of cDNA arrays. *Mol. Cell. Biol.* **23:** 2316.

Giraldez A.J., Mishima Y., Rihel J., Grocock R.J., Van Dongen S., Inoue K., Enright A.J., and Schier A.F. 2006. Zebrafish MiR-430 promotes deadenylation and clearance of maternal mRNAs. *Science* **312:** 75.

Grosshans H., Johnson T., Reinert K.L., Gerstein M., and Slack F.J. 2005. The temporal patterning microRNA *let-7* regulates several transcription factors at the larval to adult transition in *C. elegans*. *Dev. Cell* **8:** 321.

Humphreys D.T., Westman B.J., Martin D.I., and Preiss T. 2005. MicroRNAs control translation initiation by inhibiting eukaryotic initiation factor 4E/cap and poly(A) tail function. *Proc. Natl. Acad. Sci.* **102:** 16961.

Jakymiw A., Lian S., Eystathioy T., Li S., Satoh M., Hamel J.C., Fritzler M.J., and Chan E.K. 2005. Disruption of GW bodies impairs mammalian RNA interference. *Nat. Cell Biol.* **7:** 1267.

Jing Q., Huang S., Guth S., Zarubin T., Motoyama A., Chen J., Di Padova F., Lin S.C., Gram H., and Han J. 2005. Involvement of microRNA in AU-rich element-mediated mRNA instability. *Cell* **120:** 623.

Johannes G., Carter M.S., Eisen M.B., Brown P.O., and Sarnow P. 1999. Identification of eukaryotic mRNAs that are translated at reduced cap binding complex eIF4F concentrations using a cDNA microarray. *Proc. Natl. Acad. Sci.* **96:** 13118.

Johnson S.M., Grosshans H., Shingara J., Byrom M., Jarvis R., Cheng A., Labourier E., Reinert K.L., Brown D., and Slack F.J. 2005. RAS is regulated by the *let-7* microRNA family. *Cell* **120:** 635.

Lagos-Quintana M., Rauhut R., Lendeckel W., and Tuschl T. 2001. Identification of novel genes coding for small expressed RNAs. *Science* **294:** 853.

Lai E.C. 2002. Micro RNAs are complementary to 3′UTR sequence motifs that mediate negative post-transcriptional regulation. *Nat. Genet.* **30:** 363.

Lau N.C., Lim L.P., Weinstein E.G., and Bartel D.P. 2001. An abundant class of tiny RNAs with probable regulatory roles in *Caenorhabditis elegans*. *Science* **294:** 858.

Lee R.C. and Ambros V. 2001. An extensive class of small RNAs in *Caenorhabditis elegans*. *Science* **294:** 862.

Lee R.C., Feinbaum R.L., and Ambros V. 1993. The *C. elegans* heterochronic gene *lin-4* encodes small RNAs with antisense complementarity to *lin-14*. *Cell* **75:** 843.

Lewis B.P., Burge C.B., and Bartel D.P. 2005. Conserved seed pairing, often flanked by adenosines, indicates that thousands of human genes are microRNA targets. *Cell* **120:** 15.

Lewis B.P., Shih I.H., Jones-Rhoades M.W., Bartel D.P., and Burge C.B. 2003. Prediction of mammalian microRNA targets. *Cell* **115:** 787.

Lim L.P., Glasner M.E., Yekta S., Burge C.B., and Bartel D.P. 2003a. Vertebrate microRNA genes. *Science* **299:** 1540.

Lim L.P., Lau N.C., Weinstein E.G., Abdelhakim A., Yekta S., Rhoades M.W., Burge C.B., and Bartel D.P. 2003b. The microRNAs of *Caenorhabditis elegans*. *Genes Dev.* **17:** 991–1008.

Lim L.P., Lau N.C., Garrett-Engele P., Grimson A., Schelter J.M., Castle J., Bartel D.P., Linsley P.S., and Johnson J.M. 2005. Microarray analysis shows that some microRNAs downregulate large numbers of target mRNAs. *Nature* **433:** 769.

Liu J., Valencia-Sanchez M.A., Hannon G.J., and Parker R. 2005a. MicroRNA-dependent localization of targeted mRNAs to mammalian P-bodies. *Nat. Cell Biol.* **7:** 719.

Liu J., Rivas F.V., Wohlschlegel J., Yates J.R., III, Parker R., and Hannon G.J. 2005b. A role for the P-body component GW182 in microRNA function. *Nat. Cell Biol.* **7:** 1261.

Nakamoto M., Jin P., O'Donnell W.T., and Warren S.T. 2005. Physiological identification of human transcripts translationally regulated by a specific microRNA. *Hum. Mol. Genet.* **14:** 3813.

Novina C.D. and Sharp P.A. 2004. The RNAi revolution. *Nature* **430:** 161.

O'Donnell K.A., Wentzel E.A., Zeller K.I., Dang C.V., and Mendell J.T. 2005. c-Myc-regulated microRNAs modulate E2F1 expression. *Nature* **435:** 839.

Olsen P.H. and Ambros V. 1999. The *lin-4* regulatory RNA controls developmental timing in *Caenorhabditis elegans* by blocking LIN-14 protein synthesis after the initiation of translation. *Dev. Biol.* **216:** 671.

Pasquinelli A.E., Reinhart B.J., Slack F., Martindale M.Q., Kuroda M.I., Maller B., Hayward D.C., Ball E.E., Degnan B., Muller P., et al. 2000. Conservation of the sequence and temporal expression of *let-7* heterochronic regulatory RNA. *Nature* **408:** 86.

Petersen C.P., Bordeleau M.E., Pelletier J., and Sharp P.A. 2006. Short RNAs repress translation after initiation in mammalian cells. *Mol. Cell* **21:** 533.

Pillai R.S., Bhattacharyya S.N., Artus C.G., Zoller T., Cougot N., Basyuk E., Bertrand E., and Filipowicz W. 2005. Inhibition of translational initiation by Let-7 MicroRNA in human cells. *Science* **309:** 1573.

Poy M.N., Eliasson L., Krutzfeldt J., Kuwajima S., Ma X., Macdonald P.E., Pfeffer S., Tuschl T., Rajewsky N., Rorsman P., and Stoffel M. 2004. A pancreatic islet-specific microRNA regulates insulin secretion. *Nature* **432:** 226.

Preiss T., Baron-Benhamou J., Ansorge W., and Hentze M.W. 2003. Homodirectional changes in transcriptome composition and mRNA translation induced by rapamycin and heat shock. *Nat. Struct. Biol.* **10:** 1039.

Rajasekhar V.K., Viale A., Socci N.D., Wiedmann M., Hu X., and Holland E.C. 2003. Oncogenic Ras and Akt signaling contribute to glioblastoma formation by differential recruitment of existing mRNAs to polysomes. *Mol. Cell* **12:** 889.

Rajewsky N. 2006. microRNA target predictions in animals. *Nat. Genet.* (suppl. 1) **38:** S8.

Rajewsky N. and Socci N.D. 2004. Computational identification of microRNA targets. *Dev. Biol.* **267:** 529.

Reinhart B.J., Slack F.J., Basson M., Pasquinelli A.E., Bettinger J.C., Rougvie A.E., Horvitz H.R., and Ruvkun G. 2000. The 21-nucleotide *let-7* RNA regulates developmental timing in *Caenorhabditis elegans. Nature* **403:** 901.

Rigoutsos I., Huynh T., Miranda K., Tsirigos A., McHardy A., and Platt D. 2006. Short blocks from the noncoding parts of the human genome have instances within nearly all known genes and relate to biological processes. *Proc. Natl. Acad. Sci.* **103:** 6605.

Saxena S., Jonsson Z.O., and Dutta A. 2003. Small RNAs with imperfect match to endogenous mRNA repress translation. Implications for off-target activity of small inhibitory RNA in mammalian cells. *J. Biol. Chem.* **278:** 44312.

Schratt G.M., Nigh E.A., Chen W.G., Hu L., and Greenberg M.E. 2004. BDNF regulates the translation of a select group of mRNAs by a mammalian target of rapamycin-phosphatidylinositol 3-kinase-dependent pathway during neuronal development. *J. Neurosci.* **24:** 7366.

Seggerson K., Tang L., and Moss E.G. 2002. Two genetic circuits repress the *Caenorhabditis elegans* heterochronic gene *lin-28* after translation initiation. *Dev. Biol.* **243:** 215.

Sheth U. and Parker R. 2003. Decapping and decay of messenger RNA occur in cytoplasmic processing bodies. *Science* **300:** 805.

Sood P., Krek A., Zavolan M., Macino G., and Rajewsky N. 2006. Cell-type-specific signatures of microRNAs on target mRNA expression. *Proc. Natl. Acad. Sci.* **103:** 2746.

Stark A., Brennecke J., Russell R.B., and Cohen S.M. 2003. Identification of *Drosophila* microRNA targets. *PLoS Biol.* **1:** E60.

Stark A., Brennecke J., Bushati N., Russell R.B., and Cohen S.M. 2005. Animal microRNAs confer robustness to gene expression and have a significant impact on 3'UTR evolution. *Cell* **123:** 1133.

Valencia-Sanchez M.A., Liu J., Hannon G.J., and Parker R. 2006. Control of translation and mRNA degradation by miRNAs and siRNAs. *Genes Dev.* **20:** 515.

Vella M.C., Reinert K., and Slack F.J. 2004a. Architecture of a validated microRNA::target interaction. *Chem. Biol.* **11:** 1619.

Vella M.C., Choi E.Y., Lin S.Y., Reinert K., and Slack F.J. 2004b. The *C. elegans* microRNA *let-7* binds to imperfect *let-7* complementary sites from the *lin-41* 3'UTR. *Genes Dev.* **18:** 132.

Wightman B., Ha I., and Ruvkun G. 1993. Posttranscriptional regulation of the heterochronic gene *lin-14* by *lin-4* mediates temporal pattern formation in *C. elegans. Cell* **75:** 855.

Wightman B., Burglin T.R., Gatto J., Arasu P., and Ruvkun G. 1991. Negative regulatory sequences in the *lin-14* 3'-untranslated region are necessary to generate a temporal switch during *Caenorhabditis elegans* development. *Genes Dev.* **5:** 1813.

Wu L., Fan J., and Belasco J.G. 2006. MicroRNAs direct rapid deadenylation of mRNA. *Proc. Natl. Acad. Sci.* **103:** 4034.

Xie X., Lu J., Kulbokas E.J., Golub T.R., Mootha V., Lindblad-Toh K., Lander E.S., and Kellis M. 2005. Systematic discovery of regulatory motifs in human promoters and 3'UTRs by comparison of several mammals. *Nature* **434:** 338.

Yekta S., Shih I.H., and Bartel D.P. 2004. MicroRNA-directed cleavage of HOXB8 mRNA. *Science* **304:** 594.

Zamore P.D. and Haley B. 2005. Ribo-gnome: The big world of small RNAs. *Science* **309:** 1519.

Zeng Y., Yi R., and Cullen B.R. 2003. MicroRNAs and small interfering RNAs can inhibit mRNA expression by similar mechanisms. *Proc. Natl. Acad. Sci.* **100:** 9779.

Zong Q., Schummer M., Hood L., and Morris D.R. 1999. Messenger RNA translation state: The second dimension of high-throughput expression screening. *Proc. Natl. Acad. Sci.* **96:** 10632.

RNomics: Identification and Function of Small Non-Protein-coding RNAs in Model Organisms

A. HÜTTENHOFER

Innsbruck Biocenter, Division of Genomics and RNomics, Innsbruck Medical University, 6020 Innsbruck, Austria

In the recent past, our knowledge on small non-protein-coding RNAs (ncRNAs) has exponentially grown. Different approaches to identify novel ncRNAs that include computational and experimental RNomics have led to a plethora of novel ncRNAs. A picture emerges, in which ncRNAs have a variety of roles during regulation of gene expression. Thereby, many of these ncRNAs appear to function in guiding specific protein complexes to target nucleic acids. The concept of RNA guiding seems to be a widespread and very effective regulatory mechanism. In addition to guide RNAs, numerous RNAs were identified by RNomics screens, lacking known sequence and structure motifs; hence no function could be assigned to them as yet. Future challenges in the field of RNomics will include elucidation of their biological roles in the cell.

Cells from all organisms known to date contain two different kinds of RNAs: mRNAs, which are translated into proteins, and ncRNAs, which function on the level of the RNA and are not translated into proteins (Eddy 2001; Mattick 2001; Hüttenhofer et al. 2002). Sizes of ncRNAs range from very large, for example, about 17 kb as Xist RNA, to extremely small (21–23 nucleotides) as microRNAs (miRNAs). In general, the sizes of the many functional ncRNAs, known up to now, vary from about 20 to 500 nucleotides, well below the size of the majority of mRNAs.

The early view of ncRNAs was that they were relics of a primordial "RNA world" in which RNA served both as the carrier of genetic information and as the catalytic agent. The current view of the RNA world is far more complex. True catalytic RNAs (so-called "ribozymes") are in fact quite rare. Instead, most ncRNAs perform their cellular duties by a range of mechanisms that are not directly catalytic. For example, a few ncRNAs, such as SRP-RNA (Halic et al. 2004; Halic and Beckmann 2005), appear to function as obligate cofactors of catalytic protein complexes. Some ncRNAs, such as 7SK RNA (Yang et al. 2001), 6S RNA (Wassarman and Storz 2000), CsrB and CsrC RNAs (Dubey et al. 2005), and perhaps *Air* RNA (Pauler et al. 2005), act as genetic regulators by means of antagonistic competition for protein-binding sites. Others serve a structural role or act as scaffolds onto which catalytic proteins can assemble. Hence, the numerous ncRNAs might be classified according to their functions: catalytic RNAs, guide RNAs, catalytic cofactor RNAs, antisense RNAs, protein-binding site–antagonists/agonist RNAs, or templating RNAs, for example.

It has been postulated that up until now, many ncRNAs in genomes of model organisms have escaped detection and that in fact in higher eukaryotes ncRNAs outnumber protein-coding mRNAs (Mattick 2004, 2005; Mattick and Makunin 2005). This paper focuses on methods for the identification of novel ncRNA species in various model organisms in the recent past. We summarize these methods designated as "RNomics" (Filipowicz 2000; Hüttenhofer et al. 2002). Subsequently, we also try to elucidate why in the period following the "early RNA world" many of these ncRNAs still have essential roles in the current "protein world."

METHODS FOR IDENTIFICATION OF ncRNAs IN MODEL ORGANISMS

The term "experimental RNomics" has been coined for identification of novel ncRNAs (Filipowicz 2000; Hüttenhofer et al. 2001). Four different methods currently exist (Fig. 1): (1) RNA sequencing (enzymatically or chemically) as the most traditional method to reveal novel ncRNA species; (2) the parallel cloning of many ncRNA by generating specialized cDNA libraries; (3) the use of microarrays to predict ncRNAs that are expressed under a given experimental condition; and (4) "genomic SELEX" and its potential application to select ncRNA candidates from the sequence space represented by the genome of an organism of interest.

Identification of ncRNAs by Chemical or Enzymatic Sequencing

In the very early days of ncRNA research, that is, some 35–40 years ago, single ncRNA species (e.g., ribosomal RNAs, tRNAs, or viral RNAs) were selected by size separation of total RNA on denaturing gels, followed by visualization and excision of specific bands, ideally representing single ncRNA species (Fig. 1A). Thus, for its identification, the ncRNA of interest must be present in high amounts, that is, visible as a distinct band in an ethidium-bromide-stained polyacrylamide gel, exposed to UV light. Subsequently, RNAs were radiolabeled and sequenced by chemical or enzymatic sequencing methods (Sanger et al. 1965; Brownlee et al. 1972; Donis-Keller et al. 1977; Ehresmann et al. 1977; Peattie 1979).

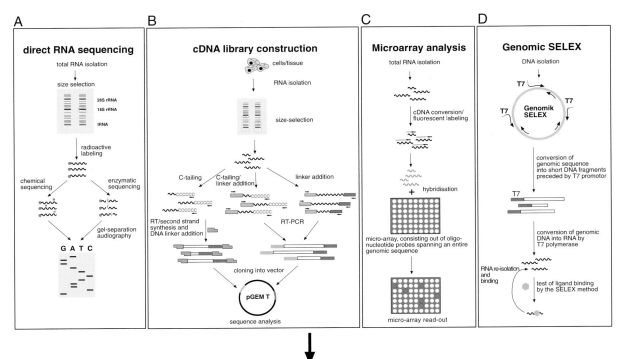

Figure 1. Experimental approaches to identify ncRNAs. Four experimental approaches (*A–D*) to identify candidates for ncRNAs in model organisms are shown. (*A*) Identification of ncRNAs by chemical or enzymatic sequencing of extracted abundant RNAs. (*B*) Identification of ncRNAs by cDNA cloning and sequencing; three different methods are indicated to reverse-transcribe ncRNAs, usually lacking poly(A) tails, into cDNAs (e.g., by C-tailing, C-tailing and linker addition, or linker addition only, followed by RT-PCR). (*C*) Identification of ncRNAs by microarray analysis. DNA oligonucleotides covering the sequence space of an entire genome are spotted onto glass slides, to which fluorescently labeled samples derived from cellular RNA are hybridized. (*D*) Identification of ncRNAs by genomic SELEX. By random priming, the sequence of a genome is converted into short PCR fragments containing a T7 promoter at their 5′ ends. Subsequently, in vitro transcription by means of T7 RNA polymerase converts this genomic sequence of an organism into RNA fragments, which can then be assayed for function, such as binding to a specific protein or small chemical ligand, by SELEX.

Identification of ncRNAs by Specialized cDNA Libraries

The second method for identification of novel ncRNA species involves the generation of cDNA libraries (Fig. 1B), in analogy with expressed sequence tag libraries (EST libraries) for identification of mRNAs (Gerhold and Caskey 1996; Ohlrogge and Benning 2000). The original mRNA cloning method is based on reverse transcription of mRNAs from an organism by an oligo (dT) primer and second-strand synthesis, resulting in a cDNA library that ideally represents all protein-coding transcripts of a genome. Compared to these conventional EST libraries, the main difference for ncRNA library approaches is the source and treatment of the cloned RNA.

Since most mRNAs are more than 500 nucleotides in length, but many ncRNAs are considerably smaller, RNAs in the size range of about 20–500 nucleotides are first isolated. This fraction is usually depleted in EST libraries as it will not be present in poly(A)$^+$ mRNA. The isolation of small-sized RNAs is achieved by size separation of total RNA (either from the entire organism at different developmental stages or from an individual organ) by denaturing polyacrylamide gel electrophoresis.

In many cases, these size- or antibody-selected RNAs lack polyadenylated tails. In general, there are three different methods to reverse-transcribe ncRNAs into cDNA as a prerequisite for cloning and sequencing (Fig. 1B): Prior to reverse transcription, RNAs are ligated to short oligonucleotide linkers (RNA or DNA) at their 5′ and 3′ ends (Hüttenhofer et al. 2004). Alternatively, at their 3′ ends, RNAs can be tailed by poly(A) polymerase, employing either ATP or CTP (Martin and Keller 1998). Reverse transcription of linker-ligated and/or tailed RNA species is followed by polymerase chain reaction (PCR), employing primers complementary to linker sequences. Subsequently, cDNA fragments are cloned into standard vector systems and sequenced.

Microarray Analysis

Microarrays have become the preferred method to monitor the levels of many transcripts in parallel and often at the whole-genome level (Fig. 1C). Microarrays, also known as DNA chips or expression arrays, are glass (or silicon) slides onto whose surface DNA probes have been printed in a grid-like arrangement. To date, single-stranded DNA oligonucleotides of 25–70 in length are the predominant type of DNA probe on commercial microarrays, although double-stranded PCR products may also serve as probes.

To analyze the entire level of cellular transcripts, samples are prepared from total RNA of an organism. The samples used for microarray hybridization can be the extracted RNA, the converted cDNA, or the cRNA; in any case, these probes will generally be labeled with fluorescent dyes such as Cy3 or Cy5. For more details on the various labeling protocols that are currently being used, see references in Stoughton (2005).

Genomic SELEX

Many ncRNAs form ribonucleoprotein particles (RNPs) at various time points in their life cycle. Such RNA-binding proteins may help an ncRNA fold into its active conformation, shield it from nucleases prior to exerting its function, or promote its annealing with target RNAs up to guiding a protein to its proper target. Other ncRNAs interact with proteins to directly regulate their activity.

The techniques discussed so far allow us to identify ncRNAs from the pool of expressed cellular RNAs after copurification with proteins; that is, by cloning, direct sequencing, or microarray analysis. Given that many such proteins bind their RNA ligands in a nanomolar range, it should also be possible to select RNA ligands from the pool of ncRNAs that an organism can possibly express even without isolating their in vivo transcripts.

This approach, termed genomic SELEX (Singer et al. 1997), is based on the in vitro generation of RNA species that are derived from a library of an organism's entire genomic DNA (Fig. 1D). The generated RNA pool will undergo successive rounds of association with a given RNA-binding protein, partitioning, and reamplification. As a result, RNA sequences that are stringently bound by the protein partner will be enriched. Once the sequence of the bound RNAs is determined, this information can be used to search for matches in the genome, and so predicted genomic regions can then be tested for the expression of unknown ncRNAs. Genomic SELEX has been successfully applied to select mRNA-binding partners of proteins (Shtatland et al. 2000; Kim et al. 2003), but studies that focused on ncRNAs have not yet been published for any organism.

Alternative Methods

Alternatively to biochemical methods, genetic and bioinformatic tools are also employed to identify ncRNAs in model organisms. Some of the first chromosomally encoded regulatory ncRNAs, for example, MicF, DsrA, and RprA of *Escherichia coli*, were discovered in the course of a genetic screen (Mizuno et al. 1983; Sledjeski and Gottesman 1995; Majdalani et al. 2001). Similarly, genetics also discovered the founding member, *lin-4* RNA, of the ever-growing class of eukaryotic miRNAs (Lee et al. 1993). For a more detailed review of genetic and biocomputational routes to ncRNA discovery, see Eddy (2002), Vogel and Sharma (2005), and Washietl et al. (2005).

FUNCTIONS OF IDENTIFIED ncRNAs

The above experimental approaches, as well as computational RNomics approaches, have identified thousands of novel ncRNA species in various model organisms from *E. coli* to *Homo sapiens* in the recent past. It thus appears that the number of ncRNAs in the present protein world is very significant and may in fact outnumber protein-coding mRNAs, especially in higher eukaryotes (Mattick 2005). So why is it that such large numbers of ncRNAs still exist—in a post-RNA-world age—and what is the function of all of these RNA species? Why have these functions not been exerted and overtaken by proteins instead?

Although for the majority of newly discovered ncRNA candidates their function is currently unknown, it appears that many of the ncRNAs with an already known function belong to just a few classes, each composed of hundreds to thousands of members, such as miRNAs or siRNAs (short interfering RNAs). These ncRNAs all seem to exhibit a common function, namely, the guiding of protein–enzyme complexes to nucleic acids; hence, they are designated as guide RNAs. By transcriptional or post-transcriptional mechanisms, these guide RNAs are able to regulate and fine-tune gene expression by interacting with either DNA or RNA.

Chimeric RNA–Protein Enzymes

Guide RNAs function as part of a catalytic RNP complex in which the RNA performs the task of substrate recognition and a protein component performs the catalysis. Because the two essential components belong to different classes of macromolecules, I refer to these RNA-guided proteins as "chimeric RNP enzymes." In general, chimeric RNP enzymes contain an unvarying protein-based enzyme portion (consisting of one or more proteins) that associates with different small guide RNAs which target the complex to its substrate by antisense complementarity (Fig. 2A).

The range of catalytic "payloads" that are guided by these RNAs is strikingly wide and includes endonucleases, polymerases, DNA-, RNA-, and histone methyltransferases, and many more. Guide RNAs belong to a few large families (see above), three of which, siRNA/miRNAs, small nucleolar RNAs (snoRNAs), and gRNAs, contain hundreds of representatives.

All guide RNA families enable an efficient form of modularity, in which multiple substrates can be processed by a single protein complex. Another type of modularity has recently been observed within the miRNA/siRNA family. Not only can multiple RNAs target a single protein catalytic complex to multiple substrates, but different catalytic enzymes can be transported to different substrates by means of similar RNP complexes (Fig. 2B). This is observed, for example, in plants, where an miRNA/siRNA-Dicer-Argonaute complex can guide either a DNA methyltransferase (Kawasaki and Taira

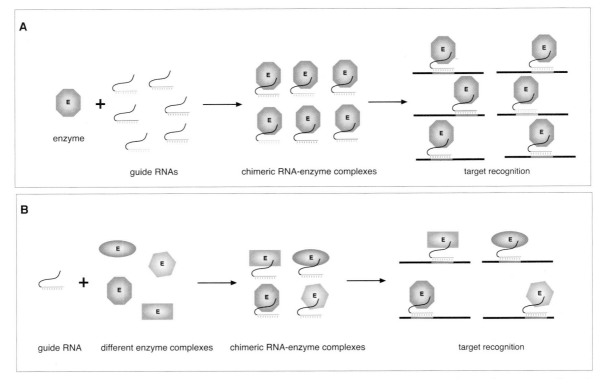

Figure 2. The concept of RNA guiding. (*A*) One non-sequence-specific enzyme complex (E), consisting of one to several proteins, binds to many different small guide RNAs, resulting in chimeric RNA–enzyme complexes that recognize their targets by Watson-Crick base-pairing and thus guide the enzymatic complex to different substrates (target recognition). (*B*) One guide RNA can recognize different enzyme complexes (E): One class of guide RNAs (such as miRNAs) might bind to different enzyme complexes (E) and thus is able to guide different enzymatic reactions.

2004; Matzke and Birchler 2005) or a histone methyltransferase (Volpe et al. 2002), or an RNA endonuclease (Meister and Tuschl 2004; Tomari and Zamore 2005). What determines the choice of protein partner remains unknown. A very similar siRNA/miRNA RNP complex guides mRNA translation inhibition in animals (Liu et al. 2004) and DNA endonucleases in *Tetrahymena* (Mochizuki and Gorovsky 2004).

Advantages of RNA Guiding

Given that RNA or DNA target recognition can also be accomplished by proteins alone, why is the RNA-guided enzyme mechanism so widely used? A possible explanation is based on the observation that an RNA-guided enzyme system requires only one (non-sequence-specific) protein for its enzymatic activity. Sequence specificity, and thereby target recognition, is accomplished by the small ncRNA component of the RNP complex. This strategy both limits the amount of the genome that must be allocated to encode the required genes and facilitates the evolution of novel targets for the complex.

Moreover, evolutionary mechanisms that generate novel targets for protein-only enzymes are necessarily more complex. This is because new genes rarely arise de novo, but rather by gene duplication, followed by mutation of the duplicated copy (Long et al. 2003). To accomplish recognition of new target sites, a sophisticated mutation mechanism would be required. This

mechanism would most likely require multiple point mutations, changing several amino acids, in order to modify the RNA-binding domain to target the new site. Since many protein mutations within RNA-binding domains would be expected to often result in loss of function, this would be a highly inefficient means of generating novel target diversity.

In contrast, RNA-guided systems avoid both the problem of requiring multiple protein enzymes to catalyze reactions that involve multiple substrates and the difficulties of evolving additional enzymes for new target sites. A single RNA-guided protein catalytic complex can perform numerous modifications or cleavages simply by associating with the appropriate guide RNA. Since guide RNA genes are generally much shorter than protein-coding genes, significant gains in genomic coding efficiency are possible. In addition, the energetic cost of synthesizing a protein molecule is much higher than that for an RNA molecule.

Furthermore, the RNA-guided enzyme system has the potential to expand its repertoire of target sites simply by duplicating the gene for the RNA guide and incorporating single nucleotide mutations within its antisense sequence. Such single-base antisense mutations will more often generate a new set of target sites and will rarely lead to loss of functionality of the RNP complex, as compared to mutations in protein genes. To achieve this flexibility of target selection with protein-only enzymes would likely require a much larger number of nucleotide substitutions, with an increased likelihood that at least one of these mutations caused a loss of function.

The General Concept of RNA Guiding Is Widespread in Biology

The first RNAs to be called guide RNAs were those found in kinetoplast mitochondria of trypanosomes, which guide the insertion or deletion of U residues into mitochondrial pre-mRNAs (Stuart et al. 2005). I suggest that the concept of guide RNAs is far more widespread than initially anticipated and can be extended to snoRNAs, si/miRNAs, and even small nuclear RNAs (snRNAs). Indeed, two additional ncRNA families have recently been identified in *Caenorhabditis elegans* (Deng et al. 2006). Whether these ncRNAs represent new guide RNA families is unknown, although it seems likely that our list of guide RNA families is still incomplete. The large families of guide RNAs thereby outnumber the relatively few representatives of catalytic ncRNAs, to which most attention has been drawn in recent years. In evolutionary terms, the concept of RNA guiding has proved to be a very powerful way of generating genetic diversity because new target sites can be generated by gene duplication of guide RNAs genes and mutation of their antisense elements.

CONCLUSIONS: CHALLENGES FOR THE FUTURE

After establishing the biological roles for some novel identified ncRNA species, such as guide RNAs (e.g., snoRNAs, miRNAs, and gRNAs), the future challenge will reside in the analysis of the function of the many other ncRNAs especially in higher eukaryotes, for which the function has not been determined, up until now. Although the number of ncRNAs in eukaryotes has been proposed to be larger than that for protein-coding genes, it still has to be demonstrated, however, whether all of these predicted ncRNA are really biologically functional. Thus, in the future, high-throughput techniques will be required to study the function of thousands of proposed ncRNA candidates in different model organisms.

The most informative approach would be the elimination of the RNA itself, or its encoding DNA gene on the genome (e.g., a knockout of the RNA gene), the latter being very time-consuming. As for mRNAs, residing in the cytoplasm of the cell, knockdown by RNA interference techniques has been proven to be a powerful high-throughput tool to study the function of encoded proteins. A similar approach could be applied to ncRNAs localized to the cytoplasm of a cell, however, not to nuclear or nucleolar localized ncRNAs or to ncRNAs located in cell organelles. For these RNA species, other techniques must be established to study their function in a high-throughput manner. Only if the functions of all candidates for ncRNAs have been established, will we know the actual number of regulatory RNA elements present in a cell.

ACKNOWLEDGMENTS

A.H. expresses his gratitude to all past and present lab members from the Division of Genomics and RNomics for their enthusiasm to work on understanding the fascinating world of noncoding RNAs.

REFERENCES

Brownlee G.G., Cartwright E., McShane T., and Williamson R. 1972. The nucleotide sequence of somatic 5 S RNA from *Xenopus laevis*. *FEBS Lett.* **25:** 8.

Deng W., Zhu X., Skogerbo G., Zhao Y., Fu Z., Wang Y., He H., Cai L., Sun H., Liu C., et al. 2006. Organization of the *Caenorhabditis elegans* small non-coding transcriptome: Genomic features, biogenesis, and expression. *Genome Res.* **16:** 20.

Donis-Keller H., Maxam A.M., and Gilbert W. 1977. Mapping adenines, guanines, and pyrimidines in RNA. *Nucleic Acids Res.* **4:** 2527.

Dubey A.K., Baker C.S., Romeo T., and Babitzke P. 2005. RNA sequence and secondary structure participate in high-affinity CsrA-RNA interaction. *RNA* **11:** 1579.

Eddy S.R. 2001. Non-coding RNA genes and the modern RNA world. *Nat. Rev. Genet.* **2:** 919.

———. 2002. Computational genomics of noncoding RNA genes. *Cell* **109:** 137.

Ehresmann C., Stiegler P., Carbon P., and Ebel J.P. 1977. Recent progress in the determination of the primary sequence of the 16 S RNA of *Escherichia coli*. *FEBS Lett.* **84:** 337.

Filipowicz W. 2000. Imprinted expression of small nucleolar RNAs in brain: Time for RNomics. *Proc. Natl. Acad. Sci.* **97:** 14035.

Gerhold D. and Caskey C.T. 1996. It's the genes! EST access to human genome content. *Bioessays* **18:** 973.

Halic M. and Beckmann R. 2005. The signal recognition particle and its interactions during protein targeting. *Curr. Opin. Struct. Biol.* **15:** 116.

Halic M., Becker T., Pool M.R., Spahn C.M., Grassucci R.A., Frank J., and Beckmann R. 2004. Structure of the signal recognition particle interacting with the elongation-arrested ribosome. *Nature* **427:** 808.

Hüttenhofer A., Brosius J., and Bachellerie J.P. 2002. RNomics: Identification and function of small, non-messenger RNAs. *Curr. Opin. Chem. Biol.* **6:** 835.

Hüttenhofer A., Cavaille J., and Bachellerie J.P. 2004. Experimental RNomics: A global approach to identifying small nuclear RNAs and their targets in different model organisms. *Methods Mol. Biol.* **265:** 409.

Hüttenhofer A., Kiefmann M., Meier-Ewert S., O'Brien J., Lehrach H., Bachellerie J.-P., and Brosius J. 2001. RNomics: An experimental approach that identifies 201 candidates for novel, small, non-messenger RNAs in mouse. *EMBO J.* **20:** 2943.

Kawasaki H. and Taira K. 2004. Induction of DNA methylation and gene silencing by short interfering RNAs in human cells. *Nature* **431:** 211.

Kim S., Shi H., Lee D.K., and Lis J.T. 2003. Specific SR protein-dependent splicing substrates identified through genomic SELEX. *Nucleic Acids Res.* **31:** 1955.

Lee R.C., Feinbaum R.L., and Ambros V. 1993. The *C. elegans* heterochronic gene *lin-4* encodes small RNAs with antisense complementarity to *lin-14*. *Cell* **75:** 843.

Liu J., Carmell M.A., Rivas F.V., Marsden C.G., Thomson J.M., Song J.J., Hammond S.M., Joshua-Tor L., and Hannon G.J. 2004. Argonaute2 is the catalytic engine of mammalian RNAi. *Science* **305:** 1437.

Long M., Betran E., Thornton K., and Wang W. 2003. The origin of new genes: Glimpses from the young and old. *Nat. Rev. Genet.* **4:** 865.

Majdalani N., Chen S., Murrow J., St. John K., and Gottesman S. 2001. Regulation of RpoS by a novel small RNA: The characterization of RprA. *Mol. Microbiol.* **39:** 1382.

Martin G. and Keller W. 1998. Tailing and 3'-end labeling of RNA with yeast poly(A) polymerase and various nucleotides. *RNA* **4:** 226.

Mattick J.S. 2001. Non-coding RNAs: The architects of eukaryotic complexity. *EMBO Rep.* **2:** 986.

———. 2004. RNA regulation: A new genetics? *Nat. Rev. Genet.* **5:** 316.

———. 2005. The functional genomics of noncoding RNA. *Science* **309:** 1527.

Mattick J.S. and Makunin I.V. 2005. Small regulatory RNAs in mammals. *Hum. Mol. Genet.* (spec. no. 1) **14:** R121.

Matzke M.A. and Birchler J.A. 2005. RNAi-mediated pathways in the nucleus. *Nat. Rev. Genet.* **6:** 24.

Meister G. and Tuschl T. 2004. Mechanisms of gene silencing by double-stranded RNA. *Nature* **431:** 343.

Mizuno T., Chou M.Y., and Inouye M. 1983. A unique mechanism regulating gene expression: Translational inhibition by a complementary RNA transcript (micRNA). *Proc. Jpn. Acad. Ser. B Phys. Biol. Sci.* **59:** 335.

Mochizuki K. and Gorovsky M.A. 2004. Small RNAs in genome rearrangement in *Tetrahymena*. *Curr. Opin. Genet. Dev.* **14:** 181.

Ohlrogge J. and Benning C. 2000. Unraveling plant metabolism by EST analysis. *Curr. Opin. Plant Biol.* **3:** 224.

Pauler F.M., Stricker S.H., Warczok K.E., and Barlow D.P. 2005. Long-range DNase I hypersensitivity mapping reveals the imprinted Igf2r and Air promoters share cis-regulatory elements. *Genome Res.* **15:** 1379.

Peattie D.A. 1979. Direct chemical method for sequencing RNA. *Proc. Natl. Acad. Sci.* **76:** 1760.

Sanger F., Brownlee G.G., and Barrell B.G. 1965. A two-dimensional fractionation procedure for radioactive nucleotides. *J. Mol. Biol.* **13:** 373.

Shtatland T., Gill S.C., Javornik B.E., Johansson H.E., Singer B.S., Uhlenbeck O.C., Zichi D.A., and Gold L. 2000. Interactions of *Escherichia coli* RNA with bacteriophage MS2 coat protein: Genomic SELEX. *Nucleic Acids Res.* **28:** E93.

Singer B.S., Shtatland T., Brown D., and Gold L. 1997. Libraries for genomic SELEX. *Nucleic Acids Res.* **25:** 781.

Sledjeski D. and Gottesman S. 1995. A small RNA acts as an antisilencer of the H-NS-silenced rcsA gene of *Escherichia coli*. *Proc. Natl. Acad. Sci.* **92:** 2003.

Stoughton R.B. 2005. Applications of DNA microarrays in biology. *Annu. Rev. Biochem.* **74:** 53.

Stuart K.D., Schnaufer A., Ernst N.L., and Panigrahi A.K. 2005. Complex management: RNA editing in trypanosomes. *Trends Biochem. Sci.* **30:** 97.

Tomari Y. and Zamore P.D. 2005. Perspective: Machines for RNAi. *Genes Dev.* **19:** 517.

Vogel J. and Sharma C.S. 2005. How to find small non-coding RNAs in bacteria. *Biol. Chem.* **386:** 1219.

Volpe T.A., Kidner C., Hall I.M., Teng G., Grewal S.I., and Martienssen R.A. 2002. Regulation of heterochromatic silencing and histone H3 lysine-9 methylation by RNAi. *Science* **297:** 1833.

Washietl S., Hofacker I.L., Lukasser M., Hüttenhofer A., and Stadler P.F. 2005. Mapping of conserved RNA secondary structures predicts thousands of functional noncoding RNAs in the human genome. *Nat. Biotechnol.* **23:** 1383.

Wassarman K.M. and Storz G. 2000. 6S RNA regulates *E. coli* RNA polymerase activity. *Cell* **101:** 613.

Yang Z., Zhu Q., Luo K., and Zhou Q. 2001. The 7SK small nuclear RNA inhibits the CDK9/cyclin T1 kinase to control transcription. *Nature* **414:** 317.

Drosophila Genome-wide RNAi Screens: Are They Delivering the Promise?

B. MATHEY-PREVOT AND N. PERRIMON

Department of Genetics, Howard Hughes Medical Institute, Harvard Medical School, Boston, Massachusetts 02115

The emergence of RNA interference (RNAi) on the heels of the successful completion of the *Drosophila* genome project was seen by many as the ace in functional genomics: Its application would quickly assign a function to all genes in this organism and help delineate the complex web of interactions or networks linking them at the systemic level. A few years wiser and a number of genome-wide *Drosophila* RNAi screens later, we reflect on the state of high-throughput RNAi screens in *Drosophila* and ask whether the initial promise was fulfilled. We review the impact that this approach has had in the field of *Drosophila* research and chart out strategies to extract maximal benefit from the application of RNAi to gene discovery and pursuit of systems biology.

The completion of the *Drosophila* genome sequence in 2000 (Adams et al. 2000) has conceptually changed the approach to functional genomics as it provided the opportunity to develop genome-wide approaches to systematically explore gene functions. Newly emerged technologies were quickly put to task to extract maximal information encrypted in the raw sequence of the *Drosophila* genome. This is best illustrated with RNA interference (RNAi), which is based on the ability of double-stranded (dsRNA), small interfering RNAs (siRNAs), or small hairpin RNAs (shRNAs) to silence a target gene through the specific destruction of that gene's mRNA (for review, see Friedman and Perrimon 2004).

In the past few years, *Drosophila* has become a premier system for systematic genome-wide cell-based RNAi high-throughput screens (RNAi HTS), largely because of two major advances. First, Clemens et al. (2000) made the seminal discovery that long dsRNAs added to the medium of *Drosophila* tissue culture cells are rapidly taken up by the cells and cause efficient knockdown of their targeted mRNAs, thus opening up the application of RNAi to cell-based assays. Second, the development of cell-based assays in *Drosophila* to a high-throughput format (Armknecht et al. 2005) coupled with the production of comprehensive *Drosophila* dsRNA libraries allowed near or full genome-scale screens to systematically interrogate the function of all genes predicted from genomic sequencing (Kiger et al. 2003; Lum et al. 2003; Boutros et al. 2004; Foley and O'Farrell 2004).

In just 3 years, several large-scale RNAi screens in *Drosophila* have been published, and the results obtained from these studies allow us to reflect on the impact that this approach has had in the field of *Drosophila* research and chart out strategies to extract maximal benefit from the application of genome-scale RNAi screens. Here, we discuss whether RNAi HTS are (1) succeeding as a functional gene discovery platform, that is, whether they allow a rapid and unbiased identification of genes involved in a specific biological processes, even in the case of pleiotropic and redundant genes, and (2) allowing us to obtain a systems biology or global picture of the functions of all genes in a given process. Because of the success of *Drosophila* genetics over the years at identifying gene functions, arguably the most interesting application of RNAi HTS is to use this nascent technology to obtain a global understanding of biological processes. In particular, RNAi HTS can be used to gain insights into the structure of signaling networks. For example, as many assays can be designed to quantitatively read the activity of pathways, e.g., using transcriptional reporters or phospho-specific antibodies, the respective contribution of every gene in the genome can be measured and used to model, in combination with other data sets, the flow of information through protein networks (Sachs et al. 2005).

RNAi HTS: THE BASICS OF THE METHODOLOGY

The RNAi HTS platform is extremely flexible and can accommodate versatile formats. Genome-scale dsRNA libraries (for a list of available *Drosophila* dsRNA libraries, see Echeverri and Perrimon 2006) can be screened in 48-(Ramet et al. 2002), 96-(Lum et al. 2003; Björklund et al. 2006), or 384-well plates in a variety of cell-based assays (Boutros et al. 2004; Agaisse et al. 2005; Baeg et al. 2005; Cherry et al. 2005; DasGupta et al. 2005; Muller et al. 2005; Nybakken et al. 2005; Philips et al. 2005; Bard et al. 2006; Gwack et al. 2006; Vig et al. 2006; Zhang et al. 2006). In addition, dsRNAs can be spotted at high density on glass slides (RNAi microarrays) and assayed in visual screens (Wheeler et al. 2004; Guertin et al. 2006), achieving even faster and cheaper means of screening large libraries (Fig. 1). Detection of phenotypes in high-throughput often relies on the use of plate reader or conventional microscopy (Armknecht et al. 2005), but it can also be based on flow cytometry (Ramet et al. 2002; Björklund et al. 2006), automated fluorometric imaging plate reader (FLIPR) (Vig et al. 2006), or high-throughput confocal microscopy (Pelkmans et al. 2005).

At the *Drosophila* RNAi Screening Center (DRSC), which we established a few years ago (see http://flyrnai/.org), screens are conducted in high-density 384-well tissue culture plates. Existing *Drosophila* cell culture

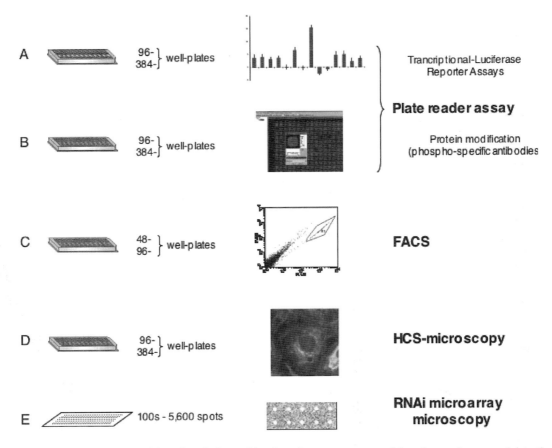

Figure 1. RNAi screening format and detection platforms. Listed are the most common cell-based assay formats and detection platforms used in *Drosophila* high-throughput RNAi screens. Genome-scale dsRNA libraries are typically prearrayed on 48-, 96-, or 384-well plates (*A–D*) or glass slides (*E*), and phenotypes are analyzed using various detection methods. (*A*) Conventional plate reader measurements include luminescence, fluorescence intensity, fluorescence polarization, time-resolved fluorescence, and absorbance detection modes. (*B*) The Aerius platform (LI-COR Biosciences) is a modified laser-based microscope that excites and scans in the far-red the emission from appropriately conjugated secondary antibodies bound to the primary, phospho-specific antibody. (*C*) Detection of phenotypes by flow cytometry using a fluorescence activated cell sorter (FACS). (*D*) Detection of phenotypes in high-content screening (HCS) microscopy approaches typically include high-throughput, automated wide-field, and, more recently, confocal microscopes. (*E*) dsRNAs microarrays spotted on glass slides seeded with cells are screened with automated microscopy. Screen examples carried out according to each format are given in the text.

lines (having distinct attributes that can be exploited in the various screens) are ideally suited for this approach; furthermore, the range of uses for these cells can be extended by either transient or stable transfection of DNA constructs before screening. In addition, we have developed and successfully implemented efficient protocols to conduct RNAi screens in primary embryonic cells, broadening the range of biological and developmental processes (e.g., neurite outgrowth and myofibrillogenesis) that can be investigated by this approach (J. Bai et al., in prep.).

The basic experimental design for screens carried out at the DRSC involves three major steps: (1) Gene-specific dsRNAs from our collection stored in 96-well plates are arrayed into 384-well assay plates using robotics. (2) Cells are uniformly and rapidly dispensed into the 384-well plates using a MultiDrop liquid dispenser. (3) After the appropriate incubation time, cells are subjected to individual treatments in a highly parallel fashion, fixed, or directly processed for the assay readout. The phenotypic output measured for each sample depends on

the assay readout: Quantitative measurements have been acquired with a plate reader, whereas qualitative measurements have been captured by automated microscopy. The incubation period with dsRNAs varies and must be optimized for specific assays/targets. In general, we have used a 3-day incubation period in our experiments. However, many RNAi effects can be detected within a day or two of treatment. If necessary, incubation with dsRNAs can be carried out for longer (up to 1 week) without deleterious cytotoxicity.

Perhaps the most important aspect of an RNAi HTS platform is the quality of the "RNAi library" to be screened. One of the primary goals at the DRSC has been to ensure that the library of dsRNAs screened is of the highest quality. Thus, over the years, a number of "upgrades" have been made based on our experience with the reagents. Our first library, "*the DRSC 10 collection*" (Boutros et al. 2002), was based on earlier annotations from BDGP/Celera and the Sanger Center, which predicted 13,672 and 20,622 genes, respectively (Adams et al. 2000; Hild et al. 2003). A direct comparison of the two predictions yielded a total of 21,306 nonredundant possible transcripts in the *Drosophila* genome, with 14,556 dsRNAs targeting annotations present in both the BDGP and Sanger Center sets and 6,750 dsRNAs targeting Sanger annotations not found in the BDGP set (Sanger-only dsRNAs). Subsequent independent analyses of whether the Sanger-only predictions represented real genes or were expressed revealed that only 10% were likely to be validated as genes containing introns (Yandell et al. 2005) and only 291 predictions were confirmed by expression by Stolc et al. (2004). Later releases of the BDGP genome annotation caused a number of revisions, including the prediction of new genes or new exons within a gene, as well as the reassignment of adjacent open reading frames (ORFs) into a single functional unit. These periodic revisions led us to update our library accordingly and to remap some of the older dsRNA to new functional units.

In addition to the upgrade due to changes in genes' annotations, the DRSC library has been updated to address the issue of off-target effects (OTEs) associated with the use of long dsRNAs. This issue emerged as an unanticipated complication in the analysis of large-scale *Drosophila* screens, likely causing the inclusion of a number of false positives among hits reported in early screens (M. Kulkarni et al. 2006). Although this issue is familiar to investigators using siRNAs in mammalian systems (where it had been recognized early on; Jackson et al. 2003), it took the *Drosophila* community by surprise, as it was widely believed that OTE was unlikely to take place in *Drosophila* because of the use of long dsRNAs. To many, the processing of long dsRNAs by Dicer into many short 21- to 23-nucleotide triggers (Hammond et al. 2000) meant that any OTE potentially associated with a particular siRNA trigger from the pool would be diluted by the specific effects of the "good" siRNAs present in excess in that pool. Although that may be true in some cases, a retrospective analysis done by the DRSC challenged this assumption. We performed a statistical analysis of the results from more than 30 DRSC genome-wide screens and asked how the various dsRNAs behaved across these screens. In particular, we examined whether dsRNA predicted to have regions of perfect homology with genes other than the intended target (using a simple string search of all possible siRNAs generated from a dsRNA against all gene sequences in *Drosophila*) led to a greater probability (than by chance alone) of causing a phenotype. In other words, there was a clear correlation between the presence of predicted off-targets in dsRNAs and their likelihood to cause a phenotype in a cell-based assay (M. Kulkarni et al. 2006). Importantly, the homology length at or above which it became problematic was 19 nucleotides rather than the initially predicted cutoff of 21 nucleotides, thus increasing the number of potentially "problematic" dsRNA reagents in our collection.

As a result of this analysis, we assembled a new dsRNA collection (the "*DRSC 2.0 collection*") to eliminate any dsRNAs predicted to have potential OTE. This was achieved by keeping all original dsRNAs from DRSC 1.0 that lacked predicted off-targets and generating 7,692 new, independently synthesized dsRNAs to replace DRSC 1.0 dsRNAs predicted to have 1 or more off-

targets. In addition, Sanger-only dsRNAs that were not detected by Stolc et al. (2004) are not represented in the DRSC 2.0 collection. As we are still learning the rules for what might constitute an offending sequence associated with OTEs, we decided to provide screeners with the ability to quickly validate the effects of dsRNAs identified in a primary screen with a second or third independent dsRNA, even if the original dsRNA had no 19-nucleotide perfect homology with any non-target genes. With this objective in mind, we are currently assembling the "*DRSC 3.0 collection*," which consists of new dsRNAs that are distinct from any dsRNA present in DRSC 2.0. These dsRNAs are devoid of predicted perfect homologies with non-target genes and correspond to every hit identified in a completed DRSC screen.

RNAi AS A FUNCTIONAL GENE DISCOVERY PLATFORM

In addition to the quality of the RNAi library, the success of RNAi HTS depends on the robustness of the cell-based assay and its applicability to high-throughput screening. Many considerations, such as signal-to-noise issues, normalization methods, choice of cell type, and specificity of the readout, should be taken into account when designing an assay. This thorough assessment is probably the most important step of RNAi HTS (for more details, see review by Echeverri and Perrimon 2006). In the context of gene discovery, a screen that leads to hundreds of positives may be considered less successful than a screen that identifies a smaller number of candidates. As such, the design of an assay should be aimed as much as possible at capturing specific features inherent to the process under study to limit the number of positives. For example, a screen for cell viability is expected to lead to hundreds of hits, whereas a screen for subcellular localization of a protein may only lead to the identification of a few hits. Altogether, assay development today is probably the most important step of the RNAi HTS area and where many sophisticated innovations will occur. Below, we describe a few screens that have been done today as a means to document the currently available technologies (Fig. 1).

Transcriptional Reporter Screens

Many assays are based on transcriptional reporters whose overall chemiluminescence or fluorescence output is rapidly measured using a plate reader (see, e.g., Boutros et al. 2004; Baeg et al. 2005; DasGupta et al. 2005; Nybakken et al. 2005; Bard et al. 2006). The generation of numerical readouts for each condition or well tested makes it possible to normalize the data and subject it to various statistical analyses. This approach was used to investigate, for example, the evolutionarily conserved Wnt/Wg signaling pathway, which regulates many aspects of metazoan development. A cell-based assay based on the "TOP-Flash" (Tcf Optimal Promoter) reporter construct was developed and optimized for high-throughput conditions in 384-well plates (DasGupta et al. 2005). The TOP-Flash construct consists of multimerized Tcf-binding sites cloned upstream of a cDNA encoding firefly luciferase.

Transfection of this construct, a Renilla luciferase normalization vector, and dsRNA into *Drosophila* cells in the presence or absence of Wg serves as the basis of the assay. A normalized readout of luciferase expression is measured under every experimental condition, with its value being directly proportional to the extent of pathway activation. The screen was performed in duplicate, and 238 hits were identified that either reduced Wg pathway activity by more than 1.5 SD or increased reporter activity by more than 3 SDs. Importantly, more than 16 of the known regulators of the Wg pathway scored in this assay, including Armadillo, Pangolin, Legless, Pygopus, Axin, CK1α, Frizzled, and Arrow. The positive and negative regulators were then systematically ordered in the pathway by several epistasis experiments to ascertain at which step in the signal transduction cascade the candidate genes have a potential function. The hits comprise many genes assignable to certain molecular complexes or biological functions, and include (1) HMG/homeodomain transcription factors, (2) kinases and phosphatases, (3) proteosomal components and ubiquitin ligases, (4) G-protein family, and (5) membrane-associated proteins. Of specific interest are some of the kinases and phosphatases, such as Cdc2 and String (Cdc25). Both have been shown previously to genetically interact with Armadillo, thus implicating them as having some role in the Wg pathway. However, their mechanisms of action in the regulation of the Wg pathway are unknown.

Antibody-based Screens

Another powerful application of antibody-based screens involves the use of phospho-specific antibodies. Provided the specificity of the phospho-antibodies is good, such screens are highly quantitative and can be performed using either a plate reader to measure overall levels of fluorescence emitted by the fluorescently coupled secondary antibody or with the Aerius platform (LI-COR Biosciences), a modified laser-based microscope that excites and scans in the far-red the emission from appropriately conjugated secondary antibodies bound to the primary, phospho-specific antibody (Fig. 1). For instance, the cellular network responsible for mitogen-activated protein kinase (MAPK) activation was investigated using a fluorescently conjugated antibody (cell signaling) that recognizes the diphosphorylated (activated) form of the single *Drosophila* ERK, Rolled. ERK activity was monitored by dpERK staining, at baseline (resting) and under stimulation by Insulin, in an SL2-derived cell line that was engineered to express yellow fluorescent protein (YFP)-tagged Rolled, as a means to normalize for total ERK protein levels using YFP fluorescence. Importantly, the kinetics of MAPK activation and the effects of known component knockdown were found to be identical in wild-type and Rolled-YFP-expressing cells; 1168 unique dsRNAs were found in the primary screen to significantly affect the level of ERK phosphorylation. Although this unbiased list was not fully validated, it included the entire *Drosophila* core pathway, and more than 60% of the candidates had identifiable human orthologs. Various criteria ranging from GO annotation consideration and evolutionary conservation were applied to filter the initial list down to 362 candidates, which were tested in secondary screens in different cell lines and under various ligand stimulations. Of those, 331 genes were validated, and greater than 85% of those could be confirmed with the use of a second or third independent dsRNA (A. Friedman and N. Perrimon, in prep.). In addition to identifying new regulators (see Table 1), data from this quantitative, unbiased approach can be integrated with other genomic and proteomic approaches (as outlined in Fig. 2) to provide a blueprint of the complex regulatory network leading to MAPK activation.

Table 1. *Drosophila* RNAi Genome-scale Screens and Gene Discovery: Selected Examples

Screen	Gene/putative function	References
Wg signaling	Evi/Wg secretion, positive regulator	Bartscherer et al. (2006)
Store-operated Ca^{2+} entry	Orai1/CRACM1/Olf186-F; modulator of CRAC-mediated current	Feske et al. (2006); Vig et al. (2006); Zhang et al. (2006)
JAK/STAT signaling	PTP61F/protein tyrosine phosphatase, negative regulator	Baeg et al. (2005); Muller et al. (2005)
Host factors involved in *Mycobacteria* infection	CD36/class-B scavenger receptor required for uptake of mycobacteria	Philips et al. (2005)
Hh signaling	Ihog/type I membrane protein binds active Hh protein and mediates its signaling	Yao et al. (2006)
Hh signaling	PP2A/multimeric protein phosphatase 2A, negative regulator	Nybakken et al. (2005)
EGFR signaling	PLC-γ/required for ER retention of cleaved Spitz during fly eye development	Schlesinger et al. (2004)
Cytokinesis inhibitors	Borr/protein involved in Aurora-B kinase pathway	Eggert et al. (2004)
MAPK pathway	dGCKIII/member of Ste20 kinase family, positive regulator	A. Friedman and N. Perrimon (in prep.)
Myofibrillogenesis	Sals/actin-binding protein regulating proper sarcomere length	J. Bai and N. Perrimon (in prep.)

This list is not meant to be exhaustive but rather highlights selective examples of genes found to participate in particular signaling pathways or biological processes investigated with unbiased RNAi screens. In addition, all screens (cited throughout the text) have identified major molecular machines including the ribosomal complex, the protein degradation machinery (proteasome, ubiquitination), the vesicular and nuclear transport machinery, and the RNA processing machinery.

RNAi phenoprints
RNAi signatures

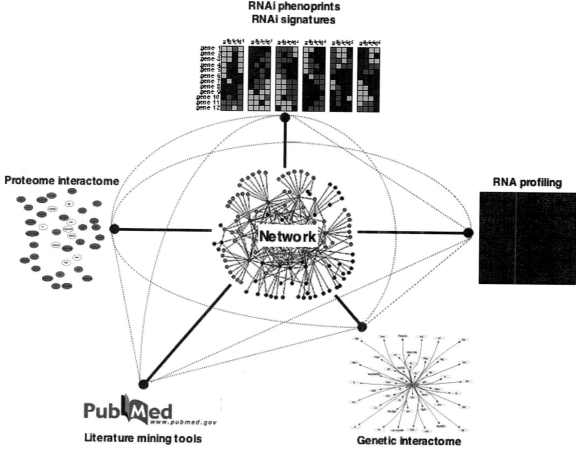

Figure 2. Network building through the integration of RNAi signatures with other data sets. RNAi HTS in *Drosophila* cells can be used to assign phenotypic signatures (referred to as phenoprints) to every *Drosophila* gene. Phenoprints can then be used to cluster genes that are functionally related, guiding functional genomics efforts to assign a biological function to uncharacterized or unknown genes based on where they cluster in RNAi screens. The RNAi signatures/phenoprints panel illustrates how 12 genes (1–12) can be functionally clustered based on the comparison of distinct phenotypes (a^n–d^n) scored in hypothetical screens (1–6). Furthermore, global correlations between phenoprints, transcriptional profiling (RNAi profiling), interactome data sets (proteome and genetic interactomes), and published literature (literature-mining tools) can be used to derive network graphs critical to data mining.

FACS-based Screens

Several large-scale RNAi screens have relied on the use of flow cytometry to follow phenotypes (Ramet et al. 2002; Björklund et al. 2006; Ulvila et al. 2006). Coupled with high-resolution imaging microscopy, this approach can be very powerful and allows multiple parameters to be analyzed simultaneously. Using this strategy, Björklund et al. (2006) set out to identify pathways regulating cell size and cell cycle progression in *Drosophila* S2 cells. RNAi-treated cells were simultaneously analyzed by fluorescence-activated cell sorter (FACS) for six distinct phenotypes (G_1 arrest, G_2/M transition, cell death, cytokinesis, and cell size in G_1 or G_2). This multiparameter analysis identified 488 candidate genes, whose gene products include cell cycle regulators, members of the ubiquitin pathway, components of vesicular and nuclear transport, and mediators of four ligand-induced signaling pathways (Wnt/Wg, p38βMAPK, FRAP/TOR, and JAK/STAT) (Björklund et al. 2006). Although profiling DNA content by FACS analysis has proved to be valuable

and informative, one slight limitation of this approach is that many *Drosophila* cell lines do not have a normal mode of chromosomes and tend to be polyploid.

Microscopy-based Screens

Arguably, the most informative cell-based assays are microscopy-based as they provide a wealth of information since specific information can be scored. Image-based screens can use not only an antibody staining, but also cellular compartments or structures (e.g., Golgi, mitochondria, nuclei, and actin filaments) that can be selectively labeled with either fluorescently labeled dyes or GFPs (green fluorescent protein) tagged with the appropriate localization tag (Kiger et al. 2003; Gwack et al. 2006). A powerful extension of this approach is exemplified in host/pathogens screens where GFP is constitutively expressed when the pathogen localizes to the endosomal compartment of a cell. This approach can be used to identify host factors highjacked by *Mycobacterium* for survival within macrophages as well as host factors involved in pathogen killing. Philips et al. (2005) developed an ex vivo assay of infection using S2 cells and GFP-expressing mycobacteria and showed that the nonpathogenic species, *M. smegmatis*, is killed by S2 cells, whereas the pathogenic species, *M. fortuitum*, is able to grow within S2 cells. During infection of S2 cells, expression of the macrophage-activated promoters *map24* and *map49* is induced, revealing that the intracellular milieu within S2 cells is similar to that within mammalian macrophages. This cell-based assay was used in a genome-wide RNAi screen to identify host factors required for entry or growth of *M. fortuitum* within S2 cells; 30 dsRNAs were reported to allow *M. fortuitum* to grow better within S2 cells, whereas 86 dsRNAs were found to disrupt infection by *M. fortuitum*. Most of the dsRNAs that disrupt infection target genes predicted to have a role in vesicular trafficking, the actin cytoskeleton, ubiquitin or proteolysis, or fatty acid metabolism. Many of these factors also disrupt phagocytosis of other pathogens, such as *Escherichia coli*. However, some appear to be uniquely required for mycobacterial infection. In particular, a member of the CD36 family of scavenger receptors was found to be required for uptake of two mycobacterial species and *Listeria monocytogenes*, but dispensable for uptake of *E. coli* and *Staphylococcus aureus*, suggesting a role in pattern recognition of distinct bacterial species (Table 1).

RNAi Microarrays

In this innovative approach, 2–3 nl of dsRNA from large dsRNA collections is arrayed at high density (upward of 5000 spots per slide) on coated glass slides. Cells are seeded onto the slide and incubated for a few days before RNAi phenotypes are scored by microscopy over the clusters of cells that have landed on each dsRNA spot. In this manner, thousands of dsRNAs can be quickly and economically screened (Wheeler et al. 2005). To demonstrate its feasibility, Wheeler et al. (2004) printed arrays with 384 preselected dsRNAs (at a density suitable to array 5600 dsRNAs on a single slide) and screened for regulation of cell number and viability; 44 dsRNAs in this set showed a reproducible phenotype. In addition to confirming the function of cell cycle regulators and apoptotic modulators, targeted by dsRNAs present in the set, these authors identified two surface receptors (InR and Pvr), as well as a number of kinases and phosphatases that affected cell number and viability, underscoring the promise of this approach for systematic synthetic lethal screens using combinations of dsRNA treatments (Wheeler et al. 2004).

ARE GENOME-WIDE RNAi SCREENS DELIVERING THE PROMISE?

RNAi HTS as a Functional Gene Discovery Platform

Clearly, as exemplified by the screens described above and others, RNAi HTS are being highly successful at gene discovery (Table 1). The success of the approach relies on an overall low rate of false negatives. Indeed, as shown in Table 2, most expected components that should score in a screen are identified in top hits. Thus, the approach is very robust at generating an enriched list of genes that are likely to be specific for a process. Follow-up experiments are then required to further validate the genes identified.

In the upcoming years, many exciting advances will take place, both in the sophistication of cell-based assay designs and in the detection of phenotypes. Assays that are more biologically relevant, such as those that use primary cells, will become favored. Either antibody-, FISH-, or Luminex-based assays relying on multiplexed endogenous readouts (Levsky et al. 2002; Pelech 2004; Sachs et al. 2005) will be preferred to the current luciferase transcriptional reporter-based assays since they capture richer information. Similarly, high-content microscopy screens that extract and quantify multiple features from each image and are carried out in different established or primary cell cultures will greatly expand our ability to probe for complex cell biological processes. In addition, improvements in image acquisition together with the development of novel molecular probes will alleviate our current limitation in addressing questions of spatial and temporal regulation of signaling pathways and cytoskeletal organization. Parallel RNAi and small-molecule screens comparing small-compound- and dsRNA-induced phenotypes will offer a powerful venue for drug target discovery (Eggert et al. 2004). Combinatorial RNAi experiments where collections of dsRNAs are screened for their ability to suppress or enhance the phenotype caused by another dsRNA or small molecules will become common, and we anticipate the need for such screens to grow up exponentially. A prevalent illustration for this kind of application will be the search for synthetic lethal phenotypes, an approach that will necessitate a very large number of experiments. Although these screens can be conducted in the 384-well plate format, the miniaturization and economy of cells and reagents intrinsic to RNAi cell microarrays (Wheeler et al. 2005) offer an ideal solution to this challenge. Other potential uses of RNAi microarrays include suppression (or enhancement) of small-compound-induced phenotype(s) as recently exploited by Guertin et al. (2006), who com-

Table 2. Low False-Negative Rate in Genome-wide RNAi Screens

| Pathway | Reference | Canonical genes | | Scoring efficiency (%) |
		expected	found	
Wg	DasGupta et al. (2005)	17	15	88
Hh	Nybakken et al. (2005)	15	12	80
JAK/STAT	Baeg et al. (2005)	6	5	83

Data reported in three published screens investigating the Wnt/Wg, Hh, and JAK/STAT pathways were used to estimate the rate of false-negative associated with genome-wide screens by tallying the number of core component genes identified in each screen and comparing it to the expected number of genes known to belong to the pathway of interest. Note that for some screens, certain core components could not be evaluated because of the design of the assay.

bined the use of the small-molecule inhibitor rapamycin and RNAi to identify TOR-regulated genes that control growth and division.

RNAi HTS as a Tool for a "Systems Biology" Approach

One of the most interesting promises of RNAi HTS is that it potentially provides a means to identify all of the parts of a network and thus could be used as the first step in a systems biology approach to understand the contribution of the genome to a biological process. Indeed, the data emerging from RNAi HTS could be used to integrate data sets generated from other "Omic" approaches (Fig. 2). The advantage of using data sets from RNAi HTS for such a purpose would be that it provides, unlike other Omic methods, direct functionality. Cell-based high-content screens (HCS) that rely on RNAi-induced cellular phenotypes are particularly well suited for this approach because they generate data sets that are rich in information. Each feature scored in such an assay is assessed independently, according to a controlled vocabulary.

The compilation of these features defines a phenotypic profile or "phenoprint," which is specifically associated with each gene knockdown (Piano et al. 2002). Using this approach, Piano et al. (2002) characterized early embryonic defects in *Caenorhabditis elegans* for 161 genes. Using time-lapse microscopy to systematically describe the defects for each gene in terms of 47 RNAi-associated phenotypes, these authors then clustered the genes into functionally related groups, an approach that can prove to be extremely powerful to functionally annotate unknown genes. However, for it to be useful in building or ordering large biological networks, RNAi HTS must generate an RNAi signature of high confidence, meaning that the rates of false positives in the RNAi HTS screens are low. Estimating the rates of false positives in RNAi screens is possibly one of the most difficult issues right now with the methodology as false positives can have originate from many sources (Echeverri and Perrimon 2006). Indeed, comparisons between related RNAi screens already performed by different groups, sometimes using different or similar RNAi libraries, are revealing poor overlap between data sets (Björklund et al. 2006; M. Kulkarni et al. 2006). The origin of the discrepancy between the studies is complex, as false positives can originate from stochastic, biological, and off-target noise. Stochastic noise refers to any experimental variation caused by random instrumentation malfunctions, plate-manufacturing defects, effectiveness of the reagents used (e.g., knockdown efficiency), as well as human error. Biological noise entails the unpredictable contribution of ill-defined biological variables to a phenotype or readout of interest that cannot easily be controlled. For instance, the state or health of a culture, passage number, or adaptation over time to certain conditions may alter how cells respond to RNAi in general or to a treatment in particular. Both stochastic and biological noise can be quantified and dealt with by performing multiple replicas of the assay and subjecting the results to strict statistical

treatments, conditions that are rarely practical when performing a primary screen. In contrast, the contribution of OTE to the rate of false positives cannot easily be accounted for unless one has prior knowledge of OTE rules, which we do not have at present. For instance, setting the limit at 19 nucleotides for perfect homology with unintended targets as definition for OTE may not be sufficient, since shorter-length homologies have been reported to lead to OTE, at least when using single siRNAs (Birmingham et al. 2006). Consequently, using multiple distinct dsRNAs targeting the same gene should help eliminate or minimize the rate of OTE, as non-overlapping dsRNAs are unlikely to have overlapping off-targets. Using two or three dsRNAs targeting the same gene is therefore highly recommended, particularly if one desires to assign a particular function to a gene based on RNAi data alone. Validation of hits in secondary screens provides an ideal opportunity to supply statistical robustness to account for stochastic and biological noise and specificity to minimize OTE. Ideally, for each gene tested in a secondary screen, one would want to array multiple copies (5–7) of two or three independent dsRNAs interspersed with mixed positive and negative control dsRNAs (13–30 total) in a screening plate. Assaying each plate in duplicate ensures that numerous data points are obtained for each dsRNA, allowing statistical significance in the measured output and providing an accurate determination of the validation rate for each gene tested.

In conclusion, RNAi HTS technology, although a successful approach for gene discovery, is not yet fully mature as a tool for system biology. Only when the rate of false positives is better understood and the quality of the data consistently reliable will RNAi signature data sets become a major player in system biology. This should be achievable in the next few years by more carefully controlled experiments and better understood reagents.

ACKNOWLEDGMENTS

We thank Jianwu Bai, Matt Booker, Adam Friedman, Lutz Kockel, Meghana Kulkarni, and Katharine Sepp for contributing ideas, material, and unpublished experimental data included in this manuscript. Work at the DRSC is supported by a grant R01 GM067761 from the NIGMS. N.P. is an Investigator of the Howard Hughes Medical Institute.

REFERENCES

Adams M.D., Celniker S.E., Holt R.A., Evans C.A., Gocayne J.D., Amanatides P.G., Scherer S.E., Li P.W., Hoskins R.A., Galle R.F., et al. 2000. The genome sequence of *Drosophila melanogaster*. *Science* **287**: 2185.

Agaisse H., Burrack L.S., Philips J., Rubin E.J., Perrimon N., and Higgins D.E. 2005. Genome-wide RNAi screen for host factors required for intracellular bacterial infection. *Science* **309**: 1248.

Armknecht S., Boutros M., Kiger A., Nybakken K., Mathey-Prevot B., and Perrimon N. 2005. High-throughput RNA interference screens in *Drosophila* tissue culture cells. *Methods Enzymol.* **392**: 55.

Baeg G.H., Zhou R., and Perrimon N. 2005. Genome-wide RNAi analysis of JAK/STAT signaling components in *Drosophila*. *Genes Dev.* **29**: 1861.

Bard F., Casano L., Mallabiabarrena A., Wallace E., Saito K.,

Kitayama H., Guizzunti G., Hu Y., Wendler F., Dasgupta R., et al. 2006. Functional genomics reveals genes involved in protein secretion and Golgi organization. *Nature* **439**: 604.

Bartscherer K., Pelte N., Ingelfinger D., and Boutros M. 2006. Secretion of Wnt ligands requires Evi, a conserved transmembrane protein. *Cell* **125**: 523.

Birmingham A., Anderson E.M., Reynolds A., Ilsley-Tyree D., Leake D., Fedorov Y., Baskerville S., Maksimova E., Robinson K., Karpilow J., et al. 2006. 3′UTR seed matches, but not overall identity, are associated with RNAi off-targets. *Nat. Methods* **3**: 199.

Björklund M., Taipale M., Varjosalo M., Saharinen J., Lahdenpera J., and Taipale J. 2006. Identification of pathways regulating cell size and cell-cycle progression by RNAi. *Nature* **439**: 1009.

Boutros M., Agaisse H., and Perrimon N. 2002. Sequential activation of signaling pathways during innate immune responses in *Drosophila*. *Dev. Cell* **3**: 711.

Boutros M., Kiger A.A., Armknecht S., Kerr K., Hild M., Koch B., Haas S.A., Paro R., and Perrimon N.; Heidelberg Fly Array Consortium. 2004. Genome-wide RNAi analysis of growth and viability in *Drosophila* cells. *Science* **303**: 832.

Cherry S., Doukas T., Armknecht S., Whelan S., Wang H., Sarnow P., and Perrimon N. 2005. Genome-wide RNAi screen reveals a specific sensitivity of IRES-containing RNA viruses to host translation inhibition. *Genes Dev.* **19**: 445.

Clemens J.C., Worby C.A., Simonson-Leff N., Muda M., Maehama T., Hemmings B.A., and Dixon J.E. 2000. Use of double-stranded RNA interference in *Drosophila* cell lines to dissect signal transduction pathways. *Proc. Natl. Acad. Sci.* **97**: 6499.

DasGupta R., Kaykas A., Moon R.T., and Perrimon N. 2005. Functional genomic analysis of the Wnt-wingless signaling pathway. *Science* **308**: 826.

Echeverri C.J. and Perrimon N. 2006. High-throughput RNAi screening in cultured cells: A user's guide. *Nat. Rev. Genet.* **7**: 373.

Eggert U.S., Kiger A.A., Richter C., Perlman Z.E., Perrimon N., Mitchison T.J., and Field C.M. 2004. Parallel chemical genetic and genome-wide RNAi screens identify cytokinesis inhibitors and targets. *PLoS Biol.* **2**: e379.

Feske S., Gwack Y., Prakriya M., Srikanth S., Puppel S.H., Tanasa B., Hogan P.G., Lewis R.S., Daly M., and Rao A. 2006. A mutation in Orai1 causes immune deficiency by abrogating CRAC channel function. *Nature* **441**: 179.

Foley E. and O'Farrell P.H. 2004. Functional dissection of an innate immune response by a genome-wide RNAi screen. *PLoS Biol.* **2**: E203.

Friedman A. and Perrimon N. 2004. Genome-wide high-throughput screens in functional genomics. *Curr. Opin. Genet. Dev.* **14**: 470.

Guertin D.A., Guntur K.V., Bell G.W., Thoreen C.C., and Sabatini D.M. 2006. Functional genomics identifies TOR-regulated genes that control growth and division. *Curr. Biol.* **16**: 958.

Gwack Y., Sharma S., Nardone J., Tanasa B., Iuga A., Srikanth S., Okamura H., Bolton D., Feske S., Hogan P.G., and Rao A. 2006. A genome-wide *Drosophila* RNAi screen identifies DYRK-family kinases as regulators of NFAT. *Nature* **441**: 646.

Hammond S.M., Bernstein E., Beach D., and Hannon G.J. 2000. An RNA-directed nuclease mediates post-transcriptional gene silencing in *Drosophila* cells. *Nature* **404**: 293.

Hild M., Beckmann B., Haas S.A., Koch B., Solovyev V., Busold C., Fellenberg K., Boutros M., Vingron M., Sauer F., et al. 2003. An integrated gene annotation and transcriptional profiling approach towards the full gene content of the *Drosophila* genome. *Genome Biol.* **5**: R3.

Jackson A.L., Bartz S.R., Schelter J., Kobayashi S.V., Burchard J., Mao M., Li B., Cavet G., and Linsley P.S. 2003. Expression profiling reveals off-target gene regulation by RNAi. *Nat. Biotechnol.* **21**: 635.

Kiger A., Baum B., Jones S., Jones M., Coulson A., Echeverri C., and Perrimon N. 2003. A functional genomic analysis of cell morphology using RNA interference. *J. Biol.* **2**: 27.

Kulkarni M.M., Booker M., Silver S.I., Friedman A., Hong P., Perrimon N., and Mathey-Prevot B. 2006. Evidence of off-target effects associated with long dsRNAs in *Drosophila* cell-based assays. *Nat. Methods* **3**: 833.

Levsky J.M., Shenoy S.M., Pezo R.C., and Singer R.H. 2002. Single-cell gene expression profiling. *Science* **297**: 836.

Lum L., Yao S., Mozer B., Rovescalli A., Von Kessler D., Nirenberg M., and Beachy P.A. 2003. Identification of Hedgehog pathway components by RNAi in *Drosophila* cultured cells. *Science* **299**: 2039.

Muller P., Kuttenkeuler D., Gesellchen V., Zeidler M.P., and Boutros M. 2005. Identification of JAK/STAT signalling components by genome-wide RNA interference. *Nature* **436**: 871.

Nybakken K., Vokes S.A., Lin T.Y., McMahon A.P., and Perrimon N. 2005. A genome-wide RNA interference screen in *Drosophila melanogaster* cells for new components of the Hh signaling pathway. *Nat. Genet.* **37**: 1323.

Pelech S. 2004. Tracking cell signaling protein expression and phosphorylation by innovative proteomic solutions. *Curr. Pharm. Biotechnol.* **5**: 69.

Pelkmans L., Fava E., Grabner H., Hannus M., Habermann B., Krausz E., and Zerial M. 2005. Genome-wide analysis of human kinases in clathrin- and caveolae/raft-mediated endocytosis. *Nature* **436**: 78.

Philips J.A., Rubin E.J., and Perrimon N. 2005. *Drosophila* RNAi screen reveals CD36 family member required for mycobacterial infection. *Science* **309**: 1251.

Piano F., Schetter A.J., Morton D.G., Gunsalus K.C., Reinke V., Kim S.K., and Kemphues K.J. 2002. Gene clustering based on RNAi phenotypes of ovary-enriched genes in *C. elegans*. *Curr. Biol.* **12**: 1959.

Ramet M., Manfruelli P., Pearson A., Mathey-Prevot B., and Ezekowitz R.A. 2002. Functional genomic analysis of phagocytosis and identification of a *Drosophila* receptor for *E. coli*. *Nature* **416**: 644.

Sachs K., Perez O., Pe'er D., Lauffenburger D.A., and Nolan G.P. 2005. Causal protein-signaling networks derived from multiparameter single-cell data. *Science* **308**: 523.

Schlesinger A., Kiger A., Perrimon N., and Shilo B.Z. 2004. Small wing PLCgamma is required for ER retention of cleaved Spitz during eye development in *Drosophila*. *Dev. Cell* **7**: 535.

Stolc V., Gauhar Z., Mason C., Halasz G., van Batenburg M.F., Rifkin S.A., Hua S., Herreman T., Tongprasit W., Barbano P.E., et al. 2004. A gene expression map for the euchromatic genome of *Drosophila melanogaster*. *Science* **306**: 655.

Ulvila J., Parikka M., Kleino A., Sormunen R., Ezekowitz R.A., Kocks C., and Ramet M. 2006. Double-stranded RNA is internalized by scavenger receptor-mediated endocytosis in *Drosophila* S2 cells. *J. Biol. Chem.* **281**: 14370.

Vig M., Peinelt C., Beck A., Koomoa D.L., Rabah D., Koblan-Huberson M., Kraft S., Turner H., Fleig A., Penner R., and Kinet J.P. 2006. CRACM1 is a plasma membrane protein essential for store-operated Ca2+ entry. *Science* **312**: 1220.

Wheeler D.B., Carpenter A.E., and Sabatini D.M. 2005. Cell microarrays and RNA interference chip away at gene function. *Nat. Genet.* (suppl.) **37**: S25.

Wheeler D.B., Bailey S.N., Guertin D.A., Carpenter A.E., Higgins C.O., and Sabatini D.M. 2004. RNAi living-cell microarrays for loss-of-function screens in *Drosophila melanogaster* cells. *Nat. Methods* **1**: 127.

Yandell M., Bailey A.M., Misra S., Shu S., Wiel C., Evans-Holm M., Celniker S.E., and Rubin G.M. 2005. A computational and experimental approach to validating annotations and gene predictions in the *Drosophila melanogaster* genome. *Proc. Natl. Acad. Sci.* **102**: 1566.

Yao S., Lum L., and Beachy P. 2006. The ihog cell-surface proteins bind Hedgehog and mediate pathway activation. *Cell* **125**: 343.

Zhang S.L., Yeromin A.V., Zhang X.H., Yu Y., Safrina O., Penna A., Roos J., Stauderman K.A., and Cahalan M.D. 2006. Genome-wide RNAi screen of Ca2+ influx identifies genes that regulate Ca2+ release-activated Ca2+ channel activity. *Proc. Natl. Acad. Sci.* **103**: 9357.

Deep Conservation of MicroRNA-target Relationships and 3'UTR Motifs in Vertebrates, Flies, and Nematodes

K. Chen* and N. Rajewsky*†

*Center for Comparative Functional Genomics, Department of Biology, New York University, New York, New York 10003; †Max Delbrück Centrum for Molecular Medicine, Berlin-Buch, 13092 Berlin, Germany

microRNAs (miRNAs) are a class of small noncoding RNAs that posttranscriptionally regulate a large fraction of genes in animal genomes. We have previously published computational miRNA target predictions in five vertebrates, six flies, and three nematodes. Here, we report a comprehensive study of the "deep" conservation of miRNA targets and conserved 3'UTR (untranslated region) motifs in general across vertebrates, flies, and nematodes. Our data indicate that although many miRNA genes and 3'UTR motifs are well-conserved, miRNA-target relationships have diverged more rapidly, and we explicitly assign each gained or lost miRNA-target relationship to one of the three clades. However, we also identify a small but significant number of deeply conserved miRNA targets and show that these are enriched for essential processes related to development. Finally, we provide lists of 3'UTR motifs that are significantly conserved, and thus likely functional, classified by their distribution in the three clades. We find hundreds of such motifs specific to each clade, dozens specific to each pair of clades, and ten shared by vertebrates, flies, and nematodes. These findings suggest that posttranscriptional control has undergone extensive rewiring during metazoan evolution and that many deeply conserved miRNA-target relationships may be vital subunits of metazoan gene regulatory networks.

The main interest of our lab is to identify and characterize gene regulatory elements and to ultimately arrive at a better understanding of gene regulatory networks. A recent focus of our lab has been gene regulation mediated by miRNAs. miRNAs are a class of small noncoding RNAs that posttranscriptionally regulate a large fraction of genes in animal genomes. To understand the function of miRNAs, it is necessary to identify and characterize their targets. We have previously published computational miRNA target predictions in five vertebrates, six flies, and three nematodes. For example, we and other groups have shown that at least 30% of all human genes are likely to be regulated by about 60 conserved vertebrate miRNAs. Moreover, in collaboration with experimental groups, we helped to determine the biological function of a few miRNAs (for a general review, see Rajewsky 2006). However, since many miRNAs are seemingly specific to certain metazoan clades, whereas others are conserved in virtually all animals, miRNAs and their targets are also an excellent system in which to study the evolution of a whole layer of gene regulation. For example, it seems possible to predict miRNA targets that are specific to a certain lineage within a metazoan clade (e.g., the *Sophophora* and *Drosophila* lineages within flies). Here, we focus on "deep" conservation of miRNA-mediated gene regulation, i.e., targets of conserved miRNAs that are shared by vertebrates, flies, and nematodes.

Although it is well known that many miRNAs are well-conserved across large evolutionary distances (Lagos-Quintana et al. 2001), the conservation of miRNA targets has only been studied in a few anecdotal cases. Notably, Pasquinelli et al. (2000) showed that *let-7* and one of its targets, *lin-41*, are broadly conserved in animals, Moss and Tang (2003) showed that *lin-28* is a conserved target of *let-7* and *lin-4* in mammals and nematodes, whereas

Floyd and Bowman (2004) and Axtell and Bartel (2005) showed that a number of miRNA-target relationships are conserved in plants.

Here, we undertake a systematic study of miRNA-target conservation among vertebrates, flies, and nematodes, using computational target predictions from our previously published PicTar algorithm (Grun et al. 2005; Krek et al. 2005; Lall et al. 2006). Recently, an independent, large-scale experimental study estimated that approximately 90% of PicTar predictions in *Drosophila* are correct at a sensitivity of 70% (Stark et al. 2005). We discovered 5 miRNA-target relationships conserved in all three clades and 264 more miRNA-target relationships conserved in two clades (for these numbers, families of paralogous miRNAs and target genes are collapsed to a single representative).

We refer to a gene that is predicted to be a target of the same miRNA in at least two clades as a "deeply conserved target." The set of such targets is significantly enriched for genes involved in essential biological processes related to development (P value $<5.8e-3$ in humans and flies) (see Materials and Methods). Among the most interesting cases, we recovered five different subunits of vacuolar ATPase as deeply conserved targets of the *miR-1* family, *lin-28* as a deeply conserved target of the *let-7* family, and *odd-skipped* as a conserved target of the *miR-8* family.

Despite the suggestive biological significance of the deeply conserved miRNA targets, since PicTar predicts a very large number of miRNA targets in each of the three genomes (9379 in humans, 3082 in *D. melanogaster*, and 2679 in *Caenorhabditis elegans*), our results imply that miRNA targets are poorly conserved overall. In principle, this result could be due to three scenarios. First, the target predictions could simply be erroneous. Second, the ancestral miRNAs could have had only a few targets and these targets have indeed been conserved, but many more targets have been gained subsequent to speciation, leading to an apparent lack of conservation of the ancestral targets. Third, the ancestral miRNA could have had many targets, and the network of miRNA-target relationships has undergone extensive rewiring during metazoan evolution. Our analysis suggests that the third scenario is most likely to be correct, and we speculate that rewiring of miRNA networks, like rewiring of transcription factor networks, may be important in bilaterian evolution and diversification (Davidson 2001). We stress the importance of analyzing miRNA conservation in at least three clades for distinguishing between the latter two evolutionary scenarios, since pairwise comparisons cannot discriminate between a gain and loss of an miRNA target.

We complement this analysis with a study of conserved 3'UTR motifs in vertebrates, flies, and nematodes, using techniques derived from Xie et al. (2005). The results of Xie et al. imply that conserved 3'UTR motifs in vertebrates are highly enriched for miRNA-binding sites, and we show that these results extend to flies and nematodes. In addition, we discovered a significant correlation between the patterns of motif conservation in these three clades, which implies that *cis*-regulatory motifs in 3'UTRs have remained well-conserved across very large evolutionary distances. We find that this correlation is strongest for human and flies and weakest for human and nematodes. We classify the hundreds of significantly conserved 3'UTR motifs that we have discovered according to their distribution in the three clades and hypothesize that many of these are likely to be functionally important *cis*-regulatory sequences.

In addition, our data are consistent with the interpretation that many of the most highly conserved miRNAs have already been discovered in these three clades, meaning that many of the remaining ones to be found are clade-specific. Of the three clades, the miRNA gene complement of *D. melanogaster* may be the most poorly sampled of the three.

Subsequent to obtaining these results, a similar computational study of miRNA target conservation in flies and nematodes appeared (Chan et al. 2005). Our work differs in several respects. First, we study three clades instead of two. This is interesting not only because we include many more species in our analysis and thus discover about three times as many potential deeply conserved regulatory relationships, but also because we can use the third clade as an outgroup in our evolutionary studies. In particular, we are able to differentiate between evolutionary scenarios two and three described above. Second, our target predictions rely on previously experimentally validated methods (Grun et al. 2005; Krek et al. 2005; Lall et al. 2006), whereas those of Chan et al. (2005) have not yet been subjected to experimental verification. For example, our fly 3'UTRs are defined based on full-length cDNAs and not artificially truncated to 500 nucleotides. As part of this work, we also attempted to predict binding sites in genes conserved across all three clades without requiring that they be aligned, exactly as in Chan et al. (2005). However, we failed to find any excess of predicted targets in real miRNAs versus randomized controls. This is in stark contrast to the strong signal-to-noise ratios observed in alignment-based target prediction methods (Lewis et al. 2003; Brennecke et al. 2005b; Grun et al. 2005; Krek et al. 2005; Lall et al. 2006). Third, our motif conservation score is defined differently and relies on methods previously used successfully for motif discovery in vertebrates (Xie et al. 2005).

MATERIALS AND METHODS

Construction of miRNA families. We clustered all miRNAs with PicTar target predictions by linking two miRNAs if they shared a nucleus (the first or second 7-mer) and applying single-linkage clustering. PicTar predictions are only generated for miRNAs conserved in all species under consideration. (For details, see Grun et al. 2005; Krek et al. 2005.)

Motif analysis. For the vertebrate alignment, we used repeat-masked UCSC alignments of the human, mouse, rat, and dog genomes downloaded from the UCSC genome browser (http://genome.ucsc.edu) as described previously by Krek et al. (2005). For the fly alignment, we used tandem-repeat-masked Mercator/MAVID alignments (Bray and Pachter 2003; http://hanuman.math.berkeley.edu/~cdewey/mercator) of *D. melanogaster*, *D. erectus*, *D. ananassae*, *D. yakuba*, and *D. pseudoobscura*, and for the nematode alignment, we used unmasked Mercator/MAVID alignments of *C. elegans*, *C. briggsae*, and *C. remanei* as described previously by Grun et al. (2005) and Lall et al. (2006). For all genes with multiple transcript variants, we kept only the transcript with the longest 3'UTR.

We experimented with different conservation scores, using as our metric the average score of all known miRNAs. We attempted to improve sensitivity by excluding very short UTRs (<200 nucleotides for vertebrates and <100 nucleotides for flies and worms), dividing by the average total count in all species, instead of the total count in just the reference species, and considering 5-, 6-, and 8-mers in the analysis, but we found that none of these performed as well as the simplest measure. When matching the motifs against miRNA sequences, we matched against the entire Rfam7.0 data set (Griffiths-Jones et al. 2003), not just the miRNAs that had PicTar predictions.

For the purposes of viewing the scatter plots (Fig. 1), we added the chicken genome to the vertebrate alignments, and for the fly alignments, we used repeat-masked UCSC alignments of *D. melanogaster*, *D. ananassae*, *D. yakuba*, *D. pseudoobscura*, *D. mojavensis*, and *D. virilis*. Visually, the additional species tend to accentuate the "arms" of the scatter plot, but otherwise, the overall patterns of conservation are not affected. In addition, to minimize noise, we eliminated all motifs that have a conserved count of zero. Intuitively, small counts induce high variance in the Z scores and tend to blur the "arms" of the scatter plots. Finally, we took the absolute value of the Z scores instead of the raw Z scores because we were concerned only with overrepresentation of conserved occurrences.

Conservation of targets. For each miRNA family, we considered the overlap between each pair of clades separately. For each clade, we took all the miRNAs in the family from that clade, as well as the union of their target sets, removing all but one copy of each set of paralogs to get a unique set of target genes for the clade. We then computed the overlap rate as the intersection of the two unique sets divided by the cardinality of the smaller of the two sets. This approach was designed to maximize sensitivity for conserved targets and does not penalize for target genes with no ortholog in the other clade.

The null model that we considered is that of sampling uniformly at random and independently without replacement from the set of all genes with orthologs in the other clade, keeping the number of unique genes (i.e., no paralogs) in the two sets constant. In this model, we assumed that 3′UTRs of homologous genes in different clades were sufficiently diverged as to be statistically independent. To compute the expected number of targets in the null model, we assumed that there were N clusters of homologs between two clades and the target sets were of size $j \leq k$. Fixing the number of targets in one clade to be j, by linearity of expectation, the number of overlaps is $k * j / N$ (recall that independence is not required for the linearity of expectation), which implies that the overlap rate is k / N. For three-way comparisons with target sets of size i, j, k, and N homologs between the three clades, $i * j * k / N^2$ so the rate is $j*k/N^2$. Another way to derive this is by taking the expectation of an appropriate hypergeometric distribution.

To compute the overall average rate, we took the unweighted average of the per-family overlap rate. An alternative is to weight the rates by the size of the smaller of the two reduced sets, thus reducing the effect of noise from small sets. When we compute the overall average rates this way, the numbers are almost exactly the same.

To identify families with significantly high conservation of targets, we computed P values for Fisher's exact test using code from the GeneMerge program (Castillo-Davis and Hartl 2003) and the multiple testing correction using the Multtest package for the R programming language from the Bioconductor project. To identify additional putative triply conserved target relationships that were missed by PicTar, we took all doubly conserved target relationships for which an orthologous miRNA and orthologous target gene existed in the third clade. We did not collapse paralogous target genes since not all of these are expected to contain the binding site. There were 265 genes that fit this category, many of which were targeted by multiple 6-mers; 73 genes did not appear in our alignments at all, leaving a total of 241 gene–6-mer pairs to check. For each of these 241 pairs, we checked each 3′UTR from the relevant species (human, mouse, rat, dog, *D. melanogaster, D. ananassae, D. pseudoobscura, D. yakuba, C. elegans, C. briggsae, C. remanei*) for the presence of the 6-mer without requiring it to be aligned, and we considered a gene to be a putative target gene if at least three of the species contained the 6-mer, one of which had to be the reference species. To compute the expected number of hits, we concatenated the lengths of the 3′UTRs (without removing repeat-masked sequence) and divided by 4^6.

In our GO term analysis, we used as our background set the set of all PicTar targets that have orthologs in one of the other two clades, and which are themselves targeted by miRNAs from multiclade families. In the evolutionary analysis, we assumed a star phylogeny and used the simple parsimony criteria to assign each target relationship conserved in either one or two clades as a gain or loss, respectively.

RESULTS

Conservation of miRNA Families

We took all human, *D. melanogaster*, and *C. elegans* miRNAs for which PicTar target predictions are available and clustered them into families of homologous miRNAs (see Materials and Methods). Our clustering method relies on the notion of a critical region for target recognition, the "nucleus," defined here to be the first or second 7-mer in the mature miRNA sequence as described by Krek et al. (2005). Recent research has established that the nucleus is the most important element of miRNA-target-binding specificity (Lewis et al. 2003; Doench and Sharp 2004; Brennecke et al. 2005b). PicTar uses this model to make most of its predictions, but it also uses the binding free energy to predict imperfect (3′ compensatory) sites. Careful examination of the families indicated that nearly all of them have very good alignments across the entire length of the mature sequence and that they are generally concordant with the human–nematode families defined by Lim et al. (2003).

The large number of families conserved in all three clades (15) is particularly striking since it is larger than the number of families conserved in any pair of clades (5 human–fly, 5 human–nematode, and 9 fly–nematode), which suggests that there is significant conservation of miRNA genes across the clades. A closer examination of the families themselves revealed that many are very well conserved at the sequence level beyond matching at the nucleus. For example, four of the five families specific to humans and flies (*miR-219, miR-7, miR-210, miR-184*) are perfectly conserved between these two clades, and in the case of *miR-7*, the genomic location of the gene inside the last intron of the *D. melanogaster* bancal gene is conserved.

As a negative control for our clustering method, we repeated our procedure with all miRNAs from humans, *D. melanogaster*, and *C. elegans*, as well as three plant species *Arabidopsis thaliana, Oryza sativa,* and *Zea mays.* We found that all animal and plant miRNAs clustered separately, with the exception of one cluster made up of a human miRNA and two *O. sativa* miRNAs, which is likely to be a coincidence. This suggests that the families as defined represent true evolutionary homologs.

CONSERVATION OF 3′UTR MOTIFS

To further investigate the evolution of miRNAs, we examined the conservation of 3′UTR motifs in these three clades, using techniques adapted from Xie et al. (2005).

We used the same 3′UTR alignments that were used for PicTar target predictions to identify 7-mers whose rate of conservation was significantly higher than that of random 7-mers. Formally, for each clade, we computed a conservation score for each 7-mer, defined as the number of conserved instances in the 3′UTR alignment divided by the total number of instances in the reference species (*Homo sapiens*, *D. melanogaster*, or *C. elegans*, respectively). We note that our analysis differs from that of Xie et al. (2005) in that they consider motifs of length 6–18 and perform clustering of similar motifs, whereas we consider only 7-mers (i.e., potential miRNA-binding sites).

For each clade, we compute a Z score for each motif based on the distribution of conservation scores, and we call a motif with a Z score >3 a highly conserved motif (HCM). In each clade, HCMs are highly enriched for miRNA-binding sites. In vertebrates, 64 of 206 HCMs correspond to binding sites; in flies, 50 of 206 HCMs correspond to binding sites; and in worms, 42 of 223 HCMs correspond to binding sites. This analysis extends the results of Xie et al. (2005) to flies and nematodes.

Recently, Bentwich et al. (2005) identified 53 candidate primate-specific miRNAs. We extracted these and computed the distribution of Z scores of their nuclei. Surprisingly, at first, the distribution of the 69 unique nuclei was not significantly different from that of all human miRNAs in a two-tailed Wilcoxon test, as would be expected if these nuclei were indeed primate-specific. However, upon closer investigation, we found that 10 of these nuclei matched to other (nonprimate-specific) human miRNAs or miRNAs in *D. melanogaster* or *C. elegans*. Excluding these k-mers from the analysis, we found that the average Z score of the remaining 59 primate-specific nuclei was 0.71 versus 1.34 for other human miRNAs and that this difference is statistically significant in a one-tailed Wilcoxon test (*P* value 0.013). This suggests that some of the primate-specific genes are paralogous to miRNAs conserved more broadly in mammals, or even in flies or nematodes.

When we considered HCMs that have a Z score >3 in more than one clade, we found that these HCMs are very highly enriched for conserved miRNAs. Of 36 HCMs conserved in both vertebrates and flies, 19 match to known miRNAs, 16 of which are conserved in both humans and *D. melanogaster*. Comparable numbers were seen in the other two-way clade comparisons (21 of 34 HCMs conserved in flies and worms matched miRNAs, of which 16 were conserved in both, and 13 of 20 HCMs conserved in humans and worms matched miRNAs, of which 10 were conserved in both). We present these results for the human–worm comparison as a scatter plot in Figure 1, along with a random control in which the motif-to-Z score assignments have been randomly permuted in all three clades (see Materials and Methods). The scatter plots for the other two pairwise comparisons and a list of all HCMs are available upon request from the authors.

The scatter plots indicate a high degree of correlation between the conservation of 3′UTR motifs in these three clades, as well as an extremely high enrichment for conserved miRNA-binding sites in the HCMs conserved in both clades. This can be visualized by comparison with randomized controls in which the Z scores for the k-mers have been randomly permuted. These control plots show very few outliers outside of the two "arms" corresponding to the clade-specific motifs, as might be expected from a scatter plot of two independent Poisson distributions, and virtually all miRNAs lie in the insignificant regime. A comparison of the human–nematode and human–fly scatter plots shows that the two "arms" are more prominent in the human–nematode plot, indicating that the motifs in these two clades are less correlated. Quantitatively, the Pearson correlation coefficients for humans versus flies, humans versus worms, and flies versus worms are 0.39, 0.12, and 0.29, respectively.

Extending this analysis to a comparison of all three clades, we found that of the ten HCMs conserved in all three clades, eight matched some miRNA and seven matched in all three clades. Two of the most well-known conserved miRNA families, *let-7* and *miR-1*, appear as HCMs conserved in all three clades. The two motifs that remain unaccounted for are GTAAATA and AGTGCCT.

GTAAATA matches to a predicted miRNA precursor sequence in *C. elegans* with an miRScanII score of 9.23 bits (Ohler et al. 2004). Only precursors of score >12.7 bits were experimentally screened in Ohler et al. (2004). It also matches the first seven bases of the 32nd highest scoring 8-mer in humans (Xie et al. 2005), which in turn matches one novel predicted miRNA precursor. The highest score of any predicted precursor sequence in *C. elegans* containing AGTGCCT is only 0.39 bits, and the motif is not found in any predicted mature miRNA sequence in humans from Xie et al. (2005).

When we examined the number of HCMs in each clade that match an miRNA from one of the other two clades, we found only 3 and 9 such HCMs for humans and *C. elegans*, respectively, but 19 for *D. melanogaster*. This suggests that many of the most highly conserved miRNAs have already been discovered in humans and *C. elegans*, whereas the *D. melanogaster* miRNA complement may be the most poorly sampled of the three. This inference is consistent with the small total number of *D. melanogaster* miRNA families as compared to worms and humans (Fig. 1), as well as the relatively low amount of effort expended on miRNA gene finding in *D. melanogaster* thus far. It is also consistent with the work of Bentwich et al. (2005), which suggests that many of the remaining human miRNAs are primate-specific. At the practical level, our results imply that miRNA gene-finding techniques relying on comparisons over moderate phylogenetic distances should continue to have success in *D. melanogaster*, whereas phylogenetic shadowing techniques using comparisons over much shorter phylogenetic distances (Berezikov et al. 2005) may be required in human and *C. elegans*.

Our motif analysis is consistent with the currently accepted view of miRNA-target-binding specificity, in that in all three clades, the second 7-mer was more conserved than the first 7-mer. In contrast, using a different measure of conservation, Chan et al. (2005) found that the first 7-mer is more conserved in flies but the second 7-mer is more conserved in worms. TargetScanS (Lewis et al. 2003), a different target prediction program for ver-

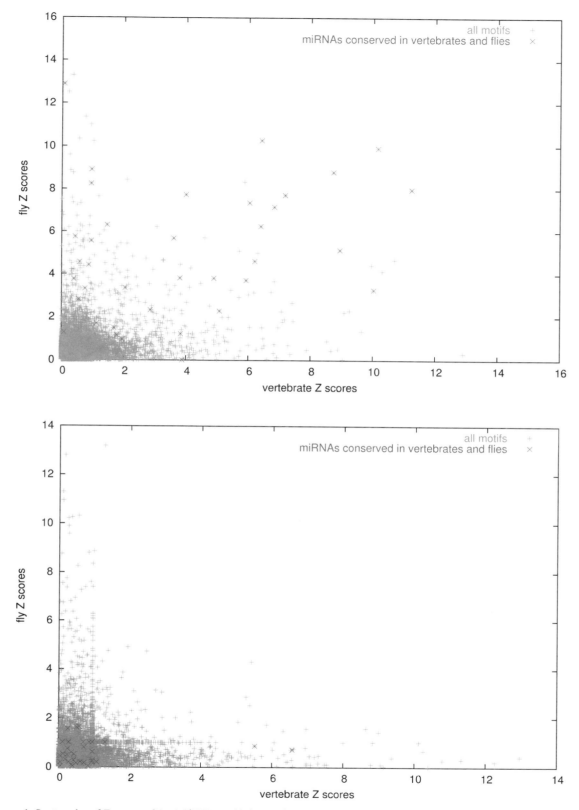

Figure 1. Scatter plot of Z scores of (*top*) 3′UTR motifs in vertebrates and flies and (*bottom*) 3′UTR motifs in vertebrates and flies with Z score-to-motif assignments randomly permuted. miRNAs found in both vertebrates and flies are marked with an X.

tebrates, uses a conserved adenosine in the first position regardless of the nucleotide at the first position in the miRNA. Although we find that an adenosine in the first position is indeed more conserved than the complementary nucleotide in humans and flies, this does not extend to nematodes. Although it is possible that the mechanism of miRNA binding is different in these clades, we have instead chosen not to use the conserved adenosine in any of our predictions.

Conservation of miRNA-target Relationships

Given the high degree of miRNA conservation observed so far, we turned next to the conservation of miRNA-target relationships. For this analysis, we used target predictions from the PicTar algorithm. Briefly, PicTar is a probabilistic algorithm that takes as input 3′UTR alignments from several related species and a set of miRNAs and computes a maximum likelihood parse of the 3′UTR sequence into binding sites and background sequence (Grun et al. 2005; Krek et al. 2005; Lall et al. 2006; http://pictar.bio.nyu.edu).

Our overall strategy was to maximize sensitivity for conserved miRNA-target relationships at the cost of specificity. To this end, we first chose the most sensitive PicTar settings available, which involved using nonrepeat-masked alignments of fewer species (human, mouse, rat, and dog for vertebrate predictions; *D. melanogaster*, *D. ananassae*, *D. pseudoobscura*, and *D. yakuba* for the fly predictions; and *C. elegans*, *C. briggsae*, and *C. remanei* for the nematode predictions). For the species used in this study, the estimated signal-to-noise ratios for the PicTar algorithm were 2.3 in humans, 2.5 in *D. melanogaster*, and 2.7 in *C. elegans*.

We downloaded pairwise sets of homologous genes for the human, *D. melanogaster*, and *C. elegans* genomes from the Inparanoid database (O'Brien et al. 2005) and augmented these with additional homologous genes from the Ensembl database (http://www.ensembl.org). We clustered all of these together into three-way homology genes using single-linkage clustering. A comparison with a previously published analysis of human and *Drosophila* targets (Grun et al. 2005) gave a very good overlap, although the homologous gene sets used were different, indicating that the quality of homology detection, even across clades as distant as these, is not a major factor in the analysis.

For each family of miRNAs, we took the union of the target sets of each of the miRNAs in a clade as the representative set. As noted previously, most of the miRNA target sets are expected to be very similar within each family since the miRNAs typically share the same nucleus, so this procedure does not inflate the number of targets much. We computed the overlap for each pair or triple of clades, counting paralogs of miRNAs or target genes only once and computing the percentage overlap as the ratio of the number of overlapping genes to the size of the smaller of the two sets of genes. We only counted genes in the representative sets that had orthologs in the other clade for the purposes of this calculation, so that miRNAs were not penalized for having many clade-specific genes.

Despite all efforts to increase sensitivity, we reached the surprising conclusion that the percentage of overlap was very low in all cases. Averaging over approximately 20 families for each pair of clades, the percentage of overlapping targets is 10% for humans and *D. melanogaster*, 11% for human and *C. elegans*, 4% for *D. melanogaster* and *C. elegans*, and 0.7% for all three clades. In a random model where the target genes are sampled randomly from the set of all genes with orthologs, we expect to see the percentage of overlapping targets to be 4% for humans and *D. melanogaster*, 5% for humans and *C. elegans*, 2% for *D. melanogaster* and *C. elegans*, and 0.09% for all three clades. Thus, we conclude that the number of conserved targets is slightly higher than random, but still small. A list of all conserved miRNA-target relationships is available by request from the authors.

For each family and pair of clades, we tested the significance of the overlap by computing a P value for Fisher's exact test and correcting for multiple hypothesis testing with the Bonferroni correction. Four families had conservation statistically greater than that expected at random (P value <0.04). Using their *D. melanogaster* representatives, these are the *miR-263b*, *miR-8*, *miR-7*, and *miR-92* families (see Materials and Methods for the other members of the families). Among the families showing significant overlap, we single out the *miR-92* family, which contains by far the largest number of conserved targets of any family (32 conserved targets between vertebrates and flies). Many of the conserved targets are important developmental transcription factors, such as *engrailed*, *mef2*, *crooked legs*, *erect wing*, *held out wings*, *spalt major*, *grain*, *E2f2*, *antennapedia*, *longitudinals lacking*, and *jun-related antigen*. GO term analysis of both the human and *D. melanogaster* target sets shows significant enrichment for transcription factors (P value <0.0001; see Materials and Methods). We point out that this family has been suggested to be important in *Drosophila* development (Leaman et al. 2005) but also note that this result has recently been put to question (Brennecke et al. 2005a).

An analysis of the conserved targets shows that they are significantly enriched for genes involved in development in flies, as compared to the set of all target genes with orthologs and which are targeted by orthologous miRNAs. These include morphogenesis (P value 2.4e-3), organ development (P value 2.0e-3), development (P value 1.8e-4), and histogenesis (P value 7.2e-3). A similar result for flies was seen previously by Grun et al. (2005). In humans, conserved targets are enriched for development (P value 5.8e-3) and amino acid phosphorylation (P value 4.3e-3). A few specific conserved miRNA-target relationships are of particular interest. The muscle-specific *miR-1* targets a subunit of vacuolar ATPase in all three clades, and four other subunits of vacuolar ATPase are conserved targets in two clades. *let-7* targets *lin-28* in both humans and worms, whereas *odd-skipped* is a conserved target of the *miR-8* family in all three clades.

We investigated the possibility that binding sites are present in 3′UTRs but not detected by PicTar, since the binding sites may have undergone rearrangements and therefore are not aligned correctly (Krutzfeld et al. 2005). We identified all miRNA-target relationships conserved in exactly two clades for which a homologous miRNA and a homologous target gene existed in the third clade. In each of these cases, we searched all the 3′UTRs of the homologous target genes in the third clade for the presence of a 6-mer-binding site (positions 2–7 from the 5′ end), without requiring that the binding site be conserved in the alignment. These requirements are significantly less strict than those implemented in the PicTar algorithm, which requires aligned 7-mer-binding sites, and are closer to the target prediction requirements of Chan et al. (2005).

After excluding those genes that do not appear in our alignments, we were left with 241 pairs of genes and k-mers to test (for this analysis, we kept all paralogous target genes, since in many cases, only a subset of paralogous genes contain a binding site). Among these, we discovered only 14 other potential triply conserved target relationships. In only two cases was a 7-mer nucleus present in all species but not conserved in the alignment. The small number of such cases suggests (1) that targets are indeed lost, and not simply missed by the algorithm, and (2) that our target prediction methods seem to be robust to rearrangements in the positions of the binding sites.

Interestingly, three of the five new *D. melanogaster* potential targets are targets of *miR-4/miR-79* and the other two are targets of *miR-8*, whereas all four of the new *C. elegans* potential targets are targets of *miR-235*. This suggests that the binding mechanism of some miRNAs may be different from that of other miRNAs (i.e., some miRNAs may require only a 6-mer nucleus or their target recognition sequence could depend on their nucleotide composition). This is potentially a valuable observation to incorporate into future target prediction algorithms.

One case of particular interest is the vacuolar-ATPase complex in which five subunits are conserved in at least two of the three clades. We failed to find binding sites in the third clade even with our relaxed requirements, or even when requiring only an unaligned 5-mer-binding site (positions 2–6). One possible reason for this is that down-regulation of a few subunits is sufficient to down-regulate the entire protein complex.

Taken together, these results demonstrate that many miRNA target sites are indeed lost over large evolutionary timescales. As previously discussed, a pairwise comparison of clades cannot distinguish between a loss or gain of an miRNA target, whereas the three-way comparison we perform here makes this distinction clear and implies that miRNA regulatory networks have undergone extensive rewiring. The results also imply that the low conservation of regulatory relationships is not due to overly stringent target prediction criteria.

CONCLUSIONS

We have performed a systematic study of the conservation of miRNAs, 3′UTR motifs, and miRNA targets in vertebrates, flies, and nematodes and have shown that although miRNAs and 3′UTR motifs are well-conserved, miRNA targets have diverged far more rapidly. We have

also argued that miRNA targets are indeed lost and gained over evolutionary timescales (as opposed to being just gained), implying a certain amount of flexibility in the network of miRNA regulation.

In this section, we argue that the lack of target conservation we observe is unlikely to be due to erroneous target predictions, but instead is likely to reflect biological reality. First, to address the possibility that our miRNA families do not represent true evolutionary homologs, we repeated our analysis on only a subset of 13 human–*Drosophila* miRNA families that contain essentially perfect alignments across the entire mature sequence and for which homology is clear and obtained exactly the same result. Second, to address the problem of false positives in the PicTar predictions, we observe that since the signal-to-noise ratios are all approximately 2, false positives affect the overlap rate by at most an expected factor of 2, which is still a small overlap.

A third and more serious problem is that of false negatives. Indeed, we estimate that 25–30% of targets are either not conserved in all the species studied or are missed due to details of the algorithm (Stark et al. 2005; N. Rajewsky, unpubl.). However, we argue that these as yet undiscovered targets would not change the overlap rate in expectation, assuming their rate of deep conservation is similar to the current set of predicted targets. In fact, because the PicTar algorithm depends in large part on strict conservation over quite a broad range of species, it is much more probable that the current set of targets is heavily biased toward the most conserved genes, and hence our overlap rate is in fact an overestimate in this regard.

The idea that the miRNAs themselves are well-conserved, whereas their targets have changed rapidly, is plausible since miRNAs typically target hundreds of genes; thus, a small change in an miRNA would therefore affect many genes, whereas the change in a single target would have a much smaller effect. We note that it is relatively easy to lose an miRNA target site (typically, it would take just one point mutation), whereas by comparison, it is more difficult to destroy a transcription-factor-binding site because these sites tend to accommodate more degeneracy. Future quantitative comparison of the evolution of miRNA-target sites, as compared to transcription-factor-binding sites, could reveal whether post-transcriptional regulatory networks mediated by miRNAs evolve faster or slower than transcriptional regulatory networks. Similarly, it will be interesting to see how the evolution of these networks relates to other types of biological networks, such as protein–protein interaction networks (Sharan et al. 2005).

The evolution of gene regulatory networks over the very large evolutionary distances we consider here has rarely been studied at the global level, largely because of the difficulties in predicting binding sites for transcription factors and posttranscriptional regulators. Because of the comparatively simple nature of miRNA-target prediction, we are able to suggest a high-level view of the evolution of gene regulatory networks in metazoans. Our results indicate that the regulators themselves (the miRNAs) appear to be well-conserved, whereas regulatory relationships between miRNAs and target genes have changed more rapidly. Nonetheless, a small core of developmentally important regulatory relationships appears to be conserved and thus may be a crucial component of the metazoan regulatory network. In addition, we identify clade-specific conserved 3'UTR motifs that may be functionally important for posttranscriptional regulation.

ACKNOWLEDGMENTS

We thank Uwe Ohler for providing us with the predicted *C. elegans* miRNA precursor sequences from the MiRscanII pipeline. This research was supported in part by the Howard Hughes Medical Institute grant through the Undergraduate Biological Sciences Education Program to New York University. In addition, K.C. thanks Lior Pachter for generous support (National Institutes of Health grant R01-HG02362-03).

REFERENCES

Axtell M.J. and Bartel D.P. 2005. Antiquity of microRNAs and their targets in land plants. *Plant Cell* **17:** 1658.

Bentwich I., Avniel A., Karov Y., Aharonov R., Gilad S., Barad O., Barzilai A., Einat P., Einav U., Meiri E., et al. 2005. Identification of hundreds of conserved and nonconserved human microRNAs. *Nat. Genet.* **37:** 766.

Berezikov E., Guryev V., van de Belt J., Wienholds E., Plasterk R.H., and Cuppen E. 2005. Phylogenetic shadowing and computational identification of human microRNA genes. *Cell* **120:** 21.

Bray N. and Pachter L. 2003. MAVID: Constrained ancestral alignment of multiple sequences. *Genome Res.* **14:** 693.

Brennecke J., Stark A., and Cohen S.M. 2005a. Not miR-ly muscular: microRNAs and muscle development. *Genes Dev.* **19:** 2261.

Brennecke J., Stark A., Russell R.B., and Cohen S.M. 2005b. Principles of microRNA-target recognition. *PloS Biol.* **3:** e85.

Castillo-Davis C.I. and Hartl D.L. 2003. GeneMerge—Postgenomic analysis, data mining, and hypothesis testing. *Bioinformatics* **19:** 891.

Chan C.S., Elemento O., and Tavazoie S. 2005. Revealing posttranscriptional regulatory elements through network-level conservation. *PLoS Comput. Biol.* **1:** e69.

Davidson E.H. 2001. *Genomic regulatory systems: Development and evolution.* Academic Press, New York.

Doench J.G. and Sharp P.A. 2004. Specificity of microRNA target selection in translational repression. *Genes Dev.* **18:** 504.

Floyd S.K. and Bowman J.L. 2004. Gene regulation: Ancient microRNA target sequences in plants. *Nature* **428:** 485.

Griffiths-Jones S., Bateman A., Marshall M., Khanna A., and Eddy S.R. 2003. Rfam: An RNA family database. *Nucleic Acids Res.* **31:** 439.

Grun D., Wang Y., Langenberger D., Gunsalus K.C., and Rajewsky N. 2005. microRNA target predictions across seven *Drosophila* species and comparison to mammalian targets. *PLoS Comput. Biol.* **1:** e13.

Krek A., Grun D., Poy M.N., Wolf R., Rosenberg L., Epstein E.J., MacMenamin P., de Piedade I., Gunsalus K.C., Stoffel M., and Rajewsky N. 2005. Combinatorial microRNA target predictions. *Nat. Genet.* **37:** 495.

Krutzfeld J., Rajewsky N., Braich R., Rajeev J.G., Tuschl T., Manoharan M., and Stoffel M. 2005. Silencing of microRNAs in vivo with "antagomirs." *Nature* **438:** 685.

Lagos-Quintana M., Rauhut R., Lendeckel W., and Tuschl T. 2001. Identification of novel genes coding for small expressed RNAs. *Science* **294:** 853.

Lall S., Grun D., Krek A., Chen K., Wang Y.L., Dewey C.N., Sood P., Colombo T., Bray N., MacMenamin P., et al. 2006. A genome-wide map of conserved microRNA targets in *C. elegans. Curr. Biol.* **16:** 460.

Leaman D., Chen P.Y., Fak J., Yalcin A., Pearce M., Unnerstall U., Marks D.S., Sander C., Tuschl T., and Gaul U. 2005. Antisense-mediated depletion reveals essential and specific functions of microRNAs in *Drosophila* development. *Cell* **121:** 1097.

Lewis B.P., Shih I.H., Jones-Rhoades M.W., Bartel D.P., and Burge C.B. 2003. Prediction of mammalian microRNA targets. *Cell* **115:** 787.

Lim L.P., Lau N.C., Weinstein E.G., Abdelhakim A., Yekta S., Rhoades M.W., Burge C.B., and Bartel D.P. 2003. The microRNAs of *Caenorhabditis elegans. Genes Dev.* **17:** 977.

Moss E.G. and Tang L. 2003. Conservation of the heterochronic regulator Lin-28, its developmental expression and microRNA complementary sites. *Dev. Biol.* **258:** 432.

O'Brien K.P., Remm M., and Sonnhammer E.L. 2005. Inparanoid: A comprehensive database of eukaryotic orthologs. *Nucleic Acids Res.* **33:** D476.

Ohler U., Yekta S., Lim L.P., Bartel D.P., and Burge C.B. 2004. Patterns of flanking sequence conservation and a characteristic upstream motif for microRNA gene identification. *RNA* **10:** 1309.

Pasquinelli A.E., Reinhart B.J., Slack F., Martindale M.Q., Kuroda M.I., Maller B., Hayward D.C., Ball E.E., Degnan B., Muller P., et al. 2000. Conservation of the sequence and temporal expression of let-7 heterochronic regulatory RNA. *Nature* **408:** 86.

Rajewsky N. 2006. microRNA target predictions in animals. *Nat. Genet.* **38:** S8.

Sharan R., Suthram S., Kelley R., Kuhn T., McCuine S., Uetz P., Sittler T., Karp R., and Ideker T. 2005. Conserved patterns of protein interaction in multiple species. *Proc. Natl. Acad. Sci.* **102:** 1972.

Stark A., Brennecke J., Bushati N., Russell R.B., and Cohen S.M. 2005. Animal microRNAs confer robustness to gene expression and have a significant impact on 3'UTR evolution. *Cell* **123:** 1133.

Xie X., Lu J., Kulbokas E.J., Golub T., Mootha V., Lindblad-Toh K., Lander E.S., and Kellis M. 2005. Systematic discovery of regulatory motifs in human promoters and 3'UTRs by comparison of several mammals. *Nature* **434:** 338.

Organ Polarity in Plants Is Specified through the Opposing Activity of Two Distinct Small Regulatory RNAs

F.T.S. Nogueira,* A.K. Sarkar,* D.H. Chitwood,*† and M.C.P. Timmermans*

*Cold Spring Harbor Laboratory, Cold Spring Harbor, New York 11724;
†Watson School of Biological Sciences, Cold Spring Harbor, New York 11724

Small RNAs and their targets form complex regulatory networks that control cellular and developmental processes in multicellular organisms. In plants, dorsoventral (adaxial/abaxial) patterning provides a unique example of a developmental process in which early patterning decisions are determined by small RNAs. A gradient of microRNA166 on the abaxial/ventral side of the incipient leaf restricts the expression of adaxial/dorsal determinants. Another class of small RNAs, the *TAS3*-derivated *trans*-acting short-interfering RNAs (ta-siRNAs), are expressed adaxially and repress the activity of abaxial factors. Loss of maize *leafbladeless1* (*lbl1*) function, a key component of the ta-siRNA biogenesis pathway, leads to misexpression of miR166 throughout the initiating leaf, implicating ta-siRNAs in the spatiotemporal regulation of miR166. The spatial restriction of ta-siRNA biogenesis components suggests that this pathway may act non-cell-autonomously in the meristem and possibly contributes to the classic meristem-borne adaxializing Sussex signal. Here, we discuss the key participants in adaxial/abaxial patterning and point out the intriguing possibility that organ polarity in plants is established by the opposing action of specific ta-siRNAs and miRNAs.

Plant shoots are characterized by indeterminate growth. The growing tip of a plant, referred to as the shoot apical meristem (SAM), contains a population of pluripotent stem cells that divide to maintain the SAM and to generate daughter cells from which lateral organs, such as leaves, arise (Fig. 1). Leaves of most seed plants are dorsoventrally flattened and differentiate distinct cell types in the upper and lower leaf surfaces to maximize the capture of sunlight and the exchange of gases that drive photosynthesis. Dorsoventral (adaxial/abaxial) polarity is specified shortly after the emergence of the leaf primordium from the meristem and is thought to reflect inherent positional differences in the developing organ relative to the SAM (Wardlaw 1940). The adaxial, dorsal side of the leaf develops in closer proximity to the tip of the SAM than the abaxial, ventral side (Fig. 1) (for review, see Bowman et al. 2002). Surgical experiments separating the incipient leaf from the remainder of the SAM result in radially symmetric, abaxialized leaves suggesting the existence of a meristem-borne signal that specifies adaxial cell fate (Sussex 1951, 1955). Although the identity of the Sussex signal remains unknown, recent studies implicate small regulatory RNAs in adaxial/abaxial patterning and raise the possibility that a mobile RNA might fulfill the requirements of a positional, polarizing signal. Here, we review the role of microRNAs (miRNAs) and *trans*-acting short interfering RNAs (ta-siRNAs) in leaf polarity and outline recent results suggesting that the opposing activity of two distinct small regulatory RNAs establishes adaxial/abaxial asymmetry in the developing leaf.

ESTABLISHMENT OF LEAF POLARITY IN *ARABIDOPSIS*

Several families of putative transcription factors play key roles in establishing adaxial/abaxial polarity in the

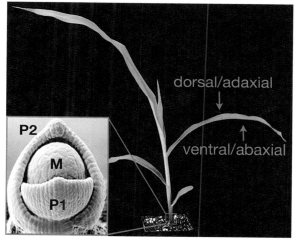

Figure 1. Leaf primordia arise on the flank of the shoot apical meristem (SAM) and establish dorsoventral (adaxial/abaxial) polarity in response to signals from the SAM. Leaves of a 2-week-old maize seedling differentiate distinct dorsal (adaxial) and ventral (abaxial) surfaces. The red box indicates the approximate position of the SAM within the surrounding leaves. The inset shows a scanning electron micrograph of a maize apex. The population of stem cells at the tip of the meristem (M) permits the reiterative development of the leaf primordia, which emerge from the flank of the SAM. The youngest leaf primordium is indicated as P1, the second youngest as P2, etc.

leaf. Members of the class III family of homeodomain-leucine zipper (HD-ZIPIII) proteins—PHABULOSA (PHB), PHAVOLUTA (PHV), and REVOLUTA (REV)—specify adaxial fate (McConnell et al. 2001; Otsuga et al. 2001; Emery et al. 2003). In contrast, the *KANADI* (*KAN*) genes, which encode transcriptional regulators belonging to the GARP family, act redundantly to promote abaxial identity (Eshed et al. 2001, 2004; Kerstetter et al. 2001). Although both the *HD-ZIPIII* and *KAN* genes are expressed evenly throughout the incipient leaf (P0), their domains of expression become restricted to the adaxial and abaxial sides of the organ, respectively, shortly after emergence of the primordium from the SAM. Constitutive *KAN* expression leads to the development of a radially symmetric, abaxialized leaf (Eshed et al. 2001, 2004; Kerstetter et al. 2001). Loss of *HD-ZIPIII* function results in a similar phenotype, suggesting that the *HD-ZIPIII* genes act, at least in part, to spatially restrict the *KAN* expression domain (Emery et al. 2003). *HD-ZIPIII* expression is in turn excluded from the abaxial side by the action of the KAN proteins. Thus, the *HD-ZIPIII* and *KAN* genes have a mutually antagonistic relationship, which may reflect their requirement to maintain a stable adaxial/abaxial boundary throughout leaf development.

In *Arabidopsis*, establishment of abaxial identity further requires the activities of members of the *YABBY* and *AUXIN RESPONSE FACTOR* (*ARF*) families (Sawa et al. 1999; Siegfried et al. 1999; Pekker et al. 2005). The *YABBY* genes *FILAMENTOUS FLOWER* (*FIL*) and *YAB3* act, at least in part, downstream of the *HD-ZIPIII* and *KAN* genes, whereas *ARF3/ETT* and *ARF4* affect organ polarity through a distinct pathway (Sawa et al. 1999; Siegfried et al. 1999; Kumaran et al. 2002; Pekker

et al. 2005). *ARF3/ETT* is expressed more broadly than *ARF4*, but both transcription factors colocalize in the abaxial domain of leaf primordia, where they act in combination with KAN proteins to promote abaxial fate (Pekker et al. 2005).

Interestingly, both the abaxial determinants *ARF3/ETT* and *ARF4*, as well as the adaxializing *HD-ZIPIII* genes, are targets for RNAi-based regulation (Fig. 2a). *HD-ZIPIII* transcripts contain complementary target sites to the nearly identical microRNAs 165 and 166 (miR165/166), which can direct cleavage of *HD-ZIPIII* mRNAs in vitro (Rhoades et al. 2002; Tang et al. 2003). *ARF3/ETT* and *ARF4* are targets of the recently discovered ta-siRNAs (Allen et al. 2005; Williams et al. 2005a). Thus, in addition to being a key developmental process, the prominent role of small regulatory RNAs in adaxial/abaxial patterning makes leaf polarity an excellent model to dissect the role of small RNAs as developmental signals.

BIOGENESIS AND FUNCTION OF miRNAs

The biogenesis and function of plant miRNAs have been extensively reviewed elsewhere (see Timmermans et al. 2004; Jones-Rhoades et al. 2006; Mallory and Vaucheret 2006). Briefly, miRNAs are processed from

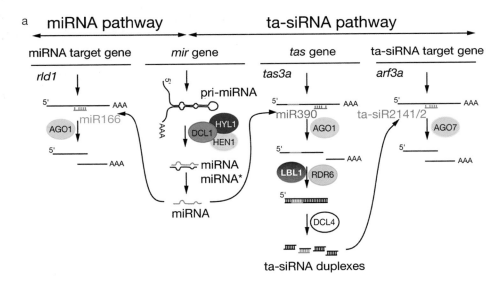

Figure 2. Plant miRNA and ta-siRNA biogenesis and function are interconnected. (*a*) *mir* genes transcribe a primary miRNA transcript (pri-miRNA), which is rapidly processed by DCL1 and HYL1. The resulting imperfect miRNA/miRNA* duplex is 3′-end methylated by HEN1 and loaded onto an AGO1-containing RISC. This complex cleaves target transcripts; *HD-ZIPIII* (*rld1*) transcripts in case of miR166 and *tas3a* mRNAs for miR390. The *tas3a* cleavage fragments are converted into double-stranded RNAs through the activities of LBL1/SGS3 and RDR6, and processed by DCL4 into 21-nucleotide double-stranded siRNAs. The *tas3a*-derived ta-siR2141/2142 acts *in trans* to cleave *arf3a* transcripts via AGO7. (*b*) Diagram of the maize *tas3a* transcript illustrating the 21-nucleotide phased processing of ta-siRNAs, which initiates from the miR390-cleavage site. Black brackets represent putative ta-siRNAs, and red brackets correspond to *Arabidopsis* ta-siR2141/ta-siR2142 siRNA homologs.

RNA polymerase II transcripts that contain a stem-loop structure (Fig. 2a). These transcripts, termed primary miRNAs (pri-miRNAs), are processed by the RNase III enzyme DICER-LIKE1 (DCL1) (Park et al. 2002; Reinhart et al. 2002). Subsequently, the approximately 21-nucleotide mature miRNA becomes incorporated into the RNA-induced silencing complex (RISC), which targets complementary transcripts for site-specific cleavage or translational repression (Chen 2004; Han et al. 2004; Vaucheret et al. 2004; Baumberger and Baulcombe 2005). Plant miRNAs and their targets frequently possess near-perfect complementarity, and this has enabled the identification of many target genes using computational approaches. Interestingly, the known plant miRNAs show a strong propensity to target transcription factors or other genes that regulate critical steps during plant development (Reinhart et al. 2002; Jones-Rhoades et al. 2006). Consequently, mutations in genes associated with miRNA biogenesis or function, such as *ARGONAUTE1* (*AGO1*), *HYPONASTIC LEAVES1* (*HYL1*), and *SERRATE* (*SE*), affect important developmental processes including adaxial/abaxial patterning (Kidner and Martienssen 2004; Grigg et al. 2005; Yu et al. 2005; Yang et al. 2006).

Elucidation of the precise developmental roles of individual miRNAs is, however, complicated by the presence of extensive redundancy. Most *MIR* genes are members of multigene families (see Bartel 2004; Jones-Rhoades et al. 2006). Whereas miRNA families in animals are small and include diverse members, plant miRNA families frequently contain many genes that can produce identical mature miRNA sequences. Family members are likely to have overlapping expression profiles and functions that buffer against the loss of any single miRNA locus, as few loss-of-function alleles of *MIR* genes have been recovered through forward-genetic screens. Dominant gain-of-function miRNA mutants, however, are more common and have revealed developmental roles for several miRNAs. For example, the *meristem enlarged1* and *jabba1-D* mutants both develop fasciated stems and enlarged meristems resulting from altered *MIR166a* and *MIR166g* expression, respectively (Kim et al. 2005; Williams et al. 2005b). Similar gain-of-function mutations for miR156, miR159, miR160, miR164, miR172, and miR319 lead to defects in vegetative and floral organ development, meristem function, and flowering (see Jones-Rhoades et al. 2006).

ROLE OF miRNA165/166 IN ADAXIAL/ABAXIAL PATTERNING

As in *Arabidopsis*, adaxial/abaxial asymmetry in maize is established through the polarized expression of *HD-ZIPIII* genes (Juarez et al. 2004b). *rolled leaf1* (*rld1*), which encodes a close homolog of *REV*, is expressed at the tip of the SAM and in a strip of cells from the center of the SAM to the site of leaf initiation (Fig. 3e). In developing leaf primordia, *rld1* is expressed on the adaxial side as well as in the vasculature. The miR165/166 target site in the *HD-ZIPIII* transcripts is conserved between monocots and dicots (Reinhart et al. 2002; Rhoades et al. 2002; Juarez et al. 2004b). Because these lineages last shared a

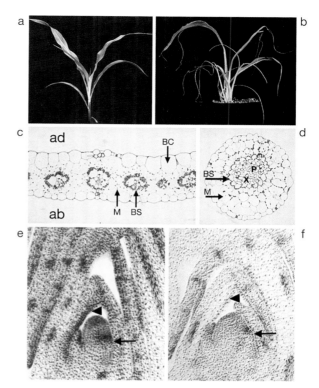

Figure 3. LBL1 is required for the specification of adaxial/dorsal fate. Compared to the wild-type maize seedling (*a*), severe *lbl1* mutants (*b*) develop thread-like, abaxialized leaves. (*c*) Transverse section through a wild-type leaf blade illustrates its dorsoventrally flattened organization, with polar veins surrounded by bundle sheath (BS) and mesophyll cells (M). Bulliform cells (BC) differentiate only in the adaxial (ad) epidermis. (*d*) Severe *lbl1* leaves are radially symmetric and comprise an irregular vascular cylinder surrounded by concentric rings of bundle sheath and mesophyll cells, and abaxial (ab) epidermis. X, xylem; P, phloem. (*e*) Longitudinal section through a wild-type apex showing *rld1* expression in the SAM, vasculature, and on the adaxial side of the incipient (*arrow*) and developing leaf primordia (*arrowhead*). (*f*) In *lbl1*, meristematic and adaxial expression of *rld1* is reduced. (Reprinted, with permission, from Timmermans et al. 1998 and Juarez et al. 2004a.)

common ancestor more than 100 million years ago, this conservation suggests an important role in plant development for the regulation of *HD-ZIPIII* genes by miR165/166. The significance of this relationship was first demonstrated by the characterization of dominant mutations in the *Arabidopsis HD-ZIPIII* genes (*phb-d*, *phv-d*, and *rev-d*) that abrogate the miR165/166 target site. Such mutations interfere with miRNA-directed transcript cleavage, leading to ectopic, abaxial *HD-ZIPIII* expression and formation of adaxialized leaves (McConnell et al. 2001; Emery et al. 2003; Tang et al. 2003). Disruption of the miRNA165/166 complementary site of maize *rld1* (*Rld1-O*) similarly results in misexpression of *rld1* on the abaxial side of developing primordia and adaxialization of the leaf (Juarez et al. 2004b).

In situ hybridization provides direct evidence that the pattern of miR165/166 expression spatially defines the expression domain of the *HD-ZIPIII* genes (Juarez et al. 2004b; Kidner and Martienssen 2004). miR165 and miR166 are expressed on the abaxial side of the developing leaf in a pattern complementary to that of the *HD-ZIPIII* genes. Interestingly, in maize, miR166 is most abundant in a group of cells below the incipient leaf, but a gradient of weaker miR166 expression extends into the abaxial side of the newly initiating primordium. In P1 primordia, higher levels of miR166 accumulate abaxially and, in older primordia, miR166 accumulates in a progressively broader domain extending adaxially and laterally (Juarez et al. 2004b). These findings suggest that miR165/166 is a highly conserved polarizing signal that specifies adaxial/abaxial polarity by restricting *HD-ZIPIII* expression to the adaxial side of the leaf in both dicots and monocots (Timmermans et al. 2004). The spatial regulation of *HD-ZIPIII* gene expression by miR165/166 may even predate the origin of angiosperm leaves. The miR165/166-directed cleavage of *HD-ZIPIII* transcripts is conserved in basal lineages of land plants, including bryophytes, lycopods, and ferns (Floyd and Bowman 2004). Organ polarity is a relatively recent landmark in plant development; thus, it is possible that the regulation of *HD-ZIPIII* genes by miR165/166 evolved as a preadaptation that was later co-opted for use in adaxial/abaxial patterning of leaves and other lateral organs (Floyd et al. 2006).

THE ta-siRNA PATHWAY

Like the miRNAs, ta-siRNAs are processed from long, RNA pol II transcripts that are not predicted to encode for proteins (Fig. 2a). Interestingly, the ta-siRNA precursors (*TAS*) are themselves targets for miRNA-directed cleavage (Allen et al. 2005). However, unlike most miRNA-directed cleavage products, *TAS* cleavage fragments are converted into double-stranded RNAs through the activities of the plant-specific Zn-finger protein SUPPRESSOR OF GENE SILENCING3 (SGS3) and RNA-DEPENDENT RNA POLYMERASE6 (RDR6), and are subsequently processed by DCL4 into 21-bp siRNAs that guide the cleavage of target mRNAs, similar to the action of miRNAs (Peragine et al. 2004; Vazquez et al. 2004; Allen et al. 2005; Gasciolli et al. 2005; Xie et al. 2005; Yoshikawa et al. 2005). Biogenesis of ta-siRNAs thus requires the activity of proteins in the miRNA pathway, such as DCL1 and AGO1, as well as SGS3, RDR6, and DCL4 (Fig. 2a).

Three gene families are known to generate ta-siRNAs in *Arabidopsis*. *TAS1* and *TAS2* are targets of miR173, whereas the production of ta-siRNAs from *TAS3* depends on miR390-mediated cleavage (Allen et al. 2005). Because DCL4 cleavage initiates at the processed end of the *TAS* precursor, ta-siRNAs are generated in a 21-nucleotide phase starting at the miRNA cleavage site (Fig. 2b). Consequently, the ta-siRNAs derived from each *TAS* locus, as well as their potential targets, can be predicted using computational approaches. The *TAS1/TAS2* loci share homology and produce related ta-siRNAs directed against a subset of pentatricopeptide repeat genes and a group of genes of unknown function (Peragine et al. 2004; Vazquez et al. 2004; Allen et al. 2005; Yoshikawa et al. 2005). Mutational analysis indicates that the *TAS1*- and

TAS2-derived ta-siRNAs have no obvious developmental role (Adenot et al. 2006). In contrast, two of the *TAS3*-derived ta-siRNAs, ta-siR2141/2, regulate the expression of *ARF2*, *ARF3/ETT*, and *ARF4*, which are known to function during shoot morphogenesis and adaxial/abaxial patterning (Allen et al. 2005; Pekker et al. 2005; Williams et al. 2005a).

TAS3 ta-siRNAs SPECIFY LEAF POLARITY THROUGH REGULATION OF miR166

Given that the abaxial determinants *ARF3/ETT* and *ARF4* are targets for *TAS3* ta-siRNAs, a role for the ta-siRNA pathway in leaf polarity can be inferred. However, the contribution of this pathway to adaxial/abaxial patterning in *Arabidopsis* is not immediately apparent. *Arabidopsis* mutants that block the biogenesis of ta-siRNAs develop no obvious leaf polarity defects (Peragine et al. 2004; Vazquez et al. 2004; Allen et al. 2005; Adenot et al. 2006). Moreover, the distribution of trichomes on leaves of such mutants, as well on plants expressing a ta-siR2141/2 insensitive allele of *ARF3/ETT*, is inconsistent with the predicted abaxializing phenotype (Peragine et al. 2004; Fahlgren et al. 2006; Hunter et al. 2006).

An essential role for the ta-siRNA pathway in adaxial/abaxial patterning became evident through the cloning and characterization of *leafbladeless1* (*lbl1*) from maize. *lbl1*, in addition to the *HD-ZIPIII* genes, is required for the specification of adaxial fate (Timmermans et al. 1998; Juarez et al. 2004a). Loss-of-function *lbl1* mutations condition an abaxialized leaf phenotype, with the most severely affected *lbl1* leaves becoming radially symmetric and fully abaxialized (Fig. 3a–d) (Timmermans et al. 1998). The recent cloning of *lbl1* showed that it encodes a homolog of the *Arabidopsis* SGS3 protein. LBL1 and the *Arabidopsis* SGS3 protein share 65% amino acid similarity overall, but the degree of sequence similarity is higher in the zinc-finger (92%) and XS domains (79%), which define the SGS3 protein family (Fig. 4) (Bateman 2002).

These findings indicate that SGS3 activity is essential for adaxial/abaxial patterning in maize, and suggest a role for ta-siRNAs in this process. miR173 and the *TAS1/TAS2* loci are not conserved between maize and *Arabidopsis*, but the maize genome includes at least one *mir390* gene and four *tas3* loci (*tas3a–tas3d*) whose transcripts are predicted targets for miR390. Most *tas3*-derived ta-siRNAs are not conserved between maize and *Arabidopsis*, but interestingly, all four maize *tas3* transcripts are predicted to yield copies of ta-siR2141/2 (Fig. 2b). The *tas3* genes, as well as *mir390*, are expressed in vegetative apices, and accordingly, ta-siR2141/2 accumulates in this tissue (F. Nogueira et al., in prep.). ta-siR2141/2 accumulation is severely reduced or abolished in *lbl1* mutants, suggestive of functional conservation between LBL1 and SGS3. As in *Arabidopsis*, of the maize *tas3*-derived ta-siRNAs, only ta-siR2141/2 has clearly identifiable candidate targets, and these include members of the *arf3* gene family (F. Nogueira et al., in prep.). 5′ RACE analysis identified *arf3a* as a direct tar-

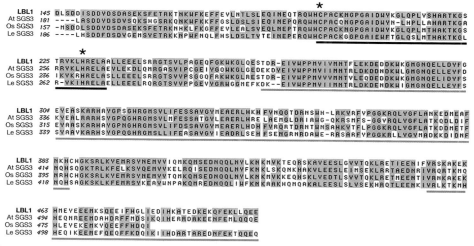

Figure 4. *lbl1* encodes a SGS3-like protein. Partial sequencing alignment of LBL1 (DQ832257) and the SGS3 proteins from *Arabidopsis thaliana* (NP_197747.1), *Oryza sativa* (AK064217.1), and *Lycopersicum esculentum* (BT013417.1). The zinc-finger domain (*black*), XS domain (*red*), and coiled coils (*blue*) characteristic of the SGS3 protein family are underlined. Asterisks mark amino acid substitutions identified in severe *lbl1* alleles.

get of ta-siR2141/2, indicating that the ta-siRNA pathway is extensively conserved between maize and *Arabidopsis*. This pathway could be even more ancient, because homologs of miR390 and *TAS3* have been identified in moss, which last shared a common ancestor with the monocots and dicots more than 400 million years ago (Arazi et al. 2005; D. Bartel, pers. comm.). Importantly, *lbl1* is required for the ta-siR2141/2-directed cleavage of *arf3a*, and thus forms an essential component of the ta-siRNA pathway. The observation that LBL1 is essential for adaxial/abaxial patterning in maize thus implies a key role for ta-siRNAs in this process.

lbl1 contributes to leaf polarity by regulating the expression of *HD-ZIPIII* genes (Juarez et al. 2004a; F. Nogueira et al., in prep.). Expression of *rld1* and its paralogs *rld2* and *phb* on the adaxial side of developing leaf primordia is reduced in *lbl1* mutants (Fig. 3e,f). Notably, *HD-ZIPIII* expression at the side of leaf initiation is altered in *lbl1* mutants, suggesting that *lbl1* and the ta-siRNA pathway affect an early step in adaxial/abaxial patterning. In situ hybridization confirms that *lbl1* is most prominently expressed in a dome of cells at the tip of the SAM that extends into the adaxial side of the initiating primordium.

Loss of LBL1 activity leads to misexpression and/or overexpression of ta-siRNA targets and downstream components. Reduced *HD-ZIPIII* expression in *lbl1*, therefore, suggests that one or more antagonists of the *HD-ZIPIII* genes are controlled by the ta-siRNA pathway. Indeed, *lbl1* affects *HD-ZIPIII* expression by regulating the spatiotemporal pattern of miR166 accumulation (F. Nogueira et al., in prep.). miR166 normally accumulates in a graded pattern on the abaxial side of incipient and young leaf primordia. However, in an *lbl1* mutant background, miR166 is ectopically expressed in a torus at the base of the SAM that broadly overlaps with the incipient leaf. Expression of miR166 in the P1 and older leaf pri-

mordia also comprises a broader domain, including both adaxial and abaxial sides. Thus, *lbl1* and the ta-siRNA pathway spatially restrict the miR166 expression domain in the SAM and developing leaf primordia, and contribute to organ polarity by setting up the abaxial specific expression gradient of miR166 in the incipient leaf. This presents the intriguing possibility that the opposing activity of two distinct small regulatory RNAs directs the early patterning decision leading to adaxial/abaxial polarity in the incipient leaf; ta-siR2141/2 defines the adaxial side of the leaf by restricting the expression domain of miR166, which in turn delineates the abaxial side by restricting expression of the adaxializing *HD-ZIPIII* genes.

The maize genome includes at least nine *mir166* loci, *mir166a* through *mir166i*. All *mir166* genes are expressed within the vegetative apex, but the spatiotemporal expression profiles vary among family members (F. Nogueira et al., unpubl.). The ta-siRNA pathway affects the accumulation of just a subset of *mir166* precursors. Transcript levels for *mir166c* and *mir166i* are increased in the meri-stem of *lbl1* as compared to wild type, whereas *mir166a* precursor levels are reduced in the mutant (F. Nogueira et al., in prep.). This suggests that the loss of ta-siRNA- directed repression of *mir166c* and *mir166i* underlies, at least partially, the ectopic miR166 accumulation in *lbl1* incipient primordia and resulting abaxialization of *lbl1* leaves.

How might the ta-siRNA pathway restrict miR166 expression? Despite our increasing understanding of miRNA biogenesis and function, little is known about the regulation of *MIR* genes themselves. ARF proteins are transcription factors that mediate auxin-dependent gene regulation through binding to specific sequence motifs within promoters of auxin-regulated genes (Ulmasov et al. 1999). Although it will require gain- or loss-of-function *arf3* mutants to assess the contribution of these transcription factors to the spatiotemporal regulation of miR166, it is not difficult to envisage that the ta-siRNA pathway may control the expression of specific mir166 family members via arf3 genes. Additionally, miRNA accumulation is likely regulated at the posttranscriptional level (Bollman et al. 2003; Yang et al. 2006). The primary mir166i transcript includes a sequence motif with modest complementarity to ta-siR2141/2. Given that the ta-siRNA pathway is expected to occur in the nucleus (Vaucheret 2006), this observation suggests that the spatiotemporal pattern of miR166 expression may be regulated directly by this ta-siRNA. This possibility is particularly intriguing because it presents a scenario in which miRNA precursors are themselves controlled by small RNAs. Moreover, leaf polarity would be specified through a cascade of direct small regulatory RNA interactions in which miR390-directed cleavage of TAS3 triggers the production of ta-siR2141/2, which in turn processes mir166i required for the polarized expression of miR166 and adaxial/abaxial patterning of the newly initiated leaf.

DIVERSE CONTRIBUTIONS OF THE TA-SIRNA PATHWAY TO LEAF POLARITY IN DISTINCT PLANT LINEAGES

Despite uncertainty regarding the precise mechanism by which the ta-siRNA pathway regulates miR166 expression, the defects observed in lbl1 clearly demonstrate a critical role for ta-siRNAs in adaxial/abaxial patterning by establishing the abaxial expression gradient of miR166 in the incipient leaf. These findings also highlight important differences between Arabidopsis and maize; namely, even though the ta-siRNA pathway is extensively conserved between these species, Arabidopsis mutants that block ta-siRNA biogenesis display no obvious leaf polarity defects (Peragine et al. 2004; Vazquez et al. 2004; Allen et al. 2005). The differential reliance on the ta-siRNA pathway for adaxial/abaxial patterning in part reflects redundancy of the pathway in Arabidopsis with ASYMMETRIC LEAVES1 (AS1) and AS2. Double mutants affecting the ta-siRNA pathway as well as AS1 or AS2 function develop weakly abaxialized leaves (Li et al. 2005; Garcia et al. 2006; Xu et al. 2006).

In addition, divergence in the nature or function of downstream targets, or the time during leaf development at which downstream targets act, could greatly influence the contribution of the ta-siRNA pathway to adaxial/abaxial patterning in different plant species. For instance, whereas the Arabidopsis ta-siRNA pathway represses FIL expression (Li et al. 2005; Garcia et al. 2006; Xu et al. 2006), members of the maize yabby gene family closely related to FIL are positively regulated by lbl1 (Juarez et al. 2004a). Additionally, although expression analysis suggests that ta-siRNAs contribute to the regulation of miR165/166 in Arabidopsis (Li et al. 2005; Xu et al. 2006), disruption of the ta-siRNA pathway affects expression of specific MIR165/166 family members only during later stages of leaf development (F. Nogueira et al., unpubl.). Distinctly, the maize ta-siRNA pathway acts foremost in the SAM to regulate miR166 expression in the newly initiated primordium, and thus directs the early decisions in adaxial/abaxial patterning (F. Nogueira et al., in prep.).

SMALL RNAs AS POTENTIAL MOBILE SIGNALS IN PLANTS

The antagonistic interaction between ta-siRNAs and miR166 highlights the complexity that can be found in small RNA regulated pathways, and illustrates the important role of small RNAs in pattern formation during development. This unique interaction also raises the question whether small RNAs contribute to the production or perception of positional information from the SAM required for adaxial/abaxial patterning. Due to their intrinsic high specificity, the idea of mobile—possibly morphogenic—small RNA signals in plants is tantalizing. However, evidence so far suggests that miRNAs act largely cell-autonomously (Parizotto et al. 2004; Alvarez et al. 2006). siRNAs generated during posttranscriptional or virus-induced gene silencing, on the other hand, are mobile, and their movement is the basis for the non-cell-autonomous nature of these silencing processes (Himber et al. 2003; Voinnet 2005). Interestingly, cell-to-cell movement is restricted to 21-bp DCL4-dependent siRNAs, whereas the 24-bp siRNAs generated by DCL3 act strictly cell-autonomously outside the phloem (Dunoyer et al. 2005). This suggests that, if produced by the correct mechanism or channeled into the correct pathway, small RNAs can traffic between cells.

In this regard, the involvement of ta-siRNAs in the specification of adaxial fate is particularly intriguing. Their biogenesis requires DCL4, presenting the possibility that this novel class of siRNAs may be able to move from cell to cell. Indeed, ectopic expression of miR166 in the lbl1 mutant also occurs outside the normal lbl1 expression domain. This nonoverlapping expression pattern, along with a role for ta-siRNAs in establishing a gradient of miR166 on the abaxial side of the incipient leaf, suggests that the ta-siRNA pathway may operate non-cell-autonomously in the SAM (F. Nogueira et al., in prep.). ta-siRNAs thus constitute a plausible component of the Sussex signal. Consistent with this notion, the most severely affected lbl1 mutant leaves resemble the abaxialized leaves that arise following the surgical separation of leaf initials from the tip of the SAM (Fig. 3d) (Sussex 1951; Timmermans et al. 1998).

CONCLUSIONS

The finding that lbl1 encodes a key component of the ta-siRNA pathway reveals an essential role for ta-siRNAs in adaxial/abaxial patterning. In addition, leaf polarity requires the activity of miR165/166. The importance of small regulatory RNAs in adaxial/abaxial polarity may reflect the need to rapidly change transcription profiles that underlie cell fate changes, or to maintain a precise balance between adaxial and abaxial determinants during primordium growth, but could also reflect a role of small RNAs acting as positional cues. The expanded uniform expression of miR166 in incipient primordia of lbl1 indicates that ta-siRNAs promote adaxial fate by restricting miR166 expression to the abaxial side of the initiating leaf. Organ polarity in

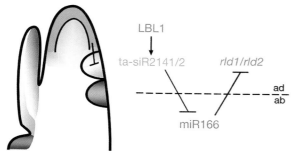

Figure 5. Organ polarity may be specified through the opposing activity of two distinct small regulatory RNAs. *lbl1* (*solid green*) is expressed in a dome of cells at the tip of the SAM, suggesting the site of *TAS3* ta-siRNA biogenesis. miR166 (*purple*) accumulates most prominently below the incipient leaf, but also in a graded pattern on the abaxial side of this primordium. ta-siR2141/2 (*pale green*) may move from its side of synthesis into the adaxial side of the initiating leaf. ta-siR2141/2 locally restricts miR166 expression, which in turn represses expression of the adaxial determinants *rld1* and *rld2* (*red*). The opposing activities of ta-siR2141/2 and miR166 thus set up polarity in the developing leaves.

plants may thus be specified through the opposing activities of ta-siR2141/2 and miR166 (Fig. 5).

The possibility that ta-siRNAs can act as mobile signals, perhaps because of their specific association with DCL4, suggests a mechanism by which ta-siR2141/2 establishes organ polarity. The local biogenesis of ta-siR2141/2 in the SAM followed by its movement through the adjacent tissue would result in a concentration gradient across the initiating leaf, which sets up a complementary gradient of miR166. A defined balance between ta-siR2141/2 and miR166 activities may specify leaf polarity. Such a balance may result through regulated accumulation of each small RNA or through their relative efficacies of cleaving target transcripts. The number of cells over which a siRNA can move is correlated directly with its abundance, suggesting a dilution of the siRNA signal away from its point source (Himber et al. 2003; Voinnet 2005). The relatively low abundance of ta-siR2141/2 (Allen et al. 2005; Williams et al. 2005a) or the involvement of an AGO7-based RISC in the *TAS3* ta-siRNA pathway could therefore be extremely significant for proper adaxial/abaxial patterning.

Although the concept of the Sussex signal has existed for more than 50 years, only recently have we started to better understand the complex genetic network that specifies adaxial/abaxial polarity. The recognition of small RNAs as early regulators of lateral organ polarity implicates them as potential polarizing signals that could perhaps contribute to the Sussex signal.

ACKNOWLEDGMENTS

The authors thank Shahinez Madi, Krista Marran, and Amanda Grieco for excellent technical assistance. This research was supported by grants from the U.S. Department of Agriculture (06-03420) and the National Science Foundation (0615752) to M.T.; F.N. was funded in part by a fellowship from the Cold Spring Harbor Laboratory Association; and D.C. is a National Science Foundation graduate research fellow and a George A. and Marjorie H. Matheson fellow.

REFERENCES

Adenot X., Elmayan T., Lauressergues D., Boutet S., Bouche N., Gasciolli V., and Vaucheret H. 2006. DRB4-dependent *TAS3* trans-acting siRNAs control leaf morphology through AGO7. *Curr. Biol.* **16:** 927.

Allen E., Xie Z., Gustafson A.M., and Carrington J.C. 2005. microRNA-directed phasing during trans-acting siRNA biogenesis in plants. *Cell* **121:** 207.

Alvarez J.P., Pekker I., Goldshmidt A., Blum E., Amsellem Z., and Eshed Y. 2006. Endogenous and synthetic microRNAs stimulate simultaneous, efficient, and localized regulation of multiple targets in diverse species. *Plant Cell* **18:** 1134.

Arazi T., Talmor-Neiman M., Stav R., Riese M., Huijser P., and Baulcombe D.C. 2005. Cloning and characterization of micro-RNAs from moss. *Plant J.* **43:** 837.

Bartel D.P. 2004. MicroRNAs: Genomics, biogenesis, mechanism, and function. *Cell* **116:** 281.

Bateman A. 2002. The SGS3 protein involved in PTGS finds a family. *BMC Bioinformatics* **3:** 21.

Baumberger N. and Baulcombe D.C. 2005. *Arabidopsis* ARG-ONAUTE1 is an RNA slicer that selectively recruits microRNAs and short interfering RNAs. *Proc. Natl. Acad. Sci.* **102:** 11928.

Bollman K.M., Aukerman M.J., Park M.Y. Hunter C., Berardini T.Z., and Poethig R.S. 2003. HASTY, the *Arabidopsis* ortholog of exportin 5/MSN5, regulates phase change and morphogenesis. *Development* **130:** 1493.

Bowman J.L., Eshed Y., and Baum S.F. 2002. Establishment of polarity in angiosperm lateral organs. *Trends Genet.* **18:** 134.

Chen X. 2004. A microRNA as a translational repressor of *APETALA2* in *Arabidopsis* flower development. *Science* **303:** 2022.

Dunoyer P., Himber C., and Voinnet O. 2005. DICER-LIKE 4 is required for RNA interference and produces the 21-nucleotide small interfering RNA component of the plant cell-to-cell silencing signal. *Nat. Genet.* **37:** 1356.

Emery J.F., Floyd S.K., Alvarez J., Eshed Y., Hawker N.P., Izhaki A., Baum S.F., and Bowman J.L. 2003. Radial patterning of *Arabidopsis* shoots by class III *HD-ZIP* and *KANADI* genes. *Curr. Biol.* **13:** 1768.

Eshed Y., Baum S.F., Perea J.V., and Bowman J.L. 2001. Establishment of polarity in lateral organs of plants. *Curr. Biol.* **11:** 1251.

Eshed Y., Izhaki A., Baum S.F., Floyd S.K., and Bowman J.L. 2004. Asymmetric leaf development and blade expansion in *Arabidopsis* are mediated by *KANADI* and *YABBY* activities. *Development* **131:** 2997.

Fahlgren N., Montgomery T.A., Howell M.D., Allen E., Dvorak S.K., Alexander A.L., and Carrington J.C. 2006. Regulation of *AUXIN RESPONSE FACTOR3* by *TAS3* ta-siRNA affects developmental timing and patterning in *Arabidopsis*. *Curr. Biol.* **16:** 939.

Floyd S.K. and Bowman J.L. 2004. Gene regulation: Ancient microRNA target sequences in plants. *Nature* **428:** 485.

Floyd S.K., Zalewski C.S., and Bowman J.L. 2006. Evolution of class III homeodomain-leucine zipper genes in streptophytes. *Genetics* **173:** 373.

Garcia D., Collier S.A., Byrne M.E., and Martienssen R.A. 2006. Specification of leaf polarity in *Arabidopsis* via the *trans*-acting siRNA pathway. *Curr. Biol.* **16:** 933.

Gasciolli V., Mallory A.C., Bartel D.P., and Vaucheret H. 2005. Partially redundant functions of *Arabidopsis* DICER-like enzymes and a role for DCL4 in producing trans-acting siRNAs. *Curr. Biol.* **15:** 1494.

Grigg S.P., Canales C., Hay A., and Tsiantis M. 2005. SER-RATE coordinates shoot meristem function and leaf axial patterning in *Arabidopsis*. *Nature* **437:** 1022.

Han J., Lee Y., Yeom K.H., Kim Y.K., Jin H., and Kim V.N. 2004. The Drosha-DGCR8 complex in primary microRNA processing. *Genes Dev.* **18:** 3016.

Himber C., Dunoyer P., Moissiard G., Ritzenthaler C., and Voinnet O. 2003. Transitivity-dependent and -independent cell-to-cell movement of RNA silencing. *EMBO J.* **22:** 4523.

Hunter C., Willmann M.R., Wu G., Yoshikawa M., de la Luz Gutierrez-Nava M., and Poethig S.R. 2006. Trans-acting siRNA-mediated repression of ETTIN and ARF4 regulates heteroblasty in *Arabidopsis*. *Development* **133:** 297.

Jones-Rhoades M.W., Bartel D.P., and Bartel B. 2006. microRNAs and their regulatory roles in plants. *Annu. Rev. Plant Biol.* **57:** 19.

Juarez M.T., Twigg R.W., and Timmermans M.C.P. 2004a. Specification of adaxial cell fate during maize leaf development. *Development* **131:** 4533.

Juarez M.T., Kui J.S., Thomas J., Heller B.A., and Timmermans M.C.P. 2004b. microRNA-mediated repression of *rolled leaf1* specifies maize leaf polarity. *Nature* **428:** 84.

Kerstetter R.A., Bollman K., Taylor R.A., Bomblies K., and Poethig R.S. 2001. *KANADI* regulates organ polarity in *Arabidopsis*. *Nature* **411:** 706.

Kidner C.A. and Martienssen R.A. 2004. Spatially restricted microRNA directs leaf polarity through ARGONAUTE1. *Nature* **428:** 81.

Kim J., Jung J.H., Reyes J.L., Kim Y.S., Kim S.Y., Chung K.S., Kim J.A., Lee M., Lee Y., Narry Kim V., et al. 2005. microRNA-directed cleavage of *ATHB15* mRNA regulates vascular development in *Arabidopsis* inflorescence stems. *Plant J.* **42:** 84.

Kumaran M.K., Bowman J.L., and Sundaresan V. 2002. *YABBY* polarity genes mediate the repression of *KNOX* homeobox genes in *Arabidopsis*. *Plant Cell* **14:** 2761.

Li H., Xu L., Wang H., Yuan Z. Cao X., Yang Z., Zhang D., Xu Y., and Huang H. 2005. The putative RNAdependent RNA polymerase RDR6 acts synergistically with ASYMMETRIC LEAVES1 and 2 to repress *BREVIPEDICELLUS* and MicroRNA165/166 in *Arabidopsis* leaf development. *Plant Cell* **17:** 2157.

Mallory A.C. and Vaucheret H. 2006. Functions of microRNAs and related small RNAs in plants. *Nat. Genet.* **38:** S31.

McConnell J.R., Emery J., Eshed Y., Bao N., Bowman J., and Barton M.K. 2001. Role of *PHABULOSA* and *PHAVOLUTA* in determining radial patterning in shoots. *Nature* **411:** 709.

Otsuga D., DeGuzman B., Prigge M.J., Drews G.N., and Clark S.E. 2001. *REVOLUTA* regulates meristem initiation at lateral positions. *Plant J.* **25:** 223.

Parizotto E.A., Dunoyer P., Rahm N., Himber C., and Voinnet O. 2004. *In vivo* investigation of the transcription, processing, endonucleolytic activity, and functional relevance of the spatial distribution of a plant miRNA. *Genes Dev.* **18:** 2237.

Park W., Li J., Song R., Messing J., and Chen X. 2002. CARPEL FACTORY, a Dicer homolog, and HEN1, a novel protein, act in microRNA metabolism in *Arabidopsis thaliana*. *Curr. Biol.* **12:** 1484.

Pekker I., Alvarez J.P., and Eshed Y. 2005. Auxin response factors mediate *Arabidopsis* organ asymmetry via modulation of KANADI activity. *Plant Cell* **17:** 2899.

Peragine A., Yoshikawa M., Wu G., Albrecht H.L., and Poethig R.S. 2004. SGS3 and SGS2/SDE1/RDR6 are required for juvenile development and the production of trans-acting siRNAs in *Arabidopsis*. *Genes Dev.* **18:** 2368.

Reinhart B.J., Weinstein E.G., Rhoades M.W., Bartel B., and Bartel D.P. 2002. MicroRNAs in plants. *Genes Dev.* **16:** 1616.

Rhoades M.W., Reinhart B.J., Lim L.P., Burge C.B., Bartel B., and Bartel D.P. 2002. Prediction of plant microRNA targets. *Cell* **110:** 513.

Sawa S., Watanabe K., Goto K., Liu Y.G., Shibata D., Kanaya E., Morita E.H., and Okada K. 1999. *FILAMENTOUS FLOWER*, a meristem and organ identity gene of *Arabidopsis*, encodes a protein with a zinc finger and HMG-related domains. *Genes Dev.* **13:** 1079.

Siegfried K.R., Eshed Y., Baum S.F., Otsuga D., Drews G.N., and Bowman J.L. 1999. Members of the *YABBY* gene family specify abaxial cell fate in *Arabidopsis*. *Development* **126:** 4117.

Sussex I.M. 1951. Experiments on the cause of dorsiventrality in leaves. *Nature* **167:** 651.

———. 1955. Morphogenesis in *Solanum tuberosum* L.: Experimental investigation of leaf dorsiventrality and orientation in the juvenile shoot. *Phytomorphology* **5:** 286.

Tang G., Reinhart B.J., Bartel D.P., and Zamore P.D. 2003. A biochemical framework for RNA silencing in plants. *Genes Dev.* **17:** 49.

Timmermans M.C.P., Juarez M.T., and Phelps-Durr T.L. 2004. A conserved microRNA signal specifies leaf polarity. *Cold Spring Harbor Symp. Quant. Biol.* **69:** 409.

Timmermans M.C.P., Schultes N.P., Jankovsky J.P., and Nelson T. 1998. *Leafbladeless1* is required for dorsoventrality of lateral organs in maize. *Development* **125:** 2813.

Ulmasov T., Hagen G., and Guilfoyle T.J. 1999. Dimerization and DNA binding of auxin response factors. *Plant J.* **19:** 309.

Vazquez F., Vaucheret H., Rajagopalan R., Lepers C., Gasciolli V., Mallory A.C., Hilbert J.L., Bartel D.P., and Crete P. 2004. Endogenous trans-acting siRNAs regulate the accumulation of *Arabidopsis* mRNAs. *Mol Cell.* **16:** 69.

Vaucheret H. 2006. Post-transcriptional small RNA pathways in plants: Mechanisms and regulation. *Genes Dev.* **20:** 759.

Vaucheret H., Vazquez F., Crete P., and Bartel D.P. 2004. The action of ARGONAUTE1 in the miRNA pathway and its regulation by the miRNA pathway are crucial for plant development. *Genes Dev.* **18:** 1187.

Voinnet O. 2005. Non-cell autonomous RNA silencing. *FEBS Lett.* **579:** 5858.

Wardlaw C.W. 1940. Experiments on organogenesis in ferns. *Growth* (suppl.) **13:** 93.

Williams L., Carles C.C., Osmont K.S., and Fletcher J.C. 2005a. A database analysis method identifies an endogenous trans-acting short-interfering RNA that targets the *Arabidopsis ARF2, ARF3,* and *ARF4* genes. *Proc. Natl. Acad. Sci.* **102:** 9703.

Williams L., Grigg S.P., Xie M., Christensen S., and Fletcher J.C. 2005b. Regulation of *Arabidopsis* shoot apical meristem and lateral organ formation by microRNA miR166g and its *AtHD-ZIP* target genes. *Development* **132:** 3657.

Xie Z., Allen E., Wilken A., and Carrington J.C. 2005. DICER-LIKE 4 functions in trans-acting small interfering RNA biogenesis and vegetative phase change in *Arabidopsis thaliana*. *Proc. Natl. Acad. Sci.* **102:** 12984.

Xu L., Yang L., Pi L., Liu Q., Ling Q., Wang H., Poethig R.S., and Huang H. 2006. Genetic interaction between the AS1-AS2 and RDR6-SGS3-AGO7 pathways for leaf morphogenesis. *Plant Cell Physiol.* **47:** 853.

Yang L., Liu Z., Lu F., Dong A., and Huang H. 2006. SERRATE is a novel nuclear regulator in primary microRNA processing in *Arabidopsis*. *Plant J.* **47:** 841.

Yoshikawa M., Peragine A., Park M.Y., and Poethig R.S. 2005. A pathway for the biogenesis of trans-acting siRNAs in *Arabidopsis*. *Genes Dev.* **19:** 2164.

Yu L., Yu X., Shen R., and He Y. 2005. HYL1 gene maintains venation and polarity of leaves. *Planta* **221:** 231.

The Function of RNAi in Plant Development

R.S. Poethig, A. Peragine, M. Yoshikawa,* C. Hunter, M. Willmann, and G. Wu

Department of Biology, University of Pennsylvania, Philadelphia, Pennsylvania 19104-6018

The morphological phenotype of mutations in genes required for posttranscriptional gene silencing (PTGS) or RNA interference (RNAi) in *Arabidopsis* demonstrates that this process is critical for normal development. One way in which RNAi contributes to gene regulation is through its involvement in the biogenesis of *trans*-acting small interfering RNAs (siRNAs). These endogenous siRNAs are derived from noncoding transcripts that are cleaved by a microRNA (miRNA) and mediate the silencing of protein-coding transcripts. Some protein-coding genes are also subject to miRNA-initiated transitive silencing. Several developmentally important transcription factors regulated by these silencing mechanisms have been identified.

Plants and animals produce two major types of 21- to 24-nucleotide RNAs: miRNAs and siRNAs (for review, see Du and Zamore 2005; Carthew 2006; Jones-Rhoades et al. 2006; Mallory and Vaucheret 2006; Valencia-Sanchez et al. 2006; Vazquez 2006). miRNAs are endogenously encoded and typically regulate genes involved in development or physiology. In contrast, siRNAs are the by-product of transcriptional or posttranscriptional silencing pathways, and in plants were thought to function primarily to suppress the expression of potentially deleterious RNAs, such as RNA derived from viruses and transposons. However, recent studies have revealed an unexpected role for RNAi in developmental gene regulation in plants and have led to the discovery of a new class of endogenous siRNAs, termed *trans*-acting siRNAs (ta-siRNAs) (Peragine et al. 2004; Vazquez et al. 2004; Allen et al. 2005; Williams et al. 2005; Yoshikawa et al. 2005). ta-siRNAs are interesting not only because of their regulatory function, but also because their biogenesis is dependent on miRNA-directed cleavage of a precursor transcript (Allen et al. 2005; Yoshikawa et al. 2005). These results not only have expanded our understanding of the normal function of RNAi in plants, but also have established a novel function for miRNA-directed transcript cleavage.

IDENTIFICATION OF TA-SIRNAS

ta-siRNAs are 21-nucleotide siRNAs that posttranscriptionally repress the expression of transcripts to which they have imperfect complementarity. ta-siRNAs were discovered independently in our laboratory (Peragine et al. 2004) and in the laboratory of H. Vaucheret (Vazquez et al. 2004). We identified this novel class of siRNAs from an analysis of mutations that affect the process of vegetative phase change in *Arabidopsis*. Like all flowering plants, *Arabidopsis* undergoes a transition from a juvenile to an adult phase of vegetative devel-

opment prior to flowering (Brink 1962; Telfer et al. 1997; Kerstetter and Poethig 1998; Baurle and Dean 2006). This transition is known as vegetative phase change, and in *Arabidopsis*, it is accompanied by changes in leaf shape and size and by the initiation of trichome production on the lower (abaxial) leaf surface (Chien and Sussex 1996; Telfer et al. 1997; Tsukaya et al. 2000). Screens for mutations that accelerate the appearance of adult vegetative traits have produced genes that fall into roughly two phenotypic classes. One class of mutations dramatically accelerates the appearance of adult vegetative traits and essentially eliminates the production of transition leaves. This class includes alleles of the miRNA export receptor *HST* (Telfer and Poethig 1998; Bollman et al. 2003), the cyclophilin 40 ortholog *SQN* (Berardini et al. 2001), and the miRNA-processing gene *SE* (Clarke et al. 1999; Prigge and Wagner 2001; Grigg et al. 2005; Yang et al. 2006). The second class of mutations includes alleles of *ZIP/AGO7*, *RDR6*, *SGS3*, and *DCL4*. Mutations of these genes have the same morphological phenotype, first described in detail for mutations of *ZIP/AGO7* (Fig. 1A,B) (Hunter et al. 2003). In contrast to the first class of mutations, which have a very pleiotropic phenotype, mutations in this second class have specific effects on leaf and flower development. These mutants precociously express a variety of traits associated with the adult phase of vegetative development, including epinastic leaf curvature, abaxial trichome production, and increased hydathode production and cause the septum of the silique to develop stigmatic tissue. Double- or triple-mutant combinations of these genes have a phenotype that is nearly identical to single mutants, implying that they operate in the same pathway (Fig. 1B) (Peragine et al. 2004).

Because two of this latter class of genes, *SGS3* and *RDR6*, had been shown to promote posttranscriptional silencing of sense transgenes in *Arabidopsis* (Dalmay et al. 2000; Mourrain et al. 2000), we reasoned that their phenotype was likely due to the up-regulation of genes that are normally silenced during the juvenile phase of development. To identify these genes, we performed a microarray analysis of RNA levels in *zip*, *sgs3*, and *rdr6* seedlings (Peragine et al. 2004). This analysis revealed two genes whose transcripts are elevated by *sgs3* and *rdr6*

---

*Present address: Department of Molecular Genetics, National Institute of Agrobiological Sciences, Tsukuba, Japan.

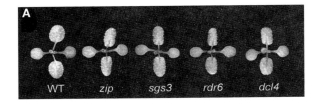

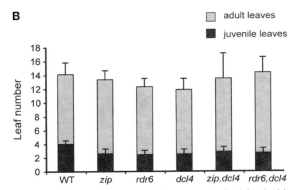

Figure 1. Phenotype of mutations in genes required for the biogenesis of ta-siRNAs in *Arabidopsis*. (*A*) 10-day-old seedlings of wild-type (WT), *zip-1*, *sgs3-11*, *rdr6-11*, and *dcl4-2*. (*B*) The number of juvenile and adult leaves in wild type and mutant plants (± S.D.). Juvenile leaves are leaves that lack abaxial trichomes. The phenotype of double mutants is nearly identical to that of single mutants. The interaction between *zip-1*, *sgs3-11*, and *rdr6-11* has been described previously (Peragine et al. 2004).

but not by *zip,* and three genes that are elevated by all three of these mutations. We were unable to identify siRNAs from these genes among known endogenous siRNAs. However, one up-regulated gene (At5g18040) had near-perfect reverse complementarity to a cloned siRNA (*siR255/siR480*) from an intergenic region of the genome. RNA ligase-mediated rapid amplification of 5′ cDNA ends (RLM-5′ RACE) revealed that At5g18040 is cleaved in the middle of this complementary region, supporting the conclusion that it is negatively regulated by *siR255/siR480*. *sgs3* and *rdr6* were found to block the production of *siR255/siR480*, accounting for the elevation of At5g18040 in these mutants. Similar results for these genes were obtained by Vazquez et al. (2004). Subsequent studies by us (Yoshikawa et al. 2005) and other workers (Allen et al. 2005; Williams et al. 2005) revealed that most of the other up-regulated genes identified in this microarray analysis are also targets of siRNAs from unrelated transcripts and showed that the elevated expression of these genes in *zip, rdr6,* and *sgs3* is correlated with the absence of these "*trans*-acting" siRNAs.

Five loci capable of producing ta-siRNAs—*TAS1a, TAS1b, TAS1c, TAS2,* and *TAS3*—have been identified in *Arabidopsis* (Fig. 2). *TAS1a, b,* and *c* are closely related and produce ta-siRNAs that target At5g18040 and members of a large family of closely related genes encoding pentatricopeptide repeat (PPR) proteins (Peragine et al. 2004; Vazquez et al. 2004; Yoshikawa et al. 2005). *TAS2* is a paralog of the *TAS1* loci, but it only produces ta-siRNAs that target PPR genes (Allen et al. 2005;

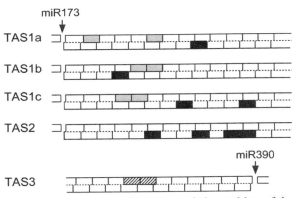

Figure 2. Sources of ta-siRNAs in *Arabidopsis*. Most of the siRNAs from these loci are in phase with a miRNA cleavage site. Only confirmed or predicted ta-siRNAs are illustrated. Targets of ta-siRNAs: (*black*) PPR genes; (*gray*) At5g18040-related genes; (*diagonal lines*) ARF2, ETT/ARF3, and ARF4.

Yoshikawa et al. 2005). *TAS1* and *TAS2* do not appear to be conserved outside *Arabidopsis thaliana*. *TAS3* regulates three members of the ARF family of auxin-related transcription factors (*ARF2, ETT/ARF3,* and *ARF4*) and is conserved throughout higher plants (Allen et al. 2005; Williams et al. 2005).

BIOGENESIS OF TA-SIRNAS

How are ta-siRNAs generated? Vazquez et al. (2004) reported that most of the 21-nucleotide siRNAs from *TAS1a* are in exact register with each other. The basis for this pattern became apparent when it was discovered that siRNAs from *TAS1, TAS2,* and *TAS3* are all in phase with a miRNA cleavage site. *TAS1* and *TAS2* are targets of *miR173,* and *TAS3* is a target of *miR390* (Allen et al. 2005; Yoshikawa et al. 2005). This immediately suggested that ta-siRNAs are derived from progressive, Dicer-mediated cleavage, starting at the miRNA target site. Evidence that miRNA-directed cleavage is indeed essential for the biogenesis of ta-siRNAs was provided by the reconstitution of *TAS1a* processing in a heterologous system (Allen et al. 2005) and by the observation that ta-siRNA biogenesis is blocked by mutations in *miR173* (Gasciolli et al. 2005) or mutations that disrupt miRNA biogenesis (Yoshikawa et al. 2005).

The specific function of *RDR6, SGS3,* and *DCL4* in the biogenesis of ta-siRNAs was determined from an analysis of the effect of mutations in these genes on the primary transcripts from *TAS1a* and *TAS2* (Fig. 3) (Yoshikawa et al. 2005). *RDR6* and *SGS3* were originally identified in screens for mutants that block the posttranscriptional silencing of transgenes (Dalmay et al. 2000; Mourrain et al. 2000). *RDR6* was proposed to be responsible for the transformation of mRNA into double-stranded RNA based on its similarity to a tomato RNA-dependent RNA polymerase and the observation that it is required for the silencing of transgenes that produce "sense" transcripts, but not transgenes that have been engineered to produce a hairpin transcript (Beclin et al. 2002). Additional evi-

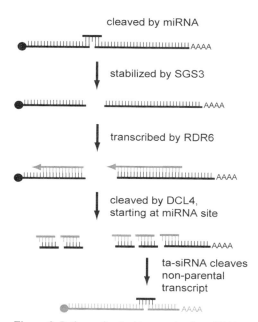

cleaved by miRNA

stabilized by SGS3

transcribed by RDR6

cleaved by DCL4, starting at miRNA site

ta-siRNA cleaves non-parental transcript

Figure 3. Pathway for the biogenesis of ta-siRNAs.

dence for this conclusion was provided by the observation that *RDR6* is required for transitivity—the production of siRNAs from an untargeted region of a transcript (Vaistij et al. 2002). *SGS3* is a plant-specific protein of unknown function. Although mutations in these genes have identical effects on transgene silencing (Beclin et al. 2002) and have identical morphological phenotypes (Peragine et al. 2004), they have opposite effects on the accumulation of the *miR173*-generated cleavage fragments of *TAS2* and *TAS1a* (Yoshikawa et al. 2005). *sgs3* decreases the accumulation of these fragments, and *rdr6* produces a dramatic increase in their accumulation. *sgs3 rdr6* double mutants have the same phenotype as *sgs3*, indicating that *SGS3* is required for the accumulation of these fragments both in the presence and in the absence of *RDR6* function. This epistatic interaction suggests that SGS3 somehow stabilizes these fragments before they are transcribed by RDR6. The subsequent cleavage of double-stranded transcripts into 21-nucleotide siRNAs is accomplished by DCL4, one of four Dicer-like proteins in *Arabidopsis* (Dunoyer et al. 2005; Gasciolli et al. 2005; Xie et al. 2005; Yoshikawa et al. 2005). The R2D2-like protein, DRB4, binds to DCL4 (Hiraguri et al. 2005) and may assist in its activity because the phenotype of a *drb4* mutation resembles *dcl4* (Adenot et al. 2006). *TAS2* and *TAS3* are also processed inefficiently by DCL2 into 22-nucleotide siRNAs and by DCL3 into 24-nucleotide siRNAs (Gasciolli et al. 2005; Bouche et al. 2006; Henderson et al. 2006).

FUNCTION OF TA-SIRNAS

The loss-of-function phenotype of genes required for RNAi in *Arabidopsis* demonstrates that this process has a role in leaf and flower development, although the subtle nature of their phenotype demonstrates that this role is

relatively limited. Most of the up-regulated genes identified by transcript profiling of *zip, rdr6,* and *sgs3* are targets of ta-siRNAs, implying that defects in this class of siRNAs largely account for the mutant phenotype of these genes. T-DNA (transferred DNA) insertions in *TAS1* and *TAS2* have no apparent morphological phenotype (Vazquez et al. 2004; Yoshikawa et al. 2005). In contrast, a T-DNA insertion in *TAS3* has a phenotype quite similar to that of *zip, rdr6, sgs3,* and *dcl4* (Adenot et al. 2006). *TAS3* produces a ta-siRNA (*tasi-ARF*) that targets members of the ARF family of auxin-related transcription factors (Allen et al. 2005; Williams et al. 2005). Two of the targets of *tasi-ARF—ETT/ARF3* and *ARF4*—are elevated in *zip, sgs3, rdr6,* and *dcl4,* and the precocious phenotype of these mutations is mimicked by transgenes that express a *tasi-ARF*-insensitive form of *ETT/ARF3* (Fahlgren et al. 2006; Hunter et al. 2006). Additional evidence that *ARF3* and *ARF4* are responsible for the phenotype of *zip, rdr6, sgs3,* and *dcl4* was provided by the isolation of loss-of-function mutations of these genes in a screen for second-site suppressors of *zip* (Hunter et al. 2006). We found that *ett* and *arf4* partially suppress the phenotype of *zip* and *rdr6,* demonstrating that the wild-type alleles of these genes are necessary for the *zip* mutant phenotype. The observation that neither single mutant is capable of completely suppressing the phenotype of *zip* suggests that both genes contribute to this phenotype, and that they have similar, but not identical, functions in leaf development. This conclusion is also supported by the observation that the level of *ARF3* mRNA necessary to replicate the phenotype of *zip* (in plants transformed with *35S::ARF3*) is considerably higher than the level of *ARF3* mRNA in *zip* mutants (Hunter et al. 2006).

The loss-of-function phenotypes of *ETT/ARF3* and *ARF4* demonstrate that these genes have several functions in leaf development. Mutations in *ETT/ARF3* were originally identified because of their effects on carpel polarity (Sessions and Zambryski 1995; Sessions et al. 1997). Subsequent studies showed that *ETT/ARF3* and *ARF4* regulate leaf polarity as well, probably in conjunction with the *KAN* family of transcription factors (Pekker et al. 2005; Hunter et al. 2006). Strong *ett/arf3* mutations simplify the shape of spongy mesophyll cells in a manner similar to that of *kan1,* and *ett/arf3 arf4* double mutants resemble *kan1 kan2* double mutants. Furthermore, *ett/arf3* is able to suppress the floral phenotype of *KAN1* overexpression (Pekker et al. 2005). Because many of the traits that change during vegetative development are polarized (e.g., trichome production and leaf curvature), it may be that temporal changes in the regulation of leaf polarity are the basis for the phase-specific expression pattern of these traits. It is unlikely that this is the sole function of *ETT/ARF3* and *ARF4,* however, because some aspects of their mutant phenotype cannot be readily explained by a loss of abaxial identity. In particular, both mutations delay abaxial trichome production and affect the length:width ratio of the lamina (Hunter et al. 2006). Because defects in the specification of abaxial identity typically result in enhanced abaxial trichome production, the trichome phenoytpe of *ett/arf3* and *arf4* is inconsistent with a defect in leaf polarity.

IS RNAi of miRNA-REGULATED GENES FUNCTIONALLY IMPORTANT?

The fact that transcript profiling of *zip, sgs3,* and *rdr6* produced only a small number of up-regulated genes suggests that RNAi does not make a major contribution to the regulation of gene expression in *Arabidopsis* (Peragine et al. 2004). However, northern analysis of several miRNA-regulated protein-coding genes has revealed that these genes are the source of siRNAs and that these siRNAs are dependent on *AGO1, DCL1, DCL3, RDR6,* and *RDR2* (Ronemus et al. 2006). Thus, at least some miRNA-regulated protein-coding genes appear to be subject to RNAi, presumably as a result of being cleaved by a miRNA. Whether this process has a significant role in the expression of these genes remains to be determined.

One miRNA-regulated gene whose expression is elevated in *zip, rdr6,* and *sgs3* is *SPL3*. *SPL3* encodes a member of the Squamosa promoter-binding family of transcription factors (Cardon et al. 1999), ten of which possess a cognate site for *miR156* (Rhoades et al. 2002). *SPL3* is particularly interesting because its mRNA increases early in vegetative development, coincident with a decrease in the level of *miR156* (Wu and Poethig 2006). This temporal change in *SPL3* expression is also observed with a GUS-SPL3 fusion protein under the transcriptional regulation of the cauliflower mosaic virus (CaMV) 35S promoter and is dependent on the presence of a functional *miR156* target site, indicating that it is likely a consequence of the decrease in *miR156*. Misexpression of *SPL3*, or its close relatives *SPL4* and *SPL5*, as a consequence of the loss of their *miR156* target site results in precocious vegetative phase change and early flowering. These results suggest that these genes have a critical role in vegetative phase change and flowering and raise the possibility that changes in the expression of *miR156* may have an important role in these transitions (Wu and Poethig 2006).

What is the function of *ZIP, SGS3,* and *RDR6* in the regulation of *SPL3*? The observation that *SPL3* is up-regulated in these mutants indicates that it is either directly or indirectly subject to RNAi. One possibility is that cleavage of *SPL3* by *miR156* makes it susceptible to transitive silencing, just as miRNA-directed cleavage initiates the production of siRNAs from the precursors of ta-siRNAs (Allen et al. 2005; Yoshikawa et al. 2005). If transitive silencing had a critical role in the activity of *miR156*, one might expect mutations in this process to have a phenotype similar to that of mutations that affect the biogenesis of *miR156*. However, *zip* and *rdr6* only cause a slight, temporally uniform increase in the abundance of *SPL3* mRNA, whereas *hst*—a mutation that dramatically reduces *miR156* levels—produces a very large and temporally variable increase in *SPL3* (Wu and Poethig 2006). Furthermore, we were unable to detect siRNAs derived from *SPL3* by northern analysis and did not find any evidence for the existence of these siRNAs in databases of endogenous small RNAs in *Arabidopsis*. Although the possibility that these siRNAs may be present at extremely low levels has not been eliminated, the existing evidence suggests that the effect of *zip, sgs3,* and *rdr6* on *SPL3*

expression may be an indirect effect of these mutations on a transcriptional regulator of *SPL3*.

FUNCTION OF RNAi IN VEGETATIVE PHASE CHANGE

The results described above demonstrate that RNAi normally promotes the expression of juvenile traits by repressing the expression of genes such as *ETT/ARF3*, *ARF4,* and *SPL3* during this phase. Therefore, one way in which phase change might be regulated is by a decrease in the activity of this silencing pathway late in shoot development. There is currently no support for this hypothesis, however. We observed no change in the abundance of *tasi-ARF* during development, nor did we detect a difference in the expression of its targets *ETT/ARF3* and *ARF4* in successive leaves (Hunter et al. 2006). Similarly, *zip* and *rdr6* cause a uniform increase in the expression of *SPL3*, rather than affecting the temporal expression pattern of this gene (Wu and Poethig 2006). These data are consistent with the observation that *sgs3* and *rdr6* block transgene silencing throughout shoot development (Dalmay et al. 2000; Mourrain et al. 2000; Peragine et al. 2004) and suggest that there is no major temporal change in the activity of the posttranscriptional gene silencing pathway during shoot growth. We conclude from these observations that RNAi regulates the sensitivity of the shoot apex to factors that regulate phase change, rather than being a component of the phase change signal.

Although there is no evidence that temporal changes in the expression of *ZIP, SGS3, RDR6,* and *DCL4* contribute to vegetative phase change, this may not be true for components of the transcriptional silencing pathway. Temporal variation in gene silencing has been described for transposons (Banks et al. 1988; Fedoroff and Banks 1988; Martienssen et al. 1990) and the *Pl-Bh* mutation (Cocciolone and Cone 1993; Hoekenga et al. 2000; Irish and McMurray 2006) in maize, as well as for transgenes in tobacco (de Carvalho et al. 1992) and *Arabidopsis* (Elmayan et al. 1998; Vaucheret et al. 2004). In general, silencing increases along the length of the shoot; genes subject to silencing are typically expressed at a higher level in organs produced early in development (cotyledons, juvenile leaves) than in organs produced later in shoot development (adult leaves, flowers). In maize, this temporal decrease in gene expression or transposon activity has been associated with increased DNA methylation (Banks et al. 1988; Fedoroff and Banks 1988; Martienssen et al. 1990; Hoekenga et al. 2000; Irish and McMurray 2006), and there is evidence that DNA methylation also increases during shoot development in *Arabidopsis* (Ruiz-Garcia et al. 2005). Whether this temporal change in gene silencing and DNA methylation represents the accumulation of changes that occur at a constant rate during shoot growth or is the result of a discrete change in the activity of a gene silencing pathway is unknown.

CONCLUSIONS

tasi-ARF and its targets are present in many flowering plants and in at least one gymnosperm (Allen et al. 2005;

Williams et al. 2005). Furthermore, *miR390*—the miRNA that is responsible for the production of *tasi-ARF*—is conserved in moss (Axtell and Bartel 2005). Although the targets of *miR390* in moss have not yet been identified, these observations indicate that ta-siRNAs appeared early in plant evolution. Given that ta-siRNAs are ancient and effective mechanisms for simultaneously regulating a large number of related and unrelated genes, one might imagine that this mechanism would have been strongly selected in the evolution of regulatory pathways. However, there is little evidence for this conclusion. Mutations that interfere with RNAi have very subtle phenotypes in *Arabidopsis*, and only a few of the known miRNA targets in *Arabidopsis* are affected by these mutations. Although additional sources of ta-siRNAs probably remain to be discovered, current evidence suggests that the number of such loci in *Arabidopsis* is not large. Why are ta-siRNAs in particular, and RNAi in general, not more widely used in developmental regulation? Did this process evolve for a purpose (e.g., virus resistance) that makes it difficult to use it for alternative purposes? This might be the case, for example, if the viral RNA substrates of RNAi and endogenous mRNAs are structurally different in some way or are shunted to different processing pathways. Or, more interestingly, does RNAi have properties that make it more useful for regulating some types of processes than others? Answers to these questions will have to await a better understanding of the mechanism of RNAi in plants and the factors that make transcripts susceptible to this process.

ACKNOWLEDGMENTS

We thank members of the Poethig lab for helpful discussions. This research was supported by a National Institutes of Health training grant to A.P., NIH postdoctoral fellowships to M.W. and C.H., and an NIH research grant to R.S.P.

REFERENCES

Adenot X., Elmayan T., Lauressergues D., Boutet S., Bouche N., Gasciolli V., and Vaucheret H. 2006. DRB4-dependent TAS3 *trans*-acting siRNAs control leaf morphology through AGO7. *Curr. Biol.* **16**: 927.

Allen E., Xie Z., Gustafson A.M., and Carrington J.C. 2005. microRNA-directed phasing during *trans*-acting siRNA biogenesis in plants. *Cell* **121**: 207.

Axtell M.J. and Bartel D.P. 2005. Antiquity of microRNAs and their targets in land plants. *Plant Cell* **17**: 1658.

Banks J.A., Masson P., and Fedoroff N. 1988. Molecular mechanisms in the developmental regulation of the maize Suppressor-mutator transposable element. *Genes Dev.* **2**: 1364.

Baurle I. and Dean C. 2006. The timing of developmental transitions in plants. *Cell* **125**: 655.

Beclin C., Boutet S., Waterhouse P., and Vaucheret H. 2002. A branched pathway for transgene-induced RNA silencing in plants. *Curr. Biol.* **12**: 684.

Berardini T.Z., Bollman K., Sun H., and Poethig R.S. 2001. Regulation of vegetative phase change in *Arabidopsis thaliana* by cyclophilin 40. *Science* **291**: 2405.

Bollman K.M., Aukerman M.J., Park M.Y., Hunter C., Berardini T.Z., and Poethig R.S. 2003. *HASTY*, the *Arabidopsis*

ortholog of *Exportin 5/MSN5*, regulates phase change and morphogenesis. *Development* **130**: 1493.

Bouche N., Lauressergues D., Gasciolli V., and Vaucheret H. 2006. An antagonistic function for *Arabidopsis* DCL2 in development and a new function for DCL4 in generating viral siRNAs. *EMBO J.* **25**: 3347.

Brink R.A. 1962. Phase change in higher plants and somatic cell heredity. *Q. Rev. Biol.* **37**: 1.

Cardon G., Hohmann S., Klein J., Nettesheim K., Saedler H., and Huijser P. 1999. Molecular characterisation of the *Arabidopsis* SBP-box genes. *Gene* **237**: 91.

Carthew R.W. 2006. Gene regulation by microRNAs. *Curr. Opin. Genet. Dev.* **16**: 203.

Chien J.C. and Sussex I.M. 1996. Differential regulation of trichome formation on the adaxial and abaxial leaf surfaces by gibberellins and photoperiod in *Arabidopsis thaliana* (L) Heynh. *Plant Physiol.* **111**: 1321.

Clarke J.H., Tack D., Findlay K., Van Montagu M., and Van Lijsebettens M. 1999. The *SERRATE* locus controls the formation of the early juvenile leaves and phase length in *Arabidopsis. Plant J.* **20**: 493.

Cocciolone S.M. and Cone K.C. 1993. *Pl-Bh*, an anthocyanin regulatory gene of maize that leads to variegated pigmentation. *Genetics* **135**: 575.

Dalmay T., Hamilton A., Rudd S., Angell S., and Baulcombe D.C. 2000. An RNA-dependent RNA polymerase gene in *Arabidopsis* is required for posttranscriptional gene silencing mediated by a transgene but not by a virus. *Cell* **101**: 543.

de Carvalho F., Gheysen G., Kushnir S., Van Montagu M., Inze D., and Castresana C. 1992. Suppression of beta-1,3-glucanase transgene expression in homozygous plants. *EMBO J.* **11**: 2595.

Du T. and Zamore P.D. 2005. microPrimer: The biogenesis and function of microRNA. *Development* **132**: 4645.

Dunoyer P., Himber C., and Voinnet O. 2005. *DICER-LIKE 4* is required for RNA interference and produces the 21-nucleotide small interfering RNA component of the plant cell-to-cell silencing signal. *Nat. Genet.* **37**: 1356.

Elmayan T., Balzergue S., Beon F., Bourdon V., Daubremet J., Guenet Y., Mourrain P., Palauqui J.C., Vernhettes S., Vialle T., et al. 1998. *Arabidopsis* mutants impaired in cosuppression. *Plant Cell* **10**: 1747.

Fahlgren N., Montgomery T.A., Howell M.D., Allen E., Dvorak S.K., Alexander A.L., and Carrington J.C. 2006. Regulation of *AUXIN RESPONSE FACTOR3* by *TAS3* ta-siRNA affects developmental timing and patterning in *Arabidopsis. Curr. Biol.* **16**: 939.

Fedoroff N.V. and Banks J.A. 1988. Is the Suppressor-mutator element controlled by a basic developmental regulatory mechanism? *Genetics* **120**: 559.

Gasciolli V., Mallory A.C., Bartel D.P., and Vaucheret H. 2005. Partially redundant functions of *Arabidopsis* DICER-like enzymes and a role for DCL4 in producing *trans*-acting siRNAs. *Curr. Biol.* **15**: 1494.

Grigg S.P., Canales C., Hay A., and Tsiantis M. 2005. *SERRATE* coordinates shoot meristem function and leaf axial patterning in *Arabidopsis. Nature* **437**: 1022.

Henderson I.R., Zhang X., Lu C., Johnson L., Meyers B.C., Green P.J. and Jacobsen S.E. 2006. Dissecting *Arabidopsis thaliana* DICER function in small RNA processing, gene silencing and DNA methylation patterning. *Nat. Genet.* **38**: 721.

Hiraguri A., Itoh R., Kondo N., Nomura Y., Aizawa D., Murai Y., Koiwa H., Seki M., Shinozaki K., and Fukuhara T. 2005. Specific interactions between Dicer-like proteins and HYL1/DRB-family dsRNA-binding proteins in *Arabidopsis thaliana. Plant Mol. Biol.* **57**: 173.

Hoekenga O.A., Muszynski M.G. and Cone K.C. 2000. Developmental patterns of chromatin structure and DNA methylation responsible for epigenetic expression of a maize regulatory gene. *Genetics* **155**: 1889.

Hunter C., Sun H., and Poethig R.S. 2003. The *Arabidopsis* heterochronic gene *ZIPPY* is an *ARGONAUTE* family member. *Curr. Biol.* **13**: 1734.

Hunter C., Willmann M.R., Wu G., Yoshikawa M., de la Luz Gutierrez-Nava M., and Poethig S.R. 2006. Trans-acting siRNA-mediated repression of *ETTIN* and *ARF4* regulates heteroblasty in *Arabidopsis*. *Development* **133:** 2973.

Irish E.E. and McMurray D. 2006. Rejuvenation by shoot apex culture recapitulates the developmental increase of methylation at the maize gene *Pl-Blotched*. *Plant Mol. Biol.* **60:** 747.

Jones-Rhoades M.W., Bartel D.P., and Bartel B. 2006. MicroRNAs and their regulatory roles in plants. *Annu. Rev. Plant Biol.* **57:** 19.

Kerstetter R.A. and Poethig R.S. 1998. The specification of leaf identity during shoot development. *Annu. Rev. Cell Dev. Biol.* **14:** 373.

Mallory A.C. and Vaucheret H. 2006. Functions of microRNAs and related small RNAs in plants. *Nat. Genet.* (suppl.) **38:** S31.

Martienssen R., Barkan A., Taylor W.C., and Freeling M. 1990. Somatically heritable switches in the DNA modification of Mu transposable elements monitored with a suppressible mutant in maize. *Genes Dev.* **4:** 331.

Mourrain P., Beclin C., Elmayan T., Feuerbach F., Godon C., Morel J.B., Jouette D., Lacombe A.M., Nikic S., Picault N., et al. 2000. *Arabidopsis SGS2* and *SGS3* genes are required for posttranscriptional gene silencing and natural virus resistance. *Cell* **101:** 533.

Pekker I., Alvarez J.P., and Eshed Y. 2005. Auxin response factors mediate *Arabidopsis* organ asymmetry via modulation of KANADI activity. *Plant Cell* **17:** 2899.

Peragine A., Yoshikawa M., Wu G., Albrecht H.L., and Poethig R.S. 2004. *SGS3* and *SGS2/SDE1/RDR6* are required for juvenile development and the production of *trans*-acting siRNAs in *Arabidopsis*. *Genes Dev.* **18:** 2368.

Prigge M.J. and Wagner D.R. 2001. The *Arabidopsis SERRATE* gene encodes a zinc-finger protein required for normal shoot development. *Plant Cell* **13:** 1263.

Rhoades M.W., Reinhart B.J., Lim L.P., Burge C.B., Bartel B., and Bartel D.P. 2002. Prediction of plant microRNA targets. *Cell* **110:** 513.

Ronemus M., Vaughn M.W., and Martienssen R. A. 2006. MicroRNA-targeted and small interfering RNA-mediated mRNA degradation is regulated by ARGONAUTE, DICER, and RNA-dependent RNA polymerase in *Arabidopsis*. *Plant Cell* **18:** 1559.

Ruiz-Garcia L., Cervera M.T., and Martinez-Zapater J.M. 2005. DNA methylation increases throughout *Arabidopsis* development. *Planta* **222:** 301.

Sessions R.A. and Zambryski P.C. 1995. *Arabidopsis* gynoecium structure in the wild and in *ettin* mutants. *Development* **121:** 1519.

Sessions A., Nemhauser J.L., McColl A., Roe J.L., Feldmann K.A., and Zambryski P.C. 1997. *ETTIN* patterns the *Arabidopsis* floral meristem and reproductive organs. *Development* **124:** 4481.

Telfer A. and Poethig R.S. 1998. *HASTY:* A gene that regulates the timing of shoot maturation in *Arabidopsis thaliana*. *Development* **125:** 1889.

Telfer A., Bollman K.M., and Poethig R.S. 1997. Phase change and the regulation of trichome distribution in *Arabidopsis thaliana*. *Development* **124:** 645.

Tsukaya H., Shoda K., Kim G.T., and Uchimiya H. 2000. Heteroblasty in *Arabidopsis thaliana* (L.) Heynh. *Planta* **210:** 536.

Vaistij F.E., Jones L., and Baulcombe D.C. 2002. Spreading of RNA targeting and DNA methylation in RNA silencing requires transcription of the target gene and a putative RNA-dependent RNA polymerase. *Plant Cell* **14:** 857.

Valencia-Sanchez M.A., Liu J., Hannon G.J., and Parker R. 2006. Control of translation and mRNA degradation by miRNAs and siRNAs. *Genes Dev.* **20:** 515.

Vaucheret H., Vazquez F., Crete P., and Bartel D.P. 2004. The action of *ARGONAUTE1* in the miRNA pathway and its regulation by the miRNA pathway are crucial for plant development. *Genes Dev.* **18:** 1187.

Vazquez F. 2006. *Arabidopsis* endogenous small RNAs: Highways and byways. *Trends Plant Sci.* **11:** 460.

Vazquez F., Vaucheret H., Rajagopalan R., Lepers C., Gasciolli V., Mallory A.C., Hilbert J.L., Bartel D.P., and Crete P. 2004. Endogenous *trans*-acting siRNAs regulate the accumulation of *Arabidopsis* mRNAs. *Mol. Cell* **16:** 69.

Williams L., Carles C.C., Osmont K.S., and Fletcher J.C. 2005. A database analysis method identifies an endogenous *trans*-acting short-interfering RNA that targets the *Arabidopsis ARF2*, *ARF3*, and *ARF4* genes. *Proc. Natl. Acad. Sci.* **102:** 9703.

Wu G. and Poethig R. S. 2006. Temporal regulation of shoot development in *Arabidopsis thaliana* by *miR156* and its target *SPL3*. *Development* **133:** 3539.

Xie Z., Allen E., Wilken A., and Carrington J.C. 2005. *DICER-LIKE 4* functions in *trans*-acting small interfering RNA biogenesis and vegetative phase change in *Arabidopsis thaliana*. *Proc. Natl. Acad. Sci.* **102:** 12984.

Yang L., Liu Z., Lu F., Dong A., and Huang H. 2006. *SERRATE* is a novel nuclear regulator in primary microRNA processing in *Arabidopsis*. *Plant J.* **47:** 841.

Yoshikawa M., Peragine A., Park M.Y., and Poethig R.S. 2005. A pathway for the biogenesis of *trans*-acting siRNAs in *Arabidopsis*. *Genes Dev.* **19:** 2164.

rasiRNAs, DNA Damage, and Embryonic Axis Specification

W.E. Theurkauf,* C. Klattenhoff,* D.P. Bratu,* N. McGinnis-Schultz,*
B.S. Koppetsch,* and H.A. Cook[†]

*Program in Molecular Medicine and Program in Cell Dynamics, University of Massachusetts Medical School,
Worcester, Massachusetts 01605; [†]Department of Biological Sciences, Wagner College, Staten Island, New York 10301

Drosophila repeat-associated small interfering RNAs (rasiRNAs) have been implicated in retrotransposon and *stellate* locus silencing. However, mutations in the rasiRNA pathway genes *armitage*, *spindle-E*, and *aubergine* disrupt embryonic axis specification, triggering defects in microtubule organization and localization of *osk* and *grk* mRNAs during oogenesis. We show that mutations in *mei-41* and *mnk*, which encode ATR and Chk2 kinases that function in DNA damage signal transduction, dramatically suppress the cytoskeletal and RNA localization defects associated with rasiRNA mutations. In contrast, *stellate* and retrotransposon silencing are not restored in *mei-41* and *mnk* double mutants. We also find that *armitage*, *aubergine*, and *spindle-E* mutations lead to germ-line-specific accumulation of γ-H2Av foci, which form at DNA double-strand breaks, and that mutations in *armi* lead to Chk2-dependent phosphorylation of Vasa, an RNA helicase required for axis specification. The *Drosophila* rasiRNA pathway thus appears to suppress DNA damage in the germ line, and mutations in this pathway block axis specification by activating an ATR/Chk2-dependent DNA damage response that disrupts microtubule polarization and RNA localization.

RNA interference (RNAi) and related processes utilize short RNAs to direct protein complexes to chromatin or RNA, triggering heterochromatin formation, transcriptional silencing, translational repression, or RNA destruction (Hannon 2002; Hutvagner and Zamore 2002; Wassenegger 2005). Mutations in genes involved in small RNA function lead to embryonic lethality in mice (Bernstein et al. 2003), disrupt embryo morphogenesis in zebra fish (Giraldez et al. 2005), and lead to defects in developmental timing in worms (Grishok et al. 2001). Mutations in RNAi-related processes also interfere with stem cell division, stem cell maintenance, and viral immunity in flies (Forstemann et al. 2005; Hatfield et al. 2005; Galiana-Arnoux et al. 2006; Wang et al. 2006), and block chromosome segregation in cultured chicken cells (Fukagawa et al. 2004) and yeast (Provost et al. 2002; Volpe et al. 2003). However, the full scope of biological functions controlled by small RNAs is only beginning to emerge, and new classes of these RNAs continue to be discovered (Aravin et al. 2006; Girard et al. 2006; Vagin et al. 2006). Moreover, the targets for most small RNAs have not been identified, and the molecular basis for the phenotypes associated with small RNA production and function are not well understood.

Mutations in the *Drosophila armitage* (*armi*), *spindle-E* (*spn-E*), and *aubergine* (*aub*) genes disrupt siRNA-guided RNA cleavage and assembly of the RNA-induced silencing complex (RISC) in ovary extracts and production of 24- to 30-nucleotide repeat-associated siRNAs (rasiRNAs), which are linked to retrotransposon and *Stellate* locus silencing (Aravin et al. 2004; Tomari et al. 2004b; Vagin et al. 2006). Strong loss-of-function mutations in these genes disrupt embryonic axis specification, triggering defects in microtubule organization and microtubule-dependent localization of mRNA and protein determinants in the developing oocyte (Cook et al. 2004).

In contrast, mutations in *argonaute-2* (*ago-2*) and *dicer-2* (*dcr-2*) that disrupt the siRNA pathway, but do not block rasiRNA production, are viable and fertile (Lee et al. 2004; Okamura et al. 2004; Tomari et al. 2004a; Deshpande et al. 2005; Vagin et al. 2006). The rasiRNA pathway thus appears to have an essential function in embryonic axis specification; however, the critical developmental targets for this pathway have not been defined.

The axes of the *Drosophila* embryo are specified through asymmetric localization of RNA and protein determinants within the oocyte, and this complex process presents a number of potential targets for rasiRNA control. Oogenesis begins in the germarium with a stem cell division that produces a cystoblast, which divides four times with incomplete cytokinesis to generate a cyst of 16 germ-line cells interconnected by ring canal junctions (Spradling 1993). The cysts reorganize and are surrounded by somatic follicle cells as they pass through the germarium. The pro-oocyte is at the posterior of cysts in region 3 of the germarium. A single microtubule organizing center (MTOC) is present at the anterior pole of the pro-oocyte, and microtubules extend from this MTOC into the 15 pro-nurse cells (Theurkauf et al. 1992). As the germ line/somatic cell complexes bud from region 3 to form stage 2 egg chambers, the MTOC re-forms at the posterior cortex of the oocyte (Theurkauf et al. 1992; Riechmann and Ephrussi 2001). The posterior MTOC directs *gurken* (*grk*) mRNA and protein to the posterior pole. Grk protein is a TGFα homolog that signals to the somatic follicle cells, which leads to posterior differentiation (Gonzalez-Reyes et al. 1995; Roth et al. 1995). During mid-oogenesis, the posterior follicle cells signal back to the oocyte, inducing a reorganization of the oocyte microtubule cytoskeleton that leads to *oskar* mRNA localization to the posterior pole and dorsal–anterior localization of Grk, which signals to the dorsal follicle cells to trigger dorsal differentiation (Neuman-Silberberg and Schupbach 1993; Gonzalez-Reyes et al. 1995; Roth et al. 1995). Microtubule polarization and Grk localization during early oogenesis thus initiate embryonic axis specification.

Mutations in the rasiRNA pathway lead to premature expression of Osk protein during early oogenesis (Cook et al. 2004), suggesting that the axis specification defects associated with these mutations may result from defects in expression of axis specification genes. However, here we show that the cytoskeletal polarization and morphogen localization defects associated with *armi* and *aub* are dramatically suppressed by null mutations in *mei-41* and *mnk*, which encode ATR and Chk2 kinases that function in DNA double-strand-break (DSB) signaling. We also show that rasiRNA pathway mutations lead to germ-line-specific accumulation of γ-H2Av foci characteristic of DNA DSBs. Significantly, the ATR/Chk2 mutations do not suppress the defects in retrotransposon and *Stellate* silencing. We therefore conclude that rasiRNA-based gene silencing is not required for axis specification, and that the critical developmental function for the *Drosophila* rasiRNA pathway is to maintain germ-line genome integrity. Mutations in the rasiRNA pathway thus lead to DNA damage, which activates an ATR and Chk2 kinase pathway that blocks axis specification by disrupting microtubule polarization and asymmetric RNA localization.

ATR AND Chk2 MUTATIONS SUPPRESS *ARMI* AND *AUB* AXIS SPECIFICATION DEFECTS

The *armi*, *spn-E*, and *aub* genes are required for production of rasiRNAs and lead to Stellate overexpression during spermatogenesis and premature Oskar protein expression during oogenesis (Aravin et al. 2001; Cook et al. 2004; Vagin et al. 2006). Mutations in these genes lead to female sterility and disrupt embryonic axis specification, suggesting that rasiRNAs control expression of genes involved in patterning the oocyte (Cook et al. 2004). However, mutations in the meiotic DSB repair

pathway also lead to axis specification defects through activation of a posttranslational DNA damage response (Ghabrial et al. 1998). In these mutants, persistent meiotic DSBs activate a damage signaling pathway that includes the ATR and Chk2 kinases, which induce the observed axis specification defects (Ghabrial and Schupbach 1999; Bartek et al. 2001; Abdu et al. 2002). These findings raised the alternative possibility that rasiRNA pathway mutations disrupt axis specification by activating the ATR and Chk2 pathway.

To genetically test the role of DNA damage signaling in the rasiRNA pathway mutant phenotype, we analyzed double mutant combinations with *mei-41* or *mnk*, which encode the *Drosophila* ATR and Chk2 homologs. We were unable to recover *mnk;spn-E* double mutants, and it is unclear whether this reflects a significant negative genetic interaction between these genes or is the result of background mutations on the *mnk* or *spn-E* chromosome. Our analyses thus focused on *armi* and *aub*, which we were able to combine with *mei-41* and *mnk*. If *armi* and *aub* mutations block axis specification through ATR/Chk2 activation, the patterning defects associated with these mutations will be suppressed in the double mutants. Initial suppression analysis focused on the dorsal appendages, which are easily scored eggshell structures that are induced through Grk signaling from the oocyte to the somatic follicle cells during mid-oogenesis (Schupbach 1987). Appendages do not form in the absence of Grk, a single appendage forms with low Grk levels, and two appendages form when signaling is normal (Gonzalez-Reyes et al. 1995; Roth et al. 1995). As shown in Table 1, *mei-41* and *mnk* dramatically suppress the appendage defects associated with *armi* and *aub*. Two appendages are present on 100% of the embryos derived from wild-type and *mei-41* females, and on 94% of the embryos derived from *mnk* single mutants (Table 1). In contrast, only 3.5% of the embryos derived from *armi*[72.1]/*armi*[1] mutant females have two dorsal appendages. Strikingly, 92% of the embryos derived from *mnk; armi*[72.1]/*armi*[1] double mutants show wild-type appendage morphology.

Table 1. *mnk* and *mei-41* Mutations Suppress Dorsal–Ventral Patterning Defects in rasiRNA Mutants

Maternal genotype	Dorsal appendage (%) phenotype			Hatch rate (%)	N
	2 (wild type)	1 (fused)	0 (absent)		
mnkP6 / mnkP6	94.1	2.3	3.6	73.9	827
mei41D3 / mei41D3	100	0	0	0	920
meiW681 / meiW68K05603	94.3	4	1.7	67.2	1281
armi72.1 / armi1	3.5	67.6	28.9	0	765
mnkP6 / mnkP6 ; armi72.1 / armi1	91.9	2.5	5.6	0	1062
mei41D3 / mei41D3 ; armi72.1 / armi1	56	38.4	5.6	0	575
meiW681 / meiW68K05603 ; armi72.1 / armi1	3.6	37.9	58.5	0	280
aubHN2 / aubQC42	47.7	40.3	12	0	1212
mnkP6, aubHN2 / mnkP6, aubQC42	97.6	2	0.4	0	296
mei41D3 / mei41D3 ; aubHN2/ aubQC42	85.2	8.6	6.2	0	859
spn-E1 / spn-E1	16.6	55.4	27.9	0	123
mei41D3 / mei41D3 ; spn-E1 / spn-E1	23.9	56.2	19.9	0	233
spn-D2 / spn-D2	34.8	54	11.2	17.3	1245
mnkP6 / mnkP6 ; spn-D2 / spn-D2	98.8	0.5	0.7	45.9	812

Two dorsal appendages are normally present at the dorsal side of a wild-type *Drosophila* egg. The mutant phenotypes are classified as weakly ventralized, which results in fusion of the dorsal appendages, and strongly ventralized, resulting in absence of dorsal appendages.

Similarly, two appendages are present on 48% of embryos derived from *aub* single mutants, and 98% of the embryos from *mnk, aub* double mutants have two appendages.

Mutations in *mei-41* also suppressed the eggshell patterning defects associated with *armi* and *aub* mutations, although suppression by *mei-41* was consistently less dramatic than suppression by *mnk*. Of the embryos from *mei-41; armi^{72.1}/armi^1* double mutants, 56% show normal appendages. The *mei-41* mutation was also less effective than *mnk* in suppressing appendage defects associated with homozygous *armi^1* (data not shown) and *aub* (Table 1), indicating that this difference is not gene- or allele-specific. Chk2 can be activated by both ATR and ATM kinases (Bartek et al. 2001; Hirao et al. 2002; Bartek and Lukas 2003), and the lower level of suppression by *mei-41*/ATR relative to *mnk*/Chk2 may therefore reflect redundant Chk2 activation by the *Drosophila* ATM homolog. However, null alleles of the *Drosophila atm* gene are lethal (Oikemus et al. 2004), making direct tests of this hypothesis difficult. Nonetheless, these initial observations suggested that the axis specification defects associated with rasiRNA pathway mutations result from activation of an ATR/Chk2 kinase DNA damage signal.

The axis specification defects associated with repair mutations are suppressed by mutations in *mei-W68*, which encodes the *Drosophila* homolog of the Spo11 nuclease that catalyzes meiotic DSB formation. In contrast, *mei-W68* has no effect on the dorsal appendage defects associated with *armi* (Table 1). Meiotic breaks thus do not appear to be the source of damage in *armi* mutations.

LOCALIZATION OF AXIS SPECIFICATION DETERMINANTS

During early oogenesis, the TGFα homolog Grk localizes to the posterior of the oocyte and signals to the overlying follicle cells, inducing posterior differentiation. During mid-oogenesis, Grk signals from the oocyte to the dorsal follicle cells to generate the dorsal–ventral axis (Gonzalez-Reyes et al. 1995; Roth et al. 1995). Mutations in *armi* and *aub* disrupt Grk protein localization at both stages, leading to posterior and dorsal–ventral axis specification defects (Cook et al. 2004). To determine whether the DNA damage signaling mutations suppress these defects in Grk localization, we analyzed the distribution of this protein by indirect immunofluorescence and laser scanning confocal microscopy (Fig. 1). For these studies, Grk protein levels within cross sections of stage 6 oocytes were measured, and an average fluorescence intensity profile for each genotype was generated (Fig. 1B, inset). In wild-type stage-6 oocytes, Grk protein accumulates near the posterior cortex (Fig. 1A, panel a). In *armi* and *aub* single mutants, in contrast, Grk protein is more uniformly distributed in the oocyte and nurse cells (Fig. 1A, c and e). However, Grk shows almost wild-type accumulation near the posterior cortex of *mnk; armi* and *mnk, aub* double mutant oocytes (Fig. 1A, d and f). The defects in dorsal–anterior localization of Grk during mid-oogenesis (Fig. 1A, panels c′ and e′) are also restored in the *mnk* double mutants (Fig. 1A, d′ and f′). Weaker suppression

is observed with *mei-41*, consistent with our analysis of the dorsal appendages (not shown).

To determine whether *mnk* and *mei-41* suppress the *armi* and *aub* induced defects in posterior morphogen localization (Cook et al. 2004), we analyzed the distribution of the pole plasm proteins Vasa (Vas) and Oskar (Osk) during mid-oogenesis. Osk localizes to the posterior in only 10% of stage-9 and stage-10 oocytes from *armi* females (2 of 23), with no detectable localization in the remaining egg chambers (Fig. 2c). In contrast, Osk shows wild-type posterior accumulation in over 80% of stage-9 and stage-10 *mnk;armi* double mutants (27 of 33; Fig. 2d). Vas localization to the posterior pole is similarly restored in the double mutants (not shown). *mei-41* leads to a less dramatic suppression of the posterior patterning defects (not shown). Osk and Vas localization are also disrupted in *aub* mutants (Fig. 2e and data not shown), and localization is restored in double mutants with *mnk* and *mei-41* (Fig. 2f). The defects in posterior and dorsal–ventral morphogen localization associated with both *armi* and *aub* thus require ATR and Chk2, which function in DNA damage signal transduction.

ATR AND CHK2 MUTATIONS RESTORE MICROTUBULE POLARIZATION

Specification of the posterior pole is initiated during early oogenesis, when the microtubule cytoskeleton reorganizes to form a polarized scaffold in the oocyte nurse cell complex. When these complexes are in the germarium, a prominent MTOC forms at the anterior pole of the oocyte, and this MTOC appears to be required for oocyte differentiation (Theurkauf et al. 1993). After cysts bud from the germarium, a posterior MTOC is established (Fig. 3a′). This asymmetric microtubule array directs Grk to the posterior pole of the oocyte, which signals to the overlying somatic follicle cells to induce posterior differentiation (Gonzalez-Reyes et al. 1995; Roth et al. 1995). Mutations in *armi* disrupt the posterior MTOC (Fig. 3b′) (Cook et al. 2004). In contrast, egg chambers double mutant for *armi* and *mnk* show near-wild-type microtubule organization at both stages (Fig. 3c′). This correlates with restoration of normal posterior localization of Grk protein (Fig. 1A, d). Egg chambers double mutant for *armi* and *mei-41* show a phenotype intermediate between the *armi* mutants and wild-type controls, consistent with partial suppression of posterior patterning defects later in oogenesis. Microtubule organization defects in *aub* are also strongly suppressed by *mnk* and more weakly suppressed by *mei-41* (data not shown). Mutations in *armi* and *aub* thus trigger Chk2-dependent defects in microtubule organization. Interestingly, DNA-damage-dependent Chk2 activation disrupts mitotic MTOCs in syncytial embryos (Takada et al. 2003). Chk2, which has an established function in control of the cell cycle machinery, thus appears to control microtubule organization during oogenesis and embryogenesis.

The axis specification defects in meiotic DSB repair mutations are also suppressed by *mnk*, suggesting that these mutations trigger Chk2-dependent defects in micro-

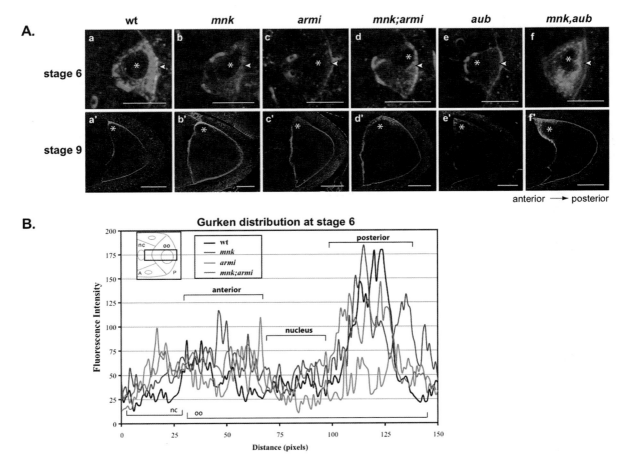

Figure 1. *mnk* restores Gurken protein localization in rasiRNA mutants. (*A*) (*a*) In wild-type stage-6 egg chambers, Gurken (Grk) protein (*green*) is tightly localized at the posterior cortex (*arrowhead*) near the oocyte nucleus (*asterisk*). (*a'*) By stage 9, Grk is localized at the dorsal anterior cortex near the oocyte nucleus (*asterisk*). Actin filaments (*red*) mark the cell boundaries. (*b,b'*) In *mnk^p6* oocytes, Grk localization is the same as in wild type. (*c,c'*) In *armi^{72.1}/armi^1* egg chambers, this localization pattern is lost, with Grk dispersed throughout the oocyte. (*d,d'*) *mnk^p6* suppresses the *armi^{72.1}/armi^1* phenotype and rescues Grk localization during early and late oogenesis. (*e*) In *aub^{QC42}/aub^{HN2}* stage-6 egg chambers, Grk localization is similar to that of *armi^{72.1}/armi^1* oocytes. (*e'*) At stage 9, Grk is localized correctly in *aub^{QC42}/aub^{HN2}*, but not at wild-type levels. (*f,f'*) In *mnk^p6*, *aub^{QC42}/ mnk^p6*, *aub^{HN2}* egg chambers, Grk localization level is restored. Images were acquired under identical conditions for either stage. Projections of three serial 0.6-μm optical sections are shown. Scale bars, 10 μm and 25 μm for stage-6 and -9 egg chambers, respectively. (*B*) Quantification of Gurken localization in stage-6 oocytes. Distribution of fluorescence was measured within a region comprising some nurse cell cytoplasm and a cross section of the oocyte (*inset*). Single Z-section images were acquired focusing on both the nucleus and the posterior cortex. The measurements for each genotype were performed post-acquisition using Image J software and averaged over 4–5 oocytes per genotype using Microsoft Excel.

tubule polarization during early oogenesis. We therefore analyzed microtubule organization in ovaries mutant for *spn-D*, which encodes a rad51C homolog required for DSB repair (Abdu et al. 2003). Mutations in *spn-D*, like mutations in *armi* and *aub*, disrupt formation of both a prominent MTOC at the anterior of stage 1 egg chambers, and of the posterior MTOC during stages 2 through 6 (Fig. 3d and d'). Significantly, these defects are suppressed in *mnk; spn-D* double mutants (Fig. 3e and e'). Therefore, both DNA repair and rasiRNA mutations lead to defects in microtubule polarization that are mediated by Chk2.

Repair mutations induce Chk2-dependent phosphorylation of Vas, a conserved RNA helicase required for posterior and dorsal–ventral patterning (Styhler et al. 1998; Ghabrial and Schupbach 1999). To determine whether rasiRNA mutations also trigger Chk2-dependent Vas phosphorylation, we probed western blots of *armi* and

mnk; armi double mutants for this conserved helicase. As shown in Figure 4, a single Vas species is observed on blots of ovary extracts from wild type and homozygous *armi^1* mutants. In contrast, a single lower electrophoretic mobility species is observed in ovaries homozygous for a stronger loss-of-function allele of *armi* (*armi^{72.1}*). Both species are observed with an inter-allelic combination (*armi^{72.1}/armi^1*). We were unable to recover *mnk; armi^{72.1}* double mutants. However, only the faster-migrating species is present in *mnk; armi^{72.1}/armi^1* double mutant extracts. Following phosphatase treatment, the lower-mobility species present in *armi* mutant extracts disappears and the faster-migrating species increases in intensity (data not shown), indicating that the lower-mobility band is a phosphorylated form of Vas. Mutations in *armi*, like meiotic DSB repair mutations, thus trigger Chk2-dependent phosphorylation of Vas.

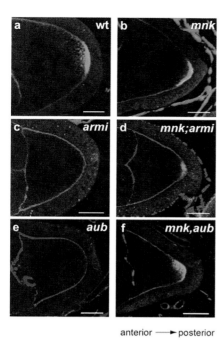

anterior ——→ posterior

Figure 2. *mnk* restores Oskar protein localization during late oogenesis in rasiRNA mutants. Egg chambers were fixed and labeled against Oskar protein (*green*) and Actin (*red*). (*a*) In wild-type stage 9–10 oocytes, Oskar (Osk) localizes tightly to the posterior cortex. (*b*) mnk^{p6} mutants have no defects in Osk localization. (*c,e*) In $armi^{72.1}/armi^1$ and aub^{QC42}/aub^{HN2} stage-9 oocytes, Osk localization is disrupted. (*d,f*) mnk^{p6} rescues the Osk localization in $armi^{72.1}/armi^1$ and aub^{QC42}/aub^{HN2} stage-9 oocytes. Egg chambers are oriented with posterior to the right. Images were acquired under identical conditions. Single optical sections are shown. Scale bar, 20 μm.

ARMI AND *AUB* MUTATIONS LEAD TO GERM-LINE γ-H2AV ACCUMULATION

The above observations indicate that the axis specification defects in *armi* and *aub* mutants require ATR and Chk2 kinases that can be activated by DNA DSBs. However, the *armi* mutation is not suppressed by *mei-W68*, suggesting that meiotic breaks are not the source of DNA damage. To determine whether the *armi* and *aub* mutations lead to DSB accumulation, we labeled mutant ovaries for the phosphorylated form of the *Drosophila* histone H2AX variant (γ-H2Av), which accumulates on chromosomes near break sites (Modesti and Kanaar 2001; Redon et al. 2002). Following chromosome breakage, *Drosophila* H2Av, like H2AX, is phosphorylated at a conserved SQ motif within an extended carboxy-terminal tail (Rogakou et al. 1998; Madigan et al. 2002). For these studies, we used an anti-phosphoprotein antibody specific for γ-H2Av (Gong et al. 2005). As shown in Figure 5a, γ-H2Av foci are normally restricted to region 2 of the germarium, where meiotic DSBs are formed (Jang et al. 2003). Consistent with earlier observations, this labeling is significantly reduced in *mei-W68* mutants, which do not initiate meiotic breaks (Fig. 5g). In *armi* and *aub* mutants, prominent γ-H2Av foci are present in germ-line cells of the germarium. However, these foci persist and increase in intensity as cysts mature and bud to form stage-2 egg chambers (Fig. 5c and e). γ-H2Av foci also persist in *spn-E* mutants, which are defective in rasiRNA function and trigger axis specification defects (data not shown).

The pattern of germ-line-specific γ-H2Av accumulation in *armi*, *aub*, and *spn-E* is similar to the pattern observed in the DNA repair mutant *spn-D*, although the

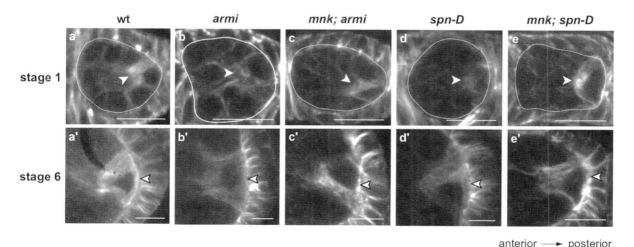

anterior ——→ posterior

Figure 3. *mnk* suppresses the microtubule organization defects caused by mutations in the rasiRNA and DNA repair pathways. (*a, a'*) Wild type, (*b, b'*) $armi^{72.1}/armi^1$, (*c, c'*) mnk^{p6}; $armi^{72.1}/armi^1$, (*d, d'*) *spn-D^2*, and (*e, e'*) mnk^{p6}; *spn-D^2* egg chambers were fixed and labeled with FITC-conjugated anti-α-tubulin antibody. (*a*) A bright MTOC is localized to the anterior pole of the oocyte in wild-type stage-1 egg chambers (*arrowhead*). (*a'*) By stage 6, the MTOC is localized along the posterior cortex (*arrowhead*). (*b, d*) In $armi^{72.1}/armi^1$ and *spn-D^2* mutant egg chambers, the stage-1 anterior MTOC and later posterior MTOC (*b', d'*) are disrupted (*arrowheads*). (*c, e*) In mnk^{p6}; $armi^{72.1}/armi^1$, and mnk^{p6}; *spn-D^2* egg chambers, the MTOC is observed at the anterior of the oocyte during stage 1 (*arrowheads*) and (*c', e'*) at the posterior of the oocyte in stage 6 (*arrowheads*). Stage-1 oocytes are outlined. Images were acquired under identical conditions. Projections of four serial 0.6-μm optical sections are shown. Posterior is oriented to the right. Scale bar, 10 μm.

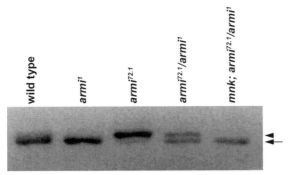

Figure 4. Severe *armi* mutations trigger *mnk*-dependent Vas phosphorylation. Western blot analysis of Vas protein in wild type, *armi¹*, *armi⁷².¹*, *armi⁷².¹/armi¹*, and *mnkᵖ⁶; armi⁷².¹/armi¹* ovary extracts. Vas from homozygous *armi⁷².¹* ovaries has a reduced electrophoretic mobility relative to Vas from wild-type ovaries. Low-mobility and wild-type-mobility forms of Vas are present in *armi⁷².¹/armi¹* ovary extracts. Only the faster-migrating form is present in *mnkᵖ⁶; armi⁷².¹/armi¹* extracts (*arrow*). Following phosphatase treatment, the lower-mobility species in *armi* mutants is lost and the faster-migrating form increases in intensity, indicating that the lower-mobility form is phosphorylated (not shown).

foci appear to arise at somewhat earlier stages in the rasiRNA mutants (Fig. 5c, e, i, j). DSB formation in *spn-D* mutants is suppressed by *mei-W68* (Abdu et al. 2002). Consistent with the observation that *mei-W68* does not suppress the dorsal–ventral axis specification defects linked to *armi* (Table 1), γ-H2Av foci persist in *mei-W68;armi* double mutants (Fig. 5h). Mutations in *armi*,

aub, and *spn-E* thus appear to trigger DNA DSBs in the germ line. Our analysis of *mei-W68;armi* double mutants strongly suggests that this is independent of meiotic DSB formation.

rasiRNA FUNCTION

The observations presented above strongly suggested that the axial patterning defects associated with *armi* and *aub* are a consequence of DNA damage signaling, and that rasiRNA-based gene silencing is not directly involved in embryonic patterning. Alternatively, the *mnk* and *mei-41* mutations could suppress the defects in rasiRNA function associated with *armi* and *aub*. If this were the case, rasiRNA-dependent silencing could be essential to axis specification. We therefore analyzed rasiRNA-dependent silencing of both the *Stellate (Ste)* gene during spermatogenesis and the *HeT-A* retrotransposon during oogenesis in single and double mutants. The *Ste* gene is repressed during spermatogenesis, apparently through mRNA turnover guided by rasiRNAs derived from the *Suppressor of Stellate* locus (Aravin et al. 2001; Gvozdev et al. 2003). Mutations in *armi* and *aub* lead to accumulation of full-length *Ste* mRNA and Stellate protein overexpression, triggering assembly of Ste crystals in mutant testes (Fig. 6A, b and c) (Stapleton et al. 2001; Aravin et al. 2004; Tomari et al. 2004b; Forstemann et al. 2005). We found that Ste crystals are present in both *mnk; armi* and *mnk, aub* double mutant testes (Fig. 6A, e and f). Ste overexpression is also linked to male sterility, and *mnk;armi* males are sterile (data not shown). *HeT-A* is a retrotransposon that contributes to

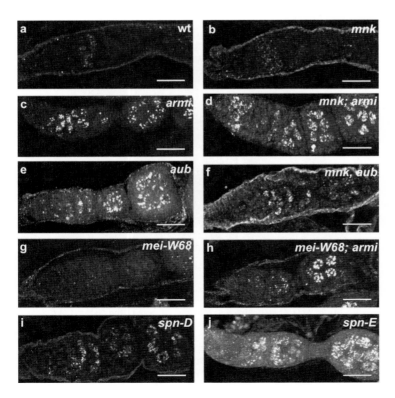

Figure 5. γ-H2Av foci accumulate in *armi* and *aub* mutant ovaries. The phosphorylated form of histone H2Av (γ-H2Av) accumulates near DSB sites. (*a, b*) In wild type and *mnkᵖ⁶* mutants, γ-H2Av foci are restricted to region 2 of the germarium, when meiotic DSBs are present. In (*c*) *armi⁷².¹/armi¹*, (*d*) *mnkᵖ⁶; armi⁷².¹/armi¹*, (*e*) *aubᵠᶜ⁴²/aubᴴᴺ²*, (*f*) *mnkᵖ⁶, aubᵠᶜ⁴²/ mnkᵖ⁶, aubᴴᴺ²*, and (*j*) *spn-E* mutant ovaries, γ-H2Av foci accumulate in germ-line cells within the germarium, and persist and increase in intensity as cysts bud from the germarium to form egg chambers. (*i*) A similar pattern is observed in ovaries mutant for *spn-D²*, which is required for DSB repair. (*g*) Mutations in *mei-W68 (mei-W68¹/mei-W68^{k05603})*, which encodes the Spo11 nuclease that initiates meiotic DSBs, suppress formation of γ-H2Av foci in region 2 of the germarium. (*h*) However, *mei-W68* does not suppress γ-H2Av focus formation in *armi* mutants (*mei-W68¹/mei-W68^{k05603}; armi⁷².¹/armi¹*). Projections of six serial 1-μm optical sections are shown. Posterior is to the right. Scale bar, 20 μm.

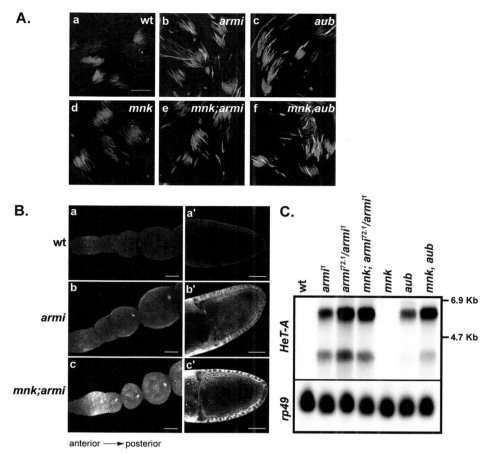

Figure 6. The *mnk* mutation does not suppress defects in rasiRNA function. (*A*) Silencing of *Stellate* locus during spermatogenesis. *Stellate* is not expressed in (*a*) wild type or (*d*) *mnk^p6* mutant testes. However, *Stellate* is overexpressed, and the protein assembles into crystals in testes from (*b*) *armi^72.1/armi^1*, (*c*) *aub^QC42/aub^HN2*, (*e*) *mnk^p6; armi^72.1/armi^1*, and (*f*) *mnk^p6, aub^QC42/ mnk^p6, aub^HN2* males. DNA (*red*) was labeled with TOTO3, and Stellate protein (*green*) was detected with anti-Stellate antibody. Projections of five serial 1-μm optical sections are shown. Scale bar, 20 μm. (*B*) FISH analysis of HeT-A retrotransposon expression. (*a, a'*) In wild-type ovaries, only background levels of HeT-A expression are detected. (*b, b'* and *c, c'*) In contrast, HeT-A is expressed at high levels in the germ-line and somatic follicle cells of early and mid-oogenesis stage *armi^72.1/armi^1* and *mnk^p6; armi^72.1/armi^1* egg chambers. Panels *a*, *b*, and *c* are projections of 12–15 serial 1.5-μm optical sections. Panels *a'*, *b'*, and *c'* are single optical sections. Posterior is oriented to the right. Scale bars, 20 μm for the left panels and 50 μm for the right panels. (*C*) Northern blot for *HeT-A*. Total ovary RNA samples were resolved on a 1% agarose-formaldehyde gel, transferred to membrane, and probed for HeT-A transcript. HeT-A transcripts are undetectable in wild-type and *mnk^p6* samples, but are abundant in RNA derived from *armi^1*, *armi^72.1/armi^1*, *mnk^p6; armi^72.1/armi^1*, *aub^QC42/aub^HN2* and *mnk^p6, aub^QC42/aub^HN2* mutant ovaries. Ribosomal protein 49 (rp49) was used as a loading control.

telomere formation in *Drosophila* (Pardue et al. 2005), and *HeT-A* expression is dramatically de-repressed in *armi*, *aub*, and *spn-E* mutant ovaries (Aravin et al. 2001; Vagin et al. 2004, 2006). *HeT-A* is not expressed at detectable levels on northern blots of wild-type or *mnk* RNAs. However, *HeT-A* transcripts are abundant in *armi* and *aub* mutants (Fig. 6C). Significantly, *HeT-A* is also overexpressed in *mnk;armi* and *mnk,aub* double mutants (Fig. 6, B and C). In fact, *HeT-A* expression is higher in the double mutants, relative to the single mutants. FISH analyses indicate that this reflects increased expression in the germ-line and somatic cells of the ovary, during both early and mid-oogenesis (Fig. 6B). Therefore, the *mnk* mutation does not suppress defects in rasiRNA-based gene silencing during spermatogenesis or oogenesis, leading us to conclude that rasiRNA-based silencing is not required for axis specification.

Taken together, these observations indicate that the critical developmental function for *armi* and *aub*, and presumably the rasiRNA pathway, is to maintain germ-line genome integrity. The embryonic patterning defects associated with these mutations, in contrast, are due to activation of a ATR/Chk2-dependent DNA damage response that disrupts structural polarization of developing oocyte.

DISCUSSION

Mutations in the *Drosophila armi, spn-E*, and *aub* genes disrupt oocyte microtubule organization and asymmetric localization of mRNAs and proteins that specify the posterior pole and dorsal–ventral axis of the oocyte and embryo. Mutations in these genes block homology-dependent RNA cleavage and RISC assembly in ovary lysates, RNAi-based gene silencing during early embryogenesis, and rasiRNA production and retrotransposon and stellate silencing. These genetic and biochemical studies suggest that siRNAs and/or rasiRNAs are required for embryonic axis specification. However, mutations in the *r2d2*, *dcr2*, and *ago 2* genes block siRNA function, but do not disrupt the rasiRNA pathway or embryonic axis specification. Defects in the rasiRNA pathway thus appear to trigger the embryonic axis specification defects associated with *armi*, *spn-E*, and *aub* mutations.

Identifying potential targets for rasiRNA control during axis specification is complex, because the functions for these RNAs are only beginning to emerge. To identify potential target rasiRNAs during axis specification, we focused on previously characterized mutations that also disrupt posterior and dorsal–ventral axis specification. We noted striking similarities between the axis specification defects produced by *armi*, *aub*, and *spn-E* and the patterning defects associated with mutations in meiotic DNA DSB repair pathway, suggesting a mechanistic link. The axis specification defects associated with DNA repair mutants result from activation of an ATR- and Chk2-dependent DNA damage signal transduction pathway. We therefore considered two alternative hypotheses: (1) The rasiRNA pathway directly controls axis specification, and ATR/Chk2 activation blocks axis specification by inhibiting this pathway. (2) Mutations that disrupt the rasiRNA pathway activate ATR and Chk2, which disrupt axis specification by targeting rasiRNA-independent processes. We show that the cytoskeletal polarization, RNA localization, and eggshell patterning defects associated with *armi* and *aub* are efficiently suppressed by *mnk* and *mei-41*, which encode Chk2 and ATR kinase components of the DNA damage signaling pathway. In addition, *armi*, *aub*, and *spn-E* mutants accumulate γ-H2Av foci characteristic of DNA breaks (Fig. 5). Finally, *armi* mutations trigger Chk2-dependent Vasa phosphorylation (Fig. 4). Significantly, HeT-A and Stellate are overexpressed in *mnk;armi* and *mnk,aub* double mutants, indicating that axis specification does not directly require rasiRNA-based silencing. Instead, the rasiRNA pathway suppresses DNA damage signaling through ATR and Chk2, which disrupt axis specification by blocking microtubule polarization and Grk signaling from the oocyte to the posterior follicle cells (Fig. 7).

The studies described here also provide new insight into the role of DNA damage signaling in axis specification. As first shown by the Schüpbach lab, meiotic DSB repair mutations trigger Chk2-dependent defects in axis specification and phosphorylation of Vasa (Abdu et al. 2002). Vasa is a putative helicase and translation initiation factor required for axis specification, and Schüpbach and colleagues have proposed that Vasa phosphorylation by Chk2 leads to defects in Grk translation that lead to the observed axial patterning defects (Abdu et al. 2002). Axis specification is initiated early in oogenesis, as the microtubule cytoskeleton reorganizes and Grk signals to the posterior follicle cells. We show that mutations in *armi* and *aub* block initial polarization of the microtubule cytoskeleton and Grk protein localization (Figs. 1 and 3). Furthermore, we find that a null allele of *vasa* does not

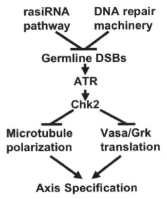

Figure 7. Model for rasiRNA control of axis specification. The rasiRNA pathway and meiotic DSB repair machinery function independently to suppress DSB in the female germ line. Mutations that disrupt either pathway activate a common DNA damage response, mediated by the ATR and Chk2 kinases. Chk2 activation blocks axis specification by disrupting microtubule organization and phosphorylating Vas, an RNA helicase required for axis specification that has been implicated in *grk* mRNA translation.

block initial polarization of the microtubule cytoskeleton, indicating that simple loss of Vasa function is not the cause of the initial cytoskeletal polarization defects in *armi*, *aub*, and *spn-E* (C. Klattenhoff and W.E. Theurkauf, unpubl.). Although Vasa is a direct or indirect target of Chk2, these data strongly suggest that Vasa phosphorylation is not the primary cause of the initial axis specification defects, at least in the rasiRNA mutants. Chk2 activation disrupts mitotic microtubule organization and spindle assembly in early embryos (Takada et al. 2003), and the findings reported here indicate that DNA damage disrupts axis specification by activating a Chk2-dependent block to microtubule organization, which in turn prevents Grk localization and signaling to the posterior follicle cells.

The role of rasiRNAs in suppressing damage signaling remains to be established. The presence of γ-H2Av foci suggests that DNA DSBs are formed in *armi*, *aub*, and *spn-E* rasiRNA pathway mutations (Fig. 5). In wild-type ovaries, γ-H2Av foci begin to accumulate in region 2 of the germarium, when the Spo11 nuclease (encoded by the *mei-W68* gene) initiates meiotic breaks (McKim and Hayashi-Hagihara 1998). The axis specification defects associated with repair mutations are efficiently suppressed by *mei-W68* mutations (Ghabrial and Schüpbach 1999; Morris and Lehmann 1999), confirming that meiotic breaks are the source of damage. However, the axis specification defects in *armi* mutants are not suppressed by *mei-W68* (Table 1), and γ-H2Av foci in *armi*, *aub*, and *spn-E* appear to arise somewhat earlier than in wild-type controls. Meiotic DSBs thus do not appear to be the primary source of DNA damage in rasiRNA pathway mutations.

Defects in heterochromatin assembly and a resulting mobilization of retrotransposons could lead to DNA damage and Chk2 activation in rasiRNA pathway mutants (Pal-Bhadra et al. 2004; Vagin et al. 2006). However, silencing of the HeT-A retrotransposon is particularly sensitive to rasiRNA pathway mutations, and this element selectively integrates at chromosome ends to form telomeres (Casacuberta and Pardue 2005; Pardue et al. 2005). The rasiRNA mutations could therefore disrupt telomere organization, leading to both loss of silencing and recognition of unprotected chromosome ends by the DSB detection machinery. If this is the case, the γ-H2Av foci in rasiRNA mutants should colocalize with telomeres. In addition, DNA damage can induce loss of retrotransposon silencing (Bradshaw and McEntee 1989; Scholes et al. 2003; Sacerdot et al. 2005). This raises the possibility that rasiRNAs function in DNA repair, and that mutations in this pathway thus lead to DNA damage that triggers the observed defects in axis specification and loss of retrotransposon silencing. The available data do not distinguish between these alternatives.

In mouse, the *piwi*-related Argonauts Miwi and Mili bind piRNAs, 30-nucleotide RNAs derived primarily from a single strand that appear to be related to rasiRNAs (Aravin et al. 2006; Girard et al. 2006; Grivna et al. 2006). Mutations in these genes disrupt spermatogenesis and lead to germ-line apoptosis (Kuramochi-Miyagawa et al. 2001; Deng and Lin 2002), which can be induced by DNA damage. We therefore speculate that the primary function for mammalian piRNAs, like *Drosophila* rasiRNAs, is to suppress DNA damage signaling.

ACKNOWLEDGMENTS

We thank Kim McKim for γ-H2Av antibody; Anne Ephrussi for anti-Oskar antibody; and Alla Sigova, Vasia Vagin, and Phil Zamore for *HeT-A* clones. We also thank Beatrice Benoit and Hanne Varmark for helpful comments on the manuscript; Phil Zamore, Vasia Vagin, Klaus Forstemann, and Yuki Tommari for stimulating discussion and sharing data prior to publication; and Beatrice Benoit for technical support. The Grk monoclonal antibody 1D12, developed by Trudi Schüpbach, was obtained from the Developmental Studies Hybridoma Bank developed under the auspices of the National Institute of Child Health and Human Development and maintained by the University of Iowa, Department of Biological Sciences, Iowa City, Iowa 52242. This work was supported by a grant to W.E.T. from the National Institute of Child Health and Human Development, National Institutes of Health (R01 HD049116).

REFERENCES

Abdu U., Brodsky M., and Schupbach T. 2002. Activation of a meiotic checkpoint during *Drosophila* oogenesis regulates the translation of Gurken through Chk2/Mnk. *Curr. Biol.* **12:** 1645.

Abdu U., Gonzalez-Reyes A., Ghabrial A., and Schupbach T. 2003. The *Drosophila* spn-D gene encodes a RAD51C-like protein that is required exclusively during meiosis. *Genetics* **165:** 197.

Aravin A., Gaidatzis D., Pfeffer S., Lagos-Quintana M., Landgraf P., Iovino N., Morris P., Brownstein M.J., Kuramochi-Miyagawa S., Nakano T., et al. 2006. A novel class of small RNAs bind to MILI protein in mouse testes. *Nature* **442:** 203.

Aravin A.A., Klenov M.S., Vagin V.V., Bantignies F., Cavalli G., and Gvozdev V.A. 2004. Dissection of a natural RNA silencing process in the *Drosophila melanogaster* germ line. *Mol. Cell. Biol.* **24:** 6742.

Aravin A.A., Naumova N.M., Tulin A.V., Vagin V.V., Rozovsky Y.M., and Gvozdev V.A. 2001. Double-stranded RNA-mediated silencing of genomic tandem repeats and transposable elements in the *D. melanogaster* germline. *Curr. Biol.* **11:** 1017.

Bartek J. and Lukas J. 2003. Chk1 and Chk2 kinases in checkpoint control and cancer. *Cancer Cell* **3:** 421.

Bartek J., Falck J., and Lukas J. 2001. CHK2 kinase - A busy messenger. *Nat. Rev. Mol. Cell Biol.* **2:** 877.

Bernstein E., Kim S.Y., Carmell M.A., Murchison E.P., Alcorn H., Li M.Z., Mills A.A., Elledge S.J., Anderson K.V., and Hannon G.J. 2003. Dicer is essential for mouse development. *Nat. Genet.* **35:** 215.

Bradshaw V.A. and McEntee K. 1989. DNA damage activates transcription and transposition of yeast Ty retrotransposons. *Mol. Gen. Genet.* **218:** 465.

Casacuberta E. and Pardue M.L. 2005. HeT-A and TART, two *Drosophila* retrotransposons with a bona fide role in chromosome structure for more than 60 million years. *Cytogenet. Genome Res.* **110:** 152.

Cook H., Koppetsch B., Wu J., and Theurkauf W. 2004. The *Drosophila* SDE3 homolog *armitage* is required for oskar mRNA silencing and embryonic axis specification. *Cell* **116:** 817.

Deng W. and Lin H. 2002. *miwi*, a murine homolog of *piwi*, encodes a cytoplasmic protein essential for spermatogenesis. *Dev. Cell* **2:** 819.

Deshpande G., Calhoun G., and Schedl P. 2005. *Drosophila* argonaute-2 is required early in embryogenesis for the assembly of centric/centromeric heterochromatin, nuclear division, nuclear migration, and germ-cell formation. *Genes Dev.* **19:** 1680.

Forstemann K., Tomari Y., Du T., Vagin V.V., Denli A.M., Bratu D.P., Klattenhoff C., Theurkauf W.E., and Zamore P.D. 2005. Normal microRNA maturation and germ-line stem cell maintenance requires Loquacious, a double-stranded RNA-binding domain protein. *PLoS Biol.* **3:** e236.

Fukagawa T., Nogami M., Yoshikawa M., Ikeno M., Okazaki T., Takami Y., Nakayama T., and Oshimura M. 2004. Dicer is essential for formation of the heterochromatin structure in vertebrate cells. *Nat. Cell Biol.* **6:** 784.

Galiana-Arnoux D., Dostert C., Schneemann A., Hoffmann J.A., and Imler J.L. 2006. Essential function in vivo for Dicer-2 in host defense against RNA viruses in *Drosophila*. *Nat. Immunol.* **7:** 590.

Ghabrial A. and Schupbach T. 1999. Activation of a meiotic checkpoint regulates translation of Gurken during *Drosophila* oogenesis. *Nat. Cell Biol.* **1:** 354.

Ghabrial A., Ray R.P., and Schupbach T. 1998. *okra* and *spindle-B* encode components of the *RAD52* DNA repair pathway and affect meiosis and patterning in *Drosophila* oogenesis. *Genes Dev.* **12:** 2711.

Giraldez A.J., Cinalli R.M., Glasner M.E., Enright A.J., Thomson J.M., Baskerville S., Hammond S.M., Bartel D.P., and Schier A.F. 2005. MicroRNAs regulate brain morphogenesis in zebrafish. *Science* **308:** 833.

Girard A., Sachidanandam R., Hannon G.J., and Carmell M.A. 2006. A germline-specific class of small RNAs binds mammalian Piwi proteins. *Nature* **442:** 199.

Gong W.J., McKim K.S., and Hawley R.S. 2005. All paired up with no place to go: Pairing, synapsis, and DSB formation in a balancer heterozygote. *PLoS Genet.* **1:** e67.

Gonzalez-Reyes A., Elliott H., and St. Johnston D. 1995. Polarization of both major body axes in *Drosophila* by gurken-torpedo signalling. *Nature* **375:** 654.

Grishok A., Pasquinelli A.E., Conte D., Li N., Parrish S., Ha I., Baillie D.L., Fire A., Ruvkun G., and Mello C.C. 2001. Genes and mechanisms related to RNA interference regulate expression of the small temporal RNAs that control *C. elegans* developmental timing. *Cell* **106:** 23.

Grivna S.T., Beyret E., Wang Z., and Lin H. 2006. A novel class of small RNAs in mouse spermatogenic cells. *Genes Dev.* **20:** 1709.

Gvozdev V.A., Aravin A.A., Abramov Y.A., Klenov M.S., Kogan G.L., Lavrov S.A., Naumova N.M., Olenkina O.M., Tulin A.V., and Vagin V.V. 2003. Stellate repeats: Targets of silencing and modules causing *cis*-inactivation and *trans*-activation. *Genetica* **117:** 239.

Hannon G.J. 2002. RNA interference. *Nature* **418:** 244.

Hatfield S.D., Shcherbata H.R., Fischer K.A., Nakahara K., Carthew R.W., and Ruohola-Baker H. 2005. Stem cell division is regulated by the microRNA pathway. *Nature* **435:** 974.

Hirao A., Cheung A., Duncan G., Girard P.M., Elia A.J., Wakeham A., Okada H., Sarkissian T., Wong J.A., Sakai T., et al. 2002. Chk2 is a tumor suppressor that regulates apoptosis in both an ataxia telangiectasia mutated (ATM)-dependent and an ATM-independent manner. *Mol. Cell. Biol.* **22:** 6521.

Hutvagner G. and Zamore P.D. 2002. A microRNA in a multiple-turnover RNAi enzyme complex. *Science* **297:** 2056.

Jang J.K., Sherizen D.E., Bhagat R., Manheim E.A., and McKim K.S. 2003. Relationship of DNA double-strand breaks to synapsis in *Drosophila*. *J. Cell Sci.* **116:** 3069.

Kuramochi-Miyagawa S., Kimura T., Yomogida K., Kuroiwa A., Tadokoro Y., Fujita Y., Sato M., Matsuda Y., and Nakano T. 2001. Two mouse *piwi*-related genes: *miwi* and *mili*. *Mech. Dev.* **108:** 121.

Lee Y.S., Nakahara K., Pham J.W., Kim K., He Z., Sontheimer E.J., and Carthew R.W. 2004. Distinct roles for *Drosophila* Dicer-1 and Dicer-2 in the siRNA/miRNA silencing pathways. *Cell* **117:** 69.

Madigan J.P., Chotkowski H.L., and Glaser R.L. 2002. DNA double-strand break-induced phosphorylation of *Drosophila* histone variant H2Av helps prevent radiation-induced apoptosis. *Nucleic Acids Res.* **30:** 3698.

McKim K.S. and Hayashi-Hagihara A. 1998. *mei-W68* in *Drosophila melanogaster* encodes a Spo11 homolog: Evidence that the mechanism for initiating meiotic recombination is conserved. *Genes Dev.* **12:** 2932.

Modesti M. and Kanaar R. 2001. DNA repair: Spot(light)s on chromatin. *Curr. Biol.* **11:** R229.

Morris J. and Lehmann R. 1999. *Drosophila* oogenesis: Versatile spn doctors. *Curr. Biol.* **9:** R55.

Neuman-Silberberg F.S. and Schupbach T. 1993. The *Drosophila* dorsoventral patterning gene gurken produces a dorsally localized RNA and encodes a TGF alpha-like protein. *Cell* **75:** 165.

Oikemus S.R., McGinnis N., Queiroz-Machado J., Tukachinsky H., Takada S., Sunkel C.E., and Brodsky M.H. 2004. *Drosophila atm/telomere fusion* is required for telomeric localization of HP1 and telomere position effect. *Genes Dev.* **18:** 1850.

Okamura K., Ishizuka A., Siomi H., and Siomi M.C. 2004. Distinct roles for Argonaute proteins in small RNA-directed RNA cleavage pathways. *Genes Dev.* **18:** 1655.

Pal-Bhadra M., Leibovitch B.A., Gandhi S.G., Rao M., Bhadra U., Birchler J.A., and Elgin S.C. 2004. Heterochromatic silencing and HP1 localization in *Drosophila* are dependent on the RNAi machinery. *Science* **303:** 669.

Pardue M.L., Rashkova S., Casacuberta E., DeBaryshe P.G., George J.A., and Traverse K.L. 2005. Two retrotransposons maintain telomeres in *Drosophila*. *Chromosome Res.* **13:** 443.

Provost P., Silverstein R.A., Dishart D., Walfridsson J., Djupedal I., Kniola B., Wright A., Samuelsson B., Radmark O., and Ekwall K. 2002. Dicer is required for chromosome segregation and gene silencing in fission yeast cells. *Proc. Natl. Acad. Sci.* **99:** 16648.

Redon C., Pilch D., Rogakou E., Sedelnikova O., Newrock K., and Bonner W. 2002. Histone H2A variants H2AX and H2AZ. *Curr. Opin. Genet. Dev.* **12:** 162.

Riechmann V. and Ephrussi A. 2001. Axis formation during *Drosophila* oogenesis. *Curr. Opin. Genet. Dev.* **11:** 374.

Rogakou E.P., Pilch D.R., Orr A.H., Ivanova V.S., and Bonner W.M. 1998. DNA double-stranded breaks induce histone H2AX phosphorylation on serine 139. *J. Biol. Chem.* **273:** 5858.

Roth S., Neuman-Silberberg F.S., Barcelo G., and Schupbach T. 1995. cornichon and the EGF receptor signaling process are necessary for both anterior-posterior and dorsal–ventral pattern formation in *Drosophila*. *Cell* **81:** 967.

Sacerdot C., Mercier G., Todeschini A.L., Dutreix M., Springer M., and Lesage P. 2005. Impact of ionizing radiation on the life cycle of *Saccharomyces cerevisiae* Ty1 retrotransposon. *Yeast* **22:** 441.

Scholes D.T., Kenny A.E., Gamache E.R., Mou Z., and Curcio M.J. 2003. Activation of a LTR-retrotransposon by telomere erosion. *Proc. Natl. Acad. Sci.* **100:** 15736.

Schupbach T. 1987. Germ line and soma cooperate during oogenesis to establish the dorsoventral pattern of egg shell and embryo in *Drosophila melanogaster*. *Cell* **49:** 699.

Spradling A.C. 1993. Germline cysts: Communes that work. *Cell* **72:** 649.

Stapleton W., Das S., and McKee B.D. 2001. A role of the *Drosophila* homeless gene in repression of Stellate in male meiosis. *Chromosoma* **110:** 228.

Styhler S., Nakamura A., Swan A., Suter B., and Lasko P. 1998. *vasa* is required for GURKEN accumulation in the oocyte, and is involved in oocyte differentiation and germline cyst development. *Development* **125:** 1569.

Takada S., Kelkar A., and Theurkauf W.E. 2003. *Drosophila* checkpoint kinase 2 couples centrosome function and spindle assembly to genomic integrity. *Cell* **113:** 87.

Theurkauf W.E., Alberts B.M., Jan Y.N., and Jongens T.A. 1993. A central role for microtubules in the differentiation of *Drosophila* oocytes. *Development* **118:** 1169.

Theurkauf W.E., Smiley S., Wong M.L., and Alberts B.M. 1992. Reorganization of the cytoskeleton during *Drosophila* oogenesis: Implications for axis specification and intercellular transport. *Development* **115:** 923.

Tomari Y., Matranga C., Haley B., Martinez N., and Zamore P.D. 2004a. A protein sensor for siRNA asymmetry. *Science* **306:** 1377.

Tomari Y., Du T., Haley B., Schwarz D., Bennett R., Cook H., Koppetsch B., Theurkauf W., and Zamore P.D. 2004b. RISC assembly defects in the *Drosophila* RNAi mutant *armitage*. *Cell* **116:** 831.

Vagin V.V., Klenov M.S., Kalmykova A., Stolyarenko A.D., Kotelnikov R.N., and Gvozdev V. 2004. The RNA interference proteins and vasa locus are involved in the silencing of retrotransposons in the female germline of *Drososophla melanogaster*. *RNA Biol.* **1:** 54.

Vagin V.V., Sigova A., Li C., Seitz H., Gvozdev V., and Zamore P.D. 2006. A distinct small RNA pathway silences selfish genetic elements in the germline. *Science* **313:** 320.

Volpe T., Schramke V., Hamilton G.L., White S.A., Teng G., Martienssen R.A., and Allshire R.C. 2003. RNA interference is required for normal centromere function in fission yeast. *Chromosome Res.* **11:** 137.

Wang X.H., Aliyari R., Li W.X., Li H.W., Kim K., Carthew R., Atkinson P., and Ding S.W. 2006. RNA interference directs innate immunity against viruses in adult *Drosophila*. *Science* **312:** 452.

Wassenegger M. 2005. The role of the RNAi machinery in heterochromatin formation. *Cell* **122:** 13.

Architecture of a MicroRNA-controlled Gene Regulatory Network That Diversifies Neuronal Cell Fates

O. HOBERT

*Howard Hughes Medical Institute, Department of Biochemistry and Molecular Biophysics,
Columbia University Medical Center, New York, New York 10032*

Individual cell types are defined by the expression of specific gene batteries. Regulatory networks that control cell-type-specific gene expression programs in the nervous system are only beginning to be understood. This paper summarizes a complex gene regulatory network, composed of several transcription factors and microRNAs (miRNAs), that controls neuronal subclass specification in the nervous system of the nematode *Caenorhabditis elegans*.

One of the key problems in developmental biology is to understand how cells within individual tissue types diversify their terminal differentiation programs. This is particularly evident in the nervous systems, where an impressive number of cells diversify from a neuronal ground state. This ground state is characterized by the expression of a generic set of neuronal function genes, such as components of the synaptic vesicle machinery, whereas neuron-specific states are characterized by a unique combinatorial expression profile of factors that define individual properties of each neuron, such as its specific morphological features or electrical properties. The human nervous system is estimated to contain 10^{11} neurons that generate 10^{14} connections, thereby illustrating the daunting nature of the task of understanding how such complex systems develop.

The diversification of individual cell types in the nervous system can be well studied in much simpler model organisms, such as the nematode *C. elegans*, which contains a nervous system of 302 cells (for a more detailed discussion, see Hobert 2006). These 302 cells fall into more than 100 different classes that can be distinguished by defined anatomical criteria (White et al. 1986). Most neuron classes can be further subdivided into subclasses (Hobert 2006). Members of a neuronal subclass are similar with regard to many properties, such as anatomical features and/or shared gene expression programs, but they can differ in very specific functional or morphological features. Examples include midline motor neuron classes, each of which is composed of individual subclasses that can only be distinguished by their specific axonal projection patterns (White et al. 1986). Other examples include bilaterally symmetric pairs of sensory neurons that have functionally diversified to express distinct classes of chemosensory receptors on the left and right side of the animal, as further discussed in this paper.

THE ASE GUSTATORY NEURONS

The ASE gustatory neuron class is one of several chemosensory neuron classes in the main head ganglia of the worm (Fig.1a) (Bargmann and Horvitz 1991). Each of these chemosensory neuron classes is composed of two bilaterally symmetric neurons that are indistinguishable by morphological criteria and in most cases examined also display the same functional properties (White et al. 1986; Hobert et al. 2002; Bergamasco and Bazzicalupo 2006). The ASE neuron class, however, is functionally lateralized in that it expresses distinct chemosensory properties on the left and right side of the animal. The ASEL (left) neuron primarily senses sodium, whereas the ASER (right) neuron primarily senses chloride and potassium (Fig.1b) (Pierce-Shimomura et al. 2001). The left/right asymmetric distribution of these chemosensory capacities correlates with the left/right asymmetric expression of a family of putative chemoreceptor genes (Fig.1b) (Yu et al. 1997; Ortiz et al. 2006).

The diversification of the anatomically symmetric ASE neurons into two functionally distinct neurons bears conceptual similarity to a poorly understood but fundamental property of most nervous systems. By anatomical and molecular criteria, most nervous systems display striking patterns of overall bilateral symmetry, yet as best exemplified in the anatomically bilaterally symmetric human brain, nervous systems display striking degrees of functional laterality (Davidson and Hugdahl 1994); that is, the left side of the brain performs tasks different from those of the right side of the brain and vice versa. How functional laterality is superimposed on a presumed bilaterally symmetric ground state is poorly understood. The ASE neuron class promises to yield insights into how the nematode *C. elegans* has solved this problem.

A GENETIC ANALYSIS OF ASE CELL FATE SPECIFICATION REVEALS A BISTABLE REGULATORY SYSTEM

Like nervous system laterality in general, ASE laterality appears to develop from a bilaterally symmetric ground state, characterized by the initially symmetric expression of genes that become restricted to either the ASEL or ASER neuron after hatching of the animal (Fig.1c) (Johnston et al. 2005). How is this switch from symmetry to asymmetry genetically programmed? The genetic amenability of *C. elegans* has enabled us to con-

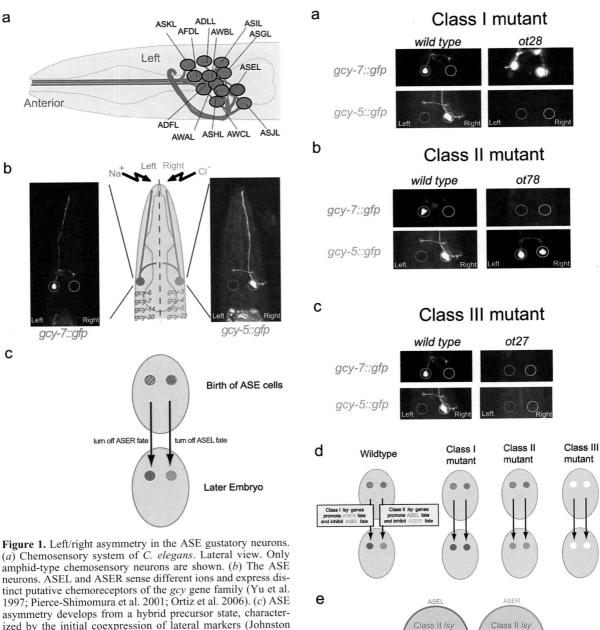

Figure 1. Left/right asymmetry in the ASE gustatory neurons. (*a*) Chemosensory system of *C. elegans*. Lateral view. Only amphid-type chemosensory neurons are shown. (*b*) The ASE neurons. ASEL and ASER sense different ions and express distinct putative chemoreceptors of the *gcy* gene family (Yu et al. 1997; Pierce-Shimomura et al. 2001; Ortiz et al. 2006). (*c*) ASE asymmetry develops from a hybrid precursor state, characterized by the initial coexpression of lateral markers (Johnston et al. 2005).

Figure 2. Mutant screen reveals several classes of ASE mutants. (*a*) Class I mutants display a "2 ASEL" phenotype. The *cog-1(ot28)* allele is shown as a representative example (Chang et al. 2003). (*Red circles*) ASEL cell position; (*blue circles*) ASER position. (*b*) Class II mutants display a "2 ASER" phenotype. The *lin-49(ot78)* allele is shown as a representative example (Chang et al. 2003). (*c*) Class III mutants display a "no ASE" phenotype. The *che-1(ot27)* allele is shown as a representative example (Chang et al. 2003). (*d*) Schematic representation of ASE fate specification from a "hybrid precursor state" (Johnston et al. 2005) to a mature L/R asymmetric state and mutant phenotypes. The progression from hybrid to asymmetric occurs in the embryo. (*e*) Class I and class II genes define a bistable system that depends on mutual inhibition.

duct large-scale genetic screens for mutants in which the left/right asymmetric expression of green fluorescent protein (GFP)-tagged terminal cell fate markers are affected ("*lsy*" phenotype for "lateral symmetry-defective"). So far, we have screened through more than 100,000 haploid genomes, which according to the estimated average mutation frequency, represents an ~50× coverage of the whole genome (Chang et al. 2003, 2004; Johnston and Hobert 2003, 2005; Johnston et al. 2006; O. Hobert et al., unpubl.). We have retrieved almost 200 mutant alleles in which distinct aspects of expression of laterally expressed *gfp* markers are affected (O. Hobert et al., unpubl.). The mutant alleles define at least 20 complementation groups. Most mutants can be classified into at least four distinct

Table 1. Identity of Genes That Affect ASE Differentiation

Gene	Mutant class	Molecular features	Expression	Instructive vs. permissive[a]	References
cog-1	Class I	Nk-type homeobox	ASER	I	Chang et al. (2003)
unc-37	Class I	WD40 domain (Groucho ortholog)	ASEL + ASER	P	Chang et al. (2003)
mir-273 family members	Class I[b]	miRNA	ASEL < ASER	I	Chang et al. (2004); O. Hobert et al. (unpubl.)
ceh-36	Class II	Otx-type homeobox	ASEL + ASER	P	Chang et al. (2003)
die-1	Class II	zinc fingers	ASEL	I	Chang et al. (2004)
lsy-2	Class II	zinc fingers	ASEL + ASER	P	Johnston and Hobert (2005)
lin-49	Class II	PHD finger, bromodomain	ASEL + ASER	P	Chang et al. (2003)
lsy-6	Class II	miRNA	ASEL	I	Johnston and Hobert (2003)
che-1	Class III	zinc fingers	ASEL + ASER	P	Chang et al. (2003)
lim-6	Class IV	LIM homeobox	ASEL	I	Hobert et al. (1999)
fozi-1	Class IV	zinc fingers	ASER	I	Johnston et al. (2006)

[a]A factor is termed "instructive" if it is not only required for the execution of a specific fate, but also sufficient, if misexpressed. In contrast, "permissive" factors are only required but not sufficient for the execution of a specific fate.
[b]This phenotype is inferred from ectopic expression analysis, not from mutant analysis.

classes (Fig. 2a–d). In class I *lsy* mutants, both ASEL and ASER express the ASEL expression profile and the ASER expression profile is lost (Fig. 2a); in class II *lsy* mutants, both ASEL and ASER express the ASER expression profile, and the ASEL expression profile is lost (Fig. 2b); in class III *lsy* mutants, all ASE-specific genes fail to be expressed (Fig. 2c); and in class IV *lsy* mutants, either the ASEL or the ASER cell expresses mixed ASEL/ASER fate characteristics. We have determined the molecular identity of some genes from each of these categories and found that they all encode gene regulatory factors, including sequence-specific DNA-binding transcription factors, general transcriptional cofactors, and miRNAs (Table 1).

The mutant phenotypes, expression pattern, and epistatic relationship of the uncovered gene regulatory factors revealed an important feature of the regulatory architecture of ASEL/R fate specification. The system classifies as a bistable system that, depending on the activity of specific regulatory factors, can exist in one of two stable states: the ASEL state or the ASER state (Johnston et al. 2005). Class I *lsy* genes control the ASER state through repression of class II *lsy* genes, which control the ASEL state through repression of class I *lsy* genes (Fig. 2e). Loss of class I *lsy* genes therefore leads to ectopic expression of class II *lsy* genes and execution of the ASEL fate in both cells, and vice versa.

THE BISTABLE SYSTEM IS CONTROLLED BY miRNAs

The class II gene *lsy-6* and the class I gene *cog-1* are two representative class I and class II genes that regulate each other's expression (Fig. 3a). We first discuss the regulation of *cog-1* by *lsy-6*, which represents a prime paradigm for one of the very few biologically validated animal miRNA/target interactions (Johnston and Hobert 2003; Ambros 2004; Carthew 2006). Mapping of the *lsy-6* locus, of which we retrieved at least four mutant alle-

les, revealed that *lsy-6* codes for a 21-nucleotide-long miRNA that binds to a single complementary site in the 3′UTR (untranslated region) of the *cog-1* homeobox gene (Fig. 3b). This interaction apparently only occurs in a single cell type, ASEL, since *cog-1* and *lsy-6* are expressed in an otherwise nonoverlapping set of neuronal and nonneuronal cells (Fig. 3c). The cell-type specificity in the overlap of a regulatory gene and its target gene is also a common theme in transcriptional regulation (see, e.g., Altun-Gultekin et al. 2001; Tsalik and Hobert 2003). The functional interaction of *lsy-6* with *cog-1* was validated using a sensor gene strategy in which the promoter of the *ceh-36* gene drives *gfp* expression in ASEL and ASER; substituting a nonregulated 3′UTR with the 3′UTR of the *cog-1* gene causes downregulation of this sensor in ASEL but not in ASER (Fig. 3d). This down-regulation depends on the presence of the *lsy-6* miRNA and the *lsy-6* complementary site in the *cog-1* 3′UTR (Fig. 3d).

Notably, at first appearance, *lsy-6* affects *cog-1* transcription, that is, the ASEL neuron, which expresses *lsy-6*, does not transcribe *cog-1* (a surprising notion since it would suggest that *lsy-6* and *cog-1* mRNA will not encounter each other); yet, removal of *lsy-6* leads to aberrant transcription of *cog-1* in ASEL (Johnston et al. 2005). We have shown that this phenomenon is explained through positive transcriptional autoregulation of *cog-1* (Johnston et al. 2005). In the presence of *lsy-6*, COG-1 protein is not produced, and since COG-1 protein autoregulates its own transcription, the *cog-1* mRNA will also disappear. Thus, a lack of overlap in the expression of miRNAs and predicted target genes, which is observed frequently (Stark et al. 2005), may therefore, at least in some cases, be a consequence of miRNA–target interaction, rather than a reflection of evolutionary divergence in transcriptional control of miRNAs and their target genes.

Contrasting proposed models that posit a prevailing theme of miRNAs as fine-tuners of gene expression or

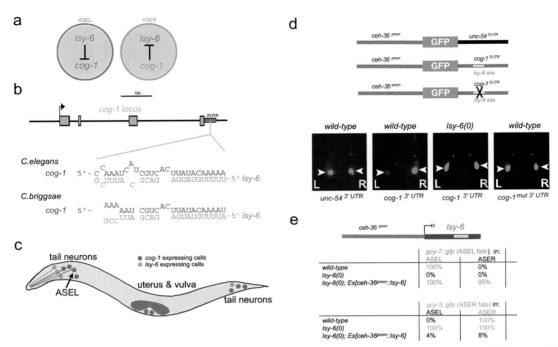

Figure 3. The *lsy-6/cog-1* interaction. (*a*) *lsy-6* and *cog-1* define two mutual inhibitory class I and class II genes. (*b*) The *cog-1* 3′UTR contains a *lsy-6* complementary site, which is phylogenetically conserved (Johnston and Hobert 2003). The seed region within the *cog-1/lsy-6* is required but not sufficient to confer *cog-1* down-regulation by *lsy-6* (Didiano and Hobert 2006). (*c*) *lsy-6* and *cog-1* expression only overlaps in ASEL. The *gfp*-tagged *cog-1* locus is expressed in the ADL, ASE, ASJ, AIA (or SMBD, SIAD, or SIAV), and PHB neuron classes, in unidentified preanal ganglion neurons, sphincter muscle, phasmid sheath cells, uterus, and the developing vulva (Palmer et al. 2002). The *gfp*-tagged *lsy-6* locus is expressed in six labial sensory neurons, ASEL and the PVQ neuron class in the tail (Johnston and Hobert 2003). (*d*) Down-regulation of the *cog-1* 3′UTR by *lsy-6* can be monitored using a sensor gene strategy (Johnston and Hobert 2003; Didiano and Hobert 2006). (*e*) *lsy-6* works as a switch. *lsy-6* is both necessary and sufficient to induce the ASEL fate (Johnston and Hobert 2003).

buffers of "genetic noise" (Bartel and Chen 2004; Stark et al. 2005; Hornstein and Shomron 2006), *lsy-6* acts as a switch in the ASE fate decision. Loss of *lsy-6* causes the ASEL neuron to switch to the execution of ASER fate, and ectopic expression of *lsy-6* in the ASER neuron switches on ASEL fate in ASER (Fig. 3e) (Johnston and Hobert 2003).

cog-1 is not only directly regulated by *lsy-6*, but is also itself required for regulation of *lsy-6* expression. This feedback mechanism is complex and involves a series of additional factors. *lsy-6* expression in ASEL is controlled by the ASEL-specific zinc finger transcription factor *die-1*, a class II *lsy* gene, retrieved from our mutant screen (Fig. 4) (Chang et al. 2004). The L/R asymmetric expression of *die-1* is controlled via the 3′UTR of *die-1*, as revealed by a sensor gene approach similar to that used for *cog-1*. The ASER-specific down-regulation of the 3′UTR of *die-1* genetically depends on *cog-1*, indicating that *cog-1* genetically activates posttranscriptional factors that negatively regulate *die-1* (Johnston et al. 2005). The *mir-273* family of miRNAs, composed of at least seven members (*mir-273, mir-51* through *mir-56*) are excellent candidates to be involved in this process. First, the *die-1* 3′UTR contains two phylogenetically conserved complementary sites to *mir-273* family members; second, some members of the family are predominantly expressed in ASER, where the *die-1* 3′UTR is down-regulated (Chang et al. 2004; O. Hobert et al., unpubl.); third, the ASER-biased expression of *mir-273*, and likely other family members as

well, is lost in *cog-1* mutants (Johnston et al. 2005); and fourth, forced expression of *mir-273* or other members of the family in ASEL induces ASER fate in ASEL (Chang et al. 2004; O. Hobert et al., unpubl.), as would be expected from a down-regulation of *die-1* expression.

lsy-6, cog-1, die-1, and *mir-273* family members therefore define a double-negative feedback loop (Fig. 5a) that provides the underlying molecular basis for the bistability of the system. A number of prominent and well-studied cell fate decisions utilize bistable feedback systems to control specific cell fate decisions. Examples include the bacteriophage λ system, which relies on mutual cross-inhibition of two repressor proteins, Cl and Cro (Fig. 5c). In a more complex example, a specific cell fate decision in the worm's developing vulva is controlled by a Notch/*lin-12*-mediated negative feedback loop between two distinct cells (Fig. 5d).

Feedback loops require the existence of specific inputs into the loop and outputs from the loop (Fig. 5e). Through genetic epistasis analysis, we determined the input and output of the ASEL/R-controlling bistable feedback loop (Johnston et al. 2005). One representative example of this type of analysis is shown in Figure 4. The key points of the epistasis analysis are that *lsy-6* requires all loop components to affect terminal differentiation markers (such as the *gcy* genes or *lim-6*), whereas, in contrast, *die-1* can exert its effect on terminal differentiation markers independently of the loop components (Fig. 4). *lsy-6* therefore provides the input into the loop, and *die-1* is the output regulator (Fig. 5a).

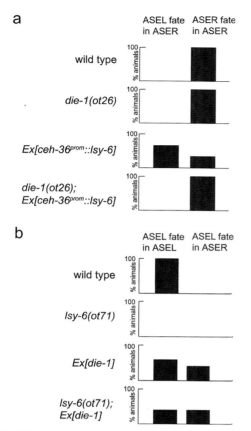

Figure 4. Determining regulatory architecture through epistasis analysis. Two examples of the genetic epistasis analysis. ASEL fate is monitored with *lim-6^prom^::gfp* and ASER fate is monitored with *gcy-5^prom^::gfp*. "Ex" indicates the forced, bilateral expression of regulatory factors from extrachromosomal arrays. (*a*) Ectopic *lsy-6* requires *die-1* to induce ASEL fate. (*b*) Ectopic *die-1* does not require *lsy-6* to induce ASEL fate. (Modified, with permission, from Johnston et al. 2005 [© National Academy of Sciences].)

The regulatory architecture downstream from *die-1* is complex. The precise quantification of mutant phenotypes as well as epistasis analysis indicates that *die-1* acts in the context of several feedforward loop motifs to affect terminal differentiation markers, such as the chemoreceptors of the *gcy* family or neuropeptides of the *flp* family (Fig. 5b). To be expressed, several of these terminal differentiation markers usually require two conditions: the presence of *die-1* and the absence of a *die-1*-repressed zinc finger transcription factor, *fozi-1*, a class IV mutant retrieved from our screen (Fig. 2d) (Johnston et al. 2006). The absence of *die-1* in ASER allows expression of this zinc finger transcription factor, thereby repressing the expression of a repressor of the ASER fate, the *lim-6* LIM homeobox gene. The sequential repression cascade that abounds in the ASEL/R regulatory architecture falls well in line with observations in the vertebrate nervous system, where the sequential repression of transcriptional repressor proteins diversifies neuronal cell fate in the spinal cord (Muhr et al. 2001).

A central feature of these sequential repression schemes is a permissive activation mechanism. This can be illustrated in the case of the ASER-specific *gcy* genes. As

shown in Figure 5a, their expression is regulated by the repression of repressor proteins, but eventually factors must exist to turn on the expression of the ASER-specific *gcy* genes. Our large-scale mutant analysis, in combination with the molecular dissection of the *cis*-regulatory architecture of ASE-expressed genes, revealed a factor that is an excellent candidate to provide such permissive activation function. In class III mutants, all ASE-expressed genes fail to be activated; all class III mutant alleles that we retrieved from our genetic screens (>20 alleles) define a single locus, *che-1*. *che-1* encodes a zinc finger transcription factor expressed in both ASEL and ASER (Chang et al. 2003; Uchida et al. 2003), which we found to bind to an experimentally determined binding site, termed the "ASE motif" that is present in all ASE-expressed genes (J. Etchberger and O. Hobert et al., unpubl.). The CHE-1 protein therefore defines a "ground state" of activation, which is modified through the sequential activity of repressors. This model also provides a mechanistic basis for the observation that directly after their birth, both the ASEL and ASER neurons coexpress genes that later become restricted to either ASEL or ASER (e.g., both *lsy-6* and *cog-1* are initially coexpressed) (Johnston et al. 2005). Presumably, these genes are first all activated by CHE-1 and then become sequentially repressed in either ASEL or ASER.

INTRODUCING THE LEFT/RIGHT BIAS

A central question that is left unanswered so far is what determines the left/right differential activity of the bistable feedback loop. Why do *lsy-6* and *die-1* "win" in ASEL, and *cog-1* and the *mir-273* family "win" in ASER? At first sight, an attractive underlying mechanism could have been some form of lateral inhibition, best studied in the case of the AC/VU cell fate decision (Fig. 5d). An apparently stochastic small difference in the level of Notch/*lin-12* activity is amplified by a feedback mechanism, so that the presumptive AC cell ultimately expresses only the *lin-12* ligand *lag-2*, whereas the presumptive VU cell expresses only *lin-12* (Greenwald 1998). Such a mechanism is, however, unlikely to exist in the ASE fate decision. First, in contrast to the AC/VU decision and other lateral inhibition phenomena, the process is not stochastic; i.e., the right cell always adopts the ASER fate and the left cell always adopts the ASEL fate. Second, laser ablation studies reveal that ASEL is not required for the adoption of the ASER fate, and ASER is not required for the adoption of the ASEL fate (Poole and Hobert 2006).

The progression of the system from a hybrid precursor state to an asymmetric state some time in late embryogenesis, after the cells are born, could, in theory, be explained by the existence of a nonautonomous signal that instructs either ASEL or ASER to become different from one another. For example, both cells could contain an intrinsic bias to one loop configuration, e.g., the ASER-promoting configuration, and a signal to ASEL could reverse the loop (e.g., by boosting *lsy-6* expression) to the opposite configuration. In this simple form, this model is also unlikely to be correct. Genetic and surgical manipulation in the early embryo rather suggest that the difference between ASEL and ASER is already predeter-

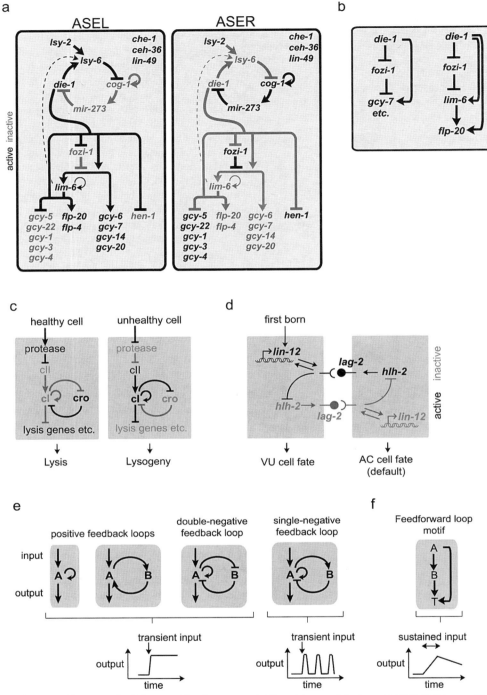

Figure 5. Regulatory architecture of cell fate specification events. (*a*) Regulatory architecture of ASEL/R fate specification. At present, it is not known where the permissively acting (i.e., ASEL/ASER-expressed) transcription factors *ceh-36* and *lin-49* fit into the gene regulatory network. The zinc finger transcription factor encoded by the *che-1* gene appears to be required for the activation of each ASEL or ASER expressed gene. (*b*) Deconvoluted network motifs extracted from panel *a* (Poole and Hobert 2006). (*c,d*) Bistable, negative feedback loops control fate decisions in other systems. Examples are the bacteriophage λ (panel *c*) (Ptashne 1992) and the AC/VU decision in *C. elegans* (panel *d*) (Greenwald 1998). (*e,f*) Network motifs and their properties. (*c,d,e,* Reprinted, with permission, from Johnston et al. 2005 [© National Academy of Sciences].)

mined at the level of very early blastomere identity, long before the ASE cells are born (Poole and Hobert 2006). One attractive possibility is that this early determination event is memorized through embryonic development until after the ASE cells are born by a chromatin-related mechanism that may bias the expression of loop components in the left cell versus the right cell. We anticipate that our ongoing genetic and molecular analyses of the many more mutants that affect ASEL/R fate specification will eventually resolve this question.

NETWORK MOTIFS

Work in simple unicellular organisms during the past few years revealed that networks of regulatory factors contain recurring wiring patterns termed "network motifs" (Lee et al. 2002; Milo et al. 2002; Shen-Orr et al. 2002). ASEL/R specification also involves several of these network motifs, namely, feedback and feedforward loops (Fig. 5) (Johnston et al. 2005, 2006). These network motifs have defined properties that may be useful to understand ASEL/R fate determination. Feedback loops have the feature of being able to amplify a transient and weak input into a robust output (Fig. 5e). Feedforward loops (Fig. 5f) are persistence detectors that measure the sustained presence of a factor A. Only if A is present long enough to activate factor B can it activate, together with factor B, the target gene T (Fig. 5f).

A working hypothesis for the ASEL/R fate determination process is therefore that an initial slight difference between ASEL and ASER, possibly determined by a signal into lineage precursors of ASEL/R, is amplified after the birth of the ASE neurons. The amplification of this input may be measured on the level of die-1 and the feedforward motifs that emanate from die-1. Only if die-1 levels have reached a certain level for a long enough time ("persistence detection") will downstream target genes be activated. Further genetic analysis will reveal additional gene regulatory factors in this network and will lead to a deeper understanding of the intricate interplay between these regulatory factors.

OPEN QUESTIONS

Our analysis has only begun to reveal the regulatory logic of ASE neuron specification. The many open questions that remain include (1) What is the molecular nature of the input into the bistable feedback loop, that is, what biases the loop into one configuration in ASEL and into the other configuration in ASER? (2) What other regulatory factors are embedded within the bistable feedback loop and within the feedforward loops that control the terminal differentiation markers, or, in other words, which of the interactions shown in Figure 5a are direct, which are indirect? (3) How is the activity of permissively acting factors, such as ceh-36, lsy-2, and lin-49, which are expressed in both ASEL and ASER (Table 1), restricted to either ASEL or ASER? We anticipate that the cloning of mutants from our large mutant collection and biochemical approaches aimed at looking at protein/nucleic acid interactions directly in the ASE neurons may answer these and other remaining questions.

CONCLUSIONS

Gene regulatory networks that control terminal cell fates can be surprisingly complex. With its attempt to provide a saturation analysis of a depth similar to the genetic analysis of early fly embryo patterning (Nüsslein-Volhard and Wieschaus 1980), it is perhaps not surprising that our large-scale genetic analysis has uncovered an intricate network of gene regulatory factors. Other cell fate decisions may be similarly controlled by complex networks, to be revealed by, for example, saturation mutant analysis. But it is also conceivable that some cell fate decisions involve more complex regulatory networks than others. This could be envisioned to be the case if one deals with diversifying the fate of cells that are largely similar to one another. Generating a group of cells with similar fates may impose regulatory constraints that may require complex changes in the regulatory wiring to further diversify individual cells within this group. The diversification of the cellular fate of ASEL and ASER is indeed a relatively recent evolutionary phenomenon since the expression of the gcy chemoreceptors is differently controlled in two related nematode species (Ortiz et al. 2006).

Given the youth of the miRNA field and the resulting paucity of experimental data on miRNA function, our studies have revealed and corroborated previous insights into miRNA function (for general reviews on miRNA function in vivo, see Ambros 2004; Carthew 2006). One theme that deserves emphasis is that gene regulation mediated by miRNAs shares many conceptual similarities with gene regulatory events mediated by transcription factors (Hobert 2004). Like transcription factors, miRNAs bind to specific cis-regulatory elements in their nucleic acid target sequences (DNA for transcription factors, RNA for miRNAs). Like cis-regulatory elements hardwired into DNA, miRNA-responsive cis-regulatory elements in mRNAs are (1) occupied in a cell-type-specific manner by trans-acting factors, i.e., cell-type-specifically expressed miRNAs and (2) they are functional only in a highly context-dependent manner (Didiano and Hobert 2006). Like transcription factors, miRNAs appear to be integrated into gene regulatory networks. They are activated by RNA polymerase-II-dependent transcription factors and control the expression of transcription factors. Like ASE-expressed transcription factors, miRNAs acting in ASE cell fate specification work as clear switches and are necessary and sufficient to induce specific cellular fates.

The so far exclusive role of miRNAs in repression of target gene expression also fits with recent themes that propose that transcription factors which determine cellular fate often act as repressors; that is, to induce a specific cell fate, one does not necessarily require a specific gene activation event, but the specific modulation of a repressed state (Muhr et al. 2001).

Since we are still in the early days of studying miRNA function, it appears premature to propose overarching, global themes of miRNA function. As the decades of work on transcription factors have taught us, large families of gene regulatory factors may easily escape an easy overall classification theme. A careful experimental analysis of miRNAs in diverse cellular contexts will ultimately reveal the full functional spectrum of this exciting class of RNA-based gene regulatory factors.

ACKNOWLEDGMENTS

The experiments presented here were conducted by a series of past and present students and postdocs. Sarah Chang and Bob Johnston identified, mapped, and cloned the genes presented here, and Bob Johnston determined the regulatory architecture of the genetic interactions.

More recent screens on ASE cell fate determination were done by Celia Antonio, Eileen Flowers, Maggie O'Meara, and Sumeet Sarin. Work on the interactions of *lsy-6* and *cog-1* was done by Dominic Didiano, studies on *che-1* were done by John Etchberger, and the early embryo work was done by Richard Poole and Bob Johnston. Thanks to members of the Hobert lab for commenting on the manuscript. This work was funded by the National Institutes of Health (2R01NS039996-05, 5R01NS050266-02) and by the Howard Hughes Medical Institute.

REFERENCES

Altun-Gultekin Z., Andachi Y., Tsalik E.L., Pilgrim D., Kohara Y., and Hobert O. 2001. A regulatory cascade of three homeobox genes, ceh-10, ttx-3 and ceh-23, controls cell fate specification of a defined interneuron class in *C. elegans*. *Development* **128**: 1951.

Ambros V. 2004. The functions of animal microRNAs. *Nature* **431**: 350.

Bargmann C.I. and Horvitz H.R. 1991. Chemosensory neurons with overlapping functions direct chemotaxis to multiple chemicals in *C. elegans. Neuron* **7**: 729.

Bartel D.P. and Chen C.Z. 2004. Micromanagers of gene expression: The potentially widespread influence of metazoan microRNAs. *Nat. Rev. Genet.* **5**: 396.

Bergamasco C. and Bazzicalupo P. 2006. Chemical sensitivity in *Caenorhabditis elegans. Cell. Mol. Life Sci.* **63**: 1510.

Carthew R.W. 2006. Gene regulation by microRNAs. *Curr. Opin. Genet. Dev.* **16**: 203.

Chang S., Johnston R.J., Jr., and Hobert O. 2003. A transcriptional regulatory cascade that controls left/right asymmetry in chemosensory neurons of *C. elegans. Genes Dev.* **17**: 2123.

Chang S., Johnston R.J., Frøkjaer-Jensen C., Lockery S., and Hobert O. 2004. MicroRNAs act sequentially and asymmetrically to control chemosensory laterality in the nematode. *Nature* **430**: 785.

Davidson R.J. and Hugdahl K., eds. 1994. *Brain asymmetry*. MIT Press, Cambridge, Massachusetts.

Didiano D. and Hobert O. 2006. Perfect seed pairing is not a generally reliable predictor for miRNA-target interactions. *Nature Struct. Mol. Biol.* **13**: 849.

Greenwald I. 1998. LIN-12/Notch signaling: Lessons from worms and flies. *Genes Dev.* **12**: 1751.

Hobert O. 2004. Common logic of transcription factor and microRNA action. *Trends Biochem. Sci.* **29**: 462.

———. 2006. Specification of the nervous system. In *Worm-Book: The online review of* C. elegans *biology*. The *C. elegans* Research Community, eds. (www.wormbook.org).

Hobert O., Johnston R.J., Jr., and Chang S. 2002. Left-right asymmetry in the nervous system: The *Caenorhabditis elegans* model. *Nat. Rev. Neurosci.* **3**: 629.

Hobert O., Tessmar K., and Ruvkun G. 1999. The *Caenorhabditis elegans* lim-6 LIM homeobox gene regulates neurite outgrowth and function of particular GABAergic neurons. *Development* **126**: 1547.

Hornstein E. and Shomron N. 2006. Canalization of development by microRNAs. *Nat. Genet.* (suppl. 1) **38**: S20.

Johnston R.J. and Hobert O. 2003. A microRNA controlling left/right neuronal asymmetry in *Caenorhabditis elegans. Nature* **426**: 845.

———. 2005. A novel *C. elegans* zinc finger transcription factor, lsy-2, required for the cell type-specific expression of the lsy-6 microRNA. *Development* **132**: 5451.

Johnston R.J., Jr., Chang S., Etchberger J.F., Ortiz C.O., and Hobert O. 2005. MicroRNAs acting in a double-negative feedback loop to control a neuronal cell fate decision. *Proc. Natl. Acad. Sci.* **102**: 12449.

Johnston R.J., Copeland J.W., Fasnacht M., Etchberger J.F., Liu J., Honig B., and Hobert O. 2006. An unusual Zn finger/FH2 domain protein controls a left/right asymmetric neuronal fate decision in *C. elegans. Development* **133**: 3317.

Lee T.I., Rinaldi N.J., Robert F., Odom D.T., Bar-Joseph Z., Gerber G.K., Hannett N.M., Harbison C.T., Thompson C.M., Simon I., et al. 2002. Transcriptional regulatory networks in *Saccharomyces cerevisiae. Science* **298**: 799.

Milo R., Shen-Orr S., Itzkovitz S., Kashtan N., Chklovskii D., and Alon U. 2002. Network motifs: Simple building blocks of complex networks. *Science* **298**: 824.

Muhr J., Andersson E., Persson M., Jessell T.M., and Ericson J. 2001. Groucho-mediated transcriptional repression establishes progenitor cell pattern and neuronal fate in the ventral neural tube. *Cell* **104**: 861.

Nüsslein-Volhard C. and Wieschaus E. 1980. Mutations affecting segment number and polarity in *Drosophila. Nature* **287**: 795.

Ortiz C.O., Etchberger J.F., Posy S.L., Frøkjær-Jensen C., Lockery S., Honig B., and Hobert O. 2006. Searching for neuronal left/right asymmetry: Genome wide analysis of nematode receptor-type guanylyl cyclases. *Genetics* **173**: 131.

Palmer R.E., Inoue T., Sherwood D.R., Jiang L.I., and Sternberg P.W. 2002. *Caenorhabditis elegans cog-1* locus encodes GTX/Nkx6.1 homeodomain proteins and regulates multiple aspects of reproductive system development. *Dev. Biol.* **252**: 202.

Pierce-Shimomura J.T., Faumont S., Gaston M.R., Pearson B.J., and Lockery S.R. 2001. The homeobox gene *lim-6* is required for distinct chemosensory representations in *C. elegans. Nature* **410**: 694.

Poole R. and Hobert O. 2006. Early embryonic programming of neuronal left/right asymmetry in *C. elegans. Curr. Biol.* (in press).

Ptashne M. 1992. *A genetic switch: Phage λ and higher organisms*. Blackwell Scientific and Cell Press, Cambridge, Massachusetts.

Shen-Orr S.S., Milo R., Mangan S., and Alon U. 2002. Network motifs in the transcriptional regulation network of *Escherichia coli. Nat. Genet.* **31**: 64.

Stark A., Brennecke J., Bushati N., Russell R.B., and Cohen S.M. 2005. Animal microRNAs confer robustness to gene expression and have a significant impact on 3'UTR evolution. *Cell* **123**: 1133.

Tsalik E.L. and Hobert O. 2003. Functional mapping of neurons that control locomotory behavior in *Caenorhabditis elegans. J. Neurobiol.* **56**: 178.

Uchida O., Nakano H., Koga M., and Ohshima Y. 2003. The *C. elegans* che-1 gene encodes a zinc finger transcription factor required for specification of the ASE chemosensory neurons. *Development* **130**: 1215.

White J.G., Southgate E., Thomson J.N., and Brenner S. 1986. The structure of the nervous system of the nematode *Caenorhabditis elegans. Philos. Trans. R. Soc. Lond. B Biol. Sci.* **314**: 1.

Yu S., Avery L., Baude E., and Garbers D.L. 1997. Guanylyl cyclase expression in specific sensory neurons: A new family of chemosensory receptors. *Proc. Natl. Acad. Sci.* **94**: 3384.

The *Caenorhabditis elegans* Argonautes ALG-1 and ALG-2: Almost Identical yet Different

B.B.J. Tops, R.H.A. Plasterk, and R.F. Ketting

Hubrecht Laboratory, Utrecht, The Netherlands

Since the discovery of the RNA interference pathway, several other small RNA pathways have been identified. These make use of the same basic machinery to generate small RNA molecules that can direct different types of (post)transcriptional silencing. The specificity for the different silencing pathways (which type of silencing a small RNA initiates) is likely accomplished by the effector molecules that bind the small RNAs: the Argonaute proteins. Two Argonaute proteins, ALG-1 and ALG-2, have been implicated in one of the silencing pathways, the microRNA (miRNA) pathway, in *Caenorhabditis elegans*. The two proteins are highly similar, and previous work suggested redundancy of the two proteins. Here, we present genetic and biochemical data that hint at individual nonredundant functions for ALG-1 and ALG-2 in the processing of precursor miRNAs to mature miRNAs.

RNA interference (RNAi) is a mechanism that degrades mRNAs in a sequence-specific manner (Fire et al. 1998). Since its discovery in *C. elegans*, identical mechanisms have been discovered in fungi, plants, flies, and mammals. One of the functions of this mechanism is to protect the genome from selfish nucleic acids, such as viruses and transposons (for review, see Plasterk 2002). Besides this posttranscriptional gene-silencing mechanism that functions by degrading the mRNAs, similar mechanisms exist that function through translational repression of mRNAs or by regulating gene silencing at the transcriptional level by heterochromatin modifications. All of these posttranscriptional and transcriptional silencing mechanisms are identical in that they require (1) small RNAs to provide the necessary sequence specificity and (2) effector molecules: Argonaute proteins that bind the small RNAs.

Depending on the type of silencing pathway, different small RNAs are distinguished. Small RNAs functioning in the RNAi pathway and directing mRNA degradation are termed small interfering RNAs (siRNAs) (Hamilton and Baulcombe 1999; Zamore et al. 2000), and small RNAs derived from RNA hairpins encoded by the genome and directing mRNA translational repression, and/or degradation, are named microRNAs (Lau et al. 2001; Ambros et al. 2003). siRNAs and miRNAs are both single-stranded RNA (ssRNA) molecules of approximately 21–23 nucleotides (depending on the organism) that are generated from longer double-stranded RNAs (dsRNAs) by the Dicer enzyme (Bernstein et al. 2001; Grishok et al. 2001; Hutvagner et al. 2001; Ketting et al. 2001). The source of the dsRNA differs for each of the silencing pathways, including viral RNA, readthrough transcripts from transposons or genome-encoded snap-back structures. In addition to these two well-studied small RNAs, other small RNA species have recently been identified: repeat-associated siRNAs (rasi's) (Aravin et al. 2003) and Piwi-associated RNAs (piRNAs) (Aravin et al. 2006; Girard et al. 2006). Neither the biogenesis nor function of these small RNAs is known.

The Argonaute proteins are the effector molecules in the different silencing pathways and are characterized by a PAZ and PIWI domain. The first is involved in binding of the small RNAs (Lingel et al. 2003; Song et al. 2003; Yan et al. 2003) and the latter resembles an RNase H motif that also binds the 5′ phosphate of the small RNAs and is required for cutting the targeted mRNA (Parker et al. 2004; Rand et al. 2004; Song et al. 2004). However, not all Argonaute proteins contain the conserved DDH motif required for this nuclease activity, so presumably not all Argonautes are cleaving mRNAs.

One of the silencing pathways is the miRNA pathway. Here we focus on miRNA function in animal systems. Long RNA transcripts are transcribed from the genome that can snap back on themselves (primary miRNA or pri-miRNA) and are processed into approximately 70-nucleotide stem-loop structures (precursors miRNA or pre-miRNA) by Drosha in the nucleus (Lee et al. 2003). The pre-miRNA serves as a template for Dicer which processes the pre-miRNA into the double-stranded miRNA of about 21 nucleotides (Bernstein et al. 2001; Grishok et al. 2001; Hutvagner et al. 2001; Ketting et al. 2001). This miRNA is subsequently unwound, and the single-stranded mature miRNA, bound by an Argonaute, can hybridize to the 3′UTR (untranslated region) of a cognate mRNA. In general, this will lead to translational inhibition without direct degradation of the messenger by the Argonaute.

One of the questions that remains is how a small RNA is directed to a specific silencing pathway (e.g., transcriptional silencing, mRNA degradation, or blocking translation). Since the small RNAs for the different silencing pathways are generated by the same proteins, the most obvious solution would be to have different effector molecules (Argonautes) for different silencing pathways, depending on the source of the dsRNA; for example, a short RNA hairpin in the case of miRNAs or a larger intermolecular dsRNA in the case of viral siRNAs.

Indeed, the genomes of most eukaryotic organisms encode several Argonautes. The *C. elegans* genome encodes 27 Argonaute family members of which some have indeed been implicated in different silencing processes. Two of these, ALG-1 and ALG-2, have been implicated in the miRNA pathway (Grishok et al. 2001). Although both Argonautes contain the DDH motif in their PIWI domain and should therefore be capable of cutting the targeted mRNA, ALG-1 and ALG-2 appear to be associated with miRNAs that block translation, although this may be the consequence of the imperfect matching between miRNA and mRNA (Hutvagner and Zamore 2002). Interestingly, although the two proteins share 88% identity at the protein level, the knockout phenotypes of the two genes differ remarkably. Mutant alleles of *alg-1(gk214 & tm492)* and *alg-1(RNAi)* animals appear to be viable, but they display a range of phenotypes resulting in very sick animals (Grishok et al. 2001; M. Jovanovic, pers. comm.). *alg-1(gk214)* may be a hypomorphic allele since it only deletes part of the amino terminus, but *alg-1(tm492)* is likely a null, because it deletes the whole PAZ domain and leads to an early stop. Additionally, a *lacZ* reporter under the translational control of the *let-7* miRNA is de-silenced in *alg-1(RNAi)* animals (Caudy et al. 2003). This is in contrast to the *alg-2(ok304)* mutant which shows minor phenotypes at best and does not de-silence the *lacZ* reporter either (Grishok et al. 2001; results not shown). *alg-1/2* double mutants are lethal (Grishok et al. 2001; this paper), indicating that these genes may act in a redundant way. However, here we describe genetic and biochemical studies which suggest that ALG-1 and ALG-2 may well carry out specific, nonredundant tasks.

MATERIALS AND METHODS

General methods. The Bristol N2 strain was used as standard wild-type strain. NL4511 and NL4517 (*alg-1::HA* and *alg-2::HA*, respectively) were generated using standard microinjection and integration techniques. VC446 and WM53 were obtained from the *Caenorhabditis* Genetics Center. Dicer assays (Ketting et al. 2001), extract preparation, western blotting (αHA, clone 12CA5 from Sigma), immunoprecipitations (αHA-matrix, clone 3F10 from Roche Applied Sciences), and size-fractionation experiments were carried out as described previously (Tops et al. 2005).

Gel-shift assay. Gel-shift assays were performed using a radioactive 5′-phosphate-labeled synthetic pre-*let-7* (*C. elegans* and human sequence) (Proligo). Gel shifts were performed for 5 minutes on ice in 10 mM Tris at pH 7.0, 10 mM $MgCl_2$, 1 mM dithiothreitol (DTT), 125 ng/μl yeast tRNA, 1 mM ATP, and 6% PEG-8000 using *C. elegans* embryonic extracts. Reactions were run on a 5% nondenaturing gel at 4°C.

GENETICS

We performed a genome-wide synthetic lethal screen for both the *alg-1(gk214)* and *alg-2(ok304)* mutant strains by RNAi, using the Ahringer RNAi library (Kamath et al. 2003; van Haaften et al. 2004), hoping to identify genes

Table 1. Genes Causing Synthetic Lethality When Knocked Down by RNAi in an *alg-1(gk214)* or *alg-2(ok304)* Genetic Background

Cosmid	Gene	Description
Synthetic lethal with *alg-1(gk214)*		
T07D3.7	*alg-2*	Argonaute involved in miRNA pathway in *C. elegans*
C25A1.5		cytochrome b_5 domain
C25A1.6		H/ACA snoRNP complex
T04D3.2	*sdz-30*	unknown
Y47G6A.19		zinc carboxypeptidase domain
F52C6.3	*phi-32*	ubiquitin-like protein
B0286.4	*ntl-2*	CCR4/NOT complex component
F44B9.7	*pqn-38*	replication factor C domain
F29B9.6	*ubc-9*	E2 ubiquitin-conjugating enzyme
Synthetic lethal with *alg-2(ok304)*		
C35C5.1	*sdc-2*	involved in *C. elegans* dosage compensation
F44A6.2	*sex-1*	involved in *C. elegans* dosage compensation
F48F7.1	*alg-1*	Argonaute involved in miRNA pathway in *C. elegans*
ZK262.8		*C. elegans*-specific

involved in miRNA biogenesis or genes involved in essential pathways regulated by miRNAs. If both proteins have similar functions and are indeed functionally redundant, we expected to find similar results for the two mutants. This is, however, not what we find. There is no overlap in the genes that cause lethality when knocked down in the *alg-1(gk214)* background and those causing lethality in the *alg-2(ok304)* background (Table 1). The genes found to be synthetic lethal with *alg-1(gk214)* must be interpreted with caution since *alg-1(gk214)* animals are already very sick. However, the genes reproducibly causing lethality in an *alg-2(ok304)* background certainly do not cause synthetic lethality in an *alg-1(gk214)* background when specifically retested. Interestingly, two of the genes identified are involved in the dosage compensation/sex determination pathway in *C. elegans* and suggest an involvement of miRNAs in the dosage compensation pathway (B.B.J. Tops et al., in prep.). The observation that there is no overlap between the genes causing synthetic lethality in *alg-1(gk214)* and *alg-2(ok304)* animals suggests that despite their high level of amino acid sequence identity, there are specific nonredundant functions for the two Argonaute proteins.

BIOCHEMISTRY

To test whether the differences between the *alg-1* and *alg-2* synthetic lethal screens could be explained by a difference in temporal or differential expression between the two Argonautes, we generated transgenic ALG-1::HA and ALG-2::HA expressing lines driven by an *alg-1* and *alg-2* promoter, respectively, and performed immunostainings on these animals. ALG-1 (result not shown) and ALG-2 are expressed in the cytoplasm from early embryogenesis (Fig. 1) to adulthood (result not shown) in most, if not all, cells in *C. elegans*, and there is no clear difference in the expression patterns of the two proteins. This appears to

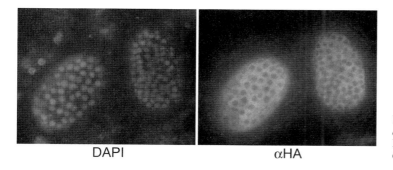

Figure 1. Subcellular localization of ALG-2. *alg-2::HA* transgenic embryos were stained with DAPI and monoclonal αHA antibodies. Only cytoplasmic staining is observed.

exclude the possibility that the observed genetic differences stem from a difference in expression pattern.

A second explanation for nonredundant functions for ALG-1 and ALG-2 could be that both Argonautes bind different subsets of miRNAs. Analysis of the miRNAs associated with ALG-1 and ALG-2 by immunoprecipitation of the HA-tagged proteins and subsequent analysis of the associated RNAs on microarrays show that there is no difference in miRNAs pulled down with ALG-1 or ALG-2 (B.B. Tops et al., in prep.), suggesting that this also is not likely to be the explanation for ALG-1- and ALG-2-specific functions.

A third explanation is that the two Argonautes associate with different proteins, resulting in different complexes with specialized functions. To test this, we size-fractionated the ALG-1- and ALG-2-containing complexes. As shown in Figure 2, in embryonic cytosolic extracts ALG-2 resides in two distinct complexes of approximately 250 and 500 kD. This is in contrast to ALG-1, which migrates in a single complex larger than 650 kD. When analyzed for *mir-40* (Fig. 2) or *mir-52* (result not shown), most of these miRNAs reside in the low-molecular-weight complex of ALG-2, and only a small fraction is comigrating with ALG-1. These data demonstrate that both Argonautes indeed reside in different complexes.

Both ALG-1 and ALG-2 have been implicated in the processing of pre-miRNAs to mature miRNAs. In the absence of one or both of the Argonaute proteins, less of the mature *lin-4* miRNA is detected in vivo, and the pre-

miRNA accumulates as it does in *dcr-1(RNAi)* animals (Grishok et al. 2001). The same is true for the *let-7* miRNA, although accumulation of the pre-miRNA is less distinct. Purified recombinant Dicer enzyme is capable of processing pre-miRNA, so biochemically, it is unlikely that ALG-1 or ALG-2 is directly involved in generating miRNAs. So why do these mutants accumulate pre-miRNAs? A possible explanation is based on experiments performed using human cell lines (Chendrimada et al. 2005; Gregory et al. 2005). From these experiments, it was concluded that the processing of pre-miRNAs into miRNAs and the loading of the miRNAs into the RNA-induced silencing complex (RISC), containing AGO2, is a coupled process. Thus, an alternative explanation for the observed accumulation of pre-miRNAs in *alg-1* and *alg-2* mutant animals is not that these animals are not capable of processing the pre-miRNA, but that they are unable to load the processed miRNA into the silencing complex and thereby unable to release the Dicer enzyme for a new round of pri-miRNA processing.

To study the processing of pre-miRNAs to miRNAs in *C. elegans* and the involvement of ALG-1 and ALG-2, we generated cytosolic extracts from *alg-1(gk214)*, *alg-2(ok304)*, and wild-type animals and tested these for in vitro pre-*let-7* processing. Wild-type extract nicely processes a synthetic pre-*let-7* into its mature form. Although we have looked carefully, we have never seen any association of in vitro processed miRNAs with proteins, either by gel-shift analysis or by size fractionation (results not shown). These experi-

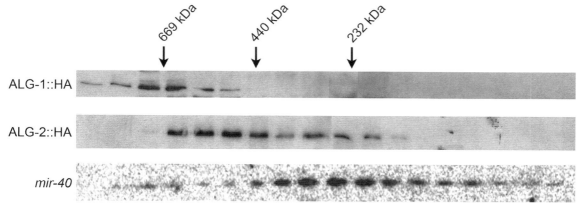

Figure 2. Size fractionation of ALG-1- and ALG-2-containing complexes. Extracts from *alg-1::HA* and *alg-2::HA* transgenic embryos were size-fractionated on a superdex200HR 10/30 column, and fractions were analyzed by western blotting using monoclonal αHA antibodies and northern blotting using *mir-40* and *mir-52* probes.

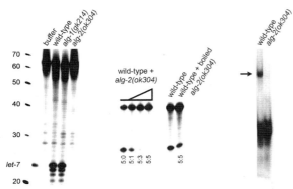

Figure 3. In vitro processing of pre-*let-7* RNA. In vitro processing of pre-*let-7* in buffer and embryonic extracts of wild-type, *alg-1(gk214)*, or *alg-2(ok304)* animals (*left*). Mixing of wild-type and *alg-2(ok304)* embryonic extracts with and without boiling of the *alg-2(ok304)* extract. Wild-type:*alg-2(ok304)* mixing ratios are indicated (*middle*). Gel-shift assay of the synthetic pre-*let-7* with wild-type and *alg-2(ok304)* extract. Arrow indicates the shifted pre-*let-7* (shifted band only contains pre-*let-7* as analyzed by denaturing urea gel) (*right*).

ments indicate that processed miRNAs are released from the Dicer complex, indicating that this in vitro system only measures pre-miRNA processing and miRNA release but not RISC loading.

Next, we analyzed the activity of our mutant extracts. Extract prepared from *alg-2(ok304)* animals is completely deficient in pre-*let-7* processing, recapitulating the observed in vivo data, although the in vitro effect is more pronounced. This is in contrast to the *alg-1(gk214)* extract that processes the pre-*let-7* normally (Fig. 3, left). We then tested whether we could reconstitute the processing activity in *alg-2(ok304)* extract by mixing in

wild-type extract. To our surprise, we could not: The *alg-2* extract has a dominant-negative effect on the processing activity in wild-type extracts (Fig. 3, middle). We could eliminate this inhibitory effect by boiling the *alg-2* extract, suggesting that the inhibitor is probably a protein and not a nucleic acid (Fig. 3, middle).

This dominant-negative effect may be explained by recent data suggesting that different silencing pathways compete for similar limiting (co)factors. Dicer was shown to interact with proteins acting in the endogenous RNAi pathway (dsRNAs generated by the host), exogenous RNAi pathway (dsRNA from the environment, e.g., viruses), and miRNA pathway. Inactivation of one pathway leads to hyperactivation of the other, suggesting that the limiting factor for these pathways is probably Dicer itself (Simmer et al. 2002; Kennedy et al. 2004; Duchaine et al. 2006; Lee et al. 2006). In extracts lacking ALG-2, cofactors normally associated with this Argonaute to generate miRNAs may still be present and bind to Dicer, somehow blocking its activity (Fig. 4). As Dicer is the limiting factor, ALG-2 mutant extract will harbor enough inhibitor to also block Dicer protein molecules from a wild-type extract. In support of this model, not only pre-miRNA processing is blocked in *alg-2* mutant extract, but also the processing of long dsRNAs (~300 nucleotides) that are typically used in RNAi is impaired (results not shown).

The observed differences between *alg-1* and *alg-2* extracts regarding pre-*let-7* processing activity may well reflect a functional difference between the two Argonaute proteins. ALG-2 may be involved in the biogenesis of miRNAs by loading the pre-miRNA in the Dicer complex and/or by facilitating the release of the processed miRNA. To test this, we performed a gel-shift assay with the synthetic pre-*let-7*. Both wild-type extract (Fig. 3, right) and *alg-1(gk214)* (result not

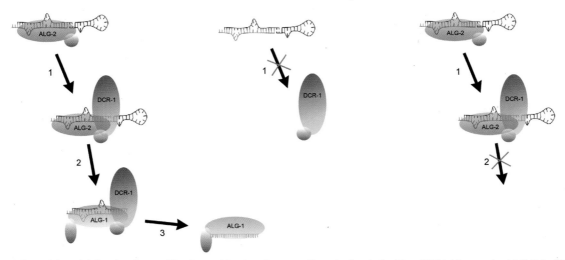

Figure 4. Model explaining the observed in vivo and in vitro data regarding *alg-1* and *alg-2* in miRNA biogenesis. ALG-2 facilitates in loading the pre-miRNA in the Dicer complex (*1*). *C. elegans* Dicer (DCR-1) processes the pre-miRNA to the double-stranded mature miRNA and ALG-2 facilitates loading the miRNA to ALG-1 (*2*). The mature miRNA/ALG-1 complex is released and can direct translational repression (*3*). In the absence of ALG-2, Dicer is unable to recognize and process the pre-miRNA. Additionally, ALG-2/Dicer-specific cofactors may bind to Dicer, preventing Dicer from associating with bona fide protein complexes, resulting in dominant inhibition of Dicer. In the absence of ALG-1, ALG-2 is unable to release its miRNA, resulting in saturation and eventually blocking of Dicer activity in vivo. In vitro stable association of ALG-2 and the double-stranded miRNA does not occur, guaranteeing continued pre-miRNA processing also in the absence of ALG-1.

shown) are capable of forming a ribonucleoprotein (RNP) complex with the pre-*let-7*, in contrast to the *alg-2(ok304)* extract that shows no affinity for pre-*let-7* at all. This suggests that ALG-2 functions in pre-miRNA processing by loading the pre-miRNA in the Dicer complex. Since ALG-2 also binds processed miRNAs, it may also facilitate the release of the processed miRNA by Dicer and/or removing the passenger strand of the miRNA, analogous to the situation in *Drosophila*, where it was shown that RISC loading depends on AGO2 slicer activity (Matranga et al. 2005; Miyoshi et al. 2005). ALG-1 could act further downstream by accepting the single-stranded miRNAs from an ALG-2 complex and inhibiting translation by binding to target mRNAs. This hypothesis fits with the observed data in that *alg-1* and *alg-2* mutant animals would both accumulate pre-*let-7* in vivo, either because Dicer is unable to form a processing complex with the pre-miRNA or because Dicer is unable to release the processed miRNA in the absence of ALG-1. In vitro, this would lead to *alg-2* mutant extracts that lack Dicer activity for the same reason as the in vivo situation (Fig. 4), but *alg-1* mutant extracts would still be capable of pre-miRNA processing since our in vitro system does not properly carry out the transfer of miRNAs from Dicer/ALG-2 to ALG-1 (RISC loading, see above). As the mature miRNAs do not remain stably associated with ALG-2 either, Dicer activity will not be inhibited.

Whatever the exact mechanism, it is likely that ALG-1 and ALG-2 have different functions. This does not exclude the fact that they can partially carry out each others function, but in a wild-type situation, both proteins most likely act at their specific step in the miRNA pathway. The *C. elegans* genome encodes 27 Argonaute proteins, and the functions of most are still unknown. However, specialized functions for different family members have been reported. Besides ALG-1 and ALG-2 that act in the miRNA pathway, RDE-1 is involved in RNAi in soma and germ line, PPW-1 seems only involved in RNAi in the germ line, and PPW-2 is involved in silencing repetitive elements: transposons and high-copy transgenic arrays (Tabara et al. 1999; Grishok et al. 2001; Tijsterman et al. 2002; Vastenhouw et al. 2003; Robert et al. 2005). Some of these specialized functions may be due to differences in temporal and/or spatial expression of the different Argonautes, but some of the differences are likely to be caused by associated proteins. Although the homology between the different family members is high, largely due to the high conservation of the PAZ and PIWI domains, the long amino-terminal tail of the Argonautes is highly diverged between different proteins and could serve as a specific site of interaction for additional proteins. Now that the similarities between the different small RNA pathways are recognized, it will be of great interest to investigate these differences.

ACKNOWLEDGMENTS

This research was supported by a ZonMW/TOP grant to R.H.A.P. and a VIDI grant to R.F.K., both supplied by the Dutch Organization for Scientific research (NWO).

REFERENCES

Ambros V., Bartel B., Bartel D.P., Burge C.B., Carrington J.C., Chen X., Dreyfuss G., Eddy S.R., Griffiths-Jones S., Marshall M., et al. 2003. A uniform system for microRNA annotation. *RNA* **9:** 277.

Aravin A.A., Lagos-Quintana M., Yalcin A., Zavolan M., Marks D., Snyder B., Gaasterland T., Meyer J., and Tuschl T. 2003. The small RNA profile during *Drosophila melanogaster* development. *Dev. Cell* **5:** 337.

Aravin A., Gaidatzis D., Pfeffer S., Lagos-Quintana M., Landgraf P., Iovino N., Morris P., Brownstein M.J., Kuramochi-Miyagawa S., Nakano T., et al. 2006. A novel class of small RNAs bind to MILI protein in mouse testes. *Nature* **442:** 203.

Bernstein E., Caudy A.A., Hammond S.M., and Hannon G.J. 2001. Role for a bidentate ribonuclease in the initiation step of RNA interference. *Nature* **409:** 363.

Caudy A.A., Ketting R.F., Hammond S.M., Denli A.M., Bathoorn A.M., Tops B.B., Silva J.M., Myers M.M., Hannon G.J., and Plasterk R.H. 2003. A micrococcal nuclease homologue in RNAi effector complexes. *Nature* **425:** 411.

Chendrimada T.P., Gregory R.I., Kumaraswamy E., Norman J., Cooch N., Nishikura K., and Shiekhattar R. 2005. TRBP recruits the Dicer complex to Ago2 for microRNA processing and gene silencing. *Nature* **436:** 740.

Duchaine T.F., Wohlschlegel J.A., Kennedy S., Bei Y., Conte D., Jr., Pang K., Brownell D.R., Harding S., Mitani S., Ruvkun G., et al. 2006. Functional proteomics reveals the biochemical niche of *C. elegans* DCR-1 in multiple small-RNA-mediated pathways. *Cell* **124:** 343.

Fire A., Xu S., Montgomery M.K., Kostas S.A., Driver S.E., and Mello C.C. 1998. Potent and specific genetic interference by double-stranded RNA in *Caenorhabditis elegans*. *Nature* **391:** 806.

Girard A., Sachidanandam R., Hannon G.J., and Carmell M.A. 2006. A germline-specific class of small RNAs binds mammalian Piwi proteins. *Nature* **442:** 199.

Gregory R.I., Chendrimada T.P., Cooch N., and Shiekhattar R. 2005. Human RISC couples microRNA biogenesis and post-transcriptional gene silencing. *Cell* **123:** 631.

Grishok A., Pasquinelli A.E., Conte D., Li N., Parrish S., Ha I., Baillie D.L., Fire A., Ruvkun G., and Mello C.C. 2001. Genes and mechanisms related to RNA interference regulate expression of the small temporal RNAs that control *C. elegans* developmental timing. *Cell* **106:** 23.

Hamilton A.J. and Baulcombe D.C. 1999. A species of small antisense RNA in posttranscriptional gene silencing in plants. *Science* **286:** 950.

Hutvagner G. and Zamore P.D. 2002. A microRNA in a multiple-turnover RNAi enzyme complex. *Science* **297:** 2056.

Hutvagner G., McLachlan J., Pasquinelli A.E., Balint E., Tuschl T., and Zamore P.D. 2001. A cellular function for the RNA-interference enzyme Dicer in the maturation of the let-7 small temporal RNA. *Science* **293:** 834.

Kamath R.S., Fraser A.G., Dong Y., Poulin G., Durbin R., Gotta M., Kanapin A., Le Bot N., Moreno S., Sohrmann M., et al. 2003. Systematic functional analysis of the *Caenorhabditis elegans* genome using RNAi. *Nature* **421:** 231.

Kennedy S., Wang D., and Ruvkun G. 2004. A conserved siRNA-degrading RNase negatively regulates RNA interference in *C. elegans*. *Nature* **427:** 645.

Ketting R.F., Fischer S.E., Bernstein E., Sijen T., Hannon G.J., and Plasterk R.H. 2001. Dicer functions in RNA interference and in synthesis of small RNA involved in developmental timing in *C. elegans*. *Genes Dev.* **15:** 2654.

Lau N.C., Lim L.P., Weinstein E.G., and Bartel D.P. 2001. An abundant class of tiny RNAs with probable regulatory roles in *Caenorhabditis elegans*. *Science* **294:** 858.

Lee R.C., Hammell C.M., and Ambros V. 2006. Interacting endogenous and exogenous RNAi pathways in *Caenorhabditis elegans*. *RNA* **12:** 589.

Lee Y., Ahn C., Han J., Choi H., Kim J., Yim J., Lee J., Provost P., Radmark O., Kim S., and Kim V.N. 2003. The nuclear RNase III Drosha initiates microRNA processing. *Nature* **425:** 415.

Lingel A., Simon B., Izaurralde E., and Sattler M. 2003. Structure and nucleic-acid binding of the *Drosophila* Argonaute 2 PAZ domain. *Nature* **426:** 465.

Matranga C., Tomari Y., Shin C., Bartel D.P., and Zamore P.D. 2005. Passenger-strand cleavage facilitates assembly of siRNA into Ago2-containing RNAi enzyme complexes. *Cell* **123:** 607.

Miyoshi K., Tsukumo H., Nagami T., Siomi H., and Siomi M.C. 2005. Slicer function of *Drosophila* Argonautes and its involvement in RISC formation. *Genes Dev.* **19:** 2837.

Parker J.S., Roe S.M., and Barford D. 2004. Crystal structure of a PIWI protein suggests mechanisms for siRNA recognition and slicer activity. *EMBO J.* **23:** 4727.

Plasterk R.H. 2002. RNA silencing: The genome's immune system. *Science* **296:** 1263.

Rand T.A., Ginalski K., Grishin N.V., and Wang X. 2004. Biochemical identification of Argonaute 2 as the sole protein required for RNA-induced silencing complex activity. *Proc. Natl. Acad. Sci.* **101:** 14385.

Robert V.J., Sijen T., van Wolfswinkel J., and Plasterk R.H. 2005. Chromatin and RNAi factors protect the *C. elegans* germline against repetitive sequences. *Genes Dev.* **19:** 782.

Simmer F., Tijsterman M., Parrish S., Koushika S.P., Nonet M.L., Fire A., Ahringer J., and Plasterk R.H. 2002. Loss of the putative RNA-directed RNA polymerase RRF-3 makes *C. elegans* hypersensitive to RNAi. *Curr. Biol.* **12:** 1317.

Song J.J., Smith S.K., Hannon G.J., and Joshua-Tor L. 2004. Crystal structure of Argonaute and its implications for RISC slicer activity. *Science* **305:** 1434.

Song J.J., Liu J., Tolia N.H., Schneiderman J., Smith S.K., Martienssen R.A., Hannon G.J., and Joshua-Tor L. 2003. The crystal structure of the Argonaute2 PAZ domain reveals an RNA binding motif in RNAi effector complexes. *Nat. Struct. Biol.* **10:** 1026.

Tabara H., Sarkissian M., Kelly W.G., Fleenor J., Grishok A., Timmons L., Fire A., and Mello C. 1999. The rde-1 gene, RNA interference, and transposon silencing in *C. elegans*. *Cell* **99:** 123.

Tijsterman M., Okihara K.L., Thijssen K., and Plasterk R.H. 2002. PPW-1, a PAZ/PIWI protein required for efficient germline RNAi, is defective in a natural isolate of *C. elegans*. *Curr. Biol.* **12:** 1535.

Tops B.B., Tabara H., Sijen T., Simmer F., Mello C.C., Plasterk R.H., and Ketting R.F. 2005. RDE-2 interacts with MUT-7 to mediate RNA interference in *Caenorhabditis elegans*. *Nucleic Acids Res.* **33:** 347.

van Haaften G., Vastenhouw N.L., Nollen E.A., Plasterk R.H., and Tijsterman M. 2004. Gene interactions in the DNA damage-response pathway identified by genome-wide RNA-interference analysis of synthetic lethality. *Proc. Natl. Acad. Sci.* **101:** 12992.

Vastenhouw N.L., Fischer S.E., Robert V.J., Thijssen K.L., Fraser A.G., Kamath R.S., Ahringer J., and Plasterk R.H. 2003. A genome-wide screen identifies 27 genes involved in transposon silencing in *C. elegans*. *Curr. Biol.* **13:** 1311.

Yan K.S., Yan S., Farooq A., Han A., Zeng L., and Zhou M.M. 2003. Structure and conserved RNA binding of the PAZ domain. *Nature* **426:** 468.

Zamore P.D., Tuschl T., Sharp P.A., and Bartel D.P. 2000. RNAi: Double-stranded RNA directs the ATP-dependent cleavage of mRNA at 21 to 23 nucleotide intervals. *Cell* **101:** 25.

MicroRNA Function and Mechanism: Insights from Zebra Fish

A.F. Schier and A.J. Giraldez

Department of Molecular and Cellular Biology, Harvard University, Cambridge, Massachusetts 02138

MicroRNAs (miRNAs) are small RNAs that bind to the 3′UTR of mRNAs. We are using zebra fish as a model system to study the developmental roles of miRNAs and to determine the mechanisms by which miRNAs regulate target mRNAs. We generated zebra fish embryos that lack the miRNA-processing enzyme Dicer. Mutant embryos are devoid of mature miRNAs and have morphogenesis defects, but differentiate multiple cell types. Injection of miR-430 miRNAs, a miRNA family expressed at the onset of zygotic transcription, rescues the early morphogenesis defects in *dicer* mutants. miR-430 accelerates the decay of hundreds of maternal mRNAs and induces the deadenylation of target mRNAs. These studies suggest that miRNAs are not obligatory components of all fate specification or signaling pathways but facilitate developmental transitions and induce the deadenylation and decay of hundreds of target mRNAs.

miRNAs are approximately 22-nucleotide small RNAs that act as posttranscriptional repressors by binding to the 3′UTR of target mRNAs (Lee et al. 1993; Reinhart et al. 2000; Bartel 2004; Kloosterman and Plasterk 2006). In animals, mature miRNAs are generated from a primary transcript (pri-miRNA) through sequential cleavage by nucleases belonging to the RNAse III family. Initially, Drosha cleaves the pri-miRNA and excises a stem-loop precursor (pre-miRNA), which is then cleaved by Dicer into an RNA duplex (Bernstein et al. 2001; Grishok et al. 2001; Hutvagner et al. 2001; Ketting et al. 2001; Knight and Bass 2001). One strand of the duplex constitutes the mature miRNA, which is incorporated into a silencing complex (RISC) and guides it to target mRNAs (Hammond et al. 2000; Hutvagner and Zamore 2002; Khvorova et al. 2003; Schwarz et al. 2003). The 5′ region of the miRNA (seed) is the main determinant of target recognition (Lai 2002; Lewis et al. 2003, 2005; Doench and Sharp 2004; Brennecke et al. 2005). Hundreds of miRNAs have been identified and thousands of targets have been predicted (Lagos-Quintana et al. 2001; Lau et al. 2001; Lee and Ambros 2001; Enright et al. 2003; Stark et al. 2003, 2005; Ambros 2004; Bartel 2004; Rehmsmeier et al. 2004; Berezikov et al. 2005; Farh et al. 2005; Krek et al. 2005; Lai 2005; Lewis et al. 2005; Miranda et al. 2006; Rajewsky 2006; Sood et al. 2006). However, the developmental roles of microRNAs are largely elusive (Alvarez-Garcia and Miska 2005; Kloosterman and Plasterk 2006), and it is still controversial how target mRNAs are repressed (Pillai 2005; Valencia-Sanchez et al. 2006). We have used the zebra fish embryo as a model system to investigate these issues.

ZEBRA FISH EMBRYOS THAT LACK MATURE miRNAs UNDERGO ABNORMAL MORPHOGENESIS

The global function of miRNAs has been unclear, and it is controversial whether miRNAs act as switches or modulators of biological processes (Bartel and Chen 2004). We therefore wished to create embryos that lack all

mature miRNAs (Fig. 1B). To this end, we used genetic and embryological manipulations to generate embryos that lacked all Dicer activity (Fig. 1B; Fig. 2) (Ciruna et al. 2002; Giraldez et al. 2005). Several lines of evidence indicated that these maternal-zygotic *dicer* mutants (MZ*dicer*) lacked mature miRNAs. For example, northern analyses detected only pre-miRNAs, and reporter genes that contain miRNA target sites in their 3′UTRs were repressed in wild type but not in MZ*dicer* mutants (Giraldez et al. 2005).

MZ*dicer* mutant embryos had severe morphogenesis defects, including abnormal gastrulation movements, impaired brain ventricle formation, and somite defects, and died on day 5 of development (Giraldez et al. 2005). Interestingly, however, different cell fates were specified, including hematopoietic, muscle, and neuronal cell types (Fig. 3B). Dorsal–ventral and anterior–posterior patterning were normal, and the major signaling pathways active in the early embryo were not grossly misregulated, including signaling by Nodal, BMP, Wnt, FGF, Hedgehog, and Retinoic Acid. Most strikingly, germ cells devoid of Dicer activity developed, renewed, and generated oocytes and sperm without apparent defects. These results indicate that mature miRNAs are not obligatory regulators of cell-fate

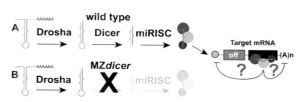

Figure 1. Maternal-zygotic *dicer* mutants do not process miRNAs. (*A*) Drosha cleaves the pri-miRNA and excises a stem-loop precursor. Dicer cleaves the hairpin to give rise to a miRNA:miRNA* duplex. One strand of this duplex is incorporated in a protein complex to form a miRNA-induced silencing complex (miRISC). The miRNA serves as a guide for this complex to the target mRNA. This leads to translational repression and accelerated mRNA degradation. (*B*) MZ*dicer* mutants lack the miRNA processing enzyme Dicer and lack mature miRNAs.

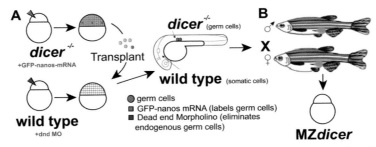

Figure 2. Germ-line replacement technique to generate MZ*dicer* mutants. (*A*) Schematic representation of the germ-line replacement technique used to generate maternal-zygotic *dicer* mutants. Depletion of host germ cells was accomplished by injection of a morpholino (Dead end morpholino; dnd MO) that blocks primordial germ cell development in host embryos, resulting in sterility. Donor germ cells labeled with GFP-nos-3′UTR were transplanted into these hosts during early embryogenesis (Ciruna et al. 2002). (*B*) Hosts containing *dicer/dicer* germ cells were raised to adulthood. These fish only harbored *dicer/dicer* PGCs. Intercrossing resulted in embryos that lacked both maternal and zygotic Dicer RNase III activity (MZ*dicer*). Additional fertile adults were generated using MZ*dicer* embryos as germ cell donors. These germ cells gave rise to mature and functional eggs and sperm, which generated embryos with a MZ*dicer* phenotype.

specification, differentiation, and signaling during early embryonic development in zebra fish. Instead, miRNA function appears to be required for the repression of mRNAs whose misexpression interferes with morphogenetic processes (Fig. 3B) (Giraldez et al. 2005).

THE MiR-430 FAMILY REGULATES MORPHOGENESIS

In vitro generated miRNA duplexes were active upon injection into MZ*dicer* mutants (Fig. 3A), suggesting that Dicer is not required for the effector steps of miRNA function (Giraldez et al. 2005). Using this injection assay, we found that members of the miR-430 family rescued the early morphogenesis defects of MZ*dicer* mutants but were not able to suppress later phenotypes (Fig. 3B–D and Fig. 4) (Giraldez et al. 2005). miR-430 expression begins after the 500-cell stage, when the zygotic genome initiates transcription. Members of the miR-430 family are the major class of miRNAs during early embryogenesis and are expressed

ubiquitously throughout embryonic development in zebra fish and *Xenopus*. After mid-embryogenesis, many other miRNAs initiate their expression (Fig. 4A) (Chen et al. 2005; Giraldez et al. 2005; Watanabe et al. 2005). Their lack likely accounts for the later defects and lethality of miR-430-injected MZ*dicer* mutants. These results suggest that miR-430 regulates mRNAs whose overexpression affects morphogenesis.

MiR-430 HAS HUNDREDS OF TARGET MRNAS

Reporter mRNAs with miR-430 target sites were efficiently degraded in wild type but not in MZ*dicer* mutants (Giraldez et al. 2006). These results led us to hypothesize that in vivo targets might also accumulate in the absence of miR-430 (Fig. 5A). We therefore performed expression profiling on wild-type embryos, MZ*dicer* mutants, and MZ*dicer* mutants injected with miR-430, leading to the identification of more than 700 mRNAs that accumulated in the absence of miR-430 (Giraldez et al. 2006). Strikingly, two-thirds of these

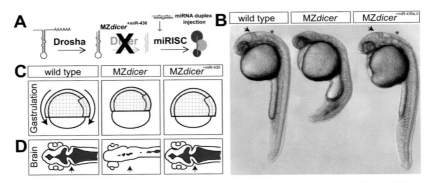

Figure 3. Abnormal morphogenesis in MZ*dicer* mutants. (*A*) Schematic representation showing that MZ*dicer* mutants lack mature miRNAs but injected processed miRNA duplexes reconstitute active miRISC. (*B*) Wild type (*left*), MZ*dicer* mutants (*center*) display morphogenesis defects in the retina, brain, trunk, and tail. MZ*dicer* injected at the one-cell stage with miR-430a+b duplex (*right*). Note the rescue of brain morphogenesis (*asterisk*), the midbrain–hindbrain boundary (*arrow*), and the trunk morphology. (*C*) Schematic representation shows the gastrulation defects in MZ*dicer* mutants with a slower epiboly compared to wild type. (*D*) Schematic representation shows a dorsal view of a 30-hr zebra fish brain in wild type, MZ*dicer*, and rescued embryos. The brain ventricles are labeled in red, and the arrow indicates the mid–hindbrain boundary.

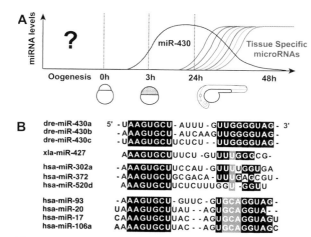

Figure 4. The miR-430 family of miRNAs. (*A*) Schematic representation showing the expression of miR-430 compared to other miRNAs in zebra fish embryos. MiR-430 starts to be expressed at the onset of zygotic transcription, is the predominant miRNA expressed during the first 18 hours of development, and decays by 48 hours postfertilization. Other miRNAs start to be expressed in a tissue-specific manner after ~18 hours postfertilization (Wienholds et al. 2005). It is currently unknown which miRNAs, if any, are expressed during oogenesis or before zygotic transcription. (*B*) Alignment of the zebra fish miR-430 family of miRNAs (*Danio rerio*, dre) with the *Xenopus laevis* miR-427 (Watanabe et al. 2005) and human miRNAs. miR-520 is a human miRNA belonging to a large miRNA cluster expressed in the placenta (Bentwich et al. 2005). miR-372 is a stem-cell miRNA that can cause germ cell tumors (Voorhoeve et al. 2006). The miR-17-20 family of miRNAs has been identified as a potential human oncogene (Ota et al. 2004; He et al. 2005).

mRNAs had 3′UTR sequences that were complementary to the miR-430 seed. Large-scale target validation using in vivo reporter assays revealed that the majority of putative targets was regulated by miR-430 in vivo (Fig. 5B). We estimate that miR-430 regulates at least 300 mRNAs. This number is likely an underestimate. First, the microarray contained only about 50% of all zebra fish genes. Second, expression profiling would not identify targets that are regulated only at the translational level. Hence, it is conceivable that miR-430 directly regulates more than 1000 different mRNAs (Giraldez et al. 2006).

MIR-430 ACCELERATES THE CLEARANCE OF MATERNAL MRNAS

Strikingly, the large majority of miR-430 targets was maternally expressed and accumulated in the absence of miR-430 (Giraldez et al. 2006). Conversely, we found that 40% of maternal mRNAs had predicted miR-430 sites. These observations suggest that miR-430 is a key regulator of maternal mRNAs and accelerates their decay (Fig. 6). The aberrant accumulation of maternal mRNAs and their prolonged translation is therefore the likely cause of the morphogenesis defects observed in the absence of miR-430.

These results establish miR-430 as a key regulator of the maternal-to-zygotic transition. This is a universal transition in animal development when the embryo initiates the transcription of its genome (zygotic phase). The preceding stage of development is driven by maternally deposited mRNAs, and the genome is silent (maternal phase). The maternal-to-zygotic transition is still poorly understood, but it coincides with the patterning of the embryo and the

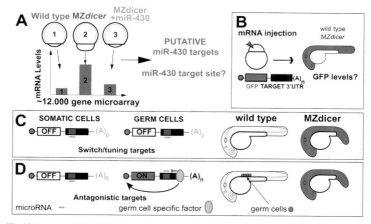

Figure 5. Regulation of miR-430 targets. (*A*) Schematic representation showing the strategy used to identify miR-430 targets in vivo using microarray analysis. Comparison of mRNA expression levels between MZ*dicer* mutants and [MZ*dicer*$^{+\text{miR-430}}$ and wild type]. Putative targets must fulfill two criteria: (1) the mRNA must be up-regulated more than 1.5-fold in MZ*dicer* compared to wild type and MZ*dicer*$^{+\text{miR-430}}$; (2) it must contain a 6-mer sequence in the 3′UTR complementary to miR-430 seed. (*B*) A fraction of the putative targets were tested using injection of a GFP reporter mRNA with the 3′UTR of the target mRNA that is wild type or mutant for the miR-430 target sites. The levels of GFP expression in wild-type and MZ*dicer* embryos were compared at 25–30 hours postfertilization. (*C*) Schematic representation showing that a target is validated when the GFP expression is repressed in wild-type but not in MZ*dicer* embryos. In case of direct regulation, mutations in the miR-430 target site abolish repression in wild type. (*D*) Schematic representation showing that some targets are regulated by miR-430 in somatic cells but not germ cells despite the presence of miR-430 in both cell types. Model showing that a germ-cell-specific factor (*orange*) binds to the 3′UTR of antagonistic targets to inhibit miRNA function or enhance translation resulting in germ-cell-specific expression of the reporter mRNA.

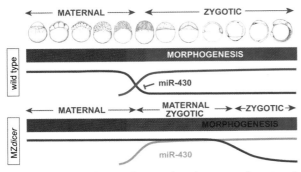

Figure 6. miR-430 accelerates the clearance of maternal mRNAs. Role of miR-430 in maternal-to-zygotic transition. In the presence of miR-430 (*blue*), a large set of maternal mRNAs (*red*) is posttranscriptionally regulated, allowing normal morphogenesis during zygotic stages. In the absence of miR-430 (*gray*), maternal mRNAs and their products accumulate and interfere with morphogenesis. (Reprinted, with permission, from Giraldez et al. 2006 [© AAAS].)

degradation of maternal mRNAs (Newport and Kirschner 1982a,b; Richter et al. 1990; Kane and Kimmel 1993; Richter 1999; Pelegri 2003; de Moor et al. 2005). Our results suggest that miR-430 transcription initiates at the maternal-to-zygotic transition to accelerate the degradation and repression of a large fraction of maternal mRNAs.

TWO CLASSES OF MiR-430 TARGETS

Reporter assays using the 3′UTR of putative miR-430 target mRNAs revealed that the large majority of targets were uniformly repressed in all embryonic cells; i.e., in germ cells and all somatic cells (Fig. 5C). This observation is consistent with the ubiquitous expression of miR-430. Strikingly, however, at least two genes, *nanos* and

tudor-like, were more susceptible to miR-430 repression in somatic cells than in germ cells (Fig. 5D) (Mishima et al. 2006). Zebra fish germ cells are set aside during blastula stages and accumulate specific mRNAs and proteins (Yoon et al. 1997; Knaut et al. 2000; Koprunner et al. 2001; Raz 2004). We found that the 3′UTRs of *nanos* and *tudor-like* were repressed in somatic cells in the presence, but not in the absence, of miR-430. In contrast, expression in germ cells was maintained even when miR-430 was present (Fig. 5D). These results suggest that some 3′UTRs can be effective miRNA targets in one cell type but not another. Such 3′UTRs might contain activating or derepressing elements that counteract the effects of the miRNA, resulting in differential, tissue-specific regulation (Fig. 4D) (Mishima et al. 2006).

MiRNAs INDUCE DEADENYLATION AND DECAY OF TARGET MRNAs

Previous studies had shown that miRNAs block the translation of target mRNAs and, in some cases, induced mRNA decay (Lee et al. 1993; Olsen and Ambros 1999; Reinhart et al. 2000; Bagga et al. 2005; Lim et al. 2005; Pillai 2005; Pillai et al. 2005). Since mRNA translation and stability depend on the poly(A) tail (Gallie 1991; de Moor et al. 2005), we tested the polyadenylation status of miR-430 targets (Fig. 7). We found that miR-430 induced the rapid deadenylation of target mRNAs (Giraldez et al. 2006; Mishima et al. 2006). Deadenylation was not an indirect effect of translational repression, because nontranslatable targets were efficiently deadenylated in the presence, but not the absence, of miR-430 (Giraldez et al. 2006; Mishima et al. 2006). Many maternal mRNAs are deadenylated in the egg and polyadenylated upon fertilization (Slater et al. 1972; Slater et al. 1973; Wilt 1973; McGrew et al. 1989; Richter 1999; de Moor et al. 2005).

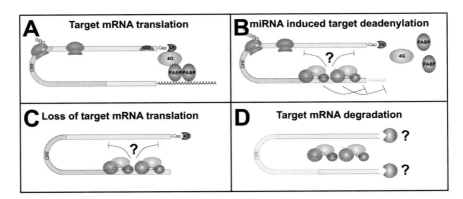

Figure 7. Model for the posttranscriptional regulation of mRNA targets by miRNAs. (*A*) Target mRNA translation: interaction between poly(A)-binding protein (PABP) on poly(A) tail with translation initiation factors eIF4G/eIF4E on Cap stimulates translation (Gallie 1991; Kapp and Lorsch 2004). (*B*) miRNA-induced target mRNA deadenylation: miRISC (Pillai 2005; Valencia-Sanchez et al. 2006) is recruited to the 3′UTR of target mRNA and accelerates deadenylation by a 3′-to-5′ exonuclease (Richter 1999; de Moor et al. 2005; Giraldez et al. 2006; Vasudevan et al. 2006; Wu et al. 2006). (*C*) Loss of target mRNA translation: the interaction between PABP and eIF4G/eIF4E is disrupted, resulting in loss of translation initiation (Gallie 1991; Kapp and Lorsch 2004; Valencia-Sanchez et al. 2006). (*D*) Target mRNA degradation: Loss of poly(A) tail results in decapping and degradation of target mRNA (Rehwinkel et al. 2005). Steps 3 and 4 are likely to occur in P-bodies (Sheth and Parker 2003; Chen et al. 2005; Ding et al. 2005; Jakymiw et al. 2005; Liu et al. 2005; Pillai 2005; Pillai et al. 2005; Rehwinkel et al. 2005; Sen and Blau 2005). The suggested role for miRNAs in accelerating mRNA deadenylation would disrupt the interaction Cap–poly(A) tail and provide a substrate for exonucleases. This could explain how miRNAs cause translational repression and mRNA decay. (Reprinted, with permission, from Giraldez et al. 2006 [© AAAS].)

miR-430 seems to revert this process by inducing the deadenylation and decay of maternal mRNAs.

CONCLUSIONS

The powerful combination of genetics, embryology, molecular biology, and bioinformatics has helped us to clarify the functions and mechanisms of miRNAs during vertebrate embryogenesis. But how general are the lessons learned from miR-430 and zebra fish embryos?

miRNAs HAVE MODULATORY ROLES

Lack of miRNAs in MZ*dicer* mutants leads to morphogenesis defects but does not lead to global disruption of fate specification or signaling. Our results therefore suggest that miRNAs have mainly modulatory roles. Indeed, miR-430 is the most abundant, if not the only, miRNA expressed during early zebra fish embryogenesis (Chen et al. 2005; Giraldez et al. 2005; Wienholds et al. 2005). Apparently, miRNAs are employed in a much more restricted fashion than are the dozens of transcription factors that specify embryonic cell fates (Schier and Talbot 2005). Similarly, loss of Dicer activity does not appear to affect zebra fish germ-cell development, suggesting that miRNAs play no or minor roles in the development of some cell types. Moreover, it is conceivable that no or few miRNAs are expressed in germ cells, despite or because of the presence of other small RNAs (Aravin et al. 2006; Girard et al. 2006; Grivna et al. 2006; Lau et al. 2006). Taken together, the results in zebra fish indicate that some cell types do not rely on microRNAs for specification and differentiation.

Studies in other systems partially support this view. For example, conditional ablation of mouse *dicer* in limbs or skin results in abnormal morphogenesis but does not lead to defects in cell-fate specification (Harfe et al. 2005; Andl et al. 2006; Yi et al. 2006). Similarly, loss of *Drosophila* miR-1, a muscle-specific miRNA, does not affect cell-fate specification, but morphogenesis during larval stages (Sokol and Ambros 2005). In contrast, studies in *Caenorhabditis elegans* have emphasized the role of the miRNAs lin-4, let-7, and lsy-6 in cell-fate decisions (Lee et al. 1993; Wightman et al. 1993; Reinhart et al. 2000; Johnston and Hobert 2003) and miR-61 in cell signaling (Yoo and Greenwald 2005). However, it is possible that the *C. elegans* data are skewed, because these miRNAs were identified in specific screens for fate defects. Our results would predict that other *C. elegans* miRNAs have more subtle roles. Conversely, it will be important to determine whether the many tissue-specific miRNAs expressed at later stages of zebra fish development have roles in fate specification or tissue homeostasis and physiology (Stark et al. 2003, 2005; Xu et al. 2003; Poy et al. 2004; Boehm and Slack 2005; Farh et al. 2005; Krutzfeldt et al. 2005; Lim et al. 2005; Wienholds et al. 2005; Hornstein and Shomron 2006; Kloosterman and Plasterk 2006; Kloosterman et al. 2006; Sood et al. 2006; Teleman et al. 2006). Ultimately, it is likely that miRNAs will be found to have a wide range of functions, but it is striking that current evidence points to modulatory or highly redundant roles.

miRNAs FACILITATE DEVELOPMENTAL TRANSITIONS

miR-430 facilitates the maternal-to-zygotic transition. Importantly, this transition is not blocked in the absence of miR-430, and the embryo does not arrest in the maternal state. Rather, the zygotic state is initiated, but mRNAs from the maternal phase are not cleared efficiently and thus interfere with subsequent development. These observations led us to suggest that miRNAs might not serve as developmental switches, but simply remove mRNAs from previous developmental states. This "spring cleaning" function ensures proper transitions from states A to B to C, and thus avoids the formation of mixed states B/C (Fig. 8). This model contrasts with *C. elegans* studies which have shown that loss of lin-4 and let-7 leads to heterochronic phenotypes, in which the animal reiterates earlier cell-fate decisions instead of initiating the next developmental stage (Ambros 1989; Lee et al. 1993; Wightman et al. 1993; Reinhart et al. 2000; Alvarez-Garcia and Miska 2005). Instead of transitioning from A to B to C, certain lineages repeat the A to B transition and do not enter C. Despite these apparent differences, miR-430, lin-4, and let-7 all act during developmental transitions, and it is possible that a specific phenotypic defect depends on the nature of the targets and the developmental context. For example, the accumulation of some targets might block the transition to the next stage, but in other systems the loss of the miRNA might cause a mixed state or delay the transition to a new state.

The role of miRNAs in developmental transitions has important implications for miRNA misregulation. For example, premature expression of a miRNA could preemptively block genes expressed at a later stage and thus

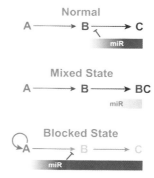

Figure 8. Model of miRNA function during development. Cell transitions from state A to state B to state C. *Normal*: miRNA is activated at the B-to-C transition and posttranscriptionally regulates mRNAs expressed in state B, thus sharpening and accelerating the transition from state B to state C (Rhoades et al. 2002; Giraldez et al. 2006). *Mixed State*: In the absence of the miRNA, gene products from state B accumulate, leading to a mixed state B/C. *Blocked State*: In disease states or upon experimental misexpression, the miRNA is expressed prematurely, resulting in the posttranscriptional regulation of genes normally expressed in state B. This prevents the transition from state A to state B, leading to the maintenance of state A. In the case of cancer, state A might correspond to a cancer stem cell (Beachy et al. 2004; He et al. 2005; Lu et al. 2005; Dalerba et al. 2006; Dews et al. 2006; Voorhoeve et al. 2006). (Reprinted, with permission, from Giraldez et al. 2006 [© AAAS].)

prevent developmental transitions (Fig. 8). This mechanism might explain how some miRNAs can function as oncogenes (He et al. 2005; Lu et al. 2005; Dews et al. 2006; Voorhoeve et al. 2006).

MiR-430 continues to be expressed during the first two days of zebra fish embryogenesis, and many of its targets are expressed during these stages (Giraldez et al. 2006). It is therefore likely that miR-430 does not only function in the maternal-to-zygotic transition, but also dampens the expression of zygotically expressed genes. Such a homeostatic role resembles the proposed roles of miRNAs in maintenance of tissue states or metabolic states (Stark et al. 2003, 2005; Xu et al. 2003; Poy et al. 2004; Boehm and Slack 2005; Farh et al. 2005; Krutzfeldt et al. 2005; Lim et al. 2005; Hornstein and Shomron 2006; Sood et al. 2006; Teleman et al. 2006).

MIRNAS HAVE HUNDREDS OF IN VIVO TARGETS

miR-430 regulates hundreds of mRNAs. These in vivo observations support cell culture and bioinformatics studies which have suggested that miRNAs have dozens of targets (Enright et al. 2003; Stark et al. 2003, 2005; Kiriakidou et al. 2004; Poy et al. 2004; Rehmsmeier et al. 2004; Krek et al. 2005; Lewis et al. 2005; Lall et al. 2006; Miranda et al. 2006). In contrast, genetic studies in *C. elegans* have identified a few key targets whose up-regulation leads to the phenotypes caused by loss of specific miRNAs (Lee et al. 1993; Wightman et al. 1993; Reinhart et al. 2000; Slack et al. 2000; Abrahante et al. 2003; Johnston and Hobert 2003; Lin et al. 2003; Abbott et al. 2005; Boehm and Slack 2005; Grosshans et al. 2005; Johnson et al. 2005; Yoo and Greenwald 2005). These two observations are not necessarily contradictory. It is conceivable that miRNAs have the capacity to regulate hundreds of targets but that the up-regulation of only a small subset of targets has severe phenotypic consequences. Detailed functional analysis of target genes is necessary to determine which and how many targets contribute to a given phenotype.

SOME MIRNA TARGETS ARE DIFFERENTIALLY REGULATED

Most miR-430 targets are uniformly susceptible to repression, but a small subset appears to be partially protected in germ cells. It is therefore conceivable that many mRNAs have acquired specific sequences that counteract the effect of miRNAs in specific tissues, times, or cell states. For example, *nanos* and *tudor-like* are miR-430 targets in somatic cells but are translated in germ cells (Mishima et al. 2006), and in tissue culture, repression of *cat-1* by miR-122 is relieved upon cellular stress (Bhattacharyya et al. 2006). Such "antagonistic targets" complement "anti-targets," which have evolved to lack miRNA-binding sites (Bartel and Chen 2004; Farh et al. 2005; Stark et al. 2005; Sood et al. 2006). Hence, mRNAs can avoid miRNA regulation by either lacking a miRNA-binding site (anti-targets) or by evolving sites that counteract the miRNA's effects (antagonistic targets). The advantage of the latter strategy is that target repression can

be regulated in a tissue-, stage-, or state-specific manner. It remains to be seen how common this mechanism is, but it might account for the observation that many predicted miRNA targets do not seem to be regulated by a given miRNA (Lewis et al. 2003; Stark et al. 2003, 2005; Giraldez et al. 2006; Miranda et al. 2006). These results also highlight the importance of in vivo validation of putative targets by testing the role of their full-length 3′UTRs.

MIRNAS REGULATE TARGET MRNAS BY DEADENYLATION

miR-430 accelerates the deadenylation of mRNA targets (Giraldez et al. 2006; Mishima et al. 2006). Further support for a role for miRNAs in deadenylation has come from cell culture studies (Wu et al. 2006) and the identification of deadenylases involved in miRNA target degradation (Behm-Ansmant et al. 2006). In some cases, however, miRNAs seem to be able to repress translation in the absence of deadenylation (Humphreys et al. 2005; Wu et al. 2006) or mRNA degradation (Pillai 2005; Petersen et al. 2006). It is therefore possible that deadenylation is one of several mechanisms that contribute to target regulation. However, since deadenylation ultimately leads to both translational repression and mRNA decay, the two main effects of miRNA action, it is tempting to speculate that deadenylation is the trigger for miRNA-mediated repression.

ACKNOWLEDGMENTS

We thank our colleagues and collaborators for their essential contributions, and Kevin Eggan and Susan Mango for helpful comments on the manuscript. A.J.G. was supported by EMBO and is currently supported by a Human Frontier Science Program fellowship. A.F.S. was an Irma T. Hirschl Trust Career Scientist and an Established Investigator of the American Heart Association. This work was also supported by grants from the National Institutes of Health (A.F.S).

REFERENCES

Abbott A.L., Alvarez-Saavedra E., Miska E.A., Lau N.C., Bartel D.P., Horvitz H.R., and Ambros V. 2005. The let-7 MicroRNA family members mir-48 mir-84 and mir-241 function together to regulate developmental timing in *Caenorhabditis elegans*. *Dev. Cell.* **9:** 403.

Abrahante J.E., Daul A.L., Li M., Volk M.L., Tennessen J.M., Miller E.A., and Rougvie A.E. 2003. The *Caenorhabditis elegans* hunchback-like gene lin-57/hbl-1 controls developmental time and is regulated by microRNAs. *Dev. Cell.* **4:** 625.

Alvarez-Garcia I. and Miska E.A. 2005. MicroRNA functions in animal development and human disease. *Development* **132:** 4653.

Ambros V. 1989. A hierarchy of regulatory genes controls a larva-to-adult developmental switch in *C. elegans*. *Cell.* **57:** 49.

———. 2004. The functions of animal microRNAs. *Nature* **431:** 350.

Andl T., Murchison E.P., Liu F., Zhang Y., Yunta-Gonzalez M., Tobias J.W., Andl C.D., Seykora J.T., Hannon G.J., and Millar S.E. 2006. The miRNA-processing enzyme dicer is essential for the morphogenesis and maintenance of hair follicles. *Curr. Biol.* **16:** 1041.

Aravin A., Gaidatzis D., Pfeffer S., Lagos-Quintana M., Landgraf P., Iovino N., Morris P., Brownstein M.J., Kuramochi-Miyagawa S., Nakano T., et al. 2006. A novel class of small RNAs bind to MILI protein in mouse testes. *Nature* **442:** 203.

Bagga S., Bracht J., Hunter S., Massirer K., Holtz J., Eachus R., and Pasquinelli A.E. 2005. Regulation by let-7 and lin-4 miRNAs results in target mRNA degradation. *Cell* **122:** 553.

Bartel D.P. 2004. MicroRNAs: Genomics, biogenesis, mechanism, and function. *Cell* **116:** 281.

Bartel D.P. and Chen C.Z. 2004. Micromanagers of gene expression: The potentially widespread influence of metazoan microRNAs. *Nat. Rev. Genet.* **5:** 396.

Beachy P.A., Karhadkar S.S., and Berman D.M. 2004. Tissue repair and stem cell renewal in carcinogenesis. *Nature* **432:** 324.

Behm-Ansmant I., Rehwinkel J., Doerks T., Stark A., Bork P., and Izaurralde E. 2006. mRNA degradation by miRNAs and GW182 requires both CCR4:NOT deadenylase and DCP1:DCP2 decapping complexes. *Genes Dev.* **20:** 1885.

Bentwich I., Avniel A., Karov Y., Aharonov R., Gilad S., Barad O., Barzilai A., Einat P., Einav U., Meiri E., et al. 2005. Identification of hundreds of conserved and nonconserved human microRNAs. *Nat. Genet.* **37:** 766.

Berezikov E., Guryev V., van de Belt J., Wienholds E., Plasterk R.H., and Cuppen E. 2005. Phylogenetic shadowing and computational identification of human microRNA genes. *Cell* **120:** 21.

Bernstein E., Caudy A.A., Hammond S.M., and Hannon G.J. 2001. Role for a bidentate ribonuclease in the initiation step of RNA interference. *Nature* **409:** 363.

Bhattacharyya S.N., Habermacher R., Martine U., Closs E.I., and Filipowicz W. 2006. Relief of microRNA-mediated translational repression in human cells subjected to stress. *Cell* **125:** 1111.

Boehm M. and Slack F. 2005. A developmental timing microRNA and its target regulate life span in *C. elegans*. *Science* **310:** 1954.

Brennecke J., Stark A., Russell R.B., and Cohen S.M. 2005. Principles of microRNA-target recognition. *PLoS Biol.* **3:** e85.

Chen P.Y., Manninga H., Slanchev K., Chien M., Russo J.J., Ju J., Sheridan R., John B., Marks D.S., Gaidatzis D., et al. 2005. The developmental miRNA profiles of zebra fish as determined by small RNA cloning. *Genes Dev.* **19:** 1288.

Ciruna B., Weidinger G., Knaut H., Thisse B., Thisse C., Raz E., and Schier A.F. 2002. Production of maternal-zygotic mutant zebra fish by germ-line replacement. *Proc. Natl. Acad. Sci.* **99:** 14919.

Dalerba P., Cho R.W., and Clarke M.F. 2006. Cancer stem cells: Models and concepts. *Annu. Rev. Med.* (in press).

de Moor C.H., Meijer H., and Lissenden S. 2005. Mechanisms of translational control by the 3′UTR in development and differentiation. *Semin. Cell Dev. Biol.* **16:** 49.

Dews M., Homayouni A., Yu D., Murphy D., Sevignani C., Wentzel E., Furth E.E., Lee W.M., Enders G.H., Mendell J.T., and Thomas-Tikhonenko A. 2006. Augmentation of tumor angiogenesis by a Myc-activated microRNA cluster. *Nat. Genet.* **38:** 1060.

Ding L., Spencer A., Morita K., and Han M. 2005. The developmental timing regulator AIN-1 interacts with miRISCs and may target the argonaute protein ALG-1 to cytoplasmic P bodies in *C. elegans*. *Mol. Cell* **19:** 437.

Doench J.G. and Sharp P.A. 2004. Specificity of microRNA target selection in translational repression. *Genes Dev.* **18:** 504.

Enright A.J., John B., Gaul U., Tuschl T., Sander C., and Marks D.S. 2003. MicroRNA targets in *Drosophila*. *Genome Biol.* **5:** R1.

Farh K.K., Grimson A., Jan C., Lewis B.P., Johnston W.K., Lim L.P., Burge C.B., and Bartel D.P. 2005. The widespread impact of mammalian MicroRNAs on mRNA repression and evolution. *Science* **310:** 1817.

Gallie D.R. 1991. The cap and poly(A) tail function synergistically to regulate mRNA translational efficiency. *Genes Dev.* **5:** 2108.

Giraldez A.J., Mishima Y., Rihel J., Grocock R.J., Van Dongen S., Inoue K., Enright A.J., and Schier A.F. 2006. Zebra fish MiR-430 promotes deadenylation and clearance of maternal mRNAs. *Science* **312:** 75.

Giraldez A.J., Cinalli R.M., Glasner M.E., Enright A.J., Thomson J.M., Baskerville S., Hammond S.M., Bartel D.P., and Schier A.F. 2005. MicroRNAs regulate brain morphogenesis in zebra fish. *Science* **308:** 833.

Girard A., Sachidanandam R., Hannon G.J., and Carmell M.A. 2006. A germline-specific class of small RNAs binds mammalian Piwi proteins. *Nature* **442:** 199.

Grishok A., Pasquinelli A.E., Conte D., Li N., Parrish S., Ha I., Baillie D.L., Fire A., Ruvkun G., and Mello C.C. 2001. Genes and mechanisms related to RNA interference regulate expression of the small temporal RNAs that control *C. elegans* developmental timing. *Cell* **106:** 23.

Grivna S.T., Beyret E., Wang Z., and Lin H. 2006. A novel class of small RNAs in mouse spermatogenic cells. *Genes Dev.* **20:** 1709.

Grosshans H., Johnson T., Reinert K.L., Gerstein M., and Slack F.J. 2005. The temporal patterning microRNA let-7 regulates several transcription factors at the larval to adult transition in *C. elegans*. *Dev. Cell* **8:** 321.

Hammond S.M., Bernstein E., Beach D., and Hannon G.J. 2000. An RNA-directed nuclease mediates post-transcriptional gene silencing in *Drosophila* cells. *Nature* **404:** 293.

Harfe B.D., McManus M.T., Mansfield J.H., Hornstein E., and Tabin C.J. 2005. The RNaseIII enzyme Dicer is required for morphogenesis but not patterning of the vertebrate limb. *Proc. Natl. Acad. Sci.* **102:** 10898.

He L., Thomson J.M., Hemann M.T., Hernando-Monge E., Mu D., Goodson S., Powers S., Cordon-Cardo C., Lowe S.W., Hannon G.J., and Hammond S.M. 2005. A microRNA polycistron as a potential human oncogene. *Nature* **435:** 828.

Hornstein E. and Shomron N. 2006. Canalization of development by microRNAs. *Nat. Genet.* (suppl.) **38:** S20.

Humphreys D.T., Westman B.J., Martin D.I., and Preiss T. 2005. MicroRNAs control translation initiation by inhibiting eukaryotic initiation factor 4E/cap and poly(A) tail function. *Proc. Natl. Acad. Sci.* **102:** 16961.

Hutvagner G. and Zamore P.D. 2002. A microRNA in a multiple-turnover RNAi enzyme complex. *Science* **297:** 2056.

Hutvagner G., McLachlan J., Pasquinelli A.E., Balint E., Tuschl T., and Zamore P.D. 2001. A cellular function for the RNA-interference enzyme Dicer in the maturation of the let-7 small temporal RNA. *Science* **293:** 834.

Jakymiw A., Lian S., Eystathioy T., Li S., Satoh M., Hamel J.C., Fritzler M.J., and Chan E.K. 2005. Disruption of GW bodies impairs mammalian RNA interference. *Nat. Cell. Biol.* **7:** 1267.

Johnson S.M., Grosshans H., Shingara J., Byrom M., Jarvis R., Cheng A., Labourier E., Reinert K.L., Brown D., and Slack F.J. 2005. RAS is regulated by the let-7 microRNA family. *Cell* **120:** 635.

Johnston R.J. and Hobert O. 2003. A microRNA controlling left/right neuronal asymmetry in *Caenorhabditis elegans*. *Nature* **426:** 845.

Kane D.A. and Kimmel C.B. 1993. The zebra fish midblastula transition. *Development* **119:** 447.

Kapp L.D. and Lorsch J.R. 2004. The molecular mechanics of eukaryotic translation. *Annu. Rev. Biochem.* **73:** 657.

Ketting R.F., Fischer S.E., Bernstein E., Sijen T., Hannon G.J., and Plasterk R.H. 2001. Dicer functions in RNA interference and in synthesis of small RNA involved in developmental timing in *C. elegans*. *Genes Dev.* **15:** 2654.

Khvorova A., Reynolds A., and Jayasena S.D. 2003. Functional siRNAs and miRNAs exhibit strand bias. *Cell* **115:** 209.

Kiriakidou M., Nelson P.T., Kouranov A., Fitziev P., Bouyioukos C., Mourelatos Z., and Hatzigeorgiou A. 2004. A combined computational-experimental approach predicts human microRNA targets. *Genes Dev.* **18:** 1165.

Kloosterman W.P. and Plasterk R.H. 2006. The diverse functions of microRNAs in animal development and disease. *Dev. Cell* **11:** 441.

Kloosterman W.P., Steiner F.A., Berezikov E., de Bruijn E., van de Belt J., Verheul M., Cuppen E., and Plasterk R.H. 2006. Cloning and expression of new microRNAs from zebra fish. *Nucleic Acids Res.* **34:** 2558.

Knaut H., Pelegri F., Bohmann K., Schwarz H., and Nüsslein-Volhard C. 2000. Zebra fish vasa RNA but not its protein is a component of the germ plasm and segregates asymmetrically before germline specification. *J. Cell Biol.* **149:** 875.

Knight S.W. and Bass B.L. 2001. A role for the RNase III enzyme DCR-1 in RNA interference and germ line development in *Caenorhabditis elegans*. *Science* **293:** 2269.

Koprunner M., Thisse C., Thisse B., and Raz E. 2001. A zebra fish nanos-related gene is essential for the development of primordial germ cells. *Genes Dev.* **15:** 2877.

Krek A., Grun D., Poy M.N., Wolf R., Rosenberg L., Epstein E.J., MacMenamin P., da Piedade I., Gunsalus K.C., Stoffel M., and Rajewsky N. 2005. Combinatorial microRNA target predictions. *Nat. Genet.* **37:** 495.

Krutzfeldt J., Rajewsky N., Braich R., Rajeev K.G., Tuschl T., Manoharan M., and Stoffel M. 2005. Silencing of microRNAs in vivo with 'antagomirs'. *Nature* **438:** 685.

Lagos-Quintana M., Rauhut R., Lendeckel W., and Tuschl T. 2001. Identification of novel genes coding for small expressed RNAs. *Science* **294:** 853.

Lai E.C. 2002. Micro RNAs are complementary to 3'UTR sequence motifs that mediate negative post-transcriptional regulation. *Nat. Genet.* **30:** 363.

———. 2005. miRNAs: Whys and wherefores of miRNA-mediated regulation. *Curr. Biol.* **15:** R458.

Lall S., Grun D., Krek A., Chen K., Wang Y.L., Dewey C.N., Sood P., Colombo T., Bray N., Macmenamin P., et al. 2006. A genome-wide map of conserved microRNA targets in *C. elegans*. *Curr. Biol.* **16:** 460.

Lau N.C., Lim L.P., Weinstein E.G., and Bartel D.P. 2001. An abundant class of tiny RNAs with probable regulatory roles in *Caenorhabditis elegans*. *Science* **294:** 858.

Lau N.C., Seto A.G., Kim J., Kuramochi-Miyagawa S., Nakano T., Bartel D.P., and Kingston R.E. 2006. Characterization of the piRNA complex from rat testes. *Science* **313:** 363.

Lee R.C. and Ambros V. 2001. An extensive class of small RNAs in *Caenorhabditis elegans*. *Science* **294:** 862.

Lee R.C., Feinbaum R.L., and Ambros V. 1993. The *C. elegans* heterochronic gene lin-4 encodes small RNAs with antisense complementarity to lin-14. *Cell* **75:** 843.

Lewis B.P., Burge C.B., and Bartel D.P. 2005. Conserved seed pairing often flanked by adenosines indicates that thousands of human genes are microRNA targets. *Cell* **120:** 15.

Lewis B.P., Shih I.H., Jones-Rhoades M.W., Bartel D.P., and Burge C.B. 2003. Prediction of mammalian microRNA targets. *Cell* **115:** 787.

Lim L.P., Lau N.C., Garrett-Engele P., Grimson A., Schelter J.M., Castle J., Bartel D.P., Linsley P.S., and Johnson J.M. 2005. Microarray analysis shows that some microRNAs downregulate large numbers of target mRNAs. *Nature* **433:** 769.

Lin S.Y., Johnson S.M., Abraham M., Vella M.C., Pasquinelli A., Gamberi C., Gottlieb E., and Slack F.J. 2003. The *C. elegans* hunchback homolog hbl-1 controls temporal patterning and is a probable microRNA target. *Dev. Cell.* **4:** 639.

Liu J., Valencia-Sanchez M.A., Hannon G.J., and Parker R. 2005. MicroRNA-dependent localization of targeted mRNAs to mammalian P-bodies. *Nat. Cell Biol.* **7:** 719.

Lu J., Getz G., Miska E.A., Alvarez-Saavedra E., Lamb J., Peck D., Sweet-Cordero A., Ebert B.L., Mak R.H., Ferrando A.A., et al. 2005. MicroRNA expression profiles classify human cancers. *Nature* **435:** 834.

McGrew L.L., Dworkin-Rastl E., Dworkin M.B., and Richter J.D. 1989. Poly(A) elongation during *Xenopus* oocyte maturation is required for translational recruitment and is mediated by a short sequence element. *Genes Dev.* **3:** 803.

Miranda K.C., Huynh T., Tay Y., Ang Y.S., Tam W.L., Thomson A.M., Lim B., and Rigoutsos I. 2006. A pattern-based method for the identification of MicroRNA binding sites and their corresponding heteroduplexes. *Cell* **126:** 1203.

Mishima Y., Giraldez A.J., Takeda Y., Fujiwara T., Sakamoto H., Schier A.F., and Inoue K. 2006. Differential regulation of germline mRNAs in soma and germ cells by zebra fish miR-430. *Curr. Biol.* **16:** 2135.

Newport J. and Kirschner M. 1982a. A major developmental transition in early *Xenopus* embryos. I. Characterization and timing of cellular changes at the midblastula stage. *Cell* **30:** 675.

———. 1982b. A major developmental transition in early *Xenopus* embryos. II. Control of the onset of transcription. *Cell* **30:** 687.

Olsen P.H. and Ambros V. 1999. The lin-4 regulatory RNA controls developmental timing in *Caenorhabditis elegans* by blocking LIN-14 protein synthesis after the initiation of translation. *Dev. Biol.* **216:** 671.

Ota A., Tagawa H., Karnan S., Tsuzuki S., Karpas A., Kira S., Yoshida Y., and Seto M. 2004. Identification and characterization of a novel gene C13orf25 as a target for 13q31-q32 amplification in malignant lymphoma. *Cancer Res.* **64:** 3087.

Pelegri F. 2003. Maternal factors in zebra fish development. *Dev. Dyn.* **228:** 535.

Petersen C.P., Bordeleau M.E., Pelletier J., and Sharp P.A. 2006. Short RNAs repress translation after initiation in mammalian cells. *Mol. Cell* **21:** 533.

Pillai R.S. 2005. MicroRNA function: Multiple mechanisms for a tiny RNA? *RNA* **11:** 1753.

Pillai R.S., Bhattacharyya S.N., Artus C.G., Zoller T., Cougot N., Basyuk E., Bertrand E., and Filipowicz W. 2005. Inhibition of translational initiation by Let-7 MicroRNA in human cells. *Science* **309:** 1573.

Poy M.N., Eliasson L., Krutzfeldt J., Kuwajima S., Ma X., Macdonald P.E., Pfeffer S., Tuschl T., Rajewsky N., Rorsman P., and Stoffel M. 2004. A pancreatic islet-specific microRNA regulates insulin secretion. *Nature* **432:** 226.

Rajewsky N. 2006. microRNA target predictions in animals. *Nat. Genet.* (suppl.) **38:** S8.

Raz E. 2004. Guidance of primordial germ cell migration. *Curr. Opin. Cell Biol.* **16:** 169.

Rehmsmeier M., Steffen P., Hochsmann M., and Giegerich R. 2004. Fast and effective prediction of microRNA/target duplexes. *RNA* **10:** 1507.

Rehwinkel J., Behm-Ansmant I., Gatfield D., and Izaurralde E. 2005. A crucial role for GW182 and the DCP1:DCP2 decapping complex in miRNA-mediated gene silencing. *RNA* **11:** 1640.

Reinhart B.J., Slack F.J., Basson M., Pasquinelli A.E., Bettinger J.C., Rougvie A.E., Horvitz H.R., and Ruvkun G. 2000. The 21-nucleotide let-7 RNA regulates developmental timing in *Caenorhabditis elegans*. *Nature* **403:** 901.

Rhoades M.W., Reinhart B.J., Lim L.P., Burge C.B., Bartel B., and Bartel D.P. 2002. Prediction of plant microRNA targets. *Cell* **110:** 513.

Richter J.D. 1999. Cytoplasmic polyadenylation in development and beyond. *Microbiol. Mol. Biol. Rev.* **63:** 446.

Richter J.D., Paris J., and McGrew L.L. 1990. Maternal mRNA expression in early development: Regulation at the 3' end. *Enzyme* **44:** 129.

Schier A.F. and Talbot W.S. 2005. Molecular genetics of axis formation in zebra fish. *Annu. Rev. Genet.* **39:** 561.

Schwarz D.S., Hutvagner G., Du T., Xu Z., Aronin N., and Zamore P.D. 2003. Asymmetry in the assembly of the RNAi enzyme complex. *Cell* **115:** 199.

Sen G.L. and Blau H.M. 2005. Argonaute 2/RISC resides in sites of mammalian mRNA decay known as cytoplasmic bodies. *Nat. Cell Biol.* **7:** 633.

Sheth U. and Parker R. 2003. Decapping and decay of messenger RNA occur in cytoplasmic processing bodies. *Science* **300:** 805.

Slack F.J., Basson M., Liu Z., Ambros V., Horvitz H.R., and Ruvkun G. 2000. The lin-41 RBCC gene acts in the *C. elegans* heterochronic pathway between the let-7 regulatory RNA and the LIN-29 transcription factor. *Mol. Cell* **5:** 659.

Slater D.W., Slater I., and Gillespie D. 1972. Post-fertilization synthesis of polyadenylic acid in sea urchin embryos. *Nature* **240:** 333.

Slater I., Gillespie D., and Slater D.W. 1973. Cytoplasmic adenylylation and processing of maternal RNA. *Proc. Natl. Acad. Sci.* **70:** 406.

Sokol N.S. and Ambros V. 2005. Mesodermally expressed *Drosophila* microRNA-1 is regulated by Twist and is required in muscles during larval growth. *Genes Dev.* **19:** 2343.

Sood P., Krek A., Zavolan M., Macino G., and Rajewsky N. 2006. Cell-type-specific signatures of microRNAs on target mRNA expression. *Proc. Natl. Acad. Sci.* **103:** 2746.

Stark A., Brennecke J., Russell R.B., and Cohen S.M. 2003. Identification of *Drosophila* MicroRNA targets. *PLoS Biol.* **1:** E60.

Stark A., Brennecke J., Bushati N., Russell R.B., and Cohen S.M. 2005. Animal MicroRNAs confer robustness to gene expression and have a significant impact on 3′UTR evolution. *Cell* **123:** 1133.

Teleman A.A., Maitra S., and Cohen S.M. 2006. *Drosophila* lacking microRNA miR-278 are defective in energy homeostasis. *Genes Dev.* **20:** 417.

Valencia-Sanchez M.A., Liu J., Hannon G.J., and Parker R. 2006. Control of translation and mRNA degradation by miRNAs and siRNAs. *Genes Dev.* **20:** 515.

Vasudevan S., Seli E., and Steitz J.A. 2006. Metazoan oocyte and early embryo development program: A progression through translation regulatory cascades. *Genes Dev.* **20:** 138.

Voorhoeve P.M., le Sage C., Schrier M., Gillis A.J., Stoop H., Nagel R., Liu Y.P., van Duijse J., Drost J., Griekspoor A., et al. 2006. A genetic screen implicates miRNA-372 and miRNA-373 as oncogenes in testicular germ cell tumors. *Cell* **124:** 1169.

Watanabe T., Takeda A., Mise K., Okuno T., Suzuki T., Minami N., and Imai H. 2005. Stage-specific expression of microRNAs during *Xenopus* development. *FEBS Lett.* **579:** 318.

Wienholds E., Kloosterman W.P., Miska E., Alvarez-Saavedra E., Berezikov E., de Bruijn E., Horvitz H.R., and Plasterk R.H. 2005. MicroRNA expression in zebra fish embryonic development. *Science* **309:** 310.

Wightman B., Ha I., and Ruvkun G. 1993. Posttranscriptional regulation of the heterochronic gene lin-14 by lin-4 mediates temporal pattern formation in *C. elegans*. *Cell* **75:** 855.

Wilt F.H. 1973. Polyadenylation of maternal RNA of sea urchin eggs after fertilization. *Proc. Natl. Acad. Sci.* **70:** 2345.

Wu L., Fan J., and Belasco J.G. 2006. MicroRNAs direct rapid deadenylation of mRNA. *Proc. Natl. Acad. Sci.* **103:** 4034.

Xu P., Vernooy S.Y., Guo M., and Hay B.A. 2003. The *Drosophila* microRNA Mir-14 suppresses cell death and is required for normal fat metabolism. *Curr. Biol.* **13:** 790.

Yi R., O'Carroll D., Pasolli H.A., Zhang Z., Dietrich F.S., Tarakhovsky A., and Fuchs E. 2006. Morphogenesis in skin is governed by discrete sets of differentially expressed microRNAs. *Nat. Genet.* **38:** 356.

Yoo A.S. and Greenwald I. 2005. LIN-12/Notch activation leads to microRNA-mediated down-regulation of Vav in *C. elegans*. *Science* **310:** 1330.

Yoon C., Kawakami K., and Hopkins N. 1997. Zebra fish vasa homologue RNA is localized to the cleavage planes of 2- and 4-cell-stage embryos and is expressed in the primordial germ cells. *Development* **124:** 3157.

MicroRNAs and Hematopoietic Differentiation

A. Fatica,* A. Rosa,* F. Fazi,† M. Ballarino,* M. Morlando,* F.G. De Angelis,*
E. Caffarelli,* C. Nervi,† and I. Bozzoni*

*Institute Pasteur Cenci-Bolognetti, Department of Genetics and Molecular Biology and I.B.P.M., University of
Rome "La Sapienza," 00185 Rome, Italy; †Department of Histology and Medical Embryology, University of Rome
"La Sapienza" and San Raffaele Bio-medical Science Park of Rome, 00128 Rome, Italy

The discovery of microRNAs (miRNAs) and of their mechanism of action has provided some very new clues on how gene expression is regulated. These studies established new concepts on how posttranscriptional control can fine-tune gene expression during differentiation and allowed the identification of new regulatory circuitries as well as factors involved therein. Because of the wealth of information available about the transcriptional and cellular networks involved in hematopoietic differentiation, the hematopoietic system is ideal for studying cell lineage specification. An interesting interplay between miRNAs and lineage-specific transcriptional factors has been found, and this can help us to understand how terminal differentiation is accomplished.

The completion of the human genome sequencing together with the development of new genome-wide approaches has provided very powerful tools for identifying gene products whose expression could be correlated with cell growth and differentiation. Nevertheless, these studies have concentrated on expression profiles of mRNAs or proteins. In the last few years, several new classes of small noncoding RNAs have been identified, and quite a lot of data have been accumulated on their functions in regulating gene expression at the posttranscriptional level, including alternative splicing, mRNA stability, and translation. These events are very relevant in cell metabolism, growth, and differentiation not only because they provide a means for expanding protein diversity and abundance, but also because they enable the cell to rapidly respond to external stimuli and to adjust the biosynthetic machineries to the different growth requirements. Inside the small noncoding RNA family, particular interest is now devoted to miRNAs. These molecules are synthesized by endogenous cellular genes and have been shown to regulate mRNA and protein abundance by controlling both the stability and the translation of the target mRNAs (Pillai et al. 2005). Since the characterization of the first miRNA (Lee et al. 1993), several hundred different miRNAs have been identified, and many data are now available on their role in regulatory circuits controlling developmental timing, cell death, cell proliferation, apoptosis, hematopoiesis, and patterning of the nervous system (Ambros 2004). The existence of numerous tissue- and developmental-stage-specific miRNAs and the evolutionary conservation of many miRNAs argue for numerous additional, yet unidentified, functions of this class of transcripts. Notably, miRNA activity has also been correlated with cancer since miRNAs with oncogenic and tumor suppressor activity have been identified and a new molecular taxonomy of human cancers based on miRNA profiling has been proposed (Caldas and Brenton 2005).

Hundreds of miRNAs operating in different organisms have been identified through cloning and genetics, and through bioinformatic methods. According to current predictions, 800 or more miRNAs operate in primates, and each miRNA may target dozens of mRNAs. Hence, it is estimated that expression of as many as 30% of human genes may be controlled by miRNAs (Lewis et al. 2005).

miRNA-coding regions have very peculiar and heterogeneous genomic organizations. Many of them are found in intronic regions and may be transcribed as part of the host gene; however, the majority are located in intergenic regions or in annotated genes but in an antisense orientation, strongly suggesting that they form independent transcription units. miRNAs have been shown to be transcribed by RNA polymerase II (RNA pol II) from transcriptional units that differ from those of protein-coding genes in that they do not possess canonical TATA boxes and are intronless. Nowadays, very little is known about their transcriptional regulation, including the factors responsible for basal and tissue-specific expression.

Some miRNAs show ubiquitous expression (Sempere et al. 2004), but others are limited to certain stages in development or to certain tissues and cell types (Lee and Ambros 2001). The study of tissue-specific miRNAs will therefore provide important clues for identifying new sets of genes and regulatory circuits involved in the control of cell-specific differentiation.

Hematopoiesis is a lifelong, highly regulated multistage process where a pluripotent self-renewing hematopoietic stem cell (HSC) gives rise to all blood cell lineages. In addition, the hematopoietic system is ideal for identifying miRNAs and regulatory factors involved in cell lineage specification, since many transcriptional and cellular networks have been already identified as crucial for differentiation. We describe here examples of how miRNAs can influence cell lineage specification and how the identification of their target mRNAs has contributed to the understanding of the molecular networks involved in the alternative control between cell growth and differentiation.

Ectopic expression and knockdown of specific miRNAs have provided powerful molecular tools able to control the switch between proliferation and differentiation, thereby also providing new potential therapeutic tools for interfering with tumorigenesis.

IDENTIFICATION OF miRNAs SPECIFICALLY EXPRESSED DURING HEMATOPOIESIS AND LEUKEMIA

Expression profiling analysis showed that most miRNAs are under the control of developmental or tissue-specific signaling. Transcriptional regulation is likely to be the major control step of miRNA expression, even if some miRNAs seem to be controlled at the posttranscriptional level (Bartel 2004; Kim 2005; Gregory and Shiekhattar 2005).

Despite their small size, miRNA expression is analyzed by standard techniques for RNA expression studies, including northern blot and quantitative polymerase chain reaction (PCR). To analyze the relative expression level of miRNAs from different tissues, large-scale cDNA cloning has been successfully utilized, even if for high-throughput analysis the most widely used method is based on microarrays (Croce and Calin 2005). However, it should be mentioned that microarray is not as quantitative as northern blot so that in most of the cases, it is not easy to exactly define the relative abundance of individual miRNAs.

In past years, several specific miRNAs differentially expressed during hematopoietic differentiation have been identified and shown to have an important role in mammalian hematopoiesis. The first case was described in mice: miR-181, miR-223, and miR-142 are differentially expressed in hematopoietic tissues, and their expression is regulated during hematopoiesis and lineage commitment (Chen et al. 2004). Notably, ectopic expression of miR-181 in murine hematopoietic progenitor cells led to an increased fraction of B-lymphoid cells in both tissue

culture differentiation assays and adult mice, suggesting a central role of miRNAs in the control of hematopoietic lineage differentiation.

Since then, many miRNA profiles have been derived by either microarray or conventional northern analysis on different hematopoietic lineages. One of the most complete miRNA expression profiles was recently obtained by a new bead-based flow cytometric method (Lu et al. 2005). In this study, the miRNA expression detected in normal tissue was globally higher than that in tumors, corroborating the idea that the miRNA expression profile reflects the differentiation state of the cell. miRNA expression was also profiled during differentiation of primary hematopoietic progenitor cells and myeloid cell lines into erythrocytes and granulocytes, respectively. As predicted, blood cell differentiation was always associated with global changes in miRNA expression (Lu et al. 2005). Analysis on specific members of the miRNA family allowed their role to be distinguished either in cell proliferation or in differentiation. In humans, miR-221 and miR-222 have been shown to be down-modulated in human CD34$^+$ cord blood progenitor cells induced toward the erythroid lineage. During early stages of erythroid differentiation that are coupled with exponential growth, Kit protein levels are high, although they decrease in terminal erythroblasts undergoing little proliferation. The expression levels of both miR-221 and miR-222 are inversely related to the amount of the Kit protein, strongly suggesting its posttranscriptional regulation during expansion of early erythroblasts (see Fig. 1). Functional studies indeed indicated that the decline of miR-221 and miR-222 unblocks Kit protein production at the translational level, leading to expansion of early erythroid precursors. The decrease of Kit protein at later differentiation stages, where little proliferation occurs, is instead due to transcriptional down-regulation (Felli et al. 2005). In agreement with these findings, treatment of CD34$^+$ progenitors with miR-221 and miR-222, via oligonucleotide transfection or lentiviral vector infection, caused impaired proliferation

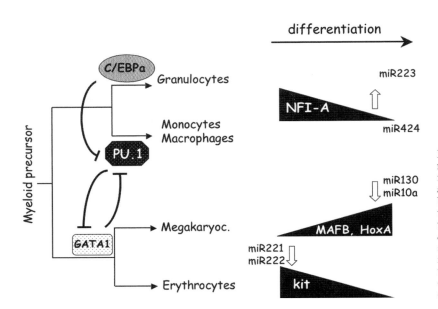

Figure 1. The transcriptional factors responsible for the commitment to the four myeloid terminal differentiation lineages are indicated. These proteins have been shown to inhibit each other's function by physical interaction. The miRNAs that undergo specific up- and down- (*open arrows*) regulation are indicated in correspondence with the effect exerted on their respective validated target. In the case of the Kit protein, its decrease at later differentiation stages is due to transcriptional down-regulation.

and accelerated differentiation of erythropoietic cultures, coupled with down-modulation of the Kit protein (Felli et al. 2005), suggesting an important role for both miRNAs in controlling the proliferation and differentiation status of early erythroblasts.

Down-regulation of miR-130 correlated instead with the up-regulation of the MAFB transcriptional factor during megakaryocytic differentiation, and the decrease of miR-10 paralleled the increase of the HOXA1 factor (Garzon et al. 2006). Interestingly, the synthesis of both MAFB and HoxA1 has been correlated with terminal commitment of megakaryocytic cells. Compared with the previous class, miR-223 was shown to increase upon granulocytic differentiation and to promote maturation of promyelocytic precursors: In fact, its ectopic expression in APL cells induced commitment to the granulocyte-specific lineage, whereas its depletion counteracted differentiation (Fazi et al. 2005). miR-223 activation correlated in time with the decrease in accumulation of the NFI-A protein while its mRNA levels remained unaltered; this, together with the experimental validation of the bona fide target site in the 3'UTR (untranslated region), indicated that NFI-A is a target of miR-223. Interestingly, the NFI-A factor was previously implicated in replication as well as in controlling changes in cell growth (Gronostajski 2000), and its knockdown was shown to enhance granulocytic differentiation (Fazi et al. 2005). NFI-A mRNA is also a target of other miRNAs up-regulated during monocyte/macrophage differentiation, such as miR-424 (Kasashima et al. 2004), therefore indicating that its repression is very important for at least two myeloid differentiation pathways.

Figure 1 summarizes a few examples of regulatory networks that correlate up- and down-regulation of miRNA expression relative to their targets during different lineages of myeloid differentiation. From the few cases studied so far, it can be suggested that those miRNAs decreasing during lineage commitment unblock the expression of key proteins required for differentiation, whereas those increasing during maturation have a negative effect on growth-promoting factors. Along this line, several lines of evidence link miRNAs to leukemias. The human miR-15a/miR-16 cluster is frequently deleted or down-regulated in patients with B-cell chronic lymphocytic leukemia (Calin et al. 2005). These miRNAs negatively regulate BCL2, which is an antiapoptotic gene that is often overexpressed in many types of human leukemias. Another miRNA, miR-155, is up-regulated in many pediatric and adult lymphomas, especially in diffuse large B-cell lymphomas, Hodgkin lymphomas, and certain types of Burkitt lymphomas (Metzler et al. 2004; Eis et al. 2005). Recently, it has been shown that miR-155 transgenic mice exhibit initially a preleukemic pre-B-cell proliferation evident in spleen and bone marrow, followed by frank B-cell malignancy. These findings indicate that the role of miR-155 is to induce polyclonal expansion, favoring the capture of secondary genetic changes for full transformation (Costinean et al. 2006).

Another example is provided by an aggressive B-cell leukemia in which Myc is translocated downstream from

miR-142, disrupting the miRNA synthesis and increasing Myc expression by the miRNA promoter. It is now clear that miRNAs may function as oncogenes and tumor suppressors (Hammond 2006); therefore, studies into miRNA expression and function might lead to an advanced understanding of the mechanisms involved in tumorigenesis.

ACTIVATORS OF MIRNA EXPRESSION IN HEMATOPOIESIS

Several transcriptional factors have been shown to have a crucial role in hematopoiesis. Important information has been obtained from overexpression and knockout experiments. Some of these factors, such as SCL/Tal1 and AML1, are common to all lineages, and their depletion affects the entire blood cell lineage. Among the factors that have lineage-specific expression patterns are GATA1, the CAAT/enhancer-binding protein C/EBPα, and PU.1. GATA1 was the first "lineage-specific" factor studied in detail and shown to be essential for erythroid and megakaryocytic lineages. PU.1 mediates lymphoid and monocyte/macrophage differentiation, and C/EBPα has a more specific function in granulopoiesis (Radomska et al. 1998). Interestingly, these factors have some functional interconnection since C/EBPα and GATA1 can inhibit PU.1 function (Radomska et al. 1998; Reddy et al. 2002), whereas increases in PU.1 lead to inhibition of GATA1 (Fig. 1).

A variety of studies in different experimental systems have provided compelling evidence indicating a very important role of C/EBPα in both growth arrest and terminal differentiation, in particular of granulocytes (Nerlov et al. 1998; Radomska et al. 1998; Wang et al. 1999; Khanna-Gupta et al. 2001; Nakajima and Ihle 2001). Although C/EBPα was known to induce granulopoiesis while suppressing monocyte differentiation, it was unclear how C/EBPα regulated this cell fate choice at the mechanistic level. In addition, mice deficient for known C/EBPα target genes did not exhibit the same block in granulocyte maturation, indicating the necessity of identifying additional C/EBPα target genes essential for myeloid cell development.

In this view, it was very interesting to find that the miR-223 promoter contained two binding sites for the C/EBPα factor and to show that this region was essential for the up-regulation of the miR in response to retinoic acid treatment. ChIP (chromatin immunoprecipitation) experiments have indeed also shown that the binding of this factor, whose synthesis and binding ability are induced upon retinoic acid treatment, correlated with differentiation and the up-regulation of miR-223 expression (Fazi et al. 2005).

A search in the miR-223 promoter also revealed the presence of an NFI-A site (Meisterernst et al. 1988; Bachurski et al. 1997) overlapping one of the two C/EBPα-binding sites. ChIP experiments demonstrated that this factor is indeed binding the miR-223 promoter but only in the absence of retinoic acid. Therefore, C/EBPα and NFI-A factors compete for the same site,

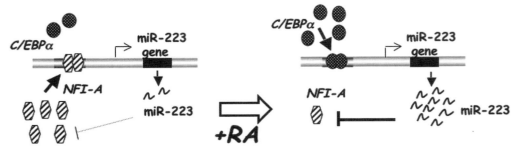

Figure 2. Schematic model of the autoregulatory loop controlling miR-223 expression. (*A*) In promyelocytic cells, NFI-A keeps low levels of miR-223 expression and, consequently, low miR-223-mediated translational repression. (*B*) Upon retinoic acid treatment, C/EBPα replaces NFI-A on the miR-223 promoter and up-regulates miR-223. This results in (1) an increase of miR-223-mediated translational repression of the NFI-A mRNA, (2) a decrease of the NF1-A protein, and (3) stimulation of granulocytic differentiation.

and their binding correlates with low (NFI-A) and high transcriptional activity of miR-223. Since NFI-A is also the target of miR-223, a very interesting autoregulatory loop appears to control granulocytic differentiation: C/EBPα activation produces the displacement of NFI-A from the miR-223 promoter and its up-regulation. In cascade, miR-223 acts by repressing NFI-A, thus subtracting it from the competition with C/EBPα and maintaining sustained levels of miR-223 expression (Fig. 2).

RNAi against either factor not only validated their activity with respect to miR-223 transcription, but also indicated their different roles in differentiation. NFI-A is required during proliferation of the precursor cells and counteracts differentiation, whereas C/EBPα is required for differentiation into granulocytes. These findings are in agreement with the roles described for C/EBPα and NFI-A. C/EBPα has been implicated in promoting differentiation and repressing genes that have the important function in stimulating growth (McKnight 2001; Tenen 2001); on the other side, NFI-A has been implicated in replication as well as in controlling changes in cell growth (Gronostajski 2000).

In analogy with the role of C/EBPα in activating the expression of lineage-specific miRNAs, it was interesting to discover that several miRNAs specifically expressed during monocyte/macrophage differentiation (Kasashima et al. 2004) contained PU.1-binding sites in their promoter. Even more striking was the fact that some of these miRNAs have among their targets the NFI-A mRNA. These findings are very interesting in view of the fact that PU.1 has been described as a major player in the commitment to this differentiation lineage and of the important role of NFI-A in repressing differentiation. Therefore, it can be concluded that in both granulocyte and monocyte/macrophage differentiation, lineage-specific transcriptional factors activate miRNAs that inhibit the expression of NFI-A, thus allowing differentiation to proceed.

NFI-A FACTOR

NFI-A was initially described as a CCAAT-box-binding transcription factor, belonging to the nuclear factor I (NFI) family of proteins (Santoro et al. 1988). The NFI family is composed of four independent genes

(NFI-A, -B, -C, and -X) and a large number of splice variants that form homodimers and heterodimers, thus creating an extensive network of possible functional dimers (Gronostajski 2000). The NFI proteins bind as dimers to the dyad symmetric consensus sequence TTGGC(N5)GCCAA; nevertheless, they could also bind very well to individual half-sites even with a reduced affinity and presumably as dimers with other proteins (Meisterernst et al. 1988; Bachurski et al. 1997).

Figure 3B shows the results of RT-PCR performed to identify the different NFI-A mRNA species expressed in HeLa cells and in the APL cell line NB4. Both cell lines display two types of NFI-A transcripts, one corresponding to the full length and the other to an isoform lacking exon 10 (black box). Skipping this exon results in changing of the reading frame, thus producing a protein variant in the carboxy-terminal portion. So far, since *trans*-activation/repression activities have been assigned to this domain, it is not known whether the two variants differ in such properties. Interestingly, both forms can undergo posttranscriptional control since they have the same 3′UTR with several target sequences for several classes of miRNAs. Figure 3A indicates that according to miRanda algorithm (John et al. 2004), five miRNA families are able to recognize the 3′UTR of NFI-A. Among them, miR-223 is specifically expressed during granulocytic differentiation, where it has been demonstrated to indeed repress NFI-A (Fazi et al. 2005). The miR-424 family is instead specifically activated in the monocyte/macrophage pathway; interestingly, the activation of miR-424 correlates with the decrease of NFI-A levels during monocyte differentiation (Fig. 3C), suggesting that it has a direct role in the down-regulation of this protein. From these observations, it can be suggested that if NFI-A repression is a general theme for differentiation, this is accomplished in the different hematopoietic lineages through activation of specific classes of miRNAs.

In conclusion, the identification of hematopoietic lineage-specific miRNAs has allowed us to unravel regulatory circuitries and factors therein involved that altogether form the complex network of interactions responsible for the commitment and maintenance of the different hematopoietic programs. The study of the regulation of these networks will also enable the identification of crucial

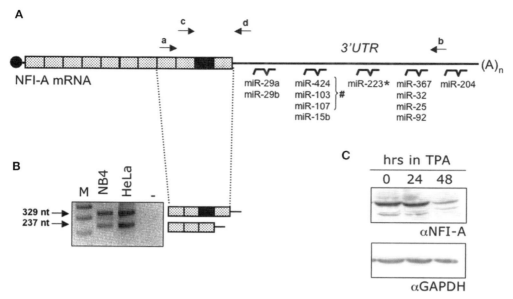

Figure 3. (*A*) Schematic representation of the NFI-A mRNA. (*Black box*) Alternatively spliced exon 10. miRNA target sites in the 3'UTR are shown. (#) miRNAs induced by TPA; (*) miRNAs induced by retinoic acid. (*B*) Nested RT-PCR on poly(A)$^+$ RNA extracted from HeLa and NB4 cell lines. An oligo(dT) was used in the retrotranscription reaction; the resulting cDNA was initially amplified with primers annealing on exon 8 and the 3'UTR (primers a and b, respectively, in panel *A*); 2 μl of a 1:100 dilution of this amplification was used for a second PCR with primers specific for exon 9 and for an inner region of the 3'UTR (primers c and d, respectively, in panel *A*). The amplification products from HeLa and NB4 cell lines were run on a 1.2% agarose gel in parallel with a negative contol (lane –). The 329-nucleotide-long amplification product corresponds to the full-length transcript, whereas that of 237 nucleotides corresponds to the transcript lacking exon 10. A schematic representation of the two products is shown on the side of the gel. (Lane M) Molecular-weight marker. (*C*) NB4 cells were treated with 16 nM TPA for the indicated times; 50 μg of protein was analyzed by western blot with anti-NFI-A antibody (*upper panel*) and anti-GAPDH antibody as a loading control (*lower panel*).

factors whose modulation can be utilized to control aberrant proliferation as in the case of several leukemias.

ACKNOWLEDGMENTS

This work was partially supported by grants from AIRC and AIRC-ROC, Sixth Research Framework Programme of the European Union, Project RIGHT (LSHB-CT-2004 005276), MURST (FIRB-p.n. RBNE015MPB and RBNE01KXC9 to I.B.), PRIN, and "Centro di eccellenza BEMM."

REFERENCES

Ambros V. 2004. The function of animal miRNAs. *Nature* **431:** 350.

Bachurski C.J., Kelly S.E., Glasser S.W., and Currier T.A. 1997. Nuclear factor I family members regulate the transcription of surfactant protein-C. *J. Biol. Chem.* **272:** 32759.

Bartel D.P. 2004. MicroRNAs: Genomics, biogenesis, mechanism, and function. *Cell* **116:** 281.

Caldas C. and Brenton J.D. 2005. Sizing up miRNAs as cancer genes. *Nat. Med.* **11:** 712.

Calin G.A., Ferracin M., Cimmino A., Di Leva G., Shimizu M., Wojcik S.E., Iorio M.V., Visone R., Sever N.I., Fabbri M., et al. 2005. A microRNA signature associated with prognosis and progression in chronic lymphocytic leukemia. *N. Engl. J. Med.* **353:** 1793.

Chen C.Z., Li L., Lodish H.F., and Bartel D.P. 2004. MicroRNAs modulate hematopoietic lineage differentiation. *Science* **303:** 83.

Costinean S., Zanesi N., Pekarsky Y., Tili E., Volinia S., Heerema N., and Croce C.M. 2006. Pre-B cell proliferation and lymphoblastic leukemia/high-grade lymphoma in Eμ-miR155 transgenic mice. *Proc. Natl. Acad. Sci.* **103:** 7024.

Croce C.M. and Calin G.A. 2005. miRNAs, cancer, and stem cell division. *Cell* **122:** 6.

Eis P.S., Tam W., Sun L., Chadburn A., Li Z., Gomez M.F., Lund E., and Dahlberg J.E. 2005. Accumulation of miR-155 and BIC RNA in human B cell lymphomas. *Proc. Natl. Acad. Sci.* **102:** 3627.

Fazi F., Rosa A., Fatica A., Gelmetti V., De Marchis M.L., Nervi C., and Bozzoni I. 2005. A minicircuitry comprised of microRNA-223 and transcription factors NFI-A and C/EBPalpha regulates human granulopoiesis. *Cell* **123:** 819.

Felli N., Fontana L., Pelosi E., Botta R., Bonci D., Fachiano F., Liuzzi F., Lulli V., Morsili O., Santoro S., et al. 2005. MicroRNAs 221 and 222 inhibit normal erythropoiesis and erythroleukemic cell growth via kit receptor down-modulation. *Proc. Natl. Acad. Sci.* **102:** 18081.

Garzon R., Pichiorri F., Palumbo T., Iuliano R., Cimmino A., Aqeilan R., Volinia S., Bhatt D., Alder H., Marcucci G., et al. 2006. MicroRNA fingerprints during human megakaryocytopoiesis. *Proc. Natl. Acad. Sci.* **103:** 5078.

Gregory R.I. and Shiekhattar R. 2005. MicroRNA biogenesis and cancer. *Cancer Res.* **65:** 3509.

Gronostajski R.M. 2000. Roles of the NF1/CTF gene family in transcription and development. *Gene* **249:** 31.

Hammond S.M. 2006. MicroRNAs as oncogenes. *Curr. Opin. Genet. Dev.* **16:** 4.

John B., Enright A.J., Aravin A., Tuschl T., Sander C., and Marks D.S. 2004. Human microRNA targets. *PLoS Biol.* **2:** e363.

Khanna-Gupta A., Zibello T., Sun H., Lekstrom-Himes J., and Berliner N. 2001. C/EBP epsilon mediates myeloid differentiation and is regulated by the CCAAT displacement protein (CDP/cut). *Proc. Natl. Acad. Sci.* **98:** 8000.

Kasashima K., Nakamura Y., and Kozu T. 2004. Altered expression profiles of microRNAs during TPA-induced differentiation of HL-60 cells. *Biochem. Biophys. Res. Commun.* **322:** 403.

Kim V.N. 2005. MicroRNA biogenesis: Coordinated cropping and dicing. *Nat. Rev. Mol. Cell Biol.* **6:** 376.

Lee R.C. and Ambros V. 2001. An extensive class of small RNAs in *Caenorhabditis elegans. Science* **294:** 862.

Lee R.C., Feinbaum R.L., and Ambros V. 1993. The *C. elegans* heterochronic gene lin-4 encodes small RNAs with antisense complementarity to lin-14. *Cell* **75:** 843.

Lewis B.P., Burge C.B., and Bartel D.P. 2005. Conserved seed pairing, often flanked by adenosines, indicates that thousands of human genes are microRNA targets. *Cell* **120:** 15.

Lu J., Getz G., Miska E.A., Alvarez-Saavedra E., Lamb J., Peck D., Sweet-Cordero A., Ebert B.L., Mak R.H., Ferrando A.A., et al. 2005. MicroRNA expression profiles classify human cancers. *Nature* **435:** 834.

McKnight S.L. 2001. McBindall—A better name for CCAAT/enhancer binding proteins? *Cell* **107:** 259.

Meisterernst M., Gander I., Rogge L., and Winnacker E.L. 1988. A quantitative analysis of nuclear factor I/DNA interactions. *Nucleic Acids Res.* **16:** 4419.

Metzler M., Wilda M., Busch K., Viehmann S., and Borkhardt A. 2004. High expression of precursor microRNA-155/BIC RNA in children with Burkitt lymphoma. *Genes Chromosomes Cancer* **39:** 167.

Nakajima H. and Ihle J.N. 2001. Granulocyte colony-stimulating factor regulates myeloid differentiation through CCAAT/enhancer-binding protein epsilon. *Blood* **98:** 897.

Nerlov C., McNagny K.M., Doderlein G., Kowenz-Leutz E., and Graf T. 1998. Distinct C/EBP functions are required for eosinophil lineage commitment and maturation. *Genes Dev.*

12: 2413.

Pillai R.S., Bhattacharyya S.N., Artus C.G., Zoller T., Cougot N., Basyuk E., Bertrand E., and Filipowicz W. 2005. Inhibition of translational initiation by Let-7 MicroRNA in human cells. *Science* **309:** 1573.

Radomska H.S., Huettner C.S., Zhang P., Cheng T., Scadden D.T., and Tenen D.G. 1998. CCAAT/enhancer binding protein alpha is a regulatory switch sufficient for induction of granulocytic development from bipotential myeloid progenitors. *Mol. Cell. Biol.* **18:** 4301.

Reddy V.A., Iwama A., Iotzova G., Schulz M., Elsasser A., Vangala R.K., Tenen D.G., Hiddemann W., and Behre G. 2002. Granulocyte inducer C/EBPalpha inactivates the myeloid master regulator PU.1: Possible role in lineage commitment decisions. *Blood* **100:** 483.

Santoro C., Mermod N., Andrews P.C., and Tjian R. 1988. A family of human CCAAT-box-binding proteins active in transcription and DNA replication: Cloning and expression of multiple cDNAs. *Nature* **334:** 218.

Sempere L.F., Freemantle S., Pitha-Rowe I., Moss E., Dmitrovsky E., and Ambros V. 2004. Expression profiling of mammalian microRNAs uncovers a subset of brain-expressed microRNAs with possible roles in murine and human neuronal differentiation. *Genome Biol.* **5:** R13.

Tenen D.G. 2001. Abnormalities of the CEBP alpha transcription factor: A major target in acute myeloid leukemia. *Leukemia* **4:** 688.

Wang X., Scott E., Sawyers C.L., and Friedman A.D. 1999. C/EBPalpha bypasses granulocyte colony-stimulating factor signals to rapidly induce PU.1 gene expression, stimulate granulocytic differentiation, and limit proliferation in 32D cl3 myeloblasts. *Blood* **94:** 560.

Expression and Suppression of Human Telomerase RNA

S. LI AND E.H. BLACKBURN

Department of Biochemistry and Biophysics, University of California, San Francisco, California 94158-2517

Telomeres are maintained by the ribonucleoprotein (RNP) enzyme telomerase, which replenishes telomeres through its unique mechanism of internal RNA-templated addition of telomeric DNA. Telomerase is active in most human cancers, typically because its core protein subunit, TERT, is up-regulated. Although the major known function of telomerase in cancer is to replenish telomeric DNA and maintain cell immortality, the regulation of the RNA component of telomerase is not well understood. In the course of investigations that have implicated telomerase RNA in key aspects of cancer progression, including metastasis, we explored some of the *cis*-acting elements affecting telomerase RNA expression and knockdown. The expression efficiency and subsequent RNA processing to produce the mature hTER differed considerably among various promoters. Together with other results, these findings establish that the crucial elements of the hTER gene affecting RNA-processing efficiency to produce the mature hTER RNA are the promoter and internal telomerase RNA-coding sequences.

THE RNA COMPONENT OF TELOMERASE: AN UNUSUAL NONCODING RNA

Elongation of human telomeres requires the enzymatic activity of the telomerase holoenzyme. In vitro, telomerase activity can be reconstituted by the catalytic subunit (hTERT) and the RNA subunit (called hTERC, hTER, or hTR). However, the routes of the RNA and protein components of telomerase from their generation to assembly into telomerase appear to be complex and likely to be highly regulated. The expression of hTERT is greatly diminished in most human adult somatic cells, with some exceptions, notably germ cells and stem cells. In contrast, the expression of hTER is readily detectable in a wide range of normal cells. The biological significance of this accumulation of telomerase RNA in human cells that lack significant levels of telomerase enzymatic activity is not well understood. Also still largely unknown are how the telomerase components are regulated, how the RNP telomerase enzyme complex is assembled, and how such assembly is controlled. We investigated the *cis*-acting elements of telomerase RNA gene constructs to explore how they affect the journey of telomerase RNA from gene to telomerase activity in human cells.

We analyzed expression and processing of human telomerase RNA in cultured transformed human cells, focusing on their dependence on *cis*-acting sequences in a variety of hTER gene constructs. Although this work was done in the course of experimenting with expression constructs designed with the goal of expressing human telomerase RNA efficiently in human tumor and other cells, the single-copy telomerase RNA gene is also an interesting small nuclear RNA (snRNA)-encoding gene for a number of reasons. First, in vertebrates, many small RNAs are encoded as multiple genes falling into gene families; single-copy snRNA genes are less common. Second, the transcriptional and processing machinery responsible for expression of telomerase RNA differs among different eukaryotic phyla. The transcription and expression of telomerase RNA have been characterized in diverse

species, including ciliates, budding yeasts, and humans (Feng et al. 1995). It has been shown that ciliate telomerase RNA is transcribed by RNA polymerase III (pol III) (Yu et al. 1990; Romero and Blackburn 1991) and that the ciliate telomerase RNA promoter functionally resembles that of the vertebrate U6/7SK promoters, and also lacks *cis*-acting control elements that lie within the RNA-coding sequence. In this way, the ciliate telomerase RNA promoter is unlike the promoters of the 5S RNA and tRNA genes transcribed by RNA pol III (Romero and Blackburn 1991; Hargrove et al. 1999). In contrast to ciliates, in budding yeast and vertebrates, telomerase RNA expression is driven by an RNA pol II promoter. Third, the processing of telomerase RNA transcripts to yield the mature telomerase RNA competent for assembly into telomerase enzyme is not well understood. Ciliate telomerase RNA has a binding site for a telomerase-specific La family protein that is involved in its biogenesis and stability (Witkin and Collins 2004). In contrast, the mature processed form of budding yeast telomerase RNA contains a binding site for the Sm protein complex (Seto et al. 1999), which is involved in the biogenesis of spliceosomal snRNAs. In budding yeast, the RNA pol-II-transcribed telomerase RNA gene products include a polyadenylated transcript. Although normally present in very low amounts, in various mutants, this polyadenylated form of telomerase RNA accumulates relative to the mature RNA (Chapon et al. 1997). In contrast to the yeast telomerase RNA, studies of human telomerase RNA suggest that its transcript is nonpolyadenylated (Feng et al. 1995). Furthermore, in contrast to both ciliate and budding yeast telomerase RNAs, hTER contains a 3′ nucleolar RNA-like box H/ACA domain (Mitchell et al. 1999). This box H/ACA domain is conserved among vertebrate telomerase RNAs, as is the presence of a small Cajal-body-specific RNA (scaRNA) sequence (Richard et al. 2003).

The detection of hTER in nucleoli (Wong et al. 2002) has further suggested that regulation of hTER biosynthesis and its incorporation into the telomerase RNP may be small nucleolar RNA (snoRNA)-like. Vertebrate snoRNAs are transcribed, and the mature snoRNA is processed, from a variety of genomic environments. The vertebrate snoRNAs that have been characterized the most extensively are encoded in intronic regions of snoRNA "host" genes, the transcripts of which often have a 5'-terminal oligopyrimidine sequence of unknown significance. Elements within the snoRNA-coding sequence itself direct snoRNA processing to release the mature snoRNA. In the case of the vertebrate box C/D snoRNAs, their processing occurs via exonucleolytic trimming, and interactions have been detected between the splicing of the host gene transcript and the snoRNA processing (Hirose and Steitz 2001; Hirose et al. 2003). A scaRNA sequence in an RNA directs that RNA to locate in Cajal bodies, nucleoplasmic bodies containing a variety of snRNAs and snoRNAs (Handwerger and Gall 2006). When its scaRNA sequence is mutated, hTER accumulates in the nucleolus (Richard et al. 2003). The expression of hTER thus may be regulated by a transcription complex characteristic of an snRNA or snoRNA promoter (Feng et al. 1995; Richard et al. 2003).

Recent work (see Kiss et al., Terns and Terns, both this volume) has identified certain Cajal body proteins that are specifically associated with hTER, as well as nucleolar proteins (Jady et al. 2004; Zhu et al. 2004). In cells lacking substantial telomerase enzymatic activity, the telomerase RNA that can accumulate in such cells is presumably stable enough to accumulate in the absence of expressed TERT protein, and potential candidates for its binding partners in these cells include the Cajal body proteins and the dyskerin complex (DKC) (Vulliamy et al. 2001; Hamma et al. 2005). Cytological analyses of telomerase components by methods that can detect labeled hTER or hTERT in cells suggest dynamic movement of telomerase RNA between different cellular compartments, including the nucleus and nucleolus (Jady et al. 2006; Tomlinson et al. 2006).

Understanding where hTER is located in its different cellular milieus, and what affects its maturation and stability, are important in order to experimentally manipulate human telomerase RNA. Therefore, we first examined the expression of human telomerase RNA from expression cassettes with various upstream and downstream sequences flanking the 451-nucleotide-long coding sequence of the mature hTER. In addition, we investigated some of the expression characteristics of ribozymes and small hairpin interfering RNA (shRNAi) constructs that affect their ability to knock down human telomerase RNA.

UPSTREAM, BUT NOT DOWNSTREAM, GENE SEQUENCES AFFECT THE EXPRESSION AND PROCESSING OF TELOMERASE RNA

We analyzed a set of hTER constructs engineered to test their utility for hTER expression in human cancer cells. The findings uncovered a number of interesting features of the telomerase RNA gene. These features are reminiscent of the expression control of other box H/ACA RNAs of vertebrates (Wang and Meier 2004).

Previous data suggested that the total amount of hTER accumulation in human tumor cells is regulated, since only low levels of hTER accumulated even when ectopically expressed from strong promoters in the context of an adenoviral, plasmid, or retroviral vector (Marusic et al. 1997; Kim et al. 2001). Furthermore, in previous work, when hTER gene transcripts were generated from retroviral or plasmid-borne cytomegalovirus (CMV) promoter-driven constructs, most of the transcript was apparently unprocessed, since the predominant product observed in tumor cell lines was the polyadenylated form of the transcript that had been produced using the polyadenylation site engineered into the construct (typically, the retroviral polyadenylation site or a viral SV40 polyadenylation site) (Kim et al. 2001). Previous data have indicated that such polyadenylated hTER transcripts are not assembled into the telomerase RNP (Feng et al. 1995; Mitchell et al. 1999). Therefore, first, we sought to identify an ideal promoter for telomerase RNA expression and correct processing by testing a variety of RNA pol II and RNA pol III promoters. The RNA pol II promoters tested were the CMV (viral), CAG, PGK (cellular protein-coding genes), and IU1 promoters. The IU1 promoter was made from one of the gene family members of the U1 snRNA gene family. U6 and tRNA RNA gene promoters were the pol III promoters chosen for testing (Fig. 1). Second, to determine the effects of downstream sequences on expression and processing efficiency, transcription termination sequences unique for IU1, U6, tRNA, and CMV were used in the expression cassettes as described previously (Bertrand et al. 1997). Flanked by these 5' and 3' sequences, two types of hTER gene constructs were tested: an hTER DNA fragment containing only the sequence of the mature 451-nucleotide hTER (nucleotides 1–451) and a 598-nucleotide long sequence including the 451-nucleotide hTER sequence plus the flanking 137 nucleotides of the 3' genomic sequence. Each of these hTER expression cassettes was engineered into a lentiviral expression vector, which was chosen for its high transduction capability and the stable and efficient expression of genes from the genomically integrated, proviral form of the lentivector-borne genes.

To monitor the expression of hTER under different promoters, we utilized the "VA13" cell line WI-38 VA13/2RA (SV40-immortalized WI-38). This SV40-transformed human fibroblast line has previously been shown to have no endogenous hTER or hTERT expression (Bryan et al. 1997). Addition of constructs expressing both these genes, hTERT and hTER, reconstitutes activity in these cells (Bryan et al. 1997). Thus, VA13 cells have served as a useful "blank slate" on which to build and study telomerase activity with engineered components in the context of a cell. We have previously shown that expression of hTER can result in stable accumulation of the hTER RNA at levels that are unchanged even when hTERT is co-overexpressed in the same cells (Li et al. 2004). Hence, the stability of hTER in VA13 cells is not dependent on the level of hTERT protein. The lentivirus carrying different hTER expression cassettes also carried a green fluorescent protein (GFP) expression element to monitor infection efficiencies (which typically produced more than 90% of the cells that were GFP-positive). The VA13 cells were infected with each lentiviral

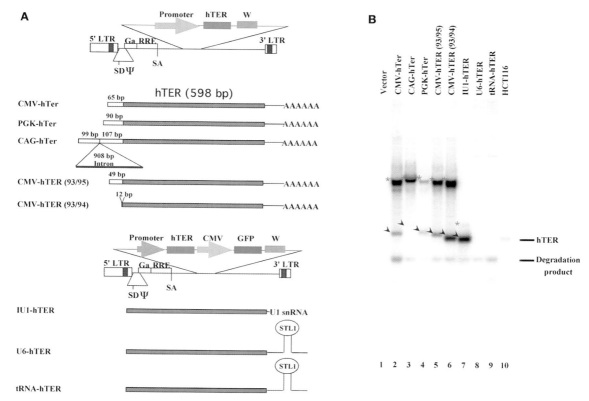

Figure 1. Expression of human telomerase RNA from lentiviral constructs with various promoters. (*A*) Lentiviral vectors: Only the relevant portion of the plasmid is illustrated. The hTER expression cassettes inserted into the lentivector were driven by various promoters as indicated. For the expression construct using the IU1, U6, and tRNA promoters, a GFP reporter expression cassette was also engineered to monitor the transduction efficiency. (W) Woodchuck hepatitis virus posttranslational regulatory element (Zufferey et al. 1999) placed at the 3′ end of GFP to enhance GFP expression. Schematic of hTER transcript from various expression cassettes: The 598-bp hTER fragment is shown (*red rectangle*). The additional sequence 5′ to hTER (*white rectangle*) is vector-derived sequence which follows the transcription initiation site. Note that the 3′ termination sequence is different due to the nature of different expression cassettes. Although a transcriptional termination site is found immediately downstream from the IU1-hTER cassette that is also utilized by RNA pol II, this termination site does not appear to cause early termination of the full-length lentiviral transcript, as a high-titer recombinant viral supernatant can be produced using these lentivectors. In contrast, when a polyadenylation signal is inserted into the viral genome in a forward orientation, the viral titer is dramatically reduced due to loss of full-length viral RNA transcript (Coffin et al. 1997). (*B*) Northern blotting analysis of telomerase RNA expressed from different promoters in VA13 cells 4 days after lentiviral infection. (*Lane 1*) Control virus; (*lane 2*) CMV-promoter-driven hTER; (*lane 3*) CAG promoter-driven hTER; (*lane 4*) PGK-promoter-driven hTER; (*lane 5*) CMV-promoter-driven hTER with shorter 5′ additional sequence; (*lane 6*) CMV-promoter-driven hTER with shortest 5′ additional sequence; (*lane 7*) IU1-promoter-driven hTER (598) and GFP reporter; (*lane 8*) U6-promoter-driven hTER and GFP reporter; (*lane 9*) tRNA-promoter-driven hTER and GFP reporter; (*lane 10*) RNA from HCT116 colon cancer cells. The asterisks indicate unprocessed hTER transcripts. The arrowheads indicate mature telomerase RNA transcripts.

construct, the unabsorbed virus was washed out of the medium after 8 hours, and at 4 days postinfection of the virus, total cellular RNA was prepared as described previously (Li et al. 2004). The expression of hTER was then detected and analyzed by northern hybridization.

Each of these RNA pol II promoter-derived expression cassettes resulted in readily detected expression of hTER, although the expression efficiency and subsequent RNA processing to produce the mature hTER differed considerably among the promoters. When hTER was expressed from the IU1 RNA pol II promoter, all of the transcript was correctly processed at the 3′ end. In contrast to hTER expressed from the IU1 promoter, the majority of hTER transcripts expressed from the CMV, CAG, and PGK promoters are polyadenylated, and only a small portion of the hTER transcript is properly processed to the 451-base mature form. Similar results were observed previously for

the CMV promoters in adenoviral plasmids and retroviral vectors (Marusic et al. 1997; Kim et al. 2001). These data suggested that transcription which directs the hTER transcript into the polyadenylation pathway may inhibit the maturation of hTER in vivo. In addition, as reported previously, the expression of endogenous and ectopically expressed WT-hTER in two human tumor cell lines—the melanoma line LOX and the bladder cancer line UM-UC3 cells—has been analyzed by northern hybridization (Li et al. 2004). Overexpression of this ectopically expressed WT-hTER was observed in both cell lines, suggesting that expression of telomerase RNA in human tumor cell lines using the IU1 promoter may bypass the regulation of total hTER accumulation. Such regulation had been suggested from previous studies in the human prostate cancer cell line LNCaP using an expression cassette with the viral CMV promoter and an SV40 polyadenylation site (Kim et al. 2001). In contrast to the pol II promoters, no hTER transcript was detectable when the U6 or tRNA promoter was used, despite delivery of the expression cassettes to more than 95% of cells, as indicated by their GFP positivity (data not shown). This result suggested that the hTER transcript derived from RNA pol III may not be stable, since the same promoter yields robust expression of ribozyme transcripts (Bertrand et al. 1997).

In addition, the dependence of the efficiency of 3'-end processing on the sequence at the 5' end of the transcript was tested. We kept the CMV promoter the same but varied the 5'-end sequence of the transcript. As shown in Figure 1 (lanes 2, 5, and 6), although the CMV construct with a 12-base insertion at the 5' end of the hTER-coding sequence was expressed at higher overall levels than that with a 49-base or 65-base insert, the ratio of the unprocessed full-length polyadenylated transcript to the processed mature-length hTER was not significantly altered (Fig. 1, lanes 2, 5, and 6). Therefore, the 5'-end sequence of the transcript did not affect the relative degrees of processing of the transcripts via polyadenylation versus 3'-end processing to form mature hTER. We next tested whether an intron in the transcript could improve the ratio of processed to unprocessed hTER transcript. The presence of an intron affects the processing pathway of mRNA transcripts; however, the CAG promoter construct did not result in high-level production to mature hTER (Fig. 1, lane 3). These findings suggest that the primary 5' determinant of relative processing of a transcript into mature hTER is the promoter, and not sequences within or downstream from hTER.

As previously reported (Li et al. 2004), two different hTER-containing fragments were engineered to be expressed from the IU1 promoter: The first one, IU1-hTER (598), contains mature hTER (nucleotides 1–451) and the flanking 137 nucleotides of 3' genomic sequence; the second one, IU1-hTER (451), contains only nucleotides 1–451 of the mature hTER. Expression levels of both hTER gene constructs were similar, and only the transcript corresponding to the mature hTER was observed in both cases (Li et al. 2004). To test whether hTER expressed from the IU1 promoter can be assembled into functional telomerase enzyme, in vitro TRAP (telomere repeat amplification protocol) assays for telomerase enzymatic activity were conducted on cell lysates from VA13 (hTert) cells infected with both of these types of hTER expression cassettes (598 or 451). Typical telomerase product ladders were detectable only in cell extracts expressing both hTERT and hTER. The results were similar with either the IU1-hTER (598) or IU1-hTER (451) expression cassette. Nucleotides 1–451, corresponding to the full-length hTER, thus appear to comprise sufficient RNA transcript information to direct hTER maturation and incorporation into a functional telomerase RNP enzyme complex (Li et al. 2004).

Finally, we also tested cassettes in which the 3'-flanking sequence immediately downstream from the coding sequence was a different sequence, unrelated to the natural hTER-flanking sequence that was used in the constructs listed above. There was no effect on the processing efficiency of varying this downstream sequence (results not shown). Thus, these results are consistent with the conclusion that the processing of the hTER transcript to produce the 3' end of the mature hTER is controlled by sequences in the coding region, consistent with the results for other vertebrate snoRNAs of the box H/ACA class (Wang and Meier 2004; Hamma et al. 2005).

A characteristic partial-length fragment of hTER was commonly seen in these ectopically expressed hTER transcripts. This fragment was mapped and found to consist of a 3'-end fragment of hTER, from which the 5' portion of the telomerase RNA, consisting of the universally conserved core of the telomerase RNA that contains the templating domain and conserved TER pseudoknot, has been removed. This 5' half of the hTER is apparently readily degraded, as it is detected only variably, and as an indistinct smear of fragments, in these experiments, leaving only the more stable 3' portion of the telomerase RNA. This remaining 3'-half fragment of hTER contains the previously identified conserved regions 4–7 of vertebrate telomerase RNA (Chen et al. 2000), which include the box H/ACA structure. Since box H/ACA snoRNAs bind the four-protein complex DKC core complex (Vulliamy et al. 2001; Hamma et al. 2005), this complex likely stabilizes this hTER fragment. A similar 3'-half degradation fragment was characterized in LNCaP human prostate cancer cells (Kim et al. 2001; and M. Kim and E.H. Blackburn, unpubl.). It is unknown whether this protected 3'-half fragment results from an endonucleolytic cleavage event that cuts off the universal core region, or from 5' to 3' exonuclease action that reaches a stopping point between the telomerase RNA universal conserved domain and the box H/ACA domain, or from combined endo- and exonucleolytic activities. Whether this consistently observed 3'-half fragment of hTER is a normal intermediate in a breakdown pathway of the endogenous telomerase RNA or has any other function is not known.

PROMOTERS DRIVING THE EXPRESSION OF RIBOZYMES OR siRNAs AFFECT THEIR EFFICIENCY IN KNOCKING DOWN HUMAN TELOMERASE RNA

For experimental purposes, we wanted to knock down telomerase RNA in human cells (Li et al. 2004). Therefore, we examined features of ribozymes or small interfering RNA gene constructs designed with the goal of optimal knockdown of hTER. A priori, the RNA knockdown agent must encounter its target RNA in the cell. However, unlike most RNAs that are commonly targeted by RNAi pathways, both experimentally and in natural settings, hTER is a noncoding, non-mRNA target. Furthermore, in contrast to the typical mRNA targets of RNA knockdown, generating the functional form of telomerase RNA in human telomerase does not appear to involve processing through the polyadenylation pathway. As described above, it is not known whether endogenous hTER is naturally subject to regulation via RNAi or by microRNA pathways. We have previously reported that the mouse telomerase RNA in a mouse melanoma cell line can be knocked down to approximately 20–30% of its normal level by hammerhead ribozyme constructs, with the ribozyme expression driven from a CMV promoter

(Nosrati et al. 2004). Similar results were observed in human cancer cell lines when ribozymes targeting human telomerase RNA were expressed from the U3B7 promoter (Bertrand et al. 1997) (data not shown). Interestingly, a similar level of knockdown was observed in these human cells regardless of whether these ribozymes were expressed from a tRNA, U6, or U3B7 promoter, even though each promoter is predicted to result in its transcript accumulating in different cellular compartments (Bertrand et al. 1997). In addition, we tested different shRNA constructs for their efficacy in knocking down the levels of hTER and telomerase activity. The shRNAs were expressed from various expression cassettes in lentiviral constructs. We found that hTER can be efficiently knocked down by an shRNA expressed from a U6 promoter (Li et al. 2004, 2005). These limited findings provisionally suggest that cellular localization of the ribozyme or shRNA transcripts may not be a major determinant of their effectiveness in hTER knock-down. In summary, telomerase RNA expression and function require trafficking within the cell via specific pathways, and further studies will be needed for a fuller understanding of the biogenesis of this RNA.

ACKNOWLEDGMENTS

The authors gratefully acknowledge support for this work provided by the UCSF Comprehensive Cancer Center Breast Cancer SPORE CA58207, grant CA96840 from the National Cancer Institute, and the Bernard Osher Foundation (to E.H.B.), and support from the Damon Runyon Cancer Research Foundation (to S.L.).

REFERENCES

Bertrand E., Castanotto D., Zhou C., Carbonnelle C., Lee N.S., Good P., Chatterjee S., Grange T., Pictet R., Kohn D., et al. 1997. The expression cassette determines the functional activity of ribozymes in mammalian cells by controlling their intracellular localization. *RNA* **3**: 75.

Bryan T.M., Marusic L., Bacchetti S., Namba M., and Reddel R.R. 1997. The telomere lengthening mechanism in telomerase-negative immortal human cells does not involve the telomerase RNA subunit. *Hum. Mol. Genet.* **6**: 921.

Chapon C., Cech T.R., and Zaug A.J. 1997. Polyadenylation of telomerase RNA in budding yeast. *RNA* **3**: 1337.

Chen J.L., Blasco M.A., and Greider C.W. 2000. Secondary structure of vertebrate telomerase RNA. *Cell* **100**: 503.

Coffin J.M., Hughes S.H., and Varmus H.E., eds. 1997. *Retroviruses*. Cold Spring Harbor Laboratory Press, Cold Spring Harbor, New York.

Feng J., Funk W.D., Wang S.S., Weinrich S.L., Avilion A.A., Chiu C.P., Adams R.R., Chang E., Allsopp R.C., Yu J., et al. 1995. The RNA component of human telomerase. *Science* **269**: 1236.

Hamma T., Reichow S.L., Varani G., and Ferre-D'Amare A.R. 2005. The Cbf5-Nop10 complex is a molecular bracket that organizes box H/ACA RNPs. *Nat. Struct. Mol. Biol.* **12**: 1101.

Handwerger K.E. and Gall J.G. 2006. Subnuclear organelles: New insights into form and function. *Trends Cell Biol.* **16**: 19.

Hargrove B.W., Bhattacharyya A., Domitrovich A.M., Kapler G.M., Kirk K., Shippen D.E., and Kunkel G.R. 1999. Identification of an essential proximal sequence element in the promoter of the telomerase RNA gene of *Tetrahymena thermophila*. *Nucleic Acids Res.* **27**: 4269.

Hirose T. and Steitz J.A. 2001. Position within the host intron is

critical for efficient processing of box C/D snoRNAs in mammalian cells. *Proc. Natl. Acad. Sci.* **98**: 12914.

Hirose T., Shu M.D., and Steitz J.A. 2003. Splicing-dependent and -independent modes of assembly for intron-encoded box C/D snoRNPs in mammalian cells. *Mol. Cell* **12**: 113.

Jady B.E., Bertrand E., and Kiss T. 2004. Human telomerase RNA and box H/ACA scaRNAs share a common Cajal body-specific localization signal. *J. Cell Biol.* **164**: 647.

Jady B.E., Richard P., Bertrand E., and Kiss T. 2006. Cell cycle-dependent recruitment of telomerase RNA and Cajal bodies to human telomeres. *Mol. Biol. Cell* **17**: 944.

Kim M.M., Rivera M.A., Botchkina I.L., Shalaby R., Thor A., and Blackburn E.H. 2001. A low threshold level of expression of mutant-template telomerase RNA is sufficient to inhibit human tumor cell growth. *Proc. Natl. Acad. Sci.* **98**: 7982.

Li S., Crothers J., Haqq C.M., and Blackburn E.H. 2005. Cellular and gene expression responses involved in the rapid growth inhibition of human cancer cells by RNA interference-mediated depletion of telomerase RNA. *J. Biol. Chem.* **280**: 23709.

Li S., Rosenberg J.E., Donjacour A.A., Botchkina I.L., Hom Y.K., Cunha G.R., and Blackburn E.H. 2004. Rapid inhibition of cancer cell growth induced by lentiviral delivery and expression of mutant-template telomerase RNA and anti-telomerase short-interfering RNA. *Cancer Res.* **64**: 4833.

Marusic L., Anton M., Tidy A., Wang P., Villeponteau B., and Bacchetti S. 1997. Reprogramming of telomerase by expression of mutant telomerase RNA template in human cells leads to altered telomeres that correlate with reduced cell viability. *Mol. Cell. Biol.* **17**: 6394.

Mitchell J.R., Cheng J., and Collins, K. 1999. A box H/ACA small nucleolar RNA-like domain at the human telomerase RNA 3' end. *Mol. Cell. Biol.* **19**: 567.

Nosrati M., Li S., Bagheri S., Ginzinger D., Blackburn E.H., Debs R.J., and Kashani-Sabet M. 2004. Antitumor activity of systemically delivered ribozymes targeting murine telomerase RNA. *Clin. Cancer Res.* **10**: 4983.

Richard P., Darzacq X., Bertrand E., Jady B.E., Verheggen C., and Kiss T. 2003. A common sequence motif determines the Cajal body-specific localization of box H/ACA scaRNAs. *EMBO J.* **22**: 4283.

Romero D.P. and Blackburn E.H. 1991. A conserved secondary structure for telomerase RNA. *Cell* **67**: 343.

Seto A.G., Zaug A.J., Sobel S.G., Wolin S.L., and Cech T.R. 1999. *Saccharomyces cerevisiae* telomerase is an Sm small nuclear ribonucleoprotein particle. *Nature* **401**: 177.

Tomlinson R.L., Ziegler T.D., Supakorndej T., Terns R.M., and Terns M.P. 2006. Cell cycle-regulated trafficking of human telomerase to telomeres. *Mol. Biol. Cell* **17**: 955.

Vulliamy T., Marrone A., Goldman F., Dearlove A., Bessler M., Mason P.J., and Dokal I. 2001. The RNA component of telomerase is mutated in autosomal dominant dyskeratosis congenita. *Nature* **413**: 432.

Wang C. and Meier U.T. 2004. Architecture and assembly of mammalian H/ACA small nucleolar and telomerase ribonucleoproteins. *EMBO J.* **23**: 1857.

Witkin K.L. and Collins K. 2004. Holoenzyme proteins required for the physiological assembly and activity of telomerase. *Genes Dev.* **18**: 1107.

Wong J.M., Kusdra L., and Collins K. 2002. Subnuclear shuttling of human telomerase induced by transformation and DNA damage. *Nat. Cell Biol.* **4**: 731.

Yu G.L., Bradley J.D., Attardi L.D., and Blackburn E.H. 1990. In vivo alteration of telomere sequences and senescence caused by mutated *Tetrahymena* telomerase RNAs. *Nature* **344**: 126.

Zhu Y., Tomlinson R.L., Lukowiak A.A., Terns R.M., and Terns M.P. 2004. Telomerase RNA accumulates in Cajal bodies in human cancer cells. *Mol. Biol. Cell* **15**: 81.

Zufferey R., Donello J.E., Trono D., and Hope T.J. 1999. Woodchuck hepatitis virus posttranscriptional regulatory element enhances expression of transgenes delivered by retroviral vectors. *J. Virol.* **73**: 2886.

RNA as a Flexible Scaffold for Proteins: Yeast Telomerase and Beyond

D.C. Zappulla and T.R. Cech

Howard Hughes Medical Institute, Department of Chemistry and Biochemistry, University of Colorado, Boulder, Colorado 80309

Yeast telomerase, the enzyme that adds a repeated DNA sequence to the ends of the chromosomes, consists of a 1157-nucleotide RNA (TLC1) plus several protein subunits: the telomerase reverse transcriptase Est2p, the regulatory subunit Est1p, the nonhomologous end-joining heterodimer Ku, and the seven Sm proteins involved in ribonucleoprotein (RNP) maturation. The RNA subunit provides the template for telomeric DNA synthesis. In addition, we have reported evidence that it serves as a flexible scaffold to tether the proteins into the complex. More generally, we consider the possibility that RNPs may be considered in three structural categories: (1) those that have specific structures determined in large part by the RNA, including RNase P, other ribozyme–protein complexes, and the ribosome; (2) those that have specific structures determined in large part by proteins, including many small nuclear RNPs (snRNPs) and small nucleolar RNPs (snoRNPs); and (3) flexible scaffolds, with no specific structure of the RNP as a whole, as exemplified by yeast telomerase. Other candidates for flexible scaffold structures are other telomerases, viral IRES (internal ribosome entry site) elements, tmRNA (transfer-messenger RNA), the SRP (signal recognition particle), and *Xist* and *roX1* RNAs that alter chromatin structure to achieve dosage compensation.

The ends of linear chromosomes cannot be replicated completely by the same enzymatic machinery that replicates the internal portions. In the absence of a special replication mechanism, chromosomes literally shrink from their ends. This DNA end-replication problem is overcome in most eukaryotes by the enzyme telomerase.

Telomerase is an RNP enzyme, and the function of one portion of its RNA subunit is gratifyingly easy to understand: It provides the template for extension of telomeric DNA (Greider and Blackburn 1989; Yu et al. 1990). The RNA itself performs other functions, the best-established example being the formation of the "template boundary," a short intramolecular helix that defines the stopping point for reverse transcription (Tzfati et al. 2000). Finally, the RNA binds a number of proteins, foremost among these being the telomerase reverse transcriptase (TERT, or Est2p in *Saccharomyces cerevisiae*). The RNA also binds a number of accessory proteins, but these are more species-specific. In yeast, they consist of Est1p, which recruits or activates telomerase at the chromosome end; the Ku heterodimer, the DNA-repair protein that has a special role in recruiting telomerase to broken chromosome ends for de novo telomere addition; and the Sm proteins, which also contribute to biogenesis of the snRNPs. Yet the function, or absence of function, of the majority of the telomerase RNA remains unknown.

At an even more fundamental level, we still do not even know how to envision the structure of the telomerase RNP. Is it a more-or-less specific three-dimensional structure formed by a precise series of RNA–RNA, RNA–protein, and protein–protein interactions, along the lines of the ribosome or RNase P, such that it could someday be crystallized? Alternatively, is it possibly a loose collection of proteins held together by RNA "strings" such that no two complexes have an identical overall shape? In the case of yeast telomerase, we describe evidence that supports the latter model: RNA as a flexible scaffold for proteins. Our interrogation of yeast telomerase has stimulated us to think more generally about the structure–function relationships in RNPs, especially those involved in catalysis, and we present those thoughts in the Conclusions.

YEAST TELOMERASE RNA SECONDARY STRUCTURE

For noncoding RNAs, knowing the nucleotide sequence is of very limited value by itself, but having a good secondary structure model provides a useful framework for investigating structure–function relationships. In the case of telomerase RNAs, secondary structures were determined for ciliated protozoan examples and showed a single-stranded template region, a nearby pseudoknot, and a long "handle" (Stem IV) whose terminal loop is essential for function (Romero and Blackburn 1991; ten Dam et al. 1991; Lingner et al. 1994; Sperger and Cech 2001). Vertebrate telomerase RNAs folded into a structure somewhat similar to that of the ciliates, followed by a 3′-terminal pair of stem-loop structures characteristic of box H/ACA snoRNAs (Mitchell et al. 1999; Chen et al. 2000).

Budding yeast telomerase RNA structures were difficult to decipher due to their unusual length (typically >1000 nucleotides) and rapid divergence, which frustrates sequence alignments and makes it difficult to apply comparative sequence analysis. The Blackburn laboratory was able to model several helices of telomerase RNAs from *Kluyveromyces* species (Tzfati et al. 2003). Furthermore, for *S. cerevisiae* telomerase RNA (TLC1), several local structures involved in binding of the Ku and Est1 proteins were established (Peterson et al. 2001; Seto et al. 2002), and a single-stranded U-rich sequence involved in Sm protein binding was identified (Seto et al. 1999). Finally, in 2004, two laboratories converged on the same secondary structure model for TLC1 RNA, as shown in Figure 1A (Dandjinou et al. 2004; Zappulla and Cech 2004).

The three long arms of this secondary structure model are unusual features for a noncoding RNA; overall, the structure looks completely different from canonical structures of rRNA, group I and group II introns, RNase P RNA, tmRNA, etc., which are highly branched structures. Furthermore, each of the long RNA arms has near its end a binding site for one of the known accessory proteins: Ku, Est1p, or Sm. The three long arms consist of double-helical regions interrupted by internal loops and bulges, features known to provide bendable joints for RNA in solution. The elements more directly involved in

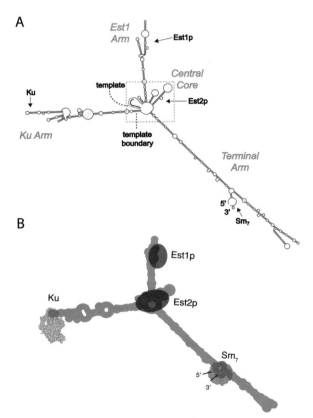

catalysis—Est2p, the RNA template, and the template boundary element—are brought together in the central core of the structure. When the protein and RNA components are represented more or less according to predicted scale, telomerase appears as small protein beads separated by long RNA arms (Fig. 1B).

FLEXIBLE SCAFFOLD MODEL

The model that the yeast telomerase RNA provides a flexible scaffold for protein binding was derived from several observations. First, large stretches of the RNA could be deleted without significant loss of function in vivo (Livengood et al. 2002). Second, most of the RNA is evolving very rapidly, as evidenced by the divergent sequences of telomerase RNAs from species in the genus *Kluyveromyces* (Tzfati et al. 2000; Seto et al. 2002) or the genus *Saccharomyces*. In fact, TLC1 RNA has only 43% sequence identity between four *Saccharomyces* species examined, compared to 82–99% for other noncoding RNAs (U1, RNase P, and 18S rRNA) and 92% for the actin mRNA open reading frame (Zappulla and Cech 2004). These observations indicate that the RNA is "flexible," meaning that it readily accommodates structural perturbations. It is likely that the RNA is also "flexible" in the sense that its long quasi-helical arms are free to twist, turn, and move in many directions, but this sort of structural flexibility is not directly addressed by our data.

TESTING THE FLEXIBLE SCAFFOLD MODEL

To provide a strong test of the flexible scaffold model, we mutated the natural binding site for the Est1p regulatory subunit on the RNA and then introduced a functional Est1p-binding element (nucleotides 524–704) at position 220, 450, or 1033 (Fig. 2A). Since mutation of the Est1p-binding site in TLC1 leads to senescence, if the repositioned Est1p site restores the essential function of Est1p in telomerase, the cell should become viable. Not only did these variant TLC1 RNAs with repositioned Est1p sites provide robust cell growth, but they maintained near wild-type telomere lengths (Zappulla and Cech 2004). Among other controls, a mutated version of the Est1p-binding site inserted at any of the same three unnatural locations did not rescue telomerase activity. Thus, it appears that as long as Est1p is tethered somewhere to the RNA, it is able to perform its function in recruiting telomerase to its site of action at the very end of the chromosome.

Given that yeast telomerase appeared to consist of proteins tethered by flexible RNA arms, we next asked whether the arm length was important. To avoid creating misfolded RNAs, the RNA secondary structure model was used to design pairs of deletions that would remove both sides of an RNA stem. First, a major portion of the terminal arm was deleted, and then deletions were made in all three major arms; in each case, the shortened "Mini-T" RNA (Fig. 2B) complemented a *tlc1Δ* mutant and telomere length was maintained at a stably shorter length. However, either the length or some other feature of the RNA arms was important for normal levels of RNA accumulation, as the Mini-T strains had substantially reduced

Figure 1. Secondary structure model of yeast telomerase RNA including protein-binding sites. (*A*) RNA structure model of Zappulla and Cech (2004), based on computer-generated RNA stability calculations and limited comparative sequence analysis and mutagenesis. Dandjinou et al. (2004) independently proposed a highly similar model. (*B*) Schematic model of the RNP with the RNA (*green*) shown approximately to scale with the proteins whose structures have been determined. Est1p and Est2p, whose detailed structures are unknown, are indicated as ovals with dimensions expected for proteins of these molecular weights. The RNA arms are shown in an extended conformation, but because the helices are interrupted by internal loops and bulges, they are expected to be highly flexible and to assume multiple conformations. Crystal structures of human Ku heterodimer (Walker et al. 2001; PDB ID 1JEQ) and archaeal Sm7 (Mura et al. 2001; PDB ID 1I8F) are shown.

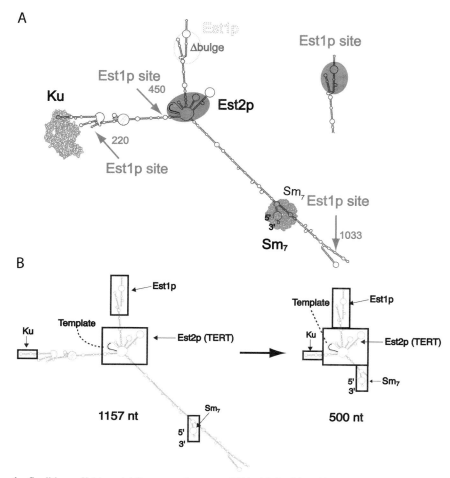

Figure 2. Testing the flexible scaffold model for yeast telomerase RNA. (*A*) Positional independence of Est1p-binding to telomerase RNA. The regulatory subunit Est1p can be relocated to any of three unnatural positions in the RNA (*red*) with retention of telomerase activity (Zappulla and Cech 2004). In these experiments, the natural Est1p-binding site (*blue*) was inactivated by deletion of a critical bulge structure. (*B*) Distance flexibility. Mini-T(500) reduces the amount of RNA separating the various protein-binding sites (*boxed*) with retention of function in vivo (Zappulla et al. 2005).

levels of TLC1 RNA. When Mini-T(500), the 500-nucleotide version of Mini-T, was integrated into its normal locus in the yeast genome, the RNA expression level was only about 5% of the wild-type level.

Mixing the Mini-T(500) yeast cells with wild type and carrying out competitive growth in liquid culture revealed a large fitness disadvantage of Mini-T, about 22% per generation (Fig. 3). However, when the same RNA was expressed from a low-copy-number centromere-containing plasmid, which gives an expression level 11–16% of wild type, the fitness was much better: only about 3% selective disadvantage per generation (Zappulla et al. 2005). Thus, it appears that most if not all of the selective disadvantage of Mini-T(500) is due to its reduced expression level rather than some functional defect. It may be that the long arms of the natural TLC1 RNA provide binding sites for general RNA-binding proteins that promote nuclear retention or RNA stability while not contributing directly to telomerase activity.

Recent work by our colleague Dr. Amy Mozdy helps illuminate why reduced TLC1 expression has such a major effect on telomere length. Wild-type haploid yeast

contain a steady-state number of about 30 TLC1 RNA molecules (Mozdy and Cech 2006). This can be compared to 32 telomeres per cell at the beginning of S phase and 64 telomeres per cell in late S phase when telomeres have been replicated. Thus, telomeres outnumber telomerase even in wild-type cells. The 5% expression level of integrated Mini-T(500) would correspond to about one molecule of TLC1 per cell. Thus, it is not surprising that the telomeres are maintained at an average length much shorter than wild type (Fig. 3A).

CONCLUSIONS

Yeast Telomerase

We have presented evidence that yeast telomerase RNA provides a flexible scaffold for tethering proteins into the complex. By "flexible," we mean that there are only loose constraints on the length, sequence, and relative orientation of the RNA arms that tether the Est1p, Ku, and Sm protein components. Given the structure of the RNA arms—short helices separated by internal loops and

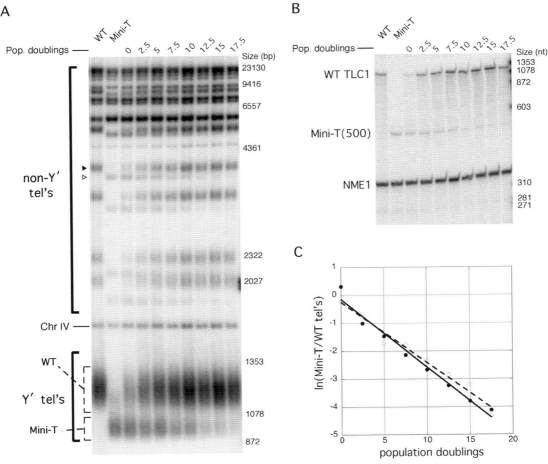

Figure 3. Yeast with Mini-T(500) integrated in the genome, replacing the wild-type *TLC1* gene, maintain shorter telomeres and show reduced fitness upon competitive growth. (*A*) Southern hybridization with a telomeric DNA probe shows the length distribution of wild-type (WT) and Mini-T telomeres. About half (17 of 32) of the yeast telomeres contain the repeated Y′ sequence just internal to the telomeric repeats, such that they are cleaved by XhoI restriction endonuclease to give a low-molecular-weight set of fragments. Non-Y′ telomeres form individual bands at higher molecular weights; one example is highlighted by a closed triangle (wild-type) and an open triangle (Mini-T). Upon continued growth in liquid medium, the cells containing WT TLC1 outcompete the Mini-T cells and take over the culture. (*B*) Northern hybridization with a probe to the region of TLC1 RNA that is shared between the WT RNA and Mini-T(500). Again, upon continued growth, the cells containing WT TLC1 outcompete the Mini-T cells. (*C*) Quantitation of the data in panel *A* and a repeat experiment (primary data not shown) indicate a fitness disadvantage of 22% per generation for Mini-T cells. Most of this fitness defect was due to reduced levels of the Mini-T(500) RNA (see text).

bulges (Fig. 1A)—they must also be flexible in the sense of allowing bending, rotation, and other dynamics. We have not tested the importance of this sort of flexibility, which could probably be done by site-specific mutagenesis, to eliminate the internal loops and bulges, followed by tests for activity (see, e.g., Nakamura et al. 1995).

We speculate that functional yeast telomerase might not even require the arms to be made of RNA. If the only purpose of the arms is to tether the proteins in a flexible manner, then the RNA arms might be replaced by protein α helices or loops, or even by polyethylene glycol. An interesting comparison concerns the scaffold proteins involved in cell signaling, which bind multiple protein kinases. Wendell Lim has asked exactly the same question of these scaffold proteins as we have asked of the telomerase RNA: "Do they simply tether components, or do they precisely orient and activate them?" (Park et al. 2003). These authors found that the yeast MAP kinase

scaffold protein Ste5 survived major rearrangements, supporting the former model: "Scaffolds are highly flexible organizing factors."

Even regarding yeast telomerase RNA, much remains unknown. The "flexible scaffold" model may break down near the active site, where it would not be surprising if the Est2p catalytic subunit had to be oriented in a specific manner relative to the RNA template that it reverse-transcribes. However, even in the central core, Chappell and Lundblad (2004) have described a collection of mutants that do not disrupt function, and they even replaced the yeast sequences in the core with the RNA pseudoknot from the ciliated protozoan *Oxytricha nova* and obtained a functional RNA. Thus, the flexible scaffold model may extend into the active-site region.

One possible explanation for the telomerase RNA of *Saccharomyces* having a structural organization different from those of other eukaryotes is the absence of RNA interference (RNAi) in the budding yeasts. Even though the long RNA arms of TLC1 are not perfect double helices, they may be close enough to attract dicer or other elements of the RNAi machinery. Thus, budding yeast telomerase RNA might have evolved without the constraint of avoiding large semihelical extensions. It is noteworthy that both U1 and U2 snRNAs from *S. cerevisiae* are also much larger than their mammalian counterparts, and, like TLC1, can be substantially reduced in size with retention of function (Igel and Ares 1988; Shuster and Guthrie 1988; Siliciano et al. 1991). The RNase P RNAs from the yeast *Candida glabrata* and the fungus *Phanerochaete chrysosporium* are also unusually large (Kachouri et al. 2005; Piccinelli et al. 2005). This supports the idea that in some yeast and other fungi, there is either selective advantage to larger size RNAs or else the absence of a selective disadvantage.

Three Classes of RNPs

In considering RNPs more generally, we quickly come to the realization that the existing data are insufficient to prove or disprove many of them as being flexible scaffold structures, in the sense that much of their RNA framework can accommodate insertions, deletions, and base changes without disruption of function. In some cases, the "accommodativeness" or malleability of the RNA can be inferred from comparative sequence analysis of functionally equivalent RNAs from divergent species. In addition, in many cases, we know something about the structural flexibility of the RNP—whether each particle has more or less the same arrangement of protein and RNA components, albeit with some conformational switching, or whether, on the other hand, proteins are loosely tethered into the complex. Here, we make the provisional assumption that structural flexibility of the RNP correlates with the ability of the RNA to accommodate insertions, deletions, and base changes in regions that serve as tethers; but in doing so, we realize that there may be cases where these features diverge. On the basis of these assumptions, we propose that RNPs may fit into three general structural categories, as follows:

1. ***RNA-determined RNP structures.*** Specific structure (or family of structures), determined in large part by the folded RNA.
   Examples: Ribosome, RNase P, and other ribozyme–protein complexes such as the yeast mitochondrial group I intron bI5 bound to Cbp2 protein, or a number of *Neurospora* group I introns bound to CYT-18 protein.
   Evidence: The RNP, and in some cases the RNA, can be crystallized. Individual ribosomal proteins often have disordered tails or internal loops, and these become ordered only upon RNP formation, so at least for the ribosome, the protein does not provide a pre-ordered template upon which the RNA folds (Brodersen et al. 2002; Klein et al. 2004). Not only do the ribosomal RNAs have conserved secondary

structures, but there are bases throughout the structures that are universally conserved among the corresponding rRNAs of all three primary kingdoms of life (Noller 1993). In *Saccharomyces* species, the sequence conservation of the rRNA far exceeds that of the telomerase RNA (99% and 43% sequence identity, respectively) (Zappulla and Cech 2004).

2. ***Protein-determined RNP structures.*** Specific structure (or family of structures), determined in large part by protein–protein interactions.
   Examples: U1, U2, U11/U12, and U5 snRNPs, box C/D, and box H/ACA snoRNPs.
   Evidence for specific structure: High-resolution cryo-electron microscopy (cryo-EM) structures of U1 and U11/U12 (Stark et al. 2001; Golas et al. 2005).
   Evidence for structure being protein-based: The majority of U2-specific proteins form stable heteromeric complexes in absence of U2 snRNA (Will and Luhrmann 2006), U5 proteins assemble in absence of snRNA (Achsel et al. 1998), and the crystal structure of Cbf5–Nop10–Gar1 complex has been determined in the absence of snoRNA (Rashid et al. 2006).

3. ***Flexible scaffold RNPs.*** Proteins are tethered into the complex by a flexible RNA scaffold, with no specific structure of the RNP as a whole.
   Example: Yeast telomerase, other candidates discussed below.
   Evidence for flexibility: Phylogenetic variation in sequence even among closely related *Saccharomyces* species, ability to delete large portions of stems and still retain function, and ability to transplant Est1p-binding element to new locations with function.
   Evidence against specific structure: The RNA does not have the properties one would expect for a specific tertiary structure, so structure would have to come from the protein; but protein–protein interactions are not strong, unlike snRNPs and snoRNPs above.
   Prediction: The RNP would not crystallize, and would be heterogeneous by EM.

Telomerases from humans and from ciliated protozoa have been extensively studied, their RNA secondary structures are well established (Romero and Blackburn 1991; Chen et al. 2000), and a specific tertiary structure has been proposed for the core of the human RNA (Theimer et al. 2005). We find it attractive to think that these much smaller telomerase RNAs may be natural Mini-T RNAs, with the protein-binding arms reduced to a minimal length so that it is now difficult to even recognize the arms. If this is the case, it should be possible to add back flexible RNA arms to separate various functional elements of these other telomerase RNAs without perturbing activity. On the other hand, these other telomerases may be assembled with protein–protein interactions having replaced some of the RNA–protein interactions seen in yeast telomerase, in which case, the RNP may have a defined three-dimensional structure.

IRES elements that permit 5′-cap-independent initiation of protein synthesis of certain viral and cellular mRNAs provide an interesting case. They bind multiple protein complexes, including the 12-protein initiation factor eIF3, as well as the small ribosomal subunit. The complex between the hepatitis C virus IRES RNA and eIF3 has some conformational flexibility, as determined by cryo-EM reconstructions (Siridechadilok et al. 2005). Both the IRES RNA and the ribosomal subunits undergo further conformational changes as translation is initiated and commences (Spahn et al. 2001; Boehringer et al. 2005). Thus, the hepatitis C IRES may be a flexible collection of proteins that becomes conformationally fixed upon binding to the ribosome. All IRESs are not identical, however. In the case of the cricket paralysis virus IRES, the RNA by itself forms a highly compact structure that interacts directly with the ribosome to achieve factor-independent translational initiation (Batey 2006).

tmRNA, named because it combines the functions of tRNA and mRNA, rescues stalled ribosomes by switching translation from a damaged mRNA to a sequence internal to the tmRNA. The tmRNA open reading frame encodes a protein degradation tag. The RNA binds several proteins, although not necessarily at the same time. These include the core subunit SmpB, EF-Tu, to recruit the RNP to the stalled ribosome, a ribonuclease that may contribute to degradation of the damaged mRNA, and several other proteins (Karzai and Sauer 2001). Certainly, it appears that the RNP must be conformationally flexible to carry out its function (Haebel et al. 2004). It is premature to know whether or not the RNP switches between discrete tertiary structures (Burks et al. 2005) or instead has no specific tertiary structure, in which case it would qualify as a flexible scaffold.

The SRP (signal recognition particle) is an RNP that recognizes the signal sequence on a nascent secretory or membrane protein as it exits the ribosome, arrests further translation, and then targets the protein to a membrane-associated receptor. The protein is then translocated into the endoplasmic reticulum or through the bacterial inner membrane. The SRP RNA has an S domain, which binds proteins that recognize the peptide signal sequence, and an Alu domain, which binds to the ribosome and causes translational arrest. These two domains are separated by a flexible hinge region comprising double-stranded RNA interspersed with internal loops and bulges (Egea et al. 2005), much like the yeast telomerase RNA "arms." The SRP may therefore be considered to be a flexible scaffold that tethers the S-domain RNA–protein complex to the Alu RNA or RNP element that binds to the ribosome. The SRP then undergoes an induced-fit conformational change and forms a discrete structure when it binds to the translating ribosome (Halic et al. 2004). The *Saccharomyces* SRP RNA is considerably longer than those of other species because of large quasi-helical extensions of the Alu domain (Van Nues and Brown 2004).

Noncoding RNAs involved in sex chromosome gene-dosage compensation may be thought of as flexible scaffolds which tether proteins that affect the transcriptional state (Wutz 2003). In mammals, the *Xist* RNA promotes heterochromatization of one of the female X chromosomes, whereas in *Drosophila*, the *roX1* and *roX2* RNAs increase transcription of the single male X chromosome.

Different domains of *Xist* RNA are responsible for transcriptional repression and association with chromatin, the latter accomplished by functionally redundant sequences dispersed through the RNA (Wutz et al. 2002). Deletion analysis has shown that multiple 10% segments of *roX1* RNA are dispensable for function, consistent with the RNA providing a flexible scaffold to bind multiple MSL (male-specific lethal) protein complexes (Stuckenholz et al. 2003).

Another intriguing candidate for a flexible scaffold is the NRON (noncoding repressor of NFAT) recently identified as a repressor of a transcription factor, NFAT (nuclear factor of activated T cells) (Willingham et al. 2005). NRON binds a number of proteins, including members of the importin β family, and appears to act as a repressor of the transcription factor NFAT. It will be important to understand something of its structure and the spatial arrangement of its protein-binding sites.

It may seem surprising that we have placed yeast telomerase into a different category from the snoRNPs. Superficially at least, they have similar features: In all cases, the RNA provides a "guide sequence" that base-pairs to the site of a chemical reaction in a nucleic acid substrate. The reaction is 2′-O-methylation in the case of the box C/D snoRNPs (Kiss-Laszlo et al. 1996), conversion of U to pseudouracil in the case of the box H/ACA snoRNPs (Ganot et al. 1997), and nucleotide addition to single-stranded DNA in the case of telomerase. In all cases, the RNA also binds a protein enzyme that catalyzes the reaction: fibrillarin or Nop1p in the case of 2′-O-methylation (Tollervey et al. 1993; Wang et al. 2000; Omer et al. 2002), Nap57/dyskerin or Cbf5p in the case of pseudo-uridylation (Zebarjadian et al. 1999), and TERT/Est2p in the case of telomerase (Lingner et al. 1997). Certainly, there is some flexibility in the box H/ACA snoRNP system: The protein trimer can accommodate about 100 different snoRNAs (Meier 2006). We place the snoRNPs in a different category from yeast telomerase because the former form a specific protein structure in the absence of the RNA, whereas stable protein–protein interactions have not been observed between Est2p and Est1p, Ku, or Sm in yeast. Time will tell whether this is a useful distinction or whether the apparent differences arise in large part because of the primitive state of our current knowledge.

ACKNOWLEDGMENTS

We thank Karen Goodrich for excellent technical assistance; Rob Batey, Quentin Vicens, and Art Zaug for helpful discussions; and Anne Stellwagen (Boston College) for an ongoing collaboration on the Ku protein. This work was supported in part by a grant from the National Institutes of Health.

REFERENCES

Achsel T., Ahrens, K., Brahms H., Teigelkamp S., and Luhrmann R. 1998. The human U5-220kD protein (hPrp8) forms a stable RNA-free complex with several U5-specific proteins, including an RNA unwindase, a homologue of ribosomal elongation factor EF-2, and a novel WD-40 protein. *Mol. Cell. Biol.* **18:** 6756.

Batey R.T. 2006. Structures of regulatory elements in mRNAs. *Curr. Opin. Struct. Biol.* **16**: 299.

Boehringer D., Thermann R., Ostareck-Lederer A., Lewis J.D., and Stark H. 2005. Structure of the hepatitis C virus IRES bound to the human 80S ribosome: Remodeling of the HCV IRES. *Structure* **13**: 1695.

Brodersen D.E., Clemons W.M., Jr., Carter A.P., Wimberly B.T., and Ramakrishnan V. 2002. Crystal structure of the 30 S ribosomal subunit from *Thermus thermophilus:* Structure of the proteins and their interactions with 16 S RNA. *J. Mol. Biol.* **316**: 725.

Burks J., Zwieb C., Muller F., Wower I., and Wower J. 2005. Comparative 3-D modeling of tmRNA. *BMC Mol. Biol.* **6**: 14.

Chappell A.S. and Lundblad V. 2004. Structural elements required for association of the *Saccharomyces cerevisiae* telomerase RNA with the Est2 reverse transcriptase. *Mol. Cell. Biol.* **24**: 7720.

Chen J.L., Blasco M.A., and Greider C.W. 2000. Secondary structure of vertebrate telomerase RNA. *Cell* **100**: 503.

Dandjinou A.T., Levesque N., Larose S., Lucier J.F., Abou Elela S., and Wellinger R.J. 2004. A phylogenetically based secondary structure for the yeast telomerase RNA. *Curr. Biol.* **14**: 1148.

Egea P.F., Stroud R.M., and Walter P. 2005. Targeting proteins to membranes: Structure of the signal recognition particle. *Curr. Opin. Struct. Biol.* **15**: 213.

Ganot P., Bortolin M.L., and Kiss T. 1997. Site-specific pseudouridine formation in preribosomal RNA is guided by small nucleolar RNAs. *Cell* **89**: 799.

Golas M.M., Sander B., Will C.L., Luhrmann R., and Stark H. 2005. Major conformational change in the complex SF3b upon integration into the spliceosomal U11/U12 di-snRNP as revealed by electron cryomicroscopy. *Mol. Cell* **17**: 869.

Greider C.W. and Blackburn E.H. 1989. A telomeric sequence in the RNA of *Tetrahymena* telomerase required for telomere repeat synthesis. *Nature* **337**: 331.

Haebel P.W., Gutmann S., and Ban N. 2004. Dial tm for rescue: tmRNA engages ribosomes stalled on defective mRNAs. *Curr. Opin. Struct. Biol.* **14**: 58.

Halic M., Becker T., Pool M.R., Spahn C.M., Grassucci R.A., Frank J., and Beckmann R. 2004. Structure of the signal recognition particle interacting with the elongation-arrested ribosome. *Nature* **427**: 808.

Igel A.H. and Ares M., Jr. 1988. Internal sequences that distinguish yeast from metazoan U2 snRNA are unnecessary for pre-mRNA splicing. *Nature* **334**: 450.

Kachouri R., Stribinskis V., Zhu Y., Ramos K.S., Westhof E., and Li Y. 2005. A surprisingly large RNase P RNA in *Candida glabrata. RNA* **11**: 1064.

Karzai A.W. and Sauer R.T. 2001. Protein factors associated with the SsrA•SmpB tagging and ribosome rescue complex. *Proc. Natl. Acad. Sci.* **98**: 3040.

Kiss-Laszlo Z., Henry Y., Bachellerie J.P., Caizergues-Ferrer M., and Kiss T. 1996. Site-specific ribose methylation of preribosomal RNA: A novel function for small nucleolar RNAs. *Cell* **85**: 1077.

Klein D.J., Moore P.B., and Steitz T.A. 2004. The roles of ribosomal proteins in the structure assembly, and evolution of the large ribosomal subunit. *J. Mol. Biol.* **340**: 141.

Lingner J., Hendrick L.L., and Cech T.R. 1994. Telomerase RNAs of different ciliates have a common secondary structure and a permuted template. *Genes Dev.* **8**: 1984.

Lingner J., Hughes T.R., Shevchenko A., Mann M., Lundblad V., and Cech T.R. 1997. Reverse transcriptase motifs in the catalytic subunit of telomerase. *Science* **276**: 561.

Livengood A.J., Zaug A.J., and Cech T.R. 2002. Essential regions of *Saccharomyces cerevisiae* telomerase RNA: Separate elements for Est1p and Est2p interaction. *Mol. Cell. Biol.* **22**: 2366.

Meier U.T. 2006. How a single protein complex accommodates many different H/ACA RNAs. *Trends Biochem. Sci.* **31**: 311.

Mitchell J.R., Cheng J., and Collins K. 1999. A box H/ACA small nucleolar RNA-like domain at the human telomerase RNA 3′ end. *Mol. Cell. Biol.* **19**: 567.

Mozdy A.D. and Cech T.R. 2006. Low abundance of telomerase in yeast: Implications for telomerase haploinsufficiency. *RNA* **12**: 1721.

Mura C., Cascio D., Sawaya M.R., and Eisenberg D.S. 2001. The crystal structure of a heptameric archaeal Sm protein: Implications for the eukaryotic snRNP core. *Proc. Natl. Acad. Sci.* **98**: 5532.

Nakamura T.M., Wang Y.H., Zaug A.J., Griffith J.D., and Cech T.R. 1995. Relative orientation of RNA helices in a group 1 ribozyme determined by helix extension electron microscopy. *EMBO J.* **14**: 4849.

Noller H.F. 1993. On the origin of the ribosome: Coevolution of subdomains of tRNA and rRNA. In *The RNA world* (ed. R.F. Gesteland and J.F. Atkins), p. 137. Cold Spring Harbor Laboratory Press, Cold Spring Harbor, New York.

Omer A.D., Ziesche S., Ebhardt H., and Dennis P.P. 2002. *In vitro* reconstitution and activity of a C/D box methylation guide ribonucleoprotein complex. *Proc. Natl. Acad. Sci.* **99**: 5289.

Park S.H., Zarrinpar A., and Lim W.A. 2003. Rewiring MAP kinase pathways using alternative scaffold assembly mechanisms. *Science* **299**: 1061.

Peterson S.E., Stellwagen A.E., Diede S.J., Singer M. S., Haimberger Z.W., Johnson C.O., Tzoneva M., and Gottschling, D.E. 2001. The function of a stem-loop in telomerase RNA is linked to the DNA repair protein Ku. *Nat. Genet.* **27**: 64.

Piccinelli P., Rosenblad M.A., and Samuelsson T. 2005. Identification and analysis of ribonuclease P and MRP RNA in a broad range of eukaryotes. *Nucleic Acids Res.* **33**: 4485.

Rashid R., Liang B., Baker D.L., Youssef O.A., He Y., Phipps K., Terns R.M., Terns M.P., and Li H. 2006. Crystal structure of a Cbf5-Nop10-Gar1 complex and implications in RNA-guided pseudouridylation and dyskeratosis congenita. *Mol. Cell* **21**: 249.

Romero D.P. and Blackburn E.H. 1991. A conserved secondary structure for telomerase RNA. *Cell* **67**: 343.

Seto A.G., Livengood A.J., Tzfati Y., Blackburn E., and Cech T.R. 2002. A bulged stem tethers Est1p to telomerase RNA in budding yeasts. *Genes Dev.* **16**: 2800.

Seto A.G., Zaug A.J., Sobel S.G., Wolin S.L., and Cech T.R. 1999. *Saccharomyces cerevisiae* telomerase is an Sm small nuclear ribonucleoprotein particle. *Nature* **401**: 177.

Shuster E.O. and Guthrie C. 1988. Two conserved domains of yeast U2 snRNA are separated by 945 nonessential nucleotides. *Cell* **55**: 41.

Siliciano P.G., Kivens W.J., and Guthrie C. 1991. More than half of yeast U1 snRNA is dispensable for growth. *Nucleic Acids Res.* **19**: 6367.

Siridechadilok B., Fraser C.S., Hall R.J., Doudna J.A., and Nogales E. 2005. Structural roles for human translation factor eIF3 in initiation of protein synthesis. *Science* **310**: 1513.

Spahn C.M., Beckmann R., Eswar N., Penczek P.A., Sali A., Blobel G., and Frank J. 2001. Structure of the 80S ribosome from *Saccharomyces cerevisiae*—tRNA-ribosome and subunit-subunit interactions. *Cell* **107**: 373.

Sperger J.M. and Cech T.R. 2001. A stem-loop of *Tetrahymena* telomerase RNA distant from the template potentiates RNA folding and telomerase activity. *Biochemistry* **40**: 7005.

Stark H., Dube P., Luhrmann R., and Kastner B. 2001. Arrangement of RNA and proteins in the spliceosomal U1 small nuclear ribonucleoprotein particle. *Nature* **409**: 539.

Stuckenholz C., Meller V.H., and Kuroda M.I. 2003. Functional redundancy within roX1, a noncoding RNA involved in dosage compensation in *Drosophila melanogaster. Genetics* **164**: 1003.

ten Dam E., van Belkum A., and Pleij K. 1991. A conserved pseudoknot in telomerase RNA. *Nucleic Acids Res.* **19**: 6951.

Theimer C.A., Blois C.A., and Feigon J. 2005. Structure of the human telomerase RNA pseudoknot reveals conserved tertiary interactions essential for function. *Mol. Cell* **17**: 671.

Tollervey D., Lehtonen H., Jansen R., Kern H., and Hurt E.C. 1993. Temperature-sensitive mutations demonstrate roles for yeast fibrillarin in pre-rRNA processing, pre-rRNA methylation, and ribosome assembly. *Cell* **72**: 443.

Tzfati Y., Fulton T.B., Roy J., and Blackburn E.H. 2000. Template boundary in a yeast telomerase specified by RNA structure. *Science* **288:** 863.

Tzfati Y., Knight Z., Roy J., and Blackburn E.H. 2003. A novel pseudoknot element is essential for the action of a yeast telomerase. *Genes Dev.* **17:** 1779.

Van Nues R.W. and Brown J.D. 2004. *Saccharomyces* SRP RNA secondary structures: A conserved S-domain and extended Alu-domain. *RNA* **10:** 75.

Walker J.R., Corpina R.A., and Goldberg J. 2001. Structure of the Ku heterodimer bound to DNA and its implications for double-strand break repair. *Nature* **412:** 607.

Wang H., Boisvert D., Kim K.K., Kim R., and Kim S.H. 2000. Crystal structure of a fibrillarin homologue from *Methanococcus jannaschii*, a hyperthermophile, at 1.6 Å resolution. *EMBO J.* **19:** 317.

Will C.L. and Luhrmann R. 2006. Spliceosome structure and function. In *The RNA world* (ed. R.F. Gesteland et al.), p.369. Cold Spring Harbor Laboratory Press, Cold Spring Harbor, New York.

Willingham A.T., Orth A.P., Batalov S., Peters E.C., Wen B.G.,

Aza-Blanc P., Hogenesch J.B., and Schultz P.G. 2005. A strategy for probing the function of noncoding RNAs finds a repressor of NFAT. *Science* **309:** 1570.

Wutz A. 2003. RNAs templating chromatin structure for dosage compensation in animals. *Bioessays* **25:** 434.

Wutz A., Rasmussen T.P., and Jaenisch R. 2002. Chromosomal silencing and localization are mediated by different domains of Xist RNA. *Nat. Genet.* **30:** 167.

Yu G.L., Bradley J.D., Attardi L.D., and Blackburn E.H. 1990. *In vivo* alteration of telomere sequences and senescence caused by mutated *Tetrahymena* telomerase RNAs. *Nature* **344:** 126.

Zappulla D.C. and Cech T.R. 2004. Yeast telomerase RNA: A flexible scaffold for protein subunits. *Proc. Natl. Acad. Sci.* **101:** 10024.

Zappulla D.C., Goodrich K., and Cech T.R. 2005. A miniature yeast telomerase RNA functions *in vivo* and reconstitutes activity *in vitro*. *Nat. Struct. Mol. Biol.* **12:** 1072.

Zebarjadian Y., King T., Fournier M.J., Clarke L., and Carbon J. 1999. Point mutations in yeast CBF5 can abolish *in vivo* pseudouridylation of rRNA. *Mol. Cell. Biol.* **19:** 7461.

Telomerase RNA Levels Limit the Telomere Length Equilibrium

C.W. Greider

Department of Molecular Biology and Genetics, Johns Hopkins
University School of Medicine, Baltimore, Maryland 21205

Small functional RNAs play essential roles in many biological processes. Regulating the level of these small RNAs can be as important as maintaining their function in cells. The telomerase RNA is maintained in cells at a steady-state level where small changes in concentration can have a profound impact on function. Cells that have half the level of the telomerase RNA cannot maintain telomeres through many cell divisions. People who are heterozygous for telomerase RNA mutations have the diseases dyskeratosis congenita and aplastic anemia, caused by short telomeres that result in loss of tissue renewal capacity. Mice heterozygous for telomerase RNA show haploinsufficiency in telomere length maintenance and also show loss of tissue renewal capacity. It is remarkable that small changes in the level of this functional RNA can have such profound effects in cells. This tight regulation highlights the importance of controlling the action of telomerase in cells.

Work over the past 20 years has shown that functional RNAs play important roles in cells. While much of the focus is on the precise mechanisms of RNA function, less attention has been focused on the importance of the level of specific functional RNAs. The RNA component of telomerase is essential for telomere maintenance. Recent experiments on the telomerase RNA have shown that changes in the level of this functional RNA have major consequences for telomere length regulation. Reduced RNA levels compromise cell viability and cause human disease. Because telomerase RNA is maintained at an equilibrium concentration well below saturation, small changes in RNA concentration can affect enzyme action. The exquisite regulation of this small RNA highlights the role of maintaining the equilibrium concentration of small functional RNAs; tight regulatory balance is likely used to regulate RNAs that play roles in other cellular processes.

TELOMERASE IS REQUIRED FOR TELOMERE LENGTH MAINTENANCE

Telomeres are the protein–DNA structures that protect chromosome ends from nucleases and recombination and distinguish these natural chromosome ends from broken DNA. When telomere function is lost, the chromosome end resembles a double-stranded DNA break and can result in chromosome end-to-end fusion and other rearrangements. Telomerase is a remarkable enzyme that maintains telomere length and thus helps assure telomere function. The normal mechanisms that replicate chromosome ends lead to a loss of telomere sequence each time the cell divides (Watson 1972; Olovnikov 1973). Telomerase overcomes this end replication problem, by adding telomere sequences back onto chromosome ends (Greider and Blackburn 1985). Telomerase enzymes from all eukaryotes contain two essential core enzyme components, a catalytic protein component, TERT, and the telomerase RNA component, or TR (Lingner et al. 1997). The telomerase RNA component contains a short, single-stranded sequence that serves as the template for the

telomeric sequences that are added onto chromosome ends by the enzyme (Greider and Blackburn 1989). In addition to the template region, the telomerase RNA structure contains regions that are essential in the catalytic function of the enzyme (Chen and Greider 2004).

TELOMERE LENGTH EQUILIBRIUM IS MAINTAINED BY REGULATING ACCESS OF TELOMERASE TO TELOMERES

Telomerase maintains telomere length by adding telomere repeats onto some, but not all, chromosome ends at each cell division. The shortest telomeres in the cell are preferentially elongated by telomerase. Thus, the length of each individual telomere is maintained as an equilibrium between shortening due to replication and lengthening by telomerase. This equilibrium is maintained within a defined set point through a series of feedback mechanisms that regulate repeat addition (Greider 1996; Smogorzewska and de Lange 2004). The specific set point for the equilibrium length is species specific; in yeast the average is around 300 bp, whereas in humans it is around 10,000 bp. Within a species, different strains may have different telomere length set points, and this set point is genetically defined (Craven and Petes 1999; Hemann and Greider 2000).

Telomere length equilibrium is regulated by at least two different mechanisms: telomere-binding proteins and modification of those proteins at the telomere. In both yeast and mammals there is a specific set of telomere proteins that bind telomeric DNA to protect the end and distinguish the telomere from a double strand break. Access of telomerase to the telomere is negatively regulated by these telomere-binding proteins (Shore 1997a; Smogorzewska and de Lange 2004; de Lange 2005). The bound telomere protein complex somehow blocks the ability of telomerase to elongate the telomere. This inhibition of telomerase elongation directly at the telomere establishes a feedback mechanism that regulates how long telomeres can get. The longer the telomere, the more telomere-binding proteins are present, the stronger the block to telomerase elongation of that specific telomere (Shore 1997b; Smogorzewska and de Lange 2004). This simple feedback system is also regulated by protein modification. Phosphorylation of specific substrates by the ATM and ATR protein kinases is required to allow telomerase access to elongate the telomere (Naito et al. 1998; Ritchie et al. 1999). In addition, the Cdk1 kinase also regulates telomere elongation (Frank et al. 2006; Vodenicharov and Wellinger 2006). The specific targets of phosphorylation and the mechanisms by which these modifications establish and maintain the length equilibrium are not yet understood in detail. This is an active area of ongoing research.

TELOMERASE IS REQUIRED IN CELLS THAT MUST DIVIDE MANY TIMES

Telomerase is essential to maintain telomere length through many cell divisions. Experimentally, deletion of either the RNA or protein component of telomerase leads to progressive telomere shortening as cells undergo multiple rounds of DNA replication (Lundblad and Szostak 1989; Singer and Gottschling 1994; Blasco et al. 1997; Lingner et al. 1997). The absence of telomerase initially has no effect on cells when telomeres are long. As the cells divide and telomeres shorten, the short telomeres trigger a DNA damage response and cell cycle arrest or apoptosis (Hemann et al. 2001a; Enomoto et al. 2002; d'Adda di Fagagna et al. 2003; IJpma and Greider 2003; Hao et al. 2004). Short telomeres are thought to trigger this DNA damage response because when telomeres become too short they no longer provide the protective function, and the chromosome end now resembles a DNA break (Hemann et al. 2000, 2001a).

Telomere maintenance is essential for cells that must divide many times. Single-celled organisms, like yeast, require telomerase for long-term growth (Lundblad and Szostak 1989). In mammals, telomerase is essential for tissues that require constant renewal. In these tissues, telomeres must be maintained specifically in stem cells that are responsible for maintaining tissue integrity. In mice, deletion of the telomerase RNA gene, mTR, causes progressive telomere shortening with increased generations of interbreeding. When telomeres are short, cell death occurs in tissues that undergo constant cell division (Lee et al. 1998; Hao et al. 2005). Short telomeres cause cell death in the testes, leading to loss of fertility (Hemann et al. 2001b). These mice also show decreased cellularity of the bone marrow and spleen and a loss of the villi in the gastrointestinal tract (Hao et al. 2005). The fact that all of the affected tissues in these mice with short telomeres are those that undergo constant turnover suggests that telomere maintenance is required for the long-term integrity of stem cells.

Telomerase is also required in cancer cells to allow for the large number of cell divisions these cells undergo during tumor initiation and growth. Telomeres are typically shorter in cancer cells than in normal cells, likely due to the large number of cell divisions that these cells have undergone. Genetic crosses between the mTR$^{-/-}$ mouse and several tumor-prone mouse models reveal that telomere shortening can limit tumor growth (Greenberg et al. 1999; Rudolph et al. 2001; Qi et al. 2003; Wong et al. 2003). The short telomeres signal through the DNA damage response pathway and limit tumor cell proliferation and thus limit tumor growth. These results support the proposals which have been made (Greider 1990; Harley et al. 1990; Shay and Wright 2002) that inhibiting telomerase may be an effective approach to cancer treatment.

THE SHORTEST TELOMERES LIMIT CELL DIVISION

The powerful ability of short telomeres to limit cell division and tissue renewal raises the question of the mechanism by which short telomeres exert this effect. Experiments with the telomerase null mice demonstrated that, at the molecular level, it is the shortest telomere that limits cell division. As the telomerase null mouse is bred for multiple generations, in the first few generations, when telomeres are long, there are no cellular or organismal phenotypes. However, in the later mTR$^{-/-}$ generations, when the telomeres are sufficiently short, loss of tissue regenerative capacity is seen. This implies that it is not the absence of telomerase itself, but rather the short telomeres, that elicits the cell growth inhibition. This effect of short telomeres is exerted by the shortest telomeres in the cell, not by an overall shortening of the telomere length distribution. When mTR heterozygous mice with long telomeres were crossed to telomerase null mice with short telomeres, the null progeny had critically short telomeres and showed cell death. In the mTR$^{+/-}$ littermates, the critically short telomeres were elongated by telomerase, but the overall length distribution was identical to that in the mTR$^{-/-}$ mice. This indicates that all of the phenotypes associated with telomere shortening are due to the shortest telomeres in the distribution, and telomerase is targeted to preferentially elongate the shortest telomeres (Hemann et al. 2001a). Thus, only a few short telomeres needed to be elongated to allow cell viability. This observation highlights the extent of fine-tuning of telomere length regulation in cells; very subtle changes in telomere length maintenance can determine whether a cell lives or dies.

HALF THE LEVEL OF TELOMERASE RNA IS NOT SUFFICIENT FOR TELOMERE MAINTENANCE

The fine-tuning of telomere length is regulated by the level of active telomerase enzyme. The importance of the level of telomerase in cells is evident from the human genetic disease dyskeratosis congenita. Mutations in the telomerase RNA gene cause autosomal dominant dyskeratosis congenita (Vulliamy et al. 2001). In this genetic disease, patients have short telomeres and most often die of bone marrow failure. This bone marrow failure likely represents a defect in the self-renewal capacity of the hematopoietic progenitor cells. The telomerase null mouse shows similar telomere shortening and loss of bone marrow regenerative capacity. Strikingly, patients with autosomal dominant dyskeratosis congenita have only one mutant allele of the telomerase RNA gene.

Autosomal dominant inheritance can be caused by a dominant interfering mutation that inactivates the product of the wild-type allele, or it can be due to haploinsufficiency, where half the level of the normal enzyme is not sufficient for an essential function.

To test whether half the level of telomerase RNA limits telomere elongation, we bred heterozygous mTR$^{+/-}$ mice. Quantitative RT-PCR showed that mTR$^{+/-}$ mice have half of the steady-state level of telomerase RNA (Hathcock et al. 2002). When mTR$^{+/-}$ mice were bred for multiple generations and maintained as heterozygotes (Fig. 1), they showed progressive telomere shortening (Hao et al. 2005). This telomere shortening in the presence of half the level of telomerase RNA implies haploinsufficiency for telomerase in telomere length maintenance. The late generation heterozygous mice that had short telomeres showed cell death in the testes and decreased tissue renewal capacity, similar to the mTR$^{-/-}$ mice with short telomeres. This loss of tissue renewal provides direct genetic evidence for haploinsufficiency in the human autosomal dominant disease dyskeratosis congenita.

The telomere shortening with increased generations of breeding in mTR$^{+/-}$ heterozygous mice implies that telomerase is maintained in cells at a very low level that is just barely enough to maintain the length equilibrium; any decrease in the level perturbs the telomere length equilibrium. The effects of limiting telomerase can last for many generations. The wild-type offspring from crosses of late-generation heterozygous mice (termed Wt* mice; Fig. 1) still had short telomeres (Hao et al.

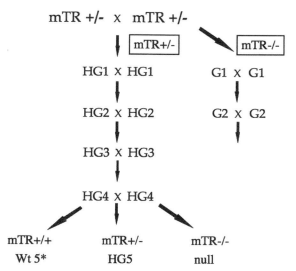

Figure 1. Breeding of telomerase heterozygous mice (see Hao et al. 2005). CAST/EiJ mice that were heterozygous for mouse telomerase RNA were bred to establish first-generation mTR$^{-/-}$ G$_1$ mice. The mTR$^{-/-}$ G$_1$ mice were bred to create second-generation mTR$^{-/-}$ G$_2$ mice (*right side*). The first-generation heterozygotes HG1 were then bred together to create second-generation heterozygotes HG2, which in turn were interbred for additional generations (*middle line*). Telomere shortening occurred with each progressive generation of heterozygote. When the late-generation heterozygotes, HG4, were bred together, they established telomerase null mTR$^{-/-}$ mice: the HG5 generation and Wt 5* mice. These Wt* mice had short telomeres compared to the parental wild-type mice that established this line.

2005). Thus, the amount of telomere elongation that occurs in germ-line cells in one generation even with wild-type levels of telomerase RNA is not sufficient to reestablish the wild-type telomere length distribution. Because short telomeres cause loss of cell viability, these wild-type mice with short telomeres showed similar loss of cell viability in testes as seen in mTR$^{+/-}$ heterozygotes and mTR$^{-/-}$ mice with short telomeres (Hao et al. 2005). Preliminary experiments suggest that, after multiple generations of breeding these Wt* mice, the telomere length is reestablished at the genetically determined wild-type telomere length set point. Thus, the limiting telomerase RNA has an effect for many generations.

As with other aspects of fundamental telomere function that are conserved, the sensitivity to telomerase RNA levels is also seen in yeast. Diploid yeast heterozygous for the telomerase RNA gene *TLC1* have half of the steady-state level of telomerase RNA. Telomere shortening also occurs in these heterozygous yeast cells as they are grown continuously as heterozygotes (Mozdy and Cech 2006). This implies that the exquisite sensitivity to telomerase RNA levels is a conserved feature of telomere length regulation.

As described above, there are multiple independent mechanisms that regulate telomere length. How these different mechanisms interact to regulate the equilibrium set point is not yet clear. The fact that telomerase specifically targets the shortest telomeres and that the shortest telomeres limit cell proliferation might suggest that when telomerase is limiting, the shortest telomeres will be repaired first. However, this mechanism appears to be insufficient to restore all telomere function when telomerase is limiting. Both patients with autosomal dominant dyskeratosis congenita and the mTR$^{+/-}$ heterozygous mice show loss of bone marrow function with half the level of telomerase, thus, short telomeres must still be causing the phenotype despite some telomere elongation. This implies that even when telomerase is targeted to the shortest telomere, a reduction in the telomerase RNA level can still overwhelm the system so that short dysfunctional telomeres are still present.

WHY IS TELOMERE ELONGATION SO TIGHTLY REGULATED?

The fact that telomere length is maintained as an equilibrium distribution might suggest that there would be a lot of leeway in telomere length regulation. However, in fact, the opposite seems to be true. The cell expends a lot of energy regulating both access of telomerase to the telomere and the level of telomerase in a cell. Why would an apparently fairly loose length regulation system in fact be so exquisitely sensitive to small perturbations?

One obvious reason an organism might regulate telomerase would be as a tumor suppressor mechanism (Greider 1990; de Lange 1995). If telomere shortening does play the role of a tumor suppressor, as has been proposed, then a tight regulation of both telomere length and telomerase may be protective against cancer. This mechanism may explain a selection for tight telomerase regulation for mammals. However, tight regulation of telomerase also occurs in yeast and likely other single-celled organisms that do not get cancer. What purpose does the tight regulation of telomerase play at the cellular level? The answer may lie in the processing of double-stranded DNA breaks. Telomerase is known to add telomeric sequences onto broken DNA ends, creating new telomeres (Haber and Thorburn 1984; Greider 1991)w; although this process is inefficient, it results in functional chromosomes. For example, there are a number of truncations of human chromosome 16 that result in α-thalassemia due to loss of the α-globin gene (Wilkie et al. 1990; Lamb et al. 1993; Viprakasit et al. 2003). There are no known essential genes distal to α-globin on chromosome 16, thus if a DNA break occurs in α-globin and telomere addition occurs, a new stable chromosome results. If telomerase were not so tightly regulated perhaps it would compete more effectively with the DNA repair machinery, and new telomeres would be added to chromosome breaks generating chromosome truncations before repair could occur. This would most often lead to loss of genes distal to the break and would be detrimental to the cell. Thus, to allow repair machinery to repair chromosome breaks, telomerase must not be allowed to access the break. The normal mechanisms that regulate telomere length through telomere-binding proteins would not offer regulation at a DNA break. Therefore, limiting the amount of telomerase in a cell may be the best way to assure it does not act where it should not.

SUMMARY

Telomerase is an essential enzyme that maintains telomere length. Despite the fact that telomere length is not precise, but rather regulated about a broad equilibrium, there is very tight regulation of both telomerase action and telomerase levels in the cell. One of the consequences of the very stringent regulation of telomerase is that having half the level of this enzyme is not sufficient to maintain telomeres, and this leads to human disease. It will be interesting to examine the processes that strictly regulate the level of telomerase in cells and the potential consequences to cells if that regulation is disturbed. These studies will give insights and possibly propose new avenues for treatment of patients with dyskeratosis congenita and other diseases caused by short telomeres.

ACKNOWLEDGMENTS

The work from the Greider lab described here was supported by National Institutes of Health grants RO1AG39383 and RO1AG027406. I thank the Greider lab members and Drs. Rachel Green, David Feldser, and Mary Armanios for critical reading of the manuscript.

REFERENCES

Blasco M.A., Lee H.-W., Hande M.P., Samper E., Lansdorp P.M., DePinho R.A., and Greider C.W. 1997. Telomere shortening and tumor formation by mouse cells lacking telomerase RNA. *Cell* **91**: 25.

Chen J.L. and Greider C.W. 2004. Telomerase RNA structure and function: Implications for dyskeratosis congenita. *Trends Biochem. Sci.* **29**: 183.

Craven R.J. and Petes T.D. 1999. Dependence of the regulation of telomere length on the type of subtelomeric repeat in the yeast *Saccharomyces cerevisiae. Genetics* **152**: 1531.

d'Adda di Fagagna F., Reaper P.M., Clay-Farrace L., Fiegler H., Carr P., Von Zglinicki T., Saretzki G., Carter N.P., and Jackson S.P. 2003. A DNA damage checkpoint response in telomere-initiated senescence. *Nature* **426**: 194.

de Lange T. 1995. Telomere dynamics and genome instability in human cancer. In *Telomeres* (ed. E.H. Blackburn and C.W. Greider), p. 265. Cold Spring Harbor Laboratory Press, Cold Spring Harbor, New York.

———. 2005. Shelterin: The protein complex that shapes and safeguards human telomeres. *Genes Dev.* **19**: 2100.

Enomoto S., Glowczewski L., and Berman J. 2002. MEC3, MEC1, and DDC2 are essential components of a telomere checkpoint pathway required for cell cycle arrest during senescence in *Saccharomyces cerevisiae. Mol. Biol. Cell* **13**: 2626.

Frank C.J., Hyde M., and Greider C.W. 2006. Regulation of telomere elongation by the cyclin-dependent kinase CDK1. *Mol. Cell* **24**: 423.

Greenberg R.A., Chin L., Femino A., Lee K.H., Gottlieb G.J., Singer R.H., Greider C.W., and DePinho R.A. 1999. Short dysfunctional telomeres impair tumorigenesis in the INK4a(delta2/3) cancer-prone mouse. *Cell* **97**: 515.

Greider C.W. 1990. Telomeres, telomerase and senescence. *Bioessays* **12**: 363.

———. 1991. Chromosome first aid. *Cell* **67**: 645.

———. 1996. Telomere length regulation. *Annu. Rev. Biochem.* **65**: 337.

Greider C.W. and Blackburn E.H. 1985. Identification of a specific telomere terminal transferase activity in *Tetrahymena* extracts. *Cell* **43**: 405.

———. 1989. A telomeric sequence in the RNA of *Tetrahymena* telomerase required for telomere repeat synthesis. *Nature* **337**: 331.

Haber J.E. and Thorburn P.C. 1984. Healing of broken linear dicentric chromosomes in yeast. *Genetics* **106**: 207.

Hao L.Y., Strong M., and Greider C.W. 2004. Phosphorylation of H2AX at short telomeres in T cells and fibroblasts. *J. Biol. Chem.* **279**: 45148.

Hao L.Y., Armanios M., Strong M.A., Karim B., Feldser D.M., Huso D., and Greider C.W. 2005. Short telomeres, even in the presence of telomerase, limit tissue renewal capacity. *Cell* **123**: 1121.

Harley C.B., Futcher A.B., and Greider C.W. 1990. Telomeres shorten during ageing of human fibroblasts. *Nature* **345**: 458.

Hathcock K.S., Hemann M.T., Opperman K.K., Strong M.A., Greider C.W., and Hodes R.J. 2002. Haploinsufficiency of mTR results in defects in telomere elongation. *Proc. Natl. Acad. Sci.* **99**: 3591.

Hemann M.T. and Greider C.W. 2000. Wild-derived inbred mouse strains have short telomeres. *Nucleic Acids Res.* **28**: 4474.

Hemann M.T., Hackett J., IJpma A., and Greider C.W. 2000. Telomere length, telomere binding proteins and DNA damage signaling. *Cold Spring Harbor Symp. Quant. Biol.* **65**: 275.

Hemann M.T., Strong M., Hao L.Y., and Greider C.W. 2001a. The shortest telomere, not average telomere length, is critical for cell viability and chromosome stability. *Cell* **107**: 67.

Hemann M.T., Rudolph L., Strong M., DePinho R.A., Chin L., and Greider C.W. 2001b. Telomere dysfunction triggers developmentally regulated germ cell apoptosis. *Mol. Biol. Cell* **12**: 2023.

IJpma A. and Greider C.W. 2003. Short telomeres induce a DNA damage response in *Saccharomyces cerevisiae. Mol. Biol. Cell* **14**: 987.

Lamb J., Harris P.C., Wilkie A.O., Wood W.G., Dauwerse J.G., and Higgs D.R. 1993. De novo truncation of chromosome 16p and healing with (TTAGGG)n in the alpha-thalassemia/mental retardation syndrome (ATR-16). *Am. J. Hum. Genet.* **52**: 668.

Lee H.-W., Blasco M.A., Gottlieb G.J., Horner J.W., Greider C.W., and DePinho R.A. 1998. Essential role of mouse telomerase in highly proliferative organs. *Nature* **392**: 569.

Lingner J., Hughes T.R., Shevchenko A., Mann M., Lundblad V., and Cech T.R. 1997. Reverse transcriptase motifs in the catalytic subunit of telomerase. *Science* **276:** 561.

Lundblad V. and Szostak J.W. 1989. A mutant with a defect in telomere elongation leads to senescence in yeast. *Cell* **57:** 633.

Mozdy A.D. and Cech T.R. 2006. Low abundance of telomerase in yeast: Implications for telomerase haploinsufficiency. *RNA* **12:** 1721.

Naito T., Matsuura A., and Ishikawa F. 1998. Circular chromosome formation in a fission yeast mutant defective in two ATM homologues. *Nat. Genet.* **20:** 203.

Olovnikov A.M. 1973. A theory of marginotomy. *J. Theor. Biol.* **41:** 181.

Qi L., Strong M.A., Karim B.O., Armanios M., Huso D.L., and Greider C.W. 2003. Short telomeres and ataxia-telangiectasia mutated deficiency cooperatively increase telomere dysfunction and suppress tumorigenesis. *Cancer Res.* **63:** 8188.

Ritchie K.B., Mallory J.C., and Petes T.D. 1999. Interactions of TLC1 (which encodes the RNA subunit of telomerase), TEL1, and MEC1 in regulating telomere length in the yeast *Saccharomyces cerevisiae. Mol. Cell. Biol.* **19:** 6065.

Rudolph K.L., Millard M., Bosenberg M.W., and DePinho R.A. 2001. Telomere dysfunction and evolution of intestinal carcinoma in mice and humans. *Nat. Genet.* **28:** 155.

Shay J.W. and Wright W.E. 2002. Telomerase: A target for cancer therapeutics. *Cancer Cell* **2:** 257.

Shore D. 1997a. Telomerase and telomere-binding proteins: Controlling the endgame. *Trends Biochem. Sci.* **22:** 233.

———. 1997b. Telomere length regulation: Getting the measure of chromosome ends. *Biol. Chem.* **378:** 591.

Singer M.S. and Gottschling D.E. 1994. TLC1: Template RNA component of *Saccharomyces cerevisiae* telomerase. *Science* **266:** 404.

Smogorzewska A. and de Lange T. 2004. Regulation of telomerase by telomeric proteins. *Annu. Rev. Biochem.* **73:** 177.

Viprakasit V., Kidd A.M., Ayyub H., Horsley S., Hughes J., and Higgs D.R. 2003. De novo deletion within the telomeric region flanking the human alpha globin locus as a cause of alpha thalassaemia. *Br. J. Haematol.* **120:** 867.

Vodenicharov M.D. and Wellinger R.J. 2006. DNA degradation at unprotected telomeres in yeast is regulated by the CDK1 (Cdc28/Clb) cell-cycle kinase. *Mol. Cell* **24:** 127.

Vulliamy T., Marrone A., Goldman F., Dearlove A., Bessler M., Mason P.J., and Dokal I. 2001. The RNA component of telomerase is mutated in autosomal dominant dyskeratosis congenita. *Nature* **413:** 432.

Watson J.D. 1972. Origin of concatameric T7 DNA. *Nat. New Biol.* **239:** 197.

Wilkie A.O., Lamb J., Harris P.C., Finney R.D., and Higgs D.R. 1990. A truncated human chromosome 16 associated with alpha thalassaemia is stabilized by addition of telomeric repeat (TTAGGG)n. *Nature* **346:** 868.

Wong K.K., Maser R.S., Bachoo R.M., Menon J., Carrasco D.R., Gu Y., Alt F.W., and DePinho R.A. 2003. Telomere dysfunction and Atm deficiency compromises organ homeostasis and accelerates ageing. *Nature* **421:** 643.

Sensing Metabolic Signals with Nascent RNA Transcripts: The T Box and S Box Riboswitches as Paradigms

T.M. HENKIN AND F.J. GRUNDY

Department of Microbiology and The RNA Group, The Ohio State University, Columbus, Ohio 43210

Recent studies in a variety of bacterial systems have revealed a number of regulatory systems in which the 5′ region of a gene plays a key role in regulation of the downstream coding sequences. These RNA regions act *in cis* to determine if the full-length transcript will be synthesized or if the coding sequence(s) will be translated. Each class of system includes an RNA element whose structure is modulated in response to a specific regulatory signal, and the signals measured can include small molecules, small RNAs (including tRNA), and physical parameters such as temperature. Multiple sets of genes can be regulated by a particular mechanism, and multiple systems of this type, each of which responds to a specific signal, can be present in a single organism. In addition, different classes of RNA elements can be found that respond to a particular signal, indicating the existence of multiple alternate solutions to the same regulatory problem. The T box and S box systems, which respond to uncharged tRNA and *S*-adenosylmethionine (SAM), respectively, provide paradigms of two systems of this type.

Direct sensing of a regulatory signal by the 5′ region of a nascent RNA (the "leader region") has recently emerged as a common mechanism for regulation of gene expression in bacteria (Grundy and Henkin 2004, 2006). In mechanisms of this type, termed "riboswitches," the regulatory signal can modulate folding of the nascent RNA transcript, which in turn can determine whether the RNA folds into the helix of an intrinsic transcriptional terminator, resulting in premature termination of transcription, or an alternate structure that allows expression of the downstream coding sequences (Fig. 1). Similar RNA rearrangements can also mediate translational regulation by sequestration of the ribosome-binding site at the start of the regulated coding sequence. Each class of riboswitch RNA recognizes its regulatory signal with high specificity, and the RNA structural elements exhibit a sensitivity to the signal that is appropriate to the in vivo pools of the effector or to the physical parameter (e.g., temperature). In addition, the RNA structural change that occurs in response to the signal must be coupled to an appropriately sensitive gene expression response. We have identified several systems of this type, including the T box system, which monitors the aminoacylation of a specific tRNA, and the S box and S_{MK} box systems, which respond directly to SAM.

Characterization of the RNA–effector interaction in these systems has provided new information about how different classes of effectors are recognized, and about the impact of these regulatory mechanisms on the cell.

THE T BOX SYSTEM: REGULATION OF AMINO ACID-RELATED GENES BY UNCHARGED tRNA

The T box system was initially uncovered by the analysis of the *Bacillus subtilis tyrS* gene, which encodes tyrosyl-tRNA synthetase. A G+C-rich helix followed by a run of U residues was identified upstream of the *tyrS*-coding sequence, leading to the prediction that regulation occurs at the level of premature termination of transcription. Expression of *tyrS* was shown to be induced when cells were grown with limiting tyrosine, and regulation was dependent on the presence of the terminator, as predicted (Henkin et al. 1992). The 5′ region of the *tyrS* gene shares a set of sequence and structural elements with a large family of aminoacyl-tRNA synthetase, amino acid biosynthesis, and transporter genes (Grundy and Henkin 1993, 1994, 2003). These features form a pattern that includes the transcriptional terminator and a competing antiterminator structure, preceded by a large stem-loop element interrupted by several bulges and internal loops (Fig. 2). One of these loops contains a single codon that corresponds to the amino acid identity of the downstream gene, so that the *tyrS* gene contains a UAC tyrosine codon, tryptophanyl genes contain a UGG tryptophan codon, etc. (Grundy and Henkin 1993). The initial pattern was recognized from a set of ten aminoacyl-tRNA synthetase genes from *Bacillus* sp., and the explosion of genome sequence

Figure 1. Regulation of gene expression by leader RNA structural rearrangements. Pairing of the RNA regions labeled C and D results in inhibition of downstream gene expression by formation of the helix of an intrinsic transcriptional terminator or a structure that sequesters the ribosome-binding site of the coding sequence. Formation of the competing B+C pairing sequesters region C, preventing transcription termination or translation inhibition. The A+B pairing, which occurs in some systems (e.g., S box) but not others (e.g., T box) competes with the B+C pairing, thereby controlling formation of the C+D pairing.

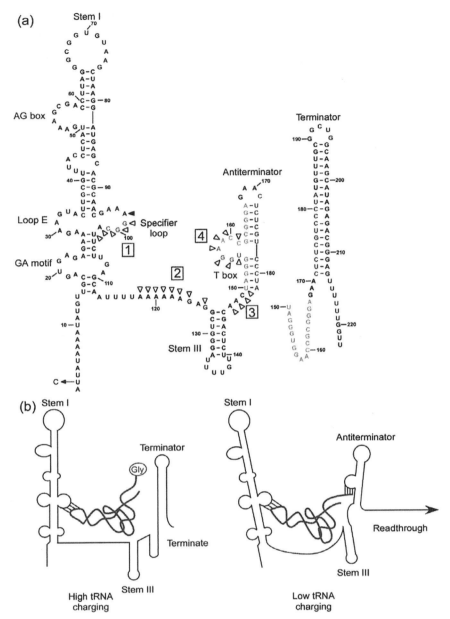

Figure 2. The T box system. (*a*) Secondary structural model of the *B. subtilis glyQS* leader RNA in the antiterminator conformation; the competing terminator conformation is shown on the right. Numbering is relative to the transcription start site. Major structural elements are labeled. Boxed numbers and arrows indicate regions that exhibit alterations in structure upon binding of the cognate tRNA^{Gly} (Yousef et al. 2005). The GGC residues in the specifier loop (in addition to the following A residue) pair with the anticodon loop of tRNA^{Gly}. (*b*) Both charged and uncharged tRNA interact with the specifier loop of the leader RNA; only uncharged tRNA can make the second interaction between the acceptor end of the tRNA and the antiterminator bulge. The second interaction stabilizes the antiterminator, preventing formation of the competing terminator helix. Conditions under which the effector tRNA is highly charged result in efficient termination and low expression of the downstream coding sequence, whereas a decrease in charging of the effector tRNA results in readthrough of the termination site and increased expression of the downstream coding sequence. Each gene responds specifically to the cognate tRNA.

information has expanded the set to more than 700 genes, all of which exhibit correspondence between the identity of the codon at the appropriate position and the predicted function of the downstream coding sequence (Merino and Yanofsky 2005; F.J. Grundy and T.M. Henkin, unpubl.). The identification of this RNA pattern suggested that amino acid limitation could be sensed via an effect on charging of the cognate tRNA.

Genetic Analysis of the *B. subtilis tyrS* Gene

The prediction that the codon was responsible for the specificity of the response to amino acid limitation was tested by mutation of the UAC tyrosine codon of the *tyrS* gene to a UUC phenylalanine codon. This single-nucleotide substitution resulted in loss of response to tyrosine availability and induction of *tyrS* expression when cells were limited for phenylalanine. Replacement of the UAC tyrosine codon with a nonsense codon resulted in transcription termination under all growth conditions, and expression was restored by introduction into the cell of a nonsense suppressor tRNA containing a compensatory anticodon mutation (Grundy and Henkin 1993). These studies established that base-pairing between the sequence at the position of the UAC codon in *tyrS* (designated the "specifier sequence") and the anticodon of the corresponding tRNA is required for readthrough of the terminator in the 5′ region of the gene and suggested that each gene containing the T box leader region pattern would respond similarly to its cognate tRNA. This prediction was subsequently tested for a variety of genes in this family.

An obvious mechanism for utilization of a codon within the 5′ region of an mRNA as a regulatory signal is provided by transcription attenuation systems similar to the *Escherichia coli trp* operon, where regulatory codons are monitored by a translating ribosome (Landick et al. 1996). In systems of this type, which rely on coupling of transcription and translation, the regulated genes are preceded by a short peptide-coding sequence that includes tandem codons for the corresponding amino acid (e.g., tryptophan codons for the *trp* biosynthesis operon). Efficient translation of the leader peptide-coding sequence requires an adequate supply of the appropriate charged tRNA (e.g., tRNA^{Trp}), and stalling of the ribosome because of limitation for specific charged tRNAs results in formation of an antiterminator structure in the nascent RNA that prevents formation of the helix of an intrinsic transcriptional terminator, therefore preventing premature termination of transcription. This type of mechanism was ruled out for the *B. subtilis tyrS* gene when introduction of a frameshift mutation, which was predicted to disrupt translation of the tyrosine codon, had no effect on *tyrS* regulation, indicating that the UAC codon does not act through translation.

Induction of *tyrS* expression by limitation for tyrosine suggested that antitermination occurs preferentially under conditions when aminoacylation of tRNA^{Tyr} is reduced, whereas termination occurs when tRNA^{Tyr} is efficiently aminoacylated. Since *tyrS* encodes the enzyme responsible for aminoacylation of tRNA^{Tyr}, this regulatory pattern would allow increased synthesis of the enzyme in response to an increase in the substrate:product ratio. Expression of an unchargeable variant of tRNA^{Tyr} in vivo resulted in efficient readthrough of the *tyrS* leader region terminator during growth in the presence of tyrosine, demonstrating that uncharged tRNA is the effector in vivo and that amino acid limitation acts through its effect on tRNA aminoacylation (Grundy et al. 1994). Nonsense suppressor tRNAs were further used to demonstrate that the presence in the cell of charged tRNA matching the specifier sequence reduces the efficiency of antitermination directed by uncharged tRNA, supporting the model that the system responds to the ratio between charged and uncharged tRNA, rather than the absolute amount of uncharged tRNA.

The molecular basis for discrimination between uncharged and charged tRNA was also established by genetic analyses. Possible base-pairing between the unpaired acceptor end of uncharged tRNA (5′-NCCA-3′) and residues within the bulge of the antiterminator element (5′-UGGN-3′) was supported by covariation between the variable position of the antiterminator (the N position in the bulge) and the tRNA discriminator position (preceding the 3′-terminal CCA). Substitution of the variable position in the *tyrS* antiterminator bulge resulted in loss of expression, despite the fact that this position is not conserved in other genes in the T box family. Restoration of antitermination by a compensatory change in tRNA^{Tyr} supported the model that base-pairing is required, and suggested that addition of an amino acid to the 3′ end of the tRNA could interfere with its interaction with the antiterminator (Grundy et al. 1994).

Extensive mutational analysis of both the *tyrS* leader sequence and tRNA^{Tyr} demonstrated that the sequence and structural elements conserved in T box leader RNAs are important for antitermination in vivo (Rollins et al. 1997; Winkler et al. 2001; Grundy et al. 2002b) and that the overall tertiary structure of the tRNA is required, although substitutions within helical domains and in the variable loop of the tRNA are tolerated (Grundy et al. 2000). Modification of the *tyrS* specifier sequence and antiterminator to permit interaction with noncognate tRNAs revealed that although certain tRNAs could be directed to interact with the *tyrS* leader RNA in vivo, other tRNAs exhibit no antitermination activity, and no tRNA is as effective as the cognate tRNA (Grundy et al. 1997). These results suggested the existence of recognition determinants in addition to the specifier sequence and the variable position of the antiterminator, but no clear pattern emerged from bioinformatics analysis of T box leaders and the corresponding tRNAs (F.J. Grundy and T.M. Henkin, unpubl.).

Biochemical Analysis of the T Box System

Although genetic analyses of the *B. subtilis tyrS* gene, and similar analyses of other T box genes, clearly demonstrated that interaction of the leader RNA with the cognate uncharged tRNA is required for antitermination, a key unanswered question was whether the tRNA–leader RNA interaction is sufficient to promote antitermination in the absence of other cellular factors. A small model RNA designed to mimic the antiterminator element was capable of specific tRNA acceptor end binding (Gerdeman et al. 2002), whereas an RNA modeled on the specifier loop region of the *B. subtilis glyQS* leader sequence could bind an anticodon stem-loop RNA (Nelson et al. 2006). However, tRNA interaction with an intact *tyrS* leader could not be demonstrated (Grundy et al. 2002b; F.J. Grundy and T.M. Henkin, unpubl.).

The *B. subtilis glyQS* leader RNA, a natural variant that lacks two large structural elements common to most T box leaders, was employed in an attempt to reproduce tRNA-dependent antitermination in vitro. A simplified transcription system utilizing purified components, including either *B. subtilis* or *E. coli* RNAP, was used to demonstrate that the *glyQS* leader region terminator is active in vitro; addition of tRNA^{Gly} generated by in vitro transcription with phage T7 RNAP is sufficient to promote efficient antitermination in the absence of other cellular components (Grundy et al. 2002a). Antitermination in vitro requires base-pairing between the tRNA anticodon and acceptor end with the *glyQS* specifier sequence and antiterminator bulge, as was observed in vivo. These studies clearly demonstrated that the nascent RNA is competent for specific tRNA binding and can discriminate between cognate and noncognate tRNAs. Mutational analysis of tRNA^{Gly} showed that the intact tRNA is required (Yousef et al. 2003), as was previously observed in vivo. tRNA^{Thr}-dependent antitermination of the more complex *B. subtilis thrS* gene in vitro was demonstrated by Putzer et al. (2002), but additional components were required. Charged tRNA (for *thrS*) or addition of an extra nucleotide at the 3′ end of the tRNA (for *glyQS*) inhibits antitermination, as predicted by the in vivo studies, and the charged tRNA (or charged tRNA mimic) acts as a competitive inhibitor (Putzer et al. 2002; Grundy et al. 2005). Transcription complexes were probed with an antisense DNA oligonucleotide complementary to sequences that form the 3′ side of the terminator helix, and binding of the oligonucleotide was monitored by cleavage with RNase H, which is specific for RNA–DNA hybrids (Yousef et al. 2005). These studies clearly demonstrated that complexes generated in the absence of tRNA or in the presence of the charged tRNA mimic are in the terminator configuration (resistant to oligonucleotide binding and RNase H cleavage), whereas complexes generated in the presence of uncharged tRNA are in the antiterminator configuration (sensitive to RNase H).

Since two positions of base-pairing between the tRNA and the leader RNA were required for antitermination, a set of transcription complexes in which transcription elongation could be transiently blocked at specific positions by binding of a DNA-binding protein to the template DNA was generated to assess the ability of uncharged tRNA^{Gly} and a charged tRNA^{Gly} mimic to interact with the nascent transcript. Early during transcription, either tRNA could interact with the transcript, and the efficiency of antitermination was dependent on the ratio of the two tRNA species; preincubation with uncharged tRNA yielded complexes that could be disrupted by addition of the charged tRNA mimic (Grundy et al. 2005). Complexes containing the entire leader RNA, including the intact antiterminator, were fully competent for interaction with uncharged tRNA, but these complexes were now resistant to challenge with the charged tRNA mimic. These results suggested that pairing between the acceptor end of the tRNA and the antiterminator results in formation of a stable complex that can no longer be disrupted by an excess of charged tRNA.

The ability of a nascent transcript extending through the antiterminator to interact with tRNA suggested the possibility that *glyQS* leader RNA synthesized in vitro by T7 RNAP could also be competent for tRNA binding. This was tested using a size exclusion filtration assay, which permitted separation of tRNA–leader RNA complexes from unbound tRNA. Binding was dependent on base-pairing with the specifier sequence and the antiterminator and required denaturation of the leader RNA and refolding in the presence of the tRNA (Yousef et al. 2005).

T Box Leader RNA Structure

Formation of leader RNA–tRNA complexes in vitro was exploited to map changes in cleavage patterns using a variety of structure-sensitive RNA cleavage agents. The *glyQS* leader RNA exhibited cleavage patterns consistent with the predicted structure, although several regions of the RNA shown as unpaired loops in the phylogenetic model were resistant to cleavage, suggesting that these regions are in fact structured (Yousef et al. 2005); these results are consistent with predictions that several unpaired domains within T box family leaders form higher-order structural domains, including kink-turn and S-turn motifs. Incubation in the presence of tRNA^{Gly} results in protection of the specifier sequence region and the antiterminator bulge, consistent with the known leader RNA–tRNA base-pairing in those regions. Protection of the anticodon loop of the tRNA, and of D19 in the D loop, was also observed. Additional regions of the leader RNA that are not known to directly interact with the tRNA were also protected in the presence of the matching uncharged tRNA, whereas no protection was observed with a mismatched tRNA, and only the specifier region was protected by the charged tRNA mimic (Fig. 1) (Yousef et al. 2005; N. Green et al., unpubl.). These results are consistent with genetic data which suggest that regions of the leader in addition to the specifier sequence and antiterminator are important for antitermination in vivo. The observation that charged tRNA confers protection only of the specifier sequence supports the model that although both uncharged and charged tRNA can make the initial interaction with the specifier sequence, only uncharged tRNA can interact with the antiterminator, and that interaction is required for stabilization of the tRNA–leader RNA complex and antitermination.

Detailed structural information is currently available only for the antiterminator domain of T box leader RNAs. The nuclear magnetic resonance (NMR) structure of a 39-nucleotide antiterminator model RNA was determined (Gerdeman et al. 2003), revealing stacking of the upper helix and the 3′ portion of the internal bulge onto the bottom helix. The bulge residues predicted to interact with the acceptor end of the tRNA appear to be highly flexible, suggesting that these residues are made available for interaction with the tRNA, allowing stabilization of the antiterminator element by tRNA binding. Alteration of a conserved C residue in the 3′ portion of the bulge resulted in an increase in mobility of the bulge region and reduction of tRNA binding to the model RNA, consistent with the deleterious effect of the corresponding substitution on antitermination in vivo and in vitro in the context of an intact leader RNA (Rollins et al. 1997; N. Green et al., unpubl.; J. Hines, pers. comm.). These results provide further support for the importance of the arrangement of the bulge residues for antiterminator function.

A key open question is the arrangement of the Stem I region that includes the specifier sequence, which is crucial for specific recognition of the anticodon loop of the cognate tRNA. Genetic studies have demonstrated the importance of conserved elements within Stem I, including the kink-turn motif below the specifier loop, the S-turn within the specifier loop, and the AG loop and GGUGNRA elements at the top (Rollins et al. 1997; Winkler et al. 2001; N. Green et al., unpubl.). A combination of genetic, biochemical, and structural biology approaches will be necessary to provide insight into the role of Stem I elements in tRNA binding.

Transcriptional versus Translational Control

Most T box family genes are found in low G+C gram-positive bacteria, but there are also a number of genes that have been identified in high G+C gram-positives as well as certain gram-negative organisms (Grundy and Henkin 2003; Merino and Yanofsky 2005). In these cases, regulation appears to operate at the level of translation initiation, rather than transcription termination, since the final helical element of the RNA does not resemble an intrinsic transcriptional terminator but instead is positioned to sequester the Shine-Dalgarno (SD) sequence, so that formation of this helix is predicted to inhibit binding of the 30S ribosomal subunit to the mRNA. T box RNAs of this type are predicted to form an alternate structure that is stabilized by binding to the cognate tRNA; in this case, the alternate structure sequesters sequences that would otherwise bind to the SD region, so that binding of uncharged tRNA allows translation of the downstream coding sequence. No systems of this type have yet been characterized biochemically, so that translational control remains to be demonstrated experimentally.

THE S BOX SYSTEM: REGULATION OF METHIONINE-RELATED GENES BY SAM

Initial analysis of the distribution of T box genes in bacterial genomes revealed that although genes involved in methionine biosynthesis were members of the T box family in certain organisms, such as *Enterococcus*, no T box genes with AUG methionine codons were uncovered in the first genomes of *Bacillus* and *Clostridium* that were available (although genes of this type were subsequently uncovered; Grundy and Henkin 2002; Rodionov et al. 2004). Multiple genes involved in biosynthesis and transport of methionine were instead found to contain leader RNA elements that fit a new pattern designated the S box (Fig. 3) (Grundy and Henkin 1998). The identification of this pattern upstream of the SAM synthetase gene in several organisms suggested that SAM, which is synthesized from methionine, was the likely final product of the regulated pathway. The role of SAM as the effector was supported by the observation that *B. subtilis* mutants with defects in SAM synthetase exhibited elevated methionine pools, whereas overexpression of SAM synthetase resulted in methionine auxotrophy (Wabiko et al. 1988; Yocum et al. 1996). Transcriptional analysis of *B. subtilis* gene expression during growth with various sulfur sources also led to the prediction that SAM is the likely effector for regulation of methionine biosynthesis (Auger et al. 2002). A combination of genetic and biochemical studies, recently capped by the three-dimensional structure of an S box RNA in complex with SAM (Montange and Batey 2006), has now established the molecular mechanism for SAM-dependent repression of S box gene expression.

Genetic Analysis of S Box RNAs

Like T box RNAs, most S box sequences include transcriptional terminators and competing antiterminators. However, in this case, the terminator helices were predicted to be less stable than the antiterminator elements,

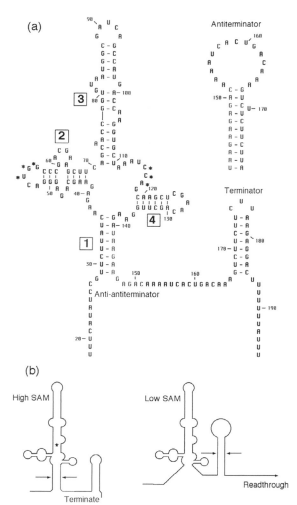

Figure 3. The S box system. (*a*) Secondary structural model of the *B. subtilis yitJ* leader RNA in the terminator conformation; the competing antiterminator conformation is shown above the terminator. The anti-antiterminator, which forms in conjunction with the terminator, sequesters sequences (*gray*) that form the 5′ side of the antiterminator. Numbering is relative to the transcription start site. Boxed numbers indicate helices 1–4 of the SAM-binding element. Asterisks indicate residues that pair to form a tertiary interaction that is facilitated by the kink-turn in the internal bulge of helix 2. (*b*) Helix 1 is stabilized by binding of SAM (asterisk within helix 3; the position of SAM has been demonstrated by crystal structure analysis of the SAM–RNA complex [Montange and Batey 2006]). Formation of helix 1 prevents formation of the antiterminator, allowing formation of the competing terminator helix and premature termination of transcription. In the absence of SAM, the RNA folds into the more stable antiterminator conformation, and transcription proceeds past the termination site, allowing expression of the downstream coding regions.

and the sequences forming the 5′ side of the antiterminator were predicted to pair with sequences further upstream, to form an anti-antiterminator element, formation of which would prevent formation of the antiterminator. These predictions were tested by mutational analysis using the *B. subtilis yitJ* gene, which encodes methylenetetrahydrofolate reductase.

Expression of *yitJ* was shown to be low when cells were grown in the presence of high methionine and was induced when cells were starved for methionine; this response was dependent on the presence of the terminator, and mutations that disrupted the antiterminator resulted in low, uninducible expression. In contrast, mutations that disrupted the anti-antiterminator resulted in high expression during growth in the presence of methionine, consistent with the prediction that this element competes with formation of the antiterminator to cause repression (Grundy and Henkin 1998). The anti-antiterminator element forms the base of a complex four-way helical junction that includes a pseudoknot, the arrangement of which is highly conserved in S box RNAs. Disruption of any of the conserved sequence or structural elements within the S box pattern (with the exception of the antiterminator) resulted in high expression during growth in methionine, indicating that these elements are required for repression (Grundy and Henkin 1998; Winkler et al. 2001; McDaniel et al. 2005). *B. subtilis* strains with elevated or reduced SAM synthetase activity in vivo exhibited alterations in S box gene expression, consistent with the model that SAM is the effector in vivo (McDaniel et al. 2003, 2006).

Biochemical Analyses of S Box RNAs

A direct interaction between SAM and S box RNAs was reported by three groups (Epshtein et al. 2003; McDaniel et al. 2003; Winkler et al. 2003). The S box leader RNA terminator element was shown to be inactive in an in vitro transcription system using purified components, and addition of SAM was sufficient to cause efficient termination in the absence of other cellular components. SAM binding resulted in rearrangement of the RNA from the antiterminator to the anti-antiterminator conformation, in agreement with the genetic studies which suggested that the anti-antiterminator element is the target for binding of the regulatory factor that promotes termination in vivo. Binding of SAM to the purified RNA was also demonstrated, and mutations that disrupt repression in vivo resulted in loss of SAM binding and loss of the SAM-dependent transcription termination in vitro. SAM binds with very high affinity to purified S box leader RNAs, and different S box genes exhibit different sensitivities to SAM both in vitro and in vivo, in a pattern consistent with the physiological role of the regulated gene products (J. Tomsic et al., unpubl.). The SAM–RNA complex is also very stable in vitro and is unlikely to dissociate within the time frame of the termination/antitermination decision.

Structural Analysis of S Box RNAs

A variety of techniques have been employed to monitor changes in S box RNA structure in response to SAM. Structural mapping using oligonucleotide binding (Epshtein et al. 2003; McDaniel et al. 2003), in-line attack (Winkler et al. 2003), and enzymatic cleavage (McDaniel et al. 2005) all showed stabilization of the anti-antiterminator element by SAM binding and multiple changes throughout the S box RNA, including stabilization of the tertiary interaction between the loop of helix 2 and the

helix 2–3 junction region (McDaniel et al. 2005). These results were recently substantiated by the crystal structure of an S box RNA in complex with SAM, which revealed a compact structure with SAM embedded in a pocket formed by the juxtaposition of helices 1 and 3 and stabilized by the tertiary interaction (Montange and Batey 2006). Like other riboswitch RNAs, the SAM-binding S box RNA exploits the entire surface of the effector molecule to mediate specific recognition.

Transcriptional versus Translational Control

As has been found for a number of other riboswitch RNAs, regulation by the S box element appears to occur predominantly at the level of premature termination of transcription in low G+C gram-positive bacteria, but it can also occur at the level of translation initiation (Grundy and Henkin 2006). In genes of this type, the final helix that is predicted to form in the presence of SAM includes the SD sequence of the downstream coding sequence, whereas in the absence of SAM, a competing helix (analogous to the antiterminator described above) prevents formation of the SD-sequestering helix. Translational control is unusual among S box genes and appears to predominate in high G+C gram-positive and in gram-negative species.

Other SAM-binding Riboswitch RNAs

Two classes of SAM-binding RNAs distinct from the S box RNAs have been uncovered. The SAM-II RNAs are found in α-proteobacteria, but their mechanism of action is unknown (Corbino et al. 2005). The S_{MK} box riboswitch is found in SAM synthetase (*metK*) genes of lactic acid bacteria including *Streptococcus* and *Enterococcus* sp., and SAM binding was shown to cause sequestration of the SD region in a pseudoknot structure that represses translation initiation in vivo (Fuchs et al. 2006). It therefore appears that there are multiple molecular mechanisms for SAM recognition by RNA and that regulation of genes involved in generation of methionine and SAM can respond to SAM pools in different ways.

CONCLUSION

The abundance of systems in which nascent RNAs act *in cis* to modulate expression of downstream coding sequences is growing rapidly. Although regulation at the levels of premature termination of transcription and translation initiation predominate so far, it is likely that more examples of other mechanisms, including effects on RNA stability, will be discovered. For some of these mechanisms, including translational regulation, it is possible that the regulatory elements can act in full-length rather than nascent transcripts, and that the elements can be located in regions other than the 5′ end of the mRNA. Uncovering novel variations on this general theme will require both new bioinformatics approaches and the ability to recognize alternative possibilities that deviate from the paradigm systems. Biophysical approaches are beginning to complement the genetics and biochemistry that have revealed so much of the basic mechanisms underlying these systems, and the emerging three-dimensional structures provide new insight into the molecular recognition properties of regulatory RNAs. The possibility that many more mechanisms of this type may operate in eukaryotes is also tantalizing.

ACKNOWLEDGMENTS

This work was supported by National Institutes of Health grants GM47823 and GM63615.

REFERENCES

Auger S., Danchin A., and Martin-Verstraete I. 2002. Global expression profile of *Bacillus subtilis* grown in the presence of sulfate or methionine. *J. Bacteriol.* **81:** 741.

Corbino K.A., Barrick J.E., Lim J., Welz R., Tucker B.J., Puskarz I., Mandal M., Rudnick N.D., and Breaker R.R. 2005. Evidence for a second class of *S*-adenosylmethionine riboswitches and other regulatory RNA motifs in alpha-proteobacteria. *Genome Biol.* **6:** R70.

Epshtein V., Mironov A.S., and Nudler E. 2003. The riboswitch-mediated control of sulfur metabolism in bacteria. *Proc. Natl. Acad. Sci.* **100:** 5052.

Fuchs R.T., Grundy F.J., and Henkin T.M. 2006. The S$_{MK}$ box is a new SAM-binding RNA for translational regulation of SAM synthetase. *Nat. Struct. Mol. Biol.* **13:** 226.

Gerdeman M.S., Henkin T.M., and Hines J.V. 2002. In vitro structure-function studies of the *Bacillus subtilis tyrS* mRNA antiterminator: Evidence for factor-independent tRNA acceptor stem binding specificity. *Nucleic Acids Res.* **30:** 1065.

———. 2003. Solution structure of the *Bacillus subtilis* T-box antiterminator RNA: Seven nucleotide bulge characterized by stacking and flexibility. *J. Mol. Biol.* **326:** 189.

Grundy F.J. and Henkin T.M. 1993. tRNA as a positive regulator of transcription antitermination in *B. subtilis. Cell* **74:** 475.

———. 1994. Conservation of a transcription antitermination mechanism in aminoacyl-tRNA synthetase and amino acid biosynthesis genes in Gram-positive bacteria. *J. Mol. Biol.* **235:** 798.

———. 1998. The S box regulon: A new global transcription termination control system for methionine and cysteine biosynthesis genes in Gram-positive bacteria. *Mol. Microbiol.* **30:** 737.

———. 2002. Synthesis of serine, glycine, cysteine and methionine. In Bacillus subtilis *and its closest relatives: From genes to cells* (ed. A.L. Sonenshein et al.), p. 245. American Society for Microbiology Press, Washington, D.C.

———. 2003. The T box and S box transcription termination control systems. *Front. Biosci.* **8:** d20.

———. 2004. Regulation of gene expression by effectors that bind to RNA. *Curr. Opin. Microbiol.* **7:** 126.

———. 2006. From ribosome to riboswitch: Control of gene expression in bacteria by RNA structural rearrangements. *Crit. Rev. Biochem. Mol. Biol.* (in press).

Grundy F.J., Rollins S.M., and Henkin T.M. 1994. Interaction between the acceptor end of tRNA and the T box stimulates antitermination in the *Bacillus subtilis tyrS* gene: A new role for the discriminator base. *J. Bacteriol.* **176:** 4518.

Grundy F.J., Winkler W.C., and Henkin T.M. 2002a. tRNA-mediated transcription antitermination *in vitro:* Codon-anticodon pairing independent of the ribosome. *Proc. Natl. Acad. Sci.* **99:** 11121.

Grundy F.J., Yousef M.R., and Henkin T.M. 2005. Monitoring uncharged tRNA during transcription of the *Bacillus subtilis glyQS* gene. *J. Mol. Biol.* **346:** 73.

Grundy F.J., Collins J.A., Rollins S.M., and Henkin T.M. 2000.

tRNA determinants for transcription antitermination of the *Bacillus subtilis tyrS* gene. *RNA* **6:** 1131.

Grundy F.J., Hodil S.E., Rollins S.M., and Henkin T.M. 1997. Specificity of tRNA-mRNA interactions in *Bacillus subtilis tyrS* antitermination. *J. Bacteriol.* **179:** 2587.

Grundy F.J., Moir T.R., Haldeman M.T., and Henkin T.M. 2002b. Sequence requirements for terminators and antiterminators in the T box transcription antitermination system: Disparity between conservation and functional requirements. *Nucleic Acids Res.* **30:** 1646.

Henkin T.M., Glass B.L., and Grundy F.J. 1992. Analysis of the *Bacillus subtilis tyrS* gene: Conservation of a regulatory sequence in multiple tRNA synthetase genes. *J. Bacteriol.* **174:** 1299.

Landick R., Turnbough C.L., and Yanofsky C. 1996. Transcription attenuation. In Escherichia coli *and* Salmonella: *Cellular and molecular biology* (ed. F.C. Neidhardt et al.), p. 1263. American Society for Microbiology, Washington, D.C.

McDaniel B.A.M., Grundy F.J., and Henkin T.M. 2005. A tertiary structural element in S box leader RNAs is required for *S*-adenosylmethionine-directed transcription termination. *Mol. Microbiol.* **57:** 1008.

McDaniel B.A.M., Grundy F.J., Artsimovitch I., and Henkin T.M. 2003. Transcription termination control of the S box system: Direct measurement of *S*-adenosylmethionine by the leader RNA. *Proc. Natl. Acad. Sci.* **100:** 3083.

McDaniel B.A., Grundy F.J., Kurlekar V.P., Tomsic J., and Henkin T.M. 2006. Identification of a mutation in the *Bacillus subtilis S*-adenosylmethionine synthetase gene that results in derepression of S-box gene expression. *J. Bacteriol.* **188:** 3674.

Merino E. and Yanofsky C. 2005. Transcription attenuation: A highly conserved regulatory strategy used by bacteria. *Trends Genet.* **21:** 260.

Montange R.K and Batey R.T. 2006. Structure of the *S*-adenosylmethionine riboswitch regulatory mRNA element. *Nature* **441:** 1172.

Nelson A.R., Henkin T.M., and Agris P.F. 2006. tRNA regulation of gene expression: Interactions of an mRNA 5′-UTR with a regulatory tRNA. *RNA* **12:** 1254.

Putzer H., Condon C., Brechemier-Baey D., Brito R., and Grunberg-Manago M. 2002. Transfer RNA-mediated antitermination in vitro. *Nucleic Acids Res.* **30:** 3026.

Rodionov D.A., Vitreschak A.G., Mironov A.A., and Gelfand M.S. 2004. Comparative genomics of the methionine metabolism in Gram-positive bacteria: A variety of regulatory systems. *Nucleic Acids Res.* **32:** 3340.

Rollins S.M., Grundy F.J., and Henkin T.M. 1997. Analysis of *cis*-acting sequence and structural elements required for antitermination of the *Bacillus subtilis tyrS* gene. *Mol. Microbiol.* **25:** 411.

Wabiko H., Ochi K., Nguyen D.M., Allen E.R., and Freese E. 1988. Genetic mapping and physiological consequences of *metE* mutations of *Bacillus subtilis. J. Bacteriol.* **170:** 2705.

Winkler W.C., Grundy F.J., Murphy B.A., and Henkin T.M. 2001. The GA motif: An RNA element common to bacterial antitermnation systems, rRNA, and eukaryotic RNAs. *RNA* **7:** 1165.

Winkler W.C., Nahvi A., Surarsan N., Barrick J.E., and Breaker R.R. 2003. An mRNA structure that controls gene expression by binding *S*-adenosylmethionine. *Nat. Struct. Biol.* **10:** 701.

Yocum R.R., Perkins J.B., Howitt C.L., and Pero J. 1996. Cloning and characterization of the *metE* gene encoding *S*-adenosylmethionine synthetase from *Bacillus subtilis. J. Bacteriol.* **178:** 4604.

Yousef M.R., Grundy F.J., and Henkin T.M. 2003. tRNA requirements for *glyQS* antitermination: A new twist on tRNA. *RNA* **9:** 1148.

———. 2005. Structural transitions induced by the interaction between tRNA^{Gly} and the *Bacillus subtilis glyQS* T box leader RNA. *J. Mol. Biol.* **349:** 273.

Genetic Control by *cis*-Acting Regulatory RNAs in *Bacillus subtilis*: General Principles and Prospects for Discovery

IRNOV, A. KERTSBURG, AND W.C. WINKLER

The University of Texas Southwestern Medical Center, Department of Biochemistry, Dallas, Texas 75390-9038

In recent years, *Bacillus subtilis*, the model organism for gram-positive bacteria, has been a focal point for study of posttranscriptional regulation. In this bacterium, more than 70 regulatory RNAs have been discovered that respond to intracellular proteins, tRNAs, and small-molecule metabolites. In total, these RNA elements are responsible for genetic control of more than 4.1% of the genome-coding capacity. This pool of RNA-based regulatory elements is now large enough that it has become a worthwhile endeavor to examine their general features and to extrapolate these simple observations to the remaining genome in an effort to predict how many more may remain unidentified. Furthermore, both metabolite- and tRNA-sensing regulatory RNAs are remarkably widespread throughout eubacteria, and it is therefore becoming increasingly clear that some of the observations for *B. subtilis* gene regulation will be generally applicable to many different species.

cis-ACTING REGULATORY RNAs IN THE MODEL BACTERIUM, *BACILLUS SUBTILIS*

Microorganisms must be capable of responding to sudden changes in their environment, such as stress conditions and nutrient limitation. To that end, microorganisms use a diverse assortment of genetic strategies for precise coordination of their genes. Although the majority of these mechanisms exert their influence over transcription initiation, a growing number have been found to regulate postinitiation processes. Just how important this posttranscriptional "layer" of gene regulation is to a given bacterium is still an open question. However, studies from *B. subtilis* and *Escherichia coli* suggest that a substantial portion of the genome is regulated by a combination of *cis*- and *trans*-acting regulatory RNAs.

cis-Acting regulatory RNAs are transcribed with the genes they regulate and are the subject of this chapter, whereas *trans*-acting regulatory RNAs are transcribed separately from their target genes and are discussed elsewhere. Considerable progress has recently been achieved in the study of *cis*-acting regulatory RNAs, helped considerably by an intensive focus on *B. subtilis* gene regulation. For example, a significant portion (>4.1%) of the *B. subtilis* genome has already been postulated to be regulated from the combined contributions of more than 70 *cis*-acting regulatory RNAs (Winkler 2005). These RNAs can be split into a few different categories: those that respond to intracellular proteins (currently 22 total), tRNAs (19 total), metabolites (currently 27 total), a collection of "orphans" that are likely to respond to metabolites (currently 7 total), and at least half a dozen less-characterized regulatory RNAs (Table 1). The ligand-binding domains for all of these regulatory RNAs are referred to as aptamers, whereas the remaining portions are involved in harnessing ligand-induced conformational

changes for genetic control. In general, protein-sensing RNAs regulate a wide variety of genes involved in carbohydrate catabolism, amino acid synthesis, and transport. tRNA-sensing RNAs, called T-box RNAs, regulate aminoacyl tRNA-synthetases, certain amino acid biosynthetic clusters, and transport genes. Metabolite-sensing RNAs primarily control expression of genes responsible for synthesis and transport of cofactors, amino sugars, amino acids, and nucleotides.

An assortment of metabolite-sensing RNAs have been discovered that respond to adenosylcobalamin, thiamine pyrophosphate (TPP), flavin mononucleotide (FMN), guanine, adenine, a precursor for queuine, lysine, glycine, and glucosamine-6-phosphate (GlcN6P). Additionally, three structurally distinct classes that sense *S*-adenosylmethionine (SAM) have been identified (for review, see Grundy and Henkin 2004a; Nudler and Mironov 2004; Winkler and Breaker 2005). With the exception of two SAM-sensing RNA elements, all of these individual classes have been identified in *B. subtilis* (see Fig. 3). Biochemical analyses suggest that these RNA structures exhibit some preorganization prior to ligand association but are conformationally modified upon binding. In general, gel-based structural probing, equilibrium dialysis-based analyses, and measurements of ligand binding by fluorescence quenching or isothermal titration calorimetry all reveal apparent K_D values that are in the low-nanomolar range. Therefore, these natural aptamers bind tightly to their target metabolites. For at least several RNA classes, recent structural data demonstrate how this is achieved (Batey 2006). X-ray crystallographic data reveal that metabolite ligands are selectively stabilized via a combination of specific hydrogen bonding, energy of base stacking, metal ion stabilization, and ionic interactions.

Table 1. Regulation of *B. subtilis* Expression by Protein-, tRNA-, and Metabolite-sensing RNAs

Regulated Gene(s)	Effector(s) ligands and mode of gene control
trpEDCFBA, pabA, trpP, ycbK Tryptophan metabolism	Trp-bound TRAP inhibits translation initiation and/or induces transcription termination (for review, see Gollnick et al. 2005).
glpFK, glpD, glpTO Glycerol metabolism	GlpP protein stabilizes an antiterminator helix (for review, see Stülke 2002).
hutPHUIGM Histidine catabolism	Histidine-bound HutP protein stabilizes an antiterminator helix (for review, see Kumar et al. 2006).
pyrR, pyrP, pyrB/C/AA/AB/K/D/F/E Pyrimidine synthesis	UMP-bound PyrR protein stabilizes an anti-antiterminator helix (Switzer et al. 1999).
ptsGHI Glucose transport	GlcT protein stabilizes an antiterminator helix; GlcT is activated by phosphorylation by HPr and inactivated due to phosphorylation by EIIGlc (for review, see Stülke 2002).
bglPH, bglS Sugar metabolism (aryl β-glucoside utilization)	LicT binds to *bglPH* and *bglS* RNAs and stabilizes an antiterminator helix when cells are exposed to aryl β-glucosides and are limited for glucose (for review, see Stülke 2002).
sacB, sacXY Sugar metabolism (sucrose utilization)	In the absence of sucrose, SacY is inactivated through phosphorylation by SacX; in its activated state, SacY stabilizes an antiterminator (for review, see Stülke 2002).
sacPA Sugar metabolism (sucrose utilization)	The SacT protein stabilizes an antiterminator helix (for review, see Stülke 2002).
rho	Rho autoregulates itself (Ingham et al. 1999).
rpsD Ribosomal protein S4	S4 is likely to associate to its own UTR for feedback repression (Grundy and Henkin 1992).
rpsJ Ribosomal protein S10	The S10 operon is likely to be bound by a ribosomal protein encoded within the operon (Li et al. 1997).
cspB, cspC Cold shock response	CspB and CspC promote translation during conditions of low temperature by associating to their 5'UTRs (Graumann and Marahiel 1999).
pyrG CTP synthesis	During conditions of low CTP reiterative transcription adds a polyG sequence that acts to stabilize an antiterminator helix (Meng et al. 2004).
cysE-cysS Cysteine synthesis	Uncharged tRNA^{Cys} binds to an intercistronic T-box RNA, instigating antitermination (Pelchat and Lapointe 1999).
leuS, pheS-pheT, serS, thrS, thrZ trpS, tyrS, tyrZ, valS, ilvBNCA-leuCBD, rtpA. proB-proA, ileS, proI, glyQ-glyS, alaS hisS-aspS, yvbW Aminoacyl synthetases, amino acid biosynthesis, and transport	Cognate uncharged tRNA binds to the 5'UTR to promote antitermination (for review, see Grundy and Henkin 2003).
yitJ, metI-metC, ykrT-ykrS, ykrW-ykrX-ykrY-ykrZ, cysH-cysP-sat-cysC-ylnD-ylnE-ylnF, yoaD-yoaC-yoaB, metE, metK, yusC-yusB-yusA, yxjG, yxjH General methionine biosynthesis, methylene tetrahydrofolate reductase, 5' methylthioadenosine recycling pathway, cysteine biosynthesis, methionine synthase, SAM synthetase, uncharacterized ABC transporters	SAM binds to the 5'UTR to promote termination (for review, see Winkler and Breaker 2005).
thiC, tenA1-thiX1-thiY1-thiz1-thiE2-thiO-thiS-thiG-thiF-thiD, ykoF-ykoE-ykoD-ykoC, yuaJ, ylmB Thiamine biosynthesis and transport, uncharacterized genes	TPP binds to the 5'UTR to promote termination (for review, see Winkler and Breaker 2005).
ypaA, ribD-ribE-ribBA-ribH Riboflavin biosynthesis and flavin transport	FMN binds to the 5'UTR to promote termination (for review, see Winkler and Breaker 2005).
yvrC-yvrB-yvrA-yvqK Unknown; similar to iron transport proteins	Adenosylcobalamin binds to the 5'UTR to promote termination (for review, see Winkler and Breaker 2005).
lysC Aspartokinase II	Lysine binds to the 5'UTR to promote termination (for review, see Winkler and Breaker 2005).
yxjA, xpt-pbuX, pbuG, purE-purK-purB-purC-purS-purQ-purL-purF-purM-purN-purH-purD Purine biosynthesis, pyrimidine nucleoside transport, xanthine permease, hypoxanthine/guanine permease	Guanine binds to the 5'UTR to promote termination (for review, see Winkler and Breaker 2005).
ydhL Uncharacterized transporter	Adenine binds to the 5'UTR to promote antitermination (for review, see Winkler and Breaker 2005).
gcvT/PA/PB Glycine efflux	Glycine binds to the 5'UTR to promote antitermination (Mandal et al. 2004).
glmS Glucosamine-6-phosphate synthesis	GlcN6P binds to the 5'UTR to promote autocatalytic self-cleavage, which in turns leads to mRNA destabilization (Winkler et al. 2004).
ykoK, ykkCDE, yxkD, ydaO, ktrAB, yybP, ykoY Transport of divalent metal ions, uncharacterized transporters, potassium transporter, unknown genes	Unknown metabolite binds to an "orphan" RNA to either promote or disrupt termination formation (Barrick et al. 2004).

Location

There are a few examples where *B. subtilis* regulatory RNAs reside within intercistronic regions of an operon. For example, regulatory RNAs that interact with proteins PyrR (pyrimidine-responsive regulator) and HutP (histidine-responsive regulator) are located between these proteins and the rest of the genes in the operon (Oda et al. 1988; Turner et al. 1994). In addition, a tRNA^Cys-sensing RNA is found between *gltX* (encoding glutamyl-tRNA synthetase) and *cysE-cysS* (encoding serine acetyltransferase and cysteinyl-tRNA synthetase, respectively) and is a site for endoribonuclease cleavage (Gagnon et al. 1994; Pelchat and Lapointe 1999). Similarly, an RNA element that binds to tryptophan-responsive attenuation protein (TRAP) is located within intercistronic spaces between folate synthesis genes *pabB* and *pabA* as well as between *rtpA* and *ycbK* (Babitzke 2004). However, the majority of *B. subtilis* regulatory RNAs are located within the 5′-untranslated regions (5′UTRs) of specific mRNA transcripts. In contrast, no *cis*-acting regulatory RNAs have been discovered within the 3′UTR portion.

On rare occasions, regulatory RNAs can occur in tandem. For example, the 5′UTR of *B. subtilis thrZ* (a cryptic threonyl-tRNA synthetase) contains three consecutive tRNA^Thr-sensing RNAs (Putzer et al. 1992). Presumably, this arrangement results in expression of *thrZ* only under conditions of extreme threonine starvation or very low levels of threonyl-tRNA synthetase. An interesting variation on this theme is exploited by glycine-sensing RNAs. These RNA structures are composed of two glycine-binding aptamers that exhibit cooperativity for glycine association; glycine binding to the first domain increases glycine binding affinity for the second domain (Mandal et al. 2004). Only upon occupation of both glycine-binding sites is expression of efflux or catabolism genes altered.

Genetic Mechanisms

In general, bacterial *cis*-acting regulatory RNAs control gene expression via transcription attenuation or translation inhibition. Both mechanistic strategies rely upon the specific orchestration of alternate base-pairing schemes (Fig. 1). For transcription attenuation, a metabolic signal is received by the aptamer domain that in turn stimulates formation of a transcription termination signal, usually in the form of an intrinsic terminator helix (responsible for rho-independent cessation of transcription). Typically, this occurs by stabilization of an anti-antiterminator helical element, thereby allowing terminator formation. It is the interchange between terminator and antiterminator helices that dictate expression levels; therefore, in the absence of the aptamer-bound signal, an alternate antiterminator helix will be formed as the default state configuration. In other instances (e.g., tRNA-sensing RNAs) (Grundy and Henkin 2003), the default conformation is formation of a transcription terminator helix. For these RNAs, ligand association promotes formation of an antiterminator helical element, thereby preventing terminator formation, often referred to as transcription antitermination. The strongest experimental

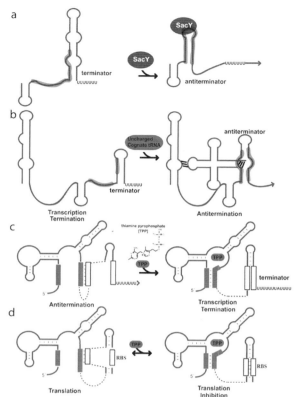

Figure 1. Genetic regulation by *cis*-acting bacterial regulatory RNAs. (*a*) Example of a protein-responsive RNA. In the absence of sucrose, SacY is inactivated through phosphorylation by SacX. In its activated state, SacY stabilizes an antiterminator helix for the *sacB* and *sacXY* transcripts (for review, see Stülke 2002). (*b*) Example of tRNA-sensing RNAs (Grundy and Henkin 2003). A decrease in amino acid levels alters intracellular tRNA charging ratios. Uncharged cognate tRNAs associate with the appropriate T-box RNA to promote transcription antitermination, thereby increasing expression of aminoacyl tRNA synthetases, biosynthesis genes, and transporters. (*c*) Example of a metabolite-sensing regulatory RNA: transcription attenuation. Binding of thiamine pyrophosphate (TPP) stimulates formation of an anti-antiterminator helix, thereby allowing formation of an intrinsic transcription terminator within the 5′UTR of a *B. subtilis* thiamine biosynthetic cluster (Mironov et al. 2002). (*d*) Example of a metabolite-sensing regulatory RNA: translation inhibition. Binding of TPP stimulates formation of a helical element that sequesters the ribosome-binding site, thereby reducing translation initiation efficiency for *E. coli thiM* transcripts (Winkler et al. 2002).

proof for transcription attenuation/antitermination-based mechanisms has derived from their reconstitution in vitro from purified components (RNA polymerase, DNA templates, ribonucleotides, and the appropriate ligand molecule). These latter experiments provide clear evidence for ligand-induced termination in the absence of additional factors, although transcription elongation factors such as NusA may participate in vivo.

A unique feature of these mechanisms is that the outcome is a single "decision" between terminator and antiterminator elements during the active process of transcription. Therefore, the "choice" of forming terminator or antiterminator elements derives from precise coordination of multiple processes including transcription kinetics, the nascent RNA folding pathway, and ligand-binding kinetics. Indeed, evidence for several different regulatory RNAs suggests that accurate harmonization of transcription and ligand-binding kinetics is important for regulation (Landick et al. 1996; Mironov et al. 2002; Yakhnin and Babitzke 2002; Zhang and Switzer 2003; Grundy and Henkin 2004b; Wickiser et al. 2005). Specifically, by synchronizing RNA polymerases on target DNA templates in vitro and studying transcript products at varying time intervals, one can map profiles for sites of transcriptional pausing and termination. Pausing signals have been found in several instances to be located near sequences required for antiterminator formation. These data suggest that institution of a transcriptional delay while the RNA polymerase footprint overlaps antiterminator or terminator elements may allow time for RNA folding and ligand binding, prior to commitment of the ultimate conformational outcome. Inclusion of NusA into this experimentation alters the profile and strength of transcriptional pausing and termination signals in vitro (Landick et al. 1996; Yakhnin and Babitzke 2002; Zhang and Switzer 2003; Grundy and Henkin 2004b). Therefore, it undoubtedly also affects transcription attenuation/antitermination mechanisms in vivo, further implicating a role for NusA in posttranscriptional processes. For FMN-sensing RNAs, these transcriptional kinetic data have been coupled with measurements of FMN association, which further support this hypothesis. Specifically, an FMN-responsive RNA (*B. subtilis ribD*) was found to be unable to reach thermodynamic equilibrium with FMN before RNA polymerase could reach the terminator sequence element. Therefore, the kinetics of transcription must be carefully tuned with the rate of FMN association to the aptamer domain. How general all of these observations are for transcription attenuation-based regulatory RNAs remains to be determined.

The second major mode of regulation is through translation inhibition. Similar to transcription attenuation, ligand association influences the thermodynamic interplay between helical pairings. However, rather than controlling terminator formation, the ribosome-binding site (RBS) is either occluded from ribosomal access or rendered more accessible upon ligand binding (Fig. 1) (see, e.g., Nou and Kadner 2000; Schlax and Worhunsky 2003; Winkler et al. 2004; Yakhnin et al. 2004). A combination of genetic and biochemical approaches has been used to demonstrate ligand-induced blocking of ribosomes for the *B. subtilis* RNAs that exert their regulatory influence via translation inhibition. This experimentation includes 30S toeprint analysis, cell-free translation, and structural probing analyses of the regulatory RNA in the presence and absence of the appropriate RNA-binding ligand. Overall, many bacterial regulatory RNAs govern gene expression by affecting the efficiency of translation initiation in this manner. Interestingly, there is a nonrandom phylogenetic distribution for transcription attenuation and translation inhibition mechanisms in eubacterial species. For reasons not yet revealed, gram-positive bacteria preferentially utilize transcription attenuation mechanisms, whereas gram-negative bacteria more often than not rely upon translation inhibition (see, e.g., Rodionov et al.

2002; Vitreschak et al. 2002). For example, more than 90% of *B. subtilis* regulatory RNAs rely on transcription attenuation-based mechanisms.

THE *GLMS* REGULATORY RNA IS A UNIQUE METABOLITE-SENSING RIBOZYME

As outlined above, there are two predominate methods for harnessing *cis*-acting regulatory RNAs for bacterial genetic control: transcription attenuation and translation inhibition. However, an alternate mechanism is used by a GlcN6P-sensing RNA identified immediately upstream of the *glmS* gene (Fig. 2) (Barrick et al. 2004). Synthesis and characterization of the RNA sequence in vitro revealed that GlcN6P promoted an autocatalytic, site-specific cleavage event near the 5′ terminus, thus demonstrating that the RNA sequence is a natural metabolite-responsive ribozyme (Winkler et al. 2004). A reciprocal relationship was observed between ribozyme self-cleavage in vitro and gene expression in vivo. Deleterious site-directed mutations of conserved positions reduced self-cleavage capability in vitro and correspondingly increased expression of a downstream *lacZ* reporter in vivo. Therefore, feedback repression of *glmS* by intracellular GlcN6P was proposed to be the regulatory function for the ribozyme (Winkler et al. 2004). Indeed, early data indicate that self-cleavage results in release of a 61-nucleotide fragment from the 5′ terminus of the approximately 1.9-kb *glmS* transcript in response to increasing levels of GlcN6P. This then promotes a significant decrease in intracellular stability for *glmS* transcripts through a process that is dependent on at least one intracellular endoribonuclease (W.C. Winkler, unpubl.). However, the complete mechanism for how self-cleavage imparts genetic control in vivo still remains to be elucidated.

Although study of the *glmS* regulatory pathway in vivo is an emerging subject, biochemical characterization of the self-cleavage reaction in vitro has progressed more rapidly. Addition of GlcN6P has been demonstrated to promote the self-cleavage reaction at least 110,000-fold over the rate of cleavage in the absence of GlcN6P (McCarthy et al. 2005). The chemical mechanism for self-cleavage appears to be similar to that of other small natural ribozymes, resulting in 5′-hydroxyl and 2′-3′-cyclic phosphate termini (Fig. 2) (Winkler et al. 2004). Recognition of the GlcN6P ligand appears to be highly specific. Even subtle differences in ligand structure, such as removal of the C6-bound phosphate or the sugar amino group, result in dramatic reduction of ligand-induced catalysis (Winkler et al. 2004; McCarthy et al. 2005). A preview into how this ligand recognition is achieved has been recently derived via nucleotide analog interference mapping (NAIM). This method employs backbone and nucleobase analogs to identify RNA functional groups essential for ligand phosphate recognition (Fig. 2) (Jansen et al. 2006). These positions cluster near the site of self-cleavage or between the P1 and P2 helices, all within the minimal core sequence. Additionally, gel-based structural probing techniques argue that the overall RNA structure is significantly preorganized prior to ligand binding and that it is only modestly altered during catalysis (Hampel and Tinsley 2006).

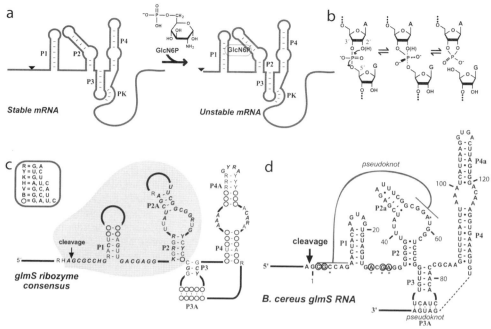

Figure 2. Genetic control by a metabolite-sensing ribozyme. (*a*) A GlcN6P sensing regulatory RNA is positioned within the 5′UTR of *glmS* transcripts for many gram-positive bacteria. GlcN6P stimulates an autocatalytic self-cleavage event at the 5′ end of the regulatory RNA, thereby releasing a short oligonucleotide and reducing mRNA stability for downstream *glmS* transcripts (W.C. Winkler, unpubl.). The details for this mechanism remain to be explored. (*b*) The proposed chemical mechanism for *glmS* ribozyme self-cleavage (Winkler et al. 2004). The 2′ hydroxyl of the −1 position acts as the attacking nucleophile with the bridging 5′ oxygen as the leaving group. The reaction proceeds via an S$_N$2-type in-line attack. (*c*) Consensus secondary structure for *glmS* ribozymes (Roth et al. 2006). Italicized nucleotides in bold type are conserved in at least 95% of representative sequences, and non-italicized letters represent greater than 80% conservation. The site of self-cleavage is indicated with an arrow. (*Gray-shaded zone*) Minimal ribozyme domain. (*d*) NAIM experimentation inferred several positions as being the likely sites for magnesium association (*black circles*) and for recognition of the ligand phosphate (*asterisks*) (Jansen et al. 2006).

Overall, the ribozyme consists of a series of four conserved stem-loop structures (P1–P4; Winkler et al. 2004) and two functionally important pseudoknot elements (Fig. 2) (Wilkinson and Been 2005; Roth et al. 2006; Soukup 2006). Neither deletion of the P3–P4 helices nor of nucleotides 5′ to the site of cleavage completely eliminates catalysis (Fig. 2) (Winkler et al. 2004). However, cleavage activity of this minimal ribozyme is approximately four orders of magnitude reduced relative to a full-length sequence when supplied with similar magnesium concentrations (Roth et al. 2006). Nonetheless, rate constants for the minimal construct are restored to almost full-length levels upon supplementation with additional magnesium ions (~100-fold). Therefore, the function of the P3–P4 region is likely to be optimization of *glmS* cleavage during physiological magnesium concentrations. However, this requirement for magnesium ions does not necessarily reflect a direct involvement of metal ions in catalysis. Indeed, magnesium could be replaced with cobalt hexammine, a structural mimic of hydrated magnesium, without significant loss in cleavage ability (Roth et al. 2006). Given the very slow rate of solvent exchange for the amine groups of cobalt hexammine, this observation suggests that only outer-shell or electrostatic interactions are required of divalent ions and that they are unlikely to directly participate in catalysis.

A common method for divalent ions to bind RNA is through interactions to nonbridging phosphate oxygens. Single-atom substitution of these positions with sulfur groups is a frequently used method that can be employed to identify these sites of backbone interactions. In many instances where phosphorothioate substitutions instigate a loss of essential magnesium ions, more thiophilic metals such as manganese can functionally substitute the missing divalents. This strategy of phosphorothioate substitution and manganese rescue has recently been employed for *glmS* ribozymes, resulting in the identification of candidate sites for magnesium association, many of which are located near the site of self-cleavage (Fig. 2) (Jansen et al. 2006).

How unusual are ribozymes in biology? To date, only nine distinct classes of naturally occurring RNA catalysts have been identified. There are four small self-cleaving RNAs in addition to *glmS* that all catalyze chain cleavage via an internal transesterification reaction. The hammerhead, hepatitis delta virus, and hairpin ribozymes all assist in processing rolling-circle replication intermediates via self-cleavage (and ligation) activities (Lilley 2003; Fedor and Williamson 2005). The Varkud satellite (VS) RNA is a mitochondrial transcript found in many natural isolates of *Neurospora* and is also thought to act in the processing of replication intermediates (Lilley 2004). The remaining known catalytic RNAs include RNase P (processing of pre-tRNAs), group I and II self-splicing introns, and the ribosome (Lilley 2003; Fedor and Williamson 2005). It remains to be determined whether catalytic RNAs are truly rare in extant organisms, or if their absence is due to a lack of intensive searching. However, the fact that only one category of metabolite-sensing regulatory RNA exploits such a mechanism for regulation suggests that during the course of evolution, transcription attenuation and translation inhibition strategies proved to be more efficient. If so, then why retain self-cleavage ability for the *glmS* RNAs? Three-dimensional resolution of structural features when the RNA is bound to the GlcN6P ligand may eventually provide the answer. In the interim, NAIM analyses have suggested that positions required for ligand phosphate recognition reside in the vicinity of the catalytic site (Jansen et al. 2006). This observation may suggest that GlcN6P binds to preorganized *glmS* ribozymes and either participates directly in the chemical mechanism of cleavage or enacts an intimate role in positioning groups at the active site, thereby functioning as a cofactor, rather than an allosteric effector. If this hypothesis proves correct, it may offer an explanation for why, throughout expansive time scales, GlcN6P binding has not been divorced from the self-cleavage reaction and replaced with transcription attenuation- or translation inhibition-based mechanisms.

METABOLITE-SENSING RNAs ARE WIDESPREAD IN BIOLOGY

One of the most fascinating aspects of metabolite- and tRNA-sensing RNAs is their widespread biological distribution (Fig. 3). In general, metabolite-sensing RNAs are enriched in gram-positive bacteria, although several classes can also be identified in gram-negative species. The metabolite ligands for these RNAs have remained unchanged throughout great evolutionary time scales (White 1976). Therefore, their respective aptamer domains would be expected to have also preserved a common structural fold during the same time period. Indeed, members of a given class of metabolite-sensing RNAs share a common structural architecture, as ascertained by gel-based structural probing and X-ray crystallography (for review, see Winkler and Breaker 2005; Batey 2006). Furthermore, many examples of metabolite-sensing RNAs have been identified in evolutionarily distant organisms. These observations together imply that members of the same RNA class that are discovered in disparate phylogenetic lineages would likely have been present prior to the divergence of the two lineages. Alternatively, the RNA sequences would had to have emerged through convergent evolution, an unlikely explanation given the vastness of potential sequence space and the significant potential for alternate structural solutions (e.g., three structurally distinct SAM sensors have now been identified).

GENERAL FEATURES OF *B. SUBTILIS* REGULATORY RNAs

The lengths of *B. subtilis* regulatory RNAs and the intergenic regions (IGRs) that host them vary among the different RNA classes (Figs. 4 and 5). The average length

for protein-, tRNA-, and metabolite-sensing RNAs is approximately 110, 196, and 326 nucleotides in length, respectively (Fig. 4). This trend is also observed for the length of IGRs that harbor these RNAs which, in total, range between 100 and 450 nucleotides in length (Figs. 4 and 5); 90% of these *cis*-acting regulatory RNAs reside in IGRs of greater than 200 nucleotides in length. In contrast, more than 70% of *B. subtilis* IGRs are less than 200 nucleotides and would therefore be unlikely to contain additional regulatory RNAs. Since most of these RNA elements in *B. subtilis* rely on transcription attenuation-based mechanisms, the region between the end of the regulatory RNA (transcription terminator) and the start of the downstream gene can be quantified. On average, this space is only 50 nucleotides in length, demonstrating that little IGR space is wasted overall. However, it should be pointed out that there is considerable variability in these data (from 1 to 186 nucleotides).

Bacterial genomes range in their nucleotide composition ratios. For example, *B. subtilis* with a genomic G + C average of 43.52% is considered a "low G + C" microorganism, whereas actinomycetes such as *Streptomyces coelicolor* exhibit substantially higher values. For organisms with reduced G + C content, it is possible that intricately structured RNAs may require a higher G + C content than the background levels, due to the increased contribution of guanine–cytosine base pairs to overall thermodynamic stability. Therefore, "islands" of higher-than-background G + C content may infer IGR locations that contain structured RNAs. The G + C % average for *B. subtilis* total genomic IGRs is 38.1. Indeed, with the exception of protein-sensing RNAs, most of the established *cis*-acting regulatory RNAs contain regions of increased G + C content (50–55%) (Fig. 5). The fact that this trend differs for protein-sensing RNAs may not be unexpected given that tRNA- and metabolite-sensing RNAs are likely to require more complex structural configurations for construction of their ligand-binding pockets. It is possible that many protein-sensing RNAs rely more on the ability of RNA-binding proteins to either recognize specific short sequences or stabilize certain small secondary structural conformations. Therefore, for some of these RNAs, it may be difficult to define which component is the "ligand" and which may be the "receptor."

An additional measure of RNA stability is its predicted folding energy. It has been observed that highly structured RNAs tend to exhibit lower predicted free energy values as compared to randomized sequence pools (Clote et al. 2005). This is in fact true for all known *cis*-acting regulatory RNAs in *B. subtilis*. The predicted folding energy for each of these RNAs is typically three standard deviations lower than that of the average of the folding energy for a pool of 1000 random sequences containing the same dinucleotide frequency (Fig. 5).

Although these observations provide some limited usefulness in the accurate characterization and description of *cis*-acting regulatory RNAs, it is possible that they could be incorporated with additional criteria for the purpose of predicting locations of novel regulatory RNAs. Importantly, this pool of more than 70 *B. subtilis* regulatory RNAs does not derive from a comprehensive, sys-

	I	II	III	IV	V	VI	VII	VIII	IX	X	XI	XII	XIII	XIV
ACTINOBACTERIA														
A. Streptomyces														
S. coelicolor	9	-	-	1	1	-	-	3	4	-	-	7	-	1
S. avermitilis	6	-	-	1	1	-	-	5	4	-	-	8	-	1
B. Corynebacterium														
C. glutamicum	-	-	-	1	-	-	-	5	-	-	-	2	-	3
C. diphteriae	2	-	-	-	-	-	-	3	-	-	-	-	-	1
C. Mycobacterium														
M. leprae	2	-	-	-	-	-	-	2	-	-	2	1	-	-
M. tuberculosis	2	-	-	-	-	-	-	2	4	-	2	1	-	1
M. avium	3	-	-	-	-	-	-	2	2	-	3	1	-	-
AQUIFICAE														
A. aeolicus	-	-	-	-	-	-	-	-	-	-	-	-	-	-
BACTEROIDETES														
P. gingivalis	5	-	-	-	-	1	-	2	-	-	-	-	-	-
CHLOROBI														
C. tepidum	1	-	-	-	1	-	-	1	-	-	-	-	-	-
CYANOBACTERIA														
P. marinus	-	-	-	-	-	-	-	1	-	-	-	-	-	-
Synechocystis sp.	-	-	-	-	-	-	-	1	-	-	-	1	1	-
DEINOCOCCUS-THERMUS														
D. radiourans	-	-	-	-	-	-	-	2	-	1	-	-	-	-
T. thermophilus	1	-	-	1	1	-	-	2	-	-	-	-	-	-
FIRMICUTES														
A. Mycoplasmatales														
M. gallisepticum	-	-	-	-	-	-	-	1	-	-	-	-	-	-
M. mobile	-	-	-	1	-	-	-	1	-	-	-	-	-	-
B. Bacillales														
S. aureus	-	2	1	2	4	-	-	2	1	1	-	-	-	1
S. epidermidis	-	2	1	2	4	-	-	5	1	1	-	-	-	1
B. subtilis	1	2	5	2	11	-	-	5	1	1	1	2	2	2
B. anthracis	1	4	6	2	17	-	-	7	2	1	1	4	-	2
B. cereus	1	4	6	2	17	-	-	7	2	1	2	4	-	2
B. halodurans	5	4	5	1	5	-	-	4	2	1	3	1	1	1
B. thuringiensis	1	4	6	2	18	-	-	7	2	1	1	4	-	2
C. Lactobacillales														
L. plantarum	-	1	-	2	-	-	1	2	-	1	1	-	-	1
L. lactis	-	1	1	2	-	-	1	2	-	-	2	-	-	-
S. mutans	-	-	-	1	-	-	1	1	1	-	-	-	-	1
S. pneumonia	-	-	1	2	-	-	1	4	1	-	-	-	-	1
E. faecalis	1	1	1	1	-	-	1	2	-	1	5	-	-	1
D. Clostridia														
C. acetobutylicum	2	3	3	2	7	-	-	5	2	1	2	2	-	-
C. tetani	-	2	1	2	4	-	-	3	3	1	-	2	-	1
C. perfringens	4	3	4	2	2	-	-	5	2	1	-	-	-	2
T. tengcongensis	2	2	1	-	3	-	-	4	4	1	1	4	1	1
PROTEOBACTERIA														
A. Alpha														
B. melitensis	3	-	-	1	-	1	-	2	2	-	-	-	-	-
A. tumefaciens	4	-	-	1	-	2	-	2	2	-	-	-	-	-
B. japonicum	5	-	-	1	-	3	-	1	2	-	-	-	-	-
R. prowazekii	-	-	-	-	-	-	-	-	-	-	-	-	-	-
B. Beta														
B. pertussis	1	-	-	1	-	-	-	1	2	-	-	-	-	1
C. violaceum	2	-	-	1	-	-	-	2	4	-	1	-	-	1
N. meningitidis	-	-	-	-	-	-	-	2	2	-	-	-	-	-
C. Delta														
D. vulgaris	2	-	-	1	-	-	-	2	2	1	-	-	-	-
G. sulfurreducens	2	-	-	-	2	-	-	2	-	-	-	1	-	1
D. Epsilon														
H. pylori	-	-	-	-	-	-	-	1	-	-	-	-	-	-
C. jejuni	-	-	-	-	-	-	-	1	-	-	-	-	-	-
E. Gamma														
H. influenzae	-	1	-	1	-	-	-	3	2	-	-	-	-	-
S. oneidensis	2	2	-	-	-	-	-	2	2	-	-	-	-	1
V. cholera	1	3	-	1	-	-	-	3	2	-	1	-	-	2
E. coli	1	1	-	1	-	-	-	3	-	-	-	-	-	2
Y .pestis	1	-	-	1	-	-	-	3	-	-	-	-	-	1
S. typhimurium	2	-	-	1	-	-	-	3	-	-	-	-	-	2
P. aeruginosa	4	-	-	1	-	-	-	1	-	-	-	-	-	2
SPIROCHAETES														
T. pallidum	-	-	-	-	-	-	-	-	-	-	-	-	-	-
THERMOTOGAE														
T. maritima	1	1	-	-	-	-	-	1	-	-	-	-	-	-

I	Adenosylcobalamin
II	Lysine
III	Purine
IV	Flavin mononucleotide
V	S-adenosylmethionine 'S box'
VI	S-adenosylmethionine 'SAM II'
VII	S-adenosylmethionine 'SAM(MK)'
VIII	Thiamine pyrophosphate
IX	Glycine
X	Glucosamine-6-phosphate
XI	*ykoK*
XII	*ydaO/yuaA* (Orphans)
XIII	*ykkC/yxkD*
XIV	*yybP/ykoY*

Figure 3. Biological distribution of metabolite-sensing regulatory RNAs. The distribution of different metabolite-sensing RNA classes is shown for representative bacterial species. Data are extracted from RFAM (Griffiths-Jones et al. 2005), except for SAM II and SAM(MK) distribution patterns, which are derived from Corbino et al. (2005) and Fuchs et al. (2006), respectively. Regulatory RNAs that are expected to sense unidentified metabolites ("orphans"; Barrick et al. 2004) are in italics and named after their downstream genes in *B. subtilis*.

tematic search but from the individual efforts of many investigators spanning several decades. Therefore, this value is still certain to be an underestimate of the true usage for regulatory RNAs in this organism. The genome of *B. subtilis* contains a total of 7334 IGRs, where an IGR is defined as the nucleotide sequence on a given genomic strand that does not overlap with any genes on the opposite strand of DNA. For this organism, there are 3178 IGRs of suitable length for housing regulatory RNAs (between 100 and 450 nucleotides in length) (Figs. 4 and 5). This represents 43% of total IGRs and the most optimistic upper limit for the total number of potential regulatory RNAs in this organism, although realistic expectations are certain to be substantially lower. Applying a G + C cutoff of more than 48% would recover virtually all tRNA- and metabolite-sensing RNAs and

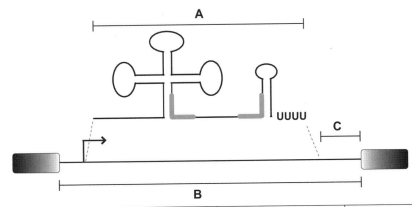

Regulatory RNA Class	A. Size of regulatory RNA				B. Size of IGR				C. Distance to downstream gene			
	Ave.	Stdev.	Max.	Min.	Ave.	Stdev.	Max.	Min.	Ave.	Stdev.	Max.	Min.
Protein-sensing RNAs	109.9	27.8	151	62	279.3	139.3	607	124	53.9	33.1	134	14
tRNA-sensing RNAs	326.2	161.1	813	217	425.6	176.6	844	248	38	19.8	84	16
Metabolite-sensing RNAs	196.2	36.6	269	107	365.8	117.6	663	105	54.8	44.5	186	1
All Classes	204.5	120.5	813	140	360.9	149.4	844	105	49.8	36.8	186	1

Figure 4. Average sizes for *B. subtilis* regulatory RNAs and the IGR that hosts them. Average lengths from the transcription start to the intrinsic transcription terminator (*A*), for the entire IGR (*B*), and from the terminator to the downstream gene (*C*) are the same as the columns in the table. (*Gray boxes*) Upstream and downstream genes.

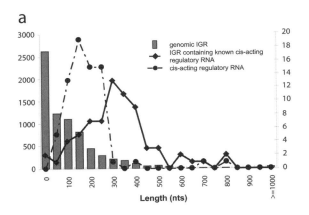

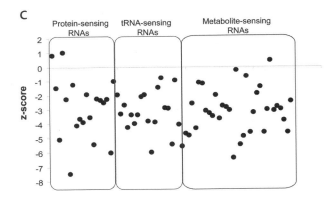

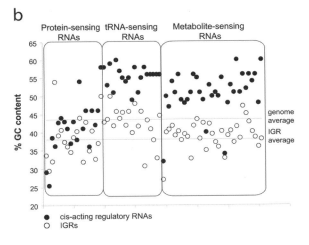

Figure 5. General features of *B. subtilis* regulatory RNAs. (*a*) Length distribution of *B. subtilis* regulatory RNAs (*circles;* right *y* axis) and the IGRs that host them (*diamonds;* right *y* axis). Length distribution for total *B. subtilis* IGRs are indicated by gray-shaded columns (left *y* axis). (*b*) Comparison of G + C % content for different classes of *B. subtilis* regulatory RNAs. The G + C % for a sliding window of 100 nucleotides (*closed circles*) is compared against the G + C % content for the entire IGR that houses known regulatory RNAs (*open circles*). The horizontal lines indicate the average G + C % values for the total genome (43.52%) and for total genomic IGRs (38.1%). (*c*) Predicted folding energies for *B. subtilis* regulatory RNAs. Each regulatory RNA sequence was randomly shuffled 1000 times while keeping the same dinucleotide frequency via the Dishuffle program (Clote et al. 2005). A predicted value for free energy of folding was determined via the RNAfold program (Vienna RNA package; Hofacker et al. 1994). The z-score was calculated for each RNA set by comparing the folding free energy of the native sequence to the average of the shuffled pool with respect to standard deviation.

reduce the total number of candidate IGRs to 454 (~6% of total IGRs). This is still almost sixfold greater than the current number of established regulatory RNAs. It is a reasonable hypothesis to expect that there are new regulatory RNAs to be discovered within this IGR pool; however, it is also obvious that additional parameters must be added before simple computational-assisted methods can accurately pinpoint novel RNAs.

Historically, identification of regulatory RNAs has been more accidental than methodical. Most were discovered from genetic and physiology-based study of downstream genes or operons. More recently, researchers have been moving toward more systematic approaches that combine computational methods with experimental verification of candidates. So far, the most successful methods for finding novel, candidate regulatory RNAs have been via comparative sequence analyses. For this, IGRs from a query genomic sequence are compared by BlastN analysis against other genomic sequences, and the resulting hits are manually examined for evidence of primary sequence and secondary structural conservation. Coupled with predictions of gene function for downstream genes, this method has been effective in identifying novel metabolite-sensing RNAs (Barrick et al. 2004; Corbino et al. 2005). The QRNA program is an example of a similar approach that minimizes the manual inspection and has been used for identifying eight new noncoding RNA (ncRNA) candidates in *Saccharomyces cerevisiae* (McCutcheon and Eddy 2003).

Predictions of thermodynamic stability have been applied as another method for finding regulatory RNAs. As explained in the previous section, this method is typically done by comparing the folding energy of native RNA against pools of randomized sequences (Workman and Krogh 1999; Rivas and Eddy 2000; Clote et al. 2005). RNAz (Washietl et al. 2005) and Dynalign (Uzilov et al. 2006) are both examples of this overall approach. RNAz has been used to predict 78 new ncRNA candidates that are conserved among humans, mice, rats, *Fugu*, and zebra fish. Dynalign is a sensitive method that can recover almost all *Escherichia coli* ncRNAs, as well as many unexplored candidates.

Candidate regulatory RNAs can also be identified by locating genetic elements that are common in either transcription attenuation or translation inhibition mechanisms. Specifically, using algorithms for identification of promoter and terminator elements, *trans*-acting regulatory RNAs have been identified in *E. coli* (Argaman et al. 2001). Additionally, the prediction of terminator and antiterminator pairings within *B. subtilis* IGRs proved to be a relatively successful approach for identification of established regulatory RNAs, as well as the prediction of numerous additional candidates (Merino and Yanofsky 2005).

Due to a high background of false positives, however, these purely computational approaches are still likely to be limited in their usefulness when applied in the absence of additional parameters. These methods and others may be bolstered in the near future by the addition of different transcriptome-based analyses. The latter methods are currently used primarily for genome annotation and expression profiling (Ruan et al. 2004). However, tag-based methods for mapping of exact 5′ and 3′ termini, as well as whole-genome tiling microarrays, may prove to be useful approaches for identifying candidate mRNAs with long UTRs, excellent starting points for discovery of regulatory RNAs (Bertone et al. 2004; Harbers and Carninci 2005). Coupling of such long UTR data sets with the other above-mentioned computational methods may, over time, prove to be the most comprehensive and accurate approach for regulatory RNA identification.

CONCLUSIONS

Regulatory RNAs convey a simple method of coupling metabolic sensing to gene control. This paper has provided a general summary of the different classes of regulatory RNAs in *B. subtilis*. This pool now consists of greater than 70 examples of protein-, tRNA-, and metabolite-sensing RNAs. Therefore, RNA-mediated genetic control can no longer be viewed as gene regulatory anomalies, given their extensive use in organisms such as *B. subtilis*. Although the list of RNAs in this bacterium is not complete, it is large enough so that it can be examined for general features. In the near future, these basic features may prove to be useful in deciding on criteria that could assist in the computational identification of novel regulatory RNAs. In general, establishment of a comprehensive catalog of *B. subtilis* *cis*-acting regulatory RNAs will help elucidate the contribution of posttranscriptional regulation to overall bacterial genetic regulation.

Note Added in Proof

During preparation of this manuscript, an example of a *B. subtilis* protein that binds a 3′UTR for apparent control of mRNA stability was reported by Serio et al. (2006). This discovery further expands the overall *B. subtilis* catalog of regulatory RNAs and hints at a novel 3′UTR-based mechanism.

ACKNOWLEDGMENTS

W.C.W. acknowledges the financial support provided by the Searle Scholars Program and the University of Texas Southwestern Medical Center Endowed Scholar Program.

REFERENCES

Argaman L., Hershberg R., Vogel J., Bejerano G., Wagner E.G., Margalit H., and Altuvia S. 2001. Novel small RNA-encoding genes in the intergenic regions of *Escherichia coli. Curr. Biol.* **11:** 941.

Babitzke P. 2004. Regulation of transcription attenuation and translation initiation by allosteric control of an RNA-binding protein: The *Bacillus subtilis* TRAP protein. *Curr. Opin. Microbiol.* **7:** 132.

Barrick J.E., Corbino K.A., Winkler W.C., Nahvi A., Mandal M., Collins J., Lee M., Roth A., Sudarsan N., Jona I., et al. 2004. New RNA motifs suggest an expanded scope for riboswicthes in bacterial genetic control. *Proc. Natl. Acad. Sci.* **101:** 6421.

Batey R.T. 2006. Structures of regulatory elements in mRNAs. *Curr. Opin. Struct. Biol.* **16:** 299.

Bertone P., Stolc V., Royce T.E., Rozowsky J.S., Urban A.E., Zhu X., Rinn J.L., Tongprasit W., Samanta M., Weissman S., et al. 2004. Global identification of human transcribed sequences with genome tiling arrays. *Science* **306:** 2242.

Clote P., Ferre F., Kranakis E., and Krizanc D. 2005. Structural RNA has lower folding energy than random RNA of the same dinucleotide frequency. *RNA* **11:** 578.

Corbino K.A., Barrick J.E., Lim J., Welz R., Tucker B.J., Puskarz I., Mandal M., Rudnick N.D., and Breaker R.R. 2005. Evidence for a second class of S-adenosylmethionine riboswitches and other regulatory RNA motifs in alpha-proteobacteria. *Genome Biol.* **6:** R70.

Fedor M.J. and Williamson J.R. 2005. The catalytic diversity of RNAs. *Nat. Rev. Mol. Cell Biol.* **6:** 399.

Fuchs R.T., Grundy F.J., and Henkin T.M. 2006. The S_{MK} box is a new SAM-binding RNA for translational regulation of SAM synthetase. *Nat. Struct. Mol. Biol.* **13:** 226.

Gagnon Y., Breton R., Putzer H., Pelchat M., Grunberg-Manago M., and Lapointe J. 1994. Clustering and co-transcription of the *Bacillus subtilis* genes encoding the aminoacyl-tRNA synthetases specific for glutamate and for cysteine and the first enzyme for cysteine biosynthesis. *J. Biol. Chem.* **269:** 7473.

Gollnick P., Babitzke P., Antson A., and Yanofsky C. 2005. Complexity in regulation of tryptophan biosynthesis in *Bacillus subtilis*. *Annu. Rev. Genet.* **39:** 47.

Graumann P. and Marahiel M.A. 1999. Cold shock proteins CspB and CspC are major stationary-phase-induced proteins in *Bacillus subtilis*. *Arch. Microbiol.* **171:** 135.

Griffiths-Jones S., Moxon S., Marshall M., Khanna A., Eddy S.R., and Bateman A. 2005. Rfam: Annotating non-coding RNAs in complete genomes. *Nucleic Acids Res.* **33:** D121.

Grundy F.J. and Henkin T.M. 1992. Characterization of the *Bacillus subtilis rpsD* regulatory target site. *J. Bacteriol.* **174:** 6763.

———. 2003. The T box and S box transcription termination control systems. *Front. Biosci.* **8:** d20.

———. 2004a. Regulation of gene expression by effectors that bind to RNA. *Curr. Opin. Microbiol.* **7:** 126.

———. 2004b. Kinetic analysis of tRNA-directed transcription antitermination of the *Bacillus subtilis glyQS* gene *in vitro*. *J. Bacteriol.* **186:** 5392.

Hampel K.J. and Tinsley M.M. 2006. Evidence for preorganization of the *glmS* ribozyme ligand binding pocket. *Biochemistry* **45:** 7861.

Harbers M. and Carninci P. 2005. Tag-based approaches for transcriptome research and genome annotation. *Nat. Methods* **2:** 495.

Hofacker I.L., Fontana W., Stadler P.F., Bonhoeffer S., Tacker M., and Schuster P. 1994. Fast folding and comparison of RNA secondary structures. *Monatsh. Chem.* **125:** 167.

Ingham C.J., Dennis J., and Furneaux P.A. 1999. Autogenous regulation of transcription termination factor Rho and the requirement for Nus factors in *Bacillus subtilis*. *Mol. Microbiol.* **31:** 651.

Jansen J.A., McCarthy T.J., Soukup G.A., and Soukup J.K. 2006. Backbone and nucleobase contacts to glucosamine-6-phosphate in the *glmS* ribozyme. *Nat. Struct. Mol. Biol.* **13:** 517.

Kumar P.K., Kumarevel T., and Mizuno H. 2006. Structural basis of HutP-mediated transcription anti-termination. *Curr. Opin. Struct. Biol.* **16:** 18.

Landick R., Turnbough C.L., and Yanofsky C. 1996. Transcription attenuation. In Escherichia coli *and* Salmonella: *Cellular and molecular biology* (F.C. Neidhardt et al.), p. 1263. ASM Press, Washington, D.C.

Li X., Lindahl L., Sha Y., and Zengel J.M. 1997. Analysis of the *Bacillus subtilis* S10 ribosomal protein gene cluster identifies two promoters that may be responsible for transcription of the entire 15-kilobase S10-*spc*-alpha cluster. *J. Bacteriol.* **179:** 7046.

Lilley D.M. 2003. The origins of RNA catalysis in ribozymes. *Trends Biochem. Sci.* **28:** 495.

———. 2004. The Varkud satellite ribozyme. *RNA* **10:** 151.

Mandal M., Lee M., Barrick J.E., Weinberg Z., Emilsson G.M.,

Ruzzo W.L., and Breaker R.R. 2004. A glycine-dependent riboswitch that uses cooperative binding to control gene expression. *Science* **306:** 275.

McCutcheon J.P. and Eddy S.R. 2003. Computational identification of non-coding RNAs in *Saccharomyces cerevisiae* by comparative genomics. *Nucleic Acids Res.* **31:** 4119.

McCarthy T.J., Plog M.A., Floy S.A., Jansen J.A., Soukup J.K., and Soukup G.A. 2005. Ligand requirements for *glmS* ribozyme self-cleavage. *Chem. Biol.* **12:** 1221.

Meng Q., Turnbough C.L., Jr., and Switzer R.L. 2004. Attenuation control of *pyrG* expression in *Bacillus subtilis* is mediated by CTP-sensitive reiterative transcription. *Proc. Natl. Acad. Sci.* **101:** 10943.

Merino E. and Yanofsky C. 2005. Transcription attenuation: A highly conserved regulatory strategy used by bacteria. *Trends Genet.* **21:** 260.

Mironov A.S., Gusarov I., Rafikov R., Lopez L.E., Shatalin K., Kreneva R.A., Perumov D.A., and Nudler E. 2002. Sensing small molecules by nascent RNA: A mechanism to control transcription in bacteria. *Cell* **111:** 747.

Nou X. and Kadner R.J. 2000. Adenosylcobalamin inhibits ribosome binding to *btuB* mRNA. *Proc. Natl. Acad. Sci.* **97:** 7190.

Nudler E. and Mironov A.S. 2004. The riboswitch control of bacterial metabolism. *Trends Biochem. Sci.* **29:** 11.

Oda M., Sugishita A., and Furukawa K. 1988. Cloning and nucleotide sequences of histidase and regulatory genes in *Bacillus subtilis hut* operon and positive regulation of the operon. *J. Bacteriol.* **170:** 3199.

Pelchat M. and Lapointe J. 1999. In vivo and in vitro processing of the *Bacillus subtilis* transcript coding for glutamyl-tRNA synthetase, serine acetyltransferase, and cysteinyl-tRNA synthetase. *RNA* **5:** 281.

Putzer H., Gendron N., and Grunberg-Manago M. 1992. Coordinate expression of the two threonyl-tRNA synthetase genes in *Bacillus subtilis:* Control by transcriptional antitermination involving a conserved regulatory sequence. *EMBO J.* **11:** 3117.

Rivas E. and Eddy S.R. 2000. Secondary structure alone is generally not statistically significant for the detection of noncoding RNAs. *Bioinformatics* **16:** 583.

Rodionov D.A., Vitreschak A.G., Mironov A.A., and Gelfand M.S. 2002. Comparative genomics of thiamin biosynthesis in procaryotes. New genes and regulatory mechanisms. *J. Biol. Chem.* **277:** 48949.

Roth A., Nahvi A., Lee M., Jona I., and Breaker R.R. 2006. Characteristics of the *glmS* ribozyme suggest only structural roles for divalent metal ions. *RNA* **12:** 607.

Ruan Y., Le Ber P., Ng H.H., and Liu E.T. 2004. Interrogating the transcriptome. *Trends Biotechnol.* **22:** 23.

Schlax P.J. and Worhunsky D.J. 2003. Translational repression mechanisms in prokaryotes. *Mol. Microbiol.* **48:** 1157.

Serio A.W., Pechter K.B., and Sonenshein A.L. 2006. *Bacillus subtilis* aconitase is required for efficient late-sporulation gene expression. *J. Bacteriol.* **188:** 6396.

Soukup G.A. 2006. Core requirements for *glmS* ribozyme self-cleavage reveal a putative pseudoknot structure. *Nucleic Acids Res.* **34:** 968.

Stülke J. 2002. Control of transcription termination in bacteria by RNA-binding proteins that modulate RNA structures. *Arch. Microbiol.* **177:** 433.

Switzer R.L., Turner R.J., and Lu Y. 1999. Regulation of the *Bacillus subtilis* pyrimidine biosynthetic operon by transcriptional attenuation: Control of gene expression by an mRNA-binding protein. *Prog. Nucleic Acid Res. Mol. Biol.* **62:** 329.

Turner R.J., Lu Y., and Switzer R.L. 1994. Regulation of the *Bacillus subtilis* pyrimidine biosynthetic (*pyr*) gene cluster by an autogenous transcriptional attenuation mechanism. *J. Bacteriol.* **176:** 3708.

Uzilov A.V., Keegan J.M., and Matthes D.H. 2006. Detection of non-coding RNAs on the basis of predicted secondary structure formation free energy change. *BMC Bioinformatics* **7:** 173.

Vitreschak A.G., Rodionov D.A., Mironov A.A., and Gelfand M.S. 2002. Regulation of riboflavin biosynthesis and transport genes in bacteria by transcriptional and translational attenuation. *Nucleic Acids Res.* **30:** 3141.

Washietl S., Hofacker I.L., and Stadler P.F. 2005. Fast and reliable prediction of noncoding RNAs. *Proc. Natl. Acad. Sci.* **102:** 2454.

White H.B. 1976. Coenzymes as fossils of an earlier metabolic state. *J. Mol. Evol.* **7:** 101.

Wickiser J.K., Winkler W.C., Breaker R.R., and Crothers D.M. 2005. The speed of RNA transcription and metabolite binding kinetics operate an FMN riboswitch. *Mol. Cell* **18:** 49.

Wilkinson S.R. and Been M.D. 2005. A pseudoknot in the 3′ non-core region of the *glmS* ribozyme enhances self-cleavage activity. *RNA* **11:** 1788.

Winkler W.C. 2005. Metabolic monitoring by bacterial mRNAs. *Arch. Microbiol.* **183:** 151.

Winkler W.C. and Breaker R.R. 2005. Regulation of bacterial gene expression by riboswitches. *Annu. Rev. Microbiol.* **59:** 487.

Winkler W., Nahvi A., and Breaker R.R. 2002. Thiamine deriva-

tives bind messenger RNAs directly to regulate bacterial gene expression. *Nature* **419:** 952.

Winkler W.C., Nahvi A., Roth A., Collins J.A., and Breaker R.R. 2004. Control of gene expression by a natural metabolite-responsive ribozyme. *Nature* **428:** 281.

Workman C. and Krogh A. 1999. No evidence that mRNA have lower folding free energies than random sequences with the same dinucleotide distribution. *Nucleic Acids Res.* **27:** 4816.

Yakhnin A.V. and Babitzke P. 2002. NusA-stimulated RNA polymerase pausing and termination participates in the *Bacillus subtilis* trp operon attenuation mechanism in vitro. *Proc. Natl. Acad. Sci.* **99:** 11067.

Yakhnin H., Zhang H., Yakhnin A.V., and Babitzke P. 2004. The *trp* RNA-binding attenuation protein of *Bacillus subtilis* regulates translation of the tryptophan transport gene *trpP* (*yhaG*) by blocking ribosome binding. *J. Bacteriol.* **186:** 278.

Zhang H. and Switzer R.L. 2003. Transcriptional pausing in the *Bacillus subtilis pyr* operon *in vitro:* A role in transcriptional attenuation? *J. Bacteriol.* **185:** 4764.

A Mg^{2+}-responding RNA That Controls the Expression of a Mg^{2+} Transporter

E.A. GROISMAN, M.J. CROMIE, Y. SHI,* AND T. LATIFI

Department of Molecular Microbiology, Washington University School of Medicine, Howard Hughes Medical Institute, St. Louis, Missouri 63110

Mg^{2+} is the most abundant divalent cation in biological systems. It is required for ATP-mediated enzymatic reactions and as a stabilizer of ribosomes and membranes. The enteric bacterium *Salmonella enterica* serovar Typhimurium harbors three Mg^{2+} transporters and a regulatory system—termed PhoP/PhoQ—whose activity is regulated by the extracytoplasmic levels of Mg^{2+}. We have determined that expression of the PhoP-activated Mg^{2+} transporter MgtA is also controlled by its 5'-untranslated region (5'UTR). The 5'UTR of the *mgtA* gene can adopt different stem-loop structures depending on the Mg^{2+} levels, which determine whether transcription reads through into the *mgtA*-coding region or stops within the 5'UTR. This makes the *mgtA* 5'UTR the first example of a cation-responding riboswitch. The initiation of *mgtA* transcription responds to extracytoplasmic Mg^{2+}, and its elongation into the coding region to cytoplasmic Mg^{2+}, which provides a singular example where the same ligand is sensed in different cellular compartments to regulate disparate steps in gene transcription. The PhoP-activated Mg^{2+} transporter MgtB is also regulated by Mg^{2+} in a strain lacking the Mg^{2+} sensor PhoQ, suggesting the presence of additional Mg^{2+}-responding devices.

All living organisms need Mg^{2+}. This is because Mg^{2+} is an essential cofactor for all ATP-mediated enzymatic reactions and because Mg^{2+} helps maintain the stability of membranes as well as of certain macromolecular complexes such as the ribosome (Reinhart 1988). Therefore, cells must preserve physiological Mg^{2+} concentrations, which might not be the same in different subcellular compartments. To achieve this task, cells need to have the means to sense the levels of Mg^{2+} as well as the ability to modify the movement of this essential cation across biological membranes and/or from storage locations.

The gram-negative bacterium *S. enterica* serovar Typhimurium harbors a Mg^{2+}-responding regulatory system (i.e., PhoP/PhoQ) (Groisman 2001), three Mg^{2+} transporters (i.e., CorA, MgtA, and MgtB) (for review, see Smith and Maguire 1998; Maguire 2006), and an integral membrane protein that is required for optimal growth in low Mg^{2+} (i.e., MgtC). The PhoP/PhoQ system consists of the PhoQ protein, a sensor for extracytoplasmic Mg^{2+}, and the transcriptional regulator PhoP (Groisman 2001). When *Salmonella* experiences low Mg^{2+}, the PhoQ protein promotes phosphorylation of the PhoP protein, which can then bind to its target promoters to modify gene transcription: PhoP-activated genes are turned on and PhoP-repressed genes are turned off. And when *Salmonella* faces high Mg^{2+}, PhoQ promotes the unphosphorylated state of the PhoP protein, which is unable to bind to its target promoters and results in no transcription of PhoP-activated genes and derepression of PhoP-repressed genes (Soncini et al. 1996; Castelli et al. 2000;

Chamnongpol and Groisman 2000; Montagne et al. 2001; Chamnongpol et al. 2003).

The CorA protein can mediate both influx and efflux of Mg^{2+} and does not exhibit sequence similarity to the MgtA or MgtB proteins, which are 50% identical to one another and solely mediate Mg^{2+} influx (for review, see Smith and Maguire 1998; Maguire 2006). MgtA and MgtB also differ from CorA in that transcription of the *mgtA* and *mgtB* genes is directly regulated by the PhoP protein in response to extracytoplasmic Mg^{2+} sensed by the PhoQ protein (Soncini et al. 1996; Yamamoto et al. 2002; Zwir et al. 2005), whereas expression of CorA responds neither to Mg^{2+} nor to the PhoP/PhoQ system (Chamnongpol and Groisman 2002). Yet, the activity of the CorA transporter is enhanced in a *phoP* mutant by an unknown mechanism (Chamnongpol and Groisman 2002).

The PhoP/PhoQ system controls expression of about 3% of the genome. This includes not only the genes encoding the Mg^{2+} transporters MgtA and MgtB, but also the genes coding for products that modify the negatively charged residues in the lipopolysaccharide (i.e., LPS), which are neutralized primarily by Mg^{2+} (Groisman et al. 1997; Groisman 2001). The LPS is the dominant molecule in the outer leaflet of the outer membrane of *Escherichia coli* and *Salmonella*, and it has been estimated that more than one-third of the total Mg^{2+} content is located in the LPS (Groisman et al. 1997). The PhoP/PhoQ system and the MgtA and CorA transporters are widely distributed within the family Enterobacteriaceae (Smith and Maguire 1995; Blanc-Potard and Groisman 1997). This is in contrast to the sporadic distribution of the MgtB and MgtC proteins. This paper discusses the discovery, characterization, and function of the first example of an Mg^{2+}-responding riboswitch, and how such RNA regulates cytoplasmic Mg^{2+} levels.

*Present address: Arizona State University, The Biodesign Institute, Center for Infectious Diseases and Vaccinology, College of Liberal Arts and Sciences, School of Life Sciences, PO Box 874501, Tempe, Arizona 85287-4501.

The 5′UTR of the *mgtA* mRNA Harbors a Mg²⁺-responding Regulatory Element

PhoP* is a variant of the PhoP protein that can promote gene transcription in the absence of the Mg²⁺ sensor PhoQ (Chamnongpol and Groisman 2000). As expected, transcription of the PhoP-activated genes *pcgL* and *pmrC* did not respond to Mg²⁺ in a *phoP* Δ*phoQ* strain. In contrast, Mg²⁺ could still modulate *mgtA* transcription in this strain. This suggested that an Mg²⁺-sensing system can regulate *mgtA* expression independently of the PhoP/PhoQ system. Consistent with this notion, Maguire's laboratory demonstrated that luciferase activity originating from a plasmid-borne *mgtA-lux* fusion was still up-regulated in media of very low Mg²⁺ in *phoP* and *phoQ* mutants (Tao et al. 1998).

The 5′UTR of the *mgtA* gene is 264 nucleotides long. Such a long 5′UTR is unusual, raising the possibility that it might be involved in the regulation of *mgtA* expression. Consistent with this notion, replacement of 100 nucleotides of the 5′UTR by an 84-bp "scar" sequence abolished Mg²⁺ regulation of *mgtA* expression in the *phoP** Δ*phoQ* strain (Cromie et al. 2006). Moreover, Mg²⁺ regulated the production of β-galactosidase in a wild-type *Salmonella* strain harboring a plasmid containing the DNA sequence corresponding to the 264-bp *mgtA* 5′UTR region cloned between a derivative of the *lac* promoter and a promoterless *lacZ* gene (Cromie et al. 2006). These results demonstrated that the 5′UTR of *mgtA* is necessary and sufficient for Mg²⁺ regulation.

Analysis of the Potential Structures That May Be Adopted by the Phylogenetically Conserved *mgtA* 5′UTR Suggests a Possible Mechanism by Which *mgtA* Expression Might Be Regulated

The M-fold program predicted that the *mgtA* 5′UTR could form several stem-loop structures including two, designated A and B, with predicted energies of –21.4 and –13.2 kcal/mole, respectively. Interestingly, an alternative stem-loop structure—termed C—could be formed by pairing of sequences originating from the right arm of stem A and the region that separates stem-loops A and B with sequences in the left arm of stem-loop B. The resulting stem has a predicted energy of –12.8 kcal/mole, which is very similar to that predicted for stem-loop B (Fig. 1A).

Examination of the microbial genome databases revealed that in several enteric species, the *mgtA* open reading frame is preceded by sequences that upon transcription could adopt stem-loop structures similar to those described above for the *Salmonella mgtA* 5′UTR. The high degree of primary and secondary structure conservation suggests that the possibility of forming the alternative stem-loop structures A plus B versus C is conserved in the *mgtA* 5′UTR of *E. coli, Citrobacter rodentium, Klebsiella pneumoniae, Erwinia chrysanthemi, Serratia marcescens,* and *Yersinia enterocolitica* (Fig. 1).

When wild-type *Salmonella* harbored a plasmid with the *mgtA* 5′UTR region from *E. coli mgtA* behind a derivative of the *lac* promoter and in front of a promoterless *lacZ*

gene, β-galactosidase activity was 15-fold higher in bacteria grown in low Mg²⁺ than in those grown in high Mg²⁺ (Cromie et al. 2006). Although this is lower than the 23-fold ratio exhibited by the isogenic strain with the plasmid harboring the *Salmonella mgtA* 5′UTR, these results demonstrate that the *E. coli mgtA* 5′UTR can also confer Mg²⁺ regulation.

Taken together with the fact that the mRNA levels for the 5′UTR and coding region of the *Salmonella mgtA* gene are differentially regulated by Mg²⁺ (Cromie et al. 2006), the analyses described above suggest a model by which *mgtA* expression might be controlled. According to this model, the *mgtA* 5′UTR may adopt different stem-loop structures, which would be determined by the cytoplasmic levels of Mg²⁺, and depending on which structure is formed (stem-loops A plus B vs. stem-loop C; Fig. 1A), transcription would stop within the 5′UTR or continue into the *mgtA*-coding region. Genetic analysis suggested that formation of stem-loops A plus B is associated with *mgtA* transcription not reaching the coding region and could be the structure favored in high Mg²⁺. In contrast, formation of stem-loop C would result in readthrough into the *mgtA*-coding region, which would be expected for a structure forming in low Mg²⁺. In sum, the 5′UTR of the *mgtA* gene appears to use an attenuation-like mechanism (Landick et al. 1996; Henkin and Yanofsky 2002) to control *mgtA* expression.

Mg²⁺ Can Modify the Structure of the *mgtA* 5′UTR

During the last few years, research carried out primarily by the Breaker, Nudler, and Henkin laboratories has demonstrated that the 5′UTR of certain bacterial transcripts can bind specific metabolites including amino acids, nucleotides, and sugars (for reviews, see Winkler 2005; Winkler and Breaker 2005). Binding of such a metabolite alters the expression of the open reading frames that follow the 5′UTR, which are typically involved in the biosynthesis or transport of the metabolite. Thus, when the levels of the metabolite build up in the cytoplasm of the bacterial cell, the 5′UTRs function as "riboswitches" that upon binding the metabolite, turn off expression of the biosynthetic or transport proteins promoting accumulation of the metabolite (Winkler 2005; Winkler and Breaker 2005).

If the *mgtA* 5′UTR functions as an Mg²⁺-responding riboswitch, one would expect that (1) it should bind Mg²⁺, (2) Mg²⁺ binding should promote the formation of particular structures in the 5′UTR thereby affecting *mgtA* expression, and (3) favoring the formation of particular structures by mutation should be associated with expression/lack of expression independently of the Mg²⁺ concentration experienced by the bacterial cell.

Enzymatic and chemical probing of the *mgtA* 5′UTR demonstrated that Mg²⁺ can alter the structure of this RNA at several places and that the 5′UTR can adopt secondary structures which are consistent with the model predicted by the M-fold program (Fig. 1A). Mg²⁺ binding to the *mgtA* 5′UTR was entirely expected given that Mg²⁺ has been known to bind to different classes of RNA molecules in different capacities (e.g., Mg²⁺ has a critical role in the

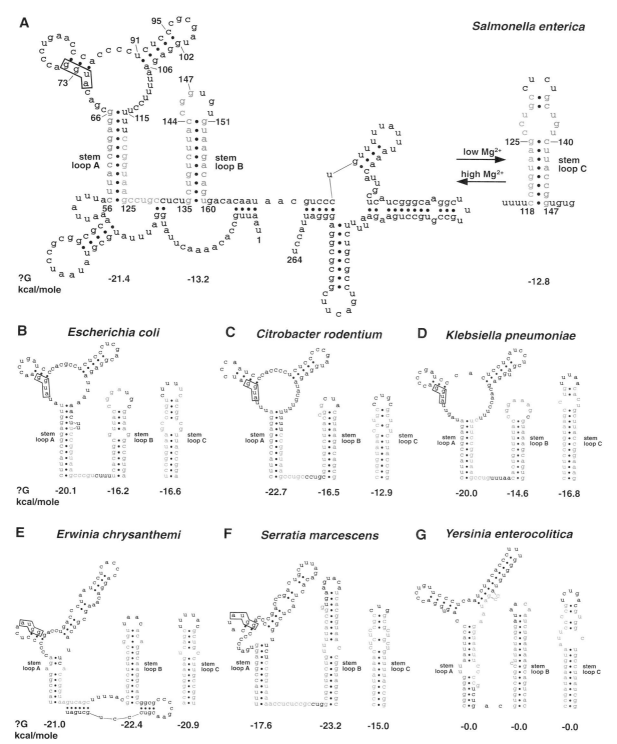

Figure 1. Predicted secondary structure of the phylogenetically conserved *mgtA* 5′UTR. (A) Schematic representation of the secondary structure of the 264-nucleotide *mgtA* 5′UTR as predicted by the M-fold program. (*Left*) Stem-loops A and B, which are postulated to form in high Mg²⁺; (*right*) stem-loop C, which is predicted to form in low Mg²⁺. Sequences in color represent regions involved in stem structures. The predicted energy for stem-loops A, B, and C is presented underneath. (*B–G*), Predicted stem-loop structures A, B, and C corresponding to the potential 5′UTRs for the *mgtA* gene from *E. coli*, *C. rodentium*, *K. pneumoniae*, *E. chrysanthemi*, *S. marscescens*, and *Y. enterocolitica*. Sequences in color represent regions involved in stem-loop structures. The predicted energy for the three stem-loops A, B, and C is presented underneath. Note conservation of AUGG sequence of loop A in all species but *Y. enterocolitica* and of the short stem-loop within loop A in *S. enterica*, *E. coli*, *C. rodentium*, and *K. pneumoniae*. Other conserved regions are shown in Cromie et al. (2006). (Reprinted, with permission, from Cromie et al. 2006 [© Elsevier].)

structure and/or enzymatic activity of several ribozymes; Hanna and Doudna 2000). Thus, we focused our analysis on those regions of the *mgtA* 5′UTR that were anticipated to be differentially affected by Mg^{2+} in ways that would affect gene expression. For example, G146 and G147 are predicted to be part of the loop in stem-loop B, which is anticipated to form in high Mg^{2+} but to be base-paired as part of stem C in low Mg^{2+} (Fig. 1A). By using RNase T1, which cleaves unpaired G residues, we detected more cleavage at both G146 and G147 at high Mg^{2+} than at low Mg^{2+}, in agreement with the prediction. Likewise, G151 is predicted to be base-paired in high Mg^{2+} but unpaired in low Mg^{2+} (Fig. 1A), and that is what the T1 cleavage treatment showed (Cromie et al. 2006). On the other hand, G149 was cleaved to a similar extent in low and high Mg^{2+} as suggested by the model (Fig. 1A). Importantly, the low and high Mg^{2+} concentrations used in this experiment correspond to those around the physiological range (Froschauer et al. 2004).

Formation of Stem-loop C Results in *mgtA* Expression

The genetic and biochemical data discussed above suggest that when stem-loop C is formed, the *mgtA*-coding region is transcribed. To test this model, we investigated the behavior of strains harboring mutant *mgtA* 5′UTRs in which the ability to form stem-loop C had been compromised. G120 is predicted to base-pair with C145 in the middle of stem C (Fig. 1A). Mutants with either the G120C or C145G single-nucleotide substitutions failed to express a reporter gene that was cloned behind the 5′UTR region (Cromie et al. 2006). This was true regardless of the Mg^{2+} concentration in which the organism was grown. On the other hand, the G120C C145G double mutant exhibited Mg^{2+}-regulated expression of the reporter gene. This supports the notion that G120 can pair with C145 and that the ability to form stem C is necessary for *mgtA* expression. Although C61 is predicted to base-pair with G120 in stem A, a strain with the C61G substitution still expressed the reporter gene in a Mg^{2+}-regulated fashion. However, instead of the tenfold ratio in transcription levels in low versus high Mg^{2+} exhibited by organisms with the wild-type *mgtA* 5′UTR, the ratio was less than fivefold. Thus, it would appear that the ability to form stem A is partially compromised in the C61G mutant, which results in more than twofold higher levels of expression in high Mg^{2+}, but not eliminated because base-stacking of G residues may still favor formation of stem A. Finally, the RNAs corresponding to the G120C and C145G single mutants exhibited aberrant T1 cleavage patterns at positions away from the mutated nucleotides, indicative that these mutations can affect the formation of structures in other regions of the 5′UTR (Cromie et al. 2006).

Looking for the Mg^{2+}-sensing Domain of the *mgtA* Riboswitch

In their most straightforward form, riboswitches can be divided into two domains: an aptamer domain that is responsible for binding a specific ligand and an expression platform domain that is responsible for altering gene expression in response to ligand binding (Winkler and Breaker 2005). In fact, RNAs can be engineered to respond to particular ligands by mixing and matching RNAs of different origins (Kim et al. 2005; Muller et al. 2006). Then, which are the regions of the 5′UTR responsible for Mg^{2+} sensing and for modulating *mgtA* expression upon Mg^{2+} binding?

As discussed above, the formation of stem-loops B and C is associated with lack of expression and expression of the *mgtA* gene, respectively. Because the predicted ΔGs for these two stem-loop structures are very similar (Fig. 1A), we would hypothesize that a different part of the 5′UTR determines which of the two structures is formed and that such part of the 5′UTR is likely involved in Mg^{2+} sensing. Stem-loop A is a prime candidate for having this role because (1) this region is transcribed before stem-loops B and C, and thus would be available to bind Mg^{2+} before the rest of the 5′UTR is made, and (2) it is conserved with other enteric species, and some of the conserved nucleotides are differentially cleaved by RNase T1 in the presence of Mg^{2+}. We targeted for mutagenesis a short stem predicted to be present in the loop-A region and found that a mutant in which the ability to form the short stem was abrogated constitutively expressed the *mgtA* gene (i.e., no longer responded to Mg^{2+}). It appears that the short stem and not the particular sequences that make up the stem is the critical element in Mg^{2+} sensing because making the compensatory mutations to restore formation of the short stem brought back Mg^{2+}-regulated expression of the *mgtA* gene. That stem-loop A may be involved in Mg^{2+} sensing is further supported by the Mg^{2+}-promoted changes in the stem-loop A structure when present by itself (Cromie et al. 2006).

The requirement for sequences in loop A for Mg^{2+} sensing does not mean that Mg^{2+} binding to this region is sufficient to trigger the conformational changes that affect *mgtA* expression. Therefore, although it is possible that loop A may constitute a Mg^{2+}-binding pocket analogous to those involved in metabolite-sensing riboswitches (Serganov et al. 2004, 2006; Montange and Batey 2006), we cannot rule out the possibility of Mg^{2+} affecting the formation of the various stem-loop structures by binding to multiple sites in the 5′UTR. In this context, it is important to note that in addition to stem-loops A, B, and C, several regions of the *Salmonella mgtA* 5′UTR show high levels of sequence identity with homologs in other species (Fig. 1A). These regions could also be involved in Mg^{2+} sensing. Alternatively or in addition, they may be conserved due to their role in regulating *mgtA* expression.

How the Mg^{2+}-responding Riboswitch May Control Expression of the MgtA Protein

We could recapitulate the Mg^{2+}-regulated transcription elongation beyond the *mgtA* 5′UTR region using an in vitro transcription system consisting of the *E. coli* RNA polymerase σ^{70} holoenzyme, linear DNA templates harboring a derivative of the *lac* promoter (Liu et al. 2004), and different concentrations of Mg^{2+} around the physiological levels (Froschauer et al. 2004). A template harboring the full-length wild-type 5′UTR of the *mgtA* gene behind the *lac* promoter derivative gave rise to two bands: a large band representing the readthrough transcript, the amount of which decreased as the Mg^{2+} concentration increased, and a small band corresponding to a truncated product of approximately 220 nucleotides. These in vitro experiments demonstrated that the Mg^{2+}-regulated expression mediated by the 5′UTR of the *mgtA* gene does not require proteins other than RNA polymerase, and they are consistent with the notion that the 5′UTR is a Mg^{2+}-responding regulatory element. Yet, it is possible that a protein may participate in the regulation of *mgtA* elongation in vivo.

There are (at least) three possible mechanisms that may explain why the *mgtA* transcript does not go beyond a certain position within the 5′UTR when bacteria experience Mg^{2+} levels that are not too low. On the one hand, the truncated product could result from RNA-mediated cleavage as demonstrated for the glucosamine-6-phosphate-responding riboswitch (Winkler et al. 2004). On the other hand, the truncated product may originate as the result of early transcription termination or due to RNA polymerase pausing at a site and then falling off the DNA template. The *mgtA* 5′UTR does not appear to function as a ribozyme because (1) the truncated product could be labeled using ^{32}pCp and T4 RNA ligase, (2) ribozymes that yield free 2′ and 3′ hydroxyl groups are typically larger than the 220-nucleotide *mgtA* 5′UTR (Jacquier 1996), (3) there was no delay in the appearance of the truncated product relative to the readthrough product as one might expect for a maturation reaction, and (4) addition of Mg^{2+} did not promote degradation of the readthrough transcript generated in vitro. Moreover, when the 5′UTR was synthesized in vitro by T7 RNA polymerase, only a single product was of the same size as that corresponding to the readthrough transcript generated with *E. coli* RNA polymerase. (The caveat with the latter result is that T7 RNA polymerase-promoted transcription of the *mgtA* templates may be too fast for the hypothetical structure having the putative ribozyme activity to form.)

Transcription attenuation mechanisms have been typically associated with the formation of transcription terminators in one of the alternative structures. However, the sequences preceding nucleotide 220 lack the typical features of Rho-independent terminators (i.e., a GC-rich RNA hairpin followed by a poly(U) sequence [Landick et al. 1996]), and the generation of a truncated product in the in vitro transcription reaction did not require the Rho protein. This raises the possibility that the truncated product may be formed as a consequence of transcription pausing followed by RNA polymerase release from the DNA template. Although more experiments are required to determine how Mg^{2+} affects transcription elongation in the *mgtA* 5′UTR, sequences beyond those corresponding to stem-loops A, B, and C appear to participate in this process because a strain harboring a derivative of the 5′UTR with the sequence corresponding to positions 1–178 (which does not include sequences beyond stem-loop B) exhibited constitutive *mgtA* expression.

A Physiological Rationale for Dual Control of *mgtA* Expression

When *Salmonella* experiences a low Mg^{2+} environment, *mgtA* is the first PhoP-activated gene that is transcribed (our unpublished results), which may reflect the organism's effort to maintain physiological levels of cytoplasmic Mg^{2+}. Once the MgtA protein is produced, it will mediate internalization of Mg^{2+}, which will eventually bind to the *mgtA* 5′UTR to shut off MgtA expression (Fig. 2). This regulatory design allows the rapid uptake of Mg^{2+} and at the same time makes MgtA an arbiter of its own expression.

The *mgtA* gene is controlled at the level of transcription initiation by the PhoP protein, which is activated in response to the extracytoplasmic levels of Mg^{2+} sensed by the PhoQ protein, and at the level of transcription elongation in response to the cytoplasmic levels of Mg^{2+} sensed by the *mgtA* 5′UTR. Thus, Mg^{2+} acts in two different compartments to control different steps in *mgtA* transcription. This dual control of the *mgtA* gene is in contrast to most genes regulated by riboswitches, which are not subjected to regulation at the level of transcription initiation (Winkler and Breaker 2003; Brantl 2004; Nudler and Mironov 2004). The dual regulation of *mgtA* expression is somewhat reminiscent of that governing expression of the *trp* biosynthetic genes of *E. coli*. In this case, the control is exerted both by the tryptophan-responding TrpR protein, a transcriptional repressor that binds to the *trp* promoter-operator region, and by a transcription attenuation mechanism that responds to charged tRNA^{Trp} (Bennett and Yanofsky 1978; Yanofsky 2004). As described for the regulation of *trp*, most of the *mgtA* regulation takes place at the level of transcription initiation (i.e., by TrpR and PhoP), and only a small fraction is controlled at the level of transcription elongation (by the leader sequence and riboswitch, respectively).

What is the purpose of the dual *mgtA* control? It may enable *Salmonella* to exert differential regulation over those determinants directly affecting cytoplasmic Mg^{2+} (i.e., the Mg^{2+} transporters MgtA and MgtB and potentially other proteins) from those mediating modifications in the bacterial cell envelope (i.e., the proteins affecting the negative charges in the LPS) when they are all under transcriptional control of the same regulatory system (i.e., PhoP/PhoQ) (Groisman 2001). Indeed, expression of the PhoP-regulated *mgtCB* operon, which encodes the inner membrane protein MgtC and the Mg^{2+} transporter MgtB, appears to have features in common with those controlling the *mgtA* gene because *mgtCB* transcription also responds to Mg^{2+} in a strain lacking the Mg^{2+} sensor PhoQ protein in a process that requires the 5′UTR preceding the *mgtC* open reading frame (our unpublished results).

Does the *mgtA*-derived 5′UTR Regulate Gene Expression by Affecting Other mRNAs *in trans*?

The Storz laboratory has demonstrated that an RNA fragment corresponding to the 5′UTR of the *mgtA* transcript accumulates in *E. coli* grown in LB media and that

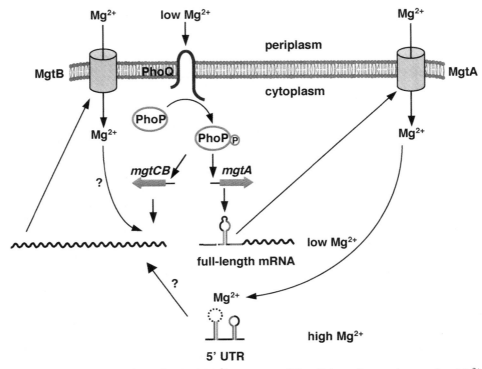

Figure 2. Model for regulation of the *Salmonella* MgtA Mg^{2+} transporter. When *Salmonella* experiences a low Mg^{2+} environment, the sensor PhoQ responds to extracytoplasmic Mg^{2+} by promoting phosphorylation of the DNA-binding protein PhoP, which binds to the *mgtA* promoter and initiates *mgtA* transcription. If cytoplasmic Mg^{2+} levels are above a certain threshold, Mg^{2+} binding to the *mgtA* 5'UTR results in the formation of stem-loop structures that promote transcription stopping within the 5'UTR. Once the cytoplasmic Mg^{2+} concentration goes below a certain level, an alternative stem-loop structure is formed, which allows readthrough into the *mgtA*-coding region. Production of the MgtA protein allows internalization of Mg^{2+}, the concentration of which will increase resulting in binding to the *mgtA* 5'UTR and shutting off of MgtA expression. The MgtB Mg^{2+} transporter is also regulated by Mg^{2+} in a process that requires the 5'UTR region preceding the *mgtC* gene. That 5'UTR region could be an Mg^{2+} sensor or it may be the target of the truncated RNA corresponding to the 220-nucleotide fragment originating from the *mgtA* 5'UTR. (Modified, with permission, from Cromie et al. 2006 [© Elsevier].)

the levels of this small RNA are lower in a strain harboring a mutation in the *hfq* gene (Kawano et al. 2005), which encodes a protein shown to bind to bacterial small RNAs exerting regulatory effects *in trans*. When we consider the fact that *mgtA* has the highest affinity of the PhoP-activated promoters (Minagawa et al. 2003) and it is the first and most highly expressed gene when the PhoP/PhoQ system is activated (our unpublished results), this raises the possibility that the *mgtA* 5'UTR might act *in trans* to regulate gene expression. The putative target mRNAs of the *mgtA* 5'UTR should be produced concurrently with the *mgtA* 5'UTR gene, and thus they are likely to be members of the PhoP regulon, but they could also be genes that are constitutively expressed. Therefore, we examined the DNA sequence between the reported transcription start site and the predicted site of translation initiation for several members of the PhoP regulon. We found that in addition to the 5'UTR of the *mgtCB* operon discussed above, several PhoP-activated genes harbor 5'UTRs that are more than 200 nucleotides long (i.e., *pagC, nagA,* and *ugtL*; Table 1). Moreover, we could identify regions of complementarity between the *mgtA* 5'UTR and the 5'UTRs of these genes. This suggests that the *mgtA* 5'UTR may have two roles: First, it may act as an Mg^{2+}-responding

riboswitch that determines whether the *mgtA*-coding region is transcribed resulting in the production of the MgtA protein. Second, it may regulate expression of other genes by acting *in trans*. Alternatively, the PhoP-regulated genes with unusually long 5'UTRs may respond to Mg^{2+} or other ligands to alter expression of the corresponding open reading frame (Fig. 2).

CONCLUSIONS

The 5'UTR of the *mgtA* gene constitutes the first example of a cation-responsive riboswitch. However, bioinformatics studies suggest that there are likely to be other cation-responding riboswitches (Barrick et al.

Table 1. PhoP-regulated Genes with Long 5'UTRs

Gene	Size of 5'UTR (in nucleotides)	Function of gene product
mgtCB	281	Mg^{2+} transport, growth in low Mg^{2+}
nagA	417	*N*-acetylglucosamine deacetylase
pagC	559	outer-membrane protein
ugtL	192	lipid-A phosphatase?

2004; Merino and Yanofsky 2005). This regulation of the Mg^{2+} transporter MgtA protein may enable *Salmonella* to exert differential control over those genes that are involved in cytoplasmic Mg^{2+} homeostasis from those involved in the modification of Mg^{2+}-binding sites in the cell envelope. Moreover, by regulating Mg^{2+} transport, MgtA may impact other riboswitches that respond to other ligands but are affected by Mg^{2+} (Yamauchi et al. 2005). The truncated *mgtA* RNA derived from the 5′UTR when organisms experience high Mg^{2+} might have the additional property of regulating genes *in trans*.

ACKNOWLEDGMENTS

Part of the research described in this paper was supported by grant AI49561 from the National Institutes of Health to E.A.G., who is an Investigator of the Howard Hughes Medical Institute.

REFERENCES

Barrick J.E., Corbino K.A., Winkler W.C., Nahvi A., Mandal M., Collins J., Lee M., Roth A., Sudarsan N., Jona I., et al. 2004. New RNA motifs suggest an expanded scope for riboswitches in bacterial genetic control. *Proc. Natl. Acad. Sci.* **101:** 6421.

Bennett G.N. and Yanofsky C. 1978. Sequence analysis of operator constitutive mutants of the tryptophan operon of *Escherichia coli*. *J. Mol. Biol.* **121:** 179.

Blanc-Potard A.B. and Groisman E.A. 1997. The *Salmonella selC* locus contains a pathogenicity island mediating intramacrophage survival. *EMBO J.* **16:** 5376.

Brantl S. 2004. Bacterial gene regulation: From transcription attenuation to riboswitches and ribozymes. *Trends Microbiol.* **12:** 473.

Castelli M.E., García Véscovi E.G., and Soncini F.C. 2000. The phosphatases activity is the target for Mg^{2+} regulation of the sensor protein PhoQ in *Salmonella*. *J. Biol. Chem.* **275:** 22948.

Chamnongpol S. and Groisman E.A. 2000. Acetyl-phosphate-dependent activation of a mutant PhoP response regulator that functions independently of its cognate sensor kinase. *J. Mol. Biol.* **300:** 291.

———. 2002. Mg^{2+} homeostasis and avoidance of metal toxicity. *Mol. Microbiol.* **44:** 561.

Chamnongpol S., Cromie M., and Groisman E.A. 2003. Mg^{2+} sensing by the Mg^{2+} sensor PhoQ of *Salmonella enterica*. *J. Mol. Biol.* **325:** 795.

Cromie M.J., Shi Y., Latifi T., and Groisman E.A. 2006. An RNA sensor for intracellular Mg^{2+}. *Cell* **125:** 71.

Froschauer E.M., Kolisek M., Dieterich F., Schweigel M., and Schweyen R.J. 2004. Fluorescence measurements of free [Mg^{2+}] by use of mag-fura 2 in *Salmonella enterica*. *FEMS Microbiol. Lett.* **237:** 49.

Groisman E.A. 2001. The pleiotropic two-component regulatory system PhoP-PhoQ. *J. Bacteriol.* **183:** 1835.

Groisman E.A., Kayser J., and Soncini F.C. 1997. Regulation of polymyxin resistance and adaptation to low-Mg^{2+} environments. *J. Bacteriol.* **179:** 7040.

Hanna R. and Doudna J.A. 2000. Metal ions in ribozyme folding and catalysis. *Curr. Opin. Chem. Biol.* **4:** 166.

Henkin T.M. and Yanofsky C. 2002. Regulation by transcription attenuation in bacteria: How RNA provides instructions for transcription termination/antitermination decisions. *Bioessays* **24:** 700.

Jacquier A. 1996. Group II introns: Elaborate ribozymes. *Biochimie* **78:** 474.

Kawano M., Reynolds A.A., Miranda-Rios J., and Storz G. 2005. Detection of 5′- and 3′-UTR-derived small RNAs and cis-encoded antisense RNAs in *Escherichia coli*. *Nucleic Acids Res.* **33:** 1040.

Kim D.S., Gusti V., Pillai S.G., and Gaur R.K. 2005. An artificial riboswitch for controlling pre-mRNA splicing. *RNA* **11:** 1667.

Landick R., Turnbough C.L., and Yanofsky C. 1996. Transcription attenuation. In Escherichia coli *and* Salmonella: *Cellular and molecular biology* (ed. F.C. Neidhardt et al.), p. 1263. ASM Press, Washington, D.C.

Liu M., Tolstorukov M., Zhurkin V., Garges S., and Adhya S. 2004. A mutant spacer sequence between −35 and −10 elements makes the P$_{lac}$ promoter hyperactive and cAMP receptor protein-independent. *Proc. Natl. Acad. Sci.* **101:** 6911.

Maguire M.E. 2006. Magnesium transporters: Properties, regulation and structure. *Front. Biosci.* **11:** 3149.

Merino E. and Yanofsky C. 2005. Transcription attenuation: A highly conserved regulatory strategy used by bacteria. *Trends Genet.* **21:** 260.

Minagawa S., Ogasawara H., Kato A., Yamamoto K., Eguchi Y., Oshima T., Mori H., Ishihama A., and Utsumi R. 2003. Identification and molecular characterization of the Mg^{2+} stimulon of *Escherichia coli*. *J. Bacteriol.* **185:** 3696.

Montagne M., Martel A., and Le Moual H. 2001. Characterization of the catalytic activities of the PhoQ histidine protein kinase of *Salmonella enterica* serovar Typhimurium. *J. Bacteriol.* **183:** 1787.

Montange R.K. and Batey R.T. 2006. Structure of the S-adenosylmethionine riboswitch regulatory mRNA element. *Nature* **441:** 1172.

Muller M., Weigand J.E., Weichenrieder O., and Suess B. 2006. Thermodynamic characterization of an engineered tetracycline-binding riboswitch. *Nucleic Acids Res.* **34:** 2607.

Nudler E. and Mironov A.S. 2004. The riboswitch control of bacterial metabolism. *Trends Biochem. Sci.* **29:** 11.

Reinhart R.A. 1988. Magnesium metabolism. A review with special reference to the relationship between intracellular content and serum levels. *Arch. Intern. Med.* **148:** 2415.

Serganov A., Polonskaia A., Phan A.T., Breaker R.R., and Patel D.J. 2006. Structural basis for gene regulation by a thiamine pyrophosphate-sensing riboswitch. *Nature* **441:** 1167.

Serganov A., Yuan Y.R., Pikovskaya O., Polonskaia A., Malinina L., Phan A.T., Hobartner C., Micura R., Breaker R.R., and Patel D.J. 2004. Structural basis for discriminative regulation of gene expression by adenine- and guanine-sensing RNAs. *Chem. Biol.* **11:** 1729.

Smith R.L. and Maguire M.E. 1995. Distribution of the CorA Mg^{2+} transport system in gram-negative bacteria. *J. Bacteriol.* **177:** 1638.

———. 1998. Microbial magnesium transport: Unusual transporters searching for identity. *Mol. Microbiol.* **28:** 217.

Soncini F.C., Garcia Vescovi E., Solomon F., and Groisman E.A. 1996. Molecular basis of the magnesium deprivation response in *Salmonella typhimurium*: Identification of PhoP-regulated genes. *J. Bacteriol.* **178:** 5092.

Tao T., Grulich P.F., Kucharski L.M., Smith R.L., and Maguire M.E. 1998. Magnesium transport in *Salmonella typhimurium*: Biphasic magnesium and time dependence of the transcription of the *mgtA* and *mgtCB* loci. *Microbiology* **144:** 655.

Winkler W.C. 2005. Riboswitches and the role of noncoding RNAs in bacterial metabolic control. *Curr. Opin. Chem. Biol.* **9:** 594.

Winkler W.C. and Breaker R.R. 2003. Genetic control by metabolite-binding riboswitches. *Chembiochem.* **4:** 1024.

———. 2005. Regulation of bacterial gene expression by riboswitches. *Annu. Rev. Microbiol.* **59:** 487.

Winkler W.C., Nahvi A., Roth A., Collins J.A., and Breaker R.R. 2004. Control of gene expression by a natural metabolite-responsive ribozyme. *Nature* **428:** 281.

Yamamoto K., Ogasawara H., Fujita N., Utsumi R., and Ishihama A. 2002. Novel mode of transcription regulation of divergently overlapping promoters by PhoP, the regulator of two-component system sensing external magnesium availability. *Mol. Microbiol.* **45:** 423.

Yamauchi T., Miyoshi D., Kubodera T., Nishimura A., Nakai S., and Sugimoto N. 2005. Roles of Mg^{2+} in TPP-dependent riboswitch. *FEBS Lett.* **579:** 2583.

Yanofsky C. 2004. The different roles of tryptophan transfer RNA in regulating trp operon expression in *E. coli* versus *B. subtilis. Trends Genet.* **20:** 367.

Zwir I., Shin D., Kato A., Nishino K., Latifi T., Solomon F., Hare J.M., Huang H., and Groisman E.A. 2005. Dissecting the PhoP regulatory network of *Escherichia coli* and *Salmonella enterica. Proc. Natl. Acad. Sci.* **102:** 2862.

Structural Studies of the Purine and SAM Binding Riboswitches

S.D. Gilbert,* R.K. Montange,* C.D. Stoddard, and R.T. Batey
Department of Chemistry and Biochemistry, University of Colorado, Boulder, Colorado 80309-0215

Riboswitches are recently discovered genetic regulatory elements found in the 5′-untranslated regions of bacterial mRNAs that act through their ability to specifically bind small-molecule metabolites. Binding of the ligand to the aptamer domain of the riboswitch is communicated to a second domain, the expression platform, which directs transcription or translation of the mRNA. To understand this process on a molecular level, structures of three of these riboswitches bound to their cognate ligands have been solved by X-ray crystallography: the purine, thiamine pyrophosphate (TPP), and S-adenosylmethionine (SAM-I) binding aptamer domains. These studies have uncovered three common themes between the otherwise different molecules. First, the natural RNA aptamers recognize directly or indirectly almost every feature of their ligand to achieve extraordinary specificity. Second, all of these RNAs use a complex tertiary architecture to establish the binding pocket. Finally, in each case, ligand binding serves to stabilize a helix that communicates the binding event to the expression platform. Here, we discuss these properties of riboswitches in the context of the purine and SAM-I riboswitches.

In mRNA, noncoding regions contain an enormous amount of information that is crucial for proper gene expression. It has long been recognized that bacterial gene regulation is controlled at the translational and transcriptional levels via proteins that bind structures and sequences in mRNA 5′UTRs (untranslated regions). The classic example of this form of regulation is that of the TRAP protein, which, in the presence of high levels of tryptophan, binds the 5′UTR, affecting the formation of regulatory stem-loop structures (Babitzke 2004). At the translational level, a number of ribosomal proteins associate with elements in the 5′UTR of their own mRNAs that are structurally similar to their rRNA-binding sites, occluding important elements for formation of the translational initiation complex (Zengel and Lindahl 1994). However, there are a number of 5′UTRs that display strong conservation in their sequence and secondary structure that could not be associated with any known protein or other regulatory factor, and thus their function and mechanism of action remained elusive.

The first of these "orphan" elements to be extensively characterized was one found by Henkin and Grundy in the 5′UTR of operons coding for genes involved in sulfur metabolism pathways (Grundy and Henkin 1998). The striking feature of this element, dubbed the S-box, was its apparent ability to fold into two mutually exclusive structures, ostensibly in the presence of a regulatory factor. It was thus postulated that the S-box uses these structures to regulate transcription through the ability of some unidentified factor to bind the 5′UTR in the presence of methionine. Almost concurrently, another RNA element was discovered, called the RFN element, that was found upstream of genes encoding proteins responsible for riboflavin synthesis in gram-positive bacteria (Gelfand et al. 1999). The RFN element, in addition to forming

mutually exclusive stem-loop structures, contains five strongly conserved helical regions and was found not only in the 5′UTR of one gene or operon across different genomes, but also within different genes and operons within one genome. Similar to the S-box, no protein factor could be identified that bound the RFN element, and so it was hypothesized that the mRNA may directly bind the reaction product of these genes, a small-molecule metabolite. Gelfand and coworkers speculated that these conserved sequences fold into structures similar in function to RNA aptamers that bind small molecules through in vitro selection methods (SELEX). Studies of a third element, the thi box, found in the 5′UTRs of thiamine biosynthetic genes, yielded further information that a small molecule (in this case, thiamine pyrophosphate) can bind RNA in the absence of a protein cofactor, thereby modulating the formation of downstream secondary structure and gene expression (Miranda-Rios et al. 2001).

In 2002, the Breaker and Nudler laboratories independently and concurrently conducted definitive studies proving this hypothesis. The first of a series of publications describing what has become known as riboswitches was an investigation of the adenosylcobalamin (vitamin B_{12}) associated element (Nahvi et al. 2002), closely followed by confirmation of the RFN (Mironov et al. 2002; Vitreschak et al. 2002), thi box (Winkler et al. 2002b), and S-box elements (McDaniel et al. 2003; Winkler et al. 2003). These mRNAs were shown to directly bind small-molecule metabolites and alter their conformation in response to the binding event using a number of biochemical techniques, most notably the "in-line" probing method pioneered by the Breaker laboratory (McDaniel et al. 2003; Winkler et al. 2003). Furthermore, it was demonstrated that mutations disrupting Watson-Crick paired helices in the conserved regions abrogated ligand binding, which could be restored by compensatory mutations that reestablished the helix. This behavior is a

---

*These two authors contributed equally to this work.

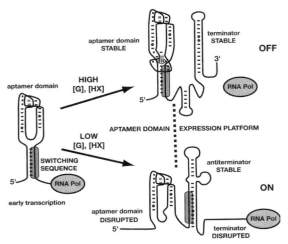

Figure 1. Schematic of transcriptional repression by the purine riboswitch. Early in transcription, the first element of a riboswitch that is formed is the aptamer domain (*left*) that binds a specific small-molecule metabolite. If a sufficient metabolite concentration is present in the cell, it binds to the aptamer and stabilizes the incorporation of a "switching sequence" into this domain (*right top*), fating the mRNA to form a rho-independent terminator, causing transcription to halt. In the absence of metabolite, the switching sequence is used by the expression platform to form an antiterminator element (*right bottom*), allowing transcription to proceed and the gene to be expressed.

hallmark of RNAs whose activity is dependent on higher-order folded structures. Finally, it was clearly shown that the binding of the ligand to the "aptamer domain" of the riboswitch is communicated to an "expression platform" (Fig. 1) comprising a secondary structural switch that is used to direct gene expression.

Bioinformatic approaches, coupled with biochemistry, have resulted in the identification of a number of other riboswitches in rapid succession (for review, see Winkler and Breaker 2005). This has been primarily achieved using the criteria that riboswitches are found in the 5′UTR of genes, that they contain conserved sequences and secondary structures that are conserved across a segment of bacterial phylogeny, and that the effector molecule is closely related to the function of the gene product in which they are found. To date, riboswitches that use thiamine pyrophosphate (Rodionov et al. 2002; Winkler et al. 2002b; Sudarsan et al. 2003a), S-adenosylmethionine (Grundy and Henkin 1998; Mandal et al. 2003; Barrick et al. 2004), adenine and guanine (Mandal et al. 2003), glycine (Mandal et al. 2004), lysine (Rodionov et al. 2003; Sudarsan et al. 2003b), glucosamine-6-phosphate (Winkler et al. 2004), flavin mononucleotide (Vitreschak et al. 2002; Winkler et al. 2002a), coenzyme B_{12} (Nahvi et al. 2002; Mandal et al. 2003), and Mg^{2+} (Cromie et al. 2006) have been identified. Currently, there are also a number of conserved secondary structures in bacterial mRNA for which there are still no identified effector molecules (Barrick et al. 2004; Corbino et al. 2005). Therefore, it is very likely that this list will continue to expand.

In this paper, we describe our efforts to determine the structure of several of these metabolite-binding RNA elements by X-ray crystallography. Through a structural

investigation of the aptamer domains of the purine and S-adenosylmethionine (S-box) riboswitches, we sought to address two fundamental questions. First, how do these mRNAs specifically bind their metabolite? Associated with this question is how these mRNAs achieve such high discrimination between their cognate ligand and a number of chemically similar metabolites also found in the cell. Second, how does the aptamer domain communicate the ligand-binding event to the expression platform? This is the crucial feature of the riboswitch that allows it to act as an effective regulator of gene expression.

PURINE RIBOSWITCH

The purine riboswitch (also known as the G-box and A-box) is a regulatory element found upstream of a number of genes involved in purine biosynthesis and transport. It comprises a series of conserved sequences spread over an approximately 60-nucleotide region (Mandal et al. 2003; Mandal and Breaker 2004b). Biochemical characterization revealed that it folds into a three-helical structure with the conserved nucleotides localizing to two terminal loops and a three-way junction (Fig. 2A). It was also noticed that a single pyrimidine residue covaried with the ligand; in guanine-responsive elements, it is always a cytosine, and in adenine-responsive elements, it is always uracil (Fig. 2a, asterisk). This observation suggested that purine specificity is dictated in large part by the ability of the nucleobase to form a Watson-Crick base pair with this pyrimidine.

Structure Determination

To determine the structure of this RNA element bound to its cognate ligand by X-ray crystallography, we used the G-box from the *xpt-pbuX* operon of *Bacillus subtilis* (Mandal et al. 2003). One of the most powerful strategies in RNA

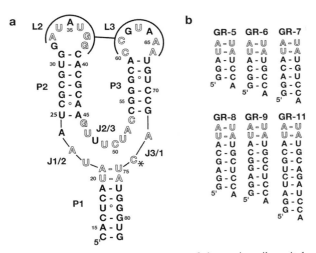

Figure 2. (*a*) Secondary structure of the purine riboswitch aptamer domain of the *xpt-pbuX* operon from *B. subtilis*. Nucleotides that are outlined are >90% conserved, and the starred nucleotide represents the pyrimidine that is cytosine in the guanine riboswitches and uridine in adenine riboswitches. (*b*) Sequence variants of the P1 helix introduced in a limited library of variants that were constructed and tested for crystallizability; construct "GR7" readily produced diffraction-quality crystals.

crystallography with respect to finding a sequence that readily crystallizes and yields high-quality diffraction data is to systematically vary peripheral elements, such as helices and loops (Ke and Doudna 2004). Unfortunately, because highly phylogenetically conserved nucleotides are interspersed throughout the RNA, the only region that could be altered without affecting the ability to bind ligand is the length of the P1 helix (Fig. 2b) (Mandal et al. 2003). Thus, in our initial survey, we used a series of constructs in which the length of this helix was between 6 and 11 bp. To circumvent any potential issues with refolding the RNA, we prepared all of the RNA using a newly developed native purification system that allowed us to rapidly purify approximately 10-mg quantities in 2–3 hours (Kieft and Batey 2004). Once prepared and bound to hypoxanthine, a soluble derivative of guanine that also binds the G-box, the RNAs were subjected to a suite of standard commercially available sparse matrices (PEG-Ion, Natrix, Crystal Screen, and Nucleic Acid Mini Screen; Hampton Research). One of the constructs (GR7, Fig. 2b) rapidly yielded crystals under several conditions in the Nucleic Acid Mini Screen (conditions 1–4). In particular, it was found that the G-box required 10–20 mM cobalt hexaammine to yield crystals that diffracted to approximately 3 Å.

Along the route toward obtaining data that would yield phase information and allow calculation of an interpretable electron density map, we encountered several issues. First was the ability to generate a crystal form that allowed us to collect quality data sets. Our initial crystal form (Fig. 3a) was of the P2 space group, with a very large unit cell that resulted in a diffraction pattern with very closely spaced reflections. Resolving these reflections for precise quantification of their volumes was almost impossible without the aid of a synchrotron X-ray source. Consequently, we sought other forms that would be more amenable to analysis with our home source. A second form was achieved by switching the growth temperature to 4°C, yielding crystals of the P3$_1$22 space group (Fig. 3b). Whereas these crystals had a smaller unit cell, analysis of the data clearly indicated that they were hemihedrally twinned, rendering them useless for obtaining phase information. Using microseeding approaches (Stura and Wilson 1990), we were able to obtain a third form, which was not twinned, but contained a C2 space group and a small unit cell. This was the form used in all subsequent work.

Obtaining phase information via the incorporation of heavy atoms is another serious issue in RNA X-ray crystallography structure determination. In protein crystallography, it is routine to use selenomethionine to incorporate a heavy atom that is suitable for obtaining experimental phases with the multiwavelength anomalous dispersion (MAD) method (Hendrickson and Ogata 1997). Although this advance has enormously facilitated protein structure determination, the same advance has not been made for RNA structure determination. Incorporation of bromine by in vitro transcription with 5-bromouridine triphosphate (Kieft et al. 2002; Zhang and Doudna 2002) or synthetically using 2′-methylseleno-substituted nucleosides has been successful (Teplova et al. 2002; Carrasco et al. 2004), but is difficult to achieve with larger RNAs (>50 nucleotides) (Hobartner et al. 2005). Another approach is to use selenomethionine-labeled U1A

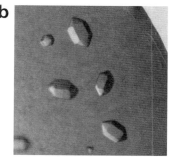

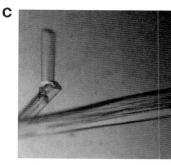

Figure 3. Crystals of the GR7 construct. The first crystal form (*a*) had one large unit cell dimension (P2 space group; a = 71.5 Å, b = 275.8 Å, c = 71.9 Å, α = γ = 90°, β = 119.6°; ~3.0 Å diffraction limit) that made crystals difficult to work with due to closely spaced reflections. The second form obtained (P3$_1$22 space group; a = b = 117.3 Å, c = 90.0 Å, α = γ = 90°, β = 120°; ~ 2.5 Å diffraction limit) were shown to be twinned using the algorithm of Yeats and coworkers. The third form (C2 space group; a = 132.3 Å, b = 35.3 Å, c = 42.2 Å, α = γ = 90°, β = 91°; ~ 1.9 Å diffraction limit) is what was used to solve the crystal structure.

bound to the RNA as a carrier of heavy atoms for phasing (Ferre-D'Amare and Doudna 2000), an approach that has been successfully applied to a number of ribozyme structures (Ferre-D'Amare et al. 1998; Rupert and Ferre-D'Amare 2001; Adams et al. 2004). However, given the predicted fold of the G-box, it was not clear how to engineer a U1A-binding site into the RNA without perturbing the ligand-binding site. After a number of heavy atom soaking trials with standard compounds such as platinum, gold, and mercury derivatives, we were unable to generate native and derivative data sets that were sufficiently isomorphous to be able to solve the phase problem using the multiple isomorphous replacement (MIR) method.

To overcome this difficulty, we turned to a recently established technique called single-wavelength anomalous dispersion (SAD) phasing (Rice et al. 2000; Dauter et al. 2002). In this method, data is only collected at a single wavelength of X-ray radiation, and the phase ambiguity is broken by visual inspection of the electron density maps generated from the two solutions. As a heavy atom, we decided to rely on cobalt hexaammine, which was a required component in the mother liquor and was likely bound to the RNA, as the crystals were a faint orange in color. The absorption peak of cobalt is very close to the wavelength of X-ray radiation from a rotating copper anode source (1.608 Å vs. 1.542 Å); at this wavelength, cobalt is a weak anomalous scatterer with 3.61 electrons contributing to the scattering. We performed an inverse-beam experiment on crystals that were grown and cryoprotected in 12 mM cobalt hexaammine (the crystals were not backsoaked to remove cobalt from the mother liquor), collecting 20° wedges with 0.5° oscillation per frame. Because the crystals showed significant mosaicity (~0.8°), we had to use D*TREK (Pflugrath 1999) to accurately merge partial reflections and integrate the volume of the entire reflection. Using SOLVE (Terwilliger 2003), we initially sought a solution containing 4 heavy atoms, which was rapidly found by the program. Further sites were found by incrementally searching for more sites, until a maximum of 10 strong sites were found. This solution was imported into crystallography and NMR system (CNS) (Brunger et al. 1998) for calculation of experimental phases, density modification using a solvent flipping routine, and calculation of an electron density map for each of the two possible solutions (the heavy-atom solution and its mirror image). Only one solution yielded a weakly interpretable map in which density for the backbone and a number of bases could be observed.

To build the model in this map of moderate quality, we chose to initially build uridine into bases that had the approximate shape of a pyrimidine, and adenine into regions that looked like purines. Using this strategy, we were able to build approximately 60% of the model. At this point, we performed one round of energy minimization and B-factor refinement on the model and used the resulting model to calculate a new phase-combined map that displayed clear electron density for all residues in the G-box and for which the sequence register could be unambiguously obtained.

Structure of the Purine Riboswitch

The architecture of the purine riboswitch is defined by two regions of extensive noncanonical base-pairing (Fig. 4) (Batey et al. 2004; Serganov et al. 2004). A distinctive side-by-side arrangement of the P2 and P3 helices is created by an unusual loop–loop (L2–L3) interaction that is formed by two sets of base quartets involving eight universally conserved nucleotides. Each quartet comprises a Watson-Crick pair between nucleotides from each loop and a noncanonical base pair docked in the minor groove of the Watson-Crick pair. These two quartets adopt an identical configuration in both the *B. subtilis xpt-pbuX* guanine riboswitch and the *Vibrio vulnificus add* adenine riboswitch. A set of two further noncanonical base pairs are found in the guanine riboswitch at the top of the L2–L3 interaction that are absent in the adenine riboswitch; nucleotides involved in these pairs are significantly less phylogenetically conserved, suggesting that they are not

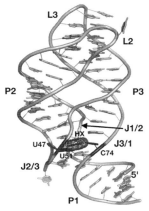

Figure 4. Global fold of the guanine riboswitch aptamer domain (PDB 1U8D). The binding pocket for the guanine analog hypoxanthine (HX, emphasized with a surface representation) is embedded within the three-way junction formed by J1/2, J2/3, and J3/1. The ligand is bound through hydrogen-bonding interactions with U51 and C74 (*dark gray*).

required for stabilization of the tertiary contact. Although this element of structure is not directly involved in ligand recognition, elimination of this feature from the RNA by conversion of each loop into a stable UUCG tetraloop completely abolishes ligand binding (Batey et al. 2004).

Ligand binding is accomplished using a complex architecture formed by the central three-way junction. This region comprises two base triples that flank each side of the central ligand-binding pocket. Ligand recognition is achieved via a base quadruple in which the purine forms a Watson-Crick pair with a pyrimidine (C74), N3 and N9 of the nucleobase are recognized by U51, and N7 by the 2′-hydroxyl group of U22 (Fig. 5a).

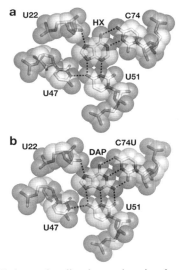

Figure 5. Hydrogen-bonding interactions involved in guanine and adenine recognition. (*a*) Hypoxanthine (HX) is bound using a series of hydrogen bonds to a pyrimidine-rich binding pocket including the ribose sugar of U22 and the bases of U47, U51, and C74 (PDB 1U8D). (*b*) Binding of the adenine analog 2,6-diaminopurine (DAP) to the guanine riboswitch containing a single point mutation C74U, demonstrating that purine specificity is completely dictated by pyrimidine 74 (PDB 2B57).

U51 is buttressed through hydrogen bonding to U47 to complete the quartet. This structure also clearly demonstrates how the purine riboswitch could accomplish a specificity swap from guanine to adenine through the change of a single nucleotide (Y74); all other contacts to the nucleobase are identical for the two ligands. Crystallization of the *xpt-pbuX* guanine riboswitch with a single C74U mutation with the adenine analog 2,6-diaminopurine directly demonstrated that the switch involves no other changes in the riboswitch architecture (Fig. 5b) (Gilbert et al. 2006).

Mechanism of Ligand Binding

This structure begs a question about the ligand–RNA interaction: How does the ligand manage to bind to a site that becomes buried within the center of the three-way junction? The essence of this problem lies in the fact that the nucleobase is approximately 98% solvent-inaccessible with no channel that would allow it entry from bulk solvent. Intrinsic to this question is, What exactly is the free state of the RNA? NMR spectroscopy and in-line probing clearly indicate that the three-way junction is conformationally flexible, lacking a single discrete structure (Mandal et al. 2003; Noeske et al. 2005). Yet, this element cannot be completely disorganized, as the ligand needs to recognize and dock with some element of the junction.

As previously noted, residues of each strand in the junction contact the ligand. Footprinting of the RNA by in-line probing yields information about the dynamic environment of each residue in the unbound state, indicating that the J3/1 strand is relatively rigid when compared with J1/2 and J2/3. The J3/1 strand presents a pyrimidine residue that we call the "specificity pyrimidine" (C74 in the *xpt-pbuX* G-box), which forms a Watson-Crick pair with the ligand. Using thermodynamic data collected from an extensive number of purine nucleobase analogs, it is clear that the hydrogen bonds between the pyrimidine and the ligand are extremely important for productive binding (Gilbert et al. 2006). A minimum of two hydrogen bonds is required in this interaction to achieve riboswitch recognition of its ligand. Furthermore, the structure indicates that the ligand does not stack upon neighboring bases in the ligand-binding pocket, underscoring the importance of hydrogen bonding in binding. Supporting this idea, when U48, a residue that is flipped into the solvent in the bound structure, is substituted with the fluorescent base 2-aminopurine, a large change in fluorescence (nearly 100%, correlated to the concentration of riboswitch in solution) is observed as the ligand is bound up by the riboswitch, consistent with a significant conformational change (Gilbert et al. 2006). From this evidence, we propose a binding mechanism in which the purine ligand is first recognized by the specificity pyrimidine C74, prompting the subsequent organization of the three-way junction.

To effectively regulate gene expression, ligand binding to the aptamer domain must be communicated to the expression platform. A consequence of the mechanism of ligand recognition that we have proposed is that ligand binding and stabilization of the P1 "communication helix" is directly and intimately coupled (Gilbert et al. 2006). Only upon recognition of the correct ligand by pyrimidine 74 will J2/3 clamp down to complete the (Purine-Y74)•U51•U47 quadruple. Concurrently, the two base triples between Watson-Crick pairs in the P1 helix and U49 and C50 of J2/3 are formed, which is instrumental in stabilizing the 3′ strand of the P1 helix ("switching sequence," Fig. 1) against incorporation into the antiterminator in the expression platform. In this way, ligand binding is highly coupled to the structural switch on the expression platform, the central feature of riboswitch function.

S-ADENOSYLMETHIONINE RIBOSWITCH

The S-box was one of the first mRNA regulatory elements described as a potential riboregulatory element. The sequence identified as the aptamer domain contains a number of phylogenetically conserved nucleotides that center within and surrounding a four-way junction motif (Fig. 6a). Biochemical and genetic analysis indicates that

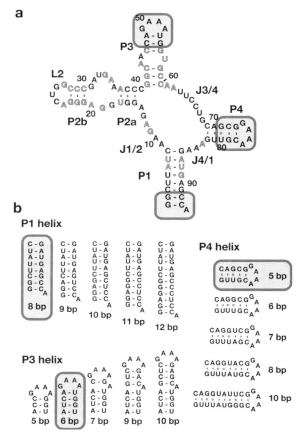

Figure 6. Secondary structure of the RNA (*a*) used for the structure determination of the SAM-I riboswitch. Nucleotides that are >95% conserved across phylogeny are depicted in gray outline. The boxed regions were the portions of sequence that are phylogenetically variable and thus systematically varied to create a library of RNAs tested for crystallizability. The library was created by varying the lengths of the P1, P3, and P4 helices (*b*), as well as changing the wild-type terminal loops of P3 and P4 to GAAA tetraloops.

many of these conserved sequence elements are critical for the formation of a kink-turn motif and a pseudoknot that are crucial for metabolite binding and gene regulation, suggesting that this RNA adopts a complex tertiary architecture.

Structure Determination

To understand how this RNA specifically binds *S*-adenosylmethionine (SAM), we solved the liganded form of the S-box. The strategy that we used to find a crystallizable RNA was an extension of that used for the guanine riboswitch. From phylogenetic alignment of over 250 sequences from the Rfam database (Griffiths-Jones et al. 2003, 2005), it was clear that the conserved nucleotides were all contained in a defined four-way junction region (Fig. 6a). From a large number of sequences, we chose to focus on one from *Thermoanaerobacter tengcongensis*, a bacterium that grows optimally at approximately 70°C, suggesting a highly stable RNA–ligand complex. Because thermophilic proteins are a popular target of crystallographers, presumably for their enhanced stability, we reasoned that their RNAs would share the same properties and thus tend to be more crystallizable. The termini of helices 1, 3, and 4 are all of variable length and sequence, indicating that they are nonessential for ligand binding and thus can be altered without affecting the RNA's SAM-binding activity. To find an appropriate RNA, we generated a library of individuals in which the length of these helices was systematically varied (Fig. 6b). Additionally, the sequence of the terminal loops of P3 and P4 were changed to GAAA, a tetraloop motif that has been shown to mediate crystal contacts in a number of RNA and RNP crystals (Pley et al. 1994; Batey et al. 2001). On the basis of these approaches, we created a library of approximately 70 variants by PCR and transcribed them into RNA using conventional methods.

Each variant was subjected to crystallization trials using four commercially available kits. These trials were allowed to incubate for 1 week at room temperature. Only constructs yielding single crystals that diffracted to <5 Å resolution in this initial screen were subjected to further trials; variants that failed this test were discarded. After investigation of a number of variants, it became clear that constructs in which the lengths of the P1, P3, and P4 helices were reduced to their minimal lengths yielded crystals in a significant number of conditions, indicating their increased propensity to crystallize as compared to RNAs with longer peripheral helices. Further screening around these initial hits using sequences with minor sequence changes yielded a construct that rapidly and reproducibly crystallized (Fig. 6b).

Again, the second consideration in our crystallization strategy was the phase problem. We used a strategy related to that used for the guanine riboswitch, relying on the ability of hexaammine compounds (cobalt, osmium, and iridium) to promiscuously, yet specifically, bind nucleic acids. In the case of the SAM riboswitch, once we found a construct that reproducibly yielded crystals, we reoptimized the crystallization conditions with iridium hexaammine. This yielded diffraction-quality crystals under conditions with 8 mM iridium hexaammine. To prepare the crystals for a MAD experiment, they were cryoprotected for 5 minutes in a solution containing 15% ethylene glycol, but no Ir(NH₃)₆Cl₃, serving also as a brief back-soak to remove excess heavy atom. These crystals yielded a clearly interpretable Patterson map, and using CNS, we were able to rapidly find a four-heavy-atom solution that yielded sufficient phase information to calculate an interpretable electron density map.

Structure of the SAM-I Riboswitch

The S-box or SAM-I motif folds into a structure consisting of two sets of coaxially stacked helices (Montange and Batey 2006). Although it is tempting to call each of these "domains" in a fashion similar to those of the group I intron, we do not know whether each of these would be able to fold in the absence of the other or the stabilizing influence of single-stranded joining regions. The first coaxial stack is between P1 and P4 with the adenosine residues of the J4/1 joining region forming a contiguous helical stretch between the two A-form helices (Fig. 7). The second stack involves P2 and P3, which is broken by a canonical kink-turn motif (Fig. 7, KT) (Klein et al. 2001) that redirects the terminal loop L2 back toward the P1/P4 stack to form a pseudoknot with J3/4. Packing of the two sets of stacks is further facilitated by extensive tertiary interactions made with the J1/2 and J3/4 stands.

Using a bipartite site created by the minor grooves of the P1 and P3 helices, the RNA binds *S*-adenosylmethionine. The ligand itself is bound in a very compact configuration (the *cis*-configuration) in which the methionine moiety

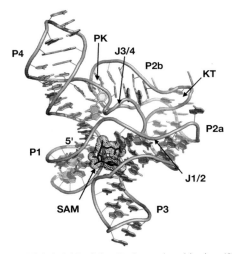

Figure 7. Global fold of the *S*-adenosylmethionine (SAM-I) riboswitch aptamer domain (PDB 2GIS). Labeling of the helices is the same as in Fig. 6a, along with the kink-turn (KT) and pseudoknot (PK) motifs. SAM is highlighted by a dotted surface representation (*dark gray*) which is found in a binding pocket formed by the minor grooves of P1 and P3 along with J1/2.

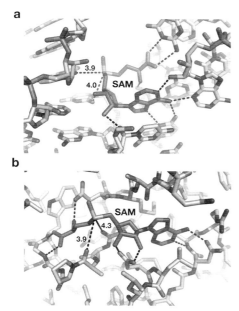

Figure 8. Comparison of the recognition of SAM by (*a*) RNA and by (*b*) the methyltransferase FtsJ (PDB 1EJ0). All hydrogen bonds shown between SAM and the RNA or protein are standard distances (2.8–3.3 Å). The electrostatic interactions between the charged sulfur moiety and carbonyl atoms in the RNA or protein are also depicted by dashes with interatomic distances (in angstroms) between the sulfur atom and the carbonyl oxygen denoted.

stacks above the adenosyl purine ring (Fig. 8a). This stacking is presumably stabilized through a pi-cation interaction between the positively charged main-chain amino group of methionine and the aromatic ring system of the adenyl moiety. This is in contrast to most proteins that bind SAM in a *trans*-configuration (Fig. 8b), presumably to position the activated methyl group for transfer to a wide variety of substrates. The *cis*-configuration of SAM creates two different faces that interact with the P1 and P3 helices in very different ways. SAM is recognized by the minor groove of P3 via extensive hydrogen bonding with the adenine ring and the main-chain atoms of the methionine moiety. It is likely that these interactions serve as the bulk of the stabilizing interactions with the RNA.

The other face of SAM comprises the ribose sugar and the sulfur moiety. The presence of the activated methyl group is sensed through an electrostatic mechanism via the positive charge on the sulfur atom. Along with the ribose sugar, the sulfur packs against the 3′ side of the P1 helix, primarily interacting with two universally conserved A-U pairs. The positively charged sulfur group is placed approximately 4 Å from two carbonyl groups (O2) of the two uracils (Fig. 8a). This electrostatic interaction results in the S-box's approximately 100-fold preference for SAM over *S*-adenosylhomocysteine (SAH), which does not possess an activated methyl group. FtsJ, a methyltransferase, uses a similar recognition strategy for discrimination between the SAM cofactor and SAH (Fig. 8b) (Bugl et al. 2000).

A further consequence of SAM binding along the P1 helix is that it facilitates a series of tertiary interactions

between the 3′ strands of the P1 and P3 helices. This small network of hydrogen bonds is clearly induced by the association of the RNA with SAM, as revealed by in-line probing. Nucleotides on the 3′ side of the P1 helix show a significant degree of protection from magnesium-induced backbone cleavage only in the presence of SAM, an indicator of ligand-induced RNA–RNA interactions. This ligand-induced tertiary structure is a critical component of SAM-dependent regulation, as these interactions serve to stabilize the 3′ side of the communication helix (P1) from being used to form an alternative secondary structure, the antiterminator helix. Thus, binding of SAM to the aptamer domain fates the expression platform to form a classic rho-independent terminator helix, causing transcription to stop.

COMMON THEMES AMONG THE RIBOSWITCH STRUCTURES

Currently, structures of purine (Batey et al. 2004; Serganov et al. 2004), *S*-adenosylmethionine (Montange and Batey 2006), and thiamine pyrophosphate (TPP) (Serganov et al. 2006; Thore et al. 2006) have been solved to atomic resolution. Each of these riboswitches is architecturally different, arising from their need to bind chemically distinct ligands. However, there are three critical similarities between the RNAs that we speculate are generalizable to all biological RNAs responsible for specifically binding small molecules.

Each of the riboswitch aptamer domains has tertiary architectural elements outside the ligand-binding pocket that serve to globally organize the RNA, in contrast to the majority of in-vitro-selected aptamers that bind small molecules. Typical in-vitro-selected aptamers, such as the ATP (Dieckmann et al. 1996; Jiang et al. 1996) and theophylline (Zimmermann et al. 1997) aptamers, bind their ligands through internal loop motifs embedded within single stem-loop structures. More complex folding involving helix–helix packing has not been structurally observed, although it likely occurs in the more complex in-vitro-selected ligase enzymes (Bartel and Szostak 1993; Bergman et al. 2004). Conversely, each riboswitch structure reveals at least one tertiary interaction via a loop–loop, pseudoknot, or loop–helix interaction in the purine, SAM, and TPP riboswitches, respectively. Disrupting mutations in these motifs result in loss of ligand binding in vitro, as well as the ability of the riboswitch to regulate gene expression in vivo. Because the global architecture of these riboswitch aptamer domains is established prior to ligand binding, this suggests that scaffolding of the binding pocket by tertiary architecture in a fashion akin to proteins is a common feature of these RNA elements. Recent studies of GTP aptamers by the Szostak group have suggested that increasing the informational complexity of the RNA is the route to greater specificity for a particular ligand (Carothers et al. 2006a, b), consistent with the observation of structurally complex biological aptamers.

A second commonality between these RNAs is the degree to which they recognize their cognate ligand. Unlike their in-vitro-selected kin, these RNAs appear to recognize the majority, if not all, of the functional groups in the ligand through direct and indirect readout mechanisms. This is exemplified by the SAM riboswitch, which recognizes its ligand through a series of hydrogen-bonding and base-stacking interactions with P3, Van der Waals, and electrostatic interactions with P1, and indirect readout to detect the presence of the methyl group and the number of methylene groups in the methionine side chain (Fig. 8a). Often the degree of recognition of ligands by in-vitro-selected aptamers is limited by the fact that their ligand was coupled to a column to effect a separation of binding and nonbinding RNAs.

Finally, and most importantly, is the coupling of ligand binding and tertiary structural changes in the aptamer domain. In each case, architectural changes in the RNA occur upon ligand binding that serves to stabilize a "communication helix," usually the P1 helix, that relays the liganded state of the aptamer domain to the expression platform. In the case of the *xpt-pbuX* guanine riboswitch, binding of the ligand to the three-way junction induces a series of two base triples between nucleotides in J2/3 and base pairs in P1. This prevents the 3′ side of the P1 helix from being used to form the antiterminator element (which would allow complete synthesis of the mRNA transcript) and instead fates the mRNA to form a rho-independent terminator element in the expression platform, aborting mRNA synthesis and shutting off gene expression (Fig. 1).

CONCLUDING REMARKS

Riboregulation by small noncoding RNAs and elements in the 5′- and 3′-UTRs of mRNAs is now understood to be a major mechanism by which gene expression is controlled. In bacteria, riboswitches control a significant number of crucial metabolic pathways at the transcriptional and translational levels via their ability to bind a variety of metabolites. In the last few years, a significant amount of effort by our group and other workers has been devoted toward uncovering the molecular basis for this form of genetic control. Despite the determination of the structures of three of the aptamer domains bound to their cognate ligand, a number of fundamental questions remain as to how productive binding is achieved and its relationship to the expression platform.

Like many forms of RNA recognition, riboswitches depend on an "induced fit" mechanism for ligand binding in which the RNA undergoes a concerted binding and local folding event (Williamson 2000; Leulliot and Varani 2001). In the case of the purine riboswitch, we have presented a model of the process by which this may occur (Gilbert et al. 2006), but this mechanism still requires further experimental evidence to fully support it. The SAM and TPP riboswitches, instead of using a single binding pocket, have a "bipartite" pocket created by the union of two helices (P1 and P3 in the case of the SAM riboswitch) (Montange and Batey 2006; Serganov et al. 2006; Thore et al. 2006). In these cases, the ligand initially docks with one of these helices, and then the second closes around the initial ligand–RNA complex to fully encapsulate the ligand. In the case of the SAM riboswitch, it is likely that

hydrogen bonding to the P3 helix precedes interactions with the P1 helix; in the case of the TPP riboswitch, the P2-J2/3-P3 coaxial stack which recognizes the pyrimidine moiety that likely initially interacts with the ligand. Detailed mechanistic analysis of these events may yield new insights into specific recognition of RNAs whose binding sites are disorganized in their free state.

Despite being locally disorganized in the free state, the binding pocket must have some structural cues that serve to guide productive ligand binding. In the case of the purine riboswitch, we have hypothesized that whereas most of the three-way junction is conformationally flexible, a critical pyrimidine residue (C/U74) is properly positioned for recognition by the purine nucleobase to initiate a productive binding process (Gilbert et al. 2006). Furthermore, we have accrued clear evidence that a number of nucleotides in and around the three-way junction have been evolutionarily selected, not because of their critical importance for the bound state, but rather to keep the free state open and receptive to ligand binding (C. Love et al., in prep.). This suggests that the structure of the unbound state is crucial for proper riboswitch function; that is, if the regulatory RNA has a tendency to misfold, the gene it controls may become constitutively activated or repressed. Thus, these RNAs are fine-tuned to avoid the "alternative conformational hell" first described by the Uhlenbeck lab that plagues a large number of RNAs in the laboratory (Uhlenbeck 1995). A comprehensive understanding of the free state of these RNAs will likely not come from NMR or X-ray crystallography, but rather from a combination of biophysical and biochemical methods that can more readily cope with dynamic states of the molecule.

A central question is, Exactly how does the RNA regulate gene expression through the interplay of the aptamer and expression platforms in the biological context? Biophysical studies by several groups have suggested that ligand binding and secondary structural rearrangements of the mRNA are slow processes (Wickiser et al. 2005a,b; Gilbert et al. 2006). If transcription by RNA polymerase proceeds through the 5′UTR at standard rate (~60–90 nucleotides/sec), then crucial steps for riboswitch function appear to be too slow for it to be an effective means of genetic control. It has been found that multiple transcriptional pause sites strategically placed within the FMN riboswitch significantly slow the rate of transcription through the regulatory element in vitro, but whether these pause sites play the same role in transcription regulation in vivo and are general to all riboswitches is unknown (Wickiser et al. 2005b).

This emphasizes the need for further structural and biophysical studies to specifically address the relationship between the aptamer domain and the expression platform. Because these domains are not tightly coupled through tertiary contacts, this will not come in the form of NMR or crystallographic structures, but rather through other biophysical techniques such as measuring force/extension curves using optical tweezers and single-molecule FRET (Onoa and Tinoco 2004). These techniques have been particularly successful at understanding RNA folding/unfolding processes, and the riboswitch is inherently an RNA folding problem. Further complicating a biophysical characterization of this process, however, is that this ligand-influenced folding event is occurring co-transcriptionally such that decisions are likely being made prior to the entire 5′UTR being synthesized. Thus, it may be useful to study subfragments of the riboswitch that represent transcriptional intermediates, such as the aptamer/antiterminator region without an intact terminator sequence. Coupled with genetic and in vivo experimental results, answers to these questions will yield a clear picture as to how biology has harnessed mRNA to be able to both carry and regulate genetic information.

ACKNOWLEDGMENTS

The authors thank the members of the Batey laboratory for support in various aspects of our work on riboswitches, in particular Sarah Mediatore, Elizabeth Pleshe, Crystal Love, and Andrea Edwards. We also are grateful for engaging discussions and feedback from the Boulder RNA community and Professor Jeffrey Kieft. This work was supported by grants from the American Cancer Society and the National Institutes of Health.

REFERENCES

Adams P.L., Stahley M.R., Kosek A.B., Wang J., and Strobel S.A. 2004. Crystal structure of a self-splicing group I intron with both exons. *Nature* **430:** 45.

Babitzke P. 2004. Regulation of transcription attenuation and translation initiation by allosteric control of an RNA-binding protein: The *Bacillus subtilis* TRAP protein. *Curr. Opin. Microbiol.* **7:** 132.

Barrick J.E., Corbino K.A., Winkler W.C., Nahvi A., Mandal M., Collins J., Lee M., Roth A., Sudarsan N., Jona I., et al. 2004. New RNA motifs suggest an expanded scope for riboswitches in bacterial genetic control. *Proc. Natl. Acad. Sci.* **101:** 6421.

Bartel D.P. and Szostak J.W. 1993. Isolation of new ribozymes from a large pool of random sequences (comment). *Science* **261:** 1411.

Batey R.T., Gilbert S.D., and Montange R.K. 2004. Structure of a natural guanine-responsive riboswitch complexed with the metabolite hypoxanthine. *Nature* **432:** 411.

Batey R.T., Sagar M.B., and Doudna J.A. 2001. Structural and energetic analysis of RNA recognition by a universally conserved protein from the signal recognition particle. *J. Mol. Biol.* **307:** 229.

Bergman N.H., Lau N.C., Lehnert V., Westhof E., and Bartel D.P. 2004. The three-dimensional architecture of the class I ligase ribozyme. *RNA* **10:** 176.

Brunger A.T., Adams P.D., Clore G.M., DeLano W.L., Gros P., Grosse-Kunstleve R.W., Jiang J.S., Kuszewski J., Nilges M., Pannu N.S., et al. 1998. Crystallography & NMR system: A new software suite for macromolecular structure determination. *Acta Crystallogr. D Biol. Crystallogr.* **54:** 905.

Bugl H., Fauman E.B., Staker B.L., Zheng F., Kushner S.R., Saper M.A., Bardwell J.C., and Jakob U. 2000. RNA methylation under heat shock control. *Mol. Cell* **6:** 349.

Carothers J.M., Oestreich S.C., and Szostak J.W. 2006a. Aptamers selected for higher-affinity binding are not more specific for the target ligand. *J. Am. Chem. Soc.* **128:** 7929.

Carothers J.M., Davis J.H., Chou J.J., and Szostak J.W. 2006b. Solution structure of an informationally complex high-affinity RNA aptamer to GTP. *RNA* **12:** 567.

Carrasco N., Buzin Y., Tyson E., Halpert E., and Huang Z. 2004. Selenium derivatization and crystallization of DNA and RNA oligonucleotides for X-ray crystallography using multiple anomalous dispersion. *Nucleic Acids Res.* **32:** 1638.

Corbino K.A., Barrick J.E., Lim J., Welz R., Tucker B.J., Puskarz I., Mandal M., Rudnick N.D., and Breaker R.R. 2005. Evidence for a second class of S-adenosylmethionine riboswitches and other regulatory RNA motifs in alpha-proteobacteria. *Genome Biol.* **6:** R70.

Cromie M.J., Shi Y., Latifi T., and Groisman E.A. 2006. An RNA sensor for intracellular Mg(2+). *Cell* **125:** 71.

Dauter Z., Dauter M., and Dodson E. 2002. Jolly SAD. *Acta Crystallogr. D Biol. Crystallogr.* **58:** 494.

Dieckmann T., Suzuki E., Nakamura G.K., and Feigon J. 1996. Solution structure of an ATP-binding RNA aptamer reveals a novel fold. *RNA* **2:** 628.

Ferre-D'Amare A.R. and Doudna J.A. 2000. Crystallization and structure determination of a hepatitis delta virus ribozyme: Use of the RNA-binding protein U1A as a crystallization module. *J. Mol. Biol.* **295:** 541.

Ferre-D'Amare A.R., Zhou K., and Doudna J.A. 1998. Crystal structure of a hepatitis delta virus ribozyme. *Nature* **395:** 567.

Gelfand M.S., Mironov A.A., Jomantas J., Kozlov Y.I., and Perumov D.A. 1999. A conserved RNA structure element involved in the regulation of bacterial riboflavin synthesis genes. *Trends Genet.* **15:** 439.

Gilbert S.D., Stoddard C.D., Wise S.J., and Batey R.T. 2006. Thermodynamic and kinetic characterization of ligand binding to the purine riboswitch aptamer domain. *J. Mol. Biol.* **359:** 754.

Griffiths-Jones S., Bateman A., Marshall M., Khanna A., and Eddy S.R. 2003. Rfam: An RNA family database. *Nucleic Acids Res.* **31:** 439.

Griffiths-Jones S., Moxon S., Marshall M., Khanna A., Eddy S.R., and Bateman A. 2005. Rfam: Annotating non-coding RNAs in complete genomes. *Nucleic Acids Res.* **33:** D121.

Grundy F.J. and Henkin T.M. 1998. The S-box regulon: A new global transcription termination control system for methionine and cysteine biosynthesis genes in gram-positive bacteria. *Mol. Microbiol.* **30:** 737.

Hendrickson W.A. and Ogata C.M. 1997. Phase determination from multiwavelength anomalous diffraction measurements. *Methods Enzymol.* **276:** 494.

Hobartner C., Rieder R., Kreutz C., Puffer B., Lang K., Polonskaia A., Serganov A., and Micura R. 2005. Syntheses of RNAs with up to 100 nucleotides containing site-specific 2′-methylseleno labels for use in X-ray crystallography. *J. Am. Chem. Soc.* **127:** 12035.

Jiang F., Kumar R.A., Jones R.A., and Patel D.J. 1996. Structural basis of RNA folding and recognition in an AMP-RNA aptamer complex. *Nature* **382:** 183.

Ke A. and Doudna J.A. 2004. Crystallization of RNA and RNA-protein complexes. *Methods* **34:** 408.

Kieft J.S. and Batey R.T. 2004. A general method for rapid and nondenaturing purification of RNAs. *RNA* **10:** 988.

Kieft J.S., Zhou K., Grech A., Jubin R., and Doudna J.A. 2002. Crystal structure of an RNA tertiary domain essential to HCV IRES-mediated translation initiation. *Nat. Struct. Biol.* **9:** 370.

Klein D.J., Schmeing T.M., Moore P.B., and Steitz T.A. 2001. The kink-turn: A new RNA secondary structure motif. *EMBO J.* **20:** 4214.

Leulliot N. and Varani G. 2001. Current topics in RNA-protein recognition: Control of specificity and biological function through induced fit and conformational capture. *Biochemistry* **40:** 7947.

Mandal M. and Breaker R.R. 2004a. Adenine riboswitches and gene activation by disruption of a transcription terminator. *Nat. Struct. Mol. Biol.* **11:** 29.

———. 2004b. Gene regulation by riboswitches. *Nat. Rev. Mol. Cell Biol.* **5:** 451.

Mandal M., Boese B., Barrick J.E., Winkler W.C., and Breaker R.R. 2003. Riboswitches control fundamental biochemical pathways in *Bacillus subtilis* and other bacteria. *Cell* **113:** 577.

Mandal M., Lee M., Barrick J.E., Weinberg Z., Emilsson G.M., Ruzzo W.L., and Breaker R.R. 2004. A glycine-dependent riboswitch that uses cooperative binding to control gene expression. *Science* **306:** 275.

McDaniel B.A., Grundy F.J., Artsimovitch I., and Henkin T.M. 2003. Transcription termination control of the S box system: Direct measurement of S-adenosylmethionine by the leader RNA. *Proc. Natl. Acad. Sci.* **100:** 3083.

Miranda-Rios J., Navarro M., and Soberon M. 2001. A conserved RNA structure (thi box) is involved in regulation of thiamin biosynthetic gene expression in bacteria. *Proc. Natl. Acad. Sci.* **98:** 9736.

Mironov A.S., Gusarov I., Rafikov R., Lopez L.E., Shatalin K., Kreneva R.A., Perumov D.A., and Nudler E. 2002. Sensing small molecules by nascent RNA: A mechanism to control transcription in bacteria. *Cell* **111:** 747.

Montange R.K. and Batey R.T. 2006. Structure of the S-adenosylmethionine riboswitch regulatory mRNA element. *Nature* **441:** 1172.

Nahvi A., Sudarsan N., Ebert M.S., Zou X., Brown K.L., and Breaker R.R. 2002. Genetic control by a metabolite binding mRNA. *Chem. Biol.* **9:** 1043.

Noeske J., Richter C., Grundl M.A., Nasiri H.R., Schwalbe H., and Wohnert J. 2005. An intermolecular base triple as the basis of ligand specificity and affinity in the guanine- and adenine-sensing riboswitch RNAs. *Proc. Natl. Acad. Sci.* **102:** 1372.

Onoa B. and Tinoco I., Jr. 2004. RNA folding and unfolding. *Curr. Opin. Struct. Biol.* **14:** 374.

Pflugrath J.W. 1999. The finer things in X-ray diffraction data collection. *Acta Crystallogr. D Biol. Crystallogr.* **55:** 1718.

Pley H.W., Flaherty K.M., and McKay D.B. 1994. Model for an RNA tertiary interaction from the structure of an intermolecular complex between a GAAA tetraloop and an RNA helix. *Nature* **372:** 111.

Rice L.M., Earnest T.N., and Brunger A.T. 2000. Single-wavelength anomalous diffraction phasing revisited. *Acta Crystallogr. D Biol. Crystallogr.* **56:** 1413.

Rodionov D.A., Vitreschak A.G., Mironov A.A., and Gelfand M.S. 2002. Comparative genomics of thiamin biosynthesis in procaryotes. New genes and regulatory mechanisms. *J. Biol. Chem.* **277:** 48949.

———. 2003. Regulation of lysine biosynthesis and transport genes in bacteria: Yet another RNA riboswitch? *Nucleic Acids Res.* **31:** 6748.

Rupert P.B. and Ferre-D'Amare A.R. 2001. Crystal structure of a hairpin ribozyme-inhibitor complex with implications for catalysis. *Nature* **410:** 780.

Serganov A., Polonskaia A., Phan A.T., Breaker R.R., and Patel D.J. 2006. Structural basis for gene regulation by a thiamine pyrophosphate-sensing riboswitch. *Nature* **441:** 1167.

Serganov A., Yuan Y.R., Pikovskaya O., Polonskaia A., Malinina L., Phan A.T., Hobartner C., Micura R., Breaker R.R., and Patel D.J. 2004. Structural basis for discriminative regulation of gene expression by adenine- and guanine-sensing mRNAs. *Chem. Biol.* **11:** 1729.

Stura E.A. and Wilson I.A. 1990. Analytical and production seeding techniques. *Methods* **1:** 38.

Sudarsan N., Barrick J.E., and Breaker R.R. 2003a. Metabolite-binding RNA domains are present in the genes of eukaryotes. *RNA* **9:** 644.

Sudarsan N., Wickiser J.K., Nakamura S., Ebert M.S., and Breaker R.R. 2003b. An mRNA structure in bacteria that controls gene expression by binding lysine. *Genes Dev.* **17:** 2688.

Teplova M., Wilds C.J., Wawrzak Z., Tereshko V., Du Q., Carrasco N., Huang Z., and Egli M. 2002. Covalent incorporation of selenium into oligonucleotides for X-ray crystal structure determination via MAD: Proof of principle. Multiwavelength anomalous dispersion. *Biochimie* **84:** 849.

Terwilliger T.C. 2003. SOLVE and RESOLVE: Automated structure solution and density modification. *Methods Enzymol.* **374:** 22.

Thore S., Leibundgut M., and Ban N. 2006. Structure of the eukaryotic thiamine pyrophosphate riboswitch with its regulatory ligand. *Science* **312:** 1208.

Uhlenbeck O.C. 1995. Keeping RNA happy. *RNA* **1:** 4.

Vitreschak A.G., Rodionov D.A., Mironov A.A., and Gelfand M.S. 2002. Regulation of riboflavin biosynthesis and transport genes in bacteria by transcriptional and translational attenuation. *Nucleic Acids Res.* **30:** 3141.

Wickiser J.K., Cheah M.T., Breaker R.R., and Crothers D.M. 2005a. The kinetics of ligand binding by an adenine-sensing riboswitch. *Biochemistry* **44:** 13404.

Wickiser J.K., Winkler W.C., Breaker R.R., and Crothers D.M. 2005b. The speed of RNA transcription and metabolite binding kinetics operate an FMN riboswitch. *Mol. Cell* **18:** 49.

Williamson J.R. 2000. Induced fit in RNA-protein recognition. *Nat. Struct. Biol.* **7:** 834.

Winkler W.C. and Breaker R.R. 2005. Regulation of bacterial gene expression by riboswitches. *Annu. Rev. Microbiol.* **59:** 487.

Winkler W.C., Cohen-Chalamish S., and Breaker R.R. 2002a. An mRNA structure that controls gene expression by binding FMN. *Proc. Natl. Acad. Sci.* **99:** 15908.

Winkler W.C., Nahvi A., and Breaker R.R. 2002b. Thiamine derivatives bind messenger RNAs directly to regulate bacterial gene expression. *Nature* **419:** 952.

Winkler W.C., Nahvi A., Roth A., Collins J.A., and Breaker R.R. 2004. Control of gene expression by a natural metabolite-responsive ribozyme. *Nature* **428:** 281.

Winkler W.C., Nahvi A., Sudarsan N., Barrick J.E., and Breaker R.R. 2003. An mRNA structure that controls gene expression by binding S-adenosylmethionine. *Nat. Struct. Biol.* **10:** 701.

Zengel J.M. and Lindahl L. 1994. Diverse mechanisms for regulating ribosomal protein synthesis in *Escherichia coli*. *Prog. Nucleic Acid Res. Mol. Biol.* **47:** 331.

Zhang L. and Doudna J.A. 2002. Structural insights into group II intron catalysis and branch-site selection. *Science* **295:** 2084.

Zimmermann G.R., Jenison R.D., Wick C.L., Simorre J.P., and Pardi A. 1997. Interlocking structural motifs mediate molecular discrimination by a theophylline-binding RNA. *Nat. Struct. Biol.* **4:** 644.

Regulating Bacterial Transcription with Small RNAs

G. Storz,* J.A. Opdyke,* and K.M. Wassarman†

*Cell Biology and Metabolism Branch, National Institute of Child Health and Human Development, Bethesda, Maryland 20892-5430; †Department of Bacteriology, University of Wisconsin, Madison, Wisconsin 53706

In recent years, the combinations of computational and molecular approaches have led to the identification of an increasing number of small, noncoding RNAs encoded by bacteria and their plasmids and phages. Most of the characterized small RNAs have been shown to operate at a posttranscriptional level, modulating mRNA stability or translation by base-pairing with the 5′ regions of the target mRNAs. However, a subset of small RNAs has been found to regulate transcription. One example is the abundant 6S RNA that has been proposed to compete for DNA binding of RNA polymerase by mimicking the open conformation of promoter DNA. Other small RNAs affect transcription termination via base-pairing interactions with sequences in the mRNA. Here, we discuss current understanding and questions regarding the roles of small RNAs in regulating transcription.

The existence of a few small, noncoding RNAs encoded on plasmids, bacteriophage, transposons, and bacterial chromosomes has been known for many decades (generally denoted small RNAs [sRNAs] in bacteria). However, the extensive role these regulatory sRNAs have has only begun to be elucidated in the last few years (for review, see Storz and Gottesman 2006). Most of the sRNAs first identified in bacterial systems were discovered fortuitously. For example, the *Escherichia coli* sRNAs MicF and DsrA were identified as regulators in multicopy screens. It was then shown that the factor necessary for the observed regulation was an RNA, rather than a protein factor. More recently, systematic approaches, both computational and experimental, have allowed the identification of almost 80 sRNAs encoded on the *E. coli* chromosome and dozens of sRNA encoded by other bacteria (for review, see Hüttenhofer and Vogel 2006). Several of the computational methods were based on searches for sequence conservation of intergenic regions between a bacterium of interest and closely related organisms. Other computational methods scanned a genome for promoter and stem-loop terminator sequences that did not span known protein-coding regions. Experimental approaches that have led to the identification of sRNAs have been direct cloning of RNAs after size fractionation as well as microarray analysis using total RNA or RNAs that coimmunoprecipitate with known RNA-binding proteins. Each of these approaches has been successful at identifying a subset of sRNAs, but no one approach has been able to identify all known sRNAs, in part due to the high diversity in sRNA conservation, biogenesis, and function.

The sRNAs characterized in *E. coli* thus far fall into three general classes based on their mode of action (for review, see Gottesman 2004; Storz et al. 2005). The first class of RNAs contains intrinsic enzymatic activity such as the *rnp* RNA that constitutes the catalytic subunit of the RNase P enzyme necessary for processing of pre-tRNAs. The second class of sRNAs functions by modulating protein activity. In some cases, this class acts by mimicking other nucleic acid structures as is illustrated by the CsrB/CsrC sRNAs, which regulate carbon utilization by controlling the activity of the major carbon storage regulator CsrA. A single CsrB RNA molecule contains 18 CsrA-binding sites. Thus, the sRNA is thought to sequester the CsrA protein away from similar binding sites on mRNAs where CsrA binding affects mRNA stability and/or translation. The third and largest class of characterized sRNAs in *E. coli* is a group of sRNAs that function to regulate target genes through the formation of base-pair interactions with mRNA. At least a third of the *E. coli* sRNAs are believed to function by this mechanism. More than half of the *E. coli* sRNAs and even higher percentages of sRNAs in other bacteria remain to be characterized; it will be interesting to see whether other modes of action will be discovered.

sRNAs that act by base-pairing with target mRNAs can be further subclassified into those that are encoded *in trans* and those RNAs that are encoded *in cis* relative to their target gene. The *trans*-encoded sRNAs regulate targets encoded at other locations on the chromosome. The complementarity between *trans*-encoded sRNAs and their target mRNAs is usually imperfect and limited in length, and many of the sRNAs are capable of regulating more than one target gene. A hallmark of *trans*-encoded sRNAs is their requirement for the RNA-binding protein Hfq for function; in fact, these sRNAs largely are found in stable complexes with Hfq. Hfq was originally identified as an *E. coli* host factor for Qβ phage replication. However, the pleiotropic phenotypes of an *hfq* mutant suggested a broader role for this protein. It is now recognized that many of the phenotypes of a strain lacking Hfq are due in part to the inability of base-pairing sRNAs to function in this mutant background. The mechanism of Hfq action is not fully understood. In vivo and in vitro studies indicate that Hfq helps facilitate base-pairing between sRNA and target mRNA molecules. In addition, Hfq is responsible for the high degree of stability of the sRNAs to which it binds; the abundance of many sRNAs is significantly reduced in *hfq* mutant strains. Further studies of Hfq are necessary to fully understand the complex role of Hfq in sRNA-mediated gene regulation.

Many of the first sRNAs to be discovered were encoded *in cis* with their target gene and were found on extrachromosomal DNA molecules such as plasmids, transposons, and bacteriophage, but recently, a small number of *cis*-encoded sRNAs expressed from the *E. coli* chromosome have been identified. By their very nature, *cis*-encoded RNAs have perfect complementarity with their target RNAs and are thought not to require Hfq for their activity.

The base-pairing between an sRNA and a target mRNA has been shown to have a number of regulatory outcomes at a posttranscriptional level. Many sRNAs can base-pair to mRNAs in regions surrounding the ribosome-binding site and inhibit translation by occluding ribosome access to this recognition sequence. An example is provided by the OxyS RNA, which base-pairs across the ribosome-binding site of *fhlA* and blocks translation. Base-pairing between sRNAs and mRNAs also has been shown to be associated with the degradation of the target mRNA and possibly the sRNA. For instance, the RyhB RNA is responsible for targeted degradation of a number of mRNAs in a manner that is dependent on the endonuclease RNase E. Although the mRNA degradation may be secondary to the block in translation, the degradation step renders the regulation irreversible. sRNA base-pairing also can activate translation of certain target genes. The 5′ region of the *rpoS* mRNA folds into a structure that occludes its own ribosome-binding site. Base-pairing between the DsrA RNA with the upstream sequences of *rpoS* allows alternative folding of the 5′ region into a structure that is accessible to the ribosome.

Although sRNAs that act posttranscriptionally, as described above, have been a focus of many studies during the past 5 years, some of the first sRNAs to be identified were found to modulate transcription, and recent results indicate that the list of sRNAs that act as transcription regulators is likely to continue to expand. Given the numerous components required for accurate transcription as well as the multiple steps involved, it is perhaps not surprising that this step in gene expression would also be subject to regulation by sRNAs. Here, we describe the growing classes of sRNAs that affect the process of transcription and discuss what remains to be learned about these sRNAs.

PROMOTER COMPETITION

E. coli 6S RNA regulates the first step in transcription, promoter recognition, through direct interaction with the general transcription machinery. Although 6S RNA (184 nucleotides) was one of the first noncoding RNAs to be discovered, its biological function remained a mystery for many years. The finding that this RNA formed a complex with RNA polymerase gave the first hints to its function (Wassarman and Storz 2000). Copurification experiments showed that the 6S RNA forms a stable complex with the housekeeping form of RNA polymerase (denoted $E\sigma^{70}$ in *E. coli*), and UV cross-linking studies revealed a direct interaction between the 6S RNA and σ^{70} within this complex. Reconstitution experiments confirmed that this interaction is strong and specific (Trotochaud and Wassarman 2005). Thus, the high specificity of 6S RNA for $E\sigma^{70}$ is most likely mediated via the σ^{70} subunit.

Phylogenetic analysis of putative 6S RNAs from divergent species revealed that 6S is present in a wide range of bacteria and contains a highly conserved secondary structure consisting of a single-stranded central region within a highly double-stranded RNA (Barrick et al. 2005; Trotochaud and Wassarman 2005). The resemblance of this structure to the conformation of DNA within an open complex during transcription initiation led to models in which RNA polymerase interacts with 6S RNA in a manner similar to its usual interactions with promoter DNA. The fact that disruption of this conserved secondary structure by mutation strongly decreases the ability of the 6S RNA to bind $E\sigma^{70}$ suggests the primary mode of recognition of 6S RNA by RNA polymerase is mediated by structural elements, rather than specific sequences (Trotochaud and Wassarman 2005).

6S RNA binding to $E\sigma^{70}$ leads to down-regulation of transcription at several σ^{70}-dependent promoters in *E. coli* (Trotochaud and Wassarman 2004). However, many promoters are not sensitive to 6S RNA, even late in stationary phase when 6S RNA levels are highest and the vast majority of $E\sigma^{70}$ is bound by 6S RNA. In addition to direct effects of 6S RNA on σ^{70}-dependent transcription, indirect regulation of transcription of promoters recognized by an alternative form of RNA polymerase ($E\sigma^S$) also has been observed. Thus, it seems likely that 6S RNA could be regulating transcription of at least two different forms of RNA polymerases using two modes of action: inhibition of $E\sigma^{70}$ by substrate competition and activation of $E\sigma^S$ by sequestration of a potent competitor ($E\sigma^{70}$). In any case, it is clear that the formation of stable 6S RNA:$E\sigma^{70}$ complexes leads to altered utilization of σ^{70} and σ^S during stationary phase. These changes in transcription allow cells to survive starvation and respond to environmental changes when nutrients are limited (Trotochaud and Wassarman 2004, 2006).

A variety of intriguing questions regarding 6S RNA action remain to be addressed. Given that only a subset of σ^{70}-dependent promoters are affected, what is the nature of the promoter specificity? Does 6S RNA affect the competition between σ^{70} and σ^S for the rest of the RNA polymerase complex? What contacts are made between 6S RNA and polymerase? Finally, how is the 6S RNA released, and how are the effects of the 6S RNA reversed?

TRANSCRIPTION TERMINATION

There are several examples of sRNAs that affect transcription termination by base-pairing. The first came from studies of the staphylococcal plasmid pT181 and the *inc18* family of streptococcal plasmids (pIP501, pAMβ1, and pSM19035), which are capable of replicating in a broad host range (Novick et al. 1989; Brantl et al. 1993; Le Chatelier et al. 1996). The replication frequencies of many plasmids are affected by negative feedback by antisense RNAs acting at different levels. In the case of pT181 and the *inc18* family of plasmids, this feedback regulation occurs by termination of mRNAs encoding replication proteins. The first system to be characterized was the pT181 plasmid (Novick et al. 1989). For this plasmid, the RepC protein is rate-limiting for replication. Two sRNAs, RNAI (~85 nucleotides) and RNAII (~150 nucleotides), are encoded on the strand opposite the *repC* mRNA. Base-pairing between either of the two sRNAs and the *repC* mRNA results in the formation of a secondary structure that leads to transcription termination (Novick et al. 1989; Brantl and Wagner 2000). In the absence of the base-pairing RNA, a different structure is formed allowing readthrough and complete synthesis of the *repC* mRNA. The antisense RNAs that modulate expression of replication proteins encoded by the *inc18* family of plasmids are thought to act in a similar fashion (Brantl et al. 1993; Brantl and Wagner 1994).

Another sRNA that likely modulates termination is encoded on the *Agrobacterium tumefaciens* Ti plasmid (Chai and Winans 2005). Again, replication frequency is regulated by an sRNA, which is encoded opposite replication genes. Increased expression of the RepE RNA (~54 nucleotides) is associated with decreased copy number, whereas decreased RepE expression results in increased copy number. RepE is encoded opposite the *repB-repC* intergenic sequence, and nuclease S1 protection analysis and assays of *repC-lacZ* fusions indicated that RepE inhibits RepC expression at the transcriptional level. As in the case for RNAI and RNAII, it is predicted that the RepE base-pairing with the *repBC* mRNA causes a secondary structural change, which results in formation of a terminator immediately upstream of *repC*.

Yet another example is provided by the *Vibrio anguillarum* virulence plasmid pJM1 (M. Stork et al., in prep.). In this case, the operon encodes four genes for ferric-siderophore transport *fatD, fatC, fatB,* and *fatA* and two genes for siderophore biosynthesis, *angR* and *angT*. It was observed that the relative level of a transcript encoding the *fatDCBA* genes is 17-fold higher than the level of the full-length transcript, and two sRNAs, RNAα and RNAβ (427 nucleotides), were found to be encoded in *cis* to the *fatDCBAangRT* operon. Again, nuclease S1 mapping and *lacZ* fusion studies suggested that the differential expression of the genes in the operon is due to transcription termination between the *fat* and *angRT* genes as a consequence of RNAβ base-pairing in this region. Exactly how base-pairing might lead to termination remains to be determined.

In the lysogenic state, bacteriophage P4 prevents the expression of its own replication genes through premature transcription termination. The phage factor responsible for efficient termination is the sRNA CI (79 nucleotides). This RNA is unlike the other examples in that the CI is processed from the leader of the "left-operon" mRNA that it regulates (Briani et al. 2000; Forti et al. 2002). The CI RNA promotes termination at one terminator in the left operon by pairing with a complementary sequence within the leader region.

The structures of the pT181 and pIP501 sRNAs and their target mRNAs as well as their base-pairing interactions have been examined in some detail (Brantl and Wagner 1994, 2000, 2002; Heidrich and Brantl 2003). The formation of the full duplex is not necessary for regulation since base-pairing intermediates preceding the formation of the full duplex are sufficient for full inhibition. It also has been shown that the control is kinetic; the

fraction of the target RNA for which termination occurs is determined by the rate of antisense RNA binding. Finally, base-pairing alone appears to be sufficient for the regulation; no requirements for additional protein factors have been observed, and antisense regulation of the pT181 and pIP501 was recapitulated in a heterologous system (Brantl and Wagner 2002). Similarly, RNAβ of the *V. anguillarum* virulence plasmid and the CI RNA of bacteriophage P4 were found to enhance termination in the absence of any additional bacterial factor in vitro (Briani et al. 2000; M. Stork et al., in prep.).

All of the sRNAs found to affect transcription termination thus far are encoded on bacterial plasmids or phage, raising the question of whether any chromosomally encoded RNAs act to modulate termination. Although independently transcribed sRNAs that have this activity have not been reported, it should be noted that the 5′-untranslated regions of an increasing number of mRNAs have been found to fold into two alternative structures that in some cases affect termination (for review, see Winkler and Breaker 2005). These so-called riboswitches have been shown to bind a variety of metabolites and even tRNAs; in the presence of the cofactor, they adopt one conformation, whereas a second conformation is adopted in the absence of the cofactor. One structure promotes termination and the other structure permits transcription readthrough.

The examples above illustrate that sRNAs can promote transcription termination and have consequences for the relative expression of genes within an operon. For these base-pairing sRNAs, a fair amount is known about the structures of the RNAs and the base-pairing interactions required for regulation. Less is known about the mechanism by which the sRNAs actually affect transcription termination. At what point does base-pairing need to occur in order for termination to occur? Does termination always occur due to the formation of an alternative structure or can duplex formation itself bring about termination as has been shown for oligonucleotides in vitro? In addition, all of the examples lead to termination. One can also imagine that base-pairing could lead to a structure that would prevent the formation of a terminator or block recognition by a termination factor such as the Rho protein, and thus would promote elongation rather than termination.

NONCODING RNA-MEDIATED TRANSCRIPTION REGULATION IN EUKARYOTES

There has also been an increase in examples of eukaryotic small RNAs (generally denoted noncoding RNAs [ncRNAs] in eukaryotes) that affect transcription (for review, see Goodrich and Kugel 2006). The biggest class is the small interfering RNAs (siRNAs) that are derived from transcripts originating from repeated sequences and act in general silencing of transcription through chromatin remodeling (for review, see Zamore and Haley 2005; Petersen et al. 2006). A different mode of transcriptional regulation is provided by the *Saccharomyces cerevisiae SRG1* RNA, which is encoded directly upstream of its regulated gene *SER3*. In this case, it is not the SRG1 RNA product that is regulatory, but rather the act of transcribing the *SRG1* gene that blocks expression of *SER3* (Martens et al. 2004, 2005). Other eukaryotic ncRNAs have been shown to affect transcription by binding to specific proteins. Examples of ncRNAs that bind and modify the activities of transcription factors include the NRSE RNA (~20-bp double-stranded neuron-restrictive silencer element RNA) and HSR1 (~2000-nucleotide heat shock RNA), which modulate the activities of the NRSF/REST repressor and HSF1 transcription factor, respectively (Kuwabara et al. 2004; Shamovsky et al. 2006). The SRA (~700-nucleotide steroid receptor RNA activator) is thought to act as a coactivator by connecting nuclear hormone receptors to the general transcription machinery (Lanz et al. 1999; Hatchell et al. 2006). The mouse B2 RNA (~178 nucleotides), which is transcribed from short interspersed repetitive elements (SINEs), has been shown to inhibit transcription initiation through directing binding to RNA polymerase during the heat shock response (Allen et al. 2004; Espinoza et al. 2004). However, in contrast to the bacterial 6S RNA, the B2 RNA does not block RNA polymerase access to DNA. Instead, the B2 RNA inhibits the initiation of RNA synthesis. The 7SK RNA (331 nucleotides) is unique in that it affects the transition from transcription initiation to elongation by sequestering the P-TEFb elongation factor in a complex with one of its inhibitors, HEX1M1 or HEX1M2 (Nguyen et al. 2001; Yang et al. 2001; Egloff et al. 2006). This inactive form of P-TEFb is not able to phosphorylate the RNA polymerase carboxy-terminal domain (CTD) as is required for promoter escape.

CONCLUSIONS

The examples described above illustrate some of the different ways in which bacterial sRNAs or eukaryotic ncRNAs can affect transcription. These general mechanisms are summarized in Figure 1. Some RNAs, such as 6S, NRSE, and B2, modulate the activity of protein complexes by binding to the proteins. Other RNAs, such as SRA and 7SK, appear to act as connectors to bring protein partners together. In contrast, plasmid-encoded RNAs such as RNAI and RepE act by changing the structure of target mRNAs. Whether these examples fully represent all the different ways that the RNAs can act on transcription remains to be seen.

The characterized bacterial sRNAs affect promoter binding and termination. Eukaryotic ncRNAs also have been shown to modulate the activities of transcriptional regulators, inhibit the initiation of RNA synthesis, and affect transcription elongation. It will be interesting to see whether transcription factors and elongation steps are also targeted by sRNAs in bacteria. Similarly, do any eukaryotic ncRNAs base-pair with mRNAs to affect termination, either directly or by altering 3′-end processing events that could lead to changes in transcription elongation and termination? One can also imagine that sRNAs or ncRNAs might be involved in the coupling of transcription with translation, transcription and RNA processing, or transcription and RNA transport. The spliceosomal U1 snRNA already has been implicated in regulation of transcription (Kwek et al. 2002).

Thousands of *cis*-encoded antisense RNAs have been reported for eukaryotic organisms. Do any of these RNAs affect transcription? In addition, although most microRNAs (miRNA) are thought to act at the level of regulation of translation and mRNA stability (for review, see Valencia-Sanchez et al. 2006), it is conceivable that they could act to create or mask binding sites for RNA-binding proteins involved in other processes as well. When gene regulation was first discovered, it was proposed that the regulators were RNA molecules (Jacob and Monod 1961). This model was superseded by the discovery of protein activators and repressors. The findings described above, however, suggest that the Jacob and Monod model should be revisited. Given that only a fraction of sRNAs and ncRNAs have been characterized, it is likely that many more examples of RNA regulators of transcription as well as unique modes of action remain to be uncovered.

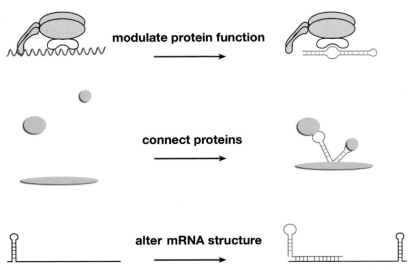

Figure 1. General mechanisms by which RNA regulators (*red*) modulate transcription.

ACKNOWLEDGMENTS

We thank R. Weisberg for comments on the manuscript.

REFERENCES

Allen T.A., Von Kaenel S., Goodrich J.A., and Kugel J.F. 2004. The SINE-encoded mouse B2 RNA represses mRNA transcription in response to heat shock. *Nat. Struct. Mol. Biol.* **11:** 816.

Barrick J.E., Sudarsan N., Weinberg Z., Ruzzo W.L., and Breaker R.R. 2005. 6S RNA is a widespread regulator of eubacterial RNA polymerase that resembles an open promoter. *RNA* **11:** 774.

Brantl S. and Wagner E.G.H. 1994. Antisense RNA-mediated transcriptional attenuation occurs faster than stable antisense/target RNA pairing: An in vitro study of plasmid pIP501. *EMBO J.* **13:** 3599.

———. 2000. Antisense RNA-mediated transcriptional attenuation: An *in vitro* study of plasmid pT181. *Mol. Microbiol.* **35:** 1469.

———. 2002. An antisense RNA-mediated transcription attenuation mechanism functions in *Escherichia coli. J. Bacteriol.* **184:** 2740.

Brantl S., Birch-Hirschfeld E., and Behnke D. 1993. RepR protein expression on plasmid pIP501 is controlled by an antisense RNA-mediated transcription attenuation mechanism. *J. Bacteriol.* **175:** 4052.

Briani F., Ghisotti D., and Dehò G. 2000. Antisense RNA-dependent transcription termination sites that modulate lysogenic development of satellite phage P4. *Mol. Microbiol.* **36:** 1124.

Chai Y. and Winans S.C. 2005. A small antisense RNA downregulates expression of an essential replicase protein of an *Agrobacterium tumefaciens* Ti plasmid. *Mol. Microbiol.* **56:** 1574.

Egloff S., Van Herreweghe E., and Kiss T. 2006. Regulation of polymerase II transcription by 7SK snRNA: Two distinct RNA elements direct P-TEFb and HEXIM1 binding. *Mol. Cell Biol.* **26:** 630.

Espinoza C.A., Allen T.A., Hieb A.R., Kugel J.F., and Goodrich J.A. 2004. B2 RNA binds directly to RNA polymerase II to repress transcript synthesis. *Nat. Struct. Mol. Biol.* **11:** 822.

Forti F., Dragoni I., Briani F., Dehò G., and Ghisotti D. 2002. Characterization of the small antisense CI RNA that regulates bacteriophage P4 immunity. *J. Mol. Biol.* **315:** 541.

Goodrich J.A. and Kugel J.F. 2006. Non-coding-RNA regulators of RNA polymerase II transcription. *Nat. Rev. Mol. Cell Biol.* **7:** 612.

Gottesman S. 2004. The small RNA regulators of *Escherichia coli:* Roles and mechanisms. *Annu. Rev. Microbiol.* **58:** 303.

Hatchell E.C., Colley S.M., Beveridge D.J., Epis M.R., Stuart L.M., Giles K.M., Redfern A.D., Miles L.E., Barker A., MacDonald L.M., et al. 2006. SLIRP, a small SRA binding protein, is a nuclear receptor corepressor. *Mol. Cell* **22:** 657.

Heidrich N. and Brantl S. 2003. Antisense-RNA mediated transcriptional attenuation: Importance of a U-turn loop structure in the target RNA of plasmid pIP501 for efficient inhibition by the antisense RNA. *J. Mol. Biol.* **333:** 917.

Hüttenhofer A. and Vogel J. 2006. Experimental approaches to identify non-coding RNAs. *Nucleic Acids Res.* **34:** 635.

Jacob F. and Monod J. 1961. Genetic regulatory mechanisms in the synthesis of proteins. *J. Mol. Biol.* **3:** 318.

Kuwabara T., Hsieh J., Nakashima K., Taira K., and Gage F.H. 2004. A small modulatory dsRNA specifies the fate of adult neural stem cells. *Cell* **116:** 779.

Kwek K.Y., Murphy S., Furger A., Thomas B., O'Gorman W., Kimura H., Proudfoot N.J., and Akoulitchev A. 2002. U1 snRNA associates with TFIIH and regulates transcriptional initiation. *Nat. Struct. Biol.* **9:** 800.

Lanz R.B., McKenna N.J., Onate S.A., Albrecht U., Wong J., Tsai S.Y., Tsai M.J., and O'Malley B.W. 1999. A steroid receptor coactivator, SRA, functions as an RNA and is present in an SRC-1 complex. *Cell* **97:** 17.

Le Chatelier E., Ehrlich S.D., and Janniere L. 1996. Counter transcript-driven attenuation system of the pAMβ1 *repE* gene. *Mol. Microbiol.* **20:** 1099.

Martens J.A., Laprade L., and Winston F. 2004. Intergenic transcription is required to repress the *Saccharomyces cerevisiae SER3* gene. *Nature* **429:** 571.

Martens J.A., Wu P.Y., and Winston F. 2005. Regulation of an intergenic transcript controls adjacent gene transcription in *Saccharomyces cerevisiae. Genes Dev.* **19:** 2695.

Novick R.P., Iordanescu S., Projan S.J., Kornblum J., and Edelman I. 1989. pT181 plasmid replication is regulated by a countertranscript driven transcriptional attenuator. *Cell* **59:** 395.

Nguyen V.T., Kiss T., Michels A.A., and Bensaude O. 2001. 7SK small nuclear RNA binds to and inhibits the activity of CDK9/cyclin T complexes. *Nature* **414:** 322.

Petersen C.P., Doench J.G., Grishok A., and Sharp P.A. 2006. The biology of short RNAs. In *The RNA world,* 3rd edition (ed. R.F. Gesteland et al.), p. 535. Cold Spring Harbor Laboratory Press, Cold Spring Harbor, New York.

Shamovsky I., Ivannikov M., Kandel E.S., Gershon D., and Nudler E. 2006. RNA-mediated response to heat shock in mammalian cells. *Nature* **440:** 556.

Storz G. and Gottesman S. 2006. Versatile roles of small RNA regulators in bacteria. In *The RNA world,* 3rd edition (ed. R.F. Gesteland et al.), p. 567. Cold Spring Harbor Laboratory Press, Cold Spring Harbor, New York.

Storz G., Altuvia S., and Wassarman K.M. 2005. An abundance of RNA regulators. *Annu. Rev. Biochem.* **74:** 199.

Trotochaud A.E. and Wassarman K.M. 2004. 6S RNA function enhances long-term cell survival. *J. Bacteriol.* **186:** 4978.

———. 2005. A highly conserved 6S RNA structure is required for regulation of transcription. *Nat. Struct. Mol. Biol.* **12:** 313.

———. 2006. 6S RNA regulation of pspF transcription leads to altered cell survival at high pH. *J. Bacteriol.* **188:** 3936.

Valencia-Sanchez M.A., Liu J., Hannon G.J., and Parker R. 2006. Control of translation and mRNA degradation by miRNAs and siRNAs. *Genes Dev.* **20:** 515.

Wassarman K.M. and Storz G. 2000. 6S RNA regulates *E. coli* RNA polymerase activity. *Cell* **101:** 613.

Winkler W.C. and Breaker R.R. 2005. Regulation of bacterial gene expression by riboswitches. *Annu. Rev. Microbiol.* **59:** 487.

Yang Z., Zhu Q., Luo K., and Zhou Q. 2001. The 7SK small nuclear RNA inhibits the CDK9/cyclin T1 kinase to control transcription. *Nature* **414:** 317.

Zamore P.D. and Haley B. 2005. Ribo-gnome: The big world of small RNAs. *Science* **309:** 1519.

Turnover and Function of Noncoding RNA Polymerase II Transcripts

M.J. Dye,* N. Gromak,* D. Haussecker,[†] S. West,* and N.J. Proudfoot*

*Sir William Dunn School of Pathology, University of Oxford, Oxford, OX1 3RE, United Kingdom;
[†]Stanford School of Medicine, Department of Pediatrics, Stanford, California 94305

In the past few years, especially since the discovery of RNA interference (RNAi), our understanding of the role of RNA in gene expression has undergone a significant transformation. This change has been brought about by growing evidence that RNA is more complex and transcription more promiscuous than has previously been thought. Many of the new transcripts are of so-called noncoding RNA (ncRNA); i.e., RNA that does not code for proteins such as mRNA, or intrinsic parts of the cellular machinery such as the highly structured RNA components of ribosomes (rRNA) and the small nuclear RNA (snRNA) components of the splicing machinery. It is becoming increasingly apparent that ncRNAs have very important roles in gene expression. This paper focuses on work from our laboratory in which we have investigated the roles and turnover of ncRNA located within the gene pre-mRNA, which we refer to as intragenic ncRNA. Also discussed are some investigations of intergenic ncRNA transcription and how these two classes of ncRNA may interrelate.

In eukaryotes, protein encoding genes are transcribed by RNA polymerase II (pol II). Typically, pol II transcription initiates within the promoter region of the gene that is encoded in the DNA template, and terminates at some point downstream from the poly(A)-addition site in the 3'-flanking region (see the diagram of the human β-globin gene; Fig. 1A). pol II transcription initiating at the promoter generates a pre-mRNA copy of this DNA template that is conventionally subdivided into the following regions: the 5'-untranslated region (5'UTR); the protein-encoding section of the pre-mRNA composed of exons interspersed with noncoding intron sequences; and the 3'UTR including the poly(A) signal. The pre-mRNA also extends beyond the poly(A)-site sequences into the 3'-flanking region up to the point of transcription termination (Fig. 1B). As 3'-flanking regions are highly unstable, few have been accurately defined. However, for the mouse β-globin (Tantravahi et al. 1993), the human β- and ε-globin (Dye and Proudfoot 2001), and the mouse serum albumin (West et al. 2006a) genes, it appears that the 3' terminus of the pre-mRNA lies between 1 and 2 kb downstream from the poly(A) site. Although the full extent of the 3'-flanking region of the majority of pre-mRNAs is unknown, it is well established that most pre-mRNAs conform to the arrangement of coding and noncoding elements seen in the β-globin gene. Different gene pre-mRNAs show great variation in terms of length and complexity. Many contain more exons and longer introns that may be spread over hundreds of kilobases of the genome (Strachan and Read 1999). Therefore, different gene pre-mRNAs may diverge greatly in the ratio of coding/noncoding sequence.

It is immediately apparent from Figure 1B that even the relatively small β-globin gene pre-mRNA is predominantly composed of ncRNA. This high ratio of ncRNA to coding RNA is typical of most pre-mRNAs. Indeed, on some gene maps, such as that of the dystrophin gene,

exons appear as tiny specks in a vast ocean of ncRNA. Some of this ncRNA is very important. The 5'UTR and 3'UTR, intronic splice signals, and poly(A)-site sequences guide essential steps in gene expression. For example, processing of the β-globin pre-mRNA by splicing out of introns and endonucleolytic cleavage followed by polyadenylation at the poly(A)-addition site yields the mature mRNA, shown in Figure 1C. Furthermore, the 5'UTR and 3'UTR are retained in the mRNA where they have important roles in nuclear export, mRNA stability, and cytoplasmic localization (Moore 2005). 3'UTRs are also involved in the control of mRNA translation which, in an estimated 30% of genes, involves interaction with another class of ncRNA, termed microRNAs (miRNAs) (Lewis et al. 2005). However, the majority of ncRNAs derived from gene introns and 3'-flanking regions are of unknown function. Much of this ncRNA has been assumed to be simply "junk" transcript, which has arisen merely as a consequence of gene evolution and as such is rapidly discarded by RNA degradation. The large energetic cost of this dominant gene transcript is apparently tolerated by mammalian nuclei. This view of ncRNA as predominantly nonfunctional transcript is now challenged by the identification of growing numbers of small regulatory RNAs, namely, small nucleolar RNAs (snoRNAs) and miRNAs (Bartel 2004; Mattick and Makunin 2006). These findings substantially alter the view that ncRNA is functionless.

In recent years, it has become apparent that for many genes, pre-mRNA processing occurs cotranscriptionally. This means that addition of the cap structure to the 5' end of the pre-mRNA, excision of introns, endonucleolytic cleavage at the poly(A) site, and even addition of the poly(A) tail occur as pol II produces the pre-mRNA copy of the DNA template (Maniatis and Tasic 2002; Proudfoot et al. 2002). Indeed, various lines of evidence indicate that pre-mRNA processing and pol II transcription are coupled

A

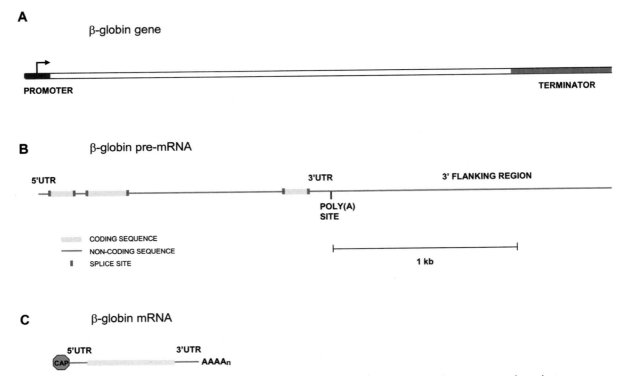

Figure 1. Structure of the human β-globin gene. (*A*) Diagram of the human β-globin gene. The promoter and terminator sequences are represented by the *green* and *red* boxes, respectively. The arrow indicates the transcription start site. (*B*) Diagram of the human β-globin pre-mRNA. Unlabeled components of the pre-mRNA are listed in the key below the diagram. (*C*) Diagram of the human β-globin mRNA.

events; in many different systems, we see that perturbation of one process may effect a change in the other. One of the earliest demonstrations of the cotranscriptional nature of pre-mRNA processing was the finding that mutation of the α-globin gene poly(A) site perturbed transcription termination in the α-globin gene 3'-flanking region (Whitelaw and Proudfoot 1986). Evidence for the coupling of transcription and pre-mRNA processing led to the generation of models showing physical linkage of components of the transcription and pre-mRNA processing apparatuses (Proudfoot et al. 2002). The transcription–RNA processing complex finds its most ambitious expression in the notion that large transcription "factories" exist as conglomerates of polymerase and processing factors at specific nuclear locations (Bartlett et al. 2006). From a purely practical viewpoint, the coupling of transcription and pre-mRNA processing provides a convenient "handle" for the experimentalist to analyze these complex processes and their potential interactions.

As mentioned above, specific roles in pre-mRNA processing and mRNA stability have been assigned to discrete intragenic ncRNA sequences. However, there is little understanding of the roles and turnover of the majority of intragenic ncRNA located in the pre-mRNA 3'-flanking region and introns. This is partly due to the difficulty of examining such rapidly metabolized RNA species. In our laboratory, we have refined and developed techniques that have enabled us to overcome some of the problems inherent in studying unstable nascent RNA. This has enabled us to make several interesting observations

concerning the biology of the 3'-flanking region and intron ncRNA transcripts discussed below.

ROLES AND TURNOVER OF 3'-FLANKING REGION NONCODING RNA TRANSCRIPTS

For several years, we have mapped nascent pre-mRNAs in the 3'-flanking regions of genes with the goal of finding sites of transcriptional termination. In particular, we have employed the nuclear run-on (NRO) technique, in which RNA molecules are radioactively end-labeled as they are transcribed by endogenous RNA polymerase in vivo. NRO analysis yields two types of data: It shows the distribution of active RNA polymerase on the DNA template of a given gene and the extent of the gene pre-mRNA. NRO analysis of 3'-flanking region transcripts of the human β-globin gene gives the pattern of hybridization signals shown in Figure 2A. Strong hybridization signals over probes 4–10 that span between 0 and 1.6 kb downstream from the poly(A) site are followed by background hybridization signals over probes A, B, and C, which lie immediately downstream from probe 10. These data show the presence of active pol II throughout the β-globin 3'-flanking region, extending up to 1.6 kb downstream from the poly(A) site. The background level hybridization signals beyond probe 10, over probes A, B, and C, indicate that pol II transcriptional termination takes place soon after this part of the 3'-flanking region has been transcribed (Dye and Proudfoot 1999). The role of specific 3'-flanking region sequences in pol II transcription

A

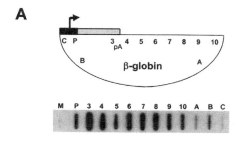

B

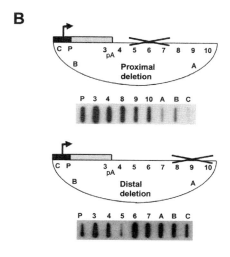

C

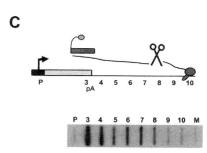

Figure 2. Nuclear Run On (NRO) analysis of β-globin gene transcripts. (*A*) NRO analysis of HeLa cells transiently transfected with the human β-globin gene construct shown in the diagram above the data panel. Characters in boldfaced type indicate the position of NRO probes. Probe M (M13 DNA) controls for background hybridization. pA indicates the position of the poly(A) site. (*B*) NRO analysis of HeLa cells transiently transfected with derivatives of the human β-globin gene construct shown in the diagrams above the data panels. (*C*) Hybrid selection NRO analysis of β-globin 3′-flanking region transcripts. (*Single wavy line*) pre-mRNA; (*filled ovals*) elongating pol II; (*box with tail*) position of the biotinylated selection probe; (*scissors*) CoTC activity.

termination was determined by deletion analysis. Although it was found that removal of the 800 bp of 3′-flanking sequence lying proximal to the β-globin poly(A) site, corresponding to probes 5, 6, and 7, has no effect on transcription termination, deletion of 3′-flanking region sequences lying more distal to the poly(A) site, corresponding to probes 8, 9, and 10, leads to a marked increase in readthrough transcription (Fig. 2B). This result shows

that 3′-flanking region sequences, located between 1 and 1.6 kb downstream from the β-globin poly(A) site, are involved in transcriptional termination (Dye and Proudfoot 2001).

We next analyzed how these 3′-flanking region sequences mediate pol II termination. An early indication of the involvement of the 3′-flanking region transcript in the termination process was derived from the experiment shown in Figure 2C. In this experiment, which employs a hybrid selection technique to measure the continuity of nascent RNA molecules, we found that β-globin pre-mRNA transcripts are not continuous throughout the 3′-flanking region (Dye and Proudfoot 1999). Detailed analysis showed that transcript discontinuity occurs at several sites in a region lying between 1 and 1.6 kb downstream from the β-globin poly(A) site. Analysis of the ε-globin gene showed a similar but weaker effect (Dye and Proudfoot 2001). These data have far-reaching implications: First, they provide strong evidence that the ncRNA generated in the β-globin 3′-flanking region transcript is targeted by an RNA cleavage activity; second, the correspondence of the site of this cotranscriptional processing activity with the termination element indicates that cotranscriptional processing of the 3′-flanking region RNA might have an important role in transcription termination; and finally, the detection of noncontinuous nascent transcripts indicates that the linear pre-mRNA, depicted in Figure 1B as extending from the start site of transcription to the termination region, does not exist in vivo. The simplest way to interpret this last point is that although the 3′-flanking region pre-mRNA does not exist in vivo as a physically continuous molecule, it does, however, exist as a continuous sequence of information, some of which undergoes cotranscriptional processing.

Cotranscriptional cleavage (CoTC) of nascent transcripts generates two highly unstable products, termed the 5′ product and the 3′ product (Fig. 3A). Detailed analysis of the 5′ product was advanced by the development, in our laboratory, of an assay termed hscRACE (hybrid selection circular rapid amplification of cDNA ends). Here, the 5′ products of CoTC were circularized using T4 RNA ligase, then reverse-transcribed and polymerase chain reaction (PCR)-amplified using specific primers in the 3′-flanking region (West et al. 2006b). DNA sequencing of the hscRACE PCR products gives the surprising result that some of the 5′ products of CoTC are oligoadenylated. Furthermore, when components of the exosome complex are depleted from the transfected cells, 5′ products of CoTC are stabilized with a concomitant increase in the length of oligoadenylate tails (West et al. 2006b). These data show that oligoadenylation of mammalian ncRNAs can serve as a marker for degradation by the exosome, as has already been described for certain ncRNAs in yeast (Fig. 3C) (Wyers et al. 2005). It will be of interest to see the extent of oligoadenylation as a general mechanism in the turnover of ncRNAs.

Whereas analysis of the 5′ product of CoTC has yielded important insights into the turnover of ncRNA, investigation of the fate of the 3′ product revealed a connection between ncRNA turnover and pol II transcription termination (West et al. 2004). In this study, it was shown that the 3′ product of CoTC in the β-globin flanking region pre-

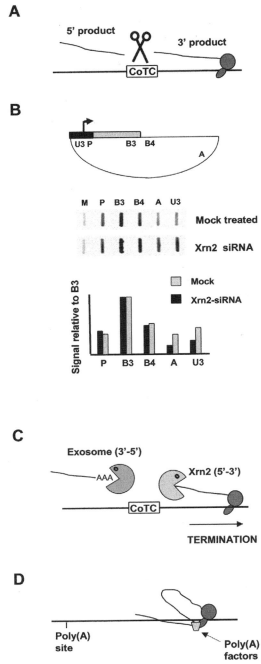

Figure 3. Turnover and roles of ncRNA in the β-globin 3′-flanking region. (*A*) Diagram of the products of CoTC in the 3′-flanking region. (*Solid black line*) DNA template; (*single wavy line*) pre-mRNA; (*scissors*) CoTC activity. (*B*) Role of exonuclease Xrn2 in pol II transcription termination. (*Upper panel*) Diagram showing the β-globin gene construct; (*middle data panel*) NRO analysis of mock-treated or Xrn2 siRNA-treated HeLa cells transiently transfected with the β-globin gene construct; (*bottom panel*) graph showing corrected NRO hybridization signals in mock-treated or Xrn2 siRNA-treated cells. (*C*) Turnover of ncRNA in the β-globin 3′-flanking region following CoTC. Oligoadenylation of the 5′ product is indicated. The "Pac Man" symbols represent 3′-5′ and 5′-3′ exonucleases. (*D*) Role of 3′-flanking region ncRNA in tethering poly(A)-site sequences to elongating pol II.

mRNA is a substrate of the nuclear 5′→3′ exonuclease, Xrn2. Furthermore, it was demonstrated that transcriptional termination of the β-globin gene was diminished when Xrn2 was depleted by transfection of Xrn2-specific siRNAs, as shown in the NRO data in Figure 3B. The decrease in transcription termination efficiency is indicated by the increase in hybridization signals (due to transcriptional readthrough) over probes A and U3. This result was also confirmed by chromatin immunoprecipitation (ChIP) measuring pol II occupancy over the β-globin gene following Xrn2 depletion (West et al. 2004). At the same time, ChIP analysis in yeast showed that the homolog of Xrn2 in *Saccharomyces cerevisiae*, the 5′→3′exonuclease Rat1protein, localizes to the 3′ end of genes and is required for pol II termination (Kim et al. 2004). In some yeast genes, transcriptional termination is entirely dependent on the poly(A) site (Birse et al. 1998). It is therefore envisaged that Rat1protein engages with the pre-mRNA at the free 5′ end generated by poly(A)-site cleavage. It is currently thought that once 5′→3′ exonuclease engages with the transcript at the free 5′ end, generated in a CoTC region or at the poly(A) site depending on the organism and specific gene, it then degrades the transcript until it "catches up" with pol II and then participates in the subsequent transcription termination process (Fig. 3C). This "torpedo" model was posited some time ago when the dependence of transcription termination on the poly(A) site was first identified (Whitelaw and Proudfoot 1986; Connelly and Manley 1988). At present, there are few details of the mechanism by which 5′→3′ exonuclease promotes transcription termination. However, the absolute requirement for a 5′-phosphate terminus, generated by cleavage at the poly(A) site or the CoTC element rather than a 5′-hydroxyl terminus generated by hammerhead ribozyme cleavage, is a clear indication that RNA turnover is involved in the termination process (West et al. 2004). To further our research in this area, we have recently developed antibodies to Xrn2 and are attempting to identify interacting protein partners to Xrn2. The generality of this termination mechanism is unknown; however, a further example of CoTC-mediated transcription termination has recently been described in the 3′-flanking region of the mouse serum albumin gene (West et al. 2006a). Whereas in the human β-actin gene it appears that, due to the effect of transcriptional pausing (see below), the initial engagement of 5′→3′exonuclease with the transcript occurs at the free 5′ end generated by cleavage at the poly(A) site (Gromak et al. 2006).

MOLECULAR TETHERING BY 3′-FLANKING REGION NONCODING RNA

Apart from being a target of CoTC, 3′-flanking region ncRNA appears to have a role in tethering of poly(A)-site sequences to pol II. We and other investigators have shown that endonucleolytic poly(A)-site cleavage of mammalian pre-mRNAs occurs some time after the poly(A)-site sequences have been transcribed (Baurén et al. 1998; Dye and Proudfoot 1999). This observation, coupled with evidence that the carboxy-terminal domain (CTD) of the large β-subunit of pol II has a role in 3′-end processing, leads us to a model in which a stable association of the poly(A) factors with the polymerase complex is established subsequent to transcription of the poly(A)-site sequences. As pol II continues transcription, the interaction of the poly(A) signal with the polymerase complex would lead to the formation of an RNA loop composed of the 3′-flanking region transcript (Fig. 3D). This association appears to be maintained at least until transcript cleavage at the poly(A) site takes place. Recent experiments in our laboratory and others have investigated the role of this structural loop in RNA 3′-end processing. It has been shown that cleavage of the nascent transcript, between the poly(A) site and termination sequences, leads to a reduction of poly(A)-site cleavage efficiency (Rigo et al. 2005). Thus, it appears that the ncRNA downstream from the poly(A) site acts as a molecular tether between the polymerase and the poly(A) signal, somehow enhancing processing at the poly(A) site.

An interesting aspect of 3′-flanking region sequences is the presence of transcriptional pause elements that have been shown to be important in altering the efficiency of 3′-end processing and subsequent transcriptional termination (Yonaha and Proudfoot 2000; Gromak et al. 2006). It has previously been assumed that pause signals operated at the DNA level and involved the recruitment of DNA-binding proteins to a specific terminator/pause sequence as is the case for pol I in both humans and yeast (Paule and White 2000). However, evidence for a similar role of DNA-binding proteins in promoting pol II termination has not been forthcoming. We therefore cannot exclude the possibility that pol II transcriptional pausing involves ncRNA. Our current understanding of transcriptional pausing is that it might increase the dwell time of the polymerase complex at a particular position, downstream from the poly(A) site, and thereby enhance processing of the pre-mRNA. Alternatively, pause sequences might act at the RNA level, stabilizing the molecular tether between the poly(A) site and pol II, thus enhancing 3′-end processing. This would in turn increase the likelihood of transcription termination.

A further potential role of ncRNA in pol II termination may relate to the occurrence of hybridization of pre-mRNA with the gene DNA template forming so-called R loops. Significant R-loop formation has been shown to impair transcription elongation (Huertas and Aquilera 2003) and may have a role in intron splicing (Li and Manley 2005). Our laboratory is currently studying the presence of R loops in candidate 3′-flanking regions of several yeast genes. It is possible that R-loop formation between ncRNA and the DNA template could induce pausing of pol II as a way of promoting pol II termination.

ROLES AND TURNOVER OF NONCODING RNA IN PRE-MRNA SPLICING

Pre-mRNA splicing is the process by which introns, largely composed of ncRNA, are excised from the pre-mRNA transcript and adjacent exons are ligated together to form a continuous mRNA (Fig. 1B,C) (Moore et al. 1993). The importance of pre-mRNA splicing in gene expression cannot be exaggerated. In particular, it provides the molecular underpinning for the generation of different mRNA isoforms and consequently different proteins from between one-half to two-thirds of all human genes by alternative splicing (Modrek and Lee 2002). Furthermore, pre-mRNA splicing has a pivotal role in the development of complex organisms. The evolution of splicing has recently been implicated as a driving force in the origin of the eukaryotic nucleus (Martin and Koonin 2006).

Although much research effort has gone into unraveling the details of splicing, there are still major gaps in our understanding of this complex process. One of these is the understanding of how splice site selection occurs. Splice site consensus sequences are remarkably simple. The splice donor (SD) site at the 5′ end of the intron is determined by a short sequence containing the conserved GU dinucleotide. The splice site at the 3′ end of the intron is more complex consisting of two sequences; the branch site that usually consists of a highly conserved A residue followed by a pyrimidine-rich tract and the splice acceptor (SA) site which is defined by a short sequence containing a conserved AG dinucleotide at the 3′ end (Proudfoot et al. 2002). Due to the length of introns and the fact that there are no obvious restraints on their internal sequence, introns often contain many of the relatively simple sequences that define splice signals. Occasionally, the arrangement of these so-called cryptic splice sites can define regions within the body of the intron as pseudoexons (Fig. 4A). However, splice site selection shows remarkable fidelity, using only the regular splice sites and generally ignoring cryptic splice sites and pseudoexons.

A critical advance in our understanding of splice site selection came with the exon definition hypothesis which suggested that identification of exons by recognition of their flanking splice sites marks them for retention in the mRNA and consequently marks the introns for removal by splicing (Berget 1995). Experimental evidence shows that exon definition is principally brought about by SR proteins bridging the exon and thus connecting splice factors binding to the flanking SA and SD splice sites (Fig. 4A). With the basic mechanism of SD-site selection understood, two further questions arise. First, how is the interaction between the chosen SD site maintained in the face of competition from cryptic SD sites within the intron? Second, especially in the case of transcription of very long introns, how is the correct SA site chosen from the many candidate SA sites that will be transcribed before it (Fig. 4A)? Recent data from ChIP analyses in yeast and mammalian cells show that splice factors are recruited to splice donor sites cotranscriptionally, possibly as soon as they emerge in the nascent transcript (Újvári and Luse 2004; Görnemann et al. 2005; Lacadie and Rosbash 2005). These data, combined with evidence for interaction of 5′ splice factors with elongating pol II and the extraordinary fidelity of SD-site selection (Maniatis and Reed 2002), give a strong indication that SD-site selection occurs on a "first come, first served" basis, through exon definition, and that a strong interaction between the SD site and the transcription complex is quickly formed. Presumably, it is this strong interaction that withstands competition from cryptic SD sites within the intron. Although SA-site selection in short introns may be based on simple sequence recognition, in long introns, this seems less likely. We suggest that the problem of splice site selection might be related to a speculative idea, concerning how pol II deals with the production of huge amounts of ncRNA when transcribing long introns (discussed below).

The classic two-step model of pre-mRNA splicing involves the formation of an intron lariat in step 1, which

A

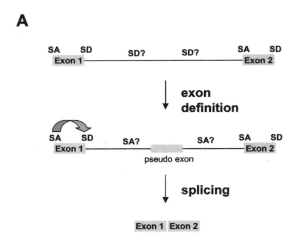

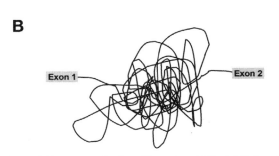

introns occur. Our study shows that cotranscriptional cleavage of a nascent intron using either the CoTC element from the β-globin gene 3′-flanking region or a fast and efficient hammerhead ribozyme has no discernable effect on steady-state mRNA level (Fig. 5A) (Dye et al. 2006). The observation that a noncontinuous pre-mRNA can be efficiently spliced gives direct evidence for the cotranscriptional establishment of a strong interaction between the splice donor site at the 5′ end of the intron mentioned above. We suggest that the establishment of this strong interaction, coupled with continued transcription, results in the formation of an RNA loop similar to that which occurs downstream from the poly(A) site (Fig. 5B). We further suggest that the loop is cotranscriptionally cleaved, allowing access of the RNA degradation machin-

B

Figure 4. Turnover and roles of intronic ncRNA. (*A*) Splice site selection and exon definition. Diagram showing position of splice sites in and around a hypothetical intron. SA and SD indicate splice acceptor site and splice donor site, respectively. SA? and SD? indicate cryptic splice sites. (*Curved arrow*) Interactions involved in exon definition. (*B*) The problem of transcribing and processing long tracts of intron sequence. (*Knotted line*) Intronic pre-mRNA.

is released concomitant with exon ligation in step 2. This model is based on experiments carried out with nuclear extracts that are optimized for splicing in vitro and electron microscopic evidence for lariat-shaped splicing intermediates in vivo (Moore et al. 1993). Further evidence for this mechanism comes from analysis of unusual highly stable intron transcripts that are exported out of the nucleus (Clement et al. 1999, 2001). However, when one considers that transcription of some introns would result in the accumulation of vast amounts of pre-mRNA near the site of transcription, it appears unlikely that splicing can only occur as outlined above. Long tracts of unspliced pre-mRNA would occupy a lot of nuclear space, and the splicing machinery would encounter difficulty in finding the correct splice site within a mass of unspliced RNA (Fig. 4B). Studies in *Drosophila* have shown that splice sites located within certain long introns are used as "stepping stones," enabling the splicing out of the intron in segments (Hatton et al. 1998). Bioinformatic analysis shows that several genes in *Drosophila* might employ this recursive splicing mechanism to process long tracts of intronic ncRNA (Burnette et al. 2005).

A recent paper from our laboratory offers an additional explanation of how the transcription and splicing of long

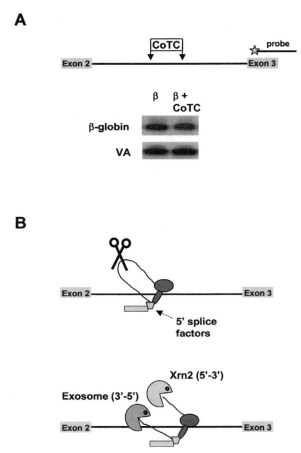

Figure 5. Cotranscriptional intron ncRNA degradation. (*A*) Cotranscriptional cleavage of intron transcripts does not influence splicing. (*Upper panel*) Diagram of intron 2 and flanking exons 2 and 3 of the β and β+CoTC globin gene constructs; (*lower panel*) nuclease S1 analysis of β and β+CoTC transcripts. The end-labeled probe used in this experiment is shown in the diagram in the upper panel. The VA signal derives from the cotransfected VA plasmid and controls for transfection efficiency and RNA recovery. (*B*) (*Upper panel*) Tethering of pre-mRNA to pol II. In the diagram (not to scale): (*solid line*) β-globin intron-2 DNA template; (*single curved line*) nascent transcript; (*gray box with black outline*) tethered exon; (*rhombus*) splice factors associated with the intron-2 splice donor site; (*scissors*) cotranscriptional cleavage of the nascent intron transcript. (*Lower panel*) Cotranscriptional intron degradation. The "Pac Man" symbols represent 3′-5′ and 5′-3′ exonucleases.

ery to remove ncRNA parts of the pre-mRNA as they are synthesized (Fig. 5B). This "exon tethering" model suggests that some introns do not act as connectors between splice sites at the RNA level and that the lariat structure may not be formed, as envisaged in the standard pre-mRNA splicing model. Having proven that pre-mRNA splicing can occur across a noncontinuous intron, we hypothesize that target sequences for an endonucleolytic cotranscriptional cleavage activity, such as that found in the 3'-flanking region of the β-globin gene, might reside in endogenous mammalian introns. Consequently, we are now searching for examples of cotranscriptional cleavage of intronic ncRNA in endogenous genes.

CoTC generates free 5' and 3' ends that are substrates of the nuclear RNA degradation machinery. We suggest that cotranscriptional degradation of intron sequences could resolve some of the problems pol II must encounter when producing very long tracts of ncRNA. Rapid intron degradation could eliminate cryptic splice sites from the intron transcript before they have a chance to compete for splicing. This hypothesis invokes another mechanism of regulating splicing. Either ncRNA sequences to be degraded are somehow marked for degradation or sequences to be protected are marked for protection. An important objection to this model would be that unrestrained RNA degradation, initiating within the intron transcript, could lead to the destruction of the entire pre-mRNA. However, our observation that splicing is completely unaffected by CoTC indicates that this is not the case. We suggest that the important splice signals, located at the 5' and 3' ends of the cotranscriptionally cleaved intron transcript, are protected from degradation, This protection could be mediated by interaction between the pre-mRNA with the transcription complex, splice factors, and hnRNPs. Binding of these factors to the transcript might hinder or slow down RNA exonuclease activity.

Introns are not simply featureless ncRNA linkers between exons, they also contain information. The importance of intronic splice signals, generally located at the 5' and 3' exon/intron junctions, in the regulation of gene expression is well established. There is also evidence for roles of other intronic sequences. Some introns contain sequences that encode small regulatory RNAs, including miRNAs and snoRNAs (Mattick and Makunin 2006) as well as important regulatory signals such as splicing enhancers and silencers (McCullough and Berget 2000; Sun and Chasin 2000; Wagner et al. 2005). Clearly, these sequences must be protected from degradation. In this regard, it is interesting to note that some miRNAs appear to be cotranscriptionally excised from their intronic RNA background (V.N. Kim, pers. comm.). It is exciting to speculate on the possibility that cotranscriptional excision of regulatory RNAs from introns could have been a driving force in the evolution of exon tethering.

In summary, our studies of 3'-flanking region and intron transcripts have shown that these intragenic ncRNAs have some very important properties. Intragenic ncRNA can act as a molecular tether connecting transcribed RNA-processing signals to elongating polymerase and thus enhancing pre-mRNA processing. They can also act as targets for cotranscriptional cleavage, which is instrumental in the process of RNA pol II tran-

scriptional termination on some gene templates. Intronic ncRNA also forms loops that may also be targets for CoTC. CoTC and subsequent transcript degradation introduces another level of complexity to the already complex process of pre-mRNA splicing. However, cotranscriptional intron degradation might also solve the dynamic and spatial problems faced by the cell in transcribing long tracts of noncoding intron sequence and might also prevent competition by cryptic splice sites.

TURNOVER AND FUNCTION OF INTERGENIC ncRNA TRANSCRIPTS

It has become apparent in recent years that transcription of ncRNA by pol II is not only confined within the boundaries of genes. Transcriptional analysis of a variety of eukaryotic genomes has revealed high levels of widespread intergenic transcription (Bertone et al. 2004; Stolc et al. 2005; Samanta et al. 2006). In general, this intergenic ncRNA is highly unstable. In the yeast *S. cerevisiae*, it has been shown that these cryptic unstable transcripts (CUTs) are stabilized when the major RNA degradation apparatus, the multisubunit exosome, is inactivated by gene knockout (Wyers et al. 2005), implicating an important role for this complex in ncRNA turnover. CUTs are of unknown function; however, it is thought that some may be involved in gene regulation processes such as that demonstrated for the *SRG1* gene transcript that regulates *SER3* gene expression by selective transcriptional interference (Martens et al. 2006).

In the yeast *Schizosaccharomyces pombe*, transcription of intergenic ncRNA is associated with epigenetic regulation of centromeric sequences (Volpe et al. 2002). It has been shown that intergenic transcripts are involved in maintaining a repressed chromatin state through RNAi pathways that cotranscriptionally recruit chromatin remodeling activities (Cam and Grewal 2004; Buhler et al. 2006). Interestingly, these transcripts are synthesized not only by pol II, but also by an RNA-dependent RNA polymerase (Rdp) activity. Similarly, in plants, intergenic transcripts synthesized by an Rdp activity and a dedicated RNA polymerase, termed pol IV, act to set up appropriate chromatin structures throughout the plant genome in response to developmental and viral cues (Herr et al. 2005). The possibility that such chromatin-associated intergenic transcription is prevalent in all eukaryotic genomes remains to be established. However, work in our laboratory, which we describe below, has shown widespread intergenic transcription across the human β-globin gene cluster that may have a similar regulatory role in the establishment of chromatin structure.

We originally detected transcription of intergenic regions in the human β-globin gene cluster, shown in the diagram in Figure 6, of erythroid and non-erythroid cells by nuclear run-on analysis following various induction treatments (Ashe et al. 1997; Plant et al. 2001). We also determined that transcription of this ncRNA likely initiates in the repetitive ERV-9 LTR (long terminal repeat) region sequence (Plant et al. 2001). These intergenic transcripts proved to be difficult to analyze due to their intrinsic instability and rapid turnover. However, in a later study, it was found that treatment of non-erythroid cells with the histone deacetylase inhibitor trichostatin A (TSA) caused widespread activation of intergenic transcription across the β-globin gene cluster which was detectable using high-amplification reverse transcriptase (RT)-PCR analysis. Figure 6 shows RT-PCR analysis of transcripts located downstream from the ERV-9 sequence. Similar results were obtained throughout the gene cluster (Haussecker and Proudfoot 2005). Interestingly, both sense and antisense transcripts were detected and, remarkably, RNAi-mediated knockdown of Dicer resulted in a considerable increase in transcript levels, suggesting that these intergenic transcripts are substrates for RNAi-related turnover, similar to that originally defined in *S. pombe* centromeres (Cam and Grewall 2004). Interestingly, genic as well as intergenic transcripts were induced by TSA treatment of HeLa cells. These genic transcripts were probably generated by transcriptional readthrough from intergenic regions. Strikingly, only the unspliced transcripts of each of the globin genes were susceptible to Dicer as shown by the relative elevation in their abundance, as compared to that of the spliced transcripts, following Dicer knockdown (Fig. 6) (Haussecker and Proudfoot 2005). Taken together, these data indicate that aberrant transcripts, whether of intergenic or unprocessed genic origin, are substrates for an RNAi-mediated degradation pathway from which, presumably, induced spliced transcripts as well as authentic genic transcripts are somehow sequestered. A connection of these results with chromatin silencing was revealed when the β-globin locus was tested for changes in chromatin structure following Dicer knockdown. As shown in Figure 7, Dicer knockdown results in an increase in histone acetylation across the

cluster, suggesting that Dicer-mediated degradation of the β-globin locus intergenic transcripts may well relate to normal repression of this cluster in non-erythroid cells.

CONCLUSIONS

We have described studies of noncoding pol II transcripts from both within and outside the boundaries of the gene. Our account of intragenic ncRNA transcripts, derived from the introns and 3′-flanking regions of protein-encoding genes, indicates that they have important roles both in connecting RNA processing signals to transcribing pol II and in serving as substrates for RNA degradation. Although a high energetic cost must be associated with the transcription and turnover of this apparently nonproductive ncRNA, it appears from our studies that the actual degradation process may have evolved in concert with the splicing, 3′-end processing, and transcriptional termination mechanisms and have an intrinsic role in them. Our studies of intergenic ncRNA transcription indicate that RNAi-related mechanisms are involved in regulating intergenic transcription in the human β-globin gene cluster and further suggest that RNAi-dependent chromatin silencing in vertebrates is not restricted to the centromeres. Furthermore, we speculate that the intergenic transcripts in the β-globin cluster might be representative of widespread intergenic ncRNA throughout the human genome and, by extrapolation, all higher eukaryotes.

Previous work from our laboratory (O'Sullivan et al. 2004) has demonstrated that transcription by pol II is associated with the formation of a gene-loop structure in which the promoter and terminator regions of the gene move into close physical contact following gene activa-

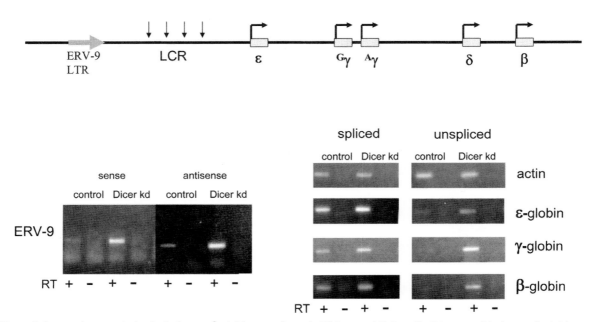

Figure 6. Intergenic transcription in the human β-globin gene cluster in TSA-treated HeLa cells. Diagram of the human β-globin gene cluster showing positions of five structural genes as well as the locus control region and ERV-9 LTR. Below the locus map, selective RT-PCR analysis is shown for the ERV-9 region as well as genic regions using primers specific for both spliced and unspliced transcripts. Nuclear RNA was isolated from TSA-induced HeLa cells with or without Dicer siRNA treatment to knockdown Dicer expression. All RT-PCRs were controlled by omitting reverse transcriptase to rule out DNA contamination (–RT). Equal amounts of nuclear RNA were used in each separate experimental panel. For ERV-9 antisense, RT-PCR was carried out using a higher number of cycles.

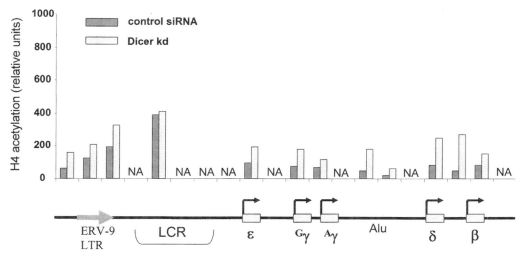

Figure 7. Changes in chromatin structure of the human β-globin gene cluster following Dicer knockdown. ChIP analysis of histone H4 acetylation across the human β-globin gene cluster in control or Dicer siRNA-treated HeLa cells. NA denotes not assayed. Positions of ChIP signals are aligned to the β-globin locus map below. For further details, see Haussecker and Proudfoot (2005).

tion. These observations have been confirmed and extended in a study showing that specific promoter and terminator factors are required for loop formation (Ansari and Hampsey 2005). Consequently, it is apparent that genes actively transcribed by pol II may possess two levels of loop conformation: gene loops and intergenic ncRNA loops (Fig. 8). This RNA and DNA loop arrangement suggests a high level of structural organization for actively expressed genes We propose that intragenic ncRNA transcription may occur within this type of gene-loop structure. In contrast, intergenic ncRNA transcription may occur outside of such a loop structure and therefore intergenic transcription can have a role in chromatin RNAi effects, rather than resulting in productive mRNA synthesis. This hypothesis requires direct testing and is the subject of ongoing research in our laboratory.

ACKNOWLEDGMENTS

We thank members of the Proudfoot lab for helpful discussion and comments. This work was supported by a Programme Grant from the Wellcome Trust to N.J.P. D.H. was supported by a postgraduate scholarship from the Darwin Trust. S.W. was supported by an EPA research studentship.

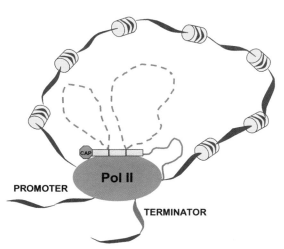

Figure 8. Gene loops and ncRNA loops may combine to define productive transcription units. Model showing the tight structural constraints that may exist to define a gene that is expressing pre-mRNA. Intergenic transcription that relates to RNAi effects may be excluded from such gene-loop regions. Diagram shows DNA associated with nucleosomes (cylinders) held at the promoter and terminator regions by the pol II complex. Tethered exons bind to the pol II complex with introns (*dashed red lines*) and 3′-flanking region (*solid red line*) looped out.

REFERENCES

Ansari A. and Hampsey M. 2005. A role for the CPF 3′-end processing machinery in RNAP II-dependent gene looping. *Genes Dev.* **19:** 2969.

Ashe H.L., Monks J., Wijgerde M., Fraser P., and Proudfoot N.J. 1997. Intergenic transcription and transinduction of the human β-globin locus. *Genes Dev.* **11:** 2494.

Bartel D.P. 2004. MicroRNAs: Genomics, biogenesis, mechanism and function. *Cell* **116:** 281.

Bartlett J., Blagojevic J., Carter D., Eskew C., Fromaget M., Job C., Shamser M., Trinidade I.F., Xu M., and Cook P.R. 2006. Specialised transcription factories. *Biochem. Soc. Symp.* **73:** 67.

Baurén G., Belikov S., and Wieslander L. 1998. Transcriptional termination in the *Balbiana ring 1* gene is closely coupled to 3′-end formation and excision of the 3′ terminal intron. *Genes Dev.* **12:** 2759.

Berget S.M. 1995. Exon recognition in vertebrate splicing. *J. Biol. Chem.* **270:** 2411.

Bertone P., Stolc V., Royce T.E., Rozowsky J.S., Urban A.E., Zhu X., Rinn J.L., Tongprasit W., Samanta M., Weismann S., et al. 2004. Global identification of human transcribed sequences with genome tiling arrays. *Science* **306:** 2242.

Birse C.E., Minvielle-Sebastia L., Lee B.A., Keller W., and Proudfoot N.J. 1998. Coupling termination of transcription to messenger RNA maturation in yeast. *Science* **280:** 298.

Buhler M., Verdel A., and Moazed D. 2006. Tethering RITS to a nascent transcript initiates RNAi- and heterochromatin-dependent gene silencing. *Cell* **125:** 873.

Burnette J.M., Miyamoto-Sato E., Schaub M.A., Conklin J., and Lopez A.J. 2005. Subdivision of large introns in *Drosophila* by recursive splicing at nonexonic elements. *Genetics* **170:** 661.

Cam H. and Grewal S.I. 2004. RNA interference and epigenetic control of heterochromatin assembly in fission yeast. *Cold Spring Harbor Symp. Quant. Biol.* **69:** 419.

Clement J.Q., Maiti S., and Wilkinson M.F. 2001. Localisation and stability of introns spliced from the Pem homeobox gene. *J. Biol. Chem.* **276:** 16919.

Clement J.Q., Qian L., Kaplinsky N., and Wilkinson M.F. 1999. The stability and fate of a spliced intron from vertebrate cells. *RNA* **5:** 206.

Connelly S. and Manley J.L. 1988. A functional mRNA polyadenylation signal is required for transcription termination by RNA polymerase II. *Genes Dev.* **2:** 440.

Dye M.J. and Proudfoot N.J. 1999. Terminal exon definition occurs cotranscriptionally and promotes termination of RNA polymerase II. *Mol. Cell* **3:** 371.

———. 2001. Multiple transcript cleavage precedes polymerase release in termination by RNA polymerase II. *Cell* **105:** 669.

Dye M.J., Gromak N., and Proudfoot N.J. 2006. Exon tethering in transcription by RNA polymerase II. *Mol. Cell* **21:** 849.

Görnemann J., Kotovic K.M., Hujer K., and Neugebauer K.M. 2005. Cotranscriptional spliceosome assembly occurs in a stepwise fashion and requires the cap binding complex. *Mol. Cell* **19:** 53.

Gromak N., West S., and Proudfoot N.J. 2006. Pause sites promote transcriptional termination of mammalian RNA polymerase II. *Mol. Cell. Biol.* **26:** 3986.

Hatton A.R., Subramaniam V., and Lopez A.J. 1998. Generation of alternative ultrabithorax isoforms and stepwise removal of a large intron by resplicing at exon-exon junctions. *Mol. Cell* **2:** 787.

Haussecker D. and Proudfoot N.J. 2005. Dicer-dependent turnover of intergenic transcripts from the human beta globin gene cluster. *Mol. Cell. Biol.* **25:** 9724.

Herr A.J., Jensen M.B., Dalmay T., and Baulcombe D.C. 2005. RNA polymerase IV directs silencing of endogenous DNA. *Science* **308:** 118.

Huertas P. and Aguilera A. 2003. Cotranscriptionally formed DNA:RNA hybrids mediate transcription elongation impairment and transcription-associated recombination. *Mol. Cell* **12:** 711.

Kim M., Krogan N.J., Vasiljeva L., Rando O.J., Nedea E., Greenblatt J.F., and Buratowski S. 2004. The yeast Rat1 exonuclease promotes transcription termination by RNA polymerase II. *Nature* **432:** 517.

Lacadie S.A. and Rosbash M. 2005. Cotranscriptional spliceosome assembly dynamics and the role of U1 snRNA:5′ss base pairing in yeast. *Mol. Cell* **19:** 65.

Lewis B.P., Burge C.B., and Bartel D.P. 2005. Conserved seed pairing, often flanked by adenosines, indicates that thousands of human genes are microRNA targets. *Cell* **120:** 15.

Li X. and Manley J.L. 2005. Inactivation of the SR protein splicing factor ASF/SF2 results in genomic instability. *Cell* **122:** 365.

Maniatis T. and Reed R. 2002. An extensive network of coupling among gene expression machines. *Nature* **416:** 499.

Maniatis T. and Tasic B. 2002. Alternative pre-mRNA splicing and proteome expansion in metazoans. *Nature* **418:** 236.

Martens J.A., Wu P.-Y., and Winston F. 2006. Regulation of an intergenic transcript controls adjacent gene transcription in *Saccharomyces cerevisiae*. *Genes Dev.* **19:** 2695.

Martin W. and Koonin E.V. 2006. Introns and the origin of nucleus-cytosol compartmentalisation. *Nature* **440:** 41.

Mattick J.S. and Makunin I.V. 2006. Non-coding RNA. *Hum. Mol. Genet.* **15:** 17.

McCullough A.J. and Berget S.M. 2000. An intronic splicing enhancer binds U1 snRNPs to enhance splicing and select 5′ splice sites. *Mol. Cell. Biol.* **20:** 9225.

Modrek B. and Lee C. 2002. A genomic view of alternative splicing. *Nat. Genet.* **30:** 13.

Moore M.J., Query C.C., and Sharp P.A. 1993. Splicing of precursors to mRNA by the spliceosome. In *The RNA world* (ed. R.F. Gesteland and J.F. Atkins), p. 303. Cold Spring Harbor Laboratory Press, Cold Spring Harbor, New York.

Moore M.J. 2005. From birth to death: The complex lives of eukaryotic mRNAs. *Science* **309:** 1514.

O'Sullivan J.M., Tan-Wong S.M., Morillon A., Lee B., Coles J., Mellro J., and Proudfoot N. 2004. Gene loops juxtapose promoters and terminators in yeast. *Nat. Genet.* **36:** 1014.

Paule M.R. and White R.J. 2000. Transcription by RNA polymerases I and III. *Nucleic Acids Res.* **28:** 1283.

Plant K.E., Routledge S.J.E., and Proudfoot N.J. 2001. Intergenic transcription in the human β-globin gene cluster. *Mol. Cell. Biol.* **21:** 6507.

Proudfoot N.J., Furger A., and Dye M.J. 2002. Integrating mRNA processing with transcription. *Cell* **108:** 501.

Rigo F., Kazerouninia A., Nag A., and Martinson H. 2005. The RNA tether from the poly(A) signal to the polymerase mediates coupling of transcription to cleavage and polyadenylation. *Mol. Cell* **20:** 733.

Samanta M.P., Tongprasit W., Sethi H., Chin C.-S., and Stolc V. 2006. Global identification of noncoding RNAs in *Saccharomyces cerevisiae* by modulating an essential RNA processing pathway. *Proc. Natl. Acad. Sci.* **103:** 4192.

Strachan T. and Read A.P. 1999. Organisation and distribution of human genes, chapter 7.2. In *Human molecular genetics*, 2nd edition. Garland Science, New York.

Stolc V., Samanta M.P., Tongprasit W., Sethi H., Liang S., Nelson D.C., Hegeman A., Nelson C., Rancour D., Bednarek S., et al. 2005. Identification of transcribed sequences in *Arabidopsis thaliana* by using high resolution genome tiling arrays. *Proc. Natl. Acad. Sci.* **102:** 4453.

Sun H. and Chasin L. 2000. Multiple splicing defects in an intronic false exon. *Mol. Cell. Biol.* **20:** 6414.

Tantravahi J., Alvira M., and Falck-Pedersen E. 1993. Characterization of the mouse β^maj-globin transcription termination region: A spacing sequence is required between the poly(A) signal sequence and multiple downstream termination elements. *Mol. Cel. Biol.* **13:** 578.

Újvári A. and Luse L.S. 2004. Newly initiated RNA encounters a factor involved in splicing immediately upon emerging from within RNA polymerase II. *J. Biol. Chem.* **279:** 49773.

Volpe T.A., Kidner C., Hall I.M., Teng G., Grewal S.I., and Martienssen R.A. 2002. Regulation of heterochromatic silencing and histone H3 lysine-9 methylation by RNAi. *Science* **297:** 1818.

Wagner E.J., Baraniak A.P., Sessions O.M., Mauger D., Moskowitz E., and Garcia-Blanco M.A. 2005. Characterization of the intronic splicing silencers flanking FGFR2 exon IIIb. *J. Biol. Chem.* **280:** 14017.

West S., Gromak N., and Proudfoot N.J. 2004. Human 5′→3′ exonuclease Xrn2 promotes transcription termination at cotranscriptional cleavage sites. *Nature* **432:** 522.

West S., Zaret K., and Proudfoot N.J. 2006a. Transcriptional termination sequences in the mouse serum albumin gene. *RNA* **12:** 655.

West S., Gromak N., Norbury C.J., and Proudfoot N.J. 2006b. Adenylation and exosome-mediated degradation of cotranscriptionally cleaved pre-messenger RNA in human cells. *Mol. Cell* **21:** 437.

Whitelaw E. and Proudfoot N. 1986. α-Thalassaemia caused by a poly(A) site mutation reveals that transcriptional termination is linked to 3′ end processing in the human α2 globin gene. *EMBO J.* **5:** 2915.

Wyers F., Rougemaile M., Badis G., Rouselle J.-C., Dufour M.-E., Boulay J., Régnault B., Devaux F., Namane A., Séraphin B., et al. 2005. Cryptic Pol II transcripts are degraded by a nuclear quality control pathway involving a new poly(A) polymerase. *Cell* **121:** 725.

Yonaha M. and Proudfoot N.J. 2000. Transcriptional termination and coupled polyadenylation in vitro. *EMBO J.* **19:** 3770.

How Does RNA Editing Affect dsRNA-mediated Gene Silencing?

B.L. BASS

Department of Biochemistry and Howard Hughes Medical Institute,
University of Utah, Salt Lake City, Utah 84112

In general, double-stranded RNA (dsRNA)-binding proteins (dsRBPs) are not sequence-specific. A dsRNA molecule in a cell will interact with any dsRBP it comes in contact with, suggesting that different dsRNA-mediated pathways intersect and affect each other. This paper analyzes evidence that the ADAR RNA editing enzymes, which act on dsRNA, affect dsRNA-mediated gene silencing pathways. Examples of how ADARs alter gene silencing pathways such as RNA interference, as well as mechanisms that allow the pathways to coexist and maintain their unique functions, are discussed.

Most of the functional groups in a double helix that allow a protein to recognize a specific base pair are in the major groove (Seeman et al. 1976), and there are many examples of proteins that enter the major groove of B-form DNA to make sequence-specific interactions (see, e.g., Wolberger 1999). However, double-stranded RNA (dsRNA) adopts an A-form helical structure with a major groove that is more narrow than that of DNA. It is difficult for a protein to enter the major groove of dsRNA to make sequence-specific contacts, and in fact, all characterized dsRNA-binding proteins (dsRBPs) bind to dsRNA of any sequence. Consistent with this, structural analyses of protein–dsRNA complexes show interactions that are predominantly in the minor groove, with major groove interactions limited to those on the exterior, involving the phosphodiester backbone (Ryter and Schultz 1998; Ramos et al. 2000; Wu et al. 2004).

Because dsRBPs are not sequence-specific, a dsRNA molecule in a cell will bind to any dsRBP it comes in contact with, and consequently, different dsRNA-mediated pathways affect each other. Researchers involved in early studies of RNA interference (RNAi) encountered this principle when they applied the RNAi protocols that worked well in flies and worms to mammalian cells (Elbashir et al. 2001). dsRNA introduced into mammalian cells encounters not only dsRBPs involved in RNAi, but also a dsRBP called PKR (Williams 2001). Upon binding to dsRNA, PKR becomes an active kinase and phosphorylates the translation initiation factor eIF2α, leading to a global shutdown of protein synthesis (Hunter et al. 1975). Introducing long dsRNA into a mammalian cell reduces expression of all mRNAs, not just the one targeted by RNAi.

Viruses capitalize on the fact that dsRBPs are not sequence-specific by synthesizing their own dsRBPs to antagonize cellular dsRNA-mediated pathways. For example, viral infection is sometimes accompanied by production of dsRNA from the viral genome (Boone et al. 1979; Maran and Mathews 1988), and to avoid the activation of PKR, viruses such as vaccinia and reovirus encode a dsRBP to compete with PKR for dsRNA (for review, see Stark et al. 1998). Not surprisingly, virus-encoded dsRBPs

and RNAs are also reported to antagonize the RNAi pathway (Li et al. 2004; Andersson et al. 2005).

The realization that dsRNA-mediated gene silencing is central to many biological processes invites the question as to how cellular dsRBPs affect this pathway. Studies are only beginning, but it seems possible that there is a finely tuned interplay between pathways involving dsRNA. This paper analyzes evidence that the dsRBPs known as adenosine deaminases that act on RNA (ADARs) affect dsRNA-mediated gene silencing.

ADAR BASICS

ADARs are RNA editing enzymes that convert adenosine (A) to inosine (I) in dsRNA (for review, see Bass 2002; Keegan et al. 2004; Valente and Nishikura 2005). The enzymes are found in all animals, where they are usually in the nucleus, and are most highly expressed in the nervous system. The A to I conversion involves a hydrolytic deamination, and, depending on the extent of base-pairing in an RNA substrate, the enzyme can selectively deaminate a specific adenosine, or nonselectively deaminate up to 50–60% of the adenosines within a double-stranded region. These two modes of selectivity can be observed in vitro with synthetic RNA, or in vivo, in the endogenous targets of ADARs (for review, see Bass 2002).

Selective deamination occurs in RNA that, although largely double-stranded, is frequently interrupted by mismatches, bulges, or loops. It is this "imperfect" helical structure that is associated with editing in codons and, as shown in Figure 1A, often forms by pairing between introns and exons. Inosine prefers to pair with cytidine, and thus, an A to I conversion can change the amino acid specified by a codon. Editing in codons results in the synthesis of multiple protein isoforms from a single encoded mRNA and serves as a mechanism of diversifying and increasing an organism's proteome. Deamination of codons typically involves the selective type of editing, which is not surprising, since nonselective editing would result in many amino acid changes and likely produce a nonfunctional protein.

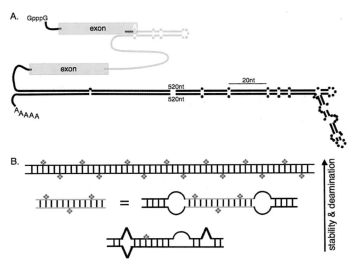

Figure 1. (*A*) The illustration depicts a pre-mRNA that contains examples of structures that are selectively (*upper*) and nonselectively (*lower*) edited. Coding exons are depicted in blue, an intron in gray, and 5′ and 3′ UTRs as black lines. Parallel lines represent base-paired regions with dots representing unpaired nucleotides. The selectively edited hairpin involves pairing between exon and intron sequences and is patterned after the R/G editing site, shown as a white A, found in certain mammalian glutamate receptor pre-mRNAs; the hairpin contains 28 bps, a loop, and mismatches as indicated (Aruscavage and Bass 2000). The nonselectively edited structure is patterned after the 3′UTR of the *C. elegans* gene, C35E7.6 (Morse et al. 2002). The edited structures are scaled relative to their actual lengths (20 nucleotides indicated), but 520 nucleotides of each strand of the 3′UTR structure were omitted as indicated. (*B*) The cartoon illustrates that the number of adenosines deaminated by an ADAR at reaction completion, or the selectivity of the enzyme, increases with the thermodynamic stability of the RNA structure. The top structure represents a long, completely base-paired dsRNA ≥ 50 bp which is deaminated nonselectively, showing 50–60% of its adenosines converted to inosines (*red diamonds*) at reaction completion. As indicated, structures that are less stable because they are shorter, or interrupted by mismatches, bulges, or loops, contain fewer inosines at the end of the reaction. Blue lines represent a specific sequence that exists as a separate molecule or between two internal loops of a longer structure.

The R/G hairpin that is edited at a specific adenosine in certain mammalian glutamate receptor pre-mRNAs is used to represent the selectively edited structure in Figure 1A (upper structure; R/G indicates that editing at the labeled A changes an arginine codon to a glycine codon). This structure is short, with a predicted free energy of –35 kcal/mole (mfold at 37°C; Mathews et al. 1999; Zuker 2003). Although selective editing usually occurs in structures that are less stable than those promoting nonselective editing, this is not always the case, as exemplified by editing of the antigenomic RNA of hepatitis delta virus (HDV; for review, see Casey 2006). The ~1700-nucleotide HDV RNA folds into a "rod-like" helical structure with a predicted free energy of –930 kcal/mol (mfold at 37°C). Despite its length and stability, the HDV RNA is edited at a specific adenosine to change an amber stop codon to a tryptophan codon (Polson et al. 1996, 1998). Importantly, although the HDV antigenomic RNA is highly base-paired, it is rare to find a complete helical turn within its structure that does not contain a mismatch, bulge, or loop. As discussed in the next section, this is key to its selective deamination.

Nonselective deamination occurs in RNA that is completely, or nearly completely, base-paired and often involves long helices containing hundreds of base pairs. In Figure 1A (lower hairpin), this type of structure is represented by a *Caenorhabditis elegans* 3′UTR which involves 1423 nucleotides that fold into a structure with a predicted free energy of –1201 kcal/mole (mfold at

20°C). This structure is heavily edited, with 42% of its 467 adenosines appearing as inosines in at least a fraction of the steady-state mRNA isolated from wild-type worms, and ~18% showing editing at all of these sites (Morse et al. 2002). As illustrated, although the nonselectively edited structures do contain mismatches, bulges, and loops, such disruptions are less frequent, and there are typically long stretches of uninterrupted, contiguous base pairs. In vivo, these types of substrates are usually found in noncoding regions of mRNAs, such as introns and untranslated regions, and are often formed by pairing between repeat elements (Morse and Bass 1999; Morse et al. 2002). Recent bioinformatic studies indicate that these types of substrates are remarkably abundant, occurring in an estimated 5% of the mRNAs encoded by our genomes (Athanasiadis et al. 2004; Kim et al. 2004; Levanon et al. 2004). As yet, the function of these structures, and the inosines within them, is unknown.

A MODEL FOR ADAR SELECTIVITY

The principles that control ADAR selectivity have been summarized previously (Bass 2002) and are briefly reviewed here using Figure 1B. As discussed above, a completely, or largely, base-paired duplex, greater than ~50 bp, is deaminated nonselectively, showing 50–60% of its adenosines deaminated at reaction completion. However, the reaction does stop, and the model is based on this observation. When AU base pairs are changed to

IU mismatches, the RNA structure becomes less stable, and the idea is that the reaction stops when there are so many IU mismatches that the RNA is too single-stranded in character to be acted on by an ADAR. Assuming there is some critical thermodynamic stability after which no further reaction can occur, it is easy to see that structures which are less stable to begin with will accommodate fewer deaminations before the reaction stops. Decreasing the length of an RNA helix, or disrupting it with mismatches, bulges, or loops, will decrease the number of deamination sites at reaction completion. At the extreme, internal loops can uncouple a long helix into a series of short helices that are each deaminated selectively (Lehmann and Bass 1999). In fact, as shown in Figure 1B, a short helix bounded by two internal loops will be deaminated at the same sites targeted when the helix exists as a separate molecule. The ability of loops to uncouple helices explains how a very long helix such as the HDV antigenomic RNA can be selectively deaminated.

DO ADARs AFFECT RNA INTERFERENCE?

Given that dsRBPs are not sequence-specific, it was proposed that ADARs might antagonize RNAi (Bass 2000), and support for this idea came from the observation that mutant *C. elegans* strains that lack ADARs exhibit aberrant gene silencing (Knight and Bass 2002). The mutant animals respond as expected when dsRNA is introduced by injection or feeding; that is, expression of the cognate mRNA is silenced at levels indistinguishable from wild type. Possibly this dsRNA never reaches the nucleus where ADARs usually reside. However, as illustrated in Figure 2, transgenic DNA, which is transcribed in the nucleus and normally expressed in a wild-type animal, is silenced in animals lacking ADARs. The silencing is dependent on factors important for the RNAi pathway, such as Dicer (Knight and Bass 2001), and is reminiscent of the cosuppression first observed in plants (for review, see Matzke and Matzke 2004). Although cosuppression of transgenic DNA is frequently observed in the germ line

of *C. elegans* (Kelly et al. 1997; for review, see Seydoux and Schedl 2001), the cosuppression observed in ADAR mutant animals is distinct because it is observed with transgenes expressed in the soma.

What is the mechanism of the somatic cosuppression observed in the ADAR mutants? In this regard, as shown in Figure 2, it is important to note that DNA introduced into *C. elegans* is covalently linked to form repetitive arrays, with genes in tandem, as well as inverted, orientation. Although in most cases transcription terminates normally, in the illustrated example giving rise to GFP mRNA, in some cases readthrough transcription of genes in an inverted orientation gives rise to dsRNA. Analysis of the wild-type animal illustrated in Figure 2 confirmed that both sense and antisense RNA were synthesized from the transgene. Because it prefers to pair with cytidine, inosine appears as guanosine in cDNA, and cDNA derived from the transgenic RNA showed the typical A to G changes indicative of editing (Knight and Bass 2002). Consistent with the fact that the sense and antisense RNAs would be completely base-paired, the RNA was nonselectively edited and contained many inosines. In contrast, but in line with the observed silencing, neither sense nor antisense RNA corresponding to the transgene sequence was detected in the ADAR mutant.

There are two obvious mechanisms by which ADARs could prevent an RNA from entering the RNAi pathway, and both may contribute to the observed silencing. First, a simple competition between ADARs and dsRBPs of the RNAi pathway might exist. Dicer and Drosha are both dsRBPs and contain the dsRNA-binding motif found in many dsRBPs (for review, see Doyle and Jantsch 2002). Furthermore, in many organisms, additional dsRBPs have been identified that facilitate the action of these enzymes (Tabara et al. 2002; Gregory et al. 2004; Han et al. 2004, 2006; Haase et al. 2005; Liu et al. 2006). In this scenario, ADARs would antagonize the RNAi pathway simply by binding dsRNA and sequestering it so that it is unavailable to dsRBPs involved in gene silencing. Of course, such a competition could occur between any dsRBP, not just ADARs, and indeed, there

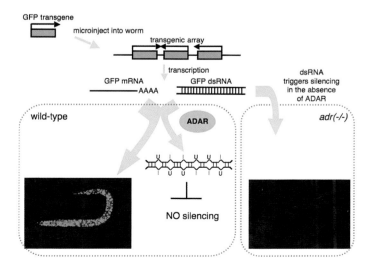

Figure 2. The cartoon illustrates the proposed pathway leading to transgene silencing in *C. elegans* strains that lack ADARs (*adr(–/–)*). Starting at the upper left corner, DNA encoding GFP (*green rectangle*) is injected into a worm, whereby it is covalently linked to form repetitive arrays that are maintained as extrachromosomal elements. In wild-type animals, GFP mRNA is expressed, as evidenced by strong GFP fluorescence (*bottom left corner*). Black arrows indicate the direction of transcription, and as shown, readthrough transcription of the repetitive array also gives rise to sense and antisense RNAs that hybridize to form dsRNA. In a wild-type animal, this dsRNA is deaminated by ADARs and thus does not lead to silencing. In animals lacking ADARs (*adr(-/-)*), the dsRNA enters the RNAi pathway, leading to silencing. Photographs of worms were taken by Jeff Habig, and strains are as described (Knight and Bass 2002).

may be a fine balance between the concentration of dsRNA and various dsRBPs.

Alternatively, the silencing phenotype of ADAR mutants may relate to the lack of editing, rather than, or in addition to, dsRNA binding. As shown in Figure 2, and discussed earlier, when ADARs edit dsRNA, AU base pairs are changed to IU mismatches, and the RNA becomes less double-stranded. Although it has not yet been demonstrated directly, by definition one would predict a dsRBP would bind less well to RNA after it has been modified by ADARs. According to this scenario, if ADARs deaminate a dsRNA before it enters the RNAi pathway, dsRBPs such as Dicer cannot bind the RNA and gene silencing cannot occur. Studies of RNA interference in vitro, using *Drosophila* extracts, support this idea. For example, if dsRNA is reacted with an ADAR before adding it to a *Drosophila* extract, it is less effective at triggering an RNAi response (Scadden and Smith 2001); production of siRNAs is reduced, and consequently, mRNA degradation is also reduced. When the dsRNA is highly edited (~50%), siRNAs are completely absent, but at intermediate amounts of editing, siRNAs are present. These authors, and others (Zamore et al. 2000), isolated siRNAs produced from edited dsRNA and found that inosine is present in the siRNA, albeit at levels lower than in the starting material, i.e., the unreacted, edited dsRNA. The latter data indicate that at intermediate amounts of editing Dicer can cleave edited dsRNA, possibly in regions that have fewer deamination sites and thus retain their overall double-stranded character.

Of course, the most interesting question in regard to the silencing observed in ADAR mutant animals is whether ADARs are involved in regulating dsRNA-mediated gene silencing of endogenous genes in vivo. So far, a specific example of this does not exist, which is not surprising, given that few genes that are natural targets of dsRNA-mediated gene silencing have been identified. Here, of course, I am not referring to genes whose expression is modulated by miRNA, of which there are many, but to systems more analogous to silencing that involves an antisense transcript, such as the *Stellate* genes in *Drosophila melanogaster* (Aravin et al. 2001, 2004), or cyclin E in *C. elegans* (Grishok and Sharp 2005). However, in support of the idea that ADARs do modulate silencing of endogenous genes, the chemotaxis defects of *C. elegans* that lack ADARs are rescued in strains that also lack components of the RNAi pathway (Tonkin and Bass 2003). Recent studies show that antisense transcription is abundant in mammalian genomes (Lehner et al. 2002; Yelin et al. 2003; Cawley et al. 2004; Chen et al. 2004; for review, see Chen et al. 2005), and it is intriguing to imagine that ADARs play a role in attenuating silencing that might result from dsRNA formed with this antisense.

When considering how ADARs might regulate silencing, it is important to note that the enzymes are very sensitive to substrate inhibition (Hough and Bass 1994). Thus, for a case where ADARs were allowing expression of a gene by deaminating its cognate dsRNA, simply increasing the levels of dsRNA would inhibit ADARs and allow silencing to take over. The inhibiting dsRNA could be generated *in cis* from the gene being silenced. Alternatively, since ADARs bind dsRNA of any sequence, a dsRNA synthesized *in trans*, at a different locus, could also inhibit ADARs and provide another way to regulate dsRNA-mediated silencing.

DO ADARs TARGET siRNA AND miRNA?

As mentioned above, long dsRNA that contains intermediate amounts of inosine is cleaved by Dicer to produce siRNAs that contain low amounts of inosine. However, because it is less stable than long dsRNA, very short dsRNA is a poor substrate for an ADAR (see prior discussion of selectivity). Thus, it seems probable that siRNA and miRNA duplexes are only rarely acted on directly by ADARs. However, it is intriguing to imagine scenarios by which ADARs might regulate the function of the slightly longer precursors of miRNAs. Again, regulation could occur because of a simple competition between ADARs and the dsRBPs required for the biogenesis and functions of small RNAs, or as a direct consequence of editing. In the latter case, an A to I change within the mature miRNA sequence would change its base-pairing properties and, thus, could actually change the mRNA that is targeted by the miRNA.

Although several studies provide "proof of principle" results, that is, they show that ADARs could in theory regulate the function of miRNAs, as yet, a definitive example of an ADAR regulating the function of an miRNA in vivo has not been found. However, it is increasingly clear that endogenous miRNA precursors are targeted by ADARs and contain inosine. Initial studies focused on specific pri-miRNAs, and although some of these were edited to a significant extent when incubated with ADARs in vitro (Yang et al. 2006) or in cells overexpressing ADARs (Luciano et al. 2004; Yang et al. 2006), only very low levels of editing were detected within endogenous miRNA precursors. For example, in 3–7% of the pri-miR-22 molecules isolated from human brain, lung, and testes, and mouse brain, at least one of the adenosines in the miRNA was edited, and in some cases, the inosines were within the mature miRNA sequence (Luciano et al. 2004). However, when considering a given adenosine within the population of pri-miR-22 molecules, the amount of editing at any particular site was extremely low, and furthermore, the editing sites were not conserved between human and mouse. Similarly, low levels of editing were observed in pri-miR-142 in mouse spleen (Yang et al. 2006).

Recently, a more systematic analysis of pri-miRNA sequences from a variety of human tissues was conducted. Six of 99 pri-miRNA sequences analyzed contained editing sites, and in contrast to the previous studies, for these 6 pri-miRNAs, specific adenosines were edited to significant levels (>10%) in a tissue-specific manner, in some cases approaching 100% editing at a single site (Blow et al. 2006). Whereas the lower levels of editing observed in the earlier studies may well represent RNA that accidentally got caught up with ADAR, the higher levels of editing in this study are more suggestive of functional editing.

Given the diverse ways metazoa regulate biological processes, it seems likely that, because ADARs *could* regulate the function of small RNAs, they probably do, and eventually a bona fide example will be discovered. In this regard, the proof-of-principle experiments are interesting because they illustrate potential mechanisms.

One idea is that editing would interfere with pri-mRNA or pre-mRNA processing, by Drosha and Dicer, respectively. Consistent with this idea, mature miR-142 levels were elevated two- to threefold in the spleen of ADAR null mice (Yang et al. 2006). This same study showed that, in vitro, pri-miRNAs containing inosine were less efficiently processed to pre-miRNAs by recombinant Drosha. This may explain why, despite the large number of mature miRNAs sequenced, no one has yet reported editing in a processed, mature miRNA sequence. Another possibility is that editing of pri- and pre-miRNAs leads to their degradation, possibly by the staphyloccoccal nuclease, Tudor-SN, a component of RISC (Caudy et al. 2003). Whereas some reports indicate that this enzyme is a non-sequence-specific nuclease that targets single-stranded RNA (Caudy et al. 2003), others observe that the enzyme stimulates cleavage of dsRNA that has three or four contiguous IU (or UI) base pairs (Scadden 2005). In fact, when HEK293 cells are transfected with a plasmid encoding pri-miR142, along with an inhibitor of Tudor-SN, an increase in edited pri-miRNA is observed (Yang et al. 2006).

Finally, another study suggests that certain ADARs may antagonize RNAi by binding to siRNA in the cytoplasm (Yang et al. 2005). Using recombinant mammalian ADARs, binding to siRNA in vitro was studied using gel-shift analyses as well as filter-binding. The interferon-inducible ADAR1p150, the only ADAR known to be present in the cytoplasm, had the highest affinity for siRNA, but ADAR1p110 and ADAR2 also bound tightly. Consistent with the idea that the ADAR1p150 antagonizes RNAi in vivo, the efficacy of RNAi in mouse embryonic fibroblasts null for ADAR1 was threefold higher than in wild-type cells or cells null for ADAR2.

YET, DsRNA-MEDIATED PATHWAYS COEXIST

Although this paper has focused on the antagonism between dsRNA-mediated pathways, it is important to note that these pathways have also evolved ways to minimize this interplay and protect their unique functions. Examples of this have been mentioned throughout this paper, and in closing, the topic will be specifically addressed.

As for all pathways in eukaryotes, one way to minimize antagonism between dsRNA-mediated pathways is to simply sequester the relevant enzymes into different cellular compartments. In the presence of interferon, a cytoplasmic form of an ADAR is induced (George and Samuel 1999), but under normal conditions, these enzymes are sequestered in the nucleus, away from PKR and Dicer, which are largely cytoplasmic (Jimenez-Garcia et al. 1993; Billy et al. 2001). As mentioned earlier, this differential compartmentalization explains why *C. elegans* that lack ADARs exhibit aberrant dsRNA-mediated silencing only when dsRNA is synthesized in

the nucleus. However, like ADARs, Drosha is in the nucleus (Lee et al. 2002), where it processes pri-miRNAs to pre-miRNAs. As discussed above, some miRNA precursors contain inosines indicating that, albeit infrequently, these molecules do interact with ADARs. But why don't more miRNA precursors contain inosine? Why aren't ADAR substrates such as the mRNAs with long, structured UTRs cleaved by Drosha in the nucleus, or Dicer in the cytoplasm, where they must be translated? Or are they?

Definitive answers to these questions await future research. However, predictions can be made by considering that, although dsRBPs are not sequence-specific, they exhibit what is sometimes called a structural specificity. Structural specificity likely plays an important role in allowing dsRBPs to coexist and carry out their unique functions. As gleaned from the earlier discussion of ADAR selectivity, an understanding of dsRBP specificity is based in the obvious: dsRBPs are just that—double-stranded RNA-binding proteins. They prefer to bind dsRNA, and although they don't recognize sequence, they recognize any structural alteration to the A-form helix. The length of a helix, whether the terminus of the helix is blunt or frayed, and whether it contains mismatches, bulges, or loops, all affect how a dsRBP recognizes a substrate. We currently have only limited information linking these observations to the actual interactions that occur between a dsRBP and its substrate. It seems likely that in some cases the structural features of the RNA will affect the physical interactions it has with a dsRBP, but it is also important to consider that anything that makes an RNA substrate more single-stranded in character will shift the double-stranded $\rightleftarrows$ single-stranded equilibrium toward the single-stranded state and affect binding.

An example of structural specificity is illustrated by returning to the discussion started at the beginning of this paper, where it was noted that long dsRNA introduced into mammalian cells not only triggers RNAi, but also activates PKR, a dsRBP that phosphorylates eIF2α to shut down all translation. A solution to the problem of how to apply RNAi techniques to mammalian cells without activating PKR came when it was determined that a specific response could be incurred using a very short RNA: the siRNA that is an intermediate of the RNAi process (Elbashir et al. 2001). Presumably, specificity is achieved because PKR binds less well to short dsRNA than at least some proteins required for RNAi (Hunter et al. 1975; Minks et al. 1979; Manche et al. 1992). The *C. elegans* dsRBP, RDE-4, which acts with Dicer in dsRNA cleavage, is an RNAi factor that shows a preference for longer dsRNA (Parker et al. 2006), and in this case, cooperativity is proposed to allow the protein to preferentially bind long dsRNA. Like PKR, ADARs do not act efficiently with very short dsRNA, consistent with the idea that mature siRNA and miRNA duplexes are not direct targets of ADARs. Thus, length appears to be one way dsRBPs gain specificity.

miRNA precursors are slightly longer, and indeed, some of these molecules are edited by ADARs. However, many are not. Even under in vitro conditions optimized for the ADAR enzyme, only 4 of 8 pri-miRNAs tested

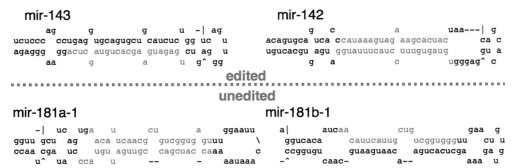

Figure 3. The lowest free-energy secondary structure is shown for four miRNA precursors, as predicted by mfold (Mathews et al. 1999; Zuker 2003). Nucleotides shown in red represent mature miRNA sequences as experimentally determined (Lagos-Quintana et al. 2002; Houbaviy et al. 2003; Poy et al. 2004). When incubated with recombinant ADAR, inosine can be detected within the two pri-miRNA sequences above the dotted line (edited), whereas inosine cannot be detected in those below the line (unedited; as reported in Yang et al. 2006).

showed inosine after incubation with recombinant ADAR (Yang et al. 2006). This may indicate that, in general, these hairpin structures have evolved so as to avoid deamination by ADARs. Figure 3 compares the predicted structures of 4 pri-mRNAs tested for their ability to be deaminated in vitro by recombinant ADAR (Yang et al. 2006). Consistent with the model of ADAR selectivity presented earlier, the predicted structures of the unedited pri-miRNAs contain more unpaired nucleotides and are less stable than those that are edited.

Although studies of the specificity of dsRBPs involved in RNAi are just beginning, it is already apparent that these dsRBPs will follow similar principles. Dicer requires a helix with a blunt terminus (Zhang et al. 2002; Vermeulen et al. 2005), whereas Drosha utilizes a frayed terminus (Han et al. 2006). Given this, it is predicted that the long structured UTRs of ADAR substrates would not be substrates for Dicer, but could be cleaved by Drosha. Similar to the ADARs, Dicer's activity is influenced by the extent of base-pairing in an RNA: Completely base-paired dsRNA is cleaved at many sites, whereas pre-miRNAs are specifically cleaved to generate the functional, mature miRNA. As evidenced from the data reviewed in this paper, dsRNA-mediated pathways clearly affect each other, but we have a long way to go in order to understand the subtleties of this interplay. Extending our current knowledge of how dsRBPs achieve specificity to a more molecular level, as well as to newly discovered dsRBPs, will no doubt lead to a more complete understanding of how dsRNA-mediated pathways intersect, yet maintain their unique functions.

ACKNOWLEDGMENTS

Thanks to members of the Bass lab for their hard work, constant questions, and excellent insight. A special thanks to lab members: to Jeff Habig, for photographs of *C. elegans*; and to Taraka Dale, Sabine Hellwig, and Heather Hundley for critically reading this manuscript. This work was supported by National Institutes of Health grants (GM067106 and GM44073). B.L.B. is an Investigator of the Howard Hughes Medical Institute.

REFERENCES

Andersson M.G., Haasnoot P.C., Xu N., Berenjian S., Berkhout B., and Akusjarvi G. 2005. Suppression of RNA interference by adenovirus virus-associated RNA. *J. Virol.* **79:** 9556.

Aravin A.A., Klenov M.S., Vagin V.V., Bantignies F., Cavalli G., and Gvozdev V.A. 2004. Dissection of a natural RNA silencing process in the *Drosophila melanogaster* germ line. *Mol. Cell. Biol.* **24:** 6742.

Aravin A.A., Naumova N.M., Tulin A.V., Vagin V.V., Rozovsky Y.M., and Gvozdev V.A. 2001. Double-stranded RNA-mediated silencing of genomic tandem repeats and transposable elements in the *D. melanogaster* germline. *Curr. Biol.* **11:** 1017.

Aruscavage P.J. and Bass B.L. 2000. A phylogenetic analysis reveals an unusual sequence conservation within introns involved in RNA editing. *RNA* **6:** 257.

Athanasiadis A., Rich A., and Maas S. 2004. Widespread A-to-I RNA editing of Alu-containing mRNAs in the human transcriptome. *PLoS Biol.* **2:** e391.

Bass B.L. 2000. Double-stranded RNA as a template for gene silencing. *Cell* **101:** 235.

———. 2002. RNA editing by adenosine deaminases that act on RNA. *Annu. Rev. Biochem.* **71:** 817.

Billy E., Brondani V., Zhang H., Muller U., and Filipowicz W. 2001. Specific interference with gene expression induced by long, double-stranded RNA in mouse embryonal teratocarcinoma cell lines. *Proc. Natl. Acad. Sci.* **98:** 14428.

Blow M.J., Grocock R.J., van Dongen S., Enright A.J., Dicks E., Futreal P.A., Wooster R., and Stratton M.R. 2006. RNA editing of human microRNAs. *Genome Biol.* **7:** R27.

Boone R.F., Parr R.P., and Moss B. 1979. Intermolecular duplexes formed from polyadenylylated vaccinia virus RNA. *J. Virol.* **30:** 365.

Casey J.L. 2006. RNA editing in hepatitis delta virus. *Curr. Top. Microbiol. Immunol.* **307:** 67.

Caudy A.A., Ketting R.F., Hammond S.M., Denli A.M., Bathoorn A.M., Tops B.B., Silva J.M., Myers M.M., Hannon G.J., and Plasterk R.H. 2003. A micrococcal nuclease homologue in RNAi effector complexes. *Nature* **425:** 411.

Cawley S., Bekiranov S., Ng H.H., Kapranov P., Sekinger E.A., Kampa D., Piccolboni A., Sementchenko V., Cheng J., Williams A.J., et al. 2004. Unbiased mapping of transcription factor binding sites along human chromosomes 21 and 22 points to widespread regulation of noncoding RNAs. *Cell* **116:** 499.

Chen J., Sun M., Hurst L.D., Carmichael G.G., and Rowley J.D. 2005. Genome-wide analysis of coordinate expression and evolution of human cis-encoded sense-antisense transcripts. *Trends Genet.* **21:** 326.

Chen J., Sun M., Kent W.J., Huang X., Xie H., Wang W., Zhou G., Shi R.Z., and Rowley J.D. 2004. Over 20% of human transcripts might form sense-antisense pairs. *Nucleic Acids Res.* **32:** 4812.

Doyle M. and Jantsch M.F. 2002. New and old roles of the double-stranded RNA-binding domain. *J. Struct. Biol.* **140:** 147.

Elbashir S.M., Harborth J., Lendeckel W., Yalcin A., Weber K., and Tuschl T. 2001. Duplexes of 21-nucleotide RNAs mediate RNA interference in cultured mammalian cells. *Nature* **411:** 494.

George C.X. and Samuel C.E. 1999. Human RNA-specific adenosine deaminase ADAR1 transcripts possess alternative exon 1 structures that initiate from different promoters, one constitutively active and the other interferon inducible. *Proc. Natl. Acad. Sci.* **96:** 4621.

Gregory R.I., Yan K.P., Amuthan G., Chendrimada T., Doratotaj B., Cooch N., and Shiekhattar R. 2004. The Microprocessor complex mediates the genesis of microRNAs. *Nature* **432:** 235.

Grishok A. and Sharp P.A. 2005. Negative regulation of nuclear divisions in *Caenorhabditis elegans* by retinoblastoma and RNA interference-related genes. *Proc. Natl. Acad. Sci.* **102:** 17360.

Haase A.D., Jaskiewicz L., Zhang H., Laine S., Sack R., Gatignol A., and Filipowicz W. 2005. TRBP, a regulator of cellular PKR and HIV-1 virus expression, interacts with Dicer and functions in RNA silencing. *EMBO Rep.* **6:** 961.

Han J., Lee Y., Yeom K.H., Kim Y.K., Jin H., and Kim V.N. 2004. The Drosha-DGCR8 complex in primary microRNA processing. *Genes Dev.* **18:** 3016.

Han J., Lee Y., Yeom K.H., Nam J.W., Heo I., Rhee J.K., Sohn S.Y., Cho Y., Zhang B.T., and Kim V.N. 2006. Molecular basis for the recognition of primary microRNAs by the Drosha-DGCR8 complex. *Cell* **125:** 887.

Houbaviy H.B., Murray M.F., and Sharp P.A. 2003. Embryonic stem cell-specific MicroRNAs. *Dev. Cell* **5:** 351.

Hough R.F. and Bass B.L. 1994. Purification of the *Xenopus laevis* double-stranded RNA adenosine deaminase. *J. Biol. Chem.* **269:** 9933.

Hunter T., Hunt T., Jackson R.J., and Robertson H.D. 1975. The characteristics of inhibition of protein synthesis by double-stranded ribonucleic acid in reticulocyte lysates. *J. Biol. Chem.* **250:** 409.

Jimenez-Garcia L.F., Green S.R., Mathews M.B., and Spector D.L. 1993. Organization of the double-stranded RNA-activated protein kinase DAI and virus-associated VA RNA$_I$ in adenovirus-2-infected HeLa cells. *J. Cell Sci.* **106:** 11.

Keegan L.P., Leroy A., Sproul D., and O'Connell M.A. 2004. Adenosine deaminases acting on RNA (ADARs): RNA-editing enzymes. *Genome Biol.* **5:** 209.

Kelly W.G., Xu S., Montgomery M.K., and Fire A. 1997. Distinct requirements for somatic and germline expression of a generally expressed *Caenorhabditis elegans* gene. *Genetics* **146:** 227.

Kim D.D., Kim T.T., Walsh T., Kobayashi Y., Matise T.C., Buyske S., and Gabriel A. 2004. Widespread RNA editing of embedded alu elements in the human transcriptome. *Genome Res.* **14:** 1719.

Knight S.W. and Bass B.L. 2001. A role for the RNase III enzyme DCR-1 in RNA interference and germ line development in *Caenorhabditis elegans*. *Science* **293:** 2269.

———. 2002. The role of RNA editing by ADARs in RNAi. *Mol. Cell* **10:** 809.

Lagos-Quintana M., Rauhut R., Yalcin A., Meyer J., Lendeckel W., and Tuschl T. 2002. Identification of tissue-specific microRNAs from mouse. *Curr. Biol.* **12:** 735.

Lee Y., Jeon K., Lee J.T., Kim S., and Kim V.N. 2002. MicroRNA maturation: Stepwise processing and subcellular localization. *EMBO J.* **21:** 4663.

Lehmann K.A. and Bass B.L. 1999. The importance of internal loops within RNA substrates of ADAR1. *J. Mol. Biol.* **291:** 1.

Lehner B., Williams G., Campbell R.D., and Sanderson C.M. 2002. Antisense transcripts in the human genome. *Trends Genet.* **18:** 63.

Levanon E.Y., Eisenberg E., Yelin R., Nemzer S., Hallegger M., Shemesh R., Fligelman Z.Y., Shoshan A., Pollock S.R., Sztybel D., et al. 2004. Systematic identification of abundant A-to-I editing sites in the human transcriptome. *Nat. Biotechnol.* **22:** 1001.

Li W.X., Li H., Lu R., Li F., Dus M., Atkinson P., Brydon E.W., Johnson K.L., Garcia-Sastre A., Ball L.A., et al. 2004. Interferon antagonist proteins of influenza and vaccinia viruses are suppressors of RNA silencing. *Proc. Natl. Acad. Sci.* **101:** 1350.

Liu X., Jiang F., Kalidas S., Smith D., and Liu Q. 2006. Dicer-2 and R2D2 coordinately bind siRNA to promote assembly of the siRISC complexes. *RNA* **12:** 1514.

Luciano D.J., Mirsky H., Vendetti N.J., and Maas S. 2004. RNA editing of a miRNA precursor. *RNA* **10:** 1174.

Manche L., Green S.R., Schmedt C., and Mathews M.B. 1992. Interactions between double-stranded RNA regulators and the protein kinase DAI. *Mol. Cell. Biol.* **12:** 5238.

Maran A. and Mathews M.B. 1988. Characterization of the double-stranded RNA implicated in the inhibition of protein synthesis in cells infected with a mutant adenovirus defective for VA RNA. *Virology* **164:** 106.

Mathews D.H., Sabina J., Zuker M., and Turner D.H. 1999. Expanded sequence dependence of thermodynamic parameters improves prediction of RNA secondary structure. *J. Mol. Biol.* **288:** 911.

Matzke M.A. and Matzke A.J. 2004. Planting the seeds of a new paradigm. *PLoS Biol.* **2:** E133.

Minks M.A., West D.K., Benvin S., and Baglioni C. 1979. Structural requirements of double-stranded RNA for the activation of 2′,5′-oligo(A) polymerase and protein kinase of interferon-treated HeLa cells. *J. Biol. Chem.* **254:** 10180.

Morse D.P. and Bass B.L. 1999. Long RNA hairpins that contain inosine are present in *Caenorhabditis elegans* poly(A)+ RNA. *Proc. Natl. Acad. Sci.* **96:** 6048.

Morse D.P., Aruscavage P.J., and Bass B.L. 2002. RNA hairpins in noncoding regions of human brain and *Caenorhabditis elegans* mRNA are edited by adenosine deaminases that act on RNA. *Proc. Natl. Acad. Sci.* **99:** 7906.

Parker G.S., Eckert D.M., and Bass B.L. 2006. RDE-4 preferentially binds long dsRNA and its dimerization is necessary for cleavage of dsRNA to siRNA. *RNA* **12:** 807.

Polson A.G., Bass B.L., and Casey J.L. 1996. RNA editing of hepatitis delta virus antigenome by dsRNA-adenosine deaminase. *Nature* **380:** 454.

Polson A.G., Ley H.L., III, Bass B.L., and Casey J.L. 1998. Hepatitis delta virus RNA editing is highly specific for the amber/W site and is suppressed by hepatitis delta antigen. *Mol. Cell. Biol.* **18:** 1919.

Poy M.N., Eliasson L., Krutzfeldt J., Kuwajima S., Ma X., Macdonald P.E., Pfeffer S., Tuschl T., Rajewsky N., Rorsman P., and Stoffel M. 2004. A pancreatic islet-specific microRNA regulates insulin secretion. *Nature* **432:** 226.

Ramos A., Grunert S., Adams J., Micklem D.R., Proctor M.R., Freund S., Bycroft M., St. Johnston D., and Varani G. 2000. RNA recognition by a Staufen double-stranded RNA-binding domain. *EMBO J.* **19:** 997.

Ryter J.M. and Schultz S.C. 1998. Molecular basis of double-stranded RNA-protein interactions: Structure of a dsRNA-binding domain complexed with dsRNA. *EMBO J.* **17:** 7505.

Scadden A.D. 2005. The RISC subunit Tudor-SN binds to hyper-edited double-stranded RNA and promotes its cleavage. *Nat. Struct. Mol. Biol.* **12:** 489.

Scadden A.D. and Smith C.W. 2001. RNAi is antagonized by A→I hyper-editing. *EMBO Rep.* **2:** 1107.

Seeman N.C., Rosenberg J.M., and Rich A. 1976. Sequence-specific recognition of double helical nucleic acids by proteins. *Proc. Natl. Acad. Sci.* **73:** 804.

Seydoux G. and Schedl T. 2001. The germline in *C. elegans*: Origins, proliferation, and silencing. *Int. Rev. Cytol.* **203:** 139.

Stark G.R., Kerr I.M., Williams B.R., Silverman R.H., and Schreiber R.D. 1998. How cells respond to interferons. *Annu. Rev. Biochem.* **67:** 227.

Tabara H., Yigit E., Siomi H., and Mello C.C. 2002. The dsRNA binding protein RDE-4 interacts with RDE-1, DCR-1, and a DExH-box helicase to direct RNAi in *C. elegans*. *Cell* **109:** 861.

Tonkin L.A. and Bass B.L. 2003. Mutations in RNAi rescue aberrant chemotaxis of ADAR mutants. *Science* **302:** 1725.

Valente L. and Nishikura K. 2005. ADAR gene family and A-to-I RNA editing: Diverse roles in posttranscriptional gene regulation. *Prog. Nucleic Acid Res. Mol. Biol.* **79:** 299.

Vermeulen A., Behlen L., Reynolds A., Wolfson A., Marshall W.S., Karpilow J., and Khvorova A. 2005. The contributions of dsRNA structure to Dicer specificity and efficiency. *RNA* **11:** 674.

Williams B.R. 2001. Signal integration via PKR. *Sci. STKE* **89:** RE2.

Wolberger C. 1999. Multiprotein-DNA complexes in transcriptional regulation. *Annu. Rev. Biophys. Biomol. Struct.* **28:** 29.

Wu H., Henras A., Chanfreau G., and Feigon J. 2004. Structural basis for recognition of the AGNN tetraloop RNA fold by the double-stranded RNA-binding domain of Rnt1p RNase III. *Proc. Natl. Acad. Sci.* **101:** 8307.

Yang W., Chendrimada T.P., Wang Q., Higuchi M., Seeburg P.H., Shiekhattar R., and Nishikura K. 2006. Modulation of microRNA processing and expression through RNA editing by ADAR deaminases. *Nat. Struct. Mol. Biol.* **13:** 13.

Yang W., Wang Q., Howell K.L., Lee J.T., Cho D.S., Murray J.M., and Nishikura K. 2005. ADAR1 RNA deaminase limits short interfering RNA efficacy in mammalian cells. *J. Biol. Chem.* **280:** 3946.

Yelin R., Dahary D., Sorek R., Levanon E.Y., Goldstein O., Shoshan A., Diber A., Biton S., Tamir Y., Khosravi R., et al. 2003. Widespread occurrence of antisense transcription in the human genome. *Nat. Biotechnol.* **21:** 379.

Zamore P.D., Tuschl T., Sharp P.A., and Bartel D.P. 2000. RNAi: Double-stranded RNA directs the ATP-dependent cleavage of mRNA at 21 to 23 nucleotide intervals. *Cell* **101:** 25.

Zhang H., Kolb F.A., Brondani V., Billy E., and Filipowicz W. 2002. Human Dicer preferentially cleaves dsRNAs at their termini without a requirement for ATP. *EMBO J.* **21:** 5875.

Zuker M. 2003. Mfold web server for nucleic acid folding and hybridization prediction. *Nucleic Acids Res.* **31:** 3406.

How a Small DNA Virus Uses dsRNA but Not RNAi to Regulate Its Life Cycle

R. Gu, Z. Zhang,* and G.G. Carmichael

*Department of Genetics and Developmental Biology, University of Connecticut Health Center,
Farmington, Connecticut 06030*

Mouse polyomavirus contains a circular DNA genome, with early and late genes transcribed from opposite strands. At early times after infection, genes encoded from the early transcription unit are predominantly expressed. After the onset of viral DNA replication, expression of genes encoded from the late transcription unit increases dramatically. At late times, late primary transcripts are inefficiently polyadenylated, leading to the generation of multigenomic RNAs that are precursors to mature mRNAs. These transcripts contain sequences complementary to the early RNAs and down-regulate early-strand gene expression by inducing RNA editing. Our recent work leads to a model where the production of the multigenomic late RNAs is also controlled by the editing of poly(A) signals, directed by overlapping primary transcripts.

Most DNA viruses exhibit striking temporal regulation of the expression of their genes. Viral genes involved in altering host functions or in regulating viral replication are expressed at early times, whereas virion structural proteins are only expressed at high levels after the onset of viral DNA replication. For some of these viruses (e.g., human adenoviruses and Simian Virus 40) (see Lewis and Manley 1985; Wiley et al. 1993), the temporal regulation of gene expression appears to be at the level of transcription initiation. Work in our laboratory has suggested that an interesting and fundamentally different type of regulation has an important role in the early-to-late switch in gene expression in cells infected with the murine polyomavirus.

Polyomavirus lytically infects mouse cells in tissue culture and is an apparently harmless passenger virus in wild mouse populations. In other rodents, it is tumorigenic and efficiently transforms rat or hamster cells in culture (Tooze 1980). Due to its small genome size and ease of manipulation, it provides us with a good model system for studying not only the molecular biology of cell transformation and tumorigenesis, but also the mechanisms of regulation of eukaryotic gene expression. We have been interested in understanding how viral RNA molecules are made and processed in infected cells. Late viral gene expression has a number of unusual features that have turned out to be useful for helping us to unravel fundamental aspects of RNA synthesis, processing, regulation, and mRNA transport from the nucleus.

GENOME ORGANIZATION

The polyoma genome is a circular DNA molecule of about 5300 bp. Our laboratory strain, 59RA, is 5327 bp (Ruley and Fried 1983). The genome is divided into "early" and "late" regions, which are expressed and regulated differently as infection proceeds (Fig. 1A) (Griffin and Fried 1975; Kamen et al. 1975, 1980a, b). The early and late transcription units extend in opposite directions around the circular genome from start sites near the unique, bidirectional origin of DNA replication (Crawford et al. 1974; Griffin and Fried 1975). Primary RNA products from the early transcription unit are alternatively spliced to yield three early mRNAs that code for the large T antigen (100 kD), the middle T antigen (56 kD), and the small T antigen (22 kD). Large T binds to sequences in or near the DNA replication origin region (Gaudray et al. 1981; Pomerantz et al. 1983; Cowie and Kamen 1984; Dilworth et al. 1984) and is involved in the initiation of DNA replication, indirectly in the autoregulation of early-strand RNA levels (Cogen 1978; Farmerie and Folk 1984; Liu and Carmichael 1993), and indirectly in the activation of high levels of expression from the late promoter (Cahill et al. 1990; Liu and Carmichael 1993). The other two early proteins are dispensable for lytic infection, but they are important for cell transformation. Late primary transcripts accumulate after the onset of DNA replication and are also spliced in alternative ways to give mRNAs that code for the three viral structural proteins VP1, VP2, and VP3.

TEMPORAL REGULATION OF GENE EXPRESSION

Gene expression during lytic infection of permissive mouse cells proceeds in a well-defined temporally regulated manner (Kamen et al. 1975; Beard et al. 1976; Piper 1979). Immediately after infection, RNA from the early transcription unit (E-RNA) begins to accumulate; however, RNA from the late transcription unit (L-RNA) accumulates more slowly. At 12–15 hours after infection, the early–late RNA ratio is about 4 to 1 (see Fig. 1B) (Kamen et al. 1975; Piper 1979; Hyde-DeRuyscher and Carmichael 1988) and in the presence

---

*Present address: Cold Spring Harbor Laboratory, Cold Spring Harbor, New York 11724.

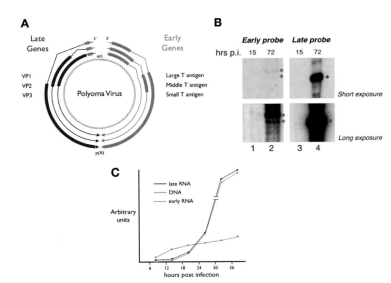

Figure 1. (*A*) The polyomavirus genome. Early and late genes are transcribed from opposite strands of the genome. (*Thick red lines*) The exons for the early genes (large, middle, and small T antigens); (*thick blue lines*) the exons for the late genes (virus capsid proteins VP1, VP2, and VP3). (*B*) An example of the early–late switch. Mouse 3T3 cells were infected with wild-type virus, and RNase protection assays were performed at 15 (an early time) and 72 (a late time) hours after infection, using probes specific for early-strand (*red stars*) and late-strand (*blue stars*) mRNAs (Hyde-DeRuyscher and Carmichael 1988). Note that at early times, late-strand RNAs are almost undetectable. (*C*) A hypothetical representation of a typical polyoma life cycle. Early-strand RNAs are seen early and accumulate somewhat at late times, but to far lower levels than late-strand RNAs, whose accumulation parallels the accumulation of viral DNA.

of DNA replication inhibitors, the ratio is 10 to 1 or higher. At 12–15 hours postinfection, viral DNA replication commences and L-RNA begins to accumulate rapidly, whereas E-RNA accumulates at a slower rate. Thus, there is a dramatic change in the relative abundances of E-RNA and L-RNA; by 24 hours postinfection, the early-to-late RNA ratio is as low as 1 to 50 (Kamen et al. 1975; Piper 1979; Hyde-DeRuyscher and Carmichael 1988). This early–late "switch" is dependent on viral DNA replication; if replication is inhibited, E-RNA accumulates to abnormally high levels with minimal accumulation of L-RNA (Cogen 1978; Heiser and Eckhart 1982; Kamen et al. 1982; Farmerie and Folk 1984; Hyde-DeRuyscher and Carmichael 1988). Figure 1B illustrates the early–late switch using an RNase protection assay, and Figure 1C presents an idealized depiction of it.

We have generated data that have uncovered unexpected mechanisms about how late genes are activated after the onset of viral DNA replication, as well as how early genes are down-regulated at late times. Furthermore, this work has led to new insights into the mechanism of action of antisense RNA in cells. It has

been commonly accepted in the field for many years that the early–late switch is the result of T antigen repression of the early promoter, coupled with a *trans*-activation of the late promoter. In fact, there is little experimental support for this notion, and it is not true. We have shown that this temporally regulated switch is *not* controlled mainly at the level of transcription initiation, but results from changes in transcription elongation and/or RNA stability (Hyde-DeRuyscher and Carmichael 1988, 1990; Liu and Carmichael 1993; Liu et al. 1994). At early times, late-strand transcripts are produced. However, these are inefficiently spliced and exported to the cytoplasm, and they have a short half-life in the nucleus (Fig. 2A) (Hyde-DeRuyscher and Carmichael 1988, 1990; Liu and Carmichael 1993; Liu et al. 1994). At late times, late nuclear RNAs are heterogeneous in size, the result of inefficient polyadenylation and transcription termination, and range from about 2.5 kb to more than 60 kb in length (Fig. 2B) (Acheson et al. 1971; Acheson 1976, 1978; Birg et al. 1977; Treisman 1980; Treisman and Kamen 1981). Most late RNA sequences never leave the nucleus as they are removed during mRNA processing and are subsequently degraded (Acheson 1976, 1984).

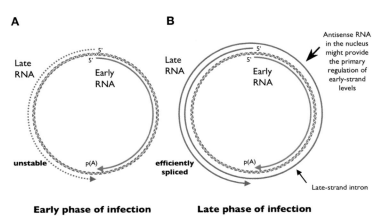

Early phase of infection   **Late phase of infection**

Figure 2. Our working model for the early–late switch. See text for details. (*A*) At early times, both strands are transcribed. However, late-strand RNAs are inefficiently spliced and are unstable in the nucleus. (*B*) After the onset of viral DNA replication, late polyadenylation/transcription termination is inefficient, leading to multigenomic primary transcripts. These are efficiently spliced, leading to mature late messages. However, the late pre-mRNAs contain intronic sequences complementary to early-strand RNAs, and this nuclear antisense RNA may down-regulate early-strand gene expression (Kumar and Carmichael 1997).

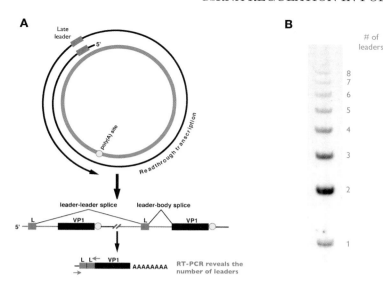

Figure 3. Processing of late pre-mRNAs. (*A*) Late pre-mRNAs contain tandem copies of the noncoding late leader exon. These exons are efficiently spliced to one another, leading to the further stabilization and processing of late mRNAs (Hyde-DeRuyscher and Carmichael 1988, 1990; Liu and Carmichael 1993). At the bottom is a diagram of late splicing. RT-PCR can be used to reveal leader-to-leader splicing. (*B*) RT-PCR assay showing that most late mRNAs have multiple tandem leader exons at their 5′ ends. Note that, consistent with the model of Fig. 2 and reported previously (Hyde-DeRuyscher and Carmichael 1990), there is a bias against single-leader mRNAs.

Late-strand pre-mRNA molecules are processed into mature mRNAs using a highly unusual pathway that involves inefficient polyadenylation and ordered splice site selection from precursors containing tandemly repeated introns and exons. Unlike early primary transcripts, late messages contain at their 5′ ends multiple tandem repeats of the 57-base noncoding late leader exon, which appears only once in the viral genome. Pre-mRNA molecules are processed by a pathway that includes the splicing of late leader exons to each other (Fig. 3A). We have shown that each class of late viral message (encoding virion structural proteins VP1, VP2, or VP3) consists of molecules with between 1 and 12 tandem leader units at their 5′ ends (Hyde-DeRuyscher and Carmichael 1990).

The life cycle of the virus appears to be connected to the processing of late pre-mRNAs, and this processing is in turn related to the inefficient use of the late polyadenylation signal. This can be seen in Figure 3B. Note that the great majority of late messages contain multiple tandem leaders at their 5′ ends. This reflects the inefficiency of late polyadenylation, which has been estimated to be about 50% each time RNA polymerase II traverses the poly(A) signal (Hyde-DeRuyscher and Carmichael 1990; Batt et al. 1994; Batt and Carmichael 1995). Interestingly, however, there is a clear bias against single-leader late mRNAs (Fig. 3B). This is consistent with the model in Figure 2A and the notion that in the absence of leader-to-leader splicing, late pre-mRNAs are relatively unstable in the nucleus. Thus, late mRNAs accumulate in response to leader-to-leader splicing, which is facilitated by inefficient polyadenylation/transcription termination. A major focus of our efforts in the recent past has been to understand how this regulation is controlled.

ANTISENSE REGULATION

An important contribution to the down-regulation of early RNA levels at late times in infection comes from the multigenomic late-strand transcripts. These RNAs are antisense to the early transcripts and may anneal with them in the nucleus, forming double-stranded RNA (dsRNA). We have developed several lines of evidence pointing to the importance of antisense regulation to the polyoma life cycle. First, mutants that express lower levels of giant late-strand RNAs always exhibit reciprocally increased early RNA levels (Adami et al. 1989; Liu and Carmichael 1993). Second, dsRNA formation in the nucleus would be expected to lead to editing by the ADAR enzyme. In the nucleus, adenosine residues in dsRNAs can be edited by ADAR to inosines by a process of hydrolytic deamination (Bass 2002). We have shown that this indeed happens in a polyomavirus infection. At late times, early-strand RNAs are extensively edited, with many transcripts exhibiting about 50% of their adenosines converted to inosines. Furthermore, these promiscuously edited RNAs are quantitatively retained in the nucleus and therefore are not translated into mutant proteins in the cytoplasm (Kumar and Carmichael 1997). Thus, RNA editing dramatically reduces the amount of cytoplasmic translatable early-strand mRNAs at late times. On the other hand, edited late-strand sequences lie within the large intron that is removed and degraded in the nucleus, so that editing does not directly affect late-strand mRNAs.

WHAT IS THE SWITCH THAT ACTIVATES LATE-STRAND SPLICING?

The polyoma early–late switch is clearly connected to the inefficient use of the late poly(A) signal. But what is the basis of this inefficiency? We have carried out many experiments to uncover the *cis*-acting sequence(s) in the polyoma late region that confers the temporal regulation of gene expression (Barrett et al. 1991; Liu and Carmichael 1993; Batt and Carmichael 1995; Huang and Carmichael 1996). Curiously, experiments aiming to uncover the "culprit" sequences that regulate the change in late polyadenylation efficiency between early and late times in infection have until recently proved frustrating. The results are summarized in Figure 4. There actually

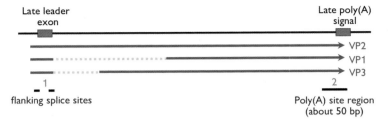

Late leader
exon

Late poly(A)
signal

VP2
VP1
VP3

1

flanking splice sites

2

Poly(A) site region
(about 50 bp)

All other regions have been replaced, with no effects on E/L switch

Figure 4. Sequences that might affect the early–late switch have been narrowed to two regions. In previous work from our laboratory (Barrett et al. 1991; Liu and Carmichael 1993; Batt and Carmichael 1995; Huang and Carmichael 1996), all late region sequences except those indicated have been removed or replaced, with no effect on the early–late switch. The splicing signals flanking the late leader exon appear to be important, as well as the poly(A) region.

appear to be no clear sequences within the late region of the virus that are essential for the early–late switch. In all of our studies, only two regions could be deleted without dramatic consequences on the switch. The first is the late leader region. Although the leader exon sequence itself is not important, its flanking splice sites must be functional (Adami et al. 1989). This most likely represents the need for leader-to-leader splicing to allow late RNA accumulation at late times.

The second region important for the early–late switch appears to include the poly(A) signal, which cannot be deleted. However, the late poly(A) region also includes elements essential for early-strand gene expression and thus indirectly for viral DNA replication. We next examined this element in greater detail. First, we asked whether the viral poly(A) signal itself is regulated. Our approach was to determine whether it can be replaced by another, unrelated element. We thus chose to mutationally inactivate the late polyadenylation signal while at the same time inserting a functional "synthetic" poly(A) signal (which obeys the known rules for the composition of an efficient processing signal; see Levitt et al. 1989) that contains some unique restriction sites. The resulting virus is called YZ-3 (Fig. 5). Note that this virus has wild-type open reading frames (ORFs) and differs from wild type only in the late poly(A) region. The early poly(A) signal is completely intact. The mutagenesis involved in the

construction of YZ-3 was difficult, as the early and late viral poly(A) signals overlap and the ORFs for viral proteins extend to almost the very end of the mRNAs (Fig. 5A). Interestingly, mutant YZ-3 is viable and grows with kinetics indistinguishable from those of our wild-type virus (Z. Liu and G. Carmichael, unpubl.). This mutant exhibits a normal early–late switch when assayed by either RNase protection assays (Fig. 5B) or reverse transcriptase–polymerase chain reaction (RT-PCR) for leader-to-leader splicing (Fig. 5C). The conclusion from these results is that the normal viral late polyadenylation signal appears to be dispensable for viral growth and for the temporal regulation of early and late gene expression. This led to the conclusion that there appears to be no identifiable element in the viral late region that contributes essentially to the early–late switch.

A NEW MODEL FOR THE EARLY–LATE SWITCH

The above results appeared to rule out the late poly(A) signal itself in the regulation of the early–late switch. However, in this region, we noticed an interesting and previously unappreciated feature of the organization of the viral genome. The early and late polyadenylation signals actually *overlap*, with the primary transcripts potentially overlapping by at least 45 bp (Fig. 6A). This is the case

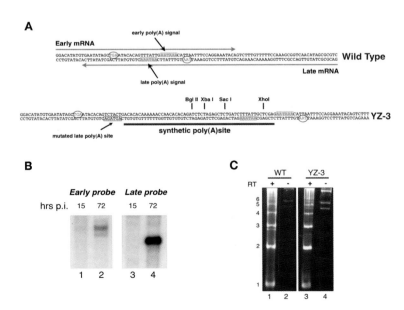

Figure 5. (*A*) Construction of a mutant virus in which the late polyadenylation signal has been replaced with a synthetic one. Mutant YZ-3 was made as described in several steps. First, a fragment spanning the polyoma wild-type late polyadenylation region was cloned into phage M13mp19. A 100-nucleotide-long synthetic DNA oligonucleotide was then used to simultaneously mutate the viral poly(A) signal and to insert a synthetic poly(A) signal containing several unique restriction enzyme cleavage sites. Following mutagenesis and confirmation by DNA sequencing, the cloned fragment was excised and reinserted into a plasmid containing the entire polyoma genome but lacking this fragment. The resulting viral genome was excised from the plasmid, recircularized by dilute ligation and transfected into mouse NIH-3T3 cells. Virus was harvested and propagated by standard methods. YZ-3 virus grows like wild-type polyoma (data not shown). (*B*) RNase protection assays as in Fig. 1B show that YZ-3 has a normal early–late switch. (*C*) RT-PCR shows that YZ-3 generates multiple tandem leaders on late mRNAs, just like wild type.

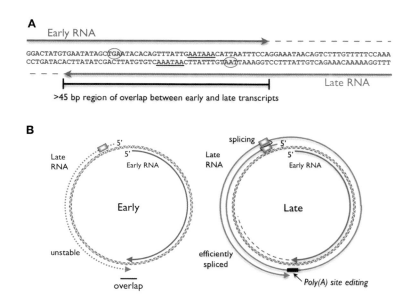

Figure 6. (*A*) The polyoma early and late polyadenylation signals overlap, with the potential for early-strand and late-strand transcripts to anneal with one another over a region of at least 45 nucleotides. (*Dotted lines*) Primary transcripts that most likely extend the region of complementarity by an even greater distance. (*B*) A new model for the early–late switch, in which poly(A)-site editing has a key role. At early times in infection, very little transcript overlap occurs, allowing early mRNAs to preferentially accumulate. At late times, overlapping transcripts become edited in the polyadenylation region, leading to a failure of the 3′-processing machinery to recognize the signals. If the late poly(A) signal is edited, transcription proceeds around the genome, with leader–leader splicing removing the large edited intron and resulting in accumulation of late mRNAs. Productive polyadenylation is thus in competition with editing. At the same time, the early poly(A) signal may also become edited. In this case, however, there is no splicing event that can remove the edited sequences, so editing leads to downregulation of gene expression.

for both wild type and YZ-3. Could this overlap be significant? Could overlapping poly(A) signals be regulated by editing? Given what we had already learned about the editing that occurs in the late phase, and that adenosines which are preferred editing targets (those preceded by A's or U's) (Polson and Bass 1994; Kumar and Carmichael 1997) are very rich in this overlap region, it seemed reasonable to hypothesize that early and late strands which overlap at their 3′ ends might serve as substrates for editing that is targeted to the polyadenylation signals. In fact, owing to the known nearest-neighbor preferences for ADAR1 editing, the sequence AAUAAA is almost perfectly suited for editing. In this model, at early times in infection, only very low levels of complementary early-strand and late-strand transcripts would be present in the nucleus, and the vast majority would not anneal with each other in their 3′ regions of complementarity. Early mRNAs would accumulate, and late pre-mRNAs would be degraded. However, after the onset of viral DNA replication, either a critical threshold concentration of sense–antisense RNAs might be reached or conditions might develop so as to facilitate annealing. Duplex regions would be edited by ADAR. This editing would presumably occur in the poly(A) regions of both early-strand and late-strand primary transcripts.

What would be the consequence of poly(A)-site editing for the viral life cycle? Interestingly, such editing might be *completely* consistent with all known results reported so far; a model for the regulation of the polyoma early–late switch by poly(A)-site editing is presented in Figure 6B. If the late poly(A) signal were to become edited at late times, it would not be recognized by the 3′-end processing machinery, and this would in turn lead to transcriptional readthrough, as is observed. Late-strand readthrough transcripts would be processed by leader-to-leader splicing and leader-to-body splicing of pre-mRNAs that eventually become polyadenylated (see Fig. 3). Splicing would salvage these RNAs and generate functional messengers

while at the same time removing the edited portions, which lay within introns. On the other hand, if the early-strand poly(A) signal were edited, early-strand readthrough would also occur, but in this case, the aberrant transcripts could not be resolved by splicing as is the case for late-strand transcripts (there is no 5′-noncoding leader exon) and would therefore be degraded in the nucleus (Maquat and Carmichael 2001). In this scenario, editing and cleavage/ polyadenylation would be in competition with one another to control the viral life cycle. In addition, poly(A)-site editing could serve opposite functions, to increase late-strand gene expression and to decrease early-strand gene expression. As the infection proceeds and complementary transcript levels rise, editing might become increasingly prevalent, leading to increasingly higher amounts of late mRNAs but fewer and fewer early mRNAs. One prediction of this new model is that at late times, but not early times, it should be possible to detect direct evidence of editing of viral poly(A) sites. Figure 7 shows one experiment where we have used an RT-PCR approach to demonstrate poly(A)-site editing at late times in polyoma

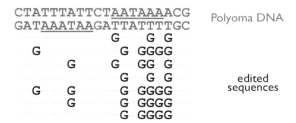

Figure 7. Poly(A) signals are efficiently edited. RT-PCR was used to amplify edited polyoma sequences as described previously (Kumar and Carmichael 1997), and seven edited sequences surrounding the polyadenylation signal (Norbury and Fried 1987) are shown. Note that the sequence AAUAAA is an exceptionally favorable target for ADAR, owing to the well-known nearest-neighbor preferences for editing (Polson and Bass 1994).

infection. In additional work to be presented elsewhere, we have carried out a number of additional experiments using other approaches to confirm the importance of poly(A)-site overlap for the viral life cycle.

CONCLUSIONS

Taken together, the experiments described above lead us to hypothesize that poly(A)-site overlap and editing is a key regulator of the polyoma early–late switch. If true, this surprising finding would identify poly(A)-site editing as a new form of gene regulation, one that has not been reported previously. Since there exist many nuclear transcripts whose 3′ ends overlap those of nearby genes expressed from the opposite DNA strand, it will be of interest to determine whether poly(A)-site editing might also contribute to the regulation of cellular gene expression.

ACKNOWLEDGMENTS

We are grateful to all members of the laboratory, past and present, who have contributed insights and comments throughout this work. In particular, we acknowledge the contributions of D. Batt, Z. Liu, R. Hyde-DeRuyscher, M. Kumar, and Y. Zhang, and the technical assistance of K. Morris. This work was supported by grants GM066816 and CA045382 from the National Institutes of Health.

REFERENCES

Acheson N. 1976. Transcription during productive infection with polyoma virus and SV40. *Cell* **8:** 1.
———. 1978. Polyoma giant RNAs contain tandem repeats of the nucleotide sequence of the entire viral genome. *Proc. Natl. Acad. Sci.* **75:** 4754.
———. 1984. Kinetics and efficiency of polyadenylation of late polyomavirus nuclear RNA: Generation of oligomeric polyadenylated RNAs and their processing into mRNA. *Mol. Cell. Biol.* **4:** 722.
Acheson N., Buetti E., Scherrer K., and Weil R. 1971. Transcription of the polyoma virus genome: Synthesis and cleavage of giant late polyoma specific RNA. *Proc. Natl. Acad. Sci.* **68:** 223.
Adami G.R., Marlor C.W., Barrett N.L., and Carmichael G.G. 1989. Leader-to-leader splicing is required for the efficient production and accumulation of polyomavirus late mRNA's. *J. Virol.* **63:** 85.
Barrett N.L., Carmichael G.G., and Luo Y. 1991. Splice site requirement for the efficient accumulation of polyoma virus late mRNAs. *Nucleic Acids Res.* **19:** 3011.
Bass B.L. 2002. RNA editing by adenosine deaminases that act on RNA. *Annu. Rev. Biochem.* **71:** 817.
Batt D.B. and Carmichael G.G. 1995. Characterization of the polyomavirus late polyadenylation signal. *Mol. Cell. Biol.* **15:** 4783.
Batt D.B., Luo Y., and Carmichael G.G. 1994. Polyadenylation and transcription termination in gene constructs containing multiple tandem polyadenylation signals. *Nucleic Acids Res.* **22:** 2811.
Beard P., Acheson N.H., and Maxwell I.H. 1976. Strand-specific transcription of polyoma virus DNA early in productive infection and in transformed cells. *J. Virol.* **17:** 20.
Birg F., Favaloro J., and Kamen R. 1977. Analysis of polyoma viral nuclear RNA by miniblot hybridization. *Proc. Natl. Acad. Sci.* **74:** 3138.
Cahill K.B., Roome A.J., and Carmichael G.G. 1990.

Replication-dependent transactivation of the polyomavirus late promoter. *J. Virol.* **64:** 992.
Cogen B. 1978. Virus-specific early RNA in 3T6 cells infected by a tsA mutant of polyoma virus. *Virology* **85:** 222.
Cowie A. and Kamen R. 1984. Multiple binding sites for polyomavirus large T antigen within regulatory sequences of polyomavirus DNA. *J. Virol.* **52:** 750.
Crawford L., Robbins A., and Nicklin P. 1974. Location of the origin and terminus of replication in polyoma virus DNA. *J. Gen. Virol.* **25:** 133.
Dilworth S., Cowie A., Kamen R., and Griffin B. 1984. DNA binding activity of polyoma virus large tumor antigen. *Proc. Natl. Acad. Sci.* **81:** 1941.
Farmerie W.G. and Folk W.R. 1984. Regulation of polyoma virus transcription by large tumor antigen. *Proc. Natl. Acad. Sci.* **81:** 6919.
Gaudray P., Tyndall C., Kamen R., and Cuzin F. 1981. The high affinity binding site on polyoma virus DNA for the viral large-T protein. *Nucleic Acids Res.* **9:** 5697.
Griffin B. and Fried M. 1975. Amplification of a specific region of the polyoma virus genome. *Nature* **256:** 175.
Heiser W.C. and Eckhart W. 1982. Polyoma virus early and late mRNAs in productively infected mouse 3T6 cells. *J. Virol.* **44:** 175.
Huang Y. and Carmichael G.G. 1996. A suboptimal 5′ splice site is a *cis*-acting determinant of nuclear export of polyomavirus late mRNAs. *Mol. Cell. Biol.* **16:** 6046.
Hyde-DeRuyscher R.P. and Carmichael G.G. 1988. Polyomavirus early-late switch is not regulated at the level of transcription initiation and is associated with changes in RNA processing. *Proc. Natl. Acad. Sci.* **85:** 8993.
———. 1990. Polyomavirus late pre-mRNA processing: DNA-replication-associated changes in leader exon multiplicity suggest a role for leader-to-leader splicing in the early-late switch. *J. Virol.* **64:** 5823.
Kamen R., Favaloro J., and Parker J. 1980a. Topography of the three late mRNA's of polyoma virus which encode the virion proteins. *J. Virol.* **33:** 637.
Kamen R., Lindstrom D.M., Shure H., and Old R.W. 1975. Virus-specific RNA in cells productively infected or transformed by polyoma virus. *Cold Spring Harbor Symp. Quant. Biol.* **39:** 187.
Kamen R., Jat P., Treisman R., Favaloro J., and Folk W.R. 1982. 5′ Termini of polyoma virus early region transcripts synthesized in vivo by wild-type virus and viable deletion mutants. *J. Mol. Biol.* **159:** 189.
Kamen R., Favaloro J., Parker J., Treisman R., Lania L., Fried M., and Mellor A. 1980b. A comparison of polyomavirus transcription in productively infected mouse cells and transformed rodent cell lines. *Cold Spring Harbor Symp. Quant. Biol.* **44:** 189.
Kumar M. and Carmichael G.G. 1997. Nuclear antisense RNA induces extensive adenosine modifications and nuclear retention of target transcripts. *Proc. Natl. Acad. Sci.* **94:** 3542.
Levitt N., Briggs D., Gil A., and Proudfoot N.J. 1989. Definition of an efficient synthetic poly(A) site. *Genes Dev.* **3:** 1019.
Lewis E.D. and Manley J.L. 1985. Control of adenovirus late promoter expression in two human cell lines. *Mol. Cell. Biol.* **9:** 2433.
Liu Z. and Carmichael G.G. 1993. Polyoma virus early-late switch: Regulation of late RNA accumulation by DNA replication. *Proc. Natl. Acad. Sci.* **90:** 8494.
Liu Z., Batt D.B., and Carmichael G.G. 1994. Targeted nuclear antisense RNA mimics natural antisense-induced degradation of polyoma virus early RNA. *Proc. Natl. Acad. Sci.* **91:** 4258.
Maquat L.E. and Carmichael G.G. 2001. Quality control of mRNA function. *Cell* **104:** 173.
Norbury C.J. and Fried M. 1987. Polyomavirus early region alternative poly(A) site: 3′-end heterogeneity and altered splicing pattern. *J. Virol.* **61:** 3754.
Piper P. 1979. Polyoma virus transcription early during productive infection of mouse 3T6 cells. *J. Mol. Biol.* **131:** 399.

Polson A.G. and Bass B.L. 1994. Preferential selection of adenosines for modification by double-stranded RNA adenosine deaminase. *EMBO J.* **13:** 5701.

Pomerantz B.J., Mueller C.R., and Hassell J.A. 1983. Polyomavirus large T antigen binds independently to multiple, unique regions on the viral genome. *J. Virol.* **47:** 600.

Ruley E. and Fried M. 1983. Sequence repeats in a polyomavirus DNA region important for gene expression. *J. Virol.* **47:** 233.

Tooze J., ed. 1980. *Molecular biology of tumour viruses*, 2nd edition: *DNA tumour viruses*. Cold Spring Harbor Laboratory, Cold Spring Harbor, New York.

Treisman R. 1980. Characterization of polyoma late mRNA leader sequences by molecular cloning and DNA sequence analysis. *Nucleic Acids Res.* **8:** 4867.

Treisman R. and Kamen R. 1981. Structure of polyoma virus late nuclear RNA. *J. Mol. Biol.* **148:** 273.

Wiley S.R., Kraus R.J., Zuo F., Murray E.E., Loritz K., and Mertz J.E. 1993. SV40 early-to-late switch involves titration of cellular transcriptional repressors. *Genes Dev.* **7:** 2206.

Regulation of Two Key Nuclear Enzymatic Activities by the 7SK Small Nuclear RNA

W.-J. He,* R. Chen,† Z. Yang,* and Q. Zhou*

*Department of Molecular and Cell Biology, University of California, Berkeley, California 94720;
†School of Life Sciences, Xiamen University, Xiamen 361005, P.R.China

7SK is a highly conserved small nuclear RNA (snRNA) in vertebrates. Since its discovery in 1968, little had been known about its function until recently, when 7SK was found to associate with the general transcription elongation factor P-TEFb. Together with the HEXIM1 protein, 7SK sequesters P-TEFb into a kinase-inactive complex, where it mediates HEXIM1's inhibition of P-TEFb. This helps maintain P-TEFb in a functional equilibrium to control transcription, cell growth, and differentiation. Although highly abundant, only a small fraction of 7SK is P-TEFb-bound. Using affinity purification, we have identified APOBEC3C as another 7SK-associated protein. As a member of the APOBEC family that functions in diverse processes through deaminating cytosine in DNA, it is unclear how APOBEC3C's activity is controlled to prevent its mutations of genomic DNA. We show that most of APOBEC3C interact with about half of nuclear 7SK, which suppresses APOBEC3C's deaminase activity and sequesters APOBEC3C in the nucleolus where it could be at a safe distance from most genomic sequences. Because the DNA substrate-binding site in APOBEC3C differs from the region for 7SK binding, 7SK does not act as a substrate competitor in inhibiting APOBEC3C. The demonstration of 7SK's suppression of yet another enzyme besides P-TEFb suggests a general role for this RNA in regulating key nuclear functions.

The 331-nucleotide-long 7SK snRNA was first described almost 40 years ago (Weinberg and Penman 1968; Zieve and Penman 1976). In the ensuing years, this RNA was gradually known to be transcribed by RNA polymerase III, highly abundant ($\sim 2 \times 10^5$ molecules/cell) and evolutionarily conserved in vertebrates (Zieve and Penman 1976; Murphy et al. 1987; Wassarman and Steitz 1991). Despite the lack of a clear demonstration of its function for many years, its high abundance and evolutionary conservation have nevertheless led to the belief that this RNA must have an important role in cell function (Murphy et al. 1987; Wassarman and Steitz 1991).

The recent years have seen the involvement of many small noncoding RNAs in the control of virtually all known biological processes. The process of RNA polymerase II (pol II)-mediated eukaryotic transcription is no exception (Goodrich and Kugel 2006). Transcription can be subdivided into several stages. The elongation stage is known to be tightly regulated and important not only for the production of full-length mRNA transcripts, but also for the coupling of transcription with major pre-mRNA processing events (Sims et al. 2004). At the center of the elongation control is the positive transcription elongation factor b (P-TEFb), which stimulates the processivity of pol II elongation and antagonizes the effects of negative elongation factors (Price 2000; Barboric and Peterlin 2005). The predominant form of P-TEFb in many human cell types consists of CDK9 and its regulatory subunit cyclin T1 (CycT1) (Price 2000; Barboric and Peterlin 2005). It phosphorylates the negative elongation factors DSIF and NELF as well as the serine-2 residues within the heptapeptide repeats ($Y_1S_2P_3T_4S_5P_6S_7$) that constitute the carboxy-terminal domain (CTD) of the largest subunit of pol II (Price 2000; Sims et al. 2004). These phosphorylation events are crucial to transform pol II from the state required for promoter-proximal pausing and capping of nascent transcripts into that for productive elongation.

It is important to point out that the kinase activity of P-TEFb is essential for the expression of a vast array of protein-coding genes (Chao and Price 2001; Shim et al. 2002). As such, P-TEFb is considered to be a general transcription factor. However, transcription of a number of genes including those of human immunodeficiency virus type 1 (HIV-1) is particularly sensitive to P-TEFb availability. In fact, HIV-1 depends on P-TEFb to such an extent that it uses the virus-encoded Tat protein and TAR RNA element, a stem-loop structure formed at the 5′ end of the nascent viral transcript, to recruit host cellular P-TEFb to the paused pol II through forming a stable TAR–Tat–P-TEFb complex (Price 2000; Barboric and Peterlin 2005).

Besides transcription, another event that takes place within the same nuclear space is DNA editing. One particular type, cytosine deamination, introduces uracil into DNA. If left uncorrected, subsequent replications could lead to the transition of guanosine to adenosine in the daughter DNA. It had previously been thought that cytosine deamination occurs spontaneously in vivo (Lindahl 1993). However, the discovery of the APOBEC family of cytosine deaminases has made it necessary to reconsider the complexity of this phenomenon and its biological implications.

The human APOBEC family consists of APOBEC1, APOBEC2, AID, and APOBEC3A to 3G (Jarmuz et al. 2002; Conticello et al. 2005). Although APOBEC1 (the prototype of this family) was initially identified as an RNA editor (Wedekind et al. 2003), more recent findings indicate that APOBEC1, APOBEC3, 3G, and AID (activation-induced deaminsase), can all act as DNA mutators by converting dC into dU in an *Escherichia* coli-based mutation assay (Harris et al. 2002). Furthermore, the purified recombinant APOBEC1, AID, and APOBEC3G can trigger deamination of dC on single-stranded DNA (ssDNA) substrates in vitro (Harris et al. 2003; Petersen-Mahrt and Neuberger 2003; Morgan et al. 2004), thus shedding new light on the intrinsic deaminase activities of these enzymes toward DNA.

Further indications that members of the APOBEC superfamily can act as ssDNA-specific mutators came from the identification of APOBEC3B, 3C, 3F, and 3G as potent antiretroviral innate immunity factors (Harris and Liddament 2004; Yu et al. 2004; Langlois et al. 2005). Although APOBEC3G and 3F are known to strongly inhibit HIV-1 and simian immunodeficiency virus (SIV) replication through converting dC to dU in the nascent reverse-transcribed minus-strand proviral DNA (Harris and Liddament 2004; Yu et al. 2004), APOBEC3B and 3C appear to function as effective inhibitors of SIV replication (Yu et al. 2004). In addition, a weak anti-HIV effect of APOBEC3C has also been documented recently (Langlois et al. 2005).

Unlike the rest of the APOBEC superfamily, which displays largely tissue-restricted expression patterns, APOBEC3C (referred to as A3C from now on) has been detected in a wide variety of human tissues as well as tumor cell lines (Jarmuz et al. 2002; Wedekind et al. 2003; Harris and Liddament 2004). Given its widespread expression and intrinsic deaminase activity toward ssDNA which occurs frequently in diverse processes ranging from DNA replication to recombination repair, to transcription, it is important to investigate how cells may suppress the potentially deleterious genomic mutations caused by A3C while directing this enzyme to its proper physiological substrate(s) and/or location.

Having discussed 7SK, the P-TEFb-mediated transcriptional elongation, and A3C-dependent cytosine deamination, what then is the relationship among these three different elements? In this paper, we first briefly review recent progress toward the elucidation of the role of 7SK in suppressing P-TEFb kinase activity as well as maintaining P-TEFb in a functional equilibrium important for transcription and the cellular decision between growth and differentiation. Second, we report an unexpected link between A3C and 7SK, which turns out to have a critical role in suppressing A3C's deaminase activity and sequesters this protein in the nucleolus, where A3C could be at a safe distance from most genomic DNA. Together, the demonstration of a 7SK-mediated suppression of two distinct nuclear enzymatic activities suggests a general role for this abundant RNA in controlling key cellular functions.

MATERIALS AND METHODS

Plasmids, deoxyoligonucleotides, and antibodies. The human A3C cDNA was amplified by reverse transcriptase–polymerase chain reaction (RT-PCR) using total HeLa RNA with the forward primer 5′-gcgaattcaatgaatccacagatcagaaacccg-3′ and reverse primer 5′-gcggtcgactcactggagactctcccgtagccttcttttcagaagtcg-3′. The PCR product was cloned into pCMV2-FLAG

(Sigma) and pECFP-C1 (BD Biosciences) to generate pAPOBEC3C-FLAG and pECFP-APOBEC3C, respectively. All plasmids were verified by sequencing.

The sequence of the deoxyoligonucleotide used in the UDG-based deamination assay was 5′-$(ATTT)_5\underline{CCC}GGG(ATT)_4$-Biotin, in which internal dC residues serving as deamination targets are underlined. In the ssDNA-binding assay, the sequences of the two DNA oligonucleotides, w/dC and no dC, were 5′-TCCCACTCCATCCAGGTCATGTTATTC-CAAATCTGTTCCA-Biotin and 5′-TGGAAGA-GATTTGGAATAAGATGAGGTGGATGGAGTGGG-A-Biotin, respectively. Finally, rabbit polyclonal anti-A3C antibodies were generated against a synthetic peptide corresponding to the carboxy-terminal 16 amino acids of human A3C.

Immunoprecipitation. FLAG-tagged human A3C (F-A3C) and its associated 7SK were immunoprecipitated with anti-FLAG agarose beads (Sigma) from nuclear extracts of transfected HeLa cells. After extensive washes with buffer D (20 mM HEPES at pH 7.9, 15% glycerol, 0.2 mM EDTA, 1 mM dithiothreitol [DTT], 0.5 mM phenylmethylsulfonyl fluoride [PMSF]) containing 0.3 M KCl and 0.2% Nonidet P-40 (NP-40), the immune complexes were eluted with the FLAG peptide. For RNase treatment, incubation of RNase A (1 μg, Roche) with immune complexes attached to the anti-FLAG beads was performed for 15 minutes at room temperature. To wash 7SK RNA off the immunoprecipitated F-A3C by high salt, buffer D containing 0.8 M KCl was used.

Immunodepletion of CDK9 or A3C from HeLa nuclear extract. Rabbit polyclonal anti-CDK9 or anti-A3C (6 μl) antibodies were incubated with 100 μl of HeLa nuclear extract (NE) in the presence of 0.2% NP-40, 2 μg/ml bovine serum albumin (BSA), 350 mM NaCl, and 2.5 μl of RNasin (40 units/μl) for 30 minutes at 4°C. The mixtures were then incubated for 30 minutes at 4°C with 15 μl of packed protein A–Sepharose beads, which were preblocked with BSA. After separating the supernatant from the beads by quick spinning, the depletion procedure was repeated twice more with fresh protein-A beads.

Immunofluorescence microscopy. HeLa cells grown on glass coverslips were permeabilized in 0.1% Triton X-100 in phosphate-buffered saline (PBS) for 30 seconds at room temperature. The permeablized cells were incubated with or without RNase A (0.1 mg/ml in PBS) for 5 minutes at 37°C followed by washing once with PBS. The cells were then fixed in freshly prepared 4% paraformaldehyde-PBS for 20 minutes at room temperature, washed three times with PBS, and blocked with 10% normal goat serum in 2% BSA-PBS for 15 minutes at room temperature. All incubations with primary antibodies (2.2 μg/ml) were performed for 1 hour at room temperature in 2% BSA-PBS plus 10% normal goat serum. After washing with PBS, cells were incubated for 45 minutes with the Alex-conjugated secondary antibody in 2% BSA-PBS plus 10% normal goat serum. After washing with PBS, cells were subjected to microscopic analysis.

FISH analysis. HeLa cells grown on glass coverslips were permeablized for 3 minutes on ice with 0.5% Triton X-100 in CSK buffer (0.1 M NaCl, 0.3 M sucrose, 10 mM PIPES at pH 6.8, 3 mM MgCl$_2$, and 1 mM PMSF). Cells were then fixed with freshly prepared 4% paraformaldehyde-PBS for 20 minutes at room temperature, repermeabilized with 0.4% Triton X-100 in PBS for 10 minutes, and washed three times with PBS. Cells were then blocked with 1× Cassein solution (Vector Laboratories, California) for 1 hour at 37°C and Avidin solution containing 200 µl of Avidin (Vector Laboratories) per milliliter of 1× Cassein solution for another 30 minutes at 37°C. After rinsing once with 1× Cassein solution, cells were prehybridized in 2× SSC/5× Denhardt's solution containing 100 µg/ml yeast tRNA for 1 hour at 37°C and then hybridized overnight at 37°C in the same solution to the biotinylated 2′-OMe RNA oligonucleotide that is antisense to 7SK (Yang et al. 2001). After hybridization, cells were washed once with 50% formamide in 2× SSC, once with 2× SSC at 37°C, and then once with PBS for 10–15 minutes each at room temperature. Cells were then blocked with 1× Cassein solution for 30 minutes at 37°C, incubated with the mouse antinucleolin monoclonal antibody for 2 hours at room temperature, and washed with 0.1% Tween-20 in PBS three times for 10 minutes each. To detect protein and hybridization signals, cells were incubated with 2 µg/ml FITC-conjugated goat anti-mouse antibody for 30 minutes and then 2 µg/ml Texas-Red-conjugated Avidin DCS (Vector Laboratories) for 15 minutes in 1× Cassein solution at room temperature. After washing four times with 0.1% Tween-20–PBS for 10 minutes each at room temperature, cells were counterstained with 0.5 µg/ml DAPI for 3–5 minutes and then rinsed with PBS once. Cells were examined by confocal microscopy.

In vitro UDG-base deaminase assay. Affinity-purified A3C protein samples were incubated for 2 hours at 37°C in 10 µl of buffer R (40 mM Tris at pH 8.0, 40 mM KCl, 50 mM NaCl, 0.5 mM EDTA, 1 mM DTT, 10% glycerol) with 3′-biotinylated deoxyoligonucleotide (0.1 pmole), which was labeled at the 5′ end with [γ-^{32}P]ATP using T4 polynucleotide kinase. Reactions were terminated by heating to 90°C for 3 minutes, and the oligonucleotide was purified by incubating with streptavidin magnetic beads (New England BioLabs) followed by washing at 72°C in buffer W (0.5 M NaCl, 2 mM EDTA, and 20 mM Tris-HCl at pH 7.5). Cytosine deamination was monitored by incubating the immobilized oligonucleotide for 30 minutes at 37°C with uracil-DNA glycosylase (1 unit, New England BioLabs) and then bringing the sample to 0.15 M in NaOH and incubating for a further 30 minutes. The oligonucleotide was then subjected to electrophoresis in 16% polyacrylamide gel electrophoresis (PAGE)-urea gel that was dried and exposed to X-ray film.

ssDNA-binding assay. F-A3C and its associated 7SK were immunoprecipitated from NE of transfected HeLa cells and washed with buffer D containing either 0.3 or 0.8 M KCl before being eluted from the anti-FLAG beads. PAGE-purified, biotin-labeled DNA oligonucleotides as

described previously (Navarro et al. 2005) were immobilized by incubating with streptavidin-coated magnetic beads. The F-A3C immunoprecipitates were incubated with 1 µg of immobilized oligonucleotides in 20 µl of buffer R containing 2 µg of poly(dI-dC) and 40 µg of BSA for 15 minutes at 30°C. The immobilized oligonucleotides were harvested using a magnetic stand and washed twice with buffer D containing 0.3 M KCl. F-A3C retained on the immobilized oligonucleotides was released by SDS-PAGE sample buffer and detected by western blotting.

RESULTS

7SK Sequesters P-TEFb into an Inactive Complex Where the CDK9 Kinase Can Be Inhibited by HEXIM1

In 2001, 32 years after its initial discovery (Weinberg and Penman 1968), 7SK was rediscovered as a P-TEFb-associated nuclear factor (Nguyen et al. 2001; Yang et al. 2001). Subsequent analyses indicate that the 7SK–P-TEFb snRNP formed in vivo also contains another protein called HEXIM1 (Michels et al. 2003, 2004; Yik et al. 2003). This protein had previously been identified as a nuclear protein whose expression could be efficiently induced when human vascular smooth-muscle cells were treated with hexamethylene bisacetamide (HMBA) (Ouchida et al. 2003), a hybrid bipolar compound known for its ability to suppress cell growth and induce terminal differentiation in many cell types (Marks et al. 1994). The current understanding based on reports from several laboratories (Michels et al. 2003; Yik et al. 2003, 2005; Blazek et al. 2005; Dulac et al. 2005; Li et al. 2005) is that the 7SK–HEXIM1–P-TEFb snRNP contains one molecule of 7SK and probably two copies each of CDK9, CycT1, and HEXIM1 (Fig. 1). Within this snRNP, HEXIM1 serves as a CDK kinase inhibitor (CKI) specific for P-TEFb (Yik et al. 2003; Michels et al. 2004). However, HEXIM1's interaction with P-TEFb and inactivation of the CDK9 kinase, mediated mainly through the carboxy-terminal half of HEXIM1, are very inefficient in the absence of 7SK (Yik et al. 2003; Michels et al. 2004). Since 7SK is able to bind both HEXIM1 and P-TEFb (Fig. 1), it is postulated to function as a molecular scaffold to mediate the HEXIM1–P-TEFb interaction (Yik et al. 2003; Michels et al. 2004).

The 7SK-binding motif in HEXIM1 has been mapped to its central arginine-rich motif (ARM) (Yik et al. 2004; Barboric et al. 2005), which also overlaps with the nuclear localization signal (NLS). Interestingly, this motif in HEXIM1 is highly homologous to and functionally interchangeable with the arginine-rich TAR RNA-binding motif in the HIV-1 Tat protein (Yik et al. 2004), which recruits P-TEFb for activated HIV transcription through forming the TAR–Tat–P-TEFb complex. This homology, together with other sequence and architectural similarities (Egloff et al. 2006) found between the 7SK–HEXIM1–P-TEFb and TAR–Tat–P-TEFb snRNPs, suggests an intriguing possibility that these two complexes may derive from a common evolutionary origin.

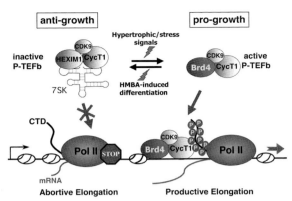

Figure 1. 7SK contributes to the maintenance of a P-TEFb functional equilibrium key for transcription and the cellular decision between growth and differentiation. In the nucleus, a major portion of P-TEFb is sequestered by 7SK and HEXIM1 into the 7SK snRNP (for simplicity, only a monomer each of P-TEFb and HEXIM1 is depicted in the snRNP), where P-TEFb's kinase activity is inhibited by HEXIM1 in a 7SK-dependent manner. Because P-TEFb within the 7SK snRNP is unable to phosphorylate the pol II CTD or associate with promoters, pol II goes into the abortive elongation mode. Treatment of HeLa cells with stress-inducing agents or cardiac myocytes with hypertrophic signals can cause a rapid disruption of the 7SK snRNP and quantitative conversion of the released P-TEFb into the Brd4-bound form. This results in the increased recruitment of P-TEFb by Brd4 to transcriptional templates and stimulation of productive elongation by pol II. On the other hand, when murine erythroleukemia cells are induced to differentiate by the treatment with HMBA, the P-TEFb equilibrium is shifted to the inactive, HEXIM1/7SK-bound state. Thus, the dynamic associations of P-TEFb with its positive and negative regulators are kept under tight cellular control in response to ever changing transcriptional demand in the cell. Since HEXIM1 is known to inhibit growth in a number of cell types, whereas Brd4 is pro-growth during mouse development, their targeting of the general transcription factor P-TEFb is expected to affect the global control of cell growth and differentiation. (Reprinted, with permission, from Zhou and Yik 2006 [©ASM].)

7SK Contributes to the Maintenance of a P-TEFb Functional Equilibrium Important for the Control of Transcription and Cell Growth

Recent reports indicate that the sequestration of P-TEFb by 7SK and HEXIM1 into a kinase-inactive complex is only half the story concerning the regulation of P-TEFb. Although approximately 50% of the total P-TEFb in log-phase HeLa cells exists in the 7SK–HEXIM1–P-TEFb snRNP, the other half has been shown to bind to the bromodomain protein Brd4 (Fig. 1) (Jang et al. 2005; Yang et al. 2005). The interaction with Brd4 is necessary to form the transcriptionally active P-TEFb and recruit P-TEFb to a chromatin template. Because Brd4 is able to interact with acetylated histones and the transcriptional Mediator complex while in complex with P-TEFb, it is proposed that the latter two serve as targets for Brd4's recruitment of P-TEFb to promoters/chromatin templates for general elongation (Fig. 1) (Jang et al. 2005; Yang et al. 2005). This mechanism may be functionally supplemented by a number of gene-specific transcriptional activators such as Tat, which are known to recruit P-TEFb to their corresponding promoters, to maximally stimulate transcriptional elongation under specific conditions (Barboric and Peterlin 2005).

It is important to point out that the two P-TEFb-containing complexes do not remain static in the cell. Rather, they undergo dynamic exchanges under a variety of conditions (Fig. 1). For example, treatment of cells with certain stress-inducing agents, particularly those that can globally interrupt transcription, such as actinomycin D, DRB (5,6-dichloro-1-β-D-ribofuranosylbenzimidazole), and UV irradiation, causes a rapid disruption of the 7SK snRNP and an enhanced formation of the Brd4–P-TEFb complex (Nguyen et al. 2001; Yang et al. 2001; Michels et al. 2003; Yik et al. 2003). Although these events could be important for stress-induced gene expression, they may also simply be an indication of a natural cellular response to those stressful events that suppress transcription and cell growth. Besides the stress-inducing agents, treatment of cardiac myocytes under conditions that cause hypertrophy has also been shown to induce the disruption of the 7SK snRNP and activation of P-TEFb (Sano et al. 2002). Because P-TEFb activity is limiting in normal heart cells, the activation of P-TEFb causes a global increase in cellular RNA and protein contents and consequently the enlargement of heart cells, leading to hypertrophy (Sano et al. 2002).

In contrast to the above conditions that shift the P-TEFb equilibrium to the active, Brd4-bound state, our recent results (N. He et al., unpubl.) indicate that persistent treatment of murine erythroleukemia cells (MELC) with HMBA, which induces terminal division and differentiation of many cell types including MELC, greatly enhances the HEXIM1 gene expression, which in turn pushes the P-TEFb equilibrium toward the inactive 7SK/HEXIM1-bound state (Fig. 1). These observations are consistent with the previous demonstrations that HEXIM1 and Brd4 are involved in the control of cell growth, albeit in opposite fashions (Houzelstein et al. 2002; Ouchida et al. 2003; Wittmann et al. 2003; Huang et al. 2004; Turano et al. 2006). Together, these data indicate that the P-TEFb equilibrium is closely linked to the intracellular transcriptional demand and proliferative state of cells and may contribute to the critical cellular decision between growth and differentiation.

Only 24% of Total Nuclear 7SK Are Associated with P-TEFb

Given the importance of 7SK in sequestering P-TEFb into the kinase-inactive 7SK snRNP, we asked what fraction of the total nuclear 7SK RNA in HeLa cells is involved in this process. To address this, we used the immobilized anti-CDK9 antibodies to quantitatively remove CDK9 and its associated factors from HeLa NE. Quantification of the amount of 7SK remaining in the depleted extract revealed that approximately 24% of this RNA was present in the 7SK–HEXIM1–P-TEFb complex (Fig. 2A).

Identification of A3C as a Major 7SK-associated Nuclear Protein

What then do the majority of 7SKs do in the cell? To examine whether 7SK may bind to other proteins besides CDK9, CycT1, and HEXIM1, we affinity-purified HeLa

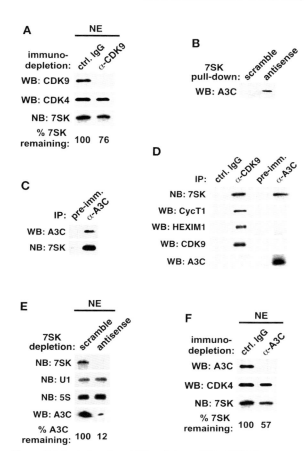

Figure 2. Most of nuclear A3C and about 43% of 7SK interact to form a complex that is different from the 7SK–HEXIM1–P-TEFb snRNP. (*A*) About 24% of nuclear 7SK is bound to CDK9. HeLa NEs were subjected to immunodepletions with the indicated antibodies and then examined by western and northern blotting for levels of CDK9, CDK4, and 7SK remaining in the depleted extracts. The 7SK levels were quantified, normalized against the CDK4 signals, and shown as percentages with the level in the control IgG-depleted NE set at 100%. (*B*) A3C coprecipitates with the affinity-purified 7SK. Western blotting was performed to detect A3C coprecipitated with 7SK, which was pulled down from HeLa NE using the biotinylated antisense 2′-OMe RNA oligonucleotide. An oligonucleotide with a scrambled sequence was used as a negative control. (*C*) 7SK coimmunoprecipitates with endogenous A3C from HeLa NE. The presence of A3C and 7SK in anti-A3C immunoprecipitates was examined by western and northern blotting, respectively. Preimmune antibodies were used as a negative control. (*D*) A3C and CDK9/cyclin T1/HEXIM1 exist in two mutually exclusive 7SK-containing complexes. HeLa NEs were subjected to immunoprecipitations with the indicated antibodies. The compositions of the purified immune complexes were examined by northern and western blotting. (*E*) About 88% of nuclear A3C is associated with 7SK. Shown are northern and western analyses of HeLa NE depleted with either the 7SK antisense or scrambled 2′-OMe RNA oligonucleotide. The A3C levels were quantified, normalized against the U1/5S signals, and displayed as percentages with the one in the scrambled oligo-depleted NE set at 100%. (*F*) About 43% of 7SK are A3C-bound. Immunodepletions and the subsequent quantification of the 7SK levels in the depleted NEs were done as described in *A*.

nuclear proteins that interact with the 7SK snRNA. The procedure used biotinylated 2′OMe oligonucleotide that was either antisense to 7SK or, as a control, contained a scrambled sequence to affinity-purify the endogenous 7SK and its associated proteins (Yang et al. 2001; Yik et

al. 2003). Analysis of these proteins by mass spectrometry revealed the presence of CDK9, CycT1, and HEXIM1 as reported previously (Yik et al. 2003). In addition, A3C was also identified as a major 7SK-associated protein. Of all the members of the APOBEC family, A3C was the only one that emerged from this purification.

Western blotting with rabbit polyclonal anti-A3C antibodies confirmed the binding of A3C to 7SK, which was precipitated from HeLa NE using the biotinylated antisense oligonucleotide but not the scrambled oligonucleotide (Fig. 2B). Reciprocally, immunoprecipitation with anti-A3C but not the preimmune antibodies derived from the same rabbit resulted in the coprecipitation of 7SK with A3C from HeLa NE as demonstrated by northern and western analyses (Fig. 2C).

A3C and CDK9/CycT1/HEXIM1 Exist in Two Mutually Exclusive 7SK-containing Complexes

Given that A3C, CDK9, CycT1, as well as HEXIM1 have all been identified as 7SK-associated proteins, we asked whether A3C and the latter three proteins, which are known to form a single snRNP (Yik et al. 2003), existed in the same or different 7SK-containing complexes. Immune complexes precipitated with either anti-CDK9 or anti-A3C antibodies from HeLa NE were analyzed by western and northern blotting (Fig. 2D). As expected, both complexes specifically harbored 7SK when compared with control precipitations performed with either irrelevant or preimmune antibodies. However, no CDK9, CycT1, or HEXIM1 was detected in the A3C-containing complex. Likewise, no A3C was present in the CDK9/CycT1/HEXIM1-containing complex. Thus, A3C existed in a separate 7SK-containing complex that was different from the previously described 7SK–HEXIM1–P-TEFb snRNP (Yik et al. 2003).

The 7SK–A3C Complex Contains Approximately 88% of Nuclear A3C and 43% of 7SK

To determine the percentage of A3C associated with 7SK in the nucleus of HeLa cells, we used the immobilized 7SK antisense 2′OMe RNA oligonucleotide to deplete 7SK from HeLa NE and examined the amount of A3C remaining in the depleted NE. Northern analysis indicated that the depletion was complete and specific, as only 7SK but not the U1 or 5S RNA was removed from NE by the antisense oligonucleotide and not by the scrambled oligonucleotide (Fig. 2E). More importantly, the depletion of 7SK reduced the nuclear level of A3C by 88%, indicating that most of the nuclear A3C were associated with 7SK in vivo.

Similarly, we also determined the percentage of 7SK associated with A3C in HeLa cells. Western analysis (Fig. 2F) indicates that A3C was specifically and quantitatively immunodepleted from HeLa NE when compared to CDK4 as a loading control. Quantification of the remaining 7SK in the depleted NE revealed that approximately 43% of this snRNA was sequestered in the 7SK–A3C complex.

7SK-dependent Localization of A3C in the Nucleolus

The above demonstration that most of nuclear A3C and a major portion of 7SK interacted with each other to form a complex prompted us to examine the subcellular localization of A3C and the role of 7SK in this process. HeLa cells expressing the native A3C were analyzed by immunofluorescence microscopy. The antigenic specificity of the polyclonal anti-A3C antibodies was further illustrated by the fact that the preimmune antibodies derived from the same rabbit produced no detectable signals in the control staining (Fig. 3A). In contrast, staining with the anti-A3C antibodies revealed a high

concentration of the protein in the nucleolus as demonstrated by its colocalization with nucleolin, a well-characterized nucleolar protein (Fig. 3A, upper panel). In addition, a very weak A3C staining could also be detected in the nucleoplasm. It is important to point out that the nucleolar staining of A3C was not an artifact of the immunofluorescence procedure, as a CFP–A3C fusion protein was also found to reside mostly in the nucleolus in live HeLa cells (Fig. 3B). Furthermore, like the situation in HeLa cells, a similar nucleolar staining for A3C was also detected in two other cell lines, 293T and CaCo-2 (data not shown).

To determine whether the nucleolar localization of A3C was 7SK RNA-dependent, we treated the detergent-permeabilized HeLa cells with RNase A followed by washing with PBS and then fixation with paraformaldehyde. After the RNase treatment, the nucleolin staining condensed into smaller areas. However, the signal intensity remained largely unchanged (Fig. 3A, lower panel), suggesting that the nucleolar structure required for nucleolin retention was not significantly affected by the destruction of RNA. In contrast, the A3C staining mostly disappeared upon RNase treatment and the subsequent wash, indicating that the nucleolar localization of A3C depended on A3C's association with the 7SK snRNA.

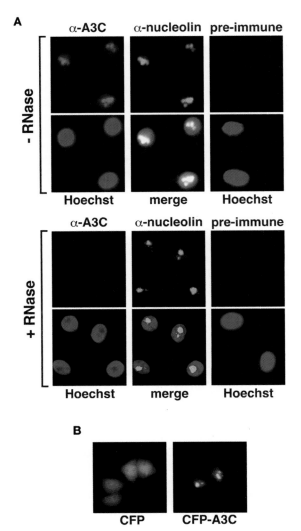

A Major Portion of Nuclear 7SK snRNA Resides in the Nucleolus

Given A3C's interaction with 7SK and its 7SK-dependent nucleolar localization, we suspected that a significant portion of cellular 7SK might also colocalize with A3C in the same subnuclear structure. To test this hypothesis, FISH (fluorescent in situ hybridization) analysis was carried out with either the scrambled or 7SK antisense RNA oligonucleotide as the hybridization probe. Whereas virtually no detectable signal was produced with the scrambled oligonucleotide, hybridization with the 7SK antisense oligonucleotide indicated that 7SK indeed was largely concentrated in the nucleolus (Fig. 4). It in fact occupied the entire nucleolus as shown by confocal microscopy, whereas nucleolin was mainly localized on the periphery of the structure. In addition to the nucleolar localization, a substantial amount of 7SK could also be detected in the nucleoplasm (Fig. 4), where it was required to mediate the interaction between HEXIM1 and P-TEFb (Michels et al. 2003; Yik et al. 2003). Taken together, our data indicate that, like A3C, a major portion of cellular 7SK existed in the nucleolus, where it could effectively target and regulate A3C.

Figure 3. A3C is largely located in the nucleolus. (*A*) 7SK-dependent nucleolar sequestration of A3C. Permeabilized HeLa cells were incubated with or without RNase A followed by washing with PBS and then fixation in paraformaldehyde. The localization of the endogenous A3C was revealed by staining with anti-A3C antibodies and the Alex546-conjugated secondary antibody. Preimmune antibodies were used as a negative control. Antinucleolin antibody was used to delineate the nucleolus. Nuclei were visualized with Hoechst 33258. (*B*) Transfected CFP–A3C is largely concentrated in the nucleolus in live HeLa cells. Cells transfected with CFP or the CFP–A3C fusion were directly photographed.

Destruction of 7SK Stimulates Cytosine Deamination in ssDNA by Affinity-purified A3C

With the demonstration of the interaction of A3C with 7SK in the nucleolus, we asked whether this interaction would affect the deaminase activity of A3C. An in vitro uracil DNA glycosylase (UDG)-based deamination assay (Petersen-Mahrt and Neuberger 2003) was performed to

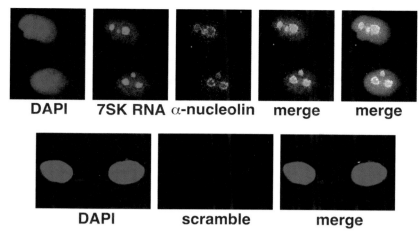

Figure 4. A major portion of 7SK snRNA reside in the nucleoli. Permeabilized and fixed HeLa cells were subjected to FISH analysis with either the scrambled or the 7SK antisense RNA oligonucleotide as a probe. Nuclei and nucleoli were visualized by staining with DAPI and antinucleolin antibody, respectively. Images were obtained with a confocal microscope.

address this question. Deamination of dC to generate dU in an ssDNA substrate was determined by treating the substrate with UDG, cleaving at the resultant abasic site with NaOH, and monitoring the ^{32}P-labeled cleavage product by gel electrophoresis. When F-A3C immunoprecipitated from transfected HeLa cells was analyzed in this assay,

the 7SK-bound F-A3C displayed very poor deaminase activity, which was only slightly above the background level as shown in control reactions containing buffer alone (Fig. 5A, compare lane 1 with lanes 3–5). However, a dramatic increase in F-A3C's deamination of the substrate (lanes 6–8) was observed when the associated 7SK was

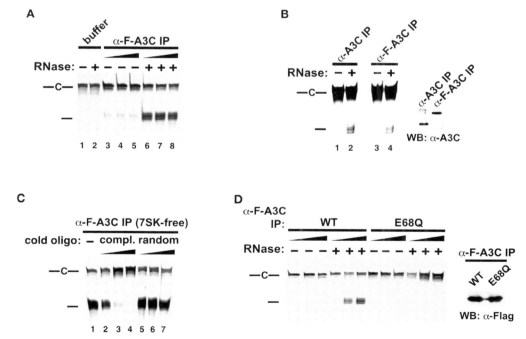

Figure 5. 7SK-dependent suppression of A3C's deaminase activity toward ssDNA. (*A*) RNase destruction of 7SK stimulates cytosine deamination of an ssDNA substrate by affinity-purified F-A3C. In vitro UDG-based deamination reactions containing either buffer alone or increasing amounts (0.5, 1.0, and 1.5 µl) of the affinity-purified 7SK–F-A3C complex were performed in the presence (+) or absence (–) of RNase A. The positions of the input dC-containing ssDNA substrate and the cleaved deaminated product are indicated. (*B*) The transiently expressed F-A3C and its endogenous counterpart display similar deaminase activities when freed from bound 7SK. Deamination by the indicated immunoprecipitates was analyzed in the presence or absence of RNase A after normalization by anti-A3C western blotting. (*C*) The deaminase activity of the 7SK-free A3C is specific for ssDNA. Deamination reactions containing labeled ssDNA substrate were performed in the presence of increasing amounts of unlabeled oligonucleotide that was either complementary to the ssDNA substrate or containing random sequences. (*D*) A catalytically inactive A3C mutant fails to deaminate cytosine even when freed from the inhibitory 7SK. Reactions containing increasing amounts of wild-type A3C or its mutant E68Q, which were normalized by anti-FLAG western blotting (*right panel*), were performed in the presence or absence of RNase A.

degraded by the addition of RNase A into the reactions. Besides the transfected F-A3C, we also analyzed the effect of 7SK degradation on endogenous A3C immunoprecipitated from HeLa NE. Indeed, both the endogenous and transfected A3C displayed a similar degree of RNase-stimulated deamination of the substrate (Fig. 5B), confirming that this stimulation was a bona fide property of native A3C in HeLa cells.

To determine whether the stimulation of A3C deaminase caused by 7SK destruction would alter the specificity of A3C toward single- or double-stranded DNA, deamination reactions containing the labeled ssDNA substrate together with increasing amounts of either a complementary DNA oligonucleotide or a control oligonucleotide with a randomized sequence were performed. The data in Figure 5C indicate that deamination of the substrate ssDNA by 7SK-free A3C was blocked in a dosage-dependent manner as the substrate began to anneal to the complementary oligonucleotide to form dsDNA. This result confirms the strict substrate preference for ssDNA by the activated, 7SK-free A3C.

Finally, the enhanced cytosine deamination of ssDNA due to 7SK degradation could be attributed entirely to the catalytic activity of A3C. The immunoprecipitated A3C mutant E68Q, which contained an amino acid substitution at its proposed Zn^{2+}-coordinated active site and therefore was catalytically inactive, displayed no detectable deaminase activity even when liberated from the bound 7SK (Fig. 5D). Taken together, these results indicate that the deaminase activity of A3C on ssDNA can be inhibited by 7SK or a 7SK-dependent factor(s) in HeLa cells.

Removal of the Bound 7SK by High Salt Stimulates A3C's Deaminase Activity

Besides the RNase-mediated destruction of 7SK, A3C could also be activated through stripping of its associated 7SK by high salt. For example, the interaction of the immunoprecipitated F-A3C with 7SK was stable in the presence of 0.3 M KCl but disrupted by washing with a buffer containing 0.8 M KCl (Fig. 6A). This situation was highly reminiscent of the salt sensitivity described previously of the interaction between

7SK and P-TEFb/HEXIM1 (Chen et al. 2004). Importantly, the 0.8 M KCl-washed anti-F-A3C immunoprecipitates were as efficient as the RNase-treated sample prepared in the presence of 0.3 M KCl in terms of their deamination of the ssDNA substrate (Fig. 6B).

Separable Binding Sites for 7SK snRNA and ssDNA on A3C

To explain the inhibition of A3C by its associated 7SK, a simple scenario could be envisioned where 7SK with plenty of single-stranded regions acted as a competitive inhibitor to prevent the binding of A3C to its substrate ssDNA. To test this hypothesis, the ability of A3C to interact with the immobilized ssDNA with or without internal dC residues was examined in the presence or absence of the associated 7SK. The data in Figure 7A indicate that the 7SK-bound F-A3C displayed a clear preference for the dC-containing ssDNA. More importantly, degradation of the bound 7SK by RNase did not significantly affect the binding of A3C to the ssDNA oligonucleotide over a broad range of A3C concentrations (Fig. 7A, compare lanes 2 and 5, and data not shown). This result suggests that the 7SK–A3C interaction that was inhibitive to A3C's activity did not interfere with the binding of A3C to its ssDNA substrate.

To further test this notion, the ssDNA-binding activities were compared between the 0.3 M KCl-washed 7SK-bound F-A3C and the high-salt-stripped 7SK-free protein. Both were equally efficient in binding to ssDNA in a dC-dependent manner (Fig. 7B). Finally, the binding of the high-salt-washed 7SK-free F-A3C to the ssDNA oligonucleotide was completed unaffected by the additions of 7SK present in total HeLa nuclear RNA at concentrations as high as 100 times over that of the ssDNA (Fig. 7C). Taken together, these results strongly implicate the existence of two different binding sites on A3C for the regulatory 7SK snRNA and ssDNA substrate. Thus, the 7SK-dependent inhibition of A3C was unlikely to be caused by the simple competition between 7SK and ssDNA for the same substrate-binding site on A3C.

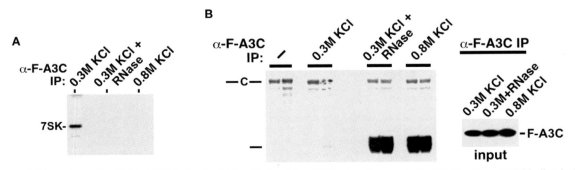

Figure 6. Disruption of the 7SK–A3C binding by high salt stimulates A3C's deaminase activity. (*A*) The 7SK–A3C binding is disrupted by high salt. The immunoprecipitated F-A3C was treated by low-salt (0.3 M KCl) washing, low-salt washing followed by RNase A incubation, or high-salt (0.8 M KCl) washing before being eluted from the FLAG beads. 7SK bound to A3C was detected by northern blotting. (*B*) A3C stripped of the bound 7SK shows enhanced deaminase activity. The affinity-purified F-A3C, which was treated under various conditions as described in *A*, was normalized by western blotting. Deaminase activities of F-A3C were tested in duplicates in UDG-based in vitro deaminase reactions as described in Fig. 5A.

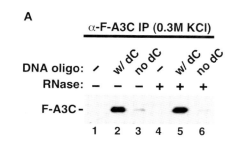

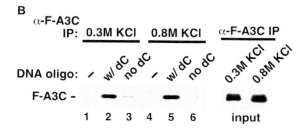

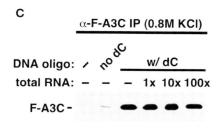

Figure 7. Association with 7SK does not interfere with A3C's binding to ssDNA substrate. (*A*) Association with 7SK does not affect A3C's binding to a dC-containing ssDNA oligonucleotide. The immunoprecipitated 7SK–F-A3C complex was washed with a buffer containing 0.3 M KCl, eluted with the FLAG peptide, and then incubated in the presence or absence of RNase A with the immobilized ssDNA either with or without deoxycytosines. After extensive washes, the bound F-A3C was detected by anti-FLAG western blotting. (*B*) Stripping 7SK from A3C with high salt does not affect A3C's ssDNA-binding ability. Immunoprecipitated F-A3C was washed by either a low-salt (0.3 M KCl) or high-salt (0.8 M KCl) buffer before elution with the FLAG peptide. The affinity-purified F-A3C was analyzed in the ssDNA-binding assay containing the indicated DNA oligonucleotide as in *A*. (*C*) Binding of 7SK-free A3C to ssDNA is unaffected by excess amounts of 7SK present in total HeLa nuclear RNA. The ssDNA-binding reactions containing constant amounts of the indicated DNA oligonucleotide and high-salt-washed 7SK-free A3C were carried out in the absence or presence of total HeLa nuclear RNA that contained 7SK at 1-, 10-, or 100-times molar excess over the DNA oligonucleotide.

DISCUSSION

Despite its high abundance and early discovery (Weinberg and Penman 1968; Zieve and Penman 1976), 7SK remains the least known noncoding small nuclear RNA in terms of its biological functions in vertebrate cells (Wassarman and Steitz 1991). Although it has recently been shown to have a scaffolding role in mediating the interaction of P-TEFb with its inhibitor HEXIM1 (Michels et al. 2003; Yik et al. 2003), only a relatively minor fraction

(about 24%; Fig. 2A) of this RNA is actually involved. Here, we report the identification of a novel 7SK-containing complex that sequesters approximately 43% of the total 7SK and 88% of A3C in HeLa cells. Unlike all other APOBEC family members whose expressions are largely tissue-restricted, A3C exists in a wide range of human tissues and tumor cell lines (Jarmuz et al. 2002; Wedekind et al. 2003). Our data indicate that the association with 7SK sequesters A3C in the nucleolus and inhibits its deaminase activity toward cytosine in ssDNA substrates.

An RNA-dependent suppression of catalytic activity is apparently not limited to A3C only. The activities of at least two APOBEC family members have been proposed to be regulated by their associated RNAs, although in neither case has the identity of the RNA involved or the exact subcellular localization where the regulation takes place been determined. In the first case, APOBEC3G has been found to reside in a large, enzymatically inactive ribonucleoprotein complex in unstimulated peripheral blood CD4[+] T cells. This complex can be converted into a smaller-size complex after treatment with RNase, which produces an enzymatically active form of APOBEC3G capable of deaminating dC on ssDNA (Chiu et al. 2005). Similarly, human AID (activation-induced deaminase) purified as a GST-fusion protein from recombinant insect cells has no measurable deaminase activity on ssDNA unless pretreated with RNase to remove the bound inhibitory RNA (Bransteitter et al. 2003). These earlier observations, together with the data presented here on A3C, suggest the likely existence of a general mechanism that employs RNA molecules to suppress the enzymatic activities of APOBEC family members.

An obvious exception to this idea is demonstrated by APOBEC1, which is the only known APOBEC protein that can also function as an RNA editor. APOBEC1 specifically deaminates cytidine 6666 in *apoB* mRNA in vivo, despite the fact that it does not specifically bind to this RNA (Maris et al. 2005). The ability of APOBEC1 to edit *apoB* mRNA is facilitated by its interaction with an auxiliary factor that binds to an 11-nucleotide mooring sequence downstream from the editing site (Mehta et al. 2000; Mehta and Driscoll 2002).

What could be the benefit for the 7SK-dependent inhibition and sequestration of the A3C deaminase in the nucleolus? We propose that the nucleolar sequestration keeps A3C at a safe distance from most genomic DNA and prevents it from binding and deaminating ssDNA that frequenetly forms in the nucleoplasm in processes ranging from DNA replication to repair, to recombination, to active transcription. This situation may be reminiscent to the control of AID activity by alteration of its subcellular localization (McBride et al. 2004; Pasqualucci et al. 2004). Responsible for somatic hypermutation and class switch recombination in activated B lymphocytes, AID is thought to act on the nontemplate ssDNA exposed by RNA polymerase during transcription (Ramiro et al. 2003). Although this event must occur in the nucleus, AID is found primarily in the cytoplasm, and a nuclear export signal (NES) located at the carboxyl terminus of AID controls the nuclear level of this protein (Ito et al.

2004). This mechanism may help prevent large amounts of AID from getting into the nucleus, where they may cause uncontrolled hypermutations in genomic DNA.

Given the relative abundance and widespread expression of A3C, it is reasonable to speculate that this protein may have an important and general role in certain key biological processes. Aside from the reverse-transcribed proviral DNA, the cellular targets of the A3C deaminase have not been identified. However, the predominantly nucleolar localizations of both A3C and 7SK, the latter of which attracts most of A3C in the cell, implicate the nucleolus as a potential site where the physiological substrates of A3C may be located. Both ribosomal RNA processing and ribosome assembly occur in this subnuclear structure. The pre-rRNA gene is the nucleolar organizer, and all of the other components of the ribosome diffuse to the newly synthesized pre-rRNA (Karpen et al. 1988). Future studies of the 7SK–A3C complex will be informative to reveal whether it has a key regulatory role in pre-rRNA transcription and processing as well as ribosomal assembly. Of particular interest is the investigation into the possibility that the deaminase activity of A3C can be activated through regulated dissociation of its inhibitory 7SK snRNA.

Exactly how the A3C-associated 7SK contributes to the inhibition of this deaminase awaits future investigation. Because the ssDNA substrate-binding site in A3C has been implicated in being different from the region required for A3C's interaction with 7SK (Fig. 7), we believe that 7SK serves as a negative regulator, rather than a simple substrate competitor in causing the inhibition. We have so far been unable to reconstitute in vitro the specific inhibition of A3C using purified protein and RNA components (data not shown). This could be due to the presence in the 7SK–A3C complex of additional yet-to-be-identified factor(s) that cooperates with 7SK to cause inhibition. If this can be confirmed, the 7SK-dependent regulation of A3C can be viewed as mechanistically similar to the 7SK-mediated control of P-TEFb. In the latter case, 7SK itself does not directly inhibit the P-TEFb kinase but rather has an intermediary role in allowing the real kinase inhibitor HEXIM1 to target and inactivate P-TEFb (Yik et al. 2003, 2004; Michels et al. 2004). Since there is so far no evidence indicating a direct inhibitory role for 7SK in inactivating either A3C or P-TEFb, we propose that at least one role of this abundant RNA is to function in a storage process to keep key nuclear enzymatic activities in check for proper cell function.

ACKNOWLEDGMENTS

We thank Vivien Lee for technical assistance. This work was supported by grants from the National Institutes of Health (AI41757) and the American Cancer Society (RSG-01-171-01-MBC) to Q.Z. and a Berkeley Scholar Fellowship to R.C.

REFERENCES

Barboric M. and Peterlin B.M. 2005. A new paradigm in eukaryotic biology: HIV Tat and the control of transcriptional elongation. *PLoS Biol.* **3**: e76.

Barboric M., Kohoutek J., Price J.P., Blazek D., Price D.H., and Peterlin B.M. 2005. Interplay between 7SK snRNA and

oppositely charged regions in HEXIM1 direct the inhibition of P-TEFb. *EMBO J.* **24**: 4291.

Blazek D., Barboric M., Kohoutek J., Oven I., and Peterlin B.M. 2005. Oligomerization of HEXIM1 via 7SK snRNA and coiled-coil region directs the inhibition of P-TEFb. *Nucleic Acids Res.* **33**: 7000.

Bransteitter R., Pham P., Scharff M.D., and Goodman M.F. 2003. Activation-induced cytidine deaminase deaminates deoxycytidine on single-stranded DNA but requires the action of RNase. *Proc. Natl. Acad. Sci.* **100**: 4102.

Chao S.-H. and Price D.H. 2001. Flavopiridol inactivates P-TEFb and blocks most RNA polymerase II transcription in vivo. *J. Biol. Chem.* **276**: 31793.

Chen R., Yang Z., and Zhou Q. 2004. Phosphorylated positive transcription elongation factor b (P-TEFb) is tagged for inhibition through association with 7SK snRNA. *J. Biol. Chem.* **279**: 4153.

Chiu Y.L., Soros V.B., Kreisberg J.F., Stopak K., Yonemoto W., and Greene W.C. 2005. Cellular APOBEC3G restricts HIV-1 infection in resting CD4+ T cells. *Nature* **435**: 108.

Conticello S.G., Thomas C.J., Petersen-Mahrt S.K., and Neuberger M.S. 2005. Evolution of the AID/APOBEC family of polynucleotide (deoxy)cytidine deaminases. *Mol. Biol. Evol.* **22**: 367.

Dulac C., Michels A.A., Fraldi A., Bonnet F., Nguyen V.T., Napolitano G., Lania L., and Bensaude O. 2005. Transcription-dependent association of multiple positive transcription elongation factor units to a HEXIM multimer. *J. Biol. Chem.* **280**: 30619.

Egloff S., Van Herreweghe E., and Kiss T. 2006. Regulation of polymerase II transcription by 7SK snRNA: Two distinct RNA elements direct P-TEFb and HEXIM1 binding. *Mol. Cell. Biol.* **26**: 630.

Goodrich J.A. and Kugel J.F. 2006. Non-coding-RNA regulators of RNA polymerase II transcription. *Nat. Rev. Mol. Cell Biol.* **7**: 612.

Harris R.S. and Liddament M.T. 2004. Retroviral restriction by APOBEC proteins. *Nat. Rev. Immunol.* **4**: 868.

Harris R.S., Petersen-Mahrt S.K., and Neuberger M.S. 2002. RNA editing enzyme APOBEC1 and some of its homologs can act as DNA mutators. *Mol. Cell* **10**: 1247.

Harris R.S., Bishop K.N., Sheehy A.M., Craig H.M., Petersen-Mahrt S.K., Watt I.N., Neuberger M.S., and Malim M.H. 2003. DNA deamination mediates innate immunity to retroviral infection. *Cell* **113**: 803.

Houzelstein D., Bullock S.L., Lynch D.E., Grigorieva E.F., Wilson V.A., and Beddington R.S. 2002. Growth and early postimplantation defects in mice deficient for the bromodomain-containing protein Brd4. *Mol. Cell. Biol.* **22**: 3794.

Huang F., Wagner M., and Siddiqui M.A. 2004. Ablation of the CLP-1 gene leads to down-regulation of the HAND1 gene and abnormality of the left ventricle of the heart and fetal death. *Mech. Dev.* **121**: 559.

Ito S., Nagaoka H., Shinkura R., Begum N., Muramatsu M., Nakata M., and Honjo T. 2004. Activation-induced cytidine deaminase shuttles between nucleus and cytoplasm like apolipoprotein B mRNA editing catalytic polypeptide 1. *Proc. Natl. Acad. Sci.* **101**: 1975.

Jang M.K., Mochizuki K., Zhou M., Jeong H.S., Brady J.N., and Ozato K. 2005. The bromodomain protein Brd4 is a positive regulatory component of P-TEFb and stimulates RNA polymerase II-dependent transcription. *Mol. Cell* **19**: 523.

Jarmuz A., Chester A., Bayliss J., Gisbourne J., Dunham I., Scott J., and Navaratnam N. 2002. An anthropoid-specific locus of orphan C to U RNA-editing enzymes on chromosome 22. *Genomics* **79**: 285.

Karpen G.H., Schaefer J.E., and Laird C.D. 1988. A *Drosophila* rRNA gene located in euchromatin is active in transcription and nucleolus formation. *Genes Dev.* **2**: 1745.

Langlois M.A., Beale R.C., Conticello S.G., and Neuberger M.S. 2005. Mutational comparison of the single-domained APOBEC3C and double-domained APOBEC3F/G antiretroviral cytidine deaminases provides insight into their DNA target site specificities. *Nucleic Acids Res.* **33**: 1913.

Li Q., Price J.P., Byers S.A., Cheng D., Peng J., and Price D.H. 2005. Analysis of the large inactive P-TEFb complex indicates that it contains one 7SK molecule, a dimer of HEXIM1 or HEXIM2, and two P-TEFb molecules containing Cdk9 phosphorylated at threonine 186. *J. Biol. Chem.* **280:** 28819.

Lindahl T. 1993. Instability and decay of the primary structure of DNA. *Nature* **362:** 709.

Maris C., Masse J., Chester A., Navaratnam N., and Allain F.H. 2005. NMR structure of the apoB mRNA stem-loop and its interaction with the C to U editing APOBEC1 complementary factor. *RNA* **11:** 173.

Marks P.A., Richon V.M., Kiyokawa H., and Rifkind R.A. 1994. Inducing differentiation of transformed cells with hybrid polar compounds: A cell cycle-dependent process. *Proc. Natl. Acad. Sci.* **91:** 10251.

McBride K.M., Barreto V., Ramiro A.R., Stavropoulos P., and Nussenzweig M.C. 2004. Somatic hypermutation is limited by CRM1-dependent nuclear export of activation-induced deaminase. *J. Exp. Med.* **199:** 1235.

Mehta A. and Driscoll D.M. 2002. Identification of domains in apobec-1 complementation factor required for RNA binding and apolipoprotein-B mRNA editing. *RNA* **8:** 69.

Mehta A., Kinter M.T., Sherman N.E., and Driscoll D.M. 2000. Molecular cloning of apobec-1 complementation factor, a novel RNA-binding protein involved in the editing of apolipoprotein B mRNA. *Mol. Cell. Biol.* **20:** 1846.

Michels A.A., Nguyen V.T., Fraldi A., Labas V., Edwards M., Bonnet F., Lania L., and Bensaude O. 2003. MAQ1 and 7SK RNA interact with CDK9/cyclin T complexes in a transcription-dependent manner. *Mol. Cell. Biol.* **23:** 4859.

Michels A.A., Fraldi A., Li Q., Adamson T.E., Bonnet F., Nguyen V.T., Sedore S.C., Price J.P., Price D.H., Lania L., and Bensaude O. 2004. Binding of the 7SK snRNA turns the HEXIM1 protein into a P-TEFb (CDK9/cyclin T) inhibitor. *EMBO J.* **23:** 2608.

Morgan H.D., Dean W., Coker H.A., Reik W., and Petersen-Mahrt S.K. 2004. Activation-induced cytidine deaminase deaminates 5-methylcytosine in DNA and is expressed in pluripotent tissues: Implications for epigenetic reprogramming. *J. Biol. Chem.* **279:** 52353.

Murphy S., Di Liegro C., and Melli M. 1987. The in vitro transcription of the 7SK RNA gene by RNA polymerase III is dependent only on the presence of an upstream promoter. *Cell* **51:** 81.

Navarro F., Bollman B., Chen H., Konig R., Yu Q., Chiles K., and Landau N.R. 2005. Complementary function of the two catalytic domains of APOBEC3G. *Virology* **333:** 374.

Nguyen V.T., Kiss T., Michels A.A., and Bensaude O. 2001. 7SK small nuclear RNA binds to and inhibits the activity of CDK9/cyclin T complexes. *Nature* **414:** 322.

Ouchida R., Kusuhara M., Shimizu N., Hisada T., Makino Y., Morimoto C., Handa H., Ohsuzu F., and Tanaka H. 2003. Suppression of NF-kappaB-dependent gene expression by a hexamethylene bisacetamide-inducible protein HEXIM1 in human vascular smooth muscle cells. *Genes Cells* **8:** 95.

Pasqualucci L., Guglielmino R., Houldsworth J., Mohr J., Aoufouchi S., Polakiewicz R., Chaganti R.S., and Dalla-Favera R. 2004. Expression of the AID protein in normal and neoplastic B cells. *Blood* **104:** 3318.

Petersen-Mahrt S.K. and Neuberger M.S. 2003. In vitro deamination of cytosine to uracil in single-stranded DNA by apolipoprotein B editing complex catalytic subunit 1 (APOBEC1). *J. Biol. Chem.* **278:** 19583.

Price D.H. 2000. P-TEFb, a cyclin-dependent kinase controlling elongation by RNA polymerase II. *Mol. Cell. Biol.* **20:** 2629.

Ramiro A.R., Stavropoulos P., Jankovic M., and Nussenzweig M.C. 2003. Transcription enhances AID-mediated cytidine deamination by exposing single-stranded DNA on the non-template strand. *Nat. Immunol.* **4:** 452.

Sano M., Abdellatif M., Oh H., Xie M., Bagella L., Giordano A., Michael L.H., DeMayo F.J., and Schneider M.D. 2002. Activation and function of cyclin T-Cdk9 (positive transcription elongation factor-b) in cardiac muscle-cell hypertrophy. *Nat. Med.* **8:** 1310.

Shim E.Y., Walker A.K., Shi Y., and Blackwell T.K. 2002. CDK-9/cyclin T (P-TEFb) is required in two postinitiation pathways for transcription in the *C. elegans* embryo. *Genes Dev.* **16:** 2135.

Sims R.J., Belotserkovskaya R., and Reinberg D. 2004. Elongation by RNA polymerase II: The short and long of it. *Genes Dev.* **18:** 2437.

Turano M., Napolitano G., Dulac C., Majello B., Bensaude O., and Lania L. 2006. Increased HEXIM1 expression during erythroleukemia and neuroblastoma cell differentiation. *J. Cell. Physiol.* **206:** 603.

Wassarman D.A. and Steitz J.A. 1991. Structural analyses of the 7SK ribonucleoprotein (RNP), the most abundant human small RNP of unknown function. *Mol. Cell. Biol.* **11:** 3432.

Wedekind J.E., Dance G.S., Sowden M.P., and Smith H.C. 2003. Messenger RNA editing in mammals: New members of the APOBEC family seeking roles in the family business. *Trends Genet.* **19:** 207.

Weinberg R.A. and Penman S. 1968. Small molecular weight monodisperse nuclear RNA. *J. Mol. Biol.* **38:** 289.

Wittmann B.M., Wang N., and Montano M.M. 2003. Identification of a novel inhibitor of breast cell growth that is down-regulated by estrogens and decreased in breast tumors. *Cancer Res.* **63:** 5151.

Yang Z., Zhu Q., Luo K., and Zhou Q. 2001. The 7SK small nuclear RNA inhibits the CDK9/cyclin T1 kinase to control transcription. *Nature* **414:** 317.

Yang Z., Yik J.H., Chen R., He N., Jang M.K., Ozato K., and Zhou Q. 2005. Recruitment of P-TEFb for stimulation of transcriptional elongation by the bromodomain protein Brd4. *Mol. Cell* **19:** 535.

Yik J.H., Chen R., Pezda A.C., and Zhou Q. 2005. Compensatory contributions of HEXIM1 and HEXIM2 in maintaining the balance of active and inactive positive transcription elongation factor b complexes for control of transcription. *J. Biol. Chem.* **280:** 16368.

Yik J.H., Chen R., Pezda A.C., Samford C.S., and Zhou Q. 2004. A human immunodeficiency virus type 1 Tat-like arginine-rich RNA-binding domain is essential for HEXIM1 to inhibit RNA polymerase II transcription through 7SK snRNA-mediated inactivation of P-TEFb. *Mol. Cell. Biol.* **24:** 5094.

Yik J.H., Chen R., Nishimura R., Jennings J.L., Link A.J., and Zhou Q. 2003. Inhibition of P-TEFb (CDK9/Cyclin T) kinase and RNA polymerase II transcription by the coordinated actions of HEXIM1 and 7SK snRNA. *Mol. Cell* **12:** 971.

Yu Q., Chen D., Konig R., Mariani R., Unutmaz D., and Landau N.R. 2004. APOBEC3B and APOBEC3C are potent inhibitors of simian immunodeficiency virus replication. *J. Biol. Chem.* **279:** 53379.

Zhou Q. and Yik J.H. 2006. The Yin and Yang of P-TEFb regulation: Implications for human immunodeficiency virus gene expression and global control of cell growth and differentiation. *Microbiol. Mol. Biol. Rev.* **70:** 646.

Zieve G. and Penman S. 1976. Small RNA species of the HeLa cell: Metabolism and subcellular localization. *Cell* **8:** 19.

The SMN Complex: An Assembly Machine for RNPs

D.J. Battle, M. Kasim, J. Yong, F. Lotti, C.-K. Lau, J. Mouaikel, Z. Zhang,
K. Han, L. Wan, and G. Dreyfuss

*Howard Hughes Medical Institute and Department of Biochemistry and Biophysics, University of Pennsylvania
School of Medicine, Philadelphia, Pennsylvania 19104-6148*

In eukaryotic cells, the biogenesis of spliceosomal small nuclear ribonucleoproteins (snRNPs) and likely other RNPs is mediated by an assemblyosome, the survival of motor neurons (SMN) complex. The SMN complex, composed of SMN and the Gemins (2–7), binds to the Sm proteins and to snRNAs and constructs the heptameric rings, the common cores of Sm proteins, on the Sm site ($AU_{5-6}G$) of the snRNAs. We have determined the specific sequence and structural features of snRNAs for binding to the SMN complex and Sm core assembly. The minimal SMN complex-binding domain in snRNAs (except U1) is composed of an Sm site and a closely adjacent 3' stem-loop. Remarkably, the specific sequence of the stem-loop is not important for SMN complex binding, but it must be located within a short distance of the 3' end of the RNA for an Sm core to assemble. This minimal snRNA-defining "snRNP code" is recognized by the SMN complex, which binds to it directly and with high affinity and assembles the Sm core. The recognition of the snRNAs is provided by Gemin5, a component of the SMN complex that directly binds the snRNP code. Gemin5 is a novel RNA-binding protein that is critical for snRNP biogenesis. Thus, the SMN complex is the identifier, as well as assembler, of the abundant class of snRNAs in cells. The function of the SMN complex, previously unanticipated because RNP biogenesis was believed to occur by self-assembly, confers stringent specificity on otherwise potentially illicit RNA–protein interactions.

RNAs exist in cells as RNPs, complexes of RNAs with RNA-binding proteins (Burd and Dreyfuss 1994; Krecic and Swanson 1999; Yong et al. 2004a). The RNA-binding proteins have essential roles in the biogenesis, function, localization, and stability of the RNAs. There are a very large number of RNA-binding proteins in eukaryotes, and they vary widely in their abundance, RNA-binding specificities, cell-type expression patterns, and biochemical properties. Given the number of cellular RNAs and the vast number of RNA-binding proteins, the problem of formation of specific RNPs is extremely complex. This is particularly important for RNPs that have intricate and highly stable multiprotein structures, exemplified by the spliceosomal snRNPs.

Spliceosomal snRNPs are the major components of the cellular mRNA splicing machinery (Will and Luhrmann 2001; Nilsen 2003). Each snRNP is composed of one or two small nuclear RNAs (snRNAs) bound to a set of RNA-binding proteins. In addition to snRNP-specific proteins, each of the major snRNAs (U1, U2, U4, and U5) is bound to a common set of seven proteins (SmB/B', SmD1, SmD2, SmD3, SmE, SmF, and SmG). These proteins form a seven-membered ring around the short, highly conserved, uridine-rich sequence on each snRNA called the Sm site (Kambach et al. 1999; Achsel et al. 2001; Stark et al. 2001). This Sm core is remarkably stable, resistant to high salt, heparin, and urea (Hamm et al. 1987; Jarmolowski and Mattaj 1993; Raker et al. 1996). Additionally, the conserved seven-nucleotide Sm site sequence (5'-AUUU/CUUG-3') is found in many RNAs. Given the high stability and remarkably slow turnover of these complexes, it is vital that cells only assemble Sm cores on proper snRNAs. Although Sm proteins will self-assemble in vitro in an ATP-independent manner on any RNA that has the short Sm site sequence,

in vivo Sm cores only assemble in an ATP-dependent manner on proper snRNAs (Kleinschmidt et al. 1989; Sumpter et al. 1992; Raker et al. 1996, 1999; Meister et al. 2001a; Pellizzoni et al. 2002a). Self-assembly of Sm cores does not occur in vivo, but rather Sm cores are assembled onto snRNAs by the SMN complex (Fischer et al. 1997; Meister et al. 2001a; Pellizzoni et al. 2002a).

The SMN protein is the product of the spinal muscular atrophy (SMA) disease gene (Lefebvre et al. 1995). Reduction of SMN results in a decreased Sm core assembly and severe motor neuron degeneration in humans (Coovert et al. 1997; Lefebvre et al. 1997; Wan et al. 2005). SMN forms a large, stable complex in the nucleus and cytoplasm of metazoan cells (Liu and Dreyfuss 1996; Paushkin et al. 2002). Besides SMN, the SMN complex contains at least six other proteins known as Gemins2–7 (Fig. 1). SMN oligomerizes via its carboxy-terminal region and is always found tightly associated with Gemin2 (Liu et al. 1997). Gemin3 (also known as DP103 or DDX20) is a DEAD-box RNA helicase (Charroux et al. 1999). Gemin4, a protein of unknown structure and function, associates with the SMN complex via Gemin3 (Charroux et al. 2000). Gemin5 is a multidomain WD-repeat-containing protein (Gubitz et al. 2002). Interestingly, Gemin6 and Gemin7 each contain a domain that has an Sm fold (Baccon et al. 2002; Pellizzoni et al. 2002b; Ma et al. 2005). Recently, another protein of unknown function, Gemin8, has been shown to be part of the SMN complex and interacts directly with Gemin6/7 (Carissimi et al. 2006).

In an ATP-dependent reaction, the SMN complex assembles Sm cores onto the Sm site of each snRNA (Fig. 2) (Fischer et al. 1997; Meister et al. 2001a; Pellizzoni et al. 2002a). snRNAs are transcribed in the nucleus by RNA polymerase II and exported to the cytoplasm for Sm core assembly by the export factor PHAX

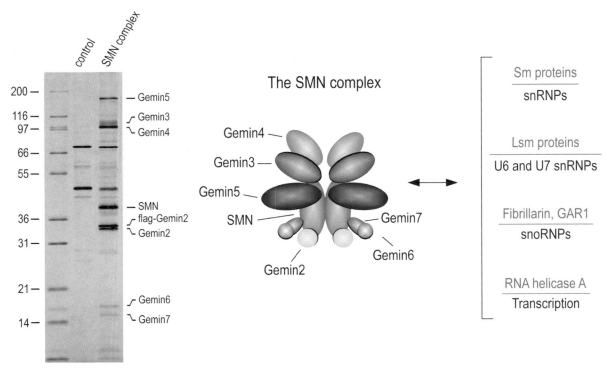

Figure 1. The SMN complex is a large, oligomeric multiprotein complex. The SMN complex contains SMN as well as at least six other proteins shown as Gemins2–7. For simplicity, the SMN complex is shown here as a dimer. Shown on the left is a silver stain of the SMN complex purified from HeLa extract. Shown on the right are several proteins that directly interact with the SMN complex, as well as the RNPs in which they function.

(Ohno et al. 2000; Segref et al. 2001). The SMN complex, presumably preloaded with Sm proteins, associates with snRNAs shortly after export and transfers the Sm proteins to the Sm site of the RNA. The 5′ cap is then hypermethylated to form the 2,2,7-trimethyl guanosine cap, and the mature snRNP is then imported into the nucleus by snurportin and importin β for final association with the snRNP-specific proteins and utilization in mRNA splicing (Mattaj and De Robertis 1985; Fischer and Luhrmann 1990; Fischer et al. 1993; Plessel et al. 1994; Huber et al. 1998).

The SMN complex interacts directly with both the Sm proteins and the snRNA. Each protein of the SMN complex, except Gemin2, directly binds to Sm proteins (Liu et al. 1997; Charroux et al. 1999, 2000; Baccon et al. 2002; Gubitz et al. 2002; Pellizzoni et al. 2002b). The Sm domains of Gemin6 and Gemin7 may bind to Sm proteins by mimicking the conserved Sm–Sm interface (Ma et al. 2005). Additionally, SmB/B′, SmD1, and SmD3 contain carboxy-terminal Arg/Gly-rich tails that are methylated by JBP1/PRMT5 in the 20S methylosome complex (Fig. 2) (Brahms et al. 2000, 2001; Friesen et al. 2001b, 2002; Meister et al. 2001b). The SMN protein itself binds via its tudor domain directly to these symmetric dimethyl-arginine tails, and methylation of these proteins enhances Sm protein binding to the SMN complex (Fig. 2) (Friesen et al. 2001a).

Independent of its interaction with Sm proteins, the SMN complex binds directly to snRNAs (Pellizzoni et al. 2002a,b; Yong et al. 2002, 2004b; Golembe et al. 2005a,b;

Battle et al. 2006). The SMN complex performs an essential function in cells by scrutinizing cellular RNAs and ensuring that Sm cores are only assembled onto proper snRNAs (Pellizzoni et al. 2002a). To accomplish this, the SMN complex must be able to recognize specific features of snRNAs in addition to the short Sm site.

HOW DOES THE SMN COMPLEX IDENTIFY snRNAs?

The snRNP Code

Recent work allowed the elucidation of specific sequence and structural features that are recognized by the SMN complex and identified snRNAs for snRNP assembly. Mapping of the minimal SMN complex interacting regions of the major spliceosomal snRNAs (U1, U2, U4, and U5) as well as some of the minor snRNAs (U11) revealed that for each of the snRNAs except U1, the minimal SMN-complex-binding region includes the Sm site, as well as at least one stem-loop immediately 3′ of the Sm site (Fig. 3) (Yong et al. 2002, 2004b). In each case, mutation of the Sm site decreases the binding of the SMN complex (Yong et al. 2004b). U1, however, is different, yet still highly specific. The Sm site of U1 is different from, and not functionally interchangeable with, the Sm site of the other snRNAs. The high-affinity SMN-complex-binding site in U1 snRNA maps to the first stem-loop near the 5′ end of the molecule (SL1), and point mutations in the loop of SL1 disrupt SMN complex binding (Fig. 3) (Yong et al. 2002).

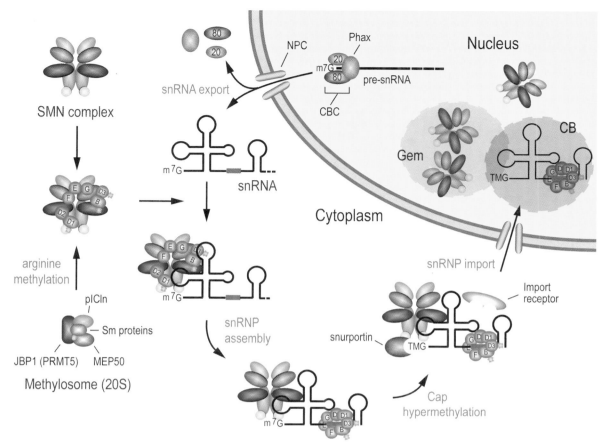

Figure 2. The snRNP assembly pathway. snRNAs are transcribed in the nucleus by RNA polymerase II, bound by the cap-binding complex (CBC) and the PHAX, and then exported to the cytoplasm. Following export, the snRNAs are bound by the SMN complex preassociated with Sm proteins that have been methylated by the 20S methylosome complex. Following Sm core assembly, 3′-end processing, and cap hypermethylation, the snRNPs are bound by snurportin and imported to the nucleus to Gems and Cajal bodies (CB).

Studies using small snRNAs encoded by *Herpesvirus saimiri* (HSURs) provided additional information that allowed a precise definition of the snRNA-binding specificity of the SMN complex. HSURs are viral RNAs whose functions are currently being investigated (Lee et al. 1988; Lee and Steitz 1990; Cook et al. 2004, 2005). The SMN complex assembles Sm cores on canonical Sm sites contained in each of these RNAs (Golembe et al. 2005b). Since these RNAs are small, contain canonical Sm sites, and are assembled into snRNPs by the SMN complex, they are ideal models for studying the details of the SMN complex–snRNA interaction. The minimal SMN-complex-binding site in each of these snRNAs again includes the Sm site and at least one stem-loop immediately 3′ of the Sm site (Fig. 3) (Golembe et al. 2005a). Extensive mutagenesis and phosphothioate interference mapping revealed that the SMN complex recognizes specific nucleotides within the Sm site. Specifically, the SMN complex recognizes the bases of the first adenosine and the first and third uridines of the Sm site (Golembe et al. 2005a). Additionally, the SMN complex senses the phosphate backbone at the positions of the first and third uridines of the Sm site, a profile that is significantly different from that of the Sm protein interaction with these

snRNAs (Golembe et al. 2005a). The SMN complex further requires the presence of the 3′ stem-loop, but the specific nucleotide sequence of the stem-loop and its length are not critical (Golembe et al. 2005a). Additionally, alkaline hydrolysis experiments demonstrated that in each case, the SMN complex forms at least one critical interaction with a region 5′ of the Sm site (Golembe et al. 2005a).

These data have allowed the determination of specific sequence and structural features that permit the SMN complex to bind an RNA and assemble them into snRNPs. Although purified Sm proteins can assemble in vitro on any RNA that contains a short stretch of uridines, in vivo Sm core assembly only occurs on RNAs that bind to the SMN complex. The SMN complex specifically recognizes the first adenosine and the first and third uridines of the Sm site. The SMN complex absolutely requires the presence of a short 7–12-bp stem-loop, although its sequence is not critical. The SMN complex requires that the 3′ end of the snRNA be within a short distance (<14 nucleotides) of the Sm site. Additionally, in all cases, the SMN complex recognizes at least one nucleotide upstream of the Sm site, although this interaction is RNA-specific and can be quite important, as in the case of U1. These features constitute a code that is read by the SMN

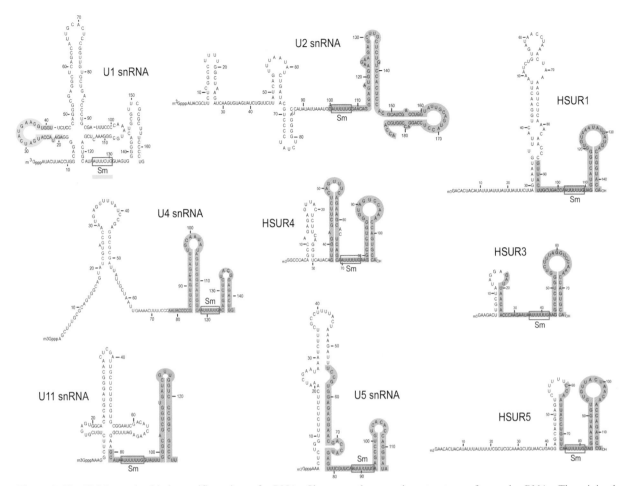

Figure 3. The SMN complex binds specific regions of snRNAs. Shown are the secondary structures of several snRNAs. The minimal SMN-complex-binding regions are highlighted in pink and yellow.

complex and determines which cellular RNAs are defined as snRNAs and assembled into snRNPs (Golembe et al. 2005a).

Gemin5 Is the snRNA-binding Protein of the SMN Complex

Several lines of evidence have demonstrated that the SMN complex itself binds snRNAs independent of Sm proteins. First, Sm proteins are readily removed from the SMN complex by simply washing the complex in high-salt buffer (Pellizzoni et al. 2002a,b; Yong et al. 2002, 2004b; Golembe et al. 2005a,b; Battle et al. 2006). These Sm-free SMN complexes fail to assemble Sm cores, but still bind snRNAs with high affinity and specificity. Second, the specificity of the Sm protein–snRNA interaction differs significantly from that of the SMN complex–snRNA interaction (Yong et al. 2002, 2004b; Golembe et al. 2005a). Although purified Sm proteins will bind any RNA containing an Sm site, the SMN complex will only bind and assemble Sm cores on RNAs that contain the additional sequence and structural features described above as the snRNP code (Golembe et al. 2005a). This left the question of which protein of the SMN complex binds and identifies snRNAs.

Gemin5 is an integral component of the SMN complex, associating with SMN in both the cytoplasm and nuclear gems of metazoan cells (Gubitz et al. 2002). Double-labeling immunofluorescence with antibodies against Gemin5 and Sm proteins shows that Gemin5 is mainly dispersed throughout the cytoplasm and concentrated in nuclear gems, whereas the majority of Sm proteins are found as assembled snRNPs in the nucleus (Fig. 4). Gemin5 is the largest component of the SMN complex, with a molecular mass of approximately 175 kD (Gubitz et al. 2002). Gemin5 is a multidomain protein containing no recognizable RNA-binding motifs. The amino-terminal domain contains 13 WD repeats, whereas the entire carboxy-terminal half of Gemin5 shows no significant sequence homology with any other known proteins (Gubitz et al. 2002). Gemin5 efficiently cross-links to snRNAs in cytoplasmic extract (Battle et al. 2006). The cross-linking is specific to snRNAs that bind to the SMN complex, and the cross-linking occurs within the SMN complex (Battle et al. 2006). Purified Gemin5 from HeLa cells, as well as recombinant Gemin5 expressed in *Escherichia coli*, directly binds to snRNAs. In all cases, the specificity of the Gemin5 interaction with snRNA matches the specificity of the full SMN complex for snRNA (Battle et al. 2006).

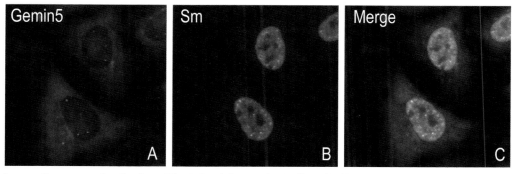

Figure 4. Immunofluorescence localization of Gemin5 and Sm proteins. Indirect double-labeling immunofluorescence on HeLa cells using anti-Gemin5 (10G11) monoclonal antibody (*A, red*) and anti-Sm (Y12) antibody (*B, green*). Combined image is shown in *C*.

Reduction of Gemin5 by RNA interference (RNAi) reduces both the ability of the SMN complex to bind snRNAs and the ability of the SMN complex to assemble Sm cores on snRNAs (Battle et al. 2006). Gemin5 therefore functions as the critical cellular factor that identifies snRNAs and allows the SMN complex to assemble them into snRNPs.

Although recognition of the snRNP code in an RNA by Gemin5 is required for Sm core assembly, it is not sufficient. Gemin5 will bind to any RNA with a snRNP code, and the position of the code within the RNA is not crucial. For example, Gemin5 will bind to an RNA construct in which the snRNP code is at the 5′ end of the molecule and significant random sequence is at the 3′ end of the molecule (Fig. 5A). In fact, the full SMN complex will bind these RNAs via Gemin5 (Golembe et al. 2005a; Battle et al. 2006). However, the SMN complex will not assemble Sm cores on an RNA that does not have the 3′ end of the RNA close in space to the Sm site (Golembe et al. 2005a). Therefore, something in the SMN complex senses the position of the 3′ end of the RNA and ensures that Sm cores are only assembled on snRNAs with proper 3′ ends (Fig. 5B).

A Role for the SMN Complex in the Biogenesis of Other RNPs

In addition to direct interactions with the components of snRNPs, snRNA, and Sm proteins, the SMN complex interacts with RNA-binding proteins that are components of other classes of RNPs, and they can therefore be considered as likely substrates of the SMN complex (see Fig. 1). These interactions suggest that the SMN complex has a central role in the biogenesis of diverse RNPs in addition to its role in snRNP biogenesis. For instance, the replication-dependent histone mRNAs are not polyadenylated at their 3′ ends, but rather are processed at their 3′ ends by the U7 snRNP. The U7 snRNA contains a single-stranded uridine-rich sequence similar to an Sm site (5′-AAUUUGUCUAG-3′). Around this site, a mixed Sm-Lsm core assembles in which SmD1 and SmD2 have been replaced by two other proteins, LSm10 and LSm11. Immunodepletion of the SMN complex from *Xenopus* egg extracts reduced U7 snRNP assembly (Pillai et al. 2003). Gemin5 does not directly interact with the U7

snRNA (Battle et al. 2006); however, the SMN complex has been reported to bind to LSm10 and LSm11 (Pillai et al. 2003; Schumperli and Pillai 2004). It is likely that the SMN complex utilizes its interaction with LSm11 to participate in the assembly of the U7 snRNP.

SMN also binds directly to fibrillarin and GAR1, constituents of box C/D and box H/ACA small nucleolar RNPs (snoRNPs), respectively (Jones et al. 2001; Pellizzoni et al. 2001a). In vitro binding assays performed with full-length and truncated forms of the protein showed that SMN interacts directly with both fibrillarin and GAR1 and that the interactions are mediated by the RG-rich domains of fibrillarin and GAR1. Co-immunoprecipitation experiments demonstrated that the SMN complex interacts with fibrillarin and GAR1 in vivo in human cells (Pellizzoni et al. 2001a). Furthermore, overexpression of a dominant-negative mutant of SMN, SMNΔN27, causes a massive reorganization of snoRNPs, pointing again to a functional interaction between snoRNPs and the SMN complex in vivo. We observed that snoRNPs are depleted from the nucleolus and accumulate in the SMNΔN27-containing structures in cells expressing this dominant-negative mutant of SMN. In these cells, transcription is inhibited in both the nucleoplasm and the nucleolus (Pellizzoni et al. 2001b), and the reorganization of snoRNPs may contribute to the inhibition of nucleolar transcription. The snoRNPs are ribonucleoprotein particles very much akin to snRNPs, and these findings therefore argue that the SMN complex very likely also has a role in the biogenesis of snoRNPs.

Another likely function of the SMN complex is suggested by the observation that the SMN complex interacts with RNA helicase A (RHA) (Pellizzoni et al. 2001b). RHA is an ATP-dependent DEAH-box RNA helicase that associates with RNA polymerase II and has been reported to play a part in transcription. Thus, the SMN complex may also have a role in transcription, specifically in the assembly of the major transcription machinery of the cell. In support of this conclusion, the overexpression of SMNΔN27 mutant leads to the inhibition of transcription in vivo, whereas wild-type SMN leads to stimulation of transcription (Pellizzoni et al. 2001b). In addition, Gemin3 has also been shown to be involved in the regulation of transcription of certain

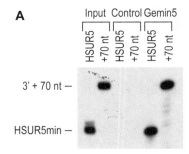

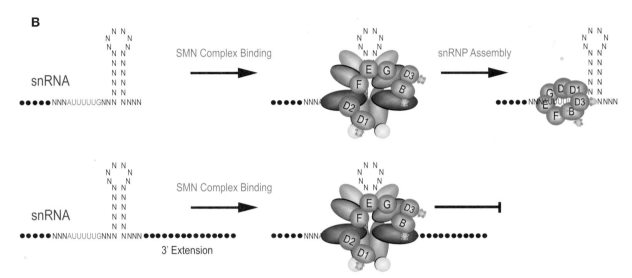

Figure 5. RNA binding by Gemin5 is necessary but not sufficient for Sm core assembly. (*A*) Gemin5 binds to an RNA containing a snRNP code derived from HSUR5 snRNA at the 5′ end of the RNA. Direct RNA-binding assay with Gemin5 and ^{32}P-labeled HSUR5 minimal SMN-complex-binding RNA or the same RNA with additional 70 nucleotides at the 3′ end was performed as described previously (Battle et al. 2006). (*B*) Gemin5 and the full SMN complex bind to RNAs containing a snRNP code followed by a long 3′ extension, but the SMN complex will only assemble Sm cores on snRNAs that have the 3′ end near the Sm site.

reporter genes by RNA polymerase II (Campbell et al. 2000; Yan et al. 2003). However, the mechanism by which SMN and its interacting partners act in these processes remains unknown.

CONCLUSION

The SMN complex has an essential role as an assemblyosome in snRNP biogenesis. This function was unexpected because RNPs in general, and Sm cores in particular, can readily form in vitro from purified Sm proteins and snRNA. In the cell, however, the potential for inaccurate Sm core assembly necessitates a specificity factor to ensure correct RNPs are formed, a function performed in eukaryotes by the SMN complex. What emerged from recent studies is the remarkable capacity of the SMN complex to identify, through Gemin5, specific RNAs as snRNAs and assemble them into snRNPs. Future studies on the functions and structure of the SMN complex should lead to a detailed picture of its mechanism of action and a better understanding of the molecular basis of SMA, and it

should pave the way for the development of therapeutic approaches to this devastating disease.

ACKNOWLEDGMENTS

We thank the members of our laboratory for helpful discussions and comments on the manuscript. We are also grateful to Stacy Grill for secretarial assistance. This work was supported by the Association Française Contre les Myopathies (A.F.M.). G.D. is an investigator of the Howard Hughes Medical Institute.

REFERENCES

Achsel T., Stark H., and Luhrmann R. 2001. The Sm domain is an ancient RNA-binding motif with oligo(U) specificity. *Proc. Natl. Acad. Sci.* **98:** 3685.

Baccon J., Pellizzoni L., Rappsilber J., Mann M., and Dreyfuss G. 2002. Identification and characterization of Gemin7, a novel component of the survival of motor neuron complex. *J. Biol. Chem.* **277:** 31957.

Battle D.J., Lau C., Wan L., Deng H., Lotti F., and Dreyfuss G. 2006. The Gemin5 protein of the SMN complex identifies snRNAs. *Mol. Cell* **23:** 273.

Brahms H., Meheus L., de Brabandere V., Fischer U., and Luhrmann R. 2001. Symmetrical dimethylation of arginine residues in spliceosomal Sm protein B/B′ and the Sm-like protein LSm4, and their interaction with the SMN protein. *RNA* **7:** 1531.

Brahms H., Raymackers J., Union A., de Keyser F., Meheus L., and Luhrmann R. 2000. The C-terminal RG dipeptide repeats of the spliceosomal Sm proteins D1 and D3 contain symmetrical dimethylarginines, which form a major B-cell epitope for anti-Sm autoantibodies. *J. Biol. Chem.* **275:** 17122.

Burd C.G. and Dreyfuss G. 1994. Conserved structures and diversity of functions of RNA-binding proteins. *Science* **265:** 615.

Campbell L., Hunter K.M., Mohaghegh P., Tinsley J.M., Brasch M.A., and Davies K.E. 2000. Direct interaction of Smn with dp103, a putative RNA helicase: A role for Smn in transcription regulation? *Hum. Mol. Genet.* **9:** 1093.

Carissimi C., Saieva L., Baccon J., Chiarella P., Maiolica A., Sawyer A., Rappsilber J., and Pellizzoni L. 2006. Gemin8 is a novel component of the survival motor neuron complex and functions in small nuclear ribonucleoprotein assembly. *J. Biol. Chem.* **281:** 8126.

Charroux B., Pellizzoni L., Perkinson R.A., Shevchenko A., Mann M., and Dreyfuss G. 1999. Gemin3: A novel DEAD box protein that interacts with SMN, the spinal muscular atrophy gene product, and is a component of gems. *J. Cell Biol.* **147:** 1181.

Charroux B., Pellizzoni L., Perkinson R.A., Yong J., Shevchenko A., Mann M., and Dreyfuss G. 2000. Gemin4. A novel component of the SMN complex that is found in both gems and nucleoli. *J. Cell Biol.* **148:** 1177.

Cook H.L., Mischo H.E., and Steitz J.A. 2004. The *Herpesvirus saimiri* small nuclear RNAs recruit AU-rich element-binding proteins but do not alter host AU-rich element-containing mRNA levels in virally transformed T cells. *Mol. Cell. Biol.* **24:** 4522.

Cook H.L., Lytle J.R., Mischo H.E., Li M.J., Rossi J.J., Silva D.P., Desrosiers R.C., and Steitz J.A. 2005. Small nuclear RNAs encoded by *Herpesvirus saimiri* upregulate the expression of genes linked to T cell activation in virally transformed T cells. *Curr. Biol.* **15:** 974.

Coovert D.D., Le T.T., McAndrew P.E., Strasswimmer J., Crawford T.O., Mendell J.R., Coulson S.E., Androphy E.J., Prior T.W., and Burghes A.H. 1997. The survival motor neuron protein in spinal muscular atrophy. *Hum. Mol. Genet.* **6:** 1205.

Fischer U. and Luhrmann R. 1990. An essential signaling role for the m3G cap in the transport of U1 snRNP to the nucleus. *Science* **249:** 786.

Fischer U., Liu Q., and Dreyfuss G. 1997. The SMN-SIP1 complex has an essential role in spliceosomal snRNP biogenesis. *Cell* **90:** 1023.

Fischer U., Sumpter V., Sekine M., Satoh T., and Luhrmann R. 1993. Nucleo-cytoplasmic transport of U snRNPs: Definition of a nuclear location signal in the Sm core domain that binds a transport receptor independently of the m3G cap. *EMBO J.* **12:** 573.

Friesen W.J., Massenet S., Paushkin S., Wyce A., and Dreyfuss G. 2001a. SMN, the product of the spinal muscular atrophy gene, binds preferentially to dimethylarginine-containing protein targets. *Mol. Cell* **7:** 1111.

Friesen W.J., Wyce A., Paushkin S., Abel L., Rappsilber J., Mann M., and Dreyfuss G. 2002. A novel WD repeat protein component of the methylosome binds Sm proteins. *J. Biol. Chem.* **277:** 8243.

Friesen W.J., Paushkin S., Wyce A., Massenet S., Pesiridis G.S., Van Duyne G., Rappsilber J., Mann M., and Dreyfuss G. 2001b. The methylosome, a 20S complex containing JBP1 and pICln, produces dimethylarginine-modified Sm proteins. *Mol. Cell. Biol.* **21:** 8289.

Golembe T.J., Yong J., and Dreyfuss G. 2005a. Specific sequence features, recognized by the SMN complex, identify snRNAs and determine their fate as snRNPs. *Mol. Cell. Biol.* **25:** 10989.

Golembe T.J., Yong J., Battle D.J., Feng W., Wan L., and Dreyfuss G. 2005b. Lymphotropic *Herpesvirus saimiri* uses the SMN complex to assemble Sm cores on its small RNAs. *Mol. Cell. Biol.* **25:** 602.

Gubitz A.K., Mourelatos Z., Abel L., Rappsilber J., Mann M., and Dreyfuss G. 2002. Gemin5, a novel WD repeat protein component of the SMN complex that binds Sm proteins. *J. Biol. Chem.* **277:** 5631.

Hamm J., Kazmaier M., and Mattaj I.W. 1987. In vitro assembly of U1 snRNPs. *EMBO J.* **6:** 3479.

Huber J., Cronshagen U., Kadokura M., Marshallsay C., Wada T., Sekine M., and Luhrmann R. 1998. Snurportin1, an m3G-cap-specific nuclear import receptor with a novel domain structure. *EMBO J.* **17:** 4114.

Jarmolowski A. and Mattaj I.W. 1993. The determinants for Sm protein binding to *Xenopus* U1 and U5 snRNAs are complex and non-identical. *EMBO J.* **12:** 223.

Jones K.W., Gorzynski K., Hales C.M., Fischer U., Badbanchi F., Terns R.M., and Terns M.P. 2001. Direct interaction of the spinal muscular atrophy disease protein SMN with the small nucleolar RNA-associated protein fibrillarin. *J. Biol. Chem.* **276:** 38645.

Kambach C., Walke S., and Nagai K. 1999. Structure and assembly of the spliceosomal small nuclear ribonucleoprotein particles. *Curr. Opin. Struct. Biol.* **9:** 222.

Kleinschmidt A.M., Patton J.R., and Pederson T. 1989. U2 small nuclear RNP assembly in vitro. *Nucleic Acids Res.* **17:** 4817.

Krecic A.M. and Swanson M.S. 1999. hnRNP complexes: Composition, structure, and function. *Curr. Opin. Cell Biol.* **11:** 363.

Lee S.I. and Steitz J.A. 1990. *Herpesvirus saimiri* U RNAs are expressed and assembled into ribonucleoprotein particles in the absence of other viral genes. *J. Virol.* **64:** 3905.

Lee S.I., Murthy S.C., Trimble J.J., Desrosiers R.C., and Steitz J.A. 1988. Four novel U RNAs are encoded by a herpesvirus. *Cell* **54:** 599.

Lefebvre S., Burlet P., Liu Q., Bertrandy S., Clermont O., Munnich A., Dreyfuss G., and Melki J. 1997. Correlation between severity and SMN protein level in spinal muscular atrophy. *Nat. Genet.* **16:** 265.

Lefebvre S., Burglen L., Reboullet S., Clermont O., Burlet P., Viollet L., Benichou B., Cruaud C., Millasseau P., and Zeviani M., et al. 1995. Identification and characterization of a spinal muscular atrophy-determining gene. *Cell* **80:** 155.

Liu Q. and Dreyfuss G. 1996. A novel nuclear structure containing the survival of motor neurons protein. *EMBO J.* **15:** 3555.

Liu Q., Fischer U., Wang F., and Dreyfuss G. 1997. The spinal muscular atrophy disease gene product, SMN, and its associated protein SIP1 are in a complex with spliceosomal snRNP proteins. *Cell* **90:** 1013.

Ma Y., Dostie J., Dreyfuss G., and Van Duyne G.D. 2005. The Gemin6-Gemin7 heterodimer from the survival of motor neurons complex has an Sm protein-like structure. *Structure* **13:** 883.

Mattaj I.W. and De Robertis E.M. 1985. Nuclear segregation of U2 snRNA requires binding of specific snRNP proteins. *Cell* **40:** 111.

Meister G., Buhler D., Pillai R., Lottspeich F., and Fischer U. 2001a. A multiprotein complex mediates the ATP-dependent assembly of spliceosomal U snRNPs. *Nat. Cell Biol.* **3:** 945.

Meister G., Eggert C., Buhler D., Brahms H., Kambach C., and Fischer U. 2001b. Methylation of Sm proteins by a complex containing PRMT5 and the putative U snRNP assembly factor pICln. *Curr. Biol.* **11:** 1990.

Nilsen T.W. 2003. The spliceosome: The most complex macromolecular machine in the cell? *Bioessays* **25:** 1147.

Ohno M., Segref A., Bachi A., Wilm M., and Mattaj I.W. 2000. PHAX, a mediator of U snRNA nuclear export whose activity is regulated by phosphorylation. *Cell* **101:** 187.

Paushkin S., Gubitz A.K., Massenet S., and Dreyfuss G. 2002. The SMN complex, an assemblyosome of ribonucleoproteins. *Curr. Opin. Cell Biol.* **14:** 305.

Pellizzoni L., Yong J., and Dreyfuss G. 2002a. Essential role for the SMN complex in the specificity of snRNP assembly. *Science* **298:** 1775.

Pellizzoni L., Baccon J., Charroux B., and Dreyfuss G. 2001a. The survival of motor neurons (SMN) protein interacts with the snoRNP proteins fibrillarin and GAR1. *Curr. Biol.* **11:** 1079.

Pellizzoni L., Baccon J., Rappsilber J., Mann M., and Dreyfuss G. 2002b. Purification of native survival of motor neurons complexes and identification of Gemin6 as a novel component. *J. Biol. Chem.* **277:** 7540.

Pellizzoni L., Charroux B., Rappsilber J., Mann M., and Dreyfuss G. 2001b. A functional interaction between the survival motor neuron complex and RNA polymerase II. *J. Cell Biol.* **152:** 75.

Pillai R.S., Grimmler M., Meister G., Will C.L., Luhrmann R., Fischer U., and Schumperli D. 2003. Unique Sm core structure of U7 snRNPs: Assembly by a specialized SMN complex and the role of a new component, Lsm11, in histone RNA processing. *Genes Dev.* **17:** 2321.

Plessel G., Fischer U., and Luhrmann R. 1994. m3G cap hypermethylation of U1 small nuclear ribonucleoprotein (snRNP) in vitro: Evidence that the U1 small nuclear RNA-(guanosine-N2)-methyltransferase is a non-snRNP cytoplasmic protein that requires a binding site on the Sm core domain. *Mol. Cell. Biol.* **14:** 4160.

Raker V.A., Plessel G., and Luhrmann R. 1996. The snRNP core assembly pathway: Identification of stable core protein heteromeric complexes and an snRNP subcore particle in vitro. *EMBO J.* **15:** 2256.

Raker V.A., Hartmuth K., Kastner B., and Luhrmann R. 1999. Spliceosomal U snRNP core assembly: Sm proteins assemble onto an Sm site RNA nonanucleotide in a specific and thermodynamically stable manner. *Mol. Cell. Biol.* **19:** 6554.

Schumperli D. and Pillai R.S. 2004. The special Sm core structure of the U7 snRNP: Far-reaching significance of a small nuclear ribonucleoprotein. *Cell. Mol. Life Sci.* **61:** 2560.

Segref A., Mattaj I.W., and Ohno M. 2001. The evolutionarily conserved region of the U snRNA export mediator PHAX is a novel RNA-binding domain that is essential for U snRNA export. *RNA* **7:** 351.

Stark H., Dube P., Luhrmann R., and Kastner B. 2001. Arrangement of RNA and proteins in the spliceosomal U1 small nuclear ribonucleoprotein particle. *Nature* **409:** 539.

Sumpter V., Kahrs A., Fischer U., Kornstadt U., and Luhrmann R. 1992. In vitro reconstitution of U1 and U2 snRNPs from isolated proteins and snRNA. *Mol. Biol. Rep.* **16:** 229.

Wan L., Battle D.J., Yong J., Gubitz A.K., Kolb S.J., Wang J., and Dreyfuss G. 2005. The survival of motor neurons protein determines the capacity for snRNP assembly: Biochemical deficiency in spinal muscular atrophy. *Mol. Cell. Biol.* **25:** 5543.

Will C.L. and Luhrmann R. 2001. Spliceosomal UsnRNP biogenesis, structure and function. *Curr. Opin. Cell Biol.* **13:** 290.

Yan X., Mouillet J.F., Ou Q., and Sadovsky Y. 2003. A novel domain within the DEAD-box protein DP103 is essential for transcriptional repression and helicase activity. *Mol. Cell. Biol.* **23:** 414.

Yong J., Pellizzoni L., and Dreyfuss G. 2002. Sequence-specific interaction of U1 snRNA with the SMN complex. *EMBO J.* **21:** 1188.

Yong J., Wan L., and Dreyfuss G. 2004a. Why do cells need an assembly machine for RNA-protein complexes? *Trends Cell Biol.* **14:** 226.

Yong J., Golembe T.J., Battle D.J., Pellizzoni L., and Dreyfuss G. 2004b. snRNAs contain specific SMN-binding domains that are essential for snRNP assembly. *Mol. Cell. Biol.* **24:** 2747.

Developing Global Insight into RNA Regulation

R.B. DARNELL

Howard Hughes Medical Institute, Laboratory of Molecular Neuro-Oncology,
The Rockefeller University, New York, New York 10021

Systematic dissection of the activity of RNA-binding proteins (RBPs) has begun to yield global insight into how they work. The paradigm we have used has been the study of Nova, a neuron-specific RBP targeted in an autoimmune neurologic disorder associated with cancer. We have developed a combination of biochemical, genetic, and bioinformatic methods to generate a global understanding of Nova's role as a splicing regulator. Genome-wide identification and validation of Nova target RNAs have yielded unexpected insights into the protein's mechanism of action, its role in neurobiology, and the unique roles RBPs have in the biology of the neuronal synapse. These studies provide us with a paradigm for understanding the role of RBPs in neurons and in disease and, more generally, with the hope that it will be feasible to develop a comprehensive understanding of posttranscriptional regulation.

RBPs IN NEURONS AND NEUROLOGIC DISEASE

RNA has greater complexity than DNA. RNA folds into complex shapes, utilizing both sequence and structure, allowing it to harbor both information content and enzymatic activity. From a single DNA template, RNA can offer combinatorial complexity (via alternative splicing of exons or RNA editing), as well as the ability to regulate protein expression in space and time (Darnell 2002). Given the need for complexity in the nervous system, where there is an intuitive mismatch between the approximately 2×10^4 primary transcripts encoded by DNA and the approximately 10^{11} neurons, each of which may harbor about 10^7 synaptic connections needing regulation, it is natural to wonder whether RNA regulation might contribute to neuronal complexity.

The first clue that neurons might have a unique system for regulating RNA metabolism came from studies of the intersection between cancer cells and neurons. Ron Evans and colleagues' interest in the study of calcitonin, a hormone expressed in the thyroid gland (and, although not as well appreciated, also in the brain) led them to study a medullary carcinoma of the thyroid tumor cell line as a model for comparative gene expression studies. These authors found that a unique transcript of the calcitonin gene was expressed in the tumor cell line and, in further studies, that this was an alternatively spliced isoform normally expressed in the brain. This isoform turned out to encode a completely different protein—calicitonin-gene-related peptide—from the primary "calicitonin" pre-mRNA transcript (Rosenfeld et al. 1983). These studies established two interesting points. First, there was the possibility that the brain may have its own special system for regulating RNA expression, able to generate unique RNA and hence protein isoforms from a single primary transcript. Second, analyzing the dysregulation of gene expression in tumor cells might paradoxically yield insight into neuron-specific gene expression.

The second insight was extended in a more systematic fashion through studies of a group of neurologic disorders termed the paraneoplastic neurologic disorders (PNDs) (Darnell and Posner 2003b, 2006). Interestingly, these disorders also manifest at the intersection of tumor and neurobiology. PND patients present with specific neurodegenerative syndromes, which can vary widely between patients, and include memory loss, blindness, cerebellar dysfunction, and motor or sensory disorders. For each set of neurologic symptoms, there are characteristic tumors present in these patients, although typically they have not yet presented clinically at the time neurologic illness sets in. These tumors have not invaded the nervous system and are typically limited in their extent of spread. A model for the pathogenesis of the disorders is based on the findings of Jerome Posner and colleagues, who in the early 1980s first found evidence of an immunologic link between cancer and neurologic syndrome (reviewed in Darnell 1996). The model proposes that tumors present in PND patients initiate the syndrome when they express proteins that are normally restricted in their expression to the nervous system. Because there is a blood–brain barrier, the immune system is able to mount what turns out to be an effective immune response to neuronal antigens expressed in peripheral tumors. This accounts for the occult nature of the tumors in these patients and the limited stage of their disorders. In fact, PNDs provide what is perhaps the best model for naturally occurring tumor immunity in humans (Darnell and Posner 2003a). Patients do not present to clinicians until some poorly understood event allows this immune response to break the immunologic blood–brain barrier and then attack those neurons that were normally expressing the neuronal antigens co-opted by the tumor. Although the details of disease pathogenesis remain under investigation (Albert and Darnell 2004), our laboratory established methods to use the high-titer antibodies in PND patients to screen expression cDNA libraries and to identify the genes encoding a number of target PND antigens (Newman et al. 1995; Darnell 1996). To date, more than a dozen such genes have been identified (Darnell and Posner 2006).

One set of PND antigens that tumor cells consistently express are neuron-specific RBPs. Two families of such proteins were discovered by using PND antisera to screen cDNA libraries—the Nova proteins (Buckanovich et al. 1993), targeted in patients harboring lung or gynecologic cancers and manifesting by neurologic symptoms of excess motor movements (paraneoplastic opsoclonus-myoclonus ataxia [POMA]), and the Hu proteins (Szabo et al. 1991). Although the functions of the Hu proteins in the brain are still incompletely understood (Musunuru and Darnell 2001), we have been able to establish that Nova regulates neuron-specific alternative splicing in an interesting subset of pre-mRNAs (Ule and Darnell 2006). Studies of the Nova proteins have established a crude template for attempting to understand RBP function on a genome-wide scale. We review here this template, which includes three main components—biochemical, genetic, and bioinformatic—that together provide a means to identify the in vivo RNAs bound by Nova and the functional understanding that results from this template.

BIOCHEMICAL UNDERPINNINGS TO UNDERSTANDING RBP FUNCTION

An essential foundation in approaching RBP function is a detailed understanding of the nature of the protein–RNA interaction. As a first step, this can be approached in an idealized setting in vitro, with the goal of yielding as detailed an understanding as possible, preferably by X-ray crystallography, of the means by which an RBP recognizes its RNA substrate.

For Nova, which we found to be an RBP harboring three KH-type RNA-binding domains (Buckanovich et al. 1996), we approached this problem by undertaking in vitro RNA selection experiments, using protocols established by Jack Szostak and colleagues (Green et al. 1991; Szostak and Ellington 1993). Two different sets of experiments were done. First, an idealized Nova target RNA was identified using long random RNA libraries and full-length recombinant Nova protein. This led to the identification of a stem-loop RNA harboring a core 4-nucleotide repeat sequence—(UCAU)$_3$—present in the loop (Fig. 1) (Buckanovich and Darnell 1997; Yang et al. 1998). Mutagenesis studies identified the CA dinucleotide as a critical invariant component of binding, with some flexibility allowed in the flanking U nucleotides and in the stem. Second, these studies were complemented by collaboration with the crystallography laboratory of Stephen Burley, who proposed solving the crystal structure of a single KH

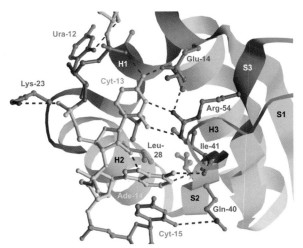

Figure 2. X-ray crystallographic structure of the Nova KH3 domain bound to RNA. The KH3 domain (*turquoise ribbon*) folds such that several side-chain amino acids (e.g., Glu-14 and Arg-54) are precisely positioned to provide appropriate hydrogen bond donor/acceptor groups to RNA (*colored stick figure*). The most precise coordination is for Cyt-13 and Ade-14 (the "CA" core of the YCAY motif), such that the hydrogen bonds are precisely those that would normally be supplied by a second-strand nucleic acid (e.g., "GT"), in this way exactly specifying the sequence. The surrounding nucleotides (Ura-12 and Cyt-15) are also restricted, but less precisely, to pyrimidine residues. (Adapted, with permission, from Lewis et al. 2000 [© Elsevier].)

domain bound to RNA. Two experiments were key to solving this structure. First, Burley's lab undertook a detailed analysis of the structure of Nova by limited proteolysis, delimiting a core protease-resistant region surrounding the KH domains that would prove to successfully form crystals. Repeating these studies in the presence of RNAs also provided a crucial result, identifying a carboxy-terminal extension on the KH domain that was protease-resistant only when Nova was bound to RNA, suggesting a role in RNA binding (Lewis et al. 1999). Second, we repeated RNA selection experiments to optimize an RNA target for crystallography with the defined, crystallizable, Nova KH3 domain (Jensen et al. 2000a). Putting these two experiments together yielded a high-resolution X-ray structure of the Nova KH3 domain bound to RNA (Fig. 2) (Lewis et al. 2000). This structure confirmed and extended our biochemistry data, demonstrating that the Nova KH domain folds to position side-chain amino acids to precisely contribute hydrogen bond donor/acceptors in the same manner a second-strand nucleic acid would, to exactly specify the CA core dinucleotide. The KH domain also delimited the flanking residues to be pyrimidines ("Y"), whereas the stem was not bound directly and appeared to function largely to keep the YCAY core element in the loop unbound to other RNA residues and thereby free for protein interaction.

GENETIC SYSTEMS AND RBP FUNCTION

The recognition of YCAY elements as the core component specified by the Nova KH domain encouraged us to search for brain transcripts that might encode

CONSENSUS   STEM- UCAUYCAUYUCAUY- STEM

Figure 1. Schematic of the consensus RNA selected with recombinant Nova fusion protein. Nova recognizes RNAs harboring a core loop element (*blue*) consisting of YCAY repeats (as determined by mutagenesis studies), presented in the context of an inverted repeat forming a stem element. (Adapted, with permission, from Buckanovich and Darnell 1997. [©ASM].)

repeats of these elements and hence by Nova targets. Our first very crude approach to this problem yielded one fortuitous hit. Using Microsoft Word as a search tool, we examined by hand intronic and exonic sequences present in a database of 350 neuronal transcripts that had been established at CSHL by Stamm and Helfman (Stamm et al. 1994). We identified the YCAY cluster in this database within an intronic sequence upstream of an alternatively spliced exon (E3A) of the inhibitory glycine receptor α2 (GlyRα2). Generation of a minigene encoding this element and the surrounding exons demonstrated that in transfected tissue culture cells, Nova was able to mediate an increase in inclusion of E3A. This led to the hypothesis that Nova might regulate alternative splicing of this, and other transcripts, in neurons.

We then crossed a biologic threshold, committing to testing this hypothesis in vivo by generating Nova null mice. These studies proved to be crucial, as they not only gave a critical test of our biochemical data, but also allowed us to develop a robust genetic template for use in further biochemical studies. Assessing GlyRα2 splicing in Nova1 KO mice, we found a consistent twofold decrease in E3A utilization (Jensen et al. 2000b), consistent with the increase mediated by excess Nova in tissue culture cells (Buckanovich and Darnell 1997).

To assay the specificity of Nova's action, we analyzed splicing of a small set of alternative exons in Nova KO mouse brain. We found that other alternatively spliced brain transcripts were unaffected by the presence or absence of Nova, indicating the specificity of Nova's action, but with one exception, the γ2L exon of the GABA$_A$ transcript (Jensen et al. 2000b). Given that Nova was targeted in a PND in which patients had an excess motor activity—interpreted by neurologists to be a defect in inhibitory motor control—the finding that two of two Nova-regulated transcripts encoded inhibitory neurotransmitter receptors was tantalizing but inconclusive, given the relatively arbitrary nature in which these transcripts had been identified. This frustration helped fuel a push to develop new methods for genome-wide identification of Nova targets, in order to test the hypothesis that Nova might regulate a restricted set of mRNAs whose functions relate to the neurology of the Nova PND syndrome. Throughout these studies, however, we followed the paradigm established above: Candidate Nova targets, identified by any method, would need to be validated in vivo in a biologically relevant genetic system (Nova KO vs. WT brain) and supported, where feasible, by biochemical studies.

BIOINFORMATICS, GENETICS, AND BIOCHEMISTRY: THE HOLISTIC APPROACH TO RBP FUNCTION

An additional advantage of identifying the GABA$_A$ transcript as a Nova target was that it was done independently of a search for specific (e.g., YCAY) binding sequences, thereby allowing us to search blindly for a necessary and sufficient element in the GABA$_A$ transcript able to mediate the action of Nova on γ2L exon inclusion. These studies revealed a core 24-nucleotide element able to confer Nova-dependent splicing on a heterologous transcript, and sequencing of this element revealed that it was full of YCAY elements (Dredge and Darnell 2003). Biochemical studies of this and one additional Nova-regulated exon (an autoregulated exon in the Nova1 transcript itself) (Dredge et al. 2005) suggested that a core cluster of three YCAY elements was critical in Nova-mediated regulation of splicing.

Taken together, these observations set the groundwork for undertaking a bioinformatic screen for Nova target transcripts (Ule et al. 2006). We developed an algorithm to score transcripts as potential Nova targets on the basis of their YCAY clusters. We used this algorithm to search a set of approximately 50 known Nova targets (including those identified by new methods described below). These studies identified YCAY clusters in Nova targets that had been identified and validated independently of their sequence composition, supporting the importance of this RNA motif in mediating Nova action. Importantly, the YCAY cluster scoring algorithm was robust enough to predict 30 RNA targets based on the presence of YCAY clusters, and these all proved to be bona fide targets when tested in Nova KO versus WT brains (Ule et al. 2006). This helped validate the algorithm itself, as well as lending further support to the significance of YCAY as a biologically relevant Nova-binding motif.

NOVA RNA MAP

These bioinformatic studies yielded an unexpected finding: The position of the YCAY cluster systematically correlated with the action of Nova on alternative exon inclusion or exclusion (Ule et al. 2006). Clusters immediately upstream or within alternative exons predicted an action of Nova to inhibit exon inclusion, whereas clusters downstream from alternative exons predicted an action to enhance exon inclusion. Thus, a very tightly defined RNA-binding map was generated, in which both the sequence and position of Nova-binding elements determined function.

Such studies clearly have mechanistic implications, suggesting that the position of Nova binding to pre-mRNA precisely relates to its effect on the splicing machinery. Thus, it was logical at this point to undertake definitive mechanistic studies to evaluate this issue. In vitro splicing assays were established, using two sets of model pre-mRNAs, those in which Nova either enhanced or inhibited exon inclusion. In both systems, we were able to demonstrate that purified Nova was able to either enhance or inhibit exon inclusion in a manner consistent with this RNA map, dependent on both the sequence and position of YCAY elements. In parallel, quantification of splicing intermediates in Nova KO brain demonstrated a direct and asymmetric action of Nova on introns harboring or proximate to YCAY-binding elements, suggesting a mechanistic model distinct from a role in exon definition, but rather one in which Nova might act locally on spliceosome assembly. Consistent with this idea, in vitro splicing assays were able to demonstrate actions of Nova on the formation of the basal machinery of the spliceosome; for example, Nova inhibition of splicing by binding at the 3′ end of an alternative exon led to inhibition of U1 small nuclear ribonucleoprotein (snRNP) binding and hence inefficient exon inclusion (Ule et al. 2006).

HOLISTIC PIONEERING: NEW METHODS TO UNDERSTAND RBP FUNCTION

An essential ingredient in developing a global understanding of RBP function is defining a robust list of biologically validated RNA targets. Several approaches have been considered by different laboratories (Blencowe 2005, 2006), although some methods have had difficulty in generating consistent results. Among these, perhaps the most widely used has been one in which RBPs of interest are immunoprecipitated, and RNA identified in the precipitates by reverse transcriptase–polymerase chain reaction (RT-PCR). Problems with this approach include precipitation (typically done under low-to-moderate stringency) of additional associated RBPs along with their RNA targets, signal:noise problems, and reassociation of RBPs with new RNA targets during immunoprecipitation (Mili and Steitz 2004).

The ability to validate Nova targets, by combining biochemical, genetic (Nova KO), and bioinformatic (YCAY cluster) analyses, gave us a base from which to try to develop new methods to overcome problems in identification of valid in vivo protein–RNA interactions. We undertook two new approaches to identifying Nova targets, one biochemically based, and a second microarray/bioinformatics based.

The first of these methods, termed CLIP (for cross-link-immunoprecipitation) (Ule et al. 2003, 2005a), takes advantage of a long-standing biochemists' trick, which is the finding that UV-B irradiation is able to induce covalent complexes between protein–nucleic acid (but not protein–protein) interactions, when contact distances are within about 1 Å. By applying UV-irradiation to acutely dissected mouse brains, we were able to covalently cross-link Nova–RNA complexes in situ. Once formed, these complexes are extremely stable (samples can be frozen and stored for future experiments), and an RBP could be rigorously purified (to homogeneity if desired). For Nova, our purification was severalfold: immunoprecipitation with a high-titer specific antibody under very stringent conditions, boiling complexes in SDS-sample buffer, running complexes on SDS-polyacrylamide gel electrophoresis (SDS-PAGE) gels, and transferring them to nitrocellulose. During this procedure, RNA is partially hydrolyzed to a modal size of approximately 50–70 nucleotides, and at the end of the purification, protein is removed by proteinase K, and RNA linkers are directionally added to the RNA, which is then PCR-amplified and sequenced. CLIP gives a snapshot of where Nova is bound in vivo; our initial studies identified binding sites present in large introns, binding sites in the vicinity of alternate exons, and binding sites in other regions (untranslated region, intergenic) and allowed us to validate seven new Nova

splicing targets that could be validated in Nova KO brain (Ule et al. 2003). Our current analysis suggests that perhaps only about 10–20% of CLIP targets harbor high YCAY cluster score sequences, with the majority of the remaining RNAs harboring lower YCAY cluster scores (J. Ule and R.B. Darnell, unpubl.). This suggests that Nova may spend a considerable amount of its time sampling a wide range of RNAs for high-affinity binding sites; further studies will be required to assess whether Nova has any biologic role in the wider range of RNAs with weaker binding sites.

A second new approach to identifying Nova target RNAs arose through a collaboration with Affymetrix, which had developed a new alternative splicing microarray that was itself in need of validation. This microarray, a prototype for more comprehensive chips now under development, harbored 40,443 perfect match and mismatch probe sets spanning alternative exons; importantly, these included probe sets for both exon-included isoforms and the corresponding exon-skipped isoform. This proved to be essential, as statistical analysis of differences in exon inclusion *or* exclusion alone, comparing Nova KO and WT brain, only yielded a predictive power of about 20%. As a result, we developed an algorithm, termed ASPIRE, in which we demanded reciprocal changes for any one putative Nova-regulated exon. By searching for such reciprocal changes in independent probe sets that measured exon inclusion *and* exon exclusion, our predictive power improved drastically. Using ASPIRE, we were able to validate 49 of 49 of our top predicted Nova-regulated exons identified in comparison of Nova KO and WT RNAs (Ule et al. 2005b).

The value of these two new methods is severalfold: First, the relatively large list of validated Nova targets provided a feed-forward data set to allow further target identification. Specifically, the ability to examine and compare known targets with control, alternatively spliced transcripts allowed the definition of the RNA map for genome-wide prediction of Nova activity. In addition, the larger set of validated targets has led us forward to be able to begin to look at Nova function with a comprehensive viewpoint.

IMPLICATIONS DERIVED FROM A GLOBAL UNDERSTANDING OF RBP ACTION

Functional Considerations

The beginning of a global understanding of the set of coregulated RNAs in the brain allows a new direction to be pursued: analysis of the functions encoded by those RNAs. From our early studies identifying GlyRα2 and GABA$_A$ transcripts as Nova targets and recognizing that they both encoded inhibitory neurotransmitter receptors, we have been aware of the possibility that Nova acts on a biologically restricted set of RNAs and that those RNAs might relate to the pathogenesis of the inhibitory motor dysfunction evident in patients with the paraneoplastic POMA syndrome. However, because these RNAs were not identified in an objective genome-wide screen, we were hesitant to make too much of these observations.

Nonetheless, we were further drawn to this observation after our first CLIP experiments. Analysis of 34 transcripts that had been identified multiple times in the 340 CLIP targets originally sequenced revealed that these too had a biologic coherence; 71% of these RNAs encoded proteins that function in the neuronal synapse (Ule et al. 2003). Since CLIP qualified as an unbiased, genome-wide screen, this provided the first compelling evidence that Nova might regulate a restricted set of transcripts in the brain. Nonetheless, the sequencing of 340 CLIP tags cannot be considered an exhaustive analysis of Nova RNA targets. Although subsequent analysis of CLIP targets has continued to strengthen our original observations (J. Ule and R.B. Darnell, unpubl.), it was a genome-wide microarray analysis that provided overwhelming evidence regarding the nature of Nova RNA targets.

Our microarray screen for Nova-regulated alternative exons interrogated 40,443 exons in 7,175 transcripts; 49 Nova-regulated transcripts were identified and validated by RT-PCR. We analyzed the biologic functions encoded by these 49 transcripts using two approaches. In the first, we used the gene-ontology (GO) annotations of encoded functions, as a means of providing an unbiased assessment in a way capable of strict statistical analysis. This analysis revealed that Nova-regulated transcripts were highly enriched (in a statistically robust manner: P <0.001, false discover rate 0.03, with 6–13-fold enrichment) in proteins that act at the cell junctions, suggesting that they were synaptic proteins (the most enriched GO categories were synapse biogenesis, synaptic transmission, cell–cell signaling, cortical actin organization beneath the membrane, cell adhesion, and extracellular matrix organization; Ule et al. 2005b).

These observations were further supported by individual annotation, searching PubMed for data regarding the biologic function of each validated Nova target. These data led to a confirmatory and more specific picture: The overwhelming majority (essentially all) of Nova targets encoded proteins functionally related to the neuronal synapse. No changes in steady-state levels were detected in the absence of Nova, indicating that Nova regulates the quality, but not the quantity, of a discrete set of synaptic proteins (Ule et al. 2005b).

This in turn suggests that Nova is likely to regulate the physiology of the synapse. Our first test of this hypothesis came from an evaluation of several Nova targets that had been repeatedly identified by CLIP, and whose encoded proteins were part of a new physiologic circuit being studied by Lily Jan and colleagues. This circuit represents a long-term inhibitory response to tetanic or coincidence (long-term potentiating [LTP]) stimuli. Jan and colleagues had found that LTP of slow inhibitory postsynaptic currents (sIPSC) were dependent on GIRK2, GABAB, and CaMKII proteins, all of which were encoded by transcripts that were Nova targets. LTP of sIPSC was therefore evaluated in Nova KO mice and was found to be completely and specifically absent (other parameters of synaptic function were normal; Huang et al. 2005). This study provided the first demonstration that identification of Nova RNA targets on a global scale

can provide specific insights into physiologic function. Moreover, this study furthered the notion that RNA regulation has important roles in modulating synaptic plasticity (hippocampal LTP), which is thought to represent the physiologic correlate of complex information processing in the brain (Ule and Darnell 2006).

Evolutionary Considerations

A second set of issues arising from developing global insight into RBP function are evolutionary considerations. In the RNA world view, RNA molecules were the first informatic and enzymatic dual-function molecules to arise in evolution. How then did proteins evolve to harness the power of RNA? This question is directly approached in considering how RBPs evolved to regulate the complexity of information at a level that is specific to RNA molecules: alternative splicing.

Our studies with Nova point out some interesting issues in considering this problem. The Nova-binding site is rather low complexity—YCAY motifs occur on average every 64 nucleotides. This motif is even simpler than the characterized transcription-factor-binding sites, which are typically in the 6–8-nucleotide range. What is the relevance of such a finding?

Such low-complexity sequences are relatively easy to evolve by mutations. And the low constraints on such evolution—the only strict requirement in Nova KH binding is a CA dinucleotide—would thereby enable a larger fraction of the genome to sample the consequences of evolving a Nova-binding site, with those reaping a benefit able to increase, over time, the density of YCAY motifs. This in turn would solidify Nova binding, as its affinity for RNA targets increases with increasing density of YCAY motifs (Buckanovich and Darnell 1997; Yang et al. 1998; Jensen et al. 2000a; Dredge and Darnell 2003; Musunuru and Darnell 2004; Dredge et al. 2005). Thus, the number of Nova-regulated targets can grow as evolution generates greater complexity in the genome; for example, through duplication of exons or genes. In fact, preliminary analysis of the evolutionary conservation of Nova-regulated exons, suggesting a growing set of Nova-regulated exons through evolution from invertebrates to chick to mammals, is consistent with this idea (J. Ule et al., unpubl.).

One corollary to the idea that Nova-binding sequences may be widely dispersed and rapidly evolving is that Nova itself is tightly fixed in evolution. This fixation is essential, as mutations altering the recognition motif of Nova would simultaneously destroy the regulation of an array of crucial alternate exon information. In fact, the Nova KH domain is extremely tightly conserved down through invertebrates (Buckanovich et al. 1993; R.B. Darnell et al., unpubl.), suggesting that the ability to bind a YCAY motif became a powerful but unalterable facet of RBP-RNA regulation early in evolution. This turns RNA regulation on its head in a way that is in harmony with the idea of the RNA world: RNA remains the powerful emerging evolutionary force, whereas the protein regulators take on roles as inert drones to mediate the regulation that RNA demands.

A second corollary is that a strict biologic coherence to transcripts harboring Nova-binding sites may be maintained and even refined through such a model. As we have observed, Nova targets are almost uncannily restricted to encoding proteins involved in synaptic biology. At the same time, it is clear that neurons (presumably through other factors) regulate alternative splicing of many other kinds of transcripts; in our analysis, the largest such groups, defined by the GO category, were transcripts encoding proteins involved in the regulation of metabolism, biosynthesis, and transcription, and yet Nova regulated no RNAs in these categories.

An interesting question for the future will be to explore the extent to which Nova might contribute to the complexity in the regulation of synaptic function between different neuronal types or even within a single neuron. It is clear that Nova-regulated exons respond to the presence of Nova in a dose-dependent manner and that different exons have a different threshold for Nova action. Thus, titration of Nova levels within a single neuron may lead to an array of actions on alternative exons, each tuned to a different degree of sensitivity to Nova levels. The correlation of Nova sensitivity with YCAY cluster scores may provide a means of evaluating this idea. Furthermore, there are many thousands of synapses within a single neuron, and the question arises as to whether Nova RNA regulation might be able to differentially modulate activity at one synapse relative to another. The exciting finding that Nova regulates LTP of sIPSC (Huang et al. 2005), together with the finding that Nova is present at neuronal synapses (R.B. Darnell et al., unpubl.), suggests that there may yet remain undiscovered dimensions to the ways in which RBPs may regulate the complexity of RNA expression.

ACKNOWLEDGMENTS

This work is an overview of the efforts of a large number of people and their experiments undertaken over the years. Although I have tried to cite the efforts of all, I would especially thank Steven Burley, Ron Buckanovich, Jennifer Darnell, Kate Dredge, Lily Jan, Kirk Jensen, Hal Lewis, Aldo Mele, Kiran Musunuru, Giovanni Stefani, and Jernej Ule for major contributions toward developing the key points in the development of the story told here. This work was supported by the National Institutes of Health (R01 NS34389 and NS40955 to R.B.D.) and the Howard Hughes Medical Institute. R.B.D. is an Investigator of the Howard Hughes Medical Institute.

REFERENCES

Albert M.L. and Darnell R.B. 2004. Paraneoplastic neurological degenerations: Keys to tumour immunity. *Nat. Rev. Cancer* **4:** 36.

Blencowe B.J. 2005. Splicing on the brain. *Nat. Genet.* **37:** 796.

———. 2006. Alternative splicing: New insights from global analyses. *Cell* **126:** 37.

Buckanovich R.J. and Darnell R.B. 1997. The neuronal RNA binding protein Nova-1 recognizes specific RNA targets in vitro and in vivo. *Mol. Cell. Biol.* **17:** 3194.

Buckanovich R.J., Posner J.B., and Darnell R.B. 1993. Nova, the paraneoplastic Ri antigen, is homologous to an RNA-binding protein and is specifically expressed in the developing motor system. *Neuron* **11:** 657.

Buckanovich R.J., Yang Y.Y., and Darnell R.B. 1996. The onconeural antigen Nova-1 is a neuron-specific RNA-binding protein, the activity of which is inhibited by paraneoplastic antibodies. *J. Neurosci.* **16:** 1114.

Darnell R.B. 1996. Onconeural antigens and the paraneoplastic neurologic disorders: At the intersection of cancer, immunity and the brain. *Proc. Natl. Acad. Sci.* **93:** 4529.

———. 2002. RNA logic in time and space. *Cell* **110:** 545.

Darnell R.B. and Posner J.B. 2003a. Observing the invisible: Successful tumor immunity in humans. *Nat. Immunol.* **4:** 201.

———. 2003b. Paraneoplastic syndromes involving the nervous system. *N. Engl. J. Med.* **349:** 1543.

———. 2006. Paraneoplastic syndromes affecting the nervous system. *Semin. Oncol.* **33:** 270.

Dredge B.K. and Darnell R.B. 2003. Nova regulates GABA(A) receptor gamma2 alternative splicing via a distal downstream UCAU-rich intronic splicing enhancer. *Mol. Cell. Biol.* **23:** 4687.

Dredge B.K., Stefani G., Engelhard C.C., and Darnell R.B. 2005. Nova autoregulation reveals dual functions in neuronal splicing. *EMBO J.* **24:** 1608.

Green R., Ellington A.D., Bartel D.P., and Szostak J.W. 1991. *In vitro* genetic analysis: Selection and amplification of rare functional nucleic acids. *Methods* **2:** 75.

Huang C.S., Shi S.H., Ule J., Ruggiu M., Barker L.A., Darnell R.B., Jan Y.N., and Jan L.Y. 2005. Common molecular pathways mediate long-term potentiation of synaptic excitation and slow synaptic inhibition. *Cell* **123:** 105.

Jensen K.B., Musunuru K., Lewis H.A., Burley S.K., and Darnell R.B. 2000a. The tetranucleotide UCAY directs the specific recognition of RNA by the Nova KH3 domain. *Proc. Natl. Acad. Sci.* **97:** 5740.

Jensen K.B., Dredge B.K., Stefani G., Zhong R., Buckanovich R.J., Okano H.J., Yang Y.Y., and Darnell R.B. 2000b. Nova-1 regulates neuron-specific alternative splicing and is essential for neuronal viability. *Neuron* **25:** 359.

Lewis H.A., Musunuru K., Jensen K.B., Edo C., Chen H., Darnell R.B., and Burley S.K. 2000. Sequence-specific RNA binding by a Nova KH domain: Implications for paraneoplastic disease and the fragile X syndrome. *Cell* **100:** 323.

Lewis H.A., Chen H., Edo C., Buckanovich R.J., Yang Y.Y., Musunuru K., Zhong R., Darnell R.B., and Burley S.K. 1999. Crystal structures of Nova-1 and Nova-2 K-homology RNA-binding domains. *Structure* **7:** 191.

Mili S. and Steitz J.A. 2004. Evidence for reassociation of RNA-binding proteins after cell lysis: Implications for the interpretation of immunoprecipitation analyses. *RNA* **10:** 1692.

Musunuru K. and Darnell R.B. 2001. Paraneoplastic neurologic disease antigens—RNA-binding proteins and signaling proteins in neuronal degeneration. *Annu. Rev. Neurosci.* **24:** 239.

———. 2004. Determination and augmentation of RNA sequence specificity of the Nova K-homology domains. *Nucleic Acids Res.* **32:** 4852.

Newman L.S., McKeever M.O., Okano H.J., and Darnell R.B. 1995. β-NAP, a cerebellar degeneration antigen, is a neuron-specific vesicle coat protein. *Cell* **82:** 773.

Rosenfeld M.G., Mermod J.J., Amara S.G., Swanson L.W., Sawchenko P.E., Rivier J., Vale W.W., and Evans R.M. 1983. Production of a novel neuropeptide encoded by the calcitonin gene via tissue-specific RNA processing. *Nature* **304:** 129.

Stamm S., Zhang M.Q., Marr T.G., and Helfman D.M. 1994. A sequence compilation and comparison of exons that are alternatively spliced in neurons. *Nucleic Acids Res.* **9:** 1515.

Szabo A., Dalmau J., Manley G., Rosenfeld M., Wong E., Henson J., Posner J.B., and Furneaux H.M. 1991. HuD, a paraneoplastic encephalomyelitis antigen contains RNA-binding domains and is homologous to Elav and sex lethal. *Cell* **67:** 325.

Szostak J.W. and Ellington A.D. 1993. In vitro selection of functional RNA sequences. In *The RNA world* (ed. R.F. Gesteland and J.F. Atkins), p. 511. Cold Spring Harbor Laboratory Press, Cold Spring Harbor, New York.

Ule J. and Darnell R.B. 2006. RNA binding proteins and the regulation of neuronal synaptic plasticity. *Curr. Opin. Neurobiol.* **16:** 102.

Ule J., Jensen K., Mele A., and Darnell R.B. 2005a. CLIP: A method for identifying protein-RNA interaction sites in living cells. *Methods* **37:** 376.

Ule J., Jensen K.B., Ruggiu M., Mele A., Ule A., and Darnell R.B. 2003. CLIP identifies Nova-regulated RNA networks in the brain. *Science* **302:** 1212.

Ule J., Stefani G., Mele A., Ruggiu M., Wang X., Taneri B., Gaasterland T., Blencowe B.J., and Darnell R.B. 2006. An RNA map predicting Nova-dependent splicing regulation. *Nature* (in press).

Ule J., Ule A., Spencer J., Williams A., Hu J.S., Cline M., Wang H., Clark T., Fraser C., Ruggiu M., et al. 2005b. Nova regulates brain-specific splicing to shape the synapse. *Nat. Genet.* **37:** 844.

Yang Y.Y.L., Yin G.L., and Darnell R.B. 1998. The neuronal RNA binding protein Nova-2 is implicated as the autoantigen targeted in POMA patients with dementia. *Proc. Natl. Acad. Sci.* **95:** 13254.

Regulation of Alternative Splicing by snoRNAs

S. Kishore and S. Stamm

University of Erlangen, Institute for Biochemistry, 91054 Erlangen, Germany

The SNURF-SNRPN locus located on chromosome 15 is maternally imprinted and generates a large transcript containing at least 148 exons. Loss of the paternal allele causes Prader-Willi syndrome (PWS). The 3′ end of the transcript harbors several evolutionarily conserved C/D box small nucleolar RNAs (snoRNAs) that are tissue-specifically expressed. With the exception of 47 copies of HBII-52 snoRNAs, none of the snoRNAs exhibit complementarity to known RNAs. Due to an 18-nucleotide sequence complementarity, HBII-52 can bind to the alternatively spliced exon Vb of the serotonin receptor 2C pre-mRNA, where it masks a splicing silencer, which results in alternative exon usage. This silencer can also be destroyed by RNA editing, which changes the amino acid sequence and appears to be independent of HBII-52. Lack of HBII-52 expression in individuals with PWS causes most likely a lack of the high-efficacy serotonin receptor, which could contribute to the disease. It is therefore possible that snoRNAs could act as versatile modulators of gene expression by modulating alternative splicing.

ALTERNATIVE SPLICING

One of the most striking results of the Human Genome Project was the demonstration that a surprisingly small number of genes generate a complex proteome. The estimated 20,000–25,000 human protein-coding genes could give rise to 100–150,000 mRNA variants as estimated by comparison of expressed sequence tags. Array analysis shows that 74% of all human genes are alternatively spliced, and a detailed array-based analysis of chromosomes 22 and 21 suggests that every protein-coding gene could undergo alternative splicing (Kampa et al. 2004). Extreme examples illustrate the potential of alternative splicing: The human neurexin 3 gene could form 1728 transcripts (Missler and Sudhof 1998) and the *Drosophila* DSCAM gene could give rise to 38016 isoforms, which is larger than the number of genes in *Drosophila* (Celotto and Graveley 2001). Unlike promoter activity that predominantly regulates the abundance of transcripts, alternative splicing influences the structure of mRNAs and their encoded proteins. As a result, it influences binding properties, intracellular localization, enzymatic activity, protein stability, and posttranslational modification of numerous gene products (Stamm et al. 2005). Alternative splicing can indirectly regulate transcript abundance. About 25–35% of alternative exons introduce frameshifts or stop codons into the pre-mRNA (Stamm et al. 2000; Lewis et al. 2003). Since approximately 75% of these exons are predicted to be subject to nonsense-mediated decay, an estimated 18–25% of transcripts will be switched off by stop codons introduced by alternative splicing and nonsense-mediated decay (Lewis et al. 2003). However, recent array analysis from mouse suggests that the actual number of regulated transcripts might be smaller than predicted (Pan et al. 2004). Finally, several proteins that regulate splice site usage shuttle between the nucleus and cytosol where they regulate translation (Sanford et al. 2004). The biological effects evoked by alternative splicing are diverse and range from a complete loss of function to subtle effects (Stamm et al. 2005). In summary, alternative splicing emerges as a key regulator for human gene expression.

WIDESPREAD EXPRESSION OF SMALL RNAs IN THE HUMAN GENOME

Recent tiling array data and detailed expression analysis showed that the expression data and gene structures collected in current databases are largely incomplete (Carninci et al. 2005). Numerous noncoding regions are transcribed into polyadenylated, stable RNAs, which are named TUF (transcript of unknown function). Within most previously well-characterized protein-coding genes, there are large numbers of noncoding transcribed fragments (transfrags) that could represent new exons or short RNAs of unknown sequence. A recent study of human gene expression using tiling arrays demonstrates that about 57% of the transfrags are not annotated in Genbank or EnsEMBL databases (Cheng et al. 2005). Transfrag sequences are not contained in previous databases since they often seem to derive from unstable RNAs, resulting in sequences that were discarded as fragmental cloning/library artifacts. Furthermore, they are not part of polyadenylated RNA and therefore escape poly(A$^+$) selection during library construction.

C/D BOX snoRNAs

C/D box snoRNAs are a group of short noncoding RNAs that have C and D boxes as characteristic sequence elements that help form the snoRNP. They reside in introns from which they are released during pre-mRNA processing of their host genes through nuclease action. A major function attributed to C/D box snoRNAs was to guide 2′-O-methylation in ribosomal, transfer, and snRNAs. The guiding activity of snoRNAs is achieved by the formation of a specific RNA:RNA duplex between the snoRNA and its target. The snoRNAs contain a region, the antisense box, that exhibits sequence complementarity to its target and forms a short, transient double strand with the target. On the target RNA, the nucleotide base-pairing with the snoRNA nucleotide positioned 5 nucleotides upstream of the snoRNA D box is methylated on the 2′-O-hydroxyl group. Several snoRNAs show complementarity toward pre-rRNA, but the rRNA is not 2'-O-methylated at the predicted positions. It was therefore proposed that snoRNAs could also function as chaperons that help correct folding in rRNA processing (Steitz and Tycowski 1995). Recently, numerous C/D box snoRNAs were discovered that show no sequence complementarity to other RNAs, suggesting that C/D box snoRNAs might have a function other than 2'-O-methylation (Filipowicz and Pogacic 2002).

C/D box snoRNAs associate with proteins to form snoRNPs that contain four evolutionarily conserved, essential proteins: fibrillarin (Nop1p), Nop56p, Nop58p, and Snu13p/15.5 kD. Fibrillarin, which exhibits amino acid sequence motifs characteristic of SAM-dependent methyltransferases, is the likely snoRNA-guided modifying enzyme, as point mutations in the methylase-like domain disrupt all rRNA methylations. Snu13p, the yeast homolog of the mammalian 15.5-kD protein, is found both in C/D box snoRNPs and in the U4/U6.U5 tri-snRNP (small nuclear ribonucleoprotein). It is likely that Snu13p/15.5 binds to an RNA structural motif (kink-turn) present in U4 snRNA that can also be formed by the C and D boxes of C/D box snoRNAs. The occurrence of Snu13p/15.5 in U4 snRNP and C/D box snoRNPs suggests a common evolutionary origin and possibly functional similarities (Watkins et al. 2000).

TISSUE-SPECIFIC snoRNAs DERIVED FROM THE SNURF-SNRPN LOCUS

The sequencing of cDNA libraries enriched for small, nonpolyadenylated RNAs resulted in the identification of snoRNAs bearing no sequence complementarity to rRNAs or snRNAs (Cavaille et al. 2000). One of these snoRNAs was HBII-52, an RNA bearing all the structural hallmarks of a typical C/D snoRNA that exhibits sequence complementarity to the alternative exon Vb of the serotonin receptor 5-HT$_{2C}$. HBII-52 is expressed from the SNURF-SNRPN locus, localized in the Prader-Willi critical region on chromosome 15. This locus appears to be one of the most complex transcriptional units in the human genome (Fig. 1A) (Runte et al. 2001). It spans more than 460 kb and contains at least 148 exons. Ten exons in the 5' part of the gene are transcribed into a bicistronic mRNA that encodes the SNURF (SmN upstream reading frame) and the SmN (small RNP in neurons) protein. The locus harbors a bipartite imprinting center (IC) that silences most maternal genes of the Prader-Willi critical region. Due to this imprinting, the SNURF-SNRPN gene is expressed only from the paternal allele. The large 3'UTR (untranslated region) of the SNURF-SNRPN locus harbors clusters of the C/D box snoRNAs, HBII-85 and HBII-52, that are present in at least 24 and 47 copies, respectively. In addition, the region harbors single copies of other C/D box snoRNAs: HBII-13, HBII-436, HBII-437, HBII-438A, and HBII-438B. The snoRNAs are flanked by noncoding exons and show a large degree of conservation between mouse and human. In contrast, their flanking, noncoding exons are only poorly conserved. In contrast to the SNURF and SmN proteins that can be detected in most tissues (Barr

et al. 1995), the snoRNAs in this locus show tissue-specific expression, indicating tissue-specific processing of the 3'UTR (Fig. 1B). Expression of HBII-52 could be detected only in brain, whereas other snoRNAs could be found in nonbrain tissues as well. The 47 snoRNAs located in the HBII-52 cluster that exhibit sequence complementarity to exon Vb of the serotonin receptor 2C mRNA (Fig. 1C) were analyzed in more detail. The corresponding mouse MBII-52 snoRNAs are expressed throughout the mouse brain. They are most abundant in the hippocampus, but absent in choroid plexus and some thalamic nuclei (Rogelj et al. 2003). The expression of MBII-52 is up-regulated during early memory consolidation in the hippocampus (Rogelj et al. 2003).

PRADER-WILLI SYNDROME

PWS is a congenital disease with an incidence of about 1 in 8,000–20,000 live births. It is disproportionately more often reported in Caucasians. It is caused by the lack of expression of genes located on chromosome 15q11-q13 that contains the SNURF-SNRPN locus. Since this region is maternally imprinted, most genes expressed are from the father's allele, and loss of their expression causes PWS. PWS shows a biphasic clinical phenotype. In the neonatal period, there is a failure to thrive, indicated by muscle hypotonia, feeding difficulties, and hypogonadism. Later, the patients are characterized by short stature and develop mild to moderate mental retardation, behavioral problems, and hyperphagia that leads to severe obesity. PWS is the most common genetic cause of marked obesity in humans, and consequences from the excess weight, such as type II diabetes, are a major complication (Butler et al. 2006). Children with PWS show low levels of growth hormone, insulin-like growth factor (IGF)-I and insulin, and elevated levels of ghrelin (Eiholzer et al. 1998a,b; Cummings et al. 2002). Subsequently, growth hormone substitution was approved for treatment of children with PWS (Carrel et al. 2006).

Genetic evidence shows that the SNURF-SNRPN locus has a major role in PWS, and its deletion causes PWS-like symptoms in mouse models (Stefan et al. 2005). However, this locus gives rise to only two proteins: SmN and SNURF. Lack of these proteins and of several other proteins (NDN, MAGEL2) located centromeric from the SNURF-SNRPN locus could be ruled out as major contributors to PWS (Ding et al. 2005). It was therefore interesting that the HBII-52 snoRNA located in the 3'UTR of SNURF-SNRPN locus exhibited an 18-nucleotide sequence complementarity to the alternative exon of the serotonin receptor 5-HT$_{2C}$.

SEROTONIN RECEPTOR 5-HT$_{2C}$

Serotonin (5-hydroxytryptamine, 5-HT) is a neurotransmitter. It acts on at least 14 different receptors that can be subdivided into seven distinct classes. With the exception of one ligand-gated ion channel, all serotonin receptors are coupled to G proteins. Alternative pre-mRNA processing largely increases the number of proteins made from these receptors. The serotonergic system

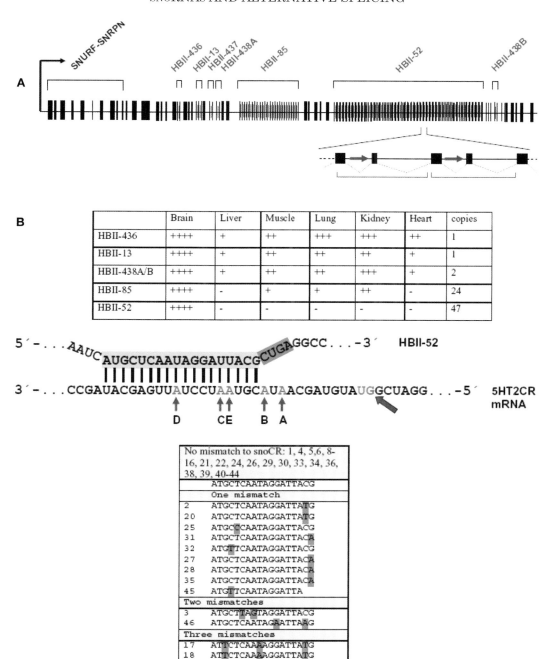

Figure 1. The SNURF-SNRPN locus. (*A*) Schematic overview of the locus. (*Black vertical lines*) Exons; (*blue vertical lines*) snoRNAs. The magnified region shows the arrangement of the snoRNAs (*black arrows*) between noncoding exons (*black boxes*). (*B*) Expression of the snoRNAs found in the locus as determined by RT-PCR and northern blot (Cavaille et al. 2000; Runte et al. 2001). The relative intensity of the snoRNAs in different tissues is indicated by plus (+) signs. (*C*) Complementarity of the 47 HBII-52 snoRNAs toward the serotonin receptor 5-HT$_{2C}$ exon Vb sequence. The base complementarity between the antisense box of the human HBII-52 snoRNA and the human 5-HT$_{2C}$ receptor is shown. Arrows indicate the A→I editing sites (*A–E*). (*Open arrow*) Proximal splice site; (*green*) D box. The table summarizes all human HBII-52 snoRNAs and their complementarity to the 5-HT$_{2C}$ exon Vb sequences. (*Green boxes*) Mismatches in the complementary regions.

interferes with numerous brain functions, including mood, sleep, sexuality, and appetite. The 5-HT$_{2C}$ (previously named 5-HT$_{1C}$) receptor (5-HT$_{2C}$R) belongs to a family of seven-transmembrane-containing G-protein-coupled receptors (GPCRs) and is mapped to human X chromosome band q24. 5-HT$_{2C}$R mRNA is widely

expressed in the central nervous system and is developmentally regulated. Mice lacking 5-HT$_{2C}$R show hyperphagia leading to obesity and develop seizures (Tecott et al. 1995). Through phospholipase-C-generated second messengers, the serotonin 2C receptor can modulate cyclic AMP accumulation, can regulate K$^+$ and Cl$^-$ channels, can both inhibit and stimulate nitric oxide (NO) levels, and can also regulate mitogenesis (Raymond et al. 2001). In summary, a broad range of psychoactive compounds including appetite suppressant, antidepressant, antipsychotic, anxiolytic, psychostimulant, and psychedelic drugs affect the 5-HT_{2C} receptor, demonstrating that the receptor has a physiological role in these processes (Giorgetti and Tecott 2004).

THE snoRNA HBII-52 REGULATES ALTERNATIVE SPLICING OF THE 5-HT_{2C} RECEPTOR

The 5-HT_{2C} receptor is located on the X chromosome and undergoes alternative splicing. Only when its alternative exon Vb is included in the mRNA can a functional receptor be made, since exon Vb skipping causes a frameshift. ExonVb is expressed throughout the brain, but it is mostly absent in the choriod plexus. Exon Vb is edited on at least five sites by ADAR2-mediated adenosin to inosine deamination. These editing events promote exon Vb inclusion (Flomen et al. 2004), generating a full-length receptor.

The antisense box of the C/D box snoRNA HBII-52 exhibits an 18-nucleotide, phylogenetically conserved sequence complementarity to the alternative exon Vb of the 5-HT_{2C} receptor. The snoRNA is located on chromosome 15 and is expressed throughout the brain, but it is absent in the choroid plexus, indicating a correlation between HBII-52 expression and exon Vb usage. We therefore analyzed the influence of HBII-52 on 5-HT_{2C}R pre-mRNA processing and found that HBII-52 promotes usage of exon Vb (Kishore and Stamm 2006). In vivo, the HBII-52 snoRNA binds transiently to exon Vb. Exon Vb contains two splicing silencer sequences that prevent inclusion of the exon in the mRNA (Wang et al. 2004) and it is therefore likely that HBII-52 blocks the action of these silencers (Kishore and Stamm 2006). The identification of these silencing elements in exon Vb also explains the earlier findings that A→I editing promotes exon Vb inclusion (Flomen et al. 2004), since the editing sites are located in the first of the two silencing elements, whose activity is destroyed by A→I editing (Kishore and Stamm 2006).

Due to the RNA editing, however, the amino acids encoded by exon Vb are changed. These amino acids are located in the intracellular loop of the receptor that couples to G proteins. Changing these amino acids through editing generates a receptor with 10–100-fold lower efficacy (Wang et al. 2000). Therefore, the most likely physiological role of the snoRNA HBII-52 is to promote inclusion of exon Vb of the 5-HT_{2C} receptor without the need of the editing that would change the receptor properties. An absence of HBII-52 would therefore reduce the expression of nonedited 5-HT_{2C} receptor mRNA. This prediction was tested in brain samples derived from individuals with PWS that do not express the HBII-52 snoRNA. In three of the four major editing sites, a reduction of the nonedited forms was observed. This strongly suggests that the absence of the HBII-52 snoRNA in individuals with PWS causes a misregulation of 5-HT_{2C} receptor mRNA isoforms (Fig. 2). Interestingly, some features of the 5-HT_{2C} knockout mice, such as hyperphagia and obesity, reflect the clinical presentation of individuals with PWS that lack the HBII-52 snoRNA. So far, HBII-52 was absent in all persons with PWS that were tested.

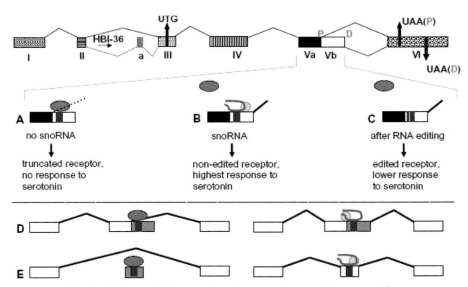

Figure 2. The working model for the influence of the HBII-52 snoRNA on the serotonin receptor. The gene structure of the serotonin receptor is shown in the top line; "a" is a novel exon that we identified. The arrow indicates HBI-36, another H/ACA box snoRNA in the receptor mRNA. (*A–C*) Regulation of the exon Vb splice site selection; (*A–C*) regulation of exon Vb. (*A*) Exon Vb contains a silencer (*brown box*) that prevents its usage, most likely by binding to a *trans*-acting factor (*red circle*). (*B*) snoRNA HBII-52 (*blue line*) can replace this silencer but needs additional sequence elements, possibly binding to other *trans*-acting factors (*yellow circle*). HBII-52 presence results therefore in exon inclusion. (*C*) Editing mutates this silencer (*striped purple box*), preventing binding of the *trans*-acting factor, which leads to exon Vb inclusion. The effect on the receptor protein is indicated below each scenario. Mechanism B is missing in patients with PWS. Silencer sequences are found in other genes misregulated in PWS. We hypothesize that in healthy individuals, HBII-52 regulates exons in these messages in a similar fashion (*D,E*).

However, two patients were described who lack the paternal HBII-52 snoRNA cluster and suffer from Angelman syndrome, but not PWS (Greger et al. 1993; Burger et al. 2002). It has not been investigated whether these patients express HBII-52, but it is possible that they either express HBII-52 from the maternal allele by an unknown mechanism or that the alternative splicing defects caused by the lack of this snoRNA are modified by a unique composition of splicing regulatory factors in these two patients.

CONCLUSIONS AND PERSPECTIVE

The results summarized here show that snoRNAs generated by the SNURF-SNRPN locus have a role in gene regulation. They are not transcriptional noise without function. An important open question is whether the regulation of alternative splicing by the HBII-52 snoRNA is a unique, special event or an example of a new regulatory function of snoRNAs. Currently, there are no other messenger or pre-mRNAs that exhibit complete complementarity to antisense elements of snoRNAs. However, a complete complementarity is not necessary in other systems that need binding of different RNAs. The complementartity of U1 snRNA to 5′ splice sites is almost always interrupted, microRNAs show only a short, degenerate complementarity to their targets, and C/D box snoRNAs functioning in ribosomal 2′-O-methylation have nonpaired regions between their antisense boxes and targets. In the HBII-52 cluster, 14 of 47 snoRNAs have between 1 and 3 mismatches toward the exon VB target sequence. If three mismatches toward a target are allowed in the 18 nucleotides of the antisense box of HBII-52, more than 1000 putative target RNAs can be identified. The exact contribution of the snoRNAs located in the SNURF-SNRPN locus to the PWS remains to be determined. If the action of these snoRNAs on their RNA targets tolerate mismatches between the target RNA and the snoRNA antisense boxes, they could act as major regulators in RNA processing and could well cause the disease. The identification of their targets and the elucidation of their exact mode of action could therefore lead to the development of new therapeutic strategies for PWS.

ACKNOWLEDGMENTS

This work was supported by the Deutsche Forschungsgemeinschaft and the BMBF.

REFERENCES

Barr J.A., Jones J., Glenister P.H., and Cattanach B.M. 1995. Ubiquitous expression and imprinting of Snrpn in the mouse. *Mamm. Genome* **6**: 405.

Burger J., Horn D., Tonnies H., Neitzel H., and Reis A. 2002. Familial interstitial 570 kbp deletion of the UBE3A gene region causing Angelman syndrome but not Prader-Willi syndrome. *Am. J. Med. Genet.* **111**: 233.

Butler M.G., Hanchett J.M., and Thompson T.E. 2006. Clinical findings and natural history of Prader-Willi syndrome. In *Management of Prader-Willi syndrome* (ed. M.G. Butler et al.), p. 3. Springer, New York.

Carninci P., Kasukawa T., Katayama S., Gough J., Frith M.C., Maeda N., Oyama R., Ravasi T., Lenhard B., Wells C., et al.

2005. The transcriptional landscape of the mammalian genome. *Science* **309**: 1559.

Carrel A.L., Lee P.D.K., and Mogul H.R. 2006. Growth hormone and Prader-Willi syndrome. In *Management of Prader-Willi syndrome* (ed. M.G. Butler et al.), p. 201. Springer, New York.

Cavaille J., Buiting K., Kiefmann M., Lalande M., Brannan C.I., Horsthemke B., Bachellerie J.P., Brosius J., and Hüttenhofer A. 2000. Identification of brain-specific and imprinted small nucleolar RNA genes exhibiting an unusual genomic organization. *Proc. Natl. Acad. Sci.* **97**: 14311.

Celotto A.M. and Graveley B.R. 2001. Alternative splicing of the *Drosophila* Dscam pre-mRNA is both temporally and spatially regulated. *Genetics* **159**: 599.

Cheng J., Kapranov P., Drenkow J., Dike S., Brubaker S., Patel S., Long J., Stern D., Tammana H., Helt G., et al. 2005. Transcriptional maps of 10 human chromosomes at 5-nucleotide resolution. *Science* **308**: 1149.

Cummings D.E., Clement K., Purnell J.Q., Vaisse C., Foster K.E., Frayo R.S., Schwartz M.W., Basdevant A., and Weigle D.S. 2002. Elevated plasma ghrelin levels in Prader Willi syndrome. *Nat. Med.* **8**: 643.

Ding F., Prints Y., Dhar M.S., Johnson D.K., Garnacho-Montero C., Nicholls R.D., and Francke U. 2005. Lack of Pwcr1/MBII-85 snoRNA is critical for neonatal lethality in Prader-Willi syndrome mouse models. *Mamm. Genome* **16**: 424.

Eiholzer U., Stutz K., Weinmann C., Torresani T., Molinari L., and Prader A. 1998a. Low insulin, IGF-I and IGFBP-3 levels in children with Prader-Labhart-Willi syndrome. *Eur. J. Pediatr.* **157**: 890.

Eiholzer U., Gisin R., Weinmann C., Kriemler S., Steinert H., Torresani T., Zachmann M., and Prader A. 1998b. Treatment with human growth hormone in patients with Prader-Labhart-Willi syndrome reduces body fat and increases muscle mass and physical performance. *Eur. J. Pediatr.* **157**: 368.

Filipowicz W. and Pogacic V. 2002. Biogenesis of small nucleolar ribonucleoproteins. *Curr. Opin. Cell Biol.* **14**: 319.

Flomen R., Knight J., Sham P., Kerwin R., and Makoff A. 2004. Evidence that RNA editing modulates splice site selection in the 5-HT2C receptor gene. *Nucleic Acids Res.* **32**: 2113.

Giorgetti M. and Tecott L.H. 2004. Contributions of 5-HT(2C) receptors to multiple actions of central serotonin systems. *Eur. J. Pharmacol.* **488**: 1.

Greger V., Woolf E., and Lalande M. 1993. Cloning of the breakpoints of a submicroscopic deletion in an Angelman syndrome patient. *Hum. Mol. Genet.* **2**: 921.

Kampa D., Cheng J., Kapranov P., Yamanaka M., Brubaker S., Cawley S., Drenkow J., Piccolboni A., Bekiranov S., Helt G., et al. 2004. Novel RNAs identified from an in-depth analysis of the transcriptome of human chromosomes 21 and 22. *Genome Res.* **14**: 331.

Kishore S. and Stamm S. 2006. The snoRNA HBII-52 regulates alternative splicing of the serotonin receptor 2C. *Science* **311**: 230.

Lewis B.P., Green R.E., and Brenner S.E. 2003. Evidence for the widespread coupling of alternative splicing and nonsense-mediated mRNA decay in humans. *Proc. Natl. Acad. Sci.* **100**: 189.

Missler M. and Sudhof T.C. 1998. Neurexins: Three genes and 1001 products. *Trends Genet.* **14**: 20.

Pan Q., Shai O., Misquitta C., Zhang W., Saltzman A.L., Mohammad N., Babak T., Siu H., Hughes T.R., Morris Q.D., et al. 2004. Revealing global regulatory features of mammalian alternative splicing using a quantitative microarray platform. *Mol. Cell* **16**: 929.

Raymond J.R., Mukhin Y.V., Gelasco A., Turner J., Collinsworth G., Gettys T.W., Grewal J.S., and Garnovskaya M.N. 2001. Multiplicity of mechanisms of serotonin receptor signal transduction. *Pharmacol. Ther.* **92**: 179.

Rogelj B., Hartmann C.E., Yeo C.H., Hunt S.P., and Giese K.P. 2003. Contextual fear conditioning regulates the expression of brain-specific small nucleolar RNAs in hippocampus. *Eur. J. Neurosci.* **18**: 3089.

Runte M., Hüttenhofer A., Gross S., Kiefmann M., Horsthemke B., and Buiting K. 2001. The IC-SNURF-SNRPN transcript serves as a host for multiple small nucleolar RNA species and as an antisense RNA for UBE3A. *Hum. Mol. Genet.* **10:** 2687.

Sanford J.R., Gray N.K., Beckmann K., and Caceres J.F. 2004. A novel role for shuttling SR proteins in mRNA translation. *Genes Dev.* **18:** 755.

Stamm S., Zhu J., Nakai K., Stoilov P., Stoss O., and Zhang M.Q. 2000. An alternative-exon database and its statistical analysis. *DNA Cell Biol.* **19:** 739.

Stamm S., Ben-Ari S., Rafalska I., Tang Y., Zhang Z., Toiber D., Thanaraj T.A., and Soreq H. 2005. Function of alternative splicing. *Gene* **344:** 1.

Stefan M., Ji H., Simmons R.A., Cummings D.E., Ahima R.S., Friedman M.I., and Nicholls R.D. 2005. Hormonal and metabolic defects in a Prader-Willi syndrome mouse model with neonatal failure to thrive. *Endocrinology* **146:** 4377.

Steitz J.A. and Tycowski K.T. 1995. Small RNA chaperones for ribosome biogenesis. *Science* **270:** 1626.

Tecott L.H., Sun L.M., Akana S.F., Strack A.M., Lowenstein D.H., Dallman M.F., and Julius D. 1995. Eating disorder and epilepsy in mice lacking 5-HT2c serotonin receptors. *Nature* **374:** 542.

Wang Q., O'Brien P.J., Chen C.-X., Cho D.-S.C., Murray J.M., and Nishikura K. 2000. Altered G protein-coupling functions of RNA editing isoform and splicing variant serotonin 2C receptors. *J. Neurochem.* **74:** 1290.

Wang Z., Rolish M.E., Yeo G., Tung V., Mawson M., and Burge C.B. 2004. Systematic identification and analysis of exonic splicing silencers. *Cell* **119:** 831.

Watkins N.J., Segault V., Charpentier B., Nottrott S., Fabrizio P., Bachi A., Wilm M., Rosbash M., Branlant C., and Luhrmann R. 2000. A common core RNP structure shared between the small nucleolar box C/D RNPs and the spliceosomal U4 snRNP. *Cell* **103:** 457.

Measuring the Rates of Transcriptional Elongation in the Female *Drosophila melanogaster* Germ Line by Nuclear Run-on

A. Sigova, V. Vagin, and P.D. Zamore

Department of Biochemistry and Molecular Pharmacology, University of Massachusetts Medical School, Worcester, Massachusetts 01605

We adapted the nuclear run-on method to measure changes in the rate of RNA polymerase II (pol II) transcription of repetitive elements and transposons in the female germ line of *Drosophila melanogaster*. Our data indicate that as little as an approximately 1.5-fold change in the rate of transcription can be detected by this method. Our nuclear run-on protocol likely measures changes in transcriptional elongation, because rates of transcription decline with time, consistent with a low rate of pol II re-initiation in the isolated nuclei. Surprisingly, we find that the retrotransposon *gypsy* and the repetitive sequence *mst40* are silenced posttranscriptionally in fly ovaries.

Eukaryotic genomes exist as chromatin (from the Greek *chroma*, meaning colored), a complex of DNA and histone proteins that facilitates efficient DNA packaging and differential regulation of gene expression. Cytologically, chromatin can be divided into heterochromatin, first identified by its dark staining, and euchromatin, which stains more lightly (Zacharias 1995). Heitz imagined euchromatin to be genetically active and rich in genes (Zacharias 1995). In contrast, he proposed heterochromatin to be genetically inert and gene-poor. Nearly eight decades later, his prescient assessment stands essentially uncorrected.

Constitutive heterochromatin (α-heterochromatin) comprises tandem arrays of repetitive DNA sequences ("satellite" sequences), punctuated occasionally by insertions of transposable elements. Constitutive heterochromatin is believed to organize specialized structures such as centromeres and telomeres that act to maintain genetic stability and ensure the segregation of chromosomes during mitosis and meiosis (Allshire et al. 1995; Nimmo et al. 1998; Bernard et al. 2001; Pidoux and Allshire 2005; Huang and Moazed 2006). Heterochromatic states are epigenetically inherited: The DNA packaging state is maintained after replication and mitosis, irrespective of the underlying DNA sequence (Elgin and Grewal 2003). At the molecular level, all heterochromatic regions contain lysine 9-methylated histone 3 and the evolutionarily conserved heterochromatic protein 1 (HP1; Swi6 in fission yeast) (Huisinga et al. 2006). Heterochromatin has other unusual properties: It replicates late in S phase, remains condensed throughout the cell cycle, resists recombination, and silences reporter genes inserted within or nearby, a property that underlies position-effect variegation (Henikoff 1992; Mahtani and Willard 1998; Copenhaver et al. 1999; Puechberty et al. 1999; Gilbert 2002; Schubeler et al. 2002). In contrast, facultative heterochromatin is transcriptionally active, but it can adopt the structural and functional characteristics of heterochromatin under special circumstances. Examples of facultative heterochromatin include the inactive X chromosome of mammals (which is

genetically identical to its active sister); the silenced rRNA genes in nucleolar dominance, a phenomenon in which rRNA genes of only one parent are active in animal and plant hybrids; and the developmentally regulated β-globin locus in animals (Weintraub et al. 1981; Forrester et al. 1989; Lewis and Pikaard 2001; Grummt and Pikaard 2003; Kim and Dean 2004; Lawrence and Pikaard 2004; Lawrence et al. 2004; Heard 2005).

Between constitutive heterochromatin and euchromatin lies β-heterochromatin, which is less condensed, largely deficient in tandem sequence arrays, filled with copies of numerous transposable element families, and contains few functional genes (Holmquist et al. 1998). Repetitive sequences account for the majority of all heterochromatic sequences in *Drosophila melanogaster*, whose genome is 30% heterochromatic, suggesting that they have a central role in heterochromatin assembly. Sequence comparison of the *D. melanogaster* fourth chromosome, which is largely heterochromatic, with the syntenic *Drosophila virilis* chromosome 6, which is euchromatic, suggests that the difference in chromatin packaging reflects the density and distribution of transposable elements (Riddle and Elgin 2006; Slawson et al. 2006).

The predominance of repetitive elements in heterochromatin suggests that these selfish genetic elements are silenced transcriptionally. Repetitive element silencing is essential for genome integrity, as their transposition is the major source of genome rearrangements (Kazazian 2004). In plants, animals, and fungi, RNA silencing has been implicated as a major defense against repetitive element transposition (Martienssen and Colot 2001; Sijen and Plasterk 2003; Kalmykova et al. 2005; Nolan et al. 2005; Savitsky et al. 2006). In the male and female germ line of *D. melanogaster*, retrotransposons and repetitive sequences are silenced by the repeat-associated small RNA (rasiRNA) pathway (Vagin et al. 2006), an RNA silencing mechanism distinct from both the RNA interference (RNAi) and microRNA (miRNA) pathways. Small silencing RNAs 24–30 nucleotides long, rasiRNAs not only are about 3–7 nucleotides longer than siRNAs and miRNAs, but they are also chemically different, in that they lack one of the 3′-terminal hydroxyl groups characteristic of animal short interfering RNAs (siRNAs) and miRNAs. Consistent with this chemical difference, rasiRNAs may not be produced by either Dicer-1, which makes *Drosophila* miRNAs, or Dicer-2, which makes siRNAs. rasiRNA-directed silencing of repetitive genetic elements requires the putative helicases Spn-E, and Armitage as well as Piwi or Aubergine, members of the Piwi subclade of Argonaute family of proteins.

Supporting the view that rasiRNAs act to silence repetitive sequence by reducing their rate of transcription, mutations in *spn-E*, *aubergine*, and *piwi* are reported to cause loss of lysine-9 methylation of histone H3 on silenced genes (Pal-Bhadra et al. 2004). Historically, heterochromatin has been regarded as transcriptionally inert, but new evidence in fission yeast suggests that heterochromatin can be transcribed by RNA pol II and then silenced by the posttranscriptional destruction of the nascent transcript, perhaps at its site of transcription (Volpe et al. 2002; Djupedal et al. 2005; Kato et al. 2005; Buhler et al. 2006).

What mechanism silences repetitive elements in the germ line of any organism is unknown. Mutations in *armi* increase the steady-state concentration of repetitive element mRNA in the fly germ line (Vagin et al. 2006). Here, we examine the transcriptional rates of the silenced and desilenced retrotransposon *gypsy* and the repetitive sequence *male-specific transcript 40* (*mst40*) in the *D. melanogaster* female germ line. Contrary to expectation, we find that loss of *gypsy* or *mst40* silencing caused by loss of the Armi protein is not accompanied by a change in the rate of transcription of either of these selfish genetic elements.

MATERIALS AND METHODS

Isolation of nuclei. The rate of transcriptional elongation was determined by nuclear run-on analysis essentially as described previously (So and Rosbash 1997). Briefly, ovaries from 50–100 females were dissected with needles into *Drosophila* Ringer's solution (182 mM KCl, 46 mM NaCl, 3 mM $CaCl_2$, 10 mM Tris-HCl at pH 7.5) and stored on ice in a 1.5-ml homogenization tube (Kontes, Vineland, New Jersey) during isolation. Isolated tissues were centrifuged at 2000g at 4°C, the supernatant discarded, and then homogenized with 20–30 strokes of the pestle in 300 µl of homogenization buffer (10 mM HEPES-KOH at pH 7.5, 10 mM KCl, 0.8 M sucrose, 1 mM EDTA, 0.5 mM dithiothreitol (DTT), and 100 µg/ml yeast tRNA (Ambion, Austin, Texas) containing one tablet of Complete Mini EDTA-free Protease Inhibitor Cocktail (Roche) for each 10 ml. An additional 200 µl of homogenization buffer was used to rinse the tube. The rinse and the homogenate were pooled and then filtered through a Bio-Spin column (Bio-Rad).

The filtrate was overlaid on a 500-µl cushion of 1 M sucrose dissolved in 10 mM HEPES (pH 7.5), 10 mM KCl, 10% glycerol, and 1 mM EDTA plus one tablet of Complete Mini EDTA-free Protease Inhibitor Cocktail (Roche) for each 10 ml and centrifuged at 10,000g for 10 minutes at 4°C. After discarding the supernatant, the nuclei were resuspended in 500 µl of nuclear resuspension (NR) buffer (40 mM HEPES-KOH at pH 8.0, 25% glycerol, 5 mM magnesium acetate, 1 mM DTT, and 0.1 mM EDTA containing for each 10 ml one tablet of Complete Mini EDTA-free Protease Inhibitor Cocktail [Roche]) and centrifuged for 2 minutes at 8000g at 4°C.

Labeling of nascent transcripts. The pellet was resuspended in 195 µl of NR buffer, and the reaction was initiated by adding 50 µl of 5x reaction mixture (5 mM magnesium acetate; 750 mM KCl; 2.5 mM each ATP, CTP, and GTP; 10 mM DTT; 1 µl of RNasin [Promega]; 50 mM of creatine phosphate; and 60 µg/ml creatine kinase) and 5 µl of [α-^{32}P]UTP (6000 Ci/mmol, 40 µCi/µl; MP Biomedicals, Irvine, California). After incubation for 30 minutes at 23°C, the reaction was stopped by adding 25 µl of RNase-free DNase (RQ1, Promega) and incubating for 5 minutes at 37°C. Proteins were digested by adding 20 µl of 15x proteinase buffer (7.5% SDS, 150 mM EDTA) and 3 µl of 10 mg/ml proteinase K and then incubating for 30 minutes at 37°C. After extraction with phenol:chloroform, the RNA was precipitated with three volumes of ethanol. The precipitate was recovered by precipitation, washed with 80% (v/v) ethanol, air-dried, and resuspended in 50 µl of H_2O. Unincorporated nucleotides were removed by three consecutive rounds of purification using mini Quick Spin Columns for RNA (Roche) according to the manufacturer's instructions and then used for hybridization as described below.

Hybridization of labeled transcript to filter-immobilized probes. Labeled RNA from isolated nuclei was hybridized to immobilized, strand-specific, in-vitro-transcribed RNA probes for firefly luciferase (*Photinus pyralis*; *Pp* luc), *roo*, *I-element*, *HeT-A*, *mst40*, *gypsy*, *rp49*, and *act5C*. RNA probes were transcribed with T7 RNA polymerase from PCR templates prepared with the oligonucleotides reported in Table 1. *Pp* luciferase was transcribed from the pGL-2 control vector (Promega) using the primers: 5′-gcg taa tac gac tca cta tag GAG AGG AAT TCA TTA TC-3′ and 5′-GAA GAG ATA GCC CTG GTT CCT G-3′. Each in-vitro-transcribed RNA probe (5 µg) was denatured in 500 µl of ice-cold 10 mM NaOH and 1 mM EDTA, transferred to Hybond-XL membrane (Amersham) using a Bio-Dot SF Microfiltration Apparatus (Bio-Rad) according to manufacturer's instructions, and immobilized on the membrane by UV irradiation (200 µjoules/cm; Stratalinker, Stratagene). The membrane was prehybridized in Church buffer (Church and Gilbert 1984) for 1 hour at 65°C and hybridization was carried out overnight at 65°C. After hybridization, membranes were washed twice with 2x SSC/0.1% (w/v) SDS for 30 minutes at 65°C and analyzed by phosphorimagery (Fuji, Tokyo, Japan).

RESULTS

Nuclear Run-on Measures Relative Rates of Transcription

Weiss (1960) demonstrated for the first time that isolated eukaryotic nuclei retain the ability to synthesize RNA. Such RNA synthesis results from transcript elongation by

Table 1. Primers Used to Generate T7 RNA Polymerase Transcription Templates for Production of Antisense RNA Probes

Element or gene	Primer sequence (5′ to 3′)	
roo	5′ primer	taatacgactcactattagggagaccacGGAGGGTTTGATTTAGGGACAGTG
	3′ primer	AGCAGAAGCAGCAACAGCAGTAG
HeT-A	5′ primer	taatacgactcactattagggagaccacGGAGAAGATCGCTGTTCTG
	3′ primer	GACACGCGAAAAGCGAAC
I-element	5′ primer	taatacgactcactattagggagaccacTTGGCCTTTAGTTTTGATGC
	3′ primer	CATCAACACAGCCCAATTGAC
act5C	5′ primer	gcgtaatacgactcactatagggTGTAGGTGGTCTCGTGGATGC
	3′ primer	GGCCACCGTGAGAAGATGAC
mst40	5′ primer	taatacgactcactattagggagaccacAACGATCGCTTGCGATCTAC
	3′ primer	TCAAATCAGACGAAGTTCAAGG
rp49	5′ primer	gcgtaatacgactcactatagggTTACTCGTTCTCTTGAGAACG
	3′ primer	GACCATCCGCCCAGCATACAG

RNA polymerases, rather than initiation of transcription after nuclear isolation (Cox 1976). These experiments led to the establishment of "nuclear run-on" as the standard method for measuring the relative rates of gene transcription (McKnight and Palmiter 1979; Swaneck et al. 1979). In a prototypical nuclear run-on experiment, intact nuclei are isolated by centrifugation and then incubated with exogenous ATP, GTP, CTP, and [α-^{32}P]UTP, incorporating radiolabeled nucleotide into transcripts initiated prior to cell lysis but elongated in the isolated nuclei. Radiolabeled nuclear RNA is isolated and then hybridized to strand-specific RNA probes immobilized on a nylon filter. The specificity of hybridization is established by comparing the hybridization signals for the genes of interest to those of control genes whose transcription is presumed to be invariant under the conditions compared (Fig. 1).

We sought to apply the nuclear run-on method to germline tissue from flies. To begin our study, we first tested whether run-on of nuclei isolated from dissected *D. melanogaster* ovaries accurately reflects changes in transcriptional rates. We compared the apparent transcriptional rates of the *RpL32* gene (commonly known as *rp49*) in nuclei isolated from ovaries of wild-type flies containing two copies of *rp49* (2x *rp49*) to that in mutant Df(3R)L127/TM6; Dp(3;1)B152/Dp(3;1)B152 flies in which one copy of the *rp49* locus on the third chromosome is deleted and each X chromosome contains an additional copy of the locus, for a total of three copies of *rp49* (hereafter, 3x *rp49*). As a control, we used the *act5C* gene. Hybridization to the firefly luciferase (*Pp Luc*) antisense transcript, whose sequence shares little similarity with any gene in *D. melanogaster*, provided a measure of the nonspecific background. The steady-state mRNA level in the ovaries of the 3x *rp49* flies was greater by a factor of 1.84 ± 0.39 (average ± standard deviation) than in the ovaries of the 2x *rp49* flies (Fig. 2A). Transcription of *rp49* in the ovaries with three copies of the gene was greater by a factor of 1.75 ± 0.24 (average ± standard deviation for three independent trials) than in the ovaries from flies bearing two copies of the *rp49* gene (Fig. 2B). We conclude that

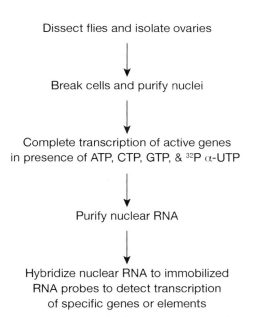

Dissect flies and isolate ovaries

Break cells and purify nuclei

Complete transcription of active genes in presence of ATP, CTP, GTP, & ^{32}P α-UTP

Purify nuclear RNA

Hybridize nuclear RNA to immobilized RNA probes to detect transcription of specific genes or elements

Figure 1. Schematic of nuclear run-on analysis.

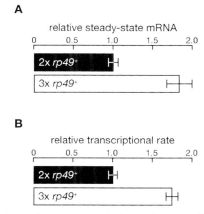

A

relative steady-state mRNA

2x *rp49*⁺

3x *rp49*⁺

B

relative transcriptional rate

2x *rp49*⁺

3x *rp49*⁺

Figure 2. Nuclear run-on analysis can detect small changes in transcriptional rates. (*A*) Relative steady-state mRNA concentration in ovaries from flies with two or three copies of the *rp49* gene. (*B*) Nuclear run-on assay in ovaries from flies with two or three copies of the *rp49* gene. The assay readily detected the approximately 1.5-fold change in the transcription of *rp49* between the wild-type *rp49/rp49* (2x *rp49*) and the Df(3R)L127/TM6; Dp(3;1)B152/Dp(3;1)B152 (3x *rp49*) flies.

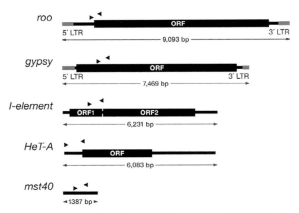

Figure 3. Position of primers used to prepare T7 RNA polymerase PCR templates to synthesize the antisense RNA probes used to detect repetitive elements.

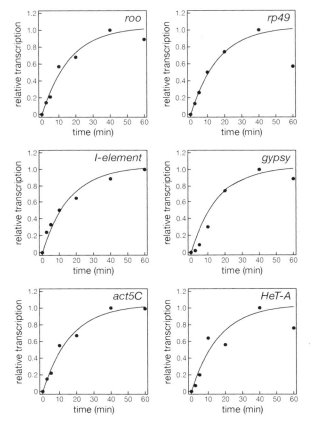

Figure 4. Nuclear run-on likely measures the rate of transcription without substantial reinitiation by RNA pol II, because transcriptional rates saturate with time.

the nuclear run-on assay readily detected an approximately 1.5-fold change in the transcriptional rate of a housekeeping gene in *D. melanogaster* ovaries.

A central assumption in this assay is that only transcripts initiated by RNA pol II prior to cell lysis are extended during the radiolabeling reaction. If this assumption is justified, the amount of [α-^{32}P]UTP incorporated in an elongating transcript should decline with time, as individual molecules of RNA pol II complete transcription but do not initiate new rounds of RNA synthesis in vitro.

Labeled transcripts were isolated and hybridized to strand-specific RNA probes for four different repetitive sequences: two long terminal repeat (LTR) retrotransposons (*roo* and *gypsy*) and two non-LTR retrotransposons (*I-element* and *HeT-A*). Probes for two single-copy genes, *act5C* and *rp49*, served as controls. The positions of the primers used to make polymerase chain reaction (PCR) templates for RNA probe transcription by T7 RNA polymerase are diagrammed in Figure 3. If no reinitiation occurs during the run-on reaction, transcription rates should saturate when there are no more unfinished transcripts left to label. We find that after about 40 minutes, transcriptional rates saturated, consistent with RNA pol II having elongated all available transcripts without reinitiating new rounds of transcription (Fig. 4).

Measuring the Rates of Repetitive Element Transcription

Hybridization-based nuclear run-on was conceived to measure the relative transcriptional rates of single-copy genes (McKnight and Palmiter 1979). Because repetitive elements are multicopy, we were concerned that their high aggregate rates of RNA synthesis might saturate the immobilized RNA probes, preventing our detecting changes in transcriptional rates. To exclude this possibility, we performed nuclear run-on using *armi* homozygous mutant ovaries, in which transposon silencing is derepressed; 100%, 50%, 25%, and 10% of the ^{32}P-radiolabeled RNA was hybridized to 5 μg of immobilized RNA probes for the transposons *roo, gypsy, I-element*, and *HeT-A*, the repetitive sequence *mst40*, and the single-copy

gene *act5C*. For all six genes, the rate of transcription decreases linearly with dilution, even for *HeT-A* retrotransposon, whose steady-state transcript levels rise approximately 170-fold in the absence of Armi (Fig. 5) (Vagin et al. 2006). We conclude that our experimental conditions can accurately measure repetitive element transcriptional rates despite their high copy number.

Repetitive Elements Are Transcribed by RNA pol II

To establish that our assay detects RNA synthesis, we conducted the run-on in the presence of three RNA chain terminators: 3'-deoxy ATP, 3'-deoxy CTP, and 3'-deoxy GTP. Consistent with RNA synthesis, run-on transcription decreased dramatically in the presence of the RNA chain terminators for all six genes examined, *act5C, rp49, gypsy, roo, I-element*, and *HeT-A* (Fig. 6).

In *D. melanogaster*, 10 μg/ml α-amanitin inhibits transcription by RNA pol II, but not by RNA pol I or III. To test whether RNA pol II predominantly transcribes repetitive elements in *D. melanogaster*, the nuclear run-on assay was conducted on nuclei preincubated with 10 μg/ml α-amanitin for 10 minutes on ice. Hybridization signals were compared with those from mock-preincubated nuclei. α-Amanitin reduced repetitive element transcription by an amount comparable to the reduction observed

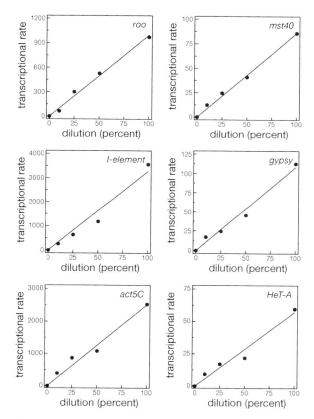

Figure 5. Nuclear run-on can detect the rates of repetitive element transcription. Dilution analysis detects no saturation of signal when registering repetitive element transcription in a nuclear run-on assay.

for the single-copy control genes *act5C* and *rp49* (Fig. 6). We conclude that in flies, repetitive elements, like protein-coding genes, are transcribed by RNA pol II.

Repetitive Elements in *D. melanogaster* Ovaries Are Silenced Posttranscriptionally

In *armi* homozygous mutants, the steady-state RNA concentration increases by a factor of three for the LTR-

retrotransposon gypsy and more than 8-fold for *mst40* (Vagin et al. 2006). *mst40* sequences are located in region 40, at the base of chromosome *2L*, close to or within the β-heterochromatin (Russell and Kaiser 1994). *gypsy* element insertions cause frequent mutations in *D. melanogaster* (Peifer and Bender 1988). Interestingly, studies of reversion of *gypsy*-induced mutations demonstrated that it is not simply the insertion of *gypsy* DNA that causes the mutant phenotype, because most *gypsy*-induced phenotypes can be suppressed by mutations in *suppressor of Hairy wing* [*su(Hw)*] (Peifer and Bender 1988). The 5´UTR (untranslated region) of *gypsy* contains an insulator sequence that binds Su(Hw) protein and is the only part of *gypsy* required to block the interaction of enhancers with the promoters they regulate. Insertion of this insulator sequence between an enhancer and a promoter by insertion of a *gypsy* element uncouples the enhancer from the promoter (Gause et al. 2001).

The *gypsy* element can only be mobilized in male and female progeny of mothers that both contain active copies of *gypsy* and are homozygous for permissive *flamenco* (*flam*) alleles (Chalvet et al. 1998; Sarot et al. 2004). What protein-coding genes, if any, reside in *flam* is unknown.

To determine the mechanism by which Armi silences *mst40* and *gypsy,* we analyzed ovaries from heterozygous and homozygous *armi* mutant females by nuclear run-on (Fig. 7). We found no significant change in transcription between *armi/+* and *armi/armi* ovaries for either *mst40* or *gypsy*. We conclude that in the female germ line of *D. melanogaster*, Armi acts to silence these two repetitive elements posttranscriptionally, rather than transcriptionally.

CONCLUSIONS

We have shown that nuclear run-on can be used to measure changes in transcriptional rates in the female *D. melanogaster* germ line. Our data indicate that as little as an approximately 1.5-fold change in the rate of transcription can be detected by this method. Our nuclear run-on protocol likely measures changes in transcriptional elongation, because rates of transcription decline with time, consistent with a low rate of pol II reinitiation in the isolated nuclei. Our data support the view that the nuclear run-on technique can be applied not only to single-copy genes, but

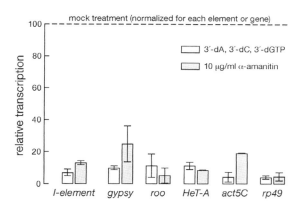

Figure 6. Repetitive element transcripts are largely RNA pol II products.

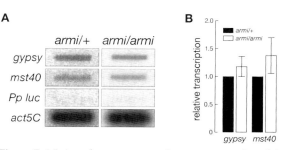

Figure 7. (*A*) A nuclear run-on experiment representative of the data used for the quantitative analysis in *B*. (*B*) The transcriptional rates of the retrotransposon *gypsy* and the repetitive sequence *mst40* do not change significantly when silencing is lost in homozygous mutant *armi* ovaries. The average (± standard deviation) relative to the transcriptional rate of *act5C* is presented.

also to multicopy transcriptional units such as repetitive elements and transposons. Surprisingly, we find that the retrotransposon *gypsy* and the repetitive sequence *mst40* are silenced posttranscriptionally in fly ovaries. Whether repetitive elements in general are silenced posttranscriptionally in the fly germ line remains to be established.

ACKNOWLEDGMENTS

We thank members of the Zamore lab for helpful discussions and comments on the manuscript and Alicia Boucher for assistance with fly husbandry. P.D.Z. is a W.M. Keck Foundation Young Scholar in Medical Research. This work was supported in part by grants from the National Institutes of Health to P.D.Z. (GM62862 and GM65236).

REFERENCES

Allshire R.C., Nimmo E.R., Ekwall K., Javerzat J.P., and Cranston G. 1995. Mutations derepressing silent centromeric domains in fission yeast disrupt chromosome segregation. *Genes Dev.* **9:** 218.

Bernard P., Maure J.F., Partridge J.F., Genier S., Javerzat J.P., and Allshire R.C. 2001. Requirement of heterochromatin for cohesion at centromeres. *Science* **294:** 2539.

Buhler M., Verdel A., and Moazed D. 2006. Tethering RITS to a nascent transcript initiates RNAi- and heterochromatin-dependent gene silencing. *Cell* **125:** 873.

Chalvet F., di Franco C., Terrinoni A., Pelisson A., Junakovic N., and Bucheton A. 1998. Potentially active copies of the gypsy retroelement are confined to the Y chromosome of some strains of *Drosophila melanogaster* possibly as the result of the female-specific effect of the flamenco gene. *J. Mol. Evol.* **46:** 437.

Church G.M. and Gilbert W. 1984. Genomic sequencing. *Proc. Natl. Acad. Sci.* **81:** 1991.

Copenhaver G.P., Nickel K., Kuromori T., Benito M.I., Kaul S., Lin X., Bevan M., Murphy G., Harris B., Parnell L.D., et al. 1999. Genetic definition and sequence analysis of *Arabidopsis* centromeres. *Science* **286:** 2468.

Cox R.F. 1976. Quantitation of elongating form A and B RNA polymerases in chick oviduct nuclei and effects of estradiol. *Cell* **7:** 455.

Djupedal I., Portoso M., Spahr H., Bonilla C., Gustafsson C.M., Allshire R.C., and Ekwall K. 2005. RNA Pol II subunit Rpb7 promotes centromeric transcription and RNAi-directed chromatin silencing. *Genes Dev.* **19:** 2301.

Elgin S.C. and Grewal S.I. 2003. Heterochromatin: Silence is golden. *Curr. Biol.* **13:** R895.

Forrester W.C., Novak U., Gelinas R., and Groudine M. 1989. Molecular analysis of the human beta-globin locus activation region. *Proc. Natl. Acad. Sci.* **86:** 5439.

Gause M., Morcillo P., and Dorsett D. 2001. Insulation of enhancer-promoter communication by a gypsy transposon insert in the *Drosophila* cut gene: Cooperation between suppressor of hairy-wing and modifier of mdg4 proteins. *Mol. Cell. Biol.* **21:** 4807.

Gilbert D.M. 2002. Replication timing and transcriptional control: Beyond cause and effect. *Curr. Opin. Cell Biol.* **14:** 377.

Grummt I. and Pikaard C.S. 2003. Epigenetic silencing of RNA polymerase I transcription. *Nat. Rev. Mol. Cell Biol.* **4:** 641.

Heard E. 2005. Delving into the diversity of facultative heterochromatin: The epigenetics of the inactive X chromosome. *Curr. Opin. Genet. Dev.* **15:** 482.

Henikoff S. 1992. Position effect and related phenomena. *Curr. Opin. Genet. Dev.* **2:** 907.

Holmquist G.P., Kapitonov V.V., and Jurka J. 1998. Mobile genetic elements, chiasmata, and the unique organization of beta-heterochromatin. *Cytogenet. Cell Genet.* **80:** 113.

Huang J. and Moazed D. 2006. Sister chromatid cohesion in silent chromatin: Each sister to her own ring. *Genes Dev.* **20:** 132.

Huisinga K.L., Brower-Toland B., and Elgin S.C. 2006. The contradictory definitions of heterochromatin: Transcription and silencing. *Chromosoma* **115:** 110.

Kalmykova A.I., Klenov M.S., and Gvozdev V.A. 2005. Argonaute protein PIWI controls mobilization of retrotransposons in the *Drosophila* male germline. *Nucleic Acids Res.* **33:** 2052.

Kato H., Goto D.B., Martienssen R.A., Urano T., Furukawa K., and Murakami Y. 2005. RNA polymerase II is required for RNAi-dependent heterochromatin assembly. *Science* **309:** 467.

Kazazian H.H.J. 2004. Mobile elements: Drivers of genome evolution. *Science* **303:** 1626.

Kim A. and Dean A. 2004. Developmental stage differences in chromatin subdomains of the beta-globin locus. *Proc. Natl. Acad. Sci.* **101:** 7028.

Lawrence R.J. and Pikaard C.S. 2004. Chromatin turn ons and turn offs of ribosomal RNA genes. *Cell Cycle* **3:** 880.

Lawrence R.J., Earley K., Pontes O., Silva M., Chen Z.J., Neves N., Viegas W., and Pikaard C.S. 2004. A concerted DNA methylation/histone methylation switch regulates rRNA gene dosage control and nucleolar dominance. *Mol. Cell* **13:** 599.

Lewis M.S. and Pikaard C.S. 2001. Restricted chromosomal silencing in nucleolar dominance. *Proc. Natl. Acad. Sci.* **98:** 14536.

Mahtani M.M. and Willard H.F. 1998. Physical and genetic mapping of the human X chromosome centromere: Repression of recombination. *Genome Res.* **8:** 100.

Martienssen R.A. and Colot V. 2001. DNA methylation and epigenetic inheritance in plants and filamentous fungi. *Science* **293:** 1070.

McKnight G.S. and Palmiter R.D. 1979. Transcriptional regulation of the ovalbumin and conalbumin genes by steroid hormones in chick oviduct. *J. Biol. Chem.* **254:** 9050.

Nimmo E.R., Pidoux A.L., Perry P.E., and Allshire R.C. 1998. Defective meiosis in telomere-silencing mutants of *Schizosaccharomyces pombe*. *Nature* **392:** 825.

Nolan T., Braccini L., Azzalin G., De Toni A., Macino G., and Cogoni C. 2005. The post-transcriptional gene silencing machinery functions independently of DNA methylation to repress a LINE1-like retrotransposon in *Neurospora crassa*. *Nucleic Acids Res.* **33:** 1564.

Pal-Bhadra M., Leibovitch B.A., Gandhi S.G., Rao M., Bhadra U., Birchler J.A., and Elgin S.C. 2004. Heterochromatic silencing and HP1 localization in *Drosophila* are dependent on the RNAi machinery. *Science* **303:** 669.

Peifer M. and Bender W. 1988. Sequences of the gypsy transposon of *Drosophila* necessary for its effects on adjacent genes. *Proc. Natl. Acad. Sci.* **85:** 9650.

Pidoux A.L. and Allshire R.C. 2005. The role of heterochromatin in centromere function. *Philos. Trans. R. Soc. Lond. B Biol. Sci.* **360:** 569.

Puechberty J., Laurent A.M., Gimenez S., Billault A., Brun-Laurent M.E., Calenda A., Marcais B., Prades C., Ioannou P., Yurov Y., and Roizes G. 1999. Genetic and physical analyses of the centromeric and pericentromeric regions of human chromosome 5: Recombination across 5cen. *Genomics* **56:** 274.

Riddle N.C. and Elgin S.C. 2006. The dot chromosome of *Drosophila*: Insights into chromatin states and their change over evolutionary time. *Chromosome Res.* **14:** 405.

Russell S.R. and Kaiser K. 1994. A *Drosophila melanogaster* chromosome 2L repeat is expressed in the male germ line. *Chromosoma* **103:** 63.

Sarot E., Payen-Groschene G., Bucheton A., and Pelisson A. 2004. Evidence for a piwi-dependent RNA silencing of the gypsy endogenous retrovirus by the *Drosophila melanogaster* flamenco gene. *Genetics* **166:** 1313.

Savitsky M., Kwon D., Georgiev P., Kalmykova A., and

Gvozdev V. 2006. Telomere elongation is under the control of the RNAi-based mechanism in the *Drosophila* germline. *Genes Dev.* **20:** 345.

Schubeler D., Scalzo D., Kooperberg C., van Steensel B., Delrow J., and Groudine M. 2002. Genome-wide DNA replication profile for *Drosophila melanogaster:* A link between transcription and replication timing. *Nat. Genet.* **32:** 438.

Sijen T. and Plasterk R.H. 2003. Transposon silencing in the *Caenorhabditis elegans* germ line by natural RNAi. *Nature* **426:** 310.

Slawson E.E., Shaffer C.D., Malone C.D., Leung W., Kellmann E., Shevchek R.B., Craig C.A., Bloom S.M., Bogenpohl J.N., Dee J., et al. 2006. Comparison of dot chromosome sequences from *D. melanogaster* and *D. virilis* reveals an enrichment of DNA transposon sequences in heterochromatic domains. *Genome Biol.* **7:** R15.

So W.V. and Rosbash M. 1997. Post-transcriptional regulation contributes to *Drosophila* clock gene mRNA cycling. *EMBO J.* **16:** 7146.

Swaneck G.E., Nordstrom J.L., Kreuzaler F., Tsai M.J., and O'Malley B.W. 1979. Effect of estrogen on gene expression in chicken oviduct: Evidence for transcriptional control of ovalbumin gene. *Proc. Natl. Acad. Sci.* **76:** 1049.

Vagin V.V., Sigova A., Li C., Seitz H., Gvozdev V., and Zamore P.D. 2006. A distinct small RNA pathway silences selfish genetic elements in the germline. *Science* **313:** 320.

Volpe T.A., Kidner C., Hall I.M., Teng G., Grewal S.I., and Martienssen R.A. 2002. Regulation of heterochromatic silencing and histone H3 lysine-9 methylation by RNAi. *Science* **297:** 1833.

Weintraub H., Larsen A., and Groudine M. 1981. Alpha-Globin-gene switching during the development of chicken embryos: Expression and chromosome structure. *Cell* **24:** 333.

Weiss S.B. 1960. Enzymatic incorporation of ribonucleoside triphosphates into the interpolynucleotide linkages of ribonucleic acid. *Proc. Natl. Acad. Sci.* **46:** 1020.

Zacharias H. 1995. Emil Heitz (1892–1965): Chloroplasts, heterochromatin, and polytene chromosomes. *Genetics* **141:** 7.

Polymorphic MicroRNA–Target Interactions: A Novel Source of Phenotypic Variation

M. GEORGES,* A. CLOP,* F. MARCQ,* H. TAKEDA,* D. PIROTTIN,* S. HIARD,* X. TORDOIR,*
F. CAIMENT,* F. MEISH,* B. BIBÉ,[†] J. BOUIX,[†] J.M. ELSEN,[†] F. EYCHENNE,[†] E. LAVILLE,[¶†]
C. LARZUL,[†] D. MILENKOVIC,[§] J. TOBIN,[‡] AND C. CHARLIER*

*Unit of Animal Genomics, Department of Animal Production, Faculty of Veterinary Medicine & CBIG,
University of Liège (B43), 4000-Liège, Belgium; [‡]Cardiovascular and Metabolic Diseases, Wyeth Research,
Cambridge, Massachusetts 02140; [†]INRA-SAGA, Institut National de la Recherche Agronomique-Station
d'Amélioration Génétique des Animaux, BP 52627-31326 Castanet-Tolosan CEDEX, France;
[¶]Station de Recherches sur la Viande, INRA, Theix, 63122 Saint-gene`s-Champanelle, France;
[§]INRA/Université de Limoges, Faculté des Sciences, 87060 Limoges Cedex, France

Studying the muscular hypertrophy of Texel sheep by forward genetics, we have identified an A-to-G transition in the 3'UTR of the GDF8 gene that reveals an illegitimate target site for microRNAs miR-1 and miR-206 that are highly expressed in skeletal muscle. This causes the down-regulation of this muscle-specific chalone and hence contributes to the muscular hypertrophy of Texel sheep. We demonstrate that polymorphisms which alter the content of putative miRNA target sites are common in human and mice, and provide evidence that both conserved and nonconserved target sites are selectively constrained. We speculate that these polymorphisms might be important mediators of phenotypic variation including disease. To facilitate studies along those lines, we have constructed a database (www.patrocles.org) listing putative polymorphic microRNA–target interactions.

Animal breeders have been conducting an unprecedented phenotype-driven screen for thousands of years, targeting behavioral, morphological, physiological, and pathological traits. This has resulted in the creation of numerous breeds that exhibit often spectacular phenotypic differences. Using the recently developed tools for genome analysis, animal geneticists are now attempting to identify the mutations that underlie this remarkable phenotypic plasticity. By doing so, they aim at contributing to a better molecular understanding of the genotype–phenotype relationship and adaptation, as well as at developing knowledge and tools allowing more proficient selection schemes (see, e.g., Andersson and Georges 2004).

One of the phenotypes that—for obvious reasons—has received considerable attention in livestock species is muscular development. Hypermuscled breeds, including Belgian Blue cattle, Piétrain pigs, and Callipyge sheep, have served as starting material for positional cloning efforts which led to the identification of structural mutations in myostatin (GDF8) (see, e.g., Grobet et al. 1997) and the calcium release channel (CRC) (Fujii et al. 1991), as well as regulatory mutation controlling the expression levels of insulin-like growth factor 2 (IGF2) (Van Laere et al. 2003) and delta-like 1 (DLK1) (Charlier et al. 2001; Freking et al. 2002; Davis et al. 2004).

We herein describe the recent positional identification of a mutation in the GDF8 3'UTR that creates an illegitimate target site for two miRNA that are highly expressed in skeletal muscle. The ensuing down-regulation of GDF8 contributes to the muscular hypertrophy of Texel sheep (Clop et al. 2006). We pursue by demonstrating that large numbers of single-nucleotide polymorphisms (SNPs), particularly in human and mice, have the potential to

affect the interaction between miRNAs and their targets and may thereby be important mediators of phenotypic variation. The corresponding information is compiled in the open access Patrocles database (Clop et al. 2006; S. Hiard et al., in prep.).

A MUTATION CREATING AN ILLEGITIMATE MIRNA TARGET SITE IN THE GDF8 3'UTR IS A QUANTITATIVE TRAIT NUCLEOTIDE AFFECTING MUSCULARITY IN SHEEP

A QTL with Major Effect on Muscularity Maps to Sheep Chromosome 2

Texel sheep are famous and hence utilized worldwide for their pronounced muscular hypertrophy. This is especially true for the hypermuscled strain of Belgian Texel. To gain a better understanding of the molecular basis of this economically important phenotype, we generated a Romanov x Belgian Texel F2 cross counting 278 individuals. Romanov sheep are characterized by a mediocre muscular development but are very fertile. F2 animals were slaughtered at a target weight of 39 kg for the males and 35 kg for the females, and 56 phenotypes quantifying carcass composition were measured on each individual. The raw phenotypes were precorrected for a number of fixed effects and covariates including sex, year, slaughter plant, and final weight. All F0, F1, and F2 animals were genotyped for 160 microsatellite markers spanning the sheep genome. Marker maps were constructed and shown to be in good agreement with published maps. The genome was scanned for quantitative trait loci (QTL) using linear regression (Haley et al. 1994), significance thresholds determined by phenotype permutation (Doerge and Churchill 1996), and confidence intervals (CI) determined by bootstrapping (Visscher et al. 1996). A highly significant effect on virtually all phenotypes measuring muscularity or fat deposition was found in the pericentromeric region of chromosome 2 (OAR2). This QTL typically explained about 25% of the breed difference, about 20% of the F_2 variance, and acted mostly in a partially recessive way.

The CI for the QTL location spanned 10 cM including the OAR2 centromere. Notably, the orthologous marker interval is known to harbor *GDF8* in the bovine. Loss-of-function mutations in the *GDF8* gene are known to have dramatic effects on muscle mass in mouse, bovine, and human (hence the alternative name myostatin), making this gene the perfect positional candidate (see, e.g., Lee 2004). We thus sequenced the coding portions from the gene after PCR amplification of the exons from genomic DNA of Texel and Romanov but did not find a single sequence difference. The same experiment was repeated using cDNA from skeletal muscle of Texel and Romanov, yielding equivalent amounts of product with identical size, sequence, and thus, coding potential. Moreover, northern blot analysis revealed a band of the expected size and with very comparable intensity in Texel and Romanov. At first glance, these results did not support a role for *GDF8* in the muscular hypertrophy of Texel sheep.

Fine-mapping Positions the QTL in the Immediate Vicinity of the *GDF8* Gene

To refine the map position of the QTL, we first increased the marker density in the CI. We then applied two fine-mapping strategies. In a first instance, we identified a F_2 ram having inherited an intact Texel chromosome from one parent, and a recombinant Texel–Romanov chromosome recombining in the CI from the other. The recombinant chromosome was of Texel descent for most of the CI (~3/4 on the OAR2p side), and of Romanov descent for the remaining quarter of the CI (on the OAR2q side). We generated 43 offspring from this ram, reared them until 35–39 kg, slaughtered them, and measured the same set of phenotypes as before. Sorting the offspring according to the paternal OAR2 homolog clearly indicated that the ram was heterozygous *Qq* for the QTL ($p = 0.0013$), thus positioning the QTL in the 1/4 portion of the CI for which the ram was heterozygous Texel/Romanov.

In a second instance, we genotyped 42 hypermuscled Belgian Texels and 108 animals from 16 other breeds for all markers available in the CI. We reasoned that selection for meatiness in Belgian Texel might have caused the near fixation of a mutation increasing muscularity, hence causing a local reduction in genetic variation ("selective sweep"). A highly significant reduction in genetic variation was observed for most of the CI. However, the signal maximized on the OAR2q side of the interval, in agreement with the results obtained with the recombinant F_2 ram. More specifically, the signature of selection was highest for marker *BULGE20*, for which the homozygosity was 89% in Texels versus 15% in controls. *BULGE20* is located less than one megabase from *GDF8*, strongly encouraging us to examine the myostatin gene more closely.

A *G* to *A* Substitution in the *GDF8* 3′UTR Is a Strong Genetic QTN Candidate

We resequenced 10.5 kb spanning the entire *GDF8* gene including 3.5 kb of upstream, and 1.9 kb of downstream, sequence for three Texel and seven control animals. This identified 20 SNPs. As expected based on the previous experiments, none of these would affect the predicted *GDF8* open reading frame. Six SNPs were located upstream of the transcription start site, two in the 5′UTR, eight in intron 1, three in intron 2, and one in the 3′UTR. However, none of them fell in a highly conserved gene segment suggestive of functionally important regulatory elements.

All 20 SNPs were genotyped on 42 hypermuscled Belgian Texels, 90 control animals (representing 11 breeds), and four rams known with virtual certainty to be heterozygous *Qq* for the QTL. The latter corresponded to the previously described "recombinant" F_2 ram, as well as to the three F_1 rams utilized to generate the Romanov × Texel F_2 population. Indeed, sorting the F_2 offspring based on the homolog inherited from the sire in the CI revealed a consistent allele substitution effect in the three F_1 sire families.

The first striking observation was the virtual monomorphism of the *GDF8* gene in Texels contrasting with the considerable polymorphism of the same gene among controls. This observation was consistent with the occurrence—in Texel—of a selective sweep driven by an advantageous *GDF8* mutation. Moreover, by virtue of the homozygosity of at least one of the four *Qq* rams, all but two (*g-2449C-G* and *g+6723G-A*) of the 20 SNPs could be excluded as being causal. More recently, the same approach applied to sheep populations in New Zealand led to the exclusion of the *g-2449C-G* SNP as well (J. McEwan, pers. comm.), leaving *g+6723G-A* as the only putative causal mutation. Finally, and consistent with the previous findings, the *g+6723G-A* SNP is the only one for which one allele (*A*) is virtually Texel-specific (allelic frequency 99% in Texel versus 1% in controls). Altogether, these observations strongly incriminated *GDF8* and the *g+6723G-A* SNP in its 3′UTR as causing the observed QTL effect.

miRNA-mediated Translation Inhibition of Mutant *GDF8* Transcripts Underlie the QTL Effect

Careful examination of the sequence context of *g+6723G-A* indicated that the *A* allele reveals one of the 3′UTR octamer motifs (AC*A*TTCCA) discovered by Xie et al. (2005) on the basis of their unusually high motif conservation score (MCS) and thought to correspond to miRNA target sites. We simulated all the possible nucleotide substitutions in the 3′UTR of the ovine *GDF8* gene (accounting for a transition rate twice as large as the transversion rate) and demonstrated that only 4.5% of these would create one of the 540 motifs described by Xie et al. (2005), whereas only 0.88% would create an octamer with an equally high MCS. This indicated that our observation was not trivial, and suggested that the *g+6726A* allele might create an illegitimate miRNA target site resulting in posttranscriptional *GDF8* down-regulation and hence in an increase in muscle mass.

ACATTCCA is showing perfect Watson-Crick base-pairing with nucleotides 1–8 of *miR-1* and *miR-206*, and near-perfect base-pairing (one G:U pair) with *miR-122*, suggesting that one or several of these miRNAs might be mediating the hypothesized effect. Using primers targeting conserved flanking sequences, we amplified and sequenced the corresponding pri-miRNA genes from sheep genomic DNA demonstrating their conservation. Using primer extension, we then demonstrated that *miR-1* and *miR-206* (but not *miR-122a*) were indeed preferentially and highly expressed in skeletal muscle. These findings are in agreement with more recent data in human and mice (see, e.g., Farh et al. 2005) and lent further support to our hypothesis.

Our model makes at least two in vivo testable predictions: (1) GDF8 protein levels should be reduced in Texel sheep when compared to controls as a result of translational inhibition of the mutant mRNA, and (2) mutant *GDF8* mRNA levels might be reduced when compared to wild-type *GDF8* mRNA levels as a result of accelerated degradation of mutant mRNA (see, e.g., Lim et al. 2005). To test the first prediction, we measured the levels of circulating GDF8 by immunoprecipitation followed by western blotting, a proven quantitative procedure that was previously successfully utilized in primates and rodents (see, e.g., Zimmers et al. 2002; Schuelke et al. 2004). A consistent threefold reduction in GDF8 levels was indeed observed in the serum of Texel sheep when compared to wild-type animals. To test the second prediction, we measured *GDF8* mRNA allelic imbalance in skeletal muscle of heterozygous *AG* individuals. Wild-type and mutant mRNA were amplified by hot-stop RT-PCR, distinguished by virtue of the *HpyCH4IV* restriction site that is present in the wild-type but not mutant allele, and quantified after denaturing acrylamide gel electrophoresis using a phosphorimager. The ratios of mutant versus wild-type intensities were compared with those obtained from *AG* genomic DNA as well as with a calibration curve, indicating a 1.5-fold reduction of mutant versus wild-type mRNA. Both in vivo predictions were thus met, in agreement with our model.

To provide direct evidence that the *g+6723G-A* SNP indeed affects the interaction between the *GDF8* 3′UTR and the miRNAs *miR-1* and *miR-206*, we generated "mutant" and "wild-type" luciferase reporter constructs endowed either with four tandem copies of an 80-bp segment of the ovine *GDF8* 3′UTR centered around the *g+6723G-A* site (pRL-4x**A** and pRL-4x**G** respectively), or with the complete ovine *GDF8* 3′UTR (pRL-3′**A** and pRL-3′**G**, respectively) in their 3′UTR. Co-transfecting COS1 cells with these constructs and pcDNA3 vectors expressing either *miR-1* or *miR-206*, or control miRNAs *miR-136* and *miR-377*, demonstrated highly specific and significant reductions in luminescence signal when transfecting mutant **A**-constructs with *miR-1* and *miR-206* expressing vectors, but not when transfecting mutant **A**-constructs with vectors expressing control miRNAs or when transfecting wild-type **G**-constructs with either of the miRNA-expressing vectors.

Taken together, our results strongly support the fact that the *g+6723G-A* SNP is indeed the causative quantitative trait nucleotide (QTN) that creates a hypomorphic allele by promoting miRNA-mediated down-regulation of *GDF8* and hence contributes to the muscular hypertrophy characteristic of Texel sheep (Fig.1).

POLYMORPHIC MIRNA–TARGET INTERACTIONS: COMMON MEDIATORS OF PHENOTYPIC VARIATION?

Common Sequence Variants Alter the Content in Putative miRNA Target Site of Thousands of Human and Mouse Genes

There is growing evidence that a substantial proportion of our genes (1/4th to 1/3rd) are regulated by miRNAs (see, e.g., Lewis et al. 2005). Thus, the ovine *g+6723G-A GDF8* mutation—rather than being a sheep idiosyncrasy—may be the prototype example of a common type of polymorphisms that make a significant contribution to phenotypic variation. That this may indeed be the case is suggested by the recent description of a *G* to *A* substitution in the 3′UTR of the human *SLITRK1* gene shown to be associated with Tourette's syndrome and thought to act by stabilizing the interaction with the coexpressed *miR-189* miRNA (Abelson et al. 2005).

To evaluate how common polymorphic miRNA–target interactions might be, we examined whether known SNPs might alter the 3′UTR content in miRNA target sites either by destroying existing sites or by creating new ones. The study was initially performed in human and mice, the two species for which the number of documented SNPs is largest. 92,967 human and 85,987 mouse SNPs, located, respectively, in 21,206 and 20,306 3′UTRs, were downloaded from Ensembl (http://www.ensembl.org/index.html). The ancestral allele was derived from the alignment with the chimpanzee for 95% of the human SNPs and from the alignment with the rat for 78% of the mouse SNPs.

miRNA target sites were defined on the basis of the collection of 540 octamers identified by Xie et al. (2005). Targets were labeled as evolutionarily conserved if the octamer was conserved across human, chimpanzee (only for mouse targets), mouse, rat (only for human targets), and dog; as nonconserved otherwise. The corresponding sequence alignments were downloaded from the UCSC genome browser (http://genome.ucsc.edu/).

We then searched for 3′UTR SNPs that would alter the target site content. Such SNPs will be hereafter referred to as pSNPs for Patrocles-SNPs. pSNPs for which the ancestral allele could be determined and that altered the target site content were classified in four groups: (1) destroying a conserved target site (DC), (2) destroying a nonconserved target site (DNC), (3) creating a nonconserved target site (CNC), or (4) shifting a target site (S). In some instances (1,235 in human; 4,278 in mouse), pSNPs were found that created a conserved site. We assumed that in these cases the allele shared by human and chimpanzee or mouse and rat was in fact a shared derived allele predating the corresponding species divergence. Hence, these pSNPs were reclassified as "DC." pSNPs changing target site content but for which the ancestral allele could not be

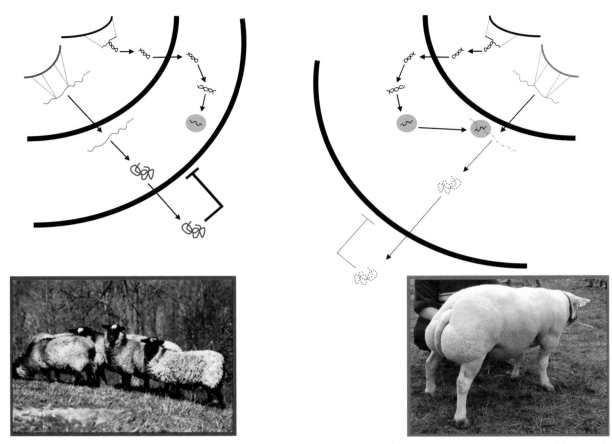

Figure 1. Schematic representation of the effect of the *g+6723G-A* mutation in the 3′UTR of the ovine *GDF8* gene. In wild-type animals (such as the Romanov sheep shown in the left picture) the *GDF8* transcripts (in *blue*) in the sarcoplasm do not interact with microRNAs *miR-1* and *miR-206* despite their high level of expression in skeletal muscle. Hence, GDF8 protein is produced at regular levels, controlling muscle growth by retro-inhibition. In mutant animals (such as the Texel sheep shown in the right picture), *GDF8* transcripts (in *red*) become targets for *miR-1* and *miR-206* as a result of the *g+6723G-A* mutation. This leads to translational inhibition of the *GDF8* transcripts, reduced GDF8 protein levels, relaxation of the retro-inhibition and, hence, enhanced muscle growth.

determined were classified as DC if one of the alleles was characterized by a conserved target site, or placed in a fifth category (polymorphic nonconserved— PNC) otherwise.

8,354 human pSNPs (i.e., 8.9%) were found to affect the 3′UTR target site content of 7,406 genes. The corresponding figures in the mouse were 7,942 SNPs in 4,582 genes. Table 1 reports the number and type of target site alterations caused by the corresponding sets of pSNPs. The total number of alterations exceeds the SNP numbers, as one SNP can have multiple effects.

Table 1. Numbers and Type of Target Site Alterations Caused by 8,354 human SNPs in the 3′UTRs of 7,406 Genes, and by 7,942 Mouse SNPs in the 3′UTR of 4,582 Genes

	Conserved	Nonconserved
Destroyed	HS: 500	HS: 4,014
	MM: 341	MM: 3,431
Created	HS: 0	HS: 4,544
	MM: 0	MM: 3,678
Polymorphic		HS: 235
		MM: 1,327
Shifted	HS: 81	
	MM: 11	

(HS) Human; (MM) mouse.

Evidence for Purifying Selection against SNPs Altering the Target Site Content of 3′UTRs

Assuming that at least part of the predicted miRNA target sites are functional, pSNPs that alter these sites will likely affect gene function and may thus not be selectively neutral, being either subject to purifying selection if deleterious or to positive selection if beneficial. To provide evidence that at least some of the putative target sites— whether conserved or not—are selectively constrained, we compared the effects on target site content of the real set of human SNPs with the effects of simulated SNP sets. The simulated SNP sets were generated by randomly changing the position of each of the 92,967 human SNPs within the same gene. For example, a SNP corresponding to a *G* to *A* transition at position x of the 3′UTR of gene y would be randomly shifted toward any of the *G* residues within the 3′UTR of gene y; a SNP corresponding to the insertion of a G residue would be randomly shifted toward any position of the corresponding 3′UTR; etc. One hundred such SNP sets were generated and subject to the same analysis as the set of real SNPs.

Figure 2 compares the distribution of the number of effects of different types obtained with the simulated and real data sets. The first noticeable observation is that the number of real pSNPs affecting target site content (8,354) is 8.4 standard deviations below the mean number of pSNPs affecting target site content across simulated data sets (9,155). This thus indicates that true SNPs are clearly "avoiding" putative miRNA target sites. This is either due to the fact that putative target sites are less prone to mutation, or more likely, that a significant proportion of putative target sites are under selective constraint. As expected, the category that is most markedly underrepresented in the real data set is the DC class: The true SNPs cause 500 such effects, which is 10 standard deviations below the average number of DC effects (783) observed with the simulated SNP sets. Conserved target sites have a higher likelihood of being genuine miRNA targets (or at least functional elements), and thus their destruction has a higher likelihood of being deleterious. More interesting is the lower than expected incidence of DNC cases with real SNPs (4,014), 8.6 standard deviations below the equivalent number obtained on average (4,595) across simulated SNP sets. This suggests that a substantial proportion of nonconserved target sites are selectively constrained and thus functional. This is somewhat unexpected, as the majority of nonconserved target sites are assumed to reside in genes that are not coexpressed with the cognate miRNA (see, e.g., Farh et al. 2005). Equally intriguing is the slightly lower than expected number of CNC cases (4,544 instead of 4,776,

differing by 3.5 standard deviations), suggesting that a nonnegligible proportion of site creations is deleterious. Note that with the simulated SNP sets, the number of CNCs generally exceeds the number of DNCs by an average of 182 units. This could be due to the fact that the putative target sites are statistically underrepresented in the analyzed 3'UTR sequences, thus, that a random mutation has a higher likelihood to create than to destroy such a site. That this difference is more pronounced with the real SNPs (530 units) probably reflects stronger purifying selection against SNPs destroying than against SNPs creating nonconserved target sites. Finally, we observed twice as many S cases as expected (81 versus 38, differing by 7.4 standard deviations). The reasons for this remain unclear.

Patrocles: The Database of Polymorphic miRNA–Target Interactions

To assist in the identification of SNPs that might affect gene function by perturbing miRNA–target interactions, we have developed a publicly accessible database that compiles information about polymorphisms that alter gene miRNA target site content in animals. The database is called Patrocles (http://www.patrocles.org), highlighting the parallel between the death of Patrocles and the down-regulation of *GDF8* in Texel as a result of the

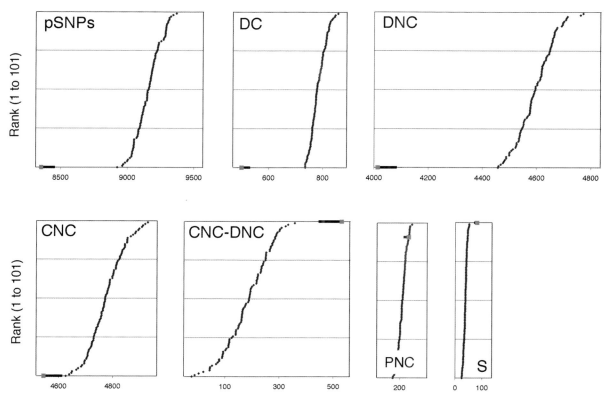

Figure 2. Distribution of the numbers of (1) SNPs altering miRNA target site content (pSNPs), (2) destructions of conserved target sites (DC), (3) destructions of nonconserved target sites (DNC), (4) creations of nonconserved target sites (CNC), (5) polymorphic target sites (PNC, ancestral allele unknown), and (6) shifted target sites (S). CNC-DNC corresponds to the difference in the number of creations versus destructions of nonconserved targets for a given SNP set. The red squares correspond to the numbers obtained with the real SNP set. The blue diamonds correspond to the numbers obtained with 100 simulated SNP sets. The black horizontal line corresponds to one standard deviation of the distribution across simulated data sets. Real and simulated data sets were ranked in ascending order.

10	126665355	1	rs11596226 T/G		CCTGTGAA M miR-27a miR-27b miR-513 GAAGCAAA M miR-686	ZRANB1	HUGO	TRABID protein (Zinc finger Ran-binding domain-containing protein 1). [Source:Uniprot/SWISSPROT;Acc:Q9UGI0]
10	126665452	1	rs10640609 AG/-	TCCAGAGA M miR-326 miR-518c*		ZRANB1	HUGO	TRABID protein (Zinc finger Ran-binding domain-containing protein 1). [Source:Uniprot/SWISSPROT;Acc:Q9UGI0]
10	126667209	-1	rs1058497 A/T	**ATACAAGA** M miR-643		CTBP2	HUGO	C-terminal-binding protein 2 (CtBP2). [Source:Uniprot/SWISSPROT;Acc:P56545]
10	126667600	-1	rs17155567 A/T rs3208601 A/T	GGGAGAAA M miR-629		CTBP2 NP_073713.1	HUGO RefSeq_peptide	C-terminal-binding protein 2 (CtBP2). [Source:Uniprot/SWISSPROT;Acc:P56545] C-terminal-binding protein 2 (CtBP2). [Source:Uniprot/SWISSPROT;Acc:P56545]
10	126667605	-1	rs17155562 G/T rs3208600 G/T	GGGAGAAA M miR-629		CTBP2 NP_073713.1	HUGO RefSeq_peptide	C-terminal-binding protein 2 (CtBP2). [Source:Uniprot/SWISSPROT;Acc:P56545] C-terminal-binding protein 2 (CtBP2). [Source:Uniprot/SWISSPROT;Acc:P56545]
10	126667705	-1	rs3208590 A/C		ACTATTAA M miR-633	CTBP2 NP_073713.1 Q5SQP8_HUMAN	HUGO RefSeq_peptide Uniprot/SPTREMBL	C-terminal-binding protein 2 (CtBP2). [Source:Uniprot/SWISSPROT;Acc:P56545] C-terminal-binding protein 2 (CtBP2). [Source:Uniprot/SWISSPROT;Acc:P56545] C-terminal-binding protein 2 (CtBP2). [Source:Uniprot/SWISSPROT;Acc:P56545]
10	127445020	-1	rs28381328 T/C	AAGGTTTA M miR-703		MMP21	HUGO	Matrix metalloproteinase-21 precursor (EC 3.4.24.-) (MMP-21). [Source:Uniprot/SWISSPROT;Acc:Q8N119]
10	127519979	1	rs17153665 A/T	ACACAACA M miR-669a CAACATTA M miR-409		NP_510869.1	RefSeq_peptide	BRCA2 and CDKN1A-interacting protein isoform BCCIPalpha [Source:RefSeq_peptide;Acc:NP_057651]

Figure 3. Screen capture showing a partial Patrocles output responding to a request for all human SNPs affecting the miRNA target site content (as defined by Lewis et al. 2005) for all genes. The output shows the chromosomal position of the SNP (*Chr.* and *Position*), the strand from which the corresponding gene is transcribed (*TrStr*), the SNP identifier (*SNP id*) as well as the two corresponding *alleles* (ancestral allele/derived allele), the target site content of the ancestral allele (*Ancestral octamers*) with indication of the type of target (X- or M-target) (*Source*) and cognate *miRNA* when known, the target site content of the derived allele (*Derived octamers*) with indication of the type of target (X- or L-target) (*Source*) and cognate *miRNA* when known, the gene identifier (*Gene Id*) with corresponding database source (*Gene source*) and gene *description.* Hyperlinks connect SNP, miRNA, and gene identifiers with the corresponding databases. Conserved target sites are marked in bold and underlined. The position of the SNP within the putative target site is highlighted in red.

acquisition of Achilles' armor in the case of Patrocles and of an illegitimate miRNA target site in the case of *GDF8*.

In its present version, the database includes data about human, mouse, dog, chimpanzee, and bovine SNPs. As more SNP data become available for other species, these will be included in the database as well.

In addition to the effect of SNPs on the 3′UTR content in octamer motifs identified by Xie et al. (2005) (X-targets), Patrocles also compiles the effect of SNPs on miRNA target site content as defined by Lewis et al. (2005) (M-targets). These are composed of a heptamer exhibiting perfect Watson-Crick (WC) complementarity with bases 2–8 ("seeds") of known miRNAs, flanked by a t1 *A* anchor on its 3′ side. This target site definition was selected to maximize specificity (Lewis et al. 2005). Known miRNAs were obtained from miRBase (http://microrna.sanger.ac.uk/sequences/index.shtml). The corresponding seeds were used across species, except for 47 miRNAs labeled as primate-specific which were only utilized to define human targets.

pSNPs destroying conserved target sites are likely to perturb gene function and hence to have phenotypic effects, including probably in some cases, to affect disease susceptibility. Equally interesting, however, are pSNPs that create target sites for miRNA that are expressed in the same tissue; i.e., SNPs creating a recognition site in an antitarget gene (Bartel and Chen 2004). The *g+6723G-A* mutation in the ovine *GDF8* gene and its effect on muscularity provide a good example of this type of pSNP. Moreover, if a pSNP destroys a nonconserved target site, it is certainly more likely to be biologically relevant if target gene and cognate miRNA have overlapping expression profiles. To facilitate the identification of truly interacting miRNA–gene pairs, we are in the process of adding information in Patrocles about the degree of coexpression of miRNA and genes. This is accomplished by confronting the experimentally determined gene expression profiles with computationally predicted miRNA expression profiles following Farh et al. (2005). The database can be queried by species, type of target site (X or L), effect of pSNP (DC, DNC, CNC, PNC, and S), pSNP identifier, gene identifier or chromosomal localization, miRNA identifier, or target site. The output provides information about the target content of the ancestral and derived alleles, including target site conservation and position of the pSNP in the target. Links are available for convenient mining of SNP and gene information. The output is available either on screen or can be downloaded as a text file (Fig. 3).

Further developments planned for the Patrocles database include the compilation of SNPs affecting the miRNA rather than the target, as well as the development of a tool to query the effect of a private SNP in a gene on putative miRNA–target interactions.

CONCLUSIONS

This work identifies a novel class of regulatory mutations that has the potential to make a significant contribution to the genetic variation for complex traits in many organisms, including disease susceptibility in man and agronomically important traits in animals and plants. The discovery of the *g+6723G-A GDF8* mutation in sheep will encourage scientists to revisit their favorite positional and physiological candidate genes with a refreshed look in their quest for causative mutations. The evolving Patrocles database provides the community with a tool that should facilitate the identification of causative pSNPs.

ACKNOWLEDGMENTS

This project was supported by grants from (1) the Walloon Ministry of Agriculture (D31/1036), (2) the "GAME" Action de Recherche Concertée from the Communauté Française de Belgique, (3) the PAI P5/25 from the Belgian SSTC (R.SSTC.0135), (4) the European Union "Callimir" STREP project, and (5) the University of Liège. Alex Clop and Haruko Takeda both benefited from E.U. Marie-Curie postdoctoral fellowships. Carole Charlier is chercheur qualifié from the FNRS.

REFERENCES

Abelson J.F., Kwan K.Y., O'Roak B.J., Baek D.Y., Stillman A.A., Morgan T.M., Mathews C.A., Pauls D.L., Rasin M.R., Gunel M., et al. 2005. Sequence variants in SLITRK1 are associated with Tourette's syndrome. *Science* **310:** 317.

Andersson L. and Georges M. 2004. Domestic animal genomics: Deciphering the genetics of complex traits. *Nat. Rev. Genet.* **5:** 202.

Bartel D.P. and Chen C.Z. 2004. Micromanagers of gene expression: The potentially widespread influence of metazoan microRNAs. *Nat. Rev. Genet.* **5:** 396.

Charlier C., Segers K., Karim L., Shay T., Gyapay G., Cockett N., and Georges M. 2001. The callipyge mutation enhances the expression of coregulated imprinted genes in cis without affecting their imprinting status. *Nat. Genet.* **27:** 367.

Clop A., Marcq F., Takeda H., Pirottin D., Tordoir X., Bibe B., Bouix J., Caiment F., Elsen J.M., Eychenne F., et al. 2006. A mutation creating a potential illegitimate microRNA target site in the myostatin gene affects muscularity in sheep. *Nat. Genet.* **38:** 813.

Davis E., Jensen C.H., Schroder H.D., Shay-Hsdfield T., Kliem A., Cockett N., Georges M., and Charlier C. 2004. Ectopic expression of DLK1 protein in skeletal muscle of padumnal heterozygotes causes the callipyge phenotype. *Curr. Biol.* **14:** 1858.

Doerge R.W. and Churchill G.A. 1996. Permutation tests for multiple loci affecting a quantitative character. *Genetics* **142:** 285.

Farh K.K., Grimson A., Jan C., Lewis B.P., Johnston W.K., Lim L.P., Burge C.B., and Bartel D.P. 2005. The widespread impact of mammalian MicroRNAs on mRNA repression and evolution. *Science* **310:** 1817.

Freking B.A., Murphy S.K., Wylie A.A., Rhodes S.J., Keele J.W., Leymaster K.A., Jirtle R.L., and Smith T.P. 2002. Identification of the single base change causing the callipyge muscle hypertrophy phenotype, the only known example of polar overdominance in mammals. *Genome Res.* **12:** 1496.

Fujii J., Otsu K., Zorzato F., de Leon S., Khanna V.K., Weiler J.E., O'Brien P.J., and MacLennan D.H. 1991. Identification of a mutation in porcine ryanodine receptor associated with malignant hyperthermia. *Science* **253:** 448.

Grobet L., Royo Martin L.J., Poncelet D., Pirottin D., Brouwers B., Riquet J., Schoeberlein A., Dunner S., Menissier F., Massabanda J., et al. 1997. A deletion in the myostatin gene causes double-muscling in cattle. *Nat. Genet.* **17:** 71.

Haley C.S., Knott S.A., and Elsen J.-M. 1994. Mapping quantitative trait loci in crosses between outbred lines using least squares. *Genetics* **136:** 1195.

Lee S.J. 2004. Regulation of muscle mass by myostatin. *Annu. Rev. Cell Dev. Biol.* **20:** 61.

Lewis B.P., Burge C.B., and Bartel D.P. 2005. Conserved seed pairing, often flanked by adenosines, indicates that thousands of human genes are microRNA targets. *Cell* **120:** 15.

Lim L.P., Lau N.C., Garrett-Engele P., Grimson A., Schelter J.M., Castle J., Bartel D.P., Linsley P.S., and Johnson J.M. 2005. Microarray analysis shows that some microRNAs downregulate large numbers of target mRNAs. *Nature* **433:** 769.

Schuelke M., Wagner K.R., Stolz L.E., Hubner C., Riebel T., Komen W., Braun T., Tobin J.F., and Lee S.J. 2004. Myostatin mutation associated with gross muscle hypertrophy in a child. *N. Engl. J. Med.* **350:** 2682.

Van Laere A.-S., Nguyen M., Braunschweig M., Nezer C., Collette C., Moreau L., Archibald A.L., Haley C.S., Buys N., Tally M., et al. 2003. Positional identification of a regulatory mutation in *IGF2* causing a major QTL effect on muscle growth in the pig. *Nature* **425:** 832.

Visscher P.M., Thompson R., and Haley C.S. 1996. Confidence intervals in QTL mapping by bootstrapping. *Genetics* **143:** 1013.

Xie X., Lu J., Kulbokas E.J., Golub T.R., Mootha V., Lindblad-Toh K., Lander E.S., and Kellis M. 2005. Systematic discovery of regulatory motifs in human promoters and 3′UTRs by comparison of several mammals. *Nature* **434:** 338.

Zimmers T.A., Davies M.V., Koniaris L.G., Haynes P., Esquela A.F., Tomkinson K.N., McPherron A.C., Wolfman N.M., and Lee S.J. 2002. Induction of cachexia in mice by systemically administered myostatin. *Science* **296:** 1486.

Expression and Function of MicroRNAs in Viruses Great and Small

C.S. Sullivan,* A. Grundhoff,* S. Tevethia,‡ R. Treisman,† J.M. Pipas,¶ and D. Ganem*

*Howard Hughes Medical Institute, Departments of Microbiology and Medicine, University of California,
San Francisco, California 94143; †Transcription Laboratory, Cancer Research UK London Research Institute,
London WC2A 3PX, United Kingdom; ‡Department of Microbiology, Pennsylvania State University Medical Center,
Hershey, Pennsylvania; ¶University of Pittsburgh, Pittsburgh, Pennsylvania

Since they employ host gene expression machinery to execute their genetic programs, it is no surprise that DNA viruses also encode miRNAs. The small size of viral genomes, and the high degree of understanding of the functions of their gene products, make them particularly favorable systems for the examination of miRNA biogenesis and function. Here we review our computational and array-based approaches for viral miRNA discovery, and we discuss the structure and function of miRNAs identified by these approaches in polyomaviruses and herpesviruses.

MicroRNAs (miRNAs) mediate posttranscriptional gene regulation in most eukaryotes and have been shown to play important regulatory roles in many cellular processes, including development, differentiation, metabolic control, apoptosis, and tumorigenesis (for review, see Du and Zamore 2005; Hammond et al. 2005; Kim 2005). In the human genome alone, more than 460 miRNAs have been identified (miRBase, http://microrna.sanger.ac.uk//sequences/index.shtml) (Griffiths-Jones 2004). MicroRNAs are derived from primary nuclear pol II transcripts (pri-miRNAs), which can be thousands of nucleotides in length. Processing of these RNAs by the nuclear microprocessor complex (which includes the enzyme Drosha) yields 60–80-nucleotide imperfect hairpins known as pre-miRNAs (Lee et al. 2002, 2003; Denli et al. 2004; Gregory et al. 2004; Han et al. 2004; Landthaler et al. 2004; Zeng et al. 2005). These pre-miRNAs are transported to the cytosol (Yi et al. 2003; Bohnsack et al. 2004; Lund et al. 2004; Zeng and Cullen 2004), where another cellular enzyme, Dicer, processes them to result in the mature approximately 22-nucleotide miRNA (Bernstein et al. 2001; Grishok et al. 2001; Hutvagner et al. 2001; Ketting et al. 2001; Chendrimada et al. 2005; Forstemann et al. 2005; Gregory et al. 2005; Jiang et al. 2005; Saito et al. 2005). The resulting miRNA enters the multiprotein RNA-induced silencing complex (RISC), where it is hypothesized to scan translating RNAs and direct their cleavage if found to have a perfect match (similar to siRNAs), or translational repression if bound to the RNA with imperfect homology (Hamilton and Baulcombe 1999; Tuschl et al. 1999; Zamore et al. 2000; Grishok et al. 2001; Hutvagner et al. 2001; Doench et al. 2003; Zeng et al. 2003). Because DNA viruses generally employ host pol II machinery to express their genes, it is expected that many such viruses will encode miRNAs—a prediction that was validated by Pfeffer et al. (2004), who first cloned miRNAs from cells infected with several herpesviruses (Pfeffer et al. 2004, 2005).

Despite rapid progress in understanding miRNA biogenesis, the functions of the vast majority of miRNAs remain unknown. The large size of the human genome and the incompletely understood nature of the events governing target recognition are the principal reasons that the targets of most cellular miRNAs are not yet known. The fact that many cellular cDNAs are of unknown function further complicates the task of deciphering the biology of many host miRNAs. In contrast, viral genomes are small, and a substantial percentage of their gene products have well-understood activities or function in known pathways. This makes viral genomes a favorable place to study the biogenesis and function of miRNAs (Sullivan and Ganem 2005). With this in mind, we have begun to search for miRNAs in DNA viruses and to capitalize on the well-understood biology of their viral progenitors to explore their biological function.

To this end, we developed a computer program, called v-miR, to screen viral genomes for inverted repeats with the properties of pre-miRNAs (as defined from a library of known human miRNAs that had been identified by cDNA cloning). On the basis of these properties, each predicted hairpin is assigned a numerical score; the higher the score, the greater the similarity of the hairpin to known pre-miRNAs (for details of the algorithm, see Grundhoff et al. 2006; the program is available to all labs on request). The program typically identifies many more hairpins than are actually involved in generating miRNAs in vivo, but it represents a useful screening tool. To evaluate the utility of the program, we decided to examine its performance on a small (5 kb), well-understood DNA virus, SV40.

SV40, a member of the polyomavirus family, causes a largely asymptomatic renal infection in its natural simian host. (However, when injected into rodents, in which it cannot efficiently complete its full replicative cycle, it causes fibrosarcomas—and it is this unnatural property, rather than its authentic biology, that has attracted the most experimental interest.) The SV40 replicative cycle is one of the best understood genetic programs in animal virology (for review, see Cole 1996). The circular 5-kb dsDNA genome directs expression of two major transcription units, early and late. Early mRNAs represent a family of spliced transcripts that encode large and small T antigens, regulatory proteins whose role in productive infection is primarily to promote viral DNA replication (for review, see Sullivan and Pipas 2002). Following T-antigen accumulation, multimers of large T bind to the viral origin of DNA replication and trigger genomic replication. Following the onset of DNA synthesis, the late transcription unit is activated, generating a series of spliced mRNAs encoding the viral structural proteins. Following accumulation of the late proteins, mature virus particles are assembled, and cell lysis releases the infectious progeny viruses. As shown in Figure 1, the early and late mRNAs are transcribed from opposite strands on the circular genome and overlap one another in the region bounded by their respective polyadenylation signals. These poly(A) signals are not 100% efficient, and longer primary transcripts resulting from readthrough of these signals have been detected in infected cells (Acheson 1978).

Figure 2A shows the readout of the v-miR program on the late strand of the SV40 genome, with hairpin scores displayed as a function of map position. The highest-scoring hairpin in Figure 2A maps just downstream from the late poly(A) site (Fig. 1). Although another equally high-scoring hairpin was identified on the early strand (not shown), only the late-strand candidate pre-miRNA was detected by northern blot analysis. When infected cells are examined with probes from the region of the hairpin, species with the mobilities of pre-miRNA and miRNAs are readily identified (Fig. 2B). As expected from their polarity and map position, they accumulate at late times after infection (Fig. 2C). Close inspection of the northerns reveals that multiple species of miRNA are visible in the 20–24-nucleotide region of the gel, and nuclease mapping confirms that both strands of the hairpin give rise to miRNAs that can be incorporated into RISC (Sullivan et al. 2005).

The location of these miRNAs indicates that they possess perfect complementarity to early (T antigen) mRNAs (Fig. 1). As such, they would be expected to trigger cleavage of those mRNAs, much as would a siRNA. Northern blotting for early mRNAs revealed the presence of a collection of small (~300 nucleotides) polyadenylated RNA fragments that accumulated preferentially at late times (Fig. 3); mapping of these fragments identified it as the probable 3′ product of miRNA-mediated cleavage of early mRNA, since their 5′ ends mapped to the regions of miRNA complementarity (not shown). To verify this, we constructed a mutant of SV40 that was incapable of generating the miRNAs. This was done by engineering multiple point mutations into the predicted pre-miRNA hairpin so as to disrupt its structure and prevent Dicer-mediated processing to the mature miRNAs. As shown in Figure 3, this mutant SV40 virus was unable to generate the predicted cleavage product following infection; as expected, mutant-infected cells accumulated enhanced levels of T-antigen mRNA and proteins. However, the mutant had no growth defect: A careful one-step growth curve reveals wild-type and mutant viruses to grow to identical titers with identical kinetics (Fig. 4). This indicates that the excess T antigen generated by the mutant serves no replicative purpose.

Why, then, has evolution selected for the production of the miRNA? One explanation relates to the fact that T antigen appears to be the major target of cytotoxic T lymphocytes (CTLs) directed against the virus. If so, then down-regulation of T-antigen synthesis might be expected to reduce susceptibility to CTL-mediated lysis. To test this, we infected simian cells bearing murine MHC-I chains with wild-type or mutant SV40, and examined susceptibility to lysis by murine CTLs directed against several epitopes of T antigen, using ^{51}Cr release assays. Figure 5 shows that cells expressing the miRNA are indeed less susceptible to lysis by CTLs; this effect could be overcome by high multiplicity of infection (not shown), suggesting that it results from reduced antigen levels, and not from some special immunomodulatory effect of the miRNA.

The SV40 miRNAs described here are conserved in all SV40 isolates, and orthologs are found in most primate polyomaviruses but are not conserved in murine polyomavirus (Py). However, examination of the murine Py sequence with v-miR (Fig. 6) reveals that the top-scoring hairpin is in a different genomic location but is also found on the late strand, 3′ to the late poly(A) site, although much farther downstream from it than the SV40 miRNA (Fig. 1). Interestingly, 25 years ago, R. Treisman, while mapping 5′ and 3′ ends of Py late RNAs, identified ends consistent with a structure identical to this hairpin, and speculated that they might have been generated by an RNase-III-like enzyme (a prescient suggestion that foreshadowed the fact that Drosha is an RNase III family

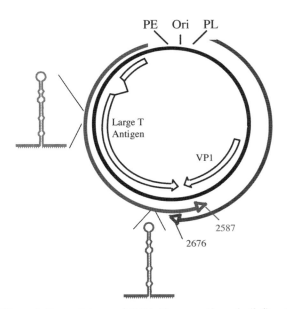

Figure 1. Transcript map of SV40. Shown are the early (*left*) and late (*right*) transcripts (*depicted as closed arrows*); coding regions are shown as open arrows. The SV40 and Py pre-miRNA hairpins are shown; both are of late polarity, but map to the central (Py) or distal (SV40) regions of the T antigen. Note that the pre-miRNAs are antisense to the early transcripts encoding T-antigen proteins. (Modified, with permission, from Sullivan et al. 2005 [Nature Publishing Group]).

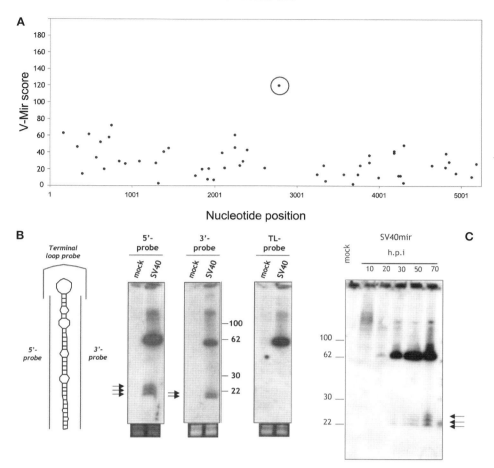

Figure 2. SV40 encodes a miRNA. (*A*) v-Mir prediction of pre-miRNAs for the SV40 genome in the late orientation. Each dot represents a candidate pre-miRNA. The vertical axis shows the v-Mir score; the higher the score, the more likely a candidate is a bona fide miRNA. Circled is the confirmed SV40-encoded pre-miRNA. (*B*) Northern blot analysis confirms SV40 encodes a miRNA. The left panel diagrams the three probes used in this figure. Arrows identify miRNAs generated from each arm of the pre-miRNA hairpin structure (5′ and 3′ probe). The control probe (TL probe) that is directed against the terminal loop only recognizes the pre-miRNA 57-nucleotide band and not the ~22-nucleotide miRNAs, demonstrating the specific processing of the stem into miRNAs. (*C*) Northern conducted on RNA harvested from cell at various times postinfection. Arrows indicate bands that correspond to miRNAs.

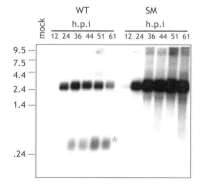

Figure 3. SV40 miRNA directs cleavage of early transcripts. Shown is northern blot of poly(A) purified RNA from cells infected with wild type (WT) of a mutant that is unable to make the miRNA (SM) at various times postinfection. The band that corresponds to the cleavage fragment is marked with an asterisk. (Modified, with permission, from Sullivan et al. 2005 [Nature Publishing Group].)

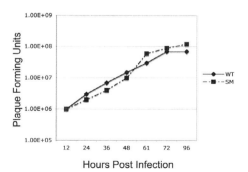

Figure 4. The SV40 miRNA mutant (SM) virus grows as well as wild type (WT) in cultured cells. Shown is a one-step growth curve of virus harvested at various times postinfection from Bsc40 monkey kidney epithelial cells that were infected at a multiplicity of infection of 5 plaque-forming units per cell. (Modified, with permission, from Sullivan et al. 2005 [Nature Publishing Group].)

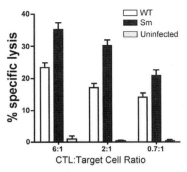

Figure 5. Cells infected with SV40 miRNA mutant (Sm) are less susceptible to cytotoxic T lymphocyte (CTL)-mediated lysis. Simian cells which express a murine class I allele were infected at a multiplicity of infection of 1 plaque-forming unit per cell with either the miRNA mutant virus (Sm, *black bars*) or wild-type virus (WT, *white bars*) at various ratios of murine CTLs (that recognize an epitope in large-T antigen) to target infected cells. (Modified, with permission, from Sullivan et al. 2005 [Nature Publishing Group].)

member) (Treisman 1981; Treisman and Kamen 1981). In 1982, Fenton and Basilico identified a fragment of early mRNA from this region that is exactly the size predicted for cleavage generated from the predicted miRNA; as expected, this fragment was detected only at late times postinfection (Fenton and Basilico 1982). We have verified that this miRNA is indeed made (C.S. Sullivan, unpubl.); together with the cleavage fragments identified by Fenton and Basilico (1982), this strongly indicates that the overall strategy of down-regulating early mRNA at late times with miRNA-directed cleavage is a conserved feature of polyomavirus biology.

The demonstrated utility of v-miR on small DNA virus genomes emboldened us to examine its ability to identify miRNAs in herpesviruses, a family of large, enveloped DNA viruses whose genomes encode 100–150 genes. We chose two herpesviruses for study—Kaposi's sarcoma-associated herpesvirus (KSHV) and Epstein-Barr virus (EBV). Both are lymphotropic DNA tumor viruses that reside in B lymphocytes and are linked to B-cell lym-

phomas; KSHV also produces the endothelial neoplasm KS. Both viruses are known to produce miRNAs. Exhaustive cloning in KSHV-infected B cells had previously identified 11 miRNAs (Cai et al. 2005; Pfeffer et al. 2005; Samols et al. 2005), and cloning from EBV-infected lymphoblastoid cells resulted in identification of 5 miRNAs (Pfeffer et al. 2004). Thus, an empiric database of identified miRNAs existed, against which we could calibrate our approach.

The large size of these viral genomes (~165 kb) indicated that v-miR would identify too many hairpins to consider using northern blotting as the secondary screen, as we had earlier done for polyomaviruses. In fact, with screening parameters (filters) similar to those used for SV40 (Fig. 1), more than 3000 hairpins were identified in KSHV alone. More advanced computational strategies and stringent filtering (based in part on the expanded number of cellular miRNAs that have been cloned) reduce this number considerably, but still leave many hairpins to screen. We therefore turned to a microarray-based approach, details of which can be found in Grundhoff et al. (2006). Briefly, for KSHV we constructed two custom arrays: (1) a "hairpin array," made up of the 3000 hairpins predicted by v-miR; and (2) a "tile array," produced by tiling across the viral genome with 50-nucleotide oligonucleotides, in 500-nucleotide steps. RNA was prepared from a KSHV-positive lymphoma cell line (BCBL-1), and from KSHV-negative BJAB cells, then differentially labeled and hybridized to the arrays. In addition, we prepared BCBL-1 RNA corresponding to the 20–25-nucleotide fraction (enriched for bona fide miRNAs, as well as containing nonspecific RNA degradation products) and to the 30–40-nucleotide fraction (a control fraction representing nonspecific degradation products alone). Again, these two preps were differentially labeled and hybridized to the arrays. Viral sequences that hybridized preferentially to infected cell RNA over uninfected cell RNA, and to the probes from the 20–25-nucleotide fraction over the 30–40-nucleotide fraction, were selected for further analysis. This involved northern blotting of BCBL-1 RNA (as compared to BJAB

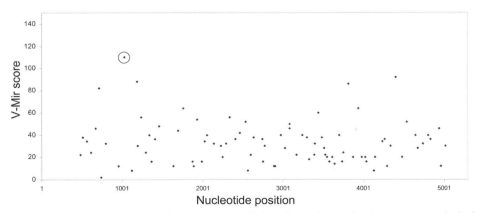

Figure 6. Polyomavirus is predicted to encode a miRNA. v-Mir prediction of pre-miRNAs for the PyV genome in the late orientation. See legend for Fig. 2A. Circled is the top-scoring predicted pre-miRNA, identical to the hairpin structure originally hypothesized by Treisman (see Treisman and Kamen 1981) to be processed by an RNase-III-like enzyme. The position of this hairpin on the genome is shown in Fig. 1.

RNA) looking for virus-specific bands of 20–25 nucleotides. Figure 7 shows the results of these analyses, for both KSHV and EBV. The KSHV experiments identified 9/11 previously known miRNAs, plus one additional species that had escaped earlier detection for trivial reasons (it harbors a cleavage site for a restriction enzyme that had been used in the cloning procedure). In EBV, 18 miRNAs were identified—a large increase over earlier cloning experiments, although most of this difference is attributable to the fact that the EBV strain used previously harbors a large deletion in a region that encodes a large cluster of miRNAs. Although the functions of all of these herpesviral miRNAs remain unknown, our results establish that v-miR can be useful in the screening of large viral genomes, when used in conjunction with additional molecular screening methods.

What does the future hold for the study of viral miRNAs? Now that good methods exist for identification of such RNAs, we can expect an avalanche of new miRNA sightings. The challenge now is to discern their function(s), a mission that will begin with identification of their molecular targets. For those miRNAs with host RNA targets, this exercise will likely be as difficult as it is for cellular miRNAs. However, we can anticipate that many viral miRNAs will have viral targets—as in the polyomaviruses. For these, not only will target identification be simpler, but divining the biological significance of the interaction should also be more straightforward, since the pathways in which many viral genes function are already

known. But this, of course, is nothing new: It is precisely these features of viral genomes—small size, limited complexity, and exploitation of host functions—that brought them (as phages) to the attention of geneticists 50 years ago, at the dawn of the age of molecular biology.

REFERENCES

Acheson N.H. 1978. Polyoma virus giant RNAs contain tandem repeats of the nucleotide sequence of the entire viral genome. *Proc. Natl. Acad. Sci.* **75:** 4754.

Bernstein E., Caudy A.A., Hammond S.M., and Hannon G.J. 2001. Role for a bidentate ribonuclease in the initiation step of RNA interference. *Nature* **409:** 363.

Bohnsack M.T., Czaplinski K., and Gorlich D. 2004. Exportin 5 is a RanGTP-dependent dsRNA-binding protein that mediates nuclear export of pre-miRNAs. *RNA* **10:** 185.

Cai X., Lu S., Zhang Z., Gonzalez C.M., Damania B., and Cullen B.R. 2005. Kaposi's sarcoma-associated herpesvirus expresses an array of viral microRNAs in latently infected cells. *Proc. Natl. Acad. Sci.* **102:** 5570.

Chendrimada T.P., Gregory R.I., Kumaraswamy E., Norman J., Cooch N., Nishikura K., and Shiekhattar R. 2005. TRBP recruits the Dicer complex to Ago2 for microRNA processing and gene silencing. *Nature* **436:** 740.

Cole C.N. 1996. Polyomaviridae: The viruses and their replication. In *Fields virology*, 3rd edition (ed. B.N. Fields et al.), p. 1997. Lippincott-Raven, Philadelphia, Pennsylvania.

Denli A.M., Tops B.B., Plasterk R.H., Ketting R.F., and Hannon G.J. 2004. Processing of primary microRNAs by the Microprocessor complex. *Nature* **432:** 231.

Doench J.G., Petersen C.P., and Sharp P.A. 2003. siRNAs can function as miRNAs. *Genes Dev.* **17:** 438.

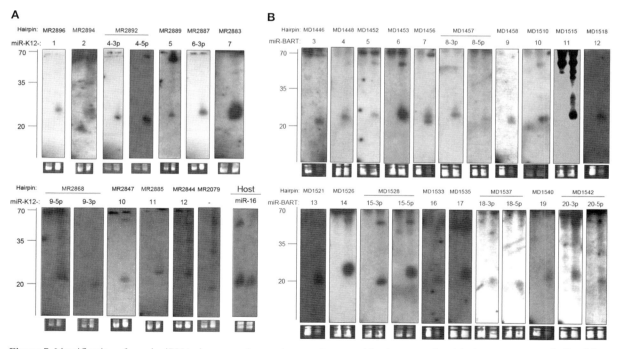

Figure 7. Identification of novel miRNAs in gamma-herpesviruses using a combined computational / microarray approach. Candidate pre-miRNAs were predicted using v-Mir, and those candidates that scored positive on microarray analysis were further validated via northern blot analysis. Northern blots using different probes identify miRNAs expressed by (*A*) Kaposi's sarcoma-associated herpesvirus (KSHV), and (*B*) Epstein-Barr virus. Candidates were considered bona fide miRNAs if they showed a distinct band around 22 nucleotides that was not detectable in RNA from uninfected cells. In each panel, RNA from uninfected (*left lane*) and infected (*right lane*) cells was probed. EtBr staining of low-MW 5S rRNA and tRNA is shown as a load control (*bottom panels*). (Modified, with permission, from Grundhoff et al. 2006 [©RNA Society].)

Du T. and Zamore P.D. 2005. microPrimer: The biogenesis and function of microRNA. *Development* **132:** 4645.

Fenton R.G. and Basilico C. 1982. Changes in the topography of early region transcription during polyoma virus lytic infection. *Proc. Natl. Acad. Sci.* **79:** 7142.

Forstemann K., Tomari Y., Du T., Vagin V.V., Denli A.M., Bratu D.P., Klattenhoff C., Theurkauf W.E., and Zamore P.D. 2005. Normal microRNA maturation and germ-line stem cell maintenance requires Loquacious, a double-stranded RNA-binding domain protein. *PLoS Biol.* **3:** e236.

Gregory R.I., Chendrimada T.P., Cooch N., and Shiekhattar R. 2005. Human RISC couples microRNA biogenesis and post-transcriptional gene silencing. *Cell* **123:** 631.

Gregory R.I., Yan K.P., Amuthan G., Chendrimada T., Doratotaj B., Cooch N., and Shiekhattar R. 2004. The Microprocessor complex mediates the genesis of microRNAs. *Nature* **432:** 235.

Griffiths-Jones S. 2004. The microRNA Registry. *Nucleic Acids Res.* **32:** D109.

Grishok A., Pasquinelli A.E., Conte D., Li N., Parrish S., Ha I., Baillie D.L., Fire A., Ruvkun G., and Mello C.C. 2001. Genes and mechanisms related to RNA interference regulate expression of the small temporal RNAs that control *C. elegans* developmental timing. *Cell* **106:** 23.

Grundhoff A., Sullivan C.S., and Ganem D. 2006. A combined computational and microarray-based approach identifies novel microRNAs encoded by human gamma-herpesviruses. *RNA* **12:** 733.

Hamilton A.J. and Baulcombe D.C. 1999. A species of small antisense RNA in posttranscriptional gene silencing in plants. *Science* **286:** 950.

Hammond S.M., Sontheimer E.J., and Carthew R.W. 2005. Dicing and slicing: The core machinery of the RNA interference pathway. *FEBS Lett.* **579:** 5822.

Han J., Lee Y., Yeom K.H., Kim Y.K., Jin H., and Kim V.N. 2004. The Drosha-DGCR8 complex in primary microRNA processing. *Genes Dev.* **18:** 3016.

Hutvagner G., McLachlan J., Pasquinelli A.E., Balint E., Tuschl T., and Zamore P.D. 2001. A cellular function for the RNA-interference enzyme Dicer in the maturation of the let-7 small temporal RNA. *Science* **293:** 834.

Jiang F., Ye X., Liu X., Fincher L., McKearin D., and Liu Q. 2005. Dicer-1 and R3D1-L catalyze microRNA maturation in *Drosophila*. *Genes Dev.* **19:** 1674.

Ketting R.F., Fischer S.E., Bernstein E., Sijen T., Hannon G.J., and Plasterk R.H. 2001. Dicer functions in RNA interference and in synthesis of small RNA involved in developmental timing in *C. elegans*. *Genes Dev.* **15:** 2654.

Kim V.N. 2005. MicroRNA biogenesis: Coordinated cropping and dicing. *Nat. Rev. Mol. Cell Biol.* **6:** 376.

Landthaler M., Yalcin A., and Tuschl T. 2004. The human DiGeorge syndrome critical region gene 8 and its *D. melanogaster* homolog are required for miRNA biogenesis. *Curr. Biol.* **14:** 2162.

Lee Y., Jeon K., Lee J.T., Kim S., and Kim V.N. 2002. MicroRNA maturation: Stepwise processing and subcellular localization. *EMBO J.* **21:** 4663.

Lee Y., Ahn C., Han J., Choi H., Kim J., Yim J., Lee J., Provost P., Radmark O., Kim S., and Kim V.N. 2003. The nuclear RNase III Drosha initiates microRNA processing. *Nature* **425:** 415.

Lund E., Guttinger S., Calado A., Dahlberg J.E., and Kutay U. 2004. Nuclear export of microRNA precursors. *Science* **303:** 95.

Pfeffer S., Zavolan M., Grasser F.A., Chien M., Russo J.J., Ju J., John B., Enright A.J., Marks D., Sander C., and Tuschl T. 2004. Identification of virus-encoded microRNAs. *Science* **304:** 734.

Pfeffer S., Sewer A., Lagos-Quintana M., Sheridan R., Sander C., Grasser F.A., van Dyk L.F., Ho C.K., Shuman S., Chien M., et al. 2005. Identification of microRNAs of the herpesvirus family. *Nat. Methods* **2:** 269.

Saito K., Ishizuka A., Siomi H., and Siomi M.C. 2005. Processing of pre-microRNAs by the Dicer-1-Loquacious complex in *Drosophila* cells. *PLoS Biol.* **3:** e235.

Samols M.A., Hu J., Skalsky R.L., and Renne R. 2005. Cloning and identification of a microRNA cluster within the latency-associated region of Kaposi's sarcoma-associated herpesvirus. *J. Virol.* **79:** 9301.

Sullivan C.S. and Ganem D. 2005. MicroRNAs and viral infection. *Mol. Cell* **20:** 3.

Sullivan C.S. and Pipas J.M. 2002. T antigens of simian virus 40: Molecular chaperones for viral replication and tumorigenesis. *Microbiol. Mol. Biol. Rev.* **66:** 179.

Sullivan C.S., Grundhoff A.T., Tevethia S., Pipas J.M., and Ganem D. 2005. SV40-encoded microRNAs regulate viral gene expression and reduce susceptibility to cytotoxic T cells. *Nature* **435:** 682.

Treisman R. 1981. "Structures of polyomavirus nuclear and cytoplasmic RNA molecules." Ph. D. thesis, University College, London.

Treisman R. and Kamen R. 1981. Structure of polyoma virus late nuclear RNA. *J Mol. Biol.* **148:** 273.

Tuschl T., Zamore P.D., Lehmann R., Bartel D.P., and Sharp P.A. 1999. Targeted mRNA degradation by double-stranded RNA in vitro. *Genes Dev.* **13:** 3191.

Yi R., Qin Y., Macara I.G., and Cullen B.R. 2003. Exportin-5 mediates the nuclear export of pre-microRNAs and short hairpin RNAs. *Genes Dev.* **17:** 3011.

Zamore P.D., Tuschl T., Sharp P.A., and Bartel D.P. 2000. RNAi: Double-stranded RNA directs the ATP-dependent cleavage of mRNA at 21 to 23 nucleotide intervals. *Cell* **101:** 25.

Zeng Y. and Cullen B.R. 2004. Structural requirements for pre-microRNA binding and nuclear export by Exportin 5. *Nucleic Acids Res.* **32:** 4776.

Zeng Y., Yi R., and Cullen B.R. 2003. MicroRNAs and small interfering RNAs can inhibit mRNA expression by similar mechanisms. *Proc. Natl. Acad. Sci.* **100:** 9779.

———. 2005. Recognition and cleavage of primary microRNA precursors by the nuclear processing enzyme Drosha. *EMBO J.* **24:** 138.

Expression and Function of MicroRNAs Encoded by Kaposi's Sarcoma-associated Herpesvirus

E. GOTTWEIN, X. CAI, AND B.R. CULLEN

Center for Virology and Department of Molecular Genetics and Microbiology, Duke University Medical Center, Durham, North Carolina 27710

microRNAs (miRNAs) are widely used by animal and plant cells to posttranscriptionally regulate cellular gene expression, but this regulatory mechanism can also be used by pathogenic viruses for the same purpose. It is now well established that numerous miRNAs are expressed by a wide range of pathogenic herpesviruses, although their mRNA targets and role in the viral life cycle remain unclear. Here, we discuss what is currently known about the expression and function of the 12 miRNAs that are expressed by the pathogenic gamma herpesvirus Kaposi's sarcoma-associated herpesvirus in latently infected human B cells.

miRNAs are a class of noncoding RNAs, generally 21 or 22 nucleotides in length, that are expressed by all animals and plants examined thus far. More than 300 miRNAs have been identified in human cells, and comparable numbers are likely to be expressed in all other vertebrate animals (Bartel 2004).

Cellular miRNAs are initially transcribed as part of one arm of an approximately 80-nucleotide stem-loop that in turn forms part of a longer capped, polyadenylated RNA (Cai et al. 2004; Lee et al. 2004). This primary miRNA (pri-miRNA) transcript is recognized by the nuclear microprocessor complex, consisting minimally of the RNase III enzyme Drosha and the RNA-binding cofactor DGCR8, which cleaves the stem of the pri-miRNA to liberate an approximately 60-nucleotide RNA hairpin—the pre-miRNA intermediate—bearing a 2-nucleotide 3′ overhang (Denli et al. 2004; Gregory et al. 2004; Han et al. 2004). Because the pri-miRNA sequences that flank the pre-miRNA hairpin are retained in the nucleus and degraded, active pri-miRNAs are not capable of simultaneously functioning as an mRNA (Cai et al. 2004). As a result, cellular miRNA hairpins are generally found in noncoding RNAs or within the introns of coding or noncoding RNA polymerase II transcripts (Bartel 2004). However, pri-miRNAs can function as mRNAs if they can be exported to the cytoplasm before they are cleaved by the microprocessor complex (Cai et al. 2004).

The pre-miRNA is exported to the cytoplasm by the karyopherin family member Exportin 5 (Yi et al. 2003), where it is recognized and bound by a second RNase III family member, Dicer. Dicer removes the terminal loop and gives rise to the miRNA duplex intermediate, an approximately 19-bp-long double-stranded RNA (dsRNA) bearing 2-nucleotide 3′ overhangs at each end (Cullen 2004). One strand of this duplex is then selectively incorporated into the RNA-induced silencing complex (RISC), where it acts as a guide RNA to direct RISC to complementary RNA sequences (Hammond et al. 2000; Martinez et al. 2002; Schwarz et al. 2002). Once

bound, RISC can inhibit mRNA function by direct cleavage of the mRNA if the degree of complementarity is high or by inhibiting the translation of the mRNA if the degree of complementarity is more limited (Doench et al. 2003; Zeng et al. 2003).

The miRNA-mediated pathway of posttranscriptional gene regulation is thought to have a key role in many aspects of differentiation and development and is active in all cell types at all stages in an animal's development (Bartel 2004). Because miRNA processing can occur efficiently with pri-miRNA precursors as short as about 200 nucleotides (Zeng and Cullen 2003), and because miRNAs are not likely to be antigenic, they also have the potential to provide a facile mechanism for viruses to down-regulate host-cell gene products that might act to limit viral replication. However, as discussed in more detail elsewhere (Cullen 2006), the fact that the pri-miRNA precursor is cleaved in the nucleus to liberate the pre-miRNA hairpin intermediate, whereas the RNA sequences flanking this hairpin structure are retained in the nucleus and degraded, suggests that viruses which have an RNA genome and/or which replicate in the cytoplasm would be unlikely to encode miRNAs. Similarly, because miRNAs act at the mRNA level, not the protein level, it seems likely that miRNAs would be most advantageous to viruses that establish long-term persistent or latent infections, during which the preexisting pool of target protein could decay, rather than viruses that have short, lytic replication cycles, which might be over before the miRNA-mediated knockdown of an mRNA would have had much effect on the level of the encoded protein. On the basis of this reasoning, it has previously been argued (Cullen 2006) that miRNAs might be prevalent in nuclear DNA viruses that establish long-term latent infections (e.g., herpesviruses), less common in nuclear DNA viruses that undergo exclusively lytic replication cycles (e.g., adenoviruses and polyomaviruses), and rare or absent in cytoplasmic DNA viruses (e.g., poxviruses) or RNA viruses (e.g., retroviruses and influenza viruses). In fact, recent evidence from a number of laboratories has confirmed the existence of numerous miRNAs in herpesviruses, has identified a single miRNA in adenovirus and SV40, and has so far failed to detect miRNAs in any RNA virus (Pfeffer et al. 2004, 2005; Cai et al. 2005, 2006; Grey et al. 2005; Sullivan et al. 2005; Grundhoff et al. 2006).

This review focuses on the miRNAs encoded by the gamma herpesvirus Kaposi's sarcoma-associated herpesvirus (KSHV), which is currently one of the best understood viruses in terms of viral miRNA expression and function. This research, together with analogous efforts focusing on a range of other pathogenic herpesviruses, suggests that miRNAs are likely to have a key role in mediating the ability of herpesviruses to maintain long-term latent infections in vivo in the face of host adaptive and innate immune responses.

KSHV EXPRESSES 12 MICRORNAs IN LATENTLY INFECTED CELLS

KSHV readily establishes latent infections in human B cells, where the virus then maintains 50–100 episomal copies of its circular viral DNA genome (Chen and Lagunoff 2005). Although KSHV encodes more than 80 proteins in its approximately 140,000-bp genome, only a small subset of these are expressed in latently infected cells (Russo et al. 1996; Sarid et al. 1998; Jenner et al. 2001). These minimally include ORF73 (LANA), which has a key role in the maintenance of the viral episome (Ye et al. 2004); ORF72 (v-cyclin), which is thought to regulate cell cycle progression in infected cells; ORF71 (V-FLIP), an antiapoptotic protein; and finally Kaposin, which appears to have oncogenic properties (McCormick and Ganem 2005). Of note, the genes encoding these four proteins are clustered together in the "latency-associated region" of the KSHV genome (Fig. 1).

On the basis of the hypothesis that viral miRNAs might be particularly advantageous to DNA viruses

during latent infection (Cullen 2006), we cDNA-cloned candidate miRNAs from the latently KSHV-infected cell line BC-1 by size selection for RNAs that were greater than 18 nucleotides and less than 25 nucleotides in size (Cai et al. 2005). This identified 167 miRNA cDNA clones of KSHV origin, and these could all be assigned to ten different miRNA stem-loop precursors designated miR-K1 to miR-K10 (Fig. 2). Work from other investigators subsequently identified two more KSHV miRNAs, miR-K11 and miR-K12, that were missed in our original analysis (Pfeffer et al. 2005; Samols et al. 2005; Grundhoff et al. 2006). All 12 KSHV miRNAs form part of one arm of an approximately 80-nucleotide RNA hairpin structure, and in several cases, the predicted "passenger" strand, that is, the strand of the miRNA duplex intermediate that is normally excluded from RISC and degraded, was also recovered (Fig. 2). This allowed confirmation of the expectation that the KSHV miRNA duplex intermediates would contain the 2-nucleotide 3' overhangs that arise from sequential processing by Drosha and Dicer.

A surprising observation to emerge from these studies was the clustering of all 12 KSHV miRNAs into an approximately 4-kb region of the approximately 140-kb KSHV genome that coincides with the KSHV "latency associated region." Specifically, 10 of the KSHV miRNAs mapped to a noncoding region located between the *Kaposin* and *ORF71* genes (Fig. 1); one miRNA, miR-K10, was actually located within an open reading frame (ORF) that forms part of the *Kaposin* gene, whereas the last miRNA, miR-K12, mapped to the 3'-untranslated region (3'UTR) of *Kaposin* (Fig. 1). Importantly, all 12 miRNAs were found to be in the same transcriptional orientation, thus suggesting that they might all be processed out of a single pri-miRNA precursor. Several cellular miRNA "clusters" have been previously identified (Bartel 2004), so this would certainly not be unprecedented.

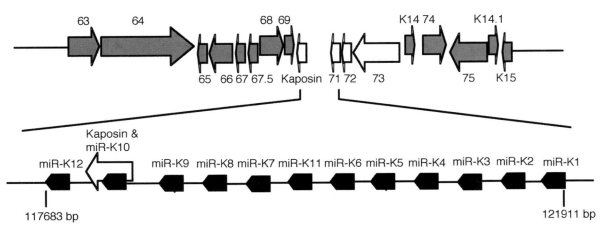

Figure 1. Genomic location of KSHV miRNAs. Schematic of a region of the KSHV genome. Known protein-coding genes, their orientation, and expression patterns are indicated. Ten of the KSHV miRNAs are arrayed between the *ORF71* and *Kaposin* genes, one (miR-K10) overlaps the Kaposin ORF, whereas miR-K12 is found in the Kaposin 3'UTR. (*White boxes*) KSHV genes expressed during latent infection; (*gray boxes*) KSHV genes expressed during lytic infection; (*black boxes*) KSHV pri-miRNA precursors. The transcriptional orientation of each gene and each pri-miRNA hairpin is indicated. (Modified, with permission, from Cai et al. 2005 [© National Academy of Sciences].)

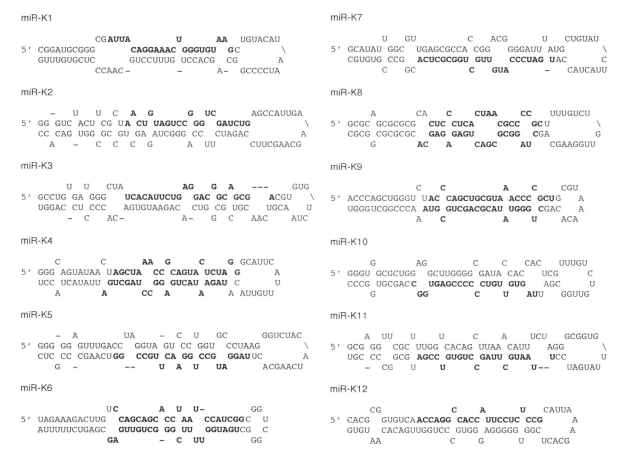

Figure 2. Predicted stem-loop structures adopted by KSHV pri-miRNAs. The miRNA sequences identified by cDNA cloning are highlighted. (Modified, with permission, from Cai et al. 2005 [© National Academy of Sciences].)

TRANSCRIPTION OF KSHV miRNAs IN LATENTLY OR LYTICALLY INFECTED CELLS

Because the latency-associated region of the KSHV genome is the only region that has been found to be consistently transcribed in all latently KSHV-infected cells (Dittmer et al. 1998; Sarid et al. 1998; Jenner et al. 2001), and because KSHV infection can induce tumorigenesis in infected immunosuppressed individuals, there has been considerable interest in the transcriptional regulation of the four genes found in this region of the KSHV genome. Building on this important earlier research (Dittmer et al. 1998; Li et al. 2002; Pearce et al. 2005), we performed a detailed analysis of the RNA transcripts that arise from this region, focusing particularly on RNAs that would be predicted to function as viral pri-miRNAs (Cai et al. 2006). This analysis was performed both in latently KSHV-infected B cells and in cells that had been induced to enter the lytic viral replication cycle by treatment with TPA.

In latently KSHV-infected B cells, we mapped two transcriptionally active promoters with imprecise cap sites located at 127,880 to 127,886 and 123,751 to 123,760, respectively (Fig. 3). The promoter located at about 127,880 was found to give rise to four distinct mRNAs, depending on which polyadenylation (pA) site and which splice sites are utilized (Cai et al. 2006). If the

pA site at 122,070 is used, then transcription from the 127,880 promoter gives rise to two spliced mRNAs that have the potential to encode the viral proteins ORF71, ORF72, and ORF73. Neither of these two mRNAs has the potential to function as a pri-miRNA. However, if the pA site at 122,070 is bypassed, which occurs about 50% of the time, then transcripts initiating at the 127,880 promoter are instead polyadenylated at the efficient PA site located at 117,436 (Fig. 3). These pre-mRNA transcripts are spliced such that the ORF71-, 72-, and 73-coding sequences are removed and the resultant mature mRNAs instead express Kaposin (Fig. 3). Importantly, the KSHV miRNAs miR-K1 to miR-K9, as well as miR-K11, are all located within the intron(s) that is spliced out of these Kaposin mRNAs. As noted above, many cellular miRNAs are also found within introns. In work published elsewhere (Cai et al. 2006), we were able to directly confirm that these KSHV miRNAs can indeed be processed out of the introns present in the Kaposin pre-mRNAs that are transcribed from the 127,880 promoter.

In addition to the 127,880 promoter, KSHV also contains a second, slightly less active, latent promoter that initiates transcription at 123,751/123,760. This promoter can again give rise to two very different mRNAs depending on whether the pA site located at 122,070 is used or ignored. If it is used, then an unspliced approximately 1.7-kb mRNA encoding ORF71 and ORF72 is expressed.

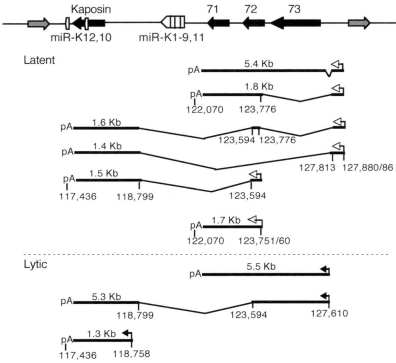

Figure 3. Transcript map of the latency-associated region of the KSHV genome. Shown is a schematic of the genes and miRNAs found in the KSHV latency-associated region as well as the pre-mRNAs and mature mRNAs that have been identified. Numbers refer to the sequence coordinates of transcription start sites, splice sites, and polyadenylation sites within the KSHV genome. Latent transcripts are shown above the dashed line and lytic transcripts below the line. Sizes of mature mRNAs are given in kilobases, and their promoters are indicated by arrows. (pA) Poly(A)-addition site. (Modified, with permission, from Cai and Cullen 2006 [© ASM].)

If it is ignored, then an approximately 1.5-kb spliced mRNA encoding Kaposin is expressed. This latter mRNA again has the potential to function as a pri-miRNA for all 12 KSHV miRNAs (Fig. 3).

Although miR-K1 to miR-K9, as well as miR-K11, are all located in the intron(s) excised from the Kaposin miRNAs described in Figure 3, two miRNAs, miR-K10 and miR-K12, are located in the Kaposin ORF and 3′UTR, respectively (see Fig. 1). As noted above, excision of either of these miRNAs in the nucleus of latently KSHV-infected cells would appear to be inconsistent with mRNA function and hence with the expression of the Kaposin proteins. Two alternative hypotheses can be advanced to explain this apparent contradiction. One possibility is that processing of the miR-K10 and miR-K12 pri-miRNAs by the microprocessor complex is inefficient, so that a substantial percentage of Kaposin mRNA is able to exit the nucleus in an intact form. Alternatively, it is possible that KSHV is able to regulate pri-miRNA processing in some way, such that miR-K10 and miR-K12 expression is only activated when necessary.

We considered the possibility that miR-K10 and miR-K12 are actually not overly important during latent infection and only become critical during lytic KSHV replication. To address this hypothesis, we sought to identify promoters located within the KSHV latency-associated region that are activated during lytic replication. It is important to note that the two latent promoters

located in this region are neither activated nor inhibited by induction of lytic KSHV replication (Dittmer et al. 1998; Cai et al. 2006).

As shown in Figure 3, we and other investigators have identified two lytic promoters in this region of the KSHV genome. One of these promoters, with a cap site located at 118,758, is an extremely powerful promoter that gives rise to high levels of about 1.3 kb that is predicted to encode Kaposin and also act as a pri-miRNA for miR-K10 and miR-K12 only (Sadler et al. 1999; Cai et al. 2006). A second lytic promoter, with a cap site located at 127,610, is a relatively weak promoter (comparable in strength to the latent 127,880 promoter) and is predicted to give rise to two mature mRNAs, again depending on whether the pA site at 122,070 is used or not (Cai et al. 2006). These are an unspliced approximately 5.5-kb mRNA potentially encoding ORF71, 72, and 73 and an approximately 5.3-kb mRNA encoding Kaposin and all 12 KSHV miRNAs.

Because the KSHV lytic 127,610 promoter is comparable in activity to the latent promoters located at 127,880 and 123,751, one would predict that induction of lytic replication would only modestly enhance the expression of the KSHV miR-K1 to miR-K9 and miR-K11 miRNAs. In contrast, because the KSHV lytic 118,758 promoter is much more active than these two latent promoters (Cai et al. 2006), one would predict that induction of lytic replication would strongly activate the expression of miR-K10 and miR-K12. As shown in Figure 4, this is

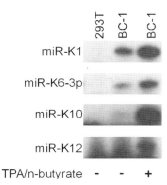

miR-K1

miR-K6-3p

miR-K10

miR-K12

TPA/n-butyrate -   -   +

Figure 4. Effect of induction of lytic KSHV replication on viral miRNA expression in BC-1 cells. BC-1 cells were either cultured normally or treated with TPA (30 ng/ml) and *n*-butyrate (300 ng/ml) for 48 hours prior to RNA isolation as described previously (Cai and Cullen 2006). The level of expression of the indicated KSHV miRNAs was then determined by primer extension analysis (Cai and Cullen 2006). Uninfected 293T cells were used as the negative control.

indeed the case. Specifically, induction of lytic KSHV replication by treatment of BC-1 cells with TPA and *n*-butyrate enhanced the expression of miR-K1 and miR-K6-3p by two- to threefold but had a far more dramatic effect on the expression of miR-K10 and miR-K12, which are actually quite difficult to detect in latently infected cells. These observations are consistent with the hypothesis that processing of miR-K10 and miR-K12 by Drosha is inefficient, so that high-level expression of these two viral miRNAs is only achieved when high-level expression of the underlying Kaposin mRNA is induced. However, it remains possible that miRNA processing

efficiency, either in general or of these two viral miRNAs in particular, is somehow enhanced upon activation of lytic KSHV replication.

KSHV MIRNAS ARE BIOLOGICALLY ACTIVE

As noted above, miRNAs function as guide RNAs for RISC. If they successfully direct RISC to an mRNA bearing a highly complementary target site, then it is expected that that mRNA will be cleaved by RISC and subsequently degraded by the action of cellular exonucleases (Bartel 2004).

To confirm that the KSHV miRNAs are indeed programming active RISC complexes in latently KSHV-infected cells, we constructed a set of two lentivirus-based indicator vectors containing, respectively, a renilla luciferase (RLuc) or a firefly luciferase (FLuc) indicator gene (Fig. 5, top) (Gottwein et al. 2006). Moreover, the RLuc indicator virus was engineered to contain two tandem target sites in its 3′UTR that are perfectly complementary to one of the KSHV miRNAs—miR-K1 to miR-K11 (Fig. 5, top). In the case of miR-K4, we constructed indicator viruses specific to KSHV miRNAs derived from both the 5′ arm (miR-K4-5p) and the 3′ arm (miR-K4-3p) of the pre-miRNA precursor as both were cDNA-cloned from latently KSHV-infected cells (Cai et al. 2005).

To confirm that the KSHV miRNAs are indeed biologically active, we prepared a viral stock of the FLuc-based control virus and a viral stock for each of the 13 RLuc-based indicator viruses by transfection of 293T cells. The FLuc viral stock was then mixed with each of the RLuc viral stocks, and the resultant mixture was used to infect

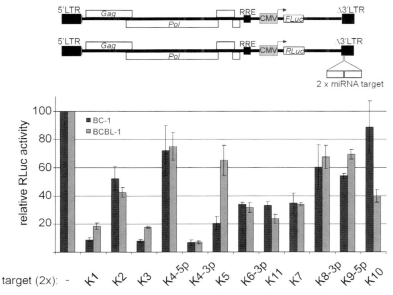

Figure 5. (*Top*) Schematic of the control vector pNL-SIN-CMV-FL, which encodes FLuc, and the indicator vector pNL-SIN-CMV-RL, which encodes RLuc. In each RLuc-based indicator construct, two perfectly complementary target sequences for one KSHV miRNA were inserted into the 3′UTR of the *Renilla luciferase* gene. (*Bottom*) BJAB, BC-1, and BCBL-1 cells were transduced with a mixture of the control and indicator viruses, and dual luciferase assays were performed. Normalized RLuc activities in KSHV-positive BC-1 and BCBL-1 cells are shown relative to those obtained in KSHV-negative BJAB cells. Values from cells transduced with the parental RLuc-expressing vector were set at 100%. (LTR) Long terminal repeat; (RRE) Rev response element; (CMV) cytomegalovirus immediate-early promoter. (Modified, with permission, from Gottwein et al. 2006 [© ASM].)

the latently KSHV-infected B-cell lines BC-1 and BCBL-1, as well as the uninfected human B-cell line BJAB. The level of RLuc and FLuc activity was then determined at 24 hours after transduction (Fig. 5, bottom). The data are presented relative to the level of RLuc activity seen in the uninfected BJAB cell line, which was set at 100, after correction for variations in the level of activity seen with the FLuc internal control indicator virus (Gottwein et al. 2006). As may be observed (Fig. 5, bottom), all the KSHV miRNAs inhibited RLuc expression to some degree. The inhibition observed was profound with miR-K1, miR-K3, and miR-K4-3p, modest with miR-K2, miR-K6-3p, miR-K11, and miR-K7, and fairly weak with miR-K8-3p, miR-K9-5p, and miR-K10.

One interesting observation was that the level of inhibition was consistent between the two KSHV-infected cell lines tested, that is, BC-1 and BCBL-1, with the exception of miR-K5 and possibly miR-K10. Analysis of the level of expression of the mature KSHV miR-K5 miRNA and of the pre-miR-K5 intermediate in fact revealed that both are expressed at much lower levels in BCBL-1 cells, where miR-K5 is less active, than in BC-1 cells (Fig. 6, top left). In contrast, all other KSHV miRNAs, including miR-K10, are expressed at comparable levels in the two cell lines (Fig. 6, top left) (Cai et al. 2005).

To try to understand the molecular mechanism underlying this discrepancy, we first sequenced the pri-miR-K5 stem-loop precursor in both BC-1 and BCBL-1 cells. As

shown in Figure 6 (top right), this analysis revealed a single nucleotide polymorphism (G→A) in BCBL-1 relative to BC-1 that is predicted to disrupt a G:C base pair in the pri-miR-K5 transcript. Of note, this change is in the passenger strand, not in the mature miR-K5 miRNA itself. We and other investigators have previously shown that efficient processing of a pri-miRNA precursor to a mature miRNA requires the entire miRNA stem-loop structure, including the basal extension of the stem that is excluded from the pre-miRNA hairpin intermediate, as well as a small amount (generally 20 nucleotides or so) of flanking single-stranded RNA sequence (Lee et al. 2003; Zeng and Cullen 2003).

To test whether this one nucleotide polymorphism in fact accounted for the poor expression of miR-K5 in BCBL-1 cells, we used PCR (polymerase chain reaction) to clone analogous 472-bp segments of the KSHV genome, centered on the pri-miR-K5 stem-loop, from both BC-1 and BCBL-1 cells and inserted these into the tetracycline-regulated pTre expression vector. These two matched KSHV miRNA expression vectors were then transfected into 293T cells, in the presence or absence of the pTet-Off plasmid and doxycycline. As expected, transfection of 293T cells with pTre-miR-K5(BC-1), in the presence of pTet-Off and the absence of doxycycline, gave rise to readily detectable levels of "pri-miR-K5," pre-miR-K5, and mature miR-K5. In contrast, in cells transfected with the very similar pTre-miR-K5(BCBL-1)

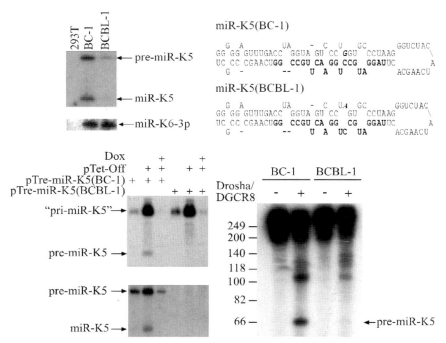

Figure 6. Differential expression of miR-K5 in BC-1 and BCBL-1 cells. (*Top left*) Northern analysis of miR-K5 (*upper panel*) and miR-K6-3p expression in total RNA preparations from 293T, BC-1, or BCBL-1 cells. (*Top right*) Predicted pre-miR-K5 stem-loop structures encoded by BC-1-derived (*upper panel*) or BCBL-1-derived (*lower panel*) KSHV. The mature miR-K5 sequence is shown in bold, and the single-nucleotide difference is shown in bold and italics. (*Bottom left*) Northern analysis of total RNA preparations from 293T cells transfected with the indicated plasmids and cultured in the presence or absence of doxycycline (Dox). The miR-K5 pri-miRNA and pre-miRNA precursors as well as mature miR-K5 are indicated. The quotation marks flanking "pri-miR-K5" signify that this RNA represents a truncated form of the authentic viral pri-miR-K5 transcript. (*Bottom right*) BC-1- or BCBL-1-derived pri-miR-K5 transcripts were incubated in vitro with recombinant Flag-Drosha/Flag-DGCR-8(+) or with control extracts prepared (–). The processed pre-miRNA is indicated by an arrow. (Modified, with permission, from Gottwein et al. 2006 [© ASM].)

vector, only the artificial pri-miR-K5 transcript was detected (Fig. 6, bottom left).

The data presented in Figure 6 (bottom left) suggested, but did not prove, that the one nucleotide polymorphism identified in Figure 6 (top right) inhibited pri-miR-K5 processing by the nuclear microprocessor complex. To confirm that this was indeed the case, we overexpressed Flag-tagged forms of human Drosha and DGCR8 in 293T cells and then recovered the recombinant Drosha–DGCR8 complex from these cells using an affinity matrix (Gottwein et al. 2006). This matrix was then incubated with a 270-nucleotide ^{32}P-labeled transcript encompassing the entire KSHV pri-miR-K5 stem-loop and adjacent flanking sequences derived from either BC-1 or BCBL-1 cells. As may be observed (Fig. 6, bottom right), the recombinant Drosha–DGCR8 complex cleaved the RNA transcript derived from BC-1 cells to give the predicted 62-nucleotide pri-miR-K5 intermediate, as well as the two approximately 100-nucleotide flanking RNA sequences. In contrast, very little Drosha processing of the analogous pri-miR-K5 RNA template derived from BCBL-1 cells was observed. We therefore conclude that the weak activity (Fig. 5, bottom) and low level of expression (Fig. 6, top left) of miR-K5 seen in BCBL-1 cells result from a single-nucleotide polymorphism (Fig. 6, top right) that inhibits processing of the pri-miR-K5 precursor by Drosha (Fig. 6, bottom right). To our knowledge, this is the first demonstration of a naturally occurring sequence polymorphism that directly perturbs miRNA processing and hence function.

CONCLUSIONS

This paper has summarized our current understanding of the expression and function of the miRNAs encoded by the pathogenic human herpesvirus KSHV. Work from our laboratory and others has identified at least 12 distinct miRNAs in the KSHV genome (Cai et al. 2005; Pfeffer et al. 2005; Grundhoff et al. 2006), all of which are expressed in latently infected cells (Fig. 2). Surprisingly, these 12 miRNAs proved to be clustered together in an approximately 4-kb region of the approximately 140-kb KSHV genome that coincided with the previously defined KSHV latency-associated region (Fig. 1) (Dittmer et al. 1998). Moreover, these 12 miRNAs were all in the same transcriptional orientation, thus suggesting that they might all derive from a single pri-miRNA precursor. In latently infected cells, this has indeed proven to be the case (Fig. 3) (Cai et al. 2006). However, in lytically infected cells, activation of a lytic promoter that is normally silent during latent infection gives rise to a distinct and very highly expressed transcript that functions both as a pri-miRNA precursor for two of the KSHV miRNAs, that is, miR-K10 and miR-K12, and as an mRNA encoding the viral Kaposin proteins (Sadler et al. 1999; Cai et al. 2006). This is an unexpected result, as miRNA processing results in the nuclear degradation of the pri-miRNA precursor so that it cannot also function as an mRNA. The most likely explanation for this paradox is that miR-K10 and miR-K12 processing is very inefficient, so that much of the Kaposin mRNA is able to reach the cytoplasm intact.

Indeed, expression of miR-K10 and miR-K12 in latently infected cells appears to be quite low, in comparison to other KSHV miRNAs, even though they all derive from processing of the same latent transcripts (Fig. 4).

The obvious and most important questions with regard to the miRNAs encoded by KSHV and other herpesviruses are, What are their function in the viral life cycle, and what are their mRNA targets? At present, the answers to these questions are not known. However, the fact that ten of the KSHV miRNAs are expressed at readily detectable and biologically active levels in latently infected cells (Fig. 5) whereas two KSHV miRNAs only appear to become fully active during lytic replication (Fig. 4) suggests that these miRNAs may have distinct roles during the viral life cycle.

ACKNOWLEDGMENTS

This research was funded by grant GM071408 from the National Institutes of Health. The authors thank Blossom Damania and Shihua Lu for their help with aspects of this research.

REFERENCES

Bartel D.P. 2004. MicroRNAs: Genomics, biogenesis, mechanism, and function. *Cell* **116:** 281.

Cai X. and Cullen B.R. 2006. Transcriptional origin of Kaposi's sarcoma-associated herpesvirus microRNAs. *J. Virol.* **80:** 2234.

Cai X., Hagedorn C.H., and Cullen B.R. 2004. Human microRNAs are processed from capped, polyadenylated transcripts that can also function as mRNAs. *RNA* **10:** 1957.

Cai X., Lu S., Zhang Z., Gonzalez C.M., Damania B., and Cullen B.R. 2005. Kaposi's sarcoma-associated herpesvirus expresses an array of viral microRNAs in latently infected cells. *Proc. Natl. Acad. Sci.* **102:** 5570.

Cai X., Schäfer A., Lu S., Bilello J.P., Desrosiers R.C., Edwards R., Raab-Traub N., and Cullen B.R. 2006. Epstein-Barr virus microRNAs are evolutionarily conserved and differentially expressed. *PLoS Pathogens* **2:** e23.

Chen L. and Lagunoff M. 2005. Establishment and maintenance of Kaposi's sarcoma-associated herpesvirus latency in B cells. *J. Virol.* **79:** 14383.

Cullen B.R. 2004. Transcription and processing of human microRNA precursors. *Mol. Cell* **16:** 861.

———. 2006. Viruses and microRNAs. *Nat. Genet.* **38:** S25.

Denli A.M., Tops B.B.J., Plasterk R.H.A., Ketting R.F., and Hannon G.J. 2004. Processing of primary microRNAs by the microprocessor complex. *Nature* **432:** 231.

Dittmer D., Lagunoff M., Renne R., Staskus K., Haase A., and Ganem D. 1998. A cluster of latently expressed genes in Kaposi's sarcoma-associated herpesvirus. *J. Virol.* **72:** 8309.

Doench J.G., Petersen C.P., and Sharp P.A. 2003. siRNAs can function as miRNAs. *Genes Dev.* **17:** 438.

Gottwein E., Cai X., and Cullen B.R. 2006. A novel assay for viral microRNA function identifies a single nucleotide polymorphism that affects Drosha processing. *J. Virol.* **80:** 5321.

Gregory R.I., Yan K.-P., Amuthan G., Chendrimada T., Doratotaj B., Cooch N., and Shickhattar R. 2004. The microprocessor complex mediates the genesis of miRNAs. *Nature* **432:** 235.

Grey F., Antoniewicz A., Allen E., Saugstad J., McShea A., Carrington J. C., and Nelson J. 2005. Identification and characterization of human cytomegalovirus-encoded microRNAs. *J. Virol.* **79:** 12095.

Grundhoff A., Sullivan C.S., and Ganem D. 2006. A combined computational and microarray-based approach identifies novel microRNAs encoded by human gamma-herpesviruses. *RNA* **12:** 1.

Hammond S.M., Bernstein E., Beach D., and Hannon G.J. 2000. An RNA-directed nuclease mediates post-transcriptional gene silencing in *Drosophila* cells. *Nature* **404:** 293.

Han J., Lee Y., Yeom K.-H., Kim Y.-K., Jin H., and Kim V.N. 2004. The Drosha-DGCR8 complex in primary microRNA processing. *Genes Dev.* **18:** 3016.

Lee Y., Kim M., Han J., Yeom K.-H., Lee S., Baek S.H., and Kim V.N. 2004. MicroRNA genes are transcribed by RNA polymerase II. *EMBO J.* **23:** 4051.

Lee Y., Ahn C., Han J., Choi H., Kim J., Yim J., Lee J., Provost P., Radmark O., Kim S., and Kim V.N. 2003. The nuclear RNase III Drosha initiates microRNA processing. *Nature* **425:** 415.

Li H., Komatsu T., Dezube B.J., and Kaye K.M. 2002. The Kaposi's sarcoma-associated herpesvirus K12 transcript from a primary effusion lymphoma contains complex repeat elements, is spliced, and initiates from a novel promoter. *J. Virol.* **76:** 11880.

Martinez J., Patkaniowska A., Urlaub H., Lührmann R., and Tuschl T. 2002. Single-stranded antisense siRNAs guide target RNA cleavage in RNAi. *Cell* **110:** 563.

McCormick C. and Ganem D. 2005. The Kaposin B protein of KSHV activates the p38/MK2 pathway and stabilizes cytokine mRNAs. *Science* **307:** 739.

Pearce M., Matsumura S., and Wilson A.C. 2005. Transcripts encoding K12, v-FLIP, v-cyclin, and the microRNA cluster of Kaposi's sarcoma-associated herpesvirus originate from a common promoter. *J. Virol.* **79:** 14457.

Pfeffer S., Zavolan M., Grässer F.A., Chien M., Russo J.J., Ju J., John B., Enright A.J., Marks D., Sander C., and Tuschl T. 2004. Identification of virus-encoded microRNAs. *Science* **304:** 734.

Pfeffer S., Sewer A., Lagos-Quintana M., Sheridan R., Sander C., Grässer F.A., van Dyk L.F., Ho C.K., Shuman S., Chien M., et al. 2005. Identification of microRNAs of the herpesvirus family. *Nat. Methods* **2:** 269.

Jenner R.G., Alba M.M., Boshoff C., and Kellam P. 2001. Kaposi's sarcoma-associated herpesvirus latent and lytic gene expression as revealed by DNA arrays. *J. Virol.* **75:** 891.

Russo J.J., Bohenzky R.A., Chien M.-C., Chen J., Yan M., Maddalena D., Parry J.P., Peruzzi D., Edelman I.S., Chang Y., and Moore P.S. 1996. Nucleotide sequence of the Kaposi sarcoma-associated herpesvirus (HHV8). *Proc. Natl. Acad. Sci.* **93:** 14862.

Sadler R., Wu L., Forghani B., Renne R., Zhong W., Herndier B., and Ganem D. 1999. A complex translational program generates multiple novel proteins from the latently expressed Kaposin (K12) locus of Kaposi's sarcoma-associated herpesvirus. *J. Virol.* **73:** 5722.

Samols M.A., Hu J., Skalsky R.L., and Renne R. 2005. Cloning and identification of a microRNA cluster within the latency-associated region of Kaposi's sarcoma-associated herpesvirus. *J. Virol.* **79:** 9301.

Sarid R., Flore O., Bohenzky R.A., Chang Y., and Moore P.S. 1998. Transcription mapping of the Kaposi's sarcoma-associated herpesvirus (human herpesvirus 8) genome in a body cavity-based lymphoma cell line (BC-1). *J. Virol.* **72:** 1005.

Schwarz D.S., Hutvágner G., Haley B., and Zamore P.D. 2002. Evidence that siRNAs function as guides, not primers, in the *Drosophila* and human RNAi pathways. *Mol. Cell* **10:** 537.

Sullivan C.S., Grundhoff A.T., Tevethia S., Pipas J.M., and Ganem D. 2005. SV40-encoded microRNAs regulate viral gene expression and reduce susceptibility to cytotoxic T cells. *Nature* **435:** 682.

Ye F.-C., Zhou F.-C., Yoo S.M., Xie J.-P., Browning P.J., and Gao S.-J. 2004. Disruption of Kaposi's sarcoma-associated herpesvirus latent nuclear antigen leads to abortive episome persistence. *J. Virol.* **78:** 11121.

Yi R., Qin Y., Macara I.G., and Cullen B.R. 2003. Exportin-5 mediates the nuclear export of pre-microRNAs and short hairpin RNAs. *Genes Dev.* **17:** 3011.

Zeng Y. and Cullen B.R. 2003. Sequence requirements for micro RNA processing and function in human cells. *RNA* **9:** 112.

Zeng Y., Yi R., and Cullen B.R. 2003. MicroRNAs and small interfering RNAs can inhibit mRNA expression by similar mechanisms. *Proc. Natl. Acad. Sci.* **100:** 9779.

Relationship between Retroviral Replication and RNA Interference Machineries

K. MOELLING, A. MATSKEVICH, AND J.-S. JUNG

Institute of Medical Virology, University of Zurich, CH-8006 Zurich, Switzerland

Small interfering RNAs (siRNAs) associated with gene silencing are cellular defense mechanisms against invading viruses. The viruses fight back by suppressors or escape mechanisms. The retroviruses developed a unique escape mechanism by disguising as DNA proviruses. An evolutionary relationship between the siRNA machinery and the replication machinery of retroviruses is likely. The RNA cleavage enzymes PIWI and RNase H proteins are structurally related. This relationship can be extended from structure to function, since the retroviral reverse transcriptase (RT)/RNase H can also cause silencing of viral RNA by siRNA. Thus, both enzymes can cleave RNA–DNA hybrids and double-stranded RNA (dsRNA) with various efficiencies shown previously and here, demonstrating that their specificities are not absolute. Other similarities may exist, for example between PAZ and the RT and between RNA-binding proteins and the viral nucleocapsid protein. Dicer has some similarities with the viral integrase, since both specifically generate dinucleotide 3′-overhanging ends. We described previously the destruction of the human immunodeficiency virus (HIV) RNA by a DNA oligonucleotide ODN (oligodeoxynucleotide). Variants of the ODN indicated high length and sequence specificities, which is reminiscent of siRNA and designated here as "siDNA." Cleavage of the viral RNA in the presence of the ODN is caused by the retroviral RT/RNase H and cellular RNase H activities. Several siRNA-mediated antiviral defense mechanisms resemble the interferon system.

We have previously designed an ODN targeted to the polypurine tract (PPT) of HIV with the idea of inhibiting HIV replication. The ODN was designed as a clamp around the PPT to inhibit the RT/RNase H during retroviral replication. The sequence consisted of an antisense strand homologous to the extended 25-nucleotide PPT. The second strand was designed on the basis of Hoogsteen base-pairing to allow a triple helix at the PPT. A linker of four nucleotides connected the strands. The PPT is the site that resists RNase H digestion during reverse transcription of the first cDNA copy, then the RT initiates at this local hybrid the second cDNA strand including cleavage by the RNase H at the ACU site, where two non-purines interrupt the PPT (Jendis et al. 1996, 1998; see Coffin et al. 1997).

The ODN, which by itself can form a hairpin-loop structure, was applied to T lymphocytes 2 hours postinfection with HIV at a low multiplicity of infection of 0.01 with ODN at 1 μM concentration without carrier. This resulted in inhibition of viral replication for 20 days without viral breakthrough (Jendis et al. 1996). Patient isolates including resistant isolates against azidothymidin (AZT) were also inhibited. A control antisense ODN was less efficient (Jendis et al. 1996, 1998). Replication of two viral isolates, HIV IIIB and BaL-1, which differ in their PPTs by 2 of 24 or 23 nucleotides, was inhibited preferentially by their homologous ODNs and to a lesser extent by their heterologous ODNs. This indicates a very high sequence specificity (Moelling et al. 2006). The PPT is highly conserved, and 1700 of 2000 isolates collected in the Los Alamos Database were identical. PPT is also conserved in SIV (simian immunodeficiency virus).

We analyzed numerous variants of the ODN by changing single nucleotides or its length. All variants were ineffective in destroying the viral RNA (Moelling et al. 2006). To understand the molecular mechanism, we performed a kinetic analysis. RT-PCR (polymerase chain reaction) analysis indicated the disappearance of the viral RNA within about 2 hours posttreatment and the absence of a detectable DNA provirus (Jendis et al. 1996; Moelling et al. 2006). Furthermore, we were unable to demonstrate a triple-helix effect. However, we could attribute the ODN-mediated molecular mechanism to destruction of the viral RNA by the retroviral RT/RNase H. We showed this by applying the ODN to purified viral particles, where the RT/RNase H was able to destroy the viral RNA and abolish viral infectivity within a few hours (Matskevich et al. 2006). The phenomenon is reminiscent of siRNA-mediated silencing and may be designated as "siDNA."

SUBSTRATE SPECIFICITIES OF RNase H AND PIWI

siRNA-mediated silencing has been shown by dsRNA, 23-mers with overhanging ends, which are generated by Dicer and processed by AGO2 and associated proteins. The cleaving PIWI domain of Argonaute was shown to be structurally related to an RNase H (Song et al. 2004; Ma et al. 2005; Yuan et al. 2005). This relationship was only deducible from structural analyses and not obvious from the primary sequences. RNases H share a characteristic conserved triad of acidic amino acids at their active centers, the DDE motif (Yang and Steitz 1995; Nowotny et al. 2005). Not only do RNases H exist in retroviral replication complexes (Moelling et al. 1971; Moelling 1975; Hansen et al. 1988), but they also exist in *Escherichia coli* and mammalian cells as RNase H1 for removal of RNA primers during DNA replication (Nowotny et al. 2005). All of these RNase H enzymes cleave RNA in RNA–DNA hybrids (Moelling et al. 1971) or at the junction of RNA–DNA, opposite to an RNA strand, for example, dur-

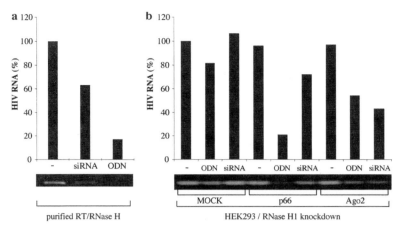

Figure 1. (*a*) Recombinant HIV RT/RNase H can cleave viral RNA in the presence of siRNA and an ODN, both targeted to the PPT of HIV RNA and measured by RT-PCR. (*b*) Cleavage of viral RNA in the presence of siRNA and ODN with cellular extracts as source of enzymes. Extracts were prepared from HEK 293 cells in which the cellular RNase H1 had been knocked down and either the p66 HIV RT/RNase H or the AGO2 protein was overexpressed. Cleavage was determined by RT-PCR.

ing HIV replication for removal of the tRNA primer after initiation of cDNA synthesis. The structural homology was, however, not shared at the functional level, because a hallmark of an RNase H is its specificity for cleaving RNA in RNA–DNA hybrids, not dsRNA as required for siRNA. Yet, both the RT/RNase H and PIWI give rise to oligonucleotides with 3′-OH and 5′-phosphate.

The PIWI domain has been shown to also cleave RNA in RNA–DNA hybrids, tested with an archaebacterial PIWI under extreme conditions, whereby the relative K_D values of single-stranded DNA to DNA–RNA hybrid to dsRNA were approximately 0.01 to 1 to 10 (Yuan et al. 2005).

We used the recombinant HIV-RT/RNase H with the ODN and siRNA and targeted them to the PPT of HIV together with HIV RNA extracted from viral particles in an in vitro reaction and then used primer sets covering the HIV-PPT RNA for analysis of cleavage of the RNA by RT-PCR analysis (Fig. 1a). As shown, RT/RNase H can perform siRNA-mediated silencing of HIV RNA, whereby the effect induced by the ODN is more efficient.

We then analyzed cells in which we had knocked down the cellular RNase H1 to more specifically determine the effect caused by the viral RT/RNase H. The cells, in which the cellular RNase H1 was knocked down, were transfected and overexpressed the HIV RT/RNase H and AGO2 (both linked to HA tags). In vitro reactions were performed with HIV RNA and ODN or siRNA, supplemented with the cellular extracts. As shown, the recombinant p66 RT/RNase H cleaved the viral RNA in the presence of ODN, and AGO2 cleaved the viral RNA in the presence of siRNA. Both enzymes also cleaved the RNA under reversed conditions, indicating that the RT/RNase H can silence RNA with low efficiency also by means of siRNA and AGO2 also by ODN (Fig. 1b). The results indicate that the specificities of the RT/RNase H and AGO2 are not absolute.

It should be noted that the HIV RT/RNase H and the murine leukemia virus (MLV) RT/RNase H have also been shown to process dsRNAs, not only RNA–DNA hybrids (Ben-Artzi et al. 1992; Hostomsky et al. 1994; Blain and Goff 1993). This activity has also been named RNase H*. Furthermore, a mutation in the active center of an archaeal

RNase H, D125N, resulted in the loss of dsRNA cleavage efficiency, but not for hybrids (Ohtani et al. 2004). In addition, stalling the RNase H by omission of the four desoxyribonucleoside triphosphates (4dNTPs), so that the RT cannot move and polymerize, increased the efficiency for cleavage of dsRNA (Götte et al. 1995). The relationship between RT/RNase H and PIWI is therefore both structural and functional, not just structural. Figure 2 summarizes these results, showing both homologous and heterologous activities. Destruction of the HIV RNA by the RT/RNase H in the presence of the "siDNA" is designated as "suicide," since a normal step in retroviral replication is mimicked by the siDNA, but prematurely, before a cDNA is made.

COMPARISON OF RETROVIRAL AND RNA INTERFERENCE MACHINERIES

Comparison of the components of the replication machinery of retroviruses suggests some further relationships besides the RNase H with the PIWI domain. Several similarities are summarized in Table 1. The RT, for example, has an attachment site for the RNA primer for initiating cDNA synthesis by the primer grip (Jacques et al. 1994). A similar function has been described for the PAZ domain of AGO2, fixing the 3′-OH end of the antisense strand in a PAZ pocket. Furthermore, AGO proteins have a binding pocket for the phosphorylated 5′ end of the antisense strand (Ma et al. 2005) with no known equivalent for the RT/RNase H.

The HIV RNase H requires RNA-binding properties similar to those of PIWI. Fusion of the RNase H to the RT domain supplies this ability. The RNase H without the RT domain has been shown to be active only in a special in situ assay by immobilizing the RNase H domain in a gel containing a radioactive hybrid (Schulze et al. 1991). Perhaps this assay may allow detection of RNase H activities of some of those AGO proteins, which appear to be inactive. Furthermore, the RT has an RNA-dependent as well as DNA-dependent DNA polymerase activity. It is not known whether it can also function as an RNA-dependent RNA polymerase. The RT also has an unwind-

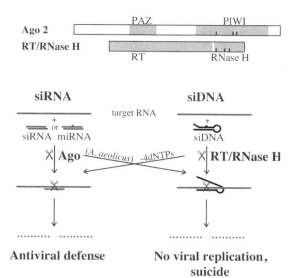

Figure 2. A relationship between siRNA components and retroviral proteins is proposed. PAZ and PIWI correspond to RT/RNase H, and Dicer corresponds to integrase (see also Table 1). PIWI of AGO2 and RT/RNase H are functionally related. This is shown by siRNA-mediated RNA silencing in parallel to "siDNA" silencing of retroviral RNA. AGO2 and RT/RNase H can also cleave the other substrates with less specificity and lower efficiency. This has been shown with the archaebacterial AGO (Yuan et al. 2005) and RT/RNase H, whereby nonspecific cleavage was increased in the absence of 4dNTPs. HIV commits suicide with the "siDNA" targeted to the PPT, which prematurely activates the viral RT/RNase H for cleavage of the viral RNA before the first cDNA is made. Thus, the viral RNA is irreversibly destroyed (Moelling et al. 2006).

ing activity (Hottiger et al. 1994), corresponding to the helicase in the RNA interference machinery. Some similarities seem to exist between the integrase of retroviruses and Dicer, both of which generate dinucleotide 3'-overhanging ends at the DNA, in the case of the DNA provirus at the two identical long terminal repeats (LTRs) and in the case of Dicer on dsRNA fragments. In addition, the integration site in the DNA of the host cell requires this type of cleavage. Both enzymes have nucleic-acid-binding domains. The integrase forms dimers for cleavage. Perhaps Dicer also dimerizes to produce 21-mers.

Table 1. Similarities of Retroviral Replication and RNA Interference Machineries

Retrovirus	siRNA
Reverse transcriptase (RT)	PAZ
RNase H	PIWI
3'-OH, 5'-phosphate oligonucleotide	3'-OH, 5'-phosphate oligonucleotide
RT (primer-grip)	PAZ pocket
RT (RNA-dependent DNA polymerase/DNA-dependent DNA polymerase)	RNA-dependent RNA polymerase
RT (DNA unwinding)	Helicase
Integrase (dinucleotides overhangs)	Dicer (dinucleotides overhangs)
Nucleocapsid (melting, RNA protection)	RNA-binding proteins, TRBP, FMRP
TRBP	TRBP
Protease	?

Furthermore, retroviruses code for a nucleocapsid (NC) protein with a bipartite basic stretch of sequences, binding to single-stranded as well as double-stranded RNA. It is a chaperone or matchmaker and straightens out hairpin-looped structures and fixes the tRNA primer to the viral RNA at the primer-binding site (PBS). It may be worth noting that the NC has, in addition to its unspecific RNA-binding properties, the ability to specifically bind and select only one tRNA from the cellular pool, which requires a sequence-specific recognition through its zinc fingers (Dannull et al. 1994). RNA-binding domains or factors are also associated with the PAZ-PIWI domains of AGO2, even though a sequence specificity has not been detected. An RNA-binding protein is the fragile X syndrome mental retardation protein (FMRP), which can associate with AGO (Caudy et al. 2002). What about the protease, another highly specific and important component in the retroviral replication machinery? Is it yet to be found in siRNA-mediated silencing? There is a protease involved in an antiviral defense mechanism, the virus fights back by a protein, which drives AGO into the proteasomal pathway in plants (D.C. Baulcombe, pers. comm.).

Structural components of retroviruses may not be related to the siRNA machinery, but what about some of the luxury functions known for HIV such as Tat, Rev, Nev, and Vif? The Tat RNA-binding protein (TRBP) is used by the virus (HIV) and the cell, where it can complex with AGO (Chendrimada et al. 2005).

RNA VIRUSES ESCAPE THE HOST DEFENSE BY SUPPRESSORS OR BY A DNA PROVIRUS

RNA viruses with their dsRNA intermediates are targets of the siRNA-mediated cellular defense, and hit back, for example, by P19 (Voinnet et al. 1999), p21 (Reed et al. 2003), or suppressor proteins counteracting AGO. Retroviruses may have escaped the siRNA-mediated antiviral defense and developed instead of suppressors a disguise as dsDNA proviruses via an RNA–DNA intermediate. They may have adopted some of the siRNA machinery for their replication machinery, specializing in tools for DNA instead of RNA. One cellular defense mechanism against an integrated DNA provirus can be by keeping it silent by DNA methylation, which in some species is mediated by small RNAs. The equivalent of this small RNA in mammals are methylases. The virus can escape this effect by fast replication.

If this also fails against viral invaders, the next level of defense is posttranscriptional gene silencing (PTGS), leading to destruction of the viral mRNA by small RNAs at noncoding untranscribed regions or inhibition of translation or initiation of translation. At this stage, DNA viruses may also be silenced. If the invader escaped even this level of defense, the ultimate rationale of the cell would be, while dying, to warn the neighboring cells by diffusion of small RNAs, thereby guaranteeing survival of an organism or a cell community, not the individual cell. Thus, diffusible small RNAs are the alarm system for bystander cells in plants. The worm *Caenorhabditis elegans* can even warn other worms in the neighborhood by secreted siRNAs. In mammals, such an effect is achieved probably not by siRNA but by interferon, a diffusible protein, which alarms bystander cells against viruses. There are other striking similarities between siRNA and the interferon response, the RNA-dependent protein kinase (PKR), which inhibits translation, and RNase L, which degrades mRNA. Thus, it appears that siRNA-mediated responses in primitive organisms have been replaced by proteins in the mammalian system.

CONCLUSIONS

The siRNA-mediated antiviral defense machinery in the RNA World may have coevolved with the retrovirus replication machinery. We and other investigators have shown a structure/function relationship between retroviral RT/RNase H and PIWI; we also pointed out the related properties of the two machineries. This analogy may raise questions or make predictions, e.g., whether the viral protease has an equivalent in the RNA interference system, yet to be discovered. Furthermore, we may not yet have determined the equivalent of the viral sequence-specific RNA-binding protein. Cellular defense against integrated DNA proviruses requires mechanisms such as siRNA-mediated methylation for silencing of the viral DNA in plants. The mammalian equivalent may occur via proteins such as methylases. Warning of neighboring cells or organisms by small RNAs may have evolved into the interferon system. Analogies to the mammalian interferon system, especially in this latter respect, are striking, whereby proteins became the mediators and may have replaced the small RNAs. In summary, some of the analogies discussed here may allow predictions for the RNA interference machinery from the retroviral replication machinery as well as from the interferon system, in particular on the role of proteins instead of siRNAs in the mammalian system.

ACKNOWLEDGMENTS

I am indebted to Dr. M. Lorger for help with some constructs and to Dr. J. Heinrich for support with the manuscript and figures.

REFERENCES

Ben-Artzi H., Zeelon E., Gorecki M., and Panet A. 1992. Double-stranded RNA-dependent RNase activity associated with human immunodeficiency virus type 1 reverse transcriptase. *Proc. Natl. Acad. Sci.* **89**: 927.

Blain S.W. and Goff S.P. 1993. Nuclease activities of Moloney murine leukemia virus reverse transcriptase. Mutants with altered substrate specificities. *J. Biol. Chem.* **268**: 23585.

Caudy A.A., Myers M., Hannon G.J., and Hammond S.M. 2002. Fragile X-related protein and VIG associate with the RNA interference machinery. *Genes Dev.* **16**: 2491.

Chendrimada T.P., Gregory R.I., Kumaraswamy E., Norman J., Cooch N., Nishikura K., and Shiekhattar R. 2005. TRBP recruits the Dicer complex to Ago2 for microRNA processing and gene silencing. *Nature* **436**: 740.

Coffin J.M., Hughes S.H., and Varmus H.E., eds. 1997. Reverse transcriptase and the generation of retroviral DNA. In *Retroviruses*, chapter 4, p. 121. Cold Spring Harbor Laboratory Press, Cold Spring Harbor, New York.

Dannull J., Surovoy A., Jung G., and Moelling K. 1994. Specific binding of HIV-1 nucleocapsid protein to PSI RNA in vitro requires N-terminal zinc finger and flanking basic amino acid residues. *EMBO J.* **13**: 1525.

Götte M., Fackler S., Hermann T., Perola E., Cellai L., Gross H.J., Le Grice S.F.J., and Heumann H. 1995. HIV-1 reverse transcriptase-associated RNase H cleaves RNA/RNA in arrested complexes: Implications for the mechanism by which RNase H discriminates between RNA/RNA and RNA/DNA. *EMBO J.* **14**: 833.

Hansen J., Schulze T., Mellert W., and Moelling K. 1988. Identification and characterization of HIV-specific RNase H by monoclonal antibody. *EMBO J.* **7**: 239.

Hostomsky Z., Hughes S.H., Goff S.P., and Le Grice S.F. 1994. Redesignation of the RNase D activity associated with retroviral reverse transcriptase as RNase H. *J. Virol.* **68**: 1970.

Hottiger M., Podust V.N., Thimmig R.L., McHenry C., and Hubscher U. 1994. Strand displacement activity of the human immunodeficiency virus type 1 reverse transcriptase heterodimer and its individual subunits. *J. Biol. Chem.* **269**: 986.

Jacques P.S., Wohrl B.M., Ottmann M., Darlix J.L., and Le Grice S.F. 1994. Mutating the "primer grip" of p66 HIV-1 reverse transcriptase implicates tryptophan-229 in template-primer utilization. *J. Biol. Chem.* **269**: 26472.

Jendis J., Strack B., and Moelling K. 1998. Inhibition of replication of drug-resistant HIV type 1 isolates by polypurine tract-specific oligodeoxynucleotide TFO A. *AIDS Res. Hum. Retrovir.* **14**: 999.

Jendis J., Strack B., Volkmann S., Boni J., and Molling K. 1996. Inhibition of replication of fresh HIV type 1 patient isolates by a polypurine tract-specific self-complementary oligodeoxynucleotide. *AIDS Res. Hum. Retrovir.* **12**: 1161.

Ma J.B., Yuan Y.R., Meister G., Pei Y., Tuschl T., and Patel D.J. 2005. Structural basis for 5′-end-specific recognition of guide RNA by the *A. fulgidus* Piwi protein. *Nature* **434**: 666.

Matskevich A.A., Ziogas A., Heinrich J., Quast S., and Moelling K. 2006. Short partially double-stranded oligodeoxynucleotide induces reverse transcriptase-mediated cleavage of HIV RNA and abrogates infectivity of virions. *AIDS Res. Hum. Retrovir.* (in press).

Moelling K. 1975. Reverse transcriptase and RNase H: Present in a murine virus and in both subunts of an avian virus. *Cold Spring Harbor Symp. Quant. Biol.* **39**: 969.

Moelling K., Abels S., Jendis J., Matskevich A., and Heinrich J. 2006. Silencing of HIV by hairpin-loop-structured DNA oligonucleotide. *FEBS Lett.* **580**: 3545.

Moelling K., Bolognesi D.P., Bauer H., Büsen W., Plassmann H.W., and Hausen P. 1971. Association of the viral reverse transcriptase with an enzyme degrading the RNA moiety of RNA-DNA hybrids. *Nat. New Biol.* **234**: 240.

Nowotny M., Gaidamakov S.A., Crouch R.J., and Yang W. 2005. Crystal structures of RNase H bound to an RNA/DNA hybrid: Substrate specificity and metal-dependent catalysis. *Cell* **121**: 1005.

Ohtani N., Yanagawa H., Tomita M., and Itaya M. 2004. Cleavage of double-stranded RNA by RNase HI from a thermoacidophilic archaeon, *Sulfolobus tokodaii* 7. *Nucleic Acids Res.* **32**: 5809.

Reed J.C., Kasschau K.D., Prokhnevsky A.I., Gopinath K., Pogue G.P., Carrington J.C., and Dolja V.V. 2003. Suppressor of RNA silencing encoded by Beet yellows virus. *Virology* **306**: 203.

Schulze T., Nawrath M., and Moelling K. 1991. Cleavage of the HIV-1 p66 reverse transcriptase/RNase H by the p9 protease in vitro generates active p15 RNase H. *Arch. Virol.* **118**: 179.

Song J.J., Smith S.K., Hannon G.J., and Joshua-Tor L. 2004. Crystal structure of Argonaute and its implications for RISC slicer activity. *Science* **305**: 1434.

Voinnet O., Pinto Y.M., and Baulcombe D.C. 1999. Suppression of gene silencing: A general strategy used by diverse DNA and RNA viruses of plants. *Proc. Natl. Acad. Sci.* **96**: 14147.

Yang W. and Steitz T.A. 1995. Recombining the structures of HIV integrase, RuvC and RNase H. *Structure* **3**: 131.

Yuan Y.R., Pei Y., Ma J.B., Kuryavyi V., Zhadina M., Meister G., Chen H.Y., Dauter Z., Tuschl T., and Patel D.J. 2005. Crystal structure of *A. aeolicus* argonaute, a site-specific DNA-guided endoribonuclease, provides insights into RISC-mediated mRNA cleavage. *Mol. Cell* **19**: 405.

Positive and Negative Modulation of Viral and Cellular mRNAs by Liver-specific MicroRNA miR-122

C.L. Jopling, K.L. Norman, and P. Sarnow

Department of Microbiology and Immunology, Stanford University School of Medicine, Stanford, California 94305

microRNAs (miRNAs) are small RNAs that in general down-regulate the intracellular abundance and translation of target mRNAs. We noted that sequestration of liver-specific miR-122 by modified antisense oligonucleotides resulted in a dramatic loss of hepatitis C virus (HCV) RNA in cultured human liver cells. A binding site for miR-122 was predicted to reside close to the 5′ end of the viral genome, and its functionality was tested by mutational analyses of the miRNA-binding site in viral RNA, resulting in reduced intracellular viral RNA abundance. Importantly, ectopic expression of miR-122 molecules that contained compensatory mutations restored viral RNA abundance, revealing a genetic interaction between miR-122 and the viral RNA genome. Studies with replication-defective viral RNAs demonstrated that miR-122 affected mRNA abundance by positively modulating RNA replication. In contrast, interaction of miR-122 with the 3′-noncoding region (3′NCR) of the cellular mRNA encoding the cationic amino acid transporter CAT-1 resulted in the down-regulation of CAT-1 protein abundance. These findings provide evidence that a specific miRNA can regulate distinct target mRNAs in both a positive and negative fashion. The positive role of miR-122 in viral replication suggests that this miRNA could be targeted for antiviral therapy.

Genetic screens in *Caenorhabditis elegans* identified the first animal miRNA, *lin-4* (Horvitz and Sulston 1980; Chalfie et al. 1981). The small *lin-4* RNA was noted to down-regulate the expression of target *lin-14* mRNA by interacting with the *lin-14* 3′-untranslated region (UTR) at multiple sites (Arasu et al. 1991; Wightman et al. 1991). The mechanism by which *lin-4* controls *lin-14* expression involves transcript stability (Bagga et al. 2005) and translation of *lin-14* mRNA (Olsen and Ambros 1999). Because Lin-14 protein levels dramatically decreased when the *lin-4* RNA was associated with *lin-14* mRNAs, it was argued that *lin-4* RNA blocked *lin-14* mRNA translation at a step following initiation. These experiments provided the first glimpse at the mechanism by which a miRNA can control expression of target mRNAs.

Since then, these small noncoding RNA molecules, approximately 22 nucleotides in length, have been detected in many eukaryotic organisms, and it is estimated that they control the expression of approximately one-fourth of all cellular mRNAs by binding to sites in the 3′NCR (Lewis et al. 2003, 2005; Bartel 2004). These small RNAs are now known as microRNAs.

Mammalian miRNA genes are encoded as monocistronic and polycistronic gene clusters and are found within intronic regions of protein-coding genes, within intronic regions of noncoding genes, and as independent transcriptional units (Lagos-Quintana et al. 2001; Lau et al. 2001; Lee et al. 2002). Transcription of miRNA genes results in the production of primary miRNA precursors that contain hairpin structures harboring the mature miRNA (Cai et al. 2004; Lee et al. 2004). To yield mature functional miRNAs, the miRNA sequence must be excised from the pri-miRNA by a maturation process that involves both nuclear and cytoplasmic cleavage events by two RNase III enzymes, Drosha and Dicer, respectively (Kim 2005). The excised RNA duplexes contain characteristic 5′ monophosphates, 3′ hydroxyl moieties, and 2-nucleotide 3′ overhangs. Next, one strand of the RNA duplex associates with a miRNA effector complex, known as miRNA-containing RNA-induced silencing complex (miRISC). Strand selection is based on the thermodynamic properties of the RNA duplex, so that the RNA strand with the weakest thermodynamic stability at its 5′ end is incorporated into miRISC (Hammond et al. 2000; Khvorova et al. 2003; Maniataki and Mourelatos 2005).

How does the miRISC complex recognize target mRNAs? Computer-assisted algorithms have predicted characteristics of miRNA-binding sites located in 3′NCRs of target mRNAs (Lewis et al. 2005). Most importantly, Watson-Crick base-pair complementarity between six consecutive nucleotides in the mRNA (the so-called seed match sequence) with corresponding nucleotides 2–7 of the miRNA (the so-called seed sequence) was noted to be essential for the formation of bona fide miRNA–mRNA complexes. In addition, the nucleotide following the seed match and its tendency to form a base-pair interaction with nucleotide 8 of the miRNA is conserved (Lewis et al. 2005). Most target recognition by a miRNA in mammalian cells involves imperfect complementarity, which leads to translational inhibition of the target mRNA (Zeng et al. 2002; Humphreys et al. 2005; Pillai et al. 2005; Petersen et al. 2006). However, some miRNAs can target mRNAs with perfect complementarity, which leads to mRNA cleavage (Yekta et al. 2004).

If one considers the abundance of miRNA molecules in cells and the minimal 6-nucleotide pairing that is needed to form specific miRNA–mRNA interactions, it is likely that viral genomes, often thousands of nucleotides in length, are targeted by miRNAs. Viruses may avoid multiplying in cells that express miRNAs with sequence complementarity to the viral genome. Alternatively, viruses could inhibit or subvert the RNA interference (RNAi) pathway to circumvent translational inhibition or enhanced turnover of viral genomes by miRNAs. Although there is much evidence for this scenario in plants and invertebrate organisms (Li and Ding 2006; Wang et al. 2006), examples of modulation of the RNAi pathway in mammalian cells have been sparse. One example was provided by primate foamy virus type 1 (PFV-1). Lecellier et al. (2005) showed that the viral RNA abundance of PFV-1 in human kidney 293T cells was enhanced in the presence of the plant tombusviral P19 protein that silences the RNAi pathway, indicating a role for the RNAi pathway in viral genome replication. Further analyses revealed a binding site for human miRNA miR-32 in the PFV-1 genome. The repressive effect of this miRNA on viral RNA amplification was confirmed when sequestration of miR-32 by antisense oligonucleotides resulted in an increase in viral RNA replication (Lecellier et al. 2005). These findings suggested that PFV-1 might encode a function that suppresses the RNAi pathway. Subsequently, it was found that ectopic expression of viral Tas protein resulted in a general inhibition of the RNAi pathway (Lecellier et al. 2005). However, it is unclear whether Tas has a role in inhibition of the RNAi pathway in cells that are normally persistently infected by PFV-1.

THE HCV RNA GENOME IS PREDICTED TO INTERACT WITH LIVER-SPECIFIC MICRORNA MIR-122

It was reported that miR-122 (Fig. 1A) was specifically expressed in the liver, where it constitutes 70% of the total miRNA population (Lagos-Quintana et al. 2002; Chang et al. 2004). As part of an ongoing project to identify

Table 1. Conservation of miR-122-binding Sites in HCV Genotypes

Genotype	5′-noncoding region
1a	UGAUGGGGGCGA**CACUCC**ACC
1b	-AUUGGGGGCGA**CACUCC**ACC
2	-AAUAGGGGCGA**CACUCC**GCC
3	—UACGAGGCGA**CACUCC**ACC
4	-UAUGAGAGCAA**CACUCC**ACC
5	-UAUUGGGGCGA**CACUCC**ACC
6	—AAUGGGGGCGA**CACUCC**ACC

Nucleotide number 1 in the seed match sequence is shown in blue, nucleotides comprising seed match sequences 2–7 are highlighted in bold, and the adenosine nucleotide at position 8 is shown in magenta.

cellular mRNA targets that can potentially be regulated by liver-specific miRNA miR-122 (Fig.1A), the abundance of miR-122 was examined by northern analysis (Fig. 1B) in a variety of liver and non-liver-derived cells. miR-122 could easily be identified in rat and human liver, in cultured human Huh7, and mouse Hepa 1-6 cells, but not in human cervical-carcinoma-derived HeLa or human liver HepG2 cells (Fig. 1B). The absence of miR-122 expression in HepG2 cells correlated with the absence of replication of hepatropic HCV in these cells (Lohmann et al. 1999).

HCV is a positive-strand RNA virus belonging to the Flaviviridae family (Bartenschlager and Lohmann 2000; Lindenbach and Rice 2003). It carries a 9.6-kb positive-strand RNA genome, with a 320-nucleotide 5′NCR, an open reading frame encoding a 3000-amino-acid polyprotein, which is subsequently proteolyzed by cellular and viral proteinases, and a shorter conserved 3′NCR (Lindenbach and Rice 2003). Because cell-culture-adapted HCV RNA can replicate in Huh7 cells but not in HepG2 cells, we examined the possibility that miR-122 regulates HCV RNA expression. To this end, we searched for sequences in the viral mRNA that could engage with the seed sequence (see above) of miR-122. Using this rule, we noted two potential binding sites for miR-122. One was located in the viral 3′NCR, but mutagenesis of this site showed that it was not functionally important (Jopling et al. 2005). The second miR-122-binding site was predicted to reside in the 5′NCR, only 21 nucleotides from the 5′ end of the viral genome (Table 1). Here, the putative seed match sequence was flanked by adenosine residues, suggesting a bona fide miRNA-binding site. Importantly, this putative seed match sequence for miR-122, including the flanking adenosine residues, was highly conserved among all viral genotypes (Table 1), with the exception of the seed match sequence in genotype 2 that lacks the anchoring adenosine at the +1 position (Table 1).

SEQUESTRATION OF MIR-122 REDUCES HCV RNA ABUNDANCE

To determine whether the predicted miR-122 interaction with the viral 5′NCR had a functional role in regulating HCV gene expression, we tested whether the accumulation of HCV RNAs would be affected when miR-122 was inactivated. Inactivation of miR-122 was accomplished after transfection of cells with 2′-O-methylated RNA oligonucleotides (122-2′OMe) with exact complementarity to

A

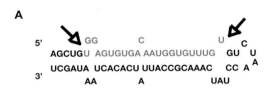

B

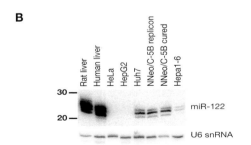

Figure 1. Tissue-specific expression of miR-122. (*A*) Predicted precursor structure of miR-122. Nucleotides highlighted in red denote the mature miR-122. The arrows indicate cleavage sites for nuclease Drosha. (*B*) Northern blot analysis of miR-122 expression in total RNA extracted from mouse and human liver, and HeLa, HepG2, Hepa 1-6, and naïve, cured and replicon Huh7 cells. Abundance of U6 snRNA was monitored as RNA loading control. An autoradiograph of the blot is shown. (Reprinted, with permission, from Jopling et al. 2005 [© AAAS].)

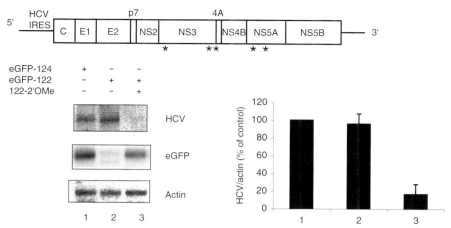

Figure 2. Sequestration of miR-122 by modified antisense RNA oligomers reduces HCV RNA abundance in cells transfected with genome-length viral RNA. (*Top*) Structure of the H77c full-length RNA, with adaptive mutations indicated by asterisks. (*Bottom*) Northern analysis of H77c RNA (HCV), eGFP, and actin RNA in Huh7 cells transfected for 5 days with the genome-length genotype 1a H77c RNA and with eGFP reporter plasmids and 122-2′OMe, as indicated. (Reprinted, with permission, from Jopling et al. 2005 [© AAAS].)

host-encoded miR-122. Such 2′OMe RNA oligonucleotides have been shown to tightly bind complementary miRNAs, leading to their functional sequestration in small RNA duplexes (Hutvagner et al. 2004; Meister et al. 2004). Functional sequestration and inactivation of miR-122 by RNA oligomers were monitored by the expression of enhanced green fluorescent protein-encoding sensor mRNAs (eGFP-122, eGFP-124) that contained sequences with perfect base complementarity to miR-122 or brain-specific miR-124 in their 3′NCRs. Due to its complete complementarity, the endogenous miR-122 should function as a small interfering RNA (siRNA) and mediate cleavage of the eGFP-122 RNA, with subsequent nucleolytic degradation. As predicted, little full-length eGFP-122 RNA was visible in cells transfected with plasmids encoding eGFP-122 (Fig. 2), although miR-124-binding-site-containing reporter mRNAs were expressed at high levels (Fig. 2). Upon transfection with 122-2′OMe, levels of eGFP-122 RNA increased, suggesting that miR-122 was sequestered.

To determine the effects of miR-122 sequestration on HCV RNA abundance, cells were transfected with full-length HCV RNAs that were synthesized by T7 RNA polymerase from a cDNA that encodes a full-length genotype 1a strain H77c (Yi and Lemon 2004). Transfection of these RNA molecules led to accumulation of viral RNA in the presence of endogenous miR-122; however, HCV RNA failed to accumulate when miR-122 was sequestered by 122-2′OMe oligomers (Fig. 2). Thus, miR-122 is required to maintain HCV RNA abundance in cultured Huh7 liver cells.

GENETIC INTERACTION BETWEEN miR-122 AND HCV

To test directly whether the putative miR-122-binding site in the viral 5′NCR was required for miR-122-mediated effects on RNA accumulation, mutations were introduced into the full-length H77c cDNA. Transfection of H77c RNAs containing substitution mutations at positions p3, p3-4, or p6 in the predicted seed match in the 5′NCR (Fig. 3A) did not lead to accumulation of detectable amounts of

viral RNA (Fig. 3B,C). However, mutation at the p1 position of the seed match sequence allowed RNA accumulation at levels similar to those of wild-type RNA (Fig. 3C). This finding is in agreement with the idea that base-pairing at the p1 position is dispensable for the formation of miRNA–mRNA complexes (Liu et al. 2003). These results argue that failure to recruit miR-122 resulted in loss of viral RNA or that the mutations had altered the overall structure of the viral RNA with subsequent effects on RNA translation, replication, or stability.

If mutations in the miR-122 seed sequence reduced RNA accumulation because of failure to bind miR-122, then ectopic expression of miR-122 duplex RNAs, containing base complementary mutations to the mutated viral genome, should restore the formation of miR-122-mutated HCV RNA complexes. Ectopic expression of wild-type miR-122 duplex RNAs did not rescue p3-, p6-, or p3-4 containing mutated viral RNAs (Fig. 3B,C), but enhanced the levels of wild-type viral RNAs (Fig. 3C), demonstrating that the introduced miR-122 RNAs were processed to functional, single-stranded miR-122 RNA molecules and that the endogenous pool of miR-122 that mediates the accumulation of viral RNA is limiting in cells. In contrast, expression of mutated miR-122 duplexes allowed accumulation of mutated viral RNAs (Fig. 3C), strongly arguing for a genetic interaction between miR-122 and the HCV genome. In addition, this result reveals that the rescue of mutated viral RNAs by mutated miR-122 RNAs must be due to a direct HCV RNA–miR-122 interaction, rather than an indirect effect via other, probably cellular, targets of miR-122.

miR-122 LIKELY REGULATES HCV RNA ABUNDANCE AT A STEP THAT OCCURS SUBSEQUENT TO VIRAL mRNA TRANSLATION

It has been assumed that miRNAs that engage in imperfect base complementarity with their target mRNAs in mammalian cells reduce the accumulation of the encoded protein either by modulating the translational

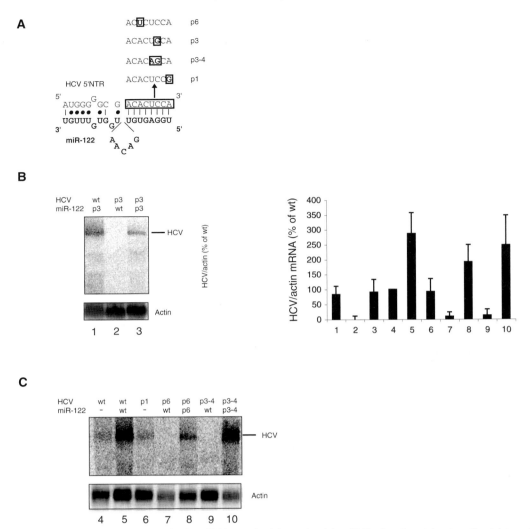

Figure 3. The predicted miR-122-binding site in HCV is required for maintaining RNA abundance due to a direct interaction with miR-122. (*A*) Positions of the mutations introduced into the H77c full-length RNA. The mutated nucleotides are enclosed in boxes. (*B,C*) Huh7 cells were transfected with synthetic duplexes corresponding to wild-type miR-122 (wt) or miR-122 with mutations in the seed complementary to the seed match mutation in the HCV genome. The duplexes were introduced into Huh7 cells 1 day prior to electroporation with wild-type H77c RNAs or mutant viral RNAs. Total RNA was harvested 5 days postelectroporation, and HCV and actin RNA levels were determined by northern blotting. Quantitation of HCV and actin RNA levels from three independent experiments and the standard deviations are shown. (Reprinted, with permission, from Jopling et al. 2005 [© AAAS].)

efficiency of the target mRNAs (Humphreys et al. 2005; Pillai et al. 2005; Petersen et al. 2006) or by degradation of mRNA (Wu et al. 2006). Thus, we examined whether miR-122 modulates translation of HCV RNA, known to occur by an unusual internal ribosome entry mechanism (Pestova et al. 1998; Ji et al. 2004; Otto and Puglisi 2004). Specifically, we monitored the accumulation of HCV core protein after transfection into Huh7 cells of in-vitro-synthesized HCV RNAs, which contained or lacked a functional miR-122-binding site. Figure 4 shows that slightly more core protein accumulated in cells transfected with wild-type (wt) than with mutant (p3) RNAs 20 hours after transfection. To test whether enhanced core production in wild-type-transfected cells reflected RNA replication, translation of replication-defective viral RNAs was examined. Results showed that wild-type and p3-mutant RNAs containing replication-lethal mutations

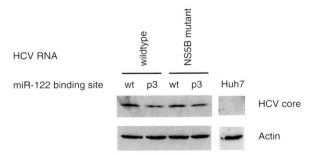

Figure 4. Mutation of the miR-122-binding site does not affect HCV mRNA translation. The p3 mutation was introduced into a replication-deficient mutant of H77c, AAG-H77, carrying amino acid changes from GDD to AAG at positions 2737 to 2739 in the viral polymerase NS5B (Yi and Lemon 2004). HCV core protein and actin expression were determined by western blotting. (Reprinted, with permission, from Jopling et al. 2005 [© AAAS].)

in the viral RNA-dependent RNA polymerase NS5B were translated with similar efficiencies (Fig. 4), suggesting that miR-122 regulates HCV RNA abundance at a step subsequent to translation, most likely at the RNA replication step. This hypothesis is supported by the fact that mutations introduced in sequences encompassing the miR-122-binding site primarily affect the replication of viral replicon RNAs (Friebe et al. 2001).

THE BINDING SITE FOR miR-122 AT THE 5′ END OF THE HCV GENOME DOWN-REGULATES GENE EXPRESSION WHEN LOCATED AT THE 3′ END OF A REPORTER mRNA

The up-regulation of HCV RNA abundance by miR-122 was surprising and unprecedented, because interactions of miRNAs with mRNAs usually lead to down-regulation of target mRNA expression. Thus, we examined whether the miR-122-binding site in the HCV genome regulates target mRNA expression when located in the 3′NCR of a reporter mRNA. Briefly, the expression of luciferase-encoding reporter mRNAs, containing HCV sequences 1–60 in their 3′NCRs, was monitored after plasmid transfection into cultured cells. In the presence of 122-2′OMe oligomers to inactivate miR-122, luciferase production was slightly up-regulated compared to random control oligomers (Fig. 5). Ectopic expression of miR-122 duplex RNAs diminished luciferase production, whereas mutant miR-122p3-4 RNA duplexes did not (Fig. 5). The finding that the miR-122-binding site in HCV 1–60 can down-regulate target gene expression when located in the 3′NCR of a reporter mRNA argues that the enhancing effect of HCV 1–60 on RNA abundance is dependent on its location in the viral genome and, likely, on the surrounding specific viral sequences.

THE LAST THREE NUCLEOTIDES IN miR-122 ARE DISPENSABLE FOR THE ENHANCEMENT OF HCV RNA ABUNDANCE

An intriguing question remains of whether any miRNA-binding site located at the 5′ end of the HCV genome enhances viral RNA abundance. So far, exchanging the binding site for the miR-122 with the binding site for the ubiquitous miR-21 yielded nonreplicating RNA molecules after transfection into Huh7 cells even when cells were supplemented with additional miR-21 (data not shown). This negative result needs to be evaluated with a caveat, because any nucleotide changes in the viral genome may affect proper folding of RNA structures or RNA sequences important for viral RNA amplification. Next, we questioned whether nucleotide sequences located 3′ of the seed sequence in miR-122 are important in up-regulating HCV RNA abundance. Because ectopic expression of wild-type miR-122 duplex RNA could further increase HCV RNA abundance (see Fig. 3C), we tested whether miR-122 duplexes containing mutations at the 3′ end of miR-122 can enhance HCV RNA abundance. Ectopic expression of miR-122 containing mutations in the 3′ six nucleotides, miR-122p18-23 (Fig. 6A), did not enhance HCV RNA abundance in transfected cells (Fig. 6B), whereas expression of miR-122 with mutations in the 3′ three nucleotides, miR-122p21-23 (Fig. 6A), enhanced HCV RNA abundance (Fig. 6B). Thus, the last three nucleotides in miR-122 are dispensable for enhancing HCV RNA abundance.

DOWN-REGULATION OF THE CATIONIC AMINO ACID TRANSPORTER CAT-1 BY miR-122

It has been suggested that miR-122 down-regulates the expression of cationic amino acid transporter CAT-1 mRNA (Chang et al. 2004). CAT-1 activity is absent in the quiescent liver, but it is increased in primary or transformed hepatocytes (Wu et al. 1994). CAT-1 mRNA contains a

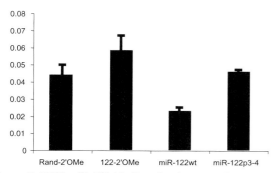

Figure 5. HCV miR-122-binding site down-regulates target gene expression when located in the 3′NCR of a reporter mRNA. The expression of luciferase plasmids in transfected cells was examined in the presence of cotransfected duplex RNAs as indicated. The activity of firefly luciferase protein normalized to Renilla luciferase activity, expressed from cotransfected plasmids, is shown.

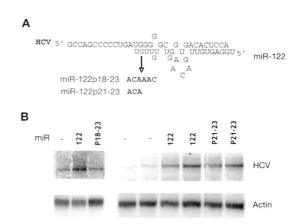

Figure 6. The last three nucleotides in miR-122 are dispensable for enhancing HCV RNA abundance. (*A*) Diagram of mutated duplex miR-122 RNAs. (*B*) HCV RNA abundance after cotransfection of H77c viral RNA and duplex mir-122 RNAs. An autoradiograph of a northern blot is shown.

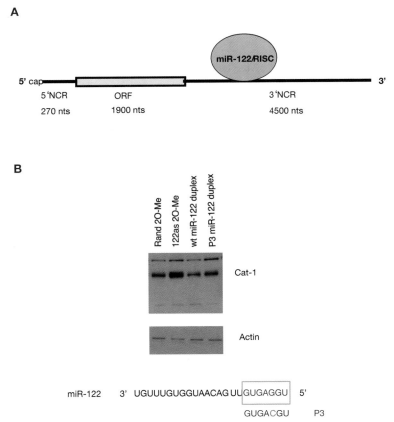

Figure 7. Down-regulation of CAT-1 protein expression by miR-122. (*A*) Diagram of the rat CAT-1 mRNA. (*B*) Effects of various antisense miR-122 oligomers and miR-122 duplexes on CAT-1 protein expression. A western blot is shown. The nucleotide sequence of wild-type and mutated p3 miR-122 is shown at the bottom of the figure.

relatively long 3'NCR (Fig. 7A) (Hatzoglou et al. 2004) that harbors several predicted binding sites for miR-122 (Chang et al. 2004), some of which are conserved across several species. We examined whether sequestration of miR-122 or ectopic expression of miR-122 duplexes affected endogenous CAT-1 protein expression in cultured rat cells. Figure 7B shows that sequestration of miR-122 by antisense-2'Ome oligomers enhanced CAT-1 expression, whereas expression of wild-type, but not mutated p3 duplex miR-122 RNAs, diminished CAT-1 protein abundance. Thus, miR-122 can down-regulate the expression of CAT-1 mRNA in cultured cells.

CONCLUSIONS

The finding that liver-specific miR-122 targets the 5'NCR of the HCV RNA genome resulting in up-regulation of intracellular RNA was surprising, because miRNAs have been found to generally bind to 3'NCRs in target mRNAs, leading to RNA turnover and repression of mRNA translation. Clearly, there does not seem to be anything unusual with the miR-122-binding site in HCV, because it down-regulates target gene expression when it resides in the 3'NCR of a reporter mRNA. The mechanism by which miR-122 up-regulates HCV RNA is unknown at present, but it likely involves a step in RNA replication. It is also possible that miR-122 affects the

localization of the viral RNA and targets it to the membraneous web structures where RNA amplification takes place (Moradpour et al. 2003, 2004). Finally, it is worth pointing out that pegylated interferon α and ribavirin therapy against HCV is frequently ineffective, particularly in patients infected with genotype 1 (Feld and Hoofnagle 2005); thus, there is a need to search for alternative antiviral targets. Sequestration of miR-122 could provide a possible antiviral tool against a rapidly evolving viral genome.

Finally, the high abundance of miR-122 in the liver raised the question of whether miR-122 targets host-cell mRNAs in this organ and what the functions of the identified targets are. Two recent studies, in which the levels of host mRNAs in the mouse liver were examined after sequestration of miR-122 revealed that inactivation of miR-122 leads to both up-regulation and down-regulation of hundreds of mRNAs, some of which contained predicted target sites in their 3'NCRs (Krützfeldt et al. 2005; Esau et al. 2006). Overall, these analyses revealed that miR-122 is involved in controlling genes encoding intermediates of the cholesterol pathway (Krützfeldt et al. 2005; Esau et al. 2006). Here, we presented data that the mRNA encoding the CAT-1 protein can be down-regulated by miR-122 in cultured rat liver cells. This finding supports the notion that CAT-1 protein expression is down-regulated in normal liver cells where miR-122 levels are high

(Hatzoglou et al. 2004). Thus, miR-122 can both up-regulate and down-regulate target mRNAs. Whether these outcomes are carried out by distinct mechanisms is an exciting area of investigation.

ACKNOWLEDGMENTS

Work performed in the authors' laboratories were supported by grants from the Wellcome Trust (C.L.J.), the Alberta Heritage Foundation for Medical Research (K.L.N.), and the National Institutes of Health (P.S.).

REFERENCES

Arasu P., Wightman B., and Ruvkun G. 1991. Temporal regulation of *lin-14* by the antagonistic action of two other heterochronic genes, *lin-4* and *lin-28*. *Genes Dev.* **5:** 1825.

Bagga S., Bracht J., Hunter S., Massirer K., Holtz J., Eachus R., and Pasquinelli A.E. 2005. Regulation by *let-7* and *lin-4* miRNAs results in target mRNA degradation. *Cell* **122:** 553.

Bartel D.P. 2004. MicroRNAs: Genomics, biogenesis, mechanism, and function. *Cell* **116:** 281.

Bartenschlager R. and Lohmann V. 2000. Replication of hepatitis C virus. *J. Gen. Virol.* **81:** 1631.

Cai X., Hagedorn C.H., and Cullen B.R. 2004. Human microRNAs are processed from capped, polyadenylated transcripts that can also function as mRNAs. *RNA* **10:** 1957.

Chalfie M., Horvitz H.R., and Sulston J.E. 1981. Mutations that lead to reiterations in the cell lineages of *C. elegans*. *Cell* **24:** 59.

Chang J.E.N., Marks D., Sander C., Lerro A., Buendia M.A., Xu C., Mason W.S., Moloshok T., Bort R., Zaret K.S., and Taylor J.E. 2004. miR-122, a mammalian liver-specific microRNA, is processed from *hcr* mRNA and may downregulate the high affinity cationic amino acid transporter CAT-1. *RNA Biol.* **1:** 106.

Esau C., Davis S., Murray S.F., Yu X.X., Pandey S.K., Pear M., Watts L., Booten S.L., Graham M., McKay R., et al. 2006. miR-122 regulation of lipid metabolism revealed by in vivo antisense targeting. *Cell Metab.* **3:** 87.

Feld J.J. and Hoofnagle J.H. 2005. Mechanism of action of interferon and ribavirin in treatment of hepatitis C. *Nature* **436:** 967.

Friebe P., Lohmann V., Krieger N., and Bartenschlager R. 2001. Sequences in the 5′ nontranslated region of hepatitis C virus required for RNA replication. *J. Virol.* **75:** 12047.

Hammond S.M., Bernstein E., Beach D., and Hannon G.J. 2000. An RNA-directed nuclease mediates post-transcriptional gene silencing in *Drosophila* cells. *Nature* **404:** 293.

Hatzoglou M., Fernandez J., Yaman I., and Closs E. 2004. Regulation of cationic amino acid transport: The story of the CAT-1 transporter. *Annu. Rev. Nutr.* **24:** 377.

Horvitz H.R. and Sulston J.E. 1980. Isolation and genetic characterization of cell-lineage mutants of the nematode *Caenorhabditis elegans*. *Genetics* **96:** 435.

Humphreys D.T., Westman B.J., Martin D.I., and Preiss T. 2005. MicroRNAs control translation initiation by inhibiting eukaryotic initiation factor 4E/cap and poly(A) tail function. *Proc. Natl. Acad. Sci.* **102:** 16961.

Hutvagner G., Simard M.J., Mello C.C., and Zamore P.D. 2004. Sequence-specific inhibition of small RNA function. *PLoS Biol.* **2:** E98.

Ji H., Fraser C.S., Yu Y., Leary J., and Doudna J.A. 2004. Coordinated assembly of human translation initiation complexes by the hepatitis C virus internal ribosome entry site RNA. *Proc. Natl. Acad. Sci.* **101:** 16990.

Jopling C.L., Yi M., Lancaster A.M., Lemon S.M., and Sarnow P. 2005. Modulation of hepatitis C virus RNA abundance by a liver-specific MicroRNA. *Science* **309:** 1577.

Khvorova A., Reynolds A., and Jayasena S.D. 2003. Functional siRNAs and miRNAs exhibit strand bias. *Cell* **115:** 209.

Kim V.N. 2005. MicroRNA biogenesis: Coordinated cropping and dicing. *Nat. Rev. Mol. Cell Biol.* **6:** 376.

Krützfeldt J., Rajewsky N., Braich R., Rajeev K.G., Tuschl T., Manoharan M., and Stoffel M. 2005. Silencing of microRNAs in vivo with "antagomirs." *Nature* **438:** 685.

Lagos-Quintana M., Rauhut R., Lendeckel W., and Tuschl T. 2001. Identification of novel genes coding for small expressed RNAs. *Science* **294:** 853.

Lagos-Quintana M., Rauhut R., Yalcin A., Meyer J., Lendeckel W., and Tuschl T. 2002. Identification of tissue-specific MicroRNAs from mouse. *Curr. Biol.* **12:** 735.

Lau N.C., Lim L.P., Weinstein E.G., and Bartel D.P. 2001. An abundant class of tiny RNAs with probable regulatory roles in *Caenorhabditis elegans*. *Science* **294:** 858.

Lecellier C.H., Dunoyer P., Arar K., Lehmann-Che J., Eyquem S., Himber C., Saib A., and Voinnet O. 2005. A cellular microRNA mediates antiviral defense in human cells. *Science* **308:** 557.

Lee Y., Jeon K., Lee J.T., Kim S., and Kim V.N. 2002. MicroRNA maturation: Stepwise processing and subcellular localization. *EMBO J.* **21:** 4663.

Lee Y., Kim M., Han J., Yeom K.H., Lee S., Baek S.H., and Kim V.N. 2004. MicroRNA genes are transcribed by RNA polymerase II. *EMBO J.* **23:** 4051.

Lewis B.P., Burge C.B., and Bartel D.P. 2005. Conserved seed pairing, often flanked by adenosines, indicates that thousands of human genes are microRNA targets. *Cell* **120:** 15.

Lewis B.P., Shih I.H., Jones-Rhoades M.W., Bartel D.P., and Burge C.B. 2003. Prediction of mammalian microRNA targets. *Cell* **115:** 787.

Li F. and Ding S.W. 2006. Virus counterdefense: Diverse strategies for evading the RNA-silencing immunity. *Annu. Rev. Microbiol.* **60:** 503.

Lindenbach B.D. and Rice C.M. 2003. Molecular biology of flaviviruses. *Adv. Virus Res.* **59:** 23.

Liu Q., Rand T.A., Kalidas S., Du F., Kim H.E., Smith D.P., and Wang X. 2003. R2D2, a bridge between the initiation and effector steps of the *Drosophila* RNAi pathway. *Science* **301:** 1921.

Lohmann V., Korner F., Koch J., Herian U., Theilmann L., and Bartenschlager R. 1999. Replication of subgenomic hepatitis C virus RNAs in a hepatoma cell line. *Science* **285:** 110.

Maniataki E. and Mourelatos Z. 2005. A human, ATP-independent, RISC assembly machine fueled by pre-miRNA. *Genes Dev.* **19:** 2979–2990.

Meister G., Landthaler M., Dorsett Y., and Tuschl T. 2004. Sequence-specific inhibition of microRNA- and siRNA-induced RNA silencing. *RNA* **10:** 544.

Moradpour D., Gosert R., Egger D., Penin F., Blum H.E., and Bienz K. 2003. Membrane association of hepatitis C virus nonstructural proteins and identification of the membrane alteration that harbors the viral replication complex. *Antivir. Res.* **60:** 103.

Moradpour D., Brass V., Bieck E., Friebe P., Gosert R., Blum H.E., Bartenschlager R., Penin F., and Lohmann V. 2004. Membrane association of the RNA-dependent RNA polymerase is essential for hepatitis C virus RNA replication. *J. Virol.* **78:** 13278.

Olsen P.H. and Ambros V. 1999. The *lin-4* regulatory RNA controls developmental timing in *Caenorhabditis elegans* by blocking LIN-14 protein synthesis after the initiation of translation. *Dev. Biol.* **216:** 671.

Otto G.A. and Puglisi J.D. 2004. The pathway of HCV IRES-mediated translation initiation. *Cell* **119:** 369.

Pestova T.V., Shatsky I.N., Fletcher S.P., Jackson R.J., and Hellen C.U.T. 1998. A prokaryotic-like mode of cytoplasmic eukaryotic ribosome binding to the initiation codon during internal translation initiation of hepatitis C and classical swine fever virus RNAs. *Genes Dev.* **12:** 67.

Petersen C.P., Bordeleau M.E., Pelletier J., and Sharp P.A. 2006. Short RNAs repress translation after initiation in mammalian cells. *Mol. Cell* **21:** 533.

Pillai R.S., Bhattacharyya S.N., Artus C.G., Zoller T., Cougot N., Basyuk E., Bertrand E., and Filipowicz W. 2005.

Inhibition of translational initiation by Let-7 microRNA in human cells. *Science* **309:** 1573.

Wang X.H., Aliyari R., Li W.X., Li H.W., Kim K., Carthew R., Atkinson P., and Ding S.W. 2006. RNA interference directs innate immunity against viruses in adult *Drosophila*. *Science* **312:** 452.

Wightman B., Burglin T.R., Gatto J., Arasu P., and Ruvkun G. 1991. Negative regulatory sequences in the lin-14 3′-untranslated region are necessary to generate a temporal switch during *Caenorhabditis elegans* development. *Genes Dev.* **5:** 1813.

Wu J.Y., Robinson D., Kung H.J., and Hatzoglou M. 1994. Hormonal regulation of the gene for the type C ecotropic retrovirus receptor in rat liver cells. *J. Virol.* **68:** 1615.

Wu L., Fan J., and Belasco J.G. 2006. MicroRNAs direct rapid deadenylation of mRNA. *Proc. Natl. Acad. Sci.* **103:** 4034.

Yekta S., Shih I.H., and Bartel D.P. 2004. MicroRNA-directed cleavage of HOXB8 mRNA. *Science* **304:** 594.

Yi M. and Lemon S.M. 2004. Adaptive mutations producing efficient replication of genotype 1a hepatitis C virus RNA in normal Huh7 cells. *J. Virol.* **78:** 7904.

Zeng Y., Wagner E.J., and Cullen B.R. 2002. Both natural and designed micro RNAs can inhibit the expression of cognate mRNAs when expressed in human cells. *Mol. Cell* **9:** 1327.

The Challenge of Viral snRNPs

N.K. Conrad, V. Fok, D. Cazalla, S. Borah, and J.A. Steitz

Department of Molecular Biophysics and Biochemistry, Howard Hughes Medical Institute,
Yale University School of Medicine, New Haven, Connecticut 06536-0812

Some gammaherpesviruses encode nuclear noncoding RNAs (ncRNAs) that assemble with host proteins. Their conservation and abundance implies that they serve important functions for the virus. This paper focuses on our studies of three classes of nuclear noncoding herpesvirus RNAs. (1) EBERs 1 and 2 are expressed by Epstein-Barr virus in latent infection of human B lymphocytes. Recent studies revealed three sites on EBER1 that associate with ribosomal protein L22. In addition, heterokaryon assays have definitively shown that both EBERs are confined to the nucleus, arguing that their contribution to viral latency is purely nuclear. (2) HSURs 1–7 are U RNAs encoded by *Herpesvirus saimiri*, which causes aggressive T-cell leukemias and lymphomas. Comparison of monkey T cells transformed with wild-type or mutant virus lacking HSURs 1 and 2 revealed significant changes in host mRNAs implicated in T-cell signaling. (3) PAN is a 1-kb polyadenylated RNA that accumulates in the nucleus of Kaposi's sarcoma-associated herpesvirus lytically infected cells. A novel element, the ENE, is essential for its high accumulation. Recent results indicate that the ENE functions to counteract poly(A)-dependent RNA degradation, which we propose contributes to nuclear surveillance of mRNA transcripts in mammalian cells. Continuing studies of these viral RNAs will provide insights into both cellular and viral gene expression.

Like cellular genomes, some—but not all—viral genomes encode small regulatory RNAs. In addition to recently discovered microRNAs (Cullen 2006), certain mammalian viruses express abundant ncRNAs of approximately 100 to 1000 nucleotides in infected cells. These viral RNAs avidly bind host proteins to become ribonucleoproteins (RNPs). Their cellular abundance and conservation among related viruses suggest important functions. Have they evolved to make the virus a better pathogen, do they serve to help the virus evade some aspect of the host's response to infection, or both?

The small viral RNAs that have been the focus of our research are all nuclear, thereby forming small nuclear RNPs (snRNPs) (Table 1). They are encoded by gammaherpesviruses that infect primates. Gammaherpesviruses are enveloped viruses with large double-stranded DNA genomes. They are lymphotropic, infecting either T or B cells, with their life cycles characterized by both lytic and latent phases. Despite the fact that the viral ncRNAs we are studying were discovered 10–25 years ago, we still lack a molecular understanding of their roles in infected cells. Recently, however, some progress on these challenging problems has been made and is reviewed here.

EBERS

Epstein-Barr virus (EBV) infection of human B lymphocytes often leads to latency (Kieff and Liebowitz 1990), a state in which most viral genes are not expressed. Among the exceptions are those for two abundant (10^6/cell) small RNAs (~170 nucleotides), EBER1 and 2 (for Epstein Barr-encoded RNA) (Table 1 and Fig. 1A). The EBERs are localized in the nucleoplasm and quantitatively bind the host La protein via the 3′ oligo(U) tracts that terminate their transcription by RNA polymerase III (Lerner et al. 1981).

EBER1 also associates with another host protein, large subunit ribosomal protein L22. This interaction results in a relocalization of the L22 protein from the nucleolus to the nucleoplasm in EBV-positive cells (Toczyski et al. 1994). Previous in vitro binding studies suggested that EBER1 stem-loops III and IV serve as binding sites for L22 (Toczyski and Steitz 1993; Dobbelstein and Shenk 1995). Recently, we carried out experiments using recombinant L22 protein in electrophoretic mobility-shift assays and found that the EBER1 molecule harbors three L22-binding sites (Fok et al. 2006b). These are the above-

Table 1. Gammaherpesviruses Small Nuclear RNAs

RNA	Virus	Abundance (copies/cell)	Length (nt)	5′ Terminus	RNA polymerase	Bound proteins
EBER1	EBV	5×10^6	167	pppA	III	La, rpL22, (PKR)[a]
EBER2	EBV	5×10^6	172	pppA	III	La
HSUR 1,2,5	HVS	10^3–10^4	114–143	$2,2,7m^3$GpppA[b]	II	Sm (HuR, hnRNP D)[d]
HSUR 3,4,6,7	HVS	10^3–10^4	75–106	$2,2,7m^3$GpppN[b]	II	Sm
PAN	KSHV	2.5×10^5	1077[c]	^{7}mGpppA[b]	II	?

Modified from Tycowski et al. (2006).
[a]In vivo binding has not been demonstrated.
[b]Caps have 5′–5′ linkages.
[c]This value excludes the poly(A) tail.
[d]Shown in vivo only for HSUR 1 (Cook et al. 2004).

A

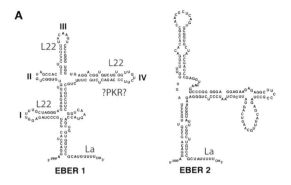

B

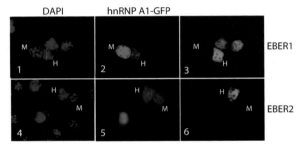

Figure 1. EBER1 and EBER2 are restricted to the nucleus. (*A*) Secondary structures of EBER1 and EBER2. The stem-loops of EBER1 are denoted with roman numerals and EBER-interacting proteins are positioned at their binding sites (Fok et al. 2006b). PKR binding at stem-loop IV has been proposed by Vuyisich et al. (2002) and McKenna et al. (2006). (*B*) Lack of EBER nucleocytoplasmic shuttling. Heterokaryons were prepared by fusing HKB5cl8 cells (Cho et al. 2002) transfected with a plasmid producing hnRNP A1-GFP and mouse NIH-3T3 cells for 6–7 hours in the presence of cycloheximide (procedures in Fok et al. 2006a). Human (*H*) and mouse (*M*) nuclei of the heterokaryons are labeled. Heterokaryons were identified by the shuttling of hnRNP A1-GFP (*green, panels 2 and 5*) into mouse nuclei, the latter distinguished by punctate DAPI staining (*panels 1 and 4*). Endogenous EBER1 (*panel 3*) and EBER2 (*panel 6*) were detected using fluorescent in situ hybridization (*yellow*).

mentioned stem-loops III and IV, and a previously undescribed site located in stem-loop I (see Fig. 1A). Interestingly, these EBER1 stem-loops do not share significant sequence similarity; hence, constructing a consensus L22-binding site awaits insights from future structural studies. The existence of multiple L22-binding sites on EBER1 inside cells has been confirmed by in vivo UV cross-linking data (Fok et al. 2006b). It will be interesting to determine how many of these regions are required to trigger L22 protein redistribution within the nucleus.

Although both EBERs are abundantly expressed in EBV-infected human B cells, they can be deleted from the viral genome with no apparent deleterious effect on the life cycle of EBV or its transformation potency (Swaminathan et al. 1991). On the other hand, the fact that no EBER-negative EBV strain has yet been isolated suggests that EBERs provide an important function(s) for the virus. Indeed, the expression of EBERs alone in EBV-negative cells promotes tumor formation in SCID mice (Komano et al. 1999; Ruf et al. 2000; Yamamoto et al. 2000). EBERs are also sufficient to induce the expression of interleukin (IL)-9 (Yang et al. 2004), IL-10 (Kitagawa et al. 2000), or insulin-like growth factor I (Iwakiri et al. 2003) in EBV-negative cell lines, leading to enhanced cell growth.

The EBERs were initially proposed to function like adenovirus VA1 RNA by inactivating the dsRNA-dependent protein kinase, PKR (Mathews and Shenk 1991), because they partially support lytic growth of a mutant adenovirus deleted for VA1 (Bhat and Thimmappaya 1983) and bind PKR in vitro (Sharp et al. 1993; Vuyisich et al. 2002; McKenna et al. 2006). However, since EBERs appear to be nuclear (Howe and Steitz 1986; Barletta et al. 1993) whereas PKR is largely cytoplasmic (Takizawa et al. 2000), it seemed unlikely that EBERs could combat this host antiviral defense by inhibiting PKR if they truly reside in a different cellular compartment.

We addressed the perplexing issue of EBER localization and its relationship to function by analyzing the potential trafficking of EBERs in EBV-transformed cells in vivo. We performed heterokaryon assays to ask whether EBERs transiently move from the nuclear to the cytoplasmic compartment (Fig. 1B). Both in this and in oocyte microinjection assays, EBER1 and 2 were confined to the nucleus, arguing that their contribution to viral latency is purely nuclear (Fok et al. 2006a). In contrast, U1 snRNA molecules in the same heterokaryons shuttled as expected from the nucleus to the cytoplasm and back (Fok et al. 2006a).

Recent results from other labs indicate that EBERs do not inhibit PKR activity in vivo when cells are challenged with various PKR stimuli (Ruf et al. 2005; Wang et al. 2005). Thus, it now seems unlikely that the physiological function of EBERs is to inhibit PKR, although their molecular activities remain unidentified. Likewise, whether their roles can be attributed to an active function of the EBER particles or to the sequestering of La, ribosomal protein L22, or some other protein partner is unknown. Perhaps a function of the EBERs is to regulate events downstream of the PKR pathway that are important for EBV growth, which may explain their ability to substitute for VA1 function (Mohr and Gluzman 1996; He et al. 1997). Our current efforts are directed at identifying host proteins in addition to La that bind EBER2 and examining changes in B-cell gene expression when EBERs are stably expressed at high levels.

HSURs

Herpesvirus saimiri (HVS) infects new-world primates and produces either lytic infection or malignant transformation of T lymphocytes. In transformed T cells, the most abundant viral transcripts are seven small RNAs (75–143 nucleotides) encoded by a cluster of genes within the viral genome (Lee et al. 1988; Albrecht and Fleckenstein 1992). HSURs (for Herpesvirus saimiri U RNAs) are synthesized by RNA polymerase II using the same distinctive promoter and 3′-end signals as cellular U RNAs. Like most

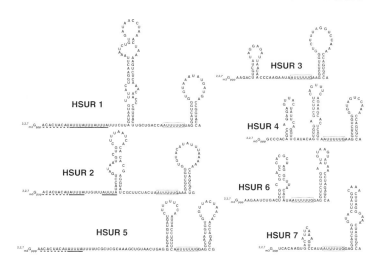

Figure 2. HSUR sequences and predicted secondary structures. HSURs and cellular Sm-class snRNAs share several characteristics; 5′-trimethylguanosine caps, 3′-terminal stem-loops, and canonical Sm protein-binding site sequences (*boxed*). Underlined are perfectly conserved sequences (*dashed lines*) and AREs (*solid lines*) present in HSURs 1, 2, and 5.

cellular U RNAs that form splicing snRNPs, HSURs traffic to the cytoplasm, where, with the aid of the SMN complex (Golembe et al. 2005), they assemble into RNP particles of the Sm class. Following their return to the nucleus, HSURs colocalize with cellular U snRNAs (Golembe et al. 2005). Interestingly, only HSURs 1 and 2 are conserved among the HVS subgroups and in their close relative *Herpesvirus ateles* (Albrecht 2000), suggesting that these two HSURs are most important to the virus.

Highly conserved sequences at the 5′ ends of HSURs 1, 2, and 5 (Fig. 2) mimic AU-rich elements (AREs) found in the 3′-untranslated regions of short-lived messages, such as those encoding cellular oncoproteins, cytokines, and lymphokines (Shaw and Kamen 1986). Thus, for many years we entertained the hypothesis that these three HSURs compete for host ARE-binding proteins, thereby enhancing the viral transformed state by modulating the levels of these important host mRNAs (Myer et al. 1992). In fact, HSUR 1 does associate in vivo with HuR and hnRNP D (Cook et al. 2004), proteins known to bind AU-rich destabilization signals and regulate mRNA stability (Fan and Steitz 1998; Loflin et al. 1999a). Moreover, the ARE has been shown to be responsible for the short half-life of HSUR 1 (Fan and Steitz 1998; Cook et al. 2004).

We subjected this model to rigorous testing by microarray and northern analyses, comparing mRNA levels in marmoset T cells transformed with wild-type versus a mutant HVS lacking HSURs 1 and 2 (Murthy et al. 1989). However, no significant differences in host ARE-containing mRNA levels were observed, refuting the hypothesis that HSURs regulate this class of host messages. Instead, HSUR 1 and 2 expression correlated with significant increases in another small set of host mRNAs, including those encoding the T-cell receptor (TCR) β and γ chains, the T-cell and natural killer (NK) cell-surface receptors, CD52 and DAP10, and intracellular proteins linked to T-cell and NK cell activation, such as SKAP55, granulysin, and NKG7. Strikingly, these host proteins are all involved in signaling pathways regulated by T-cell receptors. Genetic rescue experiments assigned this novel coordinate regulation directly to the HSURs: Up-regulation of expression of several of these genes in deletion mutant HVS-transformed cells was restored by transduction with a lentiviral vector carrying HSURs 1 and 2 (Fig. 3) (Cook et al. 2005). These are the first phenotypic changes attributable to the HSURs.

Subsequently, we have begun to probe the molecular mechanisms underlying this unexpected role for Sm snRNPs in regulating a remarkably defined and physiologically relevant set of downstream targets involved in the activation of virally transformed T cells during HVS latency. The first question we asked was whether the HSURs might be involved in transcriptional regulation of these host genes, as was recently demonstrated for the 7SK snRNP (Goodrich and Kugel 2006 and references therein). Preliminary results suggest that this is not the case and that the role of the HSUR snRNPs is instead posttranscriptional. Currently, we are testing the stability of the up-regulated host mRNAs in the presence and absence of HSURs 1 and 2. In addition, the cDNA sequence for each affected gene is being determined and compared to the marmoset genomic sequence—even though no discrepancies in northern blot patterns were noted, either small differences in splice site utilization or RNA editing of these host mRNAs could conceivably be regulated by the HSURs.

PAN RNA

The Kaposi's sarcoma-associated herpesvirus (KSHV) causes Kaposi's sarcoma in AIDS patients and other immunocompromised individuals and is also associated with two lymphoproliferative disorders (Viejo-Borbolla et al. 2004). KSHV encodes an abundant approximately 1-kb nuclear ncRNA, the polyadenylated nuclear (PAN) RNA (also known as *nut1* and T1.1; Sun et al. 1996; Zhong et al. 1996). Unlike the nuclear ncRNAs of EBV and HVS, which are present in latently infected cells, PAN RNA is expressed only in the viral lytic phase. Interestingly, PAN is not conserved: Even the closest KSHV relative, rhesus rhadinovirus, encodes no such RNA. PAN transcription by RNA polymerase II is controlled by the viral transactivator ORF50, whose activity is essential in driving the transition from viral latency to lytic phase (Song et al. 2001;

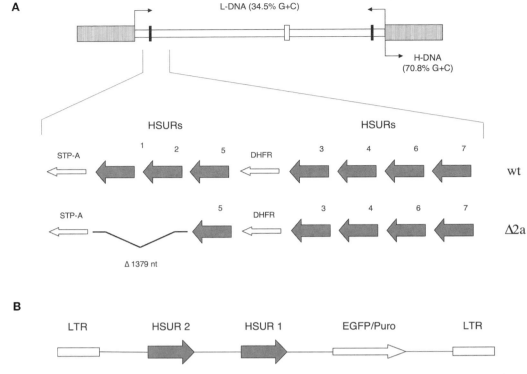

Figure 3. Genomic organization of the HSURs. (*A*) Scheme depicting the genomes of wild-type (*wt*) HVS A-11 and mutant (Δ*2a*) HVS A-11 used for in vitro transformation of peripheral blood lymphocytes from the common marmoset. The deletion of a 1379-bp region at the leftmost end of the HVS A-11 genome specifically removes the genes that code for HSURs 1 and 2 (Murthy et al. 1989). The genes for STP-A (*saimiri* transformation-associated protein of strain A) and DHFR (dihydrofolate reductase), as well as the genes encoding the other five HSURs, remain intact in the viral genome. (*B*) Diagram showing the lentiviral vector used to restore the expression of HSURs 1 and 2 in the T-cell line transformed with the mutant HVS A-11 (Cook et al. 2005). The vector (Li et al. 2003) contains the gene coding for enhanced green fluorescent protein (eGFP) under the control of the CMV promoter, which allows the sorting of GFP-expressing cells after transduction.

Chang et al. 2002). At approxiumately 2×10^5 to 3×10^5 copies per cell, PAN RNA is by far the most abundant lytic phase transcript, comprising as much as 80% of the polyadenylated RNA in lytically reactivated cell lines (Sun et al. 1996; Song et al. 2001).

The remarkable nuclear abundance of PAN RNA is achieved by high transcription rates, as well as through the action of a 79-nucleotide RNA element called the ENE, which acts posttranscriptionally and is located near the 3′ end of the transcript (Conrad and Steitz 2005). The ENE not only functions in the context of PAN RNA, but can significantly increase the nuclear abundance of an otherwise inefficiently expressed intronless β-globin transcript (Collis et al. 1990). The nuclear localization of intronless β-globin RNA containing the ENE can be overcome by tethering any of several export factors to the RNA or by inserting an intron (Conrad and Steitz 2005), suggesting that the ENE is either a weak nuclear retention element or that it stabilizes nuclear transcripts that would otherwise be degraded.

Importantly, the ability of the ENE to enhance the nuclear levels of PAN RNA is dependent on the canonical 3′-end formation machinery, arguing that the ENE either inhibits a poly(A)-dependent decay pathway or enhances 3′-end formation of intronless transcripts (Conrad and Steitz 2005). To examine the effects of the ENE on 3′cleavage in vivo, we compared the activity of the ENE to that of the process-

ing enhancer of the mouse histone H2A gene. This 101-nucleotide-long RNA element increases the abundance of intronless transcripts by enhancing 3′-end formation both in vivo and in vitro, in addition to playing a role in mRNA export (Huang and Carmichael 1997; Huang et al. 1999).

We generated several constructs to compare the activities of the ENE and the H2A element. One has the entire PAN transcribed region (Fig. 4A; WT), one deletes the 79-nucleotide ENE (and 36 nucleotides of downstream sequence) replacing it with a BglII restriction site (Δ115Bgl), and two have the 101-nucleotide H2A element inserted in the forward (H2A-F) or reverse (H2A-R) orientation into the BglII site. We then performed RNase protection assays (RPA) to examine the relative in vivo levels of properly cleaved transcripts and of uncleaved readthrough (RT) transcripts (Fig. 4) (Conrad and Steitz 2005). A representative RPA (Fig. 4B) and quantitation of the results of four independent experiments (Fig. 4C) show that the inclusion of either the ENE or the H2A element increases the abundance of cleaved transcripts approximately 5- and 8-fold, respectively (Fig. 4B, compare "cleaved" lanes 2–5; Fig. 4C, left panel). The H2A element concomitantly decreases the levels of the RT transcripts approximately 9-fold, whereas the levels of the RT transcripts are unaffected by the ENE or the H2A reverse orientation control (Fig. 4B, compare PAN-RT lanes 2–5;

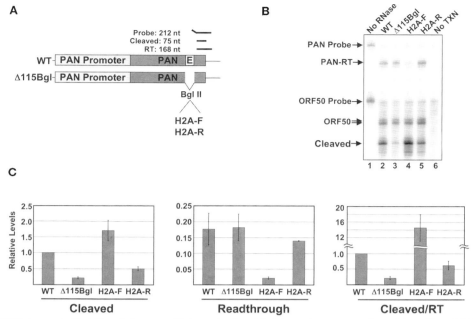

Figure 4. The ENE does not affect 3′-end cleavage efficiency in vivo. (*A*) Schematic representation of plasmids and PAN probe used for RPA. Boxed regions are the PAN promoter (*light gray*) and transcribed region (*dark gray*), whereas the lines represent vector sequence. Boxed letter E indicates the 79-nucleotide ENE in the WT construct. The ENE was replaced with a BglII site in Δ115Bgl, represented by the "V"-shaped line. The 101-nucleotide H2A element (Huang and Carmichael 1997) was inserted into the BglII site in the forward (*H2A-F*) or reverse (*H2A-R*) orientation. (*B*) Representative RPA result. Arrows demark the positions of the undigested PAN or ORF50 probes, the protected ORF50 product (a doublet), as well as the PAN-RT (readthrough) and cleaved products. (*C*) Quantitation of four independent RPA experiments with standard deviations. The left and middle panels show the levels of cleaved and RT products, respectively, normalized to the amount of WT cleaved product. Note the difference in scale. The right panel shows the cleaved-to-RT ratios for the other three constructs relative to that for WT. For clarity, the y-axis is presented as a broken scale.

Fig. 4C, middle panel). Thus, although both the ENE and the H2A element increase the cleaved-to-RT ratios (Fig. 4C, right panel), they do so by different means. These data are consistent with previous results concluding that the H2A element enhances 3′-end cleavage (Huang and Carmichael 1997; Huang et al. 1999) and that, in contrast, the ENE increases the stability of the cleaved transcript.

To further investigate a role for the ENE in stabilizing polyadenylated transcripts, we examined the effect of the ENE on transcript decay in vivo and in vitro (Conrad et al. 2006). Indeed, in an in vitro decay system (Ford et al. 1999), the ENE inhibits RNA decay by inhibiting deadenylation, the first step in RNA decay. In vivo, PAN RNA produced through a transcriptional pulse (Loflin et al. 1999b) exhibits two-component exponential decay kinetics, with some transcripts being rapidly degraded ($t_{1/2}$ ~15 min), while others decay more slowly ($t_{1/2}$ ~4 hr). Presence of the ENE shifts more PAN RNA or intronless β-globin transcripts into the slowly decaying population. Action of the ENE in vivo, as in vitro, is dependent on the poly(A) tail because the decay of non-polyadenylated transcripts (with their 3′ ends formed by U7 snRNP-dependent cleavage, which—like those of histone messages—do not undergo polyadenylation) are not affected by the presence of the ENE. These data argue that the PAN ENE inhibits rapid deadenylation-dependent RNA decay in vivo.

How does the ENE protect transcripts from deadenylation? The predicted secondary structure provides a possible answer to this question. The ENE folds into a predominantly double-stranded RNA with U-rich segments forming a central bulge (11/12 nucleotides are uridines; Fig. 5A). The structure is reminiscent of box H/ACA snoRNAs, which guide site-specific rRNA modification through hybridization with similarly bulged nucleotides in their pseudouridylation pockets (Kiss 2002). We propose that the U-rich regions of the ENE hybridize to the poly(A) tail of PAN *in cis* to protect it from the exonucleolytic attack (Fig. 5B) that initiates the decay process. In fact, we have demonstrated intramolecular interactions between the ENE and poly(A) tail both for naked transcripts and for protein-bound RNAs generated in transfected cells (Conrad et al. 2006). Most strikingly, we find that the interaction of the poly(A) tail with the ENE (Fig. 5B) is sufficient to interfere with the activity of a purified deadenylase on naked RNA in vitro. Surprisingly, this mechanism may not be unique to the ENE. The 3′UTR of the yeast *EDC1* transcript contains a U-rich sequence (36/38 nucleotides are uridines) that allows the mRNA to evade the normal cytoplasmic deadenylation machinery (Muhlrad and Parker 2005). Thus, we predict that some mammalian transcripts will likely use a mechanism similar to that of the ENE for regulation of mRNA stability. Detailed structure/function analyses are under way to define the minimal requirements for ENE activity with the hope of using bioinformatics to predict candidate ENE-like cellular sequences.

Currently, we are solidifying preliminary data that the rapid decay inhibited by the ENE comprises a surveillance

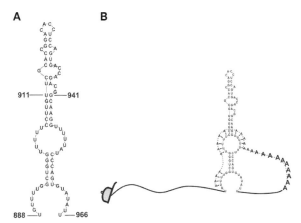

Figure 5. Predicted ENE structure and model for ENE action. (*A*) Predicted structure of the 79-nucleotide ENE. A second structure of similar free energy is also predicted, but changes are restricted to the upper part of the stem (nucleotides 911–941), while the lower stem and bulged U-rich sequences are maintained as shown. Numbers refer to the nucleotide position relative to the transcription start site as defined in Song et al. (2001). (*B*) Possible interactions between the U-rich single-stranded regions of the ENE and the poly(A) tail. G-C base pairs in the adjacent stems might engage in additional A-minor interactions (Nissen et al. 2001). Details are described in the text. The number of As depicted in the poly(A) tail is for illustration only and is not a proposal for its true length.

system for mRNA transcripts in the mammalian cell nucleus. Specifically, when transcripts are retained in the nucleus by the VSV M protein (Faria et al. 2005), their decay—whether or not they contain introns—is counteracted by the ENE. Mutation of an intron's 5′- or 3′-splice site leads to transcript decay, which likewise is overcome by the ENE (Conrad et al. 2006). The ENE thus provides a potent probe for elucidating the mechanism of nuclear RNA surveillance in mammalian cells (Moore 2002), whereas other recent advances are in yeast (Bousquet-Antonelli et al. 2000; Hilleren et al. 2001; Libri et al. 2002; Das et al. 2003; Jensen et al. 2004; Kadaba et al. 2004; Fasken and Corbett 2005; Kuai et al. 2005; LaCava et al. 2005; Milligan et al. 2005; Vanacova et al. 2005; Wyers et al. 2005).

A related, but nonoverlapping, challenge is to elucidate how PAN RNA contributes to the lytic reactivation of KSHV. We are tackling this problem using three approaches. First, we are attempting to construct a PAN knockout virus (Zhou et al. 2002). Although successful knockouts have been made for several KSHV genes (Zhou et al. 2002; Luna et al. 2004; Ye et al. 2004; Xu et al. 2005; Han and Swaminathan 2006), deleting PAN is particularly difficult because its gene partially overlaps those of two other viral transcripts. One of these overlapping transcripts encodes the K7 antiapoptotic factor (Wang et al. 2002), whereas the second is a recently identified 6.1-kb RNA called $T_{6.1}$, containing a 100-amino acid open reading frame of unknown significance (Taylor et al. 2005). The impact of these shared sequences on the interpretation of a PAN knockout phenotype is an important consideration. Second, we are integrating an inducible PAN expression cassette into the genome of a

human B-cell line to compare how host mRNA levels change in response to PAN overexpression. Perhaps PAN RNA, like HSURs 1 and 2, serves to modulate the levels of certain host transcripts in a way advantageous to the virus. Third, we are using affinity selection to capture a tagged version of PAN RNA transfected into human cells to examine which host proteins associate with PAN in vivo. Likewise, a tagged version of PAN RNA will be purified from lytically induced B cells to identify any viral proteins that selectively bind PAN. The functional consequences of the RNA–protein interactions observed represent a further challenge for investigation.

CONCLUSIONS

Synthesizing noncoding regulatory RNAs appears to be an especially effective strategy for a virus struggling to thrive in the hostile environment of a host organism. Because the mammalian immune system specializes in protein recognition, a foreign RNA molecule more readily escapes detection. Particularly during latency, when only a few of many viral genes are expressed, it would be disadvantageous to abundantly express a viral protein. Moreover, because an RNA can be designed to bind multiple host proteins to sequester or inactivate them, encoding an RNA multiplies the potential of the limited number of viral genes expressed during latency.

The best-understood small viral RNA is adenovirus VA1, which combats one branch of the interferon pathway by binding and inhibiting PKR. Adenovirus thereby overcomes the PKR-induced phosphorylation of translation initiation factor eIF2a that would otherwise shut down protein synthesis in infected cells (Mathews and Shenk 1991). Although it seems unlikely that any of the three sets of gammaherpesviruses snRNPs we are studying act exactly in this way, VA1 RNA provides a powerful paradigm for investigating the functions of other viral ncRNAs.

Another striking feature of viral regulatory RNAs is their variety, even among closely related viruses (Table 1). In the three systems we are investigating, both RNA polymerase III and RNA polymerase II are engaged by the virus to synthesize small RNAs, but in the latter case the transcripts mimic either cellular U RNAs (with distinctive caps and no poly(A)) or mRNAs with poly(A) tails. The different classes of RNPs that assemble on these RNAs could play active roles related to those of their cellular counterparts. Recent investigations have further uncovered the fact that gammaherpesviruses also encode microRNAs (Cullen 2006; Nair and Zavolan 2006), but whether these RNAs serve to regulate the translation of viral or host mRNAs is not yet known. Like all studies of viral infection, novel insights into host cell metabolism may be the most significant outcome of the investigation of viral snRNPs.

ACKNOWLEDGMENTS

We want to recognize Heidi Cook for her pioneering work leading to a phenotype for the HSURs and Hannah Mischo for helpful contributions to those studies. We thank Eleanor Marshall for generating the PAN H2A constructs and for her preliminary observations regarding the effects of the H2A element in PAN RNA. This work was supported by National Institutes of Health grant CA16038. J.A.S. is an investigator of the Howard Hughes Medical Institute. N.K.C. is funded by NIH grant T32-CA09159-29, and S.B. by an Anna Fuller predoctoral fellowship.

REFERENCES

Albrecht J.C. 2000. Primary structure of the *Herpesvirus ateles* genome. *J. Virol.* **74:** 1033.

Albrecht J.C. and Fleckenstein B. 1992. Nucleotide sequence of HSUR 6 and HSUR 7, two small RNAs of herpesvirus saimiri. *Nucleic Acids Res.* **20:** 1810.

Barletta J.M., Kingma D.W., Ling Y., Charache P., Mann R.B., and Ambinder R.F. 1993. Rapid in situ hybridization for the diagnosis of latent Epstein-Barr virus infection. *Mol. Cell. Probes* **7:** 105.

Bhat R.A. and Thimmappaya B. 1983. Two small RNAs encoded by Epstein-Barr virus can functionally substitute for the virus-associated RNAs in the lytic growth of adenovirus 5. *Proc. Natl. Acad. Sci.* **80:** 4789.

Bousquet-Antonelli C., Presutti C., and Tollervey D. 2000. Identification of a regulated pathway for nuclear pre-mRNA turnover. *Cell* **102:** 765.

Chang P.J., Shedd D., Gradoville L., Cho M.S., Chen L.W., Chang J., and Miller G. 2002. Open reading frame 50 protein of Kaposi's sarcoma-associated herpesvirus directly activates the viral PAN and K12 genes by binding to related response elements. *J. Virol.* **76:** 3168.

Cho M.S., Yee H., and Chan S. 2002. Establishment of a human somatic hybrid cell line for recombinant protein production. *J. Biomed. Sci.* **9:** 631.

Collis P., Antoniou M., and Grosveld F. 1990. Definition of the minimal requirements within the human beta-globin gene and the dominant control region for high level expression. *EMBO J.* **9:** 233.

Conrad N.K. and Steitz J.A. 2005. A Kaposi's sarcoma virus RNA element that increases the nuclear abundance of intronless transcripts. *EMBO J.* **24:** 1831.

Conrad N.K., Mili S., Shu M.-D., Marshall E.L., and Steitz J.A. 2006. Identification of a rapid mammalian deadenylation-dependent decay pathway and its inhibition by a viral RNA element. *Mol. Cell* (in press).

Cook H.L., Mischo H.E., and Steitz J.A. 2004. The *Herpesvirus saimiri* small nuclear RNAs recruit AU-rich element-binding proteins but do not alter host AU-rich element-containing mRNA levels in virally transformed T cells. *Mol. Cell. Biol.* **24:** 4522.

Cook H.L., Lytle J.R., Mischo H.E., Li M.J., Rossi J.J., Silva D.P., Desrosiers R.C., and Steitz J.A. 2005. Small nuclear RNAs encoded by *Herpesvirus saimiri* upregulate the expression of genes linked to T cell activation in virally transformed T cells. *Curr. Biol.* **15:** 974.

Cullen B.R. 2006. Viruses and microRNAs. *Nat. Genet.* (suppl.) **38:** S25.

Das B., Butler J.S., and Sherman F. 2003. Degradation of normal mRNA in the nucleus of *Saccharomyces cerevisiae*. *Mol. Cell. Biol.* **23:** 5502.

Dobbelstein M. and Shenk T. 1995. *In vitro* selection of RNA ligands for the ribosomal L22 protein associated with Epstein-Barr virus-expressed RNA by using randomized and cDNA-derived RNA libraries. *J. Virol.* **69:** 8027.

Fan X.C. and Steitz J.A. 1998. Overexpression of HuR, a nuclear-cytoplasmic shuttling protein, increases the *in vivo* stability of ARE-containing mRNAs. *EMBO J.* **17:** 3448.

Faria P.A., Chakraborty P., Levay A., Barber G.N., Ezelle H.J., Enninga J., Arana C., van Deursen J., and Fontoura B.M. 2005. VSV disrupts the Rae1/mrnp41 mRNA nuclear export pathway. *Mol. Cell* **17:** 93.

Fasken M.B. and Corbett A.H. 2005. Process or perish: Quality control in mRNA biogenesis. *Nat. Struct. Mol. Biol.* **12:** 482.

Fok V., Friend K., and Steitz J.A. 2006a. Epstein-Barr virus non-coding RNAs are confined to the nucleus, whereas their partner, the human La protein, undergoes nucleocytoplasmic shuttling. *J. Cell Biol.* **173:** 319.

Fok V., Mitton-Fry R., Grech A., and Steitz J.A. 2006b. Multiple domains of EBER 1, an Epstein-Barr virus non-coding RNA, recruit human ribosomal protein L22. *RNA* **12:** 872.

Ford L.P., Watson J., Keene J.D., and Wilusz J. 1999. ELAV proteins stabilize deadenylated intermediates in a novel *in vitro* mRNA deadenylation/degradation system. *Genes Dev.* **13:** 188.

Golembe T.J., Yong J., Battle D.J., Feng W., Wan L., and Dreyfuss G. 2005. Lymphotropic *Herpesvirus saimiri* uses the SMN complex to assemble Sm cores on its small RNAs. *Mol. Cell. Biol.* **25:** 602.

Goodrich J.A. and Kugel J.F. 2006. Non-coding-RNA regulators of RNA polymerase II transcription. *Nat. Rev. Mol. Cell Biol.* **7:** 12.

Han Z. and Swaminathan S. 2006. Kaposi's sarcoma-associated herpesvirus lytic gene ORF57 is essential for infectious virion production. *J. Virol.* **80:** 5251.

He B., Gross M., and Roizman B. 1997. The gamma(1)34.5 protein of herpes simplex virus 1 complexes with protein phosphatase 1alpha to dephosphorylate the alpha subunit of the eukaryotic translation initiation factor 2 and preclude the shut-off of protein synthesis by double-stranded RNA-activated protein kinase. *Proc. Natl. Acad. Sci.* **94:** 843.

Hilleren P., McCarthy T., Rosbash M., Parker R., and Jensen T.H. 2001. Quality control of mRNA 3´-end processing is linked to the nuclear exosome. *Nature* **413:** 538.

Howe J.G. and Steitz J.A. 1986. Localization of Epstein-Barr virus-encoded small RNAs by *in situ* hybridization. *Proc. Natl. Acad. Sci.* **83:** 9006.

Huang Y. and Carmichael G.G. 1997. The mouse histone H2a gene contains a small element that facilitates cytoplasmic accumulation of intronless gene transcripts and of unspliced HIV-1-related mRNAs. *Proc. Natl. Acad. Sci.* **94:** 10104.

Huang Y., Wimler K.M., and Carmichael G.G. 1999. Intronless mRNA transport elements may affect multiple steps of pre-mRNA processing. *EMBO J.* **18:** 1642.

Iwakiri D., Eizuru Y., Tokunaga M., and Takada K. 2003. Autocrine growth of Epstein-Barr virus-positive gastric carcinoma cells mediated by an Epstein-Barr virus-encoded small RNA. *Cancer Res.* **63:** 7062.

Jensen T.H., Boulay J., Olesen J.R., Colin J., Weyler M., and Libri D. 2004. Modulation of transcription affects mRNP quality. *Mol. Cell* **16:** 235.

Kadaba S., Krueger A., Trice T., Krecic A.M., Hinnebusch A.G., and Anderson J. 2004. Nuclear surveillance and degradation of hypomodified initiator tRNAMet in *S. cerevisiae*. *Genes Dev.* **18:** 1227.

Kieff E. and Liebowitz D. 1990. Epstein-Barr virus and its replication. In *Fields virology*, 2nd edition (ed. D.M. Knipe et al.), p. 1889. Raven Press, New York.

Kiss T. 2002. Small nucleolar RNAs: An abundant group of non-coding RNAs with diverse cellular functions. *Cell* **109:** 145.

Kitagawa N., Goto M., Kurozumi K., Maruo S., Fukayama M., Naoe T., Yasukawa M., Hino K., Suzuki T., Todo S., and Takada K. 2000. Epstein-Barr virus-encoded poly(A)⁻ RNA supports Burkitt's lymphoma growth through interleukin-10 induction. *EMBO J.* **19:** 6742.

Komano J., Maruo S., Kurozumi K., Oda T., and Takada K. 1999. Oncogenic role of Epstein-Barr virus-encoded RNAs in Burkitt's lymphoma cell line Akata. *J. Virol.* **73:** 9827.

Kuai L., Das B., and Sherman F. 2005. A nuclear degradation pathway controls the abundance of normal mRNAs in *Saccharomyces cerevisiae*. *Proc. Natl. Acad. Sci.* **102:** 13962.

LaCava J., Houseley J., Saveanu C., Petfalski E., Thompson E., Jacquier A., and Tollervey D. 2005. RNA degradation by the exosome is promoted by a nuclear polyadenylation complex. *Cell* **121:** 713.

Lee S.I., Murthy S.C., Trimble J.J., Desrosiers R.C., and Steitz J.A. 1988. Four novel U RNAs are encoded by a herpesvirus. *Cell* **54:** 599.

Lerner M.R., Andrews N.C., Miller G., and Steitz J.A. 1981. Two small RNAs encoded by Epstein-Barr virus and complexed with protein are precipitated by antibodies from patients with systemic lupus erythematosus. *Proc. Natl. Acad. Sci.* **78:** 805.

Li M.J., Bauer G., Michienzi A., Yee J.K., Lee N.S., Kim J., Li S., Castanotto D., Zaia J., and Rossi J.J. 2003. Inhibition of HIV-1 infection by lentiviral vectors expressing Pol III-promoted anti-HIV RNAs. *Mol. Ther.* **8:** 196.

Libri D., Dower K., Boulay J., Thomsen R., Rosbash M., and Jensen T.H. 2002. Interactions between mRNA export commitment, 3′-end quality control, and nuclear degradation. *Mol. Cell. Biol.* **22:** 8254.

Loflin P., Chen C.Y., and Shyu A.B. 1999a. Unraveling a cytoplasmic role for hnRNP D in the *in vivo* mRNA destabilization directed by the AU-rich element. *Genes Dev.* **13:** 1884.

Loflin P.T., Chen C.Y., Xu N., and Shyu A.B. 1999b. Transcriptional pulsing approaches for analysis of mRNA turnover in mammalian cells. *Methods* **17:** 11.

Luna R.E., Zhou F., Baghian A., Chouljenko V., Forghani B., Gao S.J., and Kousoulas K.G. 2004. Kaposi's sarcoma-associated herpesvirus glycoprotein K8.1 is dispensable for virus entry. *J. Virol.* **78:** 6389.

Mathews M.B. and Shenk T. 1991. Adenovirus virus-associated RNA and translation control. *J. Virol.* **65:** 5657.

McKenna S.A., Kim I., Liu C.W., and Puglisi J.D. 2006. Uncoupling of RNA binding and PKR kinase activation by viral inhibitor RNAs. *J. Mol. Biol.* **358:** 1270.

Milligan L., Torchet C., Allmang C., Shipman T., and Tollervey D. 2005. A nuclear surveillance pathway for mRNAs with defective polyadenylation. *Mol. Cell. Biol.* **25:** 9996.

Mohr I. and Gluzman Y. 1996. A herpesvirus genetic element which affects translation in the absence of the viral GADD34 function. *EMBO J.* **15:** 4759.

Moore M.J. 2002. Nuclear RNA turnover. *Cell* **108:** 431.

Muhlrad D. and Parker R. 2005. The yeast EDC1 mRNA undergoes deadenylation-independent decapping stimulated by Not2p, Not4p, and Not5p. *EMBO J.* **24:** 1033.

Murthy S.C., Trimble J.J., and Desrosiers R.C. 1989. Deletion mutants of herpesvirus saimiri define an open reading frame necessary for transformation. *J. Virol.* **63:** 3307.

Myer V.E., Lee S.I., and Steitz J.A. 1992. Viral small nuclear ribonucleoproteins bind a protein implicated in messenger RNA destabilization. *Proc. Natl. Acad. Sci.* **89:** 1296.

Nair V. and Zavolan M. 2006. Virus-encoded microRNAs: Novel regulators of gene expression. *Trends Microbiol.* **14:** 169.

Nissen P., Ippolito J.A., Ban N., Moore P.B., and Steitz T.A. 2001. RNA tertiary interactions in the large ribosomal subunit: The A-minor motif. *Proc. Natl. Acad. Sci.* **98:** 4899.

Ruf I.K., Lackey K.A., Warudkar S., and Sample J.T. 2005. Protection from interferon-induced apoptosis by Epstein-Barr virus small RNAs is not mediated by inhibition of PKR. *J. Virol.* **79:** 14562.

Ruf I.K., Rhyne P.W., Yang C., Cleveland J.L., and Sample J.T. 2000. Epstein-Barr virus small RNAs potentiate tumorigenicity of Burkitt lymphoma cells independently of an effect on apoptosis. *J. Virol.* **74:** 10223.

Sharp T.V., Schwemmle M., Jeffrey I., Laing K., Mellor H., Proud C.G., Hilse K., and Clemens M.J. 1993. Comparative analysis of the regulation of the interferon-inducible protein kinase PKR by Epstein-Barr virus RNAs EBER-1 and EBER-2 and adenovirus VAI RNA. *Nucleic Acids Res.* **21:** 4483.

Shaw G. and Kamen R. 1986. A conserved AU sequence from the 3′ untranslated region of GM-CSF mRNA mediates selective mRNA degradation. *Cell* **46:** 659.

Song M.J., Brown H.J., Wu T.T., and Sun R. 2001. Transcription activation of polyadenylated nuclear RNA by rta in human herpesvirus 8/Kaposi's sarcoma-associated herpesvirus. *J. Virol.* **75:** 3129.

Sun R., Lin S.F., Gradoville L., and Miller G. 1996. Polyadenylylated nuclear RNA encoded by Kaposi sarcoma-associated herpesvirus. *Proc. Natl. Acad. Sci.* **93:** 11883.

Swaminathan S., Tomkinson B., and Kieff E. 1991. Recombinant Epstein-Barr virus with small RNA (EBER)

genes deleted transforms lymphocytes and replicates *in vitro*. *Proc. Natl. Acad. Sci.* **88:** 1546.

Takizawa T., Tatematsu C., Watanabe M., Yoshida M., and Nakajima K. 2000. Three leucine-rich sequences and the N-terminal region of double-stranded RNA-activated protein kinase (PKR) are responsible for its cytoplasmic localization. *J. Biochem.* **128:** 471.

Taylor J.L., Bennett H.N., Snyder B.A., Moore P.S., and Chang Y. 2005. Transcriptional analysis of latent and inducible Kaposi's sarcoma-associated herpesvirus transcripts in the K4 to K7 region. *J. Virol.* **79:** 15099.

Toczyski D.P. and Steitz J.A. 1993. The cellular RNA-binding protein EAP recognizes a conserved stem-loop in the Epstein-Barr virus small RNA EBER 1. *Mol. Cell. Biol.* **13:** 703.

Toczyski D.P., Matera A.G., Ward D.C., and Steitz J.A. 1994. The Epstein-Barr virus (EBV) small RNA EBER1 binds and relocalizes ribosomal protein L22 in EBV-infected human B lymphocytes. *Proc. Natl. Acad. Sci.* **91:** 3463.

Tycowski K.T., Kolev N.G., Conrad N.K., Fok V., and Steitz J.A. 2006. The ever-growing world of small nuclear ribonucleoproteins. In *The RNA world*, 3rd edition (ed. R.F. Gesteland et al.), p. 327. Cold Spring Harbor Laboratory Press, Cold Spring Harbor, New York.

Vanacova S., Wolf J., Martin G., Blank D., Dettwiler S., Friedlein A., Langen H., Keith G., and Keller W. 2005. A new yeast poly(A) polymerase complex involved in RNA quality control. *PLoS Biol.* **3:** e189.

Viejo-Borbolla A., Ottinger M., and Schulz T.F. 2004. Human herpesvirus 8: Biology and role in the pathogenesis of Kaposi's sarcoma and other AIDS-related malignancies. *Curr. HIV/AIDS Rep.* **1:** 5.

Vuyisich M., Spanggord R.J., and Beal P.A. 2002. The binding site of the RNA-dependent protein kinase (PKR) on EBER1 RNA from Epstein-Barr virus. *EMBO Rep.* **3:** 622.

Wang H.W., Sharp T.V., Koumi A., Koentges G., and Boshoff C. 2002. Characterization of an anti-apoptotic glycoprotein encoded by Kaposi's sarcoma-associated herpesvirus which resembles a spliced variant of human survivin. *EMBO J.* **21:** 2602.

Wang Y., Xue S.A., Hallden G., Francis J., Yuan M., Griffin B.E., and Lemoine N.R. 2005. Virus-associated RNA I-deleted adenovirus, a potential oncolytic agent targeting EBV-associated tumors. *Cancer Res.* **65:** 1523.

Wyers F., Rougemaille M., Badis G., Rousselle J.C., Dufour M.E., Boulay J., Regnault B., Devaux F., Namane A., Seraphin B., et al. 2005. Cryptic pol II transcripts are degraded by a nuclear quality control pathway involving a new poly(A) polymerase. *Cell* **121:** 725.

Xu Y., AuCoin D.P., Huete A.R., Cei S.A., Hanson L.J., and Pari G.S. 2005. A Kaposi's sarcoma-associated herpesvirus/human herpesvirus 8 ORF50 deletion mutant is defective for reactivation of latent virus and DNA replication. *J. Virol.* **79:** 3479.

Yamamoto N., Takizawa T., Iwanaga Y., Shimizu N., and Yamamoto N. 2000. Malignant transformation of B lymphoma cell line BJAB by Epstein-Barr virus-encoded small RNAs. *FEBS Lett.* **484:** 153.

Yang L., Aozasa K., Oshimi K., and Takada K. 2004. Epstein-Barr virus (EBV)-encoded RNA promotes growth of EBV-infected T cells through interleukin-9 induction. *Cancer Res.* **64:** 5332.

Ye F.C., Zhou F.C., Yoo S.M., Xie J.P., Browning P.J., and Gao S.J. 2004. Disruption of Kaposi's sarcoma-associated herpesvirus latent nuclear antigen leads to abortive episome persistence. *J. Virol.* **78:** 11121.

Zhong W., Wang H., Herndier B., and Ganem D. 1996. Restricted expression of Kaposi sarcoma-associated herpesvirus (human herpesvirus 8) genes in Kaposi sarcoma. *Proc. Natl. Acad. Sci.* **93:** 6641.

Zhou F.C., Zhang Y.J., Deng J.H., Wang X.P., Pan H.Y., Hettler E., and Gao S.J. 2002. Efficient infection by a recombinant Kaposi's sarcoma-associated herpesvirus cloned in a bacterial artificial chromosome: Application for genetic analysis. *J. Virol.* **76:** 6185.

MSL Complex Associates with Clusters of Actively Transcribed Genes along the *Drosophila* Male X Chromosome

E. Larschan,*† A.A. Alekseyenko,*†‡ W.R. Lai,* P.J. Park,*§ and M.I. Kuroda*†‡

*Harvard-Partners Center for Genetics and Genomics, Brigham & Women's Hospital, †Department of Genetics, Harvard Medical School, ‡Howard Hughes Medical Institute, §Children's Hospital Informatics Program, Boston, Massachusetts 02115

Dosage compensation in *Drosophila* serves as a model system for understanding the targeting of chromatin-modifying complexes to their sites of action. The MSL (male-specific lethal) complex up-regulates transcription of the single male X chromosome, thereby equalizing levels of transcription of X-linked genes between the sexes. Recruitment of the MSL complex to its binding sites on the male X chromosome requires each of the MSL proteins and at least one of the two large noncoding roX RNAs. To better understand how the MSL complex specifically targets the X chromosome, we have defined the binding using high-resolution genomic tiling arrays. Our results indicate that the MSL complex largely associates with transcribed genes that are present in clusters along the X chromosome. We hypothesize that after initial recruitment of the MSL complex to the X chromosome by unknown mechanisms, nascent transcripts or chromatin marks associated with active transcription attract the MSL complex to its final targets. Defining MSL-complex-binding sites will provide a tool for understanding functions of large noncoding RNAs that have remained elusive.

Regulatory roles for noncoding RNAs are being discovered at an exciting rate, as biologists delve into a previously hidden RNA world. Although the roles of small RNAs are becoming relatively clear in many systems, mechanisms by which large RNAs regulate gene expression in the nucleus remain mysterious. Prominent examples are the RNAs involved in dosage compensation in mammals and in the fruit fly, *Drosophila melanogaster*.

Dosage compensation makes X-linked gene expression equivalent in males (XY) and females (XX). In *Drosophila*, this occurs primarily by increasing transcription of X-linked genes in males (Hamada et al. 2005; Straub et al. 2005). Two noncoding *roX* (RNA on X) RNAs, *roX1* and *roX2*, interact with five MSL proteins to associate specifically with the male X chromosome in a finely banded pattern along its length (Fig. 1). Each protein component of the MSL complex is essential for dosage compensation, whereas the *roX* RNAs are functionally redundant (Meller and Rattner 2002). The MSL complex is required for site-specific acetylation of histone H4 on lysine 16 (H4K16ac) on the X (Turner et al. 1992; Bone et al. 1994), which is likely to have a key role in up-regulation of transcription (Hilfiker et al. 1997). Recently, H4K16ac has been implicated in destabilizing higher-order chromatin structure (Shogren-Knaak et al. 2006), suggesting a mechanism by which H4K16ac might influence transcription by causing increased accessibility of the transcriptional machinery to the DNA template.

The targeting of the MSL complex to hundreds of sites along the length of the polytene X chromosome, viewed at the resolution of light microscopy, has been known for many years (Fig. 1) (Kuroda et al. 1991). More recently, it was discovered that *roX* RNAs are required for this precise targeting to the majority of

A

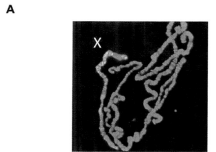

B

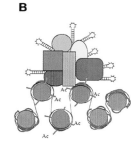

MSL1 & MSL2 core
MSL3 chromodomain
MOF H4 acetyltransferase
MLE helicase
roX1 RNA 3.7 Kb ⎱ functionally
roX2 RNA 0.5 Kb ⎰ redundant

Figure 1. MSL complex binding to the male X chromosome on polytenes and schematic of the MSL complex. (*A*) Immunostaining of the MSL3-TAP protein binding specifically to the male X chromosome. (*Blue*) DAPI staining for DNA; (*red*) MSL3-TAP immunostaining. (*B*) The MSL complex contains five protein and two noncoding RNA components. The actual organization and structure of the MSL complex are unknown. MSL1 and MSL2 components of the complex are essential for complex formation. MSL3 contains a chromodomain, a domain present in many chromatin-associated proteins. MOF is a histone acetyltransferase specific for H4 acetylated at lysine 16, and MLE is an RNA helicase. Two noncoding RNAs encoded on the X chromosome are present in the MSL complex, *roX1* (3.7 kb) and *roX2* (0.5 kb). (*A*, Adapted, with permission, from Alekseyenko et al. 2006.)

sites on the X chromosome (Meller and Rattner 2002). Signature DNA sequences that could be responsible for the X-chromosome specificity of MSL complex binding have not been identified. There is evidence for spreading from *roX* genes in *cis* (Kelley et al. 1999; Park et al. 2002; Oh et al. 2003) and recognition of X segments *in trans* (Demakova et al. 2003; Fagegaltier and Baker 2004; Oh et al. 2004), but the rules for target recognition are not known. Insertion of a strong enhancer into some ectopic positions on the X chromosome can create new cytological sites of MSL binding, suggesting that transcription can activate MSL recognition (Sass et al. 2003). However, the actual identities of direct MSL targets and their key features were largely unknown. As a first step to defining the targeting mechanism, we and other investigators have determined the genome-wide MSL-binding pattern in *Drosophila* embryos and cell lines (Alekseyenko et al. 2006; Gilfillan et al. 2006; Legube et al. 2006). Knowing the nature of these precise targets is a key first step toward understanding how MSL proteins and *roX* RNAs collaborate to bind the X chromosome and regulate X-linked genes.

CHROMATIN IP ANALYSIS OF THE MSL COMPLEX ON HIGH-RESOLUTION GENOMIC TILING ARRAYS REVEALS A LARGE SET OF COMMONLY BOUND GENES

To determine the precise locations of the MSL complex along the X chromosome, we designed genomic tiling arrays (NimbleGen) composed of 388,000 × 50-mers, spaced with 50-bp gaps along the entire nonrepetitive X chromosome (~22 Mb), and most of chromosome 2L (~19.6 Mb). We performed chromatin immunoprecipitations (ChIPs) with modifications designed to optimize our specificity including using a TAP epitope-tagged MSL3 subunit expressed from the native *msl3* promoter as our affinity reagent. Three different cell types were used for our analysis: SL2 cells (embryonic origin), Clone-8 cells (larval wing imaginal disc), and late-stage embryos (mixed cell population).

We compared binding clusters identified on the X chromosome versus 2L and found strong enrichment for the X chromosome (Alekseyenko et al. 2006). For example, in multiple analyses of Clone-8 cells, 972 binding clusters were identified over the X chromosome, whereas none were seen on chromosome 2L, confirming the chromosomal specificity of MSL binding. Biological replicates identified a strongly reproducible set of binding clusters. Furthermore, when binding patterns among different cell types were compared, a strong degree of overlap was observed (Fig. 2) (Alekseyenko et al. 2006). When we compared the lists of genes clearly bound by the MSL complex in SL2 cells, Clone-8 cells, and embryos, we found about 600 genes that were common to all three data sets. A map of MSL-binding clusters along the entire euchromatic X chromosome graphically demonstrates the conservation of MSL-binding sites in different cell types (Fig. 2).

COMPARISON WITH EXPRESSION MICROARRAYS REVEALS THAT THE MSL COMPLEX PREFERS EXPRESSED GENES, WITH STRONGER BINDING TOWARD THE 3′ END OF TRANSCRIPTION UNITS

In parallel with our ChIPs, we purified RNA from MSL3-TAP-tagged SL2 and Clone-8 cells and performed expression analyses using Affymetrix *Drosophila* microarrays. When the annotated genome was aligned with our expression and binding data, we saw a clear correlation of binding with expressed genes (e.g., red genes in Fig. 2) and not with nonexpressed genes (black in Fig. 2) or intergenic regions. When quantified, about 90% of the binding clusters were within expressed genes, whereas only 7% were within nonexpressed genes and less than 3% were in intergenic regions (Alekseyenko et al. 2006). The MSL-binding site map of the entire euchromatic X chromosome indicates that there are clusters of MSL-binding regions along the X chromosome that correspond to domains of active transcription (red genes in Fig. 2).

Furthermore, MSL binding was clearly not centered at 5′ regulatory regions, but often appeared to cover a large portion of each transcription unit. To analyze this objectively, we scaled all bound genes to the same relative length and found that binding on average was enriched over the middle and 3′ end, and away from the 5′ end (Fig. 3). This was seen in genes of all lengths, and was most evident in long genes. This pattern is clearly distinct from typical sequence-specific transcription factors, which bind to discrete target sequences generally in 5′ regulatory regions (Ren et al. 2000). The pattern is also distinct from general transcription factors thought to increase accessibility of promoter regions to RNA polymerase (Kim et al. 2005). The association of the MSL complex to bodies of genes, with stronger binding toward the 3′ end, is instead reminiscent of binding patterns for factors that regulate transcription elongation or termination (Simic et al. 2003; Carrozza et al. 2005; Keogh et al. 2005; Kizer et al. 2005; Rao et al. 2005).

Previous analyses of X-chromosome specificity relied largely on comparing the whole X chromosome to autosomes for sequences that might specify regulation by dosage compensation. With our newly identified set of precise binding sites, we focused our search for sequences that were enriched in these specific segments, when compared to autosomes or to X segments that were not bound by the MSL complex. These searches once again failed to identify unique sequence signatures that might specify MSL recognition.

ATTRACTION OF MSL COMPLEXES IS LINKED TO GENE ACTIVITY OR TO THE CHROMATIN CONTEXT OF TRANSCRIBED GENES ON THE X CHROMOSOME

Since SL2 cells, Clone-8 cells, and embryos display very similar patterns of MSL binding, it is possible that degenerate sequences have evolved on commonly expressed genes to identify them as MSL targets (Gilfillan et al. 2006; Legube et al. 2006). To test whether sequence

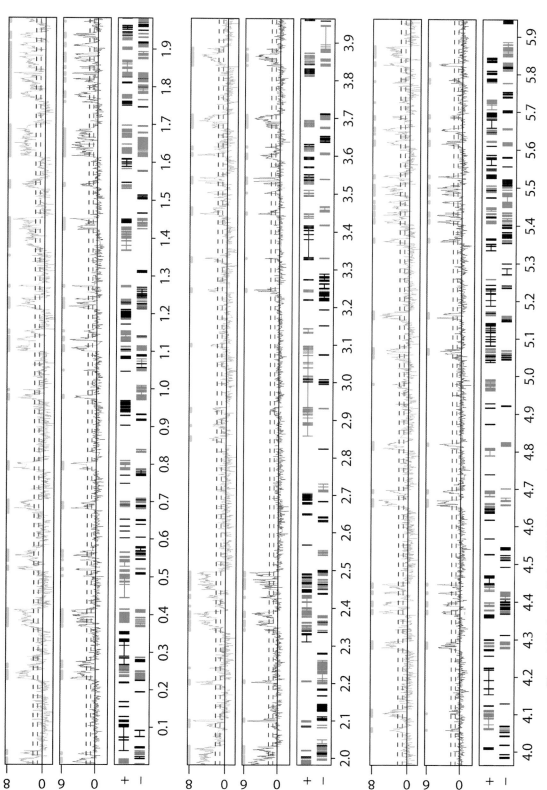

Figure 2. Binding profile for MSL3-TAP across the entire euchromatic X chromosome as determined by high-resolution ChIP on chip profile. Two ChIP on chip binding profile traces for MSL3-TAP are shown. The top profile is for Clone-8 cells (*green*) and the bottom profile is for late-stage embryos (*blue*). Below the binding profiles: (*rectangles*) exons; (*red*) transcribed gene; (*black*) nontranscribed gene as determined by Affymetrix gene expression analysis. (*Continues on following pages.*)

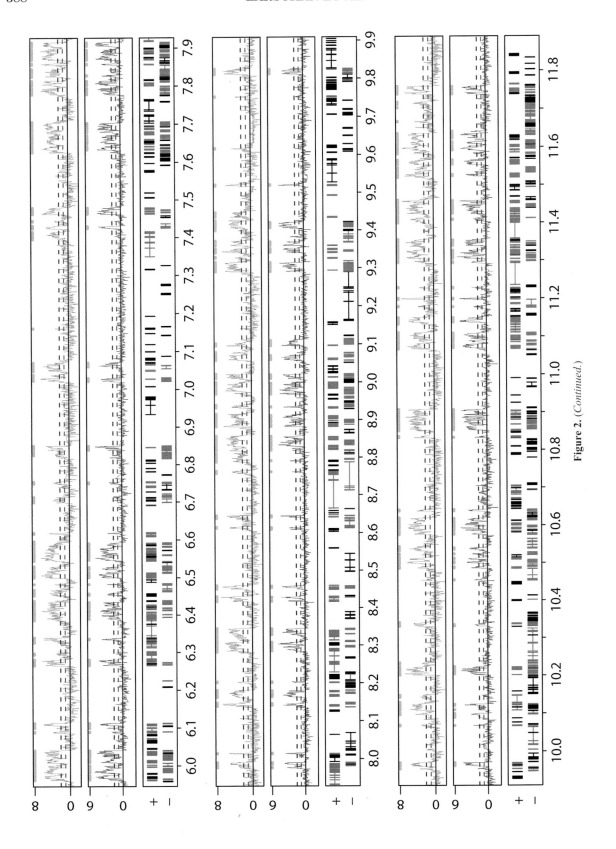

Figure 2. (*Continued.*)

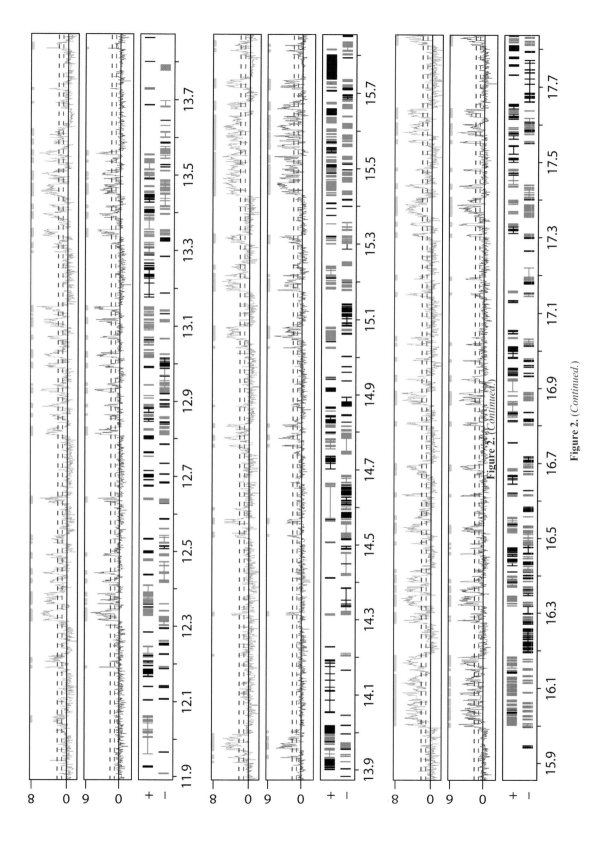

Figure 2. (*Continued.*)

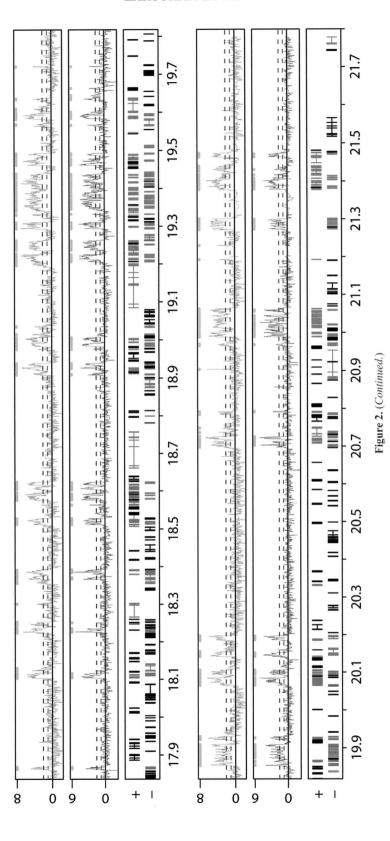

Figure 2. (*Continued.*)

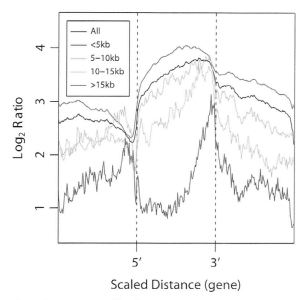

Figure 3. Average profile of MSL binding to scaled genes from different size classes. Average binding profiles for Clone-8 cells. (*Black*) Average profile for all bound genes. (*Red*) genes <5 kb; (*orange*) 5–10 kb; (*yellow*) 10–15 kb; (*green*) >15 kb. (Adapted, with permission, from Alekseyenko et al. 2006.)

then asked whether or not differential binding correlated with differential expression in the two cell types. By comparing the relative expression values for these genes in two microarray experiments for each cell type, we found that all of these genes are differentially expressed, and the differential is much more than a twofold change that could be attributed to dosage compensation. Overall, a clear correlation between differential expression of this set of genes and MSL binding is evident.

Figure 4 (left) shows two examples of genes that are bound by the MSL complex in SL2 cells (top profiles) but are not bound in Clone-8 cells (bottom profiles) and are specifically transcribed in SL2 cells. Figure 4 (right) shows two examples of genes that are specifically expressed and bound in Clone-8 cells. In each case, the gene of interest is centered below the profiles. We validated the binding and transcription levels of these candidates by real-time polymerase chain reaction (PCR) analyses for differential MSL3-TAP binding, MSL1 binding, H4K16 acetylation, and transcript level (Alekseyenko et al. 2006). The enrichment of MSL1 and site-specific acetylation of H4K16 both correlated well with differential binding of TAP-tagged MSL3 at these genes. Furthermore, real-time PCR analysis of RNA levels validated the expression microarray differences seen in the two cell types (Alekseyenko et al. 2006). Our results strongly suggest that sequence alone is not sufficient for MSL binding, because the same gene sequences can be clearly bound or clearly unbound depending on the cell type. Instead, our results suggest a model in which a majority of X-linked genes have evolved a mechanism to attract MSL complexes that is linked to gene activity or to the chromatin context of transcribed genes.

alone is sufficient for MSL recognition, we asked whether there were any genes that were bound in one cell type but unbound in the other cell type. Using strict bound/unbound criteria, we identified 14 genes that were bound in Clone-8 cells and not in SL2 cells, and 2 genes that showed the opposite pattern (Fig. 4) (Alekseyenko et al. 2006). We

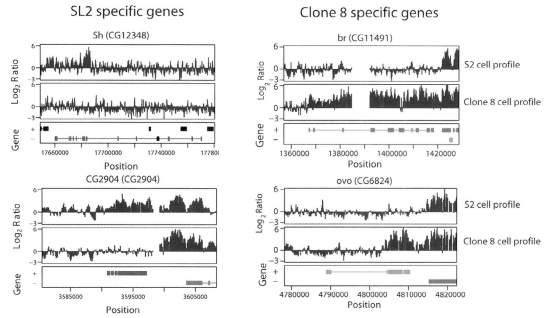

Figure 4. Sequence alone is not sufficient to specify MSL binding. Examples of ChIP-chip tiling along two genes that were bound in SL2 cells but clearly unbound in Clone-8 cells (*left*) and two genes clearly bound in Clone-8 cells but not in SL2 cells (*right*). (*Top profiles*) SL2 cells; (*bottom profiles*) Clone-8 cells. The central gene is the one of interest in each case and is specifically transcribed in the cell type in which it is bound. For *CG2904* and *br*, the whole gene appears covered by MSL binding, whereas the *Sh* and *ovo* genes show strong enrichment over the 3′ portion but not over the 5′ region. (Adapted, with permission, from Alekseyenko et al. 2006.)

CONCLUSIONS

The MSL complex performs a specialized function in flies, but its characterization is likely to have broad implications regarding the mechanism by which chromatin modification factors search for and identify active genes. We have identified the X-chromosome-specific binding pattern of the MSL complex by ChIP-microarray analysis in several distinct cell types. In all cases, we see strong enrichment for the X chromosome and not for chromosome 2L and strong enrichment over most active genes (Fig. 2). When the profiles of all bound genes are scaled to align at the 5′ and 3′ ends, we see a marked preference for the middle and 3′ ends of genes, rather than the 5′ end (Fig. 3). These data are consistent with a two-step model for MSL complex binding in which the complex first identifies the X chromosome via sequence elements or noncoding RNAs and subsequently identifies target genes by recognition of chromatin modifications or nascent mRNA transcripts (Fig. 5).

The MSL-binding profile correlates well with that of its targeted modification, H4K16ac, on selected X-linked genes (Smith et al. 2001). The skew toward the 3′ end of genes is unlike the profile of transcription initiation factors and instead reminiscent of factors that function in transcription elongation or termination. Together, our results suggest that the MSL complex is unlikely to function directly at the promoter like a typical transcription factor. An appealing idea is that an improvement of transcription elongation might improve ultimate mRNA production, perhaps by local recycling of RNA polymerase or other components of the general transcriptional machinery (Smith et al. 2001). Recent work has indicated that there can be a strong link between transcription termination and reinitiation via a looping mechanism (Ansari and Hampsey 2005). Thus, the MSL complex may be involved in promoting this transient association of the 5′ and 3′ ends. Recently, the genomic distribution of histone H3.3, a histone variant associated with transcription, showed increased enrichment on X-linked genes in *Drosophila* SL2 cells when compared to autosomal genes (Mito et al. 2005). This enrichment favors the 5′ ends of transcription units and so might reflect a stimulation of transcription initiation or elongation due to MSL action.

Our results suggest that the MSL complex targets genes predominantly in the context of active transcription. This is consistent with the predominance of the MSL complex in interband regions of polytene chromosomes (Bone et al. 1994) and with experiments in which enhancer sequences responsive to the GAL4 activator protein were able to create new MSL-binding sites that required the expression of GAL4 (Sass et al. 2003). At the same time, our results are also consistent with recent cytological comparisons of the elongating form of RNA polymerase II with the MSL pattern on the polytene X chromosome, in which colocalization was observed but was clearly incomplete (Kotlikova et al. 2006). For example, many genes that we identified as differentially transcribed between SL2 cells and Clone-8 cells were unbound in both (22% vs. 1.2% of commonly transcribed genes). This type of gene would show a lack of colocalization of

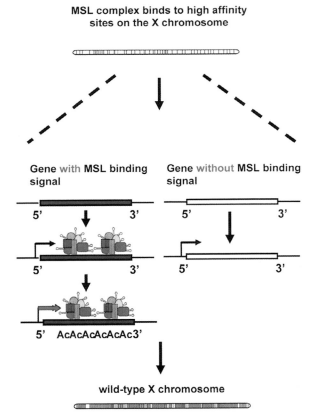

Figure 5. Model for MSL complex recruitment to expressed genes on the X chromosome. The MSL complex recognizes chromosomal regions distributed along the X chromosome based on the presence of specific sequences or noncoding RNAs. Subsequently, the MSL complex targets the 3′ end of transcribed genes by interaction with histone modifications or nascent transcripts. Next, the MSL complex acetylates histone H4 and up-regulates active genes to equalize the dosage of transcripts of X-linked genes. Possible mechanisms for up-regulation include facilitating elongation, termination, or reinitiation of transcription.

RNA polymerase and MSL complex when transcribed. Consistent with the largely invariant pattern of MSL binding seen on polytene chromosomes by Kotlikova et al. (2006), we found that the majority of MSL targets are commonly expressed genes. Differentially regulated genes may have been less likely to evolve the ability to attract the MSL complex and perhaps may have other mechanisms to compensate for dosage differences. Our results suggest intrinsic recognition of many, but not all, X-linked genes within the context of transcription.

Recognition of expressed genes makes excellent biological sense for the MSL complex in two ways. The most obvious is that only expressed genes need to be up-regulated. In this regard, it is notable that binding is independent of the absolute transcription level of individual genes, as dosage compensation must be able to operate on genes with a wide range of intrinsic expression levels. Another important reason to link binding to transcription may be to prevent the MSL complex from ectopically influencing genes that should not be expressed. When *roX* genes are inserted into P transposons and mislocalized at random positions in the genome, they attract the MSL complex, which can spread from the site of insertion into flanking chromatin (Kelley et al. 1999). In several instances, such insertions have occurred in regions where the *mini-white* reporter gene is silenced in females, but activated in males through action of the MSL complex (Kelley and Kuroda 2003). MSL action appeared to have the capacity to overcome Polycomb, HP1, and unidentified modes of silencing. Clearly, the MSL complex must normally be limited in its targeting to avoid potentially catastrophic male-specific activation of silent genes. Our results suggest that the MSL complex is excluded from clusters of nontranscribed genes present on the X chromosome (Fig. 2, black genes). Clustering of transcribed and nontranscribed genes in domains along the X chromosome is consistent with previous analysis of *Drosophila* chromosome 2L (Spellman and Rubin 2002).

How does the MSL complex locate its target genes? Studies of *roX* genes suggest that spreading in *cis* can occur from high local concentrations of the MSL complex. An interesting extension of this idea is that covering large segments of transcription units may occur by a very local spreading mechanism related to the much longer range spreading that can be seen from *roX* transgenes inserted on autosomes. Both long-range and local spreading could be the consequence of an attraction of the MSL complex to chromatin modifications that mark RNA polymerase II transcription units, such as histone H3 methylated at lysine 36 (Carrozza et al. 2005; Keogh et al. 2005; Rao et al. 2005).

In vivo and in vitro data indicate that the recruitment of the *Saccharomyces cerevisiae* Rpd3 small complex, Rpd3(S), to the 3′ end of transcribed genes requires histone H3 methylated at lysine 36 (H3K36Me) (Carrozza et al. 2005; Keogh et al. 2005; Rao et al. 2005). Rpd3(S) deacetylates histones in the wake of transcription, resetting the chromatin state and preventing abberant initiation of transcription at cryptic promoters within genes. The Eaf3 protein component of Rpd3(S) contains a chromodomain and is an MSL3 homolog. Appropriate targeting of Rpd3(S) to nucleosomes containing H3K36Me requires the Eaf3 chromodomain, suggesting that the MSL3 chromodomain may be involved in the identification of histone modifications present on transcribed genes. Distinguishing expressed genes from the bulk of the genome is likely to be an important conserved function common to many chromatin organizing and modifying activities. Future studies will examine how *roX* RNAs facilitate MSL targeting and the role of histone modifications and nascent mRNAs in MSL complex recruitment.

ACKNOWLEDGMENTS

We are grateful to members of the Kuroda lab for many helpful discussions. This work was supported by the National Institutes of Health (GM45744 to M.I.K. and GM67825 to P.J.P.), a grant to E.L., a Leukemia and Lymphoma Society Fellow (5198-05), and the Howard Hughes Medical Institute. M.I.K. is an HHMI Investigator.

REFERENCES

Ansari A. and Hampsey M. 2005. A role for the CPF 3′-end processing machinery in RNAP II-dependent gene looping. *Genes Dev.* **19:** 2969.

Alekseyenko A.A., Larschan E., Lai W.R., Park P.J., and Kuroda M.I. 2006. High-resolution ChIP-chip analysis reveals that the *Drosophila* MSL complex selectively identifies active genes on the male X chromosome. *Genes Dev.* **20:** 848.

Bone J.R., Lavender J., Richman R., Palmer M.J., Turner B.M., and Kuroda M.I. 1994. Acetylated histone H4 on the male X chromosome is associated with dosage compensation in *Drosophila*. *Genes Dev.* **8:** 96.

Carrozza M.J., Li B., Florens L., Suganuma T., Swanson S.K., Lee K.K., Shia W.J., Anderson S., Yates J., Washburn M.P., and Workman J.L. 2005. Histone H3 methylation by Set2 directs deacetylation of coding regions by Rpd3S to suppress spurious intragenic transcription. *Cell* **123:** 581.

Demakova O.V., Kotlikova I.V., Gordadze P.R., Alekseyenko A.A., Kuroda M.I., and Zhimulev I.F. 2003. The MSL complex levels are critical for its correct targeting to the chromosomes in *Drosophila melanogaster*. *Chromosoma* **112:** 103.

Fagegaltier D. and Baker B.S. 2004. X chromosome sites autonomously recruit the dosage compensation complex in *Drosophila* males. *PLoS Biol.* **2:** 1854.

Gilfillan G.D., Straub T., de Wit E., Greil F., Lamm R., van Steensel B., and Becker P.B. 2006. Chromosome-wide gene-specific targeting of the *Drosophila* dosage compensation complex. *Genes Dev.* **20:** 858.

Hamada F.N., Park P.J., Gordadze P.R., and Kuroda M.I. 2005. Global regulation of X chromosomal genes by the MSL complex in *Drosophila melanogaster*. *Genes Dev.* **19:** 2289.

Hilfiker A., Hilfiker-Kleiner D., Pannuti A., and Lucchesi J.C. 1997. mof, a putative acetyl transferase gene related to the Tip60 and MOZ human genes and to the SAS genes of yeast, is required for dosage compensation in *Drosophila*. *EMBO J.* **16:** 2054.

Kelley R.L. and Kuroda M.I. 2003. The *Drosophila roX1* RNA gene can overcome silent chromatin by recruiting the male-specific lethal dosage compensation complex. *Genetics* **164:** 565.

Kelley R.L., Meller V.H., Gordadze P.R., Roman G., Davis R.L., and Kuroda M.I. 1999. Epigenetic spreading of the *Drosophila* dosage compensation complex from *roX* RNA genes into flanking chromatin. *Cell* **98:** 513.

Keogh M.C., Kurdistani S.K., Morris S.A., Ahn S.H., Podolny V., Collins S.R., Schuldiner M., Chin K., Punna T., Thompson N.J., Boone C., Emili A., Weissman J.S., Hughes T.R., Strahl B.D., Grunstein M., Greenblatt J.F., Buratowski S., and Krogan N.J. 2005. Cotranscriptional set2 methylation of histone H3 lysine 36 recruits a repressive Rpd3 complex. *Cell* **123:** 593.

Kim T.H., Barrera L.O., Zheng M., Qu C., Singer M.A., Richmond T.A., Wu Y., Green R.D., and Ren B. 2005. A high-resolution map of active promoters in the human genome. *Nature* **436:** 876.

Kizer K.O., Phatnani H.P., Sibata Y., Hall H., Greenleaf A.L., and Strahl B.D. 2005. A novel domain in Set2 mediates RNA polymerase II interaction and couples histone H3 methylation with transcript elongation. *Mol. Cell. Biol.* **25:** 3305.

Kotlikova I.V., Demakova O.V., Semeshin V.F., Shloma V.V., Boldyreva L.V., Kuroda M.I., and Zhimulev I.F. 2006. The *Drosophila* dosage compensation complex binds to polytene chromosomes independently of developmental changes in transcription. *Genetics* **172:** 963.

Kuroda M.I., Kernan M.J., Kreber R., Ganetzky B., and Baker B.S. 1991. The maleless protein associates with the X chromosome to regulate dosage compensation in *Drosophila*. *Cell.* **66:** 935.

Legube G., McWeeney S.K., Lercher M.J., and Akhtar A. 2006. X-chromosome wide profiling of MSL-1 distribution and dosage compensation in *Drosophila*. *Genes Dev.* **20:** 871.

Meller V.H. and Rattner B.P. 2002. The roX genes encode redundant male-specific lethal transcripts required for targeting of the MSL complex. *EMBO J.* **21:** 1084.

Mito Y., Henikoff J.G., and Henikoff S. 2005. Genome-scale profiling of histone H3.3 replacement patterns. *Nat. Genet.* **37:** 1090.

Oh H., Bone J.R., and Kuroda M.I. 2004. Multiple classes of MSL binding sites target dosage compensation to the X chromosome of *Drosophila*. *Curr. Biol.* **14:** 481.

Oh H., Park Y., and Kuroda M.I. 2003. Local spreading of MSL complexes from *roX* genes on the *Drosophila* male X chromosome. *Genes Dev.* **17:** 1334.

Park Y., Kelley R.L., Oh H., Kuroda M.I., and Meller V.H. 2002. Extent of chromatin spreading determined by *roX* RNA recruitment of MSL proteins. *Science* **298:** 1620.

Rao B., Shibata Y., Strahl B.D., and Lieb J.D. 2005. Dimethylation of histone H3 at lysine 36 demarcates regulatory and nonregulatory chromatin genome-wide. *Mol. Cell. Biol.* **25:** 9447.

Ren B., Robert F., Wyrick J.J., Aparicio O., Jennings E.G., Simon I., Zeitlinger J., Schreiber J., Hannett N., Kanin E., Volkert T.L., Wilson C.J., Bell S.P., and Young R.A. 2000. Genome-wide location and function of DNA binding proteins. *Science* **290:** 2306.

Sass G.L., Pannuti A., and Lucchesi J.C. 2003. Male-specific lethal complex of *Drosophila* targets activated regions of the X chromosome for chromatin remodeling. *Proc. Natl.* *Acad. Sci.* **100:** 8287.

Shogren-Knaak M., Ishii H., Sun J.M., Pazin M.J., Davie J.R., and Peterson C.L. 2006. Histone H4-K16 acetylation controls chromatin structure and protein interactions. *Science* **311:** 844.

Simic R., Lindstrom D.L., Tran H.G., Roinick K.L., Costa P.J., Johnson A.D., Hartzog G.A., and Arndt K.M. 2003. Chromatin remodeling protein Chd1 interacts with transcription elongation factors and localizes to transcribed genes. *EMBO J.* **22:** 1846.

Smith E.R., Allis C.D., and Lucchesi J.C. 2001. Linking global histone acetylation to the transcription enhancement of X-chromosomal genes in *Drosophila* males. *J. Biol. Chem.* **276:** 31483.

Spellman P.T. and Rubin G.M. 2002. Evidence for large domains of similarly expressed genes in the *Drosophila* genome. *J. Biol.* **1:** 5.

Straub T., Gilfillan G.D., Maier V.K., and Becker P.B. 2005. The *Drosophila* MSL complex activates the transcription of target genes. *Genes Dev.* **19:** 2284.

Turner B.M., Birley A.J., and Lavender J. 1992. Histone H4 isoforms acetylated at specific lysine residues define individual chromosomes and chromatin domains in *Drosophila* polytene nuclei. *Cell* **69:** 375.

Noncoding RNAs of the H/ACA Family

M. TERNS AND R. TERNS

*Departments of Biochemistry and Molecular Biology, and Genetics, University of Georgia,
Athens Georgia 30602*

The H/ACA RNAs are an abundant family of *trans*-acting, noncoding RNAs found in eukaryotes and archaea. More than 100 H/ACA RNAs are known to exist in humans. The function of the majority of the identified H/ACA RNAs is to guide site-specific pseudouridylation of ribosomal RNA. In eukaryotes, H/ACA RNAs also mediate the processing of pre-rRNA, provide the template for telomere synthesis, and guide pseudouridylation of other classes of target RNAs (e.g., small nuclear RNAs [snRNAs]). Thus, currently, the H/ACA RNAs are known to be integrally involved in the production of both ribosomes and spliceosomes, and in the maintenance of chromosome integrity. In addition, dozens of H/ACA RNAs have been identified for which no function has yet been determined. The H/ACA RNAs select and present substrate molecules via base-pairing. All H/ACA RNAs contain conserved sequence elements (box H and box ACA) and assemble with a core set of four proteins to form functional ribonucleoprotein complexes (RNPs). Mutations in key RNA and protein components of H/ACA RNPs result in dyskeratosis congenita, a serious multisystem genetic disease. Impressive progress has been made very recently in understanding the biogenesis, trafficking, and function of H/ACA RNPs.

FUNCTIONAL DIVERSITY OF H/ACA RNPs

The H/ACA RNAs are one of the largest classes of noncoding RNAs (second only perhaps to microRNAs), with numbers of unique H/ACA RNA species exceeding 100 in mammals and plants (Brown et al. 2003; Kiss et al. 2004; Lestrade and Weber 2006). H/ACA RNAs are present in all eukaryotes and archaea, but not apparently in bacteria, and have evolved to provide a remarkably diverse set of functions. Protein translation, pre-mRNA splicing, and genome stability depend on the function of H/ACA RNAs in eukaryotes (Kiss 2002; Terns and Terns 2002; Meier 2005; Yu et al. 2005). H/ACA RNAs involved in different functions are adapted to interact with various sets of partner proteins. The H/ACA RNAs all assemble with a core set of proteins to form metabolically stable H/ACA RNA–protein complexes (RNPs). For the majority (i.e., the RNAs involved in RNA pseudouridylation), these four core proteins and the H/ACA RNA comprise a complete functional unit (Baker et al. 2005; Charpentier et al. 2005). In other cases (e.g., RNAs involved in RNA processing and telomere synthesis), additional proteins are required to make a catalytically active complex (Lubben et al. 1995; Autexier and Lue 2006). Although a catalytic role for some H/ACA RNAs has not been excluded, in the known instances, catalytic activity resides in one of the partner proteins.

H/ACA RNPs act *in trans* on multiple cellular targets to influence major cellular processes (Fig. 1). The targets are usually other cellular RNAs—rRNAs and snRNAs—but at least one DNA substrate—the vertebrate telomere—is also known. H/ACA RNAs recognize substrates by base-pairing. The H/ACA RNPs function at a variety of distinct sites, all within the nucleus, where their substrates are found. Established sites of action include nucleoli (rRNA substrates), Cajal bodies (snRNA substrates), and chromosome ends (telomere substrates).

Small Nucleolar RNPs and Ribosome Biogenesis

By far, the vast majority of H/ACA RNPs participate in ribosome biogenesis in the nucleolus and thus are called small nucleolar RNPs (snoRNPs) (or simply sRNPs in the anucleate archaea) (Fig. 1). The H/ACA snoRNPs contribute to ribosome production via at least two distinct mechanisms. The RNAs interact with specific regions of pre-rRNA and guide either site-specific nucleotide modification (conversion of uridines to pseudouridines) or endoribonucleolytic cleavage. Most H/ACA RNPs catalyze

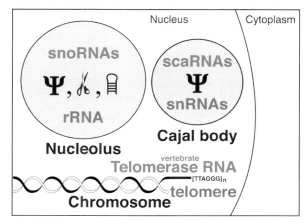

Figure 1. The molecular targets, functions, and sites of action of eukaryotic H/ACA RNAs. Distinct classes of H/ACA RNAs (*green*) act on target molecules (*red*) found in various nuclear compartments. The snoRNAs are essential for pseudouridylation (Ψ), processing (scissors), and folding (hairpin) of rRNAs transcribed within the nucleolus. scaRNAs localize to Cajal bodies to guide the pseudouridylation of snRNAs. Telomerase RNA moves to chromosome ends during S phase where it functions to guide and template telomere synthesis ($TTAGGG_n$).

rRNA pseudouridylation, producing more than 100 modifications in humans (Maden 1990). Pseudouridylation is also presumably the ancestral function, as it is common to both eukaryotes and archaea. In eukaryotes, a few H/ACA snoRNPs (U17/snR30, snR10, and perhaps E2 and E3) are required for pre-rRNA processing (Tollervey 1987; Morrissey and Tollervey 1993; Mishra and Eliceiri 1997; Atzorn et al. 2004). The processing snoRNPs are essential for the generation of mature rRNA species (e.g., 28S, 5.8S, and 18S rRNAs of vertebrates) and cell viability.

Precisely how targeted pseudouridylation of pre-rRNA aids in the assembly of active ribosomes is not yet clear. The occurrence of pseudouridine in diverse classes of noncoding RNAs (rRNA, tRNA, snRNA, and snoRNAs) and not mRNAs suggests that the modification has a role in structured RNAs. Moreover, the fact that the pseudouridines cluster to functionally important domains of the large and small rRNAs (e.g., peptidyl transferase and decoding centers) strongly implicates the modification in ribosome function (Ofengand et al. 2001; Decatur and Fournier 2003; King et al. 2003; Yu et al. 2005). It is anticipated that transient, site-specific base-pairing of H/ACA RNAs to nascent pre-rRNA molecules also contributes to productive rRNA folding (i.e., RNA chaperoning) (Bachellerie et al. 1995; Steitz and Tycowski 1995). Individually, no H/ACA pseudouridylation guide RNA has been found to be essential for cell viability. However, loss of more than one (or global loss of) pseudouridine is lethal (Decatur and Fournier 2003; Yu et al. 2005).

Small Cajal Body RNPs and Pre-mRNA Splicing

In eukaryotes, some H/ACA RNAs guide pseudouridylation of snRNAs (e.g., U1, U2, U4, and U5) that function in pre-mRNA splicing (Fig. 1) (Darzacq et al. 2002; Kiss et al. 2002; Richard et al. 2003). This relatively newly discovered class of H/ACA RNPs functions in Cajal bodies, conserved nuclear structures that appear to be important centers for a variety of RNP assembly and RNA modification reactions (Gall 2000; Cioce and Lamond 2005; Matera and Shpargel 2006). Accordingly, this class of pseudouridylation guide H/ACA RNPs are called the small Cajal body RNPs (scaRNPs) (Darzacq et al. 2002). In some cases, pseudouridylation of spliceosomal snRNAs by scaRNPs has been shown to be essential for snRNP function and pre-mRNA splicing (Yu et al. 2005).

Vertebrate Telomerase and Genome Stability

Telomerase, the enzyme that adds telomeric DNA repeats to the ends of linear chromosomes in eukaryotic cells (Greider and Blackburn 1985), is required to maintain telomere length and prevent genome instability (de Lange 2005; Hug and Lingner 2006). Vertebrate telomerase is a specialized H/ACA RNP (Fig. 1). The enzyme includes telomerase RNA (an adapted H/ACA RNA) and the four core H/ACA RNP proteins plus telomerase reverse transcriptase (TERT) (Mitchell et al. 1999b; Chen et al. 2000; Dragon et al. 2000; Pogacic et al. 2000; Wang and Meier 2004; Autexier and Lue 2006). Despite the fact

that telomerase RNA stably associates with the pseudouridine synthase component of H/ACA RNPs, there is no indication that telomerase functions in RNA pseudouridylation (Chen et al. 2000). A short sequence in the RNA provides the template for the reverse transcription of telomeric DNA repeats by TERT.

Discovery of Novel H/ACA RNAs

In recent years, large-scale "RNomic" cloning efforts and innovative computational approaches have led to the discovery of numerous H/ACA RNAs in a variety of eukaryotes and archaea (see, e.g., Brown et al. 2001; Hüttenhofer et al. 2001; Klein et al. 2002; Tang et al. 2002; Yuan et al. 2003; Kiss et al. 2004; Russell et al. 2004; Schattner et al. 2004; Liang et al. 2005; Torchet et al. 2005). The current catalog of H/ACA RNAs includes a remarkable number of "orphan" RNAs that do not exhibit recognizable complementarity to known targets (i.e., rRNAs and snRNAs). Some of the new RNAs appear to contain additional sequence elements (beyond canonical pseudouridylation guide RNA domains), suggesting that they may be involved in distinct specialized functions (like telomerase RNA). One novel trypanosomal H/ACA RNA has been found to guide pseudouridylation of the spliced leader RNA that is added *in trans* to mRNAs in these organisms (Uliel et al. 2004). This is the first known instance of mRNAs harboring pseudouridine; however, it is probable that the role of this pseudouridine relates to the function of the spliced leader RNP (which shares features with snRNPs), rather than expression of the mRNA product.

ANATOMY OF THE H/ACA RNAs

The basic unit of the H/ACA RNAs is an imperfect hairpin structure with a short single-stranded tail that harbors the signature ACA sequence element (box ACA) or the variant ANANNA (box H) (Fig. 2) (Balakin et al. 1996; Ganot et al. 1997a,b). Each hairpin unit contains an internal loop flanked by upper and lower stems. The two halves of the loop typically bear short antisense elements that form a bipartite target recognition site (pseudouridylation pocket). Substrate sequences flanking the modification site are bound by the antisense elements, and the target uridine and an adjacent nucleotide are left unpaired. The distance (~15 nucleotides) between box ACA (or box H) and the target uridine is conserved (Ganot et al. 1997a; Ni et al. 1997). The molecular basis for the "$n + 15$" spacing rule has been proposed to reflect the binding properties of the pseudouridine synthase that recognizes both the ACA element and the target uridine (Baker et al. 2005).

A common feature of archaeal H/ACA RNAs is a kink-turn (k-turn) motif located at the junction of the upper stem and apical loop. k-turns exist in a variety of RNAs where they have been found to bind a series of related proteins that induce a severe bend or "kink" of about 80–120° in the phosphodiester backbone of RNA helices (Klein et al. 2001). The k-turn motif found in the archaeal H/ACA RNAs is a relaxed variant and is bound by L7Ae

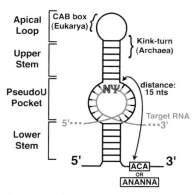

Figure 2. Anatomy of an H/ACA pseudouridylation guide RNA. The basic hairpin unit of the H/ACA RNA can be divided into an apical loop, upper stem, pseudouridylation pocket, and lower stem. In some eukaryotic H/ACA RNAs, a CAB box is found in the apical loop. A kink-turn motif is found in the upper stem of the archaeal H/ACA RNAs. The signature box H (ANANNA) or ACA motif is located immediately 3′ of the hairpin unit. Base-pairing of the target RNA (*red*) with antisense sequences in the pseudouridylation pocket positions the unpaired substrate uridine at the apex of the pocket. The approximately 15-nucleotide distance between the modified nucleotide (Ψ) and box ACA (or H) is conserved.

as detailed below (Hamma and Ferre-D'Amare 2004). There is currently no evidence for the existence of k-turns in eukaryotic H/ACA RNAs, and the eukaryotic homolog of L7Ae binds RNA without apparent sequence or structure specificity, at least in vitro (Henras et al. 2001; Wang and Meier 2004).

All H/ACA RNAs share basic architecture, but each functional class has unique features that enable distinct substrate interaction, protein binding, subcellular localization, and function (Fig. 3). The rRNA pseudouridylation guide RNAs of the nucleolus are the archetypes (and largely define the basic anatomy described above). Members of this largest H/ACA RNA family differ from one another in antisense element sequence (providing for recognition of distinct rRNA target sites) and can be composed of one, two, or three hairpin structures. All three configurations have been identified in archaea (Tang et al. 2002). Single-hairpin H/ACA RNAs exist in early diverging eukaryotes (trypanosmes) (Uliel et al. 2004), but double-hairpin molecules prevail in all other eukaryotic organisms.

The scaRNAs that guide pseudouridylation of snRNAs within the Cajal body in eukaryotes are distinguished from the snoRNAs by short sequence elements called CAB boxes (Cajal body box, consensus UGAG) found in the apical loop of each hairpin (Fig. 3). The CAB boxes are responsible for the Cajal body localization of this class of H/ACA RNAs (Richard et al. 2003). Otherwise, many snRNA pseudouridylation guide RNAs strongly resemble the canonical double-hairpin snoRNAs. In addition, however, interesting composite scaRNAs have been identified that consist of fusions of H/ACA RNAs with C/D RNAs (Jady and Kiss 2001; Darzacq et al. 2002; Kiss et al. 2002), a related but distinct family of RNAs that guide RNA ribose methylation (Terns and Terns 2002; Yu et al. 2005).

Telomerase RNA is a hybrid composed of an H/ACA domain and a specialized telomerase domain in vertebrates (Fig. 3) (Mitchell et al. 1999a; Chen et al. 2000). The 5′ half of the RNA contains a large pseudoknot domain and a short template region required to synthesize telomeric repeats. The 3′ half is a modified double-hairpin H/ACA RNA. The H/ACA domain of vertebrate telomerase RNA is essential for RNA stability, 3′-end formation, retention, and trafficking in the nucleus, as well as the function of the enzyme (Mitchell et al. 1999a; Lukowiak et al. 2001; Fu and Collins 2003; Jady et al. 2004). Similar to the scaRNAs, telomerase RNA localizes to Cajal bodies via a single CAB box in the apical loop of the 3′ hairpin (Jady et al. 2004; Zhu et al. 2004; Jady et al. 2006; Tomlinson et al. 2006). The 5′ hairpin participates in binding telomerase reverse transcriptase (Chen et al. 2000; Mitchell and Collins 2000).

Finally, the small set of H/ACA RNAs that function in rRNA processing (e.g., U17/snR30 and snR10) has unique structural features that greatly depart from those found in pseudouridylation guide RNAs (Fig. 3). U17/snR30 is the only H/ACA RNA whose sequence is conserved among diverse eukaryotes, and two conserved antisense sequence elements that are essential for pre-18S rRNA cleavages have been identified (Morrissey and Tollervey 1993). These elements are located in the 3′ hairpin on opposite strands of the equivalent of the pseudouridine pocket (Cervelli et al. 2002; Atzorn et al. 2004; Eliceiri 2006).

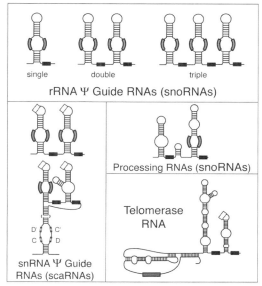

Figure 3. Structural diversity of H/ACA RNAs. H/ACA RNAs that guide pseudouridylation of rRNA (snoRNAs) are composed of one, two, or three tandemly arranged hairpin units each with an associated box H or ACA element (*black boxes*). The other classes of H/ACA RNAs harbor various specialized sequence elements. The snRNA guide RNAs (scaRNAs) can resemble double-hairpin H/ACA RNAs (*upper*) or be hybrid RNA containing both H/ACA and C/D elements (*lower*). scaRNAs and telomerase RNA contain CAB box elements (*white boxes*). Gray boxes indicate sequences involved in base-pairing with RNA or DNA substrates.

CORE PROTEIN SUBUNITS
OF H/ACA RNPs

All H/ACA RNAs associate with a set of four highly conserved proteins and function as RNP complexes (Terns and Terns 2002; Meier 2005; Yu et al. 2005). The four core proteins include the enzyme required for pseudouridine formation and three smaller proteins: Gar1 (glycine-arginine-rich), Nop10 (nucleolar protein of 10 kD), and Nhp2 (nonhistone protein) or the archaeal homolog L7Ae. The pseudouridine synthase is variably called Cbf5 (yeast and archaea), dyskerin (human), or NAP57 (rat) and is related to the TruB family of pseudouridine synthases that modify tRNA (Koonin 1996). The identity and role of the proteins in pseudouridylation and rRNA processing were established in early genetic studies performed in yeast (Girard et al. 1992; Bousquet-Antonelli et al. 1997; Lafontaine et al. 1998; Watkins et al. 1998). These defining studies paved the path for the discovery and characterization of H/ACA RNPs in other organisms.

ROLE OF H/ACA RNPS IN
DYSKERATOSIS CONGENITA

Mutations in components of H/ACA RNPs are responsible for a serious inherited disease termed dyskeratosis congenita (DC). DC patients suffer a variety of clinical symptoms (indicating involvement of several proliferating cell types) that range from skin depigmentation and nail dystrophies to life-threatening bone marrow failure, premature aging, and increased propensity for certain cancers (Meier 2003; Marrone et al. 2005; Mason et al. 2005). The X-linked form of the disease results from point mutations in the dyskerin gene that lead to single-amino-acid changes in the protein (Heiss et al. 1998). Studies employing DC patient cells and mouse models of DC found that dyskerin mutations lead to defects in rRNA modification and processing, and/or telomere shortening (depending on the mutation and system investigated) (Mitchell et al. 1999b; Mochizuki et al. 2004). However, in the case of the autosomal form of DC, it is more clear that the etiology is associated with loss of telomerase function, as the relevant mutations are found in the gene encoding telomerase RNA (Vulliamy et al. 2001; Chen and Greider 2004).

ORGANIZATION AND FUNCTION OF THE
CORE H/ACA RNP

Reconstitution of Archaeal Pseudouridylation
Guide RNPs

Functional H/ACA RNPs have recently been reconstituted from recombinant archaeal proteins and in vitro transcribed RNAs (Baker et al. 2005; Charpentier et al. 2005). The studies performed with these reconstituted systems have very significantly advanced our understanding of the assembly, organization, and function of the H/ACA RNPs. For example, this work established that the four core proteins (Cbf5, Gar1, L7Ae, and Nop10) and a guide RNA are necessary and sufficient for

efficient pseudouridylation in vitro —the first evidence that these four proteins are directly involved in the modification and that they comprise the complete set of proteins required for RNA-guided pseudouridylation. Thus, although other known pseudouridine synthases act as single-subunit enzymes to recognize substrate RNAs and modify target uridines (Ofengand et al. 2001), Cbf5 requires three protein cofactors and a guide RNA for rRNA modification.

These studies also provided substantial insight into the interactions that mediate the assembly and contribute to the functional organization of the H/ACA RNP (Fig. 4) (Baker et al. 2005; Charpentier et al. 2005). Two of the proteins, L7Ae and Cbf5, interact directly with the H/ACA guide RNA. As was established in other work, L7Ae binds the k-turn in the upper stem of archaeal H/ACA RNAs (Rozhdestvensky et al. 2003; Baker et al. 2005; Charpentier et al. 2005; Hamma et al. 2005). It is now clear that Cbf5 is the component that recognizes the signature sequence element box ACA. Interaction of Cbf5 also depends on the pseudouridylation pocket and apical loop of the H/ACA RNA (Baker et al. 2005). In addition, Cbf5 recruits the essential proteins Nop10 and Gar1 to the complex via independent protein–protein interactions with each. L7Ae does not interact with any of the other three core H/ACA RNP proteins in the absence of the guide RNA (Baker et al. 2005).

Studies with single- and double-hairpin H/ACA RNAs indicate that each hairpin nucleates the assembly of all four core H/ACA RNP proteins (Baker et al. 2005; Charpentier et al. 2005; Hamma et al. 2005). This observation is consistent with the earlier interpretation of electron microscopic images of purified yeast H/ACA RNP particles that suggested symmetric organization of core RNPs at each hairpin unit (Watkins et al. 1998).

Studies to address how the reconstituted archaeal H/ACA RNPs recognize target RNA have yielded somewhat unexpected results. Clearly, the guide RNA has a key role in this process, but simple base-pairing of the guide and substrate RNAs, although necessary, is apparently not sufficient for target interaction. Stable guide RNA/target RNA duplexes cannot be detected under the conditions that support H/ACA RNP function in vitro

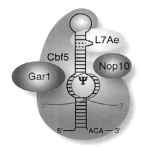

Figure 4. Organization of a pseudouridylation guide RNP. Recent studies with archaeal components indicate that L7Ae interacts with the k-turn of the H/ACA RNA; Cbf5 interacts with box ACA, the pseudouridylation pocket, and apical loop of the RNA; and Nop10 and Gar1 interact with Cbf5. (Adapted, with permission, from Baker et al. 2005.)

(Charpentier et al. 2005). However, stable duplexes are detected in the presence of Nop10 and Cbf5, indicating a role for these two proteins in substrate recognition (Charpentier et al. 2005). Experiments to systematically assess whether L7Ae and Gar1 influence target RNA association have not yet been reported. Target recognition has also been observed to require a uridine in the substrate RNA (Charpentier et al. 2005). Since the target uridine is not predicted to be involved in base-pairing with the guide RNA, this observation also supports the role of one or more proteins in target RNA binding. One possibility is that protein binding is required to remodel the guide RNA to an open conformation, making the antisense elements of the loop available for base-pairing with target RNA. As a protein that recognizes both the pseudouridylation pocket and target uridine (Baker et al. 2005), Cbf5 is an obvious candidate for a remodeling function. Nop10 could influence target interaction directly or through its known interaction with Cbf5.

Structural Characterization of H/ACA RNP Complexes

Until very recently, the three-dimensional structures of all of the four core H/ACA RNP proteins were unknown; however, an abundance of new information is now available from a series of atomic-level structures of complexes of various components of the RNP (Fig. 5) (Hamma and Ferre-D'Amare 2004; Hamma et al. 2005; Manival et al. 2006; Rashid et al. 2006; Yu 2006). This work has provided not only the structures of the individual component proteins, but also insight into the molecular basis for key interactions and functions.

Cbf5 is the heart of the H/ACA RNP complex—binding the H/ACA guide RNA and recruiting the two essential protein cofactors Nop10 and Gar1, as well as catalyzing the isomerization of the target uridine. As expected on the basis of sequence homology, the overall structure of Cbf5 is similar to that of TruB, the bacterial pseudouridine synthase that catalyzes the highly conserved pseudouridylation of U55 in the T loop of tRNAs (Hoang and Ferre-D'Amare 2001). Like TruB, Cbf5 is composed of two major domains: an amino-terminal catalytic domain and a carboxy-terminal PUA domain (Fig. 5). The PUA (pseudouridine synthase and archaeosine transglycosylase) domain is common to a variety of RNA-binding proteins (Aravind and Koonin 1999) and, as described below, is likely to be involved in interaction of Cbf5 with the H/ACA guide RNA. The catalytic domains of Cbf5 and TruB are virtually superimposable. Careful inspection reveals that all atoms important for catalysis, including a catalytic aspartate universal to all pseudouridine synthases, occupy nearly identical positions in Cbf5 and TruB. Thus, both enzymes likely use the same catalytic mechanism to isomerize uridine to pseudouridine (cleavage of the N-glycosidic bond, rotation of the uracil base, and reformation of a C-glycosidic bond).

At the same time, Cbf5 exhibits a few striking differences from TruB that can be correlated with known functional differences. For example, in place of the "thumb-loop" domain of TruB that is involved in tRNA T-loop recognition and positioning of the target uridine in the active site cleft, Cbf5 has a unique structural element, the β7/β10 hairpin loop. This loop arrives at a similar spatial location in the active site and likely has a role in recognition of the guide RNA/rRNA duplex. Another significant

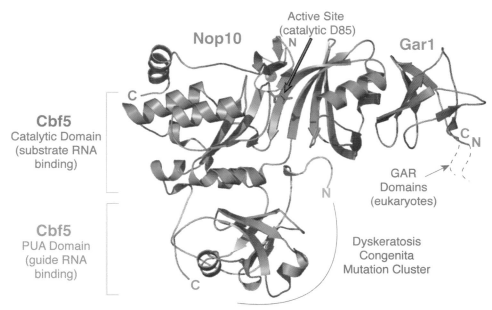

Figure 5. Structure of a heterotrimeric complex of Cbf5, Gar1, and Nop10. The pseudouridine synthase Cbf5 consists of two major domains: the catalytic domain (*green*) that contains the active site (D85), and a PUA domain (*cyan*). A long amino-terminal extension (*gold*) wraps around the PUA domain. Nop10 (*red*) and Gar1 (*blue*) interact on opposite sides of the active site of Cbf5. Modeling indicates that the mutations that cause dyskeratosis congenita cluster to a face of the PUA domain. (Adapted, with permission, from Rashid et al. 2006 [© Elsevier].)

distinction is the longer amino-terminal extension present in Cbf5. The amino terminus of Cbf5 encircles the carboxy-terminal PUA domain, thereby expanding and constraining the PUA domain relative to that of TruB. It is also noteworthy that the overall fold of the PUA domain of Cbf5 is more similar to that of the PUA domain of archaeosine transglycosylase than TruB. Archaeosine transglycosylase catalyzes the first step in an archaea-specific, tRNA modification and interacts specifically with the acceptor stem and 3′ CCA of the tRNA. It is tempting to speculate that the PUA domain of Cbf5 interacts with the lower stem and box ACA of the guide RNA in an analogous fashion; however, this is not currently known. Two patches enriched in basic residues flanking the active site and numerous residues distributed across both the catalytic and PUA domains are likely involved in RNA interaction (Hamma et al. 2005; Manival et al. 2006; Rashid et al. 2006).

Information on the precise binding sites of both Nop10 and Gar1 is available from the trimeric Cbf5/Nop10/Gar1 (Rashid et al. 2006) and dimeric Cbf5/Nop10 (Hamma et al. 2005; Manival et al. 2006) structures that have been solved (Fig. 5). As predicted from biochemical studies (Baker et al. 2005), Nop10 and Gar1 bind to distinct sites on Cbf5 (Rashid et al. 2006). The proteins interact with the catalytic domain of Cbf5 on opposite sides of the active site, suggesting that they function in catalytic activation of Cbf5 and/or interaction with the guide RNA/target RNA. The residues involved in the interactions of Nop10 and Gar1 with Cbf5 are generally conserved, indicating that assembly will be similar in archaea and eukaryotes.

Nop10 appears to be intrinsically unstructured, but is well ordered when bound to Cbf5. Nop10 consists of an amino-terminal zinc ribbon domain (in which a single zinc molecule is coordinated by four conserved cysteine residues) connected to a carboxy-terminal α-helix by a short linker region (Hamma et al. 2005; Manival et al. 2006; Rashid et al. 2006). Each of these three domains makes extensive contact with Cbf5. The amino-terminal zinc-binding domain of Nop10 is not conserved in eukaryotes, and mutations predicted to disrupt zinc binding do not appreciably affect Nop10 function in vitro (Charpentier et al. 2005). Nop10 binds adjacent to the active site of Cbf5 and interacts with critical residues in the conserved pseudouridine synthase motif I, consistent with a role in catalytic activation. The binding of Nop10 to Cbf5 enhances the positive surface potential of the complex by shielding electronegative charges of Cbf5 and by contributing electropositive charges, which could be important in RNA interaction.

Gar1 forms a six-β-barrel structure and binds Cbf5 on the opposite side of the active site from Nop10 (Rashid et al. 2006). The binding of Gar1 to Cbf5 appears to influence formation of the β7/β10 hairpin loop that is thought to be important in substrate RNA recognition within the active site of Cbf5 (discussed above). This loop is disordered in Cbf5 in the absence of Gar1 (Hamma et al. 2005).

As discussed above, L7Ae does not interact with the other H/ACA RNP proteins in the absence of the guide RNA (Baker et al. 2005). However, the details of the interaction between the L7Ae and an RNA k-turn have been captured in a co-crystal structure (Hamma and Ferre-D'Amare 2004). As has been found for other k-turn binding proteins, L7Ae induces a dramatic, approximately 120° bend in the phosphodiester backbone of the RNA and flips out a nucleotide. This L7Ae-induced kink in the guide RNA presumably influences other regions of the guide RNA (e.g., pseudouridylation pocket) to positively affect catalytic activity.

Eukaryotic Pseudouridylation Guide H/ACA RNPs

The organization of H/ACA RNPs has also been investigated in eukaryotes, and the available evidence indicates that the architecture of the pseudouridylation guide RNPs is essentially similar in archaea and eukaryotes. For example, the mammalian Cbf5 homolog (dyskerin/Nap57) also interacts with both Nop10 and Gar1 in the absence of guide RNAs (Wang and Meier 2004). Moreover, in yeast, Cbf5/Nop10/Gar1complexes were found to assemble in vivo in the absence of the L7Ae homolog (Nhp2p) and H/ACA RNAs (Henras et al. 2004).

One apparent difference in the eukaryotic and archaeal core H/ACA RNPs is in the L7Ae homologs (L7Ae in archaea and Nhp2 in eukaryotes). L7Ae interacts directly with the k-turn motif in archaeal H/ACA RNAs. However, eukaryotic H/ACA RNAs do not possess recognizable k-turns, and although Nhp2 exhibits nonspecific RNA-binding properties, it appears to be recruited to the complex via interaction with a complex of Cbf5 and Nop10 (Henras et al. 2001; Wang and Meier 2004). Nonetheless, it is quite possible that L7Ae and Nhp2 end up in similar spatial locations in the assembled RNPs and perform similar roles.

ASSEMBLY AND TRAFFICKING PATHWAYS IN EUKARYOTES

H/ACA RNP assembly and function involve dynamic and complex cellular trafficking of components in eukaryotic cells (Fig. 6). Current evidence indicates that nascent H/ACA RNAs are packaged into stable, but perhaps inactive, RNPs cotranscriptionally. It appears that the H/ACA RNP proteins are targeted to Cajal bodies for further maturation and then sent on to sites of function (i.e., nucleoli, Cajal bodies, or telomeres). These steps in assembly and trafficking appear to involve dynamic associations with non-H/ACA RNP proteins.

RNP Assembly at the Site of Transcription

Recent work in both yeast and mammalian systems indicates that RNP biogenesis begins on nascent H/ACA RNA proteins with cotranscriptional recruitment of three of the four core H/ACA RNP proteins, namely, Cbf5, Nop10, and Nhp2 (L7Ae homolog), and an assembly factor called Naf1 (Ballarino et al. 2005; Yang et al. 2005; Darzacq et al. 2006; Hoareau-Aveilla et al. 2006; Richard et al. 2006; Richard and Kiss 2006). Naf1 appears to associate with assembling H/ACA complexes as a surrogate for Gar1 that is replaced during maturation of the RNP.

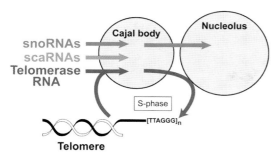

Figure 6. Intracellular trafficking of H/ACA RNAs. The major classes of H/ACA RNAs all appear to be initially targeted to Cajal bodies. scaRNAs are retained and function in Cajal bodies. snoRNAs quickly move on to nucleoli. During S phase, Cajal bodies containing telomerase RNA move to the nucleolar periphery and then appear to deliver the RNA to telomeres.

Naf1 and the core H/ACA proteins other than Gar1 are found at H/ACA genes in chromatin immunoprecipitations and a clever cell-based assay (Ballarino et al. 2005; Yang et al. 2005; Darzacq et al. 2006). Naf1 can also be found associated with the H/ACA RNPs (snoRNPs, scaRNPs, and telomerase) (Dez et al. 2002; Fatica et al. 2002; Yang et al. 2002; Hoareau-Aveilla et al. 2006). The association is with free snoRNP fractions (where immature snoRNPs would be expected) rather than snoRNPs associated with ribosome components (i.e., engaged in function) (Dez et al. 2002). In addition, the protein is not found at known sites of mature H/ACA RNP function (such as nucleoli and Cajal bodies) (Dez et al. 2002; Yang et al. 2002; Darzacq et al. 2006; Hoareau-Aveilla et al. 2006). Naf1 is essential for the accumulation of all classes of H/ACA RNAs (snoRNAs, scaRNAs, and telomerase RNA) (Dez et al. 2002; Fatica et al. 2002; Yang et al. 2002; Darzacq et al. 2006; Hoareau-Aveilla et al. 2006).

Evidence indicates that Naf1 and Gar1 bind to a common site on Cbf5 in a mutually exclusive manner. The proteins share a region of protein homology (Fatica et al. 2002), and the recent structure of the archaeal Cbf5/Gar1/Nop10 complex (Rashid et al. 2006) revealed that it is the homologous region of Gar1 that interacts with Cbf5. In vitro binding assays indicate that interaction of the two proteins with Cbf5 is mutually exclusive (Darzacq et al. 2006). The exchange of Naf1 for Gar1 may mediate the controlled transition to an active H/ACA RNP.

RNP Maturation at Cajal Bodies

As indicated above, newly assembled H/ACA RNPs may not be functional. Moreover, there is evidence that all classes of H/ACA RNAs are rapidly targeted to Cajal bodies soon after synthesis (Fig. 6). Cajal bodies are intranuclear structures that house machinery which carries out posttranscriptional RNA alterations and assembly of RNA–protein complexes (Gall 2000; Cioce and Lamond 2005; Matera and Shpargel 2006). Gar1, but not Naf1, has been detected in Cajal bodies, suggesting that this may be the site of Naf1 exchange and Gar1 incorporation into maturing H/ACA RNPs (Pogacic et al. 2000).

Several observations suggest that the survival of motor neurons (SMN) complex may function in the incorporation of Gar1 into H/ACA RNPs and other aspects of complex maturation. The SMN complex is a known snRNP assembly factor (Terns and Terns 2001; Meister et al. 2002; Gubitz et al. 2004) and is concentrated in Cajal bodies (Matera and Frey 1998). In addition, the SMN protein interacts with Gar1 both in vivo and in vitro, and via the same domains that mediate binding and assembly of snRNPs (Pellizzoni et al. 2001; Whitehead et al. 2002). SMN has also been independently linked with telomerase by the observation that SMN antibodies can immunoprecipitate telomerase complexes (Bachand et al. 2002). Finally, the recent discovery that H/ACA scaRNAs and telomerase RNA also associate with a subset of Sm proteins, the snRNP components that interact with SMN, suggests further involvement of SMN in the assembly of these RNPs (Fu and Collins 2006).

Additional proteins implicated in the early steps of H/ACA RNP biogenesis or trafficking including Shq1, Nopp140, and two putative RNA or DNA helicases called RVB1/RVB2 (also known as TIP49a/b or p50/p55) (King et al. 2001; Dez et al. 2002; Fatica et al. 2002; Wang et al. 2002; Yang et al. 2002). The available evidence indicates that these proteins are required for H/ACA RNA accumulation but are not stable components of mature H/ACA RNPs.

Delivery to Functional Destinations

In the end, mechanisms must exist to ensure that a given H/ACA RNP reaches its site of action, which may be the nucleolus (snoRNPs), Cajal body (scaRNPs), or telomere (telomerase) (Fig. 6). *cis*-Acting sequences important for the ultimate subcellular targeting of some H/ACA RNAs have been identified.

The discrete RNA elements that mediate nucleolar targeting of H/ACA snoRNAs were identified by examining the localization patterns of large panels of mutants injected in *Xenopus* oocytes (Lange et al. 1999; Narayanan et al. 1999; Ruhl et al. 2000). The critical nucleolar localization sequences include box H, box ACA, and the stem that tethers the two elements together (Narayanan et al. 1999; Lukowiak et al. 2001). These studies also demonstrated that the ability of H/ACA RNAs to base-pair with target RNAs is not critical for localization to nucleoli (Lange et al. 1999; Narayanan et al. 1999).

H/ACA snoRNAs rapidly traverse Cajal bodies en route to nucleoli, but the scaRNAs and telomerase RNA remain associated with Cajal bodies (but see more on telomerase RNA trafficking below). Cajal body localization of the scaRNAs and telomerase RNA relies on one or two copies of the CAB (Cajal body) box found in the apical loop(s) of these H/ACA RNAs (Fig. 3) (Richard et al. 2003). Both the CAB boxes and H/ACA domain (i.e., nucleolar targeting information) are required for proper Cajal body localization (Richard et al. 2003). Mutation of the CAB boxes results in loss of Cajal body localization and appearance of the RNAs in nucleoli, indicating that the CAB signal dominates the nucleolar targeting signal in these RNAs (Richard et al. 2003; Jady et al. 2004).

Recent work indicates that the association of a set of Sm proteins with scaRNAs and telomerase RNA depends on the CAB box (Fu and Collins 2006). It is not known whether the association is direct. Curiously, mutations of the CAB box of telomerase RNA that disrupt the association of the Sm proteins (and presumably Cajal body localization) do not appear to significantly affect the biogenesis, stability, or function of telomerase (Fu and Collins 2006). In addition, the Ku70/80 proteins (components of the nonhomologous end-joining DNA-repair complexes) associate both in vitro and in vivo with a 3′ fragment of telomerase RNA that includes the CAB box (Ting et al. 2005).

The activity of telomerase is restricted to S phase (Ten Hagen et al. 1990; Wright et al. 1999), and its trafficking (as well as that of its partner TERT) appears to be regulated as a function of the cell cycle (Jady et al. 2006; Tomlinson et al. 2006). Throughout most of the cell cycle, telomerase RNA is detected in Cajal bodies. However, during S phase, Cajal bodies containing the RNA mobilize to the periphery of nucleoli (early S) and then the RNA moves into foci adjacent to the Cajal bodies and finally to telomeres (peaking at mid S phase) (Jady et al. 2006; Tomlinson et al. 2006). It has been proposed that Cajal bodies directly deliver telomerase RNA to telomeres during S phase (Jady et al. 2006). Endogenous TERT protein is found in unidentified foci during most of the cell cycle and appears within nucleoli in early S phase and, like telomerase RNA, at Cajal-body-associated foci and telomeres in mid S phase (Tomlinson et al. 2006). The regulated intracellular trafficking of the key components of telomerase may be a mechanism for the control of telomerase activity.

CONCLUSIONS

The H/ACA RNPs are a diverse and abundant family of *trans*-acting RNA–protein complexes that eukaryotic cells depend on for multiple essential functions. H/ACA RNPs have apparently been present in cells for more than 2–3 billion years. The ancestral function appears to have been pseudouridylation of rRNA and ribosome biogenesis, but numerous distinct H/ACA RNA species have evolved to expand the substrates and functions of the H/ACA RNPs.

The focus of this review has been the impressive progress made in understanding the biogenesis, structure, and function of the H/ACA RNPs during the last decade. However, aspects of the expression of the H/ACA RNAs are also noteworthy. H/ACA RNAs were among the first gene products recognized to arise from introns. The vast majority of vertebrate H/ACA RNAs are encoded within introns of protein-coding pre-mRNAs and are processed out following splicing (Filipowicz and Pogacic 2002; Kiss 2002; Terns and Terns 2002). Moreover, the discovery of brain-specific mammalian H/ACA RNAs (Cavaille et al. 2000) reveals that the H/ACA RNPs likely participate in tissue-specific functions that are not yet understood.

The large number of orphan H/ACA RNAs also pre-dicts new substrates and functions to be discovered. Approximately 20% of the human H/ACA RNAs do not match known targets (Lestrade and Weber 2006). It has long been known that snoRNAs themselves contain pseudouridine residues, and it seems likely that H/ACA guide RNAs will be identified for these modifications, and perhaps also for modification of other noncoding RNAs (e.g., telomerase RNA and microRNAs).

With so much more to learn, it seems likely that the H/ACA RNPs will continue to be a lead system for understanding the biology of noncoding RNA–protein complexes for years to come.

Note Added in Proof

We want to alert the reader to an article of interest published after the writing of this review. Li and Ye (2006) have now described the structure of the H/ACA RNP including a guide RNA as well as all four essential proteins. The results support many of the hypotheses developed from the previous studies that were described above.

ACKNOWLEDGMENTS

We thank Caryn Hale and Connor Hale for assistance with figure preparation, and members of the Terns lab (past and present) who have contributed to our research on H/ACA RNPs, especially Dan Baker, Osama Youssef, Caryn Hale, Sarah Marshburn, Rebecca Tomlinson, Zhu-Hong Li, Tania Zeigler, Emem Adolf, Eladio Abreu, Tim Supakorndej, Natasha Starostina, Michael Chastkofsky, David Dy, Yusheng Zhu, Sarah Whitehead-Finch, Aarthi Narayanan, and Andrew Lukowiak. We express our sincere gratitude to Michael Adams (Department of Biochemistry and Molecular Biology, University of Georgia) for his crucial support of our endeavors to understand the archaeal H/ACA RNPs. Finally, we have enjoyed fruitful collaborations on this topic with Hong Li, Tamas Kiss, Sean Eddy, Chris Counter, Yi-Tao Yu, Joan Steitz, Louis Droogmans, and Henri Grosjean. This work was supported by National Institutes of Health grants RO1 GM54682 and RO1 CA104676 to M. Terns and R. Terns.

REFERENCES

Aravind L. and Koonin E.V. 1999. Novel predicted RNA-binding domains associated with the translation machinery. *J. Mol. Evol.* **48:** 291.

Atzorn V., Fragapane P., and Kiss T. 2004. U17/snR30 is a ubiquitous snoRNA with two conserved sequence motifs essential for 18S rRNA production. *Mol. Cell. Biol.* **24:** 1769.

Autexier C. and Lue N.F. 2006. The structure and function of telomerase reverse transcriptase. *Annu. Rev. Biochem.* **75:** 493.

Bachand F., Boisvert F.M., Cote J., Richard S., and Autexier C. 2002. The product of the survival of motor neuron (SMN) gene is a human telomerase-associated protein. *Mol. Biol. Cell* **13:** 3192.

Bachellerie J.P., Michot B., Nicoloso M., Balakin A., Ni J., and Fournier M.J. 1995. Antisense snoRNAs: A family of nucleolar RNAs with long complementarities to rRNA. *Trends Biochem. Sci.* **20:** 261.

Baker D.L., Youssef O.A., Chastkofsky M.I., Dy D.A., Terns R.M., and Terns M.P. 2005. RNA-guided RNA modification:

Functional organization of the archaeal H/ACA RNP. *Genes Dev.* **19:** 1238.

Balakin A.G., Smith L., and Fournier M.J. 1996. The RNA world of the nucleolus: Two major families of small RNAs defined by different box elements with related functions. *Cell* **86:** 823.

Ballarino M., Morlando M., Pagano F., Fatica A., and Bozzoni I. 2005. The cotranscriptional assembly of snoRNPs controls the biosynthesis of H/ACA snoRNAs in *Saccharomyces cerevisiae. Mol. Cell. Biol.* **25:** 5396.

Bousquet-Antonelli C., Henry Y., G'Elugne J.P., Caizergues-Ferrer M., and Kiss T. 1997. A small nucleolar RNP protein is required for pseudouridylation of eukaryotic ribosomal RNAs. *EMBO J.* **16:** 4770.

Brown J.W., Clark G.P., Leader D.J., Simpson C.G., and Lowe T. 2001. Multiple snoRNA gene clusters from *Arabidopsis. RNA* **7:** 1817.

Brown J.W., Echeverria M., Qu L.H., Lowe T.M., Bachellerie J.P., Hüttenhofer A., Kastenmayer J.P., Green P.J., Shaw P., and Marshall D.F. 2003. Plant snoRNA database. *Nucleic Acids Res.* **31:** 432.

Cavaille J., Buiting K., Kiefmann M., Lalande M., Brannan C.I., Horsthemke B., Bachellerie J.P., Brosius J., and Hüttenhofer A. 2000. Identification of brain-specific and imprinted small nucleolar RNA genes exhibiting an unusual genomic organization. *Proc. Natl. Acad. Sci.* **97:** 14311.

Cervelli M., Cecconi F., Giorgi M., Annesi F., Oliverio M., and Mariottini P. 2002. Comparative structure analysis of vertebrate U17 small nucleolar RNA (snoRNA). *J. Mol. Evol.* **54:** 166.

Charpentier B., Muller S., and Branlant C. 2005. Reconstitution of archaeal H/ACA small ribonucleoprotein complexes active in pseudouridylation. *Nucleic Acids Res.* **33:** 3133.

Chen J.L. and Greider C.W. 2004. Telomerase RNA structure and function: Implications for dyskeratosis congenita. *Trends Biochem Sci.* **29:** 183.

Chen J.L., Blasco M.A., and Greider C.W. 2000. Secondary structure of vertebrate telomerase RNA. *Cell* **100:** 503.

Cioce M. and Lamond A.I. 2005. Cajal bodies: A long history of discovery. *Annu. Rev. Cell Dev. Biol.* **21:** 105.

Darzacq X., Jady B.E., Verheggen C., Kiss A.M., Bertrand E., and Kiss T. 2002. Cajal body-specific small nuclear RNAs: A novel class of 2′-O- methylation and pseudouridylation guide RNAs. *EMBO J.* **21:** 2746.

Darzacq X., Kittur N., Roy S., Shav-Tal Y., Singer R.H., and Meier U.T. 2006. Stepwise RNP assembly at the site of H/ACA RNA transcription in human cells. *J. Cell Biol.* **173:** 207.

de Lange T. 2005. Shelterin: The protein complex that shapes and safeguards human telomeres. *Genes Dev.* **19:** 2100.

Decatur W.A. and Fournier M.J. 2003. RNA-guided nucleotide modification of ribosomal and other RNAs. *J. Biol. Chem.* **278:** 695.

Dez C., Noaillac-Depeyre J., Caizergues-Ferrer M., and Henry Y. 2002. Naf1p, an essential nucleoplasmic factor specifically required for accumulation of box H/ACA small nucleolar RNPs. *Mol. Cell. Biol.* **22:** 7053.

Dragon F., Pogacic V., and Filipowicz W. 2000. In vitro assembly of human H/ACA small nucleolar RNPs reveals unique features of U17 and telomerase RNAs. *Mol. Cell. Biol.* **20:** 3037.

Eliceiri G.L. 2006. The vertebrate E1/U17 small nucleolar ribonucleoprotein particle. *J. Cell. Biochem.* **98:** 486.

Fatica A., Dlakic M., and Tollervey D. 2002. Naf1p is a box H/ACA snoRNP assembly factor. *RNA* **8:** 1502.

Filipowicz W. and Pogacic V. 2002. Biogenesis of small nucleolar ribonucleoproteins. *Curr. Opin. Cell Biol.* **14:** 319.

Fu D. and Collins K. 2003. Distinct biogenesis pathways for human telomerase RNA and H/ACA small nucleolar RNAs. *Mol. Cell* **11:** 1361.

———. 2006. Human telomerase and Cajal body ribonucleoproteins share a unique specificity of Sm protein association. *Genes Dev.* **20:** 531.

Gall J.G. 2000. Cajal bodies: The first 100 years. *Annu. Rev. Cell Dev. Biol.* **16:** 273.

Ganot P., Bortolin M.L., and Kiss T. 1997a. Site-specific pseudouridine formation in preribosomal RNA is guided by small nucleolar RNAs. *Cell* **89:** 799.

Ganot P., Caizergues-Ferrer M., and Kiss T. 1997b. The family of box ACA small nucleolar RNAs is defined by an evolutionarily conserved secondary structure and ubiquitous sequence elements essential for RNA accumulation. *Genes Dev.* **11:** 941.

Girard J.P., Lehtonen H., Caizergues-Ferrer M., Amalric F., Tollervey D., and Lapeyre B. 1992. GAR1 is an essential small nucleolar RNP protein required for pre-rRNA processing in yeast. *EMBO J.* **11:** 673.

Greider C.W. and Blackburn E.H. 1985. Identification of a specific telomere terminal transferase activity in *Tetrahymena* extracts. *Cell* **43:** 405.

Gubitz A.K., Feng W., and Dreyfuss G. 2004. The SMN complex. *Exp. Cell Res.* **296:** 51.

Hamma T. and Ferre-D'Amare A.R. 2004. Structure of protein L7Ae bound to a K-turn derived from an archaeal box H/ACA sRNA at 1.8 Å resolution. *Structure* **12:** 893.

Hamma T., Reichow S.L., Varani G., and Ferre-D'Amare A.R. 2005. The Cbf5-Nop10 complex is a molecular bracket that organizes box H/ACA RNPs. *Nat. Struct. Mol. Biol.* **12:** 1101.

Heiss N.S., Knight S.W., Vulliamy T.J., Klauck S.M., Wiemann S., Mason P.J., Poustka A., and Dokal I. 1998. X-linked dyskeratosis congenita is caused by mutations in a highly conserved gene with putative nucleolar functions. *Nat. Genet.* **19:** 32.

Henras A.K., Capeyrou R., Henry Y., and Caizergues-Ferrer M. 2004. Cbf5p, the putative pseudouridine synthase of H/ACA-type snoRNPs, can form a complex with Gar1p and Nop10p in absence of Nhp2p and box H/ACA snoRNAs. *RNA* **10:** 1704.

Henras A., Dez C., Noaillac-Depeyre J., Henry Y., and Caizergues-Ferrer M. 2001. Accumulation of H/ACA snoRNPs depends on the integrity of the conserved central domain of the RNA-binding protein Nhp2p. *Nucleic Acids Res.* **29:** 2733.

Hoang C. and Ferre-D'Amare A.R. 2001. Cocrystal structure of a tRNA Psi55 pseudouridine synthase: Nucleotide flipping by an RNA-modifying enzyme. *Cell* **107:** 929.

Hoareau-Aveilla C., Bonoli M., Caizergues-Ferrer M., and Henry Y. 2006. hNaf1 is required for accumulation of human box H/ACA snoRNPs, scaRNPs, and telomerase. *RNA* **12:** 832.

Hug N. and Lingner J. 2006. Telomere length homeostasis. *Chromosoma* **115:** 413.

Hüttenhofer A., Kiefmann M., Meier-Ewert S., O'Brien J., Lehrach H., Bachellerie J.P., and Brosius J. 2001. RNomics: An experimental approach that identifies 201 candidates for novel, small, non-messenger RNAs in mouse. *EMBO J.* **20:** 2943.

Jady B.E. and Kiss T. 2001. A small nucleolar guide RNA functions both in 2′-O-ribose methylation and pseudouridylation of the U5 spliceosomal RNA. *EMBO J.* **20:** 541.

Jady B.E., Bertrand E., and Kiss T. 2004. Human telomerase RNA and box H/ACA scaRNAs share a common Cajal body-specific localization signal. *J. Cell Biol.* **164:** 647.

Jady B.E., Richard P., Bertrand E., and Kiss T. 2006. Cell cycle-dependent recruitment of telomerase RNA and Cajal bodies to human telomeres. *Mol. Biol. Cell.* **17:** 944.

King T.H., Liu B., McCully R.R., and Fournier M.J. 2003. Ribosome structure and activity are altered in cells lacking snoRNPs that form pseudouridines in the peptidyl transferase center. *Mol. Cell* **11:** 425.

King T.H., Decatur W.A., Bertrand E., Maxwell E.S., and Fournier M.J. 2001. A well-connected and conserved nucleoplasmic helicase is required for production of box C/D and H/ACA snoRNAs and localization of snoRNP proteins. *Mol. Cell. Biol.* **21:** 7731.

Kiss A.M., Jady B.E., Bertrand E., and Kiss T. 2004. Human box H/ACA pseudouridylation guide RNA machinery. *Mol. Cell. Biol.* **24:** 5797.

Kiss A.M., Jady B.E., Darzacq X., Verheggen C., Bertrand E., and Kiss T. 2002. A Cajal body-specific pseudouridylation guide RNA is composed of two box H/ACA snoRNA-like domains. *Nucleic Acids Res.* **30:** 4643.

Kiss T. 2002. Small nucleolar RNAs: An abundant group of non-coding RNAs with diverse cellular functions. *Cell* **109:** 145.

Klein D.J., Schmeing T.M., Moore P.B., and Steitz T.A. 2001. The kink-turn: A new RNA secondary structure motif. *EMBO J.* **20:** 4214.

Klein R.J., Misulovin Z., and Eddy S.R. 2002. Noncoding RNA genes identified in AT-rich hyperthermophiles. *Proc. Natl. Acad. Sci.* **99:** 7542.

Koonin E.V. 1996. Pseudouridine synthases: Four families of enzymes containing a putative uridine-binding motif also conserved in dUTPases and dCTP deaminases. *Nucleic Acids Res.* **24:** 2411.

Lafontaine D.L., Bousquet-Antonelli C., Henry Y., Caizergues-Ferrer M., and Tollervey D. 1998. The box H + ACA snoR-NAs carry Cbf5p, the putative rRNA pseudouridine synthase. *Genes Dev.* **12:** 527.

Lange T.S., Ezrokhi M., Amaldi F., and Gerbi S.A. 1999. Box H and box ACA are nucleolar localization elements of U17 small nucleolar RNA. *Mol. Biol. Cell* **10:** 3877.

Lestrade L. and Weber M.J. 2006. snoRNA-LBME-db, a comprehensive database of human H/ACA and C/D box snoR-NAs. *Nucleic Acids Res.* **34:** D158.

Li L. and Ye K. 2006. Crystal structure of an H/ACA box ribonucleoprotein particle. *Nature* **443:** 302.

Liang X.H., Uliel S., A. Hury, Barth S., Doniger T., Unger R., and Michaeli S. 2005. A genome-wide analysis of C/D and H/ACA-like small nucleolar RNAs in *Trypanosoma brucei* reveals a trypanosome-specific pattern of rRNA modification. *RNA* **11:** 619.

Lubben B., Fabrizio P., Kastner B., and Luhrmann R. 1995. Isolation and characterization of the small nucleolar ribonucleoprotein particle snR30 from *Saccharomyces cerevisiae*. *J. Biol. Chem.* **270:** 11549.

Lukowiak A.A., Narayanan A., Li Z.H., Terns R.M., and Terns M.P. 2001. The snoRNA domain of vertebrate telomerase RNA functions to localize the RNA within the nucleus. *RNA* **7:** 1833.

Maden B.E. 1990. The numerous modified nucleotides in eukaryotic ribosomal RNA. *Prog. Nucleic Acid Res. Mol. Biol.* **39:** 241.

Manival X., Charron C., Fourmann J.B., Godard F., Charpentier B., and Branlant C. 2006. Crystal structure determination and site-directed mutagenesis of the *Pyrococcus abyssi* aCBF5-aNOP10 complex reveal crucial roles of the C-terminal domains of both proteins in H/ACA sRNP activity. *Nucleic Acids Res.* **34:** 826.

Marrone A., Walne A., and Dokal I. 2005. Dyskeratosis congenita: Telomerase, telomeres and anticipation. *Curr. Opin. Genet. Dev.* **15:** 249.

Mason P.J., Wilson D.B., and Bessler M. 2005. Dyskeratosis congenita—A disease of dysfunctional telomere maintenance. *Curr. Mol. Med.* **5:** 159.

Matera A.G. and Frey M.R. 1998. Coiled bodies and gems: Janus or gemini? *Am. J. Hum. Genet.* **63:** 317.

Matera A.G. and Shpargel K.B. 2006. Pumping RNA: Nuclear bodybuilding along the RNP pipeline. *Curr. Opin. Cell Biol.* **18:** 317.

Meier U.T. 2003. Dissecting dyskeratosis. *Nat. Genet.* **33:** 116.
———. 2005. The many facets of H/ACA ribonucleoproteins. *Chromosoma* **114:** 1.

Meister G., Eggert C., and Fischer U. 2002. SMN-mediated assembly of RNPs: A complex story. *Trends Cell Biol.* **12:** 472.

Mishra R.K. and Eliceiri G.L. 1997. Three small nucleolar RNAs that are involved in ribosomal RNA precursor processing. *Proc. Natl. Acad. Sci.* **94:** 4972.

Mitchell J.R. and Collins K. 2000. Human telomerase activation requires two independent interactions between telomerase RNA and telomerase reverse transcriptase. *Mol. Cell* **6:** 361.

Mitchell J.R., Cheng J., and Collins K. 1999a. A box H/ACA small nucleolar RNA-like domain at the human telomerase RNA 3′ end. *Mol. Cell. Biol.* **19:** 567.

Mitchell J.R., E. Wood E., and Collins K. 1999b. A telomerase component is defective in the human disease dyskeratosis congenita. *Nature* **402:** 551.

Mochizuki Y., He J., Kulkarni S., Bessler M., and Mason P.J. 2004. Mouse dyskerin mutations affect accumulation of telomerase RNA and small nucleolar RNA, telomerase activity, and ribosomal RNA processing. *Proc. Natl. Acad. Sci.* **101:** 10756.

Morrissey J.P. and Tollervey D. 1993. Yeast snR30 is a small nucleolar RNA required for 18S rRNA synthesis. *Mol. Cell. Biol.* **13:** 2469.

Narayanan A., Lukowiak A., Jady B.E., Dragon F., Kiss T., Terns R.M., and Terns M.P. 1999. Nucleolar localization signals of box H/ACA small nucleolar RNAs. *EMBO J.* **18:** 5120.

Ni J., Tien A.L., and Fournier M.J. 1997. Small nucleolar RNAs direct site-specific synthesis of pseudouridine in ribosomal RNA. *Cell* **89:** 565.

Ofengand J., Malhotra A., Remme J., Gutgsell N.S., Del Campo M., Jean-Charles S., Peil L., and Kaya Y. 2001. Pseudouridines and pseudouridine synthases of the ribosome. *Cold Spring Harbor Symp. Quant. Biol.* **66:** 147.

Pellizzoni L., Baccon J., Charroux B., and Dreyfuss G. 2001. The survival of motor neurons (SMN) protein interacts with the snoRNP proteins fibrillarin and GAR1. *Curr. Biol.* **11:** 1079.

Pogacic V., Dragon F., and Filipowicz W. 2000. Human H/ACA small nucleolar RNPs and telomerase share evolutionarily conserved proteins NHP2 and NOP10. *Mol. Cell. Biol.* **20:** 9028.

Rashid R., Liang B., Baker D.L., Youssef O.A., He Y., Phipps K., Terns R.M., Terns M.P., and Li H. 2006. Crystal structure of a Cbf5-Nop10-Gar1 complex and implications in RNA-guided pseudouridylation and dyskeratosis congenita. *Mol. Cell* **21:** 249.

Richard P. and Kiss T. 2006. Integrating snoRNP assembly with mRNA biogenesis. *EMBO Rep.* **7:** 590.

Richard P., Kiss A.M., Darzacq X., and Kiss T. 2006. Cotranscriptional recognition of human intronic box H/ACA snoR-NAs occurs in a splicing-independent manner. *Mol. Cell. Biol.* **26:** 2540.

Richard P., Darzacq X., Bertrand E., Jady B.E., Verheggen C., and Kiss T. 2003. A common sequence motif determines the Cajal body-specific localization of box H/ACA scaRNAs. *EMBO J.* **22:** 4283.

Rozhdestvensky T.S., Tang T.H., Tchirkova I.V., Brosius J., Bachellerie J.P., and Hüttenhofer A. 2003. Binding of L7Ae protein to the K-turn of archaeal snoRNAs: A shared RNA binding motif for C/D and H/ACA box snoRNAs in Archaea. *Nucleic Acids Res.* **31:** 869.

Ruhl D.D., Pusateri M.E., and Eliceiri G.L. 2000. Multiple conserved segments of E1 small nucleolar RNA are involved in the formation of a ribonucleoprotein particle in frog oocytes. *Biochem. J.* **348:** 517.

Russell A.G., Schnare M.N., and Gray M.W. 2004. Pseudouridine-guide RNAs and other Cbf5p-associated RNAs in *Euglena gracilis*. *RNA* **10:** 1034.

Schattner P., Decatur W.A., Davis C.A., Ares M., Jr., Fournier M.J., and Lowe T.M. 2004. Genome-wide searching for pseudouridylation guide snoRNAs: Analysis of the *Saccharomyces cerevisiae* genome. *Nucleic Acids Res.* **32:** 4281.

Steitz J.A. and Tycowski K.T. 1995. Small RNA chaperones for ribosome biogenesis. *Science* **270:** 1626.

Tang T.H., Bachellerie J.P., Rozhdestvensky T., Bortolin M.L., Huber H., Drungowski M., Elge T., Brosius J., and Hüttenhofer A. 2002. Identification of 86 candidates for small non-messenger RNAs from the archaeon *Archaeoglobus fulgidus*. *Proc. Natl. Acad. Sci.* **99:** 7536.

Ten Hagen K.G., Gilbert D.M., Willard H.F., and Cohen S.N. 1990. Replication timing of DNA sequences associated with human centromeres and telomeres. *Mol. Cell. Biol.* **10:** 6348.

Terns M.P. and Terns R.M. 2001. Macromolecular complexes: SMN—The master assembler. *Curr. Biol.* **11:** R862.
———. 2002. Small nucleolar RNAs: Versatile trans-acting molecules of ancient evolutionary origin. *Gene Expr.* **10:** 17.

Ting N.S., Yu Y., Pohorelic B., Lees-Miller S.P., and Beattie T.L. 2005. Human Ku70/80 interacts directly with hTR, the RNA component of human telomerase. *Nucleic Acids Res.* **33:** 2090.

Tollervey D. 1987. A yeast small nuclear RNA is required for normal processing of pre-ribosomal RNA. *EMBO J.* **6:** 4169.

Tomlinson R.L., Ziegler T.D., Supakorndej T., Terns R.M., and Terns M.P. 2006. Cell cycle-regulated trafficking of human telomerase to telomeres. *Mol. Biol. Cell* **17:** 955.

Torchet C., Badis G., Devaux F., Costanzo G., Werner M., and Jacquier A. 2005. The complete set of H/ACA snoRNAs that guide rRNA pseudouridylations in *Saccharomyces cerevisiae*. *RNA* **11:** 928.

Uliel S., Liang X.H., Unger R., and Michaeli S. 2004. Small nucleolar RNAs that guide modification in trypanosomatids: Repertoire, targets, genome organisation, and unique functions. *Int. J. Parasitol.* **34:** 445.

Vulliamy T., Marrone A., Goldman F., Dearlove A., Bessler M., Mason P.J., and Dokal I. 2001. The RNA component of telomerase is mutated in autosomal dominant dyskeratosis congenita. *Nature* **413:** 432.

Wang C. and Meier U.T. 2004. Architecture and assembly of mammalian H/ACA small nucleolar and telomerase ribonucleoproteins. *EMBO J.* **23:** 1857.

Wang C., Query C.C., and Meier U.T. 2002. Immunopurified small nucleolar ribonucleoprotein particles pseudouridylate rRNA independently of their association with phosphorylated Nopp140. *Mol. Cell. Biol.* **22:** 8457.

Watkins N.J., Gottschalk A., Neubauer G., Kastner B., Fabrizio P., Mann M., and Luhrmann R. 1998. Cbf5p, a potential pseudouridine synthase, and Nhp2p, a putative RNA-binding protein, are present together with Gar1p in all H BOX/ACA-motif snoRNPs and constitute a common bipartite structure. *RNA* **4:** 1549.

Whitehead S.E., Jones K.W., Zhang X., Cheng X., Terns R.M., and Terns M.P. 2002. Determinants of the interaction of the spinal muscular atrophy disease protein SMN with the dimethylarginine-modified box H/ACA small nucleolar ribonucleoprotein GAR1. *J. Biol. Chem.* **277:** 48087.

Wright W.E., Tesmer V.M., Liao M.L., and Shay J.W. 1999. Normal human telomeres are not late replicating. *Exp. Cell Res.* **251:** 492.

Yang P.K., Rotondo G., Porras T., Legrain P., and Chanfreau G. 2002. The Shq1p.Naf1p complex is required for box H/ACA small nucleolar ribonucleoprotein particle biogenesis. *J. Biol. Chem.* **277:** 45235.

Yang P.K., Hoareau C., Froment C., Monsarrat B., Henry Y., and Chanfreau G. 2005. Cotranscriptional recruitment of the pseudouridylsynthetase Cbf5p and of the RNA binding protein Naf1p during H/ACA snoRNP assembly. *Mol. Cell. Biol.* **25:** 3295.

Yu Y.T. 2006. The most complex pseudouridylase. *Structure* **14:** 167.

Yu Y.T., Terns R.M., and Terns M.P. 2005. Mechanisms and functions of RNA-guided RNA modification. *Top. Curr. Genet.* **12:** 223.

Yuan G., Klambt C., Bachellerie J.P., Brosius J., and Hüttenhofer A. 2003. RNomics in *Drosophila melanogaster:* Identification of 66 candidates for novel non-messenger RNAs. *Nucleic Acids Res.* **31:** 2495.

Zhu Y., Tomlinson R.L., Lukowiak A.A., Terns R.M., and Terns M.P. 2004. Telomerase RNA accumulates in Cajal bodies in human cancer cells. *Mol. Biol. Cell* **15:** 81.

Biogenesis and Intranuclear Trafficking of Human Box C/D and H/ACA RNPs

T. Kiss,*† E. Fayet,* B.E. Jády,* P. Richard,* and M. Weber*

*Laboratoire de Biologie Moléculaire Eucaryote du CNRS, UMR5099, IFR109, 31062 Toulouse, France;
†Biological Research Center, Hungarian Academy of Sciences, Szeged, Hungary

Box C/D and H/ACA snoRNAs represent two abundant groups of small noncoding RNAs. The majority of box C/D and H/ACA snoRNAs function as guide RNAs in the site-specific 2′-O-methylation and pseudouridylation of rRNAs, respectively. The box C/D snoRNAs associate with fibrillarin, Nop56, Nop58, and 15.5K/NHPX proteins to form functional snoRNP particles, whereas all box H/ACA snoRNAs form complexes with the dyskerin, Nop10, Nhp2, and Gar1 snoRNP proteins. Recent studies demonstrate that the biogenesis of mammalian snoRNPs is a complex process that requires numerous *trans*-acting factors. Most vertebrate snoRNAs are posttranscriptionally processed from pre-mRNA introns, and the early steps of snoRNP assembly are physically and functionally coupled with the synthesis or splicing of the host pre-mRNA. The maturing snoRNPs follow a complicated intranuclear trafficking process that is directed by transport factors also involved in nucleocytoplasmic RNA transport. The human telomerase RNA (hTR) carries a box H/ACA RNA domain that shares a common Cajal-body-specific localization element with a subclass of box H/ACA RNAs, which direct pseudouridylation of spliceosomal snRNAs in the Cajal body. However, besides concentrating in Cajal bodies, hTR also accumulates at a small, structurally distinct subset of telomeres during S phase. This suggests that a cell-cycle-dependent, dynamic localization of hTR to telomeres may play an important regulatory role in human telomere synthesis.

The mammalian nucleus contains many small nuclear RNAs (snRNAs) that can be grouped under two main classes based on their intranuclear localization (for review, see Yu et al. 1999). Early cell fractionation experiments revealed that the abundant U1, U2, U4, U5, and U6 snRNAs localize to the nucleoplasm, whereas another abundant snRNA, U3, was found to copurify with the nucleoli and, therefore, it was designated as small nucleolar RNA (snoRNA) (Reddy et al. 1981). Later, it turned out that the U3 snoRNA represents the founding member of a major class of small noncoding RNAs that reside in the nucleolus, share the evolutionarily conserved C (RUGAUGA) and D (uCUGA) box motifs, and associate with a nucleolar protein, fibrillarin (Fig. 1) (Tyc and Steitz 1989).

In the early 1990s, three novel human snoRNAs, U17, E2, and E3, which lacked C and D boxes and did not associate with fibrillarin, were discovered (Kiss and Filipowicz 1993; Ruff et al. 1993). Later, construction and characterization of a cDNA library of human nucleolar RNAs presented further examples of mammalian snoRNAs lacking C and D boxes (Ganot et al. 1997b). In vitro structure probing experiments, followed by structural comparisons with yeast snoRNAs, revealed that all snoRNAs devoid of C and D boxes share the conserved box H (AnAnnA) and ACA motifs and fold into a common "hairpin-hinge-hairpin-tail" secondary structure, demonstrating that they constitute a novel, evolutionarily conserved class of snoRNAs (Fig. 1) (Balakin et al. 1996; Kiss et al. 1996; Ganot et al. 1997b).

In the past years, several hundreds of box C/D and H/ACA snoRNAs have been identified in a broad variety of organisms (for review, see Bachellerie et al. 2002; Hüttenhofer et al. 2005). For a compilation of vertebrate

snoRNAs, see http://www-snorna.biotoul.fr/index.php (Lestrade and Weber 2006). We learned that both box C/D and H/ACA snoRNA families show a high structural and functional conservation from humans to Archaea (for review, see Terns and Terns 2002; Omer et al. 2003; Bertrand and Fournier 2004; Tran et al. 2004). The majority of box C/D snoRNAs function as guide RNAs in the site-specific 2′-O-methylation of rRNAs (Cavaillé et al. 1996; Kiss-László et al. 1996), whereas most box H/ACA snoRNAs direct pseudouridylation of rRNAs (Fig. 1) (Ganot et al. 1997a; Ni et al. 1997). The box C/D and H/ACA modification guide snoRNAs function in the form of snoRNPs. The box C/D snoRNAs are associated with four box C/D snoRNP proteins, fibrillarin, Nop56, Nop58, and 15.5K/NHPX (Fig. 1) (Schimmang et al. 1989; Tyc and Steitz 1989; Wu et al. 1998; Lafontaine and Tollervey 1999; Newman et al. 2000; Watkins et al. 2000). The box H/ACA snoRNAs form snoRNP complexes with dyskerin, Nhp2, Nop10, and Gar1 proteins (Balakin et al. 1996; Ganot et al. 1997b; Henras et al. 1998; Watkins et al. 1998; Lyman et al. 1999). Both box C/D and H/ACA guide RNAs select the substrate nucleotides for modification through forming transient base-pairing interactions with complementary rRNA sequences. The associated snoRNP proteins, fibrillarin and dyskerin, catalyze the 2′-O-methyl transfer and the uridine-to-pseudouridine isomerization reactions, respectively (Wang et al. 2000; Hoang and Ferre-D'Amare 2001).

Besides directing rRNA modification, the box C/D and H/ACA guide RNPs also function in methylation and pseudouridylation of spliceosomal snRNAs and other cellular RNAs, including some tRNAs and mRNAs. The

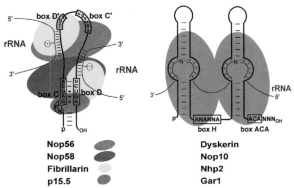

Figure 1. Schematic structure of box C/D 2′-O-methylation and box H/ACA pseudouridylation guide snoRNPs. Black lines represent snoRNA sequences. Positions and consensus sequences of the conserved C and D boxes and the related C′ and D′ boxes as well as the H and ACA boxes are shown. Blue lines indicate pre-rRNA sequences interacting with the antisense elements of box C/D and H/ACA guide RNAs. The 2′-O-methylated nucleotides located five nucleotides upstream of the D or D′ box sequences are indicated (*m*). The uridine residues selected for pseudouridylation are shown (Ψ). Binding of the 15.5K (NHPX) protein to the terminal Kink-turn motif of box C/D snoRNAs is a prerequisite for recruitment of the two highly related proteins Nop56 and Nop58 and two copies of the methyltransferase, fibrillarin (Cahill et al. 2002; Watkins et al. 2002). Organization of box H/ACA snoRNP proteins is still unknown. Electron microscopy images of purified yeast snR30 box H/ACA snoRNP particles showed a highly symmetric bipartite structure, suggesting that two sets of the four box H/ACA snoRNP core proteins dyskerin, Nop10, Nhp2, and Gar1 bind to the 5′- and 3′-hairpins of snR30 (Watkins et al. 1998).

interested reader may consult several recent reviews on the structure, function, and evolution of box C/D and H/ACA RNPs (Kiss 2001; Bachellerie et al. 2002; Filipowicz and Pogacic 2002; Terns and Terns 2002; Decatur and Fournier 2003; Bertrand and Fournier 2004; Henras et al. 2004; Dennis and Omer 2005). Here, we focus on recent advances concerning the biogenesis and intranuclear trafficking of mammalian box C/D and H/ACA snoRNPs, including the human telomerase H/ACA RNP.

SPLICING-DEPENDENT ASSEMBLY OF INTRONIC BOX C/D snoRNPs

In vertebrates, the great majority of box C/D and H/ACA snoRNAs are encoded within introns of pre-mRNAs (Fig. 2) (Leverette et al. 1992; Fragapane et al. 1993; Kiss and Filipowicz 1993; Tycowski et al. 1993). Normally, the mature intronic snoRNAs are processed from the removed and debranched host introns by 5′ to 3′ and 3′ to 5′ exonucleolytic activities (Tycowski et al. 1993; Kiss and Filipowicz 1995; Cavaillé and Bachellerie 1996; Watkins et al. 1996). Therefore, splicing of the host pre-mRNA is essential for providing linear precursor snoRNA substrates for the processing exonucleases (Ooi et al. 1998). Consistent with a splicing-mediated precursor snoRNA release, all host introns for vertebrate snoRNAs encode only a single intronic snoRNA (the one intron/one snoRNA rule). The biosynthesis of functional intronic

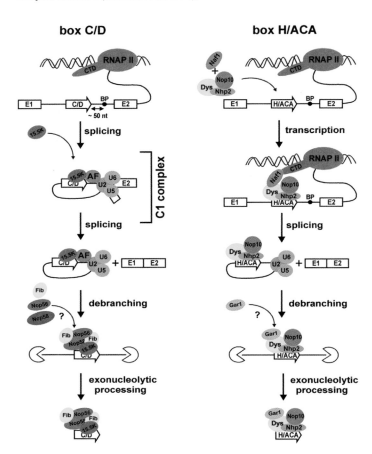

Figure 2. Models for splicing- and transcription-dependent assembly of mammalian intron-encoded snoRNPs. The great majority of mammalian box C/D and H/ACA snoRNAs are processed from pre-mRNA introns. Whereas most box C/D snoRNAs are located about 50 nucleotides upstream of the branch point (BP), box H/ACA snoRNAs possess a random location relative to the 5′ and 3′ splice sites. Recognition of intronic box C/D snoRNA sequences occurs at a relatively late step of host pre-mRNA splicing. A putative assembly factor (AF) that likely interacts with splicing factors associated with the BP region of the host pre-mRNA facilitates the recruitment of the 15.5K/NHPX box C/D snoRNP protein to the Kink-turn motif of the intronic snoRNA. Assembly of box H/ACA snoRNPs already occurs during pre-mRNA synthesis. The Naf1 protein may promote cotranscriptional recruitment of the dyskerin/Nop10/Nhp2 protein complex to the nascent H/ACA snoRNA through forming specific interactions with both dyskerin and the CTD of RNAP II. Binding of Nop56, Nop58, and fibrillarin box C/D and Gar1 box H/ACA snoRNP proteins is believed to occur in the Cajal body. For other details, see the text.

snoRNPs includes the ordered recruitment of the four box C/D (fibrillarin, Nop56, Nop58, and 15.5K/NHPX) or box H/ACA (dyskerin, Nhp2, Nop10, and Gar1) snoRNP proteins. Binding of snoRNP proteins is essential for the correct processing and metabolic stability of the mature snoRNA, since the associated snoRNP proteins define the termini of the snoRNA by protecting them from the processing exonucleases. However, normally, mammalian pre-mRNA introns are rapidly degraded after splicing (Padgett et al. 1986). The rapid intron turnover may facilitate the recycling of ribonucleotides and splicing factors bound to the removed intron lariat (Green 1991). Therefore, efficient intronic snoRNA expression might require an active mechanism that recruits snoRNP proteins to the nascent snoRNA already during synthesis or splicing of the host pre-mRNA.

The 5′ and 3′ terminal regions of box C/D snoRNAs encompassing the box C and D sequences, respectively, fold into a stem-internal loop-stem structure, called the Kink-turn (Fig. 1) (Watkins et al. 2000; Klein et al. 2001). The noncanonical G-A, A-G, and U-U base pairs formed by conserved nucleotides in the C and D box motifs are important for the establishment of a functional Kink-turn structure and for docking the 15.5K/NHPX snoRNP protein. Binding of 15.5K/NHPX induces a sharp band in the phosphodiester backbone of the two contiguous RNA stems of the Kink-turn (Vidovic et al. 2000; Szewczak et al. 2002, 2005; Watkins et al. 2002). This conformational change provides the structural requirements for the subsequent binding of Nop58, Nop56, and two copies of fibrillarin (Cahill et al. 2002). In vivo cross-linking experiments showed that one fibrillarin and Nop58 bind to the box D and C sequences in the upper stem of Kink-turn, respectively (Cahill et al. 2002). Nop56 and another copy of fibrillarin can be cross-linked to internal copies of the C and D boxes, termed the C′ and D′ boxes. Docking of these proteins is likely facilitated by protein–protein interactions, since generally the C′ and D′ boxes show poor sequence conservation (Kiss-László et al. 1998).

The Steitz group noticed that human box C/D snoRNAs possess a preferential intronic location. Namely, most box C/D snoRNAs are located about 80–90 nucleotides upstream of the 3′ splice site (Hirose and Steitz 2001). In vivo and in vitro snoRNA processing experiments confirmed that an optimal distance of about 50 nucleotides between the snoRNA coding region and the branch point of the host intron is required for efficient snoRNA processing. Increasing or decreasing the spacer length between the snoRNA and the branch point seriously compromises snoRNA accumulation (Hirose and Steitz 2001; Hirose et al. 2003). By using a coupled in vitro splicing/snoRNA processing system, the Steitz group demonstrated that recruitment of 15.5K/NHPX to box C/D intronic snoRNAs occurs specifically at the C1 splicing complex stage, indicating that 15.5K/NHPX is actively recruited to intronic box C/D snoRNAs by a splicing-dependent mechanism (Fig. 1) (Hirose et al. 2003). They proposed that a putative box C/D snoRNP assembly factor (AF) that interacts directly or indirectly with the U2 spliceosomal snRNP, or another splicing factor associated with the branch point region in the C1 splicing complex, recruits and deposits the 15.5K/NHPX protein onto the Kink-turn of box C/D snoRNAs. Identification of putative factors promoting the splicing-dependent assembly of box C/D intronic snoRNPs requires further efforts.

COUPLING BOX H/ACA snoRNP ASSEMBLY WITH POLYMERASE II TRANSCRIPTION

In contrast to box C/D snoRNAs, human H/ACA snoRNAs have no preferential intronic location relative to the 5′ or 3′ splice sites of the host introns (Richard et al. 2006; Schattner et al. 2006). Moreover, whereas most box C/D snoRNA genes are found within relatively short introns, human box H/ACA snoRNAs tend to reside within introns of longer than average length. Consistent with this, human box H/ACA snoRNAs are processed from the introns of transiently expressed natural or artificial host pre-mRNAs in a position-independent fashion, indicating that the splicing machine does not participate in the assembly of intronic box H/ACA snoRNPs (Richard et al. 2006).

Chromatin and pre-mRNA coimmunoprecipitation and in situ localization experiments demonstrated that binding of dyskerin, Nhp2, and Nop10 snoRNP proteins to intronic H/ACA snoRNAs is an early event that occurs shortly after or already during the synthesis of the host pre-mRNA (Fig. 2) (Darzacq et al. 2006; Richard et al. 2006). In vivo processing studies performed on transiently expressed artificial pre-mRNA transcripts confirmed that assembly of box H/ACA snoRNPs already occurs on the newly synthesized pre-mRNAs and demonstrated that snoRNP assembly and splice site selection, although they occur at the same time, are independent molecular events (Darzacq et al. 2006; Richard et al. 2006).

Correct and efficient expression of mammalian box H/ACA snoRNPs requires RNA polymerase (RNAP) II transcription (Richard et al. 2006). Precursor snoRNAs synthesized by RNAP III or RNAP I are either not processed or are poorly processed, and the resulting mature-sized snoRNAs fail to correctly localize in the nucleolus. This suggests that an RNAP-II-associated factor may promote the cotranscriptional assembly of box H/ACA pre-snoRNPs. In fact, expression of both yeast and mammalian H/ACA snoRNPs requires two evolutionarily conserved H/ACA-specific processing/assembly factors, Naf1 and Shq1 (Dez et al. 2002; Fatica et al. 2002; Yang et al. 2002; Hoareau-Aveilla et al. 2006). Yeast Naf1 and Shq1 can form a complex, and they can interact with the box H/ACA snoRNP core proteins, dyskerin (Cbf5p in yeast) and Nhp2. Moreover, yeast Naf1 specifically associates with the carboxy-terminal domain (CTD) of the largest subunit of RNAP II (Fatica et al. 2002). Recombinant mammalian dyskerin, Nop10, and Nhp2 are capable of forming a protein-only complex that can specifically bind to H/ACA RNAs (Wang and Meier 2004; Darzacq et al. 2006). Naf1 may recruit the dyskerin/ Nop10/Nhp2 H/ACA protein core complex to the newly synthesized intronic H/ACA pre-snoRNAs through interacting with dyskerin and CTD (Fig. 2). Consistent with its involvement in coupling H/ACA snoRNP assembly with RNAP II transcription, Naf1 specifically associates with actively transcribed H/ACA snoRNA genes, both in yeast and mammalian cells (Ballarino et al. 2005; Yang et al. 2005; Darzacq et al. 2006). It remains uncertain when and where Gar1, the last-binding H/ACA protein, binds to the maturing snoRNP. Gar1 is essential for snoRNA-directed pseudouridylation, but it is dispensable for accumulation of box H/ACA snoRNAs (Bousquet-Antonelli et al. 1997). Under in vitro conditions, Naf1 and Gar1 bind competitively to dyskerin, suggesting that Gar1 may replace Naf1 at a later stage of H/ACA snoRNP biogenesis (Darzacq et al. 2006).

MULTIPLE *TRANS*-ACTING FACTORS ARE INVOLVED IN MAMMALIAN snoRNP BIOGENESIS

Besides the box C/D and H/ACA snoRNP core proteins, several *trans*-acting protein factors have been implicated in the biogenesis of snoRNPs. Although most of the available data were obtained in the yeast *Saccharomyces cerevisiae* system, recent studies on human U3 snoRNP assembly gave new insights into the biogenesis of human box C/D snoRNPs (Verheggen et al. 2001, 2002; Boulon et al. 2004; Watkins et al. 2004). U3 is the most abundant mammalian snoRNA that is transcribed from independent genes by RNAP II, instead of being processed from pre-mRNA introns. During the nucleoplasmic biosynthesis of U3 snoRNP, the precursor U3 snoRNA that carries a short uridine-rich 3′ trailer transiently interacts with numerous non-snoRNP proteins to form a large, structurally dynamic, multiprotein processing complex that is, in fact, larger than the mature U3 snoRNP accumulating in the nucleolus (Boulon et al. 2004; Watkins et al. 2004). The proteins which transiently associate with precursor U3 include known RNA processing factors (TGS1, La, LSm4, and Rrp46), putative RNP assembly factors (Nopp140, Tip48, and Tip49), and well-characterized RNA export factors (CRM1, PHAX, Ran, and the cap binding complex [CBC]).

The La, LSm4, and Rrp46 proteins associate exclusively with the 3′-extended precursor of U3, indicating that these proteins function in the 3′ end formation of the U3 snoRNA (Watkins et al. 2004). Indeed, the La and Lsm4 proteins, the latter as a component of the Lsm2 to Lsm8 heteroheptameric complex, possess well-established functions in stabilizing and processing of mature 3′ ends of tRNAs, snRNAs, and snoRNAs (Achsel et al. 1999; Perumal and Reddy 2002; Wolin and Cedervall 2002; Beggs 2005; Maraia and Bayfield 2006). The La and Lsm proteins likely provide stability for the nascent U3 snoRNA through binding to its uridine-rich 3′ overhang. Rpr46 is an integral component of the human nuclear exosome complex which consists of multiple 3′ to 5′ exoribonucleases. Therefore, Rpr46 likely participates in the 3′ end trimming of U3 (Allmang et al. 1999; Vasiljeva and Buratowski 2006). The Lsm heteroheptameric complex associated with the 3′ terminal trailer of the precursor U3 snoRNA may recruit the exosome complex to the maturing U3 snoRNP (Fromont-Racine et al. 2000). Like all RNAP II transcripts, the nascent U3 snoRNA contains a monomethyl-G (m7G) cap structure

that is hypermethylated to mature trimethyl-G (TMG) by the TGS1 methyltransferase (Mouaikel et al. 2002).

An important question is whether the information obtained on the processing of the RNAP II-transcribed U3 snoRNA holds true for the biogenesis of intron-encoded box C/D snoRNAs. RNAi-mediated depletion experiments demonstrated that all factors implicated in U3 maturation, apart from the easily understandable exception of the TGS1 methyltransferase, are also important for the accumulation of the intron-encoded U14 snoRNA (Watkins et al. 2004). This indicates that the nuclear machines supporting the biogenesis of independently transcribed and intron-encoded box C/D snoRNPs share common structural and functional principles.

Moreover, it seems that some essential components of the mammalian "box C/D processome," namely the Nopp140, Tip48, and Tip49 proteins, also function in the biogenesis of box H/ACA snoRNPs. Nopp140 is a phosphoprotein that was found to associate with both box H/ACA and C/D snoRNPs (Meier and Blobel 1994; Isaac et al. 1998; Yang et al. 2000; Wang et al. 2002). Depletion of Nopp140 inhibits the accumulation of both box C/D and H/ACA snoRNAs (Isaac et al. 1998; Yang et al. 2000; Watkins et al. 2004). Two other "U3 processome"-associated proteins, Tip48 and Tip49 (also called p50 and p55), were first identified by using a mouse in vitro box C/D snoRNP reconstitution system (Newman et al. 2000). Tip48/p50 and Tip49/p55 are evolutionarily conserved interrelated proteins with ATPase and DNA helicase activity (Ikura et al. 2000). Since Tip48/p50 and Tip49/p55 reside in the nucleoplasm and do not associate with mature nucleolar snoRNPs, they were proposed to function in the assembly and/or nucleolar transport of box C/D snoRNPs. Indeed, genetic depletion of the yeast equivalents of Tip48 and Tip49 (Rvb1p and Rvb2p) diminished accumulation of box C/D snoRNPs and, unexpectedly, also blocked box H/ACA snoRNP production, demonstrating that Tip48/Rvb1p and Tip49/Rvb2p, together with Nopp140, function as general snoRNP biogenesis factors (King et al. 2001). Nopp140 and Tip49 associate with mature-sized TMG-capped U3 snoRNA in the nucleoplasm, suggesting that they function in the last steps of the nucleoplasmic biogenesis of snoRNPs (Watkins et al. 2004).

The survival of motor neurons protein (SMN), the protein product of a gene responsible for spinal muscular atrophy (SMA), is an essential component of a macromolecular complex required for the assembly of snRNPs (for review, see Paushkin et al. 2002; Yong et al. 2004). SMN has been reported to specifically interact in vivo and in vitro with fibrillarin and Gar1, leading to the notion that SMN is involved in the assembly of both box C/D and H/ACA snoRNPs (Jones et al. 2001; Pellizzoni et al. 2001; Terns and Terns 2001). Expression of a dominant-negative mutant of SMN caused fibrillarin, Gar1, and U3 snoRNP to accumulate outside the nucleolus (Pellizzoni et al. 2001). Depletion of HeLa SMN protein resulted in reduced levels of U3 snoRNA, further supporting an SMN function in U3 snoRNP biogenesis (Watkins et al. 2004). However, in the above experiment, loss of SMN had no significant effect on the accumulation of the U8 and U14 box C/D snoRNAs. Therefore, further studies are required to clarify the function of SMN in mammalian snoRNP biogenesis.

INTRANUCLEAR TRAFFICKING OF MATURING BOX C/D AND H/ACA snoRNPs

Whereas binding of the 15.5K/NHPX box C/D and the dyskerin, Nhp2, and Nop10 box H/ACA snoRNP core proteins to nascent intronic snoRNAs occurs in the nucleoplasm at the site of host pre-mRNA transcription and splicing, it remains largely speculative where subsequent steps of mammalian snoRNP maturation take place. Precursor snoRNAs are not detectable in the nucleolus, indicating that only fully processed snoRNAs are transported into the nucleolus (Samarsky et al. 1998; Verheggen et al. 2002). Binding of snoRNP proteins is essential not only for snoRNA processing and stability, but also for targeting mature snoRNPs into the nucleolus. Therefore, the snoRNA core structures, namely the box C/D and H/ACA elements and the neighboring helix structures, that direct snoRNP protein binding, also function as nucleolar localization signals (Lange et al. 1998, 1999; Samarsky et al. 1998; Narayanan et al. 1999a,b; Verheggen et al. 2001). A more recent study found that box C/D snoRNP proteins lack a simple, structurally well-defined nucleolar targeting motif. Instead, the four box C/D snoRNP proteins collectively constitute an efficient nucleolar localization signal (Verheggen et al. 2001).

Early snoRNA trafficking studies found that fluorescein-labeled box C/D snoRNAs injected into *Xenopus* oocyte nuclei transiently appear in the nucleoplasmic Cajal bodies before accumulating in the nucleolus (Narayanan et al. 1999a). In mammalian cells, both endogenous and ectopically expressed box C/D snoRNAs show a weak Cajal-body-specific accumulation beside massive concentration in the nucleolus (Samarsky et al. 1998; Darzacq et al. 2002). In contrast to box C/D snoRNAs, box H/ACA snoRNAs microinjected into *Xenopus* oocytes show no Cajal-body-specific accumulation, and endogenous mammalian box H/ACA snoRNAs are not detectable in the Cajal body (Narayanan et al. 1999b). However, plant box H/ACA snoRNAs and, upon overexpression, mammalian box H/ACA snoRNAs, appear in Cajal bodies (Shaw et al. 1998; Richard et al. 2003). It is therefore possible that both box C/D and H/ACA snoRNPs accumulate in Cajal bodies prior to nucleoli, but box H/ACA snoRNAs traverse the Cajal body more rapidly than box C/D snoRNAs do. Transient accumulation of box C/D and perhaps box H/ACA snoRNPs in Cajal bodies strongly suggests that some steps of snoRNP biogenesis occur in this nuclear organelle. Consistent with a function in snoRNP biogenesis, the Cajal body had long been known to contain the fibrillarin, dyskerin, and Gar1 snoRNP proteins, and the putative snoRNP assembly/processing factor, Nopp140 (Meier and Blobel 1994; Bohmann et al. 1995; Narayanan et al. 1999b).

A more recent study on the biogenesis of U3 snoRNP provided further evidence for the participation of Cajal body in snoRNP maturation (Verheggen et al. 2002). The 3′-extended precursor of human U3 snoRNA that carries a monomethyl-G primary cap accumulates both at the site of transcription and within the Cajal body, but it is excluded from the nucleolus. The precursor U3 snoRNA is localized to the transcription site and the Cajal body is associated with 15.5K/NHPX, but it is still not assembled with fibrillarin and Nop58. Importantly, the mature-sized U3 snoRNA that carries a TMG cap and associates with all C/D snoRNP proteins is also detectable in the Cajal body, although it accumulates mainly in the nucleolus. These observations indicate that association of the late-binding box C/D snoRNP proteins, final 3′ end trimming, and cap hypermethylation of the U3 snoRNA take place in the Cajal body. Consistently, the nuclear fraction of the human methyltransferase TGS1 responsible for U3 cap hypermethylation accumulates in the Cajal body.

Contrary to the fact that snoRNP maturation has no cytoplasmic phase (Terns and Dahlberg 1994; Terns et al. 1995), processing of U3 snoRNA depends on transport factors with well-established roles in the nucleocytoplasmic snRNA transport (Boulon et al. 2004; Watkins et al. 2004). Like RNAP-II-specific nascent spliceosomal snRNAs, the newly synthesized U3 precursor associates with the m7G-cap-binding complex (CBP) and PHAX (phosporylated adapter for RNA export) (Fig. 3). However, in the case of U3, binding of PHAX is not followed by recruitment of the transport factors CRM1 and Ran-GTP, which is required for the cytoplasmic transportation of nascent snRNAs (Ohno et al. 2002). In vivo depletion experiments confirmed that U3 is targeted into the Cajal body by a PHAX-dependent and CRM1-independent mechanism. Efficient targeting of U3 into the Cajal body, in addition to the m7G cap, also requires an intact terminal box C′/D motif (the Kink-turn) (Boulon et al. 2004). Since these two *cis*-acting Cajal body localization elements of U3 function synergistically, it is possible that besides CBC, additional not yet identified factors contribute to the recruitment of PHAX to the nascent U3 snoRNA, or alternatively, PHAX directly interacts with 15.5/NHPX bound to the box C′/D motif of U3 (Boulon et al. 2004). Coimmunoprecipitation experiments found PHAX to be associated with other RNAP-II-transcribed box C/D snoRNAs, U8 and U13, as well as with the human telomerase RNA that carries a box H/ACA snoRNA-like domain (see below). Moreover, depletion of PHAX inhibited the accumulation of the intron-encoded U14 box C/D snoRNA (Watkins et al. 2004). Hence, it is possible that PHAX functions in the intranuclear trafficking of both box C/D and box H/ACA snoRNPs.

In the Cajal body, the precursor U3 snoRNA is packaged with box C/D core proteins and undergoes 3′ end processing and TMG cap formation before translocation into the nucleolus (Fig. 3) (Verheggen et al. 2002). Surprisingly, depletion experiments found that targeting of mature U3 snoRNP from the Cajal body into the nucleolus requires CRM1 (Boulon et al. 2004; Watkins et al. 2004). Indeed, coimmunoprecipitation experiments confirmed that CRM1 interacts with U3 and associates mainly, if not exclusively, with mature-sized, TMG-capped U3 snoRNAs, supporting the idea that CRM1 functions in a late step of U3 snoRNP biogenesis.

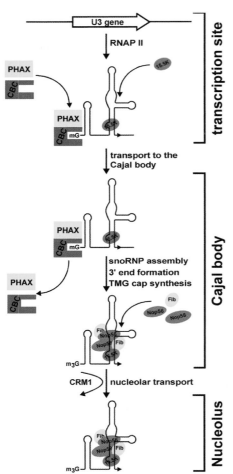

Figure 3. Model for assembly and intranuclear trafficking of mammalian U3 snoRNP. The box C′/D motif (Kink turn) of the 3′-extended nascent U3 snoRNA binds the 15.5K/NHPX protein. Binding of CBC and PHAX to the m7G monomethyl cap (mG) supports U3 transport into the Cajal body where maturation of U3 snoRNA and assembly of U3 snoRNP is completed. A transient interaction with CRM1 is essential for routing mature U3 snoRNP into the nucleolus. It seems that the mammalian intron-encoded snoRNPs follow a processing/trafficking pathway similar or identical to that described for the U3 snoRNP.

SMALL CAJAL-BODY-SPECIFIC RNPs

The Cajal body is an evolutionarily conserved, multifunctional subnuclear organelle (Ogg and Lamond 2002; Gall 2003; Cioce and Lamond 2005). The RNAP-II-specific U1, U2, U4, and U5 Sm spliceosomal snRNPs had long been known to cycle through Cajal bodies and were suspected to undergo maturation in this nuclear organelle (Carmo-Fonseca et al. 1992; Bohmann et al. 1995; Sleeman and Lamond 1999). The mature snRNAs carry numerous posttranscriptionally synthesized 2′-O-methylated nucleotides and pseudouridines (Massenet et al. 1998). During the past years, several box C/D 2′-O-methylation and box H/ACA pseudouridylation guide RNAs have been identified and demonstrated, or confidently predicted, to function in the modification of human Sm snRNAs (Lestrade and Weber 2006 and ref-

erences therein). The first guide RNA linked to Sm snRNA modification, U85, showed an unusual structural organization, because it contained both a box C/D and a box H/ACA snoRNA-like domain (Fig. 4) (Jády and Kiss 2001). The H/ACA domain of U85 is inserted into the middle of its box C/D domain. In vivo and in vitro modification experiments demonstrated that the box C/D domain of U85 directs 2′-O-methylation, whereas its box H/ACA domain guides pseudouridylation of the U5 snRNA. Later, additional box C/D–H/ACA composite guide RNAs directing Sm snRNA modification were discovered, and another guide RNA, U93, which directs pseudouridylation of the U2 snRNA, was found to contain two tandemly arranged H/ACA RNA domains (Darzacq et al. 2002; Kiss et al. 2002). Moreover, irregular structural arrangements were also observed for box C/D RNAs directing 2′-O-methylation of Sm snRNAs. These guide RNAs are significantly longer than canonical box C/D snoRNAs involved in rRNA methylation and they are frequently composed of two box C/D-like domains and carry a m7G cap, indicating that they are RNAP II transcripts (Darzacq et al. 2002; Tycowski et al. 2004).

Cell fractionation and in situ localization experiments revealed that all modification guide RNAs implicated in 2′-O-methylation and/or pseudouridylation of Sm snRNAs specifically accumulate in Cajal bodies and were therefore called small Cajal-body-specific RNAs (scaRNAs) (Darzacq et al. 2002; Kiss et al. 2002, 2004). Accumulation

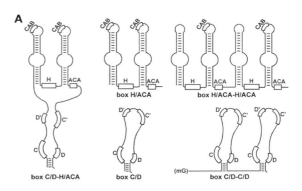

Figure 4. Human small Cajal-body-specific RNAs (scaRNAs). (*A*) Schematic structural organization of human scaRNAs. Positions of the conserved box elements are indicated. Box H/ACA scaRNAs possess a Cajal-body-specific localization element (CAB box) that is located in the terminal loop of the 5′ or 3′ hairpin. Some H/ACA scaRNAs have two CAB boxes. Elements supporting the Cajal-body-specific accumulation of box C/D scaRNAs are unknown. (*B*) Sequence conservation of the CAB box motifs of vertebrate scaRNAs. The frequencies of the four ribonucleotides G, A, U, and C in the four positions of 278 putative CAB box motifs identified in vertebrate scaRNAs are indicated by the heights of the corresponding letters.

of scaRNAs in the Cajal body instead of the nucleolus was unexpected, since they carry the box H/ACA and/or C/D motifs responsible for the nucleolar localization of snoRNAs (see above). A sequence comparison followed by mutational analysis demonstrated that the terminal loop of the 5′ or the 3′ hairpin of H/ACA scaRNAs contains a short Cajal-body-specific localization motif, the Cajal body box (CAB box) (Richard et al. 2003). In a few cases, both the 5′ and 3′ hairpins carry a CAB box motif that seems to function synergistically. Upon alteration of the CAB box motifs, the mutant H/ACA scaRNAs accumulate in the nucleolus, and authentic box H/ACA snoRNAs can be targeted into the Cajal body by inclusion of an exogenous CAB box. The original CAB box consensus (ugAG) was based on a handful of human scaRNA sequences (Richard et al. 2003). More recently, experimental and computer-based approaches identified numerous putative box H/ACA scaRNAs from a broad variety of vertebrate organisms (Lestrade and Weber 2006 and references therein). The frequency of ribonucleotides observed at each position of 278 putative CAB box motifs is shown in Figure 4B. Mutational analysis confirmed that the first two nucleotides are less critical, whereas the highly conserved final two nucleotides (AG) are absolutely essential for the Cajal-body-specific accumulation of scaRNAs. So far, no Cajal-body-specific localization element has been identified in box C/D scaRNAs that lack an H/ACA domain.

Since both box C/D and H/ACA snoRNAs may transit through Cajal bodies, it is possible that the Cajal body localization signal of scaRNAs is a retention element that prevents translocation of scaRNPs from the Cajal body to the nucleolus. Most probably, the CAB box functions through binding a specific protein factor(s) (Richard et al. 2003). Fu and Collins have recently proposed that two Sm proteins, SmB and SmD3, associate directly or indirectly with the CAB box of a subpopulation of human box H/ACA scaRNAs and the telomerase RNA that is in fact an H/ACA scaRNA (see below) (Fu and Collins 2006). However, telomerase RNA microinjected into *Xenopus* oocytes did not associate with Sm proteins and, in human HeLa cells, we failed to detect an interaction between Sm proteins and box H/ACA scaRNAs, including the human telomerase RNA (Lukowiak et al. 2001; our unpublished data). This may indicate that interaction of SmB and SmD3 proteins with scaRNAs may depend on growth conditions and/or cell lines.

HUMAN TELOMERASE IS A BOX H/ACA scaRNP

Telomerase is a ribonucleoprotein enzyme that is responsible for the synthesis of telomeric DNA at the ends of eukaryotic linear chromosomes (Collins and Mitchell 2002; Cong et al. 2002). The telomerase holoenzyme is composed of the telomerase RNA (TR) and a set of associated proteins, including the telomerase reverse transcriptase (TERT). TR provides a scaffold for binding of telomerase RNP proteins and contains the template region that is copied repeatedly by the associated TERT to produce telomeric DNA repeats (Fig. 5A). The human TR (hTR) is divided into two major structural domains (Fig.

5B). The 5′ half of hTR that carries the template sequence folds into a large pseudoknot structure that is an evolutionarily conserved feature of TRs. The 3′ terminal region of hTR possesses a box H/ACA RNA-like structure and is associated with the four box H/ACA snoRNP core proteins, dyskerin, Nhp2, Nop10, and Gar1 (Mitchell et al. 1999; Chen et al. 2000; Antal et al. 2002; Meier 2005). The H/ACA domain provides metabolic stability for human telomerase RNP and is also indispensable for the enzymatic activity of telomerase (Mitchell et al. 1999; Mitchell and Collins 2000; Martin-Rivera and Blasco 2001; Chen et al. 2002; Fu and Collins 2003).

In situ localization experiments showed that in human cancer cells, both endogenous and ectopically expressed hTRs accumulate in Cajal bodies (Jády et al. 2004; Zhu et al. 2004). Mutational analysis revealed that targeting of hTR into Cajal bodies is supported by a short signal sequence, UGAG, which is located in the terminal loop of the 3′ hairpin of the H/ACA domain of hTR, and is structurally and functionally indistinguishable from the CAB box motif of box H/ACA scaRNAs (Fig. 5B) (Jády et al. 2004). The 3′ terminal box H/ACA domain of hTR contains all the elements required for the correct expression of canonical intron-encoded box H/ACA scaRNAs, because it is efficiently processed from the second intron of the human β-globin pre-mRNA expressed in human or mouse cells and the excised RNA accumulates in Cajal bodies (our unpublished data).

However, although authentic scaRNAs are normally not detectable outside the Cajal body, hTR accumulation is not confined to this nuclear organelle. Instead, hTR has a complex and dynamic intranuclear trafficking that seems to be strictly regulated by the cell cycle. In the nuclei of HeLa S-phase cells, besides accumulating in Cajal bodies, hTR

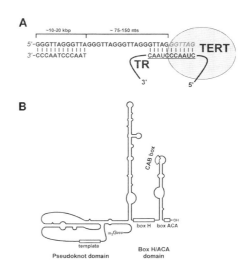

Figure 5. Structure and function of human telomerase RNA. (*A*) Synthesis of telomeric DNA by telomerase. The template sequence (*underlined*) of telomerase RNA (TR) recognizes the terminal nucleotides of the telomeric G strand. The telomerase reverse transcriptase (TERT) incorporates deoxynucleotides (*in italics*) complementary to the template sequence of TR. (*B*) Schematic structure of human telomerase RNA. Positions of the template region, the H, ACA, and CAB boxes are shown.

also concentrates at a few (most frequently 1 to 3) telomeres (Jády et al. 2006; Tomlinson et al. 2006). Accumulation of hTR at telomeres likely reflects telomere elongation events. This is supported by the facts that telomere synthesis is known to occur during S-phase and that hTERT also concentrates at a few telomeres in HeLa S-phase cells (Ten Hagen et al. 1990; Wright et al. 1999; Tomlinson et al. 2006). Telomeres accumulating hTR feature long G strands that, in marked contrast to the majority of telomeres, are accessible for in situ hybridization with a G-strand-specific fluorescent oligonucleotide probe without denaturation (Jády et al. 2006). The highly accessible G strands may represent nascent telomeric overhangs synthesized by the associated telomerase. The notion that human telomerase accumulates only at a few telomeres implies that only a small subset of human telomeres are elongated within one cell cycle, as has been demonstrated in yeast, where telomerase extends less than 10% of telomeres during every cell cycle (Teixeira et al. 2004). Therefore, intranuclear trafficking of hTR may play an important regulatory role in human telomere synthesis (Jády et al. 2006; Tomlinson et al. 2006).

The *cis-* and *trans-*acting elements supporting hTR accumulation at telomeres remain unknown. The 3′ terminal H/ACA domain of hTR alone, when expressed in HeLa cells, fails to concentrate at telomeres, suggesting that the 5′ terminal template domain is required for targeting hTR to telomeres (our unpublished data). Interestingly, more than 25% of HeLa telomeres accumulating hTR colocalize with Cajal bodies, suggesting that Cajal bodies may function in some aspects of human telomere biogenesis (Jády et al. 2006). The Cajal body is a mobile subnuclear organelle that can translocate large distances through the nucleoplasm and therefore has been implicated in the intranuclear transport and sorting of nuclear factors (Platani et al. 2000, 2002; Ogg and Lamond 2002; Cioce and Lamond 2005). In vivo imaging demonstrated that in HeLa S-phase cells, Cajal bodies moving in the interchromatin space transiently associate with telomeres with a frequency of approximately 1.8 associations per hour. This indicates that during the entire S phase, Cajal bodies interact with 5–7 telomeres (Jády et al. 2006). The biological significance of the cell-cycle-regulated interaction of Cajal bodies with telomeres remains unclear, but according to an attractive hypothesis, Cajal bodies may deliver hTR to telomeres.

CONCLUSIONS

Studies during the past years revealed that biogenesis of mammalian box C/D and H/ACA snoRNPs is a more complex process than previously anticipated. Assembly of functional box C/D and H/ACA snoRNPs requires numerous auxiliary factors, in addition to the four box C/D and H/ACA snoRNP proteins. The initial steps of the assembly of intronic snoRNPs, namely recognition of the nascent box H/ACA and C/D snoRNA sequences through recruitment of the first-binding snoRNP proteins, are actively facilitated by the RNAP II transcription complex or the pre-mRNA splicing machinery. Maturing of box C/D and

H/ACA snoRNPs requires an intranuclear trafficking of the maturing snoRNPs from the site of snoRNA synthesis to the site of the function of the mature snoRNP. Different steps of box C/D and H/ACA RNP maturation can be linked to subnuclear compartments where the maturing snoRNPs transit through. The mature box C/D and H/ACA snoRNPs involved in rRNA modification or maturation accumulate in the nucleolus, whereas box C/D and H/ACA RNPs directing spliceosomal snRNA modification reside in the Cajal body (scaRNPs). Sequestering the functionally active guide RNPs into the nucleolus or Cajal body may be an important mechanism to avoid undesired RNA modification events in the nucleoplasm. The human telomerase RNA (hTR) accumulates in Cajal bodies and shares a common Cajal body localization element with box H/ACA scaRNAs. However, hTR shows a more complex intranuclear localization pattern. In S-phase cells, hTR specifically accumulates at a few telomeres that possess long G-strand overhangs. Hence, studies on the biosynthesis of mammalian snoRNPs and scaRNPs revealed unexpected principles of the intranuclear trafficking of macromolecules and gave new insights into the molecular mechanism of the assembly of small RNPs.

ACKNOWLEDGMENTS

This work was supported by grants from la Ligue Nationale contre le Cancer and la Fondation pour la Recherche Médicale. E.F. and P.R. were supported by le Ministère l'Education Nationale, de la Recherche, et de la Technologie and la Fondation pour la Recherche Médicale, respectively.

REFERENCES

Achsel T., Brahms H., Kastner B., Bachi A., Wilm M., and Lührmann R. 1999. A doughnut-shaped heteromer of human Sm-like proteins binds to the 3′-end of U6 snRNA, thereby facilitating U4/U6 duplex formation in vitro. *EMBO J.* **18:** 5789.

Allmang C., Kufel J., Chanfreau G., Mitchell P., Petfalski E., and Tollervey D. 1999. Functions of the exosome in rRNA, snoRNA and snRNA synthesis. *EMBO J.* **18:** 5399.

Antal M., Boros E., Solymosy F., and Kiss T. 2002. Analysis of the structure of human telomerase RNA in vivo. *Nucleic Acids Res.* **30:** 912.

Bachellerie J.P., Cavaillé J., and Hüttenhofer A. 2002. The expanding snoRNA world. *Biochimie* **84:** 775.

Balakin A.G., Smith L., and Fournier M.J. 1996. The RNA world of the nucleolus: Two major families of small RNAs defined by different box elements with related functions. *Cell* **86:** 823.

Ballarino M., Morlando M., Pagano F., Fatica A., and Bozzoni I. 2005. The cotranscriptional assembly of snoRNPs controls the biosynthesis of H/ACA snoRNAs in *Saccharomyces cerevisiae*. *Mol. Cell. Biol.* **25:** 5396.

Beggs J.D. 2005. Lsm proteins and RNA processing. *Biochem. Soc. Trans.* **33:** 433.

Bertrand E. and Fournier M.J. 2004. The snoRNPs and related machines: Ancient devices that mediate maturation of rRNA and other RNAs. In *The nucleolus* (ed. M.O.J. Olson), p. 225. Kluwer Academic/Plenum Publishers, New York and Landes Bioscience, Georgetown, Texas.

Bohmann K., Ferreira J., Santama N., Weis K., and Lamond A.I. 1995. Molecular analysis of the coiled body. *J. Cell Sci.* (suppl.) **19:** 107.

Boulon S., Verheggen C., Jády B.E., Girard C., Pescia C., Paul C., Ospina J.K., Kiss T., Matera A.G., Bordonne R., and Bertrand E. 2004. PHAX and CRM1 are required sequentially to transport U3 snoRNA to nucleoli. *Mol. Cell* **16:** 777.

Bousquet-Antonelli C., Henry Y., G'Elugne J P., Caizergues-Ferrer M., and Kiss T. 1997. A small nucleolar RNP protein is required for pseudouridylation of eukaryotic ribosomal RNAs. *EMBO J.* **16:** 4770.

Cahill N.M., Friend K., Speckmann W., Li Z.H., Terns R.M., Terns M.P., and Steitz J.A. 2002. Site-specific cross-linking analyses reveal an asymmetric protein distribution for a box C/D snoRNP. *EMBO J.* **21:** 3816.

Carmo-Fonseca M., Pepperkok R., Carvalho M.T., and Lamond A.I. 1992. Transcription-dependent colocalization of the U1, U2, U4/U6, and U5 snRNPs in coiled bodies. *J. Cell Biol.* **117:** 1.

Cavaillé J. and Bachellerie J.P. 1996. Processing of fibrillarin-associated snoRNAs from pre-mRNA introns: An exonucleolytic process exclusively directed by the common stem-box terminal structure. *Biochimie* **78:** 443.

Cavaillé J., Nicoloso M., and Bachellerie J.P. 1996. Targeted ribose methylation of RNA in vivo directed by tailored antisense RNA guides. *Nature* **383:** 732.

Chen J.L., Blasco M.A., and Greider C.W. 2000. Secondary structure of vertebrate telomerase RNA. *Cell* **100:** 503.

Chen J.L., Opperman K.K., and Greider C.W. 2002. A critical stem-loop structure in the CR4-CR5 domain of mammalian telomerase RNA. *Nucleic Acids Res.* **30:** 592.

Cioce M. and Lamond A.I. 2005. Cajal bodies: A long history of discovery. *Annu. Rev. Cell Dev. Biol.* **21:** 105.

Collins K. and Mitchell J.R. 2002. Telomerase in the human organism. *Oncogene* **21:** 564.

Cong Y.S., Wright W.E., and Shay J.W. 2002. Human telomerase and its regulation. *Microbiol. Mol. Biol. Rev.* **66:** 407.

Darzacq X., Jády B.E., Verheggen C., Kiss A.M., Bertrand E., and Kiss T. 2002. Cajal body-specific small nuclear RNAs: A novel class of 2′-O-methylation and pseudouridylation guide RNAs. *EMBO J.* **21:** 2746.

Darzacq X., Kittur N., Roy S., Shav-Tal Y., Singer R.H., and Meier U.T. 2006. Stepwise RNP assembly at the site of H/ACA RNA transcription in human cells. *J. Cell Biol.* **173:** 207.

Decatur W.A. and Fournier M.J. 2003. RNA-guided nucleotide modification of ribosomal and other RNAs. *J. Biol. Chem.* **278:** 695.

Dennis P.P. and Omer A. 2005. Small non-coding RNAs in Archaea. *Curr. Opin. Microbiol.* **8:** 685.

Dez C., Noaillac-Depeyre J., Caizergues-Ferrer M., and Henry Y. 2002. Naf1p, an essential nucleoplasmic factor specifically required for accumulation of box H/ACA small nucleolar RNPs. *Mol. Cell. Biol.* **22:** 7053.

Fatica A., Dlakic M., and Tollervey D. 2002. Naf1 p is a box H/ACA snoRNP assembly factor. *RNA* **8:** 1502.

Filipowicz W. and Pogacic V. 2002. Biogenesis of small nucleolar ribonucleoproteins. *Curr. Opin. Cell Biol.* **14:** 319.

Fragapane P., Prislei S., Michienzi A., Caffarelli E., and Bozzoni I. 1993. A novel small nucleolar RNA (U16) is encoded inside a ribosomal protein intron and originates by processing of the pre-mRNA. *EMBO J.* **12:** 2921.

Fromont-Racine M., Mayes A.E., Brunet-Simon A., Rain J.C., Colley A., Dix I., Decourty L., Joly N., Ricard F., Beggs J.D., and Legrain P. 2000. Genome-wide protein interaction screens reveal functional networks involving Sm-like proteins. *Yeast* **17:** 95.

Fu D. and Collins K. 2003. Distinct biogenesis pathways for human telomerase RNA and H/ACA small nucleolar RNAs. *Mol. Cell* **11:** 1361.

———. 2006. Human telomerase and Cajal body ribonucleoproteins share a unique specificity of Sm protein association. *Genes Dev.* **20:** 531.

Gall J.G. 2003. The centennial of the Cajal body. *Nat. Rev. Mol. Cell Biol.* **4:** 975.

Ganot P., Bortolin M.L., and Kiss T. 1997a. Site-specific pseudouridine formation in preribosomal RNA is guided by small nucleolar RNAs. *Cell* **89:** 799.

Ganot P., Caizergues-Ferrer M., and Kiss T. 1997b. The family of box ACA small nucleolar RNAs is defined by an evolutionarily conserved secondary structure and ubiquitous sequence elements essential for RNA accumulation. *Genes Dev.* **11:** 941.

Green M.R. 1991. Biochemical mechanisms of constitutive and regulated pre-mRNA splicing. *Annu. Rev. Cell Biol.* **7:** 559.

Henras A.K., Dez C., and Henry Y. 2004. RNA structure and function in C/D and H/ACA s(no)RNPs. *Curr. Opin. Struct. Biol.* **14:** 335.

Henras A., Henry Y., Bousquet-Antonelli C., Noaillac-Depeyre J., Gelugne J.P., and Caizergues-Ferrer M. 1998. Nhp2p and Nop10p are essential for the function of H/ACA snoRNPs. *EMBO J.* **17:** 7078.

Hirose T. and Steitz J.A. 2001. Position within the host intron is critical for efficient processing of box C/D snoRNAs in mammalian cells. *Proc. Natl. Acad. Sci.* **98:** 12914.

Hirose T., Shu M.D., and Steitz J.A. 2003. Splicing-dependent and -independent modes of assembly for intron-encoded box C/D snoRNPs in mammalian cells. *Mol. Cell* **12:** 113.

Hoang C. and Ferre-D'Amare A.R. 2001. Cocrystal structure of a tRNA Psi55 pseudouridine synthase: Nucleotide flipping by an RNA-modifying enzyme. *Cell* **107:** 929.

Hoareau-Aveilla C., Bonoli M., Caizergues-Ferrer M., and Henry Y. 2006. hNaf1 is required for accumulation of human H/ACA snoRNPs, scaRNPs, and telomerase. *RNA* **12:** 832.

Hüttenhofer A., Schattner P., and Polacek N. 2005. Non-coding RNAs: Hope or hype? *Trends Genet.* **21:** 289.

Ikura T., Ogryzko V.V., Grigoriev M., Groisman R., Wang J., Horikoshi M., Scully R., Qin J., and Nakatani Y. 2000. Involvement of the TIP60 histone acetylase complex in DNA repair and apoptosis. *Cell* **102:** 463.

Isaac C., Yang Y., and Meier U.T. 1998. Nopp140 functions as a molecular link between the nucleolus and the coiled bodies. *J. Cell Biol.* **142:** 319.

Jády B.E. and Kiss T. 2001. A small nucleolar guide RNA functions both in 2′-O-ribose methylation and pseudouridylation of the U5 spliceosomal RNA. *EMBO J.* **20:** 541.

Jády B.E., Bertrand E., and Kiss T. 2004. Human telomerase RNA and box H/ACA scaRNAs share a common Cajal body-specific localization signal. *J. Cell Biol.* **164:** 647.

Jády B.E., Richard P., Bertrand E., and Kiss T. 2006. Cell cycle-dependent recruitment of telomerase RNA and Cajal bodies to human telomeres. *Mol. Biol. Cell* **17:** 944.

Jones K.W., Gorzynski K., Hales C.M., Fischer U., Badbanchi F., Terns R.M., and Terns M.P. 2001. Direct interaction of the spinal muscular atrophy disease protein SMN with the small nucleolar RNA-associated protein fibrillarin. *J. Biol. Chem.* **276:** 38645.

King T.H., Decatur W.A., Bertrand E., Maxwell E.S., and Fournier M.J. 2001. A well-connected and conserved nucleoplasmic helicase is required for production of box C/D and H/ACA snoRNAs and localization of snoRNP proteins. *Mol. Cell. Biol.* **21:** 7731.

Kiss T. 2001. Small nucleolar RNA-guided post-transcriptional modification of cellular RNAs. *EMBO J.* **20:** 3617.

Kiss T. and Filipowicz W. 1993. Small nucleolar RNAs encoded by introns of the human cell cycle regulatory gene RCC1. *EMBO J.* **12:** 2913.

———. 1995. Exonucleolytic processing of small nucleolar RNAs from pre-mRNA introns. *Genes Dev.* **9:** 1411.

Kiss T., Bortolin M.L., and Filipowicz W. 1996. Characterization of the intron-encoded U19 RNA, a new mammalian small nucleolar RNA that is not associated with fibrillarin. *Mol. Cell. Biol.* **16:** 1391.

Kiss A.M., Jády B.E., Bertrand E., and Kiss T. 2004. Human box H/ACA pseudouridylation guide RNA machinery. *Mol. Cell. Biol.* **24:** 5797.

Kiss A.M., Jády B.E., Darzacq X., Verheggen C., Bertrand E., and Kiss T. 2002. A Cajal body-specific pseudouridylation guide RNA is composed of two box H/ACA snoRNA-like domains. *Nucleic Acids Res.* **30:** 4643.

Kiss-László Z., Henry Y., and Kiss T. 1998. Sequence and structural elements of methylation guide snoRNAs essential for site-specific ribose methylation of pre-rRNA. *EMBO J.* **17:** 797.

Kiss-László Z., Henry Y., Bachellerie J.P., Caizergues-Ferrer M., and Kiss T. 1996. Site-specific ribose methylation of pre-ribosomal RNA: A novel function for small nucleolar RNAs. *Cell* **85:** 1077.

Klein D.J., Schmeing T.M., Moore P.B., and Steitz T.A. 2001. The kink-turn: A new RNA secondary structure motif. *EMBO J.* **20:** 4214.

Lafontaine D.L. and Tollervey D. 1999. Nop58p is a common component of the box C+D snoRNPs that is required for snoRNA stability. *RNA* **5:** 455.

Lange T.S., Borovjagin A., Maxwell E.S., and Gerbi S.A. 1998. Conserved boxes C and D are essential nucleolar localization elements of U14 and U8 snoRNAs. *EMBO J.* **17:** 3176.

Lange T.S., Ezrokhi M., Amaldi F., and Gerbi S.A. 1999. Box H and box ACA are nucleolar localization elements of U17 small nucleolar RNA. *Mol. Biol. Cell* **10:** 3877.

Lestrade L. and Weber M.J. 2006. snoRNA-LBME-db, a comprehensive database of human H/ACA and C/D box snoRNAs. *Nucleic Acids Res.* **34:** D158.

Leverette R.D., Andrews M.T., and Maxwell E.S. 1992. Mouse U14 snRNA is a processed intron of the cognate hsc70 heat shock pre-messenger RNA. *Cell* **71:** 1215.

Lukowiak A.A., Narayanan A., Li Z.H., Terns R.M., and Terns M.P. 2001. The snoRNA domain of vertebrate telomerase RNA functions to localize the RNA within the nucleus. *RNA* **7:** 1833.

Lyman S.K., Gerace L., and Baserga S.J. 1999. Human Nop5/Nop58 is a component common to the box C/D small nucleolar ribonucleoproteins. *RNA* **5:** 1597.

Maraia R.J. and Bayfield M.A. 2006. The La protein-RNA complex surfaces. *Mol. Cell* **21:** 149.

Martin-Rivera L. and Blasco M.A. 2001. Identification of functional domains and dominant negative mutations in vertebrate telomerase RNA using an in vivo reconstitution system. *J. Biol. Chem.* **276:** 5856.

Massenet S., Mougin A., and Branlant C. 1998. Posttranscriptional modification in the U small nuclear RNAs. In *Modification and editing of RNA* (ed. H. Grosjean and R. Benne), p. 201. ASM Press, Washington D.C.

Meier U.T. 2005. The many facets of H/ACA ribonucleoproteins. *Chromosoma* **114:** 1.

Meier U.T. and Blobel G. 1994. NAP57, a mammalian nucleolar protein with a putative homolog in yeast and bacteria. *J. Cell Biol.* **127:** 1505.

Mitchell J.R. and Collins K. 2000. Human telomerase activation requires two independent interactions between telomerase RNA and telomerase reverse transcriptase. *Mol. Cell* **6:** 361.

Mitchell J.R., Cheng J., and Collins K. 1999. A box H/ACA small nucleolar RNA-like domain at the human telomerase RNA 3′ end. *Mol. Cell. Biol.* **19:** 567.

Mouaikel J., Verheggen C., Bertrand E., Tazi J., and Bordonne R. 2002. Hypermethylation of the cap structure of both yeast snRNAs and snoRNAs requires a conserved methyltransferase that is localized to the nucleolus. *Mol. Cell* **9:** 891.

Narayanan A., Speckmann W., Terns R., and Terns M.P. 1999a. Role of the box C/D motif in localization of small nucleolar RNAs to coiled bodies and nucleoli. *Mol. Biol. Cell* **10:** 2131.

Narayanan A., Lukowiak A., Jády B.E., Dragon F., Kiss T., Terns R.M., and Terns M.P. 1999b. Nucleolar localization signals of box H/ACA small nucleolar RNAs. *EMBO J.* **18:** 5120.

Newman D.R., Kuhn J.F., Shanab G.M., and Maxwell E.S. 2000. Box C/D snoRNA-associated proteins: Two pairs of evolutionarily ancient proteins and possible links to replication and transcription. *RNA* **6:** 861.

Ni J., Tien A.L., and Fournier M.J. 1997. Small nucleolar RNAs direct site-specific synthesis of pseudouridine in ribosomal RNA. *Cell* **89:** 565.

Ogg S.C. and Lamond A.I. 2002. Cajal bodies and coilin—Moving towards function. *J. Cell Biol.* **159:** 17.

Ohno M., Segref A., Kuersten S., and Mattaj I.W. 2002. Identity elements used in export of mRNAs. *Mol. Cell* **9:** 659.

Omer A.D., Ziesche S., Decatur W.A., Fournier M.J., and Dennis P.P. 2003. RNA-modifying machines in archaea. *Mol. Microbiol.* **48:** 617.

Ooi S.L., Samarsky D.A., Fournier M.J., and Boeke J.D. 1998. Intronic snoRNA biosynthesis in *Saccharomyces cerevisiae* depends on the lariat-debranching enzyme: Intron length effects and activity of a precursor snoRNA. *RNA* **4:** 1096.

Padgett R.A., Grabowski P.J., Konarska M.M., Seiler S., and Sharp P.A. 1986. Splicing of messenger RNA precursors. *Annu. Rev. Biochem.* **55:** 1119.

Paushkin S., Gubitz A.K., Massenet S., and Dreyfuss G. 2002. The SMN complex, an assemblyosome of ribonucleoproteins. *Curr. Opin. Cell Biol.* **14:** 305.

Pellizzoni L., Baccon J., Charroux B., and Dreyfuss G. 2001. The survival of motor neurons (SMN) protein interacts with the snoRNP proteins fibrillarin and GAR1. *Curr. Biol.* **11:** 1079.

Perumal K. and Reddy R. 2002. The 3′ end formation in small RNAs. *Gene Expr.* **10:** 59.

Platani M., Goldberg I., Lamond A.I., and Swedlow J.R. 2002. Cajal body dynamics and association with chromatin are ATP-dependent. *Nat. Cell Biol.* **4:** 502.

Platani M., Goldberg I., Swedlow J.R., and Lamond A.I. 2000. In vivo analysis of Cajal body movement, separation, and joining in live human cells. *J. Cell Biol.* **151:** 1561.

Reddy R., Li W.Y., Henning D., Choi Y.C., Nohga K., and Busch H. 1981. Characterization and subcellular localization of 7-8 S RNAs of Novikoff hepatoma. *J. Biol. Chem.* **256:** 8452.

Richard P., Kiss A.M., Darzacq X., and Kiss T. 2006. Cotranscriptional recognition of human intronic box H/ACA snoRNAs occurs in a splicing-independent manner. *Mol. Cell. Biol.* **26:** 2540.

Richard P., Darzacq X., Bertrand E., Jády B.E., Verheggen C., and Kiss T. 2003. A common sequence motif determines the Cajal body-specific localisation of box H/ACA scaRNAs. *EMBO J.* **22:** 4283.

Ruff E.A., Rimoldi O.J., Raghu B., and Eliceiri G.L. 1993. Three small nucleolar RNAs of unique nucleotide sequences. *Proc. Natl. Acad. Sci.* **90:** 635.

Samarsky D.A., Fournier M.J., Singer R.H., and Bertrand E. 1998. The snoRNA box C/D motif directs nucleolar targeting and also couples snoRNA synthesis and localization. *EMBO J.* **17:** 3747.

Schattner P., Barberan-Soler S., and Lowe T.M. 2006. A computational screen for mammalian pseudouridylation guide H/ACA RNAs. *RNA* **12:** 15.

Schimmang T., Tollervey D., Kern H., Frank R., and Hurt E.C. 1989. A yeast nucleolar protein related to mammalian fibrillarin is associated with small nucleolar RNA and is essential for viability. *EMBO J.* **8:** 4015.

Shaw P.J., Beven A.F., Leader D.J., and Brown J.W. 1998. Localization and processing from a polycistronic precursor of novel snoRNAs in maize. *J. Cell Sci.* **111:** 2121.

Sleeman J.E. and Lamond A.I. 1999. Newly assembled snRNPs associate with coiled bodies before speckles, suggesting a nuclear snRNP maturation pathway. *Curr. Biol.* **9:** 1065.

Szewczak L.B., DeGregorio S.J., Strobel S.A., and Steitz J.A. 2002. Exclusive interaction of the 15.5 kD protein with the terminal box C/D motif of a methylation guide snoRNP. *Chem. Biol.* **9:** 1095.

Szewczak L.B., Gabrielsen J.S., Degregorio S.J., Strobel S.A., and Steitz J.A. 2005. Molecular basis for RNA kink-turn recognition by the h15.5K small RNP protein. *RNA* **11:** 1407.

Teixeira M.T., Arneric M., Sperisen P., and Lingner J. 2004. Telomere length homeostasis is achieved via a switch between telomerase- extendible and -nonextendible states. *Cell* **117:** 323.

Ten Hagen K.G., Gilbert D.M., Willard H.F., and Cohen S.N. 1990. Replication timing of DNA sequences associated with human centromeres and telomeres. *Mol. Cell. Biol.* **10:** 6348.

Terns M.P. and Dahlberg J.E. 1994. Retention and 5′ cap trimethylation of U3 snRNA in the nucleus. *Science* **264:** 959.

Terns M.P. and Terns R.M. 2001. Macromolecular complexes: SMN—The master assembler. *Curr. Biol.* **11:** R862.

———. 2002. Small nucleolar RNAs: Versatile trans-acting molecules of ancient evolutionary origin. *Gene Expr.* **10:** 17.

Terns M.P., Grimm C., Lund E., and Dahlberg J.E. 1995. A common maturation pathway for small nucleolar RNAs. *EMBO J.* **14:** 4860.

Tomlinson R.L., Ziegler T.D., Supakorndej T., Terns R.M., and Terns M.P. 2006. Cell cycle-regulated trafficking of human telomerase to telomeres. *Mol. Biol. Cell* **17:** 955.

Tran E., Brown J., and Maxwell E.S. 2004. Evolutionary origins of the RNA-guided nucleotide-modification complexes: From the primitive translation apparatus? *Trends Biochem. Sci.* **29:** 343.

Tyc K. and Steitz J.A. 1989. U3, U8 and U13 comprise a new class of mammalian snRNPs localized in the cell nucleolus. *EMBO J.* **8:** 3113.

Tycowski K.T., Aab A., and Steitz J.A. 2004. Guide RNAs with 5′ caps and novel box C/D snoRNA-like domains for modification of snRNAs in metazoa. *Curr. Biol.* **14:** 1985.

Tycowski K.T., Shu M.D., and Steitz J.A. 1993. A small nucleolar RNA is processed from an intron of the human gene encoding ribosomal protein S3. *Genes Dev.* **7:** 1176.

Vasiljeva L. and Buratowski S. 2006. Nrd1 interacts with the nuclear exosome for 3′ processing of RNA polymerase II transcripts. *Mol. Cell* **21:** 239.

Verheggen C., Lafontaine D.L., Samarsky D., Mouaikel J., Blanchard J.M., Bordonne R., and Bertrand E. 2002. Mammalian and yeast U3 snoRNPs are matured in specific and related nuclear compartments. *EMBO J.* **21:** 2736.

Verheggen C., Mouaikel J., Thiry M., Blanchard J.M., Tollervey D., Bordonne R., Lafontaine D.L., and Bertrand E. 2001. Box C/D small nucleolar RNA trafficking involves small nucleolar RNP proteins, nucleolar factors and a novel nuclear domain. *EMBO J.* **20:** 5480.

Vidovic I., Nottrott S., Hartmuth K., Luhrmann R., and Ficner R. 2000. Crystal structure of the spliceosomal 15.5kD protein bound to a U4 snRNA fragment. *Mol. Cell* **6:** 1331.

Wang C. and Meier U.T. 2004. Architecture and assembly of mammalian H/ACA small nucleolar and telomerase ribonucleoproteins. *EMBO J.* **23:** 1857.

Wang C., Query C.C., and Meier U.T. 2002. Immunopurified small nucleolar ribonucleoprotein particles pseudouridylate rRNA independently of their association with phosphorylated Nopp140. *Mol. Cell. Biol.* **22:** 8457.

Wang H., Boisvert D., Kim K.K., Kim R., and Kim S.H. 2000. Crystal structure of a fibrillarin homologue from *Methanococcus jannaschii*, a hyperthermophile, at 1.6 Å resolution. *EMBO J.* **19:** 317.

Watkins N.J., Dickmanns A., and Luhrmann R. 2002. Conserved stem II of the box C/D motif is essential for nucleolar localization and is required, along with the 15.5K protein, for the hierarchical assembly of the box C/D snoRNP. *Mol. Cell. Biol.* **22:** 8342.

Watkins N.J., Leverette R.D., Xia L., Andrews M.T., and Maxwell E.S. 1996. Elements essential for processing intronic U14 snoRNA are located at the termini of the mature snoRNA sequence and include conserved nucleotide boxes C and D. *RNA* **2:** 118.

Watkins N.J., Gottschalk A., Neubauer G., Kastner B., Fabrizio P., Mann M., and Luhrmann R. 1998. Cbf5p, a potential pseudouridine synthase, and Nhp2p, a putative RNA-binding protein, are present together with Gar1p in all H BOX/ACA-motif snoRNPs and constitute a common bipartite structure. *RNA* **4:** 1549.

Watkins N.J., Lemm I., Ingelfinger D., Schneider C., Hossbach M., Urlaub H., and Luhrmann R. 2004. Assembly and maturation of the U3 snoRNP in the nucleoplasm in a large dynamic multiprotein complex. *Mol. Cell* **16:** 789.

Watkins N.J., Segault V., Charpentier B., Nottrott S., Fabrizio P., Bachi A., Wilm M., Rosbash M., Branlant C., and Luhrmann R. 2000. A common core RNP structure shared between the small nucleoar box C/D RNPs and the spliceosomal U4 snRNP. *Cell* **103:** 457.

Wolin S.L. and Cedervall T. 2002. The La protein. *Annu. Rev. Biochem.* **71:** 375.

Wright W.E., Tesmer V.M., Liao M.L., and Shay J.W. 1999. Normal human telomeres are not late replicating. *Exp. Cell Res.* **251:** 492.

Wu P., Brockenbrough J.S., Metcalfe A.C., Chen S., and Aris J.P. 1998. Nop5p is a small nucleolar ribonucleoprotein component required for pre-18 S rRNA processing in yeast. *J. Biol. Chem.* **273:** 16453.

Yang P.K., Rotondo G., Porras T., Legrain P., and Chanfreau G. 2002. The Shq1p.Naf1p complex is required for box H/ACA small nucleolar ribonucleoprotein particle biogenesis. *J. Biol. Chem.* **277:** 45235.

Yang P.K., Hoareau C., Froment C., Monsarrat B., Henry Y., and Chanfreau G. 2005. Cotranscriptional recruitment of the pseudouridylsynthetase Cbf5p and of the RNA binding protein Naf1p during H/ACA snoRNP assembly. *Mol. Cell. Biol.* **25:** 3295.

Yang Y., Isaac C., Wang C., Dragon F., Pogacic V., and Meier U.T. 2000. Conserved composition of mammalian box H/ACA and box C/D small nucleolar ribonucleoprotein particles and their interaction with the common factor Nopp140. *Mol. Biol. Cell* **11:** 567.

Yong J., Wan L., and Dreyfuss G. 2004. Why do cells need an assembly machine for RNA-protein complexes? *Trends Cell Biol.* **14:** 226.

Yu Y.-T., Scharl E.C., Smith C.M., and Steitz J.A. 1999. The growing world of small nuclear ribonucleoproteins. In *The RNA world,* 2nd edition (ed. R.F. Gesteland et al.), p. 487. Cold Spring Harbor Laboratory Press, Cold Spring Harbor, New York.

Zhu Y., Tomlinson R.L., Lukowiak A.A., Terns R.M., and Terns M.P. 2004. Telomerase RNA accumulates in Cajal bodies in human cancer cells. *Mol. Biol. Cell* **15:** 81.

RNA and Protein Actors in X-Chromosome Inactivation

O. Masui and E. Heard

CNRS UMR 218, Institut Curie, Paris 75005, France

In female mammals, one of the two X chromosomes is converted from the active euchromatic state into inactive heterochromatin during early embryonic development. This process, known as X-chromosome inactivation, results in the transcriptional silencing of over a thousand genes and ensures dosage compensation between the sexes. Here, we discuss the possible mechanisms of action of the Xist transcript, a remarkable noncoding RNA that triggers the X-inactivation process and also seems to participate in setting up the epigenetic marks that provide the cellular memory of the inactive state. So far, no functional protein partners have been identified for Xist RNA, but different lines of evidence suggest that it may act at multiple levels, including nuclear compartmentalization, chromatin modulation, and recruitment of Polycomb group proteins.

The inactive X chromosome, or "Barr body," was first identified over 50 years ago as a heteropycnotic structure, only present in female somatic cells (Barr and Bertram 1949; Ohno and Hauschka 1960). Mary Lyon then published her seminal paper in 1961 making the link between this structure and the genetic inactivity of one of the two X chromosomes in females (Lyon 1961). She further proposed that this must be an early developmental event, which is then inherited mitotically, in order to explain the large patches of coat color mosaicism observed in female mammals heterozygous for X-linked genes.

X inactivation represents a powerful model system for studying mammalian epigenetics, as it involves differential regulation of two homologous chromosomes within the same nucleus, in a mitotically heritable but developmentally reversible manner. In placental mammals, X inactivation is initiated by a master control locus, the X-inactivation center (Xic), and the noncoding Xist transcript it produces. Xist RNA accumulates over the chromosome from which it is produced and is responsible for inducing *cis*-limited silencing of the >1000 genes on the X chromosome (Penny et al. 1996; Marahrens et al. 1998; Wutz and Jaenisch 2000). Since *Xist* was discovered in 1990, it has been the object of intense investigation; however, its regulation and its mechanism of action as a functional RNA still remain largely mysterious. Furthermore, the manner by which this silencing signal is transformed into the stable, transcriptionally repressed state that characterizes the inactive X chromosome is still poorly understood. However, several epigenetic marks, including histone modifications, Polycomb group proteins, and DNA methylation, are clearly important for the maintenance of the inactive state. Here, we summarize recent findings suggesting that Xist RNA may have multiple functions in the X-inactivation process: in the recruitment of chromatin modifiers such as Polycomb group proteins and also at the level of nuclear compartmentalization. We describe data showing that Xist RNA may participate in the formation of a silent nuclear compartment into which X-linked genes are recruited when they become inactivated. Spatial segregation from nucleoplasmic transcription factors of the X chromosome to be

inactivated may thus represent a strategy, in addition to chromatin marks, for allowing the differential treatment of the two X chromosomes within the same nucleus. These findings also provide new insights into the structure and sequence organization of the cytologically defined Barr body. We also discuss another level at which RNA may function during the X-inactivation process, which is for the recruitment of Polycomb group proteins to the inactive X chromosome. These proteins are involved in the maintenance of the inactive state, although the degree to which they are required for X chromosome inactivity may vary between lineages. Both PRC2 and PRC1 complexes appear to associate with the inactive X chromosome. The mechanisms employed for targeting these Polycomb complexes to the inactive X chromosome are only just beginning to be unraveled and appear to involve multiple strategies, including both histone modifications and RNA components. The inactive X chromosome thus provides, more than ever before, a useful model system for studying the interplay between nuclear organization, chromatin, RNA, and epigenetics.

DEVELOPMENTAL TIMING OF Xist ACTION

The early onset of *Xist* expression during development is consistent with its role in initiating X inactivation (Kay et al. 1993). In mice, there are two waves of X inactivation. The first is subject to imprinting and the second is random. During preimplantation development, the paternal *Xist* allele is expressed from the 2-cell stage, around the time of major zygotic genome activation. The maternal *Xist* allele remains inactive throughout this period due to a repressive imprint, the nature of which remains unknown, that is established in the maternal germ line (Tada et al. 2000). The early expression of the paternal *Xist* gene results in the inactivation of the paternal X chromosome from the 8-cell stage onward (Mak et al. 2004; Okamoto et al. 2004, 2005). *Xist* is essential for imprinted X inactivation, as demonstrated by the early lethality of mouse embryos with a paternally inherited deletion of the gene (Marahrens et al. 1997). The inactivity of the paternal X chromosome that is initiated in cleavage stages is maintained in extraembryonic lineages such as the trophectoderm, but in the inner cell mass of the blastocyst the paternal X is reactivated (between 3.5 and 4.5 days postcoitum [dpc]). By this stage, the repressive maternal *Xist* imprint is also lost. In this way, when the second wave of X inactivation initiates in the epiblast (around 5.5 dpc), either the paternal or maternal X chromosome will up-regulate *Xist*, and this triggers random X inactivation. Female embryonic stem (ES) cells, which are derived from the inner cell mass of blastocysts, represent a useful in vitro model system for X inactivation. Upon differentiation, *Xist* is up-regulated and accumulates on one of the two X chromosomes; this is followed by gene silencing and the appearance of a number of epigenetic marks (Fig. 1). Again, knockout studies have shown that *Xist* is essential for this random X-inactivation process (Penny et al. 1996).

Although the timing of *Xist* expression and X inactivation pointed to an early time window for its function, the exact window in which it could act was defined in a series of elegant studies using ES cells carrying an inducible Xist cDNA transgene (Wutz and Jaenisch 2000). Xist RNA-dependent silencing could only be triggered during the first 48–72 hours of differentiation. This early time window for *Xist* action suggests either that chromatin is somehow rendered refractory to Xist RNA silencing as differentiation progresses, or alternatively, that Xist RNA requires the presence of a specific factor that is only present during early development for its silencing activity. During this early time window, the inactive X is fully reversible upon arrest of *Xist* expression. However, following 72 hours of differentiation and *Xist* expression, X inactivation can no longer be reversed if the inducible Xist cDNA is turned off (Wutz and Jaenisch 2000). Thus, some form of chromosomal memory must be established on the inactive X during differentiation (Kohlmaier et al. 2004). Some of the changes that could be involved in this chromosomal memory, and that are induced following Xist RNA coating of the X chromosome in differentiating female ES cells, include a shift to asynchronous replica-

tion timing (Takagi et al. 1982), incorporation of the histone variant macro H2A (Mermoud et al. 1999; Costanzi et al. 2000), DNA (CpG) methylation (Norris et al. 1991), and a variety of histone modifications (Chaumeil et al. 2002; for review, see Heard 2004). In particular, changes in histone H3 and H4 modifications occur early on and include the loss of active, euchromatic marks such as H3K9 and H4 acetylation (Jeppesen and Turner 1993; Boggs et al. 1996; Keohane et al. 1996), as well as H3K4 di- and tri-methylation. These changes can be observed just one day after Xist RNA coating both in early mouse embryos at the 4- to 8-cell stage (Okamoto et al. 2004) and in differentiating ES cells at day 2. Marks generally associated with gene repression, such as the di-methylation of H3K9 (Heard et al. 2001; Boggs et al. 2002; Mermoud et al. 2002; Peters et al. 2002) and tri-methylation of H3K27 (Plath et al. 2003; Silva et al. 2003; Rougeulle et al. 2004), appear on the inactive X at around day 2 of ES cell differentiation. In early mouse embryos, where the temporal resolution of events is more straightforward than in differentiating ES cells, the time of onset of the repressive (H3K27me3 and H3K9me2) marks occurs later (>16-cell stage) than the loss of euchromatic histone marks on the paternal X chromosome. Polycomb group proteins associate with the inactive X chromosome during the same developmental time window as the acquisition of repressive histone marks and are responsible for some of the above histone modifications. For example, Ezh2 (PRC2 complex) is responsible for H3K27me3, and Ring 1a/b (PRC1 complex) is responsible for ubiquitination of H2AK119, as discussed below. The timing of appearance of Polycomb group proteins, following Xist RNA coating during early development (Mak et al. 2004; Okamoto et al. 2004), or after induction of an Xist cDNA (Plath et al. 2003; Kohlmeier et al. 2004), suggests that they could be recruited to the chromosome by Xist RNA itself, although this remains to be proven. However, rather than being involved in the initiation of silencing, as discussed below, these complexes may form part of the chromosomal memory of the inactive X chromosome. Thus, contrary to the initial belief that Xist RNA would only be involved in the initial silencing events of X inactivation, it now appears that it may also have a role in recruiting some of the early epigenetic marks that take over once silencing has occurred.

FUNCTIONAL DOMAINS OF Xist RNA

The functional Xist transcript is spliced, polyadenylated, and measures 17,000–19,000 nucleotides in length. Despite such features, which usually characterize messenger RNAs, it is retained in the nucleus and is untranslated. Although the overall structure of the *Xist* gene is well conserved between eutherian mammals, its sequence is remarkably poorly conserved, given its central function in controlling X inactivation. However, this is consistent with its presumed role at the RNA level and the fact that conservation may be mainly in its secondary or tertiary RNA structure, as opposed to its primary sequence. Furthermore, the most highly conserved regions in this transcript comprise a series of repeats (termed A–E, see

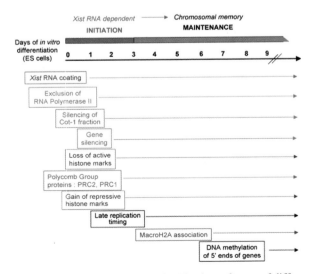

Figure 1. Kinetics of random XCI. The time of onset of different events during the onset of random X inactivation is shown.

Fig. 2A). The A repeats, located at the 5′ end of the first exon of *Xist*, are the most conserved of all. Indeed, these repeats were shown to be capable of inducing repression of X-linked genes in an in vitro assay (Allaman-Pillet et al. 2000). To define the functional regions of the Xist transcript in vivo, male ES cells containing inducible *Xist* cDNAs with different deletions were tested for their capacity to induce silencing and chromosome coating *in cis* (Wutz et al. 2002). Because the inducible transgene was located on the single X chromosome in a male cell, X inactivation induced nullisomy. Cell death was thus used as an assay for the capacity of different Xist deletions to induce inactivation. In this way, the A repeats were shown to be the only region apparently critical for *Xist*'s silencing function. A number of different regions of Xist RNA were shown to be involved in its capacity to coat a chromosome *in cis*, as well as its ability to recruit potential epigenetic marks described above, such as macroH2A and H3K27

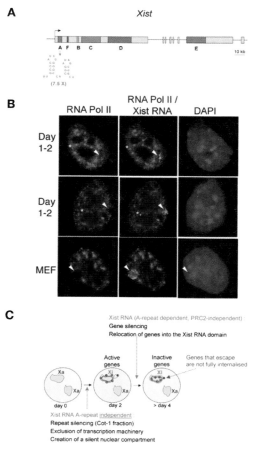

B

C

Figure 2. The *Xist* gene and it potential role in creating a silent nuclear compartment. (*A*) A map of the murine *Xist* gene is shown, indicating the conserved repeats (A–E). The sequence of the most highly conserved A repeats, involved in the gene silencing function of Xist, is shown. (*B*) Example of RNA polymerase II immunofluorescence combined with Xist RNA FISH in early differentiating ES cells (*top two panels*) and embryonic fibroblasts (*lower panel*). This shows overall exclusion of RNA pol II at the level of the domain of nuclear Xist RNA accumulation as described by Chaumeil et al. (2006). (*C*) Model for two types of Xist RNA function during the onset of X inactivation, based on Chaumeil et al. (2006).

tri-methylation (Wutz et al. 2002; Plath et al. 2003; Kohlmaier et al. 2004). The fact that a Xist transcript, mutated for the A repeats and incapable of inducing gene silencing, can still recruit Polycomb group complexes and associated H3K27me3 and H2AK119 ubiquitination, demonstrates that these chromatin changes are not sufficient for the silencing function of Xist.

The molecular mechanisms that underlie Xist RNA's capacity to induce transcriptional silencing remain unclear. One possibility is that Xist RNA acts to change chromatin in order to induce the silent state. As mentioned above, Polycomb group proteins are unlikely to be involved in this initiation step. The rapid loss of active histone modifications (H3 and H4 acetylation as well as H3K4 and H3K36 methylation) following Xist RNA coating (Chaumeil et al. 2002, 2006) suggests that Xist RNA may have a function in recruiting histone deacetylase and/or demethylase complexes. On the other hand, Xist RNA might recruit chromatin remodeling enzymes that result in the active ejection and replacement of "actively" marked histones by unmodified histones (which then become modified by Polycomb group complexes) or even histones with certain inactive marks. A role for histone variants (for example the H3.1, H3.2, and H3.3 forms of histone H3) in the initiation of silencing could also be possible, but so far remains unknown. Whether Xist RNA has a direct role in the loss of active histone marks, and whether this loss is a cause or consequence of gene silencing, are clearly areas that merit future investigation. Alternative models for Xist RNA's silencing function include the recruitment of a repressor protein or complex, although so far no such complex has been identified. Attempts to purify Xist A-repeat-specific partners identified hnRNPC (Brown and Baldry 1996), one of the most ubiquitous hnRNPs, although no link with X inactivation has yet been reported. Xist RNA may also recruit factors that induce other repressive histone marks such as H3K9me2 and H4K20me1, which might be involved in initiation and/or maintenance of X inactivation. However, the modifying enzymes involved and their link with Xist RNA remain to be found. Finally, a longstanding hypothesis is that Xist RNA could participate in the formation of a silent nuclear compartment within which the X chromosome is silenced. Recent findings suggesting that Xist RNA may indeed function in this way are described in the next section.

XIST RNA IN NUCLEAR COMPARTMENTALIZATION

A possible architectural role for Xist RNA in creating a repressive nuclear compartment or structure around the X chromosome was proposed some years ago (Clemson et al. 1996). Xist RNA is intriguingly restricted in its nuclear localization to the vicinity of the chromosome territory from which it is expressed (Brown et al. 1992; Clemson et al. 1996). This restriction does not appear to be dependent on the DNA of the chromosome itself, as the appearance of the Xist RNA domain remains unperturbed after DNase treatment (Clemson et al. 1996). It has therefore been proposed that Xist RNA may show only an indirect association with the X chromosome and a closer association with the nuclear matrix. Potential support for this has come from the finding that scaffold attachment factor A (SAF-A) is enriched on the inactive X chromosome in an RNA-dependent manner (Helbing and Fackelmayer 2003). Xist RNA might thus form a stable structure with nuclear matrix or scaffold factors, which could be important for initiation and/or maintenance of the inactive state (Fackelmayer 2005).

Our laboratory recently set out to determine whether Xist RNA might function at the level of nuclear organization, using differentiating ES cells to assess changes in nuclear organization relative to chromatin changes and gene silencing. We showed that Xist RNA chromosome coating leads to the rapid exclusion of RNA polymerase II and associated transcription factors (Fig. 2B). This represents the earliest event following Xist RNA accumulation described so far and precedes the loss of active histone marks such as H3K9 acetylation and H3K4 dimethylation (Chaumeil et al. 2006). Only subsequently do genes cease to be transcribed, as detected at the primary transcript level by RNA FISH, which allows the exact timing of transcriptional silencing to be assessed at the single-cell level. Similar findings were found in early mouse embryos: RNA pol II was found to be excluded from the Xist RNA-coated paternal X chromosome from the 4- to 8-cell stage in mouse embryos (Okamoto et al. 2004). We also found that the RNA pol II-excluded compartment within the Xist RNA accumulation is transcriptionally silent, as detected by RNA FISH using Cot-1 DNA as a probe, which detects middle repetitive elements. Although it has been reported that Cot-1 RNA detected in this way corresponds to transcribed repetitive elements present in introns, or 5′ and 3′ UTRs of genes, we have so far not found any genic loci that lie within this Cot-1 hole at early differentiation stages (Chaumeil et al. 2006; J. Chow and E. Heard, unpubl.). Thus, the rapid RNA pol II exclusion and Cot-1 silencing induced by Xist RNA only seem to affect a repeat-rich fraction at the core of the X-chromosome territory. The exact nature of the sequences present within the silent Xist RNA compartment is currently under investigation.

On the other hand, X-linked genes, which are still active at early stages of differentiation (days 1–2), were always found to be located outside, or at the periphery of this Xist RNA domain and within the nucleoplasmic RNA pol II, as might be expected given their transcriptional activity. However, upon transcriptional silencing (by day 4), most genes were found to show a significant shift in position, to a location within the Xist RNA domain. The only loci that continued to show a more external location even at late differentiation stages were the *Xist* gene, which remains active on the otherwise inactive X, and *Jarid1c*, which escapes from X inactivation in approximately 50% of differentiated cells. The external location of these genes therefore appears to be linked to their transcriptional activity. We also made the very surprising observation that Xist RNA can trigger the exclusion of RNA pol II within the Xist RNA domain, even in the absence of the A-repeat silencing region of the Xist transcript. This was an unexpected finding, because the silencing function of Xist RNA was previously believed to be solely dependent on its A-repeat region (Wutz et al. 2002). Indeed, the A-repeat-deleted Xist transcript was found to be deficient in its capacity to trigger X inactivation at the level of all the X-linked genes examined. In this case, genes remain external to the Xist RNA domain and continue to be transcribed throughout differentiation. The formation of a Xist RNA compartment, depleted of transcription factors, is therefore not in itself sufficient to induce relocation and silencing of X-linked genes. However, the Cot-1 RNA fraction of the X chromosome that lies within the Xist RNA domain was found to be silenced in an Xist A-repeat-independent fashion. This implies that Xist RNA may exert different types of silencing function—one aimed at repeats and the other at genes.

This study provided the first evidence for a new and early step in the X-inactivation process. It also suggests a novel role for Xist RNA in the formation of a silent nuclear compartment which initially comprised the more repetitive part of the X chromosome (Fig. 2C). This new function for Xist RNA is independent of its A repeats and results in the rapid exclusion of the transcription machinery from the X chromosome chosen to be inactivated. Gene repression occurs subsequently, requires Xist A-repeat action, and is accompanied by a shift from outside to inside this silent nuclear compartment. This shift is not simply a consequence of increased Xist RNA coating or increased X-chromosome compaction, as we found that Xist RNA always coats approximately 70% of the volume of the X chromosome at all stages of differentiation examined. Rather, the onset of X inactivation is accompanied by a dynamic 3D reorganization of X-linked genes. Given that the movement of genes toward the interior of the transcriptionally inert Xist RNA domain does not precede gene silencing, but rather accompanies or follows it, one explanation is that the transcriptional repression induced by the Xist A repeat containing RNA results in the capacity of genes to become internalized. In some cases, actively transcribed genes have been shown to be located in putative "transcription factories" (see Osborne et al. 2004). When Xist A-repeat-induced silencing occurs, a gene may become unleashed from such a transcription compartment and thus become more internally located by default. An alternative but not mutually exclusive explanation for gene relocation could be that the Xist transcript, and the ribonucleoprotein structure it forms, participate in "reeling" genes into the silent domain, either through local chromatin condensation or through active translocation. The fact that genes do not undergo internalization in the Xist A-repeat mutant, despite the recruitment of Polycomb group proteins, suggests that these proteins are not sufficient to induce relocation, although they may still be necessary.

REDEFINING THE BARR BODY

Our studies have provided a novel perspective on the molecular content of the cytologically defined, heteropycnotic Barr body. Although the appearance of a DAPI-dense structure at the level of the Xist RNA-coated X chromosome occurs early on in differentiation, our work suggests that this may in fact correspond to the silent, repetitive core of the X-chromosome territory that is confined within the Xist RNA domain, rather than to genes. All of the X-linked genes we have examined to date tend to be located on the more peripheral portion of the X-chromosome territory (as determined by DNA FISH using a mouse X-chromosome paint), whatever their activity status, and the relocation of X-linked genes from outside to inside the Xist RNA compartment during X inactivation actually only entails a shift in position in the order of 0.1–0.8 μm. This is also the case for the human X chromosome, which has been shown to consist of a repetitive inner core, with genes located on its outer rim (Clemson et al. 2006). It is therefore very likely that the heteropycnotic structure originally identified by Barr in 1949 (Barr and Bertram 1949) in fact corresponds mainly to the repetitive core of the inactive X. Furthermore, our analysis of the Xist A-repeat mutants suggests that this heteropycnotic structure may form even in the absence of gene silencing. The implication of this is that studies over the decades which have assessed the presence of an inactive X chromosome only at the level of presence or absence of a Barr body need to be reconsidered, as transcriptional activity of genes and the formation of a Barr body may be separable events.

However, ultrastructural analyses of the inactive X chromosome, using electron microscopy, will be required to obtain a more detailed vision of its structure and the distribution of repeats versus genes. Furthermore, whether the repetitive core of the X chromosome contains any genes whatsoever remains to be found. It would be predicted that such genes, were they to exist, might be subject to very rapid silencing in an Xist A-repeat-independent fashion. Genes that remain located at the periphery of or outside the silent Xist RNA compartment throughout differentiation, such as the *Jarid1c* gene, may provide potential insights into the process of escape from X inactivation. The external location of the *Jarid1c* gene may reflect a resistance of this locus to be internalized, despite the silencing action of the Xist A repeats. This would be consistent with the recent finding that *Jarid1c* is flanked by CTCF boundary elements, which could render it more resistant to relocation compared to other genes (Filippova et al. 2005). Intriguingly, however, escape genes on the human inactive X chromosome do not appear to be more externally located, relative to the Xist RNA domain, than those genes that are subject to X inactivation (Clemson et al. 2006).

The regions of Xist RNA, other than the A repeats, that are capable of creating a silent nuclear compartment and heteropycnotic structure remain to be defined through the analysis of further Xist mutants. It is also not yet known whether the silent nuclear compartment created by Xist RNA represents a physical or biochemical entity. Although SAF-A interacts with the inactive X chromosome in somatic cells, this has not yet been demonstrated in differentiating ES cells, nor has any evidence for a direct interaction between Xist RNA and SAF-A been reported. Accessibility assays might address the extent to which Xist RNA and its protein partners might create a physical barrier to transcription factors. It is, however, unlikely that the exclusion of the transcription machinery

by Xist RNA is a result of the recruitment of Polycomb group complexes to the X, even though this is also an Xist A-repeat-independent function. Polycomb group complexes may, however, play a role in stabilizing the repressive environment created by Xist RNA, and it will be interesting to investigate the extent to which the silent nuclear compartment defined by Xist RNA is maintained in Polycomb group mutants.

POLYCOMB GROUP PROTEINS IN X INACTIVATION

Background to the Function and Constitution of PcG Complexes

Polycomb group (PcG) genes were first identified in *Drosophila* in the 1940s (for review, see Ringrose and Paro 2004). Mutations in PcG genes lead to perturbations in the maintenance of repression of homeotic genes, which are important for body segmentation. Misregulation of *Hox* genes induces severe developmental defects in flies, and this relationship between PcG genes and *Hox* genes is well conserved from *Drosophila* through to mammals. In mammals, however, PcG genes appear to be involved in numerous other processes, including stem-cell renewal and differentiation, hematopoesis, cellular senescence, and X-chromosome inactivation. Indeed, although the PcG proteins are highly conserved throughout evolution, the homologs for each of the PcG genes in *Drosophila* are found to be expanded and to exist as gene families in mammals, which presumably confers their divergent roles (for review, see Gil et al. 2005). Both genetic and biochemical experiments have shown that most PcG proteins work in complexes that can be divided into two main classes, although these are fairly dynamic and can differ in their exact compositions (Otte and Kwaks 2003; Levine et al. 2004). One of these is termed Polycomb repressive complex 1 (PRC1) and the other is known as PRC2 (also called ESC-E(Z) complex or EED-EZH2 complex) (Fig. 3).

PRC1 was the first complex to be purified from *Drosophila* embryos (Shao et al. 1999; Saurin et al. 2001) and HeLa cells (Levine et al. 2002), and it represents a huge 1- to 2-MD protein complex with its core components being Pc, Ph, Psc, Scm, and Ring (see Fig. 3). These PRC1 proteins have multiple functions affecting chromatin structure (Levine et al. 2004), including a ubiquitin ligase activity for histone H2A K119 (Wang et al. 2004), mediated by Ring1b as the ubiquitin E3 ligase (de Napoles et al. 2004; Wang et al. 2004).

The second PcG protein complex, PRC2, was purified from both *Drosophila* embryos (Ng et al. 2000; Tie et al. 2001, 2003; Czermin et al. 2002; Müller et al. 2002) and mammalian cells (Cao et al. 2002; Kuzmichev et al. 2002) as an approximately 600-kD complex and shown to contain Esc (Eed in mammals), Su(z)12 (Suz12 in mammals), and E(z) (Ezh2 in mammals) as core components (Fig. 3). The E(z)/Ezh2 protein contains a SET (*Su*-(var)3-9, *E*(z), *Trithorax*) domain and, consistent with other SET-domain proteins, it has a histone methyltransferase (HMTase) activity. In vitro studies have shown that E(z) acts to

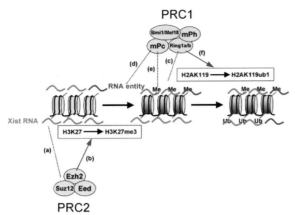

Figure 3. Schematic representation of Polycomb group complex proteins that associate with the Xist RNA-coated X chromosome. Possible relationships between PRC1/2 and chromatin changes on the inactive X chromosome are shown. Blue arrows indicate enzymatic activity for the reaction depicted by a black arrow in the rectangle. Broken lines indicate potential binding affinity. (*a*) Xist RNA leads to PRC2 complex recruitment through unknown mechanisms (Schoeftner et al. 2006); (*b*) the PRC2 protein, Ezh2, catalyzes tri-methylation of histone H3 at lysine 27 (Kuzmichev et al. 2002). Both Eed and Suz12 are required to form a stable PRC2 complex and for the HMTase activity of Ezh2. (*c*) Ring1b protein is recruited by Xist RNA in a PRC2-independent manner (Schoeftner et al. 2006). (*d*) One of the murine Polycomb homologs, Cbx7, is retained on the inactive X chromosome in an RNA-dependent manner (Bernstein et al. 2006) and through binding to H3K27me3 for which it has a high affinity. (*e*) mPc proteins have affinity for H3K27me3 (Bernstein et al. 2006). (*f*) Ring1a and Ring1b proteins catalyze mono-ubiquitylation of histone H2A at lysine 119 (de Napoles et al. 2004). Certain PRC1-associated proteins, such as Scm, have not so far been detected at the level of the inactive X chromosome, although their presence has not been formally excluded.

methylate histone H3 at K9/K27 (Cao et al. 2002; Czermin et al. 2002; Kuzmichev et al. 2002; Müller et al. 2002) and histone H1at K26 (Kuzmichev et al. 2004). In vivo studies, involving *Suz12* and *Eed* mutant embryos, confirm that Ezh2 is indeed responsible for H3K27 di- and tri-methylation (Pasini et al. 2004; Kalantry et al. 2006). In mammals, PRC2 consists of at least Ezh2, Eed, Suz12, and RbAp48 (Cao et al. 2002; Kuzmichev et al. 2002). In addition, the Eed gene can produce several isoforms of the Eed protein that are used for different substrates and in different cellular situations (Kuzmichev et al. 2004, 2005). The core components of PRC2 appear to be crucial for the stability of the complex as well as its chromatin-associated activities. Ezh2 alone does not have HMTase activity and can only act in the presence of both Eed and Suz12 (Cao and Zhang 2004; Pasini et al. 2004). In fact, in the absence of Suz12, the Ezh2 protein appears to be highly unstable (Pasini et al. 2004).

The fundamental roles of the PRC2 complex in mammals are illustrated by the embryonic lethality that mutants for most of these proteins induce. For example, mouse mutants in Eed (Faust et al. 1995; Wang et al. 2001), Ezh2 (O'Carroll et al. 2001), and Suz12 (Pasini et al. 2004) have all been reported to show early developmental lethality, around 7.5 dpc.

Evidence for a Role for PcG Complexes in X Inactivation

Several lines of evidence suggest that PcG proteins are implicated in X inactivation. Circumstantial evidence comes from immunofluorescence studies. First, several components of PRC1 and PRC2 complexes have been shown to accumulate on the inactive X chromosome during early development and ES cell differentiation, following Xist RNA coating of the X chromosome (Mak et al. 2002; de Napoles et al. 2004; Plath et al. 2004). Second, two of the histone modifications associated with PcG function, H3K27me3 and H2AK119 mono-ubiquitylation (H2Aub1), become enriched on the inactive X during a similar time window (Mak et al. 2002; Silva et al. 2003; de Napoles et al. 2004; Plath et al. 2004). Genetic evidence for an involvement of PcG proteins in X inactivation first came from the analysis of *Eed*$^{-/-}$ mutant mice (Wang et al. 2001). Female mutant mice carrying a GFP transgene on their paternally inherited X chromosome, which should be silent in extraembryonic tissues due to imprinted paternal X inactivation, showed a proportion of GFP-positive cells after 5.5 dpc. This result suggests that *Eed* mutant mice can initiate imprinted X inactivation, but cannot maintain it efficiently. No apparent effect was observed on the random X-inactivation process in the embryo-proper of Eed$^{-/-}$ mutants (Wang et al. 2001; Kalantry and Magnuson 2006), although a minor effect had been reported in one study (Silva et al. 2003). More recently, a conditional *Eed* knockout was generated in ES cells containing an inducible Xist cDNA, and the absence of Eed (and the resulting lack of Ezh2-mediated H3K27 methylation) was found to have no impact on either the initiation or the maintenance of Xist RNA-mediated silencing (Schoeftner et al. 2006). These findings are consistent with the data described above, showing that PRC2 is not involved in the initiation of X inactivation, as ES cells expressing the Xist transcript mutated for its A repeats are incapable of gene silencing, despite the recruitment of PRC2 and H3K27me3 to the X chromosome (Plath et al. 2003; Kohlmaier et al. 2004).

Taken together, these genetic studies suggest that PRC2 is unlikely to play an important role in the initiation of X inactivation and, furthermore, that it is not critical for maintenance of random X inactivation in embryonic tissues. Presumably, the participation of multiple epigenetic marks in random X inactivation, including H3K9me2 and H4K20me1 and other unknown factors, may render the requirement for PRC2 less critical. However, PRC2 does seem to play an important role in the maintenance of imprinted X inactivation in extraembryonic tissues. Further insight into this has come from a recent study showing that the defect in maintenance of the inactive state is only found in differentiated Eed$^{-/-}$ trophoblast cells (Kalantry et al. 2006). Intriguingly, in undifferentiated Eed$^{-/-}$ trophoblast stem cells, many characteristics of the inactive X, such as Xist RNA coating, PRC1 and PRC2 proteins, and associated histone modifications, are no longer detectable, despite its transcriptional inactivity (Kalantry et al. 2006). The authors conclude that PcG complexes are not necessary to maintain transcriptional silencing of the inactive X chromosome in undifferentiated stem cells. Instead, PcG proteins could be involved in the cellular memory that prevents transcriptional activation of the inactive X during differentiation.

The exact mechanism underlying PRC2's role in maintaining inactivity of the X chromosome is unknown. Drawing on parallels with other systems, it is known that a subset of PRC2 complexes contain HDAC activity in human cells (van der Vlag and Otte 1999) and in Drosophila (Tie et al. 2001, 2003). Deacetylation of histones might therefore also participate in PRC2's action on the X chromosome. Consistent with this, the accumulation of PRC2 on the Xi occurs within a similar time window to H4 hypoacetylation, although H3K9 hypoacetylation seems to be a slightly earlier event (Heard et al. 2001; Chaumeil et al. 2002; Okamoto et al. 2004). Alternatively, the H3K27me3 mark and/or other marks induced by PRC2 complexes (e.g., H1K26 methylation; Kuzmichev et al. 2004) may have a direct effect on chromatin accessibility or packaging. Yet another possibility, supported by substantial genetic and biochemical studies in different species, is that PRC2 leads to the recruitment of the PRC1 complex and that this performs the maintenance function. In vitro studies have shown that PRC1 may act at several levels to maintain silencing. These include inhibition of nucleosome remodeling mediated by SWI/SNF (Shao et al. 1999; Lavigne et al. 2004), induction of chromatin compaction (Francis et al. 2004; Lavigne et al. 2004), inhibition of transcription initiation (Dellino et al. 2004), and ubiquitinated H2A-mediated gene silencing (Wang et al. 2004).

Although PRC1 complex proteins and associated H2A ubiquitination appear to associate with the inactive X chromosome during both imprinted and random X inactivation (de Napoles et al. 2004; Fang et al. 2004; Plath et al. 2004), so far their exact role(s) in X inactivation remains unclear. Schoeftner et al. (2006) have recently shown that PRC1 recruitment by Xist RNA to the X chromosome is independent of gene silencing, similarly to PRC2. Mutant homozygous mice have been reported for some components of PRC1 such as Ring1A (del Mar Lorente et al. 2000), Ring1B (Voncken et al. 2003), Cbx2/M33 (Coré et al. 1997; Katoh-Fukui et al. 1998), Bmi1 (van der Lugt et al. 1994), Mel18 (Akasaka et al. 1996), Phc1 (Takihara et al. 1997), and Phc2 (Isono et al. 2005), which were shown to accumulate on the Xi in certain cell types. These mutants, except for Ring1B, which shows early embryonic lethality, show homeotic abnormalities consistent with the misregulation of Hox genes, although effects in the maintenance of X inactivation were not assessed. However, XX ES cells mutated for Ring1a and Ring1b have been created, and although H2A ubiquitination in the inactive X is clearly disrupted, they do not show any obvious alteration in X-inactivation status (de Napoles et al. 2004). The probable redundancy between different PRC proteins and other epigenetic marks on the X chromosome may render determination of the functional importance of PRC1 proteins in X inactivation a difficult task. Further genetic studies will be required, involving multiple knockouts and careful analyses in different embryonic and extraembryonic lineages.

Recruitment of PRC2 and PRC1 Complexes to the Inactive X Chromosome

In Drosophila, the PRC2 complex is targeted to specific regions for repression via Polycomb response elements (PREs) (Bantignies and Cavalli 2006). In mammals, DNA sequences equivalent to PREs are still being sought. In the case of X inactivation, the recruitment of PRC2 appears to be a direct consequence of Xist RNA coating, which, as mentioned earlier, might suggest that Xist itself targets PRC2 to the X chromosome. However, it cannot be excluded that PRC2 is recruited as an immediate consequence of some chromosomal change induced by Xist RNA.

In the case of the PRC1 complex, its targeting to chromatin is thought to be at least partly dependent on the presence of PRC2-induced H3K27me3 in Drosophila. The Polycomb (Pc) protein, which is a core component of PRC1, has been shown to bind to H3K27me3 with strong affinity (Fischle et al. 2003; Min et al. 2003), and this has thus been proposed as a mechanism through which the PRC2 complex could lead to the recruitment of PRC1. The domain of the Pc protein that appears to bind to the H3K27me3 mark is a conserved amino acid motif known as the chromodomain. In mammals there are five Pc homologs: Cbx2/M33/Mpc1, Cbx4/Mpc2, Cbx6, Cbx7, and Cbx8/Mpc3. All of these, except Cbx4, have been shown to have a high affinity for H3K27me3 (Bernstein et al. 2006). We recently set out to address whether Pc proteins associate with the inactive X chromosome and to gain insight into the mechanism of their recruitment. Using female mouse ES cells, we were able to detect the presence of several of these Cbx proteins on the inactive X chromosome (Bernstein et al. 2006). When fused to GFP, all Cbx proteins, except Cbx4, showed preferential accumulation on the H3K27 tri-Me enriched inactive X chromosome, following transient transfection in day 3–6 differentiating female ES cells. The accumulation was most striking for Cbx7 (Fig. 4). Using mutated forms of the Cbx7 protein, in which critical amino acids were changed within its chromodomain, we demonstrated that the association of Cbx7 with the inactive X is likely to be dependent on interactions between its chromodomain and H3K27me3. However, this interaction alone may not be sufficient for Cbx7 binding to the inactive X chromosome, as we showed that RNase treatment disrupts the X-chromosomal association of Cbx7 (Bernstein et al. 2006). This suggests that the Cbx7 protein is helped in its association with the X chromosome by an RNA entity, the nature of which remains to be defined. This could be reminiscent of previous in vitro findings showing that the chromodomain is capable of binding RNA molecules (Akhtar et al. 2000). Furthermore, several recent studies have implicated intergenic, noncoding RNA in Polycomb targeting or function (for review, see Ringrose et al. 2004). Obviously, Xist RNA is a tantalizing candidate for the RNA partner of Cbx7, but any association between Xist RNA and a Polycomb group protein remains to be demonstrated. Alternatively, the RNA involved could be of a totally different nature, similar to the RNA component associated with constitutive heterochromatin, that

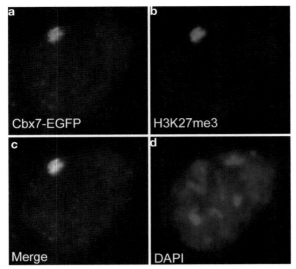

Figure 4. Preferential accumulation of Cbx7–GFP protein on the inactive X chromosome following transient transfection in female ES cells differentiated for more than 3 days. (*a*) Cbx7-EGF. (*b*) Immunofluorescence for histone H3K27me2/3. (*c*) Merged image. (*d*) DAPI.

participates in HP1 binding to chromocenters in mouse cells (Maison et al. 2002). Our study suggests that the murine Pc homolog, Cbx7, in the PRC1 complex may require both H3K27me3 and an RNA entity in order to be targeted to the X chromosome. Schoeftner et al. (2006) showed that Eed (and presumably the H3K27me3 mark) is also required for the recruitment of the canonical PRC1 proteins Mph1 and Mph2 by Xist RNA. However, they found that Eed is not required for recruitment of Ring 1b and that this protein is recruited by Xist RNA and mediates ubiquitination of histone H2A in Eed$^{-/-}$ ES cells, which lack histone H3K27me3. The implications of these findings are that PRC1 proteins may exist in more than one complex and may be recruited by independent mechanisms during X inactivation.

Taken together, these studies reveal the complexity of interactions and recruitment strategies that probably underlie targeting of the PRC1 complexes to the inactive X chromosome. The PRC2-induced H3K27me3 modification and one or more RNA entities, possibly including Xist itself, may be involved. Clearly, our understanding of the different protein and RNA partners involved in PcG targeting to the inactive X chromosome will require biochemical analyses in the future.

CONCLUSION

The X-inactivation process thus involves multiple actors, both RNA and protein. So far, Xist RNA has taken center stage. However, its partners have remained elusive and its mechanisms of action are likely to be complex. It appears to be multifunctional RNA acting at different levels during X inactivation, including its recently defined role in nuclear compartmentalization. The creation of a silent nuclear compartment by Xist RNA, independently of its gene silencing action, adds a

new dimension to its function and opens up new concepts in X inactivation. Repeat elements on the X chromosome may be treated differently from genes, and their mechanisms of silencing, although both Xist-dependent, may differ. It is, of course, tempting to speculate that RNA interference may play a role in the silencing of repeats and/or genes, as has been shown in other organisms, but this possibility remains to be explored for X inactivation. The Polycomb group proteins also appear to be intimately linked to another level of Xist RNA function, although there is increasing evidence that multiple strategies may underlie their recruitment to the X chromosome. Furthermore, unraveling their exact role in maintaining the inactive state through genetic studies is likely to be complicated by the fact that numerous other epigenetic marks, many of which are still unknown, undoubtedly participate in this process. The future will require combined efforts using genetic, biochemical, and developmental approaches to identify these marks and define their roles.

ACKNOWLEDGMENTS

The authors apologize for any omissions in references that may have been made. We thank Julie Chaumeil for discussions and helpful comments on the manuscript. The work by our group described in this review was funded by the Human Frontier Science Program, the French Ministry of Research (ACI), the European Union's Network of Excellence (Epigenome), HEROIC (Highthroughput Epigenetic Regulatory Organization in Chromatin), an Integrated Project funded by the European Union under the 6th Framework Programme (LSHG-CT-2005-018883), the Schlumberger Foundation, and the Canceropole (IDF).

REFERENCES

Akhtar A., Zink D., and Becker P.B. 2000. Chromodomains are protein-RNA interaction modules. *Nature* **407:** 405.

Akasaka T., Kanno M., Balling R., Mieza M.A., Taniguchi M., and Koseki H. 1996. A role for mel-18, a Polycomb group-related vertebrate gene, during the anteroposterior specification of the axial skeleton. *Development* **122:** 1513.

Allaman-Pillet N., Djemai A., Bonny C., and Schorderet D.F. 2000. The 5′ repeat elements of the mouse Xist gene inhibit the transcription of X-linked genes. *Gene Expr.* **9:** 93.

Bantignies F. and Cavalli G. 2006. Cellular memory and dynamic regulation of polycomb group proteins. *Curr. Opin. Cell Biol.* **18:** 275.

Barr M.L. and Bertram E.G. 1949. A morphological distinction between neurones of the male and female and the behaviour of the nucleolar satellite during accelerated nucleoprotein synthesis. *Nature* **163:** 676.

Bernstein E., Duncan E.M., Masui O., Gil J., Heard E., and Allis C.D. 2006. Mouse polycomb proteins bind differentially to methylated histone H3 and RNA and are enriched in facultative heterochromatin. *Mol. Cell. Biol.* **26:** 2560.

Boggs B.A., Connors B., Sobel R.E., Chinault A.C., and Allis C.D. 1996. Reduced levels of histone H3 acetylation on the inactive X chromosome in human females. *Chromosoma* **105:** 303.

Boggs B.A., Cheung P., Heard E., Spector D.L., Chinault A.C., and Allis C.D. 2002. Differentially methylated forms of histone H3 show unique association patterns with inactive human X chromosomes. *Nat. Genet.* **30:** 73.

Brown C.J. and Baldry S.E.L. 1996. Evidence that heteronuclear proteins interact with the XIST RNA in vitro. *Somat. Cell Mol. Genet.* **22:** 403.

Brown C.J., Hendrich B.D., Rupert J.L., Lafreniere R.G., Xing Y., Lawrence J., and Willard H.F. 1992. The human XIST gene: Analysis of a 17 kb inactive X-specific RNA that contains conserved repeats and is highly localized within the nucleus. *Cell* **71:** 527.

Cao R. and Zhang Y. 2004. SUZ12 is required for both the histone methyltransferase activity and the silencing function of the EED-EZH2 complex. *Mol. Cell* **15:** 57.

Cao R., Wang L., Wang H., Xia L., Erdjument-Bromage H., Tempst P., Jones R.S., and Zhang Y. 2002. Role of histone H3 lysine 27 methylation in Polycomb-group silencing. *Science* **298:** 1039.

Chaumeil J., Le Baccon P., Wutz A., and Heard E. 2006. A novel role for Xist RNA in the formation of a repressive nuclear compartment into which genes are recruited when silenced. *Genes Dev.* **20:** 2223.

Chaumeil J., Okamoto I., Guggiari M., and Heard E. 2002. Integrated kinetics of X chromosome inactivation in differentiating embryonic stem cells. *Cytogenet. Genome Res.* **99:** 75.

Clemson C.M., McNeil J.A., Willard H.F., and Lawrence J.B. 1996. XIST RNA paints the inactive X chromosome at interphase: Evidence for a novel RNA involved in nuclear/chromosome structure. *J. Cell Biol.* **132:** 259.

Clemson C.M., Hall L.L., Byron M., McNeil J., and Lawrence J.B. 2006. The X chromosome is organized into a gene-rich outer rim and an internal core containing silenced nongenic sequences. *Proc. Natl. Acad. Sci.* **103:** 7688.

Coré N., Bel S., Gaunt S.J., Aurrand-Lions M., Pearce J., Fisher A., and Djabali M. 1997. Altered cellular proliferation and mesoderm patterning in Polycomb-M33-deficient mice. *Development* **124:** 721.

Costanzi C., Stein P., Worrad D.M., Schultz R.M., and Pehrson J.R. 2000. Histone macroH2A1 is concentrated in the inactive X chromosome of female preimplantation mouse embryos. *Development* **127:** 2283.

Czermin B., Melfi R., McCabe D., Seitz V., Imhof A., and Pirrotta V. 2002. *Drosophila* enhancer of Zeste/ESC complexes have a histone H3 methyltransferase activity that marks chromosomal Polycomb sites. *Cell* **111:** 185.

Dellino G.I., Schwartz Y.B., Farkas G., McCabe D., Elgin S.C., and Pirrotta V. 2004. Polycomb silencing blocks transcription initiation. *Mol. Cell* **13:** 887.

del Mar Lorente M., Marcos-Gutiérrez C., Pérez C., Schoorlemmer J., Ramirez A., Magin T., and Vidal M. 2000. Loss- and gain-of-function mutations show a polycomb group function for Ring1A in mice. *Development* **127:** 5093.

de Napoles M., Mermoud J.E., Wakao R., Tang Y.A., Endoh M., Appanah R., Nesterova T.B., Silva J., Otte A.P., Vidal M., et al. 2004. Polycomb group proteins Ring1A/B link ubiquitylation of histone H2A to heritable gene silencing and X inactivation. *Dev. Cell* **7:** 663.

Fackelmayer F.O. 2005. A stable proteinaceous structure in the territory of inactive X chromosomes. *J. Biol. Chem.* **280:** 1720.

Fang J., Chen T., Chadwick B., Li E., and Zhang Y. 2004. Ring1b-mediated H2A ubiquitination associates with inactive X chromosomes and is involved in initiation of X inactivation. *J. Biol. Chem.* **279:** 52812.

Faust C., Schumacher A., Holdener B., and Magnuson T. 1995. The *Eed* mutation disrupts anterior mesoderm production in mice. *Development* **121:** 273.

Filippova G.N., Cheng M.K., Moore J.M., Truong J.-P., Hu Y.J., Nguyen D.K., Tsuchiya K.D., and Disteche C.M. 2005. Boundaries between chromosomal domains of X inactivation and escape bind CTCF and lack CpG methylation during early development. *Dev. Cell* **8:** 31.

Fischle W., Wang Y., Jacobs S.A., Kim Y., Allis C.D., and Khorasanizadeh S. 2003. Molecular basis for the discrimination of repressive methyl-lysine marks in histone H3 by Polycomb and HP1 chromodomains. *Genes Dev.* **17:** 1870.

Francis N.J., Kingston R.E., and Woodcock C.L. 2004.

Chromatin compaction by a polycomb group protein complex. *Science* **306:** 1574.

Gil J., Bernard D., and Peter G. 2005. Role of Polycomb group proteins in stem cell self-renewal and cancer. *DNA Cell Biol.* **24:** 117.

Heard E. 2004. Recent advances in X-chromosome inactivation. *Curr. Opin. Cell Biol.* **16:** 247.

Heard E., Rougeulle C., Arnaud D., Avner P., Allis C.D., and Spector D.L. 2001. Methylation of histone H3 at Lys-9 is an early mark on the X chromosome during X inactivation. *Cell* **107:** 727.

Isono K., Fujimura Y., Shinga J., Yamaki M., O-Wang J., Takihara Y., Murahashi Y., Takada Y., Mizutani-Koseki Y., and Koseki H. 2005. Mammalian polyhomeotic homologs Phc2 and Phc1 act in synergy to mediate polycomb repression of Hox genes. *Mol. Cell. Biol.* **25:** 6694.

Jeppesen P. and Turner B.M. 1993. The inactive X chromosome in female mammals is distinguished by a lack of histone H4 acetylation, a cytogenetic marker for gene expression. *Cell* **74:** 281.

Kalantry S. and Magnuson T. 2006. The Polycomb group protein EED is dispensable for the initiation of random X-chromosome inactivation. *PLoS Genet.* **2:** e66

Kalantry S., Mills K.C., Yee D., Otte A.P., Panning B., and Magnuson T. 2006. The Polycomb protein Eed protects the inactive X-chromosome from differentiation-induced reactivation. *Nat. Cell Biol.* **8:** 195.

Katoh-Fukui Y., Tsuchiya R., Shiroishi T., Nakahara Y., Hashimoto N., Noguchi K., and Higashinakagawa T. 1998. Male-to-female sex reversal in M33 mutant mice. *Nature* **393:** 688.

Kay G.F., Penny G.D., Patel D., Ashworth A., Brockdorff N., and Rastan S. 1993. Expression of Xist during mouse development suggests a role in the initiation of X chromosome inactivation. *Cell* **72:** 171.

Keohane A.M., O'Neill L.P., Belyaev N.D., Lavender J.S., and Turner B.M. 1996. X-inactivation and histone H4 acetylation in embryonic stem cells. *Dev. Biol.* **180:** 618.

Kohlmaier A., Savarese F., Lachner M., Martens J., Jenuwein T., and Wutz A. 2004. A chromosomal memory triggered by Xist regulates histone methylation in X inactivation. *PLoS Biol.* **2:** E171.

Kuzmichev A., Jenuwein T., Tempst P., and Reinberg D. 2004. Different EZH2-containing complexes target methylation of histone H1 or nucleosomal histone H3. *Mol. Cell* **14:** 183.

Kuzmichev A., Nishioka K., Erdjument-Bromage H., Tempst P., and Reinberg D. 2002. Histone methyltransferase activity associated with a human multiprotein complex containing the Enhancer of Zeste protein. *Genes Dev.* **16:** 2893.

Kuzmichev A., Margueron R., Vaquero A., Preissner T.S., Scher M., Kirmizis A., Ouyang X., Brockdorff N., Abate-Shen C., Farnham P., and Reinberg D. 2005. Composition and histone substrates of polycomb repressive group complexes change during cellular differentiation. *Proc. Natl. Acad. Sci.* **102:** 1859.

Lavigne M., Francis N.J., King I.F., and Kingston R.E. 2004. Propagation of silencing; recruitment and repression of naive chromatin in trans by polycomb repressed chromatin. *Mol. Cell* **13:** 415.

Levine S.S., King I.F., and Kingston R.E. 2004. Division of labor in polycomb group repression. *Trends Biochem. Sci.* **29:** 478.

Levine S.S., Weiss A., Erdjument-Bromage H., Shao Z., Tempst P., and Kingston R.E. 2002. The core of the polycomb repressive complex is compositionally and functionally conserved in flies and humans. *Mol. Cell. Biol.* **22:** 6070.

Lyon M.F. 1961. Gene action in the X-chromosome of the mouse (*Mus musculus* L.). *Nature* **190:** 372.

Maison C., Bailly D., Peters A.H., Quivy J.P., Roche D., Taddei A., Lachner M., Jenuwein T., and Almouzni G. 2002. Higher-order structure in pericentric heterochromatin involves a distinct pattern of histone modification and an RNA component. *Nat. Genet.* **30:** 329.

Mak W., Baxter J., Silva J., Newall A.E., Otte A.P., and Brockdorff N. 2002. Mitotically stable association of polycomb group proteins eed and enx1 with the inactive x chromosome in trophoblast stem cells. *Curr. Biol.* **12:** 1016-1020.

Mak W., Nesterova T.B., de Napoles M., Appanah R., Yamanaka S., Otte A.P., and Brockdorff N. 2004. Reactivation of the paternal X chromosome in early mouse embryos. *Science* **303:** 666.

Marahrens Y., Loring J., and Jaenisch R. 1998. Role of the Xist gene in X chromosome choosing. *Cell* **92:** 657.

Marahrens Y., Panning B., Dausman J., Strauss W., and Jaenisch R. 1997. Xist-deficient mice are defective in dosage compensation but not spermatogenesis. *Genes Dev.* **11:** 156.

Mermoud J.E., Costanzi C., Pehrson J.R., and Brockdorff N. 1999. Histone macroH2A1.2 relocates to the inactive X chromosome after initiation and propagation of X-inactivation. *J. Cell Biol.* **147:** 1399.

Mermoud J.E., Popova B., Peters A.H., Jenuwein T., and Brockdorff N. 2002. Histone H3 lysine 9 methylation occurs rapidly at the onset of random X chromosome inactivation. *Curr. Biol.* **12:** 247.

Min J., Zhang Y., and Xu R.M. 2003. Structural basis for specific binding of Polycomb chromodomain to histone H3 methylated at Lys 27. *Genes Dev.* **17:** 1823.

Müller J., Hart C.M., Francis N.J., Vargas M.L., Sengupta A., Wild B., Miller E.L., O'Connor M.B., Kingston R.E., and Simon J.A. 2002. Histone methyltransferase activity of a *Drosophila* Polycomb group repressor complex. *Cell* **111:** 197-208.

Ng J., Hart C.M., Morgan K., and Simon J.A. 2000. A *Drosophila* ESC-E(Z) protein complex is distinct from other polycomb group complexes and contains covalently modified ESC. *Mol. Cell. Biol.* **20:** 3069.

Norris D.P., Brockdorff N., and Rastan S. 1991. Methylation status of CpG-rich islands on active and inactive mouse X chromosomes. *Mamm. Genome* **1:** 78.

O'Carroll D., Erhardt S., Pagani M., Barton S.C., Surani M.A., and Jenuwein T. 2001. The polycomb-group gene Ezh2 is required for early mouse development. *Mol. Cell. Biol.* **21:** 4330.

Ohno S. and Hauschka T.S. 1960. Allocycly of the X-chromosome in tumors and normal tissues. *Cancer Res.* **20:** 541.

Okamoto I., Otte A.P., Allis C.D., Reinberg D., and Heard E. 2004. Epigenetic dynamics of imprinted X inactivation during early mouse development. *Science* **303:** 644.

Okamoto I., Arnaud D., Otte AP, Disteche C., Avner P., and Heard E. 2005. Evidence for de novo imprinted X-chromosome inactivation independent of meiotic inactivation in mice. *Nature* **438:** 369.

Osborne C.S., Chakalova L., Brown K.E., Carter D., Horton A., Debrand E., Goyenechea B., Mitchell J.A., Lopes S., Reik W., and Fraser P. 2004. Active genes dynamically colocalize to shared sites of ongoing transcription. *Nat. Genet.* **36:** 1065.

Otte A.P. and Kwaks T.H. 2003. Gene repression by Polycomb group protein complexes: A distinct complex for every occasion? *Curr. Opin. Genet. Dev.* **13:** 448.

Pasini D., Bracken A.P., Jensen M.R., Lazzerini Denchi E., and Helin K. 2004. Suz12 is essential for mouse development and for EZH2 histone methyltransferase activity. *EMBO J.* **23:** 4061.

Penny G.D., Kay G.F., Sheardown S.A., Rastan S., and Brockdorff N. 1996. The Xist gene is required *in cis* for X chromosome inactivation. *Nature* **379:** 131.

Peters A.H.F.M., Mermoud J.E., O'Carroll D., Pagani M., Schweizer D., Brockdorff N., and Jenuwein T. 2002. Histone H3 lysine 9 methylation is an epigenetic imprint of facultative heterochromatin. *Nat. Genet.* **30:** 77.

Plath K., Talbot D., Hamer K.M., Otte A.P., Yang T.P., Jaenisch R., and Panning B. 2004. Developmentally regulated alterations in Polycomb repressive complex 1 proteins on the inactive X chromosome. *J. Cell Biol.* **167:** 1025.

Plath K., Fang J., Mlynarczyk-Evans S.K., Cao R., Worringer K.A., Wang H., de la Cruz C.C., Otte A.P., Panning B., and

Zhang Y. 2003. Role of histone H3 lysine 27 methylation in X inactivation. *Science* **300:** 131.

Ringrose L. and Paro R. 2004. Epigenetic regulation of cellular memory by the Polycomb and Trithorax group proteins. *Annu. Rev. Genet.* **38:** 413.

Ringrose L., Ehret H., and Paro R. 2004. Distinct contributions of histone H3 lysine 9 and 27 methylation to locus-specific stability of Polycomb complexes. *Mol. Cell* **16:** 641.

Rougeulle C., Chaumeil J., Sarma K., Allis C.D., Reinberg D., Avner P., and Heard E. 2004. Differential histone H3 Lys-9 and Lys-27 methylation profiles on the X chromosome. *Mol. Cell. Biol.* **24:** 5475.

Saurin A.J., Shao Z., Erdjument-Bromage H., Tempst P., and Kingston R.E. 2001. A *Drosophila* Polycomb group complex includes Zeste and dTAFII proteins. *Nature* **412:** 655.

Schoeftner S., Sengupta A.K., Kubicek S., Mechtler K., Spahn L., Koseki H., Jenuwein T., and Wutz A. 2006 Recruitment of PRC1 function at the initiation of X inactivation independent of PRC2 and silencing. *EMBO J.* **25:** 3110.

Shao Z., Raible F., Mollaaghababa R., Guyon J.R., Wu C.T., Bender W., and Kingston R.E. 1999. Stabilization of chromatin structure by PRC1, a Polycomb complex. *Cell* **98:** 37.

Silva J., Mak W., Zvetkova I., Appanah R., Nesterova T.B., Webster Z., Peters A.H., Jenuwein T., Otte A.P., and Brockdorff N. 2003. Establishment of histone h3 methylation on the inactive X chromosome requires transient recruitment of Eed-Enx1 polycomb group complexes. *Dev. Cell* **4:** 481.

Tada T., Obata Y., Tada M., Goto Y., Nakatsuji N., Tan S., Kono T., and Takagi N. 2000. Imprint switching for non-random X-chromosome inactivation during mouse oocyte growth. *Development* **127:** 3101.

Takagi N., Sugawara O., and Sasaki M. 1982. Regional and temporal changes in the pattern of X-chromosome replication during the early post-implantation development of the female mouse. *Chromosoma* **85:** 275.

Takihara Y., Tomotsune D., Shirai M., Katoh-Fukui Y., Nishii K., Motaleb M.A., Nomura M., Tsuchiya R., Fujita Y., Shibata Y., et al. 1997. Targeted disruption of the mouse homolog of the *Drosophila* polyhomeotic gene leads to altered anteroposterior patterning and neural crest defects. *Development* **124:** 3673.

Tie F., Furuyama T., Prasad-Sinha J., Jane E., and Harte P.J. 2001. The *Drosophila* Polycomb Group proteins ESC and E(Z) are present in a complex containing the histone-binding protein p55 and the histone deacetylase RPD3. *Development* **128:** 275.

Tie F., Prasad-Sinha J., Birve A., Rasmuson-Lestander A., and Harte P.J. 2003. A 1-megadalton ESC/E(Z) complex from *Drosophila* that contains polycomblike and RPD3. *Mol. Cell. Biol.* **23:** 3352.

van der Vlag J. and Otte A.P. 1999. Transcriptional repression mediated by the human polycomb-group protein EED involves histone deacetylation. *Nat. Genet.* **23:** 474.

van der Lugt N.M., Domen J., Linders K., van Roon M., Robanus-Maandag E., te Riele H., van der Valk M., Deschamps J., Sofroniew M., van Lohuizen M., et al. 1994. Posterior transformation, neurological abnormalities, and severe hematopoietic defects in mice with a targeted deletion of the bmi-1 proto-oncogene. *Genes Dev.* **8:** 757.

Voncken J.W., Roelen B.A., Roefs M., de Vries S., Verhoeven E., Marino S., Deschamps J., and van Lohuizen M. 2003. Rnf2 (Ring1b) deficiency causes gastrulation arrest and cell cycle inhibition. *Proc. Natl. Acad. Sci.* **100:** 2468.

Wang H., Wang L., Erdjument-Bromage H., Vidal M., Tempst P., Jones R.S., and Zhang Y. 2004. Role of histone H2A ubiquitination in Polycomb silencing. *Nature* **431:** 873.

Wang J., Mager J., Chen Y., Schneider E., Cross J.C., Nagy A., and Magnuson T. 2001. Imprinted X inactivation maintained by a mouse Polycomb group gene. *Nat. Genet.* **28:** 371.

Wutz A. and Jaenisch R.A. 2000. A shift from reversible to irreversible X inactivation is triggered during ES cell differentiation. *Mol. Cell* **5:** 695.

Wutz A., Rasmussen T.P., and Jaenisch R.A. 2002. Chromosomal silencing and chromosomal localization are mediated by different domains of Xist RNA. *Nat. Genet.* **30:** 167.

X-Chromosome Kiss and Tell: How the Xs Go Their Separate Ways

M.C. Anguera, B.K. Sun, N. Xu, and J.T. Lee

Howard Hughes Medical Institute, Department of Molecular Biology, Massachusetts General Hospital, Department of Genetics, Harvard Medical School, Boston, Massachusetts 02114

Loci associated with noncoding RNAs have important roles in X-chromosome inactivation (XCI), the dosage compensation mechanism by which one of two X chromosomes in female cells becomes transcriptionally silenced. The Xs start out as epigenetically equivalent chromosomes, but XCI requires a cell to treat two identical X chromosomes in completely different ways: One X chromosome must remain transcriptionally active while the other becomes repressed. In the embryo of eutherian mammals, the choice to inactivate the maternal or paternal X chromosome is random. The fact that the Xs always adopt opposite fates hints at the existence of a *trans*-sensing mechanism to ensure the mutually exclusive silencing of one of the two Xs. This paper highlights recent evidence supporting a model for mutually exclusive choice that involves homologous chromosome pairing and the placement of asymmetric chromatin marks on the two Xs.

NONCODING RNAs OF THE X-INACTIVATION CENTER

XCI is the transcriptional silencing of one X chromosome in female cells in order to equalize the dosage of X-linked genes between males (XY) and females (XX) (Lyon 1961). There are two lineage-specific forms of XCI, referred to as "imprinted" and "random" XCI (Boumil and Lee 2001; Heard 2005; Lucchesi et al. 2005). The imprinted form of XCI occurs in extraembryonic tissues of eutherians and is characterized by exclusive silencing of the paternal X (Huynh and Lee 2001; Takagi and Sasaki 1975). Random XCI—where both Xs have an equal chance of undergoing inactivation—is a multistep process that occurs in the embryo proper (Avner and Heard 2001; Cohen and Lee 2002; Clerc and Avner 2003; Heard 2004). These phases have been defined genetically and consist of counting, choice, establishment of silencing, and maintenance of silencing. The counting mechanism determines the X-to-autosome ratio and inactivates one X chromosome per diploid nucleus. This is followed by a choice step where the Xs are designated to become active and inactive Xs (Xa and Xi, respectively). Transcriptional silencing of Xi begins during the establishment phase and is propagated along the chromosome. Finally, the silent Xi is preserved in new cell populations during the maintenance phase.

Both random and imprinted XCI require the X-inactivation center (*Xic*), an X-linked domain that contains a number of noncoding RNA (ncRNA) genes important for XCI (shown in Fig. 1) (Plath et al. 2002; Willard and Carrel 2001). The *Xist* gene (X-inactive specific transcript) encodes a 17-kb alternatively spliced ncRNA that accumulates *in cis* along the X chromosome designated for silencing (Borsani et al. 1991; Brockdorff et al. 1991, 1992; Brown et al. 1991, 1992; Clemson et al. 1996). This noncoding locus is essential for the silencing step (Penny et al. 1996; Marahrens et al. 1997). *Xist* expression is regulated with the help of its noncoding antisense partner, *Tsix* (Lee et al. 1999a; Lee and Lu 1999; Sado et al. 2001). The transcription of *Tsix* inhibits *Xist* expression *in cis*, effectively blocking silencing on the future Xa (Luikenhuis et al. 2001; Morey et al. 2001; Sado et al. 2001; Stavropoulos et al. 2001; Lee 2002a). *Tsix* expression is regulated by another locus that makes the ncRNA called *Xite*, located upstream of the major *Tsix* transcriptional start site. *Xite* functions in part as an enhancer of *Tsix* to ensure the persistence of *Tsix* expression during cellular differentiation (Ogawa and Lee 2003; Stavropoulos et al. 2005). In short, *Xist* silences the future Xi, whereas *Tsix* and *Xite* together designate the future Xa.

X-CHROMOSOME INACTIVATION: TWO IDENTICAL SUBSTRATES, TWO OPPOSITE OUTCOMES

Recent models suggest that these noncoding genes work together to mediate counting and choice and determine the pattern of X-inactivation in a cell-autonomous fashion. The process of XCI requires a cell to act oppositely upon two epigenetically equivalent chromosomes: As one X persists as a transcriptionally active chromosome, the other becomes globally silent. In the embryo of eutherian mammals, the choice to inactivate the maternal or paternal X is random and invariably takes place in a mutually exclusive manner. The precision with which choice is determined implies the existence of a cross-talking process or a feedback mechanism to guarantee the distinct fates of the two Xs. The loss of mutual exclusion in homozygous *Tsix* knockout mice has provided the first experimental evidence for the idea of *trans*-sensing (Lee 2002a, 2005).

Conceptually, mutually exclusive fates of the X chromosomes could be achieved in several ways (Fig. 2). One possibility is that the two X chromosomes are not really equivalent at the beginning of XCI (Fig. 2A), a possibility congruent with the imprinted status of the Xs in extraembryonic cells in which the stereotypical paternal

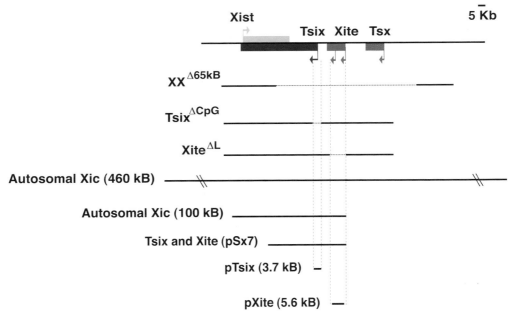

Figure 1. Map of the X-inactivation center (*Xic*) and locations of existing genetic deletions and transgene insertions. (*Dashed lines*) Genetic deletions (designated by Δ); (*solid lines*) autosomal insertions of various regions of the *Xic*. The 460-kb *Xic* autosomal transgene, encompassing 130 kb 5′ and 310 kb 3′ of *Xist*, is not drawn to scale (indicated by the diagonal cross bars).

X silencing is the rule. Although the field has yet to reach a consensus on the nature of the "imprint," differential CpG methylation within the *Xist-Tsix-Xite* regions has been implicated (Norris et al. 1994; Ariel et al. 1995; Courtier et al. 1995; Zuccotti and Monk 1995; McDonald et al. 1998; Prissette et al. 2001; Boumil et al. 2006). We note that slight differences in the methylation pattern and the lack of functional evidence thus far leave open the question of which, if any, of these marks constitute the primary imprint. Because the *Tsix* domains of differential methylation coincide with binding sites for the chromatin insulator and transcription factor, CTCF (Chao et al. 2002; Boumil et al. 2006), parallels to genomic imprinting at the *H19/Igf2* locus and *Rasgrf1* locus have frequently been drawn (Bell and Felsenfeld 2000; Hark et al. 2000; Holmgren et al. 2001; Yoon et al. 2002, 2005). At these autosomal imprinted loci, the differential binding of CTCF to differentially methylated imprinting centers appears to be of primary importance in setting up the mutually exclusive fates of the maternal and paternal chromosomes.

In a similar vein, what has been considered "random" X-inactivation could employ such a chromosome-specific mark, but that mark would be imposed zygotically rather than gametically. In a departure from conventional thought, two recent models suggest that "differential states" are already present prior to the onset of XCI (Williams and Wu 2004; Mlynarczyk-Evans et al. 2006). Although the nature of the "states" is unclear, the states are proposed to result in a situation in which both X chromosomes are not inactivated in a completely stochastic sense, but exist in predeterministic states that can alternate between the two Xs before XCI that predisposes one X to be silent at the onset of XCI. In one case, it is argued that the two active Xs of female embryonic stem (ES) cells switch

between states in which the sister chromatids are in close apposition and another in which they are farther apart (Mlynarczyk-Evans et al. 2006). The model further proposes that the configuration in which the *Xic*s are farther apart "predetermines" the future Xa. Additional characterization will be required to demonstrate if and how they are involved in X-inactivation choice.

A second possibility—one generally preferred by the field—is the concept of a limiting factor that is present in quantities sufficient for only one X chromosome (Fig. 2B) (Lyon 1972; Rastan 1983), and it is the stochastic binding of this factor, or factors to the Xs, that determines the random pattern of silencing. This putative factor could be a "negative" factor (acting to repress the *Xic*) or a "positive" factor (acting to induce it), depending on its mechanism of action. For instance, in the classic one-factor hypothesis, the X chromosomes are predisposed to inactivate by default, and the binding of the so-called "blocking factor" (BF) to one X is specifically required to block its *Xic* from initiating silencing. Conversely, if X-inactivation is actively triggered and does not occur by default, the interaction of the single positive factor—so-called competence factor (CF)—would be required to initiate inactivation. An alternative to the one-factor hypothesis proposes that two factors (one BF and one CF) are required for XCI (Lee and Lu 1999; Lee 2005). The nature of the negative/positive factor could be a unique factor, a complex of factors, or a unique privileged site within the nucleus. In any case, the limited quantity of the factors and their sole action on one or the other X are directly responsible for the "asymmetric" action on two otherwise epigenetically equivalent Xs.

A third possibility invokes direct contact between the Xs as the basis of communication and determination of

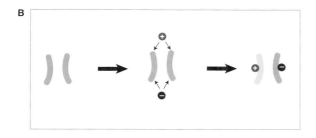

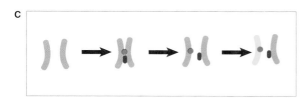

Figure 2. Conceptual models of mutually exclusive X-chromosome inactivation. (*Light blue* chromosomes) Active Xs; (*yellow* chromosomes) X chromosomes that are inactivated. X-inactivation proceeds from left to right. (*Panel A*) X chromosomes are not equivalent before the onset of XCI. Even before X-inactivation is initiated, the presence of marks or different "states" (represented by the star and the hexagon) distinguishes the two Xs. At the onset of X-inactivation, the X chromosome containing the mutually exclusive mark is silenced. This mark may or may not persist into the differentiated state. (*Panel B*) Limited positive and negative factors interact with one of the two X chromosomes. Before XCI begins, both X chromosomes are competent to become inactivated. The presence of a single positive/competence factor (*green*) or negative/blocking factor (*red*) can associate stochastically with either X. Binding of the positive factor triggers inactivation, whereas the negative factor protects the X from inactivation. (*Panel C*) X chromosomes interact with one another to coordinate mutually exclusive choice. Before XCI initiation, both X chromosomes are epigenetically equivalent. Pairing of the two Xs facilitates cross-talk for the formation of asymmetrical chromosomes, with the association of two factors (represented by a *pink circle* and *purple oval*). The asymmetry allows the cell to act uniquely upon one X, but not the other. Although schematized here as soluble factors, this asymmetry may also be chromosomally based (e.g., a different chromatin state, histone, and/or DNA modifications).

their distinct fates (Fig. 2C). *Trans*-chromosomal interaction occurs in phenomena such as transvection in *Drosophila* and *Neurospora* (Aramayo and Metzenberg 1996; Wu and Morris 1999; Chen et al. 2002; Duncan 2002; Coulthard et al. 2005; Vazquez et al. 2006) and in mammalian autosomal imprinting (LaSalle and Lalande 1995, 1996). Such *trans*-sensing mechanisms have also been proposed for X-inactivation (Marahrens 1999; Lee 2002b). In principle, physical contact between the two Xs could coordinate the silencing process and ensure that one and only one X becomes the future Xi, thereby providing a mechanism for establishing distinct fates for the two Xs.

We now highlight advances within the past year that shed light on the nature of mutually exclusive choice and the mechanism by which asymmetric marks are placed on the Xs. Intriguingly, homologous Xs do appear to come in physical contact just prior to the onset of XCI. Furthermore, the initiation of XCI is preceded by chromatin modifications unique to the *Xic* on the future Xi. Here, we propose a speculative model of early events in X-inactivation linking the physical pairing between the two X chromosomes to the establishment of *Xic* asymmetry.

THE EPHEMERAL ACT OF PAIRING

Although *trans*-sensing has long been suspected to occur at the *Xic*, experimental evidence for such interactions has, until recently, been completely lacking. The "chaotic choice" phenotype in female cells lacking *Tsix* provided the first experimental evidence for the idea of necessary cross-talk (Lee 2005). In homozygous *Tsix*[-/-] ES cells, cell differentiation results in aberrant XCI patterns in which female nuclei exhibit two Xi, one Xi, or no Xi. In recent papers, two groups independently examined whether homologous pairing of the X chromosomes might mediate *trans*-sensing. They used fluorescence in situ hybridization (FISH) to monitor the X–X distances during the various phases of XCI in female mouse ES cells, a model system that faithfully recapitulates the steps of XCI upon cell differentiation in culture (Lee et al. 1996; Panning et al. 1997; Clerc and Avner 1998; Marahrens et al. 1998). They found that the two X chromosomes transiently pair with each other during the onset of XCI, most likely just prior to *Xist* up-regulation (Bacher et al. 2006; Xu et al. 2006). Curiously, it appears that the majority of X–X pairs occur very close to the periphery of the nuclear envelope (Bacher et al. 2006), although the significance of this is presently not known, as no specific nuclear compartment has been identified.

Both groups found that the X–X associations are transient, as X–X pairing disappears during later stages of cellular differentiation and in fully differentiated somatic cells (Bacher et al. 2006; Xu et al. 2006). By asking whether pairing coincides with several chromatin changes that occur in sequence during XCI, Xu et al. (2006) observed that the association takes place in the Xist+ fraction but not in the Ezh2+ or the H3-3meK27+ subpopulation, suggesting that pairing occurs specifically in the fraction of cells that has entered the XCI pathway, but has not yet recruited the full silencing machinery. The brevity and timing of the X–X association are intriguing, as indeed the time window coincides with X chromosome counting and choice—the point at which the future Xi and Xa are designated.

How much of the X chromosome is engaged in pairing interactions and how close do the Xs actually get? By testing probes along the length of the X, from centromere to telomere, Xu et al. (2006) found that X–X pairing occurs specifically between the *Xic* regulatory regions and not between any other regions of the X chromosome. Then, biochemical analysis using the "chromosome conformation capture" (3C) technique (Dekker et al. 2002) determined with greater precision that the two *Xic* regions are in direct physical contact with each other and that this contact takes place when pairing is seen to occur by FISH analysis.

WHY DO X-CHROMOSOMES PAIR?

The *Xic* region contains genes for the noncoding *Xite*, *Tsix*, and *Xist* RNAs, the major players involved in regulating counting, choice, and silencing during XCI. Could these regulatory sequences facilitate the transient X–X interactions necessary for mutually exclusive choice? Genetic analysis shows that, interestingly, the *Xic* domains required for pairing map precisely to genes that regulate counting and choice—*Tsix* and *Xite*. An X chromosome carrying a 65-kb deletion downstream from *Xist* loses the ability to pair with its wild-type homolog (Fig. 1) (Clerc and Avner 1998; Bacher et al. 2006). Within this 65-kb region, subdeletions involving either 12.5 kb of *Xite* (Ogawa and Lee 2003) or 3.7 kb of *Tsix* (Lee and Lu 1999) are sufficient to perturb the pairing process (Xu et al. 2006). Notably, the loss of pairing in the homozygous deletion of *Tsix* specifically correlates with aberrant XCI patterns in the differentiating female ES cells: The occurrence of cells with two Xi, one Xi, and no Xi in any differentiating population implies a disruption in both counting (i.e., aberrant numbers of Xis) and mutually exclusive choice (i.e., the two- and no-Xi phenotype) (Lee 2002a, 2005; Ogawa and Lee 2003). Thus, deletions of elements necessary for X–X pairing compromise the regulation of counting and choice.

The connection between counting/choice and X–X pairing is further supported by analysis of transgenic cell lines containing multiple copies of *Tsix* or *Xite* transgenes inserted into autosomes (summarized in Fig. 3). Male ES cells carrying full-length *Xic* transgenes (100–460 kb) display novel interchromosomal association between the X and the transgene-bearing autosome (Bacher et al. 2006; Xu et al. 2006). Significantly, these male cells also display ectopic XCI (Lee et al. 1996, 1999b; Heard et al. 1999), further correlating pairing with the ability to undergo XCI. In female ES cells, the same transgenes also induce de novo X–autosome (X–A) pairing, and regions as small as 3.7 kb of *Tsix* and 5.6 kb of *Xite*—at least when multimerized—are sufficient to create new X–A pairs (Xu et al. 2006). A direct physical interaction between the X and the autosome can also be visualized by 3C analysis (Xu et al. 2006). Thus, sequences within *Tsix* and *Xite* not only are necessary, but are also sufficient to form new pairs.

The analysis of *Tsix/Xite* transgenes in a female context revealed an intriguing difference from that in a male. The presence of extra copies of *Tsix* and *Xite* in an XX context disrupts cell differentiation, whereas it has no measurable effect in XY cells (Lee 2005). The aberrant cell differentiation apparently results from the absence of *Xist* upregulation and XCI in the transgenic female cells. It was then observed that the occurrence of ectopic X–A pairs competitively inhibits the formation of endogenous X–X pairs (Xu et al. 2006). Taken together, these results suggest that X–X pairing is a prerequisite for the initiation of XCI, which in turn is required for proper cell differentiation. It is therefore hypothesized that homologous chromosome pairing is one of the earliest events of XCI and is specifically required for counting and choice,

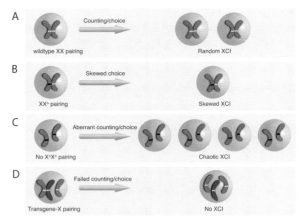

Figure 3. Summary of the effects of *Xic*, *Tsix*, and *Xite* genetic deletions and autosomal transgenes on the X–X pairing event and XCI. (*Panel A*) In wild-type XX ES cells, the homologous pairing between two *Xic* regions (designated by the *yellow regions*) on two active X chromosomes (*green* chromosomes) generates cross-talk resulting in asymmetrically "marked" chromosomes (future Xi shown in *gray*) for mutually exclusive choice. (*Panel B*) Cell lines containing either a single-copy deletion of *Xic* or *Tsix* (designated as XX$^\Delta$ cells, with the deletion in *black*) exhibit normal X–X pairing with skewed choice and inactivation on the mutated X. The single-copy deletion of *Xite* disrupts the dynamics of the X–X pairing event, also resulting in skewed inactivation of the mutated X. (*Panel C*) Homozygous deletion of *Tsix* (designated as X$^\Delta$X$^\Delta$ cells) disrupts X–X pairing and results in chaotic choice, with cells containing either 0, 1 Xi, or 2 Xi, and abnormal XCI. (*Panel D*) Cell lines containing multiple copies of either the *Xic*, *Tsix*, or *Xite* transgenes (*yellow regions*) on autosomes (*blue*) exhibit de novo X–autosome pairing and disruptions in X–X pairing. These abnormal *trans*-interactions may out-compete the normal X–X association, resulting in the presence of two Xa per cell and the failure of counting/choice and XCI.

without which the silencing mechanism of XCI cannot be called upon in female cells undergoing differentiation.

The discovery of pairing provokes many new questions regarding the mechanism of XCI. Does the pairing process involve specific DNA domains within *Tsix* and *Xite*? Does it require transcription, the ncRNA outputs of *Tsix* and *Xite*, or particular chromatin modifications to the genetic locus? Interestingly, transgenic subfragments of *Tsix* and *Xite* that are most effective at nucleating pairing carry promoter elements (Xu et al. 2006). If transcription of these ncRNAs is indeed required, then pairing, counting, and choice must be added to the list of the already diverse functions of ncRNA elements in gene regulation. Finally, what do the Xs communicate to each other while paired, and how are the distinct fates of each X decided by this mysterious act of cross-talking?

FROM THERE TO HERE: PAIRING AND ASYMMETRIC *XIC* FATES

Prior to pairing, the two X chromosomes appear to be epigenetically equivalent. When they pair and come apart again, the Xs appear to be marked for different transcriptional fates. If mutually exclusive choice arises from these *trans*-allelic interactions, the interactions must create physical differences between the two chromosomes that signal for one X to remain active and the other X to become repressed. In theory, the asymmetry could be created by many possibilities, including differential binding of a protein factor, transient localization of the Xs to a certain nuclear region, or specific "marks" and epigenetic modifications.

Recent evidence supports the idea of allele-specific differences in chromatin states. It was shown previously that the *Xist* loci exist in different chromatin environments consistent with the expression status of *Xist*: The active *Xist* gene is associated with an "open" chromatin state with increased levels of H3-K4 dimethylation. In contrast, the silent *Xist* allele exists in a "closed" chromatin state characterized by H3-K9 dimethylation and H3-K27 trimethylation (Goto and Monk 1998; Navarro et al. 2005; Sado et al. 2005). These observations—made using cells that have already undergone XCI—do not indicate whether the chromatin modifications are a cause or consequence of the asymmetrical expression of *Xist*. Additionally, *Tsix* has been recently shown to influence the local chromatin structure of the *Xist* promoter region in post-XCI cells and embryos (Navarro et al. 2005; Sado et al. 2005). These studies suggest an interplay between *Tsix* and the local chromatin environment of *Xist* that would support transcription of the locus, where *Tsix* may function to maintain an open chromatin state at *Xist*. However, because these experiments were also carried out using cells that had already undergone XCI, they contribute to the uncertainty of whether the chromatin modifications actually correlate with the monoallelic expression of *Xist* and the role of *Tsix* in this process (Sado et al. 2005). That is, could these chromatin marks be the causal

asymmetry for mutually exclusive choice and therefore predict the monoallelic expression pattern of *Xist*?

To address this question, a recent study examined the chromatin environment at *Xist* at time points before, during, and after the onset of XCI using female ES cells (Sun et al. 2006). The study used an allele-specific chromatin immunoprecipitation (ChIP) to assay the relative levels of histone modifications on the two X chromosomes of hybrid female ES cells (one X of *Mus castaneus* origin and the other of 129 origin). Three of the most studied histone modifications were examined: histone H3-K4 dimethylation, H3-K27 trimethylation, and H4 acetylation. The results are consistent with the idea that, before XCI, the two Xs are epigenetically identical because each modification was detected at relatively equal levels on both X chromosomes in wild-type female ES cells. In fibroblasts, which have already undergone XCI, the expressed *Xist* allele is enriched with H3-K4 dimethylation and H4 acetylation and reduced in H3-K27 trimethylation, consistent with the previous reports showing that the expressed *Xist* allele is in an open chromatin state (Navarro et al. 2005; Sado et al. 2005).

The situation during the onset of XCI, however, revealed an unexpected result—the presence of a transient heterochromatic environment at the *Xist* loci (Sun et al. 2006). During this phase of XCI, the authors note that *Xist* expression is up-regulated about 30-fold and that, oddly, this transcription occurs from the locus buried in "heterochromatin." This *Xist* locus contains increased H3-K27 trimethylation and a modest decrease in H4 acetylation levels at this stage of XCI, marks normally characteristic of silenced genes. The chromatin environments, summarized in Figure 4, are completely opposite

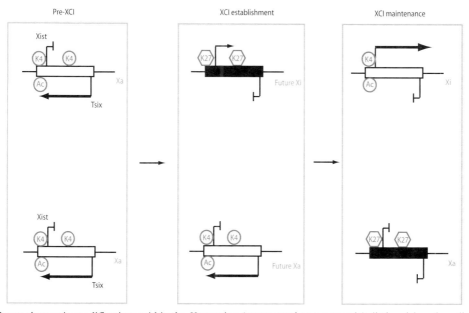

Figure 4. Different chromatin modifications within the *Xist* region (represented as a rectangle) distinguish and predict the future Xi and Xa. The *Xist* region, with modifications shown occurring at the promoter region (upstream of the start site) or within the *Xist* gene body (downstream from the start site). Before XCI, the two X chromosomes are epigenetically identical and the *Xist* region is euchromatic (*open rectangle*), with robust *Tsix* expression. (*Green*) Euchromatic modifications, with histone H3-K4 dimethylation denoted as K4 and histone H4 acetylation as Ac. The *Xist* region becomes heterochromatic (*filled rectangle*), and histone H3-K27 trimethylation is represented in red as K27. During the establishment of XCI, the two X chromosomes exhibit different chromatin modifications, with the future Xi being heterochromatic (with paradoxical *Xist* expression) and the future Xa being euchromatic. These modifications reverse during the final maintenance phase of XCI.

for the future Xa and Xi, and the marks change as XCI progresses. The *Xist* region is euchromatic on both alleles before XCI, supporting the hypothesis that the two Xs are epigenetically equivalent at the pre-XCI stage. Following counting/choice, the future Xi acquires heterochromatic features at the *Xist* region along with the onset of *Xist* expression. The future Xa allele continues to be euchromatic within the *Xist* region after the pairing and counting/choice events. During the final maintenance phase of XCI, the chromatin patterns invert, with the Xi allele exhibiting euchromatic features and increased expression of *Xist,* and the Xa allele being heterochromatic and therefore transcriptionally silent.

What is the function of *Tsix* in influencing the local chromatin environment and regulating the monoallelic expression patterns of *Xist*? The authors repeated the same allele-specific ChIP experiments using a heterozygous female mutant of *Tsix*, and found that each *Xist* allele contains different chromatin modifications corresponding to the absence of *Tsix* (Sun et al. 2006). On the *Tsix⁻* allele, the *Xist* locus is heterochromatic with decreased H3-K4 dimethylation, H4 hypoacetylation, and increased H3-K27 trimethylation. Indeed, it is this mutated X that is invariably silenced during XCI: During the onset of XCI, *Xist* becomes expressed exclusively from the *Tsix⁻* chromosome. Intriguingly, these asymmetric chromatin marks occur in the *Tsix⁺/⁻* ES cells even before induction of XCI by cell differentiation. These observations demonstrate that the chromatin marks preempt asymmetric *Xist* expression and thereby argue that the *Tsix*-driven chromatin changes are causally linked to the establishment of unequal and mutually exclusive X-chromosome fates.

Taken together, these experiments delineate a series of preemptive asymmetrical chromatin changes at *Xist* that differentiate the two X chromosomes and predict the monoallelic expression pattern of *Xist* (summarized in Fig. 4). Before the onset of XCI, the biallelic expression of *Tsix* on both X chromosomes keeps each *Xist* locus in an open, euchromatic state. During the onset of XCI, *Tsix* is silenced on one of the two X chromosomes. The loss of *Tsix* expression from that *Xic* allele influences the formation of the heterochromatic state in *Xist*, which leads paradoxically, in turn, to the transcriptional activation of that *Xist* allele. The euchromatic status of the second allele prevents the up-regulation of *Xist* on the future Xa. For reasons entirely unknown at the present time, the chromatin environment at *Xist* inverts to more conventional patterns after the establishment of XCI, in agreement with observations made by other studies in late-stage embryos and in differentiated cells (Navarro et al. 2005; Sado et al. 2005).

A number of interesting problems remain to be resolved. First, does *Xist* truly favor a repressive chromatin environment for transcription and is heterochromatin sufficient to induce *Xist* expression? If so, *Xist* may be classified as a heterochromatin-preferring gene, much like the *Light* gene of *Drosophila melanogaster*, which appears to require heterochromatin for transcriptional activation (Wakimoto and Hearn 1990; Yasuhara and Wakimoto 2006). Second, why do the chromatin states invert after

the establishment of XCI patterns, especially when *Xist* expression continues into the maintenance phase of XCI, and become constitutive in all somatic female cells? If heterochromatin is required to initiate *Xist* expression, one might expect it to be required for maintenance of expression in somatic cells. Finally, when in the XCI pathway does the heterochromatic state arise—during the pairing process or after the pairs come apart?

MODEL: CROSS-TALKING GENERATES X-ASYMMETRY FOR MUTUALLY EXCLUSIVE CHOICE

In this review, we have presented evidence supporting a model of XCI where X–X communication results in mutually exclusive choice. Our model incorporates facets of the one-factor model for X-chromosome counting/choice, which is currently widely accepted by the field (Lyon 1972; Rastan 1983). It posits the generation of protein factors, produced primarily from autosomes (but does not rule out X-linked factors) in amounts proportional to the actual number of chromosomes present. These proteins form the complex dubbed the blocking factor (BF), which binds one *Xic* and represses the initiation of silencing on that X (Lyon 1972; Rastan 1983). All remaining *Xic*s, lacking the association with the limiting BF, are induced by default to initiate silencing on the linked chromosome.

Although the single-factor (BF) model is simple and elegant, recent observations from *Tsix* and *Xite* genetic analyses have not been easy to reconcile. The alternative "two-factor" model has been proposed based on the principal observation that *Xic* mutations behave differently in the XX and XY contexts (Lee and Lu 1999; Lee 2005). First, given that *Tsix*–/+ and *Tsix*–/– mutations in female cells lead to loss of *Xist* repression, one would expect male cells lacking *Tsix* to inappropriately up-regulate *Xist* and silence its sole X. However, *Tsix*-/Y ES cells do not undergo XCI to any significant degree (Lee and Lu 1999; Sado et al. 2001), and *Tsix*-/Y mice are perfectly normal. (Note: This applies when the *Tsix* mutation is paternally inherited to avoid imprinting effects at *Tsix* [Lee 2000; Sado et al. 2001].) Along similar lines, female ES cells carrying multicopy *Xite* and *Tsix* transgenes form no Xi at all. Thus, although one could argue that one X is protected by BF, the second X is clearly not silenced "by default." Finally, in reporter assays where the *Xist* promoter drives expression of luciferase, differentiation of female cells results in increased luciferase expression only in XX and not XY cells (Sun et al. 2006).

Clearly then, XY cells lack a factor that would ordinarily be present in XX cells. The two-factor model thus incorporates the need for BF binding on the future Xa and introduces the need for a second factor—dubbed the competence factor (CF). CF is believed to play a part in inducing the inactivation of the future Xi and would be produced only in the context of supernumerary Xs. Indeed, CF is thought to consist of X-linked factors not titrated away by the fixed quantity of autosomal factors produced during the counting process. Naturally, every X in excess of one would produce one X-linked CF, enabling the silencing of all but one X (one that binds BF) in the genome.

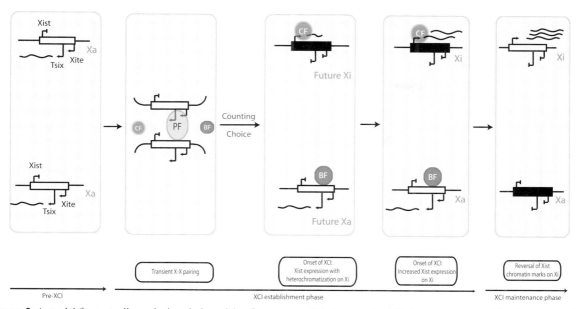

Figure 5. A model for mutually exclusive choice arising from cross-talk between paired X chromosomes, resulting in asymmetrical Xs. The two X chromosomes are both euchromatic and epigenetically identical at the pre-XCI stage. The two chromosomes are then paired together, perhaps with the help of a pairing factor (PF), to facilitate the cross-talk needed to distinguish between the chromosomes. Following the transient X–X pairing event, the future Xi and future Xa are distinguished by different chromatin modifications within the *Xist* region. Binding of a competence factor (CF) to an unknown site in this region of the future Xi assists with the transcriptional up-regulation of *Xist* despite the heterochromatic environment. The binding of a blocking factor (BF), again to an unknown site, may function to help *Tsix* keep *Xist* repressed on the euchromatic future Xa. The chromatin modifications switch for the Xi and Xa alleles during the final maintenance phase of XCI.

We hypothesize that this counting process is intimately linked to pairing, perhaps occurring simultaneously in time and space (Fig. 5). BF and CF must bind mutually exclusively. The proposed cross-talk during the paired state may provide a platform on which the asymmetric distribution of BF and CF takes place. During the pre-XCI stage, the X chromosomes are epigenetically identical and exist in random positions relative to each other in the nucleus. As XCI begins, the two Xs transiently pair via putative pairing factor(s) (PF) and they do so perhaps within a particular region in the nucleus, with closest contact points occurring at *Tsix* and *Xite*. It is envisioned that BF and CF compete to bind. Where the hypothetical BF and CF would bind is a subject of debate. Previously, it has been speculated that BF may bind *Tsix* or *Xite* of the future Xa and ensure the persistent expression of *Tsix* (Lee and Lu 1999; Lee 2005). On the other hand, CF may bind either to *Tsix/Xite* to down-regulate their Xa-promoting activities or it may bind *Xist* directly, perhaps recruited there by the newly created transient heterochromatic status of the *Tsix/Xist* domain.

By this model, cross-talking via X–X pairing eventually leads to the mutually exclusive binding of BF and CF and divergent X-fates. BF's unique binding to one *Xic* enables the monoallelic persistence of *Tsix* expression on the future Xa. The absence of BF binding and the unique binding of CF to the remaining *Xic* would then silence *Tsix* and enable the induction of *Xist*. Therefore, pairing generates the asymmetry in *Tsix* expression, which in turn dictates the asymmetric chromatin modifications and expression patterns of *Xist*.

The future promises to bring exciting new revelations pertaining to the mechanisms of counting, choice, and XCI. With no shortage of models, it seems, however, that XCI is likely to become even more complicated before the mechanisms are clarified. What we have discussed herein applies primarily to the random form of XCI. It is possible that imprinted XCI employs a significantly different mechanism of silencing, as indeed the marsupial genome appears to lack *Xist* (Duret et al. 2006; Hore et al., in prep.; L.S. Davidow et al., in prep.) . In this case, one wonders whether pairing and differential chromatin marks also direct the initiation of XCI in the marsupial. Recent studies have also introduced the contrasting viewpoint that an XX-specific CF may not be required for XCI (Morey et al. 2004; Vigneau et al. 2006). On the basis of results from a deletion of the *DXPas34* repeat in *Tsix*, the work suggests that inhibition of *Tsix* transcription (at least in the context of this deletion) has similar effects *in cis* in male and female cells, thus questioning whether XX–XY differences exist and whether they could be used as a basis for postulating a need for a CF. Yet, as discussed above, XX–XY differences do actually exist. Clearly, many unanswered questions regarding mechanisms remain and will continue to challenge the field for years to come.

ACKNOWLEDGMENTS

We thank Jennifer Erwin and Janice Yoo-Jin Ahn for critical reading of this manuscript, and all members of the Lee Laboratory for helpful discussions. M.C.A. is funded by the National Institutes of Health (GM076955-01), B.K.S. is funded by the Medical Scientist Training Program, and J.T.L. is funded by the National Institutes of Health and Howard Hughes Medical Institute.

REFERENCES

Aramayo R. and Metzenberg R.L. 1996. Meiotic transvection in fungi. *Cell* **86:** 103.

Ariel M., Robinson E., McCarrey J.R., and Cedar H. 1995. Gamete-specific methylation correlates with imprinting of the murine Xist gene. *Nat. Genet.* **9:** 312.

Avner P. and Heard E. 2001. X-chromosome inactivation: Counting, choice and initiation. *Nat. Rev. Genet.* **2:** 59.

Bacher C.P., Guggiari M., Brors B., Augui S., Clerc P., Avner P., Eils R., and Heard E. 2006. Transient colocalization of X-inactivation centers accompanies the initiation of X inactivation. *Nat. Cell Biol.* **8:** 293.

Bell A.C. and Felsenfeld G. 2000. Methylation of a CTCF-dependent boundary controls imprinted expression of the Igf2 gene. *Nature* **405:** 482.

Borsani G., Tonlorenzi R., Simmler M.C., Dandolo L., Arnaud D., Capra V., Grompe M., Pizzuti A., Muzny D., and Lawrence C., et al. 1991. Characterization of a murine gene expressed from the inactive X chromosome. *Nature* **351:** 325.

Boumil R.M. and Lee J.T. 2001. Forty years of decoding the silence in X-chromosome inactivation. *Hum. Mol. Genet.* **10:** 2225.

Boumil R.M., Ogawa Y., Sun B.K., Huynh K.D., and Lee J.T. 2006. Differential methylation of Xite and CTCF sites in Tsix mirrors the pattern of X-inactivation choice in mice. *Mol. Cell. Biol.* **26:** 2109.

Brockdorff N., Ashworth A., Kay G.F., McCabe V.M., Norris D.P., Cooper P.J., Swift S., and Rastan S. 1992. The product of the mouse Xist gene is a 15 kb inactive X-specific transcript containing no conserved ORF and located in the nucleus. *Cell* **71:** 515.

Brockdorff N., Ashworth A., Kay G.F., Cooper P., Smith S., McCabe V.M., Norris D.P., Penny G.D., Patel D., and Rastan S. 1991. Conservation of position and exclusive expression of mouse Xist from the inactive X chromosome. *Nature* **351:** 329.

Brown C.J., Ballabio A., Rupert J.L., Lafreniere R.G., Grompe M., Tonlorenzi R., and Willard H. 1991. A gene from the region of the human X inactivation center is expressed exclusively from the inactive X chromosome. *Nature* **349:** 38.

Brown C.J., Hendrich B.D., Rupert J.L., Lafreniere R.G., Xing Y., Lawrence J., and Willard H.F. 1992. The human XIST gene: Analysis of a 17 kb inactive X-specific RNA that contains conserved repeats and is highly localized within the nucleus. *Cell* **71:** 527.

Chao W., Huynh K.D., Spencer R.J., Davidow L.S., and Lee J.T. 2002. CTCF, a candidate trans-acting factor for X-inactivation choice. *Science* **295:** 345.

Chen J.L., Huisinga K.L., Viering M.M., Ou S.A., Wu C.T., and Geyer P.K. 2002. Enhancer action in trans is permitted throughout the *Drosophila* genome. *Proc. Natl. Acad. Sci.* **99:** 3723.

Clemson C.M., McNeil J.A., Willard H., and Lawrence J.B. 1996. XIST RNA paints the inactive X chromosome at interphase: Evidence for a novel RNA involved in nuclear/chromosome structure. *J. Cell Biol.* **132:** 259.

Clerc P. and Avner P. 1998. Role of the region 3′ to Xist exon 6 in the counting process of X-chromosome inactivation. *Nat. Genet.* **19:** 249.

———. 2003. Multiple elements within the Xic regulate random X inactivation in mice. *Semin. Cell Dev. Biol.* **14:** 85.

Cohen D.E. and Lee J.T. 2002. X-chromosome inactivation and the search for chromosome-wide silencers. *Curr. Opin. Genet. Dev.* **12:** 219.

Coulthard A.B., Nolan N., Bell J.B., and Hilliker A.J. 2005. Transvection at the vestigial locus of *Drosophila melanogaster*. *Genetics* **170:** 1711.

Courtier B., Heard E., and Avner P. 1995. Xce haplotypes show modified methylation in a region of the active X chromosome lying 3′ to Xist. *Proc. Natl. Acad. Sci.* **92:** 3531.

Dekker J., Rippe K., Dekker M., and Kleckner N. 2002. Capturing chromosome conformation. *Science* **295:** 1306.

Duncan I.W. 2002. Transvection effects in *Drosophila*. *Annu. Rev. Genet.* **36:** 521.

Duret L., Chureau C., Samain S., Weissenbach J., and Avner P. 2006. The Xist RNA gene evolved in eutherians by pseudogenization of a protein-coding gene. *Science* **312:** 1653.

Goto T. and Monk M. 1998. Regulation of X-chromosome inactivation in development in mice and humans. *Microbiol. Mol. Biol. Rev.* **62:** 362.

Hark A.T., Schoenherr C.J., Katz D.J., Ingram R.S., Levorse J.M., and Tilghman S.M. 2000. CTCF mediates methylation-sensitive enhancer-blocking activity at the H19/Igf2 locus. *Nature* **405:** 486.

Heard E. 2004. Recent advances in X-chromosome inactivation. *Curr. Opin. Cell Biol.* **16:** 247.

———. 2005. Delving into the diversity of facultative heterochromatin: The epigenetics of the inactive X chromosome. *Curr. Opin. Genet. Dev.* **15:** 482.

Heard E., Mongelard F., Arnaud D., and Avner P. 1999. Xist yeast artificial chromosome transgenes function as X-inactivation centers only in multicopy arrays and not as single copies. *Mol. Cell. Biol.* **19:** 3156.

Holmgren C., Kanduri C., Dell G., Ward A., Mukhopadhya R., Kanduri M., Lobanenkov V., and Ohlsson R. 2001. CpG methylation regulates the Igf2/H19 insulator. *Curr. Biol.* **11:** 1128.

Huynh K.D. and Lee J.T. 2001. Imprinted X inactivation in eutherians: A model of gametic execution and zygotic relaxation. *Curr. Opin. Cell Biol.* **13:** 690.

LaSalle J.M. and Lalande M. 1995. Domain organization of allele-specific replication within the GABRB3 gene cluster requires a biparental 15q11-13 contribution. *Nat. Genet.* **9:** 386.

———. 1996. Homologous association of oppositely imprinted chromosomal domains. *Science* **272:** 725.

Lee J.T. 2000. Disruption of imprinted X inactivation by parent-of-origin effects at Tsix. *Cell* **103:** 17.

———. 2002a. Homozygous Tsix mutant mice reveal a sex-ratio distortion and revert to random X-inactivation. *Nat. Genet.* **32:** 195.

———. 2002b. Is X-chromosome inactivation a homology effect? *Adv. Genet.* **46:** 25.

———. 2005. Regulation of X-chromosome counting by Tsix and Xite sequences. *Science* **309:** 768.

Lee J.T. and Lu N. 1999. Targeted mutagenesis of Tsix leads to nonrandom X inactivation. *Cell* **99:** 47.

Lee J.T., Davidow L.S., and Warshawsky D. 1999a. Tsix, a gene antisense to Xist at the X-inactivation centre. *Nat. Genet.* **21:** 400.

Lee J.T., Lu N., and Han Y. 1999b. Genetic analysis of the mouse X inactivation center defines an 80-kb multifunction domain. *Proc. Natl. Acad. Sci.* **96:** 3836.

Lee J.T., Strauss W.M., Dausman J.A., and Jaenisch R. 1996. A 450 kb transgene displays properties of the mammalian X-inactivation center. *Cell* **86:** 83.

Lucchesi J.C., Kelly W.G., and Panning B. 2005. Chromatin remodeling in dosage compensation. *Annu. Rev. Genet.* **39:** 615.

Luikenhuis S., Wutz A., and Jaenisch R. 2001. Antisense transcription through the Xist locus mediates Tsix function in embryonic stem cells. *Mol. Cell. Biol.* **21:** 8512.

Lyon M.F. 1961. Gene action in the X-chromosome of the mouse (*Mus musculus* L.). *Nature* **190:** 372.

———. 1972. X-chromosome inactivation and developmental patterns in mammals. *Biol. Rev. Camb. Philos. Soc.* **47:** 1.

Marahrens Y. 1999. X-inactivation by chromosomal pairing events. *Genes Dev.* **13:** 2624.

Marahrens Y., Loring J., and Jaenisch R. 1998. Role of the Xist gene in X chromosome choosing. *Cell* **92:** 657.

Marahrens Y., Panning B., Dausman J., Strauss W., and Jaenisch R. 1997. Xist-deficient mice are defective in dosage compensation but not spermatogenesis. *Genes Dev.* **11:** 156.

McDonald L.E., Paterson C.A., and Kay G.F. 1998. Bisulfite genomic sequencing-derived methylation profile of the xist gene throughout early mouse development. *Genomics* **54:** 379.

Mlynarczyk-Evans S., Royce-Tolland M., Alexander M.K., Andersen A.A., Kalantry S., Gribnau J., and Panning B. 2006. X chromosomes alternate between two states prior to random X-inactivation. *PLoS Biol.* **4:** e159.

Morey C., Arnaud D., Avner P., and Clerc P. 2001. Tsix-mediated repression of Xist accumulation is not sufficient for normal random X inactivation. *Hum. Mol. Genet.* **10:** 1403.

Morey C., Navarro P., Debrand E., Avner P., Rougeulle C., and Clerc P. 2004. The region 3′ to Xist mediates X chromosome counting and H3 Lys-4 dimethylation within the Xist gene. *EMBO J.* **23:** 594.

Navarro P., Pichard S., Ciaudo C., Avner P., and Rougeulle C. 2005. Tsix transcription across the Xist gene alters chromatin conformation without affecting Xist transcription: Implications for X-chromosome inactivation. *Genes Dev.* **19:** 1474.

Norris D.P., Patel D., Kay G.F., Penny G.D., Brockdorff N., Sheardown S.A., and Rastan S. 1994. Evidence that random and imprinted Xist expression is controlled by preemptive methylation. *Cell* **77:** 41.

Ogawa Y. and Lee J.T. 2003. Xite, X-inactivation intergenic transcription elements that regulate the probability of choice. *Mol. Cell* **11:** 731.

Panning B., Dausman J., and Jaenisch R. 1997. X chromosome inactivation is mediated by Xist RNA stabilization. *Cell* **90:** 907.

Penny G.D., Kay G.F., Sheardown S.A., Rastan S., and Brockdorff N. 1996. Requirement for Xist in X chromosome inactivation. *Nature* **379:** 131.

Plath K., Mlynarczyk-Evans S., Nusinow D.A., and Panning B. 2002. Xist RNA and the mechanism of X chromosome inactivation. *Annu. Rev. Genet.* **36:** 233.

Prissette M., El-Maarri O., Arnaud D., Walter J., and Avner P. 2001. Methylation profiles of DXPas34 during the onset of X-inactivation. *Hum. Mol. Genet.* **10:** 31.

Rastan S. 1983. Non-random X-chromosome inactivation in mouse X-autosome translocation embryos—Location of the inactivation centre. *J. Embryol. Exp. Morphol.* **78:** 1.

Sado T., Hoki Y., and Sasaki H. 2005. Tsix silences Xist through modification of chromatin structure. *Dev. Cell* **9:** 159.

Sado T., Wang Z., Sasaki H., and Li E. 2001. Regulation of imprinted X-chromosome inactivation in mice by Tsix. *Development* **128:** 1275.

Stavropoulos N., Lu N., and Lee J.T. 2001. A functional role for Tsix transcription in blocking Xist RNA accumulation but not in X-chromosome choice. *Proc. Natl. Acad. Sci.* **98:** 10232.

Stavropoulos N., Rowntree R.K., and Lee J.T. 2005. Identification of developmentally specific enhancers for Tsix in the regulation of X chromosome inactivation. *Mol. Cell. Biol.* **25:** 2757.

Sun B.K., Deaton A.M., and Lee J.T. 2006. A transient heterochromatic state in Xist preempts X inactivation choice without RNA stabilization. *Mol. Cell* **21:** 617.

Takagi N. and Sasaki M. 1975. Preferential inactivation of the paternally derived X-chromosome in the extraembryonic membranes of the mouse. *Nature* **256:** 640.

Vazquez J., Muller M., Pirrotta V., and Sedat J.W. 2006. The Mcp element mediates stable long-range chromosome-chromosome interactions in *Drosophila*. *Mol. Biol. Cell* **17:** 2158.

Vigneau S., Augui S., Navarro P., Avner P., and Clerc P. 2006. An essential role for the DXPas34 tandem repeat and Tsix transcription in the counting process of X chromosome inactivation. *Proc. Natl. Acad. Sci.* **103:** 7390.

Wakimoto B.T. and Hearn M.G. 1990. The effects of chromosome rearrangements on the expression of heterochromatic genes in chromosome 2L of *Drosophila melanogaster*. *Genetics* **125:** 141.

Willard H.F. and Carrel L. 2001. Making sense (and antisense) of the X inactivation center. *Proc. Natl. Acad. Sci.* **98:** 10025.

Williams B.R. and Wu C.T. 2004. Does random X-inactivation in mammals reflect a random choice between two X chromosomes? *Genetics* **167:** 1525.

Wu C.T. and Morris J.R. 1999. Transvection and other homology effects. *Curr. Opin. Genet. Dev.* **9:** 237.

Xu N., Tsai C.L., and Lee J.T. 2006. Transient homologous chromosome pairing marks the onset of X inactivation. *Science* **311:** 1149.

Yasuhara J.C. and Wakimoto B.T. 2006. Oxymoron no more: The expanding world of heterochromatic genes. *Trends Genet.* **22:** 330.

Yoon B.J., Herman H., Sikora A., Smith L.T., Plass C., and Soloway P.D. 2002. Regulation of DNA methylation of Rasgrf1. *Nat. Genet.* **30:** 92.

Yoon B., Herman H., Hu B., Park Y.J., Lindroth A., Bell A., West A.G., Chang Y., Stablewski A., Piel J.C., Loukinov D.I., Lobanenkov V.V., and Soloway P.D. 2005. Rasgrf1 imprinting is regulated by a CTCF-dependent methylation-sensitive enhancer blocker. *Mol. Cell. Biol.* **25:** 11184.

Zuccotti M. and Monk M. 1995. Methylation of the mouse Xist gene in sperm and eggs correlates with imprinted Xist expression and paternal X-inactivation. *Nat. Genet.* **9:** 316.

Genetic Analyses of DNA Methyltransferases in *Arabidopsis thaliana*

X. Zhang* and S.E. Jacobsen*†

*Department of Molecular, Cell and Developmental Biology, †Howard Hughes Medical Institute,
University of California, Los Angeles, California 90095

DNA methylation is a conserved epigenetic silencing mechanism that functions to suppress the proliferation of transposons and regulate the expression of endogenous genes. In plants, mutations that cause severe loss of DNA methylation result in reactivation of transposons as well as developmental abnormalities. We use the flowering plant *Arabidopsis thaliana* as a model system to study the establishment and maintenance of DNA methylation as well as its role in regulating plant development. The genetic evidence presented here suggests that methylation at CG and non-CG sites functions in a partially redundant and locus-specific manner to regulate a wide range of developmental processes. Results from recent studies also suggested that the dynamic nature of non-CG methylation, which is critically important for its regulatory function, is largely due to its complicated interactions with other epigenetic pathways such as RNAi and histone modifications. Finally, the use of genomic approaches has significantly broadened our understanding of the patterning of DNA methylation on a genome-wide scale and has led to the identification of hundreds of candidate genes that are controlled by DNA methylation.

Most eukaryotic organisms modify their genomic DNA in certain regions of the genomes through the addition of a methyl group to the C5 position of cytosine residues. This process is commonly referred to as "DNA methylation," and it is the most prevalent type of DNA modification in eukaryotes. DNA methylation is an important epigenetic silencing mechanism, as genomic regions that are densely methylated often correlate with reduced transcriptional activity. For this reason, as well as the observation that the repetitive fractions of eukaryotic genomes are often methylated, DNA methylation is generally thought to have evolved initially as a defense mechanism against transposons or other types of invading DNA (Yoder et al. 1997; Martienssen and Colot 2001). However, it has also been "domesticated" by the genome to regulate the expression of endogenous genes and became critically important for normal development. For example, null mutations in the maintenance methyltransferase DNMT1 or the de novo methyltransferase DNMT3a/b genes cause embryonic lethality in mouse, and the loss of DNMT1 function in a human cell line leads to reduced cell viability (Goll and Bestor 2005; Egger et al. 2006). Similarly, severe loss of DNA methylation in plants leads to several developmental abnormalities (Chan et al. 2005).

In mammals, DNA methylation is established and maintained mostly at CG sites through the action of the DNA methyltransferases DNMT1 and DNMT3a/b (Bird 2002; Goll and Bestor 2005). In contrast, plant genomes contain DNA methylation in all sequence contexts (Chan et al. 2005). The plant DNMT1-homolog METHYLTRANSFERASE 1 (MET1) maintains DNA methylation at CG sites (CG methylation) (Finnegan et al. 1996; Ronemus et al. 1996; Kankel et al. 2003; Saze et al. 2003), whereas the DNMT3a/3b homolog DOMAINS REARRANGED METHYLASE 1 and 2 (DRM1/2) are responsible for the de novo methylation in all sequence contexts as well as for the maintenance methylation at asymmetric CHH sites (where H = A, C, or T; CHH methylation) (Cao and Jacobsen 2002b). In addition, the plant-specific CHROMOMETHYLASE3 (CMT3) is responsible for DNA methylation at CNG sites (CNG methylation), as well as a subset of CHH sites (Bartee et al. 2001; Lindroth et al. 2001).

The added complexity of DNA methylation in plants compared to animals is reflected not only in the additional sequence contexts and the methyltransferase CMT3, but also in the intimate relationship and complicated interplay between DNA methylation and RNAi as well as histone H3 methylation. Components of the RNAi and histone methylation pathways are required for the establishment and maintenance of proper DNA methylation patterns. Conversely, some mutations that result in the partial loss of DNA methylation also affect the functions of the RNAi pathway and histone methylation (Chan et al. 2005). Thus, DNA methylation, RNAi, and histone methylation function together to maintain the proper function of plant genomes.

Here, we describe the results of our genetic and genomic analyses of DNA methylation in *Arabidopsis*. Analyses of multiple mutants that lack DNA methylation in different combinations of sequence contexts suggested that CG, CNG, and CHH methylation act in a partially redundant manner to regulate many aspects of normal plant development. We also discuss the complex interactions between DNA methylation, RNAi, and histone methylation. Finally, results from our genome-wide mapping of the sites of DNA methylation and transcriptional analysis of DNA methyltransferase mutants provided direct evidence for the critically important role of DNA methylation in silencing transposable elements and regulating endogenous genes.

DNA METHYLATION CONTROLS MANY ASPECTS OF *ARABIDOPSIS* DEVELOPMENT

Severe loss of CG methylation in mutants such as *met1* and *ddm1* resulted in developmental abnormalities, suggesting that CG methylation is important in controlling endogenous gene expression (Finnegan et al. 1996; Ronemus et al. 1996; Jeddeloh et al. 1999; Kankel et al. 2003; Saze et al. 2003). In fact, a few genes have been found to be misregulated in *met1* and are responsible for some of the developmental phenotypes. The most common type of misregulation is likely due to derepression caused by the loss of DNA methylation. For example, the hypomethylation and ectopic overexpression of the *FWA* gene in *met1* result in late flowering (Soppe et al. 2000). A second type of misregulated genes is exemplified by *SUPERMAN*. The hypermethylation and silencing of the *SUPERMAN* gene cause several floral defects such as increased number of stamens, incompletely fused carpels, and partial sterility (Jacobsen and Meyerowitz 1997). This hypermethylation process is not yet well understood but appears to be dependent on DRM1/2 (X. Zhang et al., unpubl.).

In contrast, both *cmt3* and *drm1/2* mutants were found to be phenotypically normal even after prolonged inbreeding (Bartee et al. 2001; Lindroth et al. 2001; Cao and Jacobsen 2002b). This might suggest that non-CG methylation is dispensable for normal plant development. However, it soon became clear that DRM1/2 and CMT3 acted in a partially redundant and locus-specific manner, and only in the *drm1/2 cmt3* triple mutant was the vast majority of non-CG methylation eliminated (Cao et al. 2003; Chan et al. 2006). Importantly, gross loss of non-CG methylation (but with little change in CG-methylation) leads to several developmental defects, such as twisted leaves, short stature, and partial sterility (Cao et al. 2003; Chan et al. 2006). Notably, loss of DRM1/2 and CMT3 activities in both the Landsberg *erecta* (L*er*) and the Columbia (Col) ecotypes of *Arabidopsis* led to very similar phenotypes, indicating that some aspects of the function of non-CG methylation in regulating plant development are evolutionarily conserved (Cao et al. 2003; Chan et al. 2006).

These results clearly demonstrated the requirement for DNA methylation in normal plant development. They also suggested that two different sets of genes were misregulated in *met1* and *drm1/2 cmt3*, as these two mutants displayed very different developmental abnormalities. It is therefore possible that CG methylation and non-CG methylation regulate distinct aspects of plant development. Alternatively, there may be considerable functional redundancies between CG and non-CG methylation. We therefore tested this hypothesis by constructing and analyzing mutants defective in both CG and non-CG methylation (*drm1/2 met1* and *cmt3 met1*) following the scheme shown in Figure 1A.

Both *drm1/2 met1* triple and *cmt3 met1* double mutant plants could be recovered at low frequency from *drm1/2/+ met1/+* and *cmt3/+ met1/+* parents, respectively. This result suggests that the loss of DRM1/2 and MET1 or CMT3 MET1 activities does not necessarily lead to gamete or embryo lethality. However, a severe defect in seed viability was observed for both mutants. This is more apparent in F_1 populations from *drm1/2 met1/+* and *cmt3 met1/+* parents, where instead of approximately 1/4, only about 1/40 and about 1/20 plants were found to be *drm1/2 met1* and *cmt3 met1*, respectively. Both *drm1/2 met1* and *cmt3 met1* adult plants exhibited a number of severe developmental defects, such as extremely late flowering, reduced leaf size, shorter stature, and complete sterility (Fig. 1B). In addition, the flowers of *cmt3 met1* plants displayed an *agamous*-like phenotype and developed sepals and numerous petals, but completely lacked stamens or carpels. Indeterminate flowers were also frequently observed. In contrast, typical flowers from *drm1/2 met1* plants only had two (instead of four) sepals, completely lacked petals, and developed numerous stamens and abnormal carpels. Thus, the additional and more severe defects of *drm1/2 met1* and *cmt3 met1* compared to previously isolated mutants strongly supported the notion that CG methylation and non-CG redundantly control developmentally important genes in *Arabidopsis*. It is also possible that the gross loss of DNA methylation in these multiple mutants may undermine the general structure and function of the chromosomes (e.g., chromosome segregation or heterochromatin condensation) and thus affect normal cell divisions.

To characterize the changes in DNA methylation pattern in the *drm1/2 met1* and *cmt3 met1* mutants, we performed bisulfite genomic sequencing at selected loci. To this end, we bisulfite-sequenced the retroelement *AtSN1* and an intergenic tandem repeat downstream of the *MEDEA* gene (MEA ISR), both of which have been well-characterized in wild-type and several mutant plants (Cao and Jacobsen 2002a; Zilberman et al. 2003; Henderson et al. 2006). As shown in Figure 2, virtually all methylation was eliminated from *cmt3 met1* at both loci. In contrast, in the *drm1/2 met1* mutant, methylation in all sequence contexts was undetectable at MEA ISR, but significant CNG methylation remained at *AtSN1*. Note that the DNA methylation patterns in *drm1/2 met1* and *cmt3 met1* were different from all previously described single or multiple DNA methyltransferase mutants (Fig. 2) (Cao and Jacobsen 2002a; Zilberman et al. 2003; Henderson et al. 2006). Taken together, these results revealed complex interdependence of the DRM1/2, CMT3, and MET1 activities in the establishment and maintenance of DNA methylation, and provided further support for their functional redundancies in a locus-specific manner.

We also attempted to construct the quadruple mutant *drm1/2 cmt3 met1*, in which the vast majority of methylation (if not all) should be eliminated. We were unable to isolate *drm1/2 cmt3 met1* from *drm1/2/+ cmt3/+ met1/+* parents, but *drm1/2 cmt3 met1/+* plants were readily obtained, indicating that the activities of DRM1/2, CMT3, and MET1 may be dispensable for male and/or female gamete development. However, inspection of developing siliques from *drm1/2 cmt3 met1/+* plants revealed that approximately 1/4 of the seeds aborted at very early stages (Fig. 3), which were likely to be *drm1/2 cmt3 met1*. It therefore appears that the proper methylation at some loci is critical for early embryo development.

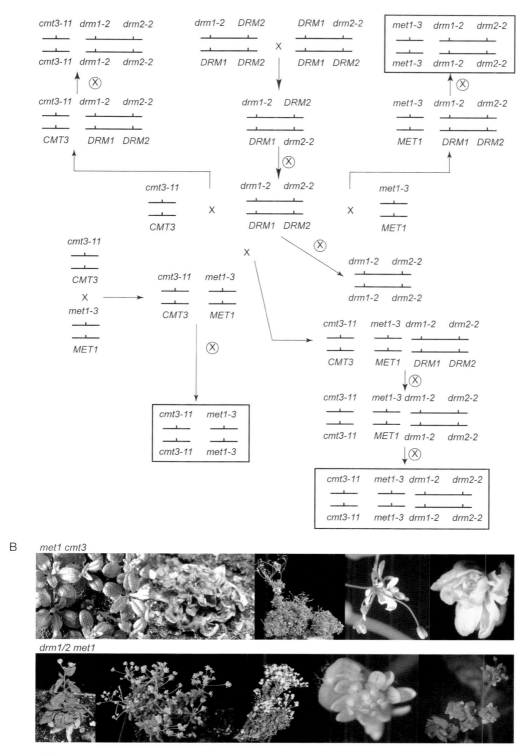

Figure 1. Partially redundant function of CG and non-CG methylation in regulating plant development. (*A*) Genetic schemes for constructing multiple mutants. (*B*) Developmental phenotypes of *cmt3 met1* (*top ro*w) and *drm1/2 met1* (*bottom row*) are significantly more severe than either *met1* or *drm1/drm2 cmt3* null mutants (Cao et al. 2003; Saze et al. 2003; Chan et al. 2006). Red arrow indicates a *met1 cmt3* plant among other plants which are not *met1 cmt3* double mutants.

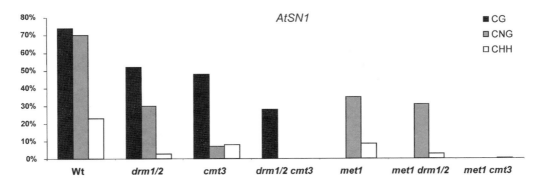

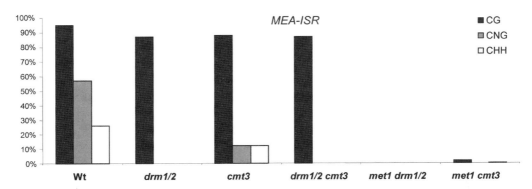

Figure 2. Severe loss of DNA methylation in *drm1/2 met1* and *met1 cmt3*. Bisulfite sequencing results of *AtSN1* and *MEA-ISR* in these two mutants are compared to those from previously described DNA methyltransferase mutants (Zilberman et al. 2003; Henderson et al. 2006). Black, gray, and open bars represent the percentage of DNA methylation at CG, CNG, and CHH sites, respectively.

In extremely rare cases (1 in ~1000), *drm1/2 cmt3 met1* plants were recovered from *drm1/2 cmt3 met1/+* parents. Such quadruple mutant plants grew very slowly, exhibited a suite of severe developmental phenotypes, and failed to flower after about 7 months (Fig. 3).

The results described above strongly suggested that the activities of MET1, DRM1/2, and CMT3 act redundantly to regulate gene expression. The severity of the phenotypes of *drm1/2 met1*, *cmt3 met1*, or *drm1/2 cmt3 met1* plants also underscored the importance of DNA methylation in regulating early embryogenesis as well as many aspects of the normal development of *Arabidopsis*. However, these results still may not have revealed the full spectrum of the

regulatory role of DNA methylation. For example, it is possible that the severe phenotypes of *drm1/2 met1*, *cmt3 met1*, or *drm1/2 cmt3 met1* could have masked the more subtle effects of the misregulation of many other genes. We derived a simple genetic scheme to test this possibility. A population of plants was created, in each of which a random subset of the chromosomes lacked MET1-dependent as well as DRM1/2- and/or CMT3-dependent methylation, whereas the rest of the chromosomes only lacked DRM1/2- and/or CMT3-dependent methylation. It has been described that during the gametogeneses of *met1/+* plants, the lack of MET1 activity in the *met1* gametes resulted in chromosomal-level loss of DNA methylation following the

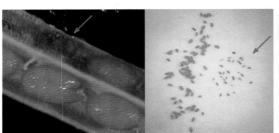 

Figure 3. DNA methylation is required for embryogenesis and normal plant development. Approximately one-fourth of the seeds from *drm1/2 cmt3 met1/+* plants abort at a very early developmental stage (*left panels*; *red arrows*). A rare *drm1/2 cmt3 met1* plant displaying very delayed and stunted development is shown in the right panels.

postmeiotic mitoses (two and three round of mitoses for male and female gametes, respectively) (Saze et al. 2003). Thus, in the mature male *met1* gamete, each chromosome has a 50% probability of being hemimethylated (25% for a female *met1* gamete), and a 50% probability of completely lacking MET1-dependent methylation (unmethylated; 75% for a female *met1* gamete). When combined with a wild-type gamete of the opposite sex to form a *met1/+* zygote, DNA methylation on the hemimethylated chromosomes is immediately restored by the introduced MET1 activity, but the unmethylated chromosomes would remain unmethylated because the methylation pattern has largely been erased. Numerous genes located on such unmethylated chromosomes would be hypomethylated and some misregulated, thus resulting in a *met1*-like phenotype (albeit milder) in *met1/+* plants. We self-pollinated *drm1/2 met1/+*, *cmt3 met1/+*, and *drm1/2 cmt3 met1/+* plants and examined the F₁ plants that were heterozygous for MET1. These plants entirely lacked DRM1/2 and/or CMT3 activities. In addition, as described above, a random subset of their chromosomes also lacked MET1-dependent methylation. A remarkably variable assortment of developmental phenotypes were observed among *drm1/2 met1/+* or *drm1/2 cmt3 met1/+* plants, ranging from mild to extremely severe (Fig. 4). This was likely due to the misregulation of different sets of genes in individual plants, because they were deficient for DRM1/2 and/or CMT3 as well as MET1-dependent methylation for different combinations of chromosomes. In addition, similar phenotypes were found to arise in multiple independent plants. This is consistent with the reproducible misregulation of specific genes controlling these phenotypes and suggests that the stochastic mutagenic effect of transposon reactivation is not a major factor. Thus, a large number of genes important for many aspects of *Arabidopsis* development are regulated by DNA methylation. Interestingly, *cmt3 met1/+* plants appeared phenotypically similar to *met1/+* and did not exhibit additional defects. Among other possibilities, it is interesting to consider that the molecular lesion responsible for the *cmt3 met1* phenotype might be recessive (similar to the *superman* epimutation in *met1*) or that the simultaneous misregulation of several genes located on different chromosomes might be required to render the phenotype.

RNAi AND HISTONE METHYLATION IN THE ESTABLISHMENT AND MAINTENANCE OF NON-CG METHYLATION

As a regulatory mechanism for gene expression and thus plant development, DNA methylation itself is likely to be highly regulated and able to respond to developmental and environmental cues. Several studies using a methylation-sensitive amplified fragment length polymorphism (AFLP) method have provided evidence that changes in DNA methylation may occur during plant development or under stress conditions (Messeguer et al. 1991; Finnegan et al. 1998; Xiong et al. 1999; Zluvova et al. 2001; Bitonti et al. 2002; Fraga et al. 2002; Ruiz-Garcia et al. 2005). MET1-mediated CG methylation is remarkably stable across multiple generations. In contrast, non-CG methylation is much more dynamic and

thus more appealing as a regulatory mechanism. This is because the establishment and maintenance of both DRM1/2-dependent CNN and CMT3-dependent CNG methylation require persistent targeting by the RNAi and histone methylation pathways, respectively.

DRM1/2 is required to establish de novo DNA methylation on an unmethylated incoming transgene (Cao and Jacobsen 2002b). To identify other upstream components of this pathway, a reverse genetic approach was undertaken where a collection of mutants defective in various epigenetic pathways were transformed with the gene *FWA*. The transgene was methylated and silenced in wild-type but not in the *drm1/2* mutant, resulting in a late-flowering phenotype (Cao and Jacobsen 2002b). Several mutants defective in the RNAi pathway, including *nrpd1a*, *nrpd1b*, *drd1*, *rdr2*, *dcl3*, and *ago4*, were found to resemble *drm1/2* in that they blocked the de novo methylation of the *FWA* transgene (Dalmay et al. 2000; Zilberman et al. 2003; Chan et al. 2004, 2006; Kanno et al. 2005; Pontier et al. 2005). Furthermore, these mutants phenocopied *drm1/2* in that they lost significant amounts of non-CG methylation at several endogenous loci tested (Hamilton et al. 2002; Chan et al. 2004, 2006; Herr et al. 2005; Kanno et al. 2005; Onodera et al. 2005; Pontier et al. 2005). These results suggested that the RNAi pathway is required for all aspects of the DRM1/2 activity. Importantly, these studies also provided a mechanism by which DRM1/2-dependent DNA methylation can be regulated in a tissue- and locus-specific manner through the activities of the RNAi pathway. Indeed, a recent deep sequencing analysis of *Arabidopsis* small RNAs revealed profound differences in the abundance and diversity of the small RNAs accumulated in different tissues (Lu et al. 2005).

Some RNAi genes such as *NRPD1a/1b* and *NRPD2a/2b* are plant-specific and have likely evolved only in the plant lineage (Herr et al. 2005; Kanno et al. 2005; Onodera et al. 2005; Pontier et al. 2005), whereas other genes such as *RDRs, DCLs*, and *ARGONAUTEs* are conserved but have undergone extensive duplications and functional diversifications in plants. As a result, the complexity of plant RNAi pathways far exceeds that of their animal counterparts. For example, the four *Arabidopsis DCL* genes are functionally specialized in the metabolism of different classes of small RNAs: DCL1 is required in the microRNA (miRNA) pathway, DCL3 is required for small interfering RNAs (siRNA), both DCL1 and DCL4 are required for *trans*-acting siRNAs (tasiRNA), and DCL1 and DCL2 are required for siRNAs derived from natural antisense transcripts (nat-siRNA) (Schauer et al. 2002; Xie et al. 2004, 2005; Borsani et al. 2005; Dunoyer et al. 2005; Gasciolli et al. 2005; Yoshikawa et al. 2005; Henderson et al. 2006). It is of particular interest to determine whether all these pathways can interact with DRM1/2 to direct DNA methylation and transcriptional gene silencing. To address this question, we constructed all pair-wise combinations of double mutants between *dcl2*, *dcl3*, and *dcl4* as well as the triple mutant *dcl2 dcl3 dcl4* (Henderson et al. 2006). Comparison of the small RNA populations from wild type and *dcl2 dcl3 dcl4* by large-scale 454 sequencing showed that DCL1 alone was not only sufficient for all aspects of miRNA metabolism, but also capable of generating 21-bp

Figure 4. Highly variable developmental phenotypes of *drm1/2 met1/+* plants produced through self-pollination of a *drm1/2 met1/+* parent.

small RNAs from some additional loci (preferentially from inverted repeats) (Henderson et al. 2006). Moreover, our results as well as several other recent studies unveiled partial but extensive functional redundancy among DCL2, DCL3, and DCL4 in processing a subset of small RNAs (Xie et al. 2004; Gasciolli et al. 2005; Henderson et al. 2006). Finally, we showed that siRNAs generated by DCL3 play a major role in directing DNA methylation, but at some loci, DCL2 and DCL4 products can also assist in this process (Henderson et al. 2006).

The specificity of RNA-directed DNA methylation (RdDM) likely resides in the primary sequences of siRNAs, and the key components that connect DRM1/2 to the RNAi pathway are the ARGONAUTE proteins. In *Arabidopsis*, AGO4 plays a major role in this process, as the *ago4* mutant displayed reduced non-CG methylation at a number of endogenous loci and failed to de novo

methylate a transgene (Zilberman et al. 2003, 2004; Chan et al. 2004). We have recently performed a detailed analysis of AGO4 with regard to its protein stability in several mutant backgrounds, its interaction with other components of the RNAi pathway, and its subnuclear localizations (Li et al. 2006; Pontes et al. 2006). Interestingly, AGO4 becomes unstable in the absence of NRPD1a, RDR2, and DCL3, but the loss of NRPD1b or DRM2 does not affect AGO4 protein level. This observation strongly suggested a hierarchical order of action, where NRPD1a, RDR2, and DCL3 function upstream of AGO4 whereas NRPD1b and DRM2 are either at the same step or downstream from AGO4 (Li et al. 2006). It is interesting to consider that such a property of AGO4 may offer a "safeguard" for the RdDM process to ensure that, in the absence of siRNAs generated by upstream components, AGO4 would not target spurious DNA methylation to inappropriate loci. Further support of the close relationship between AGO4 and NRPD1b came from the observations that they not only colocalize in the nucleus, but also physically interact with each other. Finally and most significantly, AGO4 localizes to the "Cajal body" where the processing and maturation of several ribonucleoprotein complexes take place. Taken together, these results have led us to propose a model where double-stranded RNAs produced by NPRD1a and RDR2 are processed into 24-nucleotide siRNAs by DCL3 in the Cajal body. An AGO4/NRPD1b/siRNA ribonucleoprotein complex is then assembled and relocated to the target locus through sequence homology to the siRNA, and recruits DRM1/2 for DNA methylation (Li et al. 2006).

In contrast to DRM1/2, recruitment of CMT3 for CNG methylation depends on the methylation of two lysine residues on the histone H3 amino-terminal tail: lysine 9 (K9) and lysine 27 (K27). This was initially indicated by the presence of a chromodomain in CMT3 (Lindroth et al. 2001), and experimental evidence came from two studies. First, a histone methyltransferase mutant (*kryptonite*) affecting H3 K9 methylation also affected CNG methylation at several endogenous loci (Jackson et al. 2002; Malagnac et al. 2002). Second, CMT3 was found to bind to H3 peptide methylated at both K9 and K27 in vitro, and H3 K27 methylation colocalized with H3 K9 methylation at CMT3-controlled loci (Lindroth et al. 2004). It is conceivable that alterations in H3 K9 and K27 methylation could lead to changes in CMT3-dependent CNG and CHH methylation, thus providing another mechanism by which non-CG DNA methylation can be regulated.

The studies described above identified factors required for non-CG methylation, but they did not address whether non-CG methylation can be reestablished once it is lost from an endogenous locus. Two lines of evidence suggested that such reestablishment can readily take place. First, the phenotype of *drm1/2 cmt3* is strictly recessive and co-segregates tightly with the *drm1/2 cmt3* genotype (Chan et al. 2006). Second, introducing DRM2 or CMT3 as a transgene into the *drm1/2 cmt3* mutant restored a wild-type phenotype as well as non-CG methylation at endogenous loci (Chan et al. 2006). These results were in stark contrast to *met1* or *ddm1* where the hypomethylated CG sites were generally not remethylated when crossed to wild type (Vongs et al. 1993; Finnegan et al. 1996; Kakutani et al. 1999; Saze et al. 2003). They are also in agreement with the notion that CG methylation is relatively stably maintained, whereas the elimination and reestablishment of non-CG methylation can be subject to developmental regulation.

HIGH-RESOLUTION MAPPING OF DNA METHYLATION IN THE ENTIRE *ARABIDOPSIS* GENOME

Extensive genetic, molecular, and genomic analyses in recent years using *Arabidopsis* as a model system have demonstrated the importance of DNA methylation in suppressing transposons and regulating gene expression (Chan et al. 2005). An accumulating body of data has also broadened our understanding of the establishment and maintenance of DNA methylation as well as its interactions with other epigenetic pathways. However, many important questions had remained unanswered. For example, the extent and distribution of DNA methylation had not been determined on a genome-wide scale, and very few genes were identified as being directly regulated by DNA methylation. This largely reflects technical difficulty in identifying the sites of DNA methylation in a complete eukaryotic genome in a high-throughput manner. We have recently optimized two biochemical methods to separate methylated and unmethylated DNA, followed by hybridization to high-density, whole-genome tiling arrays. This has allowed the genome-wide and high-resolution mapping of DNA methylation in *Arabidopsis* (Zhang et al. 2006).

DNA-methylated regions account for approximately 19% of the *Arabidopsis* genome and are highly enriched in centromeric and pericentromeric heterochromatin, where transposable elements and other repetitive sequences cluster. siRNA clusters, but not miRNA-producing or target genes, are also heavily methylated. Unexpectedly, we found that roughly one-third of *Arabidopsis* genes were methylated in the transcribed regions (body-methylation). Such methylation is biased toward the 3′ end of genes, largely independent of siRNAs and maintained primarily by MET1. The function of genic body-methylation remains unclear, as neither sense nor antisense transcription of body-methylated genes changes systematically when it is lost. In contrast, although the number of genes that contain DNA methylation in their promoters is relatively small, promoter-methylation appears to be more important in down-regulating gene expression level. Interestingly, a subset of promoter-methylated genes appear to have very tissue-specific expression patterns.

The vast majority of genomic regions methylated in wild type remains methylated in *drm1/2 cmt3*; a relatively small number of hypomethylated regions are generally CG-poor. This is consistent with previous results that the loss of non-CG methylation does not significantly disturb CG methylation, and suggests that in most cases non-CG methylation colocalizes with CG methylation. In contrast, only about a third of the methylated regions are still detectable in *met1*. The residual methylation is highly enriched at repetitive sequences, and is likely maintained by DRM1/2 and CMT3.

Comparison of the genes that were most significantly increased in expression in *met1* and *drm1/2 cmt3* revealed an important distinction. Most genes that were overexpressed in *met1* were pseudogenes with a pronounced heterochromatic distribution. In contrast, most up-regulated genes (69%) in *drm1/2 cmt3* had known functions and were distributed throughout euchromatin. These results provide strong evidence for the role of non-CG methylation in regulating gene expression and identified a host of candidate genes that may be directly regulated by non-CG methylation. Moreover, they are in agreement with the notion discussed earlier that CG methylation is important for the silencing of transposons and the maintenance of heterochromatin, whereas non-CG methylation plays a more central role in regulating developmentally important genes.

CONCLUSIONS

In summary, we have presented genetic evidence that CG and non-CG methylation function in a partially redundant manner to regulate many aspects of the *Arabidopsis* development. Analyses of the interactions between DNA methylation and other epigenetic pathways such as RNAi and histone methylation provided important clues as to how DNA methylation itself is likely regulated. Finally, the results from the genomic analysis have broadened our understanding of DNA methylation patterns on a genome-wide scale and have identified many candidate genes controlled by DNA methylation.

Future studies should be directed to address several outstanding questions. For example, a large number of genes were found to be misregulated in *met1* or *drm1/drm2 cmt3* (Zhang et al. 2006), yet our genetic analysis suggested that many more genes may be controlled redundantly by CG and non-CG methylation. DNA methylation and expression profiling of mutants such as *drm1/2 met1*, *cmt3 met1*, or *drm1/2 cmt3 met1* should provide candidates for such genes. Another interesting finding described here is that many promoter-methylated genes are expressed in a tissue-specific manner (Zhang et al. 2006), thus raising the possibility that demethylation may occur at their promoters in specific tissues. Future studies should determine whether a general mechanism exists that removes DNA methylation from these promoters in specific tissues, thus allowing the expression of otherwise silenced genes.

ACKNOWLEDGMENTS

We thank Dr. Simon Chan for his generous help in constructing DNA methyltransferase mutants, and members of the Jacobsen laboratory for helpful discussions and critical reading of the manuscript. Work in the Jacobsen lab is supported by National Institutes of Health grant GM60398 and a grant from the National Institutes of Health ENCODE Program HG003523. S.E.J. is an investigator of the Howard Hughes Medical Institute.

REFERENCES

Bartee L., Malagnac F., and Bender J. 2001. *Arabidopsis cmt3* chromomethylase mutations block non-CG methylation and silencing of an endogenous gene. *Genes Dev.* **15:** 1753.

Bird A. 2002. DNA methylation patterns and epigenetic memory. *Genes Dev.* **16:** 6.

Bitonti M.B., Cozza R., Chiappetta A., Giannino D., Ruffini Castiglione M., Dewitte W., Mariotti D., Van Onckelen H., and Innocenti A.M. 2002. Distinct nuclear organization, DNA methylation pattern and cytokinin distribution mark juvenile, juvenile-like and adult vegetative apical meristems in peach (*Prunus persica* (L.) Batsch). *J. Exp. Bot.* **53:** 1047.

Borsani O., Zhu J., Verslues P.E., Sunkar R., and Zhu J.K. 2005. Endogenous siRNAs derived from a pair of natural *cis*-antisense transcripts regulate salt tolerance in *Arabidopsis*. *Cell* **123:** 1279.

Cao X. and Jacobsen S.E. 2002a. Locus-specific control of asymmetric and CpNpG methylation by the DRM and CMT3 methyltransferase genes. *Proc. Natl. Acad. Sci.* (suppl. 4) **99:** 16491.

———. 2002b. Role of the *Arabidopsis* DRM methyltransferases in de novo DNA methylation and gene silencing. *Curr. Biol.* **12:** 1138.

Cao X., Aufsatz W., Zilberman D., Mette M.F., Huang M.S., Matzke M., and Jacobsen S.E. 2003. Role of the DRM and CMT3 methyltransferases in RNA-directed DNA methylation. *Curr. Biol.* **13:** 2212.

Chan S.W., Henderson I.R., and Jacobsen S.E. 2005. Gardening the genome: DNA methylation in *Arabidopsis thaliana*. *Nat. Rev. Genet.* **6:** 351.

Chan S.W.-L., Henderson I.R., Zhang X., Shah G., Chien J.S.-C., and Jacobsen S.E. 2006. RNAi, DRD1 and histone methylation actively target developmentally important non-CG DNA methylation in *Arabidopsis*. *PLoS Genet.* **2:**e83.

Chan S.W., Zilberman D., Xie Z., Johansen L.K., Carrington J.C., and Jacobsen S.E. 2004. RNA silencing genes control de novo DNA methylation. *Science* **303:** 1336.

Dalmay T., Hamilton A., Rudd S., Angell S., and Baulcombe D.C. 2000. An RNA-dependent RNA polymerase gene in *Arabidopsis* is required for posttranscriptional gene silencing mediated by a transgene but not by a virus. *Cell* **101:** 543.

Dunoyer P., Himber C., and Voinnet O. 2005. DICER-LIKE 4 is required for RNA interference and produces the 21-nucleotide small interfering RNA component of the plant cell-to-cell silencing signal. *Nat. Genet.* **37:** 1356.

Egger G., Jeong S., Escobar S.G., Cortez C.C., Li T.W., Saito Y., Yoo C.B., Jones P.A., and Liang G. 2006. Identification of DNMT1 (DNA methyltransferase 1) hypomorphs in somatic knockouts suggests an essential role for DNMT1 in cell survival. *Proc. Natl. Acad. Sci.* **103:** 14080.

Finnegan E.J., Peacock W.J., and Dennis E.S. 1996. Reduced DNA methylation in *Arabidopsis thaliana* results in abnormal plant development. *Proc. Natl. Acad. Sci.* **93:** 8449.

Finnegan E.J., Genger R.K., Peacock W.J., and Dennis E.S. 1998. DNA methylation in plants. *Annu. Rev. Plant Physiol. Plant Mol. Biol.* **49:** 223.

Fraga M.F., Canal M.J., and Rodriguez R. 2002. Phase-change related epigenetic and physiological changes in *Pinus radiata* D. Don. *Planta* **215:** 672.

Gasciolli V., Mallory A.C., Bartel D.P., and Vaucheret H. 2005. Partially redundant functions of *Arabidopsis* DICER-like enzymes and a role for DCL4 in producing *trans*-acting siRNAs. *Curr. Biol.* **15:** 1494.

Goll M.G. and Bestor T.H. 2005. Eukaryotic cytosine methyltransferases. *Annu. Rev. Biochem.* **74:** 481.

Hamilton A., Voinnet O., Chappell L., and Baulcombe D. 2002. Two classes of short interfering RNA in RNA silencing. *EMBO J.* **21:** 4671.

Henderson I.R., Zhang X., Lu C., Johnson L., Meyers B.C., Green P.J., and Jacobsen S.E. 2006. Dissecting *Arabidopsis thaliana* DICER function in small RNA processing, gene silencing and DNA methylation patterning. *Nat. Genet.* **38:** 721.

Herr A.J., Jensen M.B., Dalmay T., and Baulcombe D.C. 2005. RNA polymerase IV directs silencing of endogenous DNA. *Science* **308:** 118.

Jackson J.P., Lindroth A.M., Cao X., and Jacobsen S.E. 2002. Control of CpNpG DNA methylation by the KRYPTONITE histone H3 methyltransferase. *Nature* **416:** 556.

Jacobsen S.E. and Meyerowitz E.M. 1997. Hypermethylated *SUPERMAN* epigenetic alleles in *Arabidopsis*. *Science* **277:** 1100.

Jeddeloh J.A., Stokes T.L., and Richards E.J. 1999. Maintenance of genomic methylation requires a SWI2/SNF2-like protein. *Nat. Genet.* **22:** 94.

Kakutani T., Munakata K., Richards E.J., and Hirochika H. 1999. Meiotically and mitotically stable inheritance of DNA hypomethylation induced by *ddm1* mutation of *Arabidopsis thaliana*. *Genetics* **151:** 831.

Kankel M.W., Ramsey D.E., Stokes T.L., Flowers S.K., Haag J.R., Jeddeloh J.A., Riddle N.C., Verbsky M.L., and Richards E.J. 2003. *Arabidopsis* MET1 cytosine methyltransferase mutants. *Genetics* **163:** 1109.

Kanno T., Huettel B., Mette M.F., Aufsatz W., Jaligot E., Daxinger L., Kreil D.P., Matzke M., and Matzke A.J. 2005. Atypical RNA polymerase subunits required for RNA-directed DNA methylation. *Nat. Genet.* **37:** 761.

Li C.F., Pontes O., El-Shami M., Henderson I.R., Bernatavichute Y.V., Chan S.W., Lagrange T., Pikaard C.S., and Jacobsen S.E. 2006. An ARGONAUTE4-containing nuclear processing center colocalized with Cajal bodies in *Arabidopsis thaliana*. *Cell* **126:** 93.

Lindroth A.M., Cao X., Jackson J.P., Zilberman D., McCallum C.M., Henikoff S., and Jacobsen S.E. 2001. Requirement of *CHROMOMETHYLASE3* for maintenance of CpXpG methylation. *Science* **292:** 2077.

Lindroth A.M., Shultis D., Jasencakova Z., Fuchs J., Johnson L., Schubert D., Patnaik D., Pradhan S., Goodrich J., Schubert I., et al. 2004. Dual histone H3 methylation marks at lysines 9 and 27 required for interaction with *CHROMOMETHY-LASE3*. *EMBO J.* **23:** 4286.

Lu C., Tej S.S., Luo S., Haudenschild C.D., Meyers B.C., and Green P.J. 2005. Elucidation of the small RNA component of the transcriptome. *Science* **309:** 1567.

Malagnac F., Bartee L., and Bender J. 2002. An *Arabidopsis* SET domain protein required for maintenance but not establishment of DNA methylation. *EMBO J.* **21:** 6842.

Martienssen R.A. and Colot V. 2001. DNA methylation and epigenetic inheritance in plants and filamentous fungi. *Science* **293:** 1070.

Messeguer R., Ganal M.W., Steffens J.C., and Tanksley S.D. 1991. Characterization of the level, target sites and inheritance of cytosine methylation in tomato nuclear DNA. *Plant Mol. Biol.* **16:** 753.

Onodera Y., Haag J.R., Ream T., Nunes P.C., Pontes O., and Pikaard C.S. 2005. Plant nuclear RNA polymerase IV mediates siRNA and DNA methylation-dependent heterochromatin formation. *Cell* **120:** 613.

Pontes O., Li C.F., Nunes P.C., Haag J., Ream T., Vitins A., Jacobsen S.E., and Pikaard C.S. 2006. The *Arabidopsis* chromatin-modifying nuclear siRNA pathway involves a nucleolar RNA processing center. *Cell* **126:** 79.

Pontier D., Yahubyan G., Vega D., Bulski A., Saez-Vasquez J., Hakimi M.A., Lerbs-Mache S., Colot V., and Lagrange T. 2005. Reinforcement of silencing at transposons and highly repeated sequences requires the concerted action of two distinct RNA polymerases IV in *Arabidopsis*. *Genes Dev.* **19:** 2030.

Ronemus M.J., Galbiati M., Ticknor C., Chen J., and Dellaporta S.L. 1996. Demethylation-induced developmental pleiotropy in *Arabidopsis*. *Science* **273:** 654.

Ruiz-Garcia L., Cervera M.T., and Martinez-Zapater J.M. 2005. DNA methylation increases throughout *Arabidopsis* development. *Planta* **222:** 301.

Saze H., Scheid O.M., and Paszkowski J. 2003. Maintenance of CpG methylation is essential for epigenetic inheritance during plant gametogenesis. *Nat. Genet.* **34:** 65.

Schauer S.E., Jacobsen S.E., Meinke D.W., and Ray A. 2002. DICER-LIKE1: Blind men and elephants in *Arabidopsis* development. *Trends Plant Sci.* **7:** 487.

Soppe W.J., Jacobsen S.E., Alonso-Blanco C., Jackson J.P., Kakutani T., Koornneef M., and Peeters A.J. 2000. The late flowering phenotype of *fwa* mutants is caused by gain-of-function epigenetic alleles of a homeodomain gene. *Mol. Cell* **6:** 791.

Vongs A., Kakutani T., Martienssen R.A., and Richards E.J. 1993. *Arabidopsis thaliana* DNA methylation mutants. *Science* **260:** 1926.

Xie Z., Allen E., Wilken A., and Carrington J.C. 2005. DICER-LIKE 4 functions in *trans*-acting small interfering RNA biogenesis and vegetative phase change in *Arabidopsis thaliana*. *Proc. Natl. Acad. Sci.* **102:** 12984.

Xie Z., Johansen L.K., Gustafson A.M., Kasschau K.D., Lellis A.D., Zilberman D., Jacobsen S.E., and Carrington J.C. 2004. Genetic and functional diversification of small RNA pathways in plants. *PLoS Biol.* **2:** E104.

Xiong L.Z., Xu C.G., Saghai Maroof M.A., and Zhang Q. 1999. Patterns of cytosine methylation in an elite rice hybrid and its parental lines, detected by a methylation-sensitive amplification polymorphism technique. *Mol. Gen. Genet.* **261:** 439.

Yoder J.A., Walsh C.P., and Bestor T.H. 1997. Cytosine methylation and the ecology of intragenomic parasites. *Trends Genet.* **13:** 335.

Yoshikawa M., Peragine A., Park M.Y., and Poethig R.S. 2005. A pathway for the biogenesis of *trans*-acting siRNAs in *Arabidopsis*. *Genes Dev.* **19:** 2164.

Zhang X., Yazaki J., Sundaresan A., Cokus S., Chan S.W., Chen H., Henderson I.R., Shinn P., Pellegrini M., Jacobsen S.E., and Ecker J.R. 2006. Genome-wide high-resolution mapping and functional analysis of DNA methylation in *Arabidopsis*. *Cell* **126:** 1189.

Zilberman D., Cao X., and Jacobsen S.E. 2003. ARGONAUTE4 control of locus-specific siRNA accumulation and DNA and histone methylation. *Science* **299:** 716.

Zilberman D., Cao X., Johansen L.K., Xie Z., Carrington J.C., and Jacobsen S.E. 2004. Role of *Arabidopsis* ARGONAUTE4 in RNA-directed DNA methylation triggered by inverted repeats. *Curr. Biol.* **14:** 1214.

Zluvova J., Janousek B., and Vyskot B. 2001. Immunohistochemical study of DNA methylation dynamics during plant development. *J. Exp. Bot.* **52:** 2265.

RNA-directed DNA Methylation and Pol IVb in *Arabidopsis*

M. Matzke, T. Kanno, B. Huettel, L. Daxinger, and A.J.M. Matzke

Gregor Mendel Institute of Molecular Plant Biology, Austrian Academy of Sciences, A-1030 Vienna, Austria

Recent work in *Arabidopsis* has revealed a plant-specific RNA polymerase, pol IV, that is specialized for RNA interference (RNAi)-mediated, chromatin-based gene silencing. Two functionally diversified pol IV complexes have been identified: pol IVa is required to produce or amplify the small RNA trigger, whereas pol IVb, together with the plant-specific SWI/SNF-like chromatin remodeling factor DRD1, acts downstream from small RNA formation to induce de novo cytosine methylation of homologous DNA by an unknown mechanism. Retrotransposon long terminal repeats (LTRs) and other unannotated sequences that encode small RNAs are prime targets for DRD1/pol IVb-mediated cytosine methylation. In *drd1* and pol IVb mutants, silent LTRs in euchromatin can be derepressed, resulting in enhanced transcription of adjacent genes or intergenic regions. In addition to mediating de novo methylation, some evidence suggests that DRD1 and pol IVb are also involved in a reciprocal process of active demethylation, perhaps in conjunction with DNA glycosylase domain-containing proteins such as ROS1. We speculate that DRD1/pol IV-dependent methylation/demethylation evolved in the plant kingdom as a means to facilitate rapid, reversible changes in gene expression, which might have adaptive significance for immobile plants growing in unpredictable environments.

Several RNAi-mediated pathways operate in the nucleus to induce epigenetic modifications in a sequence-specific manner (Matzke and Birchler 2005). In a wide range of organisms, including fission yeast, *Drosophila*, and plants, the RNAi machinery is used to package tandem repeats, such as those in pericentromeric regions, into heterochromatin containing histone H3 lysine 9 (H3K9) methylation (Bernstein and Allis 2005; Kavi et al. 2005). In principle, RNAi-mediated heterochromatin assembly can occur without detectable DNA cytosine methylation, which is absent in fission yeast and adult flies. It is known, however, that RNA can guide methylation of cytosines in DNA as illustrated by the phenomenon of RNA-directed DNA methylation, which has been documented most thoroughly in plants. Despite initial reports of RNA-directed DNA methylation in human cells, recent data suggest that this process requires several plant-specific proteins—most notably, an unusual RNA polymerase IV—which has strengthened the possibility that RNA-directed DNA methylation is limited to the plant kingdom.

In this paper, we discuss genetic screens that have identified proteins needed for RNA-directed DNA methylation and transcriptional gene silencing in *Arabidopsis thaliana*, and we describe features of endogenous DNA target sequences that are methylated and silenced by these proteins. We propose a hypothetical mechanism for RNA-directed DNA methylation and speculate on a potential connection with the reciprocal process of active demethylation of 5-methylcytosine. After discussing related RNAi-mediated transcriptional gene silencing pathways in other organisms, we consider the possible importance of this silencing pathway for plants.

DISCOVERY AND CHARACTERISTICS OF RNA-DIRECTED DNA METHYLATION

Evidence for a sequence-specific signal for DNA methylation was initially obtained in transgenic plants containing multiple, unlinked copies of identical promoters, which became methylated and transcriptionally silenced in a reversible, homology-dependent manner (Matzke et al. 1989; Vaucheret 1993; Park et al. 1996). Although it was initially unclear whether DNA–DNA or RNA–DNA interactions were involved in the homologous sequence interactions that led to methylation and silencing, the discovery of RNA-directed DNA methylation in viroid-infected tobacco plants shifted the experimental focus to the latter possibility (Wassenegger et al. 1994). Further work on promoter-dependent silencing demonstrated that double-stranded RNAs containing promoter sequences could direct methylation and transcriptional silencing of homologous promoters *in trans* (Jones et al. 1999, 2001; Mette et al. 1999, 2000; Aufsatz et al. 2002a). Processing of the double-stranded RNAs to small RNAs 21–24 nucleotides in length suggested a mechanistic link to RNAi (Mette et al. 2000; Sijen et al. 2001).

Detailed analysis of sequences modified by RNA-directed DNA methylation in plants revealed two distinctive features of the modification pattern: First, methylation is largely restricted to the region of RNA–DNA sequence homology and second, methylation is acquired at cytosines in all sequence contexts (CG, CNG, and CNN, where N is A, T, or C) (Pélissier et al. 1999; Pélissier and Wassenegger 2000). This complex pattern renders DNA methylation more versatile in plants than in most other organisms that methylate their DNA. For example, plant sequences that are deficient in CG dinucleotides, which are the exclusive sites of methylation in mammals, but rich in asymmetrical CNN nucleotide groups can be methylated and potentially silenced by RNA-directed DNA methylation. In addition, there is a difference in heritability of cytosine methylation during DNA replication depending on the sequence context: Whereas symmetrical CG methylation can be maintained through the action of maintenance methyltransferases that act on hemimethylated substrates, asymmetrical CNN methylation is not efficiently maintained and needs the continuous presence of the inducing RNA (Jones et al. 2001; Aufsatz et al. 2002a). Thus, CNN methylation can be rapidly lost in dividing cells if the RNA trigger is withdrawn. Because it is poorly maintained, CNN methylation can be regarded as a measure of de novo methylation in the presence of RNA. As discussed below, the facile reversibility of CNN methylation has implications for plant promoters that are silenced by CNN methylation but unaffected by CG methylation.

PROTEINS REQUIRED FOR RNA-DIRECTED DNA METHYLATION

Forward and reverse genetics approaches in *Arabidopsis* have revealed that RNA-directed DNA methylation requires for the most part conserved DNA methyltransferases and histone modifying enzymes (Mathieu and Bender 2004; Chan et al. 2005). The DRM type of DNA methyltransferase catalyzes de novo methylation of cytosines in all sequence contexts within a region of RNA–DNA sequence homology (Cao et al. 2003). In addition, however, full de novo methylation of CG dinucleotides requires the CG-specific methyltransferase MET1 (Aufsatz et al. 2004), which might act at the de novo step in a manner similar to that of its postreplicative maintenance function on hemimethylated DNA (Fig. 1). DRM, which stands for domains rearranged methyltransferase, is homologous to the Dnmt3 group in mammals. As the name implies, however, the catalytic domains are in a different order in plants (VI, IX, X, I–V) than in mammals (I–VI, IX, X) (Cao et al. 2000). The different arrangement might extend the substrate specificity of DRM enzymes beyond CG dinucleotides, thus accounting for the extensive CNG and CNN methylation observed in plant nuclear DNA. Maintenance of symmetrical CG and CNG methylation during DNA replication is executed, respectively, by MET1, which is homologous to mammalian Dnmt1, and the plant-specific CHRO-MOMETHYLASE3 (CMT3) (Fig. 1) (Mathieu and Bender 2004; Chan et al. 2005).

Forward genetic screens have identified two histone-modifying enzymes that participate in the RNA-directed DNA methylation pathway. HDA6, which is one of 18 histone deacetylases encoded in the *Arabidopsis* genome, is needed to maintain and reinforce CG methylation induced by RNA (Aufsatz et al. 2002b). SUVH4/KYP is a H3K9 methyltransferase that contributes to maintenance of CNG methylation (Jackson et al. 2002; Malagnac et al. 2002).

RNA-directed DNA methylation requires at least three proteins of the RNAi machinery to generate and use small RNAs that guide epigenetic modifications and silencing. RDR2, one of six RNA-dependent RNA polymerases encoded in the *Arabidopsis* genome, is thought to synthesize double-stranded RNA from single-stranded primary transcripts; DCL3, one of four dicer-like proteins in *Arabidopsis*, processes double-stranded RNA to the heterochromatin-specific 24-nucleotide small RNAs; and AGO4, one of ten argonaute proteins in *Arabidopsis*, presumably binds small RNAs in nuclear effector complexes that facilitate DNA and histone methylation (Xie et al. 2004; Chan et al. 2005).

DISCOVERY OF DRD1 AND POL IVB

The identification of RNAi proteins and enzymes that catalyze epigenetic modifications in response to RNA signals provides information about components of the RNA-directed DNA methylation pathway but does not illuminate how RNA accesses the homologous target DNA to provide a substrate for de novo cytosine methylation. We have recently completed a mutant screen that has the potential to provide insight into this process. The screen exploited a transgene system based on the seed-specific α′ promoter, which was used to drive expression of a target gene encoding green fluorescent protein (GFP). A second, unlinked "silencer" complex encodes a hairpin RNA comprising α′ promoter sequences. In wild-type plants, the hairpin RNA is processed to small RNAs, which trigger methylation and silencing of the homologous target α′ promoter, resulting in GFP-negative seeds.

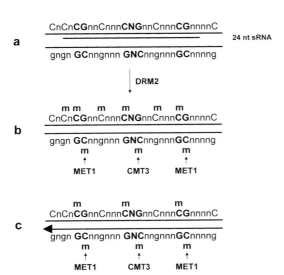

Figure 1. DNA methyltransferases cooperate to establish and maintain RNA-directed DNA methylation: a hypothetical model. (*a*) A 24-nucleotide small RNA base-pairs to the complementary DNA strand (*top*), providing a signal for DRM2 to catalyze de novo methylation (m) of cytosines in all sequence contexts within the region of RNA–DNA sequence complementarity. (*b*) The resulting hemimethylated CGs and CNGs (methylated on one strand but not the other) are recognized by MET1 and CMT3, respectively, which catalyze methylation of the opposite cytosines on the bottom DNA strand. These steps, which are not directly guided by RNA, can be regarded as "de novo" methylation because they occur on unmethylated single strands in nonreplicating DNA. The small RNA-targeted DNA is now fully methylated at CGs and CNGs, whereas CNN methylation is still limited to the top (RNA-targeted) strand. (*c*) During rounds of DNA replication (*leftward arrow*), methylation at symmetrical CGs and CNGs will be maintained by MET1 and CMT3, respectively, because they recognize the hemimethylated substrate (methylated on the parental strand, unmethylated on the daughter strand). In contrast, asymmetrical CNN methylation is eventually lost during rounds of DNA replication in the absence of the RNA trigger.

In retrospect, the choice of the α′ promoter as a target promoter was fortunate, because it turned out to be an excellent reporter of silencing by CNN methylation, which, as mentioned above, is indicative of RNA-directed de novo methylation. Moreover, the factors we identified in the mutant screen using this promoter turned out to be expressed throughout the plant, not only in seeds.

Following chemical mutagenesis of seeds doubly homozygous for the target and silencer complexes, mutants were identified by examining M2 seeds (representing the first generation in which a recessive mutation can be homozygous) for recovery of GFP activity. The screen retrieved three complementation groups, named *drd* for defective in RNA-directed DNA methylation. Subsequent map-based cloning revealed that DRD1 is a plant-specific SWI/SNF-like, putative chromatin remodeling protein (Kanno et al. 2004). DRD2 and DRD3 are subunits of a novel RNA polymerase, pol IV, which is found only in plants (Kanno et al. 2005b).

In the three *drd* mutants, the target α′ promoter displayed an unusual pattern of methylation in which most CG methylation was retained, whereas non-CG methylation was lost. This pattern indicated that the α′ promoter is not silenced by CG methylation but primarily by non-CG methylation. In particular, the heavy loss of CNN methylation in the mutants suggested defects in de novo methylation (Kanno et al. 2004). This hypothesis was tested in experiments designed to assess de novo and maintenance methylation in the *drd1* mutant. (In the context of RNA-directed DNA methylation, de novo methylation is defined as methylation that is induced on unmethylated DNA upon introduction of the RNA trigger; maintenance methylation is defined as the methylation that is retained following removal of the RNA trigger.) These experiments indeed showed that DRD1 is required for RNA-directed de novo methylation of target promoters. Surprisingly, however, experiments using the "maintenance setup" revealed that DRD1 is also required for full loss of CG methylation after segregating the silencer complex that is the source of the inducing RNA (Kanno et al. 2005a). These results led to the proposal that DRD1 facilitates dynamic regulation of DNA methylation: This putative chromatin remodeling factor is needed to direct methylation in response to RNA signals, but it is also required to fully erase methylation when the trigger RNA is withdrawn. We presume that the same applies to DRD2 and DRD3, because the methylation patterns are identical in all three *drd* mutants (Kanno et al. 2004, 2005b).

DRD2 and DRD3 correspond, respectively, to the second largest and largest subunits of pol IV, the existence of which was first inferred from the analysis of the *Arabidopsis* genome sequence (Arabidopsis Genome Initiative 2000). Accordingly, DRD2 and DRD3 have been renamed NRPD2a and NRPD1b, respectively. NRPD2a appears to be the only expressed second largest subunit of pol IV in *Arabidopsis* (Onodera et al. 2005). In contrast, NRPD1b has a homolog, NRPD1a, which was first identified as SDE4 in a forward genetic screen for silencing defective mutants (Dalmay et al. 2000; Herr et al. 2005). NRPD2a teams up with each of the two distinct largest subunits to generate two functionally diversified

pol IV complexes. Whereas both pol IV complexes are ultimately required for methylation of target DNA, they act at different steps in the pathway. pol IVa, containing NRPD1a and NRPD2a, is needed to produce or amplify the small RNA trigger. In contrast, pol IVb, containing NRPD1b and NRPD2a, acts downstream from this step to somehow use small RNA signals to guide cytosine methylation of the homologous DNA region by an unknown mechanism. Indeed, it is still unclear whether either pol IV complex transcribes to any significant extent and if so whether the template is DNA or RNA (Herr 2005; Vaucheret 2005; Vaughn and Martienssen 2005). In a later section, we present a speculative model for how the two pol IV complexes might cooperate to establish and maintain RNA-directed DNA methylation.

The distinct roles of NRPD1a and NRPD1b in RNA-directed DNA methylation can probably be attributed to differences in their carboxy-terminal domains (CTDs), which is the most divergent region between the two proteins. NRPD1b has an extended CTD that contains an additional 522 amino acids that are not present in NRPD1a. The possible function of the extended CTD in NRPD1b can be considered by comparing it to the CTD of the largest subunit of RNA polymerase II (pol II), which transcribes mRNA precursors. The CTD of pol II is unique among subunits of DNA-dependent RNA polymerase in consisting solely of different numbers of copies of a heptapeptide repeat, Tyr-Ser-Pro-Thr-Ser-Pro-Ser, in which the serine residues provide potential phosphorylation sites. The pol II CTD is dispensable for RNA polymerization and appears to serve as a recruitment platform for proteins that process mRNA (Buratowski 2003; Meinhart et al. 2005). However, it is increasingly appreciated that the CTD can also recruit proteins that catalyze epigenetic modifications. For example, in budding yeast, sequential phosphorylation of two serines leads to sequential recruitment of two histone methyltransferases, Set1 and Set2, which catalyze histone modifications characteristic of active transcription (methylation at H3K4 and H3K36) (Hampsey and Reinberg 2003). In human cells, many different proteins have been proposed to associate in a functional manner with the pol II CTD, including DNA methyltransferases (Carty and Greenleaf 2002), although there is not yet any information on these potential affiliations in plants.

Returning to the pol IV largest subunits: The extended CTD of NRPD1b, the largest subunit of pol IVb, does not consist solely of copies of a heptapeptide repeat, but it does contain ten copies of an imperfect 16-amino-acid repeat, which contains potential phosphorylation sites (Pontier et al. 2005). Intriguingly, the region containing these repeats is missing in NRPD1a (Fig. 2). It is tempting to speculate that the extended CTD of NRPD1b can recruit DNA methyltransferases, thus accounting for its involvement in RNA-directed de novo methylation. Determining whether this is the case will require further biochemical work and identification of proteins that interact with the CTD of NRPD1b.

Although only NRPD1b, and not NRPD1a, has an extended CTD containing repeats, both pol IV largest subunits have a motif found in DCL proteins (in this context, DCL stands for defective chloroplasts and leaves,

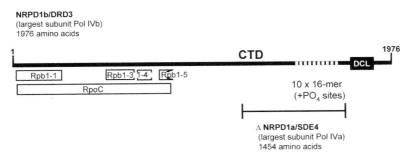

Figure 2. Largest subunits of pol IV differ in their carboxy-terminal domains. The largest subunit of pol IVb (NRPD1b/DRD3) is 1976 amino acids; the largest subunit of pol IVa (NRPD1a/SDE4) is 1454 amino acids. The size difference is due to an additional 522 amino acids in the carboxy-terminal domain (CTD) of NRPD1b. The extended CTD of NRPD1b contains ten copies of an imperfect 16-amino-acid repeat (*white dashes*) that has potential phosphorylation sites (Pontier et al. 2005). This region may provide a recruitment platform for enzymes that induce epigenetic modifications. Both NRPD1a and NRPD1b contain a motif of unknown function that is also found in proteins called defective chloroplasts and leaves (DCL). Conserved domains that might be required for RNA polymerase activity in the amino-terminal half of NRPD1b are boxed (Kanno et al. 2005b). For a discussion of differences in the active sites of pol IV subunits as compared to their counterparts in other DNA-dependent RNA polymerases, see Herr (2005).

not dicer-like) (Fig. 2). DCL proteins are found only in plants and are thought to be involved in rRNA processing (Bellaoui and Gruissem 2004; Lahmy et al. 2004). The function of the DCL motif in the two pol IV largest subunits is presently unknown, but it might be involved in mediating interactions with RNA.

ENDOGENOUS TARGETS OF DRD1 AND POL IVB

The forward genetic screens that identified DRD1, NRPD1b, and NRPD2a (Kanno et al. 2004, 2005b) as well as NPRD1a (Dalmay et al. 2000; Herr et al. 2005) were carried out using transgene systems. Although transgenes supplied convenient readouts to monitor release of gene silencing, they did not provide information about the endogenous targets of these silencing proteins in the *Arabidopsis* genome. Our initial studies on possible targets of DRD1-dependent silencing and methylation focused on sequences that are reactivated or lose methylation in DDM1 (DECREASE IN DNA METHYLATION), which is also a SWI/SNF-like protein (Jeddeloh et al. 1999); however, unlike the plant-specific DRD1, DDM1 has a mammalian homolog, Lsh (Dennis et al. 2001). DNA methylation analysis of several types of tandem repeats in *Arabidopsis*, such as 180-bp centromeric repeats, 5S rDNA repeats, and *Athila* retrotransposon arrays, supported the idea that DRD1, NRPD1b, and NRPD2a (the two subunits are collectively referred to hereafter as pol IVb) are important for CNN methylation, which is greatly reduced in *drd1* and pol IVb mutants. In contrast, predominantly CG methylation is lost from these repeated sequences in *ddm1* mutants. Thus, unlike DDM1, DRD1 and pol IVb do not appear to have a decisive role in maintaining CG methylation in tandem repeats that are packaged into heterochromatin.

To investigate the endogenous targets of the DRD1/pol IVb-dependent silencing pathway in more detail, we carried out two approaches that identified transcripts that differentially accumulate in wild-type and mutant plants: cDNA-AFLP and suppression subtractive hybridization

(SSH). These techniques do not discriminate among protein-coding and non-protein-coding regions of the genome and can be used to detect relatively modest changes in expression. The initial experiments were carried out using a *drd1* mutant. As starting material, we used polyadenylated RNA, which is enriched in pol II transcripts, isolated from seedlings.

Transcripts representing sequences that are up- or down-regulated in the *drd1* mutant compared to wild-type plants were recovered from both the cDNA-AFLP and SSH approaches. Putative targets included annotated genes and unannotated intergenic regions, many of which are present in euchromatic chromosome arms or at the periphery of pericentromeric heterochromatin. Notably, tandem repeats and transposon sequences were underrepresented in the two screens. Real-time quantitative reverse transcriptase–polymerase chain reaction (RT-PCR) was used to verify differences in expression levels of the candidate target genes in the *drd1* mutant and in other mutants in the pol IV pathway (*nrpd2a, nrpd1b, nrpd1a, rdr2*) as well as in a *met1* mutant. The validated targets could be grouped into three categories (Huettel et al. 2006; B. Huettel, unpubl.): (1) Sequences that are up-regulated in a *drd1* mutant and in pol IV mutants; these could be placed into three subcategories depending on whether reactivation was less, more, or approximately the same in a *met1* mutant; (2) sequences that are down-regulated in a *drd1* mutant and in pol IV mutants; and (3) sequences that are up-regulated in a *drd1* mutant but not consistently up-regulated in pol IV mutants. Here, we further discuss categories 1 and 2.

Category 1

Most of the candidate target sequences we identified by cDNA-AFLP and SSH analyses were up-regulated in the *drd1* mutant compared to wild-type plants. Strikingly, we identified in both approaches a pair of putative targets that are adjacent to each other in the same orientation on chromosome 5: a truncated *LINE* element and a gene encoding the ribosomal protein RPL18C. In addition to being up-regulated in a *drd1* mutant, these two transcripts were also found to be up-regulated in other mutants defective in the pol IV pathway (*nrpd1a, nrpd1b, nrpd2a,* and *rdr2*). Analysis of the intergenic region between *LINE* (long interspersed nuclear element) and the *RPL18C* gene revealed a previously unannotated solo LTR, which is a member of a small *Copia*-like retrotransposon family that we named LTRCO. The expression data suggested that the solo LTR was derepressed in the *drd1* mutant, leading to bidirectional transcriptional up-regulation of the flanking sequences to produce an antisense *LINE* transcript and a sense *RPL18C* transcript (Huettel et al. 2006).

Derepression of the solo LTR was accompanied by loss of cytosine methylation, particularly asymmetrical CNN methylation, from this element. An examination of histone modifications by chromatin immunoprecipitation (ChIP) analysis demonstrated that the solo LTR was associated with euchromatic histone modifications—H3K4 trimethylation (H3K4me3) and H3 acetylation (acetyl H3)—as well as H3K27 monomethylation (H3K27me) as the main repressive histone modification. Negligible H3K9 dimethylation (H3K9me2), which is a hallmark of constitutive heterochromatin in *Arabidopsis* (Fuchs et al. 2006), was detected. A similar histone modification pattern comprising a combination of euchromatic marks and H3K27me but little or no H3K9me2 was observed for the transgene α′ promoter. Notably, other members of the LTRCO family were not detectably reactivated in the *drd1* mutant. This may have been due to the fact that they were modified by H3K9me2 and lacked euchromatic marks. The LTRs that remained silent in the *drd1* mutant nevertheless lost CNN methylation in mutant plants. This result is significant because it demonstrates that even though the LTRs are not derepressed in a *drd1* mutant, they are targets of DRD1-dependent cytosine methylation (Huettel et al. 2006).

Consistent with the solo LTR being a target of DRD1/pol IVb-dependent silencing and methylation, 24-nucleotide small RNAs originating from this sequence could be detected on northern blots. The LTR small RNAs were present at wild-type levels in *drd1* and *nrpd1b* mutants, but absent in *nrpd2a, nrpd1a,* and *rdr2* plants. This supports the view that pol IVa (containing *nrpd1a* and *nrpd2a*) is involved in the production or amplification of small RNAs, whereas DRD1 and pol IVb (containing *nrpd1b* and *nrpd2a*) act downstream from this step to induce DNA cytosine methylation.

We extended the analysis by examining three intergenic targets—*IG1, IG2,* and *IG5*—that are up-regulated in the *drd1* and pol IVb mutants (Huettel et al. 2006). *IG2* and *IG5* are adjacent to LTRs that encode small RNAs, whereas *IG1* is part of an unannotated intergenic region that is apparently bidirectionally transcribed because it encodes 19 small RNAs of both sense and antisense polarities (Lu et al. 2005). For the most part, the up-regulated *IG1, IG2,* and *IG5* sequences have patterns of histone modification that conform to those observed for the solo LTR and transgene α′ promoter; that is, euchromatic marks and H3K27me but little or no H3K9me2. Only *IG5* had in addition detectable H3K9me2. Strikingly, however, the *IG5* sequence was considerably more up-regu-

lated in a *met1* mutant than in the *drd1* and pol IVb mutants. This finding suggests that CG methylation is more important than CNN methylation for silencing *IG5* transcription. In contrast, the solo LTR of the LTRCO family and *IG1* were not appreciably reactivated in the *met1* mutant. *IG2* represented an intermediate case that was reactivated to approximately the same degree in both the *drd1* and *met1* mutants (Huettel et al. 2006). Although the basis of the variable reactivation of different targets in *drd1* or *met1* mutants is not known, one possibility is that the sequence composition, particularly the number and density of CG dinucleotides, might have a role. Sequences that are rich in CGs and depleted in CNN nucleotide groups might reactivate more effectively in *met1* mutants than in *drd1* mutants, whereas the converse may be true for sequences rich in CNNs and deficient in CG dinucleotides.

The presence of histone modifications typical of euchromatin at many DRD1 target genes is in agreement with their location in gene-rich chromosome arms or at the boundary between euchromatin and pericentromeric heterochromatin (Huettel et al. 2006). In contrast, the LTRs of the LTRCO family that failed to reactivate in the *drd1* mutant are embedded in transposon-rich pericentromeric heterochromatin. These findings substantiate the proposal that DRD1 and pol IVb act to reversibly silence promoters in euchromatin.

Although the screens we performed to identify sequences reactivated in the *drd1* mutant were not exhaustive, the consistency of the available data allows us to draw provisional conclusions about the characteristics of sequences that are methylated and silenced by DRD1 and pol IVb. It is likely that all sequences with homologous small RNAs can be targets of DRD1/pol IVb-mediated cytosine methylation, which is characterized by methylation of cytosines in all sequence contexts. Importantly, however, only a subset of sequences targeted for methylation is reactivated in *drd1* and pol IVb mutants. Strongly reactivated targets tend to have a euchromatic character, reside in gene-rich regions, and be more sensitive to CNN methylation than to CG methylation. Further experiments using *Arabidopsis* whole-genome tiling arrays, which have recently become commercially available, will build on these studies of DRD1 and pol IVb target sequences.

LTRs as Preferential Targets for DRD1/pol IVb-mediated Methylation

Our investigation suggested that retrotransposon LTRs are frequent targets of DRD1/pol IVb-mediated cytosine methylation. Consistent with this, many LTRs in the *A. thaliana* genome encode small RNAs (Lu et al. 2005), which could guide DRD1/pol IVb-mediated methylation of cognate DNA sequences. The targeting of LTRs by a small RNA-mediated pathway in *Arabidopsis* contrasts with the situation in fission yeast. Despite an early report that LTR activity in fission yeast can be regulated by RNAi-mediated chromatin modifications and influence the expression of adjacent meiotic genes (Schramke and Allshire 2003), most of the 262 LTRs in this organism do not seem to be targets of RNAi-mediated chromatin modifications, nor do they encode small RNAs (Cam et al. 2005).

Given that our initial mutant screen scored for reactivation of a silent transgene promoter, it is not too surprising that the natural targets of the DRD1/pol IVb silencing pathway include LTRs, which are promoter/enhancer elements that can be used by pol II. The identification of the solo LTR of the LTRCO family as a derepressed regulatory element in both the SSH and cDNA-AFLP analyses suggests that this sequence might have special features which render it particularly conspicuous to the DRD1/pol IVb protein machinery. Two considerations are its length and sequence composition. Like the transgene α′ promoter, the solo LTR is several hundred base pairs in length and deficient in CG dinucleotides, having only two CGs in 376 bp. The low CG density in these sequences might favor silencing by CNN methylation, which is the repressive mark lost most consistently in *drd1* and pol IVb mutants. The relatively short length of the solo LTR might prevent stable maintenance of constitutive heterochromatin (Tran et al. 2005).

A third noteworthy feature of the solo LTR is the presence of an internal tandem repeat comprising four copies of a 36-bp monomer, which is the source of a short RNA identified by massively parallel signature sequencing (Lu et al. 2005). Because tandem repeats can ensure the continuous production of double-stranded RNAs and small RNAs by RNA-dependent RNA polymerase and Dicer, respectively (Martienssen 2003), they might provide a strong nucleation site for RNAi-mediated chromatin modifications. However, the correlation between tandem repeats in LTRs and the ability to be silenced by the DRD1/pol IVb machinery is not perfect. For example, although the *Copia*-like LTR driving transcription of *IG5* contains tandem repeats, the truncated *Athila* LTR upstream of *IG2* lacks such repeats, although direct repeats are present in full-length copies of the *Athila* LTR present elsewhere in the genome (Huettel et al. 2006; B. Huettel, unpubl.). Moreover, as illustrated by the repeat-free transgene α′ promoter, tandem repeats are not essential for DRD1/pol IVb-mediated DNA methylation when silencing is induced by a hairpin RNA that is independent of RNA-dependent RNA polymerase activity (Kanno et al. 2005b). Nevertheless, many LTRs in the *Arabidopsis* genome contain short internal tandem repeats (Table 1), so it is worth considering in future studies whether these reiterated regions increase the probability that RNAi-mediated chromatin modifications will be targeted to LTRs. We were unable to detect comparable internal tandem repeats in LTRs of fission yeast (B. Huettel, unpubl.), which might at least partially explain the resistance of these elements to RNAi-mediated chromatin modifications.

Category 2

The second category of putative DRD1 targets identified in the cDNA-AFLP and SSH analyses comprises sequences that are down-regulated in the *drd1* mutant. The most intriguing candidate from this category, which was subsequently validated by real-time RT-PCR, is

ROS1 (REPRESSOR OF SILENCING). ROS1 is a large, plant-specific protein that contains a DNA glycosylase domain. ROS1, and a related protein DEMETER, are thought to be involved in active demethylation by participating in a base-repair-type mechanism that removes 5-methylcytosine and replaces it with an unmethylated cytosine (Kapoor et al. 2005; Morales-Ruiz et al. 2006). The down-regulation of *ROS1* in the *drd1* mutant and pol IVb mutants might explain the failure to fully erase CG methylation from target promoters in these mutants (Kanno et al. 2005a,b). The observation that *ROS1* is down-regulated to an even greater extent in *nrpd1a* and *rdr2* mutants strengthens the connection of ROS1 to the entire pol IV pathway (Huettel et al. 2006). Although the nature of this connection is not yet clear, we speculate that a DRD1/pol IVb complex can act with DNA methyltransferases to induce de novo methylation and with DNA glycosylases to remove methylation, thus keeping euchromatic promoters in a potentially reversible epigenetic state (Huettel et al. 2006). In addition to passive loss of methylation (which results from a failure to maintain methylation, in particular CNN methylation, during rounds of DNA replication in the absence of the RNA trigger), active demethylation through an enzymatic activity might provide an alternate way to lose cytosine methylation in the pol IV pathway. Importantly, active demethylation can occur on non-replicating DNA, which permits epigenetic gene silencing to be reversed in stationary cells. The ability to switch gene expression states in nondividing cells by means of reversible pol-IV-mediated DNA methylation can potentially contribute to physiological adaptation in mature plant organs.

POSSIBLE MECHANISM OF DRD1/POL IV-DEPENDENT DNA METHYLATION

Our data on DRD1 and pol IVb targets suggest that this silencing pathway can reversibly silence promoters in euchromatin via cycles of de novo DNA methylation and either passive or active loss of methylation (Fig. 3). This conclusion is consistent with other proposals which suggest that pol IVb is recruited by euchromatic modifications left by another euchromatic RNA polymerase (Pontier et al. 2005) and that pol IV is important for genes that alternate between decondensed euchromatic states and condensed chromocenter-associated heterochromatic states (Onodera et al. 2005). Although there is general agreement about these basic features of DRD1/pol IVb-dependent silencing, the mechanism by which these proteins elicit de novo methylation is unknown. Here, we present a speculative model that takes into account the unique contribution of each pol IV complex.

The DRD1/pol IVb pathway appears to be particularly important for silencing transposon promoters, such as solo LTRs, in euchromatin. Silencing prevents pol II from using these promoters to initiate transcription, which could extend into flanking plant genes and have potentially deleterious consequences, such as unscheduled synthesis of antisense RNAs. To silence transposon promoters via the pol IV pathway, an essential first step is

Table 1. Tandem Repeats in the LTR Promoter Regions of *A. thaliana* Retrotransposons

Repeatmasker ID	Length (bp)	Repeat position	Repeat size (unit consensus)	Copy number
LTR/*Gypsy*				
ATGP2	1287	219–305	30	2,9
		335–409	18	4,2
ATGP2N	2033	119–212	41	2,3
		380–464	31	2,7
		796—881	30	2,9
ATGP5	779	81–132	18	2,9
ATGP8	1795	263–294	16	2
ATGP9B	478	46–165	54	2,2
ATGP9	508	27–117	18	5,1
		99–219	53	2,3
Athila2	1744	294–421	59	2,2
Athila3	1610	288–336	12	3,9
		1253–1326	38	1,9
Athila4B	1209	369–399	13	2,5
Athila6A	1888	216–329	45	2,5
		1463–1649	81	2,3
Athila7	1427	227–386	64	2,5
Athila8B	1660	1436–1473	18	2,2
LTR/*Copia*				
AtCopia11	204	8–53	22	2,2
AtCopia39	395	167–200	14	2,4
AtCopia42	1326	627–668	18	2,3
		835—936	41	2,5
AtCopia49	502	97–155	28	2,1
AtCopia54	288	62–111	20	2,5
AtCopia65A	305	229–265	17	2,2
AtCopia66	509	455–485	16	2
AtCopia79	500	168–204	18	2,1
AtCopia89	214	11–48	19	1,9
AtCopia95	841	332–408	22	3,3

Family	Sequences analyzed	Tandem repeats	Frequency	Copy number	Period size (average value)
LTR/*Gypsy*	29	12	0,41	2 to 4	35,5
LTR/*Copia*	104	10	0,1	2 to 3	21,5

Sequences and nomenclature of retrotransposons were extracted from the Repeatmasker library (URL: www.girinst.org) and searched for tandem repeats with "tandem repeat finder" using default options (URL: tandem.bu.edu/trf/trf.html). Summaries for the *Gypsy* and *Copia* families are shown at the bottom.

to generate double-stranded RNA homologous to these promoter elements (here we use an intergenic solo LTR as an example). A double-stranded RNA could either be encoded by other copies of the LTR elsewhere in the genome and act *in trans* on the target LTR, or it could be generated *in cis* at the target locus. This can occur by pol II readthrough transcription from other transposon promoters or adjacent plant genes, with the overlapping transcripts annealing to form double-stranded RNA. Alternatively, pol II synthesis of a single-stranded primary transcript that is aberrant (e.g., lacking a poly(A) tail or 5′-cap structure) can be a template for RDR2, which would synthesize the second RNA strand. Following dicing of the double-stranded RNAs by DCL3, the resulting small RNAs can guide DRD1 and pol IVb to facilitate de novo methylation of the target LTR. This step might not require extensive transcription by the DRD1/pol IVb complex but simply the establishment of an open chromatin conformation at the small RNA-targeted site that exposes DNA in this region to DNA methyltransferases. The extended CTD of NRPD1b, including the 16-amino-

acid repeats, might provide a platform for recruiting the DNA methyltransferase DRM2, which catalyzes de novo methylation of cytosines in all sequence contexts.

Once an LTR is methylated, pol II cannot initiate transcription from this promoter. However, pol IVa, which is proposed to be specialized for transcribing methylated DNA (Herr et al. 2005; Onodera et al. 2005), would be able to recognize and transcribe the methylated LTR. The LTR-containing transcript produced by pol IVa is then copied by RDR2 and diced by DCL3. The resulting small RNAs are again used by the DRD1/pol IVb machinery to maintain and even enhance methylation, which in turn would provoke increased transcription by pol IVa and elevated amounts of small RNAs, thus fueling the entire cycle. In this scenario, pol IVb-facilitated DNA methylation is restricted to regions targeted by small RNAs, whereas pol IVa transcription is confined to DNA sequences that are modified by methylation. The two pol IV complexes thus act in a concerted manner to both initiate and maintain DNA methylation that blocks pol II transcription from the target promoter.

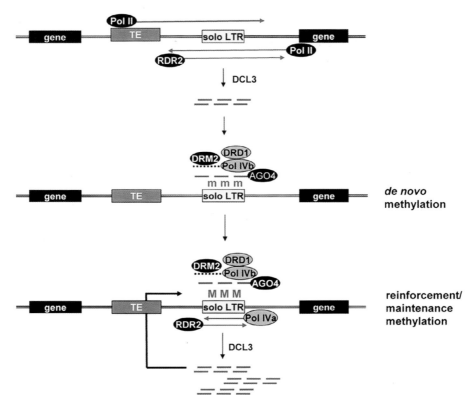

Figure 3. Hypothetical model for concerted action of pol IV complexes in RNA-directed de novo and maintenance methylation. The target is an intergenic solo LTR (*yellow*) located in euchromatin. (*Red*) RNAs. (*Top*) Double-stranded RNAs containing LTR sequences can potentially form by annealing of overlapping pol II transcripts initiating in adjacent transposon or plant gene promoters or by RDR2 activity on a single-stranded, aberrant pol II transcript. Double-stranded RNAs are processed to 24-nucleotide small RNAs by DCL3. (*Middle*) De novo methylation (small m) in response to RNA signals is catalyzed primarily by DRM2, which we speculate can interact with the repeats in the CTD of NRPD1b, the largest subunit of pol IVb (*short dotted black line*). DRD1 is a putative SWI/SNF-like factor that might facilitate opening of chromatin at the small RNA-targeted site. AGO4 presumably interacts with small RNAs via its PAZ domain. (*Bottom*) pol IVa preferentially transcribes methylated DNA, producing aberrant transcripts that are copied by RDR2. The resulting double-stranded RNA is processed by DCL3, leading to reinforcement and maintenance of methylation (large M), again by the pol IVb/DRD1/DRM2 complex. (*Green*) Components found only in plants.

REVERSIBLE RNA-DIRECTED DNA METHYLATION IS HIGHLY DEVELOPED IN PLANTS

Many components of the reversible RNA-directed DNA methylation machinery—including pol IVa, pol IVb, and DRD1—as well as asymmetrical CNN methylation and ROS1 are found only in plants. Why do plants need this silencing pathway? One reason is to silence transposons that integrate into euchromatin and/or are too short to be stably packaged into constitutive heterochromatin (Tran et al. 2005). In addition, the DRD1/pol IVb pathway might benefit plants by contributing to a genome-wide system of small RNA-regulated, transposon-derived promoters/enhancers that modulate gene expression under changeable and/or suboptimal growth conditions. Stress can induce the production of small RNAs (Sunkar and Zhu 2004) and reactivate retrotransposon LTRs to influence transcription of neighboring plant genes (Kashkush et al. 2003). These findings are consistent with a scenario in which LTRs and small RNAs are key participants in reversible DRD1/pol IVb-dependent silencing. There are more than 3000 annotated solo LTRs in the *A. thaliana* genome (Peterson-Burch

et al. 2004), many of which reside in intergenic regions of euchromatic chromosome arms. These solo LTRs could supply epigenetically controlled rheostats or switches for nearby genes, whose increased or decreased expression might have adaptive significance for the plant. If so, the DRD1/pol IV pathway might fulfill a need for rapid, reversible changes in gene expression that help sessile plants to cope with environmental challenges.

RNA-DIRECTED DNA METHYLATION IN HUMAN CELLS

The possibility that RNA-directed DNA methylation occurs in human cells was first suggested by experiments in which small RNAs homologous to either transgene or endogenous gene promoters appeared to trigger transcriptional gene silencing and promoter methylation (Kawasaki and Taira 2004; Morris et al. 2004). The most recent studies suggest, however, that promoter-directed small RNAs induce either very low levels of methylation or no detectable methylation of target promoters in human cells (Castanotto et al. 2005; Ting et al. 2005). Nevertheless, promoter-directed small RNAs appear to reliably induce transcriptional gene silencing associated with H3K9 and H3K27 methylation in these cells (Ting et al. 2005; Weinberg et al. 2005). Given that crucial components of the RNA-directed DNA methylation machinery are found only in the plant kingdom, it is not too surprising that mammals do not have an identical mechanism for eliciting small RNA-directed transcriptional gene silencing and chromatin modifications.

Although the incidence and function of RNA-mediated transcriptional silencing in mammalian cells remain to be determined, there is considerable interest in the possibility that promoter-directed small RNAs might have therapeutic value in human medicine (Morris 2006). For instance, it has been reported that small RNAs directed to the LTR promoter region of human immunodeficiency virus type 1 (HIV-1) induce prolonged transcriptional gene silencing in cell culture (Suzuki et al. 2005). There are also possible disease connections. For example, even low levels of RNA-directed DNA methylation might have implications for cancer development if rare transcriptional silencing of a tumor suppressor gene confers a growth advantage (Castanotto et al. 2005).

RNA POLYMERASE II INVOLVEMENT IN EPIGENETIC GENE SILENCING

Even though pol IV appears to be specific to plants, there is evidence that another RNA polymerase—pol II—is required for RNAi-mediated heterochromatin formation in fission yeast. Genetic screens have identified roles in this process for three pol II subunits: Rpb7, Rpb2, and Rpb1. Interestingly, the subunits are needed at different steps of the pathway, which is reminiscent of the distinct contributions of the two pol IV complexes to RNA-directed DNA methylation in plants. In fission yeast, Rpb7 is required for synthesis of the precursor of small RNAs (Djupedal et al. 2005), whereas Rpb2 is needed to generate small RNAs from the precursor RNA (Kato et al. 2005). The largest pol II subunit Rpb1, in particular, the CTD, is needed downstream from these steps to convert small RNAs into chromatin modifications (Schramke et al. 2005). An involvement of pol II in RNAi-mediated transcriptional silencing in human cells is suggested by the sensitivity of this process to α-amanitin (Weinberg et al. 2005). Thus, even though plants have evolved specialized pol IV complexes for RNA-directed DNA methylation, pol II appears to be able to carry out a related role in RNAi-mediated heterochromatin assembly in fission yeast and possibly humans. Presumably, pol IV complexes in plants are uniquely suited to dealing with the complexities of RNA-directed DNA methylation, as evidenced by their involvement in multiple stages of this process including not only de novo and maintenance methylation, but also possibly active demethylation.

CONCLUSIONS

The requirement for DNA-dependent RNA polymerases in gene silencing has been regarded as "paradoxical" (Vaughn and Martienssen 2005) and "counter-intuitive" (Huisinga et al. 2006). Yet the clear evidence for pol IV

involvement in RNAi-mediated chromatin-based silencing in plants has confirmed that these multisubunit enzymes are key players in epigenetic regulation. The elaboration of RNA-directed DNA methylation and the pol IV pathway in the plant kingdom probably reflects the need for flexible mechanisms of gene regulation that enable swift responses to an ever-changing environment. The likelihood that dispersed transposon promoters are agents of this epigenetic regulatory system validates Barbara McClintock's concept of transposons as "controlling elements" (Shapiro 2005). That the functions of pol IV complexes were first discerned in genetic analyses of transgene silencing attests to the fruitfulness of studying "foreign" DNA in plants (Matzke and Matzke 2004). Future work will address the many aspects of pol IV that remain to be clarified, including the mechanisms by which the two pol IV complexes act at discrete steps of the RNA-directed DNA methylation pathway, the complete catalog of target sequences, and the full range of functions in plant physiology and development.

ACKNOWLEDGMENTS

Financial support has been provided by the Austrian Fonds zur Förderung der wissenschaftlichen Forschung (grant numbers I26-B03 and P-16545-B12) and the European Union (contract number HPRN-CT 2002-00257).

REFERENCES

Arabidopsis Genome Initiative. 2000. Analysis of the genome sequence of the flowering plant *Arabidopsis thaliana*. *Nature* **408:** 796.

Aufsatz W., Mette M.F., Matzke A.J.M., and Matzke M. 2004. The role of MET1 in RNA-directed *de novo* and maintenance methylation of CG dinucleotides. *Plant Mol. Biol.* **54:** 793.

Aufsatz W., Mette M.F., van der Winden J., Matzke A.J.M., and Matzke M.A. 2002a. RNA-directed DNA methylation in *Arabidopsis*. *Proc. Natl. Acad. Sci.* **99:** 16499.

Aufsatz W., Mette M.F., van der Winden J., Matzke M., and Matzke A.J.M. 2002b. HDA6, a putative histone deacetylase needed to enhance DNA methylation induced by double stranded RNA. *EMBO J.* **21:** 6832.

Bellaoui M. and Gruissem W. 2004. Altered expression of the *Arabidopsis* ortholog of DCL affects normal plant development. *Planta* **219:** 819.

Bernstein E. and Allis C.D. 2005. RNA meets chromatin. *Genes Dev.* **19:** 1635.

Buratowski S. 2003. The CTD code. *Nat. Struct. Biol.* **10:** 679.

Cam H.P., Sugiyama T., Chen E.S., Chen X., FitzGerald P.C., and Grewal S.I. 2005. Comprehensive analysis of heterochromatin- and RNAi-mediated epigenetic control of the fission yeast genome. *Nat. Genet.* **37:** 809.

Cao X., Springer N.M., Muszynski M.G., Phillips R.L., Kaeppler S., and Jacobsen S.E. 2000. Conserved plant genes with similarity to mammalian *de novo* DNA methyltransferases. *Proc. Natl. Acad. Sci.* **97:** 4979.

Cao X., Aufsatz W., Zilberman D., Mette M.F., Huang M.S., Matzke M., and Jacobsen S.E. 2003. Role of the DRM and CMT3 methyltransferases in RNA-directed DNA methylation. *Curr. Biol.* **13:** 2212.

Carty S.M. and Greenleaf A.L. 2002. Hypophosphorylated C-terminal repeat domain-associating proteins in the nuclear proteome link transcription to DNA/chromatin modification and RNA processing. *Mol. Cell. Proteomics* **1:** 598.

Castanotto D., Tommasi S., Li M., Li H., Yanow S., Pfeifer G.P., and Rossi J.J. 2005. Short hairpin RNA-directed cytosine (CpG) methylation of the *RASSF1A* gene promoter in HeLa cells. *Mol. Ther.* **12:** 179.

Chan S.W.-L., Henderson I.R., and Jacobsen S.E. 2005. Gardening the genome: DNA methylation in *Arabidopsis thaliana*. *Nat. Rev. Genet.* **6:** 351.

Dalmay T., Hamilton A., Rudd S., Angell S., and Baulcombe D.C. 2000. An RNA-dependent RNA polymerase gene in *Arabidopsis* is required for posttranscriptional gene silencing mediated by a transgene but not by a virus. *Cell* **101:** 543.

Dennis K., Fan T., Geiman T., Yan Q., and Muegge K. 2001. Lsh, a member of the SNF2 family, is required for genome-wide methylation. *Genes Dev.* **15:** 2940.

Djupedal I., Portoso M., Spahr H., Bonilla C., Gustafsson C.M., Allshire R.C., and Ekwall K. 2005. RNA Pol II subunit Rpb7 promotes centromeric transcription and RNAi-directed chromatin silencing. *Genes Dev.* **19:** 2301.

Fuchs J., Demidov D., Houben A., and Schubert I. 2006. Chromosomal histone modification patterns—From conservation to diversity. *Trends Plant Sci.* **11:** 199.

Hampsey M. and Reinberg D. 2003. Tails of intrigue: Phosphorylation of RNA polymerase II mediates histone methylation. *Cell* **113:** 429.

Herr A.J. 2005. Pathways through the small RNA world of plants. *FEBS Lett.* **579:** 5879.

Herr A.J., Jensen M.B., Dalmay T., and Baulcombe D.C. 2005. RNA polymerase IV directs silencing of endogenous DNA. *Science* **308:** 118.

Huettel B., Kanno T., Daxinger L., Aufsatz W., Matzke A.J.M., and Matzke M. 2006. Endogenous targets of RNA-directed DNA methylation and Pol IV in *Arabidopsis*. *EMBO J.* **25:** 2828.

Huisinga K.L., Brower-Toland B., and Elgin S.C.R. 2006. The contradictory definitions of heterochromatin: Transcription and silencing. *Chromosoma* **115:** 110.

Jackson J.P., Lindroth A.M., Cao X., and Jacobsen S.E. 2002. Control of CpNpG DNA methylation by the KRYPTONITE histone H3 methyltransferase. *Nature* **416:** 556.

Jeddeloh J.A., Stokes T.L., and Richards E.J. 1999. Maintenance of genomic methylation requires a SWI2/SNF2-like protein. *Nat. Genet.* **22:** 94.

Jones L., Ratcliff F., and Baulcombe D.C. 2001. RNA-directed transcriptional gene silencing in plants can be inherited independently of the RNA trigger and requires Met1 for maintenance. *Curr. Biol.* **11:** 747.

Jones L., Hamilton A.J., Voinnet O., Thomas C.L., Maule A.J., and Baulcombe D.C. 1999. RNA-DNA interactions and DNA methylation in post-transcriptional gene silencing. *Plant Cell* **11:** 2291.

Kanno T., Aufsatz W., Jaligot E., Mette M.F., Matzke M., and Matzke A.J.M. 2005a. A SNF2-like protein facilitates dynamic control of DNA methylation. *EMBO Rep.* **6:** 649.

Kanno T., Mette M.F., Kreil D.P., Aufsatz W., Matzke M., and Matzke A.J.M. 2004. Involvement of putative SNF2 chromatin remodelling protein DRD1 in RNA-directed DNA methylation. *Curr. Biol.* **14:** 801.

Kanno T., Huettel B., Mette M.F., Aufsatz W., Jaligot E., Daxinger L., Kreil D.P., Matzke M., and Matzke A.J.M. 2005b. Atypical RNA polymerase subunits required for RNA-directed DNA methylation. *Nat. Genet.* **37:** 761.

Kapoor A., Agius F., and Zhu J.K. 2005. Preventing transcriptional gene silencing by active DNA demethylation. *FEBS Lett.* **579:** 5889.

Kashkush K., Feldman M., and Levy A.A. 2003. Transcriptional activation of retrotransposons alters the expression of adjacent genes in wheat. *Nat. Genet.* **33:** 102.

Kato H., Goto D.B., Martienssen R.A., Urano T., Furukawa K., and Murakami Y. 2005. RNA polymerase II is required for RNAi-dependent heterochromatin assembly. *Science* **309:** 467.

Kavi H.H., Xie W., Fernandez H.R., and Birchler J.A. 2005. Global analysis of siRNA-mediated transcriptional gene silencing. *BioEssays* **27:** 1209.

Kawasaki H. and Taira K. 2004. Induction of DNA methylation and gene silencing by short interfering RNAs in human cells. *Nature* **431:** 211.

Lahmy S., Guilleminot J., Cheng C.-M., Bechtold N., Albert S., Pelletier G., Delseny M., and Devic M. 2004. DOMINO1, a member of a small plant-specific gene family, encodes a protein essential for nuclear and nucleolar functions. *Plant J.* **39:** 809.

Lu C., Tej S.S., Luo S., Haudenschild C.D., Meyers B.C., and Green P.J. 2005. Elucidation of the small RNA component of the transcriptome. *Science* **309:** 467.

Malagnac F., Bartee L., and Bender J. 2002. An *Arabidopsis* SET domain protein required for maintenance but not establishment of DNA methylation. *EMBO J.* **21:** 6842.

Martienssen R.A. 2003. Maintenance of heterochromatin by RNA interference of tandem repeats. *Nat. Genet.* **35:** 213.

Mathieu O. and Bender J. 2004. RNA-directed DNA methylation. *J. Cell Sci.* **117:** 4861.

Matzke M.A. and Birchler J.A. 2005. RNAi-mediated pathways in the nucleus. *Nat. Rev. Genet.* **6:** 24.

Matzke M.A. and Matzke A.J.M. 2004. Planting the seeds of a new paradigm. *PLoS Biol.* **2:** E133.

Matzke M.A., Primig M., Trnovsky J., and Matzke A.J.M. 1989. Reversible methylation and inactivation of marker genes in sequentially transformed plants. *EMBO J.* **8:** 643.

Meinhart A., Kamenski T., Hoeppner S., Baumli S., and Cramer P. 2005. A structural perspective of CTD function. *Genes Dev.* **19:** 1401.

Mette M.F., van der Winden J., Matzke M.A., and Matzke A.J.M. 1999. Production of aberrant promoter transcripts contributes to methylation and silencing of unlinked homologous promoters *in trans*. *EMBO J.* **18:** 241.

Mette M.F., Aufsatz W., van der Winden J., Matzke M.A., and Matzke A.J.M. 2000. Transcriptional silencing and promoter methylation triggered by double stranded RNA. *EMBO J.* **19:** 5194.

Morales-Ruiz T., Ortega-Galisteo A.P., Ponferrada-Marin M.I., Martinez-Macias M.I., Ariza R.R., and Roldán-Aroja T. 2006. *DEMETER* and *REPRESSOR OF SILENCING 1* encode 5-methylcytosine DNA glycosylases. *Proc. Natl. Acad. Sci.* **103:** 6853.

Morris K.V. 2006. Therapeutic potential of siRNA-mediated transcriptional gene silencing. *Biotechniques* (suppl.) **2006:** 7.

Morris K.V., Chan S.W., Jacobsen S.E., and Looney D.J. 2004. Small interfering RNA-induced transcriptional gene silencing in human cells. *Science* **305:** 1289.

Onodera Y., Haag J.R., Ream T., Costa Nunes P., Pontes O., and Pikaard C.S. 2005. Plant nuclear RNA polymerase IV mediates siRNA and DNA methylation-dependent heterochromatin formation. *Cell* **120:** 613.

Park Y.D., Papp I., Moscone E.A., Iglesias V.A., Vaucheret H., Matzke A.J.M,. and Matzke M.A. 1996. Gene silencing mediated by promoter homology occurs at the level of transcription and results in meiotically heritable alterations in methylation and gene activity. *Plant J.* **9:** 183.

Pélissier T. and Wassenegger M. 2000. A DNA target of 30 bp is sufficient for RNA-directed DNA methylation. *RNA* **6:** 55.

Pélissier T., Thalmeir S., Kempe D., Sänger H.-L., and Wassenegger M. 1999. Heavy *de novo* methylation at symmetrical and non-symmetrical sites is a hallmark of RNA-directed DNA methylation. *Nucleic Acids Res.* **27:** 1625.

Peterson-Burch B.D., Nettleton D., and Voytas D.F. 2004. Genomic neighbourhoods for *Arabidopsis* retrotransposons: A role for targeted integration in the distribution of the Metaviridae. *Genome Biol.* **5:** R78.

Pontier D., Yahubyan G., Vega D., Bulski A., Saez-Vasquez J., Hakimi M.A., Lerbs-Mache S., Colot V., and Lagrange T. 2005. Reinforcement of silencing at transposons and highly repeated sequences requires the concerted action of two distinct RNA polymerases IV in *Arabidopsis*. *Genes Dev.* **19:** 2030.

Schramke V. and Allshire R. 2003. Hairpin RNAs and retrotransposon LTRs effect RNAi and chromatin-based gene silencing. *Science* **301:** 1069.

Schramke V., Sheedy D.M., Denli A.M., Bonila C., Ekwall K., Hannon G.J., and Allshire R.C. 2005. RNA-interference-directed chromatin modification coupled to RNA polymerase II transcription. *Nature* **435:** 1275.

Shapiro J.A. 2005. Retrotransposons and regulatory suites. *BioEssays* **27:** 122.

Sijen T., Vijn I., Rebocho A., van Blokland R., Roelofs D., Mol J.N.M., and Kooter J.M. 2001. Transcriptional and posttranscriptional gene silencing are mechanistically related. *Curr. Biol.* **11:** 436.

Sunkar R. and Zhu J.K. 2004. Novel and stress-regulated microRNAs and other small RNAs from *Arabidopsis*. *Plant Cell* **16:** 2001.

Suzuki K., Shijuuku T., Fukamachi T., Zaunders J., Guillemin G., Cooper D., and Kelleher A. 2005. Prolonged transcriptional silencing and CpG methylation induced by siRNAs targeted to the HIV-1 promoter region. *J. RNAi Gene Silencing* **1:** 66.

Ting A.H., Schuebel K.E., Herman J.G., and Baylin S.B. 2005. Short double-stranded RNA induces transcriptional gene silencing in human cancer cells in the absence of DNA methylation. *Nat. Genet.* **37:** 906.

Tran R., Zilberman D., de Bustos C., Ditt R.F., Henikoff J.G., Lindroth A.M., Delrow J., Boyle T., Kwong S., Bryson T.D., et al. 2005. Chromatin and siRNA pathways cooperate to maintain DNA methylation of small transposable elements in *Arabidopsis*. *Genome Biol.* **6:** R90.

Vaucheret H. 1993. Identification of a general silencer for 19S and 35S promoters in a transgenic tobacco plant: 90 bp of homology in the promoter sequence are sufficient for transinactivation. *C.R. Acad. Sci.* **316:** 1471.

———. 2005. RNA polymerase IV and transcriptional silencing. *Nat. Genet.* **37:** 659.

Vaughn M.W. and Martienssen R.A. 2005. Finding the right template: RNA Pol IV, a plant-specific RNA polymerase. *Mol. Cell* **17:** 754.

Wassenegger M., Heimes S., Riedel L., and Sänger H.-L. 1994. RNA-directed *de novo* methylation of genomic sequences in plants. *Cell* **76:** 567.

Weinberg M.S., Villeneuve L.M., Ehsani A., Aagaard L., Chen Z.X., Riggs A.D., Rossi J.J., and Morris K.V. 2005. The antisense strand of small interfering RNAs directs histone methylation and transcriptional gene silencing in human cells. *RNA* **12:** 256.

Xie Z., Johansen L.K., Gustafson A.M., Kasschau K.D., Lellis A.D., Zilberman D., Jacobsen S.E., and Carrington J.C. 2004. Genetic and functional diversification of small RNA pathways in plants. *PLoS Biol.* **2:** E104.

et al. 2003. Genome-wide insertional mutagenesis of *Arabidopsis thaliana*. *Science* **301**: 653.

Arabidopsis Genome Initiative. 2000. Analysis of the genome sequence of the flowering plant *Arabidopsis thaliana*. *Nature* **408**: 796.

Aufsatz W., Mette M.F., van der Winden J., Matzke A.J., and Matzke M. 2002. RNA-directed DNA methylation in *Arabidopsis*. *Proc. Natl. Acad. Sci.* **99**: 16499.

Baulcombe D. 2004. RNA silencing in plants. *Nature* **431**: 356.

Bernstein E., Caudy A.A., Hammond S.M., and Hannon G.J. 2001. Role for a bidentate ribonuclease in the initiation step of RNA interference. *Nature* **409**: 363.

Boeger H., Bushnell D.A., Davis R., Griesenbeck J., Lorch Y., Strattan J.S., Westover K.D., and Kornberg R.D. 2005. Structural basis of eukaryotic gene transcription. *FEBS Lett.* **579**: 899.

Bouche N., Lauressergues D., Gasciolli V., and Vaucheret H. 2006. An antagonistic function for *Arabidopsis* DCL2 in development and a new function for DCL4 in generating viral siRNAs. *EMBO J.* **25**: 3347.

Boudonck K., Dolan L., and Shaw P.J. 1999. The movement of coiled bodies visualized in living plant cells by the green fluorescent protein. *Mol. Biol. Cell* **10**: 2297.

Brodersen P. and Voinnet O. 2006. The diversity of RNA silencing pathways in plants. *Trends Genet.* **22**: 268.

Buhler M., Verdel A., and Moazed D. 2006. Tethering RITS to a nascent transcript initiates RNAi- and heterochromatin-dependent gene silencing. *Cell* **125**: 873.

Bushnell D.A. and Kornberg R.D. 2003. Complete, 12-subunit RNA polymerase II at 4.1-Å resolution: Implications for the initiation of transcription. *Proc. Natl. Acad. Sci.* **100**: 6969.

Cao X. and Jacobsen S.E. 2002. Role of the *Arabidopsis* DRM methyltransferases in de novo DNA methylation and gene silencing. *Curr. Biol.* **12**: 1138.

Cao X., Aufsatz W., Zilberman D., Mette M.F., Huang M.S., Matzke M., and Jacobsen S.E. 2003. Role of the DRM and CMT3 methyltransferases in RNA-directed DNA methylation. *Curr. Biol.* **13**: 2212.

Carmell M.A., Xuan Z., Zhang M.Q., and Hannon G.J. 2002. The Argonaute family: Tentacles that reach into RNAi, developmental control, stem cell maintenance, and tumorigenesis. *Genes Dev.* **16**: 2733.

Cioce M. and Lamond A.I. 2005. Cajal bodies: A long history of discovery. *Annu. Rev. Cell Dev. Biol.* **21**: 105.

Cramer P., Bushnell D.A., and Kornberg R.D. 2001. Structural basis of transcription: RNA polymerase II at 2.8 angstrom resolution. *Science* **292**: 1863.

Deleris A., Gallego-Bartolome J., Bao J., Kasschau K.D., Carrington J.C., and Voinnet O. 2006. Hierarchical action and inhibition of plant Dicer-like proteins in antiviral defense. *Science* **313**: 68.

Djupedal I., Portoso M., Spahr H., Bonilla C., Gustafsson C.M., Allshire R.C., and Ekwall K. 2005. RNA Pol II subunit Rpb7 promotes centromeric transcription and RNAi-directed chromatin silencing. *Genes Dev.* **19**: 2301.

Doyle O., Corden J.L., Murphy C., and Gall J.G. 2002. The distribution of RNA polymerase II largest subunit (RPB1) in the *Xenopus* germinal vesicle. *J. Struct. Biol.* **140**: 154.

Gall J.G. 2001. A role for Cajal bodies in assembly of the nuclear transcription machinery. *FEBS Lett.* **498**: 164.

Gasciolli V., Mallory A.C., Bartel D.P., and Vaucheret H. 2005. Partially redundant functions of *Arabidopsis* DICER-like enzymes and a role for DCL4 in producing trans-acting siRNAs. *Curr. Biol.* **15**: 1494.

Gnatt A.L., Cramer P., Fu J., Bushnell D.A., and Kornberg R.D. 2001. Structural basis of transcription: An RNA polymerase II elongation complex at 3.3 Å resolution. *Science* **292**: 1876.

Grewal S.I. and Moazed D. 2003. Heterochromatin and epigenetic control of gene expression. *Science* **301**: 798.

Grewal S.I. and Rice J.C. 2004. Regulation of heterochromatin by histone methylation and small RNAs. *Curr. Opin. Cell Biol.* **16**: 230.

Handwerger K.E. and Gall J.G. 2006. Subnuclear organelles: New insights into form and function. *Trends Cell Biol.* **16**: 19.

Hannon G.J. 2002. RNA interference. *Nature* **418**: 244.

Henderson I.R., Zhang X., Lu C., Johnson L., Meyers B.C., Green P.J., and Jacobsen S.E. 2006. Dissecting *Arabidopsis thaliana* DICER function in small RNA processing, gene silencing and DNA methylation patterning. *Nat. Genet.* **38**: 721.

Herr A.J., Jensen M.B., Dalmay T., and Baulcombe D.C. 2005. RNA polymerase IV directs silencing of endogenous DNA. *Science* **308**: 118.

Hunter C., Sun H., and Poethig R.S. 2003. The *Arabidopsis* heterochronic gene ZIPPY is an ARGONAUTE family member. *Curr. Biol.* **13**: 1734.

Jackson J.P., Lindroth A.M., Cao X., and Jacobsen S.E. 2002. Control of CpNpG DNA methylation by the KRYPTONITE histone H3 methyltransferase. *Nature* **416**: 556.

Jady B.E., Richard P., Bertrand E., and Kiss T. 2006. Cell cycle-dependent recruitment of telomerase RNA and Cajal bodies to human telomeres. *Mol. Biol. Cell* **17**: 944.

Kanno T., Mette M.F., Kreil D.P., Aufsatz W., Matzke M., and Matzke A.J. 2004. Involvement of putative SNF2 chromatin remodeling protein DRD1 in RNA-directed DNA methylation. *Curr. Biol.* **14**: 801.

Kanno T., Huettel B., Mette M.F., Aufsatz W., Jaligot E., Daxinger L., Kreil D.P., Matzke M., and Matzke A.J. 2005. Atypical RNA polymerase subunits required for RNA-directed DNA methylation. *Nat. Genet.* **37**: 761–765.

Kato H., Goto D.B., Martienssen R.A., Urano T., Furukawa K., and Murakami Y. 2005. RNA polymerase II is required for RNAi-dependent heterochromatin assembly. *Science* **309**: 467.

Kawasaki H. and Taira K. 2004. Induction of DNA methylation and gene silencing by short interfering RNAs in human cells. *Nature* **431**: 211.

Li C.F., Pontes O., El-Shami M., Henderson I.R., Bernatavichute Y.V., Chan S.W.-L., Lagrange T., Pikaard C.S., and Jacobsen S.E. 2006. An ARGONAUTE4-containing nuclear processing center colocalized with Cajal bodies in *Arabidopsis thaliana*. *Cell* **126**: 93.

Lippman Z. and Martienssen R. 2004. The role of RNA interference in heterochromatic silencing. *Nature* **431**: 364.

Lippman Z., May B., Yordan C., Singer T., and Martienssen R. 2003. Distinct mechanisms determine transposon inheritance and methylation via small interfering RNA and histone modification. *PLoS Biol.* **1**: E67.

Liu J.L., Murphy C., Buszczak M., Clatterbuck S., Goodman R., and Gall J.G. 2006. The *Drosophila melanogaster* Cajal body. *J. Cell Biol.* **172**: 875.

Makeyev E.V. and Bamford D.H. 2002. Cellular RNA-dependent RNA polymerase involved in posttranscriptional gene silencing has two distinct activity modes. *Mol. Cell* **10**: 1417.

Matera A.G. and Shpargel K.B. 2006. Pumping RNA: Nuclear bodybuilding along the RNP pipeline. *Curr. Opin. Cell Biol.* **18**: 317.

Minakhin L., Bhagat S., Brunning A., Campbell E.A., Darst S.A., Ebright R.H., and Severinov K. 2001. Bacterial RNA polymerase subunit omega and eukaryotic RNA polymerase subunit RPB6 are sequence, structural, and functional homologs and promote RNA polymerase assembly. *Proc. Natl. Acad. Sci.* **98**: 892.

Morgan G.T., Doyle O., Murphy C., and Gall J.G. 2000. RNA polymerase II in Cajal bodies of amphibian oocytes. *J. Struct. Biol.* **129**: 258.

Morris K.V., Chan S.W., Jacobsen S.E., and Looney D.J. 2004. Small interfering RNA-induced transcriptional gene silencing in human cells. *Science* **305**: 1289.

Motamedi M.R., Verdel A., Colmenares S.U., Gerber S.A., Gygi S.P., and Moazed D. 2004. Two RNAi complexes, RITS and RDRC, physically interact and localize to noncoding centromeric RNAs. *Cell* **119**: 789.

Murphy C., Wang Z., Roeder R.G., and Gall J.G. 2002. RNA polymerase III in Cajal bodies and lampbrush chromosomes of the *Xenopus* oocyte nucleus. *Mol. Biol. Cell* **13**: 3466.

Noma K., Sugiyama T., Cam H., Verdel A., Zofall M., Jia S., Moazed D., and Grewal S.I. 2004. RITS acts in *cis* to promote RNA interference-mediated transcriptional and post-transcriptional silencing. *Nat. Genet.* **36:** 1174.

Onodera Y., Haag J.R., Ream T., Costa Nunes P., Pontes O., and Pikaard C.S. 2005. Plant nuclear RNA polymerase IV mediates siRNA and DNA methylation-dependent heterochromatin formation. *Cell* **120:** 613.

Pal-Bhadra M., Leibovitch B.A., Gandhi S.G., Rao M., Bhadra U., Birchler J.A., and Elgin S.C. 2004. Heterochromatic silencing and HP1 localization in *Drosophila* are dependent on the RNAi machinery. *Science* **303:** 669.

Pelissier T. and Wassenegger M. 2000. A DNA target of 30 bp is sufficient for RNA-directed DNA methylation. *RNA* **6:** 55.

Peragine A., Yoshikawa M., Wu G., Albrecht H.L., and Poethig R.S. 2004. SGS3 and SGS2/SDE1/RDR6 are required for juvenile development and the production of *trans*-acting siRNAs in *Arabidopsis*. *Genes Dev.* **18:** 2368.

Pontes O., Li C.F., Costa Nunes P., Haag J., Ream T., Vitins A., Jacobsen S.E., and Park M.Y. 2006. The *Arabidopsis* chromatin-modifying nuclear siRNA pathway involves a nucleolar RNA processing center. *Cell* **126:** 79.

Pontier D., Yahubyan G., Vega D., Bulski A., Saez-Vasquez J., Hakimi M.A., Lerbs-Mache S., Colot V., and Lagrange T. 2005. Reinforcement of silencing at transposons and highly repeated sequences requires the concerted action of two distinct RNA polymerases IV in *Arabidopsis*. *Genes Dev.* **19:** 2030.

Richards E.J. and Elgin S.C. 2002. Epigenetic codes for heterochromatin formation and silencing: Rounding up the usual suspects. *Cell* **108:** 489.

Sasaki T., Shiohama A., Minoshima S., and Shimizu N. 2003. Identification of eight members of the Argonaute family in the human genome small star, filled. *Genomics* **82:** 323.

Schauer S.E., Jacobsen S.E., Meinke D.W., and Ray A. 2002. DICER-LIKE1: Blind men and elephants in *Arabidopsis* development. *Trends Plant Sci.* **7:** 487.

Schiebel W., Haas B., Marinkovic S., Klanner A., and Sanger H.L. 1993. RNA-directed RNA polymerase from tomato leaves. II. Catalytic in vitro properties. *J. Biol. Chem.* **268:** 11858.

Sentenac A. 1985. Eukaryotic RNA polymerases. *Crit. Rev. Biochem.* **18:** 31.

Shopland L.S., Byron M., Stein J.L., Lian J.B., Stein G.S., and Lawrence J.B. 2001. Replication-dependent histone gene expression is related to Cajal body (CB) association but does not require sustained CB contact. *Mol. Biol. Cell* **12:** 565.

Sleeman J.E. and Lamond A.I. 1999. Newly assembled snRNPs associate with coiled bodies before speckles, suggesting a nuclear snRNP maturation pathway. *Curr. Biol.* **9:** 1065.

Sontheimer E.J. and Carthew R.W. 2004. Molecular biology. Argonaute journeys into the heart of RISC. *Science* **305:** 1409.

Sweetser D., Nonet M., and Young R.A. 1987. Prokaryotic and eukaryotic RNA polymerases have homologous core subunits. *Proc. Natl. Acad. Sci.* **84:** 1192.

Tamaru H. and Selker E.U. 2001. A histone H3 methyltransferase controls DNA methylation in *Neurospora crassa*. *Nature* **414:** 277.

Tomlinson R.L., Ziegler T.D., Supakorndej T., Terns R.M., and Terns M.P. 2006. Cell cycle-regulated trafficking of human telomerase to telomeres. *Mol. Biol. Cell* **17:** 955.

Vaucheret H. 2006. Post-transcriptional small RNA pathways in plants: Mechanisms and regulations. *Genes Dev.* **20:** 759.

Vaucheret H., Vazquez F., Crete P., and Bartel D.P. 2004. The action of ARGONAUTE1 in the miRNA pathway and its regulation by the miRNA pathway are crucial for plant development. *Genes Dev.* **18:** 1187.

Vazquez F., Vaucheret H., Rajagopalan R., Lepers C., Gasciolli V., Mallory A.C., Hilbert J.L., Bartel D.P., and Crete P. 2004. Endogenous *trans*-acting siRNAs regulate the accumulation of *Arabidopsis* mRNAs. *Mol. Cell* **16:** 69.

Verdel A., Jia S., Gerber S., Sugiyama T., Gygi S., Grewal S.I., and Moazed D. 2004. RNAi-mediated targeting of heterochromatin by the RITS complex. *Science* **303:** 672.

Volpe T.A., Kidner C., Hall I.M., Teng G., Grewal S.I., and Martienssen R.A. 2002. Regulation of heterochromatic silencing and histone H3 lysine-9 methylation by RNAi. *Science* **297:** 1833.

Wassenegger M. 2000. RNA-directed DNA methylation. *Plant Mol. Biol.* **43:** 203.

Wassenegger M. and Krczal G. 2006. Nomenclature and functions of RNA-directed RNA polymerases. *Trends Plant Sci.* **11:** 142.

Wassenegger M., Heimes S., Riedel L., and Sanger H.L. 1994. RNA-directed de novo methylation of genomic sequences in plants. *Cell* **76:** 567.

Westover K.D., Bushnell D.A., and Kornberg R.D. 2004. Structural basis of transcription: Separation of RNA from DNA by RNA polymerase II. *Science* **303:** 1014.

Woychik N.A., Liao S.-M., Kolodziej P.A., and Young R.A. 1990. Subunits shared by eukaryotic nuclear RNA polymerases. *Genes Dev.* **4:** 313.

Xie Z., Johansen L.K., Gustafson A.M., Kasschau K.D., Lellis A.D., Zilberman D., Jacobsen S.E., and Carrington J.C. 2004. Genetic and functional diversification of small RNA pathways in plants. *PLoS Biol.* **2:** E104.

Zilberman D., Cao X., and Jacobsen S.E. 2003. ARGONAUTE4 control of locus-specific siRNA accumulation and DNA and histone methylation. *Science* **299:** 716.

Zilberman D., Cao X., Johansen L.K., Xie Z., Carrington J.C., and Jacobsen S.E. 2004. Role of *Arabidopsis* ARGONAUTE4 in RNA-directed DNA methylation triggered by inverted repeats. *Curr. Biol.* **14:** 1214.

A Paragenetic Perspective on Integration of RNA Silencing into the Epigenome and Its Role in the Biology of Higher Plants

R.A. Jorgensen, N. Doetsch, A. Müller, Q. Que, K. Gendler, and C.A. Napoli

Department of Plant Sciences, University of Arizona, Tucson, Arizona 85721-0036

We describe features of RNA silencing and associated epigenetic imprints that illustrate potential roles for RNA interference (RNAi) in maintenance and transmission of epigenetic states between cells, throughout a plant, and perhaps even across sexual generations. Three types of transgenes can trigger RNAi of homologous endogenous plant genes: (1) "sense" transgenes that overexpress translatable transcripts, (2) inverted repeat (IR) transgenes that produce double-stranded RNA (dsRNA), and (3) antisense transgenes. Each mode of RNAi produces a different characteristic developmental silencing pattern. Single-copy transgenes are sufficient for sense-RNAi and antisense-RNAi, but not inverted repeat-RNAi. A single premature termination codon dramatically attenuates sense-RNAi, but it has no effect on antisense or inverted repeat-RNAi. We report here that antisense transgenes altered by removal of nonsense codons generate silencing patterns characteristic of sense-RNAi. Duplication of a sense overexpression transgene results in two types of epigenetic events: (1) complete loss of silencing and (2) altered developmental pattern of silencing. We also report that duplicating only the transgene promoter results in complete loss of silencing, whereas duplicating only transcribed sequences produces the second class, which are vein-based patterns. We infer that the latter class is due to systemic RNA silencing signals that interact with certain epigenetic states of the transgene to imprint it with information generated at a distance elsewhere in the plant.

R. Alexander Brink (1960) proposed that chromosomes possess two functions, one genetic and the other "paragenetic," the genetic components being stable and the paragenetic components being labile and programmable in ontogeny. Brink proposed calling the stable genetic components "orthochromatin" and the labile paragenetic components "parachromatin." Unlike euchromatin and heterochromatin, which are defined as alternative states and described as residing at distinct locations in the genome, orthochromatin (which Brink equated with DNA) and parachromatin (which Brink described as chromatin–protein complexes) are intimately associated with each other throughout the genome. The function of orthochromatin is to stably maintain and transmit genetic information, whereas the function of parachromatin is to mediate gene expression and to store programmable information regarding the history of a cell lineage at individual genetic loci. Thus, the terms paragenetic and parachromatin simply refer to epigenetic states that are chromosomally based. Brink consciously chose to propose the term paragenetic rather than use the term epigenetic because of the latter's broader and less specific definition in Waddington's (1942) original usage.

Today it is well established that the molecular basis for the lability of chromatin at individual loci is chromatin remodeling complexes that establish and maintain histone modifications, DNA modifications, nucleosome structure and positioning, and higher-order chromatin states (for review, see Mellor 2006). The recent recognition that components of the RNAi machinery participate in the establishment and maintenance of (at least some) chromatin states has expanded concepts of chromatin organization to include a broader role for noncoding RNA (ncRNA) molecules than was previously understood. ncRNA is an extremely abundant and diverse class of molecules, including not only microRNA (miRNA) and its precursors, but also antisense transcripts that often "readthrough" into annotated coding sequences, opposing "normal" sense transcripts. The resulting formation of dsRNA recruits the nuclear RNAi machinery to the locus, which then precipitates heterochromatin formation, involving histone and DNA modifications (Matzke and Birchler 2005).

Here, we describe features of RNA silencing and associated epigenetic imprints at chimeric *Chalcone synthase* (*Chs*) transgene loci in petunia flowers that illustrate potential roles for RNAi in the maintenance and transmission of metastable states between cells, throughout a plant, and perhaps even across sexual generations.

SENSE-RNAi AND THE ROLE OF mRNA TRANSLATION

Three types of transgenes can trigger RNAi in plants: (1) "sense" transgenes that have been engineered to overproduce a translatable transcript, (2) inverted repeat (or "hairpin") transgenes that have been engineered to produce dsRNA transcripts, and (3) antisense transgenes, which are engineered to produce a transcript that is complementary to its target transcript (Jorgensen 2003). We refer to these three modes of transgene-induced RNAi as sense-RNAi, inverted repeat-RNAi, and antisense-RNAi, respectively. Sense-RNAi, originally known as sense cosuppression, results from attempted overexpression of a host protein (Napoli et al. 1990) and requires high levels of transcription of a translatable transcript (Que et al. 1997). A single premature termination codon (PTC) is sufficient to dramatically attenuate this mode of RNAi, indicating the importance of translation of the full-length coding region of the sense transcript. Sense-RNAi also requires a host RNA-dependent RNA polymerase (RdRP), which recognizes and copies overexpressed transcripts to produce the dsRNA molecules necessary to trigger RNAi (Beclin et al. 2002). Typical sense overexpression transgenes carry only the amino-acid-coding region of the target transcript and lack the target's 5'- and 3'-untranslated regions (UTRs), as well as introns. However, none of these elements are necessary for silencing by single-copy transgenes (Q. Que et al., unpubl.).

Sense transgenes that have not been engineered for protein overproduction (e.g., see van der Krol et al. 1990) induce RNAi only when they are integrated in the genome as inverse repeats of the T-DNA element that carries the sense transgene and the selective marker from *Agrobacterium* into the plant genome (Que et al. 1997; Stam et al. 1997). Readthrough transcription of these long inverse repeats of T-DNA is thought to produce the double-stranded "foldback" RNA that triggers RNAi (Muskens et al. 2000). Because inverse repeat integrants may also produce dsRNA corresponding to the transgene promoter, stochastic epigenetic loss of silencing frequently occurs (Stam et al. 1998).

Single-copy antisense transgenes can trigger RNAi despite possessing multiple nonsense codons in all reading frames. Instead, they are thought to trigger RNAi by pairing with complementary target transcripts to form dsRNA. Silencing of anthocyanin pigment genes in petunia flowers by sense overexpression transgenes is visually distinct from antisense silencing. Sense-RNAi occurs as a developmental silencing pattern that is centered on the fusion zones between adjacent flower petals (Fig. 1a),

whereas antisense transgenes never produce this pattern, instead producing palely pigmented flowers that may be blotchy and/or have white petal edges (Fig. 1b) (Napoli et al. 1990; Jorgensen et al. 1996; Que et al. 1998).

The inability of an antisense transgene to produce the developmental silencing pattern of sense overexpression transgenes could be due to the fact that the antisense transcript—even though it possesses a translation start codon in the same optimal translation context as the corresponding sense overexpression transcript—possesses several premature termination codons. To test this possibility, we engineered an antisense overexpression transgene by altering three nonsense codons to sense codons in the antisense strand of a 246-codon open reading frame (ORF) from a 740-bp segment of the petunia chalcone-synthase-coding region; 20% of the transformants carrying this transgene exhibited the developmental silencing pattern characteristic sense-RNAi superimposed on the typical antisense phenotype (Fig. 1c,d). A control antisense construct retaining the three nonsense codons, and thus lacking a long ORF, did not produce any flowers with the sense-RNAi pattern.

An interesting technical aside is that the "nonsense-codons-removed" translatable antisense construct was more efficient than the control antisense plant at producing plants with completely white flowers. The nonsense-codon-removed antisense approach offers a potentially useful alternative to standard sense-RNAi and inverted repeat-RNAi. Silencing target genes by standard sense-RNAi constructs has been avoided in crop genetic engineering because of the possibility that epigenetic loss of

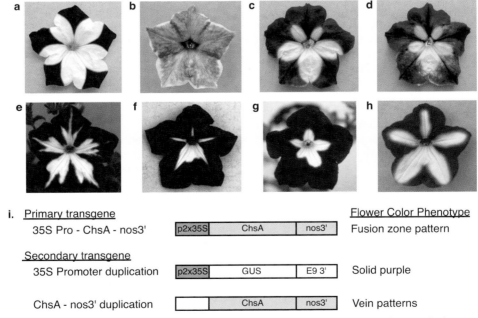

Figure 1. RNA silencing by *Chs* transgenes in petunia flowers. (*a*) Sense overexpression transgene (e.g., see "primary transgene" in *i*). (*b*) Antisense *Chs* transgene driven by 2×35S. (*c,d*) Antisense *Chs* transgene with premature termination codons removed (see text). (*e,f,g*) Three classes of vein-based patterns resulting from interaction between two transgene copies. (*h*) Vein-based pattern induced by transcribed sequence duplication (see *i*, "*ChsA–nos3'* duplication"). (*i*) *Chs* transgene constructs. 2×35S is the cauliflower mosaic virus promoter with two copies of the upstream "enhancer" region. *ChsA* refers to the coding sequence of the chalcone synthase gene. *nos3'* refers to the 3'-untranslated region of the *Agrobacterium tumefaciens nopaline synthase* (*nos*) gene, which contains polyadenylation signals. (*f,g,* Reprinted, with permission, from Jorgensen 1995 [© AAAS].)

the silent state would also result in overexpression of the target gene product by the sense transgene, as we have shown can occur with the petunia *Hf1* gene (Jorgensen et al. 2002). In contrast, inducing sense-RNAi with a translatable antisense transgene leaves no possibility of unintended overexpression of the target gene product.

SENSE-TRANSGENE DUPLICATION INDUCES NOVEL EPIGENETIC STATES THAT INTERACT WITH SYSTEMIC SILENCING SIGNALS

Duplication of a chalcone synthase transgene results in two classes of epigenetic change: (1) a complete loss of RNA silencing, resulting in uniformly purple flowers, and (2) a change in the morphological features that control the pattern of silencing, resulting in patterns of silencing based on the vasculature, i.e., vein-based patterns (Fig. 1e,f,g) (Jorgensen et al. 1996; Que and Jorgensen 1998). At least three heritably distinct types of vein-based patterns have been recognized (Jorgensen 1995; Jorgensen and Napoli 1996), indicating that multiple types of epigenetic events occur which are responsible for this class of silencing phenotypes.

By duplicating only the promoter of the overexpression transgene or by duplicating only its transcribed sequences (Fig. 1i), we found that the two classes of epigenetic changes are separable: (1) promoter duplication is responsible for the first class, loss of RNA silencing, and (2) duplication of the transcribed sequences is responsible for the second class, vein-based silencing patterns (example shown in Fig. 1h).

Interestingly, promoter duplication resulted in loss of silencing only with inverted repeat integrants of the secondary transgene (promoter-only construct). Presumably, this is because the loss of silencing is due to promoter-derived dsRNA, and transgene inverted repeats are necessary for production of this dsRNA by transcriptional readthrough and transcript foldback, as discussed above. In contrast, single-copy integrants were sufficient to induce epigenetic changes when only transcribed sequences were duplicated (resulting in vein-based silencing patterns). Importantly, the vein-based silencing pattern state can also be generated by allelic interactions, demonstrating that ectopic location of the second transgene copy is not required. It is not clear how single-copy "promoterless" *Chs* transgenes induce epigenetic events; however, it is likely that such genes are transcribed from neighboring promoters, even if at low levels, and so it is conceivable that transcripts mediate epigenetic alteration of the target transgene, perhaps in RdRP-dependent fashion. In addition, the possibility remains that a direct DNA–DNA interaction between the two transgene copies is responsible.

Paramutation

Genetic segregation of the two transgene copies by outcrossing showed that the new vein-based silencing patterns could persist in the absence of the "promoterless" *Chs* transgene; thus, these epigenetic events are paramutations, according to Brink's broader, early definition. Interestingly, segregation of inverted repeat promoter transgenes away from the target sense overexpression *Chs* transgene also resulted in some plants with vein-based patterns. We would suggest that these vein-based patterns are due to "spreading" of the epigenetic state from the promoter of the sense transgene into the adjacent transcribed sequences, accompanied by loss or reduction in DNA methylation at the promoter in the absence of the inverted repeat promoter transgene.

Paramutation at the *b* locus in maize is RNAi-mediated via an upstream array of tandem repeats (Alleman et al. 2006). Given that duplication of transcribed *Chs* sequences is sufficient to generate paramutations, as shown here, paramutation is probably not limited to transcriptional effects, nor to the presence of upstream repeats. Thus, an interesting possibility to consider is that naturally occurring alleles or paralogs of endogenous genes can interact via their transcribed regions to produce paramutation-like alterations in the chromatin of either or both gene copies.

Systemic RNA Silencing Signals

It is now well established that RNA and protein molecules can be trafficked through plasmodesmata in a regulated fashion and can even enter or leave the phloem stream via plasmodesmal connections to the phloem companion cells (Lucas and Lee 2004). This trafficking system has been referred to as the "RNA information superhighway" because it is potentially capable of long-distance transport of large numbers of informational macromolecules (Jorgensen et al. 1998). RNA silencing is also transmissible through plasmodesmata and the phloem and can spread through the tissues of an organ, as well as throughout the plant (Vaucheret 2006).

Our observation that duplication of transcribed sequences is responsible for patterns of silencing based on the vasculature suggests that these patterns are due to RNA silencing signals that emanate from the phloem of the major veins and move into the surrounding tissues via plasmodesmata. Two possibilities exist, either (1) the new epigenetic event (a paramutation) caused the transgene to create a phloem-transmissible RNA silencing signal that did not previously exist or (2) the new epigenetic state of the transgene is responsive to an RNA signal that was already present (or both). Because of the diversity of distinct, heritable epigenetic states that can arise from gene duplication (Fig. 1e,f,g), there could be (1) different types of RNA signals produced by the transgene depending on its epigenetic state, perhaps deriving from different or overlapping segments of the transcribed sequences, and/or (2) different responses, or efficiencies of response, of the transgene to the silencing signal that are determined by the epigenetic state of the transgene. Either way, it seems clear that epigenetic (paragenetic) states and systemic RNA silencing signals can interact.

RNA-DIRECTED HETEROCHROMATINIZATION IN PLANTS

Higher plants possess several novel classes of proteins that interact with the RNAi machinery in the formation of heterochromatin-like states. These include two novel types of DNA methyltransferases (chromomethyltransferase and domains-rearranged methyltransferase), a plant-specific clade of SNF2 ATPases, a large and diverse plant-specific clade of SET domain proteins related to *Drosophila Su(var)3-9* and *Schizosaccharomyces pombe clr4* and characterized by a novel plant-specific SRA-YDG domain, plant-specific RNA polymerase subunits, which comprise a novel RNA polymerase IV, and plant-specific clades of Argonaute, Dicer-like, and RdRP proteins that participate in chromatin alterations and not miRNA silencing (Matzke and Birchler 2005).

Although plants, animals, and fungi share fundamental mechanisms of chromatin modification, such as histone methylation and acetylation, DNA methylation, ATP-dependent nucleosome remodeling, and RNAi-mediated heterochromatin formation, higher plants appear to possess a more diversified complement of chromatin proteins involved in all these processes which may have been a basis for the evolution of novel plant-specific properties and functions of chromatin. Importantly, many of these novel forms of chromatin proteins are involved in RNA-mediated alteration of chromatin, suggesting that higher plants could have a special capacity for chromatin remodeling involving RNA molecules (R.A. Jorgensen et al., in prep.).

CONCLUSIONS

R.A. Brink's (1960) suggestion that chromosomes possess a paragenetic function in addition to their genetic function and that the physical nature of this paragenetic function is a variety of forms or states of chromatin that can reside at any genetic locus is especially intriguing when considered in light of two important features of higher plants: (1) the evolutionary diversification of eukaryotic chromatin that appears to involve plant-specific mechanisms based on RNA and a diversified RNAi machinery, and (2) the capacity of plants to traffic informational macromolecules (RNA and protein) cell-to-cell and systemically in a regulated manner (known as the "RNA information superhighway"). Given these capabilities, it is at least conceivable that higher plants may possess the capacity to store information at numerous genetic loci in the form of "paragenetic" chromatin states and that these states can be reprogrammed during ontogeny or in response to the environment.

As was shown here, paragenetic states appear to interact with systemic RNA silencing signals. Systemic trafficking of RNA (and protein) molecules could permit integration of paragenetic information over the whole of the organism, as well as differentiation of information states in different parts of the organism. Given the possibility of these events occurring at many thousands of genetic loci, such a system could operate as a high-capacity storage device that is reprogrammable during the life of the organism and reset each sexual generation.

If plants do possess such a system for information processing and storage that can integrate and assess information perhaps in large amounts and at both cellular and organismal levels, it would imply that plants might then be capable of making somewhat informed "decisions" to,

for instance, fine-tune gene expression states during growth and development and in physiological responses to the environment. Such a means for information processing and decision making might be considered a form of "intelligence," but it would be one that is fundamentally different from the form of intelligence that evolved separately in animals for different purposes. For instance, timescales for the responsiveness of plant "intelligence" would be much slower than that of animal intelligence, given that plants generally have no possibility of relocating in response to the environment, whether in search of nutrients or to avoid predators. Instead, the "intelligence" of plants would be expected to address much longer timescales (e.g., diurnal) than are dictated by predator–prey interactions in animals. Furthermore, in long-lived perennials, plants' "memories" of past events could theoretically last for many years, and so could the implications of their "decisions." And finally, because plants do not sequester their germ lines, they also might conceivably pass on some of their paragenetically based memories to their offspring, possibly increasing their chances of survival and reproductive success.

ACKNOWLEDGMENTS

We thank Jin Wang and Jin-Xia Wu for their excellent technical assistance. This work was funded by grants to R.A.J. from the U.S. Department of Energy, Energy Biosciences Program (Award DE-FG03-98ER20308) and the National Science Foundation Plant Genome Research Program (Award 0421679), as well as a USDA postdoctoral fellowship to N.D.

REFERENCES

Alleman M., Sidorenko L., McGinnis K., Seshadri V., Dorweiler J.E., White J., Sikkink K., and Chandler V.L. 2006. An RNA-dependent RNA polymerase is required for paramutation in maize. *Nature* **442:** 295.

Brink R.A. 1960. Paramutation and chromosome organization. *Q. Rev. Biol.* **35:** 120.

Beclin C., Boutet S., Waterhouse P., and Vaucheret H. 2002. A branched pathway for transgene-induced RNA silencing in plants. *Curr. Biol.* **12:** 684.

Jorgensen R.A. 1995. Cosuppression, flower color patterns, and metastable gene expression states. *Science* **268:** 686.

———. 2003. Sense cosuppression in plants: Past, present, and future. In *RNAi: A guide to gene silencing* (ed. G.J. Hannon), p. 5. Cold Spring Harbor Laboratory Press, Cold Spring Harbor, New York.

Jorgensen R.A. and Napoli C.A. 1996. A responsive regulatory system is revealed by sense suppression of pigment genes in *Petunia* flowers. In *Genomes: Proceedings of the 22nd Stadler Genetics Symposium* (ed. J.P. Gustafson and R.B. Flavell), p. 159. Plenum Press, New York.

Jorgensen R.A., Que Q., and Napoli C.A. 2002. Maternally controlled ovule abortion results from cosuppression of dihydroflavonol-4-reductase or flavonoid-3′, 5′-hydroxylase genes in *Petunia hybrida. Funct. Plant Biol.* **29:** 1501.

Jorgensen R.A., Atkinson R.G., Forster R.L., and Lucas W.J. 1998. An RNA-based information superhighway in plants. *Science* **279:** 1486.

Jorgensen R.A., Cluster P.D., English J., Que Q., and Napoli C.A. 1996. Chalcone synthase cosuppression phenotypes in petunia flowers: Comparison of sense vs. antisense constructs and single-copy vs. complex T-DNA sequences. *Plant Mol. Biol.* **31:** 957.

Lucas W.J. and Lee J.Y. 2004. Plasmodesmata as a supracellular control network in plants. *Nat. Rev. Mol. Cell Biol.* **5:** 712.

Matzke M.A. and Birchler J.A. 2005. RNAi-mediated pathways in the nucleus. *Nat. Rev. Genet.* **6:** 24.

Mellor J. 2006. The dynamics of chromatin remodeling at promoters. *Mol. Cell* **19:** 147.

Muskens M.W.M., Vissers A.P.A., Mol J.N.M., and Kooter J.M. 2000. Role of inverted DNA repeats in transcriptional and post-transcriptional gene silencing. *Plant Mol. Biol.* **43:** 243.

Napoli C., Lemieux C., and Jorgensen R. 1990. Introduction of a chimeric chalcone synthase gene into petunia results in reversible co-suppression of homologous genes *in trans*. *Plant Cell* **2:** 279.

Que Q. and Jorgensen R.A. 1998. Homology-based control of gene expression patterns in transgenic petunia flowers. *Dev. Genet.* **22:** 100.

Que Q., Wang H.-Y., and Jorgensen R.A. 1998. Distinct patterns of pigment suppression are produced by allelic sense and antisense chalcone synthase transgenes in petunia flowers. *Plant J.* **13:** 401.

Que Q., Wang H.-Y., English J., and Jorgensen R.A. 1997. The frequency and degree of cosuppression by sense chalcone synthase transgenes are dependent on transgene promoter strength and are reduced by premature nonsense codons in the transgene coding sequence. *Plant Cell* **9:** 1357.

Stam M., Viterbo A., Mol J.N.M., and Kooter J.M. 1998. Position-dependent methylation and transcriptional silencing of transgenes in inverted T-DNA repeats: Implications for posttranscriptional silencing of homologous host genes in plants. *Mol. Cell. Biol.* **18:** 6165.

Stam M., De Bruin R., Kenter S., Van der Hoorn R.A.L., Van Blokland R., Mol J.N.M., and Kooter J.M. 1997. Post-transcriptional silencing of chalcone synthase in *Petunia* by inverted transgene repeats. *Plant J.* **12:** 63.

van der Krol A.R., Mur L.A., Beld M., Mol J.N.M., and Stuitje A.R. 1990. Flavonoid genes in petunia: Addition of a limited number of gene copies may lead to a suppression of gene expression. *Plant Cell* **2:** 291.

Vaucheret H. 2006. Post-transcriptional small RNA pathways in plants: Mechanisms and regulations. *Genes Dev.* **20:** 759.

Waddington C.H. 1942. The epigenotype. *Endeavour* **1:** 18.

RNAi-mediated Heterochromatin Assembly in Fission Yeast

M. Zofall and S.I.S. Grewal

Laboratory of Molecular Cell Biology, National Cancer Institute, National Institutes of Health,
Bethesda, Maryland 20892

The organization of DNA into heterochromatin domains is critical for a variety of chromosomal functions, including gene silencing, recombination suppression, and chromosome segregation. In fission yeast, factors involved in the RNAi pathway such as Argonaute, Dicer, and RNA-dependent RNA polymerase are required for assembly of heterochromatin structures. The RNAi Argonaute-containing RITS complex and RNA-dependent RNA polymerase localize throughout heterochromatin domains. These factors are important components of a self-reinforcing loop mechanism operating *in cis* to process repeat transcripts into siRNAs, which involve in heterochromatin assembly. In this paper, we describe our results suggesting that slicing of repeat transcripts by the Argonaute is an important step in their conversion into siRNAs and heterochromatic silencing. Mutations in conserved residues known to be essential for slicer activity of Argonautes result in loss of siRNAs corresponding to centromeric repeats, accumulation of repeat transcripts, and defects in heterochromatin assembly. We also discuss our recent finding that heterochromatin proteins such as Swi6/HP1 serve as a platform that could recruit both silencing and anti-silencing factors to heterochromatic loci.

Eukaryotic genomes contain large tracts of repetitive DNA sequences that are often packaged in the form of condensed, inaccessible heterochromatic structures (Hall and Grewal 2003). The assembly of heterochromatic structures at different loci regulates a variety of chromosomal processes. In addition to transcriptional repression, heterochromatin prohibits recombination throughout large chromosomal domains. These repressive effects of heterochromatin are widely believed to reflect the need for organisms to prohibit the transcription and proliferation of transposable DNA elements, which, due to their ability to transpose or recombine with other elements, are a major cause of genomic instability (Birchler et al. 2000; Henikoff 2000; Hall and Grewal 2003). Heterochromatin has also been shown to play an important role in regulating gene expression, ensuring proper segregation of chromosomes, and facilitating long-range chromatin interactions between distant chromosomal regions (Bernard et al. 2001; Nonaka et al. 2002; Jia et al. 2004b; Pidoux and Allshire 2004). How heterochromatin mediates diverse cellular functions is the focus of intense research in different systems.

The study of heterochromatin assembly and its various functions is particularly amenable to genetic and biochemical analyses in the model organism fission yeast *Schizosaccharomyces pombe*. Several *trans*-acting factors involved in heterochromatin formation in fission yeast are conserved in higher eukaryotic species (Grewal and Elgin 2002; Cam and Grewal 2004) , and recent studies strongly suggest that similar mechanisms of heterochromatin assembly might operate in fission yeast and mammals (Grewal and Elgin 2002; Hall and Grewal 2003; Hiragami and Festenstein 2005; Ebert et al. 2006; Huisinga et al. 2006). In particular, an evolutionarily conserved Clr4 protein—the fission yeast homolog of mammalian Suv39h—has been shown to specifically methylate lysine 9 of histone H3 (H3K9), which serves as a binding site for recruitment of chromodomain proteins, including Swi6, a structural and functional homolog of mammalian heterochromatin protein 1 (HP1), to heterochromatic loci (Lorentz et al. 1992; Ivanova et al. 1998; Rea et al. 2000; Thon and Verhein-Hansen 2000; Bannister et al. 2001; Nakayama et al. 2001; Sadaie et al. 2004). Recent work has led to the surprising discovery of a critical role of the RNA interference (RNAi) pathway in the nucleation and assembly of heterochromatic structures at specific loci of the fission yeast genome (Hall et al. 2002; Volpe et al. 2002). Significant progress has been made in elucidating the role of small RNAs and the RNAi machinery in the assembly of heterochromatic structures (Motamedi et al. 2004; Noma et al. 2004; Verdel et al. 2004; Matzke and Birchler 2005; Sugiyama et al. 2005). These studies suggest that the RNAi machinery operates as a stable component of heterochromatin domains to promote both transcriptional and posttranscriptional silencing of heterochromatic sequences. In this paper, we describe our recent progress in understanding the mechanism of heterochromatin assembly and silencing in fission yeast.

HETEROCHROMATIC REGIONS IN THE FISSION YEAST GENOME

Fission yeast chromosomes contain large blocks of heterochromatin associated with a variety of repeat elements, in addition to small heterochromatic "islands" associated with a few meiosis-specific genes (Fig. 1) (Hall and Grewal 2003; Cam et al. 2005). Mapping of heterochromatin markers such as H3K9 methylation and Swi6 has identified extended chromosomal domains coated with heterochromatin complexes at centromeres, telomeres, and the mating-type locus (Partridge et al. 2000; Cam et al. 2005). All three heterochromatic regions share a common feature— each of these domains contains *dg/dh* repeat elements that are preferential targets of RNAi-mediated heterochromatin formation (Chikashige et al. 1989; Hahnenberger et al. 1991; Grewal and Klar 1997; Hall et al. 2002; Volpe et al. 2002; Cam and Grewal 2004; Kanoh et al. 2005).

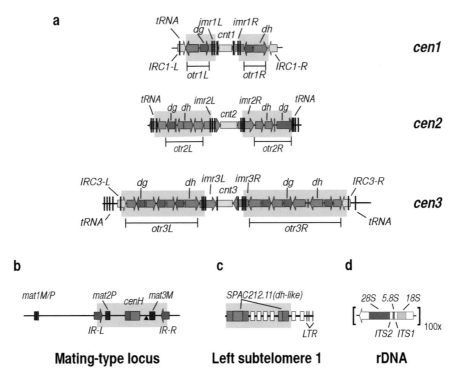

Figure 1. Heterochromatin domains in the fission yeast genome. Schematic diagrams of the major heterochromatic regions in the fission yeast genome, such as centromeres, telomeres, mating-type region, and *rDNA* locus, are shown. The regions coated with heterochromatin markers (such as H3K9me and Swi6) and RNAi machinery are shaded in pink. (*a*) Each centromere contains a unique central core (*cnt*), flanked by innermost inverted repeat (*imr*) and outer repeat region (*otr*), composed of a complex arrangement of tandem *dg* and *dh* repeats. Clusters of *tRNA* genes and/or *IRC* inverted repeats surround heterochromatin domains at centromeres. (*b*) At the mating-type region, *mat2* and *mat3* genes are located within a 20-kb heterochromatin domain exhibiting both suppression of transcription and recombination. A *cenH* element that shares strong homology with *dg* and *dh* centromeric repeats serves as an RNAi-dependent heterochromatin nucleation center. Heterochromatin can also be nucleated independently at a site between *cenH* and *mat3* (*black triangle*) that contains binding sites for Atf1/Pcr1. The heterochromatic domain is restricted by boundary elements *IR-L* and *IR-R*. (*c*) *cenH*-like elements (SPAC212.11) are also present at subtelomeric regions which display a broad distribution of heterochromatin and RNAi machinery. (*d*) Heterochromatin and RNAi factors also associate with an *rDNA* repeat, although the mechanism of heterochromatin assembly at these loci is not well understood. Small heterochromatic "islands" are also associated with certain meiosis-specific genes such as *mei4* and *ssm4* (not shown).

At centromeres, which in fission yeast are large complex structures ranging in size from 35 kb to 110 kb, tandem arrays of *dg/dh* repeat elements are important constituents of pericentromeric regions that surround the central core (*cnt*) domain (Clarke et al. 1986; Chikashige et al. 1989; Hahnenberger et al. 1991). Heterochromatin assembled at pericentromeric repeat sequences has been shown to be critical for preferential recruitment of cohesin, which is essential for proper segregation of chromosomes during cell division (Bernard et al. 2001; Nonaka et al. 2002). Distinct *trans*-acting factors interact with the *cnt* domain that primarily is enriched for factors essential for kinetochore assembly such as histone H3 variant CENP-A and Mis6 protein (Saitoh et al. 1997; Partridge et al. 2000; Takahashi et al. 2000; Bjerling and Ekwall 2002). In addition to CENP-A, low levels of histone H3 methylated at lysine 4 (H3K4me) are also present at the *cnt* domain (Cam et al. 2005), although the biological significance of H3K4me present at this site is not known.

A copy of a *dg/dh*-like element referred to as *cenH* is also present at the silent mating-type region (Grewal and Klar 1997). *cenH* is essential for establishment of hete-

rochromatin throughout a 20-kb chromosomal domain containing *mat2* and *mat3* loci (Grewal and Klar 1997; Nakayama et al. 2000; Hall et al. 2002), which serve as donors of genetic information for the active *mat1* locus during mating-type switching (Klar 1989). Detailed analyses have uncovered the existence of at least two redundant mechanisms for the initial targeting of heterochromatin to the mating-type locus. In addition to the RNAi machinery acting through the *cenH* element (Hall et al. 2002), DNA-binding factors Atf1/Pcr1 belonging to the ATF/CREB protein family mediate heterochromatin nucleation at a distinct site (Jia et al. 2004a; Kim et al. 2004). Once nucleated, heterochromatin can spread across the entire silent mating-type interval, a process that depends on Swi6, Clr4, and histone deacetylases (HDACs) (Hall et al. 2002; Noma et al. 2004; Yamada et al. 2005). Heterochromatin assembled at the mating-type locus performs multiple functions. In addition to the well-recognized function of heterochromatin in transcriptional silencing of developmentally important mating-type genes, it is also essential for recombinational suppression, which is believed to be critical for protecting the genomic integrity of the locus (Egel 1984; Klar and Bonaduce 1991; Thon et al. 1994; Grewal and Klar 1997).

Heterochromatin also mediates spreading of a protein complex, involved in mating-type switching, from an enhancer element located adjacent to the *mat3* locus to across the entire heterochromatic domain (Jia et al. 2004b). Importantly, this heterochromatin-mediated spreading occurs in a cell-type-specific manner only in *M* cells but not in *P* cells, which forms the basis for differential utilization of *mat2* or *mat3* loci during mating-type (*mat1*) switching (Jia et al. 2004b).

Sequences that share homology with centromeric repeats are also found embedded within subtelomeric regions. A partial or full-length copy of a RecQ-like gene (*SPAC212.11*) that has a *cenH*-like element in its coding region is believed to be essential for RNAi-mediated heterochromatin formation across subtelomeric regions (Cam et al. 2005; Kanoh et al. 2005; Hansen et al. 2006). Redundant mechanisms of heterochromatin nucleation are also operating at telomeres. In addition to RNAi machinery, a telomere repeat-binding protein Taz1, belonging to the TRF family (Cooper et al. 1997; Nimmo et al. 1998), is able to independently nucleate heterochromatin (Kanoh et al. 2005; Hansen et al. 2006). Although deletions of RNAi factors alone have no detectable effect on heterochromatin-mediated silencing at telomeres, due to redundancy in heterochromatin assembly mechanisms, mutations in RNAi factors result in defects in clustering of telomeres (Hall et al. 2003). This observation has led to suggestions that siRNAs promote higher-order chromatin organization by acting as a "glue" to hold dispersed repeats into a common structure (Hall et al. 2003).

Heterochromatin complexes are also found distributed at tandem *rDNA* repeats, which are also targets of RNAi machinery (Cam et al. 2005). The mechanism of heterochromatin assembly and silencing at *rDNA* in fission yeast remains to be fully explored. Recent studies suggest that RNAi and heterochromatin machineries are required for gene silencing and maintenance of genomic integrity at these loci (Thon and Verhein-Hansen 2000; Shankaranarayana et al. 2003; Bjerling et al. 2004; Cam et al. 2005).

In addition to repeat elements present at major heterochromatic regions, the fission yeast genome also contains several other repeat sequences, such as TF2 retrotransposons, WTF repeats, and solo long terminal repeats (LTRs) (Bowen et al. 2003). These repeat elements are not coated with heterochromatin components H3K9me and Swi6 (Cam et al. 2005). Therefore, it seems that RNAi and heterochromatin machineries selectively target a specialized class of repeat elements, in particular *dg* and *dh* repeat elements found at all major heterochromatic loci.

HETEROCHROMATIN DOMAIN BOUNDARIES

A remarkable feature of heterochromatin protein complexes is their ability to spread *in cis*, resulting in epigenetic silencing of adjacent loci (Grewal and Moazed 2003; Talbert and Henikoff 2006). However, in their natural chromosomal contexts, heterochromatin domains are often surrounded by boundary DNA elements that prevent inappropriate spreading of repressive chromatin into euchromatic regions (Sun and Elgin 1999; West et al. 2002). Analysis of heterochromatin distribution in the fission yeast genome has revealed the existence of boundary elements surrounding the heterochromatin domains at all three centromeres and the silent mating-type locus (Partridge et al. 2000; Noma et al. 2001; Thon et al. 2002; Cam et al. 2005; Scott et al. 2006). Identical inverted repeat (*IR*) elements surrounding the silent mating-type locus serve as heterochromatin domain boundaries (Fig. 1) (Noma et al. 2001; Thon et al. 2002). A marked decrease in H3K9me and Swi6 localization is observed coincident with the presence of *IR* elements (Noma et al. 2001). Moreover, deletions of these elements result in spreading of heterochromatin to surrounding euchromatin domains (Noma et al. 2001). A sharp decrease in H3K9me and Swi6 levels is also observed at the pericentromeric heterochromatin domain boundaries, in which case clusters of *tRNA* genes and/or inverted repeat elements, referred to as *IRCs*, have been suggested to serve as boundary elements (Fig. 1) (Partridge et al. 2000; Cam et al. 2005; Noma et al. 2006; Scott et al. 2006).

Our recent analyses suggest that *IR*s and *tRNA* genes might share a common boundary mechanism. We found that the core boundary sequences within *IR* elements contain B-boxes (Noma et al. 2006), the high-affinity binding sites for the RNA polymerase III (pol III) transcription initiation factor TFIIIC, which is known to be important for transcription of *tRNA*s (Huang et al. 2000; Huang and Maraia 2001). The binding of TFIIIC to *IR*s is essential for boundary function (Noma et al. 2006). In addition, TFIIIC binding to *tRNA* gene clusters surrounding pericentromeric regions coincides precisely with a sharp decrease in H3K9me and Swi6 (Cam et al. 2005; Noma et al. 2006). A distinctive feature of *IR* elements, however, is that TFIIIC binds to these loci without recruiting pol III. High levels of TFIIIC without pol III are also present at the so-called chromosome organizing clamps (*COC*s) that are tethered to the nuclear peripheral compartment, at which TFIIIC is concentrated into 5–10 bodies (Noma et al. 2006). According to a model, TFIIIC-bound sequences tether these loci to the nuclear periphery, thus creating a barrier against the spread of heterochromatin, in addition to facilitating higher-order genome organization.

Interestingly, TFIIIC does not bind to the *IRC* elements that surround *cen1* and *cen3*. Instead, these elements show preferential enrichment for a JmjC domain-containing antisilencing factor Epe1 (Zofall and Grewal 2006; see below), which we find is essential for transcription and barrier function of the *IRC* boundaries (Zofall and Grewal 2006).

HISTONE-MODIFYING ENZYMES AND HETEROCHROMATIC TRANSCRIPTIONAL SILENCING

Factors involved in modifications of histones have been shown to play an important role in the assembly of heterochromatic structures. In addition to the Clr4, which exists in a cullin 4 (Cul4)–Rik1-based E3 ubiquitin ligase complex (Hong et al. 2005; Horn et al. 2005; Jia et al. 2005; Thon et al. 2005), multiple HDACs (Sir2, Clr3, and Clr6) have been implicated in heterochromatin formation in fission yeast (Grewal et al. 1998; Freeman-Cook et al. 1999; Shankaranarayana et al. 2003; Ekwall 2005). Among these, Sir2 is an NAD-dependent histone deacetylase (Freeman-Cook et al. 1999; Shankaranarayana et al. 2003), whereas Clr3 and Clr6 belong to class 1 and class 2 family histone deacetylases, respectively (Grewal et al. 1998; Bjerling et al. 2002). Current evidence suggests multiple roles for HDACs in the establishment and maintenance of heterochromatic structures. Apart from setting up a histone modification pattern ("histone code") essential for localization of heterochromatin proteins, such as Swi6, HDAC proteins have been shown to be important for nucleation and spreading of heterochromatin complexes. This is best exemplified by studies at the mating-type region, in which case Clr3 HDAC cooperates with DNA-binding factors Atf1/Pcr1 to nucleate heterochromatin by mediating the targeting of Clr4 in addition to facilitating the assembly of "closed" chromatin across an entire heterochromatin domain (Jia et al. 2004a; Yamada et al. 2005).

Heterochromatin is believed to be highly condensed and inaccessible to factors involved in different aspects of DNA metabolism, including transcriptional machinery (Grewal and Elgin 2002). However, the mechanism of transcriptional silencing by heterochromatin is not fully understood. The oligomerization of Swi6/HP1 molecules bound to H3K9me via their chromoshadow domain (Brasher et al. 2000; Cowieson et al. 2000) might facilitate chromatin condensation, resulting in silencing of underlying sequences. Nonetheless, recent studies have shown that Swi6/HP1 associates with target loci in a highly dynamic fashion (Cheutin et al. 2003, 2004; Festenstein et al. 2003). Our recent work suggests that Swi6 and another chromodomain protein, Chp2, bound to H3K9me provide a loading platform for localization of HDAC Clr3, which in turn facilitates establishment of "closed" chromatin structure refractory to transcriptional machinery including RNA polymerase II (Yamada et al. 2005). In other words, chromodomain proteins bound to the H3K9me "anchor" serve as adapters for recruitment and/or spreading of regulatory proteins such as HDAC Clr3, to silence heterochromatic sequences, presumably through establishment of higher-order chromatin structures (Yamada et al. 2005).

RNAi-MEDIATED HETEROCHROMATIC SILENCING

RNAi was first described as a posttranscriptional silencing mechanism in which double-stranded RNA (dsRNA) triggers destruction of cognate RNAs (Fire et al. 1998). In this mechanism, long dsRNA is processed into siRNAs ranging in size from about 21 to 24 nucleotides by an RNase III enzyme Dicer (Hannon 2002). siRNAs are incorporated into an RNA-induced silencing complex (RISC) that, among other factors, contains Argonaute, which serves as a catalytic engine to mediate siRNA-guided degradation of target mRNA (Liu et al. 2004; Meister et al. 2004; Rand et al. 2004). In some organisms, RNA-dependent RNA polymerases (RdRPs) have been shown to be important for RNAi silencing (Meister and

Tuschl 2004). As mentioned above, studies in fission yeast have implicated RNAi machinery, including Dicer (Dcr1), Argonaute (Ago1), and RdRP (Rdp1) in heterochromatic silencing (Hall et al. 2002; Reinhart and Bartel 2002; Volpe et al. 2002). Further genetic and biochemical studies have provided significant insights into the mechanism of RNAi-mediated heterochromatin assembly. An RNAi-induced initiation of transcriptional silencing (RITS) complex has been identified (Verdel et al. 2004). RITS contains Ago1, Chp1, and a protein named Tas3. In addition, RITS also contains siRNAs derived from *dg/dh* repeats present at different heterochromatic loci (Verdel et al. 2004; Cam et al. 2005). Genome-mapping analyses have shown that components of RITS and Rdp1 are distributed throughout heterochromatic domains in a pattern almost identical to H3K9me and Swi6 distribution (Cam et al. 2005). Moreover, stable binding of RITS to chromatin depends at least in part on the binding of Chp1 chromodomain to H3K9me (Noma et al. 2004). Deletion of an H3K9-specific methyltransferase Clr4 or Chp1 chromodomain results in delocalization of RITS from heterochromatic loci and concurrent defects in processing of repeat transcripts into siRNAs (Noma et al. 2004; Cam et al. 2005). RITS is also essential for the recruitment of the RDRC complex, which contains Rdp1, the RdRP activity of which is essential for siRNA production and heterochromatin assembly (Motamedi et al. 2004; Sugiyama et al. 2005).

We have proposed that RNAi-mediated heterochromatin assembly occurs via a self-reinforcing loop mechanism (Fig. 2) (Noma et al. 2004; Sugiyama et al. 2005). According to this model, siRNAs and/or DNA-binding factors mediate initial targeting of heterochromatin factors such as Clr4 to methylate H3K9. H3K9me allows stable binding of RITS across heterochromatin domains, which in turn mediates recruitment of RDRC to facilitate processing of repeat transcripts into siRNAs by Dcr1. siRNAs produced *in cis* can feed back to target additional heterochromatin complexes. The exact mechanism by which siRNAs target chromatin modifications is not clear. It has been suggested that RITS tethered to nascent transcripts via siRNAs might mediate recruitment of chromatin-modifying activities such as Clr4 or that siRNAs directly facilitate recruitment of the Clr4 complex to heterochromatic repeats (Verdel et al 2004; Jia et al. 2005).

ARGONUATE SLICER FUNCTION IS ESSENTIAL FOR siRNA PRODUCTION AND HETEROCHROMATIC SILENCING

Argonaute proteins are defined by the presence of two conserved domains: PAZ (piwi-argonaute-zwille) and PIWI (P-element induced wimpy testis). Although the PAZ domain can be found in both Dicer and Argonaute proteins, the PIWI domain is present only in Argonaute proteins (Meister and Tuschl 2004). Recent studies have shown that the PIWI domain is structurally related to RNase H ribonucleases (Song et al. 2004; Yuan et al. 2005), and biochemical mutagenesis analyses of mammalian Argonaute 2 protein showed that the cryptic

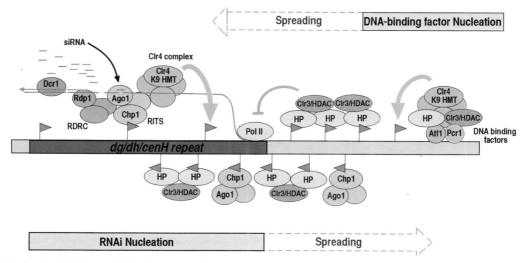

Figure 2. Redundant RNAi- and DNA-based mechanisms contribute to heterochromatin assembly. Sequence-specific DNA-binding proteins such as Atf1/Pcr1 cooperate with heterochromatin assembly factors including HDAC Clr3 to recruit Clr4 histone methyltransferase (HMT) to a specific site, and RNAi machinery targets heterochromatin factors to the *dg/dh/cenH* repeat elements. RNAi-mediated heterochromatin assembly involves processing of repeat transcripts by RITS, RDRC, and Dicer into siRNAs that are believed to mediate targeting of Clr4. Methylation of H3K9 by Clr4 (indicated by red flags) is not only necessary for RITS to stay stably associated with heterochromatic loci via Chp1 chromodomain, but it also recruits Swi6 and Chp2 (HPs), which, among other factors, mediate spreading of Clr3 HDAC implicated in limiting pol II accessibility to heterochromatic sequences. Spreading of H3K9me-Swi6 also allows RNAi machinery to spread beyond the initial nucleation site, potentially providing an opportunity to exert control over sequences incapable of nucleating heterochromatin.

RNase H domain is essential for Slicer activity of the RISC complex (Liu et al. 2004; Meister et al. 2004; Rand et al. 2004).

We sought to explore whether the endonucleolytic activity associated with the PIWI domain of Argonaute proteins is required for heterochromatic silencing. The aspartate residues 580 and 651 of a single fission yeast Ago1 protein were identified by a structure-alignment approach as a part of the highly conserved catalytic carboxylate triad "DDE," known to be conserved in all RNase H fold nucleases (Fig. 3a) (Yang and Steitz 1995; Song et al. 2004; Yuan et al. 2005). We constructed fission yeast strains carrying mutations in either of two conserved aspartate residues (i.e., D580A or D651A). Immunoblotting assay showed that mutant protein levels are comparable to that of wild-type Ago1 protein (Fig. 3c). However, we found that mutations abolished silencing of a *ura4⁺* reporter gene (*otr1::ura4⁺*) inserted at the pericentromeric repeats region of *cen1* (Fig. 3b). To test whether the loss of silencing in mutant cells is caused by defects in heterochromatin assembly at centromeres, we performed chromatin immunoprecipitation (ChIP) assays to examine the status of H3K9me and Swi6 localization at *otr1::ura4⁺*. These analyses revealed that localization of both H3K9me and Swi6 at centromeric *otr1::ura4⁺* was severely affected in both *ago1^{D580A}* and *ago1^{D651A}* mutants (Fig. 4c). On the basis of these analyses, we conclude that the catalytic activity of Ago1 protein might be essential for the heterochromatin assembly at centromeres.

We next investigated whether *ago1^{D580A}* and *ago1^{D651A}* mutations also affect silencing and heterochromatin assembly at the naturally silenced *dg/dh* centromeric

repeat elements. Reverse transcriptase-polymerase chain reaction (RT-PCR) analyses of total RNA showed that transcripts originating from the *dg* centromeric repeat elements could be readily detected in the *ago1* mutant strains (Fig. 4b). We also observed increased accumulation of transcripts derived from *cenH* in *ago1^{D580A}* and *ago1^{D651A}* mutants even though comparable levels of wild-type and mutant Ago1 proteins could be ChIPed at this site (our unpublished data). These results suggest that the processing of repeat transcripts into heterochromatic siRNAs requires the conserved catalytic aspartate residues of Ago1 PIWI domain. Indeed, although siRNAs could be readily detected in wild-type background cells, there were no detectable siRNAs present in mutant background cells (Fig. 4a). Intriguingly, we found that mutations in Ago1 had only a minor effect on H3K9me and Swi6 localization at the repeat elements (Fig. 4d), despite severely reduced levels of H3K9me and Swi6 observed at the *otr1::ura4⁺* reporter construct (Fig. 4c). Significant levels of H3K9me and Swi6 remaining at the repeat elements in *ago1* mutants are consistent with our previous results showing that an RNAi-independent pathway involving Clr3 HDAC contributes to heterochromatin nucleation by mediating targeting of Clr4 to centromeric repeats (Yamada et al. 2005). Based on these results and previous studies (Sadaie et al. 2004; Yamada et al. 2005), cells defective in siRNA production can target H3K9me to centromeric repeat elements, but siRNA production is necessary for spreading of heterochromatin complexes into the *otr1::ura4⁺* gene. How might siRNAs facilitate local spreading of heterochromatin? In light of evidence showing that assembly of heterochromatic structures requires RNAs (Maison et al. 2002; Muchardt et al. 2002; Hall et al. 2003), one pos-

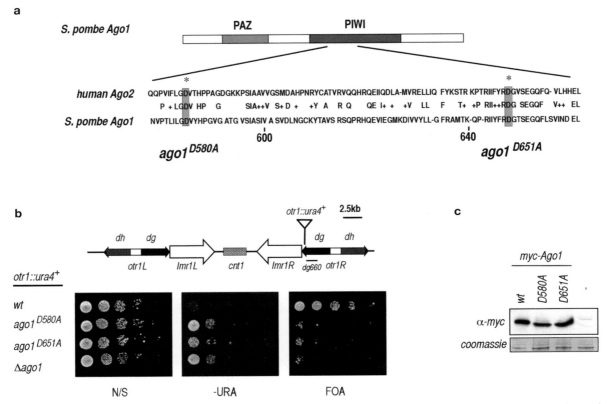

Figure 3. Mutations in conserved Ago1 residues implicated in "slicer" activity of Argonaute family proteins affect heterochromatin silencing. (*a*) Alignment of PIWI domains of human Ago2 and *S. pombe* Ago1 proteins is shown. The aspartate residues highlighted in red are known to be essential for slicing activity of human Ago2 protein. (*b*) Mutations of conserved aspartate residues (D580A and D651A) abolish heterochromatic silencing at centromeres. Effects of mutations on silencing of a *ura4*[+] reporter gene inserted at outer pericentromeric repeats (*otr1::ura4*[+]) were investigated by performing a serial dilution plating assay. Tenfold serial dilutions of indicated cultures were plated on nonselective (N/S), uracil-deficient (-ura), and counterselective (5-FOA) medium and allowed to grow at 32°C for 2 days. (*c*) Expression of wild-type and mutant Ago1 proteins was compared by western blot analysis. Whole-cell extracts prepared from cells expressing wild-type or mutant proteins tagged at the amino terminus by a myc epitope tag were examined by western blotting with anti-myc(A14) antibody (Santa Cruz Biotechnology).

sibility is that siRNAs produced *in cis* by the RNAi machinery have a structural role in organizing repeat elements into higher-order condensed structures that, in addition to mediating silencing of the repeat elements, facilitate local spreading of heterochromatin complexes.

Swi6/HP1 MEDIATES RECRUITMENT OF BOTH SILENCING AND ANTISILENCING FACTORS

RNA pol II transcribes centromeric *dg/dh* repeats even in the presence of heterochromatin (Cam et al. 2005; Djupedal et al. 2005). Since transcription of repeats is essential for generation of siRNA precursors, mutations in pol II impair RNAi-mediated heterochromatin assembly (Djupedal et al. 2005; Kato et al. 2005). However, the mechanism by which pol II gains access to sequences heavily coated with heterochromatin was not clear. Our recent work revealed that a JmjC domain-containing protein Epe1 facilitates pol II accessibility and transcription of heterochromatic repeats (Zofall and Grewal 2006). Epe1 is required for transcription of repeats specifically in a heterochromatin context. However, in mutant cells defective in heterochromatin silencing, Epe1 is dispens-

able for transcription of repeats. Thus, it seems that Epe1 is specifically required for pol II transcription in the presence of repressive heterochromatin.

Epe1 was first identified in a screen for factors that negatively regulate the integrity of heterochromatin (Ayoub et al. 2003). Whereas loss of Epe1 enhances heterochromatic silencing and results in efficient spreading of heterochromatin into euchromatic regions, its overexpression destabilizes heterochromatin that correlates with increased histone acetylation and H3K4me (Ayoub et al. 2003). We found that Epe1 is paradoxically enriched throughout heterochromatic domains (Zofall and Grewal 2006). More surprisingly, the recruitment of Epe1 to heterochromatic loci is dependent on its interaction with Swi6. Epe1 binds directly to Swi6 in vitro, and these proteins form a complex in vivo. These observations, together with our recent work showing that Swi6 recruits Clr3 HDAC to restrict pol II accessibility to heterochromatic repeats (Yamada et al. 2005), suggest that Swi6 serves as an oscillator of heterochromatic transcription by directing recruitment of both silencing and antisilencing factors, the balance of which might be critical for proper maintenance of heterochromatic structures (Fig. 5).

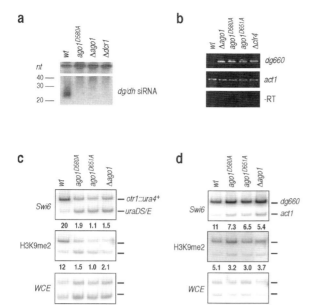

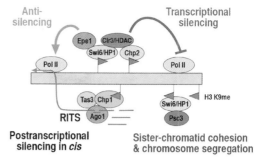

Figure 5. Binding of chromodomain proteins to H3K9me establishes a platform for recruitment of distinct activities. H3K9me nucleated by DNA-binding factors or RNAi pathway tethers chromodomain proteins Chp1, Chp2, and Swi6/HP1, which mediate targeting of distinct activities involved in heterochromatin maintenance. In addition to recruiting cohesin, which promotes proper chromosome segregation and chromatid cohesion, Swi6/HP1 targets opposing activities of HDAC Clr3 and jmjC protein Epe1. Whereas HDAC Clr3 limits pol II accessibility and transcription, Epe1 facilitates transcription and promotes pol II association with heterochromatic regions. Chromodomain protein Chp1 is a component of RITS that allows RNAi machinery to act *in cis* to process repeat transcripts into siRNAs.

Figure 4. Effects of mutations in conserved PIWI domain of Ago1 on siRNA production, repeat silencing, and heterochromatin formation. (*a*) Ago1 mutants are defective in siRNA production. 20 μg of small RNA fraction was resolved on 15% urea-PAGE gel and transferred to Hybond^{N+} membrane. Membrane was hybridized with single-stranded RNA probes transcribed with α-^{32}P-labeled UTP and hydrolyzed to lengths averaging ~50 nucleotides (Hamilton and Baulcombe 1999). (*b*) Mutations in the PIWI domain of Ago1 result in accumulation of transcripts derived from centromeric repeats. The levels of *dg* repeat transcripts were analyzed by RT-PCR using total RNA prepared from indicated strains. The same RNA samples were also used to amplify control *act1* transcript in the presence and absence of reverse transcriptase. PCR products were separated on agarose gel and were visualized by ethidium bromide staining. (*c*, *d*) Loss of silencing in Ago1 mutants correlates with defects in heterochromatin assembly. Chromatin immunoprecipitation (ChIP) analysis was used to measure levels of Swi6 or H3K9me at *dg* centromeric repeats (*d*) or *otr1::ura4+* (*c*). ChIP was performed and analyzed as described previously (Nakayama et al. 2000) using antibodies against Swi6 and dimethylated H3K9 (Upstate Biotechnology).

The mechanism by which Epe1 counteracts heterochromatic silencing is not known. Since several JmjC domain proteins have been shown to catalyze histone demethylation (Tsukada et al. 2005; Fodor et al. 2006; Klose et al. 2006; Whetstine et al. 2006), it is formally possible that Epe1 affects heterochromatin stability through removal of the repressive lysine methylation mark. However, no such activity has been detected for Epe1 (Tsukada et al. 2005). A possibility remains that Epe1 modulates chromatin via a yet undefined mechanism.

CONCLUDING REMARKS

Studies from several systems suggest an important role for RNAi machinery in assembly of repressive chromatin structures (Mochizuki and Gorovsky 2004; Chan et al. 2005; Matzke and Birchler 2005). Recent investigation into the role of RNAi in heterochromatin formation in fission

yeast has shown that RNAi factors localize throughout heterochromatin domains and that their localization across these domains is mediated by components of heterochromatin such as H3K9me and Swi6/HP1 (Noma et al. 2004; Cam et al. 2005; Sugiyama et al. 2005). These studies suggest that RNAi-mediated heterochromatin assembly occurs via a self-reinforcing loop mechanism in which chromatin-associated RITS complex serves as a core for stable binding of RNAi factors, including RdRP Rdp1, which cooperates to process repeat transcripts into siRNAs, essential for further strengthening of heterochromatic structures (Noma et al. 2004; Sugiyama et al. 2005). Rdp1 is a functional RNA-dependent RNA polymerase, and its RdRP activity is essential for generation of siRNAs (Motamedi et al. 2004; Sugiyama et al. 2005). Similarly, Dicer is required for the processing of repeat transcripts into siRNAs. However, the precise function of the Ago1 protein remained unexplored. Our analysis suggests that mutations in conserved Ago1 residues that are known to be essential for Slicer activity of Argonaute family proteins (Liu et al. 2004; Meister et al. 2004; Rand et al. 2004) severely affect processing of repeat transcripts into siRNAs, resulting in defects in heterochromatin assembly at centromeres. These results indicate that siRNA-guided cleavage of nascent repeat transcripts by Ago1 is likely an essential part of the siRNA production cycle. The cleaved transcripts might be a preferential substrate for Rdp1-mediated conversion into dsRNA or, alternatively, cleavage interferes with maturation of transcript such as polyadenylation and capping, making them preferential targets for Rdp1. Future work will address these possibilities.

An important characteristic of heterochromatic structures is their ability to exert a repressive influence on transcription of underlying sequences. It is therefore surprising that continuous transcription of heterochromatic repeats by pol II is required for RNAi-mediated heterochromatin assembly. Our work suggests that conserved heterochromatin protein Swi6/HP1 not only provides a platform for recruitment/spreading of silencing factors, but also targets antisilencing factors capable of facilitating pol II accessibility to sequences coated with heterochromatin. Therefore, proper maintenance of heterochromatin seems to require dynamic equilibrium of opposing chromatin-modifying activities such as HDACs that promote assembly of condensed higher-order structures and JmjC domain protein Epe1 capable of destabilizing these structures. Future investigations into mechanisms governing the relative influence of these activities is expected to provide insights into heterochromatin-mediated epigenetic reprogramming of target loci.

ACKNOWLEDGMENTS

We thank members of the Grewal laboratory for helpful discussions, and Hugh Cam and Ken-ichi Noma for help in manuscript preparation. Research in our laboratory is supported by the Intramural Research Program of the National Institutes of Health, National Cancer Institute.

REFERENCES

Ayoub N., Noma K., Isaac S., Kahan T., Grewal S.I., and Cohen A. 2003. A novel jmjC domain protein modulates heterochromatization in fission yeast. *Mol. Cell. Biol.* **23:** 4356.

Bannister A.J., Zegerman P., Partridge J.F., Miska E.A., Thomas J.O., Allshire R.C., and Kouzarides T. 2001. Selective recognition of methylated lysine 9 on histone H3 by the HP1 chromo domain. *Nature* **410:** 120.

Bernard P., Maure J.F., Partridge J.F., Genier S., Javerzat J.P., and Allshire R.C. 2001. Requirement of heterochromatin for cohesion at centromeres. *Science* **294:** 2539.

Birchler J.A., Bhadra M.P., and Bhadra U. 2000. Making noise about silence: Repression of repeated genes in animals. *Curr. Opin. Genet. Dev.* **10:** 211.

Bjerling P. and Ekwall K. 2002. Centromere domain organization and histone modifications. *Braz. J. Med. Biol. Res.* **35:** 499.

Bjerling P., Ekwall K., Egel R., and Thon G. 2004. A novel type of silencing factor, Clr2, is necessary for transcriptional silencing at various chromosomal locations in the fission yeast *Schizosaccharomyces pombe*. *Nucleic Acids Res.* **32:** 4421.

Bjerling P., Silverstein R.A., Thon G., Caudy A., Grewal S., and Ekwall K. 2002. Functional divergence between histone deacetylases in fission yeast by distinct cellular localization and in vivo specificity. *Mol. Cell. Biol.* **22:** 2170.

Bowen N.J., Jordan I.K., Epstein J.A., Wood V., and Levin H.L. 2003. Retrotransposons and their recognition of pol II promoters: A comprehensive survey of the transposable elements from the complete genome sequence of *Schizosaccharomyces pombe*. *Genome Res.* **13:** 1984.

Brasher S.V., Smith B.O., Fogh R.H., Nietlispach D., Thiru A., Nielsen P.R., Broadhurst R.W., Ball L.J., Murzina N.V., and Laue E.D. 2000. The structure of mouse HP1 suggests a unique mode of single peptide recognition by the shadow chromo domain dimer. *EMBO J.* **19:** 1587.

Cam H. and Grewal S.I. 2004. RNA interference and epigenetic control of heterochromatin assembly in fission yeast. *Cold Spring Harbor Symp. Quant. Biol.* **69:** 419.

Cam H., Sugiyama T., Chen E.S., Chen X., Fitzgerald P., and Grewal S.I. 2005. Comprehensive analysis of heterochromatin- and RNAi-mediated epigenetic control of the fission yeast genome. *Nat. Genet.* **37:** 809.

Chan S.W., Henderson I.R., and Jacobsen S.E. 2005. Gardening

the genome: DNA methylation in *Arabidopsis thaliana*. *Nat. Rev. Genet.* **6:** 351.

Cheutin T., Gorski S.A., May K.M., Singh P.B., and Misteli T. 2004. In vivo dynamics of Swi6 in yeast: Evidence for a stochastic model of heterochromatin. *Mol. Cell. Biol.* **24:** 3157.

Cheutin T., McNairn A.J., Jenuwein T., Gilbert D.M., Singh P.B., and Misteli T. 2003. Maintenance of stable heterochromatin domains by dynamic HP1 binding. *Science* **299:** 721.

Chikashige Y., Kinoshita N., Nakaseko Y., Matsumoto T., Murakami S., Niwa O., and Yanagida M. 1989. Composite motifs and repeat symmetry in *S. pombe* centromeres: Direct analysis by integration of NotI restriction sites. *Cell* **57:** 739.

Clarke L., Amstutz H., Fishel B., and Carbon J. 1986. Analysis of centromeric DNA in the fission yeast *Schizosaccharomyces pombe*. *Proc. Natl. Acad. Sci.* **83:** 8253.

Cooper J.P., Nimmo E.R., Allshire R.C., and Cech T.R. 1997. Regulation of telomere length and function by a Myb-domain protein in fission yeast. *Nature* **385:** 744.

Cowieson N.P., Partridge J.F., Allshire R.C., and McLaughlin P.J. 2000. Dimerisation of a chromo shadow domain and distinctions from the chromodomain as revealed by structural analysis. *Curr. Biol.* **10:** 517.

Djupedal I., Portoso M., Spahr H., Bonilla C., Gustafsson C.M., Allshire R.C., and Ekwall K. 2005. RNA Pol II subunit Rpb7 promotes centromeric transcription and RNAi-directed chromatin silencing. *Genes Dev.* **19:** 2301.

Ebert A., Lein S., Schotta G., and Reuter G. 2006. Histone modification and the control of heterochromatic gene silencing in *Drosophila*. *Chromosome Res.* **14:** 377.

Egel R. 1984. Two tightly linked silent cassettes in the mating-type region of *Schizosaccharomyces pombe*. *Curr. Genet.* **8:** 199.

Ekwall K. 2005. Genome-wide analysis of HDAC function. *Trends Genet.* **21:** 608.

Festenstein R., Pagakis S.N., Hiragami K., Lyon D., Verreault A., Sekkali B., and Kioussis D. 2003. Modulation of heterochromatin protein 1 dynamics in primary mammalian cells. *Science* **299:** 719.

Fire A., Xu S., Montgomery M.K., Kostas S.A., Driver S.E., and Mello C.C. 1998. Potent and specific genetic interference by double-stranded RNA in *Caenorhabditis elegans*. *Nature* **391:** 806.

Fodor B.D., Kubicek S., Yonezawa M., O'Sullivan R.J., Sengupta R., Perez-Burgos L., Opravil S., Mechtler K., Schotta G., and Jenuwein T. 2006. Jmjd2b antagonizes H3K9 trimethylation at pericentric heterochromatin in mammalian cells. *Genes Dev.* **20:** 1557.

Freeman-Cook L.L., Sherman J.M., Brachmann C.B., Allshire R.C., Boeke J.D., and Pillus L. 1999. The *Schizosaccharomyces pombe* hst4(+) gene is a SIR2 homologue with silencing and centromeric functions. *Mol. Biol. Cell* **10:** 3171.

Grewal S.I. and Elgin S.C. 2002. Heterochromatin: New possibilities for the inheritance of structure. *Curr. Opin. Genet. Dev.* **12:** 178.

Grewal S.I. and Klar A.J. 1997. A recombinationally repressed region between mat2 and mat3 loci shares homology to centromeric repeats and regulates directionality of mating-type switching in fission yeast. *Genetics* **146:** 1221.

Grewal S.I. and Moazed D. 2003. Heterochromatin and epigenetic control of gene expression. *Science* **301:** 798.

Grewal S.I., Bonaduce M.J., and Klar A.J. 1998. Histone deacetylase homologs regulate epigenetic inheritance of transcriptional silencing and chromosome segregation in fission yeast. *Genetics* **150:** 563.

Hahnenberger K.M., Carbon J., and Clarke L. 1991. Identification of DNA regions required for mitotic and meiotic functions within the centromere of *Schizosaccharomyces pombe* chromosome I. *Mol. Cell. Biol.* **11:** 2206.

Hall I.M. and Grewal S. I. 2003. Structure and function of heterochromatin: Implications for epigenetic gene silencing and genome organization. In *RNAi: A guide to gene silencing* (ed. G.J. Hannon), p. 205. Cold Spring Harbor Laboratory Press, Cold Spring Harbor, New York.

Hall I.M., Noma K., and Grewal S.I. 2003. RNA interference machinery regulates chromosome dynamics during mitosis and meiosis in fission yeast. *Proc. Natl. Acad. Sci.* **100:** 193.

Hall I.M., Shankaranarayana G.D., Noma K., Ayoub N., Cohen A., and Grewal S.I. 2002. Establishment and maintenance of a heterochromatin domain. *Science* **297:** 2232.

Hamilton A.J. and Baulcombe D.C. 1999. A species of small antisense RNA in posttranscriptional gene silencing in plants. *Science* **286:** 950.

Hannon G.J. 2002. RNA interference. *Nature* **418:** 244.

Hansen K.R., Ibarra P.T., and Thon G. 2006. Evolutionary-conserved telomere-linked helicase genes of fission yeast are repressed by silencing factors, RNAi components and the telomere-binding protein Taz1. *Nucleic Acids Res.* **34:** 78.

Henikoff S. 2000. Heterochromatin function in complex genomes. *Biochim. Biophys. Acta* **1470:** 1.

Hiragami K. and Festenstein R. 2005. Heterochromatin protein 1: A pervasive controlling influence. *Cell. Mol. Life Sci.* **62:** 2711.

Hong E.E., Villen J., Gerace E.L., Gygi S.P., and Moazed D. 2005. A Cullin E3 ubiquitin ligase complex associates with Rik1 and the Clr4 histone H3-K9 methyltransferase and is required for RNAi-mediated heterochromatin formation. *RNA Biol.* **2:** 106.

Horn P.J., Bastie J.N., and Peterson C.L. 2005. A Rik1-associated, cullin-dependent E3 ubiquitin ligase is essential for heterochromatin formation. *Genes Dev.* **19:** 1705.

Huang Y. and Maraia R.J. 2001. Comparison of the RNA polymerase III transcription machinery in *Schizosaccharomyces pombe, Saccharomyces cerevisiae* and human. *Nucleic Acids Res.* **29:** 2675.

Huang Y., Hamada M., and Maraia R.J. 2000. Isolation and cloning of four subunits of a fission yeast TFIIIC complex that includes an ortholog of the human regulatory protein TFIIICbeta. *J. Biol. Chem.* **275:** 31480.

Huisinga K.L., Brower-Toland B., and Elgin S.C. 2006. The contradictory definitions of heterochromatin: Transcription and silencing. *Chromosoma* **115:** 110.

Ivanova A.V., Bonaduce M.J., Ivanov S.V., and Klar A.J. 1998. The chromo and SET domains of the Clr4 protein are essential for silencing in fission yeast. *Nat. Genet.* **19:** 192.

Jia S., Kobayashi R., and Grewal S.I. 2005. Ubiquitin ligase component Cul4 associates with Clr4 histone methyltransferase to assemble heterochromatin. *Nat. Cell Biol.* **7:** 1007.

Jia S., Noma K., and Grewal S.I. 2004a. RNAi-independent heterochromatin nucleation by the stress-activated ATF/CREB family proteins. *Science* **304:** 1971.

Jia S., Yamada T., and Grewal S.I. 2004b. Heterochromatin regulates cell type-specific long-range chromatin interactions essential for directed recombination. *Cell* **119:** 469.

Kanoh J., Sadaie M., Urano T., and Ishikawa F. 2005. Telomere binding protein Taz1 establishes Swi6 heterochromatin independently of RNAi at telomeres. *Curr. Biol.* **15:** 1808.

Kato H., Goto D.B., Martienssen R.A., Urano T., Furukawa K., and Murakami Y. 2005. RNA polymerase II is required for RNAi-dependent heterochromatin assembly. *Science* **309:** 467.

Kim H.S., Choi E.S., Shin J.A., Jang Y.K., and Park S.D. 2004. Regulation of Swi6/HP1-dependent heterochromatin assembly by cooperation of components of the mitogen-activated protein kinase pathway and a histone deacetylase Clr6. *J. Biol. Chem.* **279:** 42850.

Klar A.J. 1989. The interconversion of yeast mating type: *Saccharomyces cerevisiae* and *Schizosaccharomyces pombe.* In *Mobile DNA* (ed. D.E. Berg and M.W. Howe), p. 671. American Society for Microbiology, Washington, D.C.

Klar A.J. and Bonaduce M.J. 1991. swi6, a gene required for mating-type switching, prohibits meiotic recombination in the mat2-mat3 "cold spot" of fission yeast. *Genetics* **129:** 1033.

Klose R.J., Yamane K., Bae Y., Zhang D., Erdjument-Bromage H., Tempst P., Wong J., and Zhang Y. 2006. The transcriptional repressor JHDM3A demethylates trimethyl histone H3 lysine 9 and lysine 36. *Nature* **442:** 312.

Liu J., Carmell M.A., Rivas F.V., Marsden C.G., Thomson J.M., Song J.J., Hammond S.M., Joshua-Tor L., and Hannon G.J. 2004. Argonaute2 is the catalytic engine of mammalian RNAi. *Science* **305:** 1437.

Lorentz A., Heim L., and Schmidt H. 1992. The switching gene swi6 affects recombination and gene expression in the mating-type region of *Schizosaccharomyces pombe. Mol. Gen. Genet.* **233:** 436.

Maison C., Bailly D., Peters A.H., Quivy J.P., Roche D., Taddei A., Lachner M., Jenuwein T., and Almouzni G. 2002. Higher-order structure in pericentric heterochromatin involves a distinct pattern of histone modification and an RNA component. *Nat. Genet.* **30:** 329.

Matzke M.A. and Birchler J.A. 2005. RNAi-mediated pathways in the nucleus. *Nat. Rev. Genet.* **6:** 24.

Meister G. and Tuschl T. 2004. Mechanisms of gene silencing by double-stranded RNA. *Nature* **431:** 343.

Meister G., Landthaler M., Patkaniowska A., Dorsett Y., Teng G., and Tuschl T. 2004. Human Argonaute2 mediates RNA cleavage targeted by miRNAs and siRNAs. *Mol. Cell* **15:** 185.

Mochizuki K. and Gorovsky M.A. 2004. Small RNAs in genome rearrangement in *Tetrahymena. Curr. Opin. Genet. Dev.* **14:** 181.

Motamedi M.R., Verdel A., Colmenares S.U., Gerber S.A., Gygi S.P., and Moazed D. 2004. Two RNAi complexes, RITS and RDRC, physically interact and localize to noncoding centromeric RNAs. *Cell* **119:** 789.

Muchardt C., Guilleme M., Seeler J.S., Trouche D., Dejean A., and Yaniv M. 2002. Coordinated methyl and RNA binding is required for heterochromatin localization of mammalian HP1alpha. *EMBO Rep.* **3:** 975.

Nakayama J., Klar A.J., and Grewal S.I. 2000. A chromodomain protein, Swi6, performs imprinting functions in fission yeast during mitosis and meiosis. *Cell* **101:** 307.

Nakayama J., Rice J.C., Strahl B.D., Allis C.D., and Grewal S.I. 2001. Role of histone H3 lysine 9 methylation in epigenetic control of heterochromatin assembly. *Science* **292:** 110.

Nimmo E.R., Pidoux A.L., Perry P.E., and Allshire R.C. 1998. Defective meiosis in telomere-silencing mutants of *Schizosaccharomyces pombe. Nature* **392:** 825.

Noma K., Allis C.D., and Grewal S.I. 2001. Transitions in distinct histone H3 methylation patterns at the heterochromatin domain boundaries. *Science* **293:** 1150.

Noma K., Cam H.P., Maraia R.J., and Grewal S.I. 2006. A role for TFIIIC transcription factor complex in genome organization. *Cell* **125:** 859.

Noma K., Sugiyama T., Cam H., Verdel A., Zofall M., Jia S., Moazed D., and Grewal S.I. 2004. RITS acts in cis to promote RNA interference-mediated transcriptional and posttranscriptional silencing. *Nat. Genet.* **36:** 1174.

Nonaka N., Kitajima T., Yokobayashi S., Xiao G., Yamamoto M., Grewal S.I., and Watanabe Y. 2002. Recruitment of cohesin to heterochromatic regions by Swi6/HP1 in fission yeast. *Nat. Cell Biol.* **4:** 89.

Partridge J.F., Borgstrom B., and Allshire R.C. 2000. Distinct protein interaction domains and protein spreading in a complex centromere. *Genes Dev.* **14:** 783.

Pidoux A.L. and Allshire R.C. 2004. Kinetochore and heterochromatin domains of the fission yeast centromere. *Chromosoma Res.* **12:** 521.

Rand T.A., Ginalski K., Grishin N.V., and Wang X. 2004. Biochemical identification of Argonaute 2 as the sole protein required for RNA-induced silencing complex activity. *Proc. Natl. Acad. Sci.* **101:** 14385.

Rea S., Eisenhaber F., O'Carroll D., Strahl B.D., Sun Z.W., Schmid M., Opravil S., Mechtler K., Ponting C.P., Allis C.D., and Jenuwein T. 2000. Regulation of chromatin structure by site-specific histone H3 methyltransferases. *Nature* **406:** 593.

Reinhart B.J. and Bartel D.P. 2002. Small RNAs correspond to centromere heterochromatic repeats. *Science* **297:** 1831.

Sadaie M., Iida T., Urano T., and Nakayama J. 2004. A chromodomain protein, Chp1, is required for the establishment of heterochromatin in fission yeast. *EMBO J.* **23:** 3825.

Saitoh S., Takahashi K., and Yanagida M. 1997. Mis6, a fission yeast inner centromere protein, acts during G1/S and forms specialized chromatin required for equal segregation. *Cell* **90:** 131.

Scott K.C., Merrett S.L., and Willard H.F. 2006. A heterochromatin barrier partitions the fission yeast centromere into discrete chromatin domains. *Curr. Biol.* **16:** 119.

Shankaranarayana G.D., Motamedi M.R., Moazed D., and Grewal S.I. 2003. Sir2 regulates histone H3 lysine 9 methylation and heterochromatin assembly in fission yeast. *Curr. Biol.* **13:** 1240.

Song J.J., Smith S.K., Hannon G.J., and Joshua-Tor L. 2004. Crystal structure of Argonaute and its implications for RISC slicer activity. *Science* **305:** 1434.

Sugiyama T., Cam H., Verdel A., Moazed D., and Grewal S.I. 2005. RNA-dependent RNA polymerase is an essential component of a self-enforcing loop coupling heterochromatin assembly to siRNA production. *Proc. Natl. Acad. Sci.* **102:** 152.

Sun F.L. and Elgin S.C. 1999. Putting boundaries on silence. *Cell* **99:** 459.

Takahashi K., Chen E.S., and Yanagida M. 2000. Requirement of Mis6 centromere connector for localizing a CENP-A-like protein in fission yeast. *Science* **288:** 2215.

Talbert P.B. and Henikoff S. 2006. Spreading of silent chromatin: Inaction at a distance. *Nat. Rev. Genet.* **7:** 793.

Thon G. and Verhein-Hansen J. 2000. Four chromo-domain proteins of *Schizosaccharomyces pombe* differentially repress transcription at various chromosomal locations. *Genetics* **155:** 551.

Thon G., Cohen A., and Klar A.J. 1994. Three additional linkage groups that repress transcription and meiotic recombination in the mating-type region of *Schizosaccharomyces pombe*. *Genetics* **138:** 29.

Thon G., Bjerling P., Bunner C.M., and Verhein-Hansen J. 2002. Expression-state boundaries in the mating-type region of fission yeast. *Genetics* **161:** 611.

Thon G., Hansen K.R., Altes S.P., Sidhu D., Singh G., Verhein-Hansen J., Bonaduce M.J., and Klar A.J. 2005. The Clr7 and Clr8 directionality factors and the Pcu4 cullin mediate heterochromatin formation in the fission yeast *Schizosaccharomyces pombe*. *Genetics* **171:** 1583.

Tsukada Y.I., Fang J., Erdjument-Bromage H., Warren M.E., Borchers C.H., Tempst P., and Y. Zhang. 2005. Histone demethylation by a family of JmjC domain-containing proteins. *Nature* **439:** 811.

Verdel A., Jia S., Gerber S., Sugiyama T., Gygi S., Grewal S.I., and Moazed D. 2004. RNAi-mediated targeting of heterochromatin by the RITS complex. *Science* **303:** 672.

Volpe T.A., Kidner C., Hall I.M., Teng G., Grewal S.I., and Martienssen R.A. 2002. Regulation of heterochromatic silencing and histone H3 lysine-9 methylation by RNAi. *Science* **297:** 1833.

West A.G., Gaszner M., and Felsenfeld G. 2002. Insulators: Many functions, many mechanisms. *Genes Dev.* **16:** 271.

Whetstine J.R., Nottke A., Lan F., Huarte M., Smolikov S., Chen Z., Spooner E., Li E., Zhang G., Colaiacovo M., and Shi Y. 2006. Reversal of histone lysine trimethylation by the JMJD2 family of histone demethylases. *Cell* **125:** 467.

Yamada T., Fischle W., Sugiyama T., Allis C.D., and Grewal S.I. 2005. The nucleation and maintenance of heterochromatin by a histone deacetylase in fission yeast. *Mol. Cell* **20:** 173.

Yang W. and Steitz T.A. 1995. Recombining the structures of HIV integrase, RuvC and RNase H. *Structure* **3:** 131.

Yuan Y.R., Pei Y., Ma J.B., Kuryavyi V., Zhadina M., Meister G., Chen H.Y., Dauter Z., Tuschl T., and Patel D.J. 2005. Crystal structure of *A. aeolicus* argonaute, a site-specific DNA-guided endoribonuclease, provides insights into RISC-mediated mRNA cleavage. *Mol. Cell* **19:** 405.

Zofall M. and Grewal S.I. 2006. Swi6/HP1 recruits a JmjC domain protein to facilitate transcription of heterochromatic repeats. *Mol. Cell* **22:** 681.

Slicing and Spreading of Heterochromatic Silencing by RNA Interference

S.M. LOCKE AND R.A. MARTIENSSEN

Cold Spring Harbor Laboratory and Watson School of Biological Sciences, Cold Spring Harbor, New York 11724

RNA interference (RNAi) can mediate gene silencing posttranscriptionally by target RNA cleavage, or transcriptionally by chromatin and DNA modification. Argonaute is an essential component of the RNAi machinery that displays endonucleolytic activity guided by bound small RNAs. This slicing activity has recently been shown to be required for gene silencing and spreading of histone modifications characteristic of heterochromatin in *Schizosaccharomyces pombe*. Argonaute proteins with catalytic and nucleic acid binding capacities are found to function in RNAi within both the plant and animal kingdoms. Here we review the requirement of slicing for silencing and spreading in *S. pombe*, plants, and humans.

INTRODUCTION

Heterochromatic Silencing and Position-effect Variegation

Chromosomal material is classically characterized as either euchromatic or heterochromatic, depending on condensation during interphase. Euchromatin has been shown to include regions of active transcription and is characterized by various histone modifications such as histone H3 and H4 acetylation, and histone H3 dimethylation on lysine 4 (K4me2). Heterochromatin, on the other hand, is condensed during interphase and is populated by deacetylated histones enriched in histone H3 dimethylated on K9 and K27 (Richards and Elgin 2002). It was previously thought to be transcriptionally inert.

The chromatin context of a gene can determine its expression, as illustrated by position-effect variegation (PEV) in *Drosophila melanogaster.* (Schotta et al. 2003; Pal-Bhadra et al. 2004; Talbert and Henikoff 2006). PEV is also observed in *S. pombe,* in which a reporter gene that is active in a euchromatic region becomes silenced when located within a region of heterochromatin (Allshire et al. 1994). This implies that the factors which confer transcriptional silence in heterochromatin can spread into transgenes located within its borders that would otherwise be active. In *S. pombe*, there are three main regions of heterochromatin: centromeres, telomeres, and the mating-type locus. Heterochromatin in all eukaryotes is associated with repetitive and transposable elements (TEs) which are often capable of silencing genes when juxtaposed with them (Lippman et al. 2004). Unexpectedly, these elements are a predominant source of small interfering RNA (siRNA) in both plants and fission yeast (Lippman et al. 2004), and RNA interference is required for silencing and heterochromatic modifications associated with PEV in yeast (Volpe et al. 2002), *Arabidopsis* (Zilberman et al. 2003), and *Drosophila* (Pal-Bhadra et al. 2004). Thus, RNAi mediates both transcriptional (TGS) and posttranscriptional (PTGS) gene silencing (Cerutti and Casas-Mollano 2006).

Argonaute, The Catalytic Engine of RNAi

Phylogenetic analysis indicates that the main components of RNAi, namely Dicer, Argonaute, and an RNA-directed RNA polymerase, were present in the last common ancestor of eukaryotes (Cerutti and Casas-Mollano 2006). Argonaute proteins are present in all species that exhibit RNAi and fall into two classes, Argonaute-like and Piwi-like (Cerutti and Casas-Mollano 2006). In Table 1, piwi-like proteins are indicated in bold. Argonaute is a 100-kD basic protein that consists of four domains: the amino-terminal; middle; PIWI, which is responsible for the RNase-H-like catalytic activity; and PAZ, the nucleic-acid-binding domain. The protein forms a positively charged groove between the PIWI domain and the PAZ domain located above (Song et al. 2004). The PAZ domain preferentially binds RNA in a sequence-independent manner by recognizing the 3′ ends of single-stranded RNAs (Song et al. 2003). The 5′ end of bound siRNA is free to base-pair strongly with complementary target RNA. The slice site is then determined as a distance relative to the 5′ end of the siRNA (Hall 2005). The slicing activity of *S. pombe* Ago1 is thus dependent on siRNA for both its catalytic activity and specificity. The PIWI domain contains two aspartates and a histidine that form a "DDH" catalytic motif similar to the "DDE" motif seen in RNase H (Song et al. 2004). Mg^{2+}-dependent catalytic activity breaks only one phosphodiester bond, leaving a 3′-OH and a 5′ phosphate (Martinez and Tuschl 2004). Argonaute proteins have roles in RNAi, developmental control, stem cell maintenance, and tumorigenesis (Carmell et al. 2002), as well as heterochromatic silencing, but the extent to which catalytic activity is required for each of these is unknown, particularly as Argonaute proteins without the conserved DDH sites exist in many animals (Song et al. 2004). Several recent papers have investigated this possibility in fission yeast, plants, and human cells.

Table 1. Argonaute Proteins of Yeast, Plants, and Humans

S. pombe	A. thaliana	H. sapiens
Ago1*	*AGO1* *	hAgo1
	AGO2	hAgo2*
	AGO3	hAgo3
	AGO4 *	hAgo4
	AGO5	hAgo5
	AGO6	**Hiwi**
	AGO7/ZIPPY	**Hili**
	AGO8	
	AGO9	
	PINHEAD/ZWILLE	

(*) Indicates proteins that have been tested and shown to have catalytic activity. Bold type indicates piwi-like proteins.

AGO SLICING ACTIVITY IS REQUIRED FOR SILENCING AND SPREADING IN *S. POMBE*

S. pombe contains only one Argonaute (Table 1) and one Dicer, Dcr1, making it an excellent system in which to study RNAi. Furthermore, heterochromatic regions of the genome are sharply defined and have been sequenced almost to completion (Wood et al. 2002). The *S. pombe* centromeres are composed of a central conserved region, bordered by inner and outer *dg* and *dh* repeats that are heterochromatic in nature (Fig. 1). Reporter genes inserted into these heterochromatic repeats are silenced, while the same genes inserted into a euchromatic region are expressed (Allshire et al. 1994). If Ago1, Dcr1, or the RNA-dependent RNA polymerase (Rdp1) is deleted, silencing of the reporter genes in heterochromatic repeats is no longer maintained (Volpe et al. 2002). Furthermore, RNAi regulates the silencing of centromeric repeats themselves. Levels of H3K4me2 at the centromere increase in *dcr1*, *rdp1*, and *ago1* mutants, while levels of H3K9me2 decrease, although *ago1* maintains higher levels of H3K9me2 than *dcr1* (Volpe et al. 2002). Forward and reverse transcripts from the centromeric repeats accumulate in *dcr1, rdp1*, and *ago1* while the mutant of the homolog to the mammalian heterochromatic protein 1 (HP1), *swi6*, only accumulates forward transcripts (Volpe et al. 2002). This suggests that RNAi machinery is responsible for repression of forward strand transcription whereas the reverse strand is always transcribed but is turned over rapidly into siRNA by the RNAi machinery (Volpe et al. 2002).

By making alanine substitutions for each of the conserved aspartates and the histidine required for catalytic activity, such as *ago1D650A*, the catalytic activity of Ago1, not just its structural contribution or nucleic-acid-binding capacity, has been found to be critical for RNAi-mediated silencing of a reporter gene integrated into the heterochromatic outer repeats of centromere 1 (Fig. 1) (Irvine et al. 2006). Recombinant Ago1 is capable of siRNA-directed endonucleolytic activity in vitro, but each of the mutants is impaired. *ago11D650A* cells do not accumulate detectable siRNA, suggesting that slicing is required for siRNA synthesis (Irvine et al. 2006). Instead, *ago1D650A* cells accumulate both forward and reverse transcripts of heterochromatic repeats (Irvine et al. 2006), which eliminates a possible role for Ago1 in the transcriptional step of pre-siRNA synthesis and rather suggests a role in its processing. It seems counterintuitive, but transcription of the repeats, and processing into siRNA, are necessary for reporter gene silencing. Consistent with this idea, reporter genes inserted upstream of the major heterochromatic repeat promoter were silenced 4 times less effectively than those inserted downstream (H. repeat promoter in Fig. 1), although bidirectional transcription of heterochromatic repeats ensured some level of silencing in both cases (Irvine et al. 2006).

Based on these results, a model for transcriptional silencing in *S. pombe* has been proposed (Fig. 2). Continuous transcription of heterochromatic repeats initiates from a reverse-strand promoter (Volpe et al. 2002). These transcripts are then turned over into siRNA (Reinhart and Bartel 2002). Transcription of the forward strand is silenced by the formation and maintenance of heterochromatin induced by these siRNAs (Volpe et al. 2002). RNase-H-like Ago1 cleaves the reverse-strand transcript, recruiting the RNA-directed RNA polymerase complex (RDRC) for double-stranded RNA (dsRNA) synthesis (Irvine et al. 2006). The RDRC is composed of Rdp1, helicase Hrr1, and oligoadenylate polymerase Cid12 (Motamedi et al. 2004). This complex uses aberrant transcripts, either directly or primed by siRNA, as templates for production of dsRNA (Motamedi et al. 2004). dsRNAs are then processed by Dcr1 into siRNA (Bernstein et al. 2001; Provost et al. 2002). The RNA-induced transcriptional silencing complex (RITS) is composed of Ago1 (bound to siRNA), Tas3, and the chromodomain protein Chp1 (Verdel et al. 2004). RITS associates with chromatin through binding of Chp1 to H3K9me (Partridge et al. 2002). This association, and slicing of transcript by Ago1, somehow recruit the RIK1 complex consisting of Rik1, Dos1/Clr8, Dos2/Clr7, cullin ubiquitin ligase Pcu4, and histone H3K9 methyl-

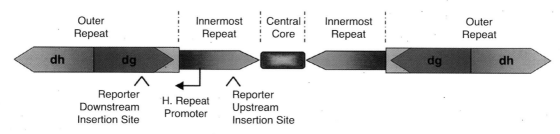

Figure 1. *S. pombe* centromere 1.

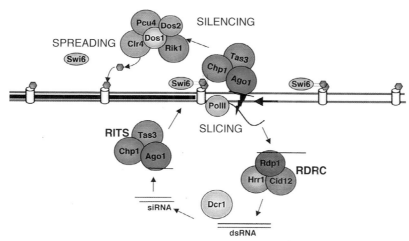

Figure 2. RNAi silencing in *S. pombe*. The symbols used are as follows: (*black arrow*) reverse promoter, (*black lightning bolt*) Ago1 endonuclease activity, (*black rectangle*) reporter transgene, (*red hexagon*) methyl group on histone H3K9.

transferase Clr4 (Horn et al. 2005; Li et al. 2005; Thon et al. 2005). Clr4 catalyzes the methylation of nearby histone H3K9 (Nakayama et al. 2001). H3K9me2 recruits Swi6, which silences transcription of the forward strand of the repeats as well as the reporter gene (Bannister et al. 2001). Swi6 is required for silencing, and as RNAi continues, H3K9 methylation and Swi6 association spread (Partridge et al. 2000; Martienssen et al. 2005; Irvine et al. 2006).

The targets of Ago1 silencing are determined by the sequence identity between the target (sliced) transcript and that of the bound guide siRNA, rather than with DNA corresponding to the promoter. This is illustrated by the presence of siRNA, and corresponding slice sites, in the repeats but not in the reporter genes embedded within them (Irvine et al. 2006). It is therefore likely that subsequent silencing of the reporter genes is first triggered by slicing and then spread by readthrough transcription by DNA-dependent RNA polymerase II (pol II) into the reporter gene (Fig. 2). Although the mechanism is still unclear, the critical role of pol II in H3K9me2 modification of the reporter genes is revealed by mutants in at least two subunits, Rpb2 and Rpb7 (Sugaya et al. 1998; Djupedal et al. 2005; Kato et al. 2005).

ARGONAUTE4 AND RNA-DIRECTED DNA METHYLATION IN PLANTS

Arabidopsis thaliana has ten Argonaute proteins (Table 1). So far, only *AGO1* and *AGO4* have been tested for and shown to have catalytic endonuclease activity (Table 1), although all ten have the conserved DDH motif (Baumberger and Baulcombe 2005). *AGO1* associates mainly with 21-nucleotide small RNAs whereas *AGO4* binds 24-nucleotide small RNAs (Qi et al. 2006), but both are capable of endonuclease activity (Baumberger and Baulcombe 2005; Qi et al. 2005), and both can bind miRNA as well as siRNA (Qi et al. 2006). *AGO1* has a major developmental role in vivo in targeting messages via miRNA (Tang et al. 2003; Kidner and Martienssen 2004) but is also required for PTGS and for associated

DNA methylation, which requires the catalytic PIWI domain (Morel et al. 2002). However, most of the small RNAs associated with both *AGO1* and *AGO4* match to heterochromatic repetitive regions, similar to *S. pombe* (Qi et al. 2006). The siRNA-mediated cleavage of target RNA suggests RNA–RNA recognition as the mechanism of specificity (Qi et al. 2006).

Unlike *S. pombe*, which has no DNA methylation, in plants, RNAi-induced silencing is often associated with DNA methylation. When target transgenes are silenced by homologous regions in viral transgenes, silencing and the associated methylation can spread from the homologous region of the target to other parts of the transgene, both upstream and downstream (Vaistij et al. 2002). This spreading depends on the RNA-dependent RNA polymerase *RDR6*, but can be subsequently inherited independent of the viral transgene in a fashion that is maintained by the maintenance methyltransferase *MET1*, a homolog of mammalian DNMT1 (Vaistij et al. 2002). This type of methylation also depends on the Argonaute protein *AGO1* (Morel et al. 2002). Thus, although histone modification was not examined in these studies, spreading of methylation resembles the situation in *S. pombe*.

On the other hand, when transgene promoter methylation is induced by homologous hairpin constructs, it is strictly limited to sequences homologous to the hairpin RNA. These sequences can be much shorter than a nucleosome, suggesting a possible RNA–DNA interaction (Matzke and Birchler 2005). In these cases, methylation is partially dependent on *AGO4* (Zilberman et al. 2003). Methylation at CpG dinucleotides by *MET1* is independent of *AGO4*, but methylation at non-CpG cytosines by the chromomethyltransferase *CMT3* and especially by the DNMT3-related methyltransferases *DRM1* and *DRM2* depends strongly on *AGO4* (Chan et al. 2004).

To investigate the role of slicing in RNA-dependent DNA methylation (RdDM), catalytic mutants of *AGO4* have been constructed and used to complement *ago4* mutants (Qi et al. 2006). A variety of endogenous methylated sequences were then examined by bisulfite sequencing. The Ac-like transposable element *SIMPLE-*

HAT has an average of 15.8% non-CpG cytosines methylated in any given molecule. This methylation disappears in *ago4-1* but is almost fully restored (8–11%) when mutants are transformed with both a wild-type and a catalytically inactive form of *AGO4* (Qi et al. 2006). In contrast, higher levels of methylation at the tandem repeat *MEA-ISR* (18.9%), which were also lost in *ago4-1*, could not be restored by catalytically inactive *AGO4* but could be restored by the wild-type form (Qi et al. 2006). Interestingly, levels of siRNA could not be fully restored by mutant *AGO4* either, indicating a requirement for *AGO4* catalytic activity. In the case of *AtMu1*, the results were ambiguous, related perhaps to the presence of a transposed copy in Landsberg *erecta* (Madlung et al. 2005), whose methylation is at least partially dependent on *AGO1* (Lippman et al. 2003).

Non-CpG methylation in epimutants at the *SUPERMAN (SUP)* locus is very low with only 21 methylated C out of 374 assayed (less than one per molecule), but has been associated with epigenetic silencing of the locus, which results in a visible phenotype (Chan et al. 2005). This level of methylation is reduced in *ago4-1* to only 17 methylated cytosines out of 765: this difference is low so that restoration by catalytically defective mutants is difficult to assess, although it was noted (Qi et al. 2006). In principle, the *superman* phenotype found in epimutants is a more sensitive assay. *superman* phenotypes are normally observed in "stabilized" strains in which silencing is maintained by a *SUP* transgene and depends on *AGO4*. This stabilization is accompanied by elevated methylation (21% mC) which is partially lost in *ago4-1* (5.5%) (Zilberman et al. 2003). In the absence of the *SUP* transgene, the *superman* phenotype is unstable, but it is even more unstable after propagation in *ago4-1* mutants (only 20–30% phenotypic plants were observed in this background) (Qi et al. 2006). The phenotype was observed at much higher frequencies among the progeny of plants transformed with each of the catalytically inactive transgenes, raising the possibility that catalytically inactive *AGO4* can partially silence *SUP* (Qi et al. 2006).

In cases where *AGO4* catalytic activity is required for DNA methylation (*MEA-ISR*), it is also required for siRNA accumulation, although the correlation with methylation is not absolute. The authors conclude that non-CpG methylation is independent of *AGO4* slicing, although there is clearly a spectrum of methylation levels promoted by catalytic mutants at different loci (Qi et al. 2006). Several explanations can be proposed. First, very low levels of catalytic activity found in the mutants may be sufficient for RdDM in some cases. Second, catalytically inactive *AGO4* might stimulate or stabilize the activity of other Argonaute proteins that retain full activity (such as *AGO1*). Third, non-catalytic *AGO4* may be sufficient for RdDM when it promotes dsRNA synthesis by *RDR2* resulting in siRNA production (Qi et al. 2006). siRNA might then be used to guide a DNA methyltransferase such as *DRM1/2*, in a mechanism completely different from histone modification in fission yeast. In this respect, it would be very interesting to examine catalytic mutants of *AGO4* for histone H3K9 methylation, and for TE silencing, as well as for DNA methylation.

AGO AND TRANSCRIPTIONAL SILENCING IN HUMAN CELLS

Seven Argonaute proteins are encoded in the human genome, of which five are Argonaute-like, but only one of those tested has been shown to be catalytically active as of yet, namely, hAgo2 (Liu et al. 2004). Mutations of the hAgo2 DDH motif eliminate endonuclease activity (Liu et al. 2004; Rivas et al. 2005). However, other Argonaute proteins are implicated in human disease. For example, Wilms' tumors often lack the chromosomal region containing hAgo3, hAgo1, and hAgo4 (Carmell et al. 2002). Also, Hiwi, a homolog to mouse *Piwi*, has been associated with testicular germ cell tumors, seminomas, and ambiguous genitalia (Carmell et al. 2002).

Human cells exhibit siRNA-induced TGS (Morris et al. 2004; Ting et al. 2005). Introduction of a siRNA targeted to the elongation factor 1 alpha (EF1A) promoter region, fused to GFP, results in a reduction of protein expression, as well as a decrease in mRNA transcript levels (Morris et al. 2004). Similar results have been observed with a siRNA targeted to the endogenous gene, containing no GFP fusion (Weinberg et al. 2006). In addition, histone H3K9 methylation increases and spreads to at least 700 bp downstream from the EFIA transcriptional start site (Weinberg et al. 2006).

It has recently been shown that Argonaute proteins are required for TGS in human cells, suggesting a conserved mechanism. siRNA directed against sequences upstream of the transcription start sites of both Huntingtin (HTT) and androgen (AR) receptors expressed in the breast cancer cell line T47D inhibit protein expression and transcript synthesis, establishing RNAi as an inducer of TGS in mammalian cells (Janowski et al. 2006). The ability of HTT-specific siRNA to silence HTT is inhibited by reducing the levels of AGO2 or AGO1, which have both been shown to associate with DNA targeted by siRNA, in T47D cells (Janowski et al. 2006).

The expression level of target genes can alter silencing effectiveness. By comparing two breast cancer cell lines in which the expression level of progesterone receptor (PR) is high in one, T47D, and 33-fold lower in another, MCF-7, it has been determined that siRNA-mediated TGS is dependent on a relatively high level of transcription (Janowski et al. 2006). The authors speculate that high levels of transcription allow access to and recognition of DNA by siRNAs (Janowski et al. 2006). However, similar observations in plants have led some to speculate that high levels of transgene expression exhaust translation cofactors, resulting in a higher concentration of substrate for RNAi (Gazzani et al. 2004). Still others have suggested that a low level of transcription simply does not produce enough aberrant RNAs for rapid and effective initiation of silencing (Melquist and Bender 2003). Whatever the reason, expression level does seem to be a variable in the success of RNAi in any system.

There has been some controversy in human studies concerning the role of siRNA-mediated silencing in histone modification. Few differences were observed in histone H3K4 or K9 modifications at the gene encoding PR when targeted by siRNA, suggesting that H3K9me2 is not required (Janowski et al. 2006). In stark contrast, the promoter region of human immunodeficiency virus-1 coreceptor CCR5 in 293T cells (a human renal epithelial cell line transformed by adenovirus E1A) shows a 14-fold increase of H3K9me2 in response to siRNAs directed to the region (Kim et al. 2006). The H3K9me2 enrichment spread to a region 100–300 bp downstream from the target site at a 7-fold increase (Kim et al. 2006), comparable to the 700 bp spreading mentioned above (Weinberg et al. 2006). AGO1 also binds the CCR5 promoter, and its association also spreads 100–300 bp downstream in response to transfection with siRNA directed against CCR5 promoter (Kim et al. 2006). Knockdown of AGO1 reduces H3K9me2 enrichment and inhibits gene silencing, suggesting that the association of AGO1 with targeted chromosomal DNA acts to recruit histone methyltransferases necessary for siRNA-mediated silencing (Kim et al. 2006). As H3K9me2 enrichment increases, AGO1 association decreases (Kim et al. 2006), suggesting that AGO1 is required for the initial boost of H3K9 methylation. As methylation increases, heterochromatin is formed and silencing occurs stably.

Argonaute Is Involved in Polycomb-mediated Silencing in Animal Cells

Polycomb genes (PcG) are regulatory factors that establish and maintain transcriptional repression by forming multimeric complexes that bind chromosomal regulatory elements (PREs) and modify chromatin structure (Pirrotta 1998; Jacobs and van Lohuizen 2002). They are absent in fission yeast, but are found in animals and in plants. Recently, a link has been described between PcG-mediated silencing in *Drosophila* and RNAi-mediated small RNA synthesis (Grimaud et al. 2006).

In the human Polycomb silencing pathway, the EZH2 histone methyltransferase methylates H3K27, which can be associated with heterochromatin (Kuzmichev et al. 2002). In HeLa cells (an immortal epithelial cell line derived from cervical cancer cells), siRNA directed against the tumor suppressor RASSF1A results in an enrichment of EZH2 and thus H3K27me3 (Kim et al. 2006). The MYT1 promoter, an endogenous target of the Polycomb silencing pathway, shows enrichment not only of EZH2 and H3K27me3, but also of AGO1 (Kim et al. 2006). Similarly, Polycomb component EZH2 was found to be enriched at the CCR5 promoter after transfection of siRNA directed against it, following the same spreading pattern as AGO1 association and H3K9me2 (Kim et al. 2006). These results implicate RNAi and AGO1 with the Polycomb pathway, a process that governs animal development. Although there are no Polycomb proteins in fission yeast, and no H3K27me3 has yet been detected, in *Drosophila*, there is some evidence that Polycomb response elements are transcribed, and that PIWI (an Argonaute homolog) mediates their association in the interphase nucleus, which is thought to be important for silencing (Grimaud et al. 2006). It remains to be seen whether these mechanisms are related.

CONCLUSIONS

The slicing function of Ago1 is necessary for transcriptional silencing and spreading of heterochromatin by RNAi

in *S. pombe* (Irvine et al. 2006). Argonaute proteins are also required for the generation of siRNA, but at least in fission yeast, siRNAs corresponding directly to the silenced promoter are not detectable, and slicing of co-transcripts that read through the silent gene seems to be sufficient. Argonaute proteins are also required for some examples of transcriptional silencing in plants and in mammalian cells. In human cells, silencing is mediated by artificial siRNA, so that a role in siRNA production can be excluded. In cases where histone H3 K9 methylation is induced, spreading of H3K9me2 and AGO1 occurs downstream from the siRNA, reminiscent of the situation in fission yeast.

However, RNAi-mediated silencing through DNA methylation (not found in fission yeast), although dependent on Argonaute, may differ in mechanism. In *Arabidopsis*, the Argonaute protein *AGO4* does not appear to require its slicing function to promote DNA methylation of at least one TE. In humans, AGO1 is required for examples of siRNA-directed DNA methylation but has yet to be demonstrated to have slicing activity. In the *Arabidopsis* study, histone modifications were not examined, but in humans, some examples of siRNA-directed silencing involve histone modification, whereas others involve DNA methylation, consistent with the idea that the two mechanisms are distinct.

In both humans and plants, it is tempting to speculate that DNA methyltransferases might be guided by siRNA themselves, whereas histone modification might be guided by the sliced target (sometimes called "aberrant RNA"). Consistent with this idea, the DNA methyltransferase DNMT3A has been shown to bind artificial siRNA (Weinberg et al. 2006) whereas siRNA have not been found in the Rik1/Clr4 complex in *S. pombe*, which may instead recognize sliced RNA products. The situation is complicated by the fact that DNA methylation can be induced by histone modification and vice versa in both humans and plants (Matzke and Birchler 2005), so that distinguishing these mechanisms may prove challenging.

ACKNOWLEDGMENTS

Work in the authors' laboratory was supported by a grant from the National Institutes of Health (GM067014) to R.M. S.M.L. is a recipient of a George A. and Marjorie H. Anderson Fellowship from the Watson School of Biological Sciences.

REFERENCES

Allshire R.C., Javerzat J.P., Redhead N.J., and Cranston G. 1994. Position effect variegation at fission yeast centromeres. *Cell* **76:** 157.

Bannister A.J., Zegerman P., Partridge J.F., Miska E.A., Thomas J.O., Allshire R.C., and Kouzarides T. 2001. Selective recognition of methylated lysine 9 on histone H3 by the HP1 chromo domain. *Nature* **410:** 120.

Baumberger N. and Baulcombe D.C. 2005. *Arabidopsis* ARGONAUTE1 is an RNA Slicer that selectively recruits microRNAs and short interfering RNAs. *Proc. Natl. Acad. Sci.* **102:** 11928.

Bernstein E., Caudy A.A., Hammond S.M., and Hannon G.J. 2001. Role for a bidentate ribonuclease in the initiation step of RNA interference. *Nature* **409:** 363.

Carmell M.A., Xuan Z., Zhang M.Q., and Hannon G.J. 2002. The Argonaute family: Tentacles that reach into RNAi, developmental control, stem cell maintenance, and tumorigenesis. *Genes Dev.* **16:** 2733.

Cerutti H. and Casas-Mollano J.A. 2006. On the origin and functions of RNA-mediated silencing: From protists to man. *Curr. Genet.* **50:** 81.

Chan S.W., Henderson I.R., and Jacobsen S.E. 2005. Gardening the genome: DNA methylation in *Arabidopsis thaliana*. *Nat. Rev. Genet.* **6:** 351.

Chan S.W., Zilberman D., Xie Z., Johansen L.K., Carrington J.C., and Jacobsen S.E. 2004. RNA silencing genes control de novo DNA methylation. *Science* **303:** 1336.

Djupedal I., Portoso M., Spahr H., Bonilla C., Gustafsson C.M., Allshire R.C., and Ekwall K. 2005. RNA Pol II subunit Rpb7 promotes centromeric transcription and RNAi-directed chromatin silencing. *Genes Dev.* **19:** 2301.

Gazzani S., Lawrenson T., Woodward C., Headon D., and Sablowski R. 2004. A link between mRNA turnover and RNA interference in *Arabidopsis*. *Science* **306:** 1046.

Grimaud C., Bantignies F., Pal-Bhadra M., Ghana P., Bhadra U., and Cavalli G. 2006. RNAi components are required for nuclear clustering of Polycomb group response elements. *Cell* **124:** 957.

Hall T.M. 2005. Structure and function of argonaute proteins. *Structure* **13:** 1403.

Horn P.J., Bastie J.N., and Peterson C.L. 2005. A Rik1-associated, cullin-dependent E3 ubiquitin ligase is essential for heterochromatin formation. *Genes Dev.* **19:** 1705.

Irvine D.V., Zaratiegui M., Tolia N.H., Goto D.B., Chitwood D.H., Vaughn M.W., Joshua-Tor L., and Martienssen R.A. 2006. Argonaute slicing is required for heterochromatic silencing and spreading. *Science* **313:** 1134.

Jacobs J.J. and van Lohuizen M. 2002. Polycomb repression: from cellular memory to cellular proliferation and cancer. *Biochim. Biophys. Acta* **1602:** 151.

Janowski B.A., Huffman K.E., Schwartz J.C., Ram R., Nordsell R., Shames D.S., Minna J.D., and Corey D.R. 2006. Involvement of AGO1 and AGO2 in mammalian transcriptional silencing. *Nat. Struct. Mol. Biol.* **13:** 787.

Kato H., Goto D.B., Martienssen R.A., Urano T., Furukawa K., and Murakami Y. 2005. RNA polymerase II is required for RNAi-dependent heterochromatin assembly. *Science* **309:** 467.

Kidner C.A. and Martienssen R.A. 2004. Spatially restricted microRNA directs leaf polarity through ARGONAUTE1. *Nature* **428:** 81.

Kim D.H., Villeneuve L.M., Morris K.V., and Rossi J.J. 2006. Argonaute-1 directs siRNA-mediated transcriptional gene silencing in human cells. *Nat. Struct. Mol. Biol.* **13:** 793.

Kuzmichev A., Nishioka K., Erdjument-Bromage H., Tempst P., and Reinberg D. 2002. Histone methyltransferase activity associated with a human multiprotein complex containing the Enhancer of Zeste protein. *Genes Dev.* **16:** 2893.

Li F., Goto D.B., Zaratiegui M., Tang X., Martienssen R., and Cande W.Z. 2005. Two novel proteins, dos1 and dos2, interact with rik1 to regulate heterochromatic RNA interference and histone modification. *Curr. Biol.* **15:** 1448.

Lippman Z., May B., Yordan C., Singer T., and Martienssen R. 2003. Distinct mechanisms determine transposon inheritance and methylation via small interfering RNA and histone modification. *PLoS Biol.* **1:** E67.

Lippman Z., Gendrel A.V., Black M., Vaughn M.W., Dedhia N., McCombie W.R., Lavine K., Mittal V., May B., Kasschau K.D., et al. 2004. Role of transposable elements in heterochromatin and epigenetic control. *Nature* **430:** 471.

Liu J., Carmell M.A., Rivas F.V., Marsden C.G., Thomson J.M., Song J.J., Hammond S.M., Joshua-Tor L., and Hannon G.J. 2004. Argonaute2 is the catalytic engine of mammalian RNAi. *Science* **305:** 1437.

Madlung A., Tyagi A.P., Watson B., Jiang H., Kagochi T., Doerge R.W., Martienssen R., and Comai L. 2005. Genomic changes in synthetic *Arabidopsis* polyploids. *Plant J.* **41:** 221.

Martienssen R.A., Zaratiegui M., and Goto D.B. 2005. RNA interference and heterochromatin in the fission yeast *Schizosaccharomyces pombe*. *Trends Genet.* **21:** 450.

Martinez J. and Tuschl T. 2004. RISC is a 5′ phosphomonoester-producing RNA endonuclease. *Genes Dev.* **18:** 975.

Matzke M.A. and Birchler J.A. 2005. RNAi-mediated pathways in the nucleus. *Nat. Rev. Genet.* **6:** 24.

Melquist S. and Bender J. 2003. Transcription from an upstream promoter controls methylation signaling from an inverted repeat of endogenous genes in *Arabidopsis*. *Genes Dev.* **17:** 2036.

Morel J.B., Godon C., Mourrain P., Beclin C., Boutet S., Feuerbach F., Proux F., and Vaucheret H. 2002. Fertile hypomorphic ARGONAUTE (ago1) mutants impaired in post-transcriptional gene silencing and virus resistance. *Plant Cell* **14:** 629.

Morris K.V., Chan S.W., Jacobsen S.E., and Looney D.J. 2004. Small interfering RNA-induced transcriptional gene silencing in human cells. *Science* **305:** 1289.

Motamedi M.R., Verdel A., Colmenares S.U., Gerber S.A., Gygi S.P., and Moazed D. 2004. Two RNAi complexes, RITS and RDRC, physically interact and localize to noncoding centromeric RNAs. *Cell* **119:** 789.

Nakayama J., Rice J.C., Strahl B.D., Allis C.D., and Grewal S.I. 2001. Role of histone H3 lysine 9 methylation in epigenetic control of heterochromatin assembly. *Science* **292:** 110.

Pal-Bhadra M., Leibovitch B.A., Gandhi S.G., Rao M., Bhadra U., Birchler J.A., and Elgin S.C. 2004. Heterochromatic silencing and HP1 localization in *Drosophila* are dependent on the RNAi machinery. *Science* **303:** 669.

Partridge J.F., Borgstrom B., and Allshire R.C. 2000. Distinct protein interaction domains and protein spreading in a complex centromere. *Genes Dev.* **14:** 783.

Partridge J.F., Scott K.S., Bannister A.J., Kouzarides T., and Allshire R.C. 2002. *cis*-acting DNA from fission yeast centromeres mediates histone H3 methylation and recruitment of silencing factors and cohesin to an ectopic site. *Curr. Biol.* **12:** 1652.

Pirrotta V. 1998. Polycombing the genome: PcG, trxG, and chromatin silencing. *Cell* **93:** 333.

Provost P., Silverstein R.A., Dishart D., Walfridsson J., Djupedal I., Kniola B., Wright A., Samuelsson B., Radmark O., and Ekwall K. 2002. Dicer is required for chromosome segregation and gene silencing in fission yeast cells. *Proc. Natl. Acad. Sci.* **99:** 16648.

Qi Y., Denli A.M., and Hannon G.J. 2005. Biochemical specialization within *Arabidopsis* RNA silencing pathways. *Mol. Cell* **19:** 421.

Qi Y., He X., Wang X.J., Kohany O., Jurka J., and Hannon G.J. 2006. Distinct catalytic and non-catalytic roles of ARGONAUTE4 in RNA-directed DNA methylation. *Nature* **443:** 1008.

Reinhart B.J. and Bartel D.P. 2002. Small RNAs correspond to centromere heterochromatic repeats. *Science* **297:** 1831.

Richards E.J. and Elgin S.C. 2002. Epigenetic codes for heterochromatin formation and silencing: Rounding up the usual suspects. *Cell* **108:** 489.

Rivas F.V., Tolia N.H., Song J.J., Aragon J.P., Liu J., Hannon G.J., and Joshua-Tor L. 2005. Purified Argonaute2 and an siRNA form recombinant human RISC. *Nat. Struct. Mol. Biol.* **12:** 340.

Schotta G., Ebert A., Dorn R., and Reuter G. 2003. Position-effect variegation and the genetic dissection of chromatin regulation in *Drosophila*. *Semin. Cell Dev. Biol.* **14:** 67.

Song J.J., Smith S.K., Hannon G.J., and Joshua-Tor L. 2004. Crystal structure of Argonaute and its implications for RISC slicer activity. *Science* **305:** 1434.

Song J.J., Liu J., Tolia N.H., Schneiderman J., Smith S.K., Martienssen R.A., Hannon G.J., and Joshua-Tor L. 2003. The crystal structure of the Argonaute2 PAZ domain reveals an RNA binding motif in RNAi effector complexes. *Nat. Struct. Biol.* **10:** 1026.

Sugaya K., Ajimura M., Tsuji H., Morimyo M., and Mita K. 1998. Alteration of the largest subunit of RNA polymerase II and its effect on chromosome stability in *Schizosaccharomyces pombe*. *Mol. Gen. Genet.* **258:** 279.

Talbert P.B. and Henikoff S. 2006. Spreading of silent chromatin: Inaction at a distance. *Nat. Rev. Genet.* **7:** 793.

Tang G., Reinhart B.J., Bartel D.P., and Zamore P.D. 2003. A biochemical framework for RNA silencing in plants. *Genes Dev.* **17:** 49.

Thon G., Hansen K.R., Altes S.P., Sidhu D., Singh D., Verhein-Hansen J., Bonaduce M.J., and Klar A.J. 2005. The Clr7 and Clr8 directionality factors and the Pcu4 cullin mediate heterochromatin formation in the fission yeast *Schizosaccharomyces pombe. Genetics* **171:** 1583.

Ting A.H., Schuebel K.E., Herman J.G., and Baylin S.B. 2005. Short double-stranded RNA induces transcriptional gene silencing in human cancer cells in the absence of DNA methylation. *Nat. Genet.* **37:** 906.

Vaistij F.E., Jones L., and Baulcombe D.C. 2002. Spreading of RNA targeting and DNA methylation in RNA silencing requires transcription of the target gene and a putative RNA-dependent RNA polymerase. *Plant Cell* **14:** 857.

Verdel A., Jia S., Gerber S., Sugiyama T., Gygi S., Grewal S.I., and Moazed D. 2004. RNAi-mediated targeting of heterochromatin by the RITS complex. *Science* **303:** 672.

Volpe T.A., Kidner C., Hall I.M., Teng G., Grewal S.I., and Martienssen R.A. 2002. Regulation of heterochromatic silencing and histone H3 lysine-9 methylation by RNAi. *Science* **297:** 1833.

Weinberg M.S., Villeneuve L.M., Ehsani A., Amarzguioui M., Aagaard L., Chen Z.X., Riggs A.D., Rossi J.J., and Morris K.V. 2006. The antisense strand of small interfering RNAs directs histone methylation and transcriptional gene silencing in human cells. *RnaNA* **12:** 256.

Wood V., Gwilliam R., Rajandream M.A., Lyne M., Lyne R., Stewart A., Sgouros J., Peat N., Hayles J., Baker S., et al. 2002. The genome sequence of *Schizosaccharomyces pombe. Nature* **415:** 871.

Zilberman D., Cao X., and Jacobsen S.E. 2003. ARGONAUTE4 control of locus-specific siRNA accumulation and DNA and histone methylation. *Science* **299:** 716.

Molecular Chaperones and Quality Control in Noncoding RNA Biogenesis

S.L. WOLIN[*†] AND E.J. WURTMANN[*]

*Departments of Cell Biology and †Molecular Biophysics and Biochemistry,
Yale University School of Medicine, New Haven, Connecticut 06536*

Although noncoding RNAs have critical roles in all cells, both the mechanisms by which these RNAs fold into functional structures and the quality control pathways that monitor correct folding are only beginning to be elucidated. Here, we discuss several proteins that likely function as molecular chaperones for noncoding RNAs and review the existing knowledge on noncoding RNA quality control. One protein, the La protein, binds many nascent noncoding RNAs in eukaryotes and is required for efficient folding of certain pre-tRNAs. In prokaryotes, the Sm-like protein Hfq is required for the function of many noncoding RNAs. Recent work in bacteria and yeast has revealed the existence of quality control systems involving polyadenylation of unstable noncoding RNAs followed by exonucleolytic degradation. In addition, the Ro protein, which is present in many animal cells and also certain bacteria, binds misfolded noncoding RNAs and is proposed to function in RNA quality control.

All cells contain an enormous variety of noncoding RNAs that have critical roles in gene expression. In the nucleus, the telomerase RNA is required to maintain chromosome ends, the 7SK RNA regulates transcription elongation, and U small nuclear RNAs (snRNAs) are key components of the spliceosome. The nucleolus contains a truly large number of small nucleolar RNAs (snoRNAs) that guide the processing and modification of the large ribosomal RNAs. In the cytoplasm, rRNAs and tRNAs are essential for protein synthesis, and the 7SL RNA-containing signal recognition particle targets newly synthesized presecretory proteins to the endoplasmic reticulum. In addition, numerous microRNAs (miRNAs) base-pair with mRNAs, regulating both stability and translation (Storz et al. 2005).

To function, many noncoding RNAs must fold into intricate structures and assemble with both proteins and other RNAs. As several studies have revealed that RNA has a tendency to become kinetically trapped in nonfunctional structures in vitro, it is likely that RNA-binding proteins assist RNA folding in cells (Herschlag 1995; Schroeder et al. 2004). RNA-binding proteins could act in several ways to facilitate formation of native RNAs. Specific RNA-binding proteins that are part of the final ribonucleoprotein complex may stabilize bound RNAs in the correct conformation. On the basis of crystallographic analyses, such a role has been proposed for many ribosomal proteins (Moore and Steitz 2003). Alternatively, similar to the molecular chaperones that function in protein folding, RNA-binding proteins that are not part of the final assembly could assist correct folding or resolve misfolded structures. Proteins that may function in this way include members of the DExH/D box family of RNA-dependent ATPases. One member of this family, *Neurospora crassa* CYT-19, disrupts alternative RNA structures that function as kinetic traps during the folding of a group I intron (Mohr et al. 2002). A related protein,

Mss116p, is required for efficient splicing of group I and group II introns in *Saccharomyces cerevisiae* (Huang et al. 2005). Another protein that may assist RNA folding through transient binding and release of nascent RNAs is the eukaryotic La protein, which binds the 3′ ends of many nascent noncoding RNAs and is often removed during 3′ maturation (Wolin and Cedervall 2002; Chakshusmathi et al. 2003). Finally, RNA-binding proteins, such as *Escherichia coli* Hfq, a member of the Sm-like family of proteins, can facilitate formation of intermolecular RNA–RNA interactions that are essential for function (Moller et al. 2002; Zhang et al. 2002).

In addition to mechanisms to assist correct folding, cells also possess RNA quality control systems to recognize and degrade misfolded and otherwise aberrant noncoding RNAs. Mutations that cause misfolding can result from gene mutations, transcriptional errors, or posttranscriptional editing events. Errors in RNA processing, failure to correctly assemble with proteins, and exposure to reactive oxygen species or other RNA-damaging agents can also result in nonfunctional RNAs. Some of the cellular mechanisms by which incorrectly folded and otherwise defective noncoding RNA molecules are recognized and degraded have been described. In both *E. coli* and yeast, degradation of several aberrant noncoding RNAs involves polyadenylation of the 3′ ends, followed by exoribonuclease digestion (Anderson 2005). However, as several yeast noncoding RNAs are degraded independently of this pathway (Alexandrov et al. 2006; Copela et al. 2006), additional quality control components remain to be uncovered.

In this paper, we examine the roles of the La and Sm-like proteins in assisting noncoding RNA biogenesis and function. We also discuss the existing information on quality control pathways for noncoding RNAs. A particular focus is the Ro protein, which binds misfolded noncoding RNAs and likely forms part of an RNA quality control pathway in animal cells.

LA STABILIZES NASCENT NONCODING RNAS AND ASSISTS PRE-TRNA FOLDING

The La protein binds to the 3′ ends of many newly synthesized noncoding RNAs, stabilizing them against exonucleases (Wolin and Cedervall 2002). Because La recognizes the sequence UUU_{OH}, which is the initial terminus of all RNAs synthesized by RNA polymerase III, RNAs bound include pre-tRNAs, pre-5S rRNAs, pre-U6 snRNA, pre-7SL RNA, and pre-RNase P RNA. As the majority of these RNAs undergo some form of 3′ processing, La is usually not bound to the mature RNAs. In the yeast *S. cerevisiae*, La also binds many noncoding RNAs synthesized by RNA polymerase II, including the spliceosomal U snRNAs and a number of snoRNAs (Wolin and Cedervall 2002; Inada and Guthrie 2004). In contrast to the RNA polymerase-III-transcribed RNAs, these RNAs are not primary transcripts, but represent partly processed RNAs that end in UUU_{OH}. As 3′ shortened forms of many nascent noncoding RNAs are detected in cells lacking La, binding by La normally stabilizes these RNAs from exonucleolytic digestion, favoring correct processing (Wolin and Cedervall 2002).

In addition to its role in protecting 3′ ends of nascent RNAs, La contributes to correct folding of certain pre-tRNAs. Although La is not essential in budding or fission yeast, La becomes essential in the presence of mutations that compromise pre-tRNA structure (Yoo and Wolin 1997; Johansson and Bystrom 2002; Chakshusmathi et al. 2003). For many of these mutations, the affected pre-tRNA is unstable without La, making it difficult to determine whether the instability is caused by misfolding, rather than the loss of end protection. However, for pre-tRNA$^{Arg}_{CCG}$, which contains a structurally fragile anticodon stem, La is required for efficient folding at low temperature or when the pre-tRNA structure is further perturbed by mutation (Fig. 1A) (Chakshusmathi et al. 2003). Chemical and enzymatic footprinting experiments suggest that in addition to binding pre-tRNA 3′ ends, La contacts and stabilizes the correctly folded anticodon stem-loop (Chakshusmathi et al. 2003).

Recent structural and biochemical studies have revealed how La protects 3′ termini from exoribonucleases. All La proteins contain three domains: an amino-terminal La motif, which adopts a winged helix fold (Alfano et al. 2004; Dong et al. 2004); a central RNA recognition motif (RRM); and a highly charged carboxyl terminus. Many La proteins also contain another RRM in the carboxyl terminus (Wolin and Cedervall 2002). Biochemical studies demonstrated that both the La motif and the central RRM are required for high-affinity binding (Goodier et al. 1997; Dong et al. 2004). A recent crystal structure of the human La motif and central RRM complexed to an oligonucleotide terminating in uridines revealed that the UUU_{OH} binds in a cleft formed by the La and RRM motifs (Teplova et al. 2006). As anticipated from mutagenesis (Dong et al. 2004), the majority of the hydrogen bonds and stacking interactions with the UUU_{OH} are from the La motif. The unusual ability of La to discriminate between RNAs ending in UUU_{OH} and UUU_p, which may help distinguish nascent RNA polymerase III transcripts from degraded RNAs, is partly due to a hydrogen bond between the 3′ OH and a conserved aspartate in the La motif (Dong et al. 2004; Teplova et al. 2006).

How might La stabilize correctly folded pre-tRNAs? As RNA footprinting experiments reveal that *S. cerevisiae* La protects the acceptor stem and the anticodon stem-loop of pre-tRNA$^{Arg}_{CCG}$ from nucleases, these regions may be in close contact with La (Chakshusmathi et al. 2003). Since the typical RNA interaction surface of the central RRM, a four-stranded antiparallel β-sheet, does not contact the 3′ uridylates (Teplova et al. 2006), it could stabilize other pre-tRNA elements. Consistent with this hypothesis, conserved residues in the RRM are necessary for maturation of several structurally impaired pre-tRNAs in *Schizosaccharomyces pombe* (Huang et al. 2006). Similarly, as the usual surface used by winged helix domain proteins to recognize double-stranded DNA is not involved in binding of La to 3′ ends (Dong et al. 2004; Teplova et al. 2006), this surface may be involved in other interactions. Finally, the carboxyl terminus, which is rich in lysines and arginines, could stabilize correctly folded RNAs. Clearly, a full description of how La stabilizes correctly folded pre-tRNAs will require a structure of the complex.

Although La is essential for viability in both mice and trypanosomes (Arhin et al. 2005; Foldynova-Trantirkova et al. 2005; Park et al. 2006), the finding that La is dispensable in two yeasts suggests that other proteins function redundantly with La in noncoding RNA biogenesis. To identify these components, our laboratory carried out genetic screens to identify mutations that cause *S. cerevisiae* La to become essential (Table 1). One class of mutations resided in essential tRNA genes, resulting in pre-tRNAs that were unstable without La (Yoo and Wolin 1997; Chakshusmathi et al. 2003). Other mutations were in proteins that, like La, contact newly synthesized noncoding RNAs. For example, La is required for viability in the presence of mutations in members of the Sm and Sm-like family of proteins (Pannone et al. 1998, 2001; Xue et al. 2000). Analyses of the mutant strains revealed that La stabilizes nascent U6 snRNAs when cells contain mutations in components of the Lsm2–Lsm8 ring (Pannone et al. 1998, 2001) and that La assists assembly of the U4 snRNA into the U4/U6 snRNP when cells contain a mutation in the snRNP core protein Smd1p (Xue et al. 2000). La is also important for growth when cells contain mutations in the arginyl tRNA synthetase or either of two tRNA modification enzymes, Trm1p and Trm61p (Calvo et al. 1999; Copela et al. 2006). Taken together, these experiments suggest that La functions redundantly with other proteins that contact nascent RNAs to stabilize these RNAs and assist their assembly into functional RNPs.

SM-LIKE PROTEIN RINGS MAY FACILITATE RNA–RNA INTERACTIONS

A family of proteins that stabilizes bound noncoding RNAs from nucleases, assists intermolecular base-pairing, and may modulate RNA structure are the Sm-like proteins. These proteins, which contain the Sm fold pre-

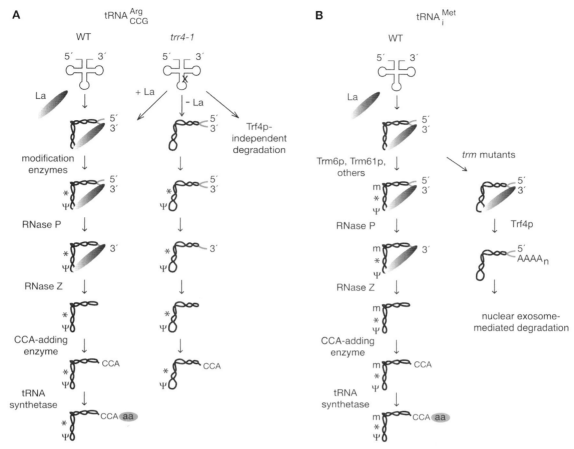

Figure 1. Biogenesis of tRNA$_{CCG}^{Arg}$ and tRNA$_i^{Met}$. (*A*) La is required for efficient folding of yeast tRNA$_{CCG}^{Arg}$. After removal of the 5′ leader by RNase P, cleavage of the 3′ trailer by an endonuclease such as RNase Z, and modification of multiple nucleotides (which occurs both before and after end maturation), tRNAs undergo CCA addition and aminoacylation. In yeast lacking La (not shown), the 3′ trailer is removed by exonucleases (Yoo and Wolin 1997). In yeast containing the *trr4-1* mutation, which disrupts the anticodon stem of tRNA$_{CCG}^{Arg}$, or during growth at low temperature, La is required to stabilize the correctly folded anticodon stem of tRNA$_{CCG}^{Arg}$. In the absence of La, the anticodon stem misfolds, and the mature tRNA is not aminoacylated (Chakshusmathi et al. 2003). In addition to the requirement for La, the *trr4-1* mutation causes a fraction of the mutant tRNA to be unstable. Degradation of the mutant tRNA does not require the Trf4 polymerase (Copela et al. 2006). (*B*) Methylation of adenosine 58 is required for stable accumulation of tRNA$_i^{Met}$. In yeast carrying mutations in either *TRM6* or *TRM61*, which encode the two subunits of the tRNA (m^{1}A58) methyltransferase (Anderson et al. 2000), the hypomethylated pre-tRNA$_i^{Met}$ undergoes polyadenylation by Trf4p and degradation by the nuclear exosome (Kadaba et al. 2004, 2006). As overexpression of *LHP1*, which encodes yeast La, results in increased levels of both pre- and mature tRNA$_i^{Met}$ in *trm6* strains (Anderson et al. 1998), binding by La may interfere with polyadenylation and/or exonucleolytic degradation. Consistent with this hypothesis, strains containing a mutation in *TRM61* and also lacking *LHP1* exhibit reduced growth and decreased tRNA$_i^{Met}$ (Calvo et al. 1999).

Table 1. Mutations That Cause La to Become Important or Essential for Growth in *S. cerevisiae*

Gene	Product	Reference
sup61$^+$	tRNA$_{CGA}^{Ser}$	Yoo and Wolin (1997)
TRT2	tRNA$_{CGU}^{Thr}$	Chakshusmathi et al. (2003)
TRR4	tRNA$_{CCG}^{Arg}$	Chakshusmathi et al. (2003)
TRM61	tRNA (m^{1}A58)methyltransferase subunit	Calvo et al. (1999)
TRM1	tRNA (m$_2^2$G)dimethyltransferase	Copela et al. (2006)
RRS1	Arginyl-tRNA synthetase	Copela et al. (2006)
SMD1	snRNP core protein Smd1p	Xue et al. (2000)
LSM5	Sm-like protein, component of U6 snRNP	Pannone et al. (2001)
LSM6	Sm-like protein, component of U6 snRNP	Pannone et al. (2001)
LSM7	Sm-like protein, component of U6 snRNP	Pannone et al. (2001)
LSM8	Sm-like protein, component of U6 snRNP	Pannone et al. (1998, 2001)

sent in the core proteins of the spliceosomal U1, U2, U4, and U5 snRNPs, associate to form six- or seven-membered rings. Sm-like proteins are evolutionarily ancient, as orthologs are present in all sequenced eukaryotic genomes, as well as in certain archaebacteria and eubacteria (Salgado-Garrido et al. 1999; Moller et al. 2002; Zhang et al. 2002). In eukaryotes, Sm-like proteins form several distinct complexes. One complex, consisting of the Lsm2–Lsm8 proteins (Lsm = like Sm), stabilizes the 3′ end of the spliceosomal U6 snRNA and assists basepairing with U4 RNA to form the U4/U6 snRNP (Pannone et al. 1998; Achsel et al. 1999; Mayes et al. 1999; Salgado-Garrido et al. 1999). A second complex, formed by Lsm1–Lsm7, binds the 3′ ends of deadenylated mRNAs and assists mRNA decapping (Tharun et al. 2000; He and Parker 2001). A third complex, containing the six proteins Lsm2–Lsm7, binds the yeast small nucleolar RNA snR5, which functions in pseudouridylation of rRNA (Fernandez et al. 2004). Two Sm-like proteins also associate with five bona fide Sm proteins to form the core of the U7 snRNP, which base-pairs with histone premRNA to direct the endonucleolytic cleavage required for mRNA 3′ end formation (Schumperli and Pillai 2004).

To date, the best-studied Sm-like protein ring is formed by E. coli Hfq. In contrast to the heteroheptameric rings found in eukaryotes, Hfq is a homohexamer (Fig. 2A). Hfq binds more than 20 noncoding RNAs in E. coli, most of which function by base-pairing with other RNAs, usually mRNAs (Zhang et al. 2003; Storz et al. 2005). For at least some of these RNAs, Hfq assists base-pairing of the noncoding RNA with its target (Moller et al. 2002; Zhang et al. 2002). Hfq also protects many bound noncoding RNAs from digestion by the endoribonuclease RNase E (Moll et al. 2003b; Zhang et al. 2003).

How does Hfq binding facilitate base-pairing of noncoding RNAs with their targets? For some mRNAs, Hfq alters the secondary structure of the mRNA, resulting in increased accessibility of the targeted sequence (Moll et al. 2003a; Geissmann and Touati 2004). For other mRNAs, no structural changes have been detected upon Hfq binding (Brescia et al. 2003; Lease and Woodson 2004). An alternative but not exclusive possibility is that Hfq accelerates base-pairing by bringing the two RNAs into proximity. Consistent with this idea, mutagenesis studies indicate that Hfq has two independent RNA-binding surfaces (Mikulecky et al. 2004). It is also possible that two Hfq rings interact with each other, bringing their individually bound RNAs together (Storz et al. 2004). A "two-ring" mechanism would be consistent with the fact that in eukaryotes, the various spliceosomal snRNAs that interact through base-pairing, such as the U4 and U6 snRNAs and the U2 and U6 snRNAs, are each bound by an Sm or Sm-like protein ring.

Do the eukaryotic Sm and Lsm proteins also modulate RNA structure and assist base-pairing? Because many of the RNAs bound by these proteins function by base-pairing with other RNAs, it has been proposed that these rings, similar to Hfq, facilitate annealing of the bound RNAs with their target sequences (Moller et al. 2002; Zhang et al. 2002). Support for this hypothesis comes from experiments showing that the Lsm2–Lsm8 complex

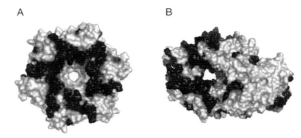

Figure 2. Hfq and Ro form rings of similar size. *Staphylococcus aureus* Hfq (*panel A*; see Schumacher et al. 2002) forms a homohexamer of 54 kD, which is similar in size to the monomeric ring formed by the *X. laevis* 60-kD Ro protein (*panel B*; see Stein et al. 2005). Surfaces with positive electrostatic potential are shown (*dark shading*). Both proteins have two RNA-binding sites. For Hfq, one binding site is around the central cavity and a second is on the distal face of the ring (Schumacher et al. 2002; Mikulecky et al. 2004). For Ro, the 3′ ends of misfolded RNAs insert into the hole, whereas helical portions bind to basic patches on the outer surface that overlap the Y RNA-binding site (Stein et al. 2005; Fuchs et al. 2006). The figure was prepared using MacPyMOL (DeLano Scientific LLC, San Carlos, California).

enhances U4:U6 annealing in vitro (Achsel et al. 1999), and that these Lsm proteins are required for multiple rounds of splicing in vitro (Verdone et al. 2004). Because both the Lsm1–Lsm7 and the Lsm2–Lsm8 complexes have recently been reconstituted from recombinant proteins (Zaric et al. 2005), information should soon be forthcoming as to whether these complexes, like Hfq, can alter RNA structure and enhance a variety of RNA–RNA interactions.

SIMILAR PATHWAYS FOR NONCODING RNA QUALITY CONTROL IN BACTERIA AND YEAST

Some of the first insights into how defective noncoding RNAs are targeted for decay came from experiments in *E. coli*. A mutant structurally unstable form of tRNA^{Trp} was found to be degraded through a pathway involving polyadenylation of the pre-tRNA by poly(A) polymerase (PAP) and degradation by the 3′ to 5′ exoribonuclease polynucleotide phosphorylase (PNPase) (Li et al. 2002). PNPase also functions with another exoribonuclease, RNase R, to degrade defective rRNAs (Cheng and Deutscher 2003). Although a requirement for polyadenylation in rRNA decay has not been reported, pre-23S rRNAs and 16S rRNAs containing short poly(A) tails have been observed in *E. coli* (Mohanty and Kushner 1999). Moreover, both overexpression of PAP and the absence of processing exoribonucleases result in accumulation of polyadenylated pre-23S rRNAs (Li et al. 1998; Mohanty and Kushner 1999), suggesting that a balance normally exists between processing and polyadenylation of precursors. In *E. coli*, polyadenylation acts as a signal for both mRNA and noncoding RNA decay and assists 3′ to 5′ exoribonucleases in initiating decay of structured RNAs by providing a single-stranded 3′ end (Deutscher 2006). Interestingly, Hfq copurifies with PAP and

PNPase and is required for the efficient polyadenylation of mRNAs that contain Rho-independent transcription terminators (Mohanty et al. 2004). Although Hfq is not required for the polyadenylation of pre-23S rRNAs, the finding that a chaperone protein is necessary to recruit PAP to some mRNAs highlights the question of whether polyadenylation of noncoding RNAs is stochastic or the result of recruitment.

A remarkably similar system for noncoding RNA quality control occurs in the yeast *S. cerevisiae*. A nuclear poly(A) polymerase, Trf4p, was found to function with the exosome, a complex of 3′ to 5′ exoribonucleases, to degrade a hypomethylated unstable pre-tRNA (Fig. 1B) (Kadaba et al. 2004). Trf4p is part of a complex called TRAMP, which also contains one of two likely RNA-binding proteins, Air1p and Air2p, and an RNA helicase, Mtr4p (LaCava et al. 2005; Vanacova et al. 2005; Wyers et al. 2005). In addition to targeting unstable pre-tRNAs for decay, TRAMP is involved in the decay of a mutant U6 snRNA, several precursors to rRNAs and snRNAs, and transcripts of intergenic regions (LaCava et al. 2005; Wyers et al. 2005; Kadaba et al. 2006). Yeast also contain a related poly(A) polymerase complex, consisting of the Trf4p-related protein Trf5p, Air1p, and Mtr4p, that has overlapping functions with the TRAMP complex (Egecioglu et al. 2006; Houseley and Tollervey 2006; Kadaba et al. 2006).

How the Trf4p and Trf5p poly(A) polymerases recognize defective RNAs is currently unknown. Because these polymerases lack the RNA-binding domain found in canonical poly(A) polymerases that act on mRNA, Air1p and Air2p are proposed to function in substrate recognition (Vanacova et al. 2005). In vitro, the TRAMP complex polyadenylates a structurally compromised mutant tRNA more efficiently than the wild-type tRNA (Vanacova et al. 2005). However, the natural tRNA targets of Trf4p polyadenylation are nascent pre-tRNAs containing both 5′ leader and 3′ trailer sequences (Kadaba et al. 2006). One possibility is that mutant pre-tRNAs that fail to be processed by the endonucleases that catalyze 5′ and 3′ maturation are preferentially targeted for decay because they accumulate as precursors. Similarly, mutant forms of U6 snRNA that are less efficient in their assembly with specific RNA-binding proteins, such as the Lsm2–Lsm8 complex, could be targets. Since the 3′ ends of many nascent noncoding RNAs are normally bound by proteins such as La, one possibility is that Trf4p may preferentially adenylate RNAs containing 3′ extensions that are not bound by proteins.

Although the Trf4p- and Trf5p-containing poly(A) polymerases function in the decay of a large number of yeast noncoding RNAs, the role of these polymerases in noncoding RNA quality control is not universal. A hypo-modified $tRNA_{AAC}^{Val}$ that is unstable at 37°C is degraded through a pathway that does not require Trf4p, Trf5p, or Rrp6p, a component of the nuclear exosome (Alexandrov et al. 2006). Similarly, two unstable mutant forms of $tRNA_{CCG}^{Arg}$ are degraded independently of Trf4p (Fig. 1A) (Copela et al. 2006). At least for $tRNA_{ACC}^{Val}$, degradation occurs at the level of the mature tRNA, rather than the pre-tRNA (Alexandrov et al. 2006). Defining these addi-tional decay pathways will likely be the focus of much effort in coming years. In addition, it will be important to elucidate how La, which binds many nascent noncoding RNAs, interfaces with Trf4p and other quality control pathways that monitor RNA integrity.

Ro BINDS MISFOLDED NONCODING RNAs AND LIKELY FUNCTIONS IN QUALITY CONTROL

A protein that likely functions in noncoding RNA quality control is the Ro 60-kD protein. This protein is found in both the nucleus and cytoplasm of many animal cells. In the cytoplasm, Ro is complexed with noncoding RNAs called Y RNAs. In the nucleus, Ro associates with mis-folded RNAs that are, at least in some cases, targeted for decay. In *Xenopus* oocyte nuclei, Ro binds a large collection of variant pre-5S rRNAs that misfold into nonfunctional structures (O'Brien and Wolin 1994; Shi et al. 1996). These variant RNAs also contain 8–10 additional nucleotides at the 3′ end, due to readthrough of the first RNA polymerase III transcription termination signal. Consistent with a general role in noncoding RNA quality control, Ro associates with variant misfolded U2 snRNAs in mouse embryonic stem cells (Chen et al. 2003).

Biochemical and structural studies have addressed how Ro recognizes its various RNA substrates. The amino-terminal two thirds of Ro consists of a series of α-helical HEAT repeats that form a ring with a central cavity (Fig. 2B) (Stein et al. 2005). The central cavity is approximately 10–15 Å in diameter and is lined with lysines and arginines. The HEAT repeat ring is clasped shut by a von Willebrand Factor A (vWFA) domain (Stein et al. 2005), which is also found in integrins and other proteins involved in cell adhesion. Within the vWFA domain is a MIDAS (metal-ion-dependent adhesion site) motif, which in integrins is a divalent-cation-dependent ligand-binding site and also transmits conformational changes that accompany ligand binding (Shimaoka et al. 2002).

A crystal structure of Ro complexed with a fragment of Y RNA revealed that Ro contains two RNA-binding sites (Fig. 2B) (Stein et al. 2005). Y RNA binds on the outer surface of the ring. A second RNA-binding site occurs within the channel, as a single-stranded RNA derived from one of the oligomers used to form the Y RNA fragment was present inside the central cavity. Biochemical experiments suggested that the 3′ ends of misfolded RNAs normally insert into the cavity and that helical portions of these RNAs bind on the outer surface. Because the binding sites of the Y RNAs and the misfolded RNAs overlap on the outer surface, Y RNAs were proposed to regulate access of Ro to other RNAs (Stein et al. 2005).

As with the Trf4p poly(A) polymerase, a major question concerns how Ro recognizes misfolded RNAs. Using misfolded pre-5S rRNA as a model, our laboratory found that Ro binds RNAs that contain both a single-stranded 3′ end and helical elements (Fuchs et al. 2006). As predicted from mutagenesis, a crystal structure of Ro bound to a fragment of misfolded pre-5S rRNA revealed that the single-stranded 3′ end binds in the central cavity, whereas an adjacent helix binds to a large basic platform on the surface of the ring. Because many of the contacts between Ro and the misfolded RNA are not sequence-specific, these experiments suggest that Ro may scavenge noncoding RNAs that fail to be bound by their specific RNA-binding proteins (Fuchs et al. 2006). Consistent with this hypothesis, the misfolded pre-5S rRNAs are poorly recognized by the 5S rRNA-binding protein TFIIIA in vivo (O'Brien and Wolin 1994).

What are the fates of the misfolded RNAs bound by Ro? In *Xenopus laevis* oocytes, the misfolded pre-5S rRNAs are inefficiently processed and eventually degraded (O'Brien and Wolin 1994). Thus, one possibility is that Ro recruits exonucleases, thus targeting associated noncoding RNAs for decay. An alternative but not exclusive possibility is that binding by Ro to the single-stranded ends of noncoding RNAs destabilizes adjacent helices, thus assisting in either degradation by exonucleases or refolding to functional forms.

CONCLUSIONS

Despite the importance of noncoding RNAs for many aspects of gene expression, much remains to be learned about both their mechanisms of folding within cells and the quality control pathways that monitor correct folding and RNP assembly. For mRNAs, it is well-established that the processes of transcription, RNP assembly, RNA processing, RNA export, and mRNA translation all interface with surveillance pathways. Far less is known as to how these steps in noncoding RNA biogenesis are monitored for formation of functional RNAs. Nonetheless, the recent identification of several components of noncoding quality control pathways has begun to yield information as to how defective noncoding RNAs and improperly assembled RNPs are recognized and handled, which should in turn result in a more complete understanding of noncoding RNA biogenesis.

ACKNOWLEDGMENTS

Work in our laboratory is supported by National Institutes of Health grants GM073863 and GM48410. E.W. is supported by a predoctoral fellowship from the National Science Foundation.

REFERENCES

Achsel T., Brahms H., Kastner B., Bachi A., Wilm M., and Luhrmann R. 1999. A doughnut-shaped heteromer of human Sm-like proteins binds to the 3′-end of U6 snRNA, thereby facilitating U4/U6 duplex formation in vitro. *EMBO J.* **18:** 5789.

Alexandrov A., Chernyakov I., Gu W., Hiley S.L., Hughes T.R., Grayhack E.J., and Phizicky E.M. 2006. Rapid tRNA decay can result from lack of nonessential modifications. *Mol. Cell* **21:** 87.

Alfano C., Sanfelice D., Babon J., Kelly G., Jacks A., Curry S., and Conte M.R. 2004. Structural analysis of cooperative RNA binding by the La motif and central RRM domain of human La protein. *Nat. Struct. Mol. Biol.* **11:** 323.

Anderson J.T. 2005. RNA turnover: Unexpected consequences of being tailed. *Curr. Biol.* **15:** R635.

Anderson J., Phan L., and Hinnebusch A.G. 2000. The Gcd10p/Gcd14p complex is the essential two-subunit tRNA(1-methyladenosine) methyltransferase of *Saccharomyces cerevisiae. Proc. Natl. Acad. Sci.* **97:** 5173.

Anderson J., Phan L., Cuesta R., Carlson B.A., Pak M., Asano K., Bjork G.R., Tamame M., and Hinnebusch A.G. 1998. The essential Gcd10p-Gcd14p nuclear complex is required for 1-methyladenosine modification and maturation of initiator methionyl-tRNA. *Genes Dev.* **12:** 3650.

Arhin G.K., Shen S., Perez I.F., Tschudi C., and Ullu E. 2005. Downregulation of the essential *Trypanosoma brucei* La protein affects accumulation of elongator methionyl-tRNA. *Mol. Biochem. Parasitol.* **144:** 104.

Brescia C.C., Mikulecky P.J., Feig A.L., and Sledjeski D.D. 2003. Identification of the Hfq-binding site on DsrA RNA: Hfq binds without altering DsrA secondary structure. *RNA* **9:** 33.

Calvo O., Cuesta R., Anderson J., Gutierrez N., Garcia-Barrio M.T., Hinnebusch A.G., and Tamame M. 1999. GCD14P, a repressor of GCN4 translation, cooperates with Gcd10p and Lhp1p in the maturation of initiator methionyl-tRNA in *Saccharomyces cerevisiae. Mol. Cell. Biol.* **19:** 4167.

Chakshusmathi G., Kim S.D., Rubinson D.A., and Wolin S.L. 2003. A La protein requirement for efficient pre-tRNA folding. *EMBO J.* **22:** 6562.

Chen X., Smith J.D., Shi H., Yang D.D., Flavell R.A., and Wolin S.L. 2003. The Ro autoantigen binds misfolded U2 small nuclear RNAs and assists mammalian cell survival after UV irradiation. *Curr. Biol.* **13:** 2206.

Cheng Z.F. and Deutscher M.P. 2003. Quality control of ribosomal RNA mediated by polynucleotide phosphorylase and RNase R. *Proc. Natl. Acad. Sci.* **100:** 6388.

Copela L.A., Chakshusmathi G., Sherrer R.L., and Wolin S.L. 2006. The La protein functions redundantly with tRNA modification enzymes to ensure tRNA structural stability. *RNA* **12:** 644.

Deutscher M.P. 2006. Degradation of RNA in bacteria: Comparison of mRNA and stable RNA. *Nucleic Acids Res.* **34:** 659.

Dong G., Chakshusmathi G., Wolin S.L., and Reinisch K.M. 2004. Structure of the La motif: A winged helix domain mediates RNA binding via a conserved aromatic patch. *EMBO J.* **23:** 1000.

Egecioglu D.E., Henras A.K., and Chanfreau G.F. 2006. Contributions of Trf4p- and Trf5p-dependent polyadenylation to the processing and degradative functions of the yeast nuclear exosome. *RNA* **12:** 26.

Fernandez C.F., Pannone B.K., Chen X., Fuchs G., and Wolin S.L. 2004. An Lsm2-Lsm7 complex in *Saccharomyces cerevisiae* associates with the small nucleolar RNA snR5. *Mol. Biol. Cell* **15:** 2842.

Foldynova-Trantirkova S., Paris Z., Sturm N.R., Campbell D.A., and Lukes J. 2005. The *Trypanosoma brucei* La protein is a candidate poly(U) shield that impacts spliced leader RNA maturation and tRNA intron removal. *Int. J. Parasitol.* **35:** 359.

Fuchs G., Stein A.J., Fu C., Reinisch K.M., and Wolin S.L. 2006. Structural and biochemical basis for misfolded RNA recognition by the Ro protein. *Nat. Struct. Mol. Biol.* **13:** 1002.

Geissmann T.A. and Touati D. 2004. Hfq, a new chaperoning role: Binding to messenger RNA determines access for small RNA regulator. *EMBO J.* **23:** 396.

Goodier J.L., Fan H., and Maraia R.J. 1997. A carboxy-terminal basic region controls RNA polymerase III transcription factor activity of human La protein. *Mol. Cell. Biol.* **17:** 5823.

He W. and Parker R. 2001. The yeast cytoplasmic LsmI/Pat1p complex protects mRNA 3′ termini from partial degradation. *Genetics* **158:** 1445.

Herschlag D. 1995. RNA chaperones and the RNA folding problem. *J. Biol. Chem.* **270:** 20871.

Houseley J. and Tollervey D. 2006. Yeast Trf5p is a nuclear poly(A) polymerase. *EMBO Rep.* **7:** 205.

Huang H.R., Rowe C.E., Mohr S., Jiang Y., Lambowitz A.M., and Perlman P.S. 2005. The splicing of yeast mitochondrial group I and group II introns requires a DEAD-box protein with RNA chaperone function. *Proc. Natl. Acad. Sci.* **102:** 163.

Huang Y., Bayfield M.A., Intine R.V., and Maraia R.J. 2006. Separate RNA-binding surfaces on the multifunctional La protein mediate distinguishable activities in tRNA maturation. *Nat. Struct. Mol. Biol.* **13:** 611.

Inada M. and Guthrie C. 2004. Identification of Lhp1p-associated RNAs by microarray analysis in *Saccharomyces cerevisiae* reveals association with coding and noncoding RNAs. *Proc. Natl. Acad. Sci.* **101:** 434.

Johansson M.J. and Bystrom A.S. 2002. Dual function of the tRNA(m(5)U54)methyltransferase in tRNA maturation. *RNA* **8:** 324.

Kadaba S., Wang X., and Anderson J.T. 2006. Nuclear RNA surveillance in *Saccharomyces cerevisiae:* Trf4p-dependent polyadenylation of nascent hypomethylated tRNA and an aberrant form of 5S rRNA. *RNA* **12:** 508.

Kadaba S., Krueger A., Trice T., Krecic A.M., Hinnebusch A.G., and Anderson J. 2004. Nuclear surveillance and degradation of hypomodified initiator tRNA[Met] in *S. cerevisiae. Genes Dev.* **18:** 1227.

LaCava J., Houseley J., Saveanu C., Petfalski E., Thompson E., Jacquier A., and Tollervey D. 2005. RNA degradation by the exosome is promoted by a nuclear polyadenylation complex. *Cell* **121:** 713.

Lease R.A. and Woodson S.A. 2004. Cycling of the Sm-like protein Hfq on the DsrA small regulatory RNA. *J. Mol. Biol.* **344:** 1211.

Li Z., Pandit S., and Deutscher M.P. 1998. Polyadenylation of stable RNA precursors in vivo. *Proc. Natl. Acad. Sci.* **95:** 12158.

Li Z., Reimers S., Pandit S., and Deutscher M.P. 2002. RNA quality control: Degradation of defective transfer RNA. *EMBO J.* **21:** 1132.

Mayes A.E., Verdone L., Legrain P., and Beggs J.D. 1999. Characterization of Sm-like proteins in yeast and their association with U6 snRNA. *EMBO J.* **18:** 4321.

Mikulecky P.J., Kaw M.K., Brescia C.C., Takach J.C., Sledjeski D.D., and Feig A.L. 2004. *Escherichia coli* Hfq has distinct interaction surfaces for DsrA, rpoS and poly(A) RNAs. *Nat. Struct. Mol. Biol.* **11:** 1206.

Mohanty B.K. and Kushner S.R. 1999. Analysis of the function of *Escherichia coli* poly(A) polymerase I in RNA metabolism. *Mol. Microbiol.* **34:** 1094.

Mohanty B.K., Maples V.F., and Kushner S.R. 2004. The Sm-like protein Hfq regulates polyadenylation dependent mRNA decay in *Escherichia coli. Mol. Microbiol.* **54:** 905.

Mohr S., Stryker J.M., and Lambowitz A.M. 2002. A DEAD-box protein functions as an ATP-dependent RNA chaperone in group I intron splicing. *Cell* **109:** 769.

Moll I., Leitsch D., Steinhauser T., and Blasi U. 2003a. RNA chaperone activity of the Sm-like Hfq protein. *EMBO Rep.* **4:** 284.

Moll I., Afonyushkin T., Vytvytska O., Kaberdin V.R., and Blasi U. 2003b. Coincident Hfq binding and RNase E cleavage sites on mRNA and small regulatory RNAs. *RNA* **9:** 1308.

Moller T., Franch T., Hojrup P., Keene D.R., Bachinger H.P., Brennan R.G., and Valentin-Hansen P. 2002. Hfq: A bacterial Sm-like protein that mediates RNA-RNA interaction. *Mol. Cell* **9:** 23.

Moore P.B. and Steitz T.A. 2003. The structural basis of large ribosomal subunit function. *Annu. Rev. Biochem.* **72:** 813.

O'Brien C.A. and Wolin S.L. 1994. A possible role for the 60 kd Ro autoantigen in a discard pathway for defective 5S ribosomal RNA precursors. *Genes Dev.* **8:** 2891.

Pannone B.K., Xue D., and Wolin S.L. 1998. A role for the yeast La protein in U6 snRNP assembly: Evidence that the La protein is a molecular chaperone for RNA polymerase III transcripts. *EMBO J.* **17:** 7442.

Pannone B.K., Kim S.D., Noe D.A., and Wolin S.L. 2001. Multiple functional interactions between components of the Lsm2-Lsm8 complex, U6 snRNA, and the yeast La protein. *Genetics* **158:** 187.

Park J.M., Kohn M.J., Bruinsma M.W., Vech C., Intine R.V., Fuhrmann S., Grinberg A., Mukherjee I., Love P.E., Ko M.S., et al. 2006. The multifunctional RNA-binding protein La is required for mouse development and for the establishment of embryonic stem cells. *Mol. Cell Biol.* **26:** 1445.

Salgado-Garrido J., Bragado-Nilsson E., Kandels-Lewis S., and Seraphin B. 1999. Sm and Sm-like proteins assemble in two related complexes of deep evolutionary origin. *EMBO J.* **18:** 3451.

Schroeder R., Barta A., and Semrad K. 2004. Strategies for RNA folding and assembly. *Nat. Rev. Mol. Cell Biol.* **5:** 908.

Schumacher M.A., Pearson R.F., Moller T., Valentin-Hansen P., and Brennan R.G. 2002. Structures of the pleiotropic translational regulator Hfq and an Hfq-RNA complex: A bacterial Sm-like protein. *EMBO J.* **21:** 3546.

Schumperli D. and Pillai R.S. 2004. The special Sm core structure of the U7 snRNP: Far-reaching significance of a small nuclear ribonucleoprotein. *Cell. Mol. Life Sci.* **61:** 2560.

Shi H., O'Brien C.A., Van Horn D.J., and Wolin S.L. 1996. A misfolded form of 5S rRNA is associated with the Ro and La autoantigens. *RNA* **2:** 769.

Shimaoka M., Takagi J., and Springer T.A. 2002. Conformational regulation of integrin structure and function. *Annu. Rev. Biophys. Biomol. Struct.* **31:** 485.

Stein A.J., Fuchs G., Fu C., Wolin S.L., and Reinisch K.M. 2005. Structural insights into RNA quality control: The Ro autoantigen binds misfolded RNAs via its central cavity. *Cell* **121:** 529.

Storz G., Altuvia S., and Wassarman K.M. 2005. An abundance of RNA regulators. *Annu. Rev. Biochem.* **74:** 199.

Storz G., Opdyke J.A., and Zhang A. 2004. Controlling mRNA stability and translation with small, noncoding RNAs. *Curr. Opin. Microbiol.* **7:** 140.

Teplova M., Yuan Y.R., Phan A.T., Malinina L., Ilin S., Teplov A., and Patel D.J. 2006. Structural basis for recognition and sequestration of UUU(OH) 3′ temini of nascent RNA polymerase III transcripts by La, a rheumatic disease autoantigen. *Mol. Cell* **21:** 75.

Tharun S., He W., Mayes A.E., Lennertz P., Beggs J.D., and Parker R. 2000. Yeast Sm-like proteins function in mRNA decapping and decay. *Nature* **404:** 515.

Vanacova S., Wolf J., Martin G., Blank D., Dettwiler S., Friedlein A., Langen H., Keith G., and Keller W. 2005. A new yeast poly(A) polymerase complex involved in RNA quality control. *PLoS Biol.* **3:** e189.

Verdone L., Galardi S., Page D., and Beggs J.D. 2004. Lsm proteins promote regeneration of pre-mRNA splicing activity. *Curr. Biol.* **14:** 1487.

Wolin S.L. and Cedervall T. 2002. The La protein. *Annu. Rev. Biochem.* **71:** 375.

Wyers F., Rougemaille M., Badis G., Rousselle J.-C., Dufour M.-E., Boulay J., Regnault B., Devaux F., Namane A., Seraphin B., et al. 2005. Cryptic pol II transcripts are degraded by a nuclear quality control pathway involving a new poly(A) polymerase. *Cell* **121:** 725.

Xue D., Rubinson D.A., Pannone B.K., Yoo C.J., and Wolin S.L. 2000. U snRNP assembly in yeast involves the La protein. *EMBO J.* **19:** 1650.

Yoo C.J. and Wolin S.L. 1997. The yeast La protein is required for the 3′ endonucleolytic cleavage that matures tRNA precursors. *Cell* **89:** 393.

Zaric B., Chami M., Remigy H., Engel A., Ballmer-Hofer K., Winkler F.K., and Kambach C. 2005. Reconstitution of two recombinant LSm protein complexes reveals aspects of their architecture, assembly and function. *J. Biol. Chem.* **22:** 16066.

Zhang A., Wassarman K.M., Ortega J., Steven A.C., and Storz G. 2002. The Sm-like Hfq protein increases OxyS RNA interaction with target mRNAs. *Mol. Cell* **9:** 11.

Zhang A., Wassarman K.M., Rosenow C., Tjaden B.C., Storz G., and Gottesman S. 2003. Global analysis of small RNA and mRNA targets of Hfq. *Mol. Microbiol.* **50:** 1111.

Stress-induced Reversal of MicroRNA Repression and mRNA P-body Localization in Human Cells

S.N. Bhattacharyya,* R. Habermacher,* U. Martine,† E.I. Closs,† and W. Filipowicz*

*Friedrich Miescher Institute for Biomedical Research, 4002 Basel, Switzerland; †Department of Pharmacology, Johannes Gutenberg University, 67, 55101 Mainz, Germany

In metazoa, microRNAs (miRNAs) imperfectly base-pair with the 3′-untranslated region (3′UTR) of mRNAs and prevent protein accumulation by either repressing translation or inducing mRNA degradation. Examples of specific mRNAs undergoing miRNA-mediated repression are numerous, but whether the repression is a reversible process remains largely unknown. Here, we show that cationic amino acid transporter 1 (CAT-1) mRNA and reporters bearing the CAT-1 3′UTR or its fragments can be relieved from the miRNA miR-122-induced inhibition in human hepatoma cells in response to different stress conditions. The derepression of CAT-1 mRNA is accompanied by its release from cytoplasmic processing bodies (P bodies) and its recruitment to polysomes, indicating that P bodies act as storage sites for mRNAs inhibited by miRNAs. The derepression requires binding of HuR, an AU-rich-element-binding ELAV family protein, to the 3′UTR of CAT-1 mRNA. We propose that proteins interacting with the 3′UTR will generally act as modifiers altering the potential of miRNAs to repress gene expression.

miRNAs are 20- to 22-nucleotide-long regulatory RNAs expressed in plants and metazoan animals. Current estimates indicate that hundreds of different miRNAs are encoded in individual genomes, with the number of human miRNAs possibly reaching 1000. Approximately 30% of all human genes are predicted to be subject to miRNA regulation. Specific functions and target mRNAs have been assigned to only a few dozen miRNAs, but it is apparent that miRNAs participate in the regulation of many different biological processes. Changes in miRNA expression are observed in human pathologies and some miRNAs were shown to act as oncogenes or tumor supressors (for review, see Ambros 2004; Bartel 2004; Wienholds and Plasterk 2005).

miRNAs regulate gene expression posttranscriptionally, by base-pairing to target mRNAs. In animals, most investigated miRNAs form imperfect hybrids with sequences in the 3′UTR, with the miRNA 5′-proximal "seed" region (positions 2–8) providing most of the pairing specificity (for review, see Filipowicz 2005; Tomari and Zamore 2005). The miRNA association results in translational repression, frequently accompanied by a considerable degradation of the mRNA by a non-RNAi mechanism (for review, see Pillai 2005; Valencia-Sanchez et al. 2006).

Argonaute (AGO) proteins are the essential and best-characterized components of miRNPs. In mammals, only one of the four AGO proteins, AGO2, is competent to catalyze cleavage of mRNA in the RNA interference (RNAi)-like mechanism (Liu et al. 2004; Meister et al. 2004). On the other hand, all four AGO proteins, AGO1–4, appear to function in miRNA repression (Liu et al. 2004; Meister et al. 2004; Pillai et al. 2004). The mechanism of translational inhibition by miRNAs is not well understood. Some natural or model miRNAs interfere with the initiation of protein synthesis (Humphreys et al. 2005; Pillai et al. 2005), although others may affect more downstream steps in translation (Olsen and Ambros 1999; Petersen et al. 2006). The

AGO proteins and repressed mRNAs are enriched in the cytoplasmic P bodies (Jakymiw et al. 2005; Liu et al. 2005; Pillai et al. 2005). The P-body association may represent a secondary event, which follows the translation inhibition step. Since mRNA catabolic enzymes also reside in P bodies, the relocation likely results in the reported mRNA degradation (Pillai 2005; Valencia-Sanchez et al. 2006).

To date, miRNAs have been primarily identified as negative regulators of expression of cellular mRNAs, and it remains unknown whether the inhibition of a specific mRNA can be effectively reversed. Clearly, the ability to disengage miRNPs from the repressed mRNA, or render them inactive, would make miRNA regulation much more dynamic and also more responsive to specific cellular needs. In this work, we present evidence that CAT-1 mRNA, which encodes the high-affinity cationic amino acid transporter and which is translationally repressed by miR-122 in Huh7 hepatoma cells, can be relieved from the miR-122 repression by subjecting Huh7 cells to different stress conditions. The derepression is accompanied by the release of CAT-1 mRNA from P bodies and its recruitment to polysomes, consistent with miR-122 inhibiting translational initiation. We provide evidence that the stress-induced up-regulation is mediated by binding of HuR, an AU-rich element (ARE)-binding protein, to the 3′UTR of CAT-1 mRNA.

RESULTS

CAT-1 mRNA as a Model to Investigate Reversibility of miRNA-mediated Repression

To investigate whether mRNA can be relieved from miRNA-mediated repression, we looked at the mRNA encoding the high-affinity cationic amino acid transporter, CAT-1, a member of the CAT (or SLC7A1-4) family of system y⁺ transporters. CAT-1, which facilitates uptake of arginine and lysine in mammalian cells, is expressed ubiquitously, but its levels vary significantly in different cells and tissues and its expression is known to undergo extensive regulation at both transcriptional and posttranscriptional levels (for review, see Hatzoglou et al. 2004). CAT-1 regulation has been studied intensively in rat C6 glioma cells, where transcription of the gene and stability and translation of the mRNA are up-regulated in response to different types of cellular stress (Yaman et al. 2002; Haztoglou et al. 2004). In human hepatoma Huh7 cells, CAT-1 mRNA activity is regulated by a liver-specific miRNA, miR-122. The human CAT-1 3′UTR contains several potential target sites for miR-122 (see Fig. 2A). Experiments performed with endogenous CAT-1 mRNA (Bhattacharyya et al. 2006) and reporters containing its 3′UTR (or fragments thereof) indicated that miR-122 has a repressive effect on translation and, in the case in some chimeric reporter mRNAs, may also cause limited destabilization of the target RNA (Chang et al. 2004; Bhattacharyya et al. 2006). Maintenance of low CAT-1 activity in liver cells is important to avoid hydrolysis of

the plasma arginine by arginase, a highly expressed enzyme in hepatocytes, which catalyzes the last step of the urea cycle (hydrolysis of arginine to ornithine and urea). However, under certain conditions, e.g., when urea cycle enzymes are down-regulated or during liver regeneration, CAT-1 expression is induced, most likely to sustain hepatocellular protein synthesis.

Amino Acid Deprivation Induces Translational Up-regulation of CAT-1 mRNA in Huh7 Cells

In cultured rat C6 glioma cells, CAT-1 protein expression increases in response to amino acid deprivation and other forms of cellular stress. Both transcriptional and posttranscriptional regulation were reported to contribute to the increase (Hatzoglou et al. 2004). We investigated the effect of amino acid starvation on expression of CAT-1 protein and mRNA in hepatoma Huh7 cells. The CAT-1 protein level increased markedly after 1 hour of starvation and then remained unchanged during several additional hours of amino acid depletion (Fig. 1A). In con-

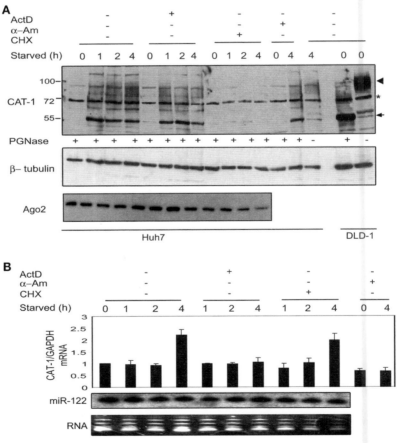

Figure 1. Starvation of Huh7 cells induces expression of CAT-1 protein independent of transcription. (*A*) Stress-induced expression of CAT-1 in Huh7 cells is a posttranscriptional event. Huh7 cells were cultured for 0–4 hours in amino-acid-depleted medium in the absence or presence of indicated inhibitors. Proteins were treated with PGNase F and analyzed by western blotting using indicated antibodies. Two lanes on the right contained protein extracts of DLD-1 cells, abundant in CAT-1. Positions of glycosylated and deglycosylated CAT-1 are indicated by an arrowhead and arrow. (*B*) Analysis of RNA isolated from Huh7 cells starved of amino acids in the presence of indicated inhibitors. (*Upper panel*) Real-time PCR quantification of CAT-1 mRNA. Normalized values are means (+/– SD) from three independent experiments, with GAPDH mRNA serving as an internal control. (*Two lower panels*) Northern analysis of miR-122 and staining of the gel with ethidium bromide. Details of methodological procedures used for experiments described in this and other figures are described in Bhattacharyya et al. (2006). (Reprinted, with permission, from Bhattacharyya et al. 2006 [© Elsevier].)

trast, the substantial increase in CAT-1 mRNA level, measured by either real-time polymerase chain reaction (PCR) (Fig. 1B) or northern blotting (not shown) (Bhattacharyya et al. 2006), was only detectable after 3–4 hours of starvation. In HepG2 hepatoma cells, which do not express miRNA miR-122, no appreciable change in either CAT-1 protein or mRNA level was observed during 4 hours of starvation (data not shown). The early induction of the CAT-1 protein in Huh7 cells was independent of RNA polymerase II transcription, since treatment with inhibitors of RNA polymerase II, either actinomycin D (ActD) or α-amanitin (α-Am), had no effect (Fig. 1A). However, the protein accumulation was inhibited by cycloheximide (CHX), an inhibitor of translational elongation (Fig. 1A). Amino acid starvation and treatment of cells with ActD or α-Am had no effect on the level of miR-122 or Ago2, essential components of miRNPs (Fig. 1A,B). However, inclusion of either inhibitor prevented the late (4-hour time point) accumulation of CAT-1 mRNA, demonstrating that they effectively inhibit RNA polymerase II transcription in Huh7 cells.

It was important to exclude the possibility that CAT-1 protein turns over rapidly in cells grown in the medium rich in amino acids, and its apparent induction by starvation is due to an increase in protein stability. To this end, we

have performed western analysis of lysates prepared from cells grown in either the presence or absence of CXH. The analysis indicated that instead of stabilizing the CAT-1 protein, the starvation increased its turnover (Bhattacharyya et al. 2006). The accelerated decay of CAT-1 in stressed cells provides a plausible explanation for the observation that steady-state levels of the protein do not continue to increase at times beyond 1 hour post-starvation. Taken together, the data indicate that starvation of Huh7 cells results in a de novo synthesis of the CAT-1 protein from the preexisting mRNA pool.

Response to Amino Acid Starvation and Other Types of Stress Is Mediated by the CAT-1 3′UTR and Involves miR-122

To test whether the translational induction described above is mediated by the CAT-1 mRNA 3′UTR, we measured the effect of starvation on the activity of different RL-cat reporters (Fig. 2A) in Huh7 cells. In RL-catA, the RL-coding region is fused to the 2.5-kb 3′UTR found in a short form of CAT-1 mRNA. In RL-catB, the 3′-proximal 1-kb region (referred to as region D) of the CAT-1 3′UTR is deleted. The remaining part contains three predicted miR-122-binding sites identified in the

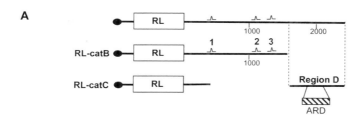

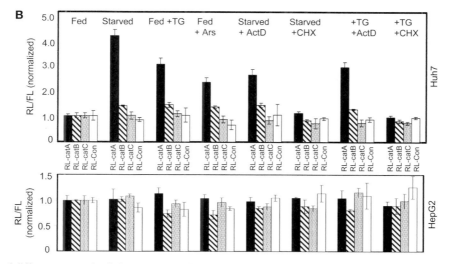

Figure 2. Effect of different types of cellular stress on activity of RL reporters in Huh7 and HepG2 cells. (*A*) Schemes of reporters bearing different segments of the CAT-1 3′UTR fused to the RL-coding region. Positions of potential miR-122-binding sites (1–3) and the approximately 1-kb region D and its AU-rich subregion ARD are indicated. (*B*) Expression of the RL-catA reporter is specifically up-regulated in stressed Huh7 cells (*upper panel*) but not HepG2 cells (*lower panel*). Cells transfected with indicated reporters were starved for 2 hours (Starved) or grown in the presence of either thapsigargin (TG; 2 hr) or arsenite (Ars; 30 min). ActD or CHX was added at the time of the shift to stress conditions. Values are normalized to activities in nonstressed (Fed) cells, which were set to 1. (Reprinted, with permission, from Bhattacharyya et al. 2006 [© Elsevier].)

2.5-kb CAT-1 3'UTR. In RL-catC, the CAT-1 3'UTR is shortened further to eliminate the region containing the miR-122 sites. Reporter activity, normalized to activity of the coexpressed firefly luciferase (FL), was tested in transfected Huh7 and HepG2 cells (Fig. 2B). In Huh7 cells, an approximately fourfold induction of RL activity was observed upon starvation of cells transfected with the reporter containing miR-122 sites, RL-catA, but not with that devoid of miRNA sites, RL-catC. Starvation increased expression of RL-catB by approximately 30% (Fig. 2B). As in the case of endogenous CAT-1 protein, the stress-induced expression of RL from RL-catA was inhibited by addition of CHX but not ActD (Fig. 2B). Likewise, other stress conditions such as the ER stress (induced by thapsigargin) or oxidative stress (induced by arsenite) stimulated expression of RL-catA approximately 2.5-fold, whereas the effect on other reporters was either minimal or absent (Fig. 2B). Amino acid starvation and treatment with thapsigargin or arsenite had no effect on the level of RL-catA mRNA, indicating that the effect was posttranscriptional (Bhattacharyya et al. 2006).

To test whether the stress-mediated activation of RL-catA indeed involves miR-122, we investigated whether responses observed in Huh7 cells can be reproduced in HepG2 cells, which do not express miR-122. Exposure of HepG2 cells to different forms of stress had no effect on expression of any RL-cat reporter (Fig. 2B, lower panel). However, when HepG2 cells were cotransfected with miR-122, up-regulation of RL-catA was clearly evident in starved cells, with activity of RL-catB and RL-catC remaining unchanged (Bhattacharyya et al. 2006).

HuR Protein Interacts with CAT-1 3'UTR and Is Essential for the Derepression

The stress-inducible RL-catA reporter differs from the reporter not undergoing activation, RL-catB, by the presence of an additional segment (region D) of the CAT-1 3'UTR (Fig. 2A). We found that region D is essential for the stress-induced relief of the miR-122-mediated repression in Huh7 cells. Moreover, this region can confer stress inducibility on a reporter inhibited by let-7 RNA, an abundant miRNA expressed in HeLa cells. The let-7-specific RL-3xBulge reporter (Pillai et al. 2005) was used to further dissect CAT-1 region D. It was found that the central part of region D, containing sequences rich in A+U and U residues and referred to hereafter as region ARD (see Fig. 2A), is essential for mediating the stress-induced reversal of miRNA repression (Bhattacharyya et al. 2006).

HuR is an AU-rich element (ARE)-binding protein, a member of the ELAV family of proteins, implicated in different aspects of posttranscriptional regulation (for review, see Brennan and Steitz 2001; Katsanou et al. 2005; Lal et al. 2005). In response to various types of cellular stress, HuR is mobilized from the nucleus to the cytosol, where it may modulate translation or increase the stability of different target mRNAs, including CAT-1 mRNA in rat glioma cells (Yaman et al. 2002). We found that HuR relocates from the nucleus to the cytosol

upon amino acid starvation also in Huh7 cells (Fig. 3A) and that the RNA-mediated depletion of HuR, using two different small interfering RNAs (siRNAs) (Fig. 3B), eliminates the RL-catA response in comparison to control cells or cells treated with control siRNA (Fig. 3C). We tested, by native gel analysis, if the ^{32}P-labeled ARD fragment, which is essential for mediating the stress-induced derepression, can interact with a recombinant GST-HuR fusion protein. As shown in Figure 3D, purified GST-HuR but not GST, formed a complex with the ARD fragment, and the complex was competed by an excess of unlabeled fragment ARD but not the ΔARD portion of region D. Additional experiments indicated that CAT-1 mRNA can be specifically immunoselected from the soluble cytoplasmic fraction of starved Huh7 cells by the anti-HuR antibody. We have also demonstrated that HuR binding per se does not cause activation of the RL reporter devoid of miR-122 sites and not repressed by the miRNA (Bhattacharyya et al. 2006).

Taken together, the above experiments indicate that HuR has a role in the stress-induced activation of CAT-1 mRNA and RL reporters undergoing repression mediated by either miR-122 or let-7 RNA.

Stress Induces HuR-dependent Relocation of CAT-1 mRNA from P bodies

mRNA reporters repressed by miRNAs were found to localize in P bodies (Liu et al. 2005; Pillai et al. 2005). We analyzed the intracellular localization of the endogenous CAT-1 mRNA and RL-cat reporters in cells grown under different conditions. In situ hybridization revealed that in nonstarved Huh7 cells, CAT-1 mRNA is concentrated in P bodies, as demonstrated by its colocalization with the P-body marker, GFP-Dcp1a (Fig. 4A). P-body enrichment of the mRNA was abolished when cells were transfected with the 2'-O-methyl oligonucleotide complementary to miR-122 but not with control anti-miR-15 oligonucleotide (Fig. 4B), indicating that the localization was dependent on miR-122. Most importantly, in Huh7 cells grown for 2 hours under amino acid deficiency, a condition that markedly increases CAT-1 protein without an effect on the mRNA level (see Fig. 1), CAT-1 mRNA was no longer detectable in P bodies (Fig. 4A). Starvation did not produce an appreciable decrease in the miR-122 signal in P bodies (Bhattacharyya et al. 2006), arguing for an effect specific for the CAT-1 mRNA and possibly only a limited number of other mRNAs among the many regulated by miR-122 in liver cells (Krutzfeldt et al. 2005). As expected, CAT-1 mRNA did not colocalize to P bodies in HepG2 cells. However, in HepG2 cells transfected with miR-122, CAT-1 mRNA was enriched in P bodies, further supporting the idea that the repression and P-body localization of CAT-1 mRNA are controlled by miR-122 (Bhattacharyya et al. 2006).

To find out whether the relocation of CAT-1 mRNA from P bodies, like the activation of its expression, also requires HuR, we studied CAT-1 distribution in Huh7 cells in which HuR protein had been knocked down by RNAi. The knockdown had no effect on the P-body

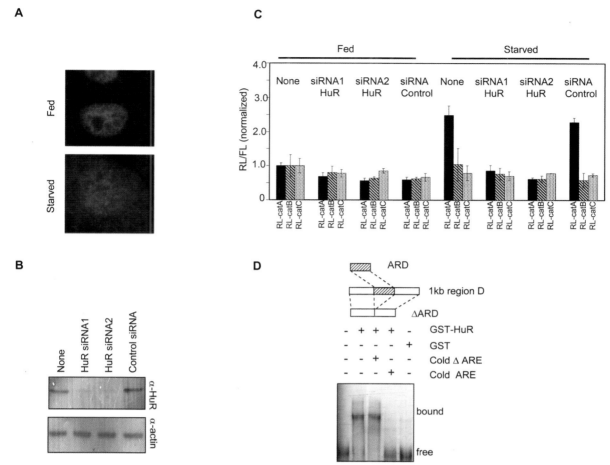

Figure 3. HuR relocates from the nucleus to the cytoplasm in stressed Huh7 cells, interacts with the CAT-1 3′UTR, and is required for the stress-induced derepression of the RL-catA reporter. (*A*) Starvation-induced relocation of HuR. Huh7 cells, either nonstarved (Fed) or starved for 2 hours (Starved) were fixed, and the localization of HuR (*red*) was determined using mouse anti-HuR monoclonal antibodies. (*B*) Two different siRNAs effectively deplete HuR. Western blot was performed 72 hours after siRNA transfection. (*C*) Knockdown of HuR prevents starvation-induced derepression of RL-catA. Huh7 cells were cotransfected with RL-cat reporters and indicated siRNAs; 72 hours after transfection, cells were transferred for an additional 2 hours to a medium either with or without amino acids. The values are means from three transfections +/– standard deviation. (*D*) Recombinant GST-HuR protein interacts with the CAT-1 ARD fragment. ^{32}P-labeled ARD RNA was incubated with either purified GST-HuR or GST alone in the absence or presence of indicated cold competitors, and complexes were analyzed on a native gel. (Reprinted, with permission, from Bhattacharyya et al. 2006 [© Elsevier].)

enrichment of CAT-1 mRNA in control cells. However, it prevented mobilization of the mRNA from these structures upon amino acid starvation. In cells transfected with nonspecific siRNA, the relocation did take place, similarly as in nontreated Huh7 cells (Fig. 4C).

We also determined the intracellular localization of different RL reporters repressed by either miR-122 or let-7 miRNA. The repressed reporters localized to P bodies, and their relocalization from these structures in response to amino acid depletion was dependent on the CAT-1 region D present downstream from the miRNA-binding sites. Reporters devoid of region D remained enriched in P bodies in both nonstarved and starved cells (Bhattacharyya et al. 2006).

In starved Huh7 cells, CAT-1 mRNA becomes extractable from cells permeabilized with digitonine, very likely as a consequence of the relocation of mRNA from P bodies to the cytosol (Pillai et al. 2005; Bhattacharyya et al. 2006). As shown in Figure 5A, the amount of CAT-1 mRNA present in the cytosol prepared from permeabilized Huh7 cells increased already after 20 minutes following the shift to the amino-acid-depleted medium. Importantly, the shift of the CAT-1 mRNA to the cytosol upon stressing the cells occurred with a kinetics similar to the appearance of HuR in this fraction and also paralleled the accumulation of CAT-1 protein (Fig. 5). Hence, the stress-induced derepression of the CAT-1 mRNA translation appears to be a rapid event.

Together, the results demonstrate that CAT-1 mRNA and RL reporters are concentrated in P bodies when repressed by the miRNA but are rapidly mobilized from these structures under conditions, including cellular stress, that preclude miRNA repression. The experiments further support a role for HuR and the CAT-1 mRNA region D in mediating the response induced by the amino acid stress. Moreover, they indicate that such a response may be a more general phenomenon, applying to different cell types and miRNA–mRNA combinations.

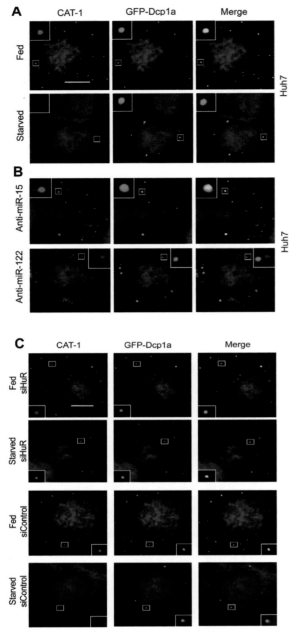

Figure 4. CAT-1 mRNA accumulation in P bodies is miR-122-dependent, and its stress-induced relocation from P-bodies requires HuR. Details of cell treatment are indicated at the left of each row. P bodies were visualized by measuring GFP-Dcp1a fluorescence (*green*), and CAT-1 by in situ hybridization with Cy3-labeled probes (*red*). DAPI (*blue*) stained the nucleus. Cells were starved for 2 hours prior to fixation. Exposure time of the red channel for starved Huh7 cells (*A*), the anti-miR-122 row (*B*), and starved siControl cells (*C*) was ten times longer than that for other images. (*Insets*) Enlargements of indicated regions. For quantification and discussion of overlap between CAT-1 mRNA and GFP-Dcp1a foci, see Bhattacharyya et al. (2006). (*A,B*) CAT-1 mRNA is mobilized from P bodies upon starvation of Huh7 cells (*A*) or upon transfection with anti-miR-122 but not anti-miR-15 oligonucleotide (*B*). Bar, 5 μm. (*C*) HuR is required for the stress-induced mobilization of CAT-1 mRNA from P bodies. Huh7 cells were cotransfected with pGFP-Dcp1a and either anti-HuR or control siRNA. After 48 hours, a fraction of the cells was starved for 2 hours. (Reprinted, with permission, from Bhattacharyya et al. 2006 [© Elsevier].)

Stress-induced Relocation of CAT-1 mRNA from P Bodies Is Accompanied by Its Entry to Polysomes

We found previously that repression of protein synthesis by *let-7* RNA in HeLa cells is accompanied by a less effective entry of target reporters to polysomes, indicative of the translation initiation block (Pillai et al. 2005). Gradient analysis of Huh7 cell extracts indicated that amino acid starvation results in an increase in the fraction of CAT-1 mRNA associated with polysomes. In contrast, β-tubulin mRNA moved toward the top of the gradient in response to starvation, consistent with a general inhibitory effect of stress on translation (Fig. 6A,B). Treatment of Huh7 cells with anti-miR-122 but not control anti-*let-7a* 2'-O-methyl oligonucleotide resulted in the CAT-1 mRNA shift to polysomes similar to that induced by starvation (Fig. 6C). Of note, a fraction of the "repressed" CAT-1 mRNA sedimented faster than the polysome-associated CAT-1 mRNA present in stressed or anti-miR-122-treated cells. Possibly, this material represents P-body aggregates containing CAT-1 mRNA.

Taken together, these data indicate that relocation of CAT-1 mRNA from P bodies and relief of the miRNA-mediated repression is accompanied by recruitment of CAT-1 mRNA to polysomes, consistent with miR-122 inhibiting translational initiation.

CONCLUSIONS

Our work demonstrates that CAT-1 mRNA, and reporters bearing the CAT-1 3'UTR or its fragments, can be relieved from miR-122-mediated repression in Huh7 cells subjected to amino acid deprivation or the ER or oxidative stress. Observations that the response to the amino acid starvation stress can be recapitulated in other cell lines by either an ectopic supply of miR-122 or the use of chimeric reporters targeted by another miRNA argue for a general importance of this type of regulation. We also demonstrate that repressed CAT-1 and reporter mRNAs accumulate in P bodies in an miR-122-dependent process and that the derepression is accompanied by the release of the mRNAs from these structures. The demonstrated stress-induced mobilization of mRNAs from P bodies provides evidence that metazoan P bodies represent sites not only of mRNA turnover, but also of storage of translationally repressed mRNAs. So far, such evidence was available for baker's yeast, an organism lacking miRNA regulation (Brengues et al. 2005). Time-course experiments measuring the appearance of CAT-1 mRNA in a soluble cytosolic fraction and the formation of CAT-1 protein suggest that reactivation of mRNAs sequestered in P bodies is a very rapid process. Hence, it is tempting to speculate that P bodies may act as general storage sites for mRNAs, which need to be quickly mobilized into polysomes under specific cellular conditions.

Although miRNAs may affect gene expression in different ways (for review, see Pillai 2005; Valencia-Sanchez et al. 2006), recent findings indicated that *let-7* RNA and some model miRNAs in mammals inhibit

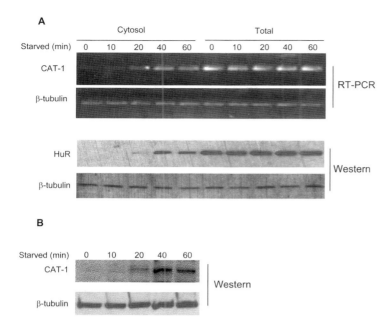

Figure 5. The stress-induced relocation of HuR protein and CAT-1 mRNA to the cytosol and the accumulation of CAT-1 protein are rapid events. (*A*) Accumulation of CAT-1 mRNA and HuR protein in the cytosol upon stressing Huh7 cells occurs with a similar kinetics. Huh7 cells were starved for amino acids for increasing time, and levels of CAT-1 mRNA (*upper panels*) and HuR protein (western; *lower panels*) were measured in the cytosolic and total extracts of starved cells. Cytosolic extracts were prepared by permeabilizing cells with digitonine (Bhattacharyya et al. 2006). (*B*) Kinetics of the CAT-1 protein accumulation in Huh7 cells subjected to amino acid deprivation stress, determined by western.

translation of reporter mRNAs at the initiation step (Humphreys et al. 2005; Pillai et al. 2005) and that repressed mRNAs localize to P bodies for either storage or degradation (Liu et al. 2005; Pillai 2005; Pillai et al. 2005; Valencia-Sanchez et al. 2006). The data presented in this work indicate that this scenario also applies to the miRNA-mediated regulation of an endogenous mRNA. The stress- or anti-miR-122-induced relocalization of CAT-1 mRNA from P bodies was accompanied by its increased association with polysomes, consistent with the miRNA inhibition acting at the initiation step of transla-

tion. Importantly, activation of CAT-1 and reporter mRNAs by exposing the cells to stress or transfecting them with anti-miR-122 oligonucleotide had no appreciable effect on the mRNA level, indicating that miR-122 in Huh7 cells controls CAT-1 mRNA mainly at the translational level and not the stability level.

Our results strongly argue for a role of HuR in the stress-induced activation of mRNAs undergoing miR-122-mediated repression. The HuR knockdown prevented both the translational activation of repressed mRNAs and their mobilization from P bodies.

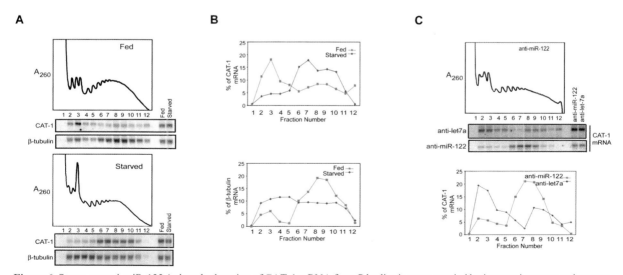

Figure 6. Stress- or anti-miR-122-induced relocation of CAT-1 mRNA from P bodies is accompanied by its recruitment to polysomes. (*A*) Distribution of CAT-1 mRNA in extracts from cells fed with amino acids (*upper panels*) or starved for amino acids (*lower panels*). RNA extracted from individual fractions was analyzed by northern blots with probes specific for CAT-1 and β-tubulin mRNAs. Two lanes at the right represent input RNA isolated from fed and starved cells. (*B*) Quantification of distribution of mRNAs analyzed in panel *A*, expressed as a percentage of total radioactivity present in each lane. (*C*) Distribution of CAT-1 mRNA in extracts from cells treated with either anti-*let-7a* or anti-miR-122 (*middle panels*). A_{260} profile of the anti-*let-7a* extract was similar to that shown at the top, and distribution of β-tubulin mRNA in both gradients was similar to that shown in panel *A*, fed cells (data not shown). Quantification of CAT-1 mRNA is at the bottom. (Reprinted, with permission, from Bhattacharyya et al. 2006 [© Elsevier].)

Moreover, a recombinant HuR interacted with the CAT-1 3'UTR fragment implicated in mediating the stimulatory effect of stress (see Fig. 3D), whereas the endogenous HuR associated with all reporter mRNAs undergoing derepression but not with their inactive variants bearing mutations in the predicted HuR-binding sites (Bhattacharyya et al. 2006). HuR is a ubiquitously expressed member of the ELAV family of proteins, which also comprises three neuronal proteins. In response to different types of cellular stress, HuR is mobilized from the nucleus to the cytosol, where it may modulate translation and/or stability of different mRNAs (for review, see Brennan and Steitz 2001; Katsanou et al. 2005; Lal et al. 2005). Our data suggest that at least some of the known effects of HuR, both translational and stability-related, may be due to the interference of HuR with the function of miRNAs, which would result in enhanced translation or stability of mRNA. HuR shuttles between the nucleus and cytoplasm, and HuR has been suggested to bind some ARE-containing mRNAs in the nucleus and chaperone them to the cytoplasm (Gallouzi and Steitz 2001; Lal et al. 2005). The in situ experiments demonstrating CAT-1 mRNA and RL reporter localization in P bodies in unstressed hepatoma and HeLa cells make it very unlikely that redistribution of mRNA between the nucleus and the cytoplasm contributes to the effects described in our work.

The demonstration that mRNAs repressed by miRNAs can depart P bodies to return to active translation indicates that P bodies are dynamic structures, exchanging their content rapidly with that of the cytosol (Andrei et al. 2005; Brengues et al. 2005). It is possible that HuR, following its relocation to the cytoplasm in stressed cells, shifts the P-body-to-cytosol equilibrium of

repressed mRNAs by binding to AREs in the 3'UTR. Whether this is accompanied by the dissociation of miRNPs from the mRNA or just prevents miRNPs from acting as effectors in the repression remains to be established (Fig. 7). It will also be interesting to study other details of HuR involvement in relieving the miRNP repression. The findings that several identified protein ligands of HuR are protein phosphatase PP2A inhibitors (Brennan et al. 2000) and that HuR can undergo methylation (Li et al. 2002) or synergize with other RNA-binding proteins (Katsanou et al. 2005) indicate that HuR is a part of an elaborated network involved in posttranscriptional regulation of gene expression.

Other examples of the reversible action of miRNAs have recently been identified in neuronal cells. In neurons, many mRNAs are transported along the dendrites as repressed mRNPs to become translated at the final destination, dendritic spines, upon synaptic activation. Such local translation is important for spine development, learning, and memory (Sutton and Schuman 2005). miRNA miR-134 is implicated in translational regulation of Limk1, a protein kinase important for spine development, in cultured rat neurons. Limk1 mRNA appears to be relieved from the miR-134-mediated repression in dendritic spines in response to extracellular stimuli, in a process involving mTOR (mammalian target of rapamycin) (Schratt et al. 2006). In *Drosophila*, stimulation of olfactory neurons, which leads to long-term memory formation, is associated with proteolysis of Armitage, a protein essential for the assembly of the RNA-induced silencing complex (RISC)/miRNP complexes. As the result of Armitage degradation, mRNAs that are normally repressed by miRNAs, including the one encoding calcium/calmodulin-dependent protein kinase II (CamKII), become

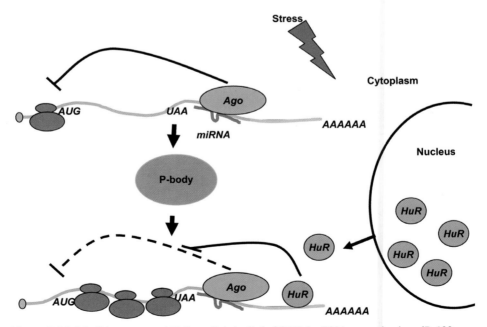

Figure 7. Model of the stress- and HuR-mediated relief of CAT-1 mRNA repression by miR-122.

effectively translated in the synapse (Ashraf et al. 2006). The regulation in *Drosophila* differs from the examples studied in mammalian cells, in that the Armitage depletion most probably indiscriminately prevents the formation of repressed mRNPs, rather than causing reactivation of specific mRNAs controlled by miRNAs.

In addition to HuR, three other ELAV proteins, HuA, HuB, and HuD, are expressed in neurons, and a role of HuD in stability and translation of some neuronal mRNAs has already been documented (Perrone-Bizzozero and Bolognani 2002). It will be interesting to find out whether, similarly to HuR in hepatoma cells, the other ELAV proteins modulate miRNA-mediated regulation in neurons. Likewise, it will be important to determine whether other classes of RNA-binding proteins interacting with the 3′UTR will act as modifiers altering the potential of miRNAs to repress gene expression.

ACKNOWLEDGMENTS

We thank C. Clayton, T. Hobman, Y. Nagamine, C. Ender, R. Pillai, C. Artus, and E. Bertrand for providing plasmids and/or antibodies. S.N.B. is a recipient of a long-term HFSP fellowship. The Friedrich Miescher Institute is supported by the Novartis Research Foundation.

REFERENCES

Andrei M.A., Ingelfinger D., Heintzmann R., Achsel T., Rivera-Pomar R., and Luhrmann R. 2005. A role for eIF4E and eIF4E-transporter in targeting mRNPs to mammalian processing bodies. *RNA* **11**: 717.

Ambros V. 2004. The functions of animal microRNAs. *Nature* **431**: 350.

Ashraf S.I., McLoon A.L., Sclarsic S.M., and Kunes S. 2006. Synaptic protein synthesis associated with memory is regulated by the RISC pathway in *Drosophila*. *Cell* **124**: 191.

Bartel D.P. 2004. MicroRNAs: Genomics, biogenesis, mechanism, and function. *Cell* **116**: 281.

Bhattacharyya S.N., Habermacher R., Martine U., Closs E.I., and Filipowicz W. 2006. Relief of microRNA-mediated translational repression in human cells subjected to stress. *Cell* **125**: 1111.

Brengues M., Teixeira D., and Parker R. 2005. Movement of eukaryotic mRNAs between polysomes and cytoplasmic processing bodies. *Science* **310**: 486.

Brennan C.M. and Steitz J.A. 2001. HuR and mRNA stability. *Cell. Mol. Life Sci.* **58**: 266.

Brennan C.M., Gallouzi I.E., and Steitz J.A. 2000. Protein ligands to HuR modulate its interaction with target mRNAs in vivo. *J. Cell Biol.* **151**: 11.

Chang J., Nicolas E., Marks D., Sander C., Lerro A., Buendia M.A., Xu C., Mason W.S., Moloshok T., Bort R., et al. 2004. miR-122, a mammalian liver-specific microRNA, is processed from *hcr* mRNA and may downregulate the high affinity cationic amino acid transporter CAT-1. *RNA Biol.* **1**: 106.

Filipowicz W. 2005. RNAi: The nuts and bolts of the RISC machine. *Cell* **122**: 17.

Gallouzi I.E. and Steitz J.A. 2001. Delineation of mRNA export pathways by the use of cell-permeable peptides. *Science* **294**: 1895.

Hatzoglou M., Fernandez J., Yaman I., and Closs E. 2004. Regulation of cationic amino acid transport: the story of the CAT-1 transporter. *Annu. Rev. Nutr.* **24**: 377.

Humphreys D.T., Westman B.J., Martin D.I., and Preiss T. 2005. MicroRNAs control translation initiation by inhibiting eukaryotic initiation factor 4E/cap and poly(A) tail function. *Proc. Natl. Acad. Sci.* **102**: 16961.

Jakymiw A., Lian S., Eystathioy T., Li S., Satoh M., Hamel J.C., Fritzler M.J., and Chan E.K. 2005. Disruption of GW bodies impairs mammalian RNA interference. *Nat. Cell Biol.* **7**: 1267.

Katsanou V., Papadaki O., Milatos S., Blackshear P.J., Anderson P., Kollias G., and Kontoyiannis D.L. 2005. HuR as a negative posttranscriptional modulator in inflammation. *Mol. Cell* **16**: 777.

Krutzfeldt J., Rajewsky N., Braich R., Rajeev K.G., Tuschl T., Manoharan M., and Stoffel M. 2005. Silencing of microRNAs in vivo with "antagomirs." *Nature* **438**: 685.

Lal A., Kawai T., Yang X., Mazan-Mamczarz K., and Gorospe M. 2005. Antiapoptotic function of RNA-binding protein HuR effected through prothymosin alpha. *EMBO J.* **24**: 1852.

Li H., Park S., Kilburn B., Jelinek M.A., Henschen-Edman A., Aswad D., Stallcup M.R., and Laird-Offringa I.A. 2002. Lipopolysaccharide-induced methylation of HuR, an mRNA-stabilizing protein, by CARM1. *J. Biol. Chem.* **277**: 44623.

Liu J., Valencia-Sanchez M.A., Hannon G.J., and Parker R. 2005. MicroRNA-dependent localization of targeted mRNAs to mammalian P-bodies. *Nat. Cell Biol.* **7**: 719.

Liu J., Carmell M.A., Rivas F.V., Marsden C.G., Thomson J.M., Song J.J., Hammond S.M., Joshua-Tor L., and Hannon G.J. 2004. Argonaute2 is the catalytic engine of mammalian RNAi. *Science* **305**: 1437.

Meister G., Landthaler M., Patkaniowska A., Dorsett Y., Teng G., and Tuschl T. 2004. Human Argonaute2 mediates RNA cleavage targeted by miRNAs and siRNAs. *Mol. Cell* **15**: 185.

Olsen P.H. and Ambros V. 1999. The *lin-4* regulatory RNA controls developmental timing in *Caenorhabditis elegans* by blocking LIN-14 protein synthesis after the initiation of translation. *Dev. Biol.* **216**: 671.

Perrone-Bizzozero N. and Bolognani F. 2002. Role of HuD and other RNA-binding proteins in neural development and plasticity. *J. Neurosci. Res.* **68**: 121.

Petersen C.P., Bordeleau M.E., Pelletier J., and Sharp P.A. 2006. Short RNAs repress translation after initiation in mammalian cells. *Mol. Cell* **21**: 533.

Pillai R.S. 2005. MicroRNA function: Multiple mechanisms for a tiny RNA? *RNA* **11**: 1753.

Pillai R.S., Artus C.G., and Filipowicz W. 2004. Tethering of human Ago proteins to mRNA mimics the miRNA-mediated repression of protein synthesis. *RNA* **10**: 1518.

Pillai R.S., Bhattacharyya S.N., Artus C.G., Zoller T., Cougot N., Basyuk E., Bertrand E., and Filipowicz W. 2005. Inhibition of translational initiation by let-7 miRNA in human cells. *Science* **309**: 1573.

Schratt G.M., Tuebing F., Nigh E.A., Kane C.G., Sabatini M.E., Kiebler M., and Greenberg M.E. 2006. A brain-specific miRNA regulates dendritic spine development. *Nature* **439**: 283.

Sutton M.A. and Schuman E.M. 2005. Local translational control in dendrites and its role in long-term synaptic plasticity. *J. Neurobiol.* **64**: 116.

Tomari Y. and Zamore P.D. 2005. Perspective: Machines for RNAi. *Genes Dev.* **19**: 517.

Valencia-Sanchez M.A., Liu J., Hannon G.J., and Parker R. 2006 Control of translation and mRNA degradation by miRNAs and siRNAs. *Genes Dev.* **20**: 515.

Wienholds E. and Plasterk R.H. 2005. MiRNA function in animal development. *FEBS Lett.* **579**: 5911.

Yaman I., Fernandez J., Sarkar B., Schneider R.J., Snider M.D., Nagy L.E., and Hatzoglou M. 2002. Nutritional control of mRNA stability is mediated by a conserved AU-rich element that binds the cytoplasmic shuttling protein HuR. *J. Biol. Chem.* **277**: 41539.

MicroRNAs Silence Gene Expression by Repressing Protein Expression and/or by Promoting mRNA Decay

I. Behm-Ansmant, J. Rehwinkel, and E. Izaurralde

MPI for Developmental Biology, D-72076 Tuebingen, Germany

MicroRNAs (miRNAs) represent a novel class of genome-encoded eukaryotic regulatory RNAs that silence gene expression posttranscriptionally. Although the proteins mediating miRNA biogenesis and function have been identified, the precise mechanism by which miRNAs regulate the expression of target mRNAs remains unclear. We summarize recent work from our laboratory demonstrating that miRNAs silence gene expression by at least two independent mechanisms: by repressing translation and/or by promoting mRNA degradation. In *Drosophila*, both mechanisms require Argonaute 1 (AGO1) and the P-body component GW182. Moreover, mRNA degradation by miRNAs is effected by the enzymes involved in general mRNA decay, including deadenylases and decapping enzymes, which also localize to P bodies. Our findings suggest a model for miRNA function in which AGO1 associates with miRNA targets through miRNA:mRNA base-pairing interactions. GW182 interacts with AGO1 and recruits deadenylases and decapping enzymes, leading to mRNA degradation. However, not all miRNA targets are degraded: Some stay in a translationally silent state, from which they may eventually be released. We propose that the final outcome of miRNA regulation (i.e., degradation vs. translational repression) is influenced by other RNA-binding proteins interacting with the targeted mRNA.

miRNAs represent a specific class of genome-encoded small RNAs that regulate gene expression posttranscriptionally (Bartel 2004; Filipowicz 2005). Hundreds of miRNAs and their targets have been identified or predicted in different organisms. They affect a broad range of biological processes including cell differentiation and proliferation, apoptosis, metabolism, and development (Bartel 2004; Filipowicz 2005).

miRNA regulatory functions are effected by the Argonaute proteins that promote decay or translational repression of mRNAs having sequences complementary to the miRNAs (Bartel 2004; Filipowicz 2005). Plant miRNAs are generally fully complementary to their targets and elicit endonucleolytic cleavage in the region of the mRNA that is base-paired with the miRNA. Endonucleolytic cleavage may also occur even when miRNAs are not fully complementary to their binding sites (Llave et al. 2002; Mallory et al. 2004; Allen et al. 2005; Guo et al. 2005; Schwab et al. 2005). This implies that plant miRNAs are similar to small interfering RNAs (siRNAs) with respect to their mode of action, although they differ on their mode of biogenesis (Bartel 2004; Filipowicz 2005).

The majority of animal miRNAs are only partially complementary to their targets and silence gene expression by mechanisms that remain elusive (Filipowicz 2005). At first, animal miRNAs were reported to repress translation without affecting mRNA levels (Lee et al. 1993; Wightman et al. 1993). More recently, several reports have shown that animal miRNAs can also induce significant degradation of target mRNAs despite imperfect mRNA–miRNA base-pairing (Bagga et al. 2005; Jing et al. 2005; Lim et al. 2005; Behm-Ansmant et al. 2006; Giraldez et al. 2006; Rehwinkel et al. 2006; Wu et al. 2006). Consistently, transcripts that are up-regulated in cells in which the miRNA pathway is inhibited (e.g., by

depletion of Dicer or Argonaute proteins) are enriched in predicted and validated miRNA targets (Giraldez et al. 2006; Rehwinkel et al. 2006). Conversely, ectopic expression of specific miRNAs in cells in which they are not normally present leads to a reduction of the levels of transcripts containing binding sites for the miRNA (Lim et al. 2005). Finally, microarrays have been used to identify miRNA targets, indicating that miRNAs do indeed affect mRNA levels (Lim et al. 2005; Giraldez et al. 2006, Rehwinkel et al. 2006). In contrast to the situation in plants, however, under most circumstances, mRNA decay by miRNAs in animal cells may not occur via endonucleolytic cleavage but rather by directing mRNAs to the general mRNA degradation machinery, and thus by accelerating their decay (Behm-Ansmant et al. 2006; Giraldez et al. 2006; Wu et al. 2006).

Several lines of evidence support the existence of a link between the miRNA pathway and general mRNA decay. First, mammalian Argonaute proteins (AGO1–AGO4), miRNAs, and miRNA targets colocalize to cytoplasmic foci known as processing bodies (P bodies) (Jakymiw et al. 2005; Liu et al. 2005a,b; Meister et al. 2005; Pillai et al. 2005; Sen and Blau 2005). P bodies are discrete cytoplasmic domains where proteins required for bulk mRNA degradation in the 5′ to 3′ direction accumulate (e.g., deadenylases, the decapping DCP1:DCP2 complex, and the 5′ to 3′ exonuclease XRN1; for review, see Valencia-Sanchez et al. 2006). Second, human AGO1 and AGO2 associate with the decapping coactivator DCP1 and with GW182, a protein with glycine-tryptophan repeats (GW repeats) which localizes to P bodies and is required for P-body integrity (Eystathioy et al. 2002, 2003; Jakymiw et al. 2005; Liu et al. 2005a,b; Meister et al. 2005; Sen and Blau 2005). Third, depletion of GW182 in human cells impairs both miRNA function and mRNA decay triggered by complementary siRNAs (Jakymiw et al. 2005;

Liu et al. 2005b; Meister et al. 2005). Similarly, miRNA function is impaired in *Drosophila* Schneider cells (S2 cells) depleted of GW182 or the decapping DCP1:DCP2 complex (Rehwinkel et al. 2005). Finally, the *Caenorhabditis elegans* protein AIN-1, which is related to GW182, is required for gene regulation by at least a subset of miRNAs (Ding et al. 2005).

Nevertheless, despite this link between miRNA-mediated silencing and mRNA degradation, there are multiple examples of miRNA targets for which translational repression is observed without detectable changes in the mRNA level (for review, see Valencia-Sanchez et al. 2006), suggesting that for these targets, regulation by miRNAs may occur independently of mRNA decay. How miRNAs regulate translation is not well understood. Two independent studies have shown that miRNAs inhibit translation initiation at an early stage involving the cap structure, as mRNAs translated via cap-independent mechanisms were shown to escape miRNA-mediated silencing (Humphreys et al. 2005; Pillai et al. 2005). Other studies have suggested that translation inhibition occurs after initiation on the basis of the observation that miRNAs and some targets remain associated with polysomes (Olsen and Ambros 1999; Kim et al. 2004; Nelson et al. 2004; Petersen et al. 2006). Moreover, Petersen et al. (2006) reported that mRNAs translated via cap-independent mechanisms are subjected to miRNA regulation. A possible explanation for these contradictory results is that miRNAs regulate gene expression by different mechanisms, although the possibility that these discrepancies are due to differences in experimental approaches cannot be excluded (see Valencia-Sanchez et al. 2006).

We have investigated the mechanism by which miRNAs silence gene expression. Here, we describe briefly the assays we use to monitor miRNA function in *Drosophila* cells, a summary of the overall results, and our current model regarding miRNA function.

miRNA-MEDIATED GENE SILENCING INVOLVES TRANSLATIONAL REPRESSION AND/OR mRNA DEGRADATION

To investigate miRNA function in *Drosophila* cells, we have generated a series of reporters in which the coding region of firefly luciferase (F-Luc) is flanked by the 3´UTRs (untranslated regions) of the *Drosophila* gene CG10011 regulated by miR-12, and of Nerfin or Vha68-1 genes, which contain binding sites for miR-9b (Fig. 1a). *Drosophila* S2 cells are transiently transfected with the F-Luc reporters, a plasmid expressing the primary miRNA transcript or the corresponding vector without insert (empty vector), and a plasmid encoding *Renilla* luciferase (R-Luc) as a transfection control. F-Luc activity is normalized to that of *Renilla* to compensate for possible differences in transfection efficiencies.

Using this assay, we have shown that expression of the F-Luc reporter having the 3´UTR of CG10011 (F-Luc-CG10011) is strongly reduced by cotransfection of a plasmid expressing miR-12, whereas expression of the firefly luciferase reporters fused to the Nerfin or Vha68-1 3´UTRs (F-Luc-Nerfin, F-Luc-Vha68-1) is inhibited by cotransfection of miR-9b (Fig. 1b) (Rehwinkel et al. 2005,

2006; Behm-Ansmant et al. 2006). In all three cases, silencing of luciferase expression by the cognate miRNAs is prevented in cells depleted of AGO1 (see Fig. 2) (Rehwinkel et al. 2005, 2006; Behm-Ansmant et al. 2006), demonstrating that silencing of these reporters occurs via the miRNA pathway. Note that the genomes of multicellular organisms encode multiple Argonaute proteins. In *Drosophila*, Argonaute paralogs have evolved specialized roles: AGO1 mediates miRNA function, whereas AGO2 catalyzes siRNA-guided endonucleolytic cleavage of mRNAs (RNA interference [RNAi]) (Okamura et al. 2004; Rand et al. 2004; Miyoshi et al. 2005).

To determine whether miRNAs silence F-Luc expression by inhibiting translation directly, or indirectly by reducing mRNA levels, we analyzed the steady-state levels of the F-Luc mRNA by northern blot and normalized them to the levels of the control R-Luc mRNA. We observed that miRNAs triggered a reduction in mRNA levels to different extents (Behm-Ansmant et al. 2006). Expression of miR-9b led to a slight decrease of the F-Luc-Nerfin mRNA levels but to a significant reduction of the F-Luc-Vha68-1 mRNA levels; miR-12 triggered a strong reduction of F-Luc-CG10011 mRNA levels (Behm-Ansmant et al. 2006).

The decrease in mRNA levels observed for the Nerfin and Vha68-1 reporters is significantly smaller than the reduction observed in F-Luc activity. Indeed, after normalizing firefly luciferase activity to the corresponding mRNA levels, we observed that miR-9b leads to a net 5-fold and 2.5-fold reduction in protein expression for the Nerfin and Vha68-1 reporters, respectively (Fig. 1c). In contrast, silencing of F-Luc-CG10011 by miR-12 can be attributed primarily to the reduction of mRNA levels (Fig. 1c). These results indicate that miRNAs silence gene expression by two mechanisms, one involving inhibition of protein expression and one a reduction of mRNA levels.

miRNAs ACCELERATE mRNA DEGRADATION

To investigate whether the reduction of mRNA levels caused by miRNAs is a consequence of increased mRNA degradation, we measured the half-lives of the reporters in the absence or presence of the cognate miRNAs following inhibition of transcription with actinomycin D. The half-lives of the F-Luc reporter having the Nerfin, Vha68-1, and CG10011 3´UTRs were about 141 minutes, 200 minutes, and longer than 300 minutes, respectively. In cells expressing the miRNAs, the half-lives of these reporters were reduced to about 38 minutes, 23 minutes, and 10 minutes (Fig. 1d). Thus, even for the Nerfin reporter, whose steady-state levels do not change dramatically in the presence of miR-9b, the degradation rate is increased.

In the presence of the miRNAs, the mRNA reporters have biphasic decay kinetics, suggesting the existence of a heterogeneous pool of mRNAs (for instance, free or bound to the miRNA) undergoing different rates of degradation. In addition, the reduction in half-lives is more marked than the changes in steady-state mRNA levels. One possible explanation for this discrepancy is that steady-state mRNA levels are determined by transcription and decay rates, and at steady state, transcription rates may compensate for decay. Alternatively, we cannot

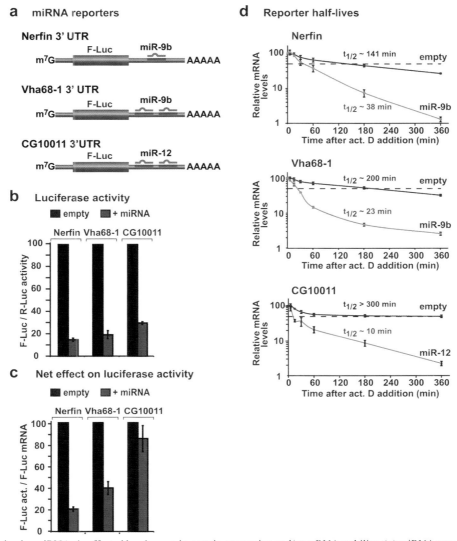

Figure 1. Silencing by miRNAs is effected by changes in protein expression and/or mRNA stability. (*a*) miRNA reporters. The F-Luc open reading frame (F-Luc) is flanked by the 3′UTRs of Nerfin, Vha68-1, or CG10011 mRNAs, which contain binding sites for miR-9b or miR-12 as indicated. (*b,c*) S2 cells were transfected with plasmids expressing miRNA reporters (Nerfin, Vha68-1, or CG10011), plasmids expressing miR-9b or miR-12 (*red bars*), or the corresponding empty vector (*black bars*). *Renilla* luciferase (R-Luc) served as a transfection control. F-Luc activity and the corresponding mRNA levels were measured and normalized to those of the *Renilla*. Normalized F-Luc activities in cells transfected with the empty vector (*black bars*) were set to 100%. In panel *c*, the normalized values of F-Luc activity shown in panel *b* are divided by the normalized mRNA levels (not shown) to estimate the net effect of miRNAs on protein expression. Error bars represent standard deviations from at least three independent experiments. (*d*) S2 cells were transfected with the miRNA reporters shown in panel *a*. Three days after transfection, cells were treated with actinomycin D and harvested at the indicated time points. The decay of the F-Luc reporters was monitored in cells coexpressing the miRNA (*pink*) or the corresponding empty vector (*blue*). RNA samples were analyzed by northern blot (not shown). The levels of F-Luc reporter transcripts normalized to those of the long-lived rp49 mRNA are plotted as a function of time. mRNA half-lives ($t_{1/2}$) are indicated.

exclude the possibility that the actinomycin D treatment is having secondary effects on mRNA stability.

MiRNAs PROMOTE DEADENYLATION AND DECAPPING

We recently reported that mRNA decay by miRNAs requires components of the CCR4:NOT deadenylase complex and the decapping DCP1:DCP2 complex (Behm-Ansmant et al. 2006). Figure 2a,b shows a comparison of the effects of depleting components of the deadenylase or decapping complexes on miRNA activity

for the Nerfin and CG10011 reporters, which are regulated mainly at the translational level and at the mRNA levels, respectively. In this representation, F-Luc activity or mRNA levels are normalized to those of *Renilla*. For each knockdown, the normalized F-Luc activities or mRNA levels measured in the absence of the miRNA are set to 100 (not shown).

For the Nerfin reporter, expression of miR-9b strongly reduces F-Luc activity but mRNA levels only slightly (Fig. 2a, black and orange bars, respectively). Both luciferase expression and mRNA levels are restored in cells depleted of AGO1 or GW182. This indicates that

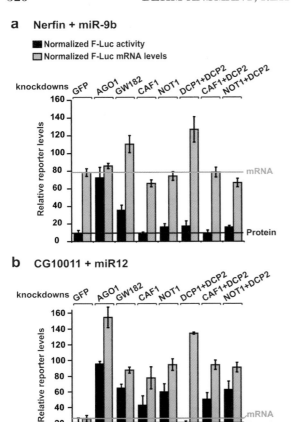

a Nerfin + miR-9b

■ Normalized F-Luc activity
■ Normalized F-Luc mRNA levels

b CG10011 + miR12

c CG1998 mRNA (K-box miRNAs)

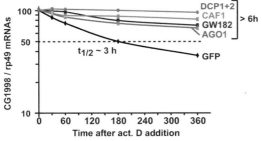

Figure 2. mRNA decay triggered by miRNAs occurs by a deadenylation and decapping mechanism requiring GW182. (*a,b*) S2 cells depleted of the proteins indicated above the panels were transfected with plasmids expressing the Nerfin or CG10011 miRNA reporters. Control cells were treated with green fluorescent protein (GFP) double-stranded RNA (dsRNA). F-Luc activity and the corresponding mRNA levels were measured and normalized to those of the *Renilla*. Normalized F-Luc activities (*black bars*) and mRNA levels (*orange bars*) in the presence of the miRNA are shown. In cells transfected with the empty vector, these values were set to 100% for each knockdown (not shown). (*Black horizontal lines*) F-Luc activity in control cells; (*orange horizontal lines*) mRNA levels in control cells. Error bars represent standard deviations from three independent experiments. (*c*) Endogenous miRNA targets are degraded by deadenylation and decapping. The decay of CG1998 mRNA (a predicted target of K-box miRNAs) was monitored in S2 cells depleted of AGO1, GW182, CAF1, or DCP1:DCP2 following inhibition of transcription by actinomycin D. Control cells were treated with GFP dsRNA. The levels of the CG1998 mRNA were analyzed by northern blot, normalized to rp49 mRNA (not shown), and plotted against time. mRNA half-lives ($t_{1/2}$) calculated from the decay curves are indicated.

GW182 also has a role in translational silencing by miRNAs, in agreement with previous studies (Jakymiw et al. 2005; Liu et al. 2005b; Meister et al. 2005; Rehwinkel et al. 2005).

There are two major deadenylase complexes in *Drosophila*: the PAN2:PAN3 and the CCR4:NOT complexes (Temme et al. 2004). Depletion of PAN2, PAN3, CAF1, or NOT1 does not have a significant effect on luciferase expression or mRNA levels of the Nerfin reporter (Fig. 2a) (data not shown). Similarly, codepletion of CAF1 or NOT1 with DCP2 does not restore luciferase expression or mRNA levels (Fig. 2a). In contrast, depletion of DCP1:DCP2 restores mRNA levels indicating that this reporter is degraded by decapping (Fig. 2a). Although mRNA levels are restored in cells depleted of decapping enzymes, luciferase expression is not. A potential explanation for this lack of restoration is provided by the analysis of mRNA levels by northern blotting. Indeed, we observe that reporter transcripts accumulating in cells depleted of the decapping enzymes are deadenylated and are thus expected to be translated less efficiently (Behm-Ansmant et al. 2006). Deadenylation is observed only in the presence of the miRNA, indicating that miRNAs trigger deadenylation. Deadenylation is, however, unlikely to be the cause of the translational repression because luciferase activity is not restored in cells depleted of deadenylases.

When we analyzed the CG10011 reporter, which is mainly regulated at the mRNA level, we observed that depletion of AGO1 or GW182 restores both luciferase expression and mRNA levels, indicating that mRNA decay is AGO1- and GW182-dependent. Depletion of CAF1 or NOT1 individually or in combination with DCP2 also restores both luciferase expression and mRNA levels, providing strong evidence for the conclusion that regulation of this reporter by miR-12 occurs at the level of mRNA degradation (Fig. 2b). Again, depletion of DCP1:DCP2 restores mRNA levels, but as observed for the Nerfin reporter, transcripts accumulating in these cells are deadenylated in the presence of the miRNA (Behm-Ansmant et al. 2006), providing a possible explanation for the partial or lack of restoration of luciferase expression.

In summary, these findings indicate that gene silencing by miRNAs involves translational repression and/or mRNA deadenylation and decapping and both require GW182. The results from our laboratory in *Drosophila* cells are remarkably consistent with recent studies in zebra fish embryos and human cells showing that miRNAs promote accelerated deadenylation of their targets (Giraldez et al. 2006; Wu et al. 2006). Moreover, Wu et al. (2006) reported that mRNAs lacking a poly(A) tail are nevertheless subjected to translational repression by miRNAs, indicating that this repression occurs by a deadenylation-independent mechanism.

A question that remains open is whether miRNA-mediated mRNA degradation is a consequence of translational repression or whether these represent two independent mechanisms by which miRNAs silence gene expression as proposed by Wu et al. (2006). As mentioned above, inhibition of translation is not always accompanied by changes in mRNA levels. Conversely, targets regulated mainly at the translational level are subjected to accelerated deadenylation, but depletion of CAF1 or NOT1 does not restore protein expression. These findings suggest that translational repression and mRNA degradation represent two independent mechanisms by which miRNAs regulate gene expression.

ENDOGENOUS miRNA REPORTERS ARE DEGRADED IN AN AGO1- AND GW182-DEPENDENT MANNER

To investigate the significance of the results obtained with the reporters for the miRNA pathway, we analyzed mRNA expression profiles in S2 cells depleted of GW182 or AGO1. We observed that depletion of these proteins leads to correlated changes in mRNA expression levels, indicating that they act in the same pathway. Furthermore, transcripts commonly up-regulated by AGO1 and GW182 are enriched in predicted and validated miRNA targets. This shows that, as for luciferase reporters, endogenous miRNA targets are regulated by GW182 (Behm-Ansmant et al. 2006).

We also analyzed the degradation rate of endogenous miRNA targets identified by expression profiling of AGO1 or GW182-depleted cells (Behm-Ansmant et al. 2006; Rehwinkel et al. 2006). The Axs and CG1998 mRNAs are predicted targets of miR-285 and K-box miRNAs, respectively (Stark et al. 2005). These mRNAs are at least three- to sixfold up-regulated in cells depleted of AGO1 or GW182. In cells depleted of AGO1, GW182, CAF1, or DCP1:DCP2, the half-lives of these mRNAs increase from approximately 3 hours to more than 6 hours, suggesting that they are degraded by deadenylation and decapping in an AGO1- and GW182-dependent manner (Fig. 2c) (Behm-Ansmant et al. 2006).

It is worth noting that for some endogenous miRNA targets, depletion of decapping enzymes or of components of the CCR4:NOT complex leads to a stronger stabilization of the mRNA than that observed in cells depleted of AGO1 or GW182. This indicates that only a fraction of these endogenous transcripts is targeted by the cognate miRNAs. Depletion of the CCR4:NOT complex or of the decapping enzymes prevents general degradation and degradation via the miRNA pathway, leading to a stronger stabilization of the transcripts. This is likely to be the case for mRNAs targeted by low-abundance miRNAs.

Together, the results obtained for the F-Luc reporters and endogenous targets demonstrate unequivocally a link between miRNA-mediated gene silencing and the machinery for general mRNA degradation. They also show that GW182 is a critical effector of miRNA function in animal cells.

GW182 BELONGS TO A CONSERVED FAMILY OF PROTEINS

GW182 belongs to a family of proteins containing a central ubiquitin-associated (UBA) domain, a carboxy-terminal RNA recognition motif (RRM), and three blocks of glycine-tryptophan repeats (referred to as amino-terminal, middle, and carboxy-terminal GW repeats). Furthermore,

a glutamine-rich (Q-rich) region is located between the UBA domain and the RRM (Fig. 3a) (Eystathioy et al. 2002; Behm-Ansmant et al. 2006).

Vertebrates contain three GW182 family members (TNRC6A/GW182, TNRC6B, and TNRC6C), but there is a single ortholog in insects and no ortholog in worms or fungi. Sequence alignment of all members of the protein family revealed the presence of two highly conserved motifs (I and II) of approximately 50 residues within the amino-terminal GW repeats (Fig. 3a). The insect proteins start exactly with motif I, whereas the vertebrate orthologs have amino-terminal extensions. Interestingly, the amino-terminal domain of GW182 proteins encompassing motif II bears similarity to the GW-like regions in

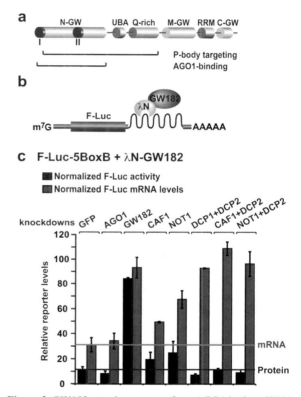

Figure 3. GW182 acts downstream from AGO1 in the miRNA pathway. (*a*) Domain organization of *Drosophila* GW182. (N-GW, M-GW, and C-GW) Amino-terminal, middle, and carboxy-terminal GW repeats, respectively; (UBA) ubiquitin-associated domain; (Q-rich) region rich in glutamine; (RRM) RNA recognition motif. Motifs I and II are indicated. The protein domains sufficient for the localization to P bodies and the interaction with AGO1 are shown below the protein outline. (*b*) Schematic representation of the F-Luc-5BoxB tethering reporter. (*c*) S2 cells depleted of the proteins indicated above the panel were transfected with the F-Luc-5BoxB reporter, a plasmid expressing *Renilla* luciferase, and vectors expressing the λN-peptide or λN-GW182. F-Luc activities (*black bars*) and mRNA levels (*green bars*) were quantitated in three independent experiments and normalized to that of the *Renilla* control. In cells expressing the λN peptide alone, these values were set to 100 for each knockdown (not shown). (*Black horizontal lines*) Normalized F-Luc activity; (*green horizontal lines*) mRNA levels in control cells. Mean values plus or minus standard deviations from three independent experiments are shown.

the *C. elegans* protein AIN-1, involved in the miRNA pathway (Ding et al. 2005). However, AIN-1 contains no UBA, Q-rich, or RRM domains, suggesting that it represents a functional analog.

C. elegans AIN-1, *Drosophila* GW182, and human TNRC6A/GW182 and TNRC6B localize to P bodies and interact with Argonaute proteins (Ding et al. 2005; Jakymiw et al. 2005; Liu et al. 2005a,b; Meister et al. 2005; Sen and Blau 2005; Behm-Ansmant et al. 2006). We have shown that it is the amino-terminal GW-repeat domain of *Drosophila* GW182 that mediates the interaction with the PIWI domain of AGO1 (Fig. 3a) (Behm-Ansmant et al. 2006). These findings suggest a conserved role for these repeats, and most likely for the highly conserved motif II, in mediating the interaction with Argonaute proteins. Interestingly, the PIWI domain adopts an RNase-H-like fold and is catalytically active, at least for a subset of Argonaute proteins including *Drosophila* AGO1 (for review, see Lingel and Sattler 2005; Miyoshi et al. 2005). It would therefore be of interest to determine whether the amino-terminal GW repeats of GW182 can modulate the catalytic activity of this domain.

Apart from the interaction with AGO1, the amino-terminal repeats and the UBA and Q-rich domains contribute to the localization of GW182 in P bodies (Fig. 3a) (Behm-Ansmant et al. 2006), which is in turn required for P-body integrity (Eystathioy et al. 2002, 2003). This suggests that GW182 may act as a molecular scaffold bringing together AGO1-containing RISCs (RNA-induced silencing complex) and mRNA decay enzymes, possibly nucleating the assembly of P bodies.

GW182 ACTS DOWNSTREAM FROM AGO1

The analyses described above combined with previous studies indicate that GW182 is recruited to miRNA targets via interactions with the Argonaute proteins (Jakymiw et al. 2005; Liu et al. 2005b; Meister et al. 2005; Behm-Ansmant et al. 2006). We used a tethering assay to begin to investigate the consequences of the recruitment of GW182 to miRNA targets. In this assay, GW182 is expressed as a fusion with the λN peptide, which binds with high affinity to five hairpins (BoxB sites) present in the 3′UTR of a luciferase reporter mRNA (F-Luc-5BoxB reporter, Fig. 3b). In this way, GW182 is directly tethered to the 3′UTR of the reporter, bypassing the requirement for AGO1 or miRNAs.

Using this assay, we have shown that tethering of GW182 to the F-Luc-5BoxB reporter silences its expression and promotes mRNA degradation (Fig. 3c) (Behm-Ansmant et al. 2006). The decrease in mRNA levels observed for the tethered mRNA does not, however, fully account for the strong reduction in F-Luc activity (Fig. 3c), indicating that GW182 silences expression of bound transcripts by a dual mechanism involving inhibition of protein expression and reduction of mRNA levels. As with miRNAs, mRNA degradation by GW182 requires the CCR4:NOT and the DCP1:DCP2 complexes. Indeed, the levels of a reporter transcript tethered to GW182 are restored in cells depleted of CAF1, NOT1, or the decapping enzymes. The restoration of mRNA levels in cells

depleted of CAF1 or NOT1 is not accompanied by a proportional increase in F-Luc expression (Fig. 3c), providing further evidence for a role of GW182 in translational regulation. Restoration of mRNA levels, but not of luciferase activity, is also observed in cells co-depleted of CAF1 and DCP2 or NOT1 and DCP2, demonstrating that GW182 promotes deadenylation and decapping of bound mRNAs (Fig. 3c). Thus, binding of GW182 appears to be a point of no return which marks transcripts as targets for the general mRNA degradation machinery.

Remarkably, GW182-mediated silencing and decay does not require AGO1 (Fig. 3c). This observation places GW182 downstream from AGO1 in the miRNA pathway. Several lines of evidence support this conclusion. First, endogenous and ectopically expressed GW182 localizes to P bodies together with decapping enzymes and the deadenylase complex (Eystathioy et al. 2002, 2003). Second, GW182 interacts with AGO1 and recruits AGO1 to P bodies when the two proteins are coexpressed (Jakymiw et al. 2005; Liu et al. 2005a,b; Meister et al. 2005; Sen and Blau 2005; Behm-Ansmant et al. 2006). Third, as mentioned above, degradation of mRNAs by GW182 or miRNAs requires the CCR4:NOT and DCP1:DCP2 complexes (Behm-Ansmant et al. 2006). Finally, miRNAs repress translation and promote mRNA degradation in a GW182-dependent manner (Jakymiw et al. 2005; Liu et al. 2005b; Meister et al. 2005; Rehwinkel et al. 2005; Behm-Ansmant et al. 2006).

CONCLUSIONS

On the basis of the results described above and the observations that GW182 associates with AGO1 and is required for miRNA-mediated gene silencing (Ding et al. 2005; Jakymiw et al. 2005; Liu et al. 2005b; Meister et al. 2005; Rehwinkel et al. 2005; Behm-Ansmant et al. 2006), we propose the following model for the mechanism of miRNA-mediated gene silencing (Fig. 4): AGO1 binds miRNA targets by means of miRNA:mRNA base-pairing interactions. AGO1 may then recruit GW182 and both proteins repress translations (Fig. 4). GW182 also marks

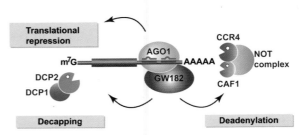

Figure 4. Model for miRNA-mediated gene silencing. miRNA-mediated gene silencing involves multiple mechanisms including translational repression, deadenylation, and decapping. In *Drosophila*, these mechanisms require AGO1 and the P-body protein GW182. Deadenylation and decapping are catalyzed by the DCP1:DCP2 decapping complex and the CCR4:NOT deadenylase complex, respectively. The mechanism by which miRNAs repress translation remains to be established. In addition, it is unclear whether silencing occurs in P bodies, although all proteins shown in this diagram have been localized to P bodies.

the transcripts as targets for decay via a deadenylation and decapping mechanism. The contribution of translational repression or mRNA degradation to gene silencing appears to differ for each miRNA:target pair and is likely to depend on the particular set of proteins bound to the 3′UTR of the mRNA. Furthermore, binding of specific RNA-binding proteins to the 3′UTR of miRNA targets can revert miRNA-mediated silencing under some physiological conditions (Bhattacharyya et al. 2006). A major challenge for future studies will be to identify how specific RNA-binding proteins influence the final outcome of miRNA regulation. Another important question that remains open is the mechanism by which miRNAs repress translation. Finally, whether silencing by miRNAs occurs in the cytosol or in cytoplasmic domains such as P bodies is currently a matter of debate that requires further investigation and a deeper understanding of the assembly, composition, and cellular function of these bodies.

ACKNOWLEDGMENTS

We are grateful to D.J. Thomas for comments on the manuscript. This study was supported by EMBO and the Human Frontier Science Program Organization (HFSPO). I.B.-A. is the recipient of a fellowship from EMBO.

REFERENCES

Allen E., Xie Z., Gustafson A.M., and Carrington J.C. 2005. microRNA-directed phasing during trans-acting siRNA biogenesis in plants. *Cell* 121: 207.

Bagga S., Bracht J., Hunter S., Massirer K., Holtz J., Eachus R., and Pasquinelli A.E. 2005. Regulation by *let-7* and *lin-4* miRNAs results in target mRNA degradation. *Cell* 122: 553.

Bartel D.P. 2004. MicroRNAs: Genomics, biogenesis, mechanism, and function. *Cell* 116: 281.

Behm-Ansmant I., Rehwinkel J., Doerks T., Stark A., Bork P., and Izaurralde E. 2006. mRNA degradation by miRNAs and GW182 requires both CCR4:NOT deadenylase and DCP1:DCP2 decapping complexes. *Genes Dev.* 20: 1885.

Bhattacharyya S.N., Habermacher R., Martine U., Closs E.I., and Filipowicz W. 2006. Relief of microRNA-mediated translational repression in human cells subjected to stress. *Cell* 125: 1111.

Ding L., Spencer A., Morita K., and Han M. 2005. The developmental timing regulator AIN-1 interacts with miRISCs and may target the argonaute protein ALG-1 to cytoplasmic P bodies in *C. elegans*. *Mol. Cell* 19: 437.

Eystathioy T., Chan E.K., Tenenbaum S.A., Keene J.D., Griffith K., and Fritzler M.J. 2002. A phosphorylated cytoplasmic autoantigen, GW182, associates with a unique population of human mRNAs within novel cytoplasmic speckles. *Mol. Biol. Cell* 13: 1338.

Eystathioy T., Jakymiw A., Chan E.K., Seraphin B., Cougot N., and Fritzler M.J. 2003. The GW182 protein colocalizes with mRNA degradation associated proteins hDcp1 and hLSm4 in cytoplasmic GW-bodies. *RNA* 9: 1171.

Filipowicz W. 2005. RNAi: The nuts and bolts of the RISC machine. *Cell* 122: 17.

Giraldez A.J., Mishima Y., Rihel J., Grocock R.J., Van Dongen S., Inoue K., Enright A.J., and Schier A.F. 2006. Zebrafish MiR-430 promotes deadenylation and clearance of maternal mRNAs. *Science* 312: 75.

Guo H.S., Xie Q., Fei J.F., and Chua N.H. 2005. MicroRNA directs mRNA cleavage of the transcription factor NAC1 to downregulate auxin signals for *Arabidopsis* lateral root development. *Plant Cell* 17: 1376.

Humphreys D.T., Westman B.J., Martin D.I., and Preiss T. 2005. MicroRNAs control translation initiation by inhibiting eukaryotic initiation factor 4E/cap and poly(A) tail function. *Proc. Natl. Acad. Sci.* 102: 16961.

Jakymiw A., Lian S., Eystathioy T., Li S., Satoh M., Hamel J.C., Fritzler M.J., and Chan E.K. 2005. Disruption of GW bodies impairs mammalian RNA interference. *Nat. Cell Biol.* 7: 1167.

Jing Q., Huang S., Guth S., Zarubin T., Motoyama A., Chen J., Di Padova F., Lin S.C., Gram H., and Han J. 2005. Involvement of microRNA in AU-rich element-mediated mRNA instability. *Cell* 120: 623.

Kim J., Krichevsky A., Grad Y., Hayes G.D., Kosik K.S., Church G.M., and Ruvkun G. 2004. Identification of many microRNAs that copurify with polyribosomes in mammalian neurons. *Proc. Natl. Acad. Sci.* 101: 360.

Lee R.C., Feinbaum R.L., and Ambros V. 1993. The *C. elegans* heterochronic gene *lin-4* encodes small RNAs with antisense complementarity to lin-14. Cell 75: 843.

Lim L.P., Lau N.C., Garrett-Engele P., Grimson A., Schelter J.M., Castle J., Bartel D.P., Linsley P.S., and Johnson J.M. 2005. Microarray analysis shows that some microRNAs downregulate large numbers of target mRNAs. *Nature* 433: 769.

Lingel A. and Sattler M. 2005. Novel modes of protein-RNA recognition in the RNAi pathway. *Curr. Opin. Struct. Biol.* 15: 107.

Llave C., Xie Z., Kasschau K.D., and Carrington J.C. 2002. Cleavage of Scarecrow-like mRNA targets directed by a class of *Arabidopsis* miRNA. *Science* 297: 2053.

Liu J., Valencia-Sanchez M.A., Hannon G.J., and Parker R. 2005a. MicroRNA-dependent localization of targeted mRNAs to mammalian P-bodies. *Nat. Cell Biol.* 7: 719.

Liu J., Rivas F.V., Wohlschlegel J., Yates J.R., Parker R., and Hannon G.J. 2005b. A role for the P-body component GW182 in microRNA function. *Nat. Cell Biol.* 7: 1261.

Mallory A.C., Reinhart B.J., Jones-Rhoades M.W., Tang G., Zamore P.D., Barton M.K., and Bartel D.P. 2004. MicroRNA control of PHABULOSA in leaf development: Importance of pairing to the microRNA 5′ region. *EMBO J.* 23: 3356.

Meister G., Landthaler M., Peters L., Chen P.Y., Urlaub H., Luhrmann R., and Tuschl T. 2005. Identification of novel Argonaute-associated proteins. *Curr. Biol.* 15: 2149.

Miyoshi K., Tsukumo H., Nagami T., Siomi H., and Siomi M.C. 2005. Slicer function of *Drosophila* Argonautes and its involvement in RISC formation. *Genes Dev.* 19: 2837.

Nelson P.T., Hatzigeorgiou A.G., and Mourelatos Z. 2004. miRNP:mRNA association in polyribosomes in a human neuronal cell line. *RNA* 10: 387.

Okamura K., Ishizuka A., Siomi H., and Siomi MC. 2004. Distinct roles for Argonaute proteins in small RNA-directed RNA cleavage pathways. *Genes Dev.* 18: 1655.

Olsen P.H. and Ambros V. 1999. The *lin-4* regulatory RNA controls developmental timing in *Caenorhabditis elegans* by blocking LIN-14 protein synthesis after the initiation of translation. *Dev. Biol.* 216: 671.

Petersen C.P., Bordeleau M.E., Pelletier J., and Sharp P.A. 2006. Short RNAs repress translation after initiation in mammalian cells. *Mol. Cell* 21: 533.

Pillai R.S., Bhattacharyya S.N., Artus C.G., Zoller T., Cougot N., Basyuk E., Bertrand E., and Filipowicz W. 2005. Inhibition of translational initiation by *let-7* microRNA in human cells. *Science* 309: 1573.

Rand T.A., Ginalski K., Grishin N.V., and Wang X. 2004. Biochemical identification of Argonaute 2 as the sole protein required for RNA-induced silencing complex activity. *Proc. Natl. Acad. Sci.* 101: 14385.

Rehwinkel J., Behm-Ansmant I., Gatfield D., and Izaurralde E. 2005. A crucial role for GW182 and the DCP1:DCP2 decapping complex in miRNA-mediated gene silencing. *RNA* 11: 1640.

Rehwinkel J., Natalin P., Stark A., Brennecke J., Cohen S.M., and Izaurralde E. 2006. Genome-wide analysis of mRNAs regulated by Drosha and Argonaute proteins in *Drosophila*. *Mol. Cell. Biol.* 26: 2965.

Schwab R., Palatnik J.F., Riester M., Schommer C., Schmid M., and Weigel D. 2005. Specific effects of microRNAs on the plant transcriptome. *Dev. Cell* **8:** 517.

Sen G.L. and Blau H.M. 2005. Argonaute 2/RISC resides in sites of mammalian mRNA decay known as cytoplasmic bodies. *Nat. Cell Biol.* **7:** 633.

Stark A., Brennecke J., Bushati N., Russell R.B., and Cohen S.M. 2005. Animal microRNAs confer robustness to gene expression and have a significant impact on 3´UTR evolution. *Cell* **123:** 1133.

Temme C., Zaessinger S., Meyer S., Simonelig M., and Wahle E. 2004. A complex containing the CCR4 and CAF1 proteins is involved in mRNA deadenylation in *Drosophila*. *EMBO J.* **23:** 2862.

Valencia-Sanchez M.A., Liu J., Hannon G.J., and Parker R. 2006. Control of translation and mRNA degradation by miRNAs and siRNAs. *Genes Dev.* **20:** 515.

Wightman B., Ha I., and Ruvkun G. 1993. Posttranscriptional regulation of the heterochronic gene *lin-14* by *lin-4* mediates temporal pattern formation in *C. elegans*. *Cell* **75:** 855.

Wu L., Fan J., and Belasco J.G. 2006. MicroRNAs direct rapid deadenylation of mRNA. *Proc. Natl. Acad. Sci.* **103:** 4034.

MicroRNAs, mRNAs, and Translation

P.A. Maroney, Y. Yu, and T.W. Nilsen

Center for RNA Molecular Biology and Department of Biochemistry, Case Western Reserve University
School of Medicine, Cleveland, Ohio 44106-4973

MicroRNAs (miRNAs) comprise a large family of regulatory molecules that repress protein production from targeted mRNAs. Although it is now clear that miRNAs exert pervasive effects on gene expression in animal cells, the mechanism(s) by which they function remains poorly understood. We have analyzed the subcellular distribution of miRNAs in actively growing HeLa cells and find that the vast majority are associated with actively translating mRNAs in polysomes. We also find that a specific miRNA-regulated mRNA (KRAS) is polysome associated and that its translation is impaired, apparently at the level of elongation. These observations are discussed in light of our current understanding of mechanism of miRNA function.

MicroRNAs (miRNAs), initially discovered in *Caenorhabditis elegans* as posttranscriptional regulators of genes involved in developmental timing, are now known to regulate the expression of the majority of genes in animals (Ambros 2004; Bartel 2004; Farh et al. 2005; Stark et al. 2005). Several hundred miRNAs have been characterized, and each miRNA has several hundred targets. Compelling informatic and experimental evidence indicates that miRNAs primarily recognize their targets via limited Watson-Crick base-pairing interactions between the 5′ end of the miRNA (nucleotides 2–8) and the 3′UTRs. Like other functional noncoding RNAs, miRNAs do not act as naked RNAs but rather exert their effects as ribonucleoproteins (RNPs); a common constituent of all miRNPs is a member of the argonaute (AGO) protein family (for review, see Bartel 2004; Pillai 2005). Although it is widely accepted that miRNA binding to a target mRNA is sufficient to elicit repression of protein production from that mRNA, there is at present no consensus as to how down-regulation takes place. Indeed, there is evidence for multiple distinct mechanisms including translational repression, enhanced mRNA degradation, or sequestration in cytoplasmic foci known as processing bodies (P-bodies). How or whether these different mechanisms are related is not yet clear. Here, we discuss recent results from our laboratory in the context of the current understanding of mechanisms of regulation by miRNAs and highlight what appear to be open questions.

MECHANISMS OF REPRESSION OF GENE EXPRESSION BY miRNAs

The earliest analyses of miRNA mechanism of action were done in *C. elegans*, and it was found that miRNA-targeted mRNAs were present when no protein encoded by those mRNAs could be detected. Moreover, the targeted mRNAs appeared to be associated with ribosomes in polysomes (Olsen and Ambros 1999; Seggerson et al. 2002). These results suggested that the block to synthesis of protein was after initiation of protein synthesis either by a direct effect on translation elongation or cotransla-

tional degradation of the newly synthesized proteins. As discussed below, these early results have not yet been explained. In subsequent analyses, it appears that much of the regulation of these initially recognized miRNA targets might be due to the degradation of the targeted mRNAs (Bagga et al. 2005), but the reasons for the discrepancies between studies conducted at different times are not yet clear. Regardless, there are now abundant examples where miRNA-mediated regulation represses protein production but does not result in lowered mRNA levels (see, e.g., O'Donnell et al. 2005).

Just as clearly, there is a significant and expanding literature which demonstrates that miRNAs can destabilize certain targeted mRNAs (Bagga et al. 2005; Lim et al. 2005; Behm-Ansmant et al. 2006; Giraldez et al. 2006; Rehwinkel et al. 2006). The fact that miRNA–mRNA interaction can result in enhanced turnover has enabled the use of mRNA microarray analysis to identify targeted miRNAs. A major question in the field is what fraction of total targets are identified by these types of experiments. There exists no analogous method to identify targets that are regulated strictly at the translational level. Furthermore, there is no straightforward way to determine whether the level of mRNA reduction corresponds to the total level of regulation; i.e., some mRNAs may be regulated both by destabilization and at the level of translation. A recent report highlights these issues; here, regulation of specific reporter constructs was observed to be mediated exclusively at the level of translation, a combination of translation and stability, or stability alone (Behm-Ansmant et al. 2006). A major challenge will be to determine how specific mRNAs are targeted for different modes of regulation.

The fact that specific regulated mRNAs appear to be regulated by distinct mechanisms is further complicated by data which suggest that each "mechanism," i.e., translational repression, or destabilization might be idiosyncratic for specific mRNAs. In this regard, there is evidence that translational repression can occur at several different steps. Two studies have demonstrated miRNA-mediated inhibition of translation at the level of initiation

(Humphreys et al. 2005; Pillai et al. 2005), and a third has provided evidence that initiation is not affected, but rather that translating ribosomes disengage prematurely (drop off) from a miRNA-regulated mRNA (Petersen et al. 2006). Although different approaches were used in these studies, it is difficult at present to rationalize the disparity in the results.

Our understanding of destabilization is somewhat clearer, but even here there are several unanswered questions. Three studies have shown that miRNA–mRNA interaction can cause deadenylation of targeted mRNAs (Behm-Ansmant et al. 2006; Giraldez et al. 2006; Wu et al. 2006). In some cases, this deadenylation leads to mRNA degradation via subsequent decapping (Behm-Ansmant et al. 2006), but in other cases, it does not. Furthermore, even in the cases where deadenylation does not lead to instability, it alone does not account for the level of translational inhibition (Behm-Ansmant et al. 2006; Wu et al. 2006).

Recently, there has been considerable interest and excitement over the potential role of processing bodies (P-bodies), also known as GW-182 bodies, in miRNA function. Originally described in budding yeast, P-bodies are cytoplasmic foci that contain a variety of factors including decapping enzymes, deadenylases, and exonucleases known to be involved in and required for mRNA decay (for review, see Valencia-Sanchez et al. 2006). In yeast, these foci can serve as a repository for untranslated mRNAs that can either be degraded or return to the translating pool (Brengues et al. 2005). The demonstration by several groups that Ago 2 interacts (directly or indirectly) with decapping enzymes led to the finding that Ago 2 localizes to these foci (Liu et al. 2005a,b; Sen and Blau 2005). A number of other provocative studies have demonstrated miRNA-dependent localization of targeted mRNAs to P-bodies (Liu et al. 2005a) and have established a role for the P-body constituent GW-182 in miRNA function (Jakymiw et al. 2005; Liu et al. 2005b; Rehwinkel et al. 2005; Behm-Ansmant et al. 2006).

Collectively, these analyses have led to the attractive hypothesis that P-bodies are central to miRNA function (for review, see Valencia-Sanchez et al. 2006). This hypothesis has the advantage of rationalizing many, but not all (see below), of the disparate observations pertaining to miRNA action. In this regard, the apparent diversity of mechanisms could result from different fates of mRNAs sequestered in P-bodies; i.e., some could be destined for decay while others could be held in stasis spatially removed from the translational machinery (P-bodies are devoid of ribosomes) (for review, see Valencia-Sanchez et al. 2006).

Although the P-body hypothesis is appealing, there are several open questions. First, what fraction of the total cellular complement of P-body-associated proteins (e.g., Ago 2, GW-182, etc.) is actually present in microscopically visible foci? Studies to date have not been quantitative. Second, is the "structural integrity" of P-bodies important for miRNA-mediated gene regulation? This has been a difficult question to address because P-bodies have resisted attempts at biochemical purification and, therefore, a definition of what constitutes a P-body has not yet

emerged. Notably, however, a recent study has shown that depletion of an LSm protein eliminates visualizable P-bodies but does not impair miRNA function (Chu and Rana 2006). Whether submicroscopic foci remain under these conditions is not clear. Third, what fraction of translationally repressed mRNAs are concentrated in P-bodies and therefore disengaged from ribosomes?

With the exception of mRNA microarray studies, which can only monitor mRNA levels, most mechanistic analyses of miRNA function have focused on specific miRNA–mRNA combinations; reporter constructs responsive to endogenous or "designer" miRNAs. To gain a different perspective, we thought that it would be informative to analyze the subcellular partitioning of bulk miRNAs. We anticipated that the "behavior" of endogenous miRNAs would be diagnostic for their mode(s) of regulation. As detailed below, these studies yielded unexpected results.

MOST miRNAs IN ACTIVELY GROWING HeLa CELLS ARE ASSOCIATED WITH TRANSLATING mRNAs

To analyze the subcellular partitioning of bulk miRNAs in HeLa cells, we prepared cytoplasmic extracts from exponentially growing cells and fractionated them on sucrose gradients. When individual fractions were analyzed for the presence of three abundant miRNAs, we were surprised to find that the vast majority co-sedimented with polysomes. By comparing the amounts of miRNAs recovered in the gradients to total miRNAs present in the cell, it was clear that most of the miRNAs were indeed present in the cytoplasmic extracts and most co-fractionated with polysomes; i.e., very few miRNAs were pelleted.

Several experiments, including dissociation of the ribosomes with EDTA, indicated that the miRNAs were polysome associated and not simply co-sedimenting with polysomes. Importantly, mild digestion of the extracts with micrococcal nuclease, which digests mRNAs in polysomes but leaves ribosomes intact, released the miRNAs such that they migrated at the top of the gradient. This experiment strongly suggested that the miRNAs were associated with polysomes because they were bound to mRNAs.

To determine whether the mRNAs associated with miRNAs were being translated, cells were treated under a variety of conditions that affect bulk protein synthesis. When cells were treated with pactamycin, which at low concentrations predominately inhibits initiation of translation (Taber et al. 1971), we observed significant accumulation of 80S ribosomes, as expected. When initiation is disrupted, elongating ribosomes finish translation and dissociate from the mRNAs. Concomitant with the inhibition of protein synthesis, mRNAs shift from heavy polysomes to lighter ones reflecting runoff of elongating ribosomes. Remarkably, the same shift from heavy to light ribosomes was observed for the miRNAs. Similarly, when cells were treated with puromycin, a drug that inhibits protein synthesis by causing premature termination of translation, similar results were obtained; i.e., mRNAs and miRNAs shifted from heavy polysomes to lighter ones. With both pactamycin and puromycin treatment, it was clear that miRNA sedimentation paralleled that of mRNAs, not bulk ribosomal RNA, thus providing additional evidence that the miRNAs we analyzed were associated with mRNAs. Furthermore, these experiments indicated that most miRNAs were associated with mRNAs that were being actively translated.

Both pactamycin and puromycin are irreversible inhibitors of protein synthesis. We wished to assess the effect of miRNA–mRNA interaction under conditions where protein synthesis could be arrested and then allowed to resume. To do this, we treated cells with hypertonic media. Hypertonic shock rapidly arrests protein synthesis by inhibiting initiation and, importantly, protein synthesis recovers upon return to isotonic media (Morley and Naegele 2002). As expected from the pactamycin results, inhibition of initiation by hypertonic shock caused disaggregation of polysomes due to ribosome runoff and a large accumulation of 80S monosomes. As a consequence of this inhibition of protein synthesis, both mRNAs and miRNAs shifted in sedimentation to lighter fractions. When cells were returned to isotonic media, there was a rapid recovery of protein synthesis and both mRNAs and miRNAs returned to heavy polysomal fractions. These results clearly indicate that most miRNA–mRNA interactions are compatible with protein synthesis and do not interfere with resumption of protein synthesis after it is arrested.

How can one interpret these results in light of other studies? First, the finding that most miRNAs are associated with polysomes may not be as surprising as it appears at first glance. In this regard, the first identified miRNA-targeted mRNAs in *C. elegans* were demonstrated to be polysome associated (Olsen and Ambros 1999; Seggerson et al. 2002), and subsequent studies have shown that the bulk of miRNAs in *Drosophila* and *C. elegans* were present in ribosomal fractions (Hammond et al. 2000, 2001; Caudy et al. 2003). Furthermore, there have been several reports of miRNAs present in polysomes (see, e.g., Kim et al. 2004; Nelson et al. 2004), and in one case, miRNA–target mRNA interaction in polysomes was documented (Nelson et al. 2004).

We believe that it is most likely that the miRNA–mRNA interactions we observe are mediated by base-pairing between the miRNAs and target mRNAs and thus are "active," but this is technically impossible to prove. Nevertheless, if we assume that the polysome-associated miRNAs are functioning as regulators, what are the implications of our results regarding mechanism? It would seem that they are in conflict with the P-body hypothesis described above, since we do not observe any accumulation of non-polyribosome-associated miRNAs. However, this may not be the case. It is possible that regulated mRNAs could exit translation and be degraded or sequestered in a P-body-component-dependent manner that involves the release of the regulatory miRNA. If this were the case, no steady-state accumulation of miRNAs in P-bodies would be expected. In this scenario, the apparent localization of Ago 2 to P-bodies (Liu et al. 2005a; Sen and Blau 2005) would be the result of tran-

sient interactions. Clearly, a quantitative assessment of the fraction of Ago 2 present in P-bodies would be helpful in determining whether this conjecture is correct. It would also be informative to determine the complement of intact mRNAs present in P-bodies and whether they are associated with miRNAs, but such analyses await the development of techniques to isolate these entities.

Regardless of the role(s) of P-body components in miRNA-mediated regulation, our results clearly indicate that miRNA–mRNA interactions are not necessarily dramatically repressive to protein synthesis. They are therefore consistent with a large number of studies which indicate that the majority of mRNAs in animal cells are miRNA targets but are still translated (see, e.g., Farh et al. 2005). Indeed, with few exceptions, the magnitude of regulation by miRNAs appears to be quite modest (see, e.g., Lewis et al. 2003; Lim et al. 2005; and see also Bartel and Chen 2004). In principle, degrees of regulation could be achieved by regulating accessibility of the miRNA to its targets. However, there is no evidence for compartmentalization of miRNAs and targets, and our data together with that of others (Kim et al. 2004; Nelson et al. 2004) suggest that regulation is primarily achieved at the level of the actively translating ribosome mRNA complex. This view is consistent with the recent demonstration that miRNA function requires the participation of Rck/p54 (Chu and Rana 2006), a DEAD box RNA helicase, whose homolog in yeast, Dhh-1, is required for translational inhibition (Coller and Parker 2005). Notably, Dhh-1 only exerts its effects on mRNAs that have engaged the translational machinery (Coller and Parker 2005).

A SPECIFIC miRNA-REGULATED mRNA IS PRESENT IN POLYSOMES AND TRANSLATIONALLY IMPAIRED

Although the experiments described above were informative regarding the behavior of bulk miRNAs, we wanted to determine how a specific miRNA-regulated mRNA partitioned within the cell. For these analyses we chose the KRAS mRNA because it has been shown to be regulated by the let-7 miRNA in HeLa cells (Johnson et al. 2005; Chu and Rana 2006). When the subcellular distribution of this mRNA was examined, it was found to cosediment with polysomes, and sedimentation was sensitive to treatment with EDTA, as expected if it were ribosome associated. More importantly, unlike the behavior of control mRNAs, the sedimentation of KRAS demonstrated reduced sensitivity to puromycin treatment and reduced sensitivity to a block to initiation of protein synthesis caused by hypertonic shock. These observations strongly suggest that miRNA-mediated inhibition of KRAS protein syntheses is due to impairment of translation elongation.

Our observations with KRAS mRNA clearly are not consistent with a block to initiation of protein synthesis (Humphreys et al. 2005; Pillai et al. 2005); if this were the case, the mRNA would not be polysome associated. They also are not compatible with enhanced rate of ribosome dropoff post-initiation (Petersen et al. 2006); if this were the case, we would have observed changes in sedimentation upon exposure to hypertonic conditions. They are also not compatible with sequestration in P-bodies because these structures lack ribosomes (Valencia-Sanchez et al. 2006). Although the behavior of KRAS does not fit well with current models of miRNA-mediated translational down-regulation, it appears to be strikingly similar to the original miRNA targets described in *C. elegans* (Olsen and Ambros 1999; Seggerson et al. 2002) which were found to be polysome associated. The behavior of KRAS mRNA also appears to be similar to that of the pal-1 mRNA in *C. elegans* which is translationally down-regulated by Gld-1. In that case, several lines of evidence suggest that translation is inhibited by the slowing or stalling of elongating ribosomes (Mootz et al. 2004). Other examples of post-initiation repression exist (Clark et al. 2000; Braat et al. 2004); whether these are miRNA-mediated is not known. It will be of considerable interest to determine whether apparently disparate cases of regulation at the level of elongation require common factors; e.g., GW182.

Our finding that KRAS is apparently down-regulated at the level of translation elongation is, as noted above, consistent with the earliest analyses of mechanism of miRNA function and it raises the possibility that other polysome-associated miRNAs also affect elongation but to a lesser extent. In this regard, a modest slowing of translation would be invisible in our analysis of bulk miRNAs.

CONCLUSIONS

It is now apparent that miRNAs can regulate protein production by targeted mRNAs in multiple ways and to widely varying extents. Given that only a few examples have been studied in any detail, it is not yet clear that all possible mechanisms have been uncovered. Furthermore, with the possible exception of enhanced turnover, the molecular details of how repression occurs are not known.

It will be important to determine how many and which mRNAs are regulated at the level of translation alone, stability, or combination of both. This will be difficult, since no technique currently allows monitoring of translation on the scale of mRNA microarray analyses. However, it may be possible to use this type of experiment to identify mRNAs that remain in polysomes when translation initiation is arrested; i.e., mRNAs highly regulated at the level of translation elongation.

Equally challenging will be defining the factors that determine the magnitude of regulation that occurs by any given mechanism. In this regard, it seems unlikely that different modes or magnitudes of regulation are specified by the miRNAs themselves. Although many potential "auxiliary" factors have been identified, it does not appear that they function in an mRNA-specific fashion (see, e.g., Caudy et al. 2002; Mourelatos et al. 2002; Meister et al. 2005). Furthermore, studies that have revealed inhibition of protein synthesis at the level of initiation (Pillai et al. 2005) and elongation (see above) used distinct 3′UTRs responsive to the same miRNA, let-7. In addition, there are numerous examples of differences in magnitude of regulation elicited by the same miRNA (see, e.g., Lewis et al. 2003; Lim et al. 2005). It thus appears that both

mechanism and magnitude of regulation are determined by the 3′UTR context in which the miRNA binds to the target mRNA. Indeed, recent analyses of a specific miRNA target in *C. elegans* provide a clear illustration of dramatic context effects (Didiano and Hobert 2006). Presumably, these effects are mediated by the presence of distinct sets of proteins bound to different 3′UTRs. Elucidating the identity of such proteins and how they interact with the miRNA regulatory machinery will clearly be important in understanding the overall impact of miRNAs on gene expression.

ACKNOWLEDGMENTS

The described work from our laboratory is currently in press in *Nature Structural and Molecular Biology* and was supported by grants from the National Institutes of Health.

REFERENCES

Ambros V. 2004. The functions of animal microRNAs. *Nature* **431:** 350.

Bagga S., Bracht J., Hunter S., Massirer K., Holtz J., Eachus R., and Pasquinelli A.E. 2005. Regulation by let-7 and lin-4 miRNAs results in target mRNA degradation. *Cell* **122:** 553.

Bartel D.P. 2004. MicroRNAs: Genomics, biogenesis, mechanism, and function. *Cell* **116:** 281.

Bartel D.P. and Chen C.-Z. 2004. Micromanagers of gene expression: The potentially widespread influence of metazoan microRNAs. *Nat. Rev. Genet.* **5:** 396.

Behm-Ansmant I., Rehwinkel J., Doerks T., Stark A., Bork P., and Izaurralde E. 2006. mRNA degradation by miRNAs and GW182 requires both CCR4:NOT deadenylase and DCP1:DCP2 decapping complexes. *Genes Dev.* **20:** 1885.

Braat A.K., Yan N., Arn E., Harrison D., and Macdonald P.M. 2004. Localization-dependent oskar protein accumulation: Control after the initiation of translation. *Dev. Cell* **7:** 125.

Brengues M., Teixeira D., and Parker R. 2005. Movement of eukaryotic mRNAs between polysomes and cytoplasmic processing bodies. *Science* **310:** 486.

Caudy A.A., Myers M., Hannon G.J., and Hammond S.M. 2002. Fragile X-related protein and VIG associate with the RNA interference machinery. *Genes Dev.* **16:** 2491.

Caudy A.A., Ketting R.F., Hammond S.M., Denli A.M., Bathoorn A.M., Tops B.B., Silva J.M., Myers M.M., Hannon G.J., and Plasterk R.H. 2003. A micrococcal nuclease homologue in RNAi effector complexes. *Nature* **425:** 411.

Chu C.-Y. and Rana T.M. 2006. Translation repression in human cells by MicroRNA-induced gene silencing requires RCK/p54. *PLoS Biol.* **4:** e210.

Clark I.E., Wyckoff D., and Gavis E.R. 2000. Synthesis of the posterior determinant Nanos is spatially restricted by a novel cotranslational regulatory mechanism. *Curr. Biol.* **10:** 1311.

Coller J. and Parker R. 2005. General translational repression by activators of mRNA decapping. *Cell* **122:** 875.

Didiano D. and Hobert O. 2006. Perfect seed pairing is not a generally reliable predictor for miRNA-target interactions. *Nat. Struct. Mol. Biol.* **13:** 849.

Farh K.K., Grimso A., Jan C., Lewis B.P., Johnston W.K., Lim L.P., Burge C.B., and Bartel D.P. 2005. The widespread impact of mammalian MicroRNAs on mRNA repression and evolution. *Science* **310:** 1817.

Giraldez A.J., Mishima Y., Rihel J., Grocock R.J., Van Dongen S., Inoue K., Enright A.J., and Schier A.F. 2006. Zebrafish MiR-430 promotes deadenylation and clearance of maternal mRNAs. *Science* **312:** 75.

Hammond S.M., Bernstein E., Beach D., and Hannon G.J. 2000. An RNA-directed nuclease mediates post-transcriptional gene silencing in *Drosophila* cells. *Nature* **404:** 293.

Hammond S.M., Boettcher S., Caudy A.A., Kobayashi R., and Hannon G.J. 2001. Argonaute2, a link between genetic and biochemical analyses of RNAi. *Science* **293:** 1146.

Humphreys D.T., Westman B.J., Martin D.I., and Preiss T. 2005. MicroRNAs control translation initiation by inhibiting eukaryotic initiation factor 4E/cap and poly(A) tail function. *Proc. Natl. Acad. Sci.* **102:** 16961.

Jakymiw A., Lian S., Eystathioy T., Li S., Satoh M., Hamel J.C., Fritzler M.J., and Chan E.K. 2005. Disruption of GW bodies impairs mammalian RNA interference. *Nat. Cell Biol.* **7:** 1267.

Johnson S.M., Grosshans H., Shingara J., Byrom M., Jarvis R., Cheng A., Labourier E., Reinert K.L., Brown D., and Slack F.J. 2005. RAS is regulated by the *let-7* microRNA family. *Cell* **120:** 635.

Kim J., Krichevsky A.M., Grad Y., Hayes G.D., Kosik K.S., Church G.M., and Ruvkun G. 2004. Identification of many microRNAs that copurify with polyribosomes in mammalian neurons. *Proc. Natl. Acad. Sci.* **101:** 360.

Lewis B.P., Shih I.-H., Jones-Rhoades M.W., Bartel D.P., and Burge C.B. 2003. Prediction of mammalian microRNA targets. *Cell* **115:** 787.

Lim L.P., Lau N.C., Garrett-Engele P., Grimson A., Schelter J.M., Castle J., Bartel D.P., Linsley P.S., and Johnson J.M. 2005. Microarray analysis shows that some microRNAs downregulate large numbers of target mRNAs. *Nature* **433:** 769.

Liu J., Valencia-Sanchez M.A., Hannon G.J., and Parker R. 2005a. MicroRNA-dependent localization of targeted mRNAs to mammalian P-bodies. *Nat. Cell Biol.* **7:** 719.

Liu J., Rivas F.V., Wohlschlegel J., Yates J.R., III, Parker R., and Hannon G.J. 2005b. A role for the P-body component GW182 in microRNA function. *Nat. Cell Biol.* **7:** 1261.

Meister G., Landthaler M., Peters L., Chen P.Y., Urlaub H., Luhrmann R., and Tuschl T. 2005. Identification of novel argonaute-associated proteins. *Curr. Biol.* **15:** 2149.

Mootz D., Ho D.M., and Hunter C.P. 2004. The STAR/Maxi-KH domain protein GLD-1 mediates a developmental switch in the translational control of *C. elegans* PAL-1. *Development* **131:** 3263.

Morley S.J. and Naegele S. 2002. Phosphorylation of eukaryotic initiation factor (eIF) 4E is not required for *de novo* protein synthesis following recovery from hypertonic stress in human kidney cells. *J. Biol. Chem.* **277:** 32855.

Mourelatos Z., Dostie J., Paushkin S., Sharma A., Charroux B., Abel L., Rappsilber J., Mann M., and Dreyfuss G. 2002. miRNPs: A novel class of ribonucleoproteins containing numerous microRNAs. *Genes Dev.* **16:** 720.

Nelson P.T., Hatzigeorgiou A.G., and Mourelatos Z. 2004. miRNP:mRNA association in polyribosomes in a human neuronal cell line. *RNA* **10:** 387.

O'Donnell K.A., Wentzel E.A., Zeller K.I., Dang C.V., and Mendell J.T. 2005. c-Myc-regulated microRNAs modulate E2F1 expression. *Nature* **435:** 839.

Olsen P.H. and Ambros V. 1999. The lin-4 regulatory RNA controls developmental timing in *Caenorhabditis elegans* by blocking LIN-14 protein synthesis after the initiation of translation. *Dev. Biol.* **216:** 671.

Petersen C.P., Bordeleau M.E., Pelletier J., and Sharp P.A. 2006. Short RNAs repress translation after initiation in mammalian cells. *Mol. Cell* **21:** 533.

Pillai R.S. 2005. MicroRNA function: Multiple mechanisms for a tiny RNA? *RNA* **11:** 1753.

Pillai R.S., Bhattacharyya S.N., Artus C.G., Zoller T., Cougot N., Basyuk E., Bertrand E., and Filipowicz W. 2005. Inhibition of translational initiation by let-7 microRNA in human cells. *Science* **309:** 1573.

Rehwinkel J., Behm-Ansmant I., Gatfield D., and Izaurralde E. 2005. A crucial role for GW182 and the DCP1:DCP2 decapping complex in miRNA-mediated gene silencing. *RNA* **11:** 1640.

Rehwinkel J., Natalin P., Stark A., Brennecke J., Cohen S.M., and Izaurralde E. 2006. Genome-wide analysis of mRNAs regulated by Drosha and Argonaute proteins in *Drosophila melanogaster*. *Mol. Cell. Biol.* **26:** 2965.

Seggerson K., Tang L., and Moss E.G. 2002. Two genetic circuits repress the *Caenorhabditis elegans* heterochronic gene lin-28 after translation initiation. *Dev. Biol.* **243:** 215.

Sen G.L. and Blau H.M. 2005. Argonaute 2/RISC resides in sites of mammalian mRNA decay known as cytoplasmic bodies. *Nat. Cell Biol.* **7:** 633-636.

Stark A., Brennecke J., Bushati N., Russell R.B., and Cohen S.M. 2005. Animal microRNAs confer robustness to gene expression and have a significant impact on 3´UTR evolution. *Cell* **123:** 1133.

Taber R., Rekosh D., and Baltimore D. 1971. Effect of pactamycin on synthesis of poliovirus proteins: A method for genetic mapping. *J. Virol.* **8:** 395.

Valencia-Sanchez M.A., Liu J., Hannon G.J., and Parker R. 2006. Control of translation and mRNA degradation by miRNAs and siRNAs. *Genes Dev.* **20:** 515.

Wu L., Fan J., and Belasco J.G. 2006. MicroRNAs direct rapid deadenylation of mRNA. *Proc. Natl. Acad. Sci.* **103:** 4034.

Regulation of Poly(A)-binding Protein through PABP-interacting Proteins

M.C. Derry, A. Yanagiya, Y. Martineau, and N. Sonenberg

Department of Biochemistry and McGill Cancer Centre, McGill University, Montréal, Québec H6G 1Y6, Canada

Translation initiation requires the participation of eukaryotic translation initiation factors (eIFs). The poly(A)-binding protein (PABP) is thought to stimulate translation by promoting mRNA circularization through simultaneous interactions with eIF4G and the 3′ poly(A) tail. PABP activity is regulated by the PABP-interacting proteins (Paips), a family of proteins consisting of Paip1, a translational stimulator, and Paip2A and Paip2B, two translational inhibitors. Paip2A controls PABP homeostasis via ubiquitination. When the cellular concentration of PABP is reduced, Paip2A becomes ubiquitinated and degraded, resulting in the relief of PABP repression. Paip1 interacts with eIF4A and eIF3, which promotes translation. The regulation of PABP activity by Paips represents the first known mechanism for controlling PABP, adding a new layer to the existing knowledge of PABP function.

Translational control is an important mechanism by which cells govern gene expression, providing a rapid response to growth and proliferation stimuli, stress, and nutrient availability. In systems with little or no transcriptional control (e.g., reticulocytes and oocytes), translation is the predominant mode of regulation of gene expression (Mathews et al. 2007). Initiation, the rate-limiting step of translation, is the main target of translational control. Translation initiation entails the recruitment of the ribosome to the mRNA, traversing of the 5′-untranslated region (5′UTR), and recognition of the initiation codon (Pestova et al. 2007). These processes are dependent on the eukaryotic translation initiation factors (eIFs). The 5′ cap structure (m7GpppN, where m is a methyl group and N is any nucleotide), which is present at the 5′ end of all nuclear-transcribed eukaryotic mRNAs, is the first mRNA structure recognized by eIFs. It is bound by the eIF4F complex, consisting of eIF4E, eIF4A, and eIF4G. eIF4E binds directly to the mRNA 5′ cap; eIF4A is an RNA helicase; and eIF4G serves as a modular scaffolding protein that binds, among other proteins, eIF4E, eIF4A, eIF3, poly(A)-binding protein (PABP), and Mnk, a serine/threonine kinase which phosphorylates eIF4E (Gingras et al. 1999; Pyronnet et al. 1999). The eIF3 complex, which contains up to 13 distinct subunits (Pestova et al. 2007), interacts with the 40S ribosomal subunit, thus serving as a link between the mRNA–eIF4F complex and the ribosome.

All eukaryotic cellular mRNAs, except those of histones, possess a poly(A) tail in their 3′UTR. Early in vitro experiments suggested a role for the poly(A) tail in translation initiation. The poly(A) tail confers a translational advantage to the mRNA in reticulocyte lysate, as addition of poly(A) RNA inhibited the translation of polyadenylated (poly(A)$^+$) mRNA (Doel and Carey 1976; Jacobson and Favreau 1983; Grossi de Sa et al. 1988). Later studies demonstrated translational stimulation by the poly(A) tail, which could not be attributed to its mRNA stabilizing effect (Munroe and Jacobson 1990a). Consistent with the importance of the poly(A) tail in translation, a positive correlation was shown between the polyadenylation state of an mRNA and translational activation during development. In many systems (e.g., *Xenopus laevis*, *Drosophila melanogaster*, mouse), the translation of a large number of maternal mRNAs is dependent on the poly(A) tail (Wickens et al. 2000).

PABP FUNCTION IN TRANSLATION

In general, PABP stimulates translation of mRNAs harboring a poly(A) tail (Sachs 2000; Kahvejian et al. 2001). PABP is an essential protein: In yeast, deletion of the Pab1 gene is lethal (Sachs et al. 1987). PABP is an abundant protein (Görlach et al. 1994) that contains, in the amino-terminal region, four phylogenetically conserved RNA recognition motifs (RRMs) (Adam et al. 1986; Sachs et al. 1987); PABP fragments containing RRMs 1 + 2 bind poly(A) RNA with an affinity similar to that of full-length PABP, whereas RRMs 3 + 4 exhibit a tenfold lower affinity for poly(A) (Burd et al. 1991; Kuhn and Pieler 1996; Deo et al. 1999).

PABP that is tethered to the 3′ end of a nonadenylated mRNA stimulates translation in *X. laevis* oocytes independently of its poly(A)-binding activity (Gray et al. 2000). A fragment containing RRMs 1 + 2 of PABP, which binds eIF4G (Imataka et al. 1998), was more effective than full-length PABP in stimulating translation (Gray et al. 2000). Fragments containing RRMs 3 + 4 or the proline-rich carboxyl terminus of PABP, termed the PABC domain, also augmented translation (Gray et al. 2000). Exogenous PABP stimulated the translation of capped poly(A)$^+$ mRNAs and, to a lesser extent, poly(A)$^-$ mRNA in yeast extracts (Otero et al. 1999) and mammalian translation systems (Kahvejian et al. 2005). These findings suggest that the mechanisms by which PABP stimulates translation are complex and may involve redundant or alternative pathways.

The PABC domain comprises a docking site for a wide range of proteins (Albrecht and Lengauer 2004). The best-characterized interactions occur with the PABP-interacting proteins (Paips), which bind to PABP and regulate its activity (Craig et al. 1998; Khaleghpour et al. 2001a; Roy et al. 2002). Other interacting proteins include the eukaryotic ribosome recycling factor (eRF3), which functions in mRNA translation termination and ribosome recycling through its interaction with PABP (Uchida et al. 2002a; Hosoda et al. 2003); deleted in azoospermia-like (DAZL) proteins, which during germ-cell development, activate silent mRNAs through binding to the 3′UTR and recruitment of PABP (Collier et al. 2005); Tob, a member of a family of proteins with antiproliferative functions (Okochi et al. 2005); and ataxin-2 homologs, which have been implicated in such cellular processes as signal transduction, embryonic development, and RNA splicing and degradation (Mangus et al. 1998; He and Parker 2000; Kiehl et al. 2000).

TRANSLATIONAL SYNERGY BETWEEN THE 5′ CAP AND THE POLY(A) TAIL

The closed-loop model for mRNA circularization was proposed more than two decades ago (Jacobson and Favreau 1983; Palatnik et al. 1984) and subsequently reiterated (Sachs and Davis 1989; Munroe and Jacobson 1990a,b; Jacobson 1996). The synergistic enhancement of translation of mRNAs that possess both a cap and a poly(A) tail further suggested a physical interaction between the two extremities of the mRNA (Gallie 1991). Electroporation of mRNAs into cells demonstrated that the translation of mRNAs was synergistically augmented by the cap and the poly(A) tail (Gallie 1991). Capped and poly(A)$^+$ mRNAs exhibited similar synergistic properties in yeast (Iizuka et al. 1994) and mammalian translation extracts (Khaleghpour et al. 2001a; Svitkin and Sonenberg 2004), indicating that the poly(A) tail plays an important role in stimulating cap-dependent translation initiation.

Proof of a direct interaction between the 5′ and 3′ ends of the mRNA was provided by the discovery of the interaction between eIF4G and PABP in yeast (Tarun and Sachs 1996) and plant systems (Le et al. 1997). In humans, a stretch of 29 amino acids in the amino terminus of eIF4G interacts with RRM 1 + 2 of PABP (Imataka et al. 1998), as in yeast PABP. However, despite its high homology with yeast PABP, human PABP does not interact with yeast eIF4G (Otero et al. 1999). The eIF4G–PABP interaction plays a critical role in *X. laevis* oocytes, since expression of an eIF4G mutant that did not bind PABP repressed translation of poly(A)$^+$ mRNAs and inhibited progesterone-induced oocyte maturation (Wakiyama et al. 2000).

MECHANISMS OF PABP-MEDIATED TRANSLATION STIMULATION

Several models have been proposed to explain the mechanism by which PABP promotes translation. First, PABP–eIF4G binding could mediate mRNA circularization, promoting the recycling of terminating ribosomes

by bridging the two ends of an mRNA, a model reinforced by the interaction between PABP and the translation termination factor eRF3 (Hoshino et al. 1999; Uchida et al. 2002a). A second model suggests that PABP stimulates 60S ribosomal subunit joining. In early experiments, mutations in a 60S ribosomal protein or in a helicase required for 60S ribosomal subunit biosynthesis partially rescued the phenotype of PABP deletion in yeast (Sachs and Davis 1989, 1990). These genetic data are consistent with biochemical experiments, in which the absence of the poly(A) tail led to a decrease in 60S ribosomal subunit joining (Munroe and Jacobson 1990a). However, other experiments support an alternative role of PABP in translation initiation, in stimulating recruitment of the 40S ribosomal subunit to the mRNA. In extracts immunodepleted of PABP, 40S ribosomal subunit recruitment was inhibited (Tarun and Sachs 1995). Consistent with these findings, in an in vitro translation system, PABP functioned as an initiation factor and stimulated both 40S initiation complex formation and 60S subunit joining (Kahvejian et al. 2005). PABP also stimulated the interaction of eIF4E with the cap structure as determined by cross-linking experiments (Kahvejian et al. 2005).

REGULATION OF PABP ACTIVITY: PABP-INTERACTING PROTEINS

In the course of searching for novel PABP-binding proteins, two novel protein partners of PABP, termed PABP-interacting proteins (Paips) were discovered: Paip1 (Craig et al. 1998) and Paip2 (Khaleghpour et al. 2001b). More recently, a homolog of Paip2 was cloned (Berlanga et al. 2006); thus, the original protein was named Paip2A and the second Paip2B.

Paip1, Paip2A, and Paip2B bind to PABP using two distinct PABP-binding motifs (PAMs; Fig. 1). PAM1 is an acidic region of approximately 25 amino acids that binds to the RRM2 in the amino terminus of PABP, whereas PAM2 is a well-defined and conserved region of approximately 15 amino acids that binds to the carboxy-terminal PABC domain of PABP (Khaleghpour et al. 2001b; Kozlov et al. 2001; Roy et al. 2002). PAM2 motifs, as well as PABC domains, have since been iden-

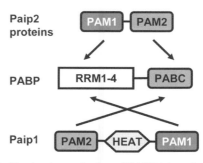

Figure 1. Structural organization of PABP interacting partners. Interactions between PABP, Paip1, and Paip2 proteins. (PAM) PABP-interacting motif; (RRM) RNA recognition motif; (PABC) PABP carboxy-terminal domain; (HEAT) heat domain (*H*untington, *e*longation factor 3, PR65/*A*, *T*OR).

tified in many different proteins of varying functions (see above), suggesting that PAM2 and PABC may play roles in protein–protein interactions in a wide range of cellular processes.

Paip1, a 75-kD protein of 479 amino acids, binds to PABP in vitro and in vivo and acts as a translational enhancer. Overexpression of Paip1 in COS-7 cells stimulated translation of a luciferase mRNA reporter (Craig et al. 1998). Deletion of the carboxyl terminus of Paip1 abrogated its ability to enhance translation. These data indicate that the PAM1 motif in Paip1 is essential for its activity (Craig et al. 1998). Amino acid sequence analysis of Paip1 revealed 25% identity and 39% similarity with the middle domain of eIF4G (Craig et al. 1998) in amino acids 619–1081 according to the new numbering system (Byrd et al. 2005). This region in eIF4G contains binding sites for eIF4A and eIF3 (Imataka and Sonenberg 1997; Morino et al. 2000). Consistent with this homology, eIF4A could be co-immunoprecipitated with Paip1 from HeLa extracts (Craig et al. 1998) and Paip1 also interacts with eIF3 (M. Derry and Y. Martineau, unpubl.) Thus, the simultaneous interactions between PABP, Paip1, eIF3, and eIF4A should facilitate the bridging of the 5′ and 3′ ends of mRNA (Fig. 2A). The existence of a complementary mode of mRNA circularization supports the importance of the circular mRNA conformation.

Paip2A, a 25-kD protein of 127 amino acids, and its recently discovered homolog Paip2B, a 25-kD protein of 123 amino acids, are antagonists of Paip1, as they inhibit the translation of poly(A)+ mRNAs (Fig. 2B). Both Paip2A and Paip2B inhibit in vitro translation of a capped poly(A)+ luciferase reporter mRNA in cell-free extracts (Khaleghpour et al. 2001a; Berlanga et al. 2006). In addition, overexpression of Paip2A or Paip2B in HeLa cells inhibited translation of a reporter mRNA (Khaleghpour et al. 2001a; Berlanga et al. 2006). Paip2 proteins inhibit the formation of 80S ribosomal complexes by competing with Paip1 for PABP binding, and by reducing the PABP–poly(A) interaction (Khaleghpour et al. 2001b). Paip2A further reduces translation by competing for PABP binding with eIF4G (Karim et al. 2006). Paip2 proteins therefore negate Paip1 activity by reducing mRNA

circularization (Fig. 2B). The PAM1 motif confers translational inhibitory activity on Paip2 proteins (Karim et al. 2006). The *D. melanogaster* homolog dPaip2 was shown to interact with dPABP, reduce dPABP binding to poly(A), and inhibit translation in vitro. Ectopic overexpression of dPaip2 in wings and wing discs resulted in a size-reduction phenotype, due to decreased cell number, whereas overexpression of dPaip2 in postreplicative tissues reduced ommatidia size in eyes and cell size in the larval fat body (Roy et al. 2004). These data demonstrate a physiological role for Paip2 proteins in regulating cell growth and proliferation.

Although no functional differences have been observed between Paip2A and Paip2B in vitro or in vivo, Paip2A and Paip2B differ in their tissue distribution in mice at both the mRNA and protein levels (Berlanga et al. 2006). These data indicate that they may function in a tissue-specific manner or may respond to different stimuli. Three Paip2B mRNAs of different lengths were identified: one of approximately 6.5 kb, corresponding to the size of the cloned cDNA, and two other species of 1.5 kb and 0.6 kb. The longest mRNA is preferentially expressed in the brain, whereas the shortest is more abundant in liver and testis (Berlanga et al. 2006). The difference between the mRNAs is confined to the 3′UTR, possibly suggesting that the longer mRNAs are controlled differently by microRNAs or *trans*-acting factors. At the protein level, both Paip2A and Paip2B are highly expressed in testis and liver (Berlanga et al. 2006). In addition, Paip2A is expressed in the brain, whereas Paip2B is mainly expressed in the pancreas (Berlanga et al. 2006), which may suggest roles in brain function and glucose homeostasis, respectively. Another difference between Paip2A and Paip2B is their level of ubiquitination, with Paip2A being more ubiquitinated (Berlanga et al. 2006). Paip2A and Paip2B diverged early during their evolution, since mammalian Paip2B is more similar to Paip2 proteins from frog, zebra fish, and salmon than to mammalian Paip2A (Berlanga et al. 2006). Both forms modulate PABP translational activity, but other distinct functions for these proteins may yet be identified. It is therefore possible that Paip2A and Paip2B diverged during evolution to accomplish different functions.

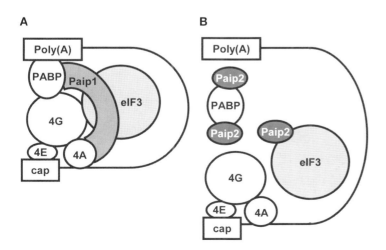

Figure 2. Model of PABP, Paip1, and Paip2 function. (*A*) Paip1 stabilizes the circularization of the mRNA by interfacing with PABP and eIF3. (*B*) Paip2 inhibits translation by reducing PABP binding to Poly(A) mRNA, Paip1, and eIF4G.

OTHER Paip FUNCTIONS

In addition to translational regulation, evidence supporting a role for Paips in stabilization of specific mRNAs continues to accumulate. Several reports suggest a role for translation in mRNA decay. Certain nucleotide sequence elements that dictate rapid mRNA decay are located within protein-coding regions and are dependent on translation (Shyu et al. 1989; Wisdom and Lee 1991). Using the *c-fos* mRNA as a model system, a role for translation in RNA turnover has been demonstrated (Schiàvi et al. 1994). Two destabilizing regions within the c-fos protein-coding region, termed protein-coding region determinants of instability (CRD), have been identified (Chen et al. 1992; Schiavi et al. 1994). Paip1 and PABP are subunits of a protein complex associated with the major CRD (mCRD) of c-fos, along with Unr, a purine-rich RNA-binding protein, hnRNP D; an AU-rich element binding protein; and NSAP1, an hnRNP R-like protein (Grosset et al. 2000). The complex stabilizes mCRD-containing mRNAs by impeding deadenylation. A bridging complex was proposed to exist between the poly(A) tail and the mCRD, which would be disrupted by ribosome transit, leading to RNA deadenylation and subsequent decay. On the basis of this report, Paip1 may be part of a decay protection complex that couples translation and mRNA decay.

Paip2A has been found in association with the 3′UTR of mRNAs known to be regulated at the level of mRNA stability. Paip2A associates with the 3′UTR of the glucose transporter GLUT5 mRNA as part of a large protein complex and is essential for the formation of this complex (Gouyon et al. 2003). Paip2A also interacts with the 3′UTR of the vascular endothelial growth factor (VEGF) mRNA. Overexpression of Paip2A led to increased stability of mRNA and increased secretion of VEGF, whereas small inhibitory RNA (siRNA)-mediated silencing of Paip2A led to decay of VEGF mRNA (Onesto et al. 2004). Surprisingly, Paip2A interacts with the VEGF mRNA in the absence of PABP (Gouyon et al. 2003) or any other proteins (Onesto et al. 2004), although Paip2A contains no known RNA-binding motif. It was suggested that in the case of VEGF mRNA, Paip2A might bind to an AU-rich region in the 3′UTR. AU-response elements (AREs) are present in the 3′UTR of many labile mRNAs and mediate their rapid degradation (Chen and Shyu 1995). Thus, Paip2A, and potentially the other Paips, may regulate mRNA stability independently of their ability to interact with and regulate PABP.

REGULATION OF Paips

Paips modulate PABP activity, but little data exist detailing how Paips are regulated, or under what circumstances. Paips are subject to ubiquitin-mediated degradation. Both Paip2A and Paip2B are ubiquitinated upon transfection into cells; Paip2A is modified to a greater extent than Paip2B, and thus is more rapidly degraded (Berlanga et al. 2006). The sequence in PABC that interacts with the PAM2 motif in the Paips is also present in the carboxyl terminus of EDD (Callaghan et al. 1998; Oughtred et al. 2002), a member of the *Homologous* to *E6-AP Carboxyl Terminus* (HECT) domain family (Huibregtse et al. 1995). These proteins function as E3 ubiquitin–protein ligases mediating ubiquitin-dependent proteolysis of specific protein targets. Thus, Paip2A is ubiquitinated upon binding to EDD through its PAM2 motif (Yoshida et al. 2006). The binding affinity of Paip2A to the EDD PABC domain is significantly weaker than to that of PABP, perhaps due to the absence of the first α-helix in the PABC domain from EDD (Deo et al. 2001; Kozlov et al. 2001); therefore, under physiological conditions, the higher affinity of Paip2A for PABP protects Paip2A from EDD-dependent proteolysis. However, upon reduction in PABP levels, for example through silencing by siRNA, Paip2A becomes free to associate with EDD and is subsequently ubiquitinated and degraded by the proteasome. As Paip2A protein levels decrease, the relative amount of free PABP is augmented, restoring overall PABP activity (Yoshida et al. 2006). The ubiquitin-mediated degradation of Paip2A constitutes the first evidence that Paip2A is posttranslationally regulated, thereby providing a mechanism for coordinately controlling protein levels of PABP and Paip2A and, consequently, modulating PABP function and translation (Fig. 3). Although Paip1 also possesses a PAM2 motif (Roy et al. 2002) and binds to the PABC domain of EDD (Deo et al. 2001), it is not degraded upon silencing of PABP (Yoshida et al. 2006). It is possible that the PAM2 motif is not sufficient for degradation of Paip1 or that an additional unknown factor is required for specific ubiquitination of Paip1.

Another potential mechanism for regulation of Paips may occur through binding to as yet unknown ligands. eIF3 has recently been identified as a new ligand for Paip1

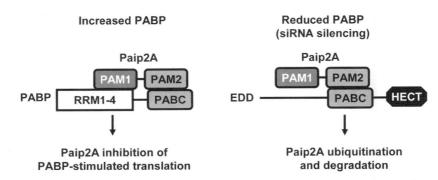

Figure 3. Model of ubiquitin-dependent degradation of Paip2A. PABP and EDD, two PABC-containing proteins, regulate the turnover of Paip2A.

(M. Derry and Y. Martineau, unpubl.). The interaction between eIF3 and Paip1 is direct and is independent of mRNA. Stimulation of cells with serum, insulin, or EGF resulted in increased eIF3–Paip1 binding, which could be reversed by wortmannin (a PI3K inhibitor), rapamycin (an mTOR inhibitor), or U0126 (a MEK1/2 inhibitor). In addition, Paip1-dependent enhancement of translation in vivo was abrogated upon co-transfection with siRNA to eIF3a. These data show that the eIF3–Paip1 interaction regulates Paip1 activity, positively correlates with increased translation in the cell, and is regulated by both the Akt-mTOR and the MEK signaling pathways.

The identification of eIF3 as a Paip1 ligand bolsters the previously proposed model (Craig et al. 1998) whereby the interaction of Paip1 with proteins, such as eIF4A, in the 5'UTR augments translation, perhaps by enhancing the activities of eIF4G or PABP or by promoting circularization. The observation that Paip1-dependent enhancement of translation is lost upon suppression of eIF3a suggests that the eIF3 may regulate Paip-dependent translational enhancement. Thus, Paip1 might be the proxy by which eIF3, and by extension the Akt/mTOR and MAPK pathways, control PABP activity.

CONCLUDING REMARKS

PABP was originally thought to enhance translation by binding to poly(A) mRNA, thus protecting it from degradation (Bernstein and Ross 1989). PABP is now known to enhance translation by direct mechanisms, including promotion of mRNA circularization through its interaction with eIF4G, ribosome recycling through its interaction with eRF3, eIF4F complex binding to the 5' cap, and 60S subunit joining (Imataka et al. 1998; Uchida et al. 2002b; Kahvejian et al. 2005). The identification of Paips as PABP binding partners adds a new layer to the existing knowledge of PABP biochemistry, representing the first known mechanism for modulating PABP translational activity. However, how Paips act to stimulate translation, their modes of regulation, and their other potential functions are still not well understood. It is becoming clear that Paips possess activity outside the previously known scope of translation initiation regulation through PABP interactions: Paips are now known to bind to other proteins, participate in other translational activities, and even regulate other cellular processes such as mRNA stability and mRNA export. Paips exist only in metazoans (Kahvejian et al. 2001) and therefore act as regulators of translation in multicellular organisms. Further research will determine in more detail the physiological role and mechanisms of regulation of Paips in cellular processes including mRNA translation, stability, and export.

ACKNOWLEDGMENTS

Y.M. was supported by a McGill Cancer Centre Post-Doctoral Fellowship. A.Y. was supported by a Postdoctoral Fellowship for Research Abroad from the Japan Society for the Promotion of Science. Work from the authors' laboratory was supported by a National Institutes of Health Public Health Service grant to N.S.

REFERENCES

Adam S.A., Nakagawa T., Swanson M.S., Woodruff T.K., and Dreyfuss G. 1986. mRNA polyadenylate-binding protein: Gene isolation and sequencing and identification of a ribonucleoprotein consensus sequence. *Mol. Cell. Biol.* **6:** 2932.

Albrecht M. and Lengauer T. 2004. Survey on the PABC recognition motif PAM2. *Biochem. Biophys. Res. Commun.* **316:** 129.

Berlanga J.J., Baass A., and Sonenberg N. 2006. Regulation of poly(A) binding protein function in translation: Characterization of the Paip2 homolog, Paip2B. *RNA* **12:** 1556.

Bernstein P. and Ross J. 1989. Poly(A), poly(A) binding protein and the regulation of mRNA stability. *Trends Biochem. Sci.* **14:** 373.

Burd C.G., Matunis E.L., and Dreyfuss G. 1991. The multiple RNA-binding domains of the mRNA poly(A)-binding protein have different RNA-binding activities. *Mol. Cell. Biol.* **11:** 3419.

Byrd M.P., Zamora M., and Lloyd R.E. 2005. Translation of eukaryotic translation initiation factor 4GI (eIF4GI) proceeds from multiple mRNAs containing a novel cap-dependent internal ribosome entry site (IRES) that is active during poliovirus infection. *J. Biol. Chem.* **280:** 18610.

Callaghan M.J., Russell A.J., Woollatt E., Sutherland G.R., Sutherland R.L., and Watts C.K. 1998. Identification of a human HECT family protein with homology to the *Drosophila* tumor suppressor gene hyperplastic discs. *Oncogene* **17:** 3479.

Chen C.Y. and Shyu A.B. 1995. AU-rich elements: Characterization and importance in mRNA degradation. *Trends Biochem. Sci.* **20:** 465.

Chen C.Y., You Y., and Shyu A.B. 1992. Two cellular proteins bind specifically to a purine-rich sequence necessary for the destabilization function of a c-fos protein-coding region determinant of mRNA instability. *Mol. Cell. Biol.* **12:** 5748.

Collier B., Gorgoni B., Loveridge C., Cooke H.J., and Gray N.K. 2005. The DAZL family proteins are PABP-binding proteins that regulate translation in germ cells. *EMBO J.* **24:** 2656.

Craig A.W., Haghighat A., Yu A.T., and Sonenberg N. 1998. Interaction of polyadenylate-binding protein with the eIF4G homologue PAIP enhances translation. *Nature* **392:** 520.

Deo R.C., Sonenberg N., and Burley S.K. 2001. X-ray structure of the human hyperplastic discs protein: An ortholog of the carboxyl terminus domain of poly(A)-binding protein. *Proc. Natl. Acad. Sci.* **98:** 4414.

Deo R.C., Bonanno J.B., Sonenberg N., and Burley S.K. 1999. Recognition of polyadenylate RNA by the poly(A)-binding protein. *Cell* **98:** 835.

Doel M.T. and Carey N.H. 1976. The translational capacity of deadenylated ovalbumin messenger RNA. *Cell* **8:** 51.

Gallie D.R. 1991. The cap and poly(A) tail function synergistically to regulate mRNA translational efficiency. *Genes Dev.* **5:** 2108.

Gingras A.-C., Gygi S.P., Raught B., Polakiewicz R.D., Abraham R.T., Hoekstra M.F., Aebersold R., and Sonenberg N. 1999. Regulation of 4E-BP1 phosphorylation: A novel two-step mechanism. *Genes Dev* **13:** 1422.

Görlach M., Burd C.G., and Dreyfuss G. 1994. The mRNA poly(A)-binding protein: Localization, abundance, and RNA-binding specificity. *Exp. Cell Res.* **211:** 400.

Gouyon F., Onesto C., Dalet V., Pages G., Leturque A., and Brot-Laroche E. 2003. Fructose modulates GLUT5 mRNA stability in differentiated Caco-2 cells: Role of cAMP-signalling pathway and PABP (polyadenylated-binding protein)-interacting protein (Paip) 2. *Biochem. J.* **375:** 167.

Gray N.K., Coller J.M., Dickson K.S., and Wickens M. 2000. Multiple portions of poly(A)-binding protein stimulate translation in vivo. *EMBO J.* **19:** 4723.

Grosset C., Chen C.Y., Xu N., Sonenberg N., Jacquemin-Sablon H., and Shyu A.B. 2000. A mechanism for translationally coupled mRNA turnover: Interaction between the poly(A) tail and a c-fos RNA coding determinant via a protein complex. *Cell* **103:** 29.

Grossi de Sa M.F., Standart N., Martins de Sa C., Akhayat O., Huesca M., and Scherrer K. 1988. The poly(A)-binding protein facilitates in vitro translation of poly(A)-rich mRNA. *Eur. J. Biochem.* **176:** 521.

He W. and Parker R. 2000. Functions of Lsm proteins in mRNA degradation and splicing. *Curr. Opin. Cell Biol.* **12:** 346.

Hoshino S., Imai M., Kobayashi T., Uchida N., and Katada T. 1999. The eukaryotic polypeptide chain releasing factor (eRF3/GSPT) carrying the translation termination signal to the 3′-Poly(A) tail of mRNA. Direct association of eRF3/GSPT with polyadenylate-binding protein. *J. Biol. Chem.* **274:** 16677.

Hosoda N., Kobayashi T., Uchida N., Funakoshi Y., Kikuchi Y., Hoshino S., and Katada T. 2003. Translation termination factor eRF3 mediates mRNA decay through the regulation of deadenylation. *J. Biol. Chem.* **278:** 38287.

Huibregtse J.M., Scheffner M., Beaudenon S., and Howley P.M. 1995. A family of proteins structurally and functionally related to the E6-AP ubiquitin-protein ligase. *Proc. Natl. Acad. Sci.* **92:** 2563.

Iizuka N., Najita L., Franzusoff A., and Sarnow P. 1994. Cap-dependent and cap-independent translation by internal initiation of mRNAs in cell extracts prepared from *Saccharomyces cerevisiae. Mol. Cell. Biol.* **14:** 7322.

Imataka H. and Sonenberg N. 1997. Human eukaryotic translation initiation factor 4G (eIF4G) possesses two separate and independent binding sites for eIF4A. *Mol. Cell. Biol.* **17:** 6940.

Imataka H., Gradi A., and Sonenberg N. 1998. A newly identified amino-terminal amino acid sequence of human eIF4G binds poly(A)-binding protein and functions in poly(A)-dependent translation. *EMBO J.* **17:** 7480.

Jacobson A. 1996. Poly(A) metabolism and translation: The closed-loop model. In *Translational control* (ed. J.W.B. Hershey et al.), p. 451. Cold Spring Harbor Laboratory Press, Cold Spring Harbor, New York.

Jacobson A. and Favreau M. 1983. Possible involvement of poly(A) in protein synthesis. *Nucleic Acids Res.* **11:** 6353.

Kahvejian A., Roy G., and Sonenberg N. 2001. The mRNA closed-loop model: The function of PABP and PABP-interacting proteins in mRNA translation. *Cold Spring Harbor Symp. Quant. Biol.* **66:** 293.

Kahvejian A., Svitkin Y.V., Sukarieh R., M'Boutchou M.N., and Sonenberg N. 2005. Mammalian poly(A)-binding protein is a eukaryotic translation initiation factor, which acts via multiple mechanisms. *Genes Dev.* **19:** 104.

Karim M.M., Svitkin Y.V., Kahvejian A., De Crescenzo G., Costa-Mattioli M., and Sonenberg N. 2006. A mechanism of translational repression by competition of Paip2 with eIF4G for poly(A) binding protein (PABP) binding. *Proc. Natl. Acad. Sci.* **103:** 9494.

Khaleghpour K., Svitkin Y.V., Craig A.W., DeMaria C.T., Deo R.C., Burley S.K., and Sonenberg N. 2001a. Translational repression by a novel partner of human poly(A) binding protein, Paip2. *Mol. Cell* **7:** 205.

Khaleghpour K., Kahvejian A., De Crescenzo G., Roy G., Svitkin Y.V., Imataka H., O'Connor-McCourt M., and Sonenberg N. 2001b. Dual interactions of the translational repressor Paip2 with poly(A) binding protein. *Mol. Cell. Biol.* **21:** 5200.

Kiehl T.R., Shibata H., and Pulst S.M. 2000. The ortholog of human ataxin-2 is essential for early embryonic patterning in *C. elegans. J. Mol. Neurosci.* **15:** 231.

Kozlov G., Trempe J.F., Khaleghpour K., Kahvejian A., Ekiel I., and Gehring K. 2001. Structure and function of the carboxyl terminus PABC domain of human poly(A)-binding protein. *Proc. Natl. Acad. Sci.* **98:** 4409.

Kuhn U. and Pieler T. 1996. *Xenopus* poly(A) binding protein: Functional domains in RNA binding and protein-protein interaction. *J. Mol. Biol.* **256:** 20.

Le H., Tanguay R.L., Balasta M.L., Wei C.C., Browning K.S., Metz A.M., Goss D.J., and Gallie D.R. 1997. Translation initiation factors eIF-iso4G and eIF-4B interact with the poly(A)-binding protein and increase its RNA binding activity. *J. Biol. Chem.* **272:** 16247.

Mangus D.A., Amrani N., and Jacobson A. 1998. Pbp1p, a factor interacting with *Saccharomyces cerevisiae* poly(A)-binding protein, regulates polyadenylation. *Mol. Cell. Biol.* **18:** 7383.

Mathews M.B., Sonenberg N., and Hershey J.W.B. 2007. Origins and principles of translational control. In *Translational control in biology and medicine* (ed. M.B. Mathews et al.), p. 1. Cold Spring Harbor Laboratory Press, Cold Spring Harbor, New York.

Morino S., Imataka H., Svitkin Y.V., Pestova T.V., and Sonenberg N. 2000. Eukaryotic translation initiation factor 4E (eIF4E) binding site and the middle one-third of eIF4GI constitute the core domain for cap-dependent translation, and the carboxyl terminus one-third functions as a modulatory region. *Mol. Cell. Biol.* **20:** 468.

Munroe D. and Jacobson A. 1990a. mRNA poly(A) tail, a 3′ enhancer of translational initiation. *Mol. Cell. Biol.* **10:** 3441.

———. 1990b. Tales of poly(A): A review. *Gene* **91:** 151.

Okochi K., Suzuki T., Inoue J., Matsuda S., and Yamamoto T. 2005. Interaction of anti-proliferative protein Tob with poly(A)-binding protein and inducible poly(A)-binding protein: Implication of Tob in translational control. *Genes Cells* **10:** 151.

Onesto C., Berra E., Grepin R., and Pages G. 2004. Poly(A)-binding protein-interacting protein 2, a strong regulator of vascular endothelial growth factor mRNA. *J. Biol. Chem.* **279:** 34217.

Otero L.J., Ashe M.P., and Sachs A.B. 1999. The yeast poly(A)-binding protein Pab1p stimulates in vitro poly(A)-dependent and cap-dependent translation by distinct mechanisms. *EMBO J.* **18:** 3153.

Oughtred R., Bedard N., Adegoke O.A., Morales C.R., Trasler J., Rajapurohitam V., and Wing S.S. 2002. Characterization of rat100, a 300-kilodalton ubiquitin-protein ligase induced in germ cells of the rat testis and similar to the *Drosophila* hyperplastic discs gene. *Endocrinology* **143:** 3740.

Palatnik C.M., Wilkins C., and Jacobson A. 1984. Translational control during early *Dictyostelium* development: Possible involvement of poly(A) sequences. *Cell* **36:** 1017.

Pestova T.V., Lorsch J.R., and Hellen C.U. 2007. The mechanism of translation initiation in eukaryotes. In *Translational control in biology and medicine* (ed. M.B. Mathews et al.), p. 87. Cold Spring Harbor Laboratory Press, Cold Spring Harbor, New York.

Pyronnet S., Imataka H., Gingras A.-C., Fukunaga R., Hunter T., and Sonenberg N. 1999. Human eukaryotic translation initiation factor 4G (eIF4G) recruits Mnk1 to phosphorylate eIF4E. *EMBO J.* **18:** 270.

Roy G., Miron M., Khaleghpour K., Lasko P., and Sonenberg N. 2004. The *Drosophila* poly(A) binding protein-interacting protein, dPaip2, is a novel effector of cell growth. *Mol. Cell. Biol.* **24:** 1143.

Roy G., De Crescenzo G., Khaleghpour K., Kahvejian A., O'Connor-McCourt M., and Sonenberg N. 2002. Paip1 interacts with poly(A) binding protein through two independent binding motifs. *Mol. Cell. Biol.* **22:** 3769.

Sachs A. 2000. Physical and functional interactions between the mRNA cap structure and the poly(A) tail. In *Translational control of gene expression* (ed. N. Sonenberg et al.), p. 447. Cold Spring Harbor Laboratory Press, Cold Spring Harbor, New York.

Sachs A.B. and Davis R.W. 1989. The poly(A) binding protein is required for poly(A) shortening and 60S ribosomal subunit-dependent translation initiation. *Cell* **58:** 857.

———. 1990. Translation initiation and ribosomal biogenesis: Involvement of a putative rRNA helicase and RPL46. *Science* **247:** 1077.

Sachs A.B., Davis R.W., and Kornberg R.D. 1987. A single domain of yeast poly(A)-binding protein is necessary and sufficient for RNA binding and cell viability. *Mol. Cell. Biol.* **7:** 3268.

Schiavi S.C., Wellington C.L., Shyu A.B., Chen C.Y., Greenberg M.E., and Belasco J.G. 1994. Multiple elements in the c-fos protein-coding region facilitate mRNA deadenylation and decay by a mechanism coupled to translation. *J. Biol. Chem.* **269:** 3441.

Shyu A.B., Greenberg M.E., and Belasco J.G. 1989. The c-fos transcript is targeted for rapid decay by two distinct mRNA degradation pathways. *Genes Dev.* **3:** 60.

Svitkin Y.V. and Sonenberg N. 2004. An efficient system for cap- and poly(A)-dependent translation in vitro. *Methods Mol. Biol.* **257:** 155.

Tarun S.Z. and Sachs A.B. 1995. A common function for mRNA 5′ and 3′ ends in translation initiation in yeast. *Genes Dev.* **9:** 2997.

———. 1996. Association of the yeast poly(A) tail binding protein with translation initiation factor eIF-4G. *EMBO J.* **15:** 7168.

Uchida N., Hoshino S., Imataka H., Sonenberg N., and Katada T. 2002a. The interaction between the eukaryotic releasing factor eRF3/GSPT and poly adenylate-binding protein is important for translation initiation. *J. Biol. Chem.* **277:** 50286.

———. 2002b. A novel role of the mammalian GSPT/eRF3 associating with poly(A)-binding protein in Cap/Poly(A)-dependent translation. *J. Biol. Chem.* **277:** 50286.

Wakiyama M., Imataka H., and Sonenberg N. 2000. Interaction of eIF4G with poly(A)-binding protein stimulates translation and is critical for *Xenopus* oocyte maturation. *Curr. Biol.* **10:** 1147.

Wickens M., Goodwin E.B., Kimble J., Strickland S., and Hentze M.W. 2000. Translational control of developmental decisions. In *Translational control of gene expression* (ed. N. Sonenberg et al.), p. 295. Cold Spring Harbor Laboratory Press, Cold Spring Harbor, New York.

Wisdom R. and Lee W. 1991. The protein-coding region of c-myc mRNA contains a sequence that specifies rapid mRNA turnover and induction by protein synthesis inhibitors. *Genes Dev.* **5:** 232.

Yoshida M., Yoshida K., Kozlov G., Lim N.S., De Crescenzo G., Pang Z., Berlanga J.J., Kahvejian A., Gehring K., Wing S.S., and Sonenberg N. 2006. Poly(A) binding protein (PABP) homeostasis is mediated by the stability of its inhibitor, Paip2. *EMBO J.* **25:** 1934.

Two Distinct Conformations of the Conserved RNA-rich Decoding Center of the Small Ribosomal Subunit Are Recognized by tRNAs and Release Factors

E.M. YOUNGMAN, L. COCHELLA, J.L. BRUNELLE, S. HE, AND R. GREEN

Howard Hughes Medical Institute, Department of Molecular Biology and Genetics, Johns Hopkins University School of Medicine, Baltimore, Maryland 21205

The mRNA is bound and poised in the decoding center of the small subunit of the ribosome where the genetic code is translated by the tRNAs, which recognize sense codons, and by the release factors, which recognize stop codons. Structural and biochemical studies have identified key universally conserved nucleotides, G530, A1492, and A1493, that are important for selection of cognate tRNA species during elongation. Here, we present evidence that these same universally conserved nucleotides are also important for interactions with the release factors, but must assume a very different structure during stop-codon recognition. These data provide mechanistic insight into how the decoding center of the ribosome has evolved to recognize distinct substrates with high fidelity, which in turn regulates the downstream chemical events of peptidyl transfer and peptide release.

The aminoacyl (A) site of the ribosome, spanning from the decoding center in the small subunit to the peptidyl transfer center in the large subunit, serves two related but distinct functions during translation. During each round of polypeptide elongation, a sense codon on the mRNA poised in the small subunit of the A site is recognized by its cognate aa-tRNA, bringing a new amino acid into the large subunit of the A site where peptidyl transfer occurs. During termination, the same site is bound by a release factor protein for recognition of a stop codon in the decoding site of the small subunit and catalysis of peptide release in the active site of the large subunit. Although peptidyl transfer and peptide release are two chemically related reactions that are catalyzed in the same active site on the ribosome, they are stimulated by the interactions that two very different "substrates" establish with the ribosome in a manner that ultimately depends on the codon sequence. It follows that these recognition events must take place with high fidelity, such that the two very different outcomes of peptidyl transfer and peptide release are carefully regulated.

Recent work has suggested that activation for peptidyl transfer is accomplished by conformational rearrangements in the active site promoted by interactions of the CCA end of the tRNA with the A loop of the peptidyl transferase center (PTC) (Schmeing et al. 2005; Brunelle et al. 2006). Although less is known about the structural and functional requirements for triggering peptide release in the PTC, a parsimonious view is that interactions with the A loop may similarly be critical for this regulated catalytic step (Caskey et al. 1971). It has also been established that in the context of an "activated" catalytic center, the universally conserved nucleotides in the PTC have no apparent catalytic role in promoting peptidyl transfer, although their identity is critical for peptide release (Youngman et al. 2004). Catalysis happens readily for the chemically facile aminolysis reaction (peptidyl

transfer), whereas the inner core of conserved PTC nucleotides has a more substantial, as yet undefined, role in the chemically more challenging hydrolytic (peptide release) reaction. These data highlight how the ribosome has evolved an active site able to execute catalysis of two distinct reactions of varying chemical difficulty, and only in the context of bound substrate.

THE DECODING CENTER AND THE ACCURACY OF TRANSLATION

What is perhaps the more remarkable layer of complexity in regulation of this active site is that the two substrates that must bind to catalyze peptidyl transfer and peptide release are very different. Catalysis of peptide bond formation involves recognition of the codon presented in the A site of the decoding center by cognate aminoacyl tRNA, whereas peptide release involves recognition of stop codons in the same decoding center by protein release factors. Thus, like the large subunit active site, the decoding center of the small subunit must perform two related functions: recognition of RNA:RNA interactions during tRNA selection and recognition of RNA:protein interactions during release factor selection. Furthermore, recognition of the codons by these distinct components takes place with high accuracy that in large part dictates the overall fidelity of the translation of the genetic code (Cochella and Green 2005b). In vivo fidelity studies suggest that the overall accuracy of translation is extremely high, on the order of 10^{-3} to 10^{-4} for codon recognition by tRNAs and on the order of 10^{-5} for stop-codon recognition by release factors (Bouadloun et al. 1983; Parker and Holtz 1984; Jorgensen et al. 1993). It follows that the task of the ribosome, tRNAs, release factors, and relevant elongation factors is to specifically recognize cognate mRNA:tRNA and mRNA:release factor interactions while discriminating against near-cognate interactions that differ by a single-nucleotide substitution in the codon.

At the simplest level, there are clearly differences in the energy of the interaction between cognate and near-cognate partners, and these differences account for some of the observed fidelity. However, at least for the case of tRNA selection, it is clear that the overall energetic differences in these interactions cannot fully explain the observed fidelity of the process in vivo. For example, it has previously been pointed out that overall differences in the G:C content of the codon:anticodon helix result in immediate discrepancies in binding energies which are not easily reconciled with the relatively uniform accuracy of protein synthesis (Ogle and Ramakrishnan 2005). Both GC- and AU-rich codon:anticodon helices are rather uniformly and accurately decoded by the protein synthesis apparatus. This apparent conundrum is rationalized by the view that it is the geometry of the codon:anticodon helix that is the critical feature recognized by the ribosome, allowing tRNA selection to occur with such high accuracy (Carter et al. 2000). This general view is well supported by kinetic studies which make the compelling argument that tRNA selection is kinetically driven by the acceleration of forward rate constants by cognate mRNA:tRNA interactions, at the expense of efficient utilization of differential off-rates of the cognate and near-cognate partners (Pape et al. 1999; Gromadski and Rodnina 2004a). More recent kinetic data have further argued that the ribosome assumes two distinct states in response to interactions in the decoding center—an "on" state and an "off" state—depending on whether correct or incorrect molecular partners are bound (Gromadski et al. 2006). These two states of the ribosome result in different downstream consequences: Cognate interactions accelerate forward rates (for GTPase activation and tRNA accommodation) and thus acceptance into the ribosome for subsequent events (peptidyl transfer), whereas near-cognate tRNAs do not accelerate these steps and thus are generally rejected from the ribosome. Although there is far less known about release factor recognition of stop versus near-stop codons, available data suggest that in this case as well, not all discrimination takes place at the level of differences in the energetics of the interaction between the mRNA and the release factor (Freistroffer et al. 2000). It seems possible that there will similarly be contributions to specificity made by the ribosome itself (e.g., the decoding center) in recognizing these precise interactions.

MOLECULAR MECHANISM OF tRNA SELECTION

Much is known about specific recognition of cognate tRNAs from high-resolution structures of the small ribosomal subunit. Early in the process of tRNA selection, docking of a cognate tRNA species with the codon of the mRNA triggers conformational rearrangements in universally conserved nucleotides in the decoding center. These changes include the rotation of a guanosine (G530) from a syn to an anti conformation by rotation around the glycosidic bond and the flipping of two adenosines (A1492 and A1493) from positions within helix 44 of 16S rRNA to extrahelical

positions (Ogle et al. 2001); we refer to this state as "flipped out." These seemingly unfavorable rearrangements are stabilized by the formation of A-minor interactions between nucleotides A1492, A1493, and G530 and the minor groove of the cognate codon:anticodon helix. These local interactions are thought in turn to favor the formation of a more global "closed" state of the ribosome observed in X-ray structures (Ogle et al. 2002). Formation of this closed state is thought to favor incorporation of the tRNA into the A site. Genetic variants that stabilize the closed state result in a decrease in the fidelity of protein synthesis (a *ram*, or ribosomal ambiguity, phenotype), whereas those that destabilize the closed state increase fidelity (known as *restrictive*) (reviewed in Ogle and Ramakrishnan 2005). Near-cognate codon:anticodon interactions, on the other hand, form irregular helices that do not stably interact with the flipped-out conserved adenosines and thus do not lead to ribosome closure and subsequent tRNA incorporation. In support of these models for tRNA selection, aminoglycoside antibiotics (such as paromomycin) known to induce miscoding during protein synthesis have been shown to bind in a pocket within helix 44 of the rRNA such that these two critical adenosines are specifically displaced from their uninduced position independent of the nature of the interaction in the decoding center (cognate vs. near-cognate) (Carter et al. 2000). Thus, paromomycin binding stabilizes the flipped-out conformation for these residues and triggers the associated downstream consequences for tRNA selection.

This summary of the kinetic and molecular determinants of high-fidelity tRNA selection by the ribosome yields an impressively detailed and consistent model for how this process takes place. Both functional and structural analyses of the related peptide release reaction lag substantially behind (Petry et al. 2005), although it is likely that these two processes are at least somewhat related. Although recognition of a specific codon by an RNA (a tRNA) and a protein factor (an RF) might depend on distinct properties of the molecular players, the downstream consequences of engagement of cognate species are clearly related: activation of the large ribosomal subunit catalytic center for the chemical reactions of aminolysis and hydrolysis. In the absence of detailed structural information about release factor engagement of the decoding center, we have taken a biochemical approach to decipher the molecular basis for the high fidelity of peptide release during protein synthesis. Here, we compare and contrast the relative contributions made by the universally conserved decoding center nucleotides to high-fidelity reading of sense codons by the tRNAs and of stop codons by the release factors.

ROLE OF CONSERVED DECODING CENTER NUCLEOTIDES IN tRNA SELECTION AND PEPTIDE RELEASE

Universally conserved nucleotides in the decoding center of the ribosome involved in specific, well-characterized molecular interactions with the codon:anticodon helix are thought to be central to the tRNA selection process (Ogle et al. 2001). We have used a site-directed mutagenesis approach to evaluate the contributions made by three key nucleotides, G530, A1492, and A1493, to decoding and peptide release. Mutations were incorporated in a plasmid-borne version of the *rrnB* operon, and the variant ribosomes were expressed in the background of wild-type ribosomes (Powers and Noller 1990). All substitutions tested at these positions resulted in a dominant-lethal growth phenotype, requiring the use of an affinity purification system for the expression and subsequent isolation of variant ribosomes. Variant 30S subunits were tagged with an MS2 RNA stem-loop inserted in the backbone of 16S rRNA that allowed for affinity purification on a glutathione matrix via association with an MS2 coat protein fused to GST (Youngman and Green 2005).

Variant ribosomes were first evaluated in two assays— GTP hydrolysis and peptidyl transfer—that report on the critical rate-determining steps in tRNA selection of GTPase activation and accommodation, respectively (Gromadski and Rodnina 2004a). For both assays, ribosomal initiation complexes were formed by loading 70S ribosome particles with a defined mRNA (with a Shine-Dalgarno sequence and an AUG start site followed by a UGG Trp codon) and an initiator tRNA (f-Met-tRNA$_f^{Met}$) using the requisite initiation factors (IFs1–3) (Cochella and Green 2005a). These ribosome complexes were then reacted with limiting amounts of a ternary complex of EF-Tu•GTP•Trp-tRNA^{Trp}, and the rates of GTP hydrolysis and peptidyl transfer were measured. All three variants tested exhibited substantial decreases in the observed rates for these reactions (up to 40-fold for A1492G in each reaction) (Fig. 1). Given that these reactions were performed under nearly satu-

rating concentrations of ribosome complexes, it is likely that the observed decreased rates reflect deficiencies in steps distinct from binding of the ternary complex to the ribosome. These data provide strong evidence for an essential role for these nucleotides in the process of tRNA selection. These data corroborate conclusions drawn from biochemical studies that probed the importance of interactions in this region by disrupting contacts within the codon:anticodon helix itself (Fahlman et al. 2006; Gromadski et al. 2006).

We next looked at the effects of substitutions in the decoding region on the related reaction of peptide release. For these experiments, ribosomal initiation complexes were formed as above, but using an mRNA that substitutes for the Trp codon a stop codon (UAG) recognized by the class I release factor RF1. The resulting ribosome complexes were then reacted with varying concentrations of RF1 and the rates of peptide release were measured as described previously (Youngman et al. 2004). Although there are defects in peptide release associated with mutations in these positions, these defects are overcome at higher concentrations of RF1 (Fig. 2). These data indicate that the predominant effect of these mutations is on RF1 binding and that the identity of these nucleotides is less important for specifying the downstream catalytic events of peptide hydrolysis.

Taken together, these two sets of experiments indicate that the identity of the decoding center nucleotides is important for the two related processes of tRNA selection and release factor recognition. The discrepancies in how these nucleotide changes affect these reactions suggest, however, that they do not have the same role in the two processes. As discussed above, structural studies indicate that all three nucleotides undergo substantial conformational rearrangements on binding cognate tRNA during tRNA selection. The simplest interpretation of our data is that changes at these positions in the rRNA affect the stability of this flipped-out state of the decoding center, thus diminishing the induced fast rates of GTPase activation and accommodation. The data for the release reaction are more complicated and show that

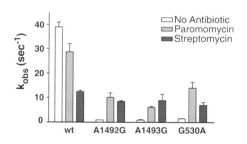

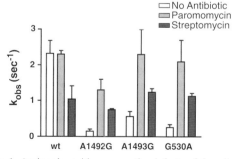

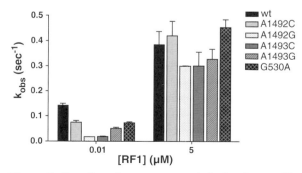

Figure 1. Aminoglycosides rescue the defects of decoding site mutant ribosomes in tRNA selection. Rate constants for GTP hydrolysis (*top*) and peptide bond formation (*bottom*) were measured at 2.25 μM ribosome complex and substoichiometric ternary complex in the absence or presence of 5 μM paromomycin or 5 μM streptomycin. Reactions were performed as described by Cochella and Green (2005a). (Reprinted, with permission, from Cochella et al. 2006 [Nature Publishing Group].)

Figure 2. Decoding site mutants are defective for peptide release at low release factor concentrations. Rate constants for peptide release were measured on wild-type and mutant ribosomes at either 10 nM ribosome complex and 10 nM RF1 (*left*) or 250 nM ribosome complex and 5 μM RF1 (*right*). Reactions were performed as described by Youngman et al. (2004), except that a high-fidelity polyamine-based buffer (Gromadski and Rodnina 2004a) was used.

reasonably mild effects of the mutations on binding of RF1 to the ribosome are overcome with higher concentrations of release factor. These data indicate that the identity of these nucleotides is not essential for catalysis per se and that their mutation may simply impose a structure on the region that is unfavorable for release factor binding. For example, specific positive interactions between the nucleotides and the RF may be disrupted (or destabilized). Alternatively, the induced state of the region (flipped-out conformation) may not be favorable for peptide release, and the wild-type identity of these nucleotides may be optimized for stable docking within helix 44 in the absence of cognate RNA:RNA duplex. In this case, mutation might destabilize the flipped-in state of the decoding center and thus allow these nucleotides to compete directly with RF binding.

EFFECTS OF AMINOGLYCOSIDE ANTIBIOTICS ON tRNA SELECTION AND PEPTIDE RELEASE

The interactions of aminoglycoside antibiotics (such as paromomycin and streptomycin) with the ribosome decoding center are well characterized structurally and based on their effects on the process of tRNA selection. These compounds are generally known for their clear effects in stimulating miscoding on the ribosome during translation. We next looked at the effects of these antibiotics on the activity of ribosomes bearing mutations in the decoding center nucleotides discussed above. As previously reported, paromomycin has little effect on GTPase activation and accommodation in wild-type ribosomes, whereas streptomycin has a mild inhibitory effect on both reactions (Pape et al. 2000; Gromadski and Rodnina 2004b). These kinetic results have been rationalized from a structural perspective based on the observation that closure of the small ribosomal subunit is stabilized by both compounds but that the extent of closure is more significant with paromomycin than with streptomycin (Carter et al. 2000). Moreover, it has been shown that when both antibiotics are supplied in the same reaction, the biochemical read-out is one characteristic of streptomycin alone, as though this compound is dominant and prevents full closure of the subunit (Gromadski and Rodnina 2004b). Not unexpectedly, both paromomycin and streptomycin substantially relieve the defects associated with the decoding center mutations, nearly to wild-type levels in the case of the GTPase activation assay (see Fig. 1). These data are consistent with the idea that mutation of the decoding center nucleotides destabilizes interactions with the codon:anticodon helix and that these defects can be suppressed by artificially inducing closure of the subunit with these compounds. Other related studies have similarly shown that these antibiotics can suppress a variety of defects in the decoding center including mismatches in the codon:anticodon helix and the loss of critical 2'-OH groups in this same region (Pape et al. 2000; Fahlman et al. 2006).

We next asked. What are the effects of these same compounds on the peptide release reaction? In striking contrast to its effects on tRNA selection, paromomycin

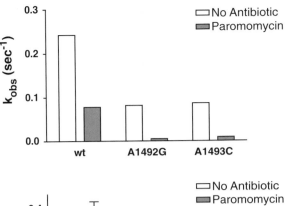

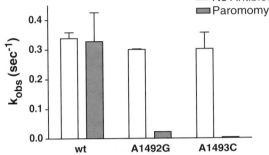

Figure 3. Paromomycin potentiates the defects of decoding site mutants in peptide release. Rate constants for peptide release were measured for wild-type and selected decoding site mutant ribosomes in the absence or presence of 50 μM paromomycin. Reactions were performed at either 50 nM ribosome complex and 50 nM RF1 (*top*) or 250 nM ribosome complex and 5 μM RF1 (*bottom*).

substantially inhibits RF1-catalyzed peptide release on wild-type ribosomes at low concentrations of RF1 (relative to its known $K_{1/2}$), although this inhibition is overcome at higher concentrations of RF1 (Fig. 3). These data suggest that paromomycin affects the release reaction by simply inhibiting release factor binding. We next looked at the effects of paromomycin on the release defects associated with mutations in the decoding center nucleotides and found that paromomycin further antagonized the defects of these mutations, rather than suppressing them (Fig. 3). A model that reconciles these results is that stabilization of the flipped-out conformation of the decoding center is in some way unfavorable for the release reaction. In the case of wild-type ribosomes, the effects of paromomycin binding can be overcome by high concentrations of the release factor. We are currently determining whether the defects of the variant ribosomes can similarly be overcome with higher release factor concentrations.

CONCLUSIONS

The experiments reported in this study provide insight into the regulated function of the universally conserved decoding center of the ribosome. This RNA-rich site is engaged by distinct classes of molecules during translation—the tRNAs and the release factors—that must both recognize a codon in the mRNA with exceptionally high fidelity and in turn must activate related downstream catalytic events in the large ribosomal subunit. In the absence of specific knowledge, it had seemed reasonable that these two processes would be highly related, albeit one RNA-based and the other protein-based, and thus would depend on similar molecular components for optimal function. What these studies have revealed is that tRNA selection and stop codon recognition during peptide release are in fact orchestrated by quite distinct mechanisms in the decoding center. Universally conserved nucleotides in the decoding center, although important at some level for both processes, are more specifically important for tRNA selection where they engage the minor groove of the codon:anticodon helix to activate efficient downstream GTPase activation and accommodation. In the release reaction, defects associated with these nucleotide substitutions are easily overcome at increasing concentrations of release factor with no residual effects on catalysis. Similarly, paromomycin has disparate effects on tRNA selection and peptide release, being overall stimulatory in the former case and overall inhibitory in the latter. In light of the well-documented effects of paromomycin on ribosome structure, we argue that tRNAs and release factors recognize distinct states of the ribosome that minimally differ in the position of A1492/A1493. Although tRNA selection depends on the flipped-out conformation of these adenosines (the "on" configuration), release factors apparently bind to the flipped-in or "off" state of the decoding center.

A number of different lines of evidence would suggest that tRNA selection and peptide release function, as they now exist, arose at different points during the evolution of the translation apparatus. Most striking among these arguments is the observation that bacterial and eukaryotic release factors are not related at the sequence level and thus apparently arose as independent solutions to the same problem (Frolova et al. 1999). In light of this, perhaps the observation that tRNA selection and peptide release depend on such distinct molecular mechanisms is less surprising. The creation of an active site that must engage a variety of distinct molecular partners is a formidable challenge, and these studies only begin to decipher how evolution has solved this problem in the protein synthesis apparatus.

ACKNOWLEDGMENTS

We thank V. Ramakrishnan and M. Rodnina for helpful discussions, the National Institutes of Health for funding, and the Howard Hughes Medical Institute for salary support.

REFERENCES

Bouadloun F., Donner D., and Kurland C.G. 1983. Codon-specific missense errors in vivo. *EMBO J.* **2:** 1351.

Brunelle J.L., Youngman E.M., Sharma D., and Green R. 2006. The interaction between C75 of tRNA and the A loop of the ribosome stimulates peptidyl transferase activity. *RNA* **12:** 33.

Carter A.P., Clemons W.M., Brodersen D.E., Morgan-Warren R.J., Wimberly B.T., and Ramakrishnan V. 2000. Functional insights from the structure of the 30S ribosomal subunit and its interactions with antibiotics. *Nature* **407:** 340.

Caskey C.T., Beaudet A.L., Scolnick E.M., and Rosman M. 1971. Hydrolysis of fMet-tRNA by peptidyl transferase. *Proc. Natl. Acad. Sci.* **68:** 3163.

Cochella L. and Green R. 2005a. An active role for tRNA in decoding beyond codon:anticodon pairing. *Science* **308:** 1178.

———. 2005b. Fidelity in protein synthesis. *Curr. Biol.* **15:** R536.

Cochella L., Brunelle J.L., and Green R. 2006. Mutational analysis reveals two independent molecular requirements during transfer RNA selection on the ribosome. *Nat. Struct. Mol. Biol.*

Fahlman R.P., Olejniczak M., and Uhlenbeck O.C. 2006. Quantitative analysis of deoxynucleotide substitutions in the codon-anticodon helix. *J. Mol. Biol.* **355:** 887.

Freistroffer D.V., Kwiatkowski M., Buckingham R.H., and Ehrenberg M. 2000. The accuracy of codon recognition by polypeptide release factors. *Proc. Natl. Acad. Sci.* **97:** 2046.

Frolova L.Y., Tsivkovskii R.Y., Sivolobova G.F., Oparina N.Y., Serpinsky O.I., Blinov V.M., Tatkov S.I., and Kisselev L.L. 1999. Mutations in the highly conserved GGQ motif of class 1 polypeptide release factors abolish ability of human eRF1 to trigger peptidyl-tRNA hydrolysis. *RNA* **5:** 1014.

Gromadski K.B. and Rodnina M.V. 2004a. Kinetic determinants of high-fidelity tRNA discrimination on the ribosome. *Mol. Cell* **13:** 191.

———. 2004b. Streptomycin interferes with conformational coupling between codon recognition and GTPase activation on the ribosome. *Nat. Struct. Mol. Biol.* **11:** 316.

Gromadski K.B., Daviter T., and Rodnina M.V. 2006. A uniform response to mismatches in codon-anticodon complexes ensures ribosomal fidelity. *Mol. Cell* **21:** 369.

Jorgensen F., Adamski F.M., Tate W.P., and Kurland C.G. 1993. Release factor-dependent false stops are infrequent in *Escherichia coli*. *J. Mol. Biol.* **230:** 41.

Ogle J.M. and Ramakrishnan V. 2005. Structural insights into translational fidelity. *Annu. Rev. Biochem.* **74:** 129.

Ogle J.M., Murphy F.V., Tarry M.J., and Ramakrishnan V. 2002. Selection of tRNA by the ribosome requires a transition from an open to a closed form. *Cell* **111:** 721.

Ogle J.M., Brodersen D.E., Clemons W.M., Jr., Tarry M.J., Carter A.P., and Ramakrishnan V. 2001. Recognition of cognate transfer RNA by the 30S ribosomal subunit. *Science* **292:** 897.

Pape T., Wintermeyer W., and Rodina M.W. 1999. Induced fit in initial selection and proofreading of aminoacyl-tRNA on the ribosome. *Eur. J. Mol. Biol.* **18:** 3800.

———. 2000. Conformational switch in the decoding region of 16S rRNA during aminoacyl-tRNA selection on the ribosome. *Nat. Struct. Biol.* **7:** 104.

Parker J. and Holtz G. 1984. Control of basal-level codon misreading in *Escherichia coli*. *Biochem. Biophys. Res. Commun.* **121:** 487.

Petry S., Brodersen D.E., Murphy F.V., Dunham C.M., Selmer M., Tarry M.J., Kelley A.C., and Ramakrishnan V. 2005. Crystal structures of the ribosome in complex with release factors RF1 and RF2 bound to a cognate stop codon. *Cell* **123:** 1255.

Powers T. and Noller H.F. 1990. Dominant lethal mutations in a conserved loop in 16S rRNA. *Proc. Natl. Acad. Sci.* **87:** 1042.

Schmeing T.M., Huang K.S., Strobel S.A., and Steitz T.A. 2005. An induced-fit mechanism to promote peptide bond formation and exclude hydrolysis of peptidyl-tRNA. *Nature* **438:** 520.

Youngman E.M. and Green R. 2005. Affinity purification of in vivo-assembled ribosomes for in vitro biochemical analysis. *Methods* **36:** 305.

Youngman E.M., Brunelle J.L., Kochaniak A.B., and Green R. 2004. The active site of the ribosome is composed of two layers of conserved nucleotides with distinct roles in peptide bond formation and peptide release. *Cell* **117:** 589.

The Expanding Universe of Noncoding RNAs

G.J. Hannon,* F.V. Rivas,* E.P. Murchison,* and J.A. Steitz[†]

*Watson School of Biological Sciences, Howard Hughes Medical Institute, Cold Spring Harbor Laboratory, Cold Spring Harbor, New York 11724; [†]Department of Molecular Biophysics and Biochemistry, Howard Hughes Medical Institute, Yale University, New Haven, Connecticut 06536

The 71st Cold Spring Harbor Symposium on Quantitative Biology celebrated the numerous and expanding roles of regulatory RNAs in systems ranging from bacteria to mammals. It was clearly evident that noncoding RNAs are undergoing a renaissance, with reports of their involvement in nearly every cellular process. Previously known classes of longer noncoding RNAs were shown to function by every possible means—acting catalytically, sensing physiological states through adoption of complex secondary and tertiary structures, or using their primary sequences for recognition of target sites. The many recently discovered classes of small noncoding RNAs, generally less than 35 nucleotides in length, most often exert their effects by guiding regulatory complexes to targets via base-pairing. With the ability to analyze the RNA products of the genome in ever greater depth, it has become clear that the universe of noncoding RNAs may extend far beyond the boundaries we had previously imagined. Thus, as much as the Symposium highlighted exciting progress in the field, it also revealed how much farther we must go to understand fully the biological impact of noncoding RNAs.

It has become customary in Summaries of Cold Spring Harbor Symposia to quote Jacob and Monod (1961). Indeed, hardly any aspect of gene expression escaped mention in their classic paper entitled "Genetic Regulatory Mechanisms in the Synthesis of Proteins" published in 1961 in the *Journal of Molecular Biology*. The possibility of a regulatory role for RNA was no exception. In the schemes presented in Figure 6 for the control of protein synthesis, not only the Messengers but also the Repressor were modeled as RNA, accompanied by the following explanations:

(1) *"The specific 'repressor' (RNA?), acting with a given operator, is synthesized by a regulator gene."*

(2) *"The chemical identification of the repressor as an RNA fraction is a logical assumption based only on the negative evidence which indicates that it is not a protein."*

The notion that RNA might use its base-pairing potential to regulate the activity of DNA was irresistible at the time, and the lasting appeal of this idea appears ever more justified with each passing year. Even though the *lac* repressor turned out to be a protein, the roster of noncoding RNAs (ncRNAs) presaged by Jacob and Monod has grown steadily. We now appreciate that ncRNAs are involved in the control of gene expression, as well as many other aspects of cellular metabolism. Beginning with the basic triumvirate of messenger RNA (mRNA), ribosomal RNA (rRNA), and transfer RNA (tRNA), new classes of RNA have materialized every few years.

The discovery of introns in 1977 seeded the realization that small nuclear RNAs (snRNAs) are building blocks of the spliceosome. snRNAs use base-pairing to recognize critical landmarks within the intron undergoing excision. Small nucleolar RNAs (snoRNAs) were later found to use sequence complementarity to guide nucleotide modification of conserved regions within eukaryotic rRNAs. This phenomenon is not restricted to eukaryotes, as snoRNAs have analogs even in Archaea. Additionally, the discovery that the RNA component of telomerase dictates the sequence added to the ends of the DNA of most eukaryotic chromosomes expanded the mechanisms by which ncRNAs act.

The demonstration of RNA catalysis in the 1980s, initially in self-splicing introns and in RNA cleavage, proved that RNA molecules—by assuming defined secondary and tertiary structures—are not confined to regulatory and structural roles in cellular machines. These findings underscored the centrality of RNA in the evolution of life and culminated with the revelation from structural studies that the ribosome is a ribozyme (Nissen et al. 2000).

The recent explosion both in the roster of ncRNAs and in our appreciation of the multifaceted regulatory potential of RNA constitutes another RNA revolution. It began in 1998, when double-stranded (ds)RNA was identified as the active principle in gene silencing in experiments conducted in *Caenorhabditis elegans* by Fire and Mello (Fire et al. 1998) and in *Trypanosoma brucei* by Ullu and Tschudi (Ngo et al. 1998). Baulcombe's laboratory then uncovered the coincident existence of short (~25 nucleotide) RNAs in plant cells undergoing posttranscriptional gene silencing (PTGS) as a result of the introduction of exogenous genes or viral infection (Hamilton and Baulcombe 1999). Links between PTGS in plants and RNA interference (RNAi) in animals were forged by the discovery of a conserved enzymatic machinery that produces small RNAs (Hammond et al. 2000; Zamore et al. 2000; Bernstein et al. 2001; Hutvagner et al. 2001) and a conserved core of the effector complexes (called RISC, for RNA-induced silencing complexes) that uses small RNAs to guide silencing of target genes (Hammond et al. 2000, 2001; Hutvagner and Zamore 2002; Martinez et al. 2002). These cellular components could exist simply to lie in wait for infecting viruses that produce dsRNAs as part of their replication cycle; indeed, small RNA pathways are critical to viral resistance in plants. However, it seemed more likely that such a flexible and adaptive machinery would have evolved to regulate naturally occurring, endogenous substrates. Already in 1993, the Ambrose and Ruvkun labs had characterized a 21-nucleotide RNA that blocks translation of the *C. elegans* lin-14 mRNA via complementarity in the 3′UTR (Lee et al. 1993; Wightman et al. 1993). This and let-7, the other "small temporal RNA," were linked to the RNAi pathway through genetic analysis of Dicer mutants (Grishok et al. 2001; Hutvagner et al. 2001; Ketting et al. 2001; Knight and Bass 2001), opening the door to the documentation of hundreds of additional novel tiny RNAs (microRNAs) present in *Drosophila*, *C. elegans*, and human cells by the Tuschl, Bartel, and Ambros labs (Lau et al. 2001; Lagos-Quintana et al. 2001; Lee and Ambros 2001).

The last several years have taught us that the RNAi pathway can regulate gene expression in an astonishing variety of ways. Genes can be silenced by modification of chromatin structure, by mRNA cleavage, and by affecting mRNA localization, polyadenylation, and productive translation (Hannon 2002; Grewal and Moazed 2003; Bartel 2004; Valencia-Sanchez et al. 2006). Small RNAs can even accomplish the ultimate form of gene silencing, marking sequences for expulsion from the genome by programmed DNA elimination (Mochizuki et al. 2002; Matzke and Birchler 2005). Clearly, gene regulation by small RNAs that are processed from dsRNA precursors by Dicer comprises a rich and varied collection of biological pathways. However, it is also evident that our explorations so far have reached only the nearest and most easily observed parts of the ncRNA universe. As the power of our observational tools increases, we begin to see glimmers of a ncRNA world that is as beautiful and complex as it is daunting and difficult to fully comprehend.

The Cold Spring Harbor Symposium on "Regulatory RNAs" was both a celebration of the richness and breadth of ncRNA function and an indicator of the many vistas that remain to be explored. It gathered scientists from a number of subfields, previously divided on the basis of organism studied, cellular process, or experimental approach. The purpose of this summary is to provide a framework for bringing together what has been learned in past decades about larger ncRNAs (generally in the size range of 60 to several hundred nucleotides) in cells from bacteria to man with the current intense focus on small RNAs (usually only 20–30 nucleotides long). Since all cellular RNA molecules exist in complexes with protein(s), deciphering how these partners contribute to the function of small RNAs is also integral to understanding their biological roles.

TODAY'S TALLY OF NCRNAS

Genome studies set the stage for thinking about the immense potential for the involvement of ncRNAs in regulating all aspects of genome organization, cellular metabolism, and organismal development. An introduction to the tremendous potential complexity of ncRNAs in eukaryotes came from Tom Gingeras (Carninci et al. 2005; Manak et al. 2006) and Mike Snyder (Bertone et al. 2004; Emanuelsson et al. 2006). Their studies collectively indicate that the transcriptome is much more extensive than previously thought. Even polling only the nonrepetitive portions of the genome in just a few cell types or developmental stages indicates that more than 85% of DNA sequences are used as templates for RNA production in mammals and flies. In fact, as many as 400,000 sites in the human genome may give rise to RNAs 20–200 nucleotides long! Additionally, many unannotated, transcribed sequences represent far distant 5′ exons of known protein-coding genes, which are used when those genes are expressed in developmentally or cell-type-specific ways. Nick Proudfoot's consideration of intragenic ncRNA regions focused attention on the possibility of co-transcriptional RNA turnover and the question of how the loop conformations of active genes may affect the fate of transcripts (West et al. 2004, 2006; Haussecker and Proudfoot 2005; Dye et al. 2006; Gromak et al. 2006).

Alex Hüttenhofer pointed out that the fraction of noncoding DNA increases across phylogeny (Hüttenhofer and Schattner 2006). In bacteria, approximately 100 small RNAs are known so far. Although this number will undoubtedly increase somewhat, it may reflect the fact that a smaller percentage of the bacterial genome, roughly 10%, constitutes noncoding regions, whereas for a typical mammalian genome, as much as 98% of the DNA may be considered non-protein-encoding.

Key emerging questions:

- How many functional classes do ncRNAs represent?

- What are their biological roles?

- What will we find in more comprehensive cell-type and tissue surveys?

- Do ncRNAs dominate the regulatory circuitry of the cell?

REGULATORY RNA MOLECULES AND SEQUENCES IN BACTERIA

Susan Gottesman introduced small RNAs in bacteria with the provocative claim, "Anything DNA can do, RNA can do better!" She categorized the known instances of regulation via RNA in bacteria as (1) interactions of mRNA 5′ leaders with metabolites or cellular milieu (such as temperature) to modify transcription or translation, (2) interactions of small RNAs with regulatory proteins to modify their activities, and (3) interactions between regulatory RNAs and mRNAs to positively or negatively affect transcription, stability, or translation (Gottesman 2005; Majdalani et al. 2005). Pairing between a regulatory RNA and its target in vivo is often facilitated by Hfq, the bacterial homolog of the Sm proteins that comprise the core of spliceosomal snRNPs (Moller et al. 2002; Zhang et al. 2002). Control of synthesis of a small RNA can be the primary regulatory step, as in the case of stress responses, including the presence of unfolded proteins in the periplasm. Gigi Storz emphasized that not only the 5′ leaders but also the 3′UTRs of mRNAs can provide recognition sites for small RNAs and discussed the importance of changes in RNA processing and stability resulting from mRNA–small RNA interactions (Storz et al. 2004, 2005).

We have long realized that alternative mRNA secondary structures induced by the passage of ribosomes can have diverse consequences. These include regulating transcription termination in the Trp and other amino acid biosynthetic operons, as well as inducing erythromycin resistance by relieving the sequestration of a ribosome-binding site (Weisblum 1995; Gollnick et al. 2005). However, additional novel examples, dubbed "riboswitches," have recently emerged (Winkler 2005; Gilbert and Batey 2006; Grundy and Henkin 2006). Tina Henkin reviewed the evidence for the T-box riboswitch that uncharged (but not charged) tRNA interacts with multiple sites in the 5′ leader RNA to tighten its structural fold, facilitating the formation of an antiterminator stem that allows transcriptional readthrough (Nelson et al. 2006). Smaller currently known ligands for riboswitches range in size from metabolites (such as amino acids, vitamins, and nucleobases) to Mg^{++}, recently identified by Eduardo Groisman as the ligand that alters transcription of the 5′UTR of the *mgtA* gene of *Salmonella* (Cromie et al. 2006). Sequence analyses of multiple riboswitches have confirmed the generality of a two-domain structure—one for ligand binding called the aptamer domain and the second referred to as the platform or expression domain. A novel type of platform domain, a self-cleaving ribozyme, was described by Wade Winkler in the 5′ leader of the *glmS* transcript in *Bacillus*.

The activity of riboswitches is generally achieved by their adopting complex secondary and tertiary structures. Sean Eddy reviewed the special challenges posed when attempting to predict and align RNA secondary structures (Dowell and Eddy 2006). Progress with a new algorithm borrowed from computational linguistics is encouraging.

Multiple insights were provided by the high-resolution 3D structures of the purine and SAM-I riboswitches bound to their cognate ligands, as presented by Rob Batey (Montange and Batey 2006). Although distinct, the structures reveal several common features: The RNA aptamer domains exploit almost every feature of their ligand to achieve extraordinary binding specificity; complex tertiary structures establish the binding pocket; and ligand binding acts to stabilize a helix that communicates to the expression domain (forcing it to assume one of two mutually exclusive alternative structures in the case of the SAM-I riboswitch).

Key emerging questions:

- Are there additional regulatory roles for small RNAs in bacteria?
- How widespread in biology are riboswitches?
- How many biological processes are regulated in this manner?
- Will all riboswitches act similarly to those already identified?
- How might we efficiently search for such regulatory RNAs and sequences?

NCRNAS PARTICIPATE IN ALMOST EVERY FUNCTION OF THE CELL NUCLEUS

The most famous ncRNAs of the nuclear compartment are the snRNA components of the spliceosome. The spliceosome comes in two flavors, the major (or U2-type) and the minor (U12-type), which has four distinct snRNAs, in the cells of higher plants and animals (Table 1) (Patel

Table 1. Noncoding Nuclear RNAs of Animals/Plants

	Location	Function
U1, U2, U4, U5, U6 snRNAs	nucleoplasm	splicing: major
U11, U12, U4$_{atac}$, U6$_{atac}$ snRNAs		minor
SL RNAs		trans
U7 snRNA	Cajal bodies	histone mRNA 3′ ends
Box C/D snoscaRNAs	nucleoli/Cajal bodies	2′-O-methylation
Box H/ACA snoscaRNAs	nucleoli/Cajal bodies	pseudouridylation
Telomerase	Cajal bodies	telomere regulation
RNase P	nucleoplasm	tRNA processing
7SK, murine B2	nucleoplasm	transcription regulation
Xist, Tsix, *roX* RNAs	nucleoplasm	dosage compensation
siRNAs in RITS (plants/fungi)	chromatin	transcriptional silencing
scnRNAs (ciliates)	macronucleus	DNA elimination

and Steitz 2003; Will and Luhrmann 2005). Some organisms (such as nematodes and trypanosomatids) trans splice, joining two distinct RNAs and releasing a Y-branched RNA comprising two intronic fragments. Although splicing itself is highly regulated (more than 70% of human gene transcripts are currently estimated to be alternatively spliced), it is not the RNA components of the spliceosome that change but the binding of proteins to the pre-mRNA that modulates spliceosome assembly. Bob Darnell used studies of the neuron-specific RNA-binding protein Nova to illustrate how a combination of biochemical, genetic, and bioinformatic tools have not only elucidated the mechanism of alternative splicing, but also identified the pre-mRNAs bound by Nova (Ule et al. 2006). The unexpected outcome is a remarkable coherence in the nature of the RNAs targeted by Nova—all encode proteins related to synapse formation or function!

The unusual role of a box C/D snoRNA (Table 1) as a splicing regulator was related by Stefan Stamm; HBII-52 snoRNA, which is not expressed in patients with Prader-Willi syndrome, base-pairs to a splicing silencer in the serotonin receptor pre-mRNA, resulting in alternative exon usage (Kishore and Stamm 2006). Gideon Dreyfuss described how the SMN (survival of motor neurons) complex of proteins, a known mediator of snRNP assembly, identifies snRNAs by the proximity of the Sm-binding site to the 3′ end and thereby discriminates against other cellular RNA targets, eliminating illicit RNA–protein interactions (Golembe et al. 2005; Battle et al. 2006).

A novel role for 7SK, an abundant nuclear RNA previously implicated in the control of transcription elongation by sequestering (in cooperation with the HEXIM1 protein) P-TEFb into a kinase-inactive complex (Yik et al. 2003), was revealed by Qiang Zhou. He reported that 7SK also binds and suppresses the cytosine deaminase activity of APOBEC3C, perhaps by inducing its sequestration in the nucleolus. The theme of RNA sequestration in the nuclear compartment as a means of regulating gene expression was echoed by David Spector, who showed that the CTN-RNA, an inosine-containing RNA that is normally retained in the nucleus, is cleaved within its 3′UTR following cellular stress. It is then transported to the cytoplasm and translated into protein (Prasanth et al. 2005). Further bioinformatic analyses have identified other nuclear retained RNAs, suggesting that we have much to understand about the function of nuclear retained stable RNA transcripts.

Versatility in the function of box H/ACA RNAs (Table 1) is also evident. Among the several hundred different H/ACA RNAs in human cells are not only pseudouridylation-guide RNAs that reside in the nucleolus (to modify rRNA) or Cajal bodies (to modify snRNAs), but also the RNA component of telomerase (Meier 2005, 2006). Although discussing different aspects of box H/ACA RNP assembly, both Tamás Kiss and Michael Terns commented on the regulated intranuclear trafficking of telomerase RNA during the cell cycle and the idea that Cajal bodies may deliver telomerase to only a fraction of telomeres during each S phase (Jady et al. 2006; Tomlinson et al. 2006).

Tom Cech brought a new perspective to the structural categorization of small RNPs based on studies of yeast telomerase. Rather than possessing a specific structure determined in large part by RNA (e.g., the ribosome) or by proteins (e.g., snRNPs), the yeast telomerase RNA serves as a flexible scaffold that tethers essential proteins together (Zappulla and Cech 2004; Zappulla et al. 2005). Liz Blackburn reported the effects of telomerase RNA knockdown or overexpression on the growth characteristics and metastatic potential of human cancer cells (Nosrati et al. 2004; Li et al. 2005). Carol Greider discussed the unanticipated generational effects of heterozygosity at the telomerase RNA locus in mice (Hao et al. 2005). The data suggest that half the level of telomerase RNA leads to telomere shortening and stem cell loss that increases in severity with each generation. Even though progeny from matings of telomerase heterozygotes may possess a wild-type genome, one generation in this wild-type state is not enough to restore telomere length, leading to "occult genetic disease!"

Large ncRNAs underlie dosage compensation in both mammals and fruit flies (Deng and Meller 2006; Heard and Disteche 2006; Rea and Akhtar 2006). In *Drosophila*, the MSL complex of five proteins in conjunction with two functionally redundant roX (RNA on X) RNAs binds and up-regulates transcription of the single male X chromosome, apparently by altering chromatin structure. Mitzi Kuroda presented results from chromatin immunoprecipitation experiments that reveal preferential binding to the middle and 3′ ends of actively transcribed genes (Alekseyenko et al. 2006); it makes sense that only active X genes should be up-regulated, but neither the specific role of the roX RNAs in this process nor the mechanisms by which MSL complexes recognize and modify chromatin only for the correct gene subset are yet understood. Perhaps, as in the case of the yeast telomerase RNA, roX RNAs act to provide a flexible scaffold that ensures delivery of just the right collection of chromatin-modifying enzymes to their target genes.

In mammals, dosage compensation is achieved by transcriptional silencing of one X chromosome in female cells. The required X inactivation center (*Xic*) contains genes for several large ncRNAs, including Xist (which accumulates on the inactive X) and its antisense partner Tsix (Heard and Disteche 2006). Jeannie Lee highlighted recent support for a model that explains the mutually exclusive silencing of one of the two X chromosomes: Inactivation occurs after transient pairing of the homologous chromosomes via their *Xic* regions from which they emerge differentially marked (Xu et al. 2006). The requirement for Tsix RNA for pairing suggests that Tsix RNA controls Xist promoter methylation, thereby silencing Xist on the opposite chromosome. Edith Heard presented studies on the mechanism by which Xist might enforce silence. She proposed that RNA polymerase II becomes excluded from an "Xist domain." Exclusion occurs before transcriptional repression of the inactive X, and prior to inactivation, expressed genes are resident at the periphery or outside the Xist domain (Chaumeil et al. 2006). Considered as a whole, the data suggest that Xist RNA participates in the formation of a silent nuclear compartment into which inactivated X-linked genes are recruited.

Key emerging questions:

- Will many alternative splicing regulators, such as Nova, act on functionally coherent sets of mRNAs?

- Will the assembly of many RNPs be chaperoned by complexes such as SMN?

- Does 7SK have yet other partners and functions than those previously reported?

- Do telomeres send signals to Cajal bodies when they get too short?

- How do roX RNA complexes recognize and activate their targets?

- How can ncRNAs like Xist create silent nuclear compartments?

- Are there other examples of such behavior in which ncRNAs target genes to nuclear substructures?

Roles for ncRNAs in Genome Organization and Heterochromatin Formation

Heterochromatin has traditionally been thought to be transcriptionally silent. The surprising finding that heterochromatin produces numerous ncRNAs, which are essential to the maintenance of the heterochromatic state, is changing our perception of heterochromatin and challenging the definition of transcriptional "silence."

Richard Jorgensen's studies have revealed a diversity of pathways that lead to RNA-induced epigenetic changes in plants and provide insight into how these integrate across tissues and generations. Genetic requirements for RNA-directed DNA methylation (RdDM) in *Arabidopsis* are being clarified through the work of Marjori Matzke, Steve Jacobsen, and Craig Pikaard. In addition to a DNA-methyltransferase, RdDM requires several components of the RNAi machinery, including DCL3 and AGO4 (Gendrel and Colot 2005). An important role is emerging for a plant-specific, atypical DNA-dependent RNA polymerase, pol IV, not only in the transcription of heterochromatic loci that serve as silencing triggers, but also downstream, during the deposition of the methyl-cytosine mark (Kanno et al. 2005a, b). Interestingly, subcellular localization studies by Steve Jacobsen and Craig Pikaard show that several steps in the RdDM pathway appear to occur in the Cajal body, implying a similarity between nuclear RISC maturation and other RNP assembly pathways (Li et al. 2006; Pontes et al. 2006).

Studies by Danesh Moazed, Shiv Grewal, and Rob Martienssen have focused on the role of RNA in regulating heterochromatin in the fission yeast *Schizosaccharomyces pombe*. In contrast to the outdated understanding of heterochromatin as a condensed state refractory to transcription, in *S. pombe*, heterochromatin-embedded "silent" transgenes retain high pol II occupancy and are transcriptionally active. Nascent transcripts from these genes are recognized by small RNAs resident in a RISC-like complex, RITS (RNA-induced transcriptional silencing), and cleaved co-transcriptionally (Irvine et al. 2006). In fact, the production of RNA from heterochromatic genes is required for the generation of small RNAs and serves as part of a feed-forward loop that is required to maintain the heterochromatic state. Thus, heterochromatin must be expressed in order to maintain "silence."

Although the details of the mechanisms by which RNA induces chromatin modifications and DNA methylation remain unclear, a series of experiments by Greg Hannon, Rob Martienssen, and Shiv Grewal have implicated the "slicer" activity of Argonaute (Ago) proteins in this process (Irvine et al. 2006; Qi et al. 2006). Interestingly, whereas Ago1 slicer activity is absolutely required for heterochromatin maintenance in *S. pombe*, in *Arabidopsis* only a subset of epigenetically controlled loci are sensitive to the loss of AGO4 cleavage activity while others are not. These seemingly divergent results can be unified by imagining that Argonaute complexes can serve two distinct roles at silenced loci. The first, which requires Argonaute's catalytic potential, is to cleave chromatin-associated nascent transcripts. By analogy with *trans*-acting siRNA generation pathways in plants, one might imagine that this cleavage triggers the production of small RNAs from the loci. Indeed, loss of small RNAs from target genes is observed upon negation of cleavage potential in both plants and yeast (Irvine et al. 2006; Qi et al. 2006). A second function, which is independent of cleavage, may be to recruit chromatin-modifying complexes that mark target loci by histone methylation. In *S. pombe* at least, such modifications also recruit RITS (Motamedi et al. 2004; Verdel et al. 2004; Buhler et al. 2006), perhaps increasing the local concentration of silencing factors and creating a micro-domain that is analogous to the silent domain created by Xist in female mammalian cells.

Key emerging questions:

- What is the functional definition of heterochromatin?

- How are certain heterochromatic sites selected for small RNA generation in plants and fungi?

- How do RISC and RITS direct histone and DNA modification?

- Do analogous small RNA-directed chromatin modification complexes exist in mammals and, if so, what is their biological function?

NCRNAS IN VIRUS-INFECTED CELLS

Nearly every regulatory strategy employed by host cells is exploited by viruses, and small RNA pathways are no exception. Table 2 summarizes the current roster of small RNAs found in virus-infected mammalian cells. The large double-stranded DNA genomes of gamma her-

Table 2. Viral Noncoding RNAs in Primates

	Virus	Function
VA 1 and 2 RNAs	Adenovirus	inactivate PKR
HSUR 1–7 snRNAs	Herpesvirus saimiri	up-regulation of host genes involved in T-cell activation
EBER 1 and 2 snRNAs	Epstein-Barr (EBV)	enhance tumorigenic potential
PAN (*polya*denylated *n*uclear) RNA	Kaposi sarcoma associated herpesvirus (KSHV)	?
Virus-encoded microRNAs	EBV	?
	KSHV	
	Polyoma virus	
	SV40	

pesviruses, such as EBV and KSHV, have emerged as the champion microRNA-encoding viral representatives (each with 10–20), although little is yet known about whether these microRNAs target viral or host-cell functions. The long-standing mystery of why a transcript called BART simply turns over in the nucleus of EBV-infected cells was elucidated by Bryan Cullen, who reported that it serves as the precursor for no fewer than 14 microRNAs (Cai et al. 2006). Joan Steitz discussed a larger (~1 kb) KSHV lytic-phase noncoding transcript called PAN, which never leaves the nucleus and has provided insights into a rapid poly(A)-dependent RNA decay pathway that may be part of a nuclear surveillance system in mammalian cells (Conrad et al. 2006).

Don Ganem uncovered the existence of a conserved cluster of SV40 microRNAs. These arise from read-through transcripts of the late region and induce RISC-mediated cleavage of early transcripts. This enables the virus to down-regulate production of T antigen, an essential early product that would otherwise become a major target of host cytotoxic T lymphocytes (Sullivan et al. 2005). A slightly different scenario was described by Gordon Carmichael for the related polyoma virus. In this case, RNA editing of dsRNA created by hybridization of the poly(A) signals at the 3′ ends of the early and late transcripts suggests an editing-versus-cleavage/polyadenylation model for the switch between early and late gene expression.

In the case of hepatitis C viral infection of liver cells, Peter Sarnow reported both positive and negative roles for a host cell microRNA (Jopling et al. 2005). miR-122 down-regulates the cationic amino acid transporter, CAT-1, by binding its 3′UTR but positively modulates viral RNA replication through complementarity to a site near the 5′ end of the viral genome. Clearly, the latter function is profoundly mysterious, as it cannot yet be linked to any of the mechanisms by which microRNAs are known to operate.

Key emerging questions:

- Did viral small RNAs evolve independently or were they originally captured from host genomes?
- Do viruses exploit microRNAs for self-regulation (e.g., of latent states) or primarily for regulation of host functions?
- Viruses often make noncanonical use of cellular components. Are microRNAs always used in the conventional way by viruses?
- Does the cellular RNAi machinery ever serve as an antiviral defense in mammals?

SMALL CYTOPLASMIC RNAs OF ANIMALS AND PLANTS

Both the cytosol and mitochondria harbor ncRNAs whose functions are well delineated (Table 3), in addition to some that have remained enigmatic for decades (e.g., vault RNAs). New insight into the role of Y RNAs, which bind a protein called Ro, was offered by Sandy Wolin. Ro's donut shape (Stein et al. 2005) allows it also to recognize misfolded ncRNAs (such as 5S and U2 snRNAs), but in a way mutually exclusive with Y RNA binding, suggesting that Y RNAs act to regulate Ro's role as part of an RNA quality control pathway in animal cells.

Table 3. Noncoding Cytoplasmic RNAs of Animals and Plants

	Location	Function
Y RNAs	cytosol	regulate Ro function
7SL (SRP)	cytosol	protein translocation to ER
MRP	mitochondria	DNA replication
Editing guide RNAs	mitochondria	U insertion/deletion in trypanosomes
Vault RNAs	vaults	?
Alu/BC1 transcripts	cytosol	?
MicroRNAs	cytosol/P bodies	repression
Multiple novel classes	?	?

Diversity of Small RNAs

Deep sequencing technology has recently revealed hidden depths of small RNA diversity in both plants and animals, with several new classes emerging at the meeting. Greg Hannon and David Bartel introduced Piwi-interacting RNAs (piRNAs), a class of small RNAs in the mammalian testes that associate with Argonaute proteins of the Piwi clade (Aravin et al. 2006; Girard et al. 2006; Lau et al. 2006). piRNAs are about 30 nucleotides long, with a very strong propensity for 5′ uracil. Although both their biogenesis and function remain elusive, piRNAs are interestingly derived from discrete genomic loci with profound strand asymmetry. Unlike microRNAs, piRNAs are not conserved, even among mammals. Yet, they are produced from syntenic regions of the mouse, rat, and human genomes.

Phil Zamore presented evidence that Piwi proteins in *Drosophila* bind repeat-associated RNAs (rasiRNAs) (Vagin et al. 2006). These 24- to 26-nucleotide-long RNAs predominantly target transposons and are biased toward antisense species. Since both piRNAs and rasiRNAs are produced in the germ line, both bind Piwi proteins, and Piwi family proteins are essential for germ-line integrity in flies and mammals, it is tempting to speculate that the two types of small RNAs are related. However, to date, lack of information concerning the biogenesis of the two small RNA classes and the function of mammalian piRNAs precludes definitive conclusions.

Why piRNAs and rasiRNAs are essential for germ-line integrity remains a mystery. However, Bill Theurkauf suggested that rasiRNAs, at least, have a role in suppressing DNA damage signaling in the germ line (Klattenhoff et al. 2007). He showed that mutations in chk2, a key damage signaling component, could suppress the effects of mutations in a *Drosophila* Piwi-family gene, Aubergine. This raises the possibility that loss of the rasiRNA pathway affects germ-line integrity by suppressing transposons, whose activity can lead to DNA damage, rather than by regulating gene expression.

David Baulcombe and Greg Hannon provided detailed characterizations of AGO1- and AGO4-bound small RNAs in *Arabidopsis* (Qi et al. 2006). Although these species were known previously, deep sequencing has allowed a new level of analysis. AGO1 binds mainly 21-nucleotide microRNAs, with a number of new candidate plant microRNAs evident in the sequenced collection. AGO4 binds 24-nucleotide RNAs, many of which depend on RNA pol IV activity and correspond to nearly every repetitive and transposable element in the *Arabidopsis* genome. Superficially, these may be considered analogs of *Drosophila* rasiRNAs, except that AGO4-bound siRNAs are produced by a Dicer-dependent mechanism whereas rasiRNAs seem to be Dicer-independent and may not even be derived from dsRNA precursors.

Detailed analyses of small RNA sequences in *C. elegans* were described by Dave Bartel. Along with numerous microRNAs and endogenous siRNAs, a new class of abundant 21-nucleotide molecules emerged (Ruby et al. 2006). They are characterized by a prominent 5′-U bias and by the presence of a highly characteristic sequence motif about 42 bp upstream of the start of the small RNA. 21U RNAs are not obviously produced by processing dsRNA precursors but instead may represent discrete transcription products from loci that cluster in the genome.

Recent advances in our understanding of other elusive small RNA classes, including plant *trans*-acting siRNAs (tasiRNAs), *Tetrahymena thermophila* scnRNAs, and *C. elegans* secondary siRNAs, were discussed in less detail in both posters and talks at the meeting.

Most deep-sequencing studies carried out to date have been specifically designed to identify small RNAs with characteristics of siRNAs and microRNAs. With a view toward genome-wide studies of RNA expression, described above, it is clear that there may still be many types of small RNAs that remain to be uncovered.

Key emerging questions:

- How many discrete classes of small RNAs exist?

- How are piRNAs and rasiRNAs produced?

- What are the functions of mammalian piRNAs and *C. elegans* 21U RNAs?

- How does biogenesis of tasiRNAs and heterochromatic siRNAs in plants and fungi relate to secondary siRNA production in worms?

MicroRNA Biogenesis

Although mechanisms that produce some of the newer classes of small RNAs remain a complete mystery, pathways leading to the production of siRNAs and microRNAs have come into increasingly sharp focus. MicroRNAs are processed from longer precursors (1–10 kb), called primary microRNAs or pri-miRNAs, that most frequently are synthesized by RNA pol II. Pri-miRNAs often encode more than one microRNA; thus, a microRNA can arise from a mono- or polycistronic primary transcript or can reside within the intron of a protein-coding gene transcript. The pri-miRNA is first reduced to a 60- to 70-nucleotide-long pre-miRNA by the Microprocessor, a nuclear multiprotein complex that contains two essential components called Drosha and DGCR8/Pasha. After transit to the cytoplasm, escorted by the export receptor Exportin 5, the pre-miRNA is further processed to approximately 22-nucleotide mature microRNA(s) by Dicer, which—like Drosha—is a member of a superfamily of enzymes that recognize and cleave dsRNAs. One or both Dicer products are then bound (probably coordinately with cleavage) by an Ago family protein to form an active complex called RISC (for review, see Du and Zamore 2005; Kim 2005). Many additional proteins have been reported to be associated with mature microRNPs.

Narry Kim addressed the nuclear events in microRNA biogenesis by asking how the Microprocessor recognizes its substrates and how and when the Microprocessor cleaves intronic microRNAs (Han et al. 2006). Studies in her lab indicate that DGCR8 recognizes a ss–dsRNA junction within the pri-miRNA adjacent to the hairpin containing the mature microRNA sequence. The Microprocessor then measures about 11 bp along the duplex, thereby positioning the active site of Drosha to cleave both strands, staggered by 2 bp. This cleavage releases the pre-miRNA, ready for secondary processing by Dicer. The processing of intron-encoded microRNAs results in production of both the microRNA and spliced mRNA. Brenda Bass discussed in more general terms the essential roles of dsRNA-binding proteins in assisting Drosha and Dicer cleavage, as well as in coordinating the progression of dsRNAs from one step to the next in the pathway of microRNA maturation (Parker et al. 2006).

The question of whether microRNAs are regulated posttranscriptionally was raised in several talks and posters. Scott Hammond's lab presented clear evidence that specific pri-miRNAs can be abundantly expressed in mammalian cells but not processed into mature microRNAs until a developmental signal releases the processing block (Thomson et al. 2006). Richard Carthew described the steps leading to RISC formation and the role of novel genes in inhibiting the production of RISC from small RNA precursors in cells of the *Drosophila* nervous system. Jim Dahlberg reported that the maturation of *Xenopus* oocytes produces a large increase in Dicer activity, indicating that microRNA biogenesis, in general, can be subject to developmental control.

Alex Schier showed that even within a developing embryo, abundantly expressed microRNAs can be prevented from acting within a privileged compartment. In zebra fish, miR-430 normally acts to promote degradation of maternal mRNAs (Giraldez et al. 2006). As expected, the maternally deposited *nanos* mRNA is targeted by miR-430 and destabilized in the soma. However, in the germ plasm, which is destined to become the germ line, *nanos* escapes the effects of miR-430 (Mishima et al. 2006). It is not yet clear whether this is accomplished by affecting RISC biogenesis within the germ compartment, by preventing RISC action either in general or with a specific microRNA, or by protecting specific targets from microRNA-mediated repression.

Key emerging questions:

- How is the diversity of microRNA precursors recognized, leading to specificity in processing?

- How are pathways of microRNA biogenesis related in plants and animals?

- To what extent is the RNAi pathway as a whole regulated in given tissues and at specific points in development?

- What levels of control are exerted on the biogenesis of individual microRNAs?

- How is control of processing an individual microRNA exerted at the posttranscriptional level?

The Structural Biology of RNA Silencing

Understanding the mechanistic basis of RNA silencing pathways provides a special challenge for structural biology. In fact, the structures of full-length and subdomains of Ago and Dicer proteins that have currently been solved to high resolution are not those with well-characterized functions, but their homologs from single-celled organisms—where it is not even clear that RNA interference pathways comparable to those in metazoans and plants exist.

Jennifer Doudna described the crystal structure of a Dicer enzyme from the parasite *Giardia* (Macrae et al. 2006). Members of the Dicer family endonucleolytically cleave dsRNAs into pieces ranging from 21 to 27 nucleotides and are critical for RNAi mediated by long dsRNAs, as well as for microRNA biogenesis. *Giardia* Dicer consists of two RNase III domains attached to a PAZ domain by a long α-helix—which appears to be the measuring device—but it lacks the amino-terminal helicase and carboxy-terminal dsRBD found in Dicers from higher eukaryotes. The structure reveals a distinct kink in the connector helix and conformational flexibility, which may allow the processing of diverse pre-miRNA substrates. An attractive idea is that different Dicers produce slightly different-length pieces, thereby steering their products into different functional pathways by handing over the resulting siRNA duplexes to different Ago family proteins via interactions between their PAZ domains and the 3′ ends of guide strands.

Three groups presented structures of Ago proteins, the core component of RISC, RITS, and perhaps other RNAi and microRNA effector complexes. Leemor Joshua-Tor recounted the seminal finding that the PIWI domain of Ago from *Pyrococcus furiosus* resembles RNase H, indicating that Ago is the component of RISC and RITS responsible for the siRNA-directed "slicer" activity (Song et al. 2004). The structure of the PIWI domain of the archaeal *Archaeoglobus fulgidus* (*Af*) Ago protein complexed with a 16-nucleotide siRNA-like duplex, described by David Barford, has revealed a profound distortion of the 5′ end of the guide RNA strand and its binding to a pocket facilitated by metal ion contacts, suggesting that the PIWI domain also participates in RNA duplex unwinding and strand selection. Importantly, the residence of the 5′ nucleotide of the guide strand in a binding pocket within the PIWI domain may explain why the first nucleotide of siRNAs and microRNAs plays little role in selection of silencing targets by RISC (Parker et al. 2005). Dinshaw Patel discussed recognition of the siRNA duplex 3′ overhang by the PAZ domains of both Dicer and Ago1 and what mutations tell us about the 5′-phosphate-binding pocket of the PIWI domain of *Af* Ago (Ma et al. 2004; Yuan et al. 2005, 2006). Clearly, future structures of siRNAs complexed with full-length Dicer and Ago proteins from organisms with well-characterized RNAi pathways will be required to illuminate the molecular details of microRNA biogenesis and function.

Translational Repression by MicroRNAs

The precise mechanisms by which microRNAs regulate the expression of their mRNA targets remain to be resolved. However, some themes did begin to emerge at the meeting. Previous studies have indicated that microRNAs can act by mechanisms ranging from mRNA cleavage (prominent in plants but still rare in mammals) to mRNA sequestration to deadenylation to direct effects on mRNA translation via inhibition of initiation or elongation.

Alex Schier recounted the elegant finding that a particular microRNA, miR-430, is expressed at the onset of zygotic transcription during zebra fish embryogenesis and induces the decay of several hundred maternal mRNAs via rapid deadenylation (Giraldez et al. 2006). Jim Dahlberg described a similar role for the microRNA pathway in clearance of maternal mRNAs in *Xenopus*, suggesting that this may be a conserved function of the RNAi pathway in vertebrates. Deadenylation is not a secondary effect of lack of translation, and some 3'UTRs are more effective microRNA targets in one cell type than another, suggesting the existence of tissue-specific regulation by microRNAs.

In a similar vein, Witek Filipowicz reported that microRNA-induced repression of translation can be reversible (Bhattacharyya et al. 2006). In human hepatoma cells, miR-122 represses CAT-1 expression, with coincident localization of the cat-1 mRNA to processing bodies (P-bodies). By subjecting cells to stress conditions, repression of cat-1 can be relieved, with coincident exit of cat-1 from P-bodies. This correlates with recognition of an AU-rich element adjacent to the miRNA target site in the 3'UTR by an ARE-binding protein, HuR. Considered together with Alex Schier's report of cellular compartments in which mRNAs are protected from miRNA-mediated repression, a picture begins to emerge in which responses to microRNAs may be modulated dynamically. This fits well with the notion that 3'UTRs may integrate many cellular signals to determine both rates of mRNA translation and rates of mRNA decay.

Phil Sharp stressed observations that mRNAs being repressed by microRNAs are found on polysomes and that some reduction (~1.5- to 2-fold) in mRNA levels often accompanies repression (Petersen et al. 2006); experiments addressing whether microRNA-mediated translational repression is localized to specific cellular site(s) concluded that it occurs in the diffuse cytoplasm rather than necessarily in discrete foci such as P-bodies or stress granules (SGs) (Leung et al. 2006). Tim Nilsen reported an analysis of the subcellular distribution of microRNAs in actively growing HeLa cells and found the vast majority to be associated with polysomes, as was a down-regulated mRNA, solidifying evidence that elongation is a likely step at which microRNA-induced translation repression is achieved (Maroney et al. 2006).

Elisa Izaurralde provided evidence that microRNAs in *Drosophila* require AGO1 and the P-body component GW182 to achieve both translational repression and mRNA degradation, which occurs via deadenylation and decapping (Behm-Ansmant et al. 2006). Importantly, Izaurralde also shed light on why such disparate mechanistic explanations for microRNA-mediated repression may have emerged from studies in different laboratories. There are at least three discrete classes of targets in *Drosophila*: Some are down-regulated at the protein level in a manner that directly reflects changes at the mRNA level; some show no detectable change at the mRNA level yet have greatly decreased protein levels, suggesting regulation solely at the level of translation; and finally, some microRNA targets are repressed at the mRNA level but also to a much greater degree at the protein level. Repression of each of these classes showed differential degrees of dependence on RISC accessory factors such as GW182 and deadenylases.

Potential molecular mechanisms underlying translational modulation by microRNAs were suggested by two talks that provided insights into the activities of other translation components—the ribosome itself and poly(A)-binding protein (PABP), which functions as an initiation factor. Rachel Green presented biochemical evidence that the decoding center of the small ribosomal subunit undergoes substrate-specific switching between distinct structural states during peptide elongation and release (Cochella and Green 2005; Cochella et al. 2007). Nahum Sonenberg summarized the multifaceted roles of PABP-interacting proteins (Paips) in regulating initiation, the rate-limiting step in translation (Cho et al. 2005; Yoshida et al. 2006). Regardless of which step(s) ultimately emerges as the target of microRNP repression, it is further possible that the direction of effects on translation might differ under different cellular conditions (e.g., during development or dependent on the cell cycle), making control by microRNAs and their associated proteins even more versatile than previously imagined.

Biological Functions of MicroRNAs

Several speakers highlighted the complexity of microRNA-mediated impacts on developmental pathways and cell-fate decisions. Marja Timmermans and Scott Poethig discussed the roles of microRNAs and tasiRNAs in the specification of organ polarity and regulation of developmental timing in plants. The Timmermans lab had previously shown that a gradient of miR-166 emanating from the abaxial side of the incipient leaf restricts the expression of adaxial determinants (Juarez et al. 2004). Follow-up studies in the maize *leafbladeless1* mutant now reveal a role for tasiRNAs in the spatiotemporal regulation of miR-166. Cloning of *leafbladeless1*, which is required for the specification of adaxial cell fate, revealed that it is the maize homolog of *Arabidopsis suppressor of gene silencing 3* (SGS3), a key component of the tasiRNA pathway. Loss of *leafbladeless1* leads to mis-expression of miR-166, suggesting that in maize, leaf polarity is established by the opposing action of two distinct types of small regulatory RNAs.

Timmermans also raised the interesting possibility that microRNAs could be mobilized to create gradients of suppression in plant tissues. In many ways, this is similar to the well-established mobility of silencing signals induced by virus infection in plants and by feeding of dsRNAs to *C. elegans*. The mechanisms underlying systemic spreading of a silencing signal in *C. elegans* were discussed by Craig Hunter. SID1 and SID2 are both transmembrane proteins; SID1 is ubiquitously expressed whereas SID2 is primarily expressed in the intestinal lumen. Paralleling their expression patterns, SID1 is involved in intracellular transport (both import and export) of long dsRNA whereas SID2 is required for uptake of dsRNA from the gut (environmental RNAi defective)(Feinberg and Hunter 2003). The biological role of these proteins is still elusive, although current experiments are ongoing regarding a potential role in viral immunity.

Scott Poethig presented studies on the juvenile-to-adult transition in *Arabidopsis*, which is accelerated in *zippy/ago7*, *sgs3*, and *RNA-dependent RNA polymerase 6 (rdr6)* mutants. These observations imply a role for RNAi in this transition. By identifying mRNAs whose abundance increases in these mutants, as well as through genetic hunts for second-site suppressors, the Poethig group identified key regulators of the transition, specifically the squamosa promoter binding protein-like family (SPL3/4/5) and the auxin-related genes ARF3 and ARF4 (Wu and Poethig 2006). The juvenile-to-adult transition is triggered when levels of SPL transcription factors increase in response to decreased levels of miR-156. ARF3 and ARF4 are targets of a tasiRNA (Hunter et al. 2006). Thus, multiple small RNA pathways converge to regulate the sensitivity of the phase-change mechanism to a temporal signal.

Oliver Hobert summarized his laboratory's work on microRNAs and gene regulatory circuits that control neuronal cell-fate specification in *C. elegans*. The interplay of microRNAs, lsy-6 and miR-273, and their respective target transcription factors, cog-1 and die-1, form a double-negative feedback loop that controls cell-fate specifi-

cation in sensory neurons (Johnston et al. 2005, 2006). On the basis of this pathway, he advanced the idea that microRNAs can function as effective on–off switches in developmental decisions, as opposed to the more familiar perception of microRNAs as fine tuners of gene expression. Irene Bozzoni echoed this important concept in her talk on microRNAs expressed in hematopoiesis and leukemia. Here, C/EBPα, a transcription factor regulating the miR-223 promoter, displaces a negative factor NFI-A, but miR-223 represses NFI-A, creating an autoregulatory loop that appears to control granulocytic differentiation (Fazi et al. 2005).

The roles of microRNAs in driving phenotypic variation, and ultimately evolution, were discussed by several investigators. In one very interesting example, Michel Georges described studies aimed at uncovering traits related to the exceptional meatiness (an economically important trait) of Texel sheep (Clop et al. 2006). Mapping of a QTL with large impact on muscle mass to the mysotatin gene revealed a G to A transition in the 3′UTR of myostatin. This change made myostatin a target for miR-1 and miR-206, both abundant microRNAs in skeletal muscle. Thus, the meatiness of Texel sheep is due to muscular hypertrophy associated with silencing of myostatin via the microRNA pathway. These results, and analysis of SNP databases, were used as support for the assertion that changes in microRNA-binding sites are an important underlying cause of phenotypic variation.

Ronald Plasterk discussed the identification of many microRNAs that are not conserved even among closely related species. In particular, certain microRNAs are found in humans but are absent from chimps (Berezikov et al. 2006). Thus, he asserted that changes in small RNA populations may help to drive adaptation on relatively short evolutionary timescales. Nikolaus Rajewsky provided support for evolutionary constraints on microRNA target sites by analyzing human SNP databases (Chen and Rajewsky 2006). He reported that negative selection is stronger on conserved predicted microRNA-binding sites in 3′UTRs than on other conserved sites. The data also predict the relevance of many non-conserved potential microRNA-binding sites, paralleling the assertions by Plasterk that small RNA regulatory pathways are a convenient way to drive evolutionary change.

Given the importance of microRNAs in nearly all aspects of cellular physiology and organismal development, it is not surprising that changes in microRNAs or their target sites are linked to disease. Frank Slack discussed the roles of microRNAs in human cancer, with a particular focus on let-7 as a noncoding tumor suppressor (Esquela-Kerscher et al. 2005; Johnson et al. 2005). Just as endogenous microRNAs can contribute to disease, it is evident that we can artificially harness the microRNA machinery to understand disease pathways. Indeed, Rene Bernards described efforts to use genome-wide, RNAi-based genetic approaches to identify new cancer targets (Bernards et al. 2006; Brummelkamp et al. 2006). Extensive and community-based efforts, described by Norbert Perrimon, are also under way in *Drosophila* to understand the connectivity of genetic pathways in that model organism (Friedman and Perrimon 2006).

Key emerging questions:

- Do microRNAs have conserved roles in regulating developmental transitions in plants and animals, and how does this relate to the convergent evolution of the microRNA pathway in these two kingdoms?

- Can microRNAs act as mobile signals, essentially behaving as morphogens?

- Is the interplay between small RNA pathways, e.g., microRNAs and tasiRNAs, unique to plants, or will similar interactions between small RNA classes be found in animals?

- What determines whether microRNAs act as rheostats or switches in gene expression pathways?

CONCLUSIONS

In retrospect, it was fitting that the 71st Cold Spring Harbor Laboratory Symposium focused on Regulatory RNAs. As we complete this review of the work presented at the meeting, admittedly a bit tardily, Craig Mello and Andy Fire have just accepted the 2006 Nobel Prize in Physiology or Medicine. Their discovery of dsRNA-induced silencing in *C. elegans* has indeed seeded a new RNA revolution. The pervasive nature and practical utility of small RNA-driven regulatory processes has highlighted to the broader community the fundamental ways in which noncoding RNAs, small and large, are intertwined with virtually every cellular process.

REFERENCES

Alekseyenko A.A., Larschan E., Lai W.R., Park P.J., and Kuroda M.I. 2006. High-resolution ChIP-chip analysis reveals that the *Drosophila* MSL complex selectively identifies active genes on the male X chromosome. *Genes Dev.* **20:** 848.

Aravin A., Gaidatzis D., Pfeffer S., Lagos-Quintana M., Landgraf P., Iovino N., Morris P., Brownstein M.J., Kuramochi-Miyagawa S., Nakano T., et al. 2006. A novel class of small RNAs bind to MILI protein in mouse testes. *Nature* **442:** 203.

Bartel D.P. 2004. MicroRNAs: Genomics, biogenesis, mechanism, and function. *Cell* **116:** 281.

Battle D.J., Lau C.K., Wan L., Deng H., Lotti F., and Dreyfuss G. 2006. The Gemin5 protein of the SMN complex identifies snRNAs. *Mol. Cell* **23:** 273.

Behm-Ansmant I., Rehwinkel J., Doerks T., Stark A., Bork P., and Izaurralde E. 2006. mRNA degradation by miRNAs and GW182 requires both CCR4:NOT deadenylase and DCP1:DCP2 decapping complexes. *Genes Dev.* **20:** 1885.

Berezikov E., Thuemmler F., van Laake L.W., Kondova I., Bontrop R., Cuppen E., and Plasterk R.H. 2006. Diversity of microRNAs in human and chimpanzee brain. *Nat. Genet.* **38:** 1375.

Bernards R., Brummelkamp T.R., and Beijersbergen R.L. 2006. shRNA libraries and their use in cancer genetics. *Nat. Methods* **3:** 701.

Bernstein E., Caudy A.A., Hammond S.M., and Hannon G.J. 2001. Role for a bidentate ribonuclease in the initiation step of RNA interference. *Nature* **409:** 363.

Bertone P., Stolc V., Royce T.E., Rozowsky J.S., Urban A.E., Zhu X., Rinn J.L., Tongprasit W., Samanta M., Weissman S., et al. 2004. Global identification of human transcribed sequences with genome tiling arrays. *Science* **306:** 2242.

Bhattacharyya S.N., Habermacher R., Martine U., Closs E.I., and Filipowicz W. 2006. Relief of microRNA-mediated translational repression in human cells subjected to stress. *Cell* **125:** 1111.

Brummelkamp T.R., Fabius A.W., Mullenders J., Madiredjo M., Velds A., Kerkhoven R.M., Bernards R., and Beijersbergen R.L. 2006. An shRNA barcode screen provides insight into cancer cell vulnerability to MDM2 inhibitors. *Nat. Chem. Biol.* **2:** 202.

Buhler M., Verdel A., and Moazed D. 2006. Tethering RITS to a nascent transcript initiates RNAi- and heterochromatin-dependent gene silencing. *Cell* **125:** 873.

Cai X., Schafer A., Lu S., Bilello J.P., Desrosiers R.C., Edwards R., Raab-Traub N., and Cullen B.R. 2006. Epstein-Barr virus microRNAs are evolutionarily conserved and differentially expressed. *PLoS Pathog.* **2:** e23.

Carninci P., Kasukawa T., Katayama S., Gough J., Frith M.C., Maeda N., Oyama R., Ravasi T., Lenhard B., Wells C., et al. 2005. The transcriptional landscape of the mammalian genome. *Science* **309:** 1559.

Chaumeil J., Le Baccon P., Wutz A., and Heard E. 2006. A novel role for Xist RNA in the formation of a repressive nuclear compartment into which genes are recruited when silenced. *Genes Dev.* **20:** 2223.

Chen K. and Rajewsky N. 2006. Natural selection on human microRNA binding sites inferred from SNP data. *Nat. Genet.* **38:** 1452.

Cho P.F., Poulin F., Cho-Park Y.A., Cho-Park I.B., Chicoine J.D., Lasko P., and Sonenberg N. 2005. A new paradigm for translational control: Inhibition via 5′-3′ mRNA tethering by Bicoid and the eIF4E cognate 4EHP. *Cell* **121:** 411.

Clop A., Marcq F., Takeda H., Pirottin D., Tordoir X., Bibe B., Bouix J., Caiment F., Elsen J.M., Eychenne F., et al. 2006. A mutation creating a potential illegitimate microRNA target site in the myostatin gene affects muscularity in sheep. *Nat. Genet.* **38:** 813.

Cochella L. and Green R. 2005. An active role for tRNA in decoding beyond codon:anticodon pairing. *Science* **308:** 1178.

Cochella L., Brunelle J.L., and Green R. 2007. Mutational analysis reveals two independent molecular requirements during transfer RNA selection on the ribosome. *Nat. Struct. Mol. Biol.* **14:** 30.

Conrad N.K., Mili S., Marshall E.L., Shu M.D., and Steitz J.A. 2006. Identification of a rapid mammalian deadenylation-dependent decay pathway and its inhibition by a viral RNA element. *Mol. Cell* **24:** 943.

Cromie M.J., Shi Y., Latifi T., and Groisman E.A. 2006. An RNA sensor for intracellular Mg(2+). *Cell* **125:** 71.

Deng X. and Meller V.H. 2006. Non-coding RNA in fly dosage compensation. *Trends Biochem. Sci.* **31:** 526.

Dowell R.D. and Eddy S.R. 2006. Efficient pairwise RNA structure prediction and alignment using sequence alignment constraints. *BMC Bioinformatics* **7:** 400.

Du T. and Zamore P.D. 2005. microPrimer: The biogenesis and function of microRNA. *Development* **132:** 4645.

Dye M.J., Gromak N., and Proudfoot N.J. 2006. Exon tethering in transcription by RNA polymerase II. *Mol. Cell* **21:** 849.

Emanuelsson O., Nagalakshmi U., Zheng D., Rozowsky J.S., Urban A.E., Du J., Lian Z., Stolc V., Weissman S., Snyder M., and Gerstein M. 2006. Assessing the performance of different high-density tiling microarray strategies for mapping transcribed regions of the human genome. *Genome Res.* (in press).

Esquela-Kerscher A., Johnson S.M., Bai L., Saito K., Partridge J., Reinert K.L., and Slack F.J. 2005. Post-embryonic expression of *C. elegans* microRNAs belonging to the lin-4 and let-7 families in the hypodermis and the reproductive system. *Dev. Dyn.* **234:** 868.

Fazi F., Rosa A., Fatica A., Gelmetti V., De Marchis M.L., Nervi C., and Bozzoni I. 2005. A minicircuitry comprised of microRNA-223 and transcription factors NFI-A and C/EBPalpha regulates human granulopoiesis. *Cell* **123:** 819.

Feinberg E.H. and Hunter C.P. 2003. Transport of dsRNA into cells by the transmembrane protein SID-1. *Science* **301**: 1545.

Fire A., Xu S., Montgomery M.K., Kostas S.A., Driver S.E., and Mello C.C. 1998. Potent and specific genetic interference by double-stranded RNA in *Caenorhabditis elegans*. *Nature* **391**: 806.

Friedman A. and Perrimon N. 2006. A functional RNAi screen for regulators of receptor tyrosine kinase and ERK signalling. *Nature* **444**: 230.

Gendrel A.V. and Colot V. 2005. *Arabidopsis* epigenetics: When RNA meets chromatin. *Curr. Opin. Plant Biol.* **8**: 142.

Gilbert S.D. and Batey R.T. 2006. Riboswitches: Fold and function. *Chem. Biol.* **13**: 805.

Giraldez A.J., Mishima Y., Rihel J., Grocock R.J., Van Dongen S., Inoue K., Enright A.J., and Schier A.F. 2006. Zebrafish MiR-430 promotes deadenylation and clearance of maternal mRNAs. *Science* **312**: 75.

Girard A., Sachidanandam R., Hannon G.J., and Carmell M.A. 2006. A germline-specific class of small RNAs binds mammalian Piwi proteins. *Nature* **442**: 199.

Golembe T.J., Yong J., and Dreyfuss G. 2005. Specific sequence features, recognized by the SMN complex, identify snRNAs and determine their fate as snRNPs. *Mol. Cell. Biol.* **25**: 10989.

Gollnick P., Babitzke P., Antson A., and Yanofsky C. 2005. Complexity in regulation of tryptophan biosynthesis in *Bacillus subtilis*. *Annu. Rev. Genet.* **39**: 47.

Gottesman S. 2005. Micros for microbes: Non-coding regulatory RNAs in bacteria. *Trends Genet.* **21**: 399.

Grewal S.I. and Moazed D. 2003. Heterochromatin and epigenetic control of gene expression. *Science* **301**: 798.

Grishok A., Pasquinelli A.E., Conte D., Li N., Parrish S., Ha I., Baillie D.L., Fire A., Ruvkun G., and Mello C.C. 2001. Genes and mechanisms related to RNA interference regulate expression of the small temporal RNAs that control *C. elegans* developmental timing. *Cell* **106**: 23.

Gromak N., West S., and Proudfoot N.J. 2006. Pause sites promote transcriptional termination of mammalian RNA polymerase II. *Mol. Cell. Biol.* **26**: 3986.

Grundy F.J. and Henkin T.M. 2006. From ribosome to riboswitch: Control of gene expression in bacteria by RNA structural rearrangements. *Crit. Rev. Biochem. Mol. Biol.* **41**: 329.

Hamilton A.J. and Baulcombe D.C. 1999. A species of small antisense RNA in posttranscriptional gene silencing in plants. *Science* **286**: 950.

Hammond S.M., Bernstein E., Beach D., and Hannon G.J. 2000. An RNA-directed nuclease mediates post-transcriptional gene silencing in *Drosophila* cells. *Nature* **404**: 293.

Hammond S.M., Boettcher S., Caudy A.A., Kobayashi R., and Hannon G.J. 2001. Argonaute2, a link between genetic and biochemical analyses of RNAi. *Science* **293**: 1146.

Han J., Lee Y., Yeom K.H., Nam J.W., Heo I., Rhee J.K., Sohn S.Y., Cho Y., Zhang B.T., and Kim V.N. 2006. Molecular basis for the recognition of primary microRNAs by the Drosha-DGCR8 complex. *Cell* **125**: 887.

Hannon G.J. 2002. RNA interference. *Nature* **418**: 244.

Hao L.Y., Armanios M., Strong M.A., Karim B., Feldser D.M., Huso D., and Greider C.W. 2005. Short telomeres, even in the presence of telomerase, limit tissue renewal capacity. *Cell* **123**: 1121.

Haussecker D. and Proudfoot N.J. 2005. Dicer-dependent turnover of intergenic transcripts from the human beta-globin gene cluster. *Mol. Cell. Biol.* **25**: 9724.

Heard E. and Disteche C.M. 2006. Dosage compensation in mammals: Fine-tuning the expression of the X chromosome. *Genes Dev.* **20**: 1848.

Hunter C., Willmann M.R., Wu G., Yoshikawa M., de la Luz Gutierrez-Nava M., and Poethig S.R. 2006. Trans-acting siRNA-mediated repression of ETTIN and ARF4 regulates heteroblasty in *Arabidopsis*. *Development* **133**: 2973.

Hüttenhofer A. and Schattner P. 2006. The principles of guiding by RNA: Chimeric RNA-protein enzymes. *Nat. Rev. Genet.* **7**: 475.

Hutvagner G. and Zamore P.D. 2002. A microRNA in a multiple-turnover RNAi enzyme complex. *Science* **297**: 2056.

Hutvagner G., McLachlan J., Pasquinelli A.E., Balint E., Tuschl T., and Zamore P.D. 2001. A cellular function for the RNA-interference enzyme Dicer in the maturation of the let-7 small temporal RNA. *Science* **293**: 834.

Irvine D.V., Zaratiegui M., Tolia N.H., Goto D.B., Chitwood D.H., Vaughn M.W., Joshua-Tor L., and Martienssen R.A. 2006. Argonaute slicing is required for heterochromatic silencing and spreading. *Science* **313**: 1134.

Jacob F. and Monod J. 1961. Genetic regulatory mechanisms in the synthesis of proteins. *J. Mol. Biol.* **3**: 318.

Jady B.E., Richard P., Bertrand E., and Kiss T. 2006. Cell cycle-dependent recruitment of telomerase RNA and Cajal bodies to human telomeres. *Mol. Biol. Cell* **17**: 944.

Johnson S.M., Grosshans H., Shingara J., Byrom M., Jarvis R., Cheng A., Labourier E., Reinert K.L., Brown D., and Slack F.J. 2005. RAS is regulated by the let-7 microRNA family. *Cell* **120**: 635.

Johnston R.J., Jr., Chang S., Etchberger J.F., Ortiz C.O., and Hobert O. 2005. MicroRNAs acting in a double-negative feedback loop to control a neuronal cell fate decision. *Proc. Natl. Acad. Sci.* **102**: 12449.

Johnston R.J., Jr., Copeland J.W., Fasnacht M., Etchberger J.F., Liu J., Honig B., and Hobert O. 2006. An unusual Zn-finger/FH2 domain protein controls a left/right asymmetric neuronal fate decision in *C. elegans*. *Development* **133**: 3317.

Jopling C.L., Yi M., Lancaster A.M., Lemon S.M., and Sarnow P. 2005. Modulation of hepatitis C virus RNA abundance by a liver-specific MicroRNA. *Science* **309**: 1577.

Juarez M.T., Kui J.S., Thomas J., Heller B.A., and Timmermans M.C. 2004. microRNA-mediated repression of rolled leaf1 specifies maize leaf polarity. *Nature* **428**: 84.

Kanno T., Aufsatz W., Jaligot E., Mette M.F., Matzke M., and Matzke A.J. 2005a. A SNF2-like protein facilitates dynamic control of DNA methylation. *EMBO Rep.* **6**: 649.

Kanno T., Huettel B., Mette M.F., Aufsatz W., Jaligot E., Daxinger L., Kreil D.P., Matzke M., and Matzke A.J. 2005b. Atypical RNA polymerase subunits required for RNA-directed DNA methylation. *Nat. Genet.* **37**: 761.

Ketting R.F., Fischer S.E., Bernstein E., Sijen T., Hannon G.J., and Plasterk R.H. 2001. Dicer functions in RNA interference and in synthesis of small RNA involved in developmental timing in *C. elegans*. *Genes Dev.* **15**: 2654.

Kim V.N. 2005. MicroRNA biogenesis: Coordinated cropping and dicing. *Nat. Rev. Mol. Cell Biol.* **6**: 376.

Kishore S. and Stamm S. 2006. The snoRNA HBII-52 regulates alternative splicing of the serotonin receptor 2C. *Science* **311**: 230.

Klattenhoff C., Bratu D.P., McGinnis-Schultz N., Koppetsch B.S., Cook H.A., and Theurkauf W.E. 2007. *Drosophila* rasiRNA pathway mutations disrupt embryonic axis specification through activation of an ATR/Chk2 DNA damage response. *Dev. Cell* **12**: 45.

Knight S.W. and Bass B.L. 2001. A role for the RNase III enzyme DCR-1 in RNA interference and germ line development in *Caenorhabditis elegans*. *Science* **293**: 2269.

Lagos-Quintana M., Rauhut R., Lendeckel W., and Tuschl T. 2001. Identification of novel genes coding for small expressed RNAs. *Science* **294**: 853.

Lau N.C., Lim L.P., Weinstein E.G., and Bartel D.P. 2001. An abundant class of tiny RNAs with probable regulatory roles in *Caenorhabditis elegans*. *Science* **294**: 858.

Lau N.C., Seto A.G., Kim J., Kuramochi-Miyagawa S., Nakano T., Bartel D.P., and Kingston R.E. 2006. Characterization of the piRNA complex from rat testes. *Science* **313**: 363.

Lee R.C. and Ambros V. 2001. An extensive class of small RNAs in *Caenorhabditis elegans*. *Science* **294**: 862.

Lee R.C., Feinbaum R.L., and Ambros V. 1993. The *C. elegans* heterochronic gene lin-4 encodes small RNAs with antisense complementarity to lin-14. *Cell* **75**: 843.

Leung A.K., Calabrese J.M., and Sharp P.A. 2006. Quantitative analysis of Argonaute protein reveals microRNA-dependent localization to stress granules. *Proc. Natl. Acad. Sci.* **103:** 18125.

Li C.F., Pontes O., El-Shami M., Henderson I.R., Bernatavichute Y.V., Chan S.W., Lagrange T., Pikaard C.S., and Jacobsen S.E. 2006. An ARGONAUTE4-containing nuclear processing center colocalized with Cajal bodies in *Arabidopsis thaliana*. *Cell* **126:** 93.

Li S., Crothers J., Haqq C.M., and Blackburn E.H. 2005. Cellular and gene expression responses involved in the rapid growth inhibition of human cancer cells by RNA interference-mediated depletion of telomerase RNA. *J. Biol. Chem.* **280:** 23709.

Ma J.B., Ye K., and Patel D.J. 2004. Structural basis for overhang-specific small interfering RNA recognition by the PAZ domain. *Nature* **429:** 318.

Macrae I.J., Zhou K., Li F., Repic A., Brooks A.N., Cande W.Z., Adams P.D., and Doudna J.A. 2006. Structural basis for double-stranded RNA processing by Dicer. *Science* **311:** 195.

Majdalani N., Vanderpool C.K., and Gottesman S. 2005. Bacterial small RNA regulators. *Crit. Rev. Biochem. Mol. Biol.* **40:** 93.

Manak J.R., Dike S., Sementchenko V., Kapranov P., Biemar F., Long J., Cheng J., Bell I., Ghosh S., Piccolboni A., and Gingeras T.R. 2006. Biological function of unannotated transcription during the early development of *Drosophila melanogaster*. *Nat. Genet.* **38:** 1151.

Maroney P.A., Yu Y., Fisher J., and Nilsen T.W. 2006. Evidence that microRNAs are associated with translating messenger RNAs in human cells. *Nat. Struct. Mol. Biol.* **13:** 1102.

Martinez J., Patkaniowska A., Urlaub H., Luhrmann R., and Tuschl T. 2002. Single-stranded antisense siRNAs guide target RNA cleavage in RNAi. *Cell* **110:** 563.

Matzke M.A. and Birchler J.A. 2005. RNAi-mediated pathways in the nucleus. *Nat. Rev. Genet.* **6:** 24.

Meier U.T. 2005. The many facets of H/ACA ribonucleoproteins. *Chromosoma* **114:** 1.

———. 2006. How a single protein complex accommodates many different H/ACA RNAs. *Trends Biochem. Sci.* **31:** 311.

Mishima Y., Giraldez A.J., Takeda Y., Fujiwara T., Sakamoto H., Schier A.F., and Inoue K. 2006. Differential regulation of germline mRNAs in soma and germ cells by zebrafish miR-430. *Curr. Biol.* **16:** 2135.

Mochizuki K., Fine N.A., Fujisawa T., and Gorovsky M.A. 2002. Analysis of a piwi-related gene implicates small RNAs in genome rearrangement in tetrahymena. *Cell* **110:** 689.

Moller T., Franch T., Hojrup P., Keene D.R., Bachinger H.P., Brennan R.G., and Valentin-Hansen P. 2002. Hfq: A bacterial Sm-like protein that mediates RNA-RNA interaction. *Mol. Cell* **9:** 23.

Montange R.K. and Batey R.T. 2006. Structure of the S-adenosylmethionine riboswitch regulatory mRNA element. *Nature* **441:** 1172.

Motamedi M.R., Verdel A., Colmenares S.U., Gerber S.A., Gygi S.P., and Moazed D. 2004. Two RNAi complexes, RITS and RDRC, physically interact and localize to noncoding centromeric RNAs. *Cell* **119:** 789.

Nelson A.R., Henkin T.M., and Agris P.F. 2006. tRNA regulation of gene expression: Interactions of an mRNA 5′-UTR with a regulatory tRNA. *RNA* **12:** 1254.

Ngo H., Tschudi C., Gull K., and Ullu E. 1998. Double-stranded RNA induces mRNA degradation in *Trypanosoma brucei*. *Proc. Natl. Acad. Sci.* **95:** 14687.

Nissen P., Hansen J., Ban N., Moore P.B., and Steitz T.A. 2000. The structural basis of ribosome activity in peptide bond synthesis. *Science* **289:** 920.

Nosrati M., Li S., Bagheri S., Ginzinger D., Blackburn E.H., Debs R.J., and Kashani-Sabet M. 2004. Antitumor activity of systemically delivered ribozymes targeting murine telomerase RNA. *Clin. Cancer Res.* **10:** 4983.

Parker G.S., Eckert D.M., and Bass B.L. 2006. RDE-4 preferentially binds long dsRNA and its dimerization is necessary for cleavage of dsRNA to siRNA. *RNA* **12:** 807.

Parker J.S., Roe S.M., and Barford D. 2005. Structural insights into mRNA recognition from a PIWI domain-siRNA guide complex. *Nature* **434:** 663.

Patel A.A. and Steitz J.A. 2003. Splicing double: Insights from the second spliceosome. *Nat. Rev. Mol. Cell Biol.* **4:** 960.

Petersen C.P., Bordeleau M.E., Pelletier J., and Sharp P.A. 2006. Short RNAs repress translation after initiation in mammalian cells. *Mol. Cell* **21:** 533.

Pontes O., Li C.F., Nunes P.C., Haag J., Ream T., Vitins A., Jacobsen S.E., and Pikaard C.S. 2006. The *Arabidopsis* chromatin-modifying nuclear siRNA pathway involves a nucleolar RNA processing center. *Cell* **126:** 79.

Prasanth K.V., Prasanth S.G., Xuan Z., Hearn S., Freier S.M., Bennett C.F., Zhang M.Q., and Spector D.L. 2005. Regulating gene expression through RNA nuclear retention. *Cell* **123:** 249.

Qi Y., He X., Wang X.J., Kohany O., Jurka J., and Hannon G.J. 2006. Distinct catalytic and non-catalytic roles of ARGONAUTE4 in RNA-directed DNA methylation. *Nature* **443:** 1008.

Rea S. and Akhtar A. 2006. MSL proteins and the regulation of gene expression. *Curr. Top. Microbiol. Immunol.* **310:** 117.

Ruby J.G., Jan C., Player C., Axtell M.J., Lee W., Nusbaum C., Ge H., and Bartel D.P. 2006. Large-scale sequencing reveals 21U-RNAs and additional microRNAs and endogenous siRNAs in *C. elegans*. *Cell* **127:** 1193.

Song J.J., Smith S.K., Hannon G.J., and Joshua-Tor L. 2004. Crystal structure of Argonaute and its implications for RISC slicer activity. *Science* **305:** 1434.

Stein A.J., Fuchs G., Fu C., Wolin S.L., and Reinisch K.M. 2005. Structural insights into RNA quality control: The Ro autoantigen binds misfolded RNAs via its central cavity. *Cell* **121:** 529.

Storz G., Altuvia S., and Wassarman K.M. 2005. An abundance of RNA regulators. *Annu. Rev. Biochem.* **74:** 199.

Storz G., Opdyke J.A., and Zhang A. 2004. Controlling mRNA stability and translation with small, noncoding RNAs. *Curr. Opin. Microbiol.* **7:** 140.

Sullivan C.S., Grundhoff A.T., Tevethia S., Pipas J.M., and Ganem D. 2005. SV40-encoded microRNAs regulate viral gene expression and reduce susceptibility to cytotoxic T cells. *Nature* **435:** 682.

Thomson J.M., Newman M., Parker J.S., Morin-Kensicki E.M., Wright T., and Hammond S.M. 2006. Extensive post-transcriptional regulation of microRNAs and its implications for cancer. *Genes Dev.* **20:** 2202.

Tomlinson R.L., Ziegler T.D., Supakorndej T., Terns R.M., and Terns M.P. 2006. Cell cycle-regulated trafficking of human telomerase to telomeres. *Mol. Biol. Cell* **17:** 955.

Ule J., Stefani G., Mele A., Ruggiu M., Wang X., Taneri B., Gaasterland T., Blencowe B.J., and Darnell R.B. 2006. An RNA map predicting Nova-dependent splicing regulation. *Nature* **444:** 580.

Vagin V.V., Sigova A., Li C., Seitz H., Gvozdev V., and Zamore P.D. 2006. A distinct small RNA pathway silences selfish genetic elements in the germline. *Science* **313:** 320.

Valencia-Sanchez M.A., Liu J., Hannon G.J., and Parker R. 2006. Control of translation and mRNA degradation by miRNAs and siRNAs. *Genes Dev.* **20:** 515.

Verdel A., Jia S., Gerber S., Sugiyama T., Gygi S., Grewal S.I., and Moazed D. 2004. RNAi-mediated targeting of heterochromatin by the RITS complex. *Science* **303:** 672.

Weisblum B. 1995. Erythromycin resistance by ribosome modification. *Antimicrob. Agents Chemother.* **39:** 577.

West S., Gromak N., and Proudfoot N.J. 2004. Human 5′ → 3′ exonuclease Xrn2 promotes transcription termination at cotranscriptional cleavage sites. *Nature* **432:** 522

West S., Gromak N., Norbury C.J., and Proudfoot N.J. 2006. Adenylation and exosome-mediated degradation of cotranscriptionally cleaved pre-messenger RNA in human cells. *Mol. Cell* **21:** 437.

Wightman B., Ha I., and Ruvkun G. 1993. Posttranscriptional regulation of the heterochronic gene lin-14 by lin-4 mediates temporal pattern formation in *C. elegans*. *Cell* **75:** 855.

Will C.L. and Luhrmann R. 2005. Splicing of a rare class of introns by the U12-dependent spliceosome. *Biol. Chem.* **386:** 713.

Winkler W.C. 2005. Riboswitches and the role of noncoding RNAs in bacterial metabolic control. *Curr. Opin. Chem. Biol.* **9:** 594.

Wu G. and Poethig R.S. 2006. Temporal regulation of shoot development in *Arabidopsis thaliana* by miR156 and its target SPL3. *Development* **133:** 3539.

Xu N., Tsai C.L., and Lee J.T. 2006. Transient homologous chromosome pairing marks the onset of X inactivation. *Science* **311:** 1149.

Yik J.H., Chen R., Nishimura R., Jennings J.L., Link A.J., and Zhou Q. 2003. Inhibition of P-TEFb (CDK9/Cyclin T) kinase and RNA polymerase II transcription by the coordinated actions of HEXIM1 and 7SK snRNA. *Mol. Cell* **12:** 971.

Yoshida M., Yoshida K., Kozlov G., Lim N.S., De Crescenzo G., Pang Z., Berlanga J.J., Kahvejian A., Gehring K., Wing S.S., and Sonenberg N. 2006. Poly(A) binding protein (PABP) homeostasis is mediated by the stability of its inhibitor, Paip2. *EMBO J.* **25:** 1934.

Yuan Y.R., Pei Y., Chen H.Y., Tuschl T., and Patel D.J. 2006. A potential protein-RNA recognition event along the RISC-loading pathway from the structure of *A. aeolicus* argonaute with externally bound siRNA. *Structure* **14:** 1557.

Yuan Y.R., Pei Y., Ma J.B., Kuryavyi V., Zhadina M., Meister G., Chen H.Y., Dauter Z., Tuschl T., and Patel D.J. 2005. Crystal structure of *A. aeolicus* argonaute, a site-specific DNA-guided endoribonuclease, provides insights into RISC-mediated mRNA cleavage. *Mol. Cell* **19:** 405.

Zamore P.D., Tuschl T., Sharp P.A., and Bartel D.P. 2000. RNAi: Double-stranded RNA directs the ATP-dependent cleavage of mRNA at 21 to 23 nucleotide intervals. *Cell* **101:** 25.

Zappulla D.C. and Cech T.R. 2004. Yeast telomerase RNA: A flexible scaffold for protein subunits. *Proc. Natl. Acad. Sci.* **101:** 10024.

Zappulla D.C., Goodrich K., and Cech T.R. 2005. A miniature yeast telomerase RNA functions in vivo and reconstitutes activity in vitro. *Nat. Struct. Mol. Biol.* **12:** 1072.

Zhang A., Wassarman K.M., Ortega J., Steven A.C., and Storz G. 2002. The Sm-like Hfq protein increases OxyS RNA interaction with target mRNAs. *Mol. Cell* **9:** 11.

Author Index

Subject Index